W9-BZW-492

To the Student

As authors our highest goal has been to write a solid mathematics textbook that will truly help you succeed in your mathematics course. After years of refining our textbooks, we have now developed a feature that we think will help ensure your success in mathematics. It is called the ***How Am I Doing? Guide to Math Success.***

The ***How Am I Doing? Guide to Math Success*** shows how you can effectively use this textbook to succeed in your mathematics course. This clear path for you to follow is based upon how our successful students have utilized the textbook in the past. Here is how it works:

EXAMPLES and PRACTICE PROBLEMS: When you study an Example, you should immediately do the Practice Problem that follows to make sure you understand each step in solving a particular problem. The worked-out solution to every Practice Problem can be found in the back of the text starting at page SP-1 so you can check your work and receive immediate guidance in case you need to review.

EXERCISE SETS—Practice, Practice, Practice: You learn math by *doing* math. The best way to learn math is to *practice, practice, practice*. The exercise sets provide the opportunity for this practice. Be sure that you complete every exercise your instructor assigns as homework. In addition, check your answers to the odd-numbered exercises in the back of the text to see whether you have correctly solved each problem.

HOW AM I DOING? MID-CHAPTER REVIEW: This feature allows you to check if you understand the important concepts covered to that point in a particular chapter. Many students find that halfway through a chapter is the point of greatest need because so many different types of problems have been covered. This review covers each of the types of problems from the first half of the chapter. Do these problems and check your answers at the back of the text. If you need to review any of these problems, simply refer back to the section and objective indicated next to the answer. Before you go further into the chapter, it is important to understand what has been learned so far.

HOW AM I DOING? CHAPTER TEST: This test (found at the end of every chapter) provides you with an excellent opportunity to both practice and review for any test you will take in class. Take this test to see how much of the chapter you have mastered. By checking your answers, you can once again refer back to the section and the objective of any exercise you want to review further. This allows you to see at once what has been learned and what still needs more study as you prepare for your test or exam.

HOW AM I DOING? CHAPTER TEST PREP VIDEO CD: If you need to review any of the exercises from the *How Am I Doing? Chapter Test,* this video CD found at the back of the text provides worked-out solutions to each of the exercises. Simply insert the CD into a computer and watch a math instructor solve each of the Chapter Test exercises in detail. By reviewing these problems, you can study through any points of difficulty and better prepare yourself for your upcoming test or exam.

These steps provide a clear path you can follow in order to successfully complete your math course. More importantly, the ***How Am I Doing? Guide to Math Success*** is a tool to help you achieve an understanding of mathematics. We encourage you to take advantage of this new feature.

John Tobey
Jeffrey Slater
North Shore Community College

MORE TOOLS FOR SUCCESS

In addition to the ***How Am I Doing? Guide to Math Success,*** your Tobey/Slater textbook is filled with other tools and features to help you succeed in your mathematics course. These include:

Blueprint for Problem Solving
The Mathematics Blueprint for Problem Solving provides you with a consistent outline to organize your approach to problem solving. You will not need to use the blueprint to solve every problem, but it is available when you are faced with a problem with which you are not familiar, or when you are trying to figure out where to begin solving a problem.

Chapter Organizers
The key concepts and mathematical procedures covered in each chapter are reviewed at the end of the chapter in a unique Chapter Organizer. This device not only lists the key concepts and methods, but provides a completely worked-out example for each type of problem. The Chapter Organizer should be used in conjunction with the ***How Am I Doing? Chapter Test*** and the ***How Am I Doing? Chapter Test Prep Video CD*** as a study aid to help you prepare for tests.

Developing Your Study Skills
These notes appear throughout the text to provide you with suggestions and techniques for improving your study skills and succeeding in your math course.

RESOURCES FOR SUCCESS

In addition to the textbook, Prentice Hall offers a wide range of materials to help you succeed in your mathematics course. These include:

Student Study Pack
Includes the *Student Solutions Manual* (fully worked-out solutions to odd-numbered exercises), access to the *Prentice Hall Tutor Center,* and the CD *Lecture Series Videos* that accompany the text. The *Student Study Pack* is available free when packaged with a new textbook.

MyMathLab
MyMathLab offers the entire textbook online with links to video clips and practice exercises in addition to tutorial exercises, homework, and tests. *MyMathLab* also offers a personalized Study Plan for each student based on student test results. The Study Plan links directly to unlimited tutorial exercises for the areas you need to study and re-test so you can practice until you have mastered the skills and concepts. *MyMathLab* is available free when packaged with a new textbook.

Annotated Instructor's Edition

Basic College Mathematics

FIFTH EDITION

John Tobey

North Shore Community College
Danvers, Massachusetts

Jeffrey Slater

North Shore Community College
Danvers, Massachusetts

This Annotated Instructor's Edition contains teaching tips that appear in the margin of the text pages. In addition, the answers to each exercise, chapter test, and cumulative test are displayed in blue ink next to the exercise or problem. Otherwise this Annotated Instructor's Edition is identical to your students' textbooks. When ordering the text for your students, be sure to use the ISBN for the student text, which is 0-13-149057-5.

Upper Saddle River, NJ 07458

This work is protected by United States copyright laws and is provided solely for the use of instructors in teaching their courses and assessing student learning. Dissemination or sale of any part of this work (including on the World Wide Web) will destroy the integrity of the work and is not permitted. The work and materials from it should never be made available to students except by instructors using the accompanying text in their classes. All recipients of this work are expected to abide by these restrictions and to honor the intended pedagogical purposes and the needs of other instructors who rely on these materials.

Senior Acquisitions Editor: *Paul Murphy*
Project Managers: *Elaine Page and Dawn Nuttall*
Editor in Chief: *Christine Hoag*
Production Editor: *Lynn Savino Wendel*
Vice President/Director of Production and Manufacturing: *David W. Riccardi*
Senior Managing Editor: *Linda Mihatov Behrens*
Executive Managing Editor: *Kathleen Schiaparelli*
Assistant Manufacturing Manager/Buyer: *Michael Bell*
Manufacturing Manager: *Trudy Pisciotti*
Executive Marketing Manager: *Eilish Collins Main*
Development Editor: *Tony Palermino*
Marketing Project Manager: *Barbara Herbst*
Marketing Assistant: *Annett Uebel*
Editor in Chief, Development: *Carol Trueheart*
Media Project Manager, Developmental Math: *Audra J. Walsh*
Media Production Editor: *Zachary Hubert*
Art Director: *Jonathan Boylan*
Interior and Cover Designer: *Susan Anderson*
Editorial Assistant: *Mary Burket*
Art Editor: *Thomas Benfatti*
Director of Creative Services: *Paul Belfanti*
Director, Image Resource Center: *Melinda Reo*
Manager, Rights and Permissions: *Zina Arabia*
Manager, Visual Research: *Beth Brenzel*
Manager, Cover Visual Research & Permissions: *Karen Sanatar*
Image Permission Coordinator: *Cynthia Vincenti*
Photo Researcher: *Kathy Ringrose*
Art Studio: *Scientific Illustrators*
Compositor: *Interactive Composition Corporation*
Cover Photo Credits: © *David Muir/Masterfile*

Photo credits appear on page P-1, which constitutes a continuation of the copyright page.

© 2005, 2002, 1998, 1995, 1991 by Pearson Education, Inc.
Pearson Prentice Hall
Pearson Education, Inc.
Upper Saddle River, New Jersey 07458

All rights reserved. No part of this book may be reproduced, in any form or by any means, without permission in writing from the publisher.

Pearson Prentice Hall® is a trademark of Pearson Education, Inc.

Printed in the United States of America

10 9 8 7 6 5 4 3 2

ISBN 0-13-149060-5 (AIE)
ISBN 0-13-149057-5 (Student Edition)

Pearson Education Ltd., London
Pearson Education Australia Pty. Limited, Sydney
Pearson Education Singapore Pte. Ltd.
Pearson Education North Asia, Ltd, Hong Kong
Pearson Education Canada, Ltd., Toronto
Pearson Educación de Mexico, S.A., de C.V.
Pearson Education, Japan, Tokyo
Pearson Education Malaysia, Pte. Ltd.

This book is dedicated to Nancy Tobey
A loving wife for thirty-eight years,
An outstanding mother of three children,
A joyful and thankful grandmother,
A dedicated but retired elementary teacher,
A true friend

Contents

CHAPTER 1

Whole Numbers 1

CHAPTER 2

Fractions 109

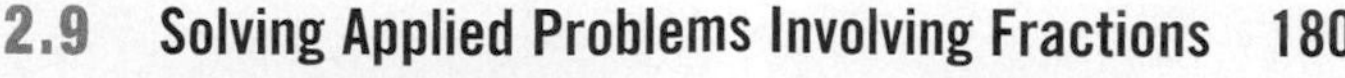

CHAPTER 3
Decimals 203

CHAPTER 4
Ratio and Proportion 269

CHAPTER 5

Percent 311

CHAPTER 6

Measurement 373

CHAPTER 7

Geometry 425

CHAPTER 8

Statistics 525

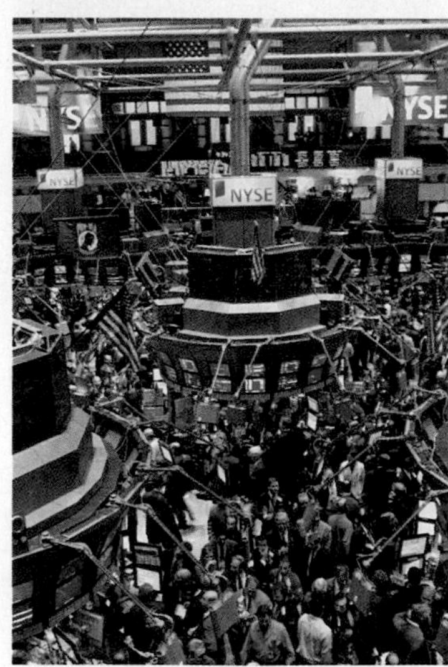

CHAPTER 9

Signed Numbers 571

CHAPTER 10

Introduction to Algebra 621

Preface

TO THE INSTRUCTOR

We share a partnership with you. For over thirty-three years we have taught mathematics courses at North Shore Community College. Each semester we join you in the daily task of sharing the knowledge of mathematics with students who often struggle with this subject. We enjoy teaching and helping students—and we are confident that you share these joys with us.

Mathematics instructors and students face many challenges today. *Basic College Mathematics* was written with these needs in mind. This textbook explains mathematics slowly, clearly, and in a way that is relevant to everyday life for the college student. As with previous editions, special attention has been given to problem solving in the fifth edition. This text is written to help students organize the information in any problem-solving situation, to reduce anxiety, and to provide a guide that enables students to become confident problem solvers.

One of the hallmark characteristics of *Basic College Mathematics* that makes the text easy to learn and teach from is the building-block organization. Each section is written to stand on its own, and each homework set is completely self-testing. Exercises are paired and graded and are of varying levels and types to ensure that all skills and concepts are covered. As a result, the text offers students an effective and proven learning program suitable for a variety of course formats—including lecture-based classes; discussion-oriented classes; distance learning centers; modular, self-paced courses; mathematics laboratories; and computer-supported centers. The book has been written to be especially helpful in online classes. The authors teach at least one course online each semester.

Basic College Mathematics is the first text in a series that includes the following:

Tobey/Slater, *Basic College Mathematics,* Fifth Edition
Tobey/Slater, *Essentials of Basic College Mathematics*
Blair/Tobey/Slater, *Prealgebra,* Third Edition
Tobey/Slater, *Beginning Algebra,* Sixth Edition
Tobey/Slater/Blair, *Beginning Algebra: Early Graphing*
Tobey/Slater, *Intermediate Algebra,* Fifth Edition
Tobey/Slater/Blair, *Beginning and Intermediate Algebra,* Second Edition

We have visited and listened to teachers across the country and have incorporated a number of suggestions into this edition to help you with the particular learning delivery system at your school. The following pages describe the key continuing features and changes in the fifth edition.

NEW! AND IMPROVED FEATURES IN THE FIFTH EDITION

How Am I Doing? Guide to Math Success

The ***How Am I Doing? Guide to Math Success*** shows how students can effectively use this textbook to succeed in their mathematics course. This clear path for them to follow is based upon how successful students have utilized this textbook in the past.

EXAMPLES and PRACTICE PROBLEMS The design of the text has been enhanced so the Examples and Practice Problems are clearly connected in a cohesive unit. This has been done to encourage students to immediately do the Practice Problem associated with each Example to make sure they understand each step in solving a particular problem. The worked-out solution to every Practice Problem can be found in the back of the text starting at page SP-1 so students can check their work and receive immediate guidance in case they need to review.

Enhanced Exercise Sets

- First, *each* exercise set has been:
  - enhanced to have a **better progression from easy to medium to challenging** problems, with appropriate quantities of each.
  - restructured to have **better matching of even and odd** problems.
  - rewritten to ensure that **all concepts are fully represented** with every example from the section covered by a group of exercises.
- In addition, the **number and frequency of *Mixed Practice* problems have been increased,** when appropriate, throughout the exercise sets. These *Mixed Practice* problems require the students to identify the type of problem and the best method they should use to solve it. **More *Verbal and Writing Skills* exercises have also been added,** where appropriate, to allow students more time to interact with new concepts and to explain them fully in their own words.
- Lastly, throughout the text the **application exercises have been updated and labeled** to better indicate the scope and relevance of each real-world and real-data problem at-a-glance. These applications relate to everyday life, global issues beyond the borders of the United States, and other academic disciplines. Many include source citations. Roughly 30 percent of the applications have been contributed by actual students based on scenarios they have encountered in their home or work lives.

How Am I Doing? Mid-Chapter Review

This feature allows you to check if your students understand the important concepts covered to that point in a particular chapter. Many students find that halfway through a chapter is the point of greatest need because so many different types of problems have been covered. This review covers each of the types of problems from the first half of the chapter. Specific section and objective references are provided with each answer to indicate where a student should look for further review.

How Am I Doing? Chapter Test

This test (found at the end of every chapter) provides your students with an excellent opportunity to both practice and review for any test they will take in class. Encourage your students to take this test to see how much of the material they have mastered and then check their answers in the back of the text. Specific section and objective references are provided with each answer to indicate where a student should look for further review.

How Am I Doing? Chapter Test Prep Video CD

If students need to review any of the exercises from the ***How Am I Doing? Chapter Test,*** this video CD found at the back of the text provides worked-out solutions to each of the exercises. Students can simply insert the CD into a computer and watch a math instructor solve each of the Chapter Test exercises in detail. By reviewing these problems, students can study through any points of difficulty and better prepare for an upcoming test or exam.

How Am I Doing? Chapter Test & TestGen

For this new edition of the text, we have provided a new file in each chapter of the *TestGen* program specific to each Chapter Test in the text. The *TestGen* Chapter Test file provides algorithms specific to the problems in the corresponding textbook Chapter Test. This provides a convenient way to create similar tests for practice or assessment purposes.

Appendix A: Consumer Finance Applications

This new appendix has been added to provide in-depth coverage of practical day-to-day finance decisions which today's students face. The topics in this appendix include: *Balancing a Checking Account, Purchasing a Home*, and *Determining the Best Deal when Purchasing a Vehicle.*

KEY FEATURES IN THE FIFTH EDITION

Developing Problem-Solving Abilities

We are committed as authors to producing a textbook that emphasizes mathematical reasoning and problem-solving techniques as recommended by AMATYC, NCTM, AMS, NADE, MAA, and other bodies. To this end, the problem sets are built on a wealth of real-life and real-data applications. Unique problems have been developed and incorporated into the exercise sets that help train students in data interpretation, mental mathematics, estimation, geometry and graphing, number sense, critical thinking, and decision making.

Mathematics Blueprint for Problem Solving

The successful Mathematics Blueprint for Problem Solving strengthens problem-solving skills by providing a consistent and interactive outline to help students organize their approach to problem solving. Once students fill in the blueprint, they can refer back to their plan as they do what is needed to solve the problem. Because of its flexibility, this feature can be used with single-step problems, multistep problems, applications, and nonroutine problems that require problem solving strategies. Students will not need to use the blueprint to solve every problem. It is available for those faced with a problem with which they are not familiar, to alleviate anxiety, to show them where to begin, and to assist them in the steps of reasoning.

Developing Your Study Skills

This highly successful feature has been expanded in the new edition. The boxed notes are integrated throughout the text to provide students with techniques for improving their study skills and succeeding in math courses.

Putting Your Skills to Work

This highly successful feature has been revised and expanded in the fifth edition. There are ten new Putting Your Skills to Work applications in the new edition. These nonroutine application problems challenge students to synthesize the knowledge they have gained and apply it to a totally new area. Each problem is specifically arranged for independent and cooperative learning or group investigation of mathematical problems that pique student interest. Students are given the opportunity to help one another discover mathematical solutions to extended problems. The investigations feature open-ended questions and extrapolation of data to areas beyond what is normally covered in such a course.

Increased Integration and Emphasis on Geometry

Due to the emphasis on geometry on many statewide exams, geometry problems are integrated throughout the text. The new edition contains over 20% more geometry problems. Additionally, examples and exercises that incorporate a principle of geometry are marked with a triangle icon for easy identification.

Graphs, Charts, and Tables

When students encounter mathematics in real-world publications, they often encounter data represented in a graph, chart, or table and are asked to make a reasonable conclusion based on the data presented. This emphasis on graphical interpretation is a continuing trend with today's expanding technology. The number of mathematical problems based on charts, graphs, and tables has been significantly increased in this edition. Students are asked to make simple interpretations, to solve medium-level problems, and to investigate challenging applied problems based on the data shown in a chart, graph, or table.

MASTERING MATHEMATICAL CONCEPTS

Text features that develop the mastery of concepts include the following:

Learning Objectives

Concise learning objectives listed at the beginning of each section allow students to preview the goals of that section.

To Think About

These critical thinking questions may follow examples in the text and appear in the exercise sets. They extend the concept being taught, providing the opportunity for all students to stretch their minds, to look for patterns, and to make conclusions based on their previous experience. The number of these exercises has significantly increased in this edition.

Cumulative Review

Almost every exercise set concludes with a section of cumulative review problems. These problems review topics previously covered, and are designed to assist students in retaining the material. Many additional applied problems have been added to the cumulative review sections.

Calculator Problems

Calculator boxes are placed in the margin of the text to alert students to a scientific calculator application. In the exercise section a scientific calculator icon is used to indicate problems that are designed for solving with a calculator. There is also instruction on how to use a scientific calculator in the appendix.

REVIEWING MATHEMATICAL CONCEPTS

At the end of each chapter we have included problems and tests to provide your students with several different formats to help them review and reinforce the ideas that they have learned. This assists them not only with that specific chapter, but reviews previously covered topics as well.

Chapter Organizers

The concepts and mathematical procedures covered are reviewed at the end of each chapter in a unique chapter organizer. This device has been extremely popular with faculty and students alike. It not only lists concepts and methods, but provides a completely worked-out example for each type of problem. Students find that preparing a similar chapter organizer on their own in higher-level math courses becomes an invaluable way to master the content of a chapter of material.

Chapter Review Problems

These problems are grouped by section as a quick refresher at the end of the chapter. They can also be used by the student as a quiz of the chapter material.

How Am I Doing? Chapter Tests

Found at the end of the chapter, the chapter test is a representative review of the material from that particular chapter that simulates an actual testing format. This provides the students with a gauge to their preparedness for the actual examination.

Cumulative Tests

At the end of each chapter is a cumulative test. One-half of the content of each cumulative test is based on the math skills learned in previous chapters. By completing these tests for each chapter, the students build confidence that they have mastered not only the contents of the chapter but those of previous chapters as well.

RESOURCES FOR THE INSTRUCTOR

Printed Resources

Annotated Instructor's Edition (ISBN: 0-13-149060-5)

- Complete student text
- Answers appear in place on the same text page as exercises
- **Teaching Tips** placed in the margin at key points where students historically need extra help.
- Answers to all exercises in the section sets, mid-chapter reviews, chapter reviews, chapter tests, cumulative tests, and practice final

Instructor's Solutions Manual (ISBN: 0-13-149058-3)

- Detailed step-by-step solutions to the even-numbered section exercises
- Solutions to every exercise (odd and even) in the mid-chapter reviews, chapter reviews, chapter tests, cumulative tests, and practice final
- Solution methods reflect those emphasized in the text

Instructor's Resource Manual with Tests (ISBN: 0-13-149059-1)

- New! One *Mini-Lecture* is provided per section. These include key learning objectives, classroom examples, and teaching notes.
- New! One *Skill Builder* is provided per section. These include concept rules, explained examples, and extra problems for students. All answers included.
- New! Two *Activities* per chapter provide short group activities in a convenient ready-to-use handout format. All answers included.

- Twenty *Additional Exercises* are provided per section for added test exercises or worksheets. All answers included.
- *Tests* provide additional suggested testing materials:
  - two *Chapter Pretests* per chapter (1 free response, 1 multiple choice)
  - six *Chapter Tests* per chapter (3 free response, 3 multiple choice)
  - two *Cumulative Tests* per even-numbered chapter (1 free response, 1 multiple choice)
  - two *Final Exams* (1 free response, 1 multiple choice)
  - answers to all items

Media Resources

TestGen with QuizMaster (Windows/Macintosh) (ISBN: 0-13-149064-8)

- Algorithmically driven, text-specific testing program—covers all objectives of the text
- New! Chapter Test file for each chapter provides algorithms specific to exercises in each *How Am I Doing? Chapter Test* from the text
- Edit and add own questions with the built-in question editor, which also allows you to create graphs, import graphics, and insert math notation
- Create a nearly unlimited number of tests and worksheets as well as assorted reports and summaries
- Networkable for administering tests and capturing grades online or on a local area network

CD Lecture Series Videos—Lab Pack (ISBN: 0-13-153061-5)

- Organized by section, contain problem solving techniques and examples from the textbook
- Step-by-step solutions to selected exercises from each textbook section marked with a video icon
- Convenient anytime access to video tutorial support when provided for students as part of the Student Study Pack

New! MyMathLab (instructor)

MyMathLab is an all-in-one, online tutorial, homework, assessment, course management tool with the following features.

- **Powered by CourseCompass™**—Pearson Education's online teaching and learning environment
- **Tutorial powered by *MathXL*®**—our online homework, tutorial, and assessment system
- Rich and flexible set of course materials, **featuring free-response exercises** algorithmically generated for unlimited practice and mastery.
- **The entire textbook online with links to multimedia resources**—video clips, practice exercises, and animations—that are correlated to the textbook examples and exercises.
- **Homework and test managers** to select and assign online exercises correlated directly to the text
- **A personalized Study Plan** generated based on student test results. The Study Plan links directly to unlimited tutorial exercises for the areas students need to study and re-test, so they can practice until they have mastered the skills and concepts.

- **Easy to use tracking** in the *MyMathLab* gradebook of all of the online homework, tests, and tutorial work.
- **Import** *TestGen* **tests**
- **Online gradebook**—designed specifically for mathematics—automatically tracks students' homework and test results and provides grade control.
- http://www.mymathlab.com

New! MathXL®

MathXL® is a powerful online homework, tutorial, and assessment system. With *MathXL* instructors can:

- Create, edit, and assign online homework and tests using algorithmically generated exercises correlated at the objective level to your textbook.
- Track student work in *MathXL*'s online gradebook.

RESOURCES FOR THE STUDENT

New! Student Study Pack (ISBN: 0-13-154604-X)

Includes the *Student Solutions Manual,* access to the *Prentice Hall Tutor Center,* and the *CD Lecture Series Videos* that accompany the text.

Printed Resources

Student Solutions Manual (ISBN: 0-13-149061-3)

- Solutions to all odd-numbered section exercises
- Solutions to every (even and odd) exercises found in the mid-chapter reviews, chapter reviews, chapter tests, and cumulative reviews
- Solution methods reflect those emphasized in the text
- Ask your bookstore about ordering

MEDIA RESOURCES

New! Chapter Test Prep Video CD (ISBN: 0-13-149070-2)

Provides step-by-step video solutions to each problem in each *How Am I Doing? Chapter Test* in the textbook. Packaged free with the text, inside the back cover.

New! MyMathLab (student)

MyMathLab is an all-in-one, online tutorial, homework, assessment, course management tool with the following student features.

- **The entire textbook online with links to multimedia resources**—video clips, practice exercises, and animations—that are correlated to the textbook examples and exercises.
- **Online tutorial, homework, and tests**
- **A personalized Study Plan** generated based on student test results. The Study Plan links directly to unlimited tutorial exercises for the areas students need to study and re-test, so they can practice until they have mastered the skills and concepts.
- http://www.mymathlab.com

New! MathXL®

MathXL® is a powerful online homework, tutorial, and assessment system. With *MathXL* students can:

- Take chapter tests and receive a personalized study plan based on their test results.
- See diagnosed weaknesses and link directly to tutorial exercises for the objectives they need to study and retest.
- Access supplemental animations and video clips directly from selected exercises.

New! MathXL® Tutorials on CD (ISBN: 0-13-153064-X)

This interactive tutorial CD-ROM provides:

- Algorithmically-generated practice exercises correlated at the objective level.
- Practice exercises accompanied by an example and a guided solution.
- Tutorial video clips within the exercise to help students visualize concepts.
- Easy-to-use tracking of student activity and scores and printed summaries of students' progress.

New! Interact Math www.interactmath.com

The power of the *MathXL* text-specific tutorial exercises available for unlimited practice online, without an access code, and without tracking capabilities.

Prentice Hall Tutor Center www.prenhall.com/tutorcenter (ISBN: 0-13-064604-0)

- Free tutorial support via phone, fax or email staffed by developmental math faculty.
- Available Sunday—Thursday 5pm EST to midnight—5 days a week, 7 hours a day
- Accessed through a registration number that may be bundled with a new text as part of the Student Study Pack or purchased separately with a used book. Comes automatically within *MyMathLab*.

ACKNOWLEDGMENTS

This book is the product of many years of work and many contributions from faculty and students across the country. We would like to thank the many reviewers and participants in focus groups and special meetings with the authors in preparation of previous editions.

Our deep appreciation to each of the following:

George J. Apostolopoulos, *DeVry Institute of Technology*

Katherine Barringer, *Central Virginia Community College*

Sohrab Bakhtyari, *St. Petersburg Junior College–Clearwater*

Christine R. Bauman, *Clark College* at Larch

Rita Beaver, *Valencia Community College*

Jamie Blair, *Orange Coast College*

Larry Blevins, *Tyler Junior College*

Vernon Bridges, *Durham Technical Community College*
Connie Buller, *Metropolitan Community College*
Oscar Caballero III, *Laredo Community College*
Brenda Callis, *Rappahannock Community College*
Joan P. Capps, *Raritan Valley Community College*
Robert Christie, *Miami-Dade Community College*
Nelson Collins, *Joliet Junior College*
Mike Contino, *California State University at Heyward*
Callie Jo Daniels, *St. Charles County Community College*
Ky Davis, *Muskingum Area Technical College*
Judy Dechene, *Fitchburg State University*
Floyd L. Downs, *Arizona State University*
Barbara Edwards, *Portland State University*
Disa Enegren, *Rose State College*
Janice F. Gahan-Rech, *University of Nebraska at Omaha*
Colin Godfrey, *University of Massachusetts, Boston*
Nancy Graham, *Rose State College*
Mary Beth Headlee, *Manatee Community College*
Sharon Louvier, *Lee College*
Doug Mace, *Baker College*
Carl Mancuso, *William Paterson College*
James A. Matovina, *Community College of Southern Nevada*
Janet McLaughlin, *Montclair State College*
Beverly Meyers, *Jefferson College*
Wayne L. Miller, *Lee College*
Gloria Mills, *Tarrant County Junior College*
Norman Mittman, *Northeastern Illinois University*
Marcia Mollé, *Metropolitan Community College*
Jody E. Murphy, *Lee College*
Katrina Nichols, *Delta College*
Leticia M. Oropesa, *University of Miami*
Sandra Orr, *West Virginia State Community College*
Jim Osborn, *Baker College*
Linda Padilla, *Joliet Junior College*
Catherine Panik, *Manatee Community College—South Campus*
Gary Phillips, *Clark College*
Elizabeth A. Polen, *County College of Morris*
Joel Rappaport, *Miami Dade Community College*
Ronald Ruemmler, *Middlesex County College*
Dennis Runde, *Manatee Community College*
Sally Search, *Tallahassee Community College*
Richard Sturgeon, *University of Southern Maine*
Ara B. Sullenberger, *Tarrant County Community College*
Margie Thrall, *Manatee Community College*

Michael Trappuzanno, *Arizona State University*
Cora S. West, *Florida Community College* at Jacksonville
Jerry Wisnieski, *Des Moines Community College*

In addition, we want to thank the following individuals for providing splendid insight and suggestions for this new edition.

John Akutagawa, *Heald Business College*
Gopa Bhowmick, *Mississippi Gulf Coast College*
Jon Blakely, *College of the Sequoias*
Matt Bourez, *College of the Sequoias*
Jared Burch, *College of the Sequoias*
Yen-Phi (Faye) Dang, *Joliet Junior College*
Naomi Gibbs, *Pitt Community College*
Laura Huerta, *Laredo Community College*
Joe Karnowski, *Norwalk Community College*
Carolyn Krause, *Delaware Technical and Community College*
Kay Kriewald, *Laredo Community College*
Douglas Lewis, *Yakima Valley Community College*
Luanne Lundberg, *Clark College*
Maria Mendez, *Laredo Community College*
Henri Onuigbo, *Wayne County Community College*
Jack Roberts, *Ivy Tech State, Sellersburg*
Cindy Satriano, *Albuquerque Technical Vocational Institute*
Jeffrey Simmons, *Ivy Tech State, Ft. Wayne*
Brad Sullivan, *Community College of Denver*
Bettie Truitt, *Black Hawk College*
Jacquelyne Wing, *Angelina College*

We have been greatly helped by a supportive group of colleagues who not only teach at North Shore Community College but have also provided a number of ideas as well as extensive help on all of our mathematics books. Also, a special word of thanks to Hank Harmeling, Tom Rourke, Wally Hersey, Bob McDonald, Judy Carter, Bob Campbell, Rick Ponticelli, Russ Sullivan, Kathy LeBlanc, Lora Connelly, Sharyn Sharaf, Donna Stefano, and Nancy Tufo. Joan Peabody has done an excellent job of typing various materials for the manuscript and her help is gratefully acknowledged. Jenny Crawford provided new problems, new ideas, and great mental energy. She greatly assisted us with error checking. Her excellent help was much appreciated. A special word of thanks to Richard Semmler and Ron Salzman for their excellent work in accuracy reviewing page proofs.

Each textbook is a combination of ideas, writing, and revisions from the authors and wise editorial direction and assistance from the editors. We want to thank our Prentice Hall editor, Paul Murphy, for his helpful insight and perspective on each phase of the revision of the textbook. Paul is a man of ideas, wisdom, and energy. He has that rare ability to look at the big picture and consider all the possibilities and then make a wise plan to accomplish the goal in the best possible way. He has found the secret of how to listen to a multitude of ideas and distill them down to a few excellent concepts upon which to improve a textbook. It has been a joy to work with him in the planning, writing, and revising of the new fifth edition of *Basic College Mathematics*. Elaine Page, our project manager, provided daily support and encouragement as the book progressed. Elaine is a woman of insight, organization, and helpful suggestions. She has an amazing ability to take hundreds of pages of ideas and condense them into a manageable collection of concrete suggestions. From time to time, as we rolled up our sleeves and went to work on the actual revision and preparation of the manuscript for production, we found that we had much to do and we asked Elaine for help. In every case, she came to our assistance and we accomplished the task together. Tony Palermino, our developmental editor, sifted through mountains of material and offered excellent suggestions for improvement and change.

For the final phases of production of this book, Dawn Nuttall has been our project manager. Dawn is a woman with remarkable perspective and thoughtful analysis. She possesses a wonderful ability to see through problems and find solutions that not only solve the difficulty at hand, but prevent further difficulties from occurring at some later point. When we encountered a wall blocking the way, Dawn always found a clear path to take us around the obstacle. It has been a delight to work with Dawn.

Our mathematics production editor, Lynn Savino Wendel, has been a wonderful help as we finalized the manuscript and checked over the page proofs. Lynn kept things moving on schedule and adjusted tasks in order to make production more efficient. Lynn knows textbook production like the back of her hand and her knowledge proved most helpful throughout the entire process.

Nancy Tobey retired from teaching and joined the team as our administrative assistant. Mailing, editing, photocopying, collating, and taping were cheerfully done each day. A special thanks goes to Nancy. We could not have finished the book without you.

Book writing is impossible for us without the loyal support of our families. Our deepest thanks and love to Nancy, Johnny, Melissa, Marcia, Shelley, Rusty, and Abby. Your understanding, your love and help, and your patience have been a source of great encouragement. Finally, we thank God for the strength and energy to write and the opportunity to help others through this textbook.

We have spent more than 33 years teaching mathematics. Each teaching day, we find that our greatest joy is helping students learn. We take a personal interest in ensuring that each student has a good learning experience in taking this course. If you have some personal comments, suggestions, or ideas for future editions of this textbook, please write to us at:

Prof. John Tobey and Prof. Jeffrey Slater
Prentice Hall Publishing
Office of the College Mathematics Editor
Room 300
75 Arlington Street
Boston, MA 02116

or e-mail us at

jtobey@northshore.edu

We wish you success in this course and in your future life!

John Tobey
Jeffrey Slater

Diagnostic Pretest: Basic College Mathematics

1. 4373
2. 26
3. 1128
4. 920 tons of sand
5. $\frac{29}{35}$
6. $8\frac{1}{4}$
7. $\frac{5}{6}$
8. 30 miles per gallon
9. 15.67542
10. 3.4
11. 22.625 centimeters
12. 41.2 miles

Chapter 1

1. Add. 3846 + 527

2. Divide. $58\overline{)1508}$

3. Subtract.
$$\begin{array}{r} 12{,}807 \\ -\ 11{,}679 \\ \hline \end{array}$$

4. The highway department used 115 truckloads of sand. Each truck held 8 tons of sand. How many tons of sand were used?

Chapter 2

5. Add. $\frac{3}{7} + \frac{2}{5}$

6. Multiply and simplify. $3\frac{3}{4} \times 2\frac{1}{5}$

7. Subtract. $2\frac{1}{6} - 1\frac{1}{3}$

8. Mike's car traveled 237 miles on $7\frac{9}{10}$ gallons of gas. How many miles per gallon did he achieve?

Chapter 3

9. Multiply.
$$\begin{array}{r} 51.06 \\ \times\ 0.307 \\ \hline \end{array}$$

10. Divide. $0.026\overline{)0.0884}$

11. The copper pipe was 24.375 centimeters long. Paula had to shorten it by cutting off 1.75 centimeters. How long will the copper pipe be when it is shortened?

12. Russ bicycled 20.5 miles on Monday, 5.8 miles on Tuesday, and 14.9 miles on Wednesday. How many miles did he bicycle on those three days?

Chapter 4

Solve each proportion problem. Round to the nearest tenth if necessary.

13. $\frac{3}{7} = \frac{n}{24}$ **14.** $\frac{0.5}{0.8} = \frac{220}{n}$

15. Wally's Landscape earned \$600 for mowing lawns at 25 houses last week. At that rate, how much would he earn for doing 45 houses?

16. Two cities that are actually 300 miles apart appear to be 8 inches apart on the road map. How many miles apart are two cities that appear to be 6 inches apart on the map?

Chapter 5

Round to the nearest tenth if necessary.

17. Change to a percent: $\frac{3}{8}$ **18.** 138% of 5600 is what number?

19. At Mountainview College 53% of the students are women. There are 2067 women at the college. How many students are at the college?

20. At a manufacturing plant it was discovered that 9 out of every 3000 parts made were defective. What percent of the parts are defective?

Chapter 6

21. 15 qt = ________ gal **22.** 3 cm = ________ meter

23. 1.56 tons = ________ lb **24.** 4900 kg = ________ milligrams

Chapter 7

Round to the nearest hundredth when necessary. Use $\pi \approx 3.14$ when necessary.

▲ **25.** Find the area of a triangle with a base of 34 meters and an altitude of 23 meters.

▲ **26.** Find the cost to install carpet in a circular area with a radius of 5 yards at a cost of \$35 per square yard.

▲ **27.** In a right triangle the longest side is 15 meters and the shortest side is 9 meters. What is the length of the other side of the triangle?

▲ **28.** How many pounds of fertilizer can be placed in a cylindrical tank that is 4 feet tall and has a radius of 5 feet if one cubic foot of fertilizer weighs 70 pounds?

13. $n = 10.3$

14. $n = 352$

15. \$1080

16. 225 miles

17. 3.75%

18. 7728

19. 3900 students

20. 0.3% are defective

21. 3.75

22. 0.03

23. 3120

24. 4,900,000,000

25. 391 square meters

26. \$2747.50

27. 12 meters

28. 21,980 pounds

29. 500

30. 100

31. 2001

32. 625 cars per quarter

33. −15

34. 12

35. $-\frac{9}{10}$

36. −18

37. $3x + 24y$

38. $x = 6$

39. $x = \frac{1}{3}$

40. The width is 21 meters. The length is 46 meters.

Chapter 8

The following double bar graph indicates the sale of Dodge Neons for Westover County as reported by the district sales managers. Use this graph to answer questions 29–32.

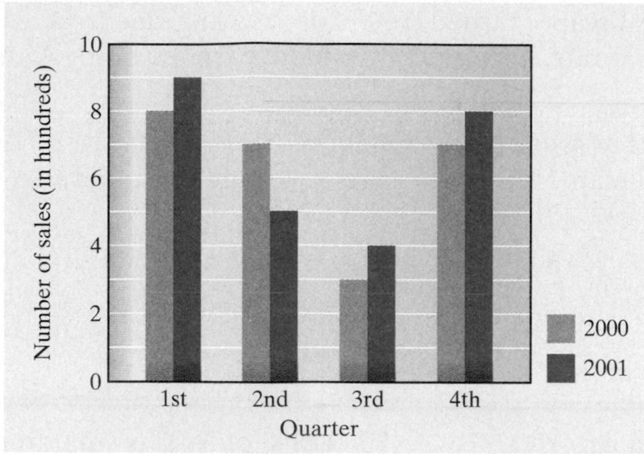

29. How many Dodge Neons were sold in the second quarter of 2001?

30. How many more Dodge Neons were sold in the fourth quarter of 2001 than were sold in the fourth quarter of 2000?

31. In which year were more Dodge Neons sold, in 2000 or 2001?

32. What is the *mean* number of Dodge Neons sold per quarter in 2000?

Chapter 9

Perform the following operations.

33. $-5 + (-2) + (-8)$

34. $-8 - (-20)$

35. $\left(-\frac{3}{4}\right) \div \left(\frac{5}{6}\right)$

36. $(-3)(2)(-1)(-3)$

Chapter 10

Simplify.

37. $9(x + y) - 3(2x - 5y)$

In exercises 38–39, solve for x.

38. $3x - 7 = 5x - 19$

39. $2(x - 3) + 4x = -2(3x + 1)$

▲ **40.** A rectangle has a perimeter of 134 meters. The length of the rectangle is 4 meters longer than double the width of the rectangle. What is the length and the width of the rectangle?

CHAPTER

1

Although most people in the United States can count on clean drinking water, that is not the case for people in many other countries of the world. Drilling wells for people who cannot get clean drinking water has become a very high priority for world relief organizations. Can you use your mathematics to keep track of the number of new wells being drilled? Turn to page 97 to find out.

Whole Numbers

1.1 UNDERSTANDING WHOLE NUMBERS

Student Learning Objectives

After studying this section, you will be able to:

1. Write numbers in expanded form.
2. Write whole numbers in standard notation.
3. Write a word name for a number and write a number for a word name.
4. Read numbers in tables.

1 Writing Numbers in Expanded Form

To count a number of objects or to answer the question "How many?" we use a set of numbers called **whole numbers.** These whole numbers are as follows.

0, 1, 2, 3, 4, 5, 6, 7, 8, 9, 10, 11, 12, 13, 14, 15, . . .

There is no largest whole number. The three dots . . . indicate that the set of whole numbers goes on indefinitely. Our number system is based on tens and ones and is called the **decimal system** (or the **base 10 system**). The numbers 0, 1, 2, 3, 4, 5, 6, 7, 8, 9 are called **digits.** The position, or placement, of the digits in the number tells the value of the digits. For examplc, in the number 521, the "5" means 5 hundreds (500). In the number 54, the "5" means 5 tens (50).

521 — 5 means 5 hundreds or 500

54 — 5 means 5 tens or 50

For this reason, our number system is called a **place-value system.**

Consider the number 5643. We will use a place-value chart to illustrate the value of each digit in the number 5643.

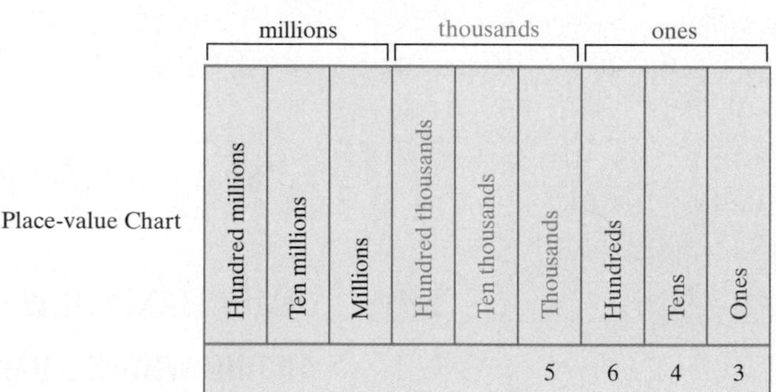

The value of the number is 5 thousands, 6 hundreds, 4 tens, 3 ones.

The place-value chart shows the value of each place, from ones on the right to hundred millions on the left. When we write very large numbers, we place a comma after every group of three digits called a **period,** moving from right to left. This makes the number easier to read. It is usually agreed that a four-digit number does not have a comma, but that numbers with five or more digits do. So 32,000 would be written with a comma but 7000 would not.

To show the value of each digit in a number, we sometimes write the number in expanded notation. For example, 56,327 is 5 ten thousands, 6 thousands, 3 hundreds, 2 tens, and 7 ones. In **expanded notation,** this is

$$50{,}000 + 6000 + 300 + 20 + 7.$$

EXAMPLE 1 Write each number in expanded notation.

(a) 2378 **(b)** 538,271 **(c)** 980,340,654

Solution

(a) Sometimes it helps to say the number to yourself.

	two thousand		three hundred		seventy		eight
2378 =	2000	+	300	+	70	+	8

(b)

Expanded notation: 538,271 = 500,000 + 30,000 + 8000 + 200 + 70 + 1

(c) When 0 is used as a placeholder, you do not include it in the expanded form.

Expanded notation: 980,340,654 = 900,000,000 + 80,000,000 + 300,000 + 40,000 + 600 + 50 + 4

Practice Problem 1 Write each number in expanded notation.

(a) 3182 **(b)** 520,890 **(c)** 709,680,059

NOTE TO STUDENT: Fully worked-out solutions to all of the Practice Problems can be found at the back of the text starting at page SP-1

2 Writing Whole Numbers in Standard Notation

The number as you usually see it is called the **standard notation.** 980,340,654 is the standard notation for the number nine hundred eighty million, three hundred forty thousand, six hundred fifty-four.

EXAMPLE 2 Write each number in standard notation.

(a) 500 + 30 + 8

(b) 300,000 + 7000 + 40 + 7

Solution

(a) 538

(b) Be careful to keep track of the place value of each digit. You may need to use 0 as a placeholder.

3 hundred thousand

300,000 + 7000 + 40 + 7 = 307,047

7 thousand

We needed to use 0 in the ten thousands place and in the hundreds place.

Teaching Tip You can remind students that a few ancient cultures actually avoided the use of zero by never writing numbers like 40. Instead they would make the number larger by writing 41 or smaller by writing 39. Needless to say, such a culture had a hard time developing effective records for business transactions.

Practice Problem 2 Write each number in standard notation.

(a) 400 + 90 + 2 **(b)** 80,000 + 400 + 20 + 7

EXAMPLE 3 Last year the population of Central City was 1,509,637. In the number 1,509,637

(a) How many ten thousands are there? **(b)** How many tens are there?
(c) What is the value of the digit 5? **(d)** In what place is the digit 6?

Solution The place-value chart will help you identify the value of each place.

(a) Look at the digit in the ten thousands place. There are 0 ten thousands.
(b) Look at the digit in the tens place. There are 3 tens.
(c) The digit 5 is in the hundred thousands place. The value of the digit is 5 hundred thousand or 500,000.
(d) The digit 6 is in the hundreds place.

NOTE TO STUDENT: Fully worked-out solutions to all of the Practice Problems can be found at the back of the text starting at page SP-1

Practice Problem 3 The campus library has 904,759 books.

(a) What digit tells the number of hundreds?
(b) What digit tells the number of hundred thousands?
(c) What is the value of the digit 4?
(d) What is the value of the digit 9? Why does this question have two answers?

3 Writing Word Names for Numbers and Numbers for Word Names

A number has the same *value* no matter how we write it. For example, "a million dollars" means the same as "$1,000,000." In fact, any number in our number system can be written in several ways or forms:

- Standard notation 521
- Expanded notation 500 + 20 + 1
- Word name five hundred twenty-one

You may want to write a number in any of these ways. To write a check, you need to use both standard notation and words.

Teaching Tip Additional coverage on balancing a checkbook can be found in the Consumer Finance Appendix.

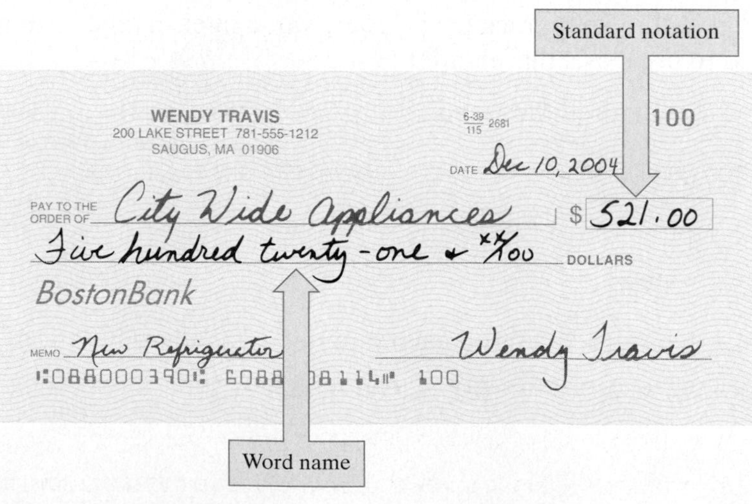

To write a word name, start from the left. Name the number in each period, followed by the name of the period, and a comma. The last period name, "ones" is not used.

EXAMPLE 4 Write a word name for 364,128,957.

Solution

Place-value Chart

Billions			Millions			Thousands			Ones		
			3	6	4	1	2	8	9	5	7
Hundreds	Tens	Ones	Hundreds	Tens	Ones	Hundreds	Tens	Ones	Hundreds	Tens	Ones

We want to write a word name for 364, 128, 957.

three hundred sixty-four million, (364)

one hundred twenty-eight thousand, (128)

nine hundred fifty-seven (957)

The answer is three hundred sixty-four million, one hundred twenty-eight thousand, nine hundred fifty-seven.

Practice Problem 4 Write a word name for 267,358,981.

EXAMPLE 5 Write the word name for each number.

(a) 1695 **(b)** 200,470 **(c)** 7,003,038

Solution Look at the place-value chart if you need help identifying the place for each digit.

(a) To help us, we will put in the optional comma: 1,695.

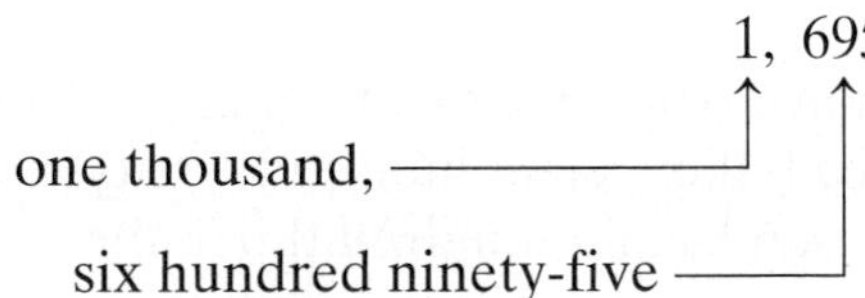

The word name is one thousand, six hundred ninety-five.

(b) 200, 470

two hundred thousand, (200)

four hundred seventy (470)

The word name is two hundred thousand, four hundred seventy.

Teaching Tip Students may wonder why they need to write the word name for a number. Remind them that they need to do this when making out a check or withdrawal slip for a passbook savings account. Sometimes banks will delay processing a check if the incorrect word name for the number is written on it.

(c)

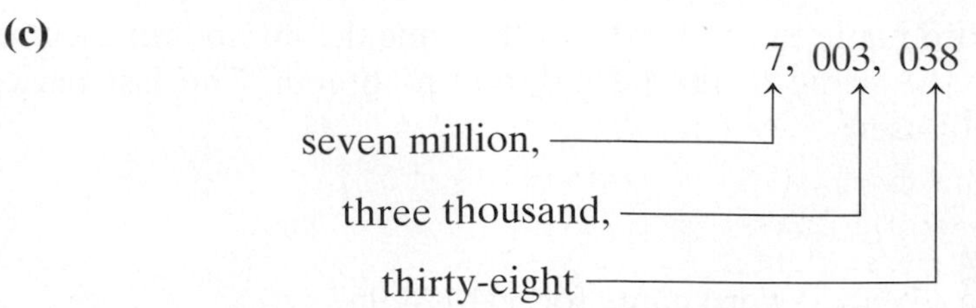

The word name is seven million, three thousand, thirty-eight.

Practice Problem 5 Write the word name for each number.

(a) 2736 **(b)** 980,306 **(c)** 12,000,021

NOTE TO STUDENT: Fully worked-out solutions to all of the Practice Problems can be found at the back of the text starting at page SP-1

CAUTION: DO NOT USE THE WORD AND FOR WHOLE NUMBERS. Many people use the word *and* when giving the word name for a whole number. For example, you might hear someone say the number 34,507 as "thirty-four thousand, five hundred *and* seven." However, this is not technically correct. In mathematics we do NOT use the word *and* when writing word names for whole numbers. In Chapter 3 we will use the word *and* to represent the decimal point. For example, 59.76 will have the word name "fifty-nine *and* seventy-six hundredths."

Very large numbers are used to measure quantities in some disciplines, such as distance in astronomy and the national debt in macroeconomics. We can extend the place-value chart to include these large numbers.

The national debt for the United States as of November 22, 2003, was $6,923,886,720,833. This number is indicated in the following place-value chart.

Place-value Chart														
Trillions			Billions			Millions			Thousands			Ones		
		6	9	2	3	8	8	6	7	2	0	8	3	3

EXAMPLE 6 Write the number for the national debt for the United States as of November 22, 2003, in the amount of $6,923,886,720,833 using a word name.

Solution The national debt on November 22, 2003 was six trillion, nine hundred twenty-three billion, eight hundred eighty-six million, seven hundred twenty thousand, eight hundred thirty-three dollars.

Practice Problem 6 As of January 1, 2004, the estimated population of the world was 6,393,646,525. Write this world population using a word name.

Occasionally you may want to write a word name as a number.

EXAMPLE 7 Write each number in standard notation.

(a) twenty-six thousand, eight hundred sixty-four

(b) two billion, three hundred eighty-six million, five hundred forty-seven thousand, one hundred ninety

Solution

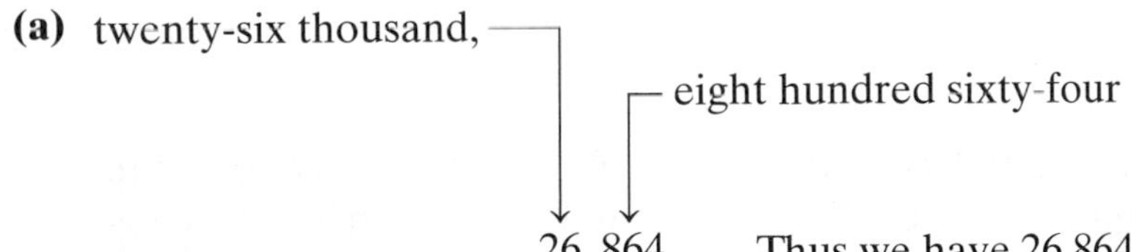

Thus we have 26,864.

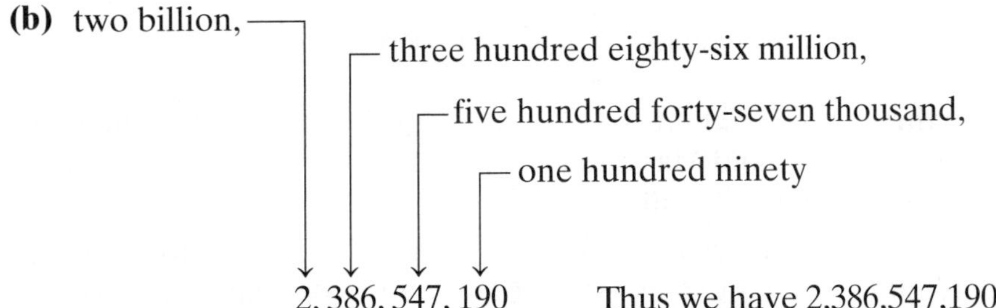

Thus we have 2,386,547,190.

Practice Problem 7 Write in standard notation.

(a) eight hundred three **(b)** thirty thousand, two hundred twenty-nine

4 Reading Numbers in Tables

Sometimes numbers found in charts and tables are abbreviated. Look at the chart below from the U.S. Bureau of the Census. Notice that the second line tells us the numbers represent thousands. To understand what these numbers mean, think "thousands." If the number 23 appears under 1740 for New Hampshire, the 23 represents 23 thousand. 23 thousand is 23,000. Note that census figures for some colonies are not available for certain years.

Estimated population of the American colonies from 1650 to 1780 (in thousands)

	Maine	New Hampshire	Vermont	Plymouth & Massachusetts	Rhode Island	Connecticut
1650	1	1	★	16	1	4
1670	★	2	★	35	2	13
1690	★	4	★	57	4	22
1700	★	5	★	56	6	26
1720	★	9	★	91	12	59
1740	★	23	★	152	25	90
1750	★	28	★	188	33	111
1770	31	62	10	235	58	184
1780	49	88	48	269	53	207

EXAMPLE 8 Refer to the chart from the previous page to answer the following questions. Write each number in standard notation.

(a) What was the estimated population of Maine in 1780?

(b) What was the estimated population of Plymouth and Massachusetts in 1720?

(c) What was the estimated population of Rhode Island in 1700?

Solution

(a) To read the chart, first look for Maine on the left. Read across to the column under 1780. The number is 49. In this chart 49 means 49 thousands.

$$49 \text{ thousands} \Rightarrow 49{,}000$$

(b) Read the line of the chart for Plymouth and Massachusetts. The number for Plymouth and Massachusetts under 1720 is 91. This means 91 thousands. We will write this as 91,000.

(c) Read the line of the chart for Rhode Island. The number for Rhode Island under 1700 is 6. This means 6 thousands. We will write this as 6000.

TO THINK ABOUT: Interpreting Data in a Table Why do you think Plymouth and Massachusetts had the largest population for the years shown in the table?

***NOTE TO STUDENT:** Fully worked-out solutions to all of the Practice Problems can be found at the back of the text starting at page SP-1*

Practice Problem 8 Refer to the chart from the previous page to answer the following questions. Write each number in standard notation.

(a) What was the estimated population of Connecticut in 1670?

(b) What was the estimated population of New Hampshire in 1780?

(c) What was the estimated population of Vermont in 1770?

Developing Your Study Skills

Class Participation

People learn mathematics through active participation, not through observation from the sidelines. If you want to do well in this course, get involved in all course activities. If you are in a traditional mathematics class, sit near the front where you can see and hear well, where your focus is on the material being covered in class. Ask questions, be ready to contribute toward solutions, and take part in all classroom activities. Your contributions are valuable to the class and to yourself. Class participation requires an investment of yourself in the learning process, which you will find pays huge dividends.

If you are in an online class or nontraditional class, be sure to e-mail the teacher or talk to the tutor on duty. Ask questions. Think about the concepts. Make your mind interact with the textbook. Be mentally involved. This active mental interaction is the key to your success.

1.1 Exercises

Student Solutions Manual | CD/Video | PH Math Tutor Center | MathXL®Tutorials on CD | MathXL® | MyMathLab® | Interactmath.com

Write each number in expanded notation.

1. 6731
6000 + 700 + 30 + 1

2. 9519
9000 + 500 + 10 + 9

3. 108,276
100,000 + 8000 + 200 + 70 + 6

4. 350,765
300,000 + 50,000 + 700 + 60 + 5

5. 23,761,345
20,000,000 + 3,000,000 + 700,000 + 60,000 + 1000 + 300 + 40 + 5

6. 46,198,253
40,000,000 + 6,000,000 + 100,000 + 90,000 + 8000 + 200 + 50 + 3

7. 103,260,768
100,000,000 + 3,000,000 + 200,000 + 60,000 + 700 + 60 + 8

8. 820,310,574
800,000,000 + 20,000,000 + 300,000 + 10,000 + 500 + 70 + 4

Write each number in standard notation.

9. 600 + 70 + 1
671

10. 500 + 90 + 6
596

11. 9000 + 800 + 60 + 3
9863

12. 7000 + 600 + 50 + 2
7652

13. 40,000 + 800 + 80 + 5
40,885

14. 60,000 + 7000 + 200 + 4
67,204

15. 700,000 + 6000 + 200
706,200

16. 900,000 + 50,000 + 40 + 7
950,047

Verbal and Writing Skills

17. In the number 56,782
(a) What digit tells the number of hundreds?
7
(b) What is the value of the digit 5?
50,000

18. In the number 318,172
(a) What digit tells the number of thousands?
8
(b) What is the value of the digit 3?
300,000

19. In the number 1,214,847
(a) What digit tells the number of hundred thousands?
2
(b) What is the value of the digit?
200,000

20. In the number 6,789,345
(a) What digit tells the number of thousands?
9
(b) What is the value of the digit?
9000

Write a word name for each number.

21. 53
fifty-three

22. 46
forty-six

23. 9304
nine thousand, three hundred four

24. 7606
seven thousand, six hundred six

25. 36,118
thirty-six thousand, one hundred eighteen

26. 55,742
fifty-five thousand, seven hundred forty-two

27. 105,261
one hundred five thousand, two hundred sixty-one

28. 370,258
three hundred seventy thousand, two hundred fifty-eight

29. 14,203,326
fourteen million, two hundred three thousand, three hundred twenty-six

30. 68,089,213
sixty-eight million, eighty-nine thousand, two hundred thirteen

31. 4,302,156,200
four billion, three hundred two million, one hundred fifty-six thousand, two hundred

32. 7,436,210,400
seven billion, four hundred thirty-six million, two hundred ten thousand, four hundred

Write each number in standard notation.

33. one thousand, five hundred sixty-one
1561

34. three thousand, one hundred eighty-nine
3189

35. twenty-seven thousand, three hundred eighty-two
27,382

36. ninety-two thousand, four hundred four
92,404

37. one hundred million, seventy-nine thousand, eight hundred twenty-six
100,079,826

38. four hundred fifty million, three hundred thousand, two hundred forty-nine
450,300,249

Applications

When writing a check, a person must write the word name for the dollar amount of the check.

39. ***Personal Finance*** Alex bought new equipment for his laboratory for $1965. What word name should he write on the check?
one thousand, nine hundred sixty-five

40. ***Personal Finance*** Alex later bought a new personal computer for $6383. What word name should he write on the check?
six thousand, three hundred eighty-three

In exercises 41–44, use the following chart prepared with data from the U.S. Bureau of the Census. Notice that the second line tells us that the numbers represent millions. These values are only approximate values representing numbers written to the nearest million. They are not exact census figures.

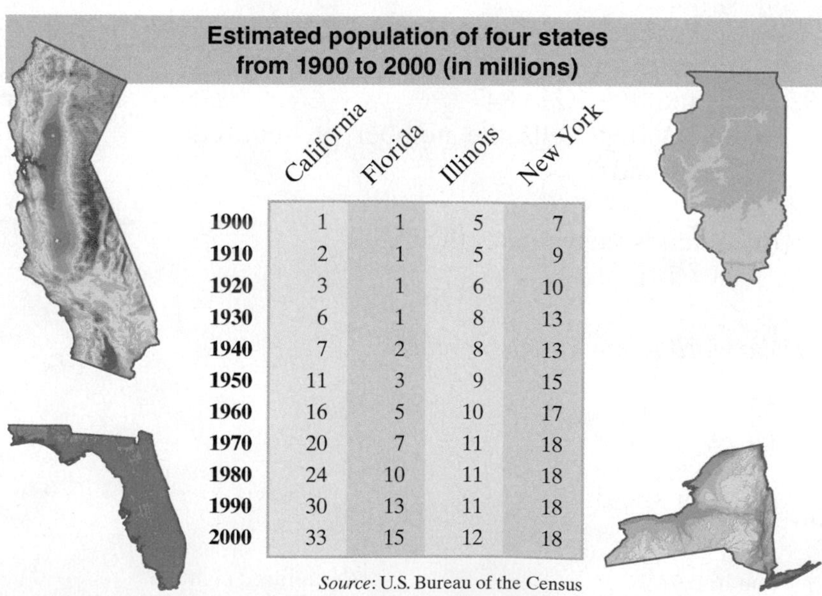

Estimated population of four states from 1900 to 2000 (in millions)

	California	Florida	Illinois	New York
1900	1	1	5	7
1910	2	1	5	9
1920	3	1	6	10
1930	6	1	8	13
1940	7	2	8	13
1950	11	3	9	15
1960	16	5	10	17
1970	20	7	11	18
1980	24	10	11	18
1990	30	13	11	18
2000	33	15	12	18

Source: U.S. Bureau of the Census

41. ***Historical Analysis*** What was the estimated population of New York in 1910?
9 million or 9,000,000

42. ***Historical Analysis*** What was the estimated population of Florida in 1970?
7 million or 7,000,000

43. ***Historical Analysis*** What was the estimated population of California in 2000?
33 million or 33,000,000

44. ***Historical Analysis*** What was the estimated population of Illinois in 1940?
8 million or 8,000,000

In exercises 45–48, use the following chart:

Number of Flights and Passengers for Selected Airlines in 1999, 2000, and 2005 (in thousands)

	1999		2000		2005**	
Airline	Flights*	Passengers	Flights*	Passengers	Flights	Passengers
American	740	72,567	791	77,185	780	75,300
Continental	428	40,059	423	40,989	401	39,520
Delta	930	101,843	922	101,809	900	98,360
Northwest	552	50,441	565	52,566	448	49,690

*Includes passenger and freight flights
**Estimated
Source: Bureau of Transportation Statistics

45. ***Airline Travel*** How many flights did Delta have in 1999?
930,000

46. ***Airline Travel*** How many passengers flew on American flights in 2000?
77,185,000

47. ***Airline Travel*** How many passengers flew on Northwest flights in 2000?
52,566,000

48. ***Airline Travel*** How many flights will Continental have in 2005?
401,000

49. ***Physics*** The speed of light is approximately 29,979,250,000 centimeters per second.
(a) What digit tells the number of ten thousands?
5
(b) What digit tells the number of ten billions?
2

50. ***Earth Science*** The radius of the earth is approximately 637,814,000 centimeters.
(a) What digit tells the number of thousands?
4
(b) What digit tells the number of millions?
7

51. ***Historical Analysis*** In 2000 849,807 immigrants came to the United States from other countries.
(a) Which digit tells the number of hundred thousands?
8
(b) Which digit tells the number of thousands?
9
(*Source:* U.S. Immigration and Naturalization Service.)

52. ***Historical Analysis*** The world's population is expected to reach 7,900,000,000 by the year 2020, according to the U.S. Bureau of the Census.
(a) Which digit tells the number of hundred millions?
9
(b) Which digit tells the number of billions?
7

53. Write in standard notation: six hundred thirteen trillion, one billion, thirty-three million, two hundred eight thousand, three.
613,001,033,208,003

54. Write in standard notation: nine hundred fourteen trillion, two billion, fifty-two million, four hundred nine thousand, six.
914,002,052,409,006

To Think About

55. Write a word name for 3,682,968,009,931,960,747. (*Hint:* The digit 1 followed by 18 zeros represents the number *1 quintillion.* 1 followed by 15 zeros represents the number *1 quadrillion.*)
three quintillion, six hundred eighty-two quadrillion, nine hundred sixty-eight trillion, nine billion, nine hundred thirty-one million, nine hundred sixty thousand, seven hundred forty-seven

56. The number 50,000,000,000,000,000,000 is represented on some scientific calculators as 5 E 19. We will cover this in more detail in a later chapter. However, for the present we can see that this is a convenient notation that allows us to record very large whole numbers. Note that this number (50 quintillion) is a 5 followed by 19 zeros. Write in standard form the number that would be represented on a calculator as 6 E 22.
60,000,000,000,000,000,000,000

57. Think about the discussion in exercise 56. If the number 4 E 20 represented on a scientific calculator was divided by 2, what number would be the result? Write your answer in standard form.
You would obtain 2 E 20. This is 200,000,000,000,000,000,000 in standard form.

58. Consider all the whole numbers between 200 and 800 that contain the digit 6. How many such numbers are there?
195

1.2 Adding Whole Numbers

1 Mastering Basic Addition Facts

We see the addition process time and time again. Carpenters add to find the amount of lumber they need for a job. Auto mechanics add to make sure they have enough parts in the inventory. Bank tellers add to get cash totals.

What is addition? We do addition when we put sets of objects together.

■■■■■ ■■■■■■■ ■■■■■■■■■■■■

5 objects + 7 objects = 12 objects

$$5 + 7 = 12$$

Usually when we add numbers, we put one number under the other in a column. The numbers being added are called **addends.** The result is called the **sum.**

Suppose that we have four pencils in the car and we bring three more pencils from home. How many pencils do we have with us now? We add 4 and 3 to obtain a value of 7. In this case, the numbers 4 and 3 are the addends and the answer 7 is the sum.

$$\begin{array}{rl} 4 & \text{addend} \\ +\,3 & \text{addend} \\ \hline 7 & \text{sum} \end{array}$$

Think about what we do when we add 0 to another number. We are not making a change, so whenever we add zero to another number, that number will be the sum. Since this is always true, this is called a *property.* Since the sum is identical to the number added to zero, this is called the **identity property of zero.**

EXAMPLE 1 Add.

(a) 8 + 5 **(b)** 3 + 7 **(c)** 9 + 0

Solution

(a) $\begin{array}{r} 8 \\ +\,5 \\ \hline 13 \end{array}$ **(b)** $\begin{array}{r} 3 \\ +\,7 \\ \hline 10 \end{array}$ **(c)** $\begin{array}{r} 9 \\ +\,0 \\ \hline 9 \end{array}$ ← *Note:* When we add zero to any other number, that number is the sum.

Practice Problem 1 Add.

(a) $\begin{array}{r} 7 \\ +\,5 \\ \hline \end{array}$ **(b)** $\begin{array}{r} 9 \\ +\,4 \\ \hline \end{array}$ **(c)** $\begin{array}{r} 3 \\ +\,0 \\ \hline \end{array}$

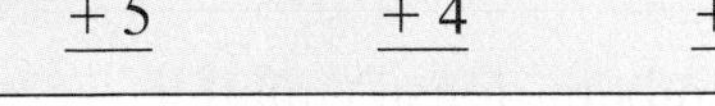

The following table shows the basic addition facts. You should know these facts. If any of the answers don't come to you quickly, now is the time to learn them. To check your knowledge try Exercises 1.2, exercises 3 and 4.

Student Learning Objectives

After studying this section, you will be able to:

1. **Master basic addition facts.**
2. **Add several single-digit numbers.**
3. **Add several-digit numbers when carrying is not needed.**
4. **Add several-digit numbers when carrying is needed.**
5. **Review the properties of addition.**
6. **Apply addition to real-life situations.**

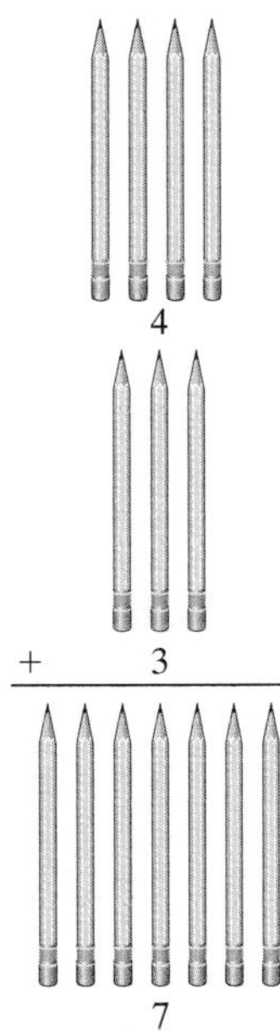

***NOTE TO STUDENT:** Fully worked-out solutions to all of the Practice Problems can be found at the back of the text starting at page SP-1*

Teaching Tip Some students will find that there are certain number facts that they do not know. For example, some students may not remember that $7 + 8 = 15$ but rather will remember that $7 + 7 = 14$ and then add one. Stress the fact that now is the time to learn all the basic addition facts by mastering the content of this addition table. Some students may need to make up flash cards of addition facts in order to master them or to improve their speed in mental addition.

Basic Addition Facts

+	0	1	2	3	4	5	6	7	8	9
0	0	1	2	3	4	5	6	7	8	9
1	1	2	3	4	5	6	7	8	9	10
2	2	3	4	5	6	7	8	9	10	11
3	3	4	5	6	7	8	9	10	11	12
4	4	5	6	7	8	9	10	11	12	13
5	5	6	7	8	9	10	11	12	13	14
6	6	7	8	9	10	11	12	13	14	15
7	7	8	9	10	11	12	13	14	15	16
8	8	9	10	11	12	13	14	15	16	17
9	9	10	11	12	13	14	15	16	17	18

To use the table to find the sum $4 + 7$, read across the top of the table to the 4 column, and then read down the left to the 7 row. The box where the 4 and 7 meet is 11, which means that $4 + 7 = 11$. Now read across the top to the 7 column and down the left to the 4 row. The box where these numbers meet is also 11. We can see that the order in which we add the numbers does not change the sum. $4 + 7 = 11$, and $7 + 4 = 11$. We call this the **commutative property of addition.**

This property does not hold true for everything in our lives. When you put on your socks and then your shoes, the result is not the same as if you put on your shoes first and then your socks! Can you think of any other examples where changing the order in which you add things would change the result?

2 Adding Several Single-Digit Numbers

If more than two numbers are to be added, we usually add from the first number to the next number and mentally note the sum. Then we add that sum to the next number, and so on.

EXAMPLE 2 Add. $3 + 4 + 8 + 2 + 5$

Solution We rewrite the addition problem in a column format.

Mentally, we do these steps.

$$\begin{array}{r} 3 \\ 4 \\ 8 \\ 2 \\ +5 \\ \hline 22 \end{array} \quad 3 + 4 = 7 \quad 7 + 8 = 15 \quad 15 + 2 = 17 \quad 17 + 5 = 22$$

Practice Problem 2 Add. $7 + 6 + 5 + 8 + 2$

NOTE TO STUDENT: Fully worked-out solutions to all of the Practice Problems can be found at the back of the text starting at page SP-1

Because the order in which we add numbers doesn't matter, we can choose to add from the top down, from the bottom up, or in any other way. One shortcut is to add first any numbers that will give a sum of 10, or 20, or 30, and so on.

EXAMPLE 3 Add.

$$\begin{array}{r} 3 \\ 4 \\ 8 \\ 2 \\ +\,6 \\ \hline \end{array}$$

Solution We mentally group the numbers into tens.

$$\begin{array}{l} 3 \\ 4 \\ 8 \\ 2 \\ 6 \end{array}$$

8 + 2 = 10 → 4 + 6 = 10

The sum is 10 + 10 + 3 or 23.

Practice Problem 3 Add. 1 + 7 + 2 + 9 + 3

3 Adding Several-Digit Numbers When Carrying Is Not Needed

Of course, many numbers that we need to add have more than one digit. In such cases, we must be careful to first add the digits in the ones column, then the digits in the tens column, then those in the hundreds column, and so on. Notice that we move from *right to left*.

EXAMPLE 4 Add. 4304 + 5163

Solution

$$\begin{array}{r} 4\ 3\ 0\ 4 \\ +\,5\ 1\ 6\ 3 \\ \hline 9\ 4\ 6\ 7 \end{array}$$

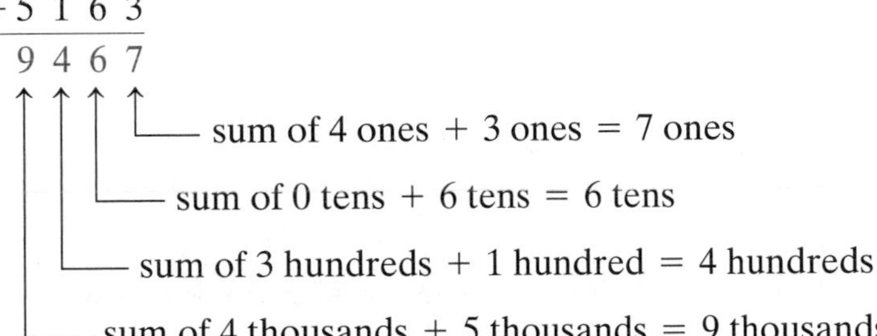

Practice Problem 4 Add.

$$\begin{array}{r} 8246 \\ +\,1702 \\ \hline \end{array}$$

4 Adding Several-Digit Numbers When Carrying Is Needed

When you add several whole numbers, often the sum in a column is greater than 9. However, we can only use *one* digit in any one place. What do we do with a two-digit sum? Look at the following example.

EXAMPLE 5 Add. 45 + 37

Solution

$$\begin{array}{r} {}^{1} \\ 4\ 5 \\ +3\ 7 \\ \hline 2 \end{array}$$

5 ones and 7 ones = 12.
We rename 12 in expanded notation: 1 ten + 2 ones.
We place the 2 ones in the ones column.
We carry the 1 ten over to the tens column.

Note: Placing the 1 in the next column is often called "carrying the one."

$$\begin{array}{r} {}^{1} \\ 4\ 5 \\ +3\ 7 \\ \hline 8\ 2 \end{array}$$

Now we can add the digits in the tens column.

Thus, 45 + 37 = 82.

Practice Problem 5 Add. $\begin{array}{r} 56 \\ +36 \\ \hline \end{array}$

NOTE TO STUDENT: ***Fully worked-out solutions to all of the Practice Problems can be found at the back of the text starting at page SP-1***

Often you must use carrying several times by bringing the left digit into the next column to the left.

Teaching Tip Remind students that when they carry a digit such as in Example 6, they may write down the digit they are carrying. Some students were probably criticized in elementary school for showing the carrying step. In college, students should feel free to write down the carrying step if it is needed. Of course, if students can do that part in their heads, there is no need to write down the carrying digit.

EXAMPLE 6 Add. 257 + 688 + 94

Solution

Thousands Column, Hundreds Column, Tens Column, Ones Column

$$\begin{array}{r} {}^{2\ 1} \\ 2\ 5\ 7 \\ 6\ 8\ 8 \\ +\ \ 9\ 4 \\ \hline 1\ 0\ 3\ 9 \end{array}$$

In the ones column we add 7 + 8 + 4 = 19. Because 19 is 1 ten and 9 ones, we place 9 in the ones column and carry 1 to the top of the tens column.

In the tens column we add 1 + 5 + 8 + 9 = 23. Because 23 tens is 2 hundreds and 3 tens, we place the 3 in the tens column and carry 2 to the top of the hundreds column.

In the hundreds column we add 2 + 2 + 6 = 10 hundreds. Because 10 hundreds is 1 thousand and 0 hundreds, we place the 0 in the hundreds column and place the 1 in the thousands column.

Practice Problem 6 Add. 789 + 63 + 297

We can add numbers in more than one way. To add $5 + 3 + 7$ we can first add the 5 and 3. We do this by using parentheses to show the first operation to be done. This shows us that $5 + 3$ is to be grouped together.

$$\begin{aligned} 5 + 3 + 7 &= (5 + 3) + 7 = 15 \\ &= \quad 8 \quad + 7 = 15 \end{aligned}$$

We could add the 3 and 7 first. We use parentheses to show that we group $3 + 7$ together and that we will add these two numbers first.

$$\begin{aligned} 5 + 3 + 7 &= 5 + (3 + 7) = 15 \\ &= 5 + \quad 10 \quad = 15 \end{aligned}$$

The way we group numbers to be added does not change the sum. This property is called the **associative property of addition.**

5 Reviewing the Properties of Addition

Look again at the three properties of addition we have discussed in this section.

1. **Associative Property of Addition** When we add three numbers, we can group them in any way.
$$\begin{aligned} (8 + 2) + 6 &= 8 + (2 + 6) \\ 10 + 6 &= 8 + 8 \\ 16 &= 16 \end{aligned}$$
2. **Commutative Property of Addition** Two numbers can be added in either order with the same result.
$$\begin{aligned} 5 + 12 &= 12 + 5 \\ 17 &= 17 \end{aligned}$$
3. **Identity Property of Zero** When zero is added to a number, the sum is that number.
$$\begin{aligned} 8 + 0 &= 8 \\ 0 + 5 &= 5 \end{aligned}$$

Because of the commutative and associative properties of addition, we can check our addition by adding the numbers in the opposite order.

Teaching Tip Some students lack confidence that they will be able to find their own errors. As a classroom activity, have students add $258 + 167 + 879$. Then have them add $879 + 167 + 258$. The sum is 1304. If you ask students how many of them made an error and detected it by adding the numbers in the opposite order and getting a different answer, there will usually be several students in the class who raise their hands.

EXAMPLE 7

(a) Add the numbers. $39 + 7284 + 3132$

(b) Check by reversing the order of addition.

Solution

(a)
$$\begin{array}{r} \scriptstyle 11 \\ 39 \\ 7284 \\ +3132 \\ \hline 10{,}455 \end{array}$$
Addition

(b)
$$\begin{array}{r} \scriptstyle 11 \\ 3132 \\ 7284 \\ +\quad 39 \\ \hline 10{,}455 \end{array}$$
Check by reversing the order.

The sum is the same in each case.

Practice Problem 7

(a) Add.
$$\begin{array}{r} 127 \\ 9876 \\ +\ 342 \\ \hline \end{array}$$

(b) Check by reversing the order.
$$\begin{array}{r} 342 \\ 9876 \\ +\ 127 \\ \hline \end{array}$$

 Applying Addition to Real-Life Situations

We use addition in all kinds of situations. There are several key words in word problems that imply addition. For example, it may be stated that there are 12 math books, 9 chemistry books, and 8 biology books on a book shelf. To find the *total* number of books implies that we add the numbers 12 + 9 + 8. Other key words are *how much, how many,* and *all.*

Sometimes a problem will have more information than you will need to answer the question. If you have too much information, to solve the problem you will need to separate out the facts that are not important. The following three steps are involved in the problem-solving process.

Step 1 Understand the problem.
Step 2 Calculate and state the answer.
Step 3 Check.

We may not write all of these steps down, but they are the steps we use to solve all problems.

EXAMPLE 8 The bookkeeper for Smithville Trucking was examining the following data for the company checking account.

Monday:	\$23,416 was deposited and \$17,389 was debited.
Tuesday:	\$44,823 was deposited and \$34,089 was debited.
Wednesday:	\$16,213 was deposited and \$20,057 was debited.

What was the total of all deposits during this period?

Solution

Step 1 *Understand the problem.*

Total implies that we will use addition. Since we don't need to know about the debits to answer this question, we use only the *deposit* amounts.

Step 2 *Calculate and state the answer.*

		11 1
Monday:	\$23,416 was deposited.	23,416
Tuesday:	\$44,823 was deposited.	44,823
Wednesday:	\$16,213 was deposited.	+ 16,213
		84,452

A total of \$84,452 was deposited on those three days.

Step 3 *Check.*

You may add the numbers in reverse order to check. We leave the check up to you.

***NOTE TO STUDENT:** Fully worked-out solutions to all of the Practice Problems can be found at the back of the text starting at page SP-1*

Practice Problem 8 North University has 23,413 men and 18,316 women. South University has 19,316 men and 24,789 women. East University has 20,078 men and 22,965 women. What is the total enrollment of *women* at the three universities?

EXAMPLE 9 Mr. Ortiz has a rectangular field whose length is 400 feet and whose width is 200 feet. What is the total number of feet of fence that would be required to fence in the field?

Solution

1. *Understand the problem.*

To help us to get a picture of what the field looks like, we will draw a diagram.

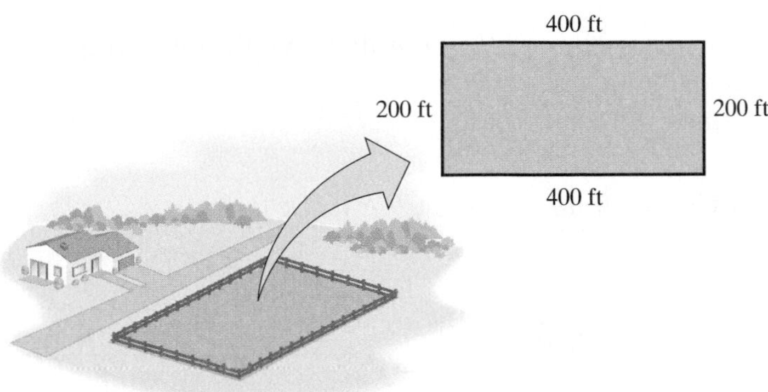

Note that ft is the abbreviation for feet. ft means feet.

2. *Calculate and state the answer.*

Since the fence will be along each side of the field, we add the lengths all around the field.

$$\begin{array}{r} 200 \\ 400 \\ 200 \\ +\,400 \\ \hline 1200 \end{array}$$

The amount of fence that would be required is 1200 feet.

3. *Check.*

Regroup the addends and add.

$$\begin{array}{r} 200 \\ 200 \\ 400 \\ +\,400 \\ \hline 1200 \end{array} \checkmark$$

Practice Problem 9 In Vermont, Gretchen fenced the rectangular field on which her sheep graze. The length of the field is 2000 feet and the width of the field is 1000 feet. What is the perimeter of the field? (*Hint:* The "distance around" an object [such as a field] is called the *perimeter.*)

Teaching Tip Stress the idea of finding the perimeter of an object by adding up the lengths of all the sides. You may want to give an example of a four-sided field that has four different lengths to stress that this concept works for figures other than rectangles.

Developing Your Study Skills

Getting Organized for an Exam

Studying adequately for an exam requires careful preparation. Begin early so that you will be able to spread your review over several days. Even though you may still be learning new material at this time, you can be reviewing concepts previously learned in the chapter. Giving yourself plenty of time for review will take the pressure off. You need this time to process what you have learned and to tie concepts together.

Adequate preparation enables you to feel confident and to think clearly with less tension and anxiety.

1.2 EXERCISES

Student Solutions Manual

CD/ Video

PH Math Tutor Center

MathXL®Tutorials on CD

MathXL®

MyMathLab®

Interactmath.com

Verbal and Writing Skills

1. Explain in your own words. Answers may vary. Samples are below.

(a) the commutative property of addition

You can change the order of the addends without changing the sum.

(b) the associative property of addition

You can group the addends in any way without changing the sum.

2. When zero is added to any number, it does not change that number. Why do you think this is called the identity property of zero?

When zero is added to any number, the sum is identical to that number.

Complete the addition facts for each table. Strive for total accuracy, but work quickly. Allow a maximum of five minutes for each table.

3.

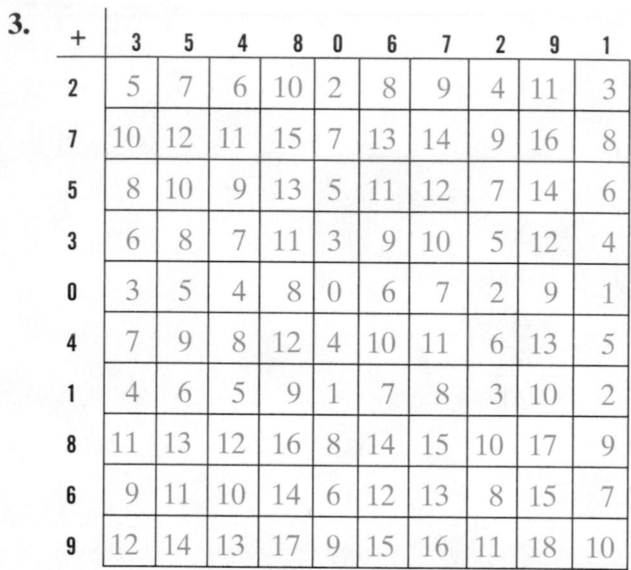

+	3	5	4	8	0	6	7	2	9	1
2	5	7	6	10	2	8	9	4	11	3
7	10	12	11	15	7	13	14	9	16	8
5	8	10	9	13	5	11	12	7	14	6
3	6	8	7	11	3	9	10	5	12	4
0	3	5	4	8	0	6	7	2	9	1
4	7	9	8	12	4	10	11	6	13	5
1	4	6	5	9	1	7	8	3	10	2
8	11	13	12	16	8	14	15	10	17	9
6	9	11	10	14	6	12	13	8	15	7
9	12	14	13	17	9	15	16	11	18	10

4.

+	1	6	5	3	0	9	4	7	2	8
3	4	9	8	6	3	12	7	10	5	11
9	10	15	14	12	9	18	13	16	11	17
4	5	10	9	7	4	13	8	11	6	12
0	1	6	5	3	0	9	4	7	2	8
2	3	8	7	5	2	11	6	9	4	10
7	8	13	12	10	7	16	11	14	9	15
8	9	14	13	11	8	17	12	15	10	16
1	2	7	6	4	1	10	5	8	3	9
6	7	12	11	9	6	15	10	13	8	14
5	6	11	10	8	5	14	9	12	7	13

Add.

5. $\begin{array}{r} 4 \\ 2 \\ 8 \\ +9 \\ \hline 23 \end{array}$

6. $\begin{array}{r} 4 \\ 6 \\ 2 \\ +7 \\ \hline 19 \end{array}$

7. $\begin{array}{r} 2 \\ 6 \\ 7 \\ 8 \\ +3 \\ \hline 26 \end{array}$

8. $\begin{array}{r} 1 \\ 5 \\ 5 \\ 9 \\ +9 \\ \hline 29 \end{array}$

9. $\begin{array}{r} 18 \\ 36 \\ +\ 3 \\ \hline 57 \end{array}$

10. $\begin{array}{r} 63 \\ 11 \\ +\ 6 \\ \hline 80 \end{array}$

11. $\begin{array}{r} 63 \\ 24 \\ +12 \\ \hline 99 \end{array}$

12. $\begin{array}{r} 54 \\ 21 \\ +23 \\ \hline 98 \end{array}$

13. $\begin{array}{r} 2847 \\ 1634 \\ +\ \ 98 \\ \hline 4579 \end{array}$

14. $\begin{array}{r} 5519 \\ 1392 \\ +\ 955 \\ \hline 7866 \end{array}$

15. $\begin{array}{r} 5631 \\ 2344 \\ +2019 \\ \hline 9994 \end{array}$

16. $\begin{array}{r} 5017 \\ 2984 \\ +1328 \\ \hline 9329 \end{array}$

17. $\begin{array}{r} 8235 \\ +5626 \\ \hline 13{,}861 \end{array}$

18. $\begin{array}{r} 6753 \\ +3265 \\ \hline 10{,}018 \end{array}$

19. $\begin{array}{r} 62{,}504 \\ +54{,}736 \\ \hline 117{,}240 \end{array}$

20. $\begin{array}{r} 83{,}596 \\ +56{,}384 \\ \hline 139{,}980 \end{array}$

Add from the top. Then check by adding in the reverse order.

21. 36, 41, 25, 6, + 13 = 121

22. 24, 39, 16, 14, + 9 = 102

23. 207, 15, 3, 57, + 861 = 1143

24. 426, 39, 6, 52, + 802 = 1325

Add.

25. 85 + 256 + 55 + 9734 = 10,130

26. 582 + 1674 + 336 + 8458 = 11,050

27. 1,362,214 + 7,002,316 + 3,214,896 = 11,579,426

28. 4,002,983 + 2,134,702 + 3,592,001 = 9,729,686

29. 837,241,000 + 298,039,240 = 1,135,280,240

30. 982,306,000 + 583,215,320 = 1,565,521,320

31. 516,208 + 24,317 + 1,763,295 = 2,303,820

32. 32,500 + 763,420 + 2,837,667 = 3,633,587

33. 75 + 132 + 25 + 51 283

34. 502 + 80 + 20 + 43 645

35. 15,216 + 485 + 5208 20,909

36. 26,002 + 599 + 3500 30,101

Applications

37. ***Consumer Mathematics*** Nadine went furniture shopping for her new home. She bought a couch for \$345, a table for \$288, and a lamp for \$74. What was the total cost of the items Nadine purchased? \$707

38. ***Consumer Mathematics*** Emily sold her handmade jewelry at three art shows during the month of July. She made \$236 at the first show, \$315 at the second, and \$275 at the third. What was the total amount of money Emily made in July? \$826

39. ***Consumer Mathematics*** Arlen sells used cars. Last week he sold cars for \$4550, \$9200, and \$6875. How much was the total amount of the three cars? \$20,625

40. ***Personal Finance*** The taxes on Jack's house two years ago were \$4658. Last year they were \$5222. This year they are \$6027. What is the total amount for three years? \$15,907

▲ **41.** ***Geometry*** Nate wants to put a fence around his backyard. The sketch below indicates the length of each side of the yard. What is the total number of feet of fence he needs for his backyard? 468 feet

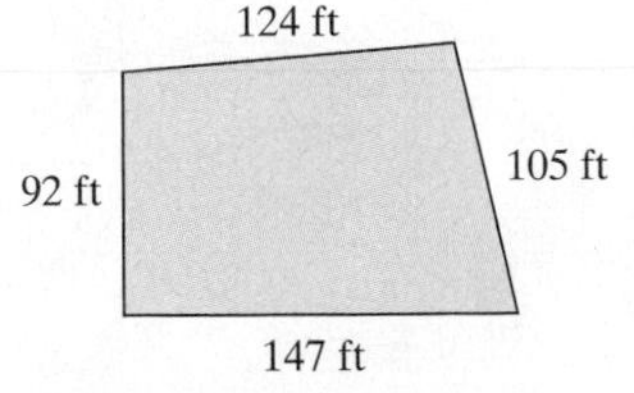

▲ **42.** ***Geometry*** Jessica has a field with the length of each side as labeled on the sketch. What is the total number of feet of fence that would be required to fence in the field? (Find the perimeter of the field.) 2335 feet

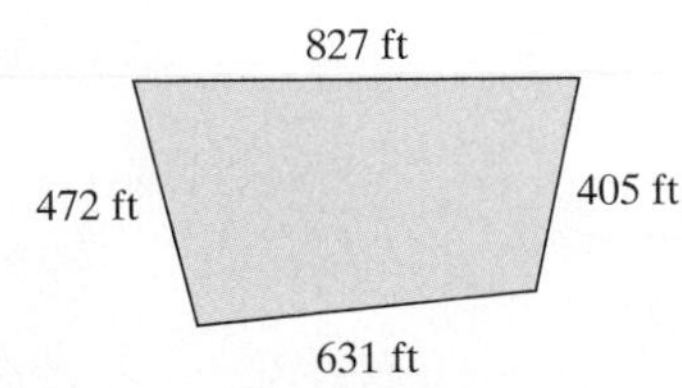

▲ **43.** ***Geography*** The Pacific Ocean, the world's largest, has an area of 64,000,000 square miles. The Atlantic Ocean has an area of 31,800,000 square miles. The Indian Ocean has an area of 25,300,000 square miles. What is the total area for these oceans? 121,100,000 square miles

▲ **44.** ***Geography*** The Arctic Ocean has an area of 5,400,000 square miles. The Mediterranean Sea has an area of 1,100,000 square miles. The Caribbean Sea has an area of 1,000,000 square miles. What is the total area for these bodies of water? 7,500,000 square miles

45. ***Geography*** As of 2003, the three most populated counties in Arizona are Maricopa with 3,072,149 people, Pima with 843,746, and Pinal with 179,727. What is the total population of these three counties? 4,095,622

▲ **46.** ***Geography*** Lake Huron, which is bordered by the United States and Canada, measures 23,010 square miles. Lake Tanganyika, which borders Tanzania and Zaire, measures 12,700 square miles. Lake Baikal in Russia measures 12,162 square miles. What is the total area for these lakes? 47,872 square miles

In exercises 47–48, be sure you understand the problem and then choose the numbers you need in order to answer each question. Then solve the problem.

47. ***Education*** The admissions department of a competitive university is reviewing applications to see whether students are *eligible* or *ineligible* for student aid. On Monday, 415 were found eligible and 27 ineligible. On Tuesday, 364 were found eligible and 68 ineligible. On Wednesday, 159 were found eligible and 102 ineligible. On Thursday, 196 were found eligible and 61 ineligible.

(a) How many students were eligible for student aid over the four days? 1134 students

(b) How many students were considered in all? 1392 students

48. ***Manufacturing*** The quality control division of a motorcycle company classifies the final assembled bike as *passing* or *failing* final inspection. In January, 14,311 vehicles passed whereas 56 failed. In February, 11,077 passed and 158 failed. In March, 12,580 passed and 97 failed.

(a) How many motorcycles passed the inspection during the three months? 37,968 motorcycles

(b) How many motorcycles were assembled during the three months in all? 38,279 motorcycles

Use the following facts to solve exercises 49 and 50. It is 87 miles from Springfield to Weston. It is 17 miles from Weston to Boston. Driving directly, it is 98 miles from Springfield to Boston. It is 21 miles from Boston to Hamilton.

49. ***Geography*** If Melissa drives from Springfield to Weston, then from Weston to Boston, and finally directly home to Springfield, how many miles does she drive? 202 miles

50. ***Geography*** If Marcia drives from Hamilton to Boston, then from Boston to Weston, and then from Weston to Springfield, how many miles does she drive? 125 miles

▲ **51.** ***Geometry*** Walter Swensen is examining the fences of a farm in Caribou, Maine. The shape of one field is in the shape of a four-sided figure with no sides equal. The field is enclosed with 2387 feet of wooden rail fence. The first side is 568 feet long, while the second side is 682 feet long. The third side is 703 feet long. How long is the fourth side? 434 feet

▲ **52.** ***Geometry*** Carlos Sontera is walking to examine the fences of a ranch in El Paso, Texas. The field he is examining is in the shape of a rectangle. The perimeter of the rectangle is 3456 feet. One side of the rectangle is 930 feet long. How long are the other sides? (*Hint:* The opposite sides of a rectangle are equal.) Two sides are 930 feet long and two sides are 798 feet long.

53. ***Personal Finance*** Answer using the information in the following Western University expense chart for the current academic year.

Western University Yearly Expenses	In-State Student, U.S. Citizen	Out-of-State Student, U.S. Citizen	Foreign Student
Tuition	$3640	$5276	$8352
Room	1926	2437	2855
Board	1753	1840	1840

How much is the total cost for tuition, room, and board for
(a) an out-of-state U.S. citizen? $9553
(b) an in-state U.S. citizen? $7319
(c) a foreign student? $13,047

To Think About

In exercises 54–55, add.

54. 2,368,521,788 + 5,721,368,701 + 4,027,399,206
12,117,289,695

55. 89 + 166 + 23 + 45 + 72 + 190 + 203 + 77 + 18 + 93 + 46 + 73 + 66
1161

56. What would happen if addition were not commutative?
Answers may vary. A sample is: You could not add the addends in reverse order to check the addition.

57. What would happen if addition were not associative?
Answers may vary. A sample is: You could not group the addends in groups that sum to 10s to make column addition easier.

Cumulative Review

Write the word name for each number.

58. 76,208,941
seventy-six million, two hundred eight thousand, nine hundred forty-one

59. 121,000,374
one hundred twenty-one million, three hundred seventy-four

Write each number in standard notation.

60. eight million, seven hundred twenty-four thousand, three hundred ninety-six
8,724,396

61. nine million, fifty-one thousand, seven hundred nineteen
9,051,719

62. twenty-eight million, three hundred eighty-seven thousand, eighteen
28,387,018

1.3 Subtracting Whole Numbers

Student Learning Objectives

After studying this section, you will be able to:

1. **Master basic subtraction facts.**
2. **Subtract whole numbers when borrowing is not necessary.**
3. **Subtract whole numbers when borrowing is necessary.**
4. **Check the answer to a subtraction problem.**
5. **Apply subtraction to real-life situations.**

1 Mastering Basic Subtraction Facts

Subtraction is used day after day in the business world. The owner of a bakery placed an ad for his cakes in a local newspaper to see if this might increase his profits. To learn how many cakes had been sold, at closing time he subtracted the number of cakes remaining from the number of cakes the bakery had when it opened. To figure his profits, he subtracted his costs (including the cost of the ad) from his sales. Finally, to see if the ad paid off, he subtracted the profits he usually made in that period from the profits after advertising. He needed subtraction to see whether it paid to advertise.

What is subtraction? We do subtraction when we take objects away from a group. If you have 12 objects and take away 3 of them, 9 objects remain.

12 objects − 3 objects = 9 objects

$12 - 3 = 9$

If you earn \$400 per month, but have \$100 taken out for taxes, how much do you have left?

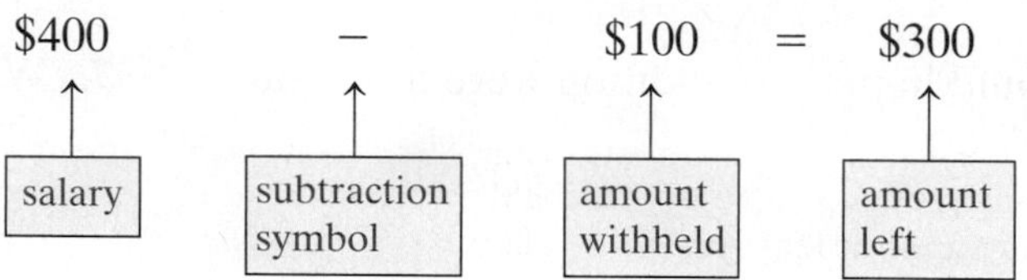

We can use addition to help with a subtraction problem.

To subtract: $200 - 196 =$ what number

We can think: $196 +$ what number $= 200$

Usually when we subtract numbers, we put one number under the other in a column. When we subtract one number from another, the answer is called the **difference.**

$$\begin{array}{r}9\\-2\\\hline 7\end{array}\qquad\begin{array}{r}8\\-3\\\hline 5\end{array}\qquad\begin{array}{r}12\\-\ 6\\\hline 6\end{array}\qquad\begin{array}{r}17\\-\ 9\\\hline 8\end{array}$$

Each of these is called the difference of the two numbers.

The other two parts of a subtraction problem have labels, although you will not often come across them. The number being subtracted is called the **subtrahend.** The number being subtracted from is called the **minuend.**

$$\begin{array}{rl}17 & \text{minuend}\\-\ 9 & \text{subtrahend}\\\hline 8 & \text{difference}\end{array}$$

In this case; the number 17 is called the *minuend.* The number 9 is called the *subtrahend.* The number 8 is called the *difference.*

QUICK RECALL OF SUBTRACTION FACTS It is helpful if you can subtract quickly. See if you can do Example 1 correctly in 15 seconds or less. Repeat again with Practice Problem 1. Strive to obtain all answers correctly in 15 seconds or less.

EXAMPLE 1 Subtract.

(a) 8 − 2 **(b)** 13 − 5 **(c)** 12 − 4

(d) 15 − 8 **(e)** 16 − 0

Solution

(a) $\begin{array}{r} 8 \\ -2 \\ \hline 6 \end{array}$ **(b)** $\begin{array}{r} 13 \\ -5 \\ \hline 8 \end{array}$ **(c)** $\begin{array}{r} 12 \\ -4 \\ \hline 8 \end{array}$

(d) $\begin{array}{r} 15 \\ -8 \\ \hline 7 \end{array}$ **(e)** $\begin{array}{r} 16 \\ -0 \\ \hline 16 \end{array}$

Practice Problem 1 Subtract.

(a) $\begin{array}{r} 9 \\ -6 \\ \hline \end{array}$ **(b)** $\begin{array}{r} 12 \\ -5 \\ \hline \end{array}$ **(c)** $\begin{array}{r} 17 \\ -8 \\ \hline \end{array}$ **(d)** $\begin{array}{r} 14 \\ -0 \\ \hline \end{array}$ **(e)** $\begin{array}{r} 18 \\ -9 \\ \hline \end{array}$

NOTE TO STUDENT: Fully worked-out solutions to all of the Practice Problems can be found at the back of the text starting at page SP-1

2 Subtracting Whole Numbers when Borrowing Is Not Necessary

When we subtract numbers with more than two digits, in order to keep track of our work, we line up the ones column, the tens column, the hundreds column, and so on. Note that we begin with the ones column, and move from right to left.

EXAMPLE 2 Subtract. 9867 − 3725

Solution

$$\begin{array}{r} 9\,8\,6\,7 \\ -3\,7\,2\,5 \\ \hline 6\,1\,4\,2 \end{array}$$

7 ones − 5 ones = 2 ones

6 tens − 2 tens = 4 tens

8 hundreds − 7 hundreds = 1 hundred

9 thousands − 3 thousands = 6 thousands

Practice Problem 2 Subtract. 7695 − 3481

 Subtracting Whole Numbers when Borrowing Is Necessary

In the subtraction that we have looked at so far, each digit in the upper number (the minuend) has been greater than the digit in the lower number (the subtrahend) for each place value. Many times, however, a digit in the lower number is greater than the digit in the upper number for that place value.

$$\begin{array}{r} 42 \\ -\ 28 \\ \hline \end{array}$$

The digit in the ones place in the lower number, the 8 of 28, is greater than the number in the ones place in the upper number, the 2 of 42. To subtract, we must *rename* 42, using place values. This is called **borrowing.**

EXAMPLE 3 Subtract. 42 − 28

Solution

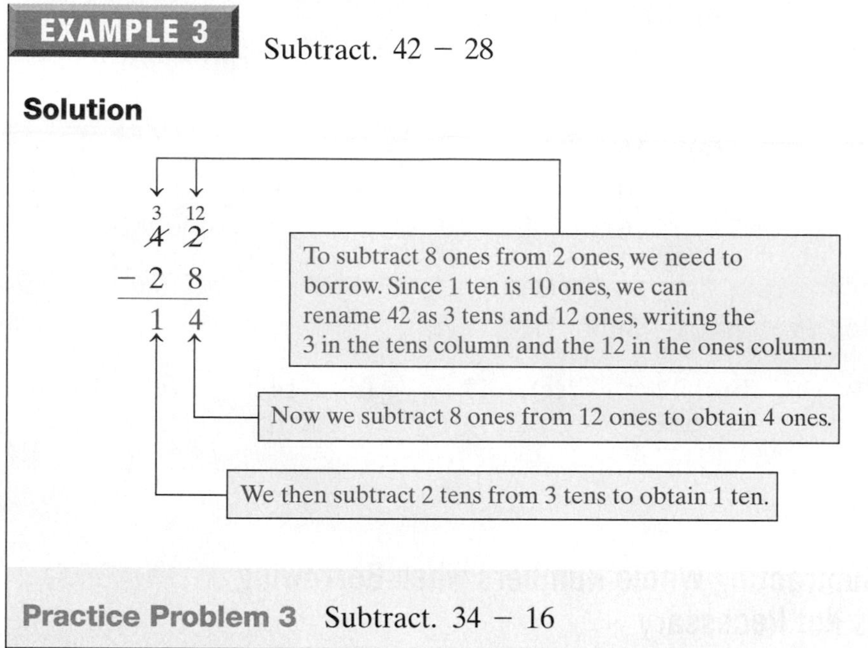

3 12
4 2
− 2 8
1 4
To subtract 8 ones from 2 ones, we need to borrow. Since 1 ten is 10 ones, we can rename 42 as 3 tens and 12 ones, writing the 3 in the tens column and the 12 in the ones column.
Now we subtract 8 ones from 12 ones to obtain 4 ones.
We then subtract 2 tens from 3 tens to obtain 1 ten.

***NOTE TO STUDENT:** Fully worked-out solutions to all of the Practice Problems can be found at the back of the text starting at page SP-1*

Practice Problem 3 Subtract. 34 − 16

EXAMPLE 4 Subtract. 864 − 548

Solution

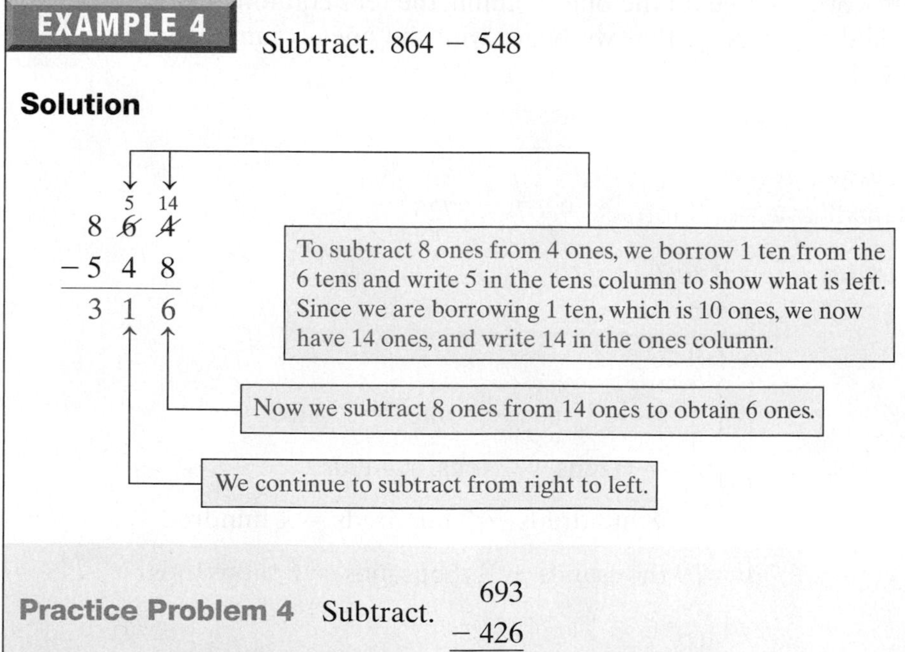

5 14
8 6 4
− 5 4 8
3 1 6
To subtract 8 ones from 4 ones, we borrow 1 ten from the 6 tens and write 5 in the tens column to show what is left. Since we are borrowing 1 ten, which is 10 ones, we now have 14 ones, and write 14 in the ones column.
Now we subtract 8 ones from 14 ones to obtain 6 ones.
We continue to subtract from right to left.

Practice Problem 4 Subtract.
$$\begin{array}{r} 693 \\ -\ 426 \\ \hline \end{array}$$

EXAMPLE 5 Subtract. 8040 − 6375

Solution

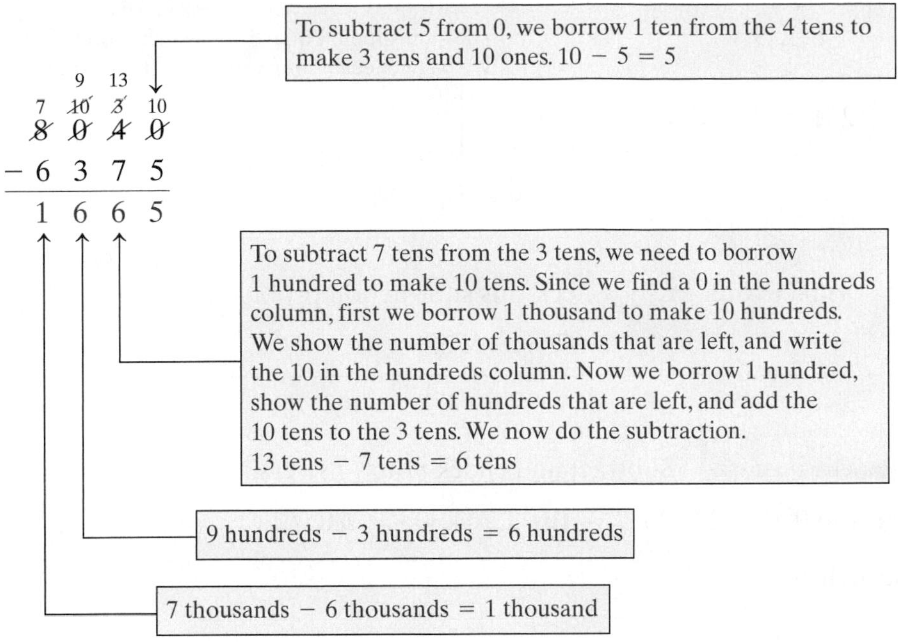

To subtract 5 from 0, we borrow 1 ten from the 4 tens to make 3 tens and 10 ones. 10 − 5 = 5
9 13
7 10 3 10
8 0 4 0
− 6 3 7 5
1 6 6 5
To subtract 7 tens from the 3 tens, we need to borrow 1 hundred to make 10 tens. Since we find a 0 in the hundreds column, first we borrow 1 thousand to make 10 hundreds. We show the number of thousands that are left, and write the 10 in the hundreds column. Now we borrow 1 hundred, show the number of hundreds that are left, and add the 10 tens to the 3 tens. We now do the subtraction. 13 tens − 7 tens = 6 tens
9 hundreds − 3 hundreds = 6 hundreds
7 thousands − 6 thousands = 1 thousand

Teaching Tip After discussing Example 5 or a similar problem, have the students subtract 6087 − 5793 as a classroom activity. If they did not obtain the correct answer of 294, they probably made an error in borrowing. Show the correct steps for borrowing in this case.

Practice Problem 5 Subtract. 9070 − 5886

EXAMPLE 6 Subtract.

(a) 9521 − 943 **(b)** 40,000 − 29,056

Solution

(a)
```
    14 11
  8  4  1 11
  9  5  2  1
 -   9  4  3
  8  5  7  8
```

(b)
```
  3  9   9  9 10
  4  0,  0  0  0
 -2  9,  0  5  6
  1  0,  9  4  4
```

Practice Problem 6 Subtract.

(a)
```
  8964
 - 985
```

(b)
```
  50,000
 -32,508
```

4 Checking the Answer to a Subtraction Problem

We observe that when 9 − 7 = 2 it follows that 7 + 2 = 9. Each subtraction problem is equivalent to a corresponding addition problem. This gives us a convenient way to check our answers to subtraction.

Teaching Tip Remind students that addition can usually be done in a lot less time than subtraction. Therefore, the checking step of adding to verify the subtraction is easier and takes less time than the original problem.

EXAMPLE 7 Check this subtraction problem.

$$5829 - 3647 = 2182$$

Solution

$$\begin{array}{r} 5829 \\ -3647 \\ \hline 2182 \end{array} \quad \text{then} \quad \begin{array}{r} 3647 \\ +2182 \\ \hline 5829 \end{array}$$

The sum should equal 5829, which it does. We have checked our work, and it is correct.

Practice Problem 7 Check this subtraction problem.

$$9763 - 5732 = 4031$$

NOTE TO STUDENT: Fully worked-out solutions to all of the Practice Problems can be found at the back of the text starting at page SP-1

EXAMPLE 8 Subtract and check your answers.

(a) 156,000 − 29,326 **(b)** 1,264,308 − 1,057,612

Solution

(a)

$$\begin{array}{r} 156{,}000 \\ -\ 29{,}326 \\ \hline 126{,}674 \end{array} \qquad \begin{array}{r} 29{,}326 \\ +\ 126{,}674 \\ \hline 156{,}000 \end{array}$$

It checks.

(b)

$$\begin{array}{r} 1{,}264{,}308 \\ -\ 1{,}057{,}612 \\ \hline 206{,}696 \end{array} \qquad \begin{array}{r} 1{,}057{,}612 \\ +\ 206{,}696 \\ \hline 1{,}264{,}308 \end{array}$$

It checks.

Practice Problem 8 Subtract and check your answers.

(a) $\begin{array}{r} 284{,}000 \\ -\ 96{,}327 \\ \hline \end{array}$ **(b)** $\begin{array}{r} 8{,}526{,}024 \\ -6{,}397{,}518 \\ \hline \end{array}$

Teaching Tip Taking the time to emphasize the idea of a variable in very simple terms in such problems as $10 = 4 + x$ will make the use of variables in later chapters much easier for the students to learn.

Subtraction can be used to solve word problems. Some problems can be expressed (and solved) with an **equation.** An equation is a number sentence with an equal sign, such as

$$10 = 4 + x$$

Here we use the letter x to represent a number we do not know. When we write $10 = 4 + x$, we are stating that 10 is equal to 4 added to some other number. Since $10 - 4 = 6$, we would assume that the number is 6. If we substitute 6 for x in the equation, we have two values that are the same.

$10 = 4 + x$

$10 = 4 + 6$ Substitute 6 for x.

$10 = 10$ Both sides of the equation are the same.

We can write an equation when one of the addends is not known, then use subtraction to solve for the unknown.

EXAMPLE 9 The librarian knows that he has eight world atlases and that five of them are in full color. How many are not in full color?

Solution We represent the number that we don't know as x and write an equation, or mathematical sentence.

$$8 = 5 + x$$

To solve an equation means to find those values that will make the equation true. We solve this equation by reasoning and by a knowledge of the relationship between addition and subtraction.

$$8 = 5 + x \text{ is equivalent to } 8 - 5 = x$$

We know that $8 - 5 = 3$. Then $x = 3$. We can check the answer by substituting 3 for x in the original equation.

$$8 = 5 + x$$

$$8 = 5 + 3 \quad \text{True} \checkmark$$

We see that $x = 3$ checks, so our answer is correct. There are three atlases not in full color.

Practice Problem 9 Form an equation for each of the following problems. Solve the equation in order to answer the question.

(a) The Salem Harbormaster's daily log noted that seventeen fishing vessels left the harbor yesterday during daylight hours. Walter was at the harbor all morning and saw twelve fishing vessels leave in the morning. How many vessels left in the afternoon? (Assume that sunset was at 6 P.M.)

(b) The Appalachian Mountain Club noted that twenty-two hikers left to climb Mount Washington during the morning. By 4 P.M., ten of them had returned. How many of the hikers were still on the mountain?

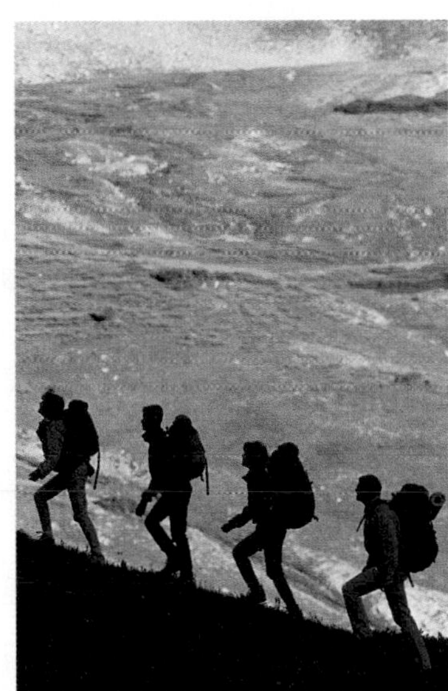

5 Applying Subtraction to Real-Life Situations

We use subtraction in all kinds of situations. There are several key words in word problems that imply subtraction. Words that involve comparison, such as *how much more, how much greater,* or how much a quantity *increased* or *decreased,* all imply subtraction. The *difference* between two numbers implies subtraction.

EXAMPLE 10 Look at the following population table.

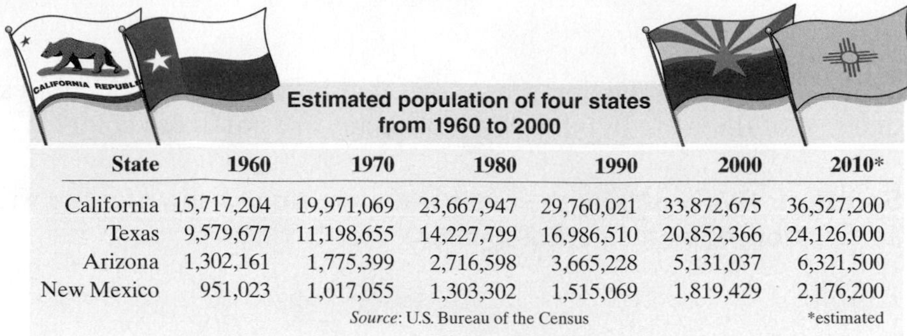

Estimated population of four states from 1960 to 2000

State	1960	1970	1980	1990	2000	2010*
California	15,717,204	19,971,069	23,667,947	29,760,021	33,872,675	36,527,200
Texas	9,579,677	11,198,655	14,227,799	16,986,510	20,852,366	24,126,000
Arizona	1,302,161	1,775,399	2,716,598	3,665,228	5,131,037	6,321,500
New Mexico	951,023	1,017,055	1,303,302	1,515,069	1,819,429	2,176,200

Source: U.S. Bureau of the Census *estimated

(a) In 1980, how much greater was the population of Texas than that of Arizona?

(b) How much did the population of California increase from 1960 to 2000?

(c) How much greater was the population of California in 1990 than that of the other three states combined?

Solution

(a)

14,227,799	1980 population of Texas
− 2,716,598	1980 population of Arizona
11,511,201	difference

The population of Texas was greater by 11,511,201.

(b)

33,872,675	2000 population of California
− 15,717,204	1960 population of California
18,155,471	difference

The population of California increased by 18,155,471 in those 40 years.

(c) First we need to find the total population in 1990 of Texas, Arizona, and New Mexico.

16,986,510	1990 population of Texas
3,665,228	1990 population of Arizona
+ 1,515,069	1990 population of New Mexico
22,166,807	

We use subtraction to compare this total with the population of California.

29,760,021	1990 population of California
− 22,166,807	
7,593,214	

The population in California in 1990 was 7,593,214 more than the population of the other three states combined.

Practice Problem 10

(a) In 1980, how much greater was the population of California than the population of Texas?

(b) How much did the population of Texas increase from 1960 to 1970?

NOTE TO STUDENT: Fully worked out solutions to all of the Practice Problems can be found at the back of the text starting at page SP-1

EXAMPLE 11 The number of real estate transfers in several towns during the years 2002 to 2004 is given in the following bar graph.

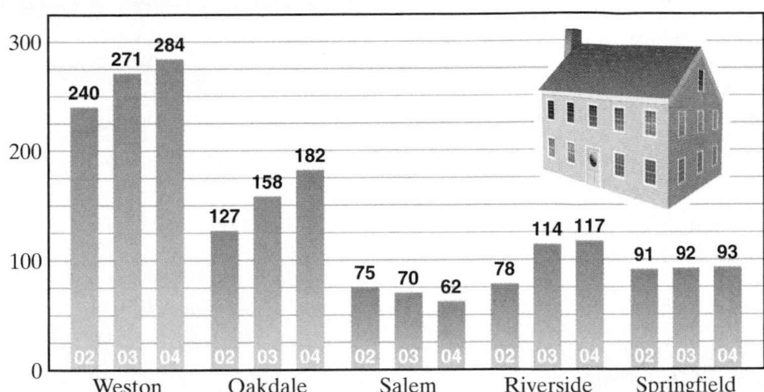

(a) What was the increase in homes sold in Weston from 2003 to 2004?

(b) What was the decrease in homes sold in Salem from 2002 to 2004?

(c) Between what two years did Oakdale have the greatest increase in sales?

Solution

(a) From the labels on the bar graph we see that 284 homes were sold in 2004 in Weston and 271 homes were sold in 2003. Thus the increase can be found by subtracting $284 - 271 = 13$. There was an increase of 13 homes sold in Weston from 2003 to 2004.

(b) In 2002, 75 homes were sold in Salem. In 2004, 62 homes were sold in Salem. The decrease in the number of homes sold is $75 - 62 = 13$. There was a decrease of 13 homes sold in Salem from 2002 to 2004.

(c) Here we will need to make two calculations in order to decide where the greatest increase occurs.

158	2003 sales	182	2004 sales
− 127	2002 sales	− 158	2003 sales
31	Sales increase from 2002 to 2003	24	Sales increase from 2003 to 2004

The greatest increase in sales in Oakdale occurred from 2002 to 2003.

Practice Problem 11 Based on the preceding bar graph, answer the following questions.

(a) What was the increase in homes sold in Riverside from 2002 to 2003?

(b) How many more homes were sold in Springfield in 2002 than in Riverside in 2002?

(c) Between what two years did Weston have the greatest increase in sales?

1.3 Exercises

Student Solutions Manual

CD/ Video

PH Math Tutor Center

MathXL®Tutorials on CD

MathXL®

MyMathLab®

Interactmath.con

Verbal and Writing Skills

1. Explain how you can check a subtraction problem.

In subtraction the minuend minus the subtrahead equals the difference. To check the problem we add the subtrahead and the difference to see if we get the minuend. If we do, the answer is correct.

2. Explain how you use borrowing to calculate $107 - 88$.

Since there are not enough ones to subtract 8 ones from 7 ones, we borrow. This means that we change the 1 hundred to an equivalent 10 tens. From the 10 tens we borrow one, making it 9 tens and 10 ones. Now we have 7 ones and 10 ones or 17 ones. 17 ones subtract 8 ones is nine ones, and 9 tens subtract 8 tens is 1 ten. Thus $107 - 88 = 19$.

3. Explain what number should be used to replace the question mark in the subtraction equation $32?5 - 1683 = 1592$.

We know that $1683 + 1592 = 32?5$. Therefore if we add 8 tens and 9 tens we get 17 tens, which is 1 hundred and 7 tens. Thus the ? should be replaced by 7.

4. Explain what steps need to be done to calculate 7 feet − 11 inches.

In subtraction we can subtract only numbers representing the same unit. Thus we need to change 7 feet to a number that measures inches. Since 1 foot equals 12 inches, 7 feet equals 84 inches. Now we subtract: 84 inches − 11 inches = 73 inches.

Try to do exercises 5–20 in one minute or less with no errors.

Subtract.

5. $\begin{array}{r} 8 \\ -3 \\ \hline 5 \end{array}$	**6.** $\begin{array}{r} 17 \\ -8 \\ \hline 9 \end{array}$	**7.** $\begin{array}{r} 15 \\ -9 \\ \hline 6 \end{array}$	**8.** $\begin{array}{r} 14 \\ -5 \\ \hline 9 \end{array}$
9. $\begin{array}{r} 16 \\ -0 \\ \hline 16 \end{array}$	**10.** $\begin{array}{r} 17 \\ -9 \\ \hline 8 \end{array}$	**11.** $\begin{array}{r} 18 \\ -9 \\ \hline 9 \end{array}$	**12.** $\begin{array}{r} 12 \\ -7 \\ \hline 5 \end{array}$
13. $\begin{array}{r} 11 \\ -4 \\ \hline 7 \end{array}$	**14.** $\begin{array}{r} 15 \\ -8 \\ \hline 7 \end{array}$	**15.** $\begin{array}{r} 13 \\ -7 \\ \hline 6 \end{array}$	**16.** $\begin{array}{r} 16 \\ -9 \\ \hline 7 \end{array}$
17. $\begin{array}{r} 11 \\ -8 \\ \hline 3 \end{array}$	**18.** $\begin{array}{r} 10 \\ -7 \\ \hline 3 \end{array}$	**19.** $\begin{array}{r} 15 \\ -6 \\ \hline 9 \end{array}$	**20.** $\begin{array}{r} 12 \\ -5 \\ \hline 7 \end{array}$

Subtract. Check your answers by adding.

21. $\begin{array}{r} 47 \\ -26 \\ \hline 21 \end{array}$ $\begin{array}{r} 26 \\ +21 \\ \hline 47 \end{array}$	**22.** $\begin{array}{r} 96 \\ -51 \\ \hline 45 \end{array}$ $\begin{array}{r} 51 \\ +45 \\ \hline 96 \end{array}$	**23.** $\begin{array}{r} 85 \\ -73 \\ \hline 12 \end{array}$ $\begin{array}{r} 73 \\ +12 \\ \hline 85 \end{array}$	**24.** $\begin{array}{r} 77 \\ -36 \\ \hline 41 \end{array}$ $\begin{array}{r} 36 \\ +41 \\ \hline 77 \end{array}$
25. $\begin{array}{r} 189 \\ -65 \\ \hline 124 \end{array}$ $\begin{array}{r} 65 \\ +124 \\ \hline 189 \end{array}$	**26.** $\begin{array}{r} 189 \\ -65 \\ \hline 124 \end{array}$ $\begin{array}{r} 65 \\ +124 \\ \hline 189 \end{array}$	**27.** $\begin{array}{r} 869 \\ -548 \\ \hline 321 \end{array}$ $\begin{array}{r} 548 \\ +321 \\ \hline 869 \end{array}$	**28.** $\begin{array}{r} 621 \\ -320 \\ \hline 301 \end{array}$ $\begin{array}{r} 320 \\ +301 \\ \hline 621 \end{array}$
29. $\begin{array}{r} 2893 \\ -572 \\ \hline 2321 \end{array}$ $\begin{array}{r} 572 \\ +2321 \\ \hline 2893 \end{array}$	**30.** $\begin{array}{r} 5780 \\ -530 \\ \hline 5250 \end{array}$ $\begin{array}{r} 530 \\ +5250 \\ \hline 5780 \end{array}$	**31.** $\begin{array}{r} 155{,}835 \\ -12{,}600 \\ \hline 143{,}235 \end{array}$ $\begin{array}{r} 12{,}600 \\ +143{,}235 \\ \hline 155{,}835 \end{array}$	**32.** $\begin{array}{r} 240{,}995 \\ -30{,}981 \\ \hline 210{,}014 \end{array}$ $\begin{array}{r} 30{,}981 \\ +210{,}014 \\ \hline 240{,}995 \end{array}$
33. $\begin{array}{r} 986{,}302 \\ -433{,}201 \\ \hline 553{,}101 \end{array}$ $\begin{array}{r} 433{,}201 \\ +553{,}101 \\ \hline 986{,}302 \end{array}$	**34.** $\begin{array}{r} 807{,}965 \\ -304{,}214 \\ \hline 503{,}751 \end{array}$ $\begin{array}{r} 304{,}214 \\ +503{,}751 \\ \hline 807{,}965 \end{array}$		

Check each subtraction. If the problem has not been done correctly, find the correct answer.

35. 129 − 19 = 110; check: 19 + 110 = 129. Correct

36. 186 − 45 = 141; check: 45 + 141 = 186. Correct

37. 8596 − 3215 = 5781; check: 3215 + 5781 = 8996. Incorrect. Correct answer: 5381

38. 9956 − 7254 = 2702; check: 7254 + 2702 = 9956. Correct

39. 6030 − 5020 = 1020; check: 5020 + 1020 = 6040. Incorrect. Correct answer: 1010

40. 7890 − 3200 = 7670; check: 3200 + 7670 = 10876. Incorrect. Correct answer: 4690

41. 47,869 − 33,846 = 13,023; check: 33,846 + 13,023 = 46,869. Incorrect. Correct answer: 14,023

42. 99,583 − 41,181 = 58,402; check: 41,181 + 58,402 = 99,583. Correct

Subtract. Use borrowing if necessary.

43. 98 − 52 = 46

44. 86 − 33 = 53

45. 174 − 82 = 92

46. 136 − 95 = 41

47. 647 − 263 = 384

48. 706 − 435 = 271

49. 955 − 237 = 718

50. 861 − 345 = 516

51. 30,000 − 19,370 = 10,630

52. 50,000 − 36,670 = 13,330

53. 152,000 − 117,908 = 34,092

54. 361,000 − 121,520 = 239,480

55. 45,312 − 37,865 = 7447

56. 64,381 − 29,997 = 34,384

57. 2,378,862 − 1,469,932 = 908,930

58. 3,554,830 − 1,710,913 = 1,843,917

Solve.

59. $x + 14 = 19$ $x = 5$

60. $x + 35 = 50$ $x = 15$

61. $57 = x + 28$ $x = 29$

62. $25 = x + 18$ $x = 7$

63. $100 + x = 127$ $x = 27$

64. $86 + x = 120$ $x = 34$

Applications

65. ***Current Events*** This year, for the college graduation ceremonies, the senior class voted for "Power of Example to Humanity." Martin Luther King received 765 votes, John F. Kennedy received 960 votes, and Mother Teresa received 778 votes. How many more votes did John F. Kennedy receive than Mother Teresa?
182 votes

960 − 778 = 182

66. ***Geography*** Three of the highest elevations on Earth are Mt. Everest in Asia, which is 8846 meters high; Mt. Aconcagua in Argentina, which is 6960 meters high; and Mt. McKinley in the United States (Alaska), which is 6194 meters high. How much higher is Mt. Everest than Mt. Aconcagua?
1886 meters

67. ***Population Trends*** In 2001, the population of Ireland was approximately 3,841,268. In the same year, the population of Portugal was approximately 10,066,523. How much less than the population of Portugal was the population of Ireland in 2001?
6,225,255

68. ***Geography*** The Nile River, the longest river in the world, is approximately 22,070,400 feet long. The Yangtze Kiang River, which is the longest river in China, is approximately 19,018,560 feet long. How much longer is the Nile River than the Yangtze Kiang River?
3,051,840 feet

69. ***Personal Finance*** Michaela's gross pay on her last paycheck was \$1280. Her deductions totaled \$318 and she deposited \$200 into her savings account. She put the remaining amount into her checking account to pay bills. How much did Michaela put into her checking account?
\$762

70. ***Personal Finance*** Adam earned \$3450 last summer at his construction job. He owed his brother \$375 and saved \$2300 to pay for his college tuition. He used the remaining amount as a down payment for a car. How much did Adam have for the down payment?
\$775

Population Trends *In answering exercises 71–78, consider the following population table.*

	1960	1970	1980	1990	2000	2010*
Illinois	10,081,158	11,110,285	11,427,409	11,430,602	12,051,683	13,216,340
Michigan	7,823,194	8,881,826	9,262,044	9,295,297	9,679,052	9,769,131
Indiana	4,662,498	5,195,392	5,490,212	5,544,159	6,045,521	6,178,300
Minnesota	3,413,864	3,806,103	4,075,970	4,375,099	4,830,784	5,263,820

Source: U.S. Census Bureau

*estimated

71. How much did the population of Minnesota increase from 1960 to 2000?
1,416,920 people

72. How much did the population of Michigan increase from 1960 to 2000?
1,855,858 people

73. In 1960, how much greater was the population of Illinois than the populations of Indiana and Minnesota combined?
2,004,796 people

74. In 2000, how much greater was the population of Illinois than the populations of Indiana and Minnesota combined?
1,175,378 people

75. How much did the population of Illinois increase from 1970 to 1990?
320,317 people

76. How much did the population of Michigan increase from 1970 to 1990?
413,471 people

77. Compare your answers to exercises 75 and 76. How much greater was the population increase of Michigan than the population increase of Illinois from 1970 to 1990?
93,154 people

78. In 2010, what will be the difference in population between the state with the highest population and the state with the lowest population?
7,952,520 people

Real Estate *The number of real estate transfers in several towns during the years 2002 to 2004 is given in the following bar graph. Use the bar graph to answer exercises 79–86. The figures in the bar graph reflect sales of single-family detached homes only.*

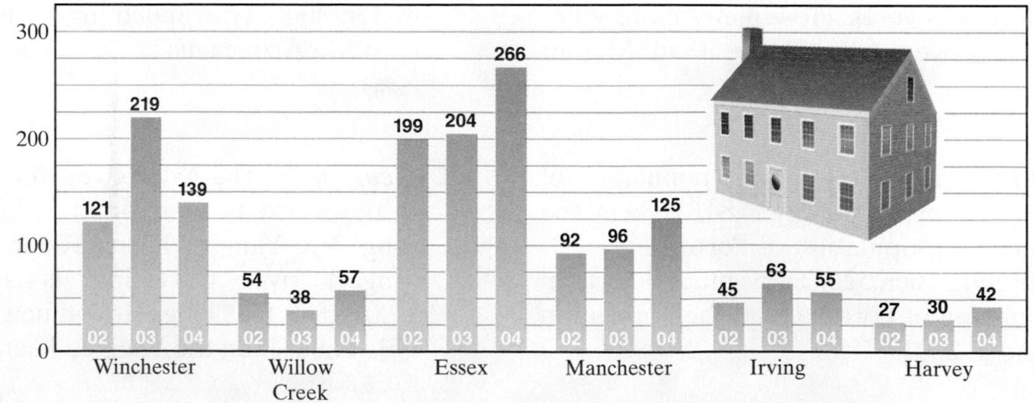

79. What was the increase in the number of homes sold in Manchester from 2003 to 2004?
29 homes

80. What was the increase in the number of homes sold in Irving from 2002 to 2003?
18 homes

81. What was the decrease in the number of homes sold in Winchester from 2003 to 2004?
80 homes

82. What was the decrease in the number of homes sold in Willow Creek from 2002 to 2003?
16 homes

83. Between what two years did the greatest change occur in the number of homes sold in Irving?
between 2002 and 2003

84. Between what two years did the greatest change occur in the number of homes sold in Winchester?
between 2002 and 2003

85. A real estate agent was trying to determine which two towns were closest to having the same number of sales of homes in 2002. What two towns should she select?
Willow Creek and Irving

86. A real estate agent was trying to determine which two towns were closest to having the same number of sales of homes in 2003. What two towns should she select?
Willow Creek and Harvey

To Think About

87. In general, subtraction is not commutative. If a and b are whole numbers, $a - b \neq b - a$. For what types of numbers would it be true that $a - b = b - a$?
It is true if a and b represent the same number, for example, if $a = 10$ and $b = 10$.

88. In general, subtraction is not associative. For example, $8 - (4 - 3) \neq (8 - 4) - 3$. In general, $a - (b - c) \neq (a - b) - c$. Can you find some numbers a, b, c for which $a - (b - c) = (a - b) - c$? (Remember, do operations inside the parentheses first.)
It is true for all a and b if $c = 0$, for example, if $a = 5$, $b = 3$, and $c = 0$.

89. ***Consumer Mathematics*** Walter Swensen wants to replace some of the fences on a farm in Caribou, Maine. The wooden rail fence costs about \$60 for wood and \$50 labor to install a fence that is 12 feet long. His son estimated he would need 276 feet of new fence. However, when he measured it he realized he would only need 216 new feet of fence. What is the difference in cost of his son's estimate versus his estimate with regard to how many feet of fence are needed?
\$550

90. ***Consumer Mathematics*** Carlos Sontera is replacing some barbed-wire fence on a ranch in El Paso, Texas. The barbed wire and poles for 12 feet of fence cost about \$80. The labor cost to install 12 feet of fence is about \$40. A ranch hand reported that 300 new feet of fence were needed. However, when Carlos actually rode out there and measured it, he found that only 228 new feet of fence were needed. What is the difference in cost of the ranch hand's estimate versus Carlos's estimate of how many feet of fence are needed?
\$720

Cumulative Review

91. Write in standard notation: eight million, four hundred sixty-six thousand, eighty-four
8,466,084

92. Write a word name for 296,308.
two hundred ninety-six thousand, three hundred eight

93. Add. $25 + 75 + 80 + 20 + 18$
218

94. Add.
$$\begin{array}{r} 278{,}563 \\ +\ 896{,}187 \\ \hline \end{array}$$
1,174,750

1.4 MULTIPLYING WHOLE NUMBERS

Student Learning Objectives

After studying this section, you will be able to:

1. **Master basic multiplication facts.**
2. **Multiply a single-digit number by a several-digit number.**
3. **Multiply a whole number by a power of 10.**
4. **Multiply a several-digit number by a several-digit number.**
5. **Use the properties of multiplication to perform calculations.**
6. **Apply multiplication to real-life situations.**

1 Mastering Basic Multiplication Facts

Like subtraction, multiplication is related to addition. Suppose that the pastry chef at the Gourmet Restaurant bakes croissants on a sheet that holds four croissants across, with room for three rows. How many croissants does the sheet hold?

We can add $4 + 4 + 4$ to get the total, or we can use a shortcut: three rows of four is the same as 3 times 4, which equals 12. This is **multiplication,** a shortcut for repeated addition.

The numbers that we multiply are called **factors.** The answer is called the **product.** For now, we will use $\times$ to show multiplication. 3×4 is read "three times four."

$$\underbrace{3}_{\text{factor}} \times \underbrace{4}_{\text{factor}} = \underbrace{12}_{\text{product}} \qquad \begin{array}{rl} 3 & \text{factor} \\ \times\ 4 & \text{factor} \\ \hline 12 & \text{product} \end{array}$$

Your skill in multiplication depends on how well you know the basic multiplication facts. Look at the table on page 39. You should learn these facts well enough to quickly and correctly give the products of any two factors in the table. To check your knowledge, try Exercises 1.4, exercises 3 and 4.

Study the table to see if you can discover any properties of multiplication. What do you see as results when you multiply zero by any number? When you multiply any number times zero, the result is zero. That is the **multiplication property of zero.**

$$2 \times 0 = 0 \qquad 5 \times 0 = 0 \qquad 0 \times 6 = 0 \qquad 0 \times 0 = 0$$

You may recall that zero plays a special role in addition. Zero is the *identity element* for addition. When we add any number to zero, that number does not change. Is there an identity element for multiplication? Look at the table. What is the identity element for multiplication? Do you see that it is 1? The **identity element for multiplication** is 1.

$$5 \times 1 = 5 \qquad 1 \times 5 = 5$$

What other properties of addition hold for multiplication? Is multiplication commutative? Does the order in which you multiply two numbers change the results? Find the product of 3×4. Then find the product of 4×3.

$$3 \times 4 = 12$$
$$4 \times 3 = 12$$

The **commutative property of multiplication** tells us that when we multiply two numbers, changing the order of the numbers gives the same result.

Teaching Tip Remind students that they may need to practice or relearn facts that they cannot instantly recall from the multiplication table. As with the addition table, some students will need to make and use multiplication flash cards in order to master the basic multiplication facts.

Basic Multiplication Facts

×	0	1	2	3	4	5	6	7	8	9	10	11	12
0	0	0	0	0	0	0	0	0	0	0	0	0	0
1	0	1	2	3	4	5	6	7	8	9	10	11	12
2	0	2	4	6	8	10	12	14	16	18	20	22	24
3	0	3	6	9	12	15	18	21	24	27	30	33	36
4	0	4	8	12	16	20	24	28	32	36	40	44	48
5	0	5	10	15	20	25	30	35	40	45	50	55	60
6	0	6	12	18	24	30	36	42	48	54	60	66	72
7	0	7	14	21	28	35	42	49	56	63	70	77	84
8	0	8	16	24	32	40	48	56	64	72	80	88	96
9	0	9	18	27	36	45	54	63	72	81	90	99	108
10	0	10	20	30	40	50	60	70	80	90	100	110	120
11	0	11	22	33	44	55	66	77	88	99	110	121	132
12	0	12	24	36	48	60	72	84	96	108	120	132	144

QUICK RECALL OF MULTIPLICATION FACTS It is helpful if you can multiply quickly. See if you can do Example 1 correctly in 15 seconds or less. Repeat again with Practice Problem 1. Strive to obtain all answers correctly in 15 seconds or less.

EXAMPLE 1 Multiply.

(a) 5×7 **(b)** 8×9 **(c)** 6×8

(d) 9×3 **(e)** 7×8

Solution

(a) $\begin{array}{r} 7 \\ \times\ 5 \\ \hline 35 \end{array}$ **(b)** $\begin{array}{r} 9 \\ \times\ 8 \\ \hline 72 \end{array}$ **(c)** $\begin{array}{r} 8 \\ \times\ 6 \\ \hline 48 \end{array}$

(d) $\begin{array}{r} 3 \\ \times\ 9 \\ \hline 27 \end{array}$ **(e)** $\begin{array}{r} 8 \\ \times\ 7 \\ \hline 56 \end{array}$

Practice Problem 1 Multiply.

(a) $\begin{array}{r} 8 \\ \times\ 8 \\ \hline \end{array}$ **(b)** $\begin{array}{r} 7 \\ \times\ 6 \\ \hline \end{array}$ **(c)** $\begin{array}{r} 5 \\ \times\ 8 \\ \hline \end{array}$

(d) $\begin{array}{r} 9 \\ \times\ 7 \\ \hline \end{array}$ **(e)** $\begin{array}{r} 9 \\ \times\ 9 \\ \hline \end{array}$

NOTE TO STUDENT: Fully worked-out solutions to all of the Practice Problems can be found at the back of the text starting at page SP-1

2 Multiplying a Single-Digit Number by a Several-Digit Number

EXAMPLE 2 Multiply. 4312×2

Solution We first multiply the ones column, then the tens column, and so on, moving right to left.

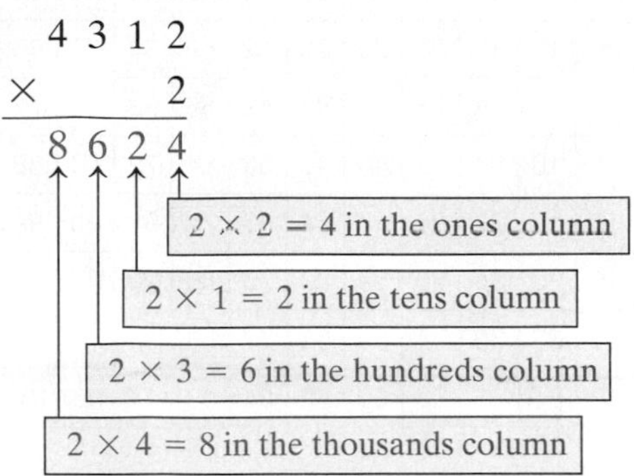

NOTE TO STUDENT: Fully worked-out solutions to all of the Practice Problems can be found at the back of the text starting at page SP-1

Practice Problem 2 Multiply. 3021×3

Usually, we will have to carry one digit of the result of some of the multiplication into the next left-hand column.

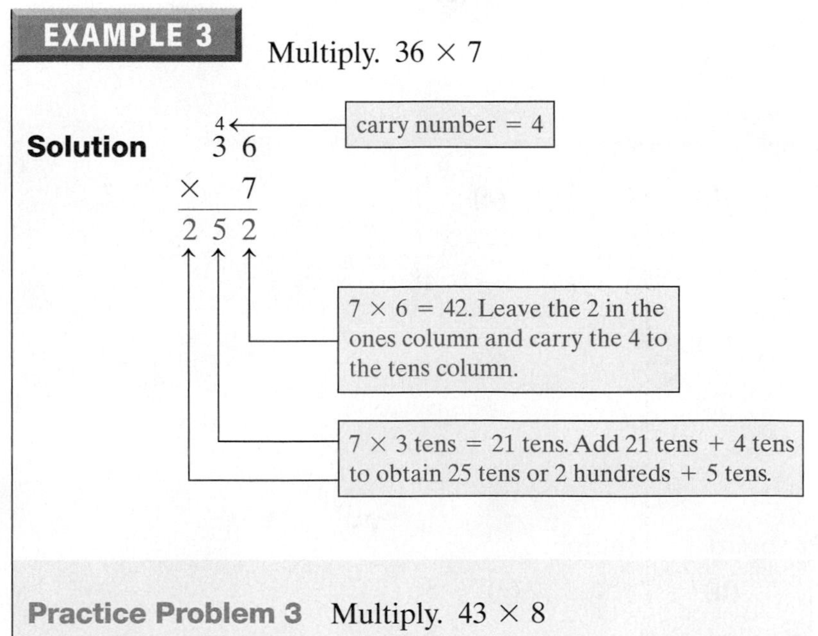

4
3 6
× 7
2 5 2
carry number = 4
7 × 6 = 42. Leave the 2 in the ones column and carry the 4 to the tens column.
7 × 3 tens = 21 tens. Add 21 tens + 4 tens to obtain 25 tens or 2 hundreds + 5 tens.

EXAMPLE 3 Multiply. 36×7

Solution

Practice Problem 3 Multiply. 43×8

EXAMPLE 4 Multiply. 359 × 9

Solution

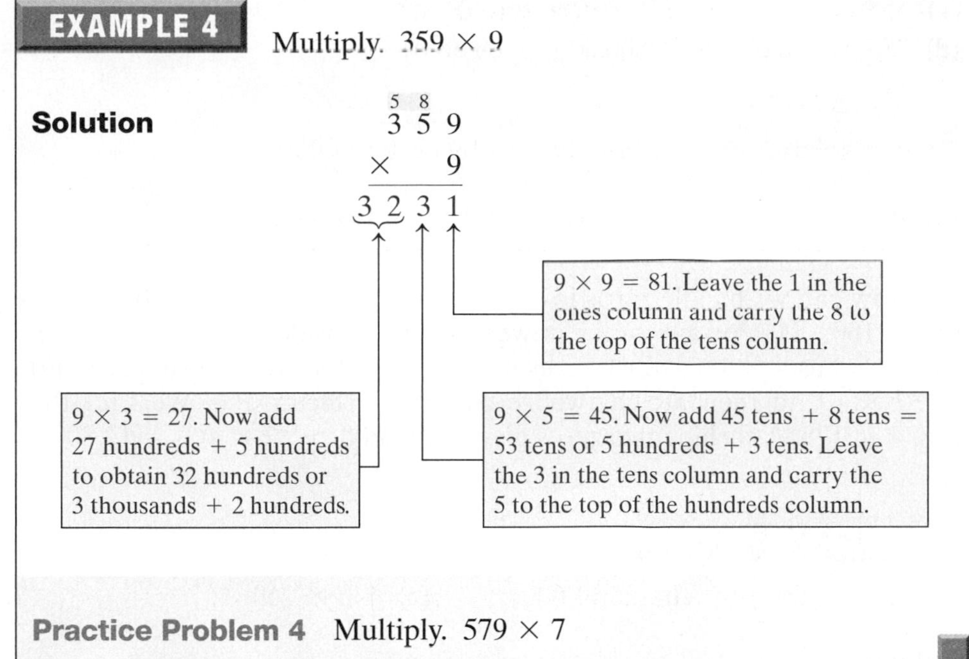

5 8
3 5 9
× 9
3 2 3 1
9 × 9 = 81. Leave the 1 in the ones column and carry the 8 to the top of the tens column.
9 × 3 = 27. Now add 27 hundreds + 5 hundreds to obtain 32 hundreds or 3 thousands + 2 hundreds.
9 × 5 = 45. Now add 45 tens + 8 tens = 53 tens or 5 hundreds + 3 tens. Leave the 3 in the tens column and carry the 5 to the top of the hundreds column.

Practice Problem 4 Multiply. 579 × 7

3 Multiplying a Whole Number by a Power of 10

Observe what happens when a number is multiplied by 10, 100, 1000, 10,000, and so on.

	one zero			two zeros	
56 × 10	=	560	56 × 100	=	5600
	three zeros			four zeros	
56 × 1000	=	56,000	56 × 10,000	=	560,000

A **power of 10** is a whole number that begins with 1 and ends in one or more zeros. The numbers 10, 100, 1000, 10,000, and so on are powers of 10.

To multiply a whole number by a power of 10:

1. Count the number of zeros in the power of 10.
2. Attach that number of zeros to the right side of the other whole number to obtain the answer.

EXAMPLE 5 Multiply 358 by each number.

(a) 10 **(b)** 100 **(c)** 1000 **(d)** 100,000

Solution

(a) 358 × 10 = 3580 (one zero) **(b)** 358 × 100 = 35,800 (two zeros)

(c) 358 × 1000 = 358,000 (three zeros)
(d) 358 × 100,000 = 35,800,000 (five zeros)

***NOTE TO STUDENT:** Fully worked-out solutions to all of the Practice Problems can be found at the back of the text starting at page SP-1*

Practice Problem 5 Multiply 1267 by each number.

(a) 10 **(b)** 1000 **(c)** 10,000 **(d)** 1,000,000

How can we handle zeros in multiplication involving a number that is not 10, 100, 1000, or any other power of 10? Consider 32 × 400. We can rewrite 400 as 4 × 100, which gives us 32 × 4 × 100. We can simply multiply 32 × 4 and then attach two zeros for the factor 100. We find that 32 × 4 = 128. Attaching two zeros gives us 12,800, or 32 × 400 = 12,800.

EXAMPLE 6 Multiply.

(a) 12 × 3000 **(b)** 25 × 600 **(c)** 430 × 260

Solution

(a) 12 × 3000 = 12 × 3 × 1000 = 36 × 1000 = 36,000
(b) 25 × 600 = 25 × 6 × 100 = 150 × 100 = 15,000
(c) 430 × 260 = 43 × 26 × 10 × 10 = 1118 × 100 = 111,800

Teaching Tip You may need to do several problems of the type 345 × 30 and 2300 × 50 until students grasp the shortcut of counting the number of end zeros and then adding them at the end. Show them that to multiply 2300 × 50, you merely multiply 23 × 5 = 115 and then add the three zeros to obtain the final answer 2300 × 50 = 115,000.

Practice Problem 6 Multiply.

(a) 9 × 60,000 **(b)** 15 × 400 **(c)** 270 × 800

Multiplying a Several-Digit Number by a Several-Digit Number

EXAMPLE 7 Multiply. 234 × 21

Solution We can consider 21 as 2 tens (20) and 1 one (1). First we multiply 234 by 1.

We also multiply 234 × 20. This gives us two **partial products.**

```
  234       234
×   1     ×  20
  234      4680
```

Now we combine these two operations together by adding the two partial products to reach the final product, which is the solution.

```
   234
×   21
   234 ←—— Multiply 234 × 1.
  4680 ←—— Multiply 234 × 20.
  4914 ←—— Add the two partial products.
```

Practice Problem 7 Multiply. 323×32

EXAMPLE 8 Multiply. 671×35

Solution

```
    6 7 1
  ×   3 5
  -------
  3 3 5 5  ←—— First multiply 671 × 5.
2 0 1 3(0) ←—— Now multiply 671 × 30.
---------
2 3 4 8 5  ←—— Now add the two partial products.
```

Note: We could omit zero on this line and leave the ones place blank.

Practice Problem 8 Multiply. 385×69

EXAMPLE 9 Multiply. 14×20

Solution

```
                          14
                        × 20
                        ----
                           0 ←—— Multiply 14 by 0.
                         280 ←—— Multiply 14 by 2 tens.
                         ---
Now add the ——————→      280     Place 28 with the 8 in the tens column.
partial products.                To line up the digits for adding, we can
                                 insert a 0 in the ones column.
```

Notice that you will also get this result if you multiply $14 \times 2 = 28$ and then attach a zero to multiply it by 10: 280.

Practice Problem 9 Multiply. 34×20

EXAMPLE 10 Multiply. 120×40

Solution

```
                      120
                     × 40
                     ----
                        0 ←—— Multiply 120 × 0.
                     4800 ←—— Multiply 120 by 4 tens.
                     ----
Now add the ——————→  4800     The answer is 480 tens. We place the 0 of the 480
partial products.             in the tens column. To line up the digits for adding,
                              we can insert a 0 in the ones column.
```

Notice that this result is the same as $12 \times 4 = 48$ with two zeros attached: 4800.

Practice Problem 10 Multiply. 130×50

EXAMPLE 11 Multiply. 684×763

Solution

$$\begin{array}{r} 684 \\ \times 763 \\ \hline 2052 \\ 4104 \\ 4788 \\ \hline 521892 \end{array}$$

$2052 \leftarrow$ Multiply 684×3.
$4104 \leftarrow$ Multiply 684×60. Note that we omit the final zero.
$4788 \leftarrow$ Multiply 684×700. Note that we omit the final two zeros.

***NOTE TO STUDENT:** Fully worked-out solutions to all of the Practice Problems can be found at the back of the text starting at page SP-1*

Practice Problem 11 Multiply. 923×675

5 Using the Properties of Multiplication to Perform Calculations

When we add three numbers, we use the associative property. Recall that the associative property allows us to group the three numbers in different ways. Thus to add $9 + 7 + 3$, we can group the numbers as $9 + (7 + 3)$ because it is easier to find the sum. $9 + (7 + 3) = 9 + 10 = 19$. We can demonstrate that multiplication is also associative.

Is this true? $2 \times (5 \times 3) = (2 \times 5) \times 3$

$$2 \times (15) = (10) \times 3$$

$$30 = 30$$

The final product is the same in both cases.

The way we group numbers to be multiplied does not change the product. This property is called the **associative property of multiplication.**

EXAMPLE 12 Multiply. $14 \times 2 \times 5$

Solution Since we can group any two numbers together, let's take advantage of multiplying by 10.

$$14 \times 2 \times 5 = 14 \times (2 \times 5) = 14 \times 10 = 140$$

Practice Problem 12 Multiply. $25 \times 4 \times 17$

For convenience, we list the properties of multiplication that we have discussed in this section.

1. **Associative Property of Multiplication.** When we multiply three numbers, the multiplication can be grouped in any way.

$$(7 \times 3) \times 2 = 7 \times (3 \times 2)$$

$$21 \times 2 = 7 \times 6$$

$$42 = 42$$

2. **Commutative Property of Multiplication.** Two numbers can be multiplied in either order with the same result.

$$9 \times 8 = 8 \times 9$$
$$72 = 72$$

3. **Identity Property of One.** When one is multiplied by a number, the result is that number.

$$7 \times 1 = 7$$
$$1 \times 15 = 15$$

4. **Multiplication Property of Zero.** The product of any number and zero yields zero as a result.

$$0 \times 14 = 0$$
$$2 \times 0 = 0$$

Sometimes you can use several properties in one problem to make the calculation easier.

EXAMPLE 13 Multiply. $7 \times 20 \times 5 \times 6$

Solution

$$\begin{aligned} 7 \times 20 \times 5 \times 6 &= 7 \times (20 \times 5) \times 6 && \text{Associative property} \\ &= 7 \times 6 \times (20 \times 5) && \text{Commutative property} \\ &= 42 \times 100 \\ &= 4200 \end{aligned}$$

Practice Problem 13 Multiply. $8 \times 4 \times 3 \times 25$

Thus far we have discussed the properties of addition and the properties of multiplication. There is one more property that links both operations.

Before we discuss that property, we will illustrate several different ways of showing multiplication. The following are all the ways to show "3 times 4."

3×4	(3)(4)	3(4) (3)4	$3 \cdot 4$	3 * 4
with an ×	with two sets of parentheses	with a single set of parentheses	with a dot	with a star

We will use parentheses to mean multiplication when we use the **distributive property.**

SIDELIGHT: The Distributive Property

Why does our method of multiplying several-digit numbers work? Why can we say that 234×21 is the same as $234 \times 1 + 234 \times 20$?

The *distributive property of multiplication over addition* allows us to distribute the multiplication and then add the results. To illustrate, 234×21 can be written as $234(20 + 1)$. By the distributive property

$$\begin{aligned} 234(20 + 1) &= (234 \times 20) + (234 \times 1) \\ &= 4680 + 234 \\ &= 4914 \end{aligned}$$

This is what we actually do when we multiply.

$$\begin{array}{r} 234 \\ \times\ 21 \\ \hline 234 \\ 4680 \\ \hline 4914 \end{array}$$

DISTRIBUTIVE PROPERTY OF MULTIPLICATION OVER ADDITION

Multiplication can be distributed over addition without changing the result.

$$5 \times (10 + 2) = (5 \times 10) + (5 \times 2)$$

6 Applying Multiplication to Real-Life Situations

To use multiplication in word problems, the number of items or the value of each item must be the same. Recall that multiplication is a quick way to do repeated addition where *each addend is the same.* In the beginning of the section we showed three rows of four croissants to illustrate 3×4. The number of croissants in each row was the same, 4. Look at another example. If we had six nickels, we could use multiplication to find the total value of the coins because the value of each nickel is the same, 5¢. Since $6 \times 5 = 30$, six nickels are worth 30¢.

In the following example the word *average* is used. The word *average* has several different meanings. In this example, we are told that the *average annual salary* of an employee at Software Associates is $21,342. This means that we can calculate the total payroll as if each employee made $21,342 even though we know that the president probably makes more than any other employee.

EXAMPLE 14 The average annual salary of an employee at Software Associates is $21,342. There are 38 employees. What is the annual payroll?

Solution

$$\begin{array}{r} \$21{,}342 \\ \times \quad 38 \\ \hline 170\ 736 \\ 640\ 26 \\ \hline 810{,}996 \end{array}$$

The total annual payroll is $810,996.

Practice Problem 14 The average cost of a new car sold last year at Westover Chevrolet was $17,348. The dealership sold 378 cars. What were the total sales of cars at the dealership last year?

***NOTE TO STUDENT:** Fully worked-out solutions to all of the Practice Problems can be found at the back of the text starting at page SP-1*

Another useful application of multiplication is area. The following example involves the area of a rectangle.

▲ **EXAMPLE 15** What is the area of a rectangular hallway that measures 4 feet wide and 8 feet long?

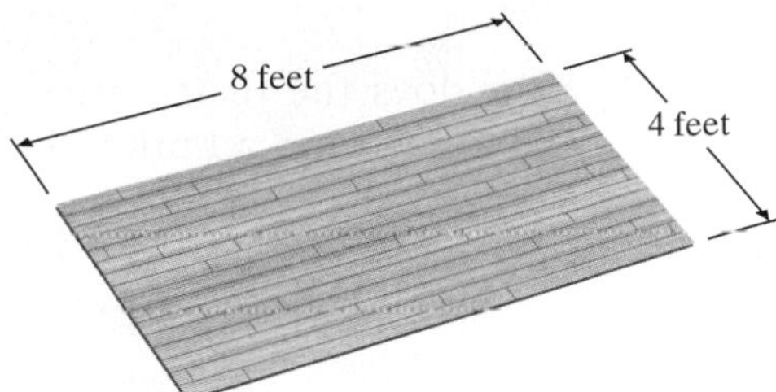

Solution The **area** of a rectangle is the product of the length times the width. Thus for this hallway

$$\text{Area} = 8 \text{ feet} \times 4 \text{ feet} = 32 \text{ square feet.}$$

The area of the hallway is 32 square feet.

Note: All measurements for area are given in square units such as square feet, square meters, square yards, and so on.

▲ **Practice Problem 15** What is the area of a rectangular rug that measures 5 yards by 7 yards? ■

Developing Your Study Skills

Why Is Homework Necessary?

Mathematics involves mastering a set of skills that you learn by practicing, not by watching someone else do it. Your instructor may make solving a mathematics problem look very easy, but for you to learn the necessary skills, you must practice them over and over again, just as your instructor once had to. There is no other way. Learning mathematics is like learning to play a musical instrument, to type, or to play a sport. No matter how much you watch someone else do it, no matter how many books you read on "how to" do it, no matter how easy it seems to be, the key to success is practice on a regular basis.

Homework provides this practice. The amount of practice needed varies for each individual, but usually students need to do most or all of the exercises provided at the end of each section in the text. The more exercises you do, the better you get. Some exercises in a set are more difficult than others, and some stress different concepts. Only by working all the exercises will you cover the full range of difficulty.

1.4 Exercises

Student Solutions Manual

CD/ Video

PH Math Tutor Center

MathXL®Tutorials on CD
MathXL®

MyMathLab®

Interactmath.com

Verbal and Writing Skills

1. Explain in your own words.

Answers may vary. Samples follow.

(a) the commutative property of multiplication

You can change the order of the factors without changing the product.

(b) the associative property of multiplication

You can group the factors in any way without changing the product.

2. How does the distributive property of multiplication over addition help us to multiply 4×13?

You can write 13 as $10 + 3$ and distribute 4 over the addition. $4 \times (10 + 3) = (4 \times 10) + (4 \times 3)$

Complete the multiplication facts for each table. Strive for total accuracy, but work quickly. (Allow a maximum of six minutes for each table.)

3.

x	6	2	3	8	0	5	7	9	12	4
5	30	10	15	40	0	25	35	45	60	20
7	42	14	21	56	0	35	49	63	84	28
1	6	2	3	8	0	5	7	9	12	4
0	0	0	0	0	0	0	0	0	0	0
6	36	12	18	48	0	30	42	54	72	24
2	12	4	6	16	0	10	14	18	24	8
3	18	6	9	24	0	15	21	27	36	12
8	48	16	24	64	0	40	56	72	96	32
4	24	8	12	32	0	20	28	36	48	16
9	54	18	27	72	0	45	63	81	108	36

4.

x	2	7	0	5	3	4	8	12	6	9
1	2	7	0	5	3	4	8	12	6	9
6	12	42	0	30	18	24	48	72	36	54
5	10	35	0	25	15	20	40	60	30	45
3	6	21	0	15	9	12	24	36	18	27
0	0	0	0	0	0	0	0	0	0	0
9	18	63	0	45	27	36	72	108	54	81
4	8	28	0	20	12	16	32	48	24	36
7	14	49	0	35	21	28	56	84	42	63
2	4	14	0	10	6	8	16	24	12	18
8	16	56	0	40	24	32	64	96	48	72

Multiply.

5. $\begin{array}{r} 32 \\ \times\ 3 \\ \hline 96 \end{array}$

6. $\begin{array}{r} 21 \\ \times\ 4 \\ \hline 84 \end{array}$

7. $\begin{array}{r} 14 \\ \times\ 5 \\ \hline 70 \end{array}$

8. $\begin{array}{r} 13 \\ \times\ 4 \\ \hline 52 \end{array}$

9. $\begin{array}{r} 87 \\ \times\ 6 \\ \hline 522 \end{array}$

10. $\begin{array}{r} 95 \\ \times\ 7 \\ \hline 665 \end{array}$

11. $\begin{array}{r} 231 \\ \times\ 3 \\ \hline 693 \end{array}$

12. $\begin{array}{r} 313 \\ \times\ 3 \\ \hline 939 \end{array}$

13. $\begin{array}{r} 429 \\ \times\ 8 \\ \hline 3432 \end{array}$

14. $\begin{array}{r} 526 \\ \times\ 9 \\ \hline 4734 \end{array}$

15. $\begin{array}{r} 6102 \\ \times\ 3 \\ \hline 18{,}306 \end{array}$

16. $\begin{array}{r} 5203 \\ \times\ 2 \\ \hline 10{,}406 \end{array}$

17. $\begin{array}{r} 12{,}203 \\ \times\ 3 \\ \hline 36{,}609 \end{array}$

18. $\begin{array}{r} 31{,}206 \\ \times\ 3 \\ \hline 93{,}618 \end{array}$

19. $\begin{array}{r} 5218 \\ \times\ 6 \\ \hline 31{,}308 \end{array}$

20. $\begin{array}{r} 3215 \\ \times\ 6 \\ \hline 19{,}290 \end{array}$

21. $\begin{array}{r} 12{,}526 \\ \times\ 8 \\ \hline 100{,}208 \end{array}$

22. $\begin{array}{r} 48{,}761 \\ \times\ 7 \\ \hline 341{,}327 \end{array}$

23. $\begin{array}{r} 235{,}702 \\ \times\ 4 \\ \hline 942{,}808 \end{array}$

24. $\begin{array}{r} 127{,}054 \\ \times\ 6 \\ \hline 762{,}324 \end{array}$

Multiply by powers of 10.

25. $\begin{array}{r} 156 \\ \times\ 10 \\ \hline 1560 \end{array}$

26. $\begin{array}{r} 278 \\ \times\ 10 \\ \hline 2780 \end{array}$

27. $\begin{array}{r} 27{,}158 \\ \times\ 100 \\ \hline 2{,}715{,}800 \end{array}$

28. $\begin{array}{r} 89{,}361 \\ \times\ 100 \\ \hline 8{,}936{,}100 \end{array}$

29. 482 ×1000 482,000	**30.** 579 ×1000 579,000	**31.** 37,256 × 10,000 372,560,000	**32.** 614,260 × 10,000 6,142,600,000

Multiply by multiples of 10.

33. 423 × 20 8460	**34.** 134 × 20 2680	**35.** 2120 × 30 63,600	**36.** 4230 × 20 84,600
37. 14,000 × 4000 56,000,000	**38.** 62,000 × 3000 186,000,000		

Multiply.

39. 514 × 12 1028 514 6168	**40.** 432 × 13 1296 432 5616	**41.** 146 × 54 584 730 7884	**42.** 163 × 35 815 489 5705
43. 89 ×64 356 534 5696	**44.** 68 ×49 612 272 3332	**45.** 607 × 25 3 035 12 14 15,175	**46.** 783 × 82 1566 6264 64,206
47. 659 × 67 4613 3954 44,153	**48.** 915 × 47 6 405 36 60 43,005	**49.** 912 × 76 5 472 63 84 69,312	**50.** 498 × 39 4 482 14 94 19,422
51. 5123 × 29 46 107 102 46 148,567	**52.** 1268 × 38 10 144 38 04 48,184	**53.** 9053 × 91 9 053 814 77 823,823	**54.** 3078 × 72 6 156 215 46 221,616
55. 5536 × 224 22 144 110 72 1107 2 1,240,064	**56.** 4127 × 415 20 635 41 27 1650 8 1,712,705	**57.** 678 ×132 1 356 20 34 67 8 89,496	**58.** 392 ×187 2 744 31 36 39 2 73,304

Mixed Practice

59. 2076 × 105 10 380 00 00 207 6 217,980	**60.** 5092 × 302 10 184 00 00 1 527 6 1,537,784	**61.** 3561 × 403 10 683 00 00 1 424 4 1,435,083	**62.** 2074 ×1003 6 222 2 074 2,080,222

63. $\begin{array}{r} 12{,}000 \\ \times \quad 60 \\ \hline \end{array}$
720,000

64. $\begin{array}{r} 8100 \\ \times \ 30 \\ \hline \end{array}$
243,000

65. $\begin{array}{r} 250 \\ \times \ 40 \\ \hline \end{array}$
10,000

66. $\begin{array}{r} 302 \\ \times \ 30 \\ \hline \end{array}$
9060

67. $\begin{array}{r} 302 \\ \times 300 \\ \hline \end{array}$
90,600

68. $\begin{array}{r} 3000 \\ \times \ 302 \\ \hline \end{array}$
906,000

69. $7 \cdot 2 \cdot 5$
70

70. $8 \cdot 3 \cdot 2$
48

71. $11 \cdot 7 \cdot 4$
308

72. $15 \cdot 4 \cdot 4$
240

73. $576 \cdot 32$
18,432

74. $485 \cdot 68$
32,980

75. $5 \cdot 8 \cdot 4 \cdot 10$
1600

76. $11 \cdot 3 \cdot 5 \cdot 4$
660

77. What is x if $x = 8 \cdot 7 \cdot 6 \cdot 0$?
$x = 0$

78. What is x if $x = 3 \cdot 12 \cdot 0 \cdot 5$?
$x = 0$

Applications

▲ **79.** ***Geometry*** Find the area of a driveway that is 38 feet long and 20 feet wide.
760 square feet

▲ **80.** ***Geometry*** Find the area of a rectangular computer chip that is 17 millimeters long and 9 millimeters wide.
153 square millimeters

▲ **81.** ***Consumer Mathematics*** Don Williams and his wife want to put down new carpet in the living room and the hallway of their house. The living room measures 12 feet by 14 feet. The hallway measures 9 feet by 3 feet. If the living room and the hallway are rectangular in shape, how many square feet of new carpet do Don and his wife need?
195 square feet

▲ **82.** ***Wildlife Management*** Robert Tobey in Copper Center, Alaska, wants to put a field under helicopter surveillance because of a roving pack of wolves that are destroying other wildlife in the area. The field consists of two rectangular regions. The first one is 4 miles by 5 miles. The second one is 12 miles by 8 miles. How many square miles does he want to place under surveillance?
116 square miles

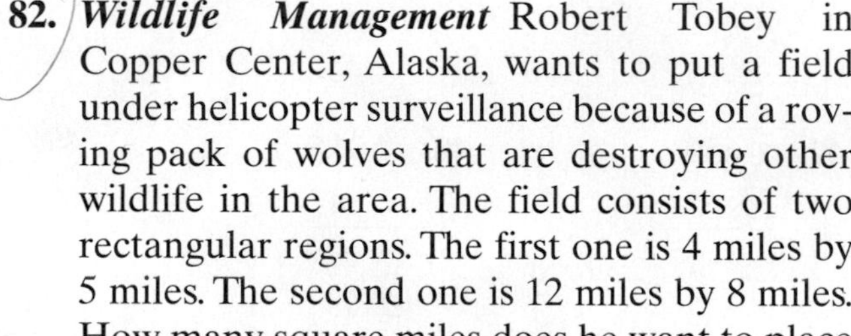

83. ***Business Decisions*** The student commons food supply needs to purchase espresso coffee. Find the cost of purchasing 240 pounds of espresso coffee beans at \$5 per pound.
\$1200

84. ***Business Decisions*** The music department of Wheaton College wishes to purchase 345 sets of headphones at the music supply store at a cost of \$8 each. What will be the total amount of the purchase?
\$2760

85. ***Personal Finance*** Helen pays \$266 per month for her car payment on her new Honda Civic. What is her automobile payment cost for a one-year period?
\$3192

86. ***Personal Finance*** A company rents a Ford Escort for a salesman at \$276 per month for eight months. What is the cost for the car rental during this time?
\$2208

87. ***Environmental Studies*** Marcos has a Toyota Corolla that gets 34 miles per gallon during highway driving. Approximately how far can he travel if he has 18 gallons of gas in the tank?
612 miles

88. ***Environmental Studies*** Cheryl has a subcompact car that gets 48 miles per gallon on highway driving. Approximately how far can she travel if she has 12 gallons of gas in the tank?
576 miles

89. ***Personal Finance*** Cameron worked 14 weeks during the summer as a house painter. He earned \$485 per week. What is the total amount Cameron earned during the summer?
\$6790

90. ***Personal Finance*** Each time Robyn gets paid, \$117 is deducted for taxes. If Robyn gets paid twice a month, how much does Robyn pay in taxes in one year?
\$2808

91. ***International Relations*** The country of Haiti has an average per capita (per person) income of \$1070. If the approximate population of Haiti is 6,890,000, what is the approximate total yearly income of the entire country?
\$7,372,300,000

92. ***International Relations*** The country of the Netherlands (Holland) has an approximate population of 15,800,000. The average per capita (per person) income is \$22,000. What is the approximate total yearly income of the entire country?
\$347,600,000,000

To Think About

Use the following information to answer exercises 93–96. There are 98 puppies in a room, with an assortment of black and white ears and paws. 18 puppies have totally black ears and 2 white paws; 26 puppies have 1 black ear and 4 white paws, and 54 puppies have no black ears and 1 white paw.

93. How many black paws are in the room?
198

94. How many white paws are in the room?
194

95. How many black ears are in the room?
62

96. How many white ears are in the room?
134

In exercises 97–100, find the value of x in each equation.

97. $5(x) = 40$
$x = 8$

98. $7(x) = 56$
$x = 8$

99. $72 = 8(x)$
$x = 9$

100. $63 = 9(x)$
$x = 7$

101. Would the distributive property of multiplication be true for roman numerals such as (XII) × (IV)? Why or why not?
No, it would not always be true. In our number system $62 = 60 + 2$. But in roman numerals IV ≠ I + V. The digit system in roman numerals involves subtraction. Thus (XII) × (IV) ≠ (XII × I) + (XII × V).

102. We saw that multiplication is distributive over addition. Is it distributive over subtraction? Why or why not? Give examples.
yes, $5 \times (8 - 3) = 5 \times 8 - 5 \times 3$,
$a \times (b - c) = (a \times b) - (a \times c)$

Cumulative Review

103. Subtract.

$$\begin{array}{r} 34{,}084 \\ -\ 27{,}328 \\ \hline \end{array}$$

6,756

104. Add.

$$\begin{array}{r} 263 \\ 27 \\ 891 \\ 5 \\ +\ 63 \\ \hline \end{array}$$

1249

105. ***Personal Finance*** Albert Gachet earns $156 per week. He has withheld weekly $12 for taxes, $2 for dues, and $3 for retirement. How much is left?

$139

106. ***Personal Finance*** Mrs. May Washington retired at $743 per month. Her cost-of-living raise increased her payment to $802 per month. What was the increase?

$59

107. ***Population Studies*** The population of Paynesville in 1990 was 32,176. In 2002, the population had increased to 34,005. By how many people did the net population increase?

1829 people

108. ***International Relations*** In 1990, the gross national product of Spain was $458,791,650,200. In 2000, it was $595,178,930,600. How much did the gross national product of Spain increase from 1990 to 2000? (*Source*: The World Bank, Washington, D.C.)

$136,387,280,400

1.5 Dividing Whole Numbers

1 Mastering Basic Division Facts

Suppose that we have eight quarters and want to divide them into two equal piles. We would discover that each pile contains four quarters.

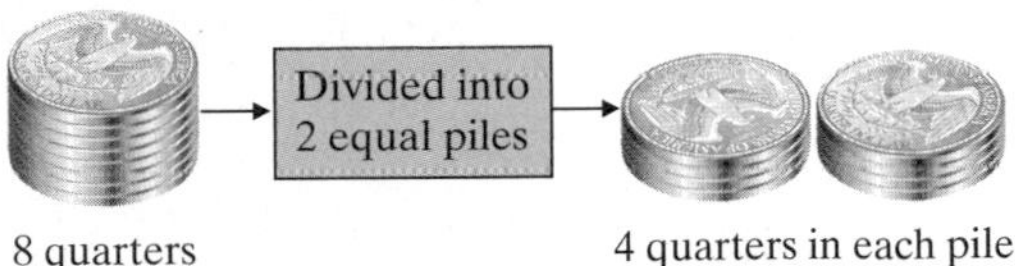

In mathematics we would express this thought by saying that

$$8 \div 2 = 4.$$

We know that this answer is right because two piles of four quarters is the same dollar amount as eight quarters. In other words, we know that $8 \div 2 = 4$ because $2 \times 4 = 8$. These two mathematical sentences are called **related sentences.** The division sentence $8 \div 2 = 4$ is related to the multiplication sentence $2 \times 4 = 8$.

In fact, in mathematics we usually define **division** in terms of multiplication. The answer to the division problem $12 \div 3$ is that number which when multiplied by 3 yields 12. Thus

$$12 \div 3 = 4 \text{ because } 3 \times 4 = 12.$$

Suppose that a surplus of \$30 in the French Club budget at the end of the year is to be equally divided among the five club members. We would want to divide the \$30 into five equal parts. We would write $30 \div 5 = 6$ because $5 \times 6 = 30$. Thus each of the five people would get \$6 in this situation.

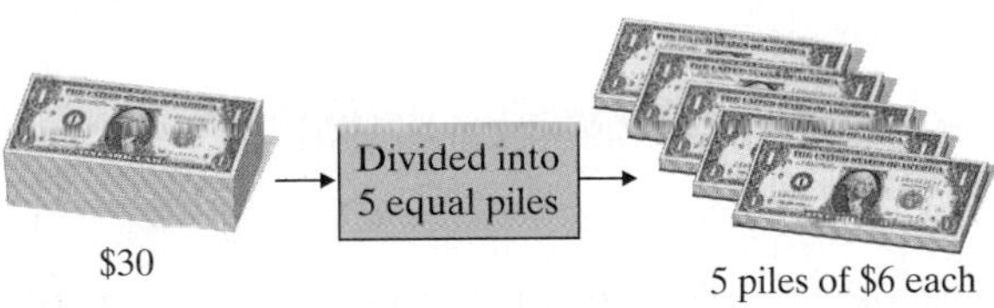

As a mathematical sentence, $30 \div 5 = 6$.
The division problem $30 \div 5 = 6$ could also be written $\frac{30}{5} = 6$ or $5\overline{)30}$ with quotient 6.

When referring to division, we sometimes use the words **divisor, dividend,** and **quotient** to identify the three parts.

$$\begin{array}{r} \text{quotient} \\ \text{divisor}\overline{)\text{dividend}} \end{array}$$

With $30 \div 5 = 6$, 30 is the dividend, 5 is the divisor, and 6 is the quotient.

$$\begin{array}{r} 6 \leftarrow \text{quotient} \\ \text{divisor} \rightarrow 5\overline{)30} \leftarrow \text{dividend} \end{array}$$

So the quotient is the answer to a division problem. It is important that you be able to do short problems involving basic division facts quickly.

Student Learning Objectives

After studying this section, you will be able to:

1. **Master basic division facts.**
2. **Perform division by a one-digit number.**
3. **Perform division by a two- or three-digit number.**
4. **Apply division to real-life situations.**

Teaching Tip Throughout all mathematics courses from basic mathematics to calculus, the words *divisor, dividend*, and *quotient* are used extensively. Stress to students that in any given division problem they need to be able to recognize which number is the divisor, which is the dividend, and which is the quotient.

EXAMPLE 1 Divide.

(a) $12 \div 4$ **(b)** $81 \div 9$ **(c)** $56 \div 8$ **(d)** $54 \div 6$

Solution

(a) $4\overline{)12}$ = 3 **(b)** $9\overline{)81}$ = 9 **(c)** $8\overline{)56}$ = 7 **(d)** $6\overline{)54}$ = 9

Practice Problem 1 Divide.

(a) $36 \div 4$ **(b)** $25 \div 5$ **(c)** $72 \div 9$ **(d)** $30 \div 6$

NOTE TO STUDENT: Fully worked-out solutions to all of the Practice Problems can be found at the back of the text starting at page SP-1

Zero can be divided by any nonzero number, but division by zero is not possible. Why is this?

Suppose that we could divide by zero. Then $7 \div 0 =$ some number. Let us represent "some number" by the letter *a*.

$$\text{If } 7 \div 0 = a, \text{ then } 7 = 0 \times a,$$

because every division problem has a related multiplication problem. But zero times any number is zero, $0 \times a = 0$. Thus

$$7 = 0 \times a = 0.$$

That is, $7 = 0$, which we know is not true. Therefore, our assumption that $7 \div 0 = a$ is wrong. Thus we conclude that we cannot divide by zero. Mathematicians state this by saying, "Division by zero is **undefined.**"

It is helpful to remember the following basic concepts:

DIVISION PROBLEMS INVOLVING THE NUMBER 1 AND THE NUMBER 0

1. Any nonzero number divided by itself is $1 (7 \div 7 = 1)$.
2. Any number divided by 1 remains unchanged $(29 \div 1 = 29)$.
3. Zero may be divided by any nonzero number; the result is always zero $(0 \div 4 = 0)$.
4. Zero can never be the divisor in a division problem ($3 \div 0$ is undefined).

Teaching Tip It is critical that all students master division with zero. They need to know that zero can be divided by any nonzero number, but that division by zero is never allowed. A simple example usually helps them to see the logic of the situation. A student club with profits of \$400 and 10 members could distribute 400 divided by 10 = \$40 to each member. A student club with profits of \$0 and 10 members would only be able to distribute 0 divided by 10 = \$0 to each member. A student club with profits of \$400 and 0 members could not distribute any money to each member. 400 divided by zero cannot be done. Therefore division by zero is **undefined.**

EXAMPLE 2 Divide, if possible. If not possible, state why.

(a) $8 \div 8$ **(b)** $9 \div 1$ **(c)** $0 \div 6$ **(d)** $20 \div 0$

Solution

(a) $\frac{8}{8} = 1$ Any number divided by itself is 1.

(b) $\frac{9}{1} = 9$ Any number divided by 1 remains unchanged.

(c) $\frac{0}{6} = 0$ Zero divided by any nonzero number is zero.

(d) $\frac{20}{0}$ cannot be done Division by zero is undefined.

Practice Problem 2 Divide, if possible.

(a) $7 \div 1$ **(b)** $\frac{9}{9}$ **(c)** $\frac{0}{5}$ **(d)** $12 \div 0$

2 Performing Division by a One-Digit Number

Our accuracy with division is improved if we have a checking procedure. For each division fact, there is a related multiplication fact.

$$\text{If } 20 \div 4 = 5, \text{ then } 20 = 4 \times 5.$$
$$\text{If } 36 \div 9 = 4, \text{ then } 36 = 9 \times 4.$$

We will often use multiplication to check our answers.

When two numbers do not divide exactly, a number called the **remainder** is left over. For example, 13 cannot be divided exactly by 2. The number 1 is left over. We call this 1 the *remainder.*

$$\begin{array}{r} 6 \\ 2\overline{)13} \\ \underline{12} \\ 1 \leftarrow \text{remainder} \end{array}$$

Thus $13 \div 2 = 6$ with a remainder of 1. We can abbreviate this answer as

6 R 1.

To check this division, we multiply $2 \times 6 = 12$ and add the remainder: $12 + 1 = 13$. That is, $2 \times 6 + 1 = 13$. The result will be the dividend if the division was done correctly. The following box shows you how to check a division that has a remainder.

$$(\text{divisor} \times \text{quotient}) + \text{remainder} = \text{dividend}$$

Teaching Tip Be sure to go over in detail how to check a division problem if there is a remainder. A few students will find this idea totally new and will need to see a few examples worked out before they understand the concept.

EXAMPLE 3 Divide. $33 \div 4$. Check your answer.

Solution

$$\begin{array}{rl} 8 & \rightarrow \text{How many times can 4 be divided into 33? 8.} \\ 4\overline{)33} & \\ \underline{32} & \leftarrow \text{What is } 8 \times 4\text{? 32.} \\ 1 & \leftarrow \text{32 subtracted from 33 is 1.} \end{array}$$

The answer is 8 with a remainder of 1. We abbreviate 8 R 1.

CHECK.

$$\begin{array}{rl} 8 & \\ \underline{\times\ 4} & \\ 32 & \text{Multiply. } 8 \times 4 = 32. \\ \underline{1} & \text{Add the remainder. } 32 + 1 = 33. \\ 33 & \text{Because the dividend is 33, the answer is correct.} \end{array}$$

Practice Problem 3 Divide. $45 \div 6$. Check your answer.

EXAMPLE 4 Divide. $158 \div 5$. Check your answer.

Solution

```
   31
5)158     5 divided into 15?  3.
  15  ←   What is 3 × 5?  15.
   08 ←   15 subtract 15?  0. Bring down 8.
    5 ←   5 divided into 8?  1. What is 1 × 5?  5.
    3 ←   8 subtract 5?  3.
```

The answer is 31 R 3.

CHECK.

```
   31
 ×  5
  155   Multiply. 31 × 5 = 155.
    3   Add the remainder 3.
  158   Because the dividend is 158, the answer is correct.
```

NOTE TO STUDENT: Fully worked-out solutions to all of the Practice Problems can be found at the back of the text starting at page SP-1

Practice Problem 4 Divide. $129 \div 6$. Check your answer.

Teaching Tip Some students do not find it necessary when doing Example 5 to show the steps of long division. They merely record it mentally. Remind students that if they can divide accurately by a single digit without writing out these steps, they are not obligated to show the steps of long division.

EXAMPLE 5 Divide. $3672 \div 7$

Solution

```
   524
7)3672    How many times can 7 be divided into 36?  5.
  35   ←  What is 5 × 7?  35.
   17  ←  36 subtract 35?  1. Bring down 7.
   14  ←  7 divided into 17?  2. What is 2 × 7?  14.
    32 ←  17 subtract 14?  3. Bring down 2.
    28 ←  7 divided into 32?  4. What is 4 × 7?  28.
     4 ←  32 subtract 28?  4.
```

The answer is 524 R 4.

Practice Problem 5 Divide. $4237 \div 8$

3 Performing Division by a Two- or Three-Digit Number

When the divisor has more than one digit, an estimation technique may help. Figure how many times the first digit of the divisor goes into the first two digits of the dividend. Try this answer as the first number in the quotient.

EXAMPLE 6 Divide. $283 \div 41$

Solution

First guess:

```
     7
41)283
   287 too large
```

How many times can the first digit of the divisor (4) be divided into the first two digits of the dividend (28)? 7. We try the answer 7 as the first number of the quotient. We multiply $7 \times 41 = 287$. We see that 287 is larger than 283.

Second guess:

```
      6
41)283     Because 7 is slightly too large, we try 6.
   246 ←   6 × 41?   246.
    37 ←   283 subtract 246?   37.
```

The answer is 6 R 37. (Note that the remainder must always be less than the divisor.)

Practice Problem 6 Divide. 229 ÷ 32

EXAMPLE 7 Divide. 33,897 ÷ 56

Solution

First guess:

```
       60
56)33897     How many times can 33 be divided by 5?   6.
   336   ←   What is 6 × 56?   336.
    29       338 subtract 336?   2. Bring down 9.
             56 cannot be divided into 29. Write 0 in quotient.
```

Second set of steps:

```
       605
56)33897
   336        Bring down 7.
     297      How many times can 5 be divided into 29?   5.
     280 ←    What is 5 × 56?   280.   Subtract 297 − 280.
      17      Remainder is 17.
```

The answer is 605 R 17.

Practice Problem 7 Divide. 42,183 ÷ 33

EXAMPLE 8 Divide. 5629 ÷ 134

Solution

```
       42
134)5629    How many times does 134 divide into 562?
    536     We guess by saying that 1 divides into 5 five times, but
            this is too large. (5 × 134 = 670!)
     269    So we try 4. What is 4 × 134?   536.
     268    Subtract 562 − 536. We obtain 26. Bring down 9.
       1    How many times does 134 divide into 269?
            We guess by saying that 1 divided into 2 goes 2 times.
            What is 2 × 134?   268. Subtract 269 − 268.
            The remainder is 1.
```

The answer is 42 R 1.

Practice Problem 8 Divide. 3227 ÷ 128

4 Applying Division to Real-Life Situations

When you solve a word problem that requires division, you will be given the total number and asked to calculate the number of items in each group or to calculate the number of groups. In the beginning of this section we showed eight quarters (the total number) and we divided them into two equal piles (the number of groups). Division was used to find how many quarters were in each pile (the number in each group). That is, $8 \div 2 = 4$. There were four quarters in each pile.

Let's look at another example. Suppose that $30 is to be divided equally among the members of a group. If each person receives $6, how many people are in the group? We use division, $30 \div 6 = 5$, to find that there are five people in the group.

EXAMPLE 9 City Service Realty just purchased nine identical computers for the real estate agents in the office. The total cost for the nine computers was $25,848. What was the cost of one computer? Check your answer.

Solution To find the cost of one computer, we need to divide the total cost by 9. Thus we will calculate $25{,}848 \div 9$.

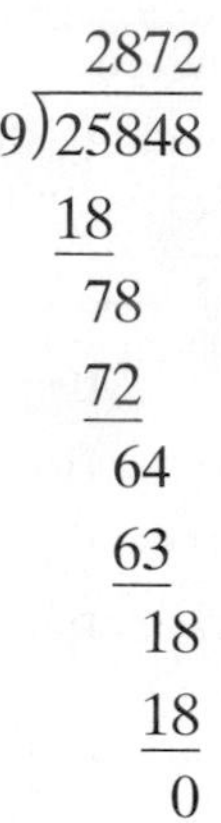

$$\begin{array}{r} 2872 \\ 9\overline{)25848} \\ \underline{18} \\ 78 \\ \underline{72} \\ 64 \\ \underline{63} \\ 18 \\ \underline{18} \\ 0 \end{array}$$

Therefore, the cost of one computer is $2872. In order to check our work we will need to see if nine computers each costing $2872 will in fact result in a total of $25,848. We use multiplication to check division.

$$\begin{array}{r} 2872 \\ \times \quad 9 \\ \hline 25848 \end{array} \checkmark$$

We did obtain 25,848. Our answer is correct.

Practice Problem 9 The Dallas police department purchased seven used identical police cars at a total cost of $117,964. Find the cost of one used car. Check your answer.

***NOTE TO STUDENT:** Fully worked-out solutions to all of the Practice Problems can be found at the back of the text starting at page SP-1*

In the following example you will see the word *average* used as it applies to division. The problem states that a car traveled 1144 miles in 22 hours. The problem asks you to find the average speed in miles per hour. This means that we will treat the problem as if the speed of the car were the same during each hour of the trip. We will use division to solve.

EXAMPLE 10 A car traveled from California to Texas, a distance of 1144 miles, in 22 hours. What was the average speed in miles per hour?

Solution When doing distance problems, it is helpful to remember that distance ÷ time = rate. We need to divide 1144 miles by 22 hours to obtain the rate or speed in miles per hour.

$$\begin{array}{r} 52 \\ 22\overline{)1144} \\ \underline{110} \\ 44 \\ \underline{44} \\ 0 \end{array}$$

The car traveled an average of 52 miles per hour.

Practice Problem 10 An airplane traveled 5138 miles in 14 hours. What was the average speed in miles per hour?

Developing Your Study Skills

Taking Notes in Class

An important part of mathematics studying is taking notes. In order to take meaningful notes, you must be an active listener. Keep your mind on what the instructor is saying, and be ready with questions whenever you do not understand something.

If you have previewed the lesson material, you will be prepared to take good notes. The important concepts will seem somewhat familiar. You will have a better idea of what needs to be written down. If you frantically try to write all that the instructor says or copy all the examples done in class, you may find your notes nearly worthless when you look at them at home. You may find that you are unable to make sense of what you have written.

Write down *important* ideas and examples as the instructor lectures, making sure that you are listening and following the logic. Include any helpful hints or suggestions that your instructor gives you or refers to in your text. You will be amazed at how easily you will forget these if you do not write them down. Try to review your notes the *same day* sometime after class. You will find the material in your notes easier to understand if you have attended class within the last few hours.

Successful note taking requires active listening and processing. Stay alert in class. You will realize the advantages of taking your own notes over copying those of someone else.

1.5 Exercises

Student Solutions Manual

CD/Video

PH Math Tutor Center

MathXL®Tutorials on CD

MathXL®

MyMathLab®

Interactmath.com

Verbal and Writing Skills

1. Explain in your own words what happens when you

(a) divide a nonzero number by itself.
When you divide a nonzero number by itself, the result is 1.

(b) divide a number by 1.
When you divide a number by 1, the result is that number.

(c) divide zero by a nonzero number.
When you divide zero by a nonzero number, the result is zero.

(d) divide a nonzero number by 0.
You cannot divide a number by 0. Division by zero is undefined.

Divide. See if you can work exercises 2–30 in three minutes or less.

2. $5\overline{)35}$ 7
3. $6\overline{)42}$ 7
4. $4\overline{)32}$ 8
5. $8\overline{)24}$ 3
6. $9\overline{)27}$ 3
7. $8\overline{)40}$ 5

8. $7\overline{)56}$ 8
9. $9\overline{)36}$ 4
10. $4\overline{)12}$ 3
11. $7\overline{)21}$ 3
12. $9\overline{)81}$ 9
13. $8\overline{)56}$ 7

14. $6\overline{)54}$ 9
15. $7\overline{)63}$ 9
16. $4\overline{)28}$ 7
17. $8\overline{)72}$ 9
18. $8\overline{)64}$ 8
19. $9\overline{)63}$ 7

20. $9\overline{)72}$ 8
21. $6\overline{)24}$ 4
22. $1\overline{)8}$ 8
23. $10\overline{)0}$ 0
24. $7\overline{)0}$ 0
25. $9 \div 0$ undefined

26. $12 \div 0$ undefined
27. $\frac{0}{8}$ 0
28. $\frac{0}{7}$ 0
29. $6 \div 6$ 1
30. $5 \div 5$ 1

Divide. In exercises 31–42, check your answer.

31. $29 \div 6$
$6\overline{)29}$ = 4 R 5
24
5
Check: 4 × 6 = 24; 24 + 5 = 29

32. $42 \div 8$
$8\overline{)42}$ = 5 R 2
40
2
Check: 5 × 8 = 40; 40 + 2 = 42

33. $76 \div 8$
$8\overline{)76}$ = 9 R 4
72
4
Check: 9 × 8 = 72; 72 + 4 = 76

34. $75 \div 9$
$9\overline{)75}$ = 8 R 3
72
3
Check: 8 × 9 = 72; 72 + 3 = 75

35. $128 \div 5$
$5\overline{)128}$ = 25 R 3
10
28
25
3
Check: 25 × 5 = 125; 125 + 3 = 128

36. $6\overline{)103}$ = 17 R 1
6
43
42
1
Check: 17 × 6 = 102; 102 + 1 = 103

37. $9\overline{)196}$ = 21 R 7
18
16
9
7
Check: 21 × 9 = 189; 189 + 7 = 196

38. $8\overline{)424}$ = 53
40
24
24
0
Check: 53 × 8 = 424

39. $9\overline{)288}$ = 32
27
18
18
0
Check: 32 × 9 = 288

40. $7\overline{)294}$ = 42
28
14
14
0
Check: 42 × 7 = 294

41. $5\overline{)185}$ = 37
15
35
35
0
Check: 37 × 5 = 185

42. $8\overline{)224}$ = 28
16
64
64
0
Check: 28 × 8 = 224

43. $4\overline{)1289}$ = 322 R 1
12
8
8
9
8
1

44. $3\overline{)758}$ = 252 R 2
6
15
15
8
6
2

45. $6\overline{)763}$ = 127 R 1
6
16
12
43
42
1

46. 57 R 4
7)403
35
53
49
4

47. 753
8)6024
56
42
40
24
24
0

48. 254
9)2286
18
48
45
36
36
0

49. 1122 R 1
3)3367
3
3
3
6
6
7
6
1

50. 1347 R 4
6)8086
6
20
18
28
24
46
42
4

51. 2056 R 2
8)16450
16
45
40
50
48
2

52. 3021 R 1
6)18,127
18
12
12
7
6
1

53. 2562 R 3
5)12,813
10
28
25
31
30
13
10
3

54. 4027 R 7
8)32,223
32
22
16
63
56
7

55. 185 ÷ 6
30 R 5
6)185
18
5
0
5

56. 202 ÷ 5
40 R 2
5)202
20
2
0
2

57. 267 ÷ 52
5 R 7
52)267
260
7

58. 321 ÷ 53
6 R 3
53)321
318
3

59. 427 ÷ 61
7
61)427
427
0

Mixed Practice

60. 6
72)432
432
0

61. 418 R 8
12)5024
48
22
12
104
96
8

62. 523 R 11
13)6810
65
31
26
50
39
11

63. 48 R 12
30)1452
120
252
240
12

64. 28 R 5
40)1125
80
325
320
5

65. 327
8)2616
24
21
16
56
56
0

66. 426
9)3834
36
23
18
54
54
0

67. 210 R 8
36)7568
72
36
36
8
0
8

68. 110 R 7
32)3527
32
32
32
7
0
7

69. 14 R 2
182)2550
182
730
728
2

70. 104 R 6
19)1982
19
82
76
6

71. 4 R 4
174)700
696
4

72. 7
128)896
896
0

73. 16
132)2112
132
792
792
0

74. 18 R 2
254)4574
254
2034
2032
2

Solve.

75. $518 \div 14 = x$. What is the value of x? $x = 37$

76. $1572 \div 131 = x$. What is the value of x? $x = 12$

Applications

77. ***Sports*** A *run* in skiing is going from the top of the ski lift to the bottom. If over seven days, 431,851 runs were made, what was the average number of ski runs per day?
61,693 runs

78. ***Farming*** Western Saddle Stable uses 21,900 pounds of feed per year to feed its 30 horses. How much does each horse eat per year?
730 pounds

79. ***Personal Finance*** For their wedding, Alan and Keisha paid $15 for each of their guest's dinner. The total bill was $2310. How many guests did they have at their wedding?
154 guests

80. ***Business Finances*** A telethon raised $3,677,880 over 15 hours. What was the average amount of money raised per hour?
$245,192

81. ***Business Finances*** A horse and carriage company in New York City bought seven new carriages at exactly the same price each. The total bill was $147,371. How much did each carriage cost?
$21,053

82. ***Real Estate*** A group of eight friends invested the same amount each in a beach property that sold for $369,432. How much did each friend pay?
$46,179

83. ***Personal Finance*** Lin wants to pay off her total of mortgage plus interest of $182,100 in 15 years. How much will she need to pay each year?
$12,140

84. ***Business Operations*** A new sorting machine sorts 26 letters per minute. The machine is fed 884 letters. How many minutes will it take to sort the letters?
34 minutes

85. ***Business Planning*** The 2nd Avenue Delicatessen is making bagel sandwiches for a New York City Marathon party. The sandwich maker has 360 bagel halves, 340 slices of turkey, and 330 slices of swiss cheese. If he needs to make sandwiches each consisting of two bagel halves, two slices of turkey, and two slices of swiss cheese, what is the greatest number of sandwiches he can make?
165 sandwiches

▲ **86.** ***Geometry*** Ace Landscaping is mowing a rectangular lawn that has an area of 2652 square feet. The company keeps a record of all lawn mowed in terms of length, width, square feet, and number of minutes it takes to mow the lawn. The width of the lawn is 34 feet. However, the page that lists the length of the lawn is soiled and the number cannot be read. Determine the length of the lawn.
78 feet

87. ***Business Management*** Dick Wightman is managing a company that is manufacturing and shipping modular homes in Canada. He has a truck that has made the trip from Toronto, Ontario to Halifax, Nova Scotia 12 times and has made the return run from Halifax to Toronto 12 times. The distance from Toronto to Halifax is 1742 kilometers.

(a) How many kilometers has the truck traveled on these 12 trips from Toronto to Halifax and back? 41,808 kilometers

(b) If Dick wants to limit the truck to a total of 50,000 kilometers driven this year, how many more kilometers can the truck be driven? 8192 kilometers

88. ***Space Travel*** The space shuttle has recently gone through a number of repairs and improvements. NASA recently approved the use of a shuttle control panel that has an area of 3526 square centimeters. The control panel is rectangular. The width of the panel is 43 centimeters. What is the length of the panel?

82 centimeters

To Think About

89. Division is not commutative. For example, $12 \div 4 \neq 4 \div 12$. If $a \div b = b \div a$, what must be true of the numbers a and b besides the fact that $b \neq 0$ and $a \neq 0$?

a and b must represent the same number. For example, if $a = 12$, then $b = 12$.

90. You can think of division as repeated subtraction. Show how $874 \div 138$ is related to repeated subtraction.

$$\begin{array}{rl} 874 & \\ -138 & \text{first time} \\ \hline 736 & \\ -138 & \text{second time} \\ \hline 598 & \\ -138 & \text{third time} \\ \hline 460 & \\ -138 & \text{fourth time} \\ \hline 322 & \\ -138 & \text{fifth time} \\ \hline 184 & \\ -138 & \text{sixth time} \\ \hline 46 & \end{array}$$

$$\begin{array}{r} 6 \text{ R } 46 \\ 138\overline{)874} \\ \underline{828} \\ 46 \end{array}$$

Cumulative Review

Solve.

91.
$$\begin{array}{r} 108 \\ \times\ 50 \\ \hline 5400 \end{array}$$

92.
$$\begin{array}{r} 7162 \\ \times\ 145 \\ \hline 35\ 810 \\ 286\ 48\ \ \\ 716\ 2\ \ \ \ \\ \hline 1{,}038{,}490 \end{array}$$

93. $316{,}214 + 89{,}981$

$$\begin{array}{r} 316{,}214 \\ +\ 89{,}981 \\ \hline 406{,}195 \end{array}$$

94. $1{,}360{,}000 - 1{,}293{,}156$

$$\begin{array}{r} 1{,}360{,}000 \\ -\ 1{,}293{,}156 \\ \hline 66{,}844 \end{array}$$

How Am I Doing? Sections 1.1–1.5

How are you doing with your homework assignments in Sections 1.1 to 1.5? Do you feel you have mastered the material so far? Do you understand the concepts you have covered? Before you go further in the textbook, take some time to do each of the following problems.

1.1

1. Write in words. 78,310,436

2. Write in expanded notation. 38,247

3. Write in standard notation. five million, sixty-four thousand, one hundred twenty-two

Use the following table to answer questions 4 and 5.

Public High School Graduates (in thousands)

Year	Graduates
1980	2,747
1995	2,273
2000	2,583

Source: U.S. Department of Education

4. How many public school graduates were there in 1980?

5. How many public school graduates were there in 2000?

1.2 *Add.*

6. $13 + 31 + 88 + 43 + 69$

7. $28{,}318 + 5{,}039 + 17{,}213$

8. $7148 + 500 + 19 + 7062$

1.3 *Subtract.*

9. $6439 - 2689$

10. $100{,}450 - 24{,}139$

11. $45{,}861{,}413 - 43{,}879{,}761$

1.4 *Multiply.*

12. $9 \times 6 \times 1 \times 2$

13. $3200 \times 40 \times 10$

14. 2658×7

15. 91×74

16. 365×908

1.5 *Divide. If there is a remainder, be sure to state it as part of the answer.*

17. $8\overline{)84{,}840}$

18. $7\overline{)51{,}633}$

19. $76\overline{)1976}$

20. $42\overline{)5838}$

Now turn to page SA-2 for the answer to each of these problems. Each answer also includes a reference to the objective in which the problem is first taught. If you missed any of these problems, you should stop and review the Examples and Practice Problems in the referenced objective. A little review now will help you master the material in the upcoming sections of the text.

1. Seventy-eight million, three hundred ten thousand, four hundred thirty-six

2. 30,000 + 8000 + 200 + 40 + 7

3. 5,064,122

4. 2,747,000

5. 2,583,000

6. 244

7. 50,570

8. 14,729

9. 3750

10. 76,311

11. 1,981,652

12. 108

13. 1,280,000

14. 18,606

15. 6734

16. 331,420

17. 10,605

18. 7376 R 1

19. 26

20. 139

1.6 Exponents and the Order of Operations

Student Learning Objectives

After studying this section, you will be able to:

1. **Evaluate expressions with whole-number exponents.**
2. **Perform several arithmetic operations in the proper order.**

1 Evaluating Expressions with Whole-Number Exponents

Sometimes a simple math idea comes "disguised" in technical language. For example, an **exponent** is just a "shorthand" number that saves writing multiplication of the same numbers.

10^3 The exponent 3 means $10 \times 10 \times 10$ (which takes longer to write).

The product 5×5 can be written as 5^2. The small number 2 is called the *exponent*. The exponent tells us how many factors are in the multiplication. The number 5 is called the **base.** The base is the number that is multiplied.

$$3 \times 3 \times 3 \times 3 = 3^4$$

(4 ← exponent; 3 ← base)

In 3^4 the base is 3 and the exponent is 4. (The 4 is sometimes called the *superscript.*) 3^4 is read as "three to the fourth power."

Teaching Tip Students who have never used exponents before often write the exponent with the same-size digit as the base. Remind them that exponents are written with a smaller digit than the base.

EXAMPLE 1 Write each product in exponent form.

(a) $15 \times 15 \times 15$ **(b)** $7 \times 7 \times 7 \times 7 \times 7$

Solution

(a) $15 \times 15 \times 15 = 15^3$ **(b)** $7 \times 7 \times 7 \times 7 \times 7 = 7^5$

Practice Problem 1 Write each product in exponent form.

(a) $12 \times 12 \times 12 \times 12$ **(b)** $2 \times 2 \times 2 \times 2 \times 2 \times 2$

NOTE TO STUDENT: Fully worked-out solutions to all of the Practice Problems can be found at the back of the text starting at page SP-1

EXAMPLE 2 Find the value of each expression.

(a) 3^3 **(b)** 7^2 **(c)** 2^5 **(d)** 1^8

Solution

(a) To find the value of 3^3, multiply the base 3 by itself 3 times.

$$3^3 = 3 \times 3 \times 3 = 27$$

(b) To find the value of 7^2, multiply the base 7 by itself 2 times.

$$7^2 = 7 \times 7 = 49$$

(c) $2^5 = 2 \times 2 \times 2 \times 2 \times 2 = 32$

(d) $1^8 = 1 \times 1 \times 1 \times 1 \times 1 \times 1 \times 1 \times 1 = 1$

Practice Problem 2 Find the value of each expression.

(a) 12^2 **(b)** 6^3 **(c)** 2^6 **(d)** 1^{10}

If a whole number does not have a visible exponent, the exponent is understood to be 1. Thus

$$3 = 3^1 \qquad \text{or} \qquad 10 = 10^1.$$

Large numbers are often expressed as a power of 10.

$10^1 = 10 = 1$ ten	$10^4 = 10{,}000 = 1$ ten thousand
$10^2 = 100 = 1$ hundred	$10^5 = 100{,}000 = 1$ hundred thousand
$10^3 = 1000 = 1$ thousand	$10^6 = 1{,}000{,}000 = 1$ million

What does it mean to have an exponent of zero? What is 10^0? Any whole number that is not zero can be raised to the zero power. The result is 1. Thus $10^0 = 1$, $3^0 = 1$, $5^0 = 1$, and so on. Why is this? Let's reexamine the powers of 10. As we go down one line at a time, notice the pattern that occurs.

As we move down one line, we decrease the exponent by 1.

$$10^5 = 100{,}000$$
$$10^4 = 10{,}000$$
$$10^3 = 1000$$
$$10^2 = 100$$
$$10^1 = 10$$
$$10^0 = 1$$

As we move down one line, we divide the previous number by 10.

Therefore, we present the following definition.

> For any whole number a other than zero, $a^0 = 1$.

If numbers with exponents are added to other numbers, it is first necessary to **evaluate**, or find the value of, the number that is raised to a power. Then we may combine the results with another number.

EXAMPLE 3 Find the value of each expression.

(a) $3^4 + 2^3$ **(b)** $5^3 + 7^0$ **(c)** $6^3 + 6$

Solution

(a) $3^4 + 2^3 = (3)(3)(3)(3) + (2)(2)(2) = 81 + 8 = 89$
(b) $5^3 + 7^0 = (5)(5)(5) + 1 = 125 + 1 = 126$
(c) $6^3 + 6 = (6)(6)(6) + 6 = 216 + 6 = 222$

NOTE TO STUDENT: Fully worked-out solutions to all of the Practice Problems can be found at the back of the text starting at page SP-1

Practice Problem 3 Find the value of each expression.

(a) $7^3 + 8^2$ **(b)** $9^2 + 6^0$ **(c)** $5^4 + 5$

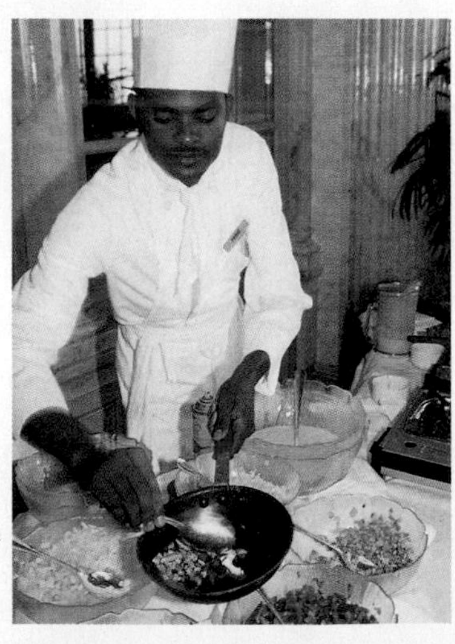

2 Performing Several Arithmetic Operations in the Proper Order

Sometimes the order in which we do things is not important. The order in which chefs hang up their pots and pans probably does not matter. The order in which they add and mix the elements in preparing food, however, makes all the difference in the world! If various cooks follow a recipe, though, they will get similar results. The recipe assures that the results will be consistent. It shows the **order of operations.**

In mathematics the order of operations is a list of priorities for working with the numbers in computational problems. This mathematical "recipe" tells how to handle certain indefinite computations. For example, how does a person solve $5 + 3 \times 2$?

A problem such as $5 + 3 \times 2$ sometimes causes students difficulty. Some people think $(5 + 3) \times 2 = 8 \times 2 = 16$. Some people think $5 + (3 \times 2) = 5 + 6 = 11$. Only one answer is right, 11. To obtain the right answer, follow the steps outlined in the following box.

Teaching Tip Students often ask why it is necessary to learn this order of operations. Explain that scientific calculators and computers follow this order of operations in performing complex calculations. It is therefore necessary for people using computers or scientific calculators to know this or they may misinterpret how the problem can be done using one of these devices. It is equally necessary in algebra courses and any higher math courses taken in college. This given order of operations is assumed in all math classes.

ORDER OF OPERATIONS

In the absence of grouping symbols:

Do first **1.** Simplify any expressions with exponents.
↕ **2.** Multiply or divide from left to right.
Do last **3.** Add or subtract from left to right.

EXAMPLE 4 Evaluate. $3^2 + 5 - 4 \times 2$

Solution

$3^2 + 5 - 4 \times 2 = 9 + 5 - 4 \times 2$	Evaluate the expression with exponents.
$= 9 + 5 - 8$	Multiply from left to right.
$= 14 - 8$	Add from left to right.
$= 6$	Subtract.

Practice Problem 4 Evaluate. $7 + 4^3 \times 3$

EXAMPLE 5 Evaluate. $5 + 12 \div 2 - 4 + 3 \times 6$

Solution There are no numbers to raise to a power, so we first do any multiplication or division in order from *left to right*.

$5 + 12 \div 2 - 4 + 3 \times 6$	Multiply or divide from left to right.
$= 5 + 6 - 4 + 3 \times 6$	Divide.
$= 5 + 6 - 4 + 18$	Multiply. Add or subtract from left to right.
$= 11 - 4 + 18$	Add.
$= 7 + 18$	Subtract.
$= 25$	Add.

Teaching Tip Students often try to do all the multiplication and then all the division in a given problem. Stress the fact that both multiplication and division have equal priority. They are to be done in order as you move from left to right.

Practice Problem 5 Evaluate. $37 - 20 \div 5 + 2 - 3 \times 4$

EXAMPLE 6 Evaluate. $2^3 + 3^2 - 7 \times 2$

Solution

$2^3 + 3^2 - 7 \times 2 = 8 + 9 - 7 \times 2$	Evaluate exponent expressions $2^3 = 8$ and $3^2 = 9$.
$= 8 + 9 - 14$	Multiply.
$= 17 - 14$	Add.
$= 3$	Subtract.

Practice Problem 6 Evaluate. $4^3 - 2 + 3^2$

You can change the order in which you compute by using grouping symbols. Place the numbers you want to calculate first within parentheses. This tells you to do those calculations first.

ORDER OF OPERATIONS

With grouping symbols:

Do first **1.** Perform operations inside the parentheses.
↕ **2.** Simplify any expressions with exponents.
3. Multiply or divide from left to right.
Do last **4.** Add or subtract from left to right.

EXAMPLE 7 Evaluate. $2 \times (7 + 5) \div 4 + 3 - 6$

Solution First, we combine numbers inside the parentheses by adding the 7 to the 5. Next, because multiplication and division have equal priority, we work from left to right doing whichever of these operations comes first.

$2 \times (7 + 5) \div 4 + 3 - 6$

$= 2 \times 12 \div 4 + 3 - 6$	Parentheses.
$= 24 \div 4 + 3 - 6$	Multiply.
$= 6 + 3 - 6$	Divide.
$= 9 - 6$	Add.
$= 3$	Subtract.

NOTE TO STUDENT: Fully worked-out solutions to all of the Practice Problems can be found at the back of the text starting at page SP-1

Practice Problem 7 Evaluate. $(17 + 7) \div 6 \times 2 + 7 \times 3 - 4$

EXAMPLE 8 Evaluate. $4^3 + 18 \div 3 - 2^4 - 3 \times (8 - 6)$

Solution

$4^3 + 18 \div 3 - 2^4 - 3 \times (8 - 6)$

$= 4^3 + 18 \div 3 - 2^4 - 3 \times 2$	Work inside the parentheses.
$= 64 + 18 \div 3 - 16 - 3 \times 2$	Evaluate exponents.
$= 64 + 6 - 16 - 3 \times 2$	Divide.
$= 64 + 6 - 16 - 6$	Multiply.
$= 70 - 16 - 6$	Add.
$= 54 - 6$	Subtract.
$= 48$	Subtract.

Practice Problem 8 Evaluate. $5^2 - 6 \div 2 + 3^4 + 7 \times (12 - 10)$

1.6 EXERCISES

Student Solutions Manual

CD/ Video

PH Math Tutor Center

MathXL®Tutorials on CD

MathXL®

MyMathLab®
interactmath.com

Verbal and Writing Skills

1. Explain what the expression 5^3 means. Evaluate 5^3.
   5^3 means $5 \times 5 \times 5$. $5^3 = 125$.

2. In exponent notation, the ___exponent___ tells how many times to multiply the base.

3. In exponent notation, the ___base___ is the number that is multiplied.

4. 10^5 is read as ___10 to the 5th power___.

5. Explain the order in which we perform mathematical operations to ensure consistency.
   To ensure consistency we
   1. perform operations inside parentheses
   2. simplify any expressions with exponents
   3. multiply or divide from left to right
   4. add or subtract from left to right

6. Use the order of operations to solve $12 \times 5 + 3 \times 5 + 7 \times 5$. Is this the same as $5(12 + 3 + 7)$? Why or why not?
   110 yes; because of the distributive property.

Write each number in exponent form.

7. $6 \times 6 \times 6 \times 6$
   6^4

8. $2 \times 2 \times 2 \times 2 \times 2$
   2^5

9. $5 \times 5 \times 5 \times 5 \times 5 \times 5$
   5^6

10. $8 \times 8 \times 8 \times 8 \times 8 \times 8 \times 8$
    8^7

11. $8 \times 8 \times 8 \times 8$
    8^4

12. $1 \times 1 \times 1 \times 1 \times 1 \times 1 \times 1$
    1^7

13. 9
    9^1

14. 27
    27^1

Find the value of each expression.

15. 2^4
    16

16. 3^3
    27

17. 4^3
    64

18. 5^2
    25

19. 6^2
    36

20. 10^3
    1000

21. 10^4
    10,000

22. 1^{20}
    1

23. 1^{17}
    1

24. 2^5
    32

25. 2^6
    64

26. 4^2
    16

27. 3^5
    243

28. 12^2
    144

29. 15^2
    225

30. 3^4
    81

31. 7^3
    343

32. 5^4
    625

33. 4^4
    256

34. 2^7
    128

35. 9^0
    1

36. 8^0
    1

37. 25^2
    625

38. 20^3
    8000

39. 10^6
    1,000,000

40. 8^1
    8

41. 13^2
    169

42. 11^2
    121

43. 9^1
    9

44. 14^2
    196

45. 7^4
    2401

46. 5^3
    125

47. $2^4 + 1^8$
    $16 + 1 = 17$

48. $7^0 + 4^3$
    $1 + 64 = 65$

49. $6^3 + 3^2$
    $216 + 9 = 225$

50. $7^3 + 4^2$
    $343 + 16 = 359$

51. $8^3 + 8$
    $512 + 8 = 520$

52. $9^2 + 9$
    $81 + 9 = 90$

Work each exercise, using the correct order of operations.

53. $7 \times 8 - 4$
$56 - 4 = 52$

54. $9 \times 3 + 10$
$27 + 10 = 37$

55. $3 \times 9 - 10 \div 2$
$27 - 5 = 22$

56. $4 \times 6 - 24 \div 4$
$24 - 6 = 18$

57. $48 \div 2^3 + 4$
$48 \div 8 + 4$
$= 6 + 4 = 10$

58. $4^3 \div 4 - 11$
$64 \div 4 - 11$
$= 16 - 11 = 5$

59. $3 \times 10^2 - 50$
$3 \times 100 - 50$
$= 300 - 50 = 250$

60. $2 \times 12^2 - 80$
$2 \times 144 - 80$
$= 288 - 80 = 208$

61. $10^2 + 3 \times (8 - 3)$
$100 + 3 \times 5$
$= 100 + 15 = 115$

62. $4^3 - 5 \times (9 + 1)$
$64 - 5 \times 10$
$= 64 - 50 = 14$

63. $(400 \div 20) \div 20$
$20 \div 20 = 1$

64. $(600 \div 30) \div 20$
$20 \div 20 = 1$

65. $950 \div (25 \div 5)$
$950 \div 5 = 190$

66. $875 \div (35 \div 7)$
$875 \div 5 = 175$

67. $(12)(5) - (12 + 5)$
$60 - 17 = 43$

68. $(3)(60) - (60 + 3)$
$180 - 63 = 117$

69. $3^2 + 4^2 \div 2^2$
$9 + 16 \div 4 = 9 + 4 = 13$

70. $7^2 + 9^2 \div 3^2$
$49 + 81 \div 9 = 49 + 9 = 58$

71. $(6)(7) - (12 - 8) \div 4$
$42 - 4 \div 4 = 42 - 1 = 41$

72. $(8)(9) - (15 - 5) \div 5$
$72 - 10 \div 5 = 72 - 2 = 70$

73. $100 - 3^2 \times 4$
$100 - 9 \times 4 = 100 - 36 = 64$

74. $130 - 4^2 \times 5$
$130 - 16 \times 5 = 130 - 80 = 50$

75. $5^2 + 2^2 + 3^3$
$25 + 4 + 27 = 56$

76. $2^3 + 3^2 + 4^3$
$8 + 9 + 64 = 81$

77. $72 \div 9 \times 3 \times 1 \div 2$
$8 \times 3 \times 1 \div 2 = 24 \div 2 = 12$

78. $120 \div 30 \times 2 \times 5 \div 8$
$4 \times 2 \times 5 \div 8 = 40 \div 8 = 5$

79. $12^2 - 6 \times 3 \times 4 \times 0$
$144 - 0 = 144$

80. $14^2 - 5 \times 2 \times 3 \times 0$
$196 - 0 = 196$

Mixed Practice

Work each exercise, using the correct order of operations

81. $4^2 \times 6 \div 3$
$16 \times 6 \div 3$
$= 96 \div 3 = 32$

82. $7^2 \times 3 \div 3$
$49 \times 3 \div 3$
$= 147 \div 3 = 49$

83. $16 - (27 \div 9) \times 4 + 1$
$16 - 3 \times 4 + 1$
$= 16 - 12 + 1 = 5$

84. $12 + (36 \div 6) \times 2 - 10$
$12 + 6 \times 2 - 10$
$= 12 + 12 - 10 = 14$

85. $3 + 3^2 \times 6 + 4$
$3 + 9 \times 6 + 4$
$= 3 + 54 + 4 = 61$

86. $5 + 4^3 \times 2 + 7$
$5 + 64 \times 2 + 7$
$= 5 + 128 + 7 = 140$

87. $32 \div 2 \times (3 - 1)^4$
$32 \div 2 \times (2)^4$
$= 32 \div 2 \times 16 = 16 \times 16 = 256$

88. $24 \div 3 \times (5 - 3)^2$
$24 \div 3 \times (2)^2$
$= 24 \div 3 \times 4 = 8 \times 4 = 32$

89. $3^2 \times 6 \div 9 + 4 \times 3$
$9 \times 6 \div 9 + 4 \times 3$
$= 6 + 12 = 18$

90. $5^2 \times 3 \div 25 + 7 \times 6$
$25 \times 3 \div 25 + 7 \times 6$
$= 3 + 42 = 45$

91. $6^2 + 3^4$
$36 + 81 = 117$

92. $5^3 + 4^2$
$125 + 16 = 141$

93. $1200 - 2^3(3) \div 6$

$1200 - 8(3) \div 6 = 1200 - 4 = 1196$

94. $2150 - 3^4(2) \div 9$

$2150 - 81(2) \div 9 = 2150 - 18 = 2132$

95. $250 \div 5 + 20 - 3^2$

$50 + 20 - 9 = 61$

96. $440 \div 20 - 10 + 2^5$

$22 - 10 + 32 = 44$

97. $250 \div (5 + 20) - 3^2$

$250 \div 25 - 9 = 10 - 9 = 1$

98. $440 \div (20 - 10) + 2^5$

$440 \div 10 + 32 = 44 + 32 = 76$

99. $2 \times 3 + (11 - 5)^2 - 4 \times 5$

$6 + (6)^2 - 20$
$= 6 + 36 - 20 = 22$

100. $(6 + 4)^2 \times 2 - 7 \times 7 + 2 \times 4$

$(10)^2 \times 2 - 49 + 8$
$= 100 \times 2 - 49 + 8 = 159$

To Think About

101. ***Astronomy*** The earth rotates once every 23 hours, 56 minutes, 4 seconds. How many seconds is that?

86,164 seconds

102. ***Astronomy*** The planet Saturn rotates once every 10 hours, 12 minutes. How many minutes is that? How many seconds?

612 minutes; 36,720 seconds

Cumulative Review

103. In the number 2,038,754

(a) What digit tells the number of ten thousands? 3

(b) What is the value of the digit 2?

2,000,000

104. Write in standard notation. two hundred million, seven hundred sixty-five thousand, nine hundred nine 200,765,909

105. Write in words. 261,763,002

two hundred sixty-one million, seven hundred sixty-three thousand, two

▲ **106.** ***Geometry*** New Boston High School has an athletic field that needs to be enclosed by fencing. The rectangular field is 250 feet wide and 480 feet long. How many feet of fencing are needed to surround the field? Grass needs to be planted for a new playing field for next year. What is the area in square feet of the amount of grass that must be planted?

1460 feet; 120,000 square feet

1.7 Rounding and Estimating

Student Learning Objectives

After studying this section, you will be able to:

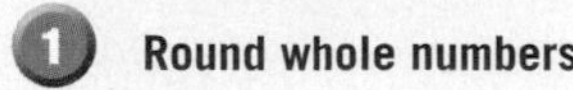

1. Round whole numbers.

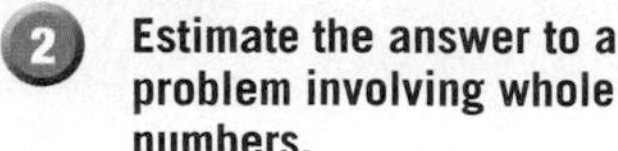

2. Estimate the answer to a problem involving whole numbers.

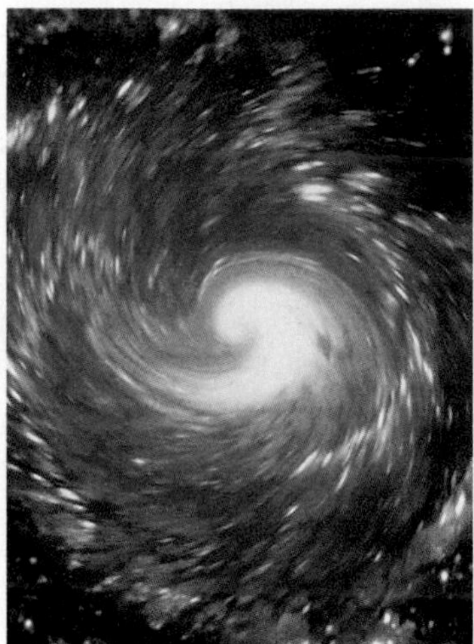

1 Rounding Whole Numbers

Large numbers are often expressed to the nearest hundred or to the nearest thousand, because an approximate number is "good enough" for certain uses.

Distances from the earth to other galaxies are measured in light-years. Although light really travels at 5,865,696,000,000 miles a year, we usually **round** this number to the nearest trillion and say it travels at 6,000,000,000,000 miles a year. To round a number, we first determine the place we are rounding to—in this case, trillion. Then we find which value is closest to the number that we are rounding. In this case, the number we want to round is closer to 6 trillion than to 5 trillion. How do we know the number is closer to 6 trillion than to 5 trillion?

To see which is the closest value, we may picture a **number line,** where whole numbers are represented by points on a line. To show how to use the number line in rounding, we will round 368 to the nearest hundred. 368 is between 300 and 400. When we round off, we pick the hundred 368 is "closest to." We draw a number line to show 300 and 400. We also show the point midway between 300 and 400 to help us to determine which hundred 368 is closest to.

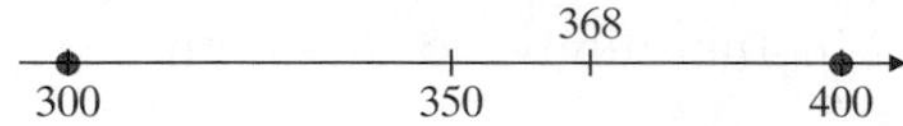

We find that the number 368 is closer to 400 than to 300, so we round 368 *up to* 400.

Let's look at another example. We will round 129 to the nearest hundred. 129 is between 100 and 200. We show this on the number line. We include the midpoint 150 as a guide.

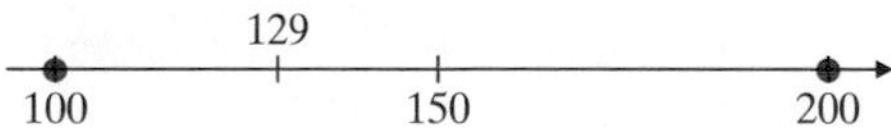

We find that the number 129 is closer to 100 than to 200, so we round 129 *down to* 100.

This leads us to the following simple rule for rounding.

Teaching Tip These rules for rounding off are used uniformly in science and mathematics classes and applications. However, a particular business or bank may employ a different rule. Inform students that if rounding off is done in a business, a different rule may be employed.

ROUNDING A WHOLE NUMBER

1. If the first digit to the right of the round-off place is
   - **(a)** *less than 5,* we make no change to the digit in the round-off place. (We know it is closer to the smaller number, so we round down.)
   - **(b)** *5 or more,* we increase the digit in the round-off place by 1. (We know it is closer to the larger number, so we round up.)
2. Then we replace the digits to the right of the round-off place by zeros.

EXAMPLE 1 Round 37,843 to the nearest thousand.

Solution

3 7, 8 4 3 According to the directions, the thousands will be the round-off place. We locate the thousands place.

↓

3 7, (8) 4 3 We see that the first digit to the right of the round-off place is 8, which is 5 or more. We increase the thousands digit by 1, and replace all digits to the right by zero.

↓

3 8, 0 0 0

We have rounded 37,843 to the nearest thousand: 38,000. This means that 37,843 is closer to 38,000 than to 37,000.

Practice Problem 1 Round 65,528 to the nearest thousand.

NOTE TO STUDENT: Fully worked-out solutions to all of the Practice Problems can be found at the back of the text starting at page SP-1

EXAMPLE 2 Round 2,445,360 to the nearest hundred thousand.

Solution

2, 4 4 5, 3 6 0 Locate the hundred thousands round-off place.

↓

2, 4 (4) 5, 3 6 0 The first digit to the right of this is less than 5, so round down. Do not change the hundred thousands digit.

↓

2, 4 0 0, 0 0 0 Replace all digits to the right by zero.

Practice Problem 2 Round 172,963 to the nearest ten thousand.

EXAMPLE 3 Round as indicated.

(a) 561,328 to the nearest ten **(b)** 3,798,152 to the nearest hundred
(c) 51,362,523 to the nearest million

Solution

(a) ↓ First locate the digit in the tens place.

561,328 The digit to the right of the tens is greater than 5.

561,330 Round up.

561,328 rounded to the nearest ten is 561,330.

(b) 3,798,152 The digit to the right of the hundreds is 5.

3,798,200 Round up.

3,798,152 rounded to the nearest hundred is 3,798,200.

(c) 51,362,523 The digit to the right of the millions is less than 5.

51,000,000 Round down.

51,362,523 rounded to the nearest million is 51,000,000.

Practice Problem 3 Round as indicated.

(a) 53,282 to the nearest ten **(b)** 164,485 to the nearest thousand
(c) 1,365,273 to the nearest hundred thousand

Teaching Tip As a classroom activity, ask students to round off 56,489 to the nearest thousand. They will obtain the correct answer 56,000. Then ask them what is wrong with this reasoning: Rounded to the nearest hundred, 56,489 becomes 56,500. Then rounding 56,500 to the nearest thousand gives 57,000. Why is this approach not correct? Usually students will point out the error in the reasoning. If not, draw a number line showing the location of each number.

EXAMPLE 4 Round 763,571.

(a) To the nearest thousand
(b) To the nearest ten thousand
(c) To the nearest million

Solution

(a) 763,571 = 764,000 to the nearest thousand. The digit to the right of the thousands is 5. We rounded up.

↓

(b) 763,571 = 760,000 to the nearest ten thousand. The digit to the right of the ten thousands is less than 5. We rounded down.

(c) 763,571 does not have any digits for millions. If it helps, you can think of this number as 0,763,571. Since the digit to the right of the millions place is 7, we round up to obtain one million or 1,000,000.

NOTE TO STUDENT: Fully worked-out solutions to all of the Practice Problems can be found at the back of the text starting at page SP-1

Practice Problem 4 Round 935,682 as indicated.

(a) To the nearest thousand
(b) To the nearest hundred thousand
(c) To the nearest million

EXAMPLE 5 Astronomers use the parsec as a measurement of distance. One parsec is approximately 30,900,000,000,000 kilometers. Round 1 parsec to the nearest trillion kilometers.

↓

Solution 30,900,000,000,000 km is 31,000,000,000,000 km or 31 trillion km to the nearest trillion kilometers.

Practice Problem 5 One light-year is approximately 9,460,000,000,000,000 meters. Round to the nearest hundred trillion meters.

2 Estimating the Answer to a Problem Involving Whole Numbers

Often we need to quickly check the answer of a calculation to be reasonably assured that the answer is correct. If you expected your bill to be "around $40" for the groceries you had selected and the cashier's total came to $41.89, you would probably be confident that the bill is correct and pay it. If, however, the cashier rang up a bill of $367, you would not just assume that it is correct. You would know an error had been made. If the cashier's total came to $60, you might not be certain, but you would probably suspect an error and check the calculation.

In mathematics we often **estimate,** or determine the approximate value of a calculation, if we need to do a quick check. There are many ways to estimate, but in this book we will use one simple principle of estimation. We use the symbol $\approx$ to mean **approximately equal to.**

PRINCIPLE OF ESTIMATION

1. Round the numbers so that there is one nonzero digit in each number.
2. Perform the calculation with the rounded numbers.

EXAMPLE 6 Estimate the sum. 163 + 237 + 846 + 922

Solution We first determine where to round each number in our problem to leave only one nonzero digit in each. In this case, we round all numbers to the nearest hundred. Then we perform the calculation with the rounded numbers.

Actual Sum	*Estimated Sum*
163	200
237	200
846	800
+ 922	+ 900
	2100

We estimate the answer to be 2100. We say the sum $\approx$ 2100. If we calculate using the exact numbers, we obtain a sum of 2168, so our estimate is quite close to the actual sum.

Practice Problem 6 Estimate the sum. 3456 + 9876 + 5421 + 1278

When we use the principle of estimation, we will not always round each number in a problem to the same place.

EXAMPLE 7 Phil and Melissa bought their first car last week. The selling price of this compact car was $8980. The dealer preparation charge was $289 and the sales tax was $449. Estimate the total cost of the car that Phil and Melissa had to pay.

Solution We round each number to have only one nonzero digit, and add the rounded numbers.

8980	9000
289	300
+ 449	+ 400
	9700

The total cost ≈ \$9700. (The exact answer is \$9718, so we see that our answer is quite close.)

NOTE TO STUDENT: Fully worked-out solutions to all of the Practice Problems can be found at the back of the text starting at page SP-1

Practice Problem 7 Greg and Marcia purchased a new sofa for \$697, plus \$35 sales tax. The store also charged them \$19 to deliver the sofa. Estimate their total cost.

Now we turn to a case where an estimate can help us discover an error.

EXAMPLE 8 Roberto added together four numbers and obtained the following result. Estimate the sum and determine if the answer seems reasonable.

$$12{,}456 + 17{,}976 + 18{,}452 + 32{,}128 \stackrel{?}{=} 61{,}012$$

Solution We round each number so that there is one nonzero digit. In this case, we round them all to the nearest ten thousand.

$$\begin{array}{rr} 12{,}456 & 10{,}000 \\ 17{,}976 & 20{,}000 \\ 18{,}452 & 20{,}000 \\ \underline{+\,32{,}128} & \underline{+\,30{,}000} \\ & 80{,}000 \end{array}$$

Our estimate is 80,000.

This is significantly different from 61,012, so we would suspect that an error has been made. In fact, Roberto did make an error. The exact sum is actually 81,012!

Practice Problem 8 Ming did the following calculation. Estimate to see if her sum appears to be correct or incorrect.

$$11{,}849 + 14{,}376 + 16{,}982 + 58{,}151 = 81{,}358$$

Next we look at a subtraction example where estimation is used.

EXAMPLE 9 The profit from Techno Industries for the first quarter of the year was \$642,987,000. The profit for the second quarter was \$238,890,000. Estimate how much less the profit was for the second quarter than for the first quarter.

Solution We round each number so that there is one nonzero digit. Then we subtract, using the two rounded numbers.

$$\begin{array}{rr} 642{,}987{,}000 & 600{,}000{,}000 \\ \underline{-\,238{,}890{,}000} & \underline{-\,200{,}000{,}000} \\ & 400{,}000{,}000 \end{array}$$

We estimate that the profit was \$400,000,000 less for the second quarter.

Practice Problem 9 The 2001 population of Florida was 16,397,426. The 2001 population of California was 34,501,728. Estimate how many more people lived in California in 2001 than in Florida.

We also use this principle to estimate results of multiplication and division.

EXAMPLE 10 Estimate the product. $56{,}789 \times 529$

Solution We round each number so that there is one nonzero digit. Then we multiply the rounded numbers to obtain our estimate.

$$\begin{array}{r} 56{,}789 \\ \times \quad 529 \\ \hline \end{array} \qquad \begin{array}{r} 60{,}000 \\ \times \quad 500 \\ \hline 30{,}000{,}000 \end{array}$$

Therefore the product is $\approx 30{,}000{,}000$. (This is reasonably close to the exact answer of 30,041,381.)

Practice Problem 10 Estimate the product. 8945×7317

EXAMPLE 11 Estimate the answer for the following division problem.

$$23\overline{)148{,}902}$$

Solution We round each number to a number with one nonzero digit. Then we perform the division, using the two rounded numbers.

$$23\overline{)148{,}902} \qquad \begin{array}{r} 5000 \\ 20\overline{)100{,}000} \end{array}$$

Our estimate is 5000. (The exact answer is 6474. We see that our estimate is "in the ballpark" but is not very close to the exact answer. Remember, an estimate is just a rough approximation of the exact answer.)

Practice Problem 11 Estimate the answer for the following division problem.

$$39\overline{)75{,}342}$$

Not all division estimates come out so easily. In some cases, you may need to carry out a long-division problem of several steps just to obtain the estimate.

EXAMPLE 12 John and Stephanie drove their car a distance of 778 miles. They used 25 gallons of gas. Estimate how many miles they can travel on 1 gallon of gas.

Solution In order to solve this problem, we need to divide 778 by 25 to obtain the number of miles John and Stephanie get with 1 gallon of gas.

We round each number to a number with one nonzero digit and then perform the division, using the rounded numbers.

$$25\overline{)778} \qquad \begin{array}{r} 26 \\ 30\overline{)800} \\ \underline{60} \\ 200 \\ \underline{180} \\ 20 \end{array} \text{ Remainder}$$

We obtain an answer of 26 with a remainder of 20. For our estimate we will use the whole number 27. Thus we estimate that the number of miles their car obtained on 1 gallon of gas was 27 miles. (This is reasonably close to the exact answer, which is just slightly more than 31 miles per gallon of gas.)

***NOTE TO STUDENT:** Fully worked-out solutions to all of the Practice Problems can be found at the back of the text starting at page SP-1*

Practice Problem 12 The highway department purchased 58 identical trucks at a total cost of $1,864,584. Estimate the cost for one truck.

Developing Your Study Skills

How To Do Homework

Set aside time each day for your homework assignments. Make a weekly schedule and write down the times each day you will devote to doing math homework. Two hours spent studying outside class for each hour in class is usual for college courses. You may need more than that for mathematics.

Before beginning to solve your homework exercises, read your textbook very carefully. Expect to spend much more time reading a few pages of a mathematics textbook than several pages of another text. Read for complete understanding, not just for the general idea.

As you begin your homework assignments, read the directions carefully. You need to understand what is being asked. Concentrate on each exercise, taking time to solve it accurately. Rushing through your work usually errors. Check your answers with those given in the back of the textbook. If your answer is incorrect, check to see that you are doing the right problem. Redo the problem, watching for errors. If it is still wrong, check with a friend. Perhaps the two of you can figure out where you are going wrong.

Work on your assignments every day and do as many exercises as it takes for you to know what you are doing. Begin by doing all the exercises that have been assigned. If there are more available in that section of your text, then do more. When you think you have done enough exercises to understand fully the topic at hand, do a few more to be sure. This may mean that you do many more exercises than the instructor assigns, but you can never practice mathematics too much. Practice improves your skills and increases your accuracy, speed, competence, and confidence.

Also, check the examples in the textbook or in your notes for a similar exercise. Can this one be solved in the same way? Give it some thought. You may want to leave it for a while by taking a break or doing a different exercise. But come back later and try again. If you are still unable to figure it out, ask your instructor for help during office hours or in class.

1.7 EXERCISES

Student Solutions Manual | CD/Video | PH Math Tutor Center | MathXL®Tutorials on CD | MathXL® | MyMathLab® | Interactmath.con

Verbal and Writing Skills

1. Explain the rule for rounding and provide examples.

Locate the rounding place. If the digit to the right of the rounding place is 5 or greater than 5, round up. If the digit to the right of the rounding place is less than 5, round down. *Note:* Examples provided by students will vary. Check for accuracy.

2. What happens when you round 98 to the nearest ten?

Since the digit to the right of tens is greater than 5, you round up. When you round up 9 tens, it becomes 10 tens or 100.

Round to the nearest ten.

3. 83
80

4. 45
50

5. 65
70

6. 57
60

7. 168
170

8. 132
130

9. 7438
7440

10. 2834
2830

11. 1672
1670

12. 7865
7870

Round to the nearest hundred.

13. 247
200

14. 661
700

15. 2781
2800

16. 1258
1300

17. 7692
7700

18. 1643
1600

Round to the nearest thousand.

19. 7621
8000

20. 3754
4000

21. 1672
2000

22. 515
1000

23. 27,863
28,000

24. 94,489
94,000

Applications

25. ***History*** The worst death rate from an earthquake was in Shaanxi, China, in 1556. That earthquake killed an estimated 832,400 people. Round this number to the nearest hundred thousand.

800,000

26. ***Astronomy*** One light year (the distance light travels in 1 year) measures 5,878,612,843,000 miles. Round this figure to the nearest hundred million.

5,878,600,000,000 miles

27. ***Astronomy*** The Hubble Space Telescope's *Guide Star Catalogue* lists 15,169,873 stars. Round this figure to the nearest million.

15,000,000 stars

28. ***Geography*** The point of highest elevation in the world is Mt. Everest in the country of Nepal. Mt. Everest is 29,028 feet above sea level. Round this figure to the nearest ten thousand.

30,000 feet

29. ***Native American Studies*** The total Native American population living on the Navajo and Trust Lands in the Arizona/New Mexico/Utah area numbered 163,298 in 2000. Round this figure to

(a) the nearest thousand.
163,000

(b) the nearest hundred.
163,300

30. ***Population Studies*** In June 2003, the population of the United States was projected to be 291,293,143. Round this figure to

(a) the nearest thousand.
291,293,000

(b) the nearest hundred.
291,293,100

▲ **31.** ***Geography*** The total area of mainland China is 3,705,392 square miles, or, 9,596,960 square kilometers. For *both* square miles and square kilometers, round this figure to

(a) the nearest hundred thousand.
3,700,000 square miles, 9,600,000 square kilometers

(b) the nearest ten thousand.
3,710,000 square miles, 9,600,000 square kilometers

▲ **32.** ***Geography*** The area of the Pacific Ocean is 165,384,000 square kilometers. Round this figure to

(a) the nearest hundred thousand.
165,400,000

(b) the nearest ten thousand.
165,380,000

Use the principle of estimation to find an estimate for each calculation.

33. $613 + 252 + 137$
600 + 300 + 100 = 1000

34. $871 + 365 + 341$
900 + 400 + 300 = 1600

35. $42 + 69 + 95 + 18$
40 + 70 + 100 + 20 = 230

36. $62 + 27 + 54 + 98$
60 + 30 + 50 + 100 = 240

37. $158{,}270 + 53{,}441 + 8701$
200,000 + 50,000 + 9000 = 259,000

38. $238{,}271 + 77{,}304 + 9551$
200,000 + 80,000 + 10,000 = 290,000

39. $567{,}984 - 129{,}562$
600,000 − 100,000 = 500,000

40. $975{,}935 - 593{,}228$
1,000,000 − 600,000 = 400,000

41. $831{,}201 - 74{,}244$
800,000 − 80,000 = 720,000

42. $382{,}140 - 56{,}117$
400,000 − 60,000 = 340,000

43. $33{,}261{,}378 - 18{,}199{,}276$
30,000,000 − 20,000,000 = 10,000,000

44. $89{,}263{,}000 - 54{,}198{,}635$
90,000,000 − 50,000,000 = 40,000,000

45. 47×62
60 × 50 = 3000

46. 43×95
100 × 40 = 4000

47. 1324×8
1000 × 8 = 8000

48. 5926×3
6000 × 3 = 18,000

49. $631{,}540 \times 312$
600,000 × 300 = 180,000,000

50. $374{,}193 \times 193$
400,000 × 200 = 80,000,000

51. $5782 \div 33$
$30\overline{)6000}$ = 200

52. $9691 \div 38$
$40\overline{)10{,}000}$ = 250

53. $156{,}721 \div 42$
$40\overline{)200{,}000}$ = 5,000

54. $581{,}361 \div 28$
$30\overline{)600{,}000}$ = 20,000

55. $3{,}885{,}720 \div 831$
$800\overline{)4{,}000{,}000}$ = 5,000

56. $12{,}447{,}312 \div 497$
$500\overline{)10{,}000{,}000}$ = 20,000

Estimate the result of each calculation. Some results are correct and some are incorrect. Which results appear to be correct? Which results appear to be incorrect?

57.

	Estimate
361	400
522	500
873	900
+ 164	+ 200
1320	2000

Incorrect

58.

	Estimate
476	500
124	100
516	500
+ 389	+ 400
1505	1500

Correct

59.

	Estimate
97,635	100,000
52,123	50,000
+ 41,986	+ 40,000
291,744	190,000

Incorrect

60.

	Estimate
26,181	30,000
47,998	50,000
+ 63,271	+ 60,000
137,450	140,000

Correct

61. 302,360 − 89,518 = 212,842

300,000 − 90,000 = 210,000
Correct

62. 735,128 − 116,733 = 518,395

700,000 − 100,000 = 600,000
Incorrect

63. 78,126,345 − 48,972,103 = 19,154,242

80,000,000 − 50,000,000 = 30,000,000
Incorrect

64. 42,765,317 − 29,318,274 = 23,447,043

40,000,000 − 30,000,000 = 10,000,000
Incorrect

65. 216 × 24 = 6184

200 × 20 = 4000
Incorrect

66. 578 × 32 = 10,496

600 × 30 = 18,000
Incorrect

67. 5896 × 72 = 424,512

6000 × 70 = 420,000
Correct

68. 8076 × 89 = 718,764

8000 × 90 = 720,000
Correct

69. $36\overline{)82,116}$ = 2281

$40\overline{)80,000}$ = 2000
Correct

70. $52\overline{)28,912}$ = 556

$50\overline{)30,000}$ = 600
Correct

71. $423\overline{)161,163}$ = 381

$400\overline{)200,000}$ = 500
Correct

72. $781\overline{)477,972}$ = 612

$800\overline{)500,000}$ = 625
4800
2000
1600
4000
4000
Correct

Applications

▲ **73.** ***Geometry*** Larry and Nella are planning an outdoor wedding by the ocean. There is a beautiful meadow that is 35 feet wide and 62 feet long. Estimate the number of square feet in the meadow.
2400 square feet

▲ **74.** ***Geometry*** A huge restaurant in New York City is 43 yards wide and 112 yards long. Estimate the number of square yards in the restaurant.
4000 square yards

75. ***International Relations*** The populations of three large German cities are Hamburg with 1,714,962 people, Berlin with 3,395,739, and Bonn with 300,822. Estimate the total population of the three cities.
5,300,000

76. ***Financial Management*** The highway departments in four towns in northwestern New York had the following budgets for snow removal for the year: $329,560, $672,940, $199,734, and $567,087. Estimate the total amount that the four towns spend for snow removal in one year.
$1,800,000

77. ***Business Management*** The local pizzeria makes 267 pizzas on an average day. Estimate how many pizzas were made in the last 134 days.
30,000 pizzas

78. ***Personal Finance*** Darcy makes \$68 for each shift she works. She is scheduled for 33 shifts during the next two months. Estimate how much she will earn in the next two months.
\$2100

79. ***Transportation*** U.S. air travel increased from 457,000,000 passengers in 1990 to 656,000,000 in 2000. Round each figure to the nearest ten million. Then estimate the increase.
200,000,000 passengers

80. ***Sports*** In 1960 the average attendance at a Patriots game (officially called the Boston Patriots at that time) was 25,783. In 2003 the average attendance at a New England Patriots game was 46,924. Estimate the increase in attendance over this time period.
20,000

▲ **81.** ***Geography*** The largest state of the United States is Alaska, with a land area of 586,412 square miles. The second largest state is Texas, with an area of 267,339 square miles. Round each figure to the nearest ten thousand. Then estimate how many square miles larger Alaska is than Texas.
320,000 square miles

▲ **82.** ***Geography*** The largest state of Mexico in land area is Chihuahua, with a land area of 244,938 square kilometers. The second largest state of Mexico in land area is Sonora, with a land area of 182,052 square kilometers. Round each figure to the nearest thousand. Then estimate how many square kilometers larger is Chihuahua than Sonora.
63,000 square kilometers

To Think About

83. ***Space Travel*** A space probe travels at 23,560 miles per hour for a distance of 7,824,560,000 miles.

(a) How many *hours* will it take the space probe to travel that distance? (Estimate.)
400,000 hours

(b) How many *days* will it take the space probe to travel that distance? (Estimate.)
20,000 days

84. ***Space Travel*** A space probe travels at 28,367 miles per hour for a distance of 9,348,487,000 miles.

(a) Estimate the number of *hours* it will take the space probe to travel that distance.
300,000 hours

(b) Estimate the number of *days* it will take the space probe to travel that distance.
15,000 days

Cumulative Review

Evaluate.

85. $26 \times 3 + 20 \div 4$
83

86. $5^2 + 3^2 - (17 - 10)$
27

87. $3 \times (16 \div 4) + 8 \times 2$
28

88. $126 + 4 - (20 \div 5)^3$
66

89. $\begin{array}{r} 5489 \\ \times\ \ 67 \\ \hline \end{array}$
367,763

90. $52\overline{)4524}$
87

1.8 Solving Applied Problems Involving Whole Numbers

1 Solving Problems Involving One Operation

When a builder constructs a new home or office building, he or she often has a *blueprint*. This accurate drawing shows the basic structure of the building. It also shows the dimensions of the structure to be built. This blueprint serves as a useful reference throughout the construction process.

Student Learning Objectives

After studying this section, you will be able to:

1. **Use the Mathematics Blueprint to solve problems involving one operation.**
2. **Use the Mathematics Blueprint to solve problems involving more than one operation.**

Similarly, when solving applied problems, it is helpful to have a "mathematics blueprint." This is a simple way to organize the information provided in the word problem. You record the facts you need to use and specify what you are solving for. You also record any other information that you feel will be helpful. We will use a Mathematics Blueprint for Problem Solving in the following situation.

Sometimes people feel totally lost when trying to solve a word problem. They sometimes say, "Where do I begin?" or "How in the world do you do this?" When you have this type of feeling, it sometimes helps to have a formal strategy or plan. Here is a plan you may find helpful:

1. *Understand the problem.*
- **(a)** Read the problem carefully.
- **(b)** Draw a picture if this helps you see the relationships more clearly.
- **(c)** Fill in the Mathematics Blueprint so that you have the facts and a method of proceeding in this situation.

2. *Solve and state the answer.*
- **(a)** Perform the calculations.
- **(b)** State the answer, including the unit of measure.

3. *Check.*
- **(a)** Estimate the answer.
- **(b)** Compare the exact answer with the estimate to see if your answer is reasonable.

Now exactly what does the Mathematics Blueprint for Problem Solving look like? It is a simple sheet of paper with four columns. Each column tells you something to do.

Gather the Facts—Find the numbers that you will need to use in your calculations.

What Am I Asked to Do?—Are you finding an area, a volume, a cost, the total number of people? What is it that you need to find?

How Do I Proceed?—Do you need to add items together? Do you need to multiply or divide? What types of calculations are required?

Key Points to Remember—Write down things you might forget. The length is in feet. The area is in square feet. We need the total number of something, not the intermediate totals. Whatever you need to help you, write it down in this column.

Mathematics Blueprint for Problem Solving

Gather the Facts	What Am I Asked to Do?	How Do I Proceed?	Key Points to Remember

EXAMPLE 1 Gerald made deposits of \$317, \$512, \$84, and \$161 into his checking account. He also made out checks for \$100 and \$125. What was the total of his deposits?

Solution

1. ***Understand the problem.*** First we read over the problem carefully and fill in the Mathematics Blueprint.

Mathematics Blueprint for Problem Solving

Gather the Facts	What Am I Asked to Do?	How Do I Proceed?	Key Points to Remember
We need only deposits—not checks. The **deposits** are \$317, \$512, \$84, and \$161.	Find the total of Gerald's four deposits.	I must add the four deposits to obtain the total.	Watch out! Don't use the **checks** of \$100 and \$125 in the calculation. We only want the total of the **deposits**.

2. ***Solve and state the answer.*** We need to *add* to find the sum of the deposits.

$$\begin{array}{r} 317 \\ 512 \\ 84 \\ +161 \\ \hline 1074 \end{array}$$

The total of the four deposits is $1074.

3. ***Check.*** Reread the problem. Be sure you have answered the question that was asked. Did it ask for the total of the deposits? Yes. ✓

Is the calculation correct? You can use estimation to check. Here we round each of the deposits so that we have one nonzero digit.

$$\begin{array}{rr} 317 & 300 \\ 512 & 500 \\ 84 & 80 \\ +161 & +200 \\ \hline & 1080 \end{array}$$

Our estimate is $1080. $1074 is close to our estimated answer of $1080. Our answer is reasonable. ✓

Thus we conclude that the total of the four deposits is $1074.

Practice Problem 1 Use the Mathematics Blueprint to solve the following problem. Diane's paycheck shows deductions of $135 for federal taxes, $28 for state taxes, $13 for FICA, and $34 for health insurance. Her gross pay (amount before deductions) is $1352. What is the total amount that is taken out of Diane's paycheck?

***NOTE TO STUDENT:** Fully worked-out solutions to all of the Practice Problems can be found at the back of the text starting at page SP-1*

Mathematics Blueprint for Problem Solving

Gather the Facts	What Am I Asked to Do?	How Do I Proceed?	Key Points to Remember

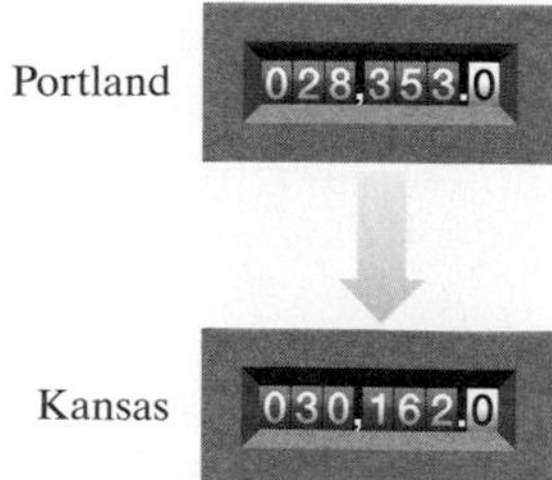

EXAMPLE 2 Theofilos looked at his odometer before he began his trip from Portland, Oregon, to Kansas City, Kansas. He checked his odometer again when he arrived in Kansas City. The two readings are shown in the figure. How many miles did Theofilos travel?

Solution

1. ***Understand the problem.*** Determine what information is given.
   The mileage reading before the trip began and when the trip was over.
   What do you need to find?
   The number of miles traveled.

Mathematics Blueprint for Problem Solving

Gather the Facts	What Am I Asked to Do?	How Do I Proceed?	Key Points to Remember
At the start of the trip, the odometer read 28,353 miles. At the end of the trip, the odometer read 30,162 miles.	Find out how many miles Theofilos traveled.	I must subtract the two mileage readings.	Subtract the mileage at the start of the trip from the mileage at the end of the trip.

2. ***Solve and state the answer.*** We need to subtract the two mileage readings to find the difference in the number of miles. This will give us the number of miles the car traveled on this trip alone.

   $30{,}162 - 28{,}353 = 1809$ The trip totaled 1809 miles.

3. ***Check.*** We estimate and compare the estimate with the preceding answer.

   $$\begin{array}{lrcr}\text{Kansas City} & 30{,}162 & \longrightarrow & 30{,}000 \\ \text{Portland} & 28{,}353 & \longrightarrow & \underline{28{,}000} \\ & & & 2{,}000\end{array}$$
   We subtract our rounded values.

   Our estimate is 2000 miles. We compare this estimate with our answer. Our answer is reasonable. ✓

2000 Presidential Race, Popular Votes

Candidate	Number of Votes
Bush (R)	50,456,002
Gore (D)	50,999,897
Nader	2,882,955

Source: Federal Election Commission

Practice Problem 2 The table on the right shows the results of the 2000 presidential race in the United States. By how many popular votes did the Democratic candidate beat the Republican candidate in that year? Why did the Democratic candidate not win the election in that year?

Mathematics Blueprint for Problem Solving

Gather the Facts	What Am I Asked to Do?	How Do I Proceed?	Key Points to Remember

***NOTE TO STUDENT:** Fully worked-out solutions to all of the Practice Problems can be found at the back of the text starting at page SP-1*

EXAMPLE 3 One horsepower is the power needed to lift 550 pounds a distance of 1 foot in 1 second. How many pounds can be lifted 1 foot in 1 second by 7 horsepower?

Solution

1. ***Understand the problem.*** Simplify the problem. If 1 horsepower can lift 550 pounds, how many pounds can be lifted by 7 horsepower? We draw and label a diagram.

We use the mathematics blueprint to organize the information.

Mathematics Blueprint for Problem Solving

Gather the Facts	What Am I Asked to Do?	How Do I Proceed?	Key Points to Remember
One horsepower will lift 550 pounds.	Find how many pounds can be lifted by 7 horsepower.	I need to multiply 550 by 7.	I do not use the information about moving one foot in one second.

2. ***Solve and state the answer.*** To solve the problem we multiply the 7 horsepower by 550 pounds for each horsepower.

$$\begin{array}{r} 550 \\ \times\ 7 \\ \hline 3850 \end{array}$$

We find that 7 horsepower moves 3850 pounds 1 foot in 1 second. We include 1 foot in 1 second in our answer because it is part of the unit of measure.

3. ***Check.*** We estimate our answer. We round 550 to 600 pounds.

$$600 \times 7 = 4200 \text{ pounds}$$

Our estimate is 4200 pounds. Our calculations in step 2 gave us 3850. Is this reasonable? This answer is close to our estimate. Our answer is reasonable. ✓

Practice Problem 3 In a measure of liquid capacity, 1 gallon is 1024 fluid drams. How many fluid drams would be in 9 gallons?

Mathematics Blueprint for Problem Solving

Gather the Facts	What Am I Asked to Do?	How Do I Proceed?	Key Points to Remember

EXAMPLE 4 Laura can type 35 words per minute. She has to type an English theme that has 5180 words. How many minutes will it take her to type the theme? How many hours and how many minutes will it take her to type the theme?

Solution

1. ***Understand the problem.*** We draw a picture. Each "package" of 1 minute is 35 words. We want to know how many packages make up 5180 words.

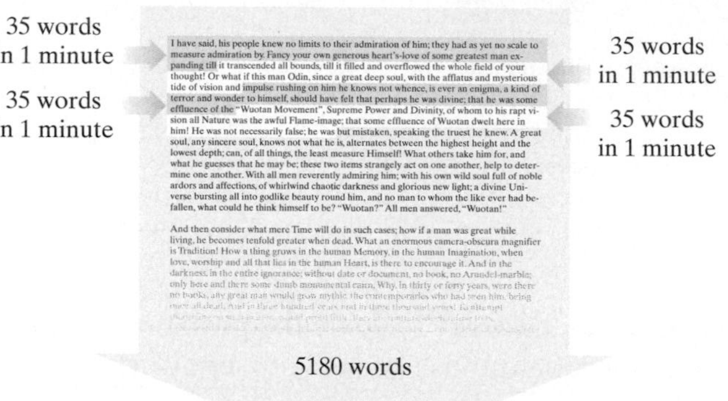

We use the Mathematics Blueprint to organize the information.

Mathematics Blueprint for Problem Solving

Gather the Facts	What Am I Asked to Do?	How Do I Proceed?	Key Points to Remember
Laura can type 35 words per minute. She must type a paper with 5180 words.	Find out how many 35-word units are in 5180 words.	I need to divide 5180 by 35.	In converting minutes to hours, I will use the fact that 1 hour = 60 minutes.

2. ***Solve and state the answer.***

$$\begin{array}{r} 148 \\ 35\overline{)5180} \\ \underline{35} \\ 168 \\ \underline{140} \\ 280 \\ \underline{280} \\ 0 \end{array}$$

It will take 148 minutes.

We will change this answer to hours and minutes. Since 60 minutes = 1 hour, we divide 148 by 60. The quotient will tell us how many hours. The remainder will tell us how many minutes.

$$\begin{array}{r} 2 \text{ R } 28 \\ 60\overline{)148} \\ \underline{-120} \\ 28 \end{array}$$

Laura can type the theme in 148 minutes or 2 hours, 28 minutes.

3. ***Check.*** The theme has 5180 words; she can type 35 words per minute. 5180 words is approximately 5000 words.

5180 words → 5000 words rounded to nearest thousand. $\quad 40\overline{)5000}$ with quotient 125

35 words per minute → 40 words per minute rounded to nearest ten. We divide our estimated values.

Our estimate is 125 minutes. This is close to our calculated answer. Our answer is reasonable. ✓

Practice Problem 4 Donna bought 45 shares of stock for $1620. How much did the stock cost her per share?

NOTE TO STUDENT: Fully worked-out solutions to all of the Practice Problems can be found at the back of the text starting at page SP-1

Mathematics Blueprint for Problem Solving

Gather the Facts	What Am I Asked to Do?	How Do I Proceed?	Key Points to Remember

 Solving Problems Involving More Than One Operation

Sometimes a chart, table, or bill of sale can be used to help us organize the data in an applied problem. In such cases, a blueprint may not be needed.

EXAMPLE 5 Cleanway Rent-A-Car bought four used luxury sedans at $21,000 each, three compact sedans at $14,000 each, and seven subcompact sedans at $8000 each. What was the total cost of the purchase?

Solution

1. ***Understand the problem.*** We will make an imaginary bill of sale to help us to visualize the problem.
2. ***Solve and state the answer.*** We do the calculation and enter the results in the bill of sale.

Car Fleet Sales, Inc. Hamilton, Massachusetts

Customer: *Cleanway Rent-A-Car*			
Quantity	**Type of Car**	**Cost per Car**	**Amount for This Type of Car**
4	Luxury sedans	$21,000	$84,000 ($4 \times \$21,000 = \$84,000$)
3	Compact sedans	$14,000	$42,000 ($3 \times \$14,000 = \$42,000$)
7	Subcompact sedans	$8,000	$56,000 ($7 \times \$8,000 = \$56,000$)
		TOTAL	$182,000 (sum of the three amounts)

The total cost of all 14 cars is $182,000.

3. ***Check.*** You may use estimation to check. The check is left to the student.

NOTE TO STUDENT: Fully worked-out solutions to all of the Practice Problems can be found at the back of the text starting at page SP-1

Practice Problem 5 Anderson Dining Commons purchased 50 tables at $200 each, 180 chairs at $40 each, and six moving carts at $65 each. What was the cost of the total purchase?

EXAMPLE 6 Dawn had a balance of $410 in her checking account last month. She made deposits of $46, $18, $150, $379, and $22. She made out checks for $316, $400, and $89. What is her balance?

Solution

1. ***Understand the problem.*** We want to *add* to get a total of all deposits and *add* to get a total of all checks.

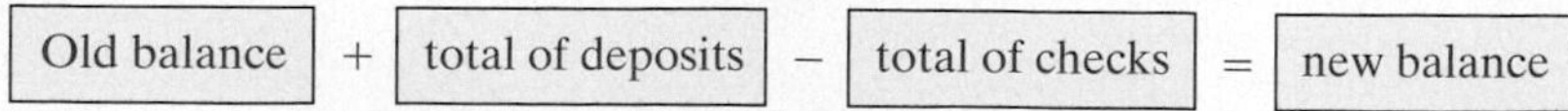

Mathematics Blueprint for Problem Solving

Gather the Facts	What Am I Asked to Do?	How Do I Proceed?	Key Points to Remember
Old balance: \$410. New deposits: \$46, \$18, \$150, \$379, and \$22. New checks: \$316, \$400, and \$89.	Find the amount of money in the checking account after deposits are made and checks are withdrawn.	**(a)** I need to calculate the total of the deposits and the total of the checks. **(b)** I add the total deposits to the old balance. **(c)** Then I subtract the total of the checks from that result.	Deposits are added to a checking account. Checks are subtracted from a checking account.

2. *Solve and state the answer.*

First we find the total sum of deposits:

$$\begin{array}{r} 46 \\ 18 \\ 150 \\ 379 \\ +\ 22 \\ \hline \$615 \end{array}$$

Then the total sum of checks:

$$\begin{array}{r} 316 \\ 400 \\ +\ 89 \\ \hline \$805 \end{array}$$

Add the deposits to the old balance and subtract the amount of the checks.

Old balance	410
+ total deposits	+ 615
	1025
− total checks	− 805
New balance	220

The new balance of the checking account is \$220.

3. *Check.* Work backward. You can add the total checks to the new balance and then subtract the total deposits. The result should be the old balance. Try it.

$$\begin{array}{r} 410 \\ -\ 615 \\ \hline 1025 \\ +\ 805 \\ \hline 220 \end{array}$$

410 Old balance ✓

↑ Work backward.

Practice Problem 6 Last month Bridget had \$498 in a savings account. She made two deposits: one for \$607 and one for \$163. The bank credited her with \$36 interest. Since last month, she has made four withdrawals: \$19, \$158, \$582, and \$74. What is her balance this month?

Mathematics Blueprint for Problem Solving

Gather the Facts	What Am I Asked to Do?	How Do I Proceed?	Key Points to Remember

EXAMPLE 7 When Lorenzo began his car trip, his gas tank was full and the odometer read 76,358 miles. He ended his trip at 76,668 miles and filled the gas tank with 10 gallons of gas. How many miles per gallon did he get with his car?

Solution

1. ***Understand the problem.***

Mathematics Blueprint for Problem Solving

Gather the Facts	What Am I Asked to Do?	How Do I Proceed?	Key Points to Remember
Odometer reading at end of trip: 76,668 miles. Odometer reading at start of trip: 76,358 miles. Used on trip: 10 gallons of gas	Find the number of miles per gallon that the car obtained on the trip.	**(a)** I need to subtract the two odometer readings to obtain the number of miles traveled. **(b)** I divide the number of miles driven by the number of gallons of gas used to get the number of miles obtained per gallon of gas.	The gas tank was full at the beginning of the trip. 10 gallons fills the tank at the end of the trip.

2. ***Solve and state the answer.*** First we subtract the odometer readings to obtain the miles traveled.

$$\begin{array}{r} 76{,}668 \\ -\ 76{,}358 \\ \hline 310 \end{array}$$

The trip was 310 miles.

Next we divide the miles driven by the number of gallons.

$$\begin{array}{r} 31 \\ 10\overline{)310} \\ \underline{30} \\ 10 \\ \underline{10} \\ 0 \end{array}$$

Thus Lorenzo obtained 31 miles per gallon on the trip.

3. ***Check.*** We do not want to round to one nonzero digit here, because, if we do, the result will be zero when we subtract. Thus we will round to the nearest hundred for the values of mileage.

$$76{,}668 \longrightarrow 76{,}700$$
$$76{,}358 \longrightarrow 76{,}400$$

Now we subtract the estimated values.

$$\begin{array}{r} 76{,}700 \\ -\ 76{,}400 \\ \hline 300 \end{array}$$

Thus we estimate the trip to be 300 miles.
Then we divide.

$$\begin{array}{r} 30 \\ 10\overline{)300} \end{array}$$

We obtain 30 miles per gallon for our estimate. This is very close to our calculated value of 31 miles per gallon. ✓

Practice Problem 7 Deidre took a car trip with a full tank of gas. Her trip began with the odometer at 50,698 and ended at 51,118 miles. She then filled the tank with 12 gallons of gas. How many miles per gallon did her car get on the trip?

NOTE TO STUDENT: Fully worked-out solutions to all of the Practice Problems can be found at the back of the text starting at page SP-1

Mathematics Blueprint for Problem Solving

Gather the Facts	What Am I Asked to Do?	How Do I Proceed?	Key Points to Remember

If you attend a traditional mathematics class that meets one or more times each week:

Developing Your Study Skills

Class Attendance

You will want to get started in the right direction by choosing to attend class every day, beginning with the first day of class. Statistics show that class attendance and good grades go together. Classroom activities are designed to enhance learning, and therefore you must be in class to benefit from them. Each day vital information and explanations are given that can help you understand concepts. Do not be deceived into thinking that you can just find out from a friend what went on in class. There is no good substitute for firsthand experience. Give yourself a push in the right direction by developing the habit of going to class every day.

If you are enrolled in an online mathematics class, a self-paced mathematics class taught in a math lab, or some other type of nontraditional class:

Developing Your Study Skills

Keeping Yourself on Schedule

In a class where you determine your own pace, you will need to commit yourself to keeping on a schedule. Follow the suggested pace provided in your course materials. Keep all your class materials organized and review them often to be sure you are doing everything that you should. If you discipline yourself to follow the suggested course schedule for the first six weeks, you will likely succeed in the class. Professor Tobey and Professor Slater both teach online mathematics courses. They have found that students usually succeed in the course as long as they do every suggested activity for the first six weeks. Make sure you succeed! Keep yourself on schedule!

1.8 Exercises

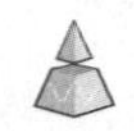

Student Solutions Manual | CD/Video | PH Math Tutor Center | MathXL®Tutorials on CD | MathXL® | MyMathLab® | Interactmath.con

Applications

You may want to use the Mathematics Blueprint for Problem Solving to help you to solve the word problems in exercises 1–34.

1. ***Real Estate*** Donna and Miguel want to buy a cabin for \$31,500. After repairs, the total cost will be \$40,300. How much will the repairs cost?
\$8800

▲ **2.** ***Geography*** China has a total area of 9,596,960 square kilometers. Bodies of water account for 270,550 square kilometers. How many square kilometers of land does China have?
9,326,410 square kilometers

3. ***Business Management*** Paula is organizing a large two-day convention. Bert's Bagels is providing the breakfast bagels. If Paula orders 120 bakers' dozen, how many bagels is that? (There are 13 in a bakers' dozen.) 1560 bagels

4. ***Business Management*** There are 144 pencils in a gross. Mr. Jim Weston ordered 14 gross of pencils for the office. How many pencils did he order?
2016 pencils

5. A 16-ounce can of beets costs 96¢. What is the unit cost of the beets? (How much do the beets cost per ounce?)
6¢ per ounce

6. ***Consumer Affairs*** A 14-ounce can of chicken soup costs 98¢. What is the unit cost of the soup? (How much does the soup cost per ounce?)
7¢ per ounce

7. ***Sports*** Kimberly began running 3 years ago. She has spent \$832 on 13 pairs of running shoes during this time. How much on average did each pair of shoes cost?
\$64

8. ***Wildlife Management*** There are approximately 50,000 bison living in the United States. If Northwest Trek, the animal preserve located in Mt. Rainier National Park, has 103 bison, how many bison are living elsewhere?
49,897

9. ***Business Management*** Sergio can bake 60 muffins in one hour. A large company ordered 300 muffins for a company breakfast. How many hours will Sergio need to fill the order? How many minutes is this?
5 hours; 300 minutes

▲ **10.** ***Geometry*** If a new amusement park covers 43 acres and there are 44,010 square feet in 1 acre, how many square feet of land does the amusement park cover?
1,892,430 square feet

11. ***Business Management*** A games arcade has recently opened in a West Chicago neighborhood. The owners were nervous about whether it would be a success. Fortunately, the gross revenues over the last four weeks were \$7356, \$3257, \$4777, and \$4992. What was the gross revenue over these four weeks for the arcade?
\$20,382

12. ***International Relations*** The two largest cities in Saudi Arabia are Riyadh, the capital, with 1,250,000 people, and Jeddah, with 900,000 people. What is the difference in population between these two cities?
350,000 people

13. ***Wildlife Management*** The Federal Nigeria game preserve has 24,111 animals, 327 full-time staff, and 793 volunteers. What is the total of these three groups? How many more volunteers are there than full-time staff?
25,231; 466

14. ***Geography*** The longest rivers in the world are the Nile River, the Amazon River, and the Mississippi River. Their lengths are 4132 miles, 3915 miles, and 3741 miles, respectively. How many total miles do these three rivers run? What is the difference in the lengths of the Nile and the Mississippi?
11,788 miles; 391 miles

15. ***World History*** Every 60 minutes, the world population increases by 100,000 people. How many people will be born during the next 480 minutes?

800,000 people

16. ***Personal Finance*** Roberto had \$2158 in his savings account six months ago. In the last six months he made four deposits: \$156, \$238, \$1119, and \$866. The bank deposited \$136 in interest over the six-month period. How much does he have in the savings account at present?

\$4673

In exercises 17–34, more than one type of operation is required.

17. ***Sports*** Carmen gives golf lessons every Saturday. She charges \$15 for adults, \$9 for children, and \$5 for club rental. Last Saturday she taught six adults and eight children, and six people needed to rent clubs. How much money did Carmen make on that day?

\$192

18. ***Business Management*** Whale Watch Excursions charges \$10 for adults, \$6 for children, and \$7 for senior citizens. On the last trip of the day, there were five adults, seven children, and three senior citizens. How much money did the company make on this trip?

\$113

19. ***Personal Finance*** Wei Mai Lee had a balance in her checking account of \$61. During the last few months, she has made deposits of \$385, \$945, \$732, and \$144. She wrote checks against her account for \$223, \$29, \$98, and \$435. When all the deposits are recorded and all the checks clear, what balance will she have in her checking account?

\$1482

20. ***Film Studies*** The 2003 Cannes Film Festival had 16,392 entries from 70 countries. The United States sent 2382 entries, Germany sent 1485 and the United Kingdom sent 1466. How many entries were sent by the other 67 countries?

11,059 entries

21. ***Real Estate*** Diana owns 85 acres of forest land in Oregon. She rents it to a timber grower for \$250 per acre per year. Her property taxes are \$57 per acre. How much profit does she make on the land each year?

\$16,405

22. ***Real Estate*** Todd owns 13 acres of commercially zoned land in the city of Columbus, Ohio. He rents it to a construction company for \$12,350 per acre per year. His property taxes to the city are \$7362 per acre per year. How much profit does he make on the land each year?

\$64,844

23. ***Environmental Studies*** Hanna wants to determine the miles-per-gallon rating of her Chevrolet Cavalier. She filled the tank when the odometer read 14,926 miles. She then drove her car on a trip. At the end of the trip, the odometer read 15,276 miles. It took 14 gallons to fill the tank. How many miles per gallon does her car deliver?

25 miles per gallon

24. ***Environmental Studies*** Gary wants to determine the miles-per-gallon rating of his Geo Metro. He filled the tank when the odometer read 28,862 miles. After ten days, the odometer read 29,438 miles and the tank required 18 gallons to be filled. How many miles per gallon did Gary's car achieve?

32 miles per gallon

25. ***Forestry*** A beautiful piece of land in the Wilmot Nature Preserve has three times as many oak trees as birches, two times as many maples as oaks, and seven times as many pine trees as maples. If there are 18 birches on the land, how many of each of the other trees are there? How many trees are there in all?

There are 54 oak trees, 108 maple trees, and 756 pine trees. In total there are 936 trees.

26. ***Business Management*** The Cool Coffee Lounge in Albuquerque, New Mexico, has 27 tables, and each table has either two or four chairs. If there are a total of 94 chairs accompanying the 27 tables, how many tables have four chairs? How many tables have two chairs?

20 tables have 4 chairs;
7 tables have 2 chairs.

Use the following list to answer exercises 27–30.

Education

The following is a partial list of the primary home languages of students attending Public School 139 in Queens, New York in the school year 2002–2003.

Language	Number of Students
Russian	226
English	183
Spanish	174
Mandarin	53
Cantonese	44
Korean	40
Hindi	29
Chinese, other dialects	21
Filipino	12
Hebrew	9
Indonesian	8
Romanian	8
Urdu	8
Dari/Farsi/Persian	7
Albanian	6
Arabic	6
Bulgarian	5
Gujarati	4

Source: Office of the Superintendent of Schools, Queens, New York.

27. How many students speak Mandarin, Cantonese, or other Chinese dialects as the primary language in their homes?
118

28. How many students speak Korean, Hindi, or Filipino as the primary language in their homes?
81

29. How many more students speak Spanish or Russian rather than English as the primary language in their homes?
217

30. How many more students speak Indonesian or Romanian rather than Albanian as the primary language in their homes?
10

Use the following bar graph to answer exercises 31–34.

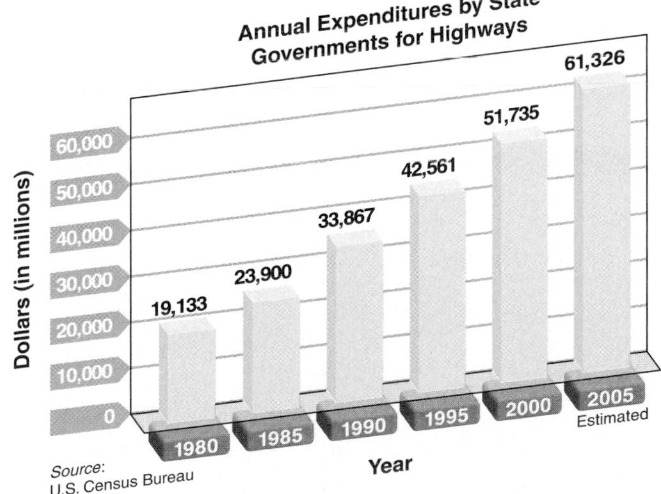

Government Finances

31. How many more dollars were spent by state governments for highways in 1990 than in 1980? $14,734,000,000

32. How many more dollars were spent by state governments for highways in 2000 than in 1985?
$27,835,000,000

33. If the exact same dollar increase occurs between 2000 and 2010 as occurred between 1990 and 2000, what will the expenditures by state governments for highways be in 2010?
$69,603,000,000

34. If the amount of money expended by state governments for highways remained constant for the years 2000 to 2003, how much money was spent for highways during that four-year period?
$206,940,000,000

Cumulative Review

35. Evaluate. 7^3

343

36. Perform in the proper order.

$3 \times 2^3 + 15 \div 3 - 4 \times 2$

$3 \times 8 + 15 \div 3 - 4 \times 2 = 24 + 5 - 8 = 21$

37. Multiply. 126×38

4788

38. Divide. $12\overline{)3096}$

258

39. Add. $96 + 123 + 57 + 526$

802

40. Subtract. $509{,}263 - 485{,}978$

23,285

41. Round to the nearest thousand. 526,195,726

526,196,000

42. Write this number in standard notation. Three billion, four hundred million, six hundred three thousand, twenty-five.

3,400,603,025

Putting Your Skills to Work

The Mathematics of Clean Water

Clean, refreshing drinking water. It is something we often take for granted. However, in many countries of the world it is a rare thing. Some villages have never had clean drinking water and the result has been sickness and infections. A second problem often related is a lack of any kind of sewerage facilities. Often there are no public or private bathrooms or toilets of any kind in rural villages. The result is a serious contamination of any possible drinking water as well as water used for cooking and cleaning.

World Vision is a relief organization that has been working to provide clean drinking water and improved sanitary conditions in countries where these are significant needs. Here are some statistics that show what changes they have been making in the country of Ghana.

These graphs refer to three phases of work done in Ghana. Phase I was during the period October 1985 to September 1990, when World Vision constructed wells with some financial support from USAID. Phase II was during the period October 1990 to September 1995, when World Vision constructed wells with financial support from the Conrad N. Hilton Foundation. Phase III was during the period October 1995 to September 2003, when World Vision constructed wells with financial support from the Conrad N. Hilton Foundation.

Problems for Individual Investigation and Analysis

1. How many more wells were dug during Phase II than Phase I? 74
2. How many wells in total were dug during the three phases? 1821

Problems for Group Investigation and Cooperative Study

3. How many new people were helped by the construction of the wells in Phase II? 330,000
4. How many new people were helped by the construction of the wells in Phase III? 350,000
5. If 455 wells were drilled in Phase I and this benefited 450,000 people, approximately how many people benefited from each well? Round your answer to the nearest whole number. 989
6. If 529 wells were drilled in Phase II and this benefited 780,000 people, approximately how many people benefited from each well? Round your answer to the nearest whole number. 1474

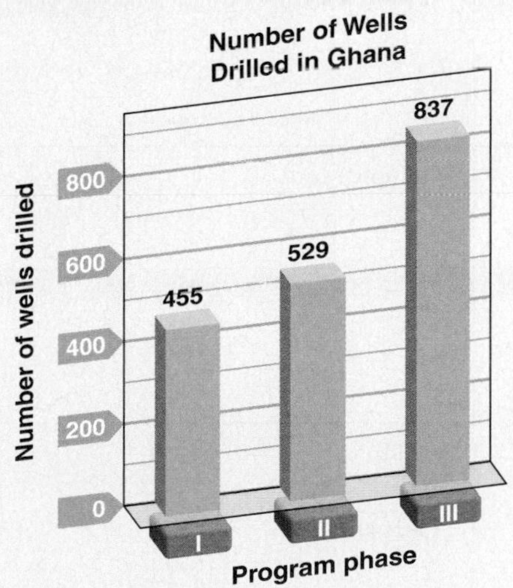

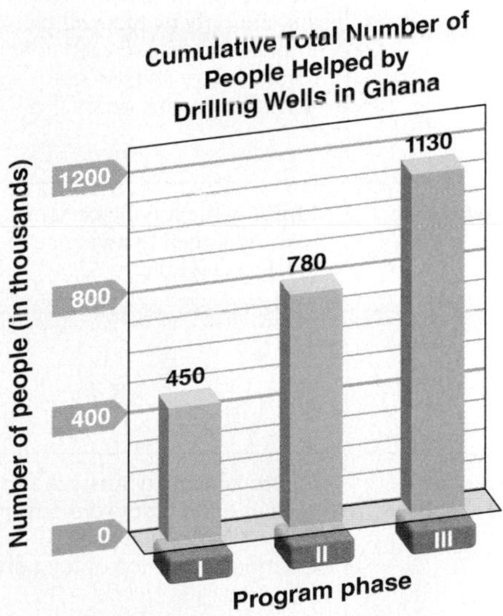

Source: World Vision

Chapter 1 Organizer

Topic	Procedure	Examples
Place value of numbers, p. 2.	Each digit has a value depending on location. millions | hundred thousands | ten thousands | thousands | hundreds | tens | ones	In the number 2,896,341, what place value does 9 have? ten thousands
Writing expanded notation, p. 2.	Take the number of each digit and multiply it by one, ten, hundred, thousand, . . . according to its place.	Write in expanded notation. 46,235 40,000 + 6000 + 200 + 30 + 5
Writing whole numbers in words, p. 4.	Take the number in each period and indicate if they are (millions) (thousands) (ones) XXX, XXX, XXX	Write in words. 134,718,216. one hundred thirty-four million, seven hundred eighteen thousand, two hundred sixteen
Adding whole numbers, p. 13.	Starting with the right column, add each column separately. If a two-digit sum occurs, "carry" the first digit over to the next column to the left.	Add. 2 1 (carries) 2 5 8 3 6 7 2 9 1 + 4 5 3 1 3 6 9
Subtracting whole numbers, p. 26.	Starting with the right column, subtract each column separately. If necessary, borrow a unit from column to the left and bring it to right as a "10."	Subtract. 13 6 3 12 (borrows, with 7, 4, 2 crossed out) 1 6,7 4 2 − 1 2,3 9 5 4,3 4 7
Multiplying several factors, p. 37.	Keep multiplying from left to right. Take each product and multiply by the next factor to the right. Continue until all factors are used once. (Since multiplication is commutative and associative, the factors can be multiplied in any order.)	Multiply. $2 \times 9 \times 7 \times 6 \times 3 = 18 \times 7 \times 6 \times 3$ $= 126 \times 6 \times 3$ $= 756 \times 3$ $= 2268$
Multiplying several-digit numbers, p. 40.	Multiply the top factor by the ones digit, then by the tens digit, then by the hundreds digit. Add the partial products together.	Multiply. 5 6 7 × 2 3 8 4 5 3 6 1 7 0 1 1 1 3 4 1 3 4, 9 4 6
Dividing by a two- or three-digit number, p. 54.	Figure how many times the first digit of the divisor goes into the first two digits of the dividend. To try this answer, multiply it back to see if it is too large or small. Continue each step of long division until finished.	Divide. 589 238)140,182 1190 2118 1904 2142 2142 0

Topic	Procedure	Examples
Exponent form, p. 63.	To show in short form the repeated multiplication of the same number, write the number being multiplied. (This is the base.) Write in smaller print above the line the number of times it appears as a factor. (This is the exponent.) To evaluate the exponent form, write the factor the number of times shown in the exponent. Then multiply.	Write in exponent form. $10 \times 10 \times 10 \times 10 \times 10 \times 10 \times 10 \times 10$ 10^8 Evaluate. 6^3 $6 \times 6 \times 6 = 216$
Order of operations, p. 64.	**1.** Perform operations inside parentheses. **2.** First raise to a power. **3.** Then do multiplication and division in order from left to right. **4.** Then do addition and subtraction in order from left to right.	Evaluate. $2^3 + 16 \div 4^2 \times 5 - 3$ Raise to a power first. $8 + 16 \div 16 \times 5 - 3$ Then do multiplication or division from left to right. $8 + 1 \times 5 - 3$ $8 + 5 - 3$ Then do addition and subtraction. $13 - 3 = 10$
Rounding, p. 70.	**1.** If the first digit to the right of the round-off place is less than 5, the digit in the round-off place is unchanged. **2.** If the first digit to the right of the round-off place is 5 or more, the digit in the round-off place is increased by 1. **3.** Digits to the right of the round-off place are replaced by zeros.	Round to the nearest hundred. 56,743 ↓ 5 6,7④3 The digit 4 is less than 5. 56,700 Round to the nearest thousand. 128,517 ↙ 1 2 8, 5 1 7 The digit 5 is obviously 5 or greater. We increase the thousands digit by 1. 129,000
Estimating the answer to a calculation, p. 72.	**1.** Round each number so that there is one nonzero digit. **2.** Perform the calculation with the rounded numbers.	Estimate the answer. $45{,}780 \times 9453$ First we round. $50{,}000 \times 9000$ Then we multiply. 50,000 × 9,000 450,000,000 We estimate the answer to be 450,000,000.

Procedure for Solving Applied Problems

Using the Mathematics Blueprint for Problem Solving, p. 81

In solving an applied problem, students may find it helpful to complete the following steps. You will not use all the steps all the time. Choose the steps that best fit the conditions of the problem.

1. Understand the problem.

- **(a)** Read the problem carefully.
- **(b)** Draw a picture if this helps you to visualize the situation. Think about what facts you are given and what you are asked to find.
- **(c)** Use the Mathematics Blueprint for Problem Solving to organize your work. Follow these four parts.
  - **1.** Gather the facts (Write down specific values given in the problem.)
  - **2.** What am I asked to do? (Identify what you must obtain for an answer.)
  - **3.** How do I proceed? (What calculations need to be done.)
  - **4.** Key points to remember. (Record any facts, warnings, formulas, or concepts you think will be important as you solve the problem.)

2. Solve and state the answer.

- **(a)** Perform the necessary calculations.
- **(b)** State the answer, including the unit of measure.

3. Check.

- **(a)** Estimate the answer to the problem. Compare this estimate to the calculated value. Is your answer reasonable?
- **(b)** Repeat your calculations.
- **(c)** Work backward from your answer. Do you arrive at the original conditions of the problem?

EXAMPLE

The Manchester highway department has just purchased two pickup trucks and three dump trucks. The cost of a pickup truck is $17,920. The cost of a dump truck is $48,670. What was the cost to purchase these five trucks?

1. ***Understand the problem.***

Mathematics Blueprint for Problem Solving

Gather the Facts	What Am I Asked to Do?	How Do I Proceed?	Key Points to Remember
Buy 2 pickup trucks 3 dump trucks Cost Pickup: $17,920 Dump: $48,670	Find the total cost of the 5 trucks.	Find the cost of 2 pickup trucks. Find the cost of 3 dump trucks. Add to get final cost of all 5 trucks.	Multiply 2 times pickup truck cost. Multiply 3 times dump truck cost.

2. ***Solve and state the answer.***

Calculate cost of pickup trucks

$$\begin{array}{r} \$17{,}920 \\ \times\ 2 \\ \hline \$35{,}840 \end{array}$$

Calculate cost of dump trucks

$$\begin{array}{r} \$48{,}670 \\ \times\ 3 \\ \hline \$146{,}010 \end{array}$$

Find total cost. $\$35{,}840 + \$146{,}010 = \$181{,}850$

The total cost of the five trucks is $181,850.

3. ***Check.*** Estimate cost of pickup trucks

$$20{,}000 \times 2 = 40{,}000$$

Estimate cost of dump trucks

$$50{,}000 \times 3 = 150{,}000$$

Total estimate

$$40{,}000 + 150{,}000 = 190{,}000$$

This is close to our calculated answer of $181,850. We determine that our answer is reasonable. ✓

Chapter 1 Review Problems

If you have trouble with a particular type of exercise, review the examples in the section indicated for that group of exercises. Answers to all exercises are located in the answer key.

Section 1.1

Write in words.

1. 376
three hundred seventy-six

2. 15,802
fifteen thousand, eight hundred two

3. 109,276
one hundred nine thousand, two hundred seventy-six

4. 423,576,055
four hundred twenty-three million, five hundred seventy-six thousand, fifty-five

Write in expanded notation.

5. 4364
4000 + 300 + 60 + 4

6. 27,986
20,000 + 7000 + 900 + 80 + 6

7. 42,166,037
40,000,000 + 2,000,000 + 100,000 + 60,000 + 6000 + 30 + 7

8. 1,305,128
1,000,000 + 300,000 + 5000 + 100 + 20 + 8

Write in standard notation.

9. nine hundred twenty-four
924

10. five thousand three hundred two
5302

11. one million, three hundred twenty-eight thousand, eight hundred twenty-eight
1,328,828

12. forty-five million, ninety-two thousand, six hundred fifty-one
45,092,651

Section 1.2

Add.

13. 76 + 39
115

14. 148 + 152
300

15. 127 + 563
690

16. 12 + 28 + 34 + 76
150

17.
$$\begin{array}{r} 123 \\ 61 \\ 9 \\ 84 \\ +\,123 \\ \hline \end{array}$$
400

18.
$$\begin{array}{r} 937 \\ 405 \\ +\,256 \\ \hline \end{array}$$
1598

19.
$$\begin{array}{r} 226 \\ 134 \\ +\,647 \\ \hline \end{array}$$
1007

20.
$$\begin{array}{r} 28{,}364 \\ +\,97{,}059 \\ \hline \end{array}$$
125,423

21.
$$\begin{array}{r} 1356 \\ 2892 \\ 561 \\ 89 \\ +\,9805 \\ \hline \end{array}$$
14,703

22.
$$\begin{array}{r} 26 \\ 503 \\ 935 \\ 1257 \\ +\,7861 \\ \hline \end{array}$$
10,582

Section 1.3

Subtract.

23. $\begin{array}{r} 36 \\ -19 \\ \hline \end{array}$ 17

24. $\begin{array}{r} 54 \\ -48 \\ \hline \end{array}$ 6

25. $\begin{array}{r} 126 \\ -99 \\ \hline \end{array}$ 27

26. $\begin{array}{r} 543 \\ -372 \\ \hline \end{array}$ 171

27. $\begin{array}{r} 1296 \\ -1137 \\ \hline \end{array}$ 159

28. $\begin{array}{r} 9000 \\ -5833 \\ \hline \end{array}$ 3167

29. $\begin{array}{r} 201{,}010 \\ -137{,}864 \\ \hline \end{array}$ 63,146

30. $\begin{array}{r} 101{,}300 \\ -98{,}274 \\ \hline \end{array}$ 3026

31. $\begin{array}{r} 6{,}325{,}034 \\ -89{,}023 \\ \hline \end{array}$ 6,236,011

32. $\begin{array}{r} 5{,}412{,}022 \\ -79{,}031 \\ \hline \end{array}$ 5,332,991

Section 1.4

Multiply.

33. $8 \times 1 \times 9 \times 2$ 144

34. $7 \times 6 \times 0 \times 4$ 0

35. $3 \cdot 4 \cdot 2 \cdot 2 \cdot 5$ 240

36. $1 \cdot 3 \cdot 10 \cdot 5 \cdot 2$ 300

37. 621×100 62,100

38. $84{,}312 \times 1000$ 84,312,000

39. $832 \times 100{,}000$ 83,200,000

40. $563 \times 1{,}000{,}000$ 563,000,000

41. $\begin{array}{r} 58 \\ \times 32 \\ \hline \end{array}$ 1856

42. $\begin{array}{r} 73 \\ \times 24 \\ \hline \end{array}$ 1752

43. $\begin{array}{r} 150 \\ \times 27 \\ \hline \end{array}$ 4050

44. $\begin{array}{r} 360 \\ \times 38 \\ \hline \end{array}$ 13,680

45. $\begin{array}{r} 709 \\ \times 36 \\ \hline \end{array}$ 25,524

46. $\begin{array}{r} 502 \\ \times 48 \\ \hline \end{array}$ 24,096

47. $\begin{array}{r} 123 \\ \times 714 \\ \hline \end{array}$ 87,822

48. $\begin{array}{r} 431 \\ \times 623 \\ \hline \end{array}$ 268,513

49. $\begin{array}{r} 1782 \\ \times 305 \\ \hline \end{array}$ 543,510

50. $\begin{array}{r} 2057 \\ \times 124 \\ \hline \end{array}$ 255,068

51. $\begin{array}{r} 300 \\ \times 500 \\ \hline \end{array}$ 150,000

52. $\begin{array}{r} 400 \\ \times 600 \\ \hline \end{array}$ 240,000

53. $\begin{array}{r} 1200 \\ \times 6000 \\ \hline \end{array}$ 7,200,000

54. $\begin{array}{r} 2500 \\ \times 3000 \\ \hline \end{array}$ 7,500,000

55. $\begin{array}{r} 100{,}000 \\ \times 20{,}000 \\ \hline \end{array}$ 2,000,000,000

56. $\begin{array}{r} 300{,}000 \\ \times 40{,}000 \\ \hline \end{array}$ 12,000,000,000

Section 1.5

Divide, if possible.

57. $20 \div 10$ 2

58. $40 \div 8$ 5

59. $0 \div 8$ 0

60. $12 \div 1$ 12

61. $7 \div 1$
7

62. $0 \div 5$
0

63. $\frac{49}{7}$
7

64. $\frac{42}{6}$
7

65. $\frac{5}{0}$
undefined

66. $\frac{24}{6}$
4

67. $\frac{56}{8}$
7

68. $\frac{48}{8}$
6

Divide. Be sure to indicate the remainder, if one exists.

69. $6\overline{)750}$ 125

70. $7\overline{)875}$ 125

71. $5\overline{)1290}$ 258

72. $4\overline{)1236}$ 309

73. $3\overline{)77,622}$ 25,874

74. $8\overline{)24,512}$ 3064

75. $6\overline{)221,748}$ 36,958

76. $5\overline{)184,605}$ 36,921

77. $8\overline{)127,890}$ 15,986 R 2

78. $7\overline{)250,485}$ 35,783 R 4

79. $67\overline{)490}$ 7 R 21

80. $72\overline{)325}$ 4 R 37

81. $21\overline{)666}$ 31 R 15

82. $22\overline{)319}$ 14 R 11

83. $68\overline{)2614}$ 38 R 30

84. $53\overline{)3202}$ 60 R 22

85. $45\overline{)4275}$ 95

86. $35\overline{)9030}$ 258

87. $132\overline{)7128}$ 54

88. $204\overline{)3876}$ 19

Section 1.6

Write in exponent form.

89. 13×13
13^2

90. $21 \times 21 \times 21$
21^3

91. $8 \times 8 \times 8 \times 8 \times 8$
8^5

92. $10 \times 10 \times 10 \times 10 \times 10 \times 10$
10^6

Evaluate.

93. 2^6
64

94. 3^4
81

95. 2^7
128

96. 5^3
125

97. 7^2
49

98. 9^2
81

99. 6^3
216

100. 4^3
64

Perform each operation in proper order.

101. $7 + 2 \times 3 - 5$
8

102. $6 \times 2 - 4 + 3$
11

103. $2^5 + 4 - (5 + 3^2)$
22

104. $4^3 + 20 \div (2 + 2^3)$
66

105. $3^3 \times 4 - 6 \div 6$
107

106. $20 \div 20 + 5^2 \times 3$
76

107. $2^3 \times 5 \div 8 + 3 \times 4$
17

108. $2^3 + 4 \times 5 - 32 \div (1 + 3)^2$
26

109. $6 \times 3 + 3 \times 5^2 - 63 \div (5 - 2)^2$
86

Section 1.7

Round to the nearest ten.

110. 1275
1280

111. 5895
5900

112. 15,305
15,310

113. 42,644
42,640

In exercises 114–117, round to the nearest thousand.

114. 12,350
12,000

115. 22,986
23,000

116. 675,800
676,000

117. 202,498
202,000

118. Round to the nearest hundred thousand. 4,649,320
4,600,000

119. Round to the nearest ten thousand. 9,995,312
10,000,000

Use the principle of estimation to find an estimate for each calculation.

120. $589 + 622 + 933 + 864$

$$\begin{array}{r} 600 \\ 600 \\ 900 \\ +\,900 \\ \hline 3000 \end{array}$$

121. $25{,}981 + 7347 + 68{,}125$

$$\begin{array}{r} 30{,}000 \\ 7000 \\ +\,70{,}000 \\ \hline 107{,}000 \end{array}$$

122. $4{,}326{,}171 - 2{,}916{,}788$

$$\begin{array}{r} 4{,}000{,}000 \\ -\,3{,}000{,}000 \\ \hline 1{,}000{,}000 \end{array}$$

123. $29{,}378 - 17{,}924$

$$\begin{array}{r} 30{,}000 \\ -\,20{,}000 \\ \hline 10{,}000 \end{array}$$

124. 1463×5982

$$\begin{array}{r} 1000 \\ \times\,6000 \\ \hline 6{,}000{,}000 \end{array}$$

125. $2{,}965{,}372 \times 893$

$$\begin{array}{r} 3{,}000{,}000 \\ \times\quad 900 \\ \hline 2{,}700{,}000{,}000 \end{array}$$

126. $83{,}421 \div 24$

$$\begin{array}{r} 4{,}000 \\ 20\overline{)80{,}000} \end{array}$$

127. $7{,}963{,}127 \div 378$

$$\begin{array}{r} 20{,}000 \\ 400\overline{)8{,}000{,}000} \end{array}$$

Section 1.8

Solve.

128. ***Consumer Decisions*** For their wedding reception, Steven and Tanya bought 18 twelve-packs of soda. How many cans of soda were there? 216 cans

129. ***Computer Applications*** Ward can type 25 words per minute on his computer. He typed for seven minutes at that speed. How many words did he type? 175 words

130. ***Travel*** In June, 2462 people visited the Renaissance Festival. There were 1997 visitors in July, and 2561 in August. How many people visited the festival during these three months? 7020 people

131. ***Farming*** Applepickers, Inc. bought a truck for \$26,300, a car for \$14,520, and a minivan for \$18,650. What was the total purchase price? \$59,470

132. ***Aviation*** A plane was flying at 14,630 feet. It flew over a mountain 4329 feet high. How many feet was it from the plane to the top of the mountain? 10,301 feet

133. ***Personal Finance*** Roberta was billed \$11,658 for tuition. She received a \$4630 grant. How much did she have to pay after the grant was deducted? \$7028

134. ***Travel*** The expedition cost a total of \$32,544 for 24 paying passengers, who shared the cost equally. What was the cost per passenger? \$1356

135. ***Business Management*** Middlebury College ordered 112 dormitory beds for \$8288. What was the cost per bed? \$74

136. ***Personal Finance*** Melissa's savings account balance last month was \$810. The bank added \$24 interest. Melissa deposited \$105, \$36, and \$177. She made withdrawals of \$18, \$145, \$250, and \$461. What will be her balance this month? \$278

137. ***Environmental Studies*** Ali began a trip on a full tank of gas with the car odometer at 56,320 miles. He ended the trip at 56,720 miles and added 16 gallons of gas to refill the tank. How many miles per gallon did he get on the trip? 25 miles per gallon

138. ***Business Management*** The maintenance group bought three lawn mowers at \$279, four power drills at \$61, and two riding tractors at \$1980. What was the total purchase price for these items? \$5041

139. ***Business Management*** Anita is opening a new café in town. She bought 15 tables at \$65 each, 60 chairs for \$12 each and eight ceiling fans for \$42 each. What was the total purchase price for these items? \$2031

Environmental Protection *Use the following bar graph to answer exercises 140–142.*

140. How many more tons of solid waste were recovered and recycled in 1995 than in 1980? 40,500,000 tons

141. What was the greatest increase in tons of solid waste recovered and recycled in a five-year period?

21,400,000 tons, from 1990 to 1995

142. If the exact same increase in the number of tons recovered occurs from 2000 to 2010 as occurred from 1990 to 2000, how many tons of solid waste will be recovered and recycled in 2010?

93,400,000 tons

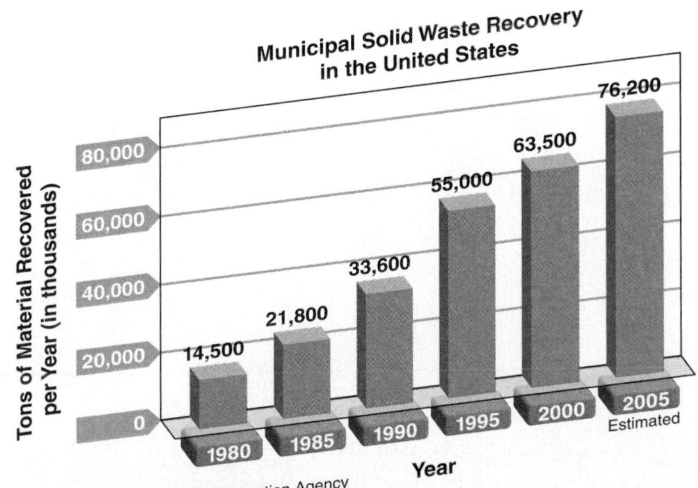

Mixed Practice

Perform each calculation.

143. $205 + 36 + 1983 + 60$
2284

144. $56{,}793 - 48{,}926$
7867

145. 396×28
11,088

146. $37\overline{)4773}$
129

147. Evaluate. $4 \times 12 - (12 + 9) + 2^3 \div 4$
29

148. ***Personal Finance*** Michael Evans has \$3000 in his checking account. He buys 3 computers at \$699 each and 2 printers at \$78 each. How much does he have remaining after the purchases? \$747

▲ **149.** ***Geometry*** Milton is building a rectangular patio in his backyard. The patio measures 22 feet by 15 feet.

(a) How many square feet is the patio?
330 square feet

(b) If Milton wanted to fence in the patio, how many feet of fence would he need?
74 feet

How Am I Doing? Chapter 1 Test

Remember to use your Chapter Test Prep Video CD to see the worked-out solutions to the test problems you want to review.

Note to Instructor: The Chapter 1 Test file in the TestGen program provides algorithms specifically matched to these problems so you can easily replicate this test for additional practice or assessment purposes.

Write the answers.

1. Write in words. 44,007,635

2. Write in expanded notation. 26,859

3. Write in standard notation. three million, five hundred eighty-one thousand, seventy-six

Add.

4. $\begin{array}{r} 189 \\ 26 \\ 12 \\ 528 \\ +\ 76 \\ \hline \end{array}$

5. $\begin{array}{r} 763 \\ 220 \\ +508 \\ \hline \end{array}$

6. $\begin{array}{r} 135{,}484 \\ 2{,}376 \\ 81{,}004 \\ +100{,}113 \\ \hline \end{array}$

Subtract.

7. $\begin{array}{r} 8961 \\ -\ 894 \\ \hline \end{array}$

8. $\begin{array}{r} 501{,}760 \\ -328{,}902 \\ \hline \end{array}$

9. $\begin{array}{r} 18{,}400{,}100 \\ -13{,}174{,}332 \\ \hline \end{array}$

Multiply.

10. $1 \times 6 \times 9 \times 7$

11. $\begin{array}{r} 45 \\ \times 96 \\ \hline \end{array}$

12. $\begin{array}{r} 326 \\ \times 592 \\ \hline \end{array}$

13. $\begin{array}{r} 18{,}491 \\ \times\quad 7 \\ \hline \end{array}$

In problems 14–16, divide. If there is a remainder, be sure to state it as part of your answer.

14. $5\overline{)15{,}071}$

15. $6\overline{)14148}$

16. $37\overline{)13{,}024}$

17. Write in exponent form. $14 \times 14 \times 14$

18. Evaluate. 2^6

1. forty-four million, seven thousand, six hundred thirty-five
2. 20,000 + 6000 + 800 + 50 + 9
3. 3,581,076
4. 831
5. 1491
6. 318,977
7. 8067
8. 172,858
9. 5,225,768
10. 378
11. 4320
12. 192,992
13. 129,437
14. 3014 R 1
15. 2358
16. 352
17. 14^3
18. 64

19. 23

20. 50

21. 79

22. 94,800

23. 6,460,000

24. 5,300,000

25. 150,000,000,000

26. 16,000

27. $2148

28. 467 feet

29. $127

30. $292

31. 748,000 square feet

32. 46 feet

In problems 19–21, perform each operation in proper order.

19. $5 + 6^2 - 2 \times (9 - 6)^2$

20. $2^4 + 3^3 + 28 \div 4$

21. $4 \times 6 + 3^3 \times 2 + 23 \div 23$

22. Round to the nearest hundred. 94,768

23. Round to the nearest ten thousand. 6,462,431

24. Round to the nearest hundred thousand. 5,278,963

Estimate the answer.

25. $4{,}867{,}010 \times 27{,}058$

26. $1423 + 3298 + 4103 + 7614$

Solve.

27. A cruise for 15 people costs $32,220. If each person paid the same amount, how much will it cost each individual?

28. The river is 602 feet wide at Big Bend Corner. A boy is in the shallow water, 135 feet from the shore. How far is the boy from the other side of the river?

29. At the bookstore, Hector bought three notebooks at $2 each, one textbook for $45, two lamps at $21 each, and two sweatshirts at $17 each. What was his total bill?

30. Patricia is looking at her checkbook. She had a balance last month of $31. She deposited $902 and $399. She made out checks for $885, $103, $26, $17, and $9. What will be her new balance?

▲ **31.** The runway at Beverly Airport needs to be resurfaced. The rectangular runway is 6800 feet long and 110 feet wide. What is the area of the runway that needs to be resurfaced?

▲ **32.** Nancy Tobey planted a vegetable garden in the backyard. However, the deer and raccoons have been stealing all the vegetables. She asked John to fence in the garden. The rectangular garden measures 8 feet by 15 feet. How many feet of fence should John purchase if he wants to enclose the garden?

CHAPTER

2

In early America a person delivering supplies from town to town often had to make his or her own calculations to find the total distance that must be traveled. Many of the old dirt roads were marked with signs indicating a distance in fractions. Could you make these kind of calculations? Turn to page 191 to find out.

Fractions

2.1 Understanding Fractions

Student Learning Objectives

After studying this section, you will be able to:

1. Use a fraction to represent part of a whole.
2. Draw a sketch to illustrate a fraction.
3. Use fractions to represent real-life situations.

1 Using a Fraction to Represent Part of a Whole

In Chapter 1 we studied whole numbers. In this chapter we will study a fractional part of a whole number. One way to represent parts of a whole is with **fractions.** The word *fraction* (like the word *fracture*) suggests that something is being broken. In mathematics, fractions represent the part that is "broken off" from a whole. The whole can be a single object (like a whole pie) or a group (the employees of a company). Here are some examples.

Single object

$$\frac{1}{3}$$

The whole is the pie on the left. The fraction $\frac{1}{3}$ represents the shaded part of the pie, 1 of 3 pieces. $\frac{1}{3}$ is read "one-third."

A group: ACE company employs 150 men, 200 women.

$$\frac{150}{350}$$

The whole is the company of 350 people (150 men plus 200 women). The fraction $\frac{150}{350}$ represents that part of the company consisting of men.

Recipe: Applesauce
4 apples
1/2 cup sugar
1 teaspoon cinnamon

The whole is 1 whole cup of sugar. This recipe calls for $\frac{1}{2}$ cup of sugar. Notice that in many real-life situations $\frac{1}{2}$ is written as 1/2.

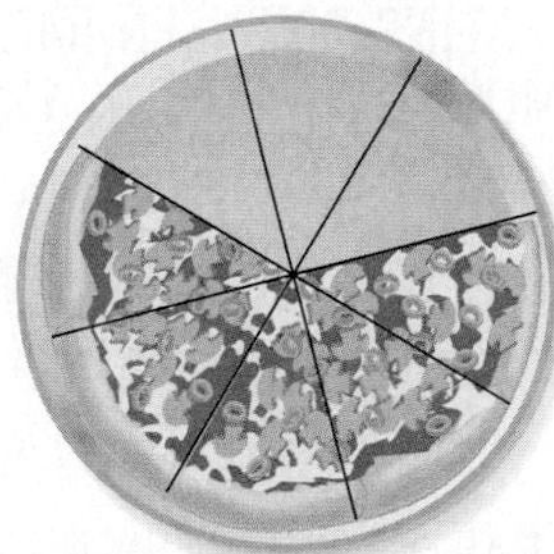

When we say "$\frac{3}{8}$ of a pizza has been eaten," we mean 3 of 8 equal parts of a pizza have been eaten. (See the figure.) When we write the fraction $\frac{3}{8}$, the number on the top, 3, is the **numerator,** and the number on the bottom, 8, is the **denominator.**

Teaching Tip Stress to students that they must know the names of the two parts of the fraction: the numerator and the denominator. You can tell them that an easy way to remember which is to recall that the *D*enominator is *D*own at the bottom of the fraction.

The numerator specifies how many parts → 3
The denominator specifies the total number of parts → 8

When we say, "$\frac{2}{3}$ of the marbles are red," we mean 2 marbles out of a total of 3 are red marbles.

Part we are interested in → 2 numerator
Total number in the group → 3 denominator

EXAMPLE 1 Use a fraction to represent the shaded or completed part of the whole shown.

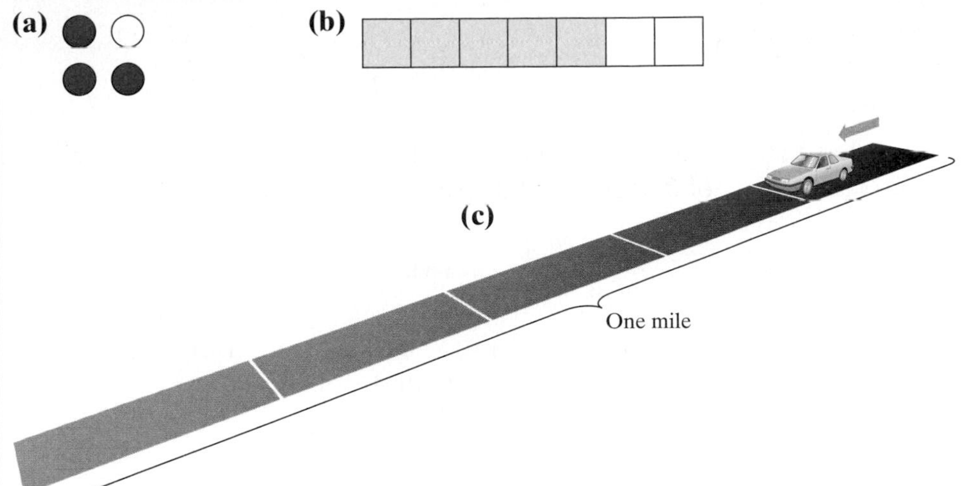

Solution

(a) Three out of four circles are red. The fraction is $\frac{3}{4}$.

(b) Five out of seven equal parts are shaded. The fraction is $\frac{5}{7}$.

(c) The mile is divided into five equal parts. The car has traveled 1 part out of 5 of the one mile distance. The fraction is $\frac{1}{5}$.

Practice Problem 1 Use a fraction to represent the shaded part of the whole.

(a) 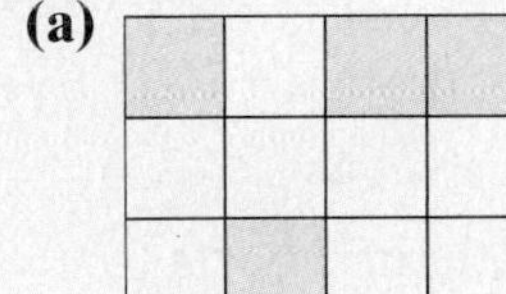

(b) 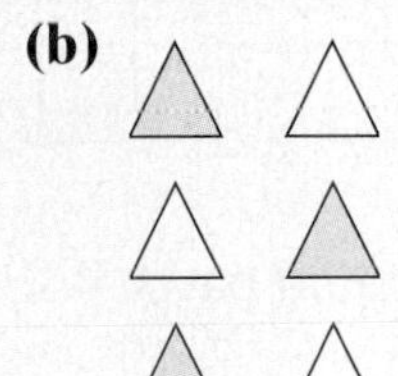

(c) 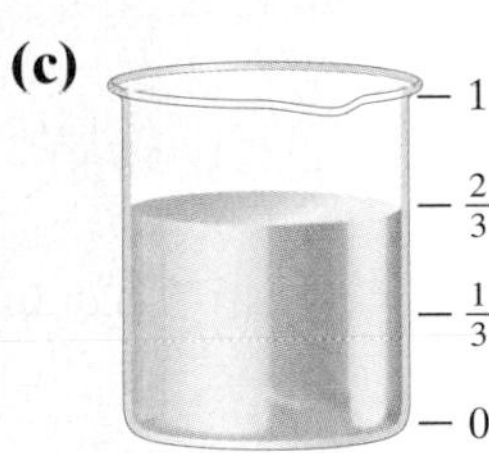

***NOTE TO STUDENT:** Fully worked-out solutions to all of the Practice Problems can be found at the back of the text starting at page SP-1*

We can also think of a fraction as a division problem.

$$\frac{1}{3} = 1 \div 3 \quad \text{and} \quad 1 \div 3 = \frac{1}{3}$$

The division way of looking at fractions asks the question:

What is the result of dividing one whole into three equal parts?

Thus we can say the fraction $\frac{a}{b}$ means the same as $a \div b$. However, special care must be taken with the number 0.

Suppose that we had four equal parts and we wanted to take none of them. We would want $\frac{0}{4}$ of the parts. Since $\frac{0}{4} = 0 \div 4 = 0$, we see that $\frac{0}{4} = 0$. Any fraction with a 0 numerator equals zero.

$$\frac{0}{8} = 0 \qquad \frac{0}{5} = 0 \qquad \frac{0}{13} = 0$$

What happens when zero is in the denominator? $\frac{4}{0}$ means 4 out of 0 parts. Taking 4 out of 0 does not make sense. We say $\frac{4}{0}$ is **undefined.**

$\frac{3}{0}, \frac{7}{0}, \frac{4}{0}$ are **undefined.**

We cannot have a fraction with 0 in the denominator. Since $\frac{4}{0} = 4 \div 0$, we say division by zero is *undefined.* We cannot divide by 0.

2 Drawing a Sketch to Illustrate a Fraction

Drawing a sketch of a mathematical situation is a powerful problem-solving technique. The picture often reveals information not always apparent in the words.

EXAMPLE 2 Draw a sketch to illustrate.

(a) $\frac{7}{11}$ of an object **(b)** $\frac{2}{9}$ of a group

Solution

(a) The easiest figure to draw is a rectangular bar.

We divide the bar into 11 equal parts. We then shade in 7 parts to show $\frac{7}{11}$.

(b) We draw 9 circles of equal size to represent a group of 9.

We shade in 2 of the 9 circles to show $\frac{2}{9}$.

Practice Problem 2 Draw a sketch to illustrate.

(a) $\frac{4}{5}$ of an object **(b)** $\frac{3}{7}$ of a group

***NOTE TO STUDENT:** Fully worked-out solutions to all of the Practice Problems can be found at the back of the text starting at page SP-1*

Recall these facts about division problems involving the number 1 and the number 0.

DIVISION INVOLVING THE NUMBER 1 AND THE NUMBER 0

1. Any nonzero number divided by itself is 1.

$$\frac{7}{7} = 1$$

2. Any number divided by 1 remains unchanged. $\frac{29}{1} = 29$

3. Zero may be divided by any nonzero number; the result is always zero.

$$\frac{0}{4} = 0$$

4. Division by zero is undefined. $\frac{3}{0}$ is undefined

3 Using Fractions to Represent Real-Life Situations

Several real-life situations can be described using fractions.

EXAMPLE 3 Use a fraction to describe each situation.

(a) A baseball player gets a hit 5 out of 12 times at bat.

(b) There are 156 men and 185 women taking psychology this semester. Describe the part of the class that consists of women.

(c) Robert Tobey found in the Alaska moose count that five-eighths of the moose observed were female.

Solution

(a) The baseball player got a hit $\frac{5}{12}$ of his times at bat.

(b) The total class is $156 + 185 = 341$. The fractional part that is women is 185 out of 341. Thus $\frac{185}{341}$ of the class is women.

156 men	185 women

Total class 341 students

(c) Five-eighths of the moose observed were female. The fraction is $\frac{5}{8}$.

Practice Problem 3 Use a fraction to describe each situation.

(a) 9 out of the 17 players on the basketball team are on the dean's list.

(b) The senior class has 382 men and 351 women. Describe the part of the class consisting of men.

(c) John needed seven-eighths of a yard of material.

EXAMPLE 4 Wanda made 13 calls, out of which she made five sales. Albert made 17 calls, out of which he made six sales. Write a fraction that describes for both people together the number of calls in which a sale was made compared with the total number of calls.

Solution There are $5 + 6 = 11$ calls in which a sale was made.

There were $13 + 17 = 30$ total calls.

Thus $\frac{11}{30}$ of the calls resulted in a sale.

Practice Problem 4 An inspector found that one out of seven belts was defective. She also found that two out of nine shirts were defective. Write a fraction that describes what part of all the objects examined were defective.

Developing Your Study Skills

Previewing New Material

Part of your study time each day should consist of looking ahead to those sections in your text that are to be covered the following day. You do not necessarily have to study and learn the material on your own, but if you survey the concepts, terminology, diagrams, and examples, the new ideas will seem more familiar to you when the instructor presents them. You can take note of concepts that appear confusing or difficult and be ready to listen carefully for your instructor's explanations. You can be prepared to ask the questions that will increase your understanding. Previewing new material enables you to see what is coming and prepares you to be ready to absorb it.

2.1 EXERCISES

Student Solutions Manual | CD/ Video | PH Math Tutor Center | MathXL®Tutorials on CD | MathXL® | MyMathLab® | Interactmath.com

Verbal and Writing Skills

1. A ___fraction___ can be used to represent part of a whole or part of a group.

2. In a fraction, the ___numerator___ tells the number of parts we are interested in.

3. In a fraction, the ___denominator___ tells the total number of parts in the whole or in the group.

4. Describe a real-life situation that involves fractions.
Answers will vary. An example is: I was late 3 out of 5 times last week. I was late 3/5 of the time.

Name the numerator and the denominator in each fraction.

5. $\frac{3}{5}$ N: 3 D: 5

6. $\frac{9}{11}$ N: 9 D: 11

7. $\frac{7}{8}$ N: 7 D: 8

8. $\frac{9}{10}$ N: 9 D: 10

9. $\frac{1}{17}$ N: 1 D: 17

10. $\frac{1}{15}$ N: 1 D: 15

In exercises 11–30, use a fraction to represent the shaded part of the object or the shaded portion of the set of objects.

11. 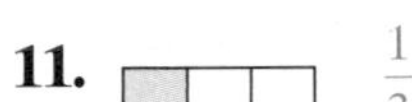 $\frac{1}{3}$

12. $\frac{1}{2}$

13. 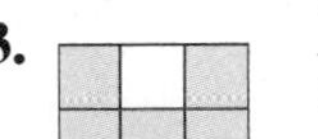 $\frac{7}{9}$

14. $\frac{5}{6}$

15. 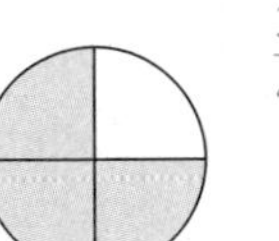 $\frac{3}{4}$

16. 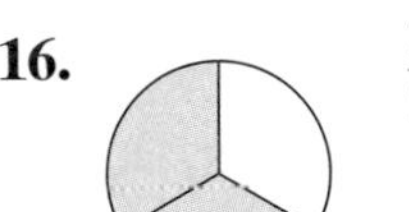 $\frac{2}{3}$

17. 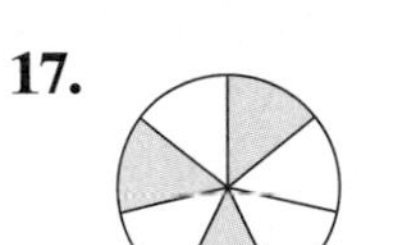 $\frac{3}{7}$

18. 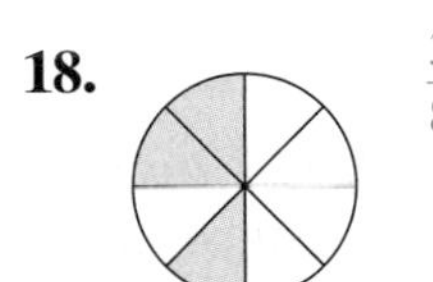 $\frac{3}{8}$

19.  $\frac{2}{5}$

20.  $\frac{1}{4}$

21. 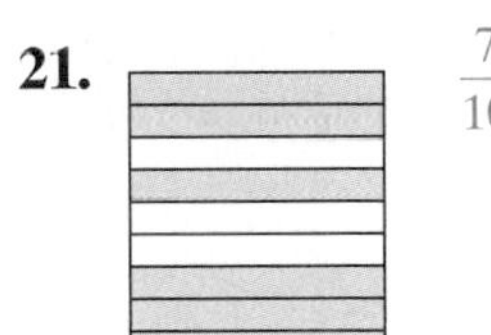 $\frac{7}{10}$

22. 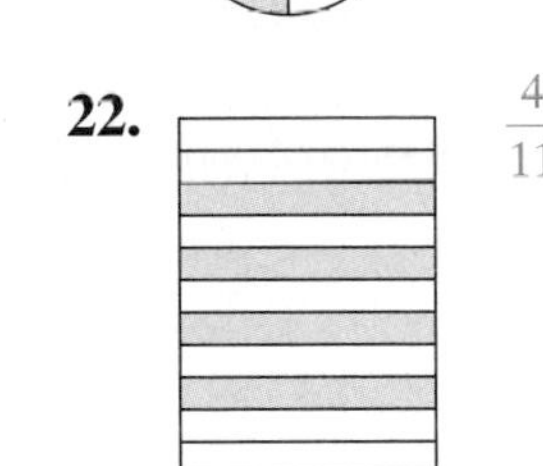 $\frac{4}{11}$

23. 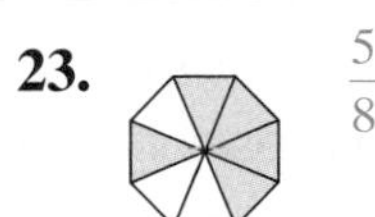 $\frac{5}{8}$

24. 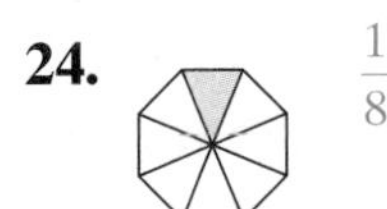 $\frac{1}{8}$

25.  $\frac{4}{7}$

26. 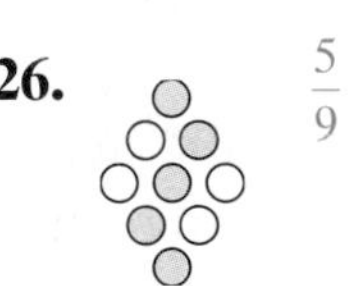 $\frac{5}{9}$

27. $\frac{7}{8}$

28. 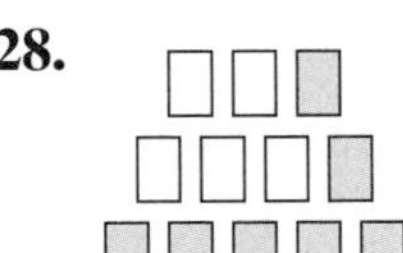 $\frac{7}{12}$

29.  $\frac{3}{5}$

30. 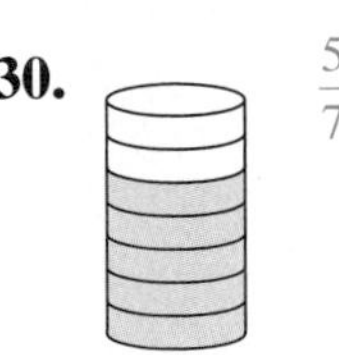 $\frac{5}{7}$

Draw a sketch to illustrate each fractional part. Object used to represent fractional parts may vary. Samples are given.

31. $\frac{1}{5}$ of an object

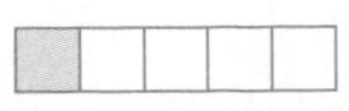

32. $\frac{3}{7}$ of an object

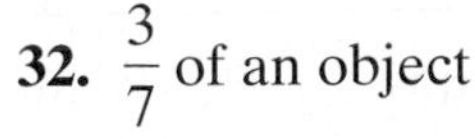

33. $\frac{3}{8}$ of an object

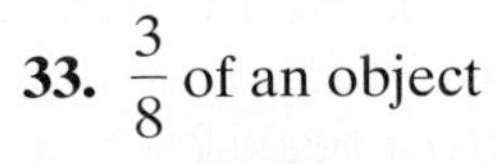

34. $\frac{5}{12}$ of an object

35. $\frac{7}{10}$ of an object

36. $\frac{5}{9}$ of an object

Applications

37. ***Making Wreaths*** Cecilia made 95 holiday wreaths to sell this year. Thirty-one of them were decorated with small silver bells, while the others had ribbon on them. What fractional part of the wreaths were made with silver bells?

$\frac{31}{95}$

38. ***Sales Tax*** The total purchase amount was 83¢, of which 5¢ was sales tax. What fractional part of the total purchase price was sales tax?

$\frac{5}{83}$

39. ***Personal Finance*** Lance bought a 100-CD jukebox for $750. Part of it was paid for with the $209 he earned parking cars for the valet service at a local wedding reception hall. What fractional part of the jukebox was paid for by his weekend earnings?

$\frac{209}{750}$

40. ***Test Taking*** A mathematics class was given 55 minutes to complete their first test. Marion finished in 46 minutes. What fractional part of the time allowed did Marion use?

$\frac{46}{55}$

41. ***Political Campaigns*** The Democratic National Committee fundraising event served 122 chicken dinners and 89 roast beef dinners to its contributors. What fractional part of the guests ate roast beef?

$\frac{89}{211}$

42. ***Education*** Bridgeton Community College has 78 full-time instructors and 31 part-time instructors. What fractional part of the faculty are part time?

$\frac{31}{109}$

43. ***Physical Fitness*** At Gold's Gym one day, 9 people were riding stationary bikes, 7 people were using rowing machines, and 13 people were running on treadmills. What fractional part of the people were using rowing machines?

$\frac{7}{29}$

44. ***Animal Shelters*** At the local animal shelter there are 12 puppies, 25 adult dogs, 14 kittens, and 31 adult cats. What fractional part of the animals are either puppies or adult dogs?

$\frac{37}{82}$

45. ***Picnic*** The picnic table held two bowls of corn, three bowls of potato salad, four bowls of baked beans, and five bowls of ribs. What fractional part of the bowls on the table contains either ribs or beans?

$\frac{9}{14}$

46. ***Music Collection*** A box of compact discs contains 5 classical CDs, 6 jazz CDs, 4 sound tracks, and 24 blues CDs. What fractional part of the total CDs is either jazz or blues?

$\frac{30}{39}$

47. ***Manufacturing*** The West Peabody Engine Company manufactured two items last week: 101 engines and 94 lawn mowers. It was discovered that 19 engines and 3 lawn mowers were defective. Of the engines that were not defective, 40 were properly constructed but 42 were not of the highest quality. Of the lawn mowers that were not defective, 50 were properly constructed but 41 were not of the highest quality.

(a) What fractional part of all items manufactured was of the highest quality?

$\frac{90}{195}$

(b) What fractional part of all items manufactured was defective?

$\frac{22}{195}$

48. ***Tour Bus*** A Chicago tour bus held 25 women and 33 men. 12 women wore jeans. 19 men wore jeans. In the group of 25 women, a subgroup of 8 women wore sandals. In the group of 19 men, a subgroup of 10 wore sandals.

(a) What fractional part of the people on the bus wore jeans?

$\frac{31}{58}$

(b) What fractional part of the women on the bus wore sandals?

$\frac{8}{25}$

To Think About

49. Illustrate a real-life example of the fraction $\frac{0}{6}$.

The amount of money each of six business owners gets if the business has a profit of $0.

50. What happens when we try to illustrate a real-life example of the fraction $\frac{6}{0}$? Why?

We cannot do it. Division by zero is undefined.

Cumulative Review

51. Add.

$$\begin{array}{r} 18 \\ 27 \\ 34 \\ 16 \\ 125 \\ +\ 21 \\ \hline 241 \end{array}$$

52. Subtract.

$$\begin{array}{r} 56{,}203 \\ -\ 42{,}987 \\ \hline 13{,}216 \end{array}$$

53. Multiply.

$$\begin{array}{r} 3178 \\ \times\ 46 \\ \hline 19068 \\ 12712 \\ \hline 146{,}188 \end{array}$$

54. Divide. $24\overline{)30{,}196}$

1258 R 4

$$\begin{array}{r} 24 \\ \hline 61 \\ 48 \\ \hline 139 \\ 120 \\ \hline 196 \\ 192 \\ \hline 4 \end{array}$$

55. ***Fund-Raiser*** The annual marching band fund-raiser for Jefferson Valley High School has collected 2004 books. 282 are science fiction, 866 are novels, 42 are cookbooks, 317 are history books, 102 are biographies, 99 are foreign language books, and 115 are textbooks. The remaining books are reference books, such as encyclopedias and dictionaries. How many reference books are there? 181

2.2 SIMPLIFYING FRACTIONS

Student Learning Objectives

After studying this section, you will be able to:

1. Write a number as a product of prime factors.
2. Reduce a fraction to lowest terms.
3. Determine whether two fractions are equal.

Teaching Tip Tell students that the ancient Greek mathematician Eratosthenes developed a method of finding prime numbers called the Sieve of Eratosthenes. This method is still used today with some slight modification by computers that generate lists of prime numbers. Suggest that they look up the topic of the Sieve of Eratosthenes in an encyclopedia if they are interested.

Teaching Tip The text here lists only three commonly used tests of divisibility. Ask students if they know any other similar tests for divisibility by some number other than 2, 3, or 5. If they cannot think of any, show them one or more of the following:

- A number is divisible by 4 if the number formed by its last two digits is divisible by 4. For example, 456,716 is divisible by 4.
- A number is divisible by 6 if it is divisible by both 2 and 3.
- A number is divisible by 8 if the number formed by its last three digits is divisible by 8. For example, 26,963,984 is divisible by 8 since 984 is divisible by 8.
- A number is divisible by 9 if the sum of its digits is divisible by 9. Thus 1,428,714 is divisible by 9 since $1 + 4 + 2 + 8 + 7 + 1 + 4$ is 27, which is divisible by 9.

1 Writing a Number as a Product of Prime Factors

A **prime number** is a whole number greater than 1 that cannot be evenly divided except by itself and 1. If you examine all the whole numbers from 1 to 50, you will find 15 prime numbers.

THE FIRST 15 PRIME NUMBERS

2, 3, 5, 7, 11, 13, 17, 19, 23, 29, 31, 37, 41, 43, 47

A **composite number** is a whole number greater than 1 that can be divided by whole numbers other than 1 and itself. The number 12 is a composite number.

$$12 = 2 \times 6 \qquad \text{and} \qquad 12 = 3 \times 4$$

The number 1 is neither a prime nor a composite number. The number 0 is neither a prime nor a composite number.

Recall that factors are numbers that are multiplied together. Prime factors are prime numbers. To check to see if a number is prime or composite, simply divide the smaller primes (such as 2, 3, 5, 7, 11, ...) into the given number. If the number can be divided exactly without a remainder by one of the smaller primes, it is a composite and not a prime.

Some students find the following rules helpful when deciding if a number can be divided by 2, 3, or 5.

DIVISIBILITY TESTS

1. A number is divisible by 2 if the last digit is 0, 2, 4, 6, or 8.
2. A number is divisible by 3 if the sum of the digits is divisible by 3.
3. A number is divisible by 5 if the last digit is 0 or 5.

To illustrate:

1. 478 is divisible by 2 since it ends in 8.
2. 531 is divisible by 3 since when we add the digits of 531 $(5 + 3 + 1)$, we get 9, which is divisible by 3.
3. 985 is divisible by 5 since it ends in 5.

EXAMPLE 1 Write each whole number as the product of prime factors.

(a) 12 **(b)** 60 **(c)** 168

Solution

(a) To start, write 12 as the product of any two factors. We will write 12 as 4×3.

$12 = 4 \times 3$ Now check whether the factors are prime. If not, factor these.

$2 \times 2 \times 3$

$12 = 2 \times 2 \times 3$ Now all factors are prime, so 12 is completely factored.

Instead of writing $2 \times 2 \times 3$, we can write $2^2 \times 3$.
Note: To start, we could write 12 as 2×6. Begin this way and follow the preceding steps. Is the product of prime factors the same? Will this always be true?

(b) We follow the same steps as in (a).

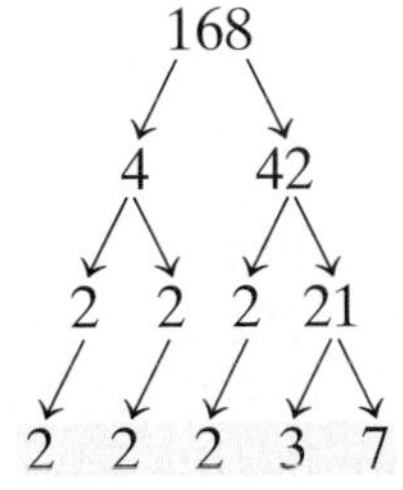

Check that all factors are prime.

Instead of writing $2 \times 2 \times 3 \times 5$, we can write $2^2 \times 3 \times 5$.
Note that in the final answer the prime factors are listed in order from least to greatest.

(c) Some students like to use a **factor tree** to help write a number as a product of prime factors as illustrated below.

168
4 42
2 2 2 21
2 2 2 3 7

$$168 = 2 \times 2 \times 2 \times 3 \times 7$$

or $$168 = 2^3 \times 3 \times 7$$

Practice Problem 1 Write each whole number as a product of primes.
(a) 18 **(b)** 72 **(c)** 400

NOTE TO STUDENT: Fully worked-out solutions to all of the Practice Problems can be found at the back of the text starting at page SP-1

Suppose we started Example 1 (c) by writing $168 = 14 \times 12$. Would we get the same answer? Would our answer be correct? Let's compare.
Again we will use a factor tree.

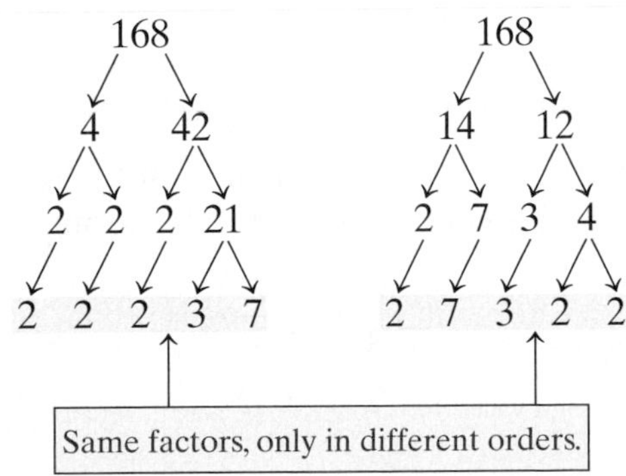

Thus $168 = 2 \times 2 \times 2 \times 3 \times 7$
or $= 2^3 \times 3 \times 7$.

The order of prime factors is not important because multiplication is commutative. No matter how we start, when we factor a composite number, we always get exactly the same prime factors.

THE FUNDAMENTAL THEOREM OF ARITHMETIC

Every composite number can be written in exactly one way as a product of prime numbers.

We have seen this in our Solution to Example 1(c).

You will be able to check this theorem again in Exercises 2.2, exercises 7–26. Writing a number as a product of prime factors is also called **prime factorization.**

2 Reducing a Fraction to Lowest Terms

You know that $5 + 2$ and $3 + 4$ are two ways to write the same number. We say they are *equivalent* because they are *equal* to the same *value.* They are both ways of writing the value 7.

Like whole numbers, fractions can be written in more than one way. For example, $\frac{2}{4}$ and $\frac{1}{2}$ are two ways to write the same number. The value of the fractions is the same. When we use fractions, we often need to write them in another form. If we make the numerator and denominator smaller, we *simplify* the fractions.

Compare the two fractions in the drawings on the left. In each picture the shaded part is the same size. The fractions $\frac{3}{4}$ and $\frac{6}{8}$ are called **equivalent fractions.** The fraction $\frac{3}{4}$ is in **simplest form.** To see how we can change $\frac{6}{8}$ to $\frac{3}{4}$, we look at a property of the number 1.

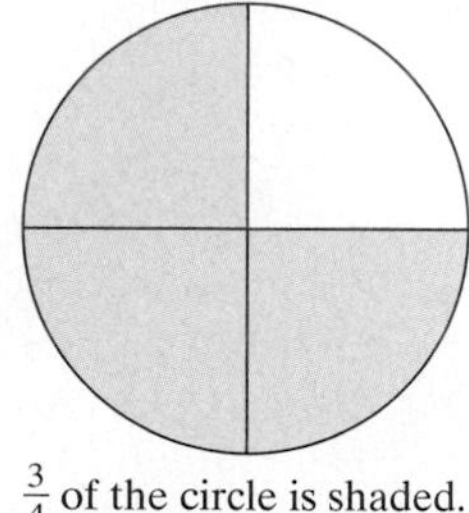

$\frac{3}{4}$ of the circle is shaded.

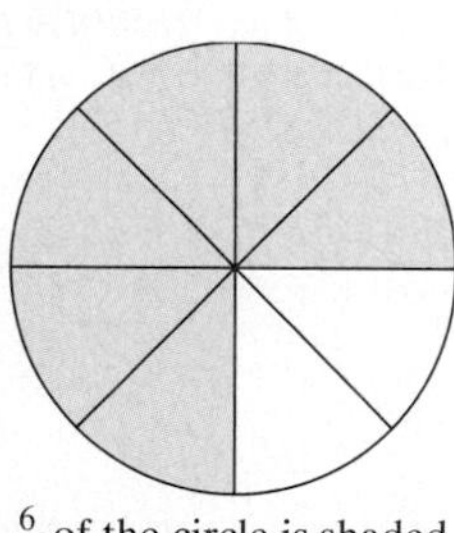

$\frac{6}{8}$ of the circle is shaded.

Any nonzero number divided by itself is 1.

$$\frac{5}{5} = \frac{17}{17} = \frac{c}{c} = 1$$

Thus, if we multiply a fraction by $\frac{5}{5}$ or $\frac{17}{17}$ or $\frac{c}{c}$ (remember, c cannot be zero), the value of the fraction is unchanged because we are multiplying by a form of 1. We can use this rule to show that $\frac{3}{4}$ and $\frac{6}{8}$ are equivalent.

$$\frac{3}{4} \times \frac{2}{2} = \frac{6}{8}$$

In general, if b and c are not zero,

$$\frac{a}{b} = \frac{a \times c}{b \times c}.$$

To reduce a fraction, we find a **common factor** in the numerator and in the denominator and divide it out. In the fraction $\frac{6}{8}$, the common factor is 2.

$$\frac{6}{8} = \frac{3 \times 2}{4 \times 2} = \frac{3}{4}$$

$$\frac{6}{8} = \frac{3}{4}$$

For all fractions (where a, b, and c are not zero), if c is a common factor,

$$\frac{a}{b} = \frac{a \div c}{b \div c}.$$

A fraction is called **simplified, reduced,** or **in lowest terms** if the numerator and the denominator have only 1 *as a common factor.*

EXAMPLE 2 Simplify (write in lowest terms).

(a) $\dfrac{15}{25}$ (b) $\dfrac{42}{56}$

Solution

(a) $\dfrac{15}{25} = \dfrac{15 \div 5}{25 \div 5} = \dfrac{3}{5}$ The greatest common factor is 5. Divide the numerator and the denominator by 5.

(b) $\dfrac{42}{56} = \dfrac{42 \div 14}{56 \div 14} = \dfrac{3}{4}$ The greatest common factor is 14. Divide the numerator and the denominator by 14.

Perhaps 14 was not the first common factor you thought of. Perhaps you did see the common factor 2. Divide out 2. Then look for another common factor, 7. Now divide out 7.

$$\frac{42}{56} = \frac{42 \div 2}{56 \div 2} = \frac{21}{28} = \frac{21 \div 7}{21 \div 7} = \frac{3}{4}$$

If we do not see large factors at first, sometimes we can simplify a fraction by dividing both numerator and denominator by a smaller common factor several times, until no common factors are left.

Practice Problem 2 Simplify by dividing out common factors.

(a) $\dfrac{30}{42}$ (b) $\dfrac{60}{132}$

***NOTE TO STUDENT:** Fully worked-out solutions to all of the Practice Problems can be found at the back of the text starting at page SP-1*

A second method to reduce or simplify fractions is called the *method of prime factors*. We factor the numerator and the denominator into prime numbers. We then divide the numerator and the denominator by any common prime factors.

EXAMPLE 3 Simplify the fractions by the method of prime factors.

(a) $\dfrac{35}{42}$ (b) $\dfrac{22}{110}$

Solution

(a) $\dfrac{35}{42} = \dfrac{5 \times 7}{2 \times 3 \times 7}$ We factor 35 and 42 into prime factors. The common prime factor is 7.

$= \dfrac{5 \times \overset{1}{\cancel{7}}}{2 \times 3 \times \underset{1}{\cancel{7}}}$ Now we divide out 7.

$= \dfrac{5 \times 1}{2 \times 3 \times 1} = \dfrac{5}{6}$ We multiply the factors in the numerator and denominator to write the reduced or simplified form.

Thus $\dfrac{35}{42} = \dfrac{5}{6}$, and $\dfrac{5}{6}$ is the simplified form.

(b) $\dfrac{22}{110} = \dfrac{2 \times 11}{2 \times 5 \times 11} = \dfrac{\overset{1}{\cancel{2}} \times \overset{1}{\cancel{11}}}{\underset{1}{\cancel{2}} \times 5 \times \underset{1}{\cancel{11}}} = \dfrac{1}{5}$

***NOTE TO STUDENT:** Fully worked-out solutions to all of the Practice Problems can be found at the back of the text starting at page SP-1*

Practice Problem 3 Simplify the fractions by the method of prime factors.

(a) $\dfrac{120}{135}$ **(b)** $\dfrac{715}{880}$

3 Determining Whether Two Fractions Are Equal

After we simplify, how can we check that a reduced fraction is *equivalent* to the original fraction? If two fractions are equal, their diagonal products or **cross products** are equal. This is called the **equality test for fractions.** If $\frac{3}{4} = \frac{6}{8}$, then

$$\frac{3}{4} \stackrel{?}{=} \frac{6}{8} \quad \begin{array}{l} \longrightarrow 4 \times 6 = 24 \longleftarrow \\ \longrightarrow 3 \times 8 = 24 \longleftarrow \end{array} \quad \boxed{\text{Products are equal.}}$$

If two fractions are unequal (we use the symbol $\neq$), their *cross* products are unequal. If $\dfrac{5}{6} \neq \dfrac{6}{7}$, then

$$\frac{5}{6} \stackrel{?}{=} \frac{6}{7} \quad \begin{array}{l} \longrightarrow 6 \times 6 = 36 \longleftarrow \\ \longrightarrow 5 \times 7 = 35 \longleftarrow \end{array} \quad \boxed{\text{Products are not equal.}}$$

Since $36 \neq 35$, we know that $\dfrac{5}{6} \neq \dfrac{6}{7}$. The test can be described in this way.

EQUALITY TEST FOR FRACTIONS

For any two fractions where a, b, c, and d are whole numbers and $b \neq 0, d \neq 0$, if $\dfrac{a}{b} = \dfrac{c}{d}$, then $a \times d = b \times c$.

EXAMPLE 4 Are these fractions equal? Use the equality test.

(a) $\dfrac{2}{11} \stackrel{?}{=} \dfrac{18}{99}$ **(b)** $\dfrac{3}{16} \stackrel{?}{=} \dfrac{12}{62}$

Solution

(a)

$$\frac{2}{11} \stackrel{?}{=} \frac{18}{99} \quad \begin{array}{l} \longrightarrow 11 \times 18 = 198 \longleftarrow \\ \longrightarrow 2 \times 99 = 198 \longleftarrow \end{array} \quad \boxed{\text{Products are equal.}}$$

Since $198 = 198$, we know that $\dfrac{2}{11} = \dfrac{18}{99}$.

(b) $\frac{3}{16} \stackrel{?}{=} \frac{12}{62}$ $\longrightarrow 16 \times 12 = 192$ $\longleftarrow$ Products are not equal.
$\longrightarrow 3 \times 62 = 186$ $\longleftarrow$

Since $192 \neq 186$, we know that $\frac{3}{16} \neq \frac{12}{62}$.

Practice Problem 4 Test whether the following fractions are equal.

(a) $\frac{84}{108} \stackrel{?}{=} \frac{7}{9}$ **(b)** $\frac{3}{7} \stackrel{?}{=} \frac{79}{182}$

2.2 EXERCISES

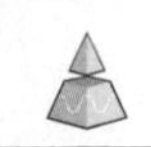

Student Solutions Manual | CD/ Video | PH Math Tutor Center | MathXL®Tutorials on CD | MathXL® | MyMathLab® | Interactmath.com

Verbal and Writing Skills

1. Which of these whole numbers are prime?
4, 12, 11, 15, 6, 19, 1, 41, 38, 24, 5, 46 11, 19, 41, 5

2. A prime number is a whole number greater than 1 that cannot be evenly __divided__ except by itself and 1.

3. A __composite__ __number__ is a whole number greater than 1 that can be divided by whole numbers other than itself and 1.

4. Every composite number can be written in exactly one way as a __product__ of __prime__ numbers.

5. Give an example of a composite number written as a product of primes. $56 = 2 \times 2 \times 2 \times 7$

6. Give an example of equivalent (equal) fractions. $\frac{23}{135} = \frac{46}{270}$

Write each number as a product of prime factors.

7. 15
3×5

8. 9
3×3

9. 35
5×7

10. 8
2^3

11. 49
7^2

12. 25
5^2

13. 64
2^6

14. 81
3^4

15. 55
5×11

16. 42
$2 \times 3 \times 7$

17. 63
$3^2 \times 7$

18. 36
$2^2 \times 3^2$

19. 75
3×5^2

20. 125
5^3

21. 54
2×3^3

22. 99
$3^2 \times 11$

23. 120
$2^3 \times 3 \times 5$

24. 135
$3^3 \times 5$

25. 184
$2^3 \times 23$

26. 216
$2^3 \times 3^3$

Determine which of these whole numbers are prime. If a number is composite, write it as the product of prime factors.

27. 47
prime

28. 31
prime

29. 57
3×19

30. 51
3×17

31. 67
prime

32. 71
prime

33. 62
2×31

34. 91
7×13

35. 89
prime

36. 97
prime

37. 127
prime

38. 119
7×17

39. 121
11×11

40. 95
5×19

41. 129
3×43

42. 143
11×13

Reduce each fraction by finding a common factor in the numerator and in the denominator and dividing by the common factor.

43. $\frac{18}{27}$

$\frac{18 \div 9}{27 \div 9} = \frac{2}{3}$

44. $\frac{16}{24}$

$\frac{16 \div 8}{24 \div 8} = \frac{2}{3}$

45. $\frac{32}{48}$

$\frac{32 \div 16}{48 \div 16} = \frac{2}{3}$

46. $\frac{30}{42}$

$\frac{30 \div 6}{42 \div 6} = \frac{5}{7}$

47. $\frac{30}{48}$

$\frac{30 \div 6}{48 \div 6} = \frac{5}{8}$

48. $\frac{48}{64}$

$\frac{48 \div 16}{64 \div 16} = \frac{3}{4}$

49. $\frac{210}{310}$

$\frac{210 \div 10}{310 \div 10} = \frac{21}{31}$

50. $\frac{110}{140}$

$\frac{110 \div 10}{140 \div 10} = \frac{11}{14}$

Reduce each fraction by the method of prime factors.

51. $\frac{3}{15}$

$\frac{3 \times 1}{3 \times 5} = \frac{1}{5}$

52. $\frac{7}{21}$

$\frac{7 \times 1}{7 \times 3} = \frac{1}{3}$

53. $\frac{66}{88}$

$\frac{2 \times 3 \times 11}{2 \times 2 \times 2 \times 11} = \frac{3}{4}$

54. $\frac{42}{56}$

$\frac{2 \times 3 \times 7}{2 \times 2 \times 2 \times 7} = \frac{3}{4}$

55. $\frac{30}{45}$

$\frac{2 \times 3 \times 5}{3 \times 3 \times 5} = \frac{2}{3}$

56. $\frac{65}{91}$

$\frac{5 \times 13}{7 \times 13} = \frac{5}{7}$

57. $\frac{27}{45}$

$\frac{3 \times 3 \times 3}{3 \times 3 \times 5} = \frac{3}{5}$

58. $\frac{28}{42}$

$\frac{2 \times 2 \times 7}{2 \times 3 \times 7} = \frac{2}{3}$

Mixed Practice

Reduce each fraction by any method.

59. $\frac{33}{36}$

$\frac{3 \times 11}{3 \times 12} = \frac{11}{12}$

60. $\frac{40}{96}$

$\frac{8 \times 5}{8 \times 12} = \frac{5}{12}$

61. $\frac{63}{108}$

$\frac{9 \times 7}{9 \times 12} = \frac{7}{12}$

62. $\frac{72}{132}$

$\frac{6 \times 12}{11 \times 12} = \frac{6}{11}$

63. $\frac{88}{121}$

$\frac{11 \times 8}{11 \times 11} = \frac{8}{11}$

64. $\frac{165}{180}$

$\frac{15 \times 11}{15 \times 12} = \frac{11}{12}$

65. $\frac{150}{200}$

$\frac{3 \times 50}{4 \times 50} = \frac{3}{4}$

66. $\frac{200}{300}$

$\frac{2 \times 100}{3 \times 100} = \frac{2}{3}$

67. $\frac{220}{260}$

$\frac{11 \times 20}{13 \times 20} = \frac{11}{13}$

68. $\frac{210}{390}$

$\frac{30 \times 7}{30 \times 13} = \frac{7}{13}$

Are these fractions equal? Why or why not?

69. $\frac{3}{11} \stackrel{?}{=} \frac{9}{33}$

$3 \times 33 \stackrel{?}{=} 11 \times 9$
$99 = 99$
yes

70. $\frac{4}{15} \stackrel{?}{=} \frac{12}{45}$

$4 \times 45 \stackrel{?}{=} 15 \times 12$
$180 = 180$
yes

71. $\frac{12}{40} \stackrel{?}{=} \frac{3}{13}$

$12 \times 13 \stackrel{?}{=} 40 \times 3$
$156 \neq 120$
no

72. $\frac{24}{72} \stackrel{?}{=} \frac{15}{45}$

$24 \times 45 \stackrel{?}{=} 72 \times 15$
$1080 = 1080$
yes

Are these fractions equal? Why or why not?

73. $\frac{23}{27} \stackrel{?}{=} \frac{92}{107}$

$23 \times 107 \stackrel{?}{=} 27 \times 92$
$2461 \neq 2484$
no

74. $\frac{70}{120} \stackrel{?}{=} \frac{41}{73}$

$70 \times 73 \stackrel{?}{=} 120 \times 41$
$5110 \neq 4920$
no

75. $\frac{27}{57} \stackrel{?}{=} \frac{45}{95}$

$27 \times 95 \stackrel{?}{=} 57 \times 45$
$2565 = 2565$
yes

76. $\frac{18}{24} \stackrel{?}{=} \frac{23}{28}$

$18 \times 28 \stackrel{?}{=} 24 \times 23$
$504 \neq 552$
no

77. $\frac{65}{70} \stackrel{?}{=} \frac{13}{14}$

$65 \times 14 \stackrel{?}{=} 70 \times 13$
$910 = 910$
yes

78. $\frac{98}{182} \stackrel{?}{=} \frac{7}{13}$

$98 \times 13 \stackrel{?}{=} 182 \times 7$
$1274 = 1274$
yes

Applications

Reduce the fractions in your answers.

79. ***Pizza Delivery*** Pizza Palace made 128 deliveries on Saturday night. The manager found that 32 of the deliveries were of more than one pizza. He wanted to study the deliveries that consisted of just one pizza. What fractional part of the deliveries were of just one pizza?

$\frac{3}{4}$

80. ***Medical Students*** Medical students frequently work long hours. Susan worked a 16-hour shift, spending 12 hours in the emergency room and 4 hours in surgery. What fractional part of her shift was she in the emergency room? What fractional part of her shift was she in surgery?

$\frac{3}{4}$ of her shift in the emergency room

$\frac{1}{4}$ of her shift in surgery

81. ***Teaching*** Professor Holbert found that 15 out of 95 students in the Introduction to Psychology class failed the midsemester exam. What fractional part of the class passed the exam?

$\frac{16}{19}$

82. ***Wireless Communications*** William works for a wireless communications company that makes beepers and mobile phones. He inspected 315 beepers and found that 20 were defective. What fractional part of the beepers were not defective?

$\frac{59}{63}$

83. ***Personal Finance*** Seth made \$17,500 during the past six months. He spent \$5000 of his earnings on a European vacation. What fractional part of his earnings did he spend on his vacation?

$\frac{2}{7}$

84. ***Real Estate*** Monique's sister and her husband have been working two jobs each to put a down payment on a plot of land where they plan to build their house. The purchase price is \$42,500. They have saved \$5500. What fractional part of the cost of the land have they saved?

$\frac{11}{85}$

Education *The following data was compiled on the students attending day classes at North Shore Community College.*

Number of Students	Daily Distance Traveled from Home to College (miles)	Length of Commute
1100	0–6	Very short
1700	7–12	Short
900	13–18	Medium
500	19–24	Long
300	More than 24	Very long

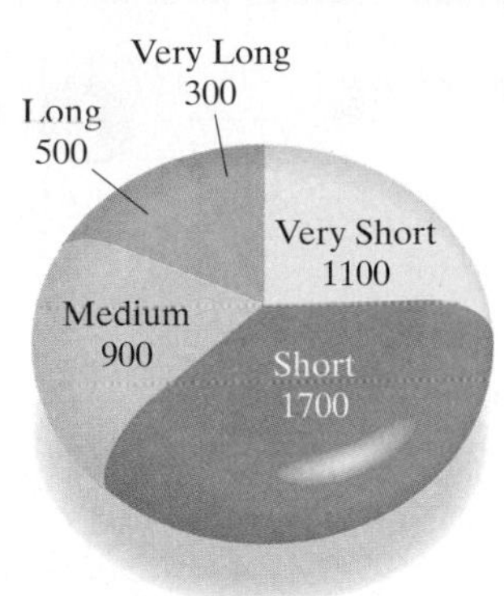

The number of students with each type of commute is displayed in the circle graph to the right.

Answer exercises 85–88 based on the preceding data. Reduce all fractions in your answers.

85. What fractional part of the student body has a short daily commute to the college?
$\frac{17}{45}$

86. What fractional part of the student body has a medium daily commute to the college?
$\frac{1}{5}$

87. What fractional part of the student body has a long or very long daily commute to the college?
$\frac{8}{45}$

88. What fractional part of the student body has a daily commute to the college that is considered less than long?
$\frac{37}{45}$

Cumulative Review

89. Multiply. 386×425
164,050

90. Divide. $15{,}552 \div 12$
1296

91. 3200×300
960,000

92. ***Charities*** In 2001 the total income of the Salvation Army was $2,792,800,000. That same year the total income of the YMCA of the USA was $3,987,500,000. How much greater was the income of the YMCA than the Salvation Army? (*Source*: Internal Revenue Service)
$1,194,700,000

2.3 Converting Between Improper Fractions and Mixed Numbers

Student Learning Objectives

After studying this section, you will be able to:

1. **Change a mixed number to an improper fraction.**
2. **Change an improper fraction to a mixed number.**
3. **Reduce a mixed number or an improper fraction, to lowest terms.**

1 Changing a Mixed Number to an Improper Fraction

We have names for different kinds of fractions. If the value of a fraction is less than 1, we say the fraction is proper.

$$\frac{3}{5}, \frac{5}{7}, \frac{1}{8} \quad \text{are called } \textbf{proper fractions.}$$

Notice that the numerator is less than the denominator. If the numerator is less than the denominator, the fraction is a proper fraction.

If the value of a fraction is greater than or equal to 1, the quantity can be written as an improper fraction or as a mixed number.

Suppose that we have 1 whole pizza and $\frac{1}{6}$ of a pizza. We could write this as $1\frac{1}{6}$. $1\frac{1}{6}$ is called a mixed number. A **mixed number** is the sum of a whole number greater than zero and a proper fraction. The notation $1\frac{1}{6}$ actually means $1 + \frac{1}{6}$ The plus sign is not usually shown.

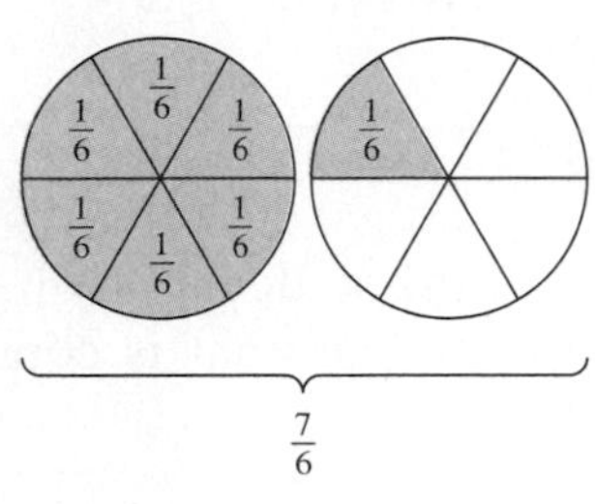

Another way of writing $1\frac{1}{6}$ pizza is to write $\frac{7}{6}$ pizza. $\frac{7}{6}$ is called an improper fraction. Notice that the numerator is greater than the denominator. If the numerator is greater than or equal to the denominator, the fraction is an improper fraction.

$$\frac{7}{6}, \frac{6}{6}, \frac{5}{4}, \frac{8}{3}, \frac{2}{2} \quad \text{are } \textbf{improper fractions.}$$

The following chart will help you visualize these different fractions and their names.

Because in some cases improper fractions are easier to add, subtract, multiply, and divide than mixed numbers, we often change mixed numbers to improper fractions when we perform calculations with them.

Value Less Than 1	Value Equal To 1	Value Greater Than 1
Proper Fraction	Improper Fraction	Improper Fraction or Mixed Number
$\frac{3}{4}$	$\frac{4}{4}$	$\frac{5}{4}$ or $1\frac{1}{4}$
$\frac{7}{8}$	$\frac{8}{8}$	$\frac{17}{8}$ or $2\frac{1}{8}$
$\frac{3}{100}$	$\frac{100}{100}$	$\frac{109}{100}$ or $1\frac{9}{100}$

CHANGING A MIXED NUMBER TO AN IMPROPER FRACTION

1. Multiply the whole number by the denominator of the fraction.
2. Add the numerator of the fraction to the product found in step 1.
3. Write the sum found in step 2 over the denominator of the fraction.

Teaching Tip Some students have been "drilled" in elementary school that improper fractions are "wrong" since they are not simplified. You may need to explain that there is nothing "improper" or incorrect about improper fractions. Some mathematical problems are best left with improper fractions as answers. In mathematics we will find several places where improper fractions are needed and are appropriate.

EXAMPLE 1 Change each mixed number to an improper fraction.

(a) $3\frac{2}{5}$ **(b)** $5\frac{4}{9}$ **(c)** $18\frac{3}{5}$

Solution

Multiply the whole number by the denominator. Add the numerator to the product. Write the sum over the denominator.

(a) $3\frac{2}{5} = \frac{3 \times 5 + 2}{5} = \frac{15 + 2}{5} = \frac{17}{5}$

(b) $5\frac{4}{9} = \frac{9 \times 5 + 4}{9} = \frac{45 + 4}{9} = \frac{49}{9}$

(c) $18\frac{3}{5} = \frac{5 \times 18 + 3}{5} = \frac{90 + 3}{5} = \frac{93}{5}$

Practice Problem 1 Change the mixed numbers to improper fractions.

(a) $4\frac{3}{7}$ **(b)** $6\frac{2}{3}$ **(c)** $19\frac{4}{7}$

NOTE TO STUDENT: Fully worked-out solutions to all of the Practice Problems can be found at the back of the text starting at page SP-1

2 Changing an Improper Fraction to a Mixed Number

We often need to change an improper fraction to a mixed number.

CHANGING AN IMPROPER FRACTION TO A MIXED NUMBER

1. Divide the numerator by the denominator.
2. Write the quotient followed by the fraction with the remainder over the denominator.

$$\text{quotient}\,\frac{\text{remainder}}{\text{denominator}}$$

EXAMPLE 2 Write each improper fraction as a mixed number.

(a) $\frac{13}{5}$ **(b)** $\frac{29}{7}$ **(c)** $\frac{105}{31}$ **(d)** $\frac{85}{17}$

Solution

(a) We divide the denominator 5 into 13.

$$\begin{array}{r} 2 \\ 5\overline{)13} \\ \underline{10} \\ 3 \end{array} \begin{array}{l} \leftarrow \text{quotient} \\ \\ \\ \leftarrow \text{remainder} \end{array}$$

The answer is in the form quotient $\dfrac{\text{remainder}}{\text{denominator}}$.

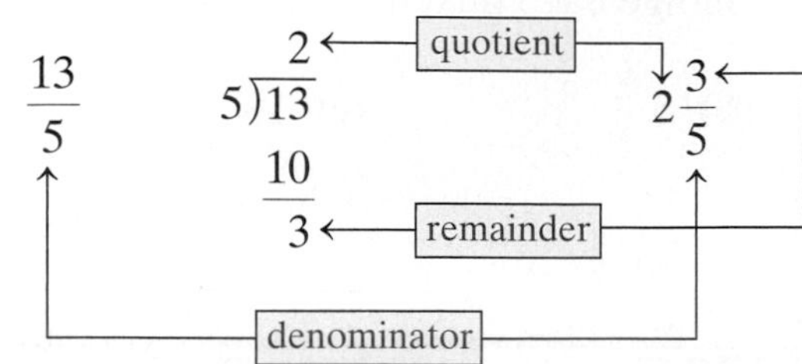

Thus $\dfrac{13}{5} = 2\dfrac{3}{5}$.

(b) $\begin{array}{r} 4 \\ 7\overline{)29} \\ \underline{28} \\ 1 \end{array}$ $\dfrac{29}{7} = 4\dfrac{1}{7}$

(c) $\begin{array}{r} 3 \\ 31\overline{)105} \\ \underline{93} \\ 12 \end{array}$ $\dfrac{105}{31} = 3\dfrac{12}{31}$

(d) $\begin{array}{r} 5 \\ 17\overline{)85} \\ \underline{85} \\ 0 \end{array}$ The remainder is 0, so $\dfrac{85}{17} = 5$, a whole number.

NOTE TO STUDENT: Fully worked-out solutions to all of the Practice Problems can be found at the back of the text starting at page SP-1

Practice Problem 2 Write as a mixed number or a whole number.

(a) $\dfrac{17}{4}$ **(b)** $\dfrac{36}{5}$ **(c)** $\dfrac{116}{27}$ **(d)** $\dfrac{91}{13}$

3 Reducing a Mixed Number or an Improper Fraction to Lowest Terms

Mixed numbers and improper fractions may need to be reduced if they are not in simplest form. Recall that we write the fraction in terms of prime factors. Then we look for common factors in the numerator and the denominator of the fraction. Then we divide the numerator and the denominator by the common factor.

EXAMPLE 3 Reduce the improper fraction. $\dfrac{22}{8}$

Solution

$$\frac{22}{8} = \frac{\overset{1}{\cancel{2}} \times 11}{\underset{1}{\cancel{2}} \times 2 \times 2} = \frac{11}{4}$$

Practice Problem 3 Reduce the improper fraction. $\dfrac{51}{15}$

EXAMPLE 4 Reduce the mixed number. $4\frac{21}{28}$

Solution We cannot reduce the whole number 4, only the fraction $\frac{21}{28}$.

$$\frac{21}{28} = \frac{3 \times \overset{1}{\cancel{7}}}{4 \times \underset{1}{\cancel{7}}} = \frac{3}{4}$$

Therefore, $4\frac{21}{28} = 4\frac{3}{4}$.

Practice Problem 4 Reduce the mixed number. $3\frac{16}{80}$

If an improper fraction contains a very large numerator and denominator, it is best to change the fraction to a mixed number before reducing.

EXAMPLE 5 Reduce $\frac{945}{567}$ by first changing to a mixed number.

Solution

$$\begin{array}{r} 1 \\ 567\overline{)945} \\ \underline{567} \\ 378 \end{array} \qquad \text{so } \frac{945}{567} = 1\frac{378}{567}$$

To reduce the fraction we write

$$\frac{378}{567} = \frac{3 \times 3 \times 3 \times 2 \times 7}{3 \times 3 \times 3 \times 3 \times 7} = \frac{\overset{1}{\cancel{3}} \times \overset{1}{\cancel{3}} \times \overset{1}{\cancel{3}} \times 2 \times \overset{1}{\cancel{7}}}{\underset{1}{\cancel{3}} \times \underset{1}{\cancel{3}} \times \underset{1}{\cancel{3}} \times 3 \times \underset{1}{\cancel{7}}} = \frac{2}{3}$$

So $\frac{945}{567} = 1\frac{378}{567} = 1\frac{2}{3}$.

Problems like Example 5 can be done in several different ways. It is not necessary to follow these exact steps when reducing this fraction.

Practice Problem 5 Reduce $\frac{1001}{572}$ by first changing to a mixed number.

Teaching Tip After explaining Example 5 or some similar problem, have the group do the following problem as a class activity. Have half the class reduce the fraction $\frac{663}{255}$ using this method of changing to a mixed number first. Have the remainder of the class reduce the fraction as an improper fraction. The answer is $2\frac{3}{5}$. Record which half of the class scored the greatest number of correct answers first. (As long as you divide the class into two groups of roughly equal ability, the improper fraction method will usually win.)

TO THINK ABOUT: When a Denominator Is Prime A student concluded that just by looking at the denominator he could tell that the fraction $\frac{1655}{97}$ cannot be reduced unless $1655 \div 97$ is a whole number. How did he come to that conclusion?

Note that 97 is a prime number. The only factors of 97 are 97 and 1. Therefore, *any* fraction with 97 in the denominator can be reduced only if 97 is a factor of the numerator. Since $1655 \div 97$ is not a whole number (see the following division), it is therefore impossible to reduce $\frac{1655}{97}$.

$$\begin{array}{r} 17 \\ 97\overline{)1655} \\ \underline{97} \\ 685 \\ \underline{679} \\ 6 \end{array}$$

You may explore this idea in Exercises 2.3, exercises 83 and 84.

2.3 Exercises

Student Solutions Manual | CD/Video | PH Math Tutor Center | MathXL®Tutorials on CD | MathXL® | MyMathLab® | Interactmath.com

Verbal and Writing Skills

1. Describe in your own words how to change a mixed number to an improper fraction.

(a) Multiply the whole number by the denominator of the fraction.
(b) Add the numerator of the fraction to the product formed in step (a).
(c) Write the sum found in step (b) over the denominator of the fraction.

2. Describe in your own words how to change an improper fraction to a mixed number.

(a) Divide the numerator by the denominator.
(b) Write the quotient followed by the fraction with the remainder over the denominator.

Change each mixed number to an improper fraction.

3. $4\frac{2}{3}$ $\frac{14}{3}$ **4.** $3\frac{5}{6}$ $\frac{23}{6}$ **5.** $2\frac{3}{7}$ $\frac{17}{7}$ **6.** $3\frac{3}{8}$ $\frac{27}{8}$ **7.** $9\frac{2}{9}$ $\frac{83}{9}$ **8.** $8\frac{3}{8}$ $\frac{67}{8}$

9. $10\frac{2}{3}$ $\frac{32}{3}$ **10.** $15\frac{3}{4}$ $\frac{63}{4}$ **11.** $21\frac{2}{3}$ $\frac{65}{3}$ **12.** $13\frac{1}{3}$ $\frac{40}{3}$ **13.** $9\frac{1}{6}$ $\frac{55}{6}$ **14.** $41\frac{1}{2}$ $\frac{83}{2}$

15. $20\frac{1}{6}$ $\frac{121}{6}$ **16.** $6\frac{6}{7}$ $\frac{48}{7}$ **17.** $10\frac{11}{12}$ $\frac{131}{12}$ **18.** $13\frac{5}{7}$ $\frac{96}{7}$ **19.** $7\frac{9}{10}$ $\frac{79}{10}$ **20.** $4\frac{1}{50}$ $\frac{201}{50}$

21. $8\frac{1}{25}$ $\frac{201}{25}$ **22.** $12\frac{5}{6}$ $\frac{77}{6}$ **23.** $5\frac{5}{12}$ $\frac{65}{12}$ **24.** $207\frac{2}{3}$ $\frac{623}{3}$ **25.** $164\frac{2}{3}$ $\frac{494}{3}$ **26.** $33\frac{1}{3}$ $\frac{100}{3}$

27. $8\frac{11}{15}$ $\frac{131}{15}$ **28.** $5\frac{19}{20}$ $\frac{119}{20}$ **29.** $5\frac{13}{25}$ $\frac{138}{25}$ **30.** $6\frac{18}{19}$ $\frac{132}{19}$

Change each improper fraction to a mixed number or a whole number.

31. $\frac{4}{3}$ $1\frac{1}{3}$ **32.** $\frac{13}{4}$ $3\frac{1}{4}$ **33.** $\frac{11}{4}$ $2\frac{3}{4}$ **34.** $\frac{9}{5}$ $1\frac{4}{5}$ **35.** $\frac{15}{6}$ $2\frac{1}{2}$ **36.** $\frac{23}{6}$ $3\frac{5}{6}$

37. $\frac{27}{8}$ $3\frac{3}{8}$ **38.** $\frac{48}{16}$ 3 **39.** $\frac{60}{12}$ 5 **40.** $\frac{42}{13}$ $3\frac{3}{13}$ **41.** $\frac{86}{9}$ $9\frac{5}{9}$ **42.** $\frac{47}{2}$ $23\frac{1}{2}$

43. $\frac{70}{3}$ $23\frac{1}{3}$ **44.** $\frac{54}{17}$ $3\frac{3}{17}$ **45.** $\frac{51}{16}$ $3\frac{3}{16}$ **46.** $\frac{19}{3}$ $6\frac{1}{3}$ **47.** $\frac{28}{3}$ $9\frac{1}{3}$ **48.** $\frac{100}{3}$ $33\frac{1}{3}$

49. $\frac{35}{2}$ $17\frac{1}{2}$

50. $\frac{132}{11}$ 12

51. $\frac{91}{7}$ 13

52. $\frac{183}{7}$ $26\frac{1}{7}$

53. $\frac{210}{15}$ 14

54. $\frac{196}{9}$ $21\frac{7}{9}$

55. $\frac{102}{17}$ 6

56. $\frac{104}{8}$ 13

57. $\frac{403}{11}$ $36\frac{7}{11}$

58. $\frac{212}{9}$ $23\frac{5}{9}$

Reduce each mixed number.

59. $2\frac{9}{12}$ $2\frac{3}{4}$

60. $2\frac{10}{15}$ $2\frac{2}{3}$

61. $4\frac{11}{66}$ $4\frac{1}{6}$

62. $3\frac{15}{90}$ $3\frac{1}{6}$

63. $15\frac{18}{72}$ $15\frac{1}{4}$

64. $10\frac{15}{75}$ $10\frac{1}{5}$

Reduce each improper fraction.

65. $\frac{24}{6}$ 4

66. $\frac{36}{4}$ 9

67. $\frac{36}{15}$ $\frac{12}{5}$

68. $\frac{63}{45}$ $\frac{7}{5}$

69. $\frac{91}{14}$ $\frac{13}{2}$

70. $\frac{143}{13}$ 11

Change to a mixed number and reduce.

71. $\frac{340}{126}$ $2\frac{88}{126} = 2\frac{44}{63}$

72. $\frac{390}{360}$ $1\frac{30}{360} = 1\frac{1}{12}$

73. $\frac{580}{280}$ $2\frac{20}{280} = 2\frac{1}{14}$

74. $\frac{764}{328}$ $2\frac{108}{328} = 2\frac{27}{82}$

75. $\frac{508}{296}$ $1\frac{212}{296} = 1\frac{53}{74}$

76. $\frac{2150}{1000}$ $2\frac{150}{1000} = 2\frac{3}{20}$

Applications

77. ***Banner Display*** The Science Museum is hanging banners all over the building to commemorate the Apollo astronauts. The art department is using $360\frac{2}{3}$ yards of starry-sky parachute fabric. Change this number to an improper fraction.

$\frac{1082}{3}$ yards

78. ***Sculpture*** For the Northwestern University alumni homecoming, the students studying sculpture have made a giant replica of the school using $244\frac{3}{4}$ pounds of clay. Change this number to an improper fraction.

$\frac{979}{4}$ pounds

79. ***Environmental Studies*** A Cape Cod cranberry bog was contaminated by waste from abandoned oil storage tanks at Otis Air Force Base. Damage was done to $\frac{151}{3}$ acres of land. Write this as a mixed number.

$50\frac{1}{3}$ acres

80. ***Painting*** To paint the walls of the new gymnasium, the custodian took measurements. Based on his calculations, he determined he needed $\frac{361}{3}$ gallons of paint. Write this as a mixed number.

$120\frac{1}{3}$

81. ***Cooking*** The cafeteria workers at Ipswich High School cafeteria used $\frac{1131}{8}$ pounds of flour while cooking for the students last week. Write this as a mixed number.

$141\frac{3}{8}$ pounds

82. ***Shelf Construction*** The new Danvers Main Building at North Shore Community Colleges had several new offices for the faculty and staff. Shelving was constructed for these offices. A total of $\frac{1373}{8}$ feet of shelving was used in the construction. Write this as a mixed number.

$171\frac{5}{8}$ feet

To Think About

83. Can $\frac{5687}{101}$ be reduced? Why or why not?

no; 101 is prime and is not a factor of 5687

84. Can $\frac{9810}{157}$ be reduced? Why or why not?

no; 157 is prime and not a factor of 9810

Cumulative Review

85. Subtract. 1,398,210 − 1,137,963

260,247

86. Estimate the answer.
78,964 × 229,350

16,000,000,000

87. Estimate the answer.
328,515 ÷ 966

300

88. ***Textbook Shipment*** Each semester college textbooks are often shipped to the bookstore in cardboard boxes that contain 24 textbooks. If 893 copies of *Basic College Mathematics* are shipped to the bookstore, how many full cartons are needed for the shipment? How many books are there in the carton that is not full?

37 full cartons are needed. There are 5 books in the carton that is not full.

2.4 Multiplying Fractions and Mixed Numbers

Student Learning Objectives

After studying this section, you will be able to:

1. **Multiply two fractions that are proper or improper.**
2. **Multiply a whole number by a fraction.**
3. **Multiply mixed numbers.**

1 Multiplying Two Fractions That Are Proper or Improper

FUDGE SQUARES

Ingredients:

2 cups sugar	1/4 teaspoon salt
4 oz chocolate	1 teaspoon vanilla
1/2 cup butter	1 cup all-purpose flour
4 eggs	1 cup nutmeats

Suppose you want to make an amount equal to half of what the recipe shown will produce. You would multiply the measure given for each ingredient by $\frac{1}{2}$.

$\frac{1}{2}$ of 2 cups sugar	$\frac{1}{2}$ of $\frac{1}{4}$ teaspoon salt
$\frac{1}{2}$ of 4 oz chocolate	$\frac{1}{2}$ of 1 teaspoon vanilla
$\frac{1}{2}$ of $\frac{1}{2}$ cup butter	$\frac{1}{2}$ of 1 cup all-purpose flour
$\frac{1}{2}$ of 4 eggs	$\frac{1}{2}$ of 1 cup nutmeats

We often use multiplication of fractions to describe taking a fractional part of something. To find $\frac{1}{2}$ of $\frac{3}{7}$, we multiply

$$\frac{1}{2} \times \frac{3}{7} = \frac{3}{14}.$$

We begin with a bar that is $\frac{3}{7}$ shaded. To find $\frac{1}{2}$ of $\frac{3}{7}$ we divide the bar in half and take $\frac{1}{2}$ of the shaded section. $\frac{1}{2}$ of $\frac{3}{7}$ yields 3 out of 14 squares.

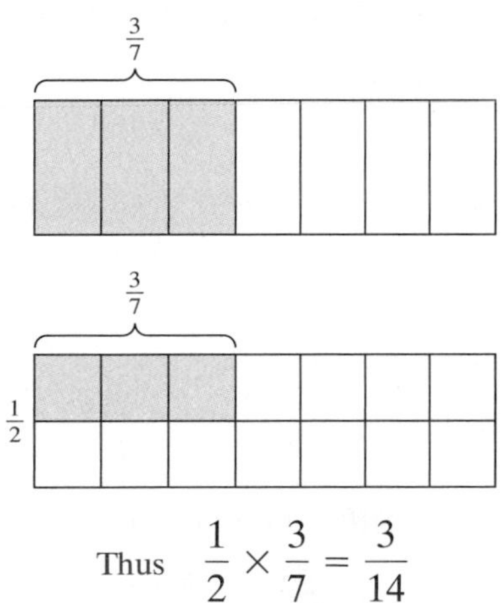

Thus $\frac{1}{2} \times \frac{3}{7} = \frac{3}{14}$

When you multiply two proper fractions together, you get a smaller fraction.

To multiply two fractions, we multiply the numerators and multiply the denominators.

$$\frac{2}{3} \times \frac{5}{7} = \frac{10}{21} \begin{array}{l} \leftarrow 2 \times 5 = 10 \\ \leftarrow 3 \times 7 = 21 \end{array}$$

MULTIPLICATION OF FRACTIONS

In general, for all positive whole numbers a, b, c, and d,

$$\frac{a}{b} \times \frac{c}{d} = \frac{a \times c}{b \times d}.$$

EXAMPLE 1 Multiply.

(a) $\frac{3}{8} \times \frac{5}{7}$ **(b)** $\frac{1}{11} \times \frac{2}{13}$

Solution

(a) $\frac{3}{8} \times \frac{5}{7} = \frac{3 \times 5}{8 \times 7} = \frac{15}{56}$ **(b)** $\frac{1}{11} \times \frac{2}{13} = \frac{1 \times 2}{11 \times 13} = \frac{2}{143}$

***NOTE TO STUDENT:** Fully worked-out solutions to all of the Practice Problems can be found at the back of the text starting at page SP-1*

Practice Problem 1 Multiply.

(a) $\frac{6}{7} \times \frac{3}{13}$ **(b)** $\frac{1}{5} \times \frac{11}{12}$

Some products may be reduced. $\frac{12}{35} \times \frac{25}{18} = \frac{300}{630} = \frac{10}{21}$

By simplifying before multiplication, the reducing can be done more easily. For a multiplication problem, a factor in the numerator can be paired with a common factor in the denominator of the same or a different fraction. We can begin by finding the prime factors in the numerators and denominators. We then divide numerator and denominator by their common prime factors.

EXAMPLE 2 Simplify first and then multiply. $\frac{12}{35} \times \frac{25}{18}$

Solution

$$\frac{12}{35} \times \frac{25}{18} = \frac{2 \cdot 2 \cdot 3}{5 \cdot 7} \times \frac{5 \cdot 5}{3 \cdot 3 \cdot 2}$$ First we find the prime factors.

$$= \frac{2 \cdot 2 \cdot 3 \cdot 5 \cdot 5}{5 \cdot 7 \cdot 3 \cdot 3 \cdot 2}$$ Write the product as one fraction.

$$= \frac{\overset{1}{\cancel{2}} \cdot 2 \cdot \overset{1}{\cancel{3}} \cdot \overset{1}{\cancel{5}} \cdot 5}{\underset{1}{\cancel{2}} \cdot \underset{1}{\cancel{3}} \cdot 3 \cdot \underset{1}{\cancel{5}} \cdot 7}$$ Arrange the factors in order and divide the numerator and denominator by the same numbers.

$$= \frac{10}{21}$$ Multiply the remaining factors.

Practice Problem 2 Simplify first and then multiply.

$$\frac{55}{72} \times \frac{16}{33}$$

Note: Although finding the prime factors of the numerators and denominators will help you avoid errors, you can also begin these problems by dividing the numerators and denominators by larger common factors. This method will be used for the remainder of the exercises in this section of the text.

2 Multiplying a Whole Number by a Fraction

When multiplying a fraction by a whole number, it is more convenient to express the whole number as a fraction with a denominator of 1. We know that $5 = \frac{5}{1}$, $7 = \frac{7}{1}$, and so on.

EXAMPLE 3 Multiply.

(a) $5 \times \frac{3}{8}$ **(b)** $\frac{22}{7} \times 14$

Solution

(a) $5 \times \frac{3}{8} = \frac{5}{1} \times \frac{3}{8} = \frac{15}{8}$ or $1\frac{7}{8}$ **(b)** $\frac{22}{7} \times 14 = \frac{22}{\cancel{7}_1} \times \frac{\cancel{14}^2}{1} = \frac{44}{1} = 44$

Teaching Tip This skill of multiplying a whole number by a fraction is very useful in everyday life. After you have explained Examples 3 and 4, ask the class to find the answer to this problem: "There are 15,041 students at the university. Approximately $\frac{3}{13}$ of the students are over the age of 25. How many students are over age 25?" The correct answer is 3471 students.

Practice Problem 3 Multiply.

(a) $7 \times \frac{5}{13}$ **(b)** $\frac{13}{16} \times 8$

EXAMPLE 4 Mr. and Mrs. Jones found that $\frac{2}{7}$ of their income went to pay federal income taxes. Last year they earned \$37,100. How much did they pay in taxes?

Solution We need to find $\frac{2}{7}$ of \$37,100. So we must multiply $\frac{2}{7} \times 37{,}100$.

$$\frac{2}{\cancel{7}_1} \times \cancel{37{,}100}^{5300} = \frac{2}{1} \times 5300 = 10{,}600$$

They paid \$10,600 in federal income taxes.

▲ **Practice Problem 4** Fred and Linda own 98,400 square feet of land. They found that $\frac{3}{8}$ of the land is in a wetland area and cannot be used for building. How many square feet of land are in the wetland area?

3 Multiplying Mixed Numbers

To multiply a fraction by a mixed number or to multiply two mixed numbers, first change each mixed number to an improper fraction.

EXAMPLE 5 Multiply.

(a) $\frac{5}{7} \times 3\frac{1}{4}$ **(b)** $20\frac{2}{5} \times 6\frac{2}{3}$ **(c)** $\frac{3}{4} \times 1\frac{1}{2} \times \frac{4}{7}$ **(d)** $4\frac{1}{3} \times 2\frac{1}{4}$

Solution

(a) $\frac{5}{7} \times 3\frac{1}{4} = \frac{5}{7} \times \frac{13}{4} = \frac{65}{28}$ or $2\frac{9}{28}$

(b) $20\frac{2}{5} \times 6\frac{2}{3} = \frac{\cancel{102}^{34}}{\cancel{5}_1} \times \frac{\cancel{20}^{4}}{\cancel{3}_1} = \frac{136}{1} = 136$

(c) $\frac{3}{4} \times 1\frac{1}{2} \times \frac{4}{7} = \frac{3}{\cancel{4}_1} \times \frac{3}{2} \times \frac{\cancel{4}^1}{7} = \frac{9}{14}$

(d) $4\frac{1}{3} \times 2\frac{1}{4} = \frac{13}{\cancel{3}_1} \times \frac{\cancel{9}^3}{4} = \frac{39}{4}$ or $9\frac{3}{4}$

NOTE TO STUDENT: Fully worked-out solutions to all of the Practice Problems can be found at the back of the text starting at page SP-1

Practice Problem 5 Multiply.

(a) $2\frac{1}{6} \times \frac{4}{7}$ **(b)** $10\frac{2}{3} \times 13\frac{1}{2}$ **(c)** $\frac{3}{5} \times 1\frac{1}{3} \times \frac{5}{8}$ **(d)** $3\frac{1}{5} \times 2\frac{1}{2}$

Teaching Tip Remind students that the area of both a rectangle and a square can be obtained by multiplying the length times the width or, in the case of the square, by multiplying the length of one side by itself. As a class activity, have them find the area of a square that measures $14\frac{1}{5}$ miles on each side. The answer is $201\frac{16}{25}$ square miles.

EXAMPLE 6 Find the area in square miles of a rectangle with width $1\frac{1}{3}$ miles and length $12\frac{1}{4}$ miles.

Length = $12\frac{1}{4}$ miles

Width = $1\frac{1}{3}$ miles

Solution We find the area of a rectangle by multiplying the width times the length.

$$1\frac{1}{3} \times 12\frac{1}{4} = \frac{\cancel{4}^1}{3} \times \frac{49}{\cancel{4}_1} = \frac{49}{3} \text{ or } 16\frac{1}{3}$$

The area is $16\frac{1}{3}$ square miles.

Practice Problem 6 Find the area in square meters of a rectangle with width $1\frac{1}{5}$ meters and length $4\frac{5}{6}$ meters.

EXAMPLE 7 Find the value of x if

$$\frac{3}{7} \cdot x = \frac{15}{42}.$$

Solution The variable x represents a fraction. We know that 3 times one number equals 15 and 7 times another equals 42.

Since $3 \cdot 5 = 15$
and $7 \cdot 6 = 42$ we know that $\frac{3}{7} \cdot \frac{5}{6} = \frac{15}{42}$.

Therefore, $x = \frac{5}{6}$.

Practice Problem 7 Find the value of x if $\frac{8}{9} \cdot x = \frac{80}{81}$.

2.4 EXERCISES

Student Solutions Manual

CD/ Video

PH Math Tutor Center

MathXL®Tutorials on CD

MathXL®

MyMathLab®

Interactmath.com

Multiply. Make sure all fractions are simplified in the final answer.

1. $\frac{3}{5}\times\frac{7}{11}$ — $\frac{21}{55}$

2. $\frac{1}{8}\times\frac{5}{11}$ — $\frac{5}{88}$

3. $\frac{3}{4}\times\frac{5}{13}$ — $\frac{15}{52}$

4. $\frac{4}{7}\times\frac{3}{5}$ — $\frac{12}{35}$

5. $\frac{6}{5}\times\frac{10}{12}$ — 1

6. $\frac{7}{8}\times\frac{16}{21}$ — $\frac{7}{8}\times\frac{16}{21}=\frac{2}{3}$

7. $\frac{7}{36}\times\frac{30}{9}$ — $\frac{7}{36}\times\frac{30}{9}=\frac{35}{54}$

8. $\frac{22}{45}\times\frac{5}{11}$ — $\frac{22}{45}\times\frac{5}{11}=\frac{2}{9}$

9. $\frac{15}{28}\times\frac{7}{9}$ — $\frac{15}{28}\times\frac{7}{9}=\frac{5}{12}$

10. $\frac{5}{24}\times\frac{18}{15}$ — $\frac{5}{24}\times\frac{18}{15}=\frac{1}{4}$

11. $\frac{9}{10}\times\frac{35}{12}$ — $\frac{9}{10}\times\frac{35}{12}=\frac{21}{8}$ or $2\frac{5}{8}$

12. $\frac{12}{17}\times\frac{3}{24}$ — $\frac{12}{17}\times\frac{3}{24}=\frac{3}{34}$

13. $8\times\frac{3}{7}$ — $\frac{8}{1}\times\frac{3}{7}=\frac{24}{7}$ or $3\frac{3}{7}$

14. $\frac{8}{9}\times 6$ — $\frac{8}{9}\times\frac{6}{1}=\frac{16}{3}$ or $5\frac{1}{3}$

15. $\frac{5}{12}\times 8$ — $\frac{5}{12}\times\frac{8}{1}=\frac{10}{3}$ or $3\frac{1}{3}$

16. $5\times\frac{7}{25}$ — $\frac{5}{1}\times\frac{7}{25}=\frac{7}{5}$ or $1\frac{2}{5}$

17. $\frac{4}{9}\times\frac{3}{7}\times\frac{7}{8}$ — $\frac{4}{9}\times\frac{3}{7}\times\frac{7}{8}=\frac{1}{6}$

18. $\frac{8}{7}\times\frac{5}{12}\times\frac{3}{10}$ — $\frac{8}{7}\times\frac{5}{12}\times\frac{3}{10}=\frac{1}{7}$

19. $\frac{5}{4}\times\frac{9}{10}\times\frac{8}{3}$ — $\frac{5}{4}\times\frac{9}{10}\times\frac{8}{3}=3$

20. $\frac{5}{7}\times\frac{15}{2}\times\frac{28}{15}$ — $\frac{5}{7}\times\frac{15}{2}\times\frac{28}{15}=10$

Multiply. Change any mixed number to an improper fraction before multiplying.

21. $2\frac{3}{4}\times\frac{8}{9}$ — $\frac{11}{4}\times\frac{8}{9}=\frac{22}{9}$ or $2\frac{4}{9}$

22. $\frac{5}{6}\times 3\frac{3}{5}$ — $\frac{5}{6}\times\frac{18}{5}=3$

23. $10\times 3\frac{1}{10}$ — $\frac{10}{1}\times\frac{31}{10}=31$

24. $12\times 5\frac{7}{12}$ — $\frac{12}{1}\times\frac{67}{12}=67$

25. $1\frac{3}{16}\times 0$ — 0

26. $0\times 6\frac{2}{3}$ — 0

27. $3\frac{7}{8}\times 1$ — $3\frac{7}{8}$

28. $\frac{5}{5}\times 11\frac{5}{7}$ — $11\frac{5}{7}$

29. $1\frac{1}{4}\times 3\frac{2}{3}$ — $\frac{5}{4}\times\frac{11}{3}=\frac{55}{12}$ or $4\frac{7}{12}$

30. $2\frac{3}{5}\times 1\frac{4}{7}$ — $\frac{13}{5}\times\frac{11}{7}=\frac{143}{35}$ or $4\frac{3}{35}$

31. $2\frac{3}{10}\times\frac{3}{5}$ — $\frac{23}{10}\times\frac{3}{5}=\frac{69}{50}$ or $1\frac{19}{50}$

32. $4\frac{3}{5}\times\frac{1}{10}$ — $\frac{23}{5}\times\frac{1}{10}=\frac{23}{50}$

33. $4\frac{1}{5}\times 12\frac{2}{9}$ — $\frac{21}{5}\times\frac{110}{9}=\frac{154}{3}$ or $51\frac{1}{3}$

34. $5\frac{1}{4}\times 10\frac{3}{7}$ — $\frac{21}{4}\times\frac{73}{7}=\frac{219}{4}$ or $54\frac{3}{4}$

35. $6\frac{2}{5}\times\frac{1}{4}$ — $\frac{32}{5}\times\frac{1}{4}=\frac{8}{5}$ or $1\frac{3}{5}$

36. $\frac{8}{9}\times 4\frac{1}{11}$ — $\frac{8}{9}\times\frac{45}{11}=\frac{40}{11}$ or $3\frac{7}{11}$

Mixed Practice

Multiply. Make sure all fractions are simplified in the final answer.

37. $\frac{11}{15} \times \frac{35}{33}$

$\frac{7}{9}$

38. $\frac{14}{17} \times \frac{34}{42}$

$\frac{2}{3}$

39. $3\frac{1}{4} \times 4\frac{2}{3}$

$15\frac{1}{6}$ or $\frac{91}{6}$

40. $1\frac{5}{8} \times 3\frac{3}{4}$

$6\frac{3}{32}$ or $\frac{195}{32}$

Solve for x.

41. $\frac{2}{7} \cdot x = \frac{18}{35}$

$x = \frac{9}{5}$

42. $\frac{5}{8} \cdot x = \frac{35}{88}$

$x = \frac{7}{11}$

43. $\frac{7}{13} \cdot x = \frac{56}{117}$

$x = \frac{8}{9}$

44. $x \cdot \frac{11}{15} = \frac{77}{225}$

$x = \frac{7}{15}$

Applications

▲ **45.** ***Geometry*** A spy is running from his captors in a forest that is $8\frac{3}{4}$ miles long and $4\frac{1}{3}$ miles wide. Find the area of the forest where he is hiding. (*Hint:* The area of a rectangle is the product of the length times the width.)

$37\frac{11}{12}$ square miles

▲ **46.** ***Geometry*** An area in the Midwest is a designated tornado danger zone. The land is $22\frac{5}{8}$ miles long and $16\frac{1}{2}$ miles wide. Find the area of the tornado danger zone. (*Hint:* The area of a rectangle is the product of the length times the width.)

$373\frac{5}{16}$ square miles

47. ***Airplane Travel*** A Lear jet airplane has 360 gallons of fuel. The plane averages $4\frac{1}{3}$ miles per gallon. How far can the plane go?

1560 miles

48. ***Land Vehicle Travel*** A jeep has $11\frac{1}{6}$ gallons of gas. The jeep averages 12 miles per gallon. How far will the jeep be able to go on what is in the tank?

134 miles

49. ***Cooking*** A recipe from Nanette's French cookbook for a scalloped potato tart requires $90\frac{1}{2}$ grams of grated cheese. How many grams of cheese would she need if she made one tart for each of her 18 cousins?

1629 grams

▲ **50.** ***Geometry*** The dormitory rooms in Selkirk Hall are being carpeted. Each room requires $20\frac{1}{2}$ square feet of carpet. If there are 30 rooms, how much carpet is needed?

615 square feet

51. ***Education*** The principal of Dexter High School estimate that $\frac{1}{18}$ of the senior class will not graduate. If there are 396 seniors, how many are not expected to graduate?

22 students

52. ***Boat Travel*** The propeller on the Ipswich River Cruise Boat turns 320 revolutions per minute. How fast would it turn at $\frac{3}{4}$ of that speed?

240 revolutions per minute

53. ***Job Search*** Carlos has sent his résumé to 12,064 companies through an Internet job search service. If $\frac{1}{32}$ of the companies e-mail him with an invitation for an interview, how many companies will he have heard from?

377 companies

54. ***Car Purchase*** Russ purchased a new Buick LeSabre for \$26,500. After one year the car was worth $\frac{4}{5}$ of the purchase price. What was the car worth after one year?

\$21,200

55. ***Jogging*** Mary jogged $4\frac{1}{4}$ miles per hour for $1\frac{1}{3}$ hours. During $\frac{1}{3}$ of her jogging time, she was jogging in the rain. How many miles did she jog in the rain?

$1\frac{8}{9}$ miles

56. ***College Students*** There were 1340 students at the Beverly campus of North Shore Community College during the spring 2003 semester. The registrar discovered that $\frac{2}{5}$ of these students live in the city of Beverly. He further discovered that $\frac{1}{4}$ of the students living in Beverly attend classes only on Monday, Wednesday, and Friday. How many students at the Beverly campus live in the city of Beverly and attend classes only on Monday, Wednesday, and Friday?

134 students

▲ 57. ***Geometry*** Reggie has 120 square feet to plant a garden. He plans to plant vegetables in two-thirds of the garden. One-half of this area will contain green peppers.

(a) What fractional portion of the garden will contain green peppers?

$\frac{1}{3}$ of the garden

(b) How many square feet will contain green peppers?

40 square feet

58. ***Personal Finance*** When Sue turned 13, her wealthy grandparents gave her an allowance and put $\frac{3}{5}$ of the money into a college fund. When she turned 18, her grandparents surprised her by telling her that $\frac{3}{8}$ of *that* money was put into a long-term fund.

(a) What portion of the original allowance went into the long-term fund?

$\frac{9}{40}$ of the allowance

(b) What amount was put into the long-term fund if the allowance was $150 per month?

$2025

To Think About

59. When we multiply two fractions, we look for opportunities to divide a numerator and a denominator by the same number. Why do we bother with that step? Why don't we just multiply the two numerators and the two denominators?

The step of dividing the numerator and denominator by the same number allows us to work with smaller numbers when we do the multiplication. Also, this allows us to avoid the step of having to simplify the fraction in the final answer.

60. Suppose there is an unknown fraction that has *not* been simplified (it is not reduced). You multiply this unknown fraction by $\frac{2}{5}$ and you obtain a simplified answer of $\frac{6}{35}$. How many possible values could this unknown fraction be? Give at least three possible answers.

There are an infinite number of answers. Any fraction that can be simplified to $\frac{3}{7}$ would be a correct answer. Thus three possible answers to this problem are $\frac{6}{14}$, $\frac{9}{21}$, or $\frac{12}{28}$.

Cumulative Review

61. ***Toll Bridge*** A total of 16,399 cars used a toll bridge in January (31 days). What is the average number of cars using the bridge in one day?

529 cars

62. ***Sales*** The Office of Investors Services has 15,456 calls made per month by the sales personnel. There are 42 sales personnel in the office. What is the average number of calls made per month by one salesperson?

368 calls

63. ***Commuter Driving*** Gerald commutes 21 miles roundtrip between home and work. If he works 240 days a year, how many miles does he drive between work and home in one year?

5040 miles

64. ***Jet Travel*** At cruising speed a new commercial jet uses 12,360 gallons of fuel per hour. How many gallons will be used in 14 hours of flying time?

173,040 gallons

2.5 DIVIDING FRACTIONS AND MIXED NUMBERS

Student Learning Objectives

After studying this section, you will be able to:

1. **Divide two proper or improper fractions.**
2. **Divide a whole number and a fraction.**
3. **Divide mixed numbers.**

1 Dividing Two Proper or Improper Fractions

Why would you divide fractions? Consider this problem.

- A copper pipe that is $\frac{3}{4}$ of a foot long is to be cut into $\frac{1}{4}$-foot pieces. How many pieces will there be?

To find how many $\frac{1}{4}$'s are in $\frac{3}{4}$, we divide $\frac{3}{4} \div \frac{1}{4}$. We draw a sketch.

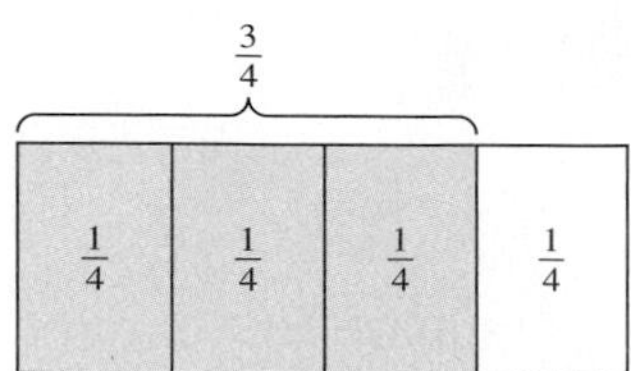

Notice that there are three $\frac{1}{4}$'s in $\frac{3}{4}$.

How do we divide two fractions? We **invert** the second fraction and multiply.

$$\frac{3}{4} \div \frac{1}{4} = \frac{3}{\cancel{4}_1} \times \frac{\cancel{4}^1}{1} = \frac{3}{1} = 3$$

When we invert a fraction, we interchange the numerator and the denominator. If we invert $\frac{5}{9}$, we obtain $\frac{9}{5}$. If we invert $\frac{6}{1}$, we obtain $\frac{1}{6}$. Numbers such as $\frac{5}{9}$ and $\frac{9}{5}$ are called **reciprocals** of each other.

Teaching Tip Tell students that in dividing two fractions it is always the second fraction that is inverted. Inform them that it is a very common mistake for students to erroneously invert the first fraction. If they have trouble remembering which one to flip, tell them to think of the following suggestion from a student who was a cook. "You cannot flip a one-sided pancake. It has to have two sides to flip. You cannot flip fraction number one. It has to be fraction two to flip it." This rule seems silly, but no student who has learned it ever inverts the wrong fraction!

RULE FOR DIVISION OF FRACTIONS

To divide two fractions, we invert the second fraction and multiply.

$$\frac{a}{b} \div \frac{c}{d} = \frac{a}{b} \times \frac{d}{c}$$

(when b, c, and d are not zero).

EXAMPLE 1 Divide.

(a) $\frac{3}{11} \div \frac{2}{5}$ **(b)** $\frac{5}{8} \div \frac{25}{16}$

Solution

(a) $\frac{3}{11} \div \frac{2}{5} = \frac{3}{11} \times \frac{5}{2} = \frac{15}{22}$

(b) $\frac{5}{8} \div \frac{25}{16} = \frac{\cancel{5}^1}{\cancel{8}_1} \times \frac{\cancel{16}^2}{\cancel{25}_5} = \frac{2}{5}$

Practice Problem 1 Divide.

(a) $\frac{7}{13} \div \frac{3}{4}$ **(b)** $\frac{16}{35} \div \frac{24}{25}$

NOTE TO STUDENT: Fully worked-out solutions to all of the Practice Problems can be found at the back of the text starting at page SP-1

2 Dividing a Whole Number and a Fraction

When dividing with whole numbers, it is helpful to remember that for any whole number a, $a = \frac{a}{1}$.

EXAMPLE 2 Divide.

(a) $\frac{3}{7} \div 2$ **(b)** $5 \div \frac{10}{13}$

Solution

(a) $\frac{3}{7} \div 2 = \frac{3}{7} \div \frac{2}{1} = \frac{3}{7} \times \frac{1}{2} = \frac{3}{14}$

(b) $5 \div \frac{10}{13} = \frac{5}{1} \div \frac{10}{13} = \frac{\overset{1}{\cancel{5}}}{1} \times \frac{13}{\underset{2}{\cancel{10}}} = \frac{13}{2}$ or $6\frac{1}{2}$

Practice Problem 2 Divide.

(a) $\frac{3}{17} \div 6$ **(b)** $14 \div \frac{7}{15}$

EXAMPLE 3 Divide, if possible.

(a) $\frac{23}{25} \div 1$ **(b)** $1 \div \frac{7}{5}$

(c) $0 \div \frac{4}{9}$ **(d)** $\frac{3}{17} \div 0$

Solution

(a) $\frac{23}{25} \div 1 = \frac{23}{25} \times \frac{1}{1} = \frac{23}{25}$

(b) $1 \div \frac{7}{5} = \frac{1}{1} \times \frac{5}{7} = \frac{5}{7}$

(c) $0 \div \frac{4}{9} = \frac{0}{1} \times \frac{9}{4} = \frac{0}{4} = 0$ Zero divided by any nonzero number is zero.

(d) $\frac{3}{17} \div 0$ Division by zero is undefined.

Practice Problem 3 Divide, if possible.

(a) $1 \div \frac{11}{13}$ **(b)** $\frac{14}{17} \div 1$ **(c)** $\frac{3}{11} \div 0$ **(d)** $0 \div \frac{9}{16}$

Teaching Tip This is a good time to stress the concept of division by zero and dividing zero by a nonzero number. After covering Example 3 or one like it, have the students do the following eight problems quickly.

1. $\frac{4}{5} \div 3$ **2.** $\frac{5}{6} \div 0$

3. $\frac{8}{9} \div 1$ **4.** $\frac{12}{17} \div 4$

5. $0 \div \frac{14}{17}$ **6.** $1 \div \frac{2}{3}$

7. $\frac{14}{19} \div 0$ **8.** $0 \div \frac{5}{8}$

In order, the correct answers are $\frac{4}{15}$, undefined, $\frac{8}{9}$, $\frac{3}{17}$, 0, $\frac{3}{2}$, undefined, 0. Students find their most common mistakes are with division problems involving zero. Remind them that if they master this concept now, they will avoid a lot of problems later.

SIDELIGHT: Invert and Multiply

Why do we divide by inverting the second fraction and multiplying? What is really going on when we do this? We are actually multiplying by 1. To see why, consider the following.

$$\frac{3}{7} \div \frac{2}{3} = \frac{\frac{3}{7}}{\frac{2}{3}}$$ We write the division by using another fraction bar.

$$= \frac{\frac{3}{7}}{\frac{2}{3}} \times 1$$ Any fraction can be multiplied by 1 without changing the value of the fraction. This is the fundamental rule of fractions.

$$= \frac{\frac{3}{7}}{\frac{2}{3}} \times \frac{\frac{3}{2}}{\frac{3}{2}}$$ Any nonzero number divided by itself equals 1.

$$= \frac{\frac{3}{7} \times \frac{3}{2}}{\frac{2}{3} \times \frac{3}{2}}$$ Definition of multiplication of fractions.

$$= \frac{\frac{3}{7} \times \frac{3}{2}}{1} = \frac{3}{7} \times \frac{3}{2}$$ Any number can be written as a fraction with a denominator of 1 without changing its value.

Thus

$$\frac{3}{7} \div \frac{2}{3} = \frac{3}{7} \times \frac{3}{2} = \frac{9}{14}.$$

3 Dividing Mixed Numbers

If one or more mixed numbers are involved in the division, they should be converted to improper fractions first.

EXAMPLE 4 Divide.

(a) $3\frac{7}{15} \div 1\frac{1}{25}$ **(b)** $\frac{3}{5} \div 2\frac{1}{7}$

Solution

(a) $3\frac{7}{15} \div 1\frac{1}{25} = \frac{52}{15} \div \frac{26}{25} = \frac{\overset{2}{\cancel{52}}}{\underset{3}{\cancel{15}}} \times \frac{\overset{5}{\cancel{25}}}{\underset{1}{\cancel{26}}} = \frac{10}{3}$ or $3\frac{1}{3}$

(b) $\frac{3}{5} \div 2\frac{1}{7} = \frac{3}{5} \div \frac{15}{7} = \frac{\overset{1}{\cancel{3}}}{5} \times \frac{7}{\underset{5}{\cancel{15}}} = \frac{7}{25}$

Practice Problem 4 Divide. **(a)** $1\frac{1}{5} \div \frac{7}{10}$ **(b)** $2\frac{1}{4} \div 1\frac{7}{8}$

***NOTE TO STUDENT:** Fully worked-out solutions to all of the Practice Problems can be found at the back of the text starting at page SP-1*

The division of two fractions may be indicated by a wide fraction bar.

EXAMPLE 5 Divide.

(a) $\dfrac{10\frac{2}{9}}{2\frac{1}{3}}$ **(b)** $\dfrac{1\frac{1}{15}}{3\frac{1}{3}}$

Solution

(a) $\dfrac{10\frac{2}{9}}{2\frac{1}{3}} = 10\frac{2}{9} \div 2\frac{1}{3} = \frac{92}{9} \div \frac{7}{3} = \frac{92}{\cancel{9}_{3}} \times \frac{\cancel{3}^{1}}{7} = \frac{92}{21}$ or $4\frac{8}{21}$

(b) $\dfrac{1\frac{1}{15}}{3\frac{1}{3}} = 1\frac{1}{15} \div 3\frac{1}{3} = \frac{16}{15} \div \frac{10}{3} = \frac{\cancel{16}^{8}}{\cancel{15}_{5}} \times \frac{\cancel{3}^{1}}{\cancel{10}_{5}} = \frac{8}{25}$

Practice Problem 5 Divide.

(a) $\dfrac{5\frac{2}{3}}{7}$ **(b)** $\dfrac{1\frac{2}{5}}{2\frac{1}{3}}$

Some students may find Example 6 difficult at first. Read it slowly and carefully. It may be necessary to read it several times before it becomes clear.

EXAMPLE 6 Find the value of x if $x \div \frac{8}{7} = \frac{21}{40}$.

Solution First we will change the division problem to an equivalent multiplication problem.

$$x \div \frac{8}{7} = \frac{21}{40}$$
$$x \cdot \frac{7}{8} = \frac{21}{40}$$

x represents a fraction.

In the numerator, we want to know what times 7 equals 21. In the denominator, we want to know what times 8 equals 40.

$$\frac{3}{5} \cdot \frac{7}{8} = \frac{21}{40}$$

Thus $x = \frac{3}{5}$.

Practice Problem 6 Find the value of x if $x \div \frac{3}{2} = \frac{22}{36}$.

EXAMPLE 7 There are 117 milligrams of cholesterol in $4\frac{1}{3}$ cups of milk. How much cholesterol is in 1 cup of milk?

Solution We want to divide the 117 by $4\frac{1}{3}$ to find out how much is in 1 cup.

$$117 \div 4\frac{1}{3} = 117 \div \frac{13}{3} = \frac{\overset{9}{\cancel{117}}}{1} \times \frac{3}{\underset{1}{\cancel{13}}} = \frac{27}{1} = 27$$

Thus there are 27 milligrams of cholesterol in 1 cup of milk.

NOTE TO STUDENT: Fully worked-out solutions to all of the Practice Problems can be found at the back of the text starting at page SP-1

Practice Problem 7 A copper pipe that is $19\frac{1}{4}$ feet long will be cut into 14 equal pieces. How long will each piece be?

Teaching Tip Students have trouble identifying which number to divide into which. Ask them to solve the following problem. "A farmer has $117\frac{1}{3}$ tons of grain. He wants to store it in 12 equal piles. How many tons should be placed in each pile?" Go over carefully why we decide to divide $117\frac{1}{3}$ by 12. This will usually help build the students' confidence before they attempt to do the word problems in this section.

Developing Your Study Skills

Why Is Review Necessary?

You master a course in mathematics by learning the concepts one step at a time. There are basic concepts like addition, subtraction, multiplication, and division of whole numbers that are considered the foundation upon which all of mathematics is built. These must be mastered first. Then the study of mathematics is built step by step upon this foundation, each step supporting the next. The process is a carefully designed procedure, and so no steps can be skipped. A student of mathematics needs to realize the importance of this building process to succeed.

Because learning new concepts depends on those previously learned, students often need to take time to review. The reviewing process will strengthen the understanding and application of concepts that are weak due to lack of mastery or passage of time. Review at the right time on the right concepts can strengthen previously learned skills and make progress possible.

Timely, periodic review of previously learned mathematical concepts is absolutely necessary in order to master new concepts. You may have forgotten a concept or grown a bit rusty in applying it. Reviewing is the answer. Make use of any review sections in your textbook, whether they are assigned or not. Look back to previous chapters whenever you have forgotten how to do something. Study the examples and practice some exercises to refresh your understanding.

Be sure that you understand and can perform the computations of each new concept. This will enable you to be able to move successfully on to the next ones.

2.5 Exercises

Student Solutions Manual | CD/ Video | PH Math Tutor Center | MathXL®Tutorials on CD | MathXL® | MyMathLab® | Interactmath.com

Make sure all fractions are simplified in the final answer.

Verbal and Writing Skills

1. In your own words explain how to remember that when you divide two fractions you invert the *second* fraction and multiply by the first. How can you be sure that you don't invert the *first* fraction by mistake?

Think of a simple problem like $3 \div \frac{1}{2}$. One way to think of it is, how many $\frac{1}{2}$'s can be placed in 3? For example, how many $\frac{1}{2}$-pound rocks could be put in a bag that holds 3 pounds of rocks? The answer is 6. If we inverted the first fraction by mistake, we would have $\frac{1}{3} \cdot \frac{1}{2} = \frac{1}{6}$. We know that is wrong since there are obviously several $\frac{1}{2}$-pound rocks in a bag that holds 3 pounds of rocks. The answer $\frac{1}{6}$ would make no sense.

2. Explain why $2 \div \frac{1}{3}$ is a larger number than $2 \div \frac{1}{2}$.

One way to think of it is to imagine how many $\frac{1}{3}$-pound rocks could be put in a bag that holds 2 pounds of rocks and then imagine how many $\frac{1}{2}$-pound rocks could be put in the same bag. The number of $\frac{1}{3}$-pound rocks would be larger. Therefore $2 \div \frac{1}{3}$ is a larger number.

Divide, if possible.

3. $\frac{7}{8} \div \frac{2}{3}$ — $\frac{7}{8} \times \frac{3}{2} = \frac{21}{16}$ or $1\frac{5}{16}$

4. $\frac{3}{13} \div \frac{9}{26}$ — $\frac{3}{13} \times \frac{26}{9} = \frac{2}{3}$

5. $\frac{2}{3} \div \frac{4}{27}$ — $\frac{2}{3} \times \frac{27}{4} = \frac{9}{2}$ or $4\frac{1}{2}$

6. $\frac{5}{16} \div \frac{3}{8}$ — $\frac{5}{16} \times \frac{8}{3} = \frac{5}{6}$

7. $\frac{5}{9} \div \frac{10}{27}$ — $\frac{5}{9} \times \frac{27}{10} = \frac{3}{2}$ or $1\frac{1}{2}$

8. $\frac{8}{15} \div \frac{24}{35}$ — $\frac{8}{15} \times \frac{35}{24} = \frac{7}{9}$

9. $\frac{2}{9} \div \frac{1}{6}$ — $\frac{2}{9} \times \frac{6}{1} = \frac{4}{3}$ or $1\frac{1}{3}$

10. $\frac{3}{4} \div \frac{2}{3}$ — $\frac{3}{4} \times \frac{3}{2} = \frac{9}{8}$ or $1\frac{1}{8}$

11. $\frac{4}{15} \div \frac{4}{15}$ — $\frac{4}{15} \times \frac{15}{4} = 1$

12. $\frac{2}{7} \div \frac{2}{7}$ — $\frac{2}{7} \times \frac{7}{2} = 1$

13. $\frac{3}{7} \div \frac{7}{3}$ — $\frac{3}{7} \times \frac{3}{7} = \frac{9}{49}$

14. $\frac{11}{12} \div \frac{1}{5}$ — $\frac{11}{12} \times \frac{5}{1} = \frac{55}{12}$ or $4\frac{7}{12}$

15. $\frac{4}{5} \div 1$ — $\frac{4}{5} \times 1 = \frac{4}{5}$

16. $1 \div \frac{3}{7}$ — $1 \times \frac{7}{3} = \frac{7}{3}$ or $2\frac{1}{3}$

17. $\frac{3}{11} \div 4$ — $\frac{3}{11} \times \frac{1}{4} = \frac{3}{44}$

18. $2 \div \frac{7}{8}$ — $\frac{2}{1} \times \frac{8}{7} = \frac{16}{7}$ or $2\frac{2}{7}$

19. $1 \div \frac{7}{27}$ — $1 \times \frac{27}{7} = \frac{27}{7}$ or $3\frac{6}{7}$

20. $\frac{9}{16} \div 1$ — $\frac{9}{16} \times 1 = \frac{9}{16}$

21. $0 \div \frac{3}{17}$ — $0 \times \frac{17}{3} = 0$

22. $0 \div \frac{5}{16}$ — $0 \times \frac{16}{5} = 0$

23. $\frac{18}{19} \div 0$ — undefined

24. $\frac{24}{29} \div 0$ — undefined

25. $8 \div \frac{4}{5}$ — $\frac{8}{1} \times \frac{5}{4} = 10$

26. $16 \div \frac{8}{11}$ — $\frac{16}{1} \times \frac{11}{8} = 22$

27. $\frac{7}{8} \div 4$ — $\frac{7}{8} \times \frac{1}{4} = \frac{7}{32}$

28. $\frac{5}{6} \div 12$ — $\frac{5}{6} \times \frac{1}{12} = \frac{5}{72}$

29. $\frac{9}{16} \div \frac{3}{4}$ — $\frac{9}{16} \times \frac{4}{3} = \frac{3}{4}$

30. $\frac{3}{4} \div \frac{9}{16}$ — $\frac{3}{4} \times \frac{16}{9} = \frac{4}{3}$ or $1\frac{1}{3}$

31. $3\frac{1}{4} \div 2\frac{1}{4}$ — $\frac{13}{4} \times \frac{4}{9} = \frac{13}{9}$ or $1\frac{4}{9}$

32. $2\frac{2}{3} \div 4\frac{1}{3}$ — $\frac{8}{3} \times \frac{3}{13} = \frac{8}{13}$

33. $6\frac{2}{5} \div 3\frac{1}{5}$

$\frac{32}{5} \times \frac{5}{16} = 2$

34. $9\frac{1}{3} \div 3\frac{1}{9}$

$\frac{28}{3} \times \frac{9}{28} = 3$

35. $6000 \div \frac{6}{5}$

$\frac{6000}{1} \times \frac{5}{6} = 5000$

36. $8000 \div \frac{4}{7}$

$\frac{8000}{1} \times \frac{7}{4} = 14{,}000$

37. $\dfrac{\frac{4}{5}}{200}$

$\frac{4}{5} \times \frac{1}{200} = \frac{1}{250}$

38. $\dfrac{\frac{5}{9}}{100}$

$\frac{5}{9} \times \frac{1}{100} = \frac{1}{180}$

39. $\dfrac{\frac{5}{8}}{\frac{25}{7}}$

$\frac{5}{8} \times \frac{7}{25} = \frac{7}{40}$

40. $\dfrac{\frac{3}{16}}{\frac{5}{8}}$

$\frac{3}{16} \times \frac{8}{5} = \frac{3}{10}$

Mixed Practice

Multiply or divide.

41. $3\frac{1}{5} \div \frac{3}{10}$

$\frac{16}{5} \times \frac{10}{3} = \frac{32}{3}$ or $10\frac{2}{3}$

42. $2\frac{1}{8} \div \frac{1}{4}$

$\frac{17}{8} \times \frac{4}{1} = \frac{17}{2}$ or $8\frac{1}{2}$

43. $2\frac{1}{3} \times \frac{1}{6}$

$\frac{7}{3} \times \frac{1}{6} = \frac{7}{18}$

44. $6\frac{1}{2} \times \frac{1}{3}$

$\frac{13}{2} \times \frac{1}{3} = \frac{13}{6}$ or $2\frac{1}{6}$

45. $5\frac{1}{4} \div 2\frac{5}{8}$

$\frac{21}{4} \times \frac{8}{21} = 2$

46. $1\frac{2}{9} \div 4\frac{1}{3}$

$\frac{11}{9} \times \frac{3}{13} = \frac{11}{39}$

47. $5 \div 1\frac{1}{4}$

$\frac{5}{1} \times \frac{4}{5} = 4$

48. $7 \div 1\frac{2}{5}$

$\frac{7}{1} \times \frac{5}{7} = 5$

49. $12\frac{1}{2} \div 5\frac{5}{6}$

$\frac{25}{2} \times \frac{6}{35} = \frac{15}{7}$ or $2\frac{1}{7}$

50. $14\frac{2}{3} \div 3\frac{1}{2}$

$\frac{44}{3} \times \frac{2}{7} = \frac{88}{21}$ or $4\frac{4}{21}$

51. $8\frac{1}{4} \div 2\frac{3}{4}$

$\frac{33}{4} \times \frac{4}{11} = 3$

52. $2\frac{3}{8} \div 5\frac{3}{7}$

$\frac{19}{8} \times \frac{7}{38} = \frac{7}{16}$

53. $3\frac{1}{2} \times \frac{9}{16}$

$\frac{7}{2} \times \frac{9}{16} = \frac{63}{32}$ or $1\frac{31}{32}$

54. $1\frac{1}{8} \times \frac{3}{8}$

$\frac{9}{8} \times \frac{3}{8} = \frac{27}{64}$

55. $3\frac{3}{4} \div 9$

$\frac{15}{4} \times \frac{1}{9} = \frac{5}{12}$

56. $5\frac{5}{6} \div 7$

$\frac{35}{6} \times \frac{1}{7} = \frac{5}{6}$

57. $\dfrac{5}{3\frac{1}{6}}$

$\frac{5}{1} \times \frac{6}{19} = \frac{30}{19}$ or $1\frac{11}{19}$

58. $\dfrac{8}{2\frac{1}{2}}$

$\frac{8}{1} \times \frac{2}{5} = \frac{16}{5}$ or $3\frac{1}{5}$

59. $\dfrac{0}{4\frac{3}{8}}$

$0 \times \frac{8}{35} = 0$

60. $\dfrac{5\frac{2}{5}}{0}$

undefined

61. $\dfrac{\frac{7}{12}}{3\frac{2}{3}}$

$\frac{7}{12} \times \frac{3}{11} = \frac{7}{44}$

62. $\dfrac{\frac{9}{10}}{3\frac{3}{5}}$

$\frac{9}{10} \times \frac{5}{18} = \frac{1}{4}$

63. $3\frac{3}{5} \times 2\frac{1}{3}$

$\frac{18}{5} \times \frac{7}{3} = \frac{42}{5}$ or $8\frac{2}{5}$

64. $4\frac{2}{3} \times 5\frac{1}{7}$

$\frac{14}{3} \times \frac{36}{7} = 24$

Review Example 6. Then find the value of x in each of the following.

65. $x \div \frac{4}{3} = \frac{21}{20}$ $x = \frac{7}{5}$

66. $x \div \frac{2}{5} = \frac{15}{16}$ $x = \frac{3}{8}$

67. $x \div \frac{9}{5} = \frac{20}{63}$ $x = \frac{4}{7}$

68. $x \div \frac{7}{3} = \frac{9}{28}$ $x = \frac{3}{4}$

Applications

Answer each question.

69. ***Leather Factory*** A leather factory in Morocco tans leather. In order to make the leather soft, it has to soak in a vat of uric acid and other ingredients. The main holding tank holds $20\frac{1}{4}$ gallons of the tanning mixture. If the mixture is distributed evenly into nine vats of equal size for the different colored leathers, how much will each vat hold?

$2\frac{1}{4}$ gallons

70. ***Marine Biology*** A specially protected stretch of beach bordering the Great Barrier Reef in Australia is used for marine biology and ecological research. The beach, which is $7\frac{1}{2}$ miles long, has been broken up into 20 equal segments for comparison purposes. How long is each segment of the beach?

$\frac{3}{8}$ mile

71. ***Vehicle Travel*** Bruce drove in a snowstorm to get to his favorite mountain to do some snowboarding. He traveled 125 miles in $3\frac{1}{3}$ hours. What was his average speed (in miles per hour)?

$37\frac{1}{2}$ miles per hour

72. ***Vehicle Travel*** Roberto drove his truck to Cedarville, a distance of 200 miles, in $4\frac{1}{6}$ hours. What was his average speed (in miles per hour)?

48 miles per hour

73. ***Cooking*** The school cafeteria is making hamburgers for the annual Senior Day Festival. The cooks have decided that because hamburger shrinks on the grill, they will allow $\frac{2}{3}$ pound of meat for each student. If the kitchen has $38\frac{2}{3}$ pounds of meat, how many students will be fed?

58 students

74. ***Making Flags*** The Patriotic Flag Company is making flags that need $3\frac{1}{8}$ yards of fabric each. The warehouse has $87\frac{1}{2}$ yards of flag fabric available. How many flags can the factory make?

28 flags

75. ***Cooking*** A coffee pot that holds 150 cups of coffee is being used at a company meeting. Each large Styrofoam cup holds $1\frac{1}{2}$ cups of coffee. How many large Styrofoam cups can be filled?

100 large Styrofoam cups

76. ***Medicine Dosage*** A small bottle of eye drops contains 16 milliliters. If the recommended use is $\frac{2}{3}$ milliliter, how many times can a person use the drops before the bottle is empty?

24 times

77. ***Time Capsule*** In 1899, a time capsule was placed behind a steel wall measuring $4\frac{3}{4}$ inches thick. On December 22, 1999, a special drill was used to bore through the wall and extricate the time capsule. The drill could move only $\frac{5}{6}$ inch at a time. How many drill attempts did it take to reach the other side of the steel wall?

It took six drill attempts.

78. ***Ink Production*** Imagination Ink supplies different colored inks for highlighter pens. Vat 1 has yellow ink, holds 150 gallons, and is $\frac{4}{5}$ full. Vat 2 has green ink, holds 50 gallons, and is $\frac{5}{8}$ full. One gallon of ink will fill 1200 pens. How many pens can be filled with the existing ink from Vats 1 and 2?

181,500 pens

To Think About

When multiplying or dividing mixed numbers it is wise to estimate your answer by rounding each mixed number to the nearest whole number.

79. Estimate your answer to $14\frac{2}{3} \div 5\frac{1}{6}$ by rounding each mixed number to the nearest whole number. Then find the exact answer. How close was your estimate?

We estimate by dividing $15 \div 5$, which is 3. The exact value is $2\frac{26}{31}$. Our estimate is very close. It is off by only $\frac{5}{31}$.

80. Estimate your answer to $18\frac{1}{4} \times 27\frac{1}{2}$ by rounding each mixed number to the nearest whole number. Then find the exact answer. How close was your estimate?

We estimate by multiplying 18×28 to obtain 504. The exact value is $501\frac{7}{8}$. Our estimate is very close. It is off by only $2\frac{1}{8}$.

Cumulative Review

81. Write in words. 39,576,304

thirty-nine million, five hundred seventy-six thousand, three hundred four.

82. Write in expanded form. 509,270

$500{,}000 + 9000 + 200 + 70$

83. Add. $126 + 34 + 9 + 891 + 12 + 27$

1099

84. Write in standard notation. eighty-seven million, five hundred ninety-five thousand, six hundred thirty-one

87,595,631

How Am I Doing? Sections 2.1–2.5

How are you doing with your homework assignments in Sections 2.1 to 2.5? Do you feel you have mastered the material so far? Do you understand the concepts you have covered? Before you go further in the textbook, take some time to do each of the following problems.

2.1

1. Use a fraction to represent the shaded part of the object.

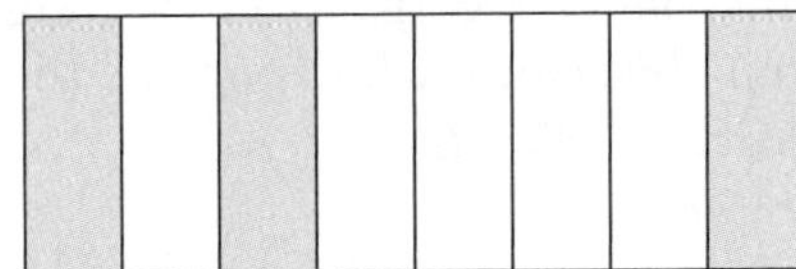

2. Frederich University had 3500 students inside the state, 2600 students outside the state but inside the country, and 800 students from outside the country. Write a fraction that describes the part of the student body from outside the country. Reduce the fraction.

3. An inspector checked 124 CD players. Of these, 5 were defective. Write a fraction that describes the part that was defective.

2.2

Reduce each fraction.

4. $\dfrac{3}{18}$ **5.** $\dfrac{13}{39}$ **6.** $\dfrac{16}{112}$ **7.** $\dfrac{175}{200}$ **8.** $\dfrac{44}{121}$

2.3

Change to an improper fraction.

9. $3\dfrac{2}{3}$ **10.** $6\dfrac{1}{9}$

Change to a mixed number.

11. $\dfrac{97}{4}$ **12.** $\dfrac{29}{5}$ **13.** $\dfrac{36}{17}$

2.4

Multiply.

14. $\dfrac{5}{11} \times \dfrac{1}{4}$ **15.** $\dfrac{3}{7} \times \dfrac{14}{9}$ **16.** $12\dfrac{1}{3} \times 5\dfrac{1}{2}$

2.5

Divide.

17. $\dfrac{3}{7} \div \dfrac{3}{7}$ **18.** $\dfrac{7}{16} \div \dfrac{7}{8}$ **19.** $6\dfrac{4}{7} \div 1\dfrac{5}{21}$ **20.** $8 \div \dfrac{12}{7}$

Now turn to page SA-5 for the answer to each of these problems. Each answer also includes a reference to the objective in which the problem is first taught. If you missed any of these problems, you should stop and review the Examples and Practice Problems in the referenced objective. A little review now will help you master the material in the upcoming sections of the text.

1. $\frac{3}{8}$
2. $\frac{8}{69}$
3. $\frac{5}{124}$
4. $\frac{1}{6}$
5. $\frac{1}{3}$
6. $\frac{1}{7}$
7. $\frac{7}{8}$
8. $\frac{4}{11}$
9. $\frac{11}{3}$
10. $\frac{55}{9}$
11. $24\frac{1}{4}$
12. $5\frac{4}{5}$
13. $2\frac{2}{17}$
14. $\frac{5}{44}$
15. $\frac{2}{3}$
16. $67\frac{5}{6}$ or $\frac{407}{6}$
17. 1
18. $\frac{1}{2}$
19. $5\frac{4}{13}$ or $\frac{69}{13}$
20. $4\frac{2}{3}$ or $\frac{14}{3}$

Note to Instructor: Fractions can be difficult for students to master. For those instructors preferring to test midway through this difficult chapter, this unique mid-chapter test has been provided for student practice.

Test on Sections 2.1–2.5

1. $\frac{23}{32}$
2. $\frac{85}{113}$
3. $\frac{1}{2}$
4. $\frac{7}{15}$
5. $\frac{4}{11}$
6. $\frac{25}{31}$
7. $\frac{3}{4}$
8. $\frac{7}{3}$ or $2\frac{1}{3}$
9. $\frac{43}{12}$
10. $\frac{33}{8}$
11. $6\frac{3}{7}$
12. $8\frac{1}{4}$
13. $\frac{21}{88}$
14. $\frac{9}{7}$ or $1\frac{2}{7}$
15. 15
16. $16\frac{1}{2}$ or $\frac{33}{2}$
17. $13\frac{5}{12}$ or $\frac{161}{12}$
18. $4\frac{16}{21}$ or $\frac{100}{21}$
19. $\frac{16}{21}$
20. $5\frac{1}{3}$ or $\frac{16}{3}$
21. 7
22. $11\frac{1}{5}$ or $\frac{56}{5}$

Solve. Make sure all fractions are simplified in the final answer.

1. Maria scored 23 out of 32 problems correct on the math final exam. Write a fraction that describes what part of the exam she completed correctly.

2. Carlos inspected the boxes that were shipped from the central warehouse. He found that 340 were the correct weight and 112 were not. Write a fraction that describes what part of the total number of the boxes were at the correct weight.

Reduce each fraction.

3. $\frac{19}{38}$

4. $\frac{35}{75}$

5. $\frac{24}{66}$

6. $\frac{125}{155}$

7. $\frac{39}{52}$

8. $\frac{84}{36}$

Change each mixed number to an improper fraction.

9. $3\frac{7}{12}$

10. $4\frac{1}{8}$

Change each improper fraction to a mixed number.

11. $\frac{45}{7}$

12. $\frac{33}{4}$

Multiply.

13. $\frac{3}{8} \times \frac{7}{11}$

14. $\frac{15}{7} \times \frac{3}{5}$

15. $18 \times \frac{5}{6}$

16. $\frac{3}{8} \times 44$

17. $2\frac{1}{3} \times 5\frac{3}{4}$

18. $1\frac{3}{7} \times 3\frac{1}{3}$

Divide.

19. $\frac{4}{7} \div \frac{3}{4}$

20. $\frac{8}{9} \div \frac{1}{6}$

21. $5\frac{1}{4} \div \frac{3}{4}$

22. $5\frac{3}{5} \div \frac{1}{2}$

Mixed Practice

Perform the indicated operations. Simplify your answers.

23. $2\frac{1}{4} \times 3\frac{1}{2}$

24. $6 \times 2\frac{1}{3}$

25. $5 \div 1\frac{7}{8}$

26. $5\frac{3}{4} \div 2$

27. $\frac{13}{20} \div \frac{4}{5}$

28. $\frac{4}{7} \div 8$

29. $\frac{9}{22} \times \frac{11}{16}$

30. $\frac{14}{25} \times \frac{65}{42}$

Solve. Simplify your answer.

▲ **31.** A garden measures $5\frac{1}{4}$ feet by $8\frac{3}{4}$ feet. What is the area of the garden in square feet?

32. A recipe for two loaves of bread calls for $2\frac{2}{3}$ cups of flour. Lexi wants to make $1\frac{1}{2}$ times as much bread. How many cups of flour will she need?

33. Lisa drove $62\frac{1}{2}$ miles to visit a friend. Three-fourths of her trip was on the highway. How many miles did she drive on the highway?

34. The butcher prepared $12\frac{3}{8}$ pounds of lean ground round. He placed it in packages that held $\frac{3}{4}$ of a pound. How many full packages did he have? How much lean ground round was left over?

35. The college computer center has 136 computers. Samuel found that $\frac{3}{8}$ of them have Windows 2000 installed on them. How many computers have Windows 2000 installed on them?

36. The fire department has finished inspecting $\frac{3}{5}$ of the homes in the city to determine if the smoke detectors in these homes are functioning properly. 12,000 homes have been inspected. How many homes still need to be inspected?

37. Yung Kim was paid \$132 last week at his part-time job. He was paid $\$8\frac{1}{4}$ per hour. How many hours did he work last week?

38. The Outdoor Shop is making some custom tents that are very light but totally waterproof. Each tent requires $8\frac{1}{4}$ yards of cloth. How many tents can be made from $56\frac{1}{2}$ yards of cloth? How much cloth will be left over?

39. A container of vanilla-flavored syrup holds $32\frac{4}{5}$ ounces. Nate uses $\frac{4}{5}$ ounce every morning in his coffee. How many days will it take Nate to use up the container?

23. $7\frac{7}{8}$ or $\frac{63}{8}$

24. 14

25. $2\frac{2}{3}$ or $\frac{8}{3}$

26. $2\frac{7}{8}$ or $\frac{23}{8}$

27. $\frac{13}{16}$

28. $\frac{1}{14}$

29. $\frac{9}{32}$

30. $\frac{13}{15}$

31. $45\frac{15}{16}$ square feet

32. 4 cups

33. $46\frac{7}{8}$ miles

34. 16 full packages; $\frac{3}{8}$ lb left over

35. 51 computers

36. 8000 homes

37. 16 hours

38. 6 tents, 7 yards left over

39. 41 days

2.6 The Least Common Denominator and Creating Equivalent Fractions

Student Learning Objectives

After studying this section, you will be able to:

1. Find the least common multiple (LCM) of two numbers.
2. Find the least common denominator (LCD) given two or three fractions.
3. Create equivalent fractions with a least common denominator.

1 Finding the Least Common Multiple (LCM) of Two Numbers

The idea of a multiple of a number is fairly straightforward.
The **multiples** of a number are the products of that number and the numbers 1, 2, 3, 4, 5, 6, 7, . . .

For example, the multiples of 4 are 4, 8, 12, 16, 20, 24, 28, . . .
The multiples of 5 are 5, 10, 15, 20, 25, 30, 35, . . .

The **least common multiple,** or **LCM,** of two natural numbers is the smallest number that is a multiple of both.

EXAMPLE 1 Find the least common multiple of 10 and 12.

Solution

The multiples of 10 are 10, 20, 30, 40, 50, 60, 70, . . .
The multiples of 12 are 12, 24, 36, 48, 60, 72, 84, . . .

The first multiple that appears on both lists is the least common multiple. Thus the number 60 is the least common multiple of 10 and 12.

NOTE TO STUDENT: Fully worked-out solutions to all of the Practice Problems can be found at the back of the text starting at page SP-1

Practice Problem 1 Find the least common multiple of 14 and 21.

EXAMPLE 2 Find the least common multiple of 6 and 8.

Solution

The multiples of 6 are 6, 12, 18, 24, 30, 36, 42, . . .
The multiples of 8 are 8, 16, 24, 32, 40, 48, 56, . . .

The first multiple that appears on both lists is the least common multiple. Thus the number 24 is the least common multiple of 6 and 8.

Practice Problem 2 Find the least common multiple of 10 and 15.

Now of course we can do the problem immediately if the larger number is a multiple of the smaller number. In such cases the larger number is the least common multiple.

EXAMPLE 3 Find the least common multiple of 7 and 35.

Solution Because $7 \times 5 = 35$, therefore 35 is a multiple of 7.
So we can state immediately that the least common multiple of 7 and 35 is 35.

Practice Problem 3 Find the least common multiple of 6 and 54.

Finding the Least Common Denominator (LCD) Given Two or Three Fractions

We need some way to determine which of two fractions is larger. Suppose that Marcia and Melissa each have some leftover pizza.

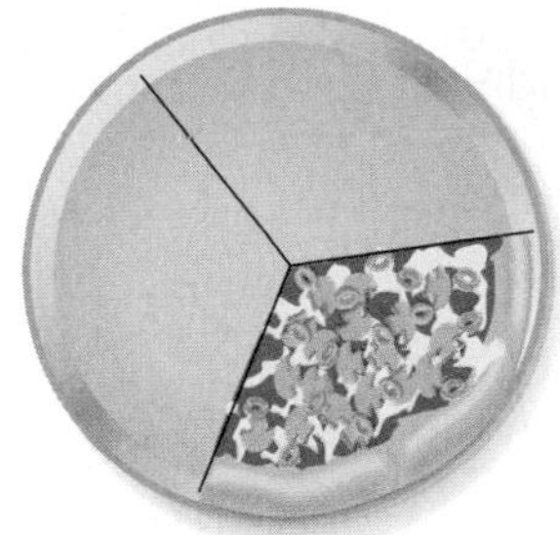

Marcia's Pizza

$\frac{1}{3}$ of a pizza left

Melissa's Pizza

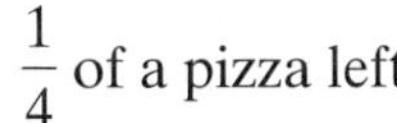

$\frac{1}{4}$ of a pizza left

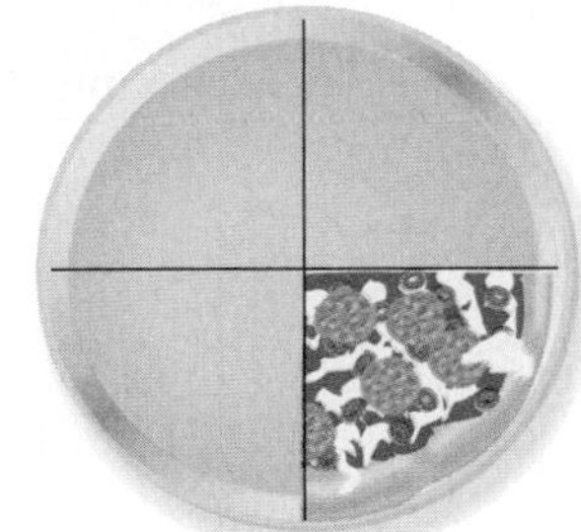

Who has more pizza left? How much more? Comparing the amounts of pizza left would be easy if each pizza had been cut into equal-sized pieces. If the original pizzas had each been cut into 12 pieces, we would be able to see that Marcia had $\frac{1}{12}$ of a pizza more than Melissa had.

Marcia's Pizza

(We know that $\frac{4}{12} = \frac{1}{3}$ by reducing.)

Melissa's Pizza

(We know that $\frac{3}{12} = \frac{1}{4}$ by reducing.)

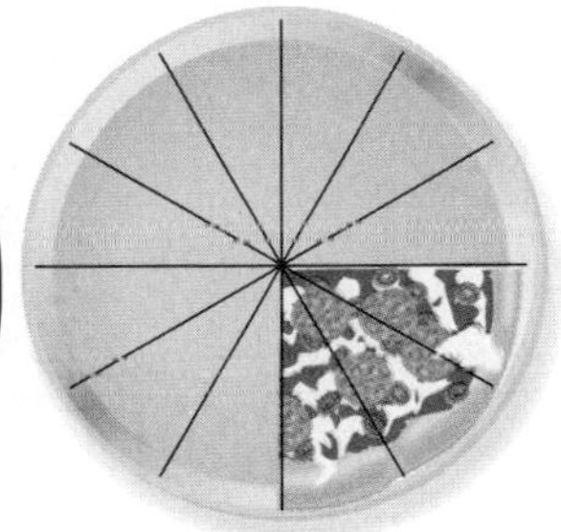

The denominator 12 appears in the fractions $\frac{4}{12}$ and $\frac{3}{12}$. We call the smallest denominator that allows us to compare fractions directly the *least common denominator,* abbreviated LCD. The number 12 is the least common denominator for the fractions $\frac{1}{3}$ and $\frac{1}{4}$.

Notice that 12 is the least common multiple of 3 and 4.

DEFINITION

The **least common denominator (LCD)** of two or more fractions is the smallest number that can be divided evenly by each of the fractions' denominators.

How does this relate to least common multiples? The LCD of two fractions is the least common multiple of the two denominators.

In some problems you may be able to guess the LCD quite quickly. With practice, you can often find the LCD mentally. For example, you now know that if the denominators of two fractions are 3 and 4, the LCD is 12. For the fractions $\frac{1}{2}$ and $\frac{1}{4}$, the LCD is 4; for the fractions $\frac{1}{3}$ and $\frac{1}{6}$, the LCD is 6. We can see that if the denominator of one fraction divides without remainder into the denominator of another, the LCD of the two fractions is the larger of the denominators.

EXAMPLE 4 Determine the LCD for each pair of fractions.

(a) $\frac{7}{15}$ and $\frac{4}{5}$ **(b)** $\frac{2}{3}$ and $\frac{5}{27}$

Solution

(a) Since 5 can be divided into 15, the LCD of $\frac{7}{15}$ and $\frac{4}{5}$ is 15. (Notice that the least common multiple of 5 and 15 is 15.)

(b) Since 3 can be divided into 27, the LCD of $\frac{2}{3}$ and $\frac{5}{27}$ is 27. (Notice that the least common multiple of 3 and 27 is 27.)

Practice Problem 4 Determine the LCD for each pair of fractions.

(a) $\frac{3}{4}$ and $\frac{11}{12}$ **(b)** $\frac{1}{7}$ and $\frac{8}{35}$

NOTE TO STUDENT: Fully worked-out solutions to all of the Practice Problems can be found at the back of the text starting at page SP-1

In a few cases, the LCD is the product of the two denominators.

EXAMPLE 5 Find the LCD for $\frac{1}{4}$ and $\frac{3}{5}$.

Solution We see that $4 \times 5 = 20$. Also, 20 is the *smallest* number that can be divided without remainder by 4 and by 5. We know this because the least common multiple of 4 and 5 is 20. So the LCD $= 20$.

Practice Problem 5 Find the LCD for $\frac{3}{7}$ and $\frac{5}{6}$.

In cases where the LCD is not obvious, the following procedure will help us find the LCD.

THREE-STEP PROCEDURE FOR FINDING THE LEAST COMMON DENOMINATOR

1. Write each denominator as the product of prime factors.
2. List all the prime factors that appear in either product.
3. Form a product of those prime factors, using each factor the greatest number of times it appears in any one denominator.

Teaching Tip Stress the fact that not all students will approach these problems the same way. Some students were taught in school to find the LCD, others to find the GCF, others the LCM. If a student wishes to find the LCD in a way different from the one presented in this book, that is fine as long as the student can obtain correct answers.

EXAMPLE 6 Find the LCD by the three-step procedure.

(a) $\frac{5}{6}$ and $\frac{4}{15}$ **(b)** $\frac{7}{18}$ and $\frac{7}{30}$ **(c)** $\frac{10}{27}$ and $\frac{5}{18}$

Solution

(a) Step 1 Write each denominator as a product of prime factors.

$$6 = 2 \times 3 \qquad 15 = 5 \times 3$$

Step 2 The LCD will contain the factors 2, 3, and 5.

$$6 = 2 \times 3 \qquad 15 = 5 \times 3$$

Step 3 $\text{LCD} = 2 \times 3 \times 5$ We form a product.

$$= 30$$

(b) Step 1 Write each denominator as a product of prime factors.

$$18 = 2 \times 9 = 2 \times 3 \times 3$$
$$30 = 3 \times 10 = 2 \times 3 \times 5$$

Step 2 The LCD will be a product containing 2, 3, and 5.

Step 3 The LCD will contain the factor 3 twice since it occurs twice in the denominator 18.

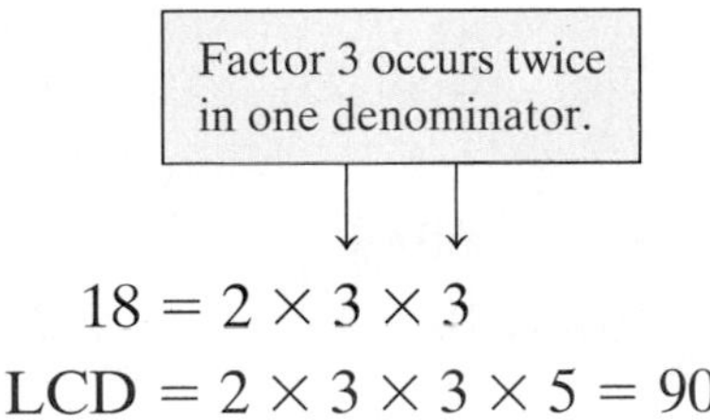
Factor 3 occurs twice
in one denominator.

$$18 = 2 \times 3 \times 3$$
$$\text{LCD} = 2 \times 3 \times 3 \times 5 = 90$$

(c) Write each denominator as a product of prime factors.

$$27 = 3 \times 3 \times 3 \qquad 18 = 3 \times 3 \times 2$$

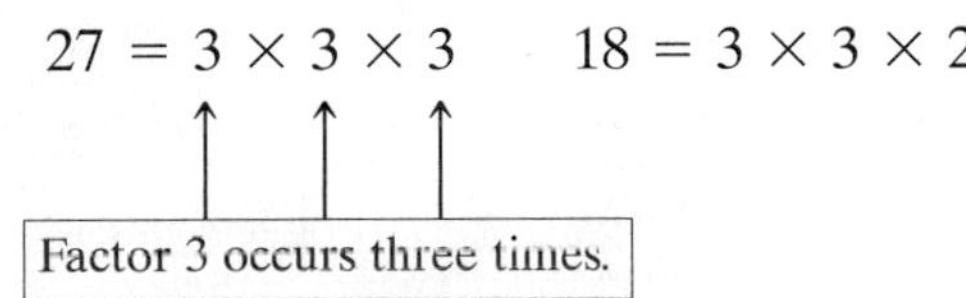
Factor 3 occurs three times.

The LCD will contain the factor 2 once but the factor 3 three times.

$$\text{LCD} = 2 \times 3 \times 3 \times 3 = 54$$

Practice Problem 6 Find the LCD for each pair of fractions.

(a) $\frac{3}{14}$ and $\frac{1}{10}$ **(b)** $\frac{1}{15}$ and $\frac{7}{50}$ **(c)** $\frac{3}{16}$ and $\frac{5}{12}$

A similar procedure can be used for three fractions.

EXAMPLE 7 Find the LCD of $\frac{7}{12}$, $\frac{1}{15}$, and $\frac{11}{30}$.

Solution

$$12 = 2 \times 2 \times 3$$
$$15 = 3 \times 5$$
$$30 = 2 \times 3 \times 5$$
$$\text{LCD} = 2 \times 2 \times 3 \times 5$$
$$= 60$$

Teaching Tip Ask the students to find the LCD for the fractions $\frac{3}{7}, \frac{4}{21}, \frac{7}{24}$. The correct answer is LCD = 168.

Practice Problem 7 Find the LCD of $\frac{3}{49}, \frac{5}{21}$, and $\frac{6}{7}$.

3 Creating Equivalent Fractions with a Least Common Denominator

We cannot add fractions with unlike denominators. To change denominators, we must (1) find the LCD and (2) build up the addends—the fractions being added—into equivalent fractions that have the LCD as the denominator. We know now how to find the LCD. Let's look at how we build up fractions. We know, for example, that

$$\frac{1}{2} = \frac{2}{4} = \frac{50}{100} \qquad \frac{1}{4} = \frac{25}{100} \qquad \text{and} \qquad \frac{3}{4} = \frac{75}{100}.$$

In these cases, we have mentally multiplied the given fraction by 1, in the form of a certain number, c, in the numerator and that same number, c, in the denominator.

$$\frac{1}{2} \times \frac{c}{c} = \frac{2}{4} \qquad \text{Here } c = 2, \frac{2}{2} = 1.$$

$$\frac{1}{2} \times \frac{c}{c} = \frac{50}{100} \qquad \text{Here } c = 50, \frac{50}{50} = 1.$$

This property is called the *building fraction property*.

BUILDING FRACTION PROPERTY

For whole numbers a, b, and c where $b \neq 0, c \neq 0$,

$$\frac{a}{b} = \frac{a}{b} \times 1 = \frac{a}{b} \times \frac{c}{c} = \frac{a \times c}{b \times c}.$$

EXAMPLE 8 Build each fraction to an equivalent fraction with the LCD.

(a) $\frac{3}{4}$, LCD = 28 **(b)** $\frac{4}{5}$, LCD = 45 **(c)** $\frac{1}{3}$ and $\frac{4}{5}$, LCD = 15

Solution

(a) $\frac{3}{4} \times \frac{c}{c} = \frac{?}{28}$ We know that $4 \times 7 = 28$, so the value c that we multiply numerator and denominator by is 7.

$$\frac{3}{4} \times \frac{7}{7} = \frac{21}{28}$$

(b) $\frac{4}{5} \times \frac{c}{c} = \frac{?}{45}$ We know that $5 \times 9 = 45$, so $c = 9$.

$$\frac{4}{5} \times \frac{9}{9} = \frac{36}{45}$$

(c) $\frac{1}{3} = \frac{?}{15}$ We know that $3 \times 5 = 15$, so we multiply numerator and denominator by 5.

$$\frac{1}{3} \times \frac{5}{5} = \frac{5}{15}$$

$\frac{4}{5} = \frac{?}{15}$ We know that $5 \times 3 = 15$, so we multiply numerator and denominator by 3.

$$\frac{4}{5} \times \frac{3}{3} = \frac{12}{15}$$

Thus $\frac{1}{3} = \frac{5}{15}$ and $\frac{4}{5} = \frac{12}{15}$.

Practice Problem 8 Build each fraction to an equivalent fraction with the LCD.

(a) $\frac{3}{5}$, LCD = 40 **(b)** $\frac{7}{11}$, LCD = 44 **(c)** $\frac{2}{7}$ and $\frac{3}{4}$, LCD = 28

NOTE TO STUDENT: Fully worked-out solutions to all of the Practice Problems can be found at the back of the text starting at page SP-1

EXAMPLE 9

(a) Find the LCD of $\frac{1}{32}$ and $\frac{7}{48}$.

(b) Build the fractions to equivalent fractions that have the LCD as their denominators.

Solution

(a) First we find the prime factors of 32 and 48.

$$32 = 2 \times 2 \times 2 \times 2 \times 2$$
$$48 = 2 \times 2 \times 2 \times 2 \times 3$$

Thus the LCD will require a factor of 2 five times and a factor of 3 one time.

$$\text{LCD} = 2 \times 2 \times 2 \times 2 \times 2 \times 3 = 96$$

(b) $\frac{1}{32} = \frac{?}{96}$ Since $32 \times 3 = 96$ we multiply by the fraction $\frac{3}{3}$.

$$\frac{1}{32} = \frac{1}{32} \times \frac{3}{3} = \frac{3}{96}$$

$\frac{7}{48} = \frac{?}{96}$ Since $48 \times 2 = 96$, we multiply by the fraction $\frac{2}{2}$.

$$\frac{7}{48} = \frac{7}{48} \times \frac{2}{2} = \frac{14}{96}$$

Teaching Tip Call students' attention to the fact that some people have learned how to find the LCD of a fraction by using least common multiples. This is a good time to remind students that a good mathematician knows many ways to solve the same problem. It helps to be able to think of two different approaches to solving problems in real life as well as in math.

Practice Problem 9

(a) Find the LCD of $\frac{3}{20}$ and $\frac{11}{15}$.

(b) Build the fractions to equivalent fractions that have the LCD as their denominators.

EXAMPLE 10

(a) Find the LCD of $\frac{2}{125}$ and $\frac{8}{75}$.

(b) Build the fractions to equivalent fractions that have the LCD as their denominators.

Solution

(a) First we find the prime factors of 125 and 75.

$$125 = 5 \times 5 \times 5$$
$$75 = 5 \times 5 \times 3$$

Thus the LCD will require a factor of 5 three times and a factor of 3 one time.

$$\text{LCD} = 5 \times 5 \times 5 \times 3 = 375$$

(b) $\frac{2}{125} = \frac{?}{375}$. Since $125 \times 3 = 375$, we multiply by the fraction $\frac{3}{3}$.

$$\frac{2}{125} = \frac{2}{125} \times \boxed{\frac{3}{3}} = \frac{6}{375}$$

$\frac{8}{75} = \frac{?}{375}$ Since $75 \times 5 = 375$, we multiply by the fraction $\frac{5}{5}$.

$$\frac{8}{75} = \frac{8}{75} \times \boxed{\frac{5}{5}} = \frac{40}{375}$$

NOTE TO STUDENT: Fully worked-out solutions to all of the Practice Problems can be found at the back of the text starting at page SP-1

Practice Problem 10

(a) Find the LCD of $\frac{5}{64}$ and $\frac{3}{80}$.

(b) Build the fractions to equivalent fractions that have the LCD as their denominators.

2.6 EXERCISES

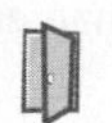
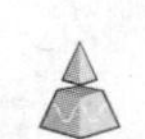

Student Solutions Manual | CD/Video | PH Math Tutor Center | MathXL®Tutorials on CD | MathXL® | MyMathLab® | Interactmath.com

Find the least common multiple (LCM) for each pair of numbers.

1. 8 and 12
24

2. 6 and 9
18

3. 20 and 50
100

4. 22 and 55
110

5. 12 and 15
60

6. 18 and 30
90

7. 9 and 36
36

8. 8 and 72
72

9. 21 and 49
147

10. 25 and 35
175

Find the LCD for each pair of fractions.

11. $\frac{1}{5}$ and $\frac{3}{10}$
LCD = 10

12. $\frac{3}{8}$ and $\frac{5}{16}$
LCD = 16

13. $\frac{3}{7}$ and $\frac{1}{4}$
LCD = 28

14. $\frac{5}{6}$ and $\frac{3}{5}$
LCD = 30

15. $\frac{2}{5}$ and $\frac{3}{7}$
LCD = 35

16. $\frac{1}{16}$ and $\frac{2}{3}$
LCD = 48

17. $\frac{1}{6}$ and $\frac{5}{9}$
$9 = 3 \times 3$
$6 = 2 \times 3$
LCD = 18

18. $\frac{1}{4}$ and $\frac{3}{14}$
$4 = 2 \times 2$
$14 = 2 \times 7$
LCD = 28

19. $\frac{7}{12}$ and $\frac{14}{15}$
$12 = 2 \times 2 \times 3$
$15 = 3 \times 5$
LCD = 60

20. $\frac{7}{15}$ and $\frac{9}{25}$
$15 = 3 \times 5$
$25 = 5 \times 5$
LCD = 75

21. $\frac{7}{32}$ and $\frac{3}{4}$
LCD = 32

22. $\frac{2}{11}$ and $\frac{1}{44}$
LCD = 44

23. $\frac{5}{10}$ and $\frac{11}{45}$
$10 = 2 \times 5$
$45 = 3 \times 3 \times 5$
LCD = 90

24. $\frac{13}{20}$ and $\frac{17}{30}$
$20 = 2 \times 2 \times 5$
$30 = 2 \times 3 \times 5$
LCD = 60

25. $\frac{7}{12}$ and $\frac{7}{30}$
$12 = 2 \times 2 \times 3$
$30 = 2 \times 3 \times 5$
LCD = 60

26. $\frac{5}{6}$ and $\frac{7}{15}$
$6 = 2 \times 3$
$15 = 3 \times 5$
LCD = 30

27. $\frac{5}{21}$ and $\frac{8}{35}$
$21 = 3 \times 7$
$35 = 5 \times 7$
LCD = 105

28. $\frac{1}{20}$ and $\frac{5}{70}$
$20 = 2 \times 2 \times 5$
$70 = 2 \times 5 \times 7$
LCD = 140

29. $\frac{5}{18}$ and $\frac{17}{45}$
$18 = 2 \times 3 \times 3$
$45 = 3 \times 3 \times 5$
LCD = 90

30. $\frac{1}{24}$ and $\frac{7}{40}$
$40 = 2 \times 2 \times 2 \times 5$
$24 = 2 \times 2 \times 2 \times 3$
LCD = 120

Find the LCD for each set of three fractions.

31. $\frac{2}{3}, \frac{1}{2}, \frac{5}{6}$
LCD = 6

32. $\frac{1}{5}, \frac{1}{3}, \frac{7}{10}$
LCD = 30

33. $\frac{5}{6}, \frac{1}{10}, \frac{3}{4}$
$6 = 2 \times 3$
$10 = 2 \times 5$
$4 = 2 \times 2$
LCD = 60

34. $\frac{5}{12}, \frac{3}{16}, \frac{1}{4}$
$12 = 2 \times 2 \times 3$
$16 = 2 \times 2 \times 2 \times 2$
$4 = 2 \times 2$
LCD = 48

35. $\frac{5}{11}, \frac{7}{12}, \frac{1}{6}$
$11 = 11$
$12 = 2 \times 2 \times 3$
$6 = 2 \times 3$
LCD = 132

36. $\frac{11}{16}, \frac{3}{20}, \frac{2}{5}$
$16 = 2 \times 2 \times 2 \times 2$
$20 = 2 \times 2 \times 5$
$5 = 5$
LCD = 80

37. $\frac{7}{12}, \frac{1}{21}, \frac{3}{14}$
$12 = 2 \times 2 \times 3$
$21 = 3 \times 7$
$14 = 2 \times 7$
LCD = 84

38. $\frac{1}{30}, \frac{3}{40}, \frac{7}{8}$
$30 = 2 \times 3 \times 5$
$40 = 2 \times 2 \times 2 \times 5$
$8 = 2 \times 2 \times 2$
LCD = 120

39. $\frac{7}{15}, \frac{11}{12}, \frac{7}{8}$

$15 = 3 \times 5$
$12 = 2 \times 2 \times 3$
$8 = 2 \times 2 \times 2$
LCD = 120

40. $\frac{5}{36}, \frac{2}{48}, \frac{1}{24}$

$36 = 2 \times 2 \times 3 \times 3$
$48 = 2 \times 2 \times 2 \times 2 \times 3$
$24 = 2 \times 2 \times 2 \times 3$
LCD = 144

Build each fraction to an equivalent fraction with the denominator specified. State the numerator.

41. $\frac{1}{3} = \frac{?}{9}$ 3

42. $\frac{1}{5} = \frac{?}{35}$ 7

43. $\frac{5}{7} = \frac{?}{49}$ 35

44. $\frac{7}{9} = \frac{?}{81}$ 63

45. $\frac{4}{11} = \frac{?}{55}$ 20

46. $\frac{2}{13} = \frac{?}{39}$ 6

47. $\frac{7}{24} = \frac{?}{48}$ 14

48. $\frac{3}{50} = \frac{?}{100}$ 6

49. $\frac{8}{9} = \frac{?}{108}$ 96

50. $\frac{6}{7} = \frac{?}{147}$ 126

51. $\frac{7}{20} = \frac{?}{180}$ 63

52. $\frac{15}{32} = \frac{?}{192}$ 90

The LCD of each pair of fractions is listed. Build each fraction to an equivalent fraction that has the LCD as the denominator.

53. LCD = 36, $\frac{7}{12}$ and $\frac{5}{9}$

$\frac{21}{36}$ and $\frac{20}{36}$

54. LCD = 20, $\frac{9}{10}$ and $\frac{3}{4}$

$\frac{18}{20}$ and $\frac{15}{20}$

55. LCD = 80, $\frac{5}{16}$ and $\frac{17}{20}$

$\frac{25}{80}$ and $\frac{68}{80}$

56. LCD = 72, $\frac{5}{24}$ and $\frac{7}{36}$

$\frac{15}{72}$ and $\frac{14}{72}$

57. LCD = 20, $\frac{9}{10}$ and $\frac{19}{20}$

$\frac{18}{20}$ and $\frac{19}{20}$

58. LCD = 240, $\frac{13}{30}$ and $\frac{41}{80}$

$\frac{104}{240}$ and $\frac{123}{240}$

Find the LCD. Build up the fractions to equivalent fractions having the LCD as the denominator.

59. $\frac{2}{5}$ and $\frac{9}{35}$

LCD = 35
$\frac{14}{35}$ and $\frac{9}{35}$

60. $\frac{7}{9}$ and $\frac{35}{54}$

LCD = 54
$\frac{42}{54}$ and $\frac{35}{54}$

61. $\frac{5}{24}$ and $\frac{3}{8}$

LCD = 24
$\frac{5}{24}$ and $\frac{9}{24}$

62. $\frac{19}{42}$ and $\frac{6}{7}$

LCD = 42
$\frac{19}{42}$ and $\frac{36}{42}$

63. $\frac{7}{10}$ and $\frac{5}{6}$

LCD = 30
$\frac{21}{30}$ and $\frac{25}{30}$

64. $\frac{9}{12}$ and $\frac{13}{18}$

LCD = 36
$\frac{27}{36}$ and $\frac{26}{36}$

65. $\frac{4}{15}$ and $\frac{5}{12}$

LCD = 60
$\frac{16}{60}$ and $\frac{25}{60}$

66. $\frac{9}{10}$ and $\frac{3}{25}$

LCD = 50
$\frac{45}{50}$ and $\frac{6}{50}$

67. $\frac{5}{18}, \frac{11}{36}, \frac{7}{12}$

LCD = 36
$\frac{10}{36}, \frac{11}{36}, \frac{21}{36}$

68. $\frac{1}{30}, \frac{7}{15}, \frac{1}{45}$

LCD = 90
$\frac{3}{90}, \frac{42}{90}, \frac{2}{90}$

69. $\frac{3}{56}, \frac{7}{8}, \frac{5}{7}$

LCD = 56
$\frac{3}{56}, \frac{49}{56}, \frac{40}{56}$

70. $\frac{5}{9}, \frac{1}{6}, \frac{3}{54}$

LCD = 54
$\frac{30}{54}, \frac{9}{54}, \frac{3}{54}$

71. $\frac{5}{63}, \frac{4}{21}, \frac{8}{9}$

LCD = 63
$\frac{5}{63}, \frac{12}{63}, \frac{56}{63}$

72. $\frac{3}{8}, \frac{5}{14}, \frac{13}{16}$

LCD = 112
$\frac{42}{112}, \frac{40}{112}, \frac{91}{112}$

Applications

73. ***Door Repair*** Suppose that you wish to compare the lengths of the three portions of the given stainless steel bolt that came out of a door.

(a) What is the LCD for the three fractions?
LCD = 16

(b) Build up each fraction to an equivalent fraction that has the LCD as a denominator.
$\frac{3}{16}, \frac{12}{16}, \frac{6}{16}$

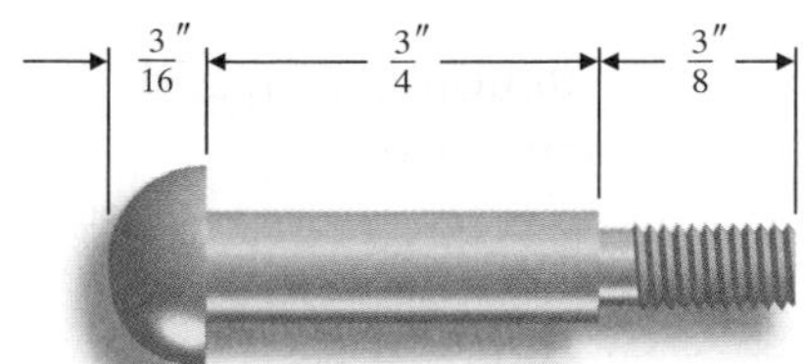

74. ***Plant Growth*** Suppose that you want to prepare a report on the growth of a plant. The total height of the plant in the pot is recorded for each week of a three-week experiment.

(a) What is the LCD for the three fractions?
LCD = 96

(b) Build up each fraction to an equivalent fraction that has the LCD for a denominator.
$\frac{15}{96}, \frac{80}{96}, \frac{84}{96}$

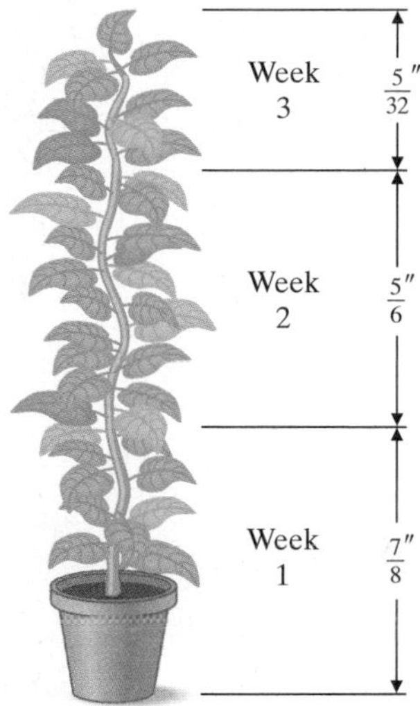

Cumulative Review

75. Divide. $32\overline{)5699}$ 178 R 3

76. Multiply. 1438×22 31,636

77. Evaluate. $(5 - 3)^2 + 4 \times 6 - 3$ 25

Personal Income *Use the following bar graph for exercises 78–83.*

78. What five-year period saw the largest increase in the poverty level for a family of four?
between 1980 and 1985

79. What five-year period saw the smallest increase in the poverty level for a family of four?
between 1995 and 2000

80. If a husband earned $8100 a year in 1995, what would his wife have had to earn for the family income to be above the poverty level?
more than $7469 a year

81. If a wife earned $9700 a year in 2000, what would her husband have had to earn for the family income to be above the poverty level?
more than $7903 a year

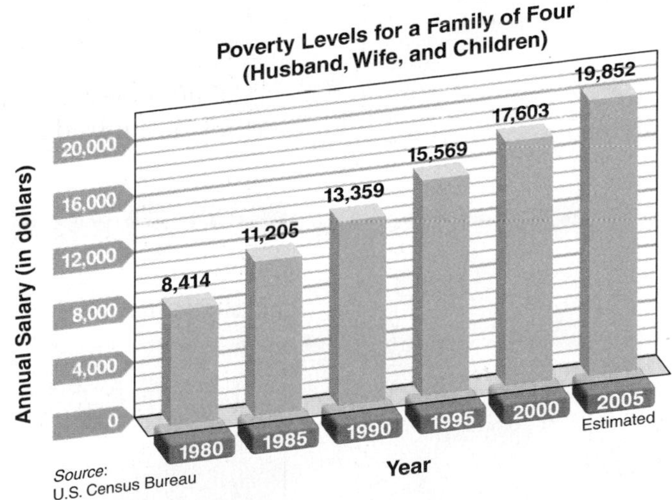

82. If the same increase in the poverty level occurs between 2005 and 2015 as between 1995 and 2005, what would be the expected poverty level for a family of four in 2015?
$24,135

83. If the same increase in the poverty level occurs between 2000 and 2010 as between 1990 and 2000, what would be the expected poverty level for a family of four in 2010?
$21,847

2.7 Adding and Subtracting Fractions

Student Learning Objectives

After studying this section, you will be able to:

1. **Add and subtract fractions with a common denominator.**
2. **Add and subtract fractions with different denominators.**

1 Adding and Subtracting Fractions with a Common Denominator

You must have common denominators (denominators that are alike) to add or subtract fractions.

If your problem has fractions without a common denominator or if it has mixed numbers, you must use what you already know about changing the form of each fraction (how the fraction looks). Only after all the fractions have a common denominator can you add or subtract.

An important distinction: You must have common denominators to add or subtract fractions, but you need not have common denominators to multiply or divide fractions.

To add two fractions that have the same denominator, add the numerators and write the sum over the common denominator.

To illustrate we use $\frac{1}{5} + \frac{2}{5} = \frac{3}{5}$. The figure shows that $\frac{1}{5} + \frac{2}{5} = \frac{3}{5}$.

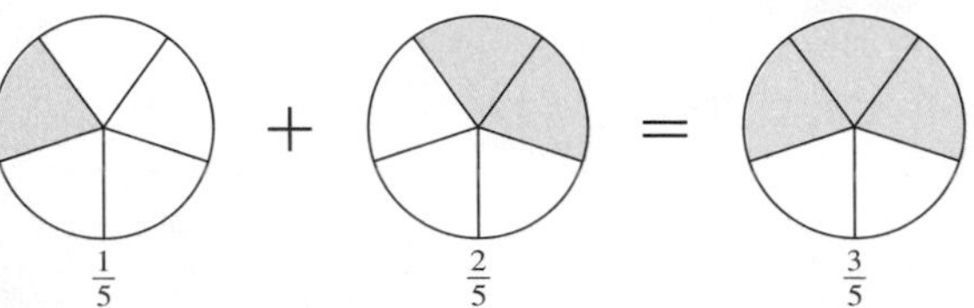

EXAMPLE 1 Add. $\frac{5}{13} + \frac{7}{13}$

Solution

$$\frac{5}{13} + \frac{7}{13} = \frac{12}{13}$$

Practice Problem 1 Add. $\frac{3}{17} + \frac{12}{17}$

***NOTE TO STUDENT:** Fully worked-out solutions to all of the Practice Problems can be found at the back of the text starting at page SP-1*

The answer may need to be reduced. Sometimes the answer may be written as a mixed number.

EXAMPLE 2 Add. **(a)** $\frac{4}{9} + \frac{2}{9}$ **(b)** $\frac{5}{7} + \frac{6}{7}$

Solution

(a) $\frac{4}{9} + \frac{2}{9} = \frac{6}{9} = \frac{2}{3}$ **(b)** $\frac{5}{7} + \frac{6}{7} = \frac{11}{7}$ or $1\frac{4}{7}$

Practice Problem 2 Add. **(a)** $\frac{1}{12} + \frac{5}{12}$ **(b)** $\frac{13}{15} + \frac{7}{15}$

A similar rule is followed for subtraction, except that the numerators are subtracted and the result placed over a common denominator. Be sure to reduce all answers when possible.

EXAMPLE 3 Subtract. **(a)** $\dfrac{5}{13} - \dfrac{4}{13}$ **(b)** $\dfrac{17}{20} - \dfrac{3}{20}$

Solution

(a) $\dfrac{5}{13} - \dfrac{4}{13} = \dfrac{1}{13}$ **(b)** $\dfrac{17}{20} - \dfrac{3}{20} = \dfrac{14}{20} = \dfrac{7}{10}$

Practice Problem 3 Subtract. **(a)** $\dfrac{5}{19} - \dfrac{2}{19}$ **(b)** $\dfrac{21}{25} - \dfrac{6}{25}$

2 Adding and Subtracting Fractions with Different Denominators

If the two fractions do not have a common denominator, we follow the procedure in Section 2.6: Find the LCD and then build up each fraction so that its denominator is the LCD.

EXAMPLE 4 Add. $\dfrac{7}{12} + \dfrac{1}{4}$

Solution The LCD is 12. The fraction $\frac{7}{12}$ already has the least common denominator.

$$\begin{array}{rr} \dfrac{7}{12} \qquad\quad = & \boxed{\dfrac{7}{12}} \\ +\dfrac{1}{4} \times \dfrac{3}{3} = + & \boxed{\dfrac{3}{12}} \\ \hline & \boxed{\dfrac{10}{12}} \end{array}$$

We will need to reduce this fraction. Then we will have

$$\frac{7}{12} + \frac{1}{4} = \frac{7}{12} + \frac{3}{12} = \frac{10}{12} = \frac{5}{6}.$$

It is very important to remember to reduce our final answer.

Practice Problem 4 Add. $\dfrac{2}{15} + \dfrac{2}{5}$

EXAMPLE 5 Add. $\dfrac{7}{20} + \dfrac{4}{15}$

Solution LCD = 60.

$$\frac{7}{20} \times \frac{3}{3} = \frac{21}{60} \qquad \frac{4}{15} \times \frac{4}{4} = \frac{16}{60}$$

Thus

$$\frac{7}{20} + \frac{4}{15} = \frac{21}{60} + \frac{16}{60} = \frac{37}{60}$$

Practice Problem 5 Add. $\dfrac{5}{12} + \dfrac{5}{16}$

A similar procedure holds for the addition of three or more fractions.

EXAMPLE 6 Add. $\frac{3}{8} + \frac{5}{6} + \frac{1}{4}$

Solution LCD = 24.

$$\frac{3}{8} \times \frac{3}{3} = \frac{9}{24} \qquad \frac{5}{6} \times \frac{4}{4} = \frac{20}{24} \qquad \frac{1}{4} \times \frac{6}{6} = \frac{6}{24}$$

$$\frac{3}{8} + \frac{5}{6} + \frac{1}{4} = \frac{9}{24} + \frac{20}{24} + \frac{6}{24} = \frac{35}{24} \quad \text{or} \quad 1\frac{11}{24}$$

NOTE TO STUDENT: Fully worked-out solutions to all of the Practice Problems can be found at the back of the text starting at page SP-1

Practice Problem 6 Add. $\frac{3}{16} + \frac{1}{8} + \frac{1}{12}$

Teaching Tip The ancient Egyptians, during the time of the building of the pyramids, were able to do several types of problems involving large fractions. However, the Egyptian mathematicians did not invent a way to find the LCD for any two fractions, but only certain fractional denominators. Thus they would probably be unable to solve problems like Example 7 because they could not find the LCD of 25 and 35.

EXAMPLE 7 Subtract. $\frac{17}{25} - \frac{3}{35}$

Solution LCD = 175.

$$\frac{17}{25} \times \frac{7}{7} = \frac{119}{175} \qquad \frac{3}{35} \times \frac{5}{5} = \frac{15}{175}$$

Thus

$$\frac{17}{25} - \frac{3}{35} = \frac{119}{175} - \frac{15}{175} = \frac{104}{175}.$$

Practice Problem 7 Subtract. $\frac{9}{48} - \frac{5}{32}$

▲ **EXAMPLE 8** John and Stephanie have a house on $\frac{7}{8}$ acre of land. They have $\frac{1}{3}$ acre of land planted with grass. How much of the land is not planted with grass?

Solution

1. ***Understand the problem.*** Draw a picture.

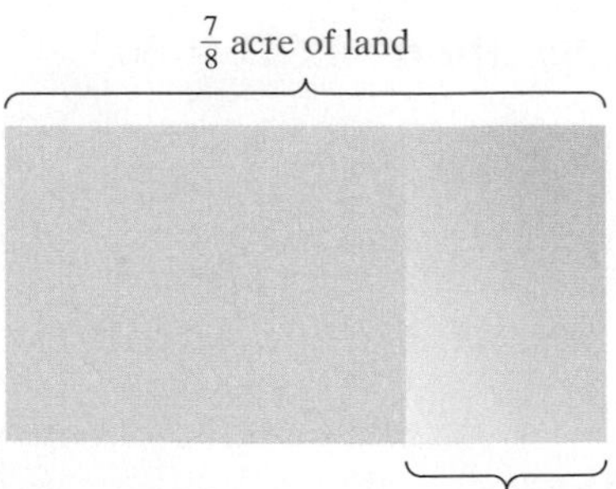

We need to subtract. $\frac{7}{8} - \frac{1}{3}$

2. *Solve and state the answer.* The LCD is 24.

$$\frac{7}{8} \times \frac{3}{3} = \frac{21}{24} \qquad \frac{1}{3} \times \frac{8}{8} = \frac{8}{24}$$

$$\frac{7}{8} - \frac{1}{3} = \frac{21}{24} - \frac{8}{24} = \frac{13}{24}$$

We conclude that $\frac{13}{24}$ acre of the land is not planted with grass.

3. *Check.* The check is left to the student.

Practice Problem 8 Leon had $\frac{9}{10}$ gallon of cleaning fluid in the garage. He used $\frac{1}{4}$ gallon to clean the garage floor. How much cleaning fluid is left? ■

Some students may find Example 9 difficult. Read it slowly and carefully.

EXAMPLE 9 Find the value of x in the equation $x + \frac{5}{6} = \frac{9}{10}$. Reduce your answer.

Solution The LCD for the two fractions $\frac{5}{6}$ and $\frac{9}{10}$ is 30.

$$\frac{5}{6} \times \frac{5}{5} = \frac{25}{30} \qquad \frac{9}{10} \times \frac{3}{3} = \frac{27}{30}$$

Thus we can write the equation in the equivalent form.

$$x + \frac{25}{30} = \frac{27}{30}$$

The denominators are the same. Look at the numerators. We must add 2 to 25 to get 27.

$$\frac{2}{30} + \frac{25}{30} = \frac{27}{30}$$

So $x = \frac{2}{30}$ and we reduce the fraction to obtain $x = \frac{1}{15}$.

Practice Problem 9 Find the value of x in the equation $x + \frac{3}{10} = \frac{23}{25}$. ■

ALTERNATE METHOD: Multiply the Denominators as a Common Denominator In all the problems in this section so far, we have combined two fractions by first finding the least common denominator. However, there is an alternate approach. You are only required to find a common denominator, not necessarily the least common denominator. One way to quickly find a common denominator of two fractions is to multiply the two denominators. However, if you use this method, the numbers will usually be larger and you will usually need to simplify the fraction in your final answer.

EXAMPLE 10 Add $\frac{11}{12} + \frac{13}{30}$ by using the product of the two denominators as a common denominator.

Solution Using this method we just multiply the numerator and denominator of each fraction by the denominator of the other fraction. Thus no steps are needed to determine what to multiply by.

$$\frac{11}{12} \times \frac{30}{30} = \frac{330}{360} \qquad \frac{13}{30} \times \frac{12}{12} = \frac{156}{360}$$

$$\text{Thus} \quad \frac{11}{12} + \frac{13}{30} = \frac{330}{360} + \frac{156}{360} = \frac{486}{360}$$

We must reduce the fraction: $\frac{486}{360} = \frac{27}{20}$ or $1\frac{7}{20}$

***NOTE TO STUDENT:** Fully worked-out solutions to all of the Practice Problems can be found at the back of the text starting at page SP-1*

Practice Problem 10 Add $\frac{15}{16} + \frac{3}{40}$ by using the product of the two denominators as a common denominator.

Some students find this alternate method helpful because you do not have to find the LCD or the number each fraction must be multiplied by. Other students find this alternate method more difficult because of errors encountered when working with large numbers in reducing the final answer. You are encouraged to try a couple of the homework exercises by this method and make up your own mind.

2.7 Exercises

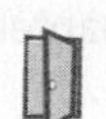

Add or subtract. Simplify all answers.

1. $\frac{5}{9} + \frac{2}{9}$ $\frac{7}{9}$

2. $\frac{5}{8} + \frac{2}{8}$ $\frac{7}{8}$

3. $\frac{7}{18} + \frac{15}{18}$ $\frac{22}{18} = \frac{11}{9} = 1\frac{2}{9}$

4. $\frac{11}{25} + \frac{17}{25}$ $\frac{28}{25} = 1\frac{3}{25}$

5. $\frac{5}{24} - \frac{3}{24}$ $\frac{2}{24} = \frac{1}{12}$

6. $\frac{21}{23} - \frac{1}{23}$ $\frac{20}{23}$

7. $\frac{53}{88} - \frac{19}{88}$ $\frac{34}{88} = \frac{17}{44}$

8. $\frac{103}{110} - \frac{3}{110}$ $\frac{100}{110} = \frac{10}{11}$

Add or subtract. Simplify all answers.

9. $\frac{1}{3} + \frac{1}{2}$ $\frac{2}{6} + \frac{3}{6} = \frac{5}{6}$

10. $\frac{1}{4} + \frac{1}{3}$ $\frac{3}{12} + \frac{4}{12} = \frac{7}{12}$

11. $\frac{3}{10} + \frac{3}{20}$ $\frac{6}{20} + \frac{3}{20} = \frac{9}{20}$

12. $\frac{4}{9} + \frac{1}{6}$ $\frac{8}{18} + \frac{3}{18} = \frac{11}{18}$

13. $\frac{1}{8} + \frac{3}{4}$ $\frac{1}{8} + \frac{6}{8} = \frac{7}{8}$

14. $\frac{5}{16} + \frac{1}{2}$ $\frac{5}{16} + \frac{8}{16} = \frac{13}{16}$

15. $\frac{4}{5} + \frac{7}{20}$ $\frac{16}{20} + \frac{7}{20} = \frac{23}{20}$ or $1\frac{3}{20}$

16. $\frac{2}{3} + \frac{4}{7}$ $\frac{14}{21} + \frac{12}{21} = \frac{26}{21}$ or $1\frac{5}{21}$

17. $\frac{3}{10} + \frac{7}{100}$ $\frac{30}{100} + \frac{7}{100} = \frac{37}{100}$

18. $\frac{13}{100} + \frac{7}{10}$ $\frac{13}{100} + \frac{70}{100} = \frac{83}{100}$

19. $\frac{3}{25} + \frac{1}{35}$ $\frac{21}{175} + \frac{5}{175} = \frac{26}{175}$

20. $\frac{3}{15} + \frac{1}{25}$ $\frac{15}{75} + \frac{3}{75} = \frac{18}{75} = \frac{6}{25}$

21. $\frac{7}{8} + \frac{5}{12}$ $\frac{21}{24} + \frac{10}{24} = \frac{31}{24}$ or $1\frac{7}{24}$

22. $\frac{5}{6} + \frac{7}{8}$ $\frac{20}{24} + \frac{21}{24} = \frac{41}{24}$ or $1\frac{17}{24}$

23. $\frac{3}{8} + \frac{3}{10}$ $\frac{15}{40} + \frac{12}{40} = \frac{27}{40}$

24. $\frac{12}{35} + \frac{1}{10}$ $\frac{24}{70} + \frac{7}{70} = \frac{31}{70}$

25. $\frac{5}{12} - \frac{1}{6}$ $\frac{5}{12} - \frac{2}{12} = \frac{3}{12} = \frac{1}{4}$

26. $\frac{37}{20} - \frac{2}{5}$ $\frac{37}{20} - \frac{8}{20} = \frac{29}{20}$ or $1\frac{9}{20}$

27. $\frac{3}{7} - \frac{1}{5}$ $\frac{15}{35} - \frac{7}{35} = \frac{8}{35}$

28. $\frac{7}{8} - \frac{5}{6}$ $\frac{21}{24} - \frac{20}{24} = \frac{1}{24}$

29. $\frac{5}{9} - \frac{5}{36}$ $\frac{20}{36} - \frac{5}{36} = \frac{15}{36} = \frac{5}{12}$

30. $\frac{9}{10} - \frac{1}{15}$ $\frac{27}{30} - \frac{2}{30} = \frac{25}{30} = \frac{5}{6}$

31. $\frac{5}{12} - \frac{7}{30}$ $\frac{25}{60} - \frac{14}{60} = \frac{11}{60}$

32. $\frac{9}{24} - \frac{3}{8}$ $\frac{9}{24} - \frac{9}{24} = 0$

33. $\frac{11}{12} - \frac{2}{3}$ $\frac{11}{12} - \frac{8}{12} = \frac{3}{12} = \frac{1}{4}$

34. $\frac{7}{10} - \frac{2}{5}$ $\frac{7}{10} - \frac{4}{10} = \frac{3}{10}$

35. $\frac{17}{21} - \frac{1}{7}$ $\frac{17}{21} - \frac{3}{21} = \frac{14}{21} = \frac{2}{3}$

36. $\frac{20}{25} - \frac{4}{5}$ $\frac{20}{25} - \frac{20}{25} = 0$

37. $\frac{7}{24} - \frac{1}{6}$

$\frac{7}{24} - \frac{4}{24} = \frac{3}{24} = \frac{1}{8}$

38. $\frac{2}{5} - \frac{2}{15}$

$\frac{6}{15} - \frac{2}{15} = \frac{4}{15}$

39. $\frac{10}{16} - \frac{5}{8}$

$\frac{10}{16} - \frac{10}{16} = 0$

40. $\frac{5}{6} - \frac{10}{12}$

$\frac{10}{12} - \frac{10}{12} = 0$

41. $\frac{23}{36} - \frac{2}{9}$

$\frac{23}{36} - \frac{8}{36} = \frac{15}{36} = \frac{5}{12}$

42. $\frac{2}{3} - \frac{1}{16}$

$\frac{32}{48} - \frac{3}{48} = \frac{29}{48}$

43. $\frac{1}{2} + \frac{2}{7} + \frac{3}{14}$

$\frac{7}{14} + \frac{4}{14} + \frac{3}{14} = \frac{14}{14} = 1$

44. $\frac{7}{8} + \frac{5}{6} + \frac{7}{24}$

$\frac{21}{24} + \frac{20}{24} + \frac{7}{24} = \frac{48}{24} = 2$

45. $\frac{5}{30} + \frac{3}{40} + \frac{1}{8}$

$\frac{20}{120} + \frac{9}{120} + \frac{15}{120} = \frac{44}{120} = \frac{11}{30}$

46. $\frac{1}{12} + \frac{3}{14} + \frac{4}{21}$

$\frac{7}{84} + \frac{18}{84} + \frac{16}{84} = \frac{41}{84}$

47. $\frac{7}{30} + \frac{2}{5} + \frac{5}{6}$

$\frac{7}{30} + \frac{12}{30} + \frac{25}{30} = \frac{44}{30} = \frac{22}{15} = 1\frac{7}{15}$

48. $\frac{1}{12} + \frac{5}{36} + \frac{32}{36}$

$\frac{3}{36} + \frac{5}{36} + \frac{32}{36} = \frac{40}{36} = \frac{10}{9} = 1\frac{1}{9}$

Study Example 9 carefully. Then find the value of x in each equation.

49. $x + \frac{1}{7} = \frac{5}{14}$

$x = \frac{3}{14}$

50. $x + \frac{1}{8} = \frac{7}{16}$

$x = \frac{5}{16}$

51. $x + \frac{2}{3} = \frac{9}{11}$

$x = \frac{5}{33}$

52. $x + \frac{3}{4} = \frac{17}{18}$

$x = \frac{7}{36}$

53. $x - \frac{1}{5} = \frac{4}{12}$

$x = \frac{8}{15}$

54. $x - \frac{2}{3} = \frac{1}{8}$

$x = \frac{19}{24}$

Applications

55. ***Cooking*** Rita is baking a cake for a dinner party. The recipe calls for $\frac{2}{3}$ cup sugar for the frosting and $\frac{3}{4}$ cup sugar for the cake. How many total cups of sugar does she need?

$1\frac{5}{12}$ cups

56. ***Fitness Training*** Kia is training for a short triathlon. On Monday she swam $\frac{1}{4}$ mile and ran $\frac{5}{6}$ mile. On Tuesday she swam $\frac{1}{2}$ mile and ran $\frac{3}{4}$ mile. How many miles has she swum so far this week? How many miles has she run so far?

$\frac{3}{4}$ mile; $1\frac{7}{12}$ miles

57. ***Food Purchase*** Laurie purchased $\frac{2}{3}$ pound of bananas and $\frac{5}{6}$ pound of seedless grapes. How many pounds of fruit did she purchase?

$1\frac{1}{2}$ pounds

58. ***Automobile Maintenance*** Mandy purchased two new steel-belted all-weather radial tires for her car. The tread depth on the new tires measures $\frac{11}{32}$ of an inch. The dealer told her that when the tires have worn down and their tread depth measures $\frac{1}{8}$ of an inch, she should replace the worn tires with new ones. How much will the tread depth decrease over the useful life of the tire? $\frac{7}{32}$ of an inch

59. ***Power Outage*** Travis typed $\frac{11}{12}$ of his book report on his computer. Then he printed out $\frac{3}{5}$ of his book report on his computer printer. Suddenly, there was a power outage, and he discovered that he hadn't saved his book report before the power went off. What fractional part of the book report was lost when the power failed?

$\frac{19}{60}$ of the book report

60. ***Childcare*** An infant's father knows that straight apple juice is too strong for his daughter. Her bottle is $\frac{1}{2}$ full, and he adds $\frac{1}{3}$ of a bottle of water to dilute the apple juice.

(a) How much is there to drink in the bottle after this addition?

$\frac{5}{6}$ of a bottle

(b) If she drinks $\frac{2}{5}$ of the bottle, how much is left? $\frac{13}{30}$ of a bottle is left.

61. ***Food Purchase*** While he was at the grocery store, Raymond purchased a box of candy for himself. On the way back to the dorm he ate $\frac{1}{4}$ of the candy. As he was putting away the groceries he ate $\frac{1}{2}$ of what was left. There are now six chocolates left in the box. How many chocolates were in the box to begin with?

16 chocolates

62. ***Gasoline Station*** On Saturday morning, the gasoline storage tank at Dusty's station was $\frac{11}{12}$ full. During Saturday and Sunday the attendants pumped out $\frac{2}{3}$ of what was is the tank on Saturday morning. How much remained in the tank when they opened Monday morning?

$\frac{11}{36}$ tank

63. ***Business Management*** The manager at Fit Factory Health Club was going through his files for 2004 and discovered that only $\frac{9}{14}$ of the members actually used the club. When he checked the numbers from the previous year of 2003, he found that $\frac{5}{6}$ of the members had used the club. What fractional part of the membership represents the decrease in club usage?

$\frac{4}{21}$ of the membership

64. ***Art*** Consuela is a collage artist. She assembles different objects on canvases to present a message to the viewer. She is working on a collage that is attached to a dart board. She places the center of a computer chip $\frac{1}{3}$ inch from the center bull's-eye. Then, continuing in a straight line from the bull's-eye, she places the center of a tea bag $\frac{5}{9}$ inch from the center of the computer chip. Continuing in a straight line, the last item she uses is a chewing gum wrapper. The center of the wrapper is placed $\frac{3}{4}$ inch from the center of the tea bag. How far from the bull's-eye is the center of the chewing gum wrapper? $1\frac{23}{36}$ inches

Cumulative Review

65. Reduce to lowest terms. $\dfrac{15}{85}$

$\dfrac{3}{17}$

66. Reduce to lowest terms. $\dfrac{27}{207}$

$\dfrac{3}{23}$

67. Change to a mixed number. $\dfrac{125}{14}$

$8\dfrac{13}{14}$

68. Change to an improper fraction. $14\dfrac{3}{7}$

$\dfrac{101}{7}$

69. Multiply. $5\dfrac{1}{2} \times 4\dfrac{3}{4}$

$26\dfrac{1}{8}$

70. Divide. $4\dfrac{1}{3} \div 1\dfrac{1}{2}$

$2\dfrac{8}{9}$

2.8 Adding and Subtracting Mixed Numbers and the Order of Operations

Student Learning Objectives

After studying this section, you will be able to:

1. **Add mixed numbers.**
2. **Subtract mixed numbers.**
3. **Evaluate fractional expressions using the order of operations.**

1 Adding Mixed Numbers

When adding mixed numbers, it is best to add the fractions together and then add the whole numbers together.

EXAMPLE 1 Add. $3\frac{1}{8} + 2\frac{5}{8}$

Solution

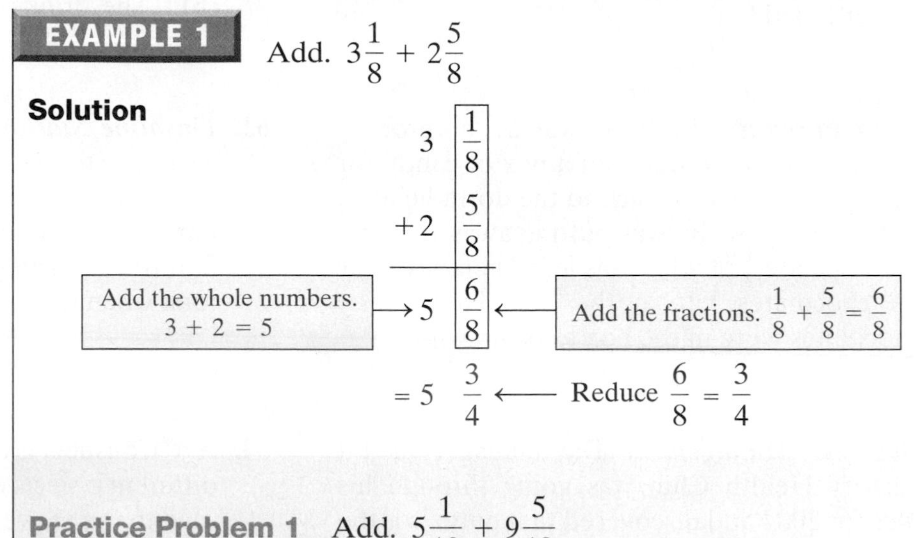

$3\ \frac{1}{8}$
$+2\ \frac{5}{8}$
Add the whole numbers. 3 + 2 = 5
$\rightarrow 5\ \frac{6}{8} \leftarrow$
Add the fractions. $\frac{1}{8} + \frac{5}{8} = \frac{6}{8}$
$= 5\ \frac{3}{4} \leftarrow$ Reduce $\frac{6}{8} = \frac{3}{4}$

NOTE TO STUDENT: Fully worked-out solutions to all of the Practice Problems can be found at the back of the text starting at page SP-1

Practice Problem 1 Add. $5\frac{1}{12} + 9\frac{5}{12}$

If the fraction portions of the mixed numbers do not have a common denominator, we must build up the fraction parts to obtain a common denominator before adding.

EXAMPLE 2 Add. $1\frac{2}{7} + 5\frac{1}{3}$

Solution The LCD of $\frac{2}{7}$ and $\frac{1}{3}$ is 21.

$$\frac{2}{7} \times \frac{3}{3} = \frac{6}{21} \qquad \frac{1}{3} \times \frac{7}{7} = \frac{7}{21}$$

Thus $1\frac{2}{7} + 5\frac{1}{3} = 1\frac{6}{21} + 5\frac{7}{21}$.

$$1\frac{2}{7} = 1\ \frac{6}{21}$$

$$+5\frac{1}{3} = +5\ \frac{7}{21}$$

Add the whole numbers. 1 + 5 $\rightarrow 6\ \frac{13}{21} \leftarrow$ Add the fractions. $\frac{6}{21} + \frac{7}{21}$

Practice Problem 2 Add. $6\frac{1}{4} + 2\frac{2}{5}$

If the sum of the fractions is an improper fraction, we convert it to a mixed number and add the whole numbers together.

EXAMPLE 3 Add. $6\frac{5}{6} + 4\frac{3}{8}$

Solution The LCD of $\frac{5}{6}$ and $\frac{3}{8}$ is 24.

$$
\begin{array}{rcl}
6\ \boxed{\frac{5}{6}\times\frac{4}{4}} & = & 6\ \boxed{\frac{20}{24}} \\
+4\ \boxed{\frac{3}{8}\times\frac{3}{3}} & = & +4\ \boxed{\frac{9}{24}} \\
\hline
\text{Add the whole numbers.} \rightarrow & & 10\ \boxed{\frac{29}{24}} \leftarrow \text{Add the fractions.}
\end{array}
$$

$= 10 + \boxed{1\frac{5}{24}}$ Since $\frac{29}{24} = 1\frac{5}{24}$

$= 11\frac{5}{24}$ We add the whole numbers $10 + 1 = 11$.

Practice Problem 3 Add. $7\frac{1}{4} + 3\frac{5}{6}$

2 Subtracting Mixed Numbers

Subtracting mixed numbers is like adding.

EXAMPLE 4 Subtract. $8\frac{5}{7} - 5\frac{5}{14}$

Solution The LCD of $\frac{5}{7}$ and $\frac{5}{14}$ is 14.

$$
\begin{array}{rcl}
8\ \boxed{\frac{5}{7}\times\frac{2}{2}} & = & 8\frac{10}{14} \\
-5\frac{5}{14} & = & -5\frac{5}{14} \\
\hline
\end{array}
$$

Subtract the whole numbers. $\rightarrow 3\frac{5}{14} \leftarrow$ Subtract the fractions.

Practice Problem 4 Subtract. $12\frac{5}{6} - 7\frac{5}{12}$

Teaching Tip Studying and understanding Example 4 or one similar to it will be most critical for the students. You may want to add a similar example on the board. We suggest working out the problem $7\frac{1}{6} - 3\frac{16}{21}$. Be sure the students see how to change this to the problem $7\frac{7}{42} - 3\frac{32}{42}$ before finally writing the expression $6\frac{49}{42} - 3\frac{32}{42} = 3\frac{17}{42}$.

Sometimes we must borrow before we can subtract.

EXAMPLE 5 Subtract.

(a) $9\frac{1}{4} - 6\frac{5}{14}$ **(b)** $15 - 9\frac{3}{16}$

Solution This example is fairly challenging. Read through each step carefully. Be sure to have paper and pencil handy and see if you can verify each step.

(a) The LCD of $\frac{1}{4}$ and $\frac{5}{14}$ is 28.

$$\begin{array}{rcl} 9\boxed{\frac{1}{4}\times\frac{7}{7}} & = & 9\frac{7}{28} \\ -6\boxed{\frac{5}{14}\times\frac{2}{2}} & = & -6\frac{10}{28} \\ \hline \end{array}$$

We cannot subtract $\frac{7}{28}-\frac{10}{28}$, so we will need to borrow.

We borrow 1 from 9 to obtain $9\frac{7}{28} = 8 + 1\frac{7}{28} = 8 + \frac{35}{28} = 8\frac{35}{28}$

We write this as

$$\begin{array}{rcl} 9\frac{7}{28} & = & 8\frac{35}{28} \\ -6\frac{10}{28} & = & -6\frac{10}{28} \\ \hline & & 2\frac{25}{28} \end{array}$$

$\boxed{8-6=2} \rightarrow 2\frac{25}{28} \leftarrow \boxed{\frac{35}{28}-\frac{10}{28}=\frac{25}{28}}$

(b) The LCD = 16.

$$\begin{array}{rcl} 15 & = & 14\frac{16}{16} \\ -\ 9\frac{3}{16} & = & -\ 9\frac{3}{16} \\ \hline & & 5\frac{13}{16} \end{array}$$

We borrow 1 from 15 to obtain $15 = 14 + 1 = 14 + \frac{16}{16} = 14\frac{16}{16}$

$\boxed{14-9=5} \rightarrow 5\frac{13}{16} \leftarrow \boxed{\frac{16}{16}-\frac{3}{16}=\frac{13}{16}}$

***NOTE TO STUDENT:** Fully worked-out solutions to all of the Practice Problems can be found at the back of the text starting at page SP-1*

Practice Problem 5 Subtract. **(a)** $9\frac{1}{8} - 3\frac{2}{3}$ **(b)** $18 - 6\frac{7}{18}$

EXAMPLE 6 A plumber had a pipe $5\frac{3}{16}$ inches long for a fitting under the sink. He needed a pipe that was $3\frac{7}{8}$ inches long, so he cut the pipe down. How much of the pipe did he cut off?

Solution We will need to subtract $5\frac{3}{16} - 3\frac{7}{8}$ to find the length that was cut off.

$$\begin{array}{rcl} 5\frac{3}{16} & = & 5\frac{3}{16} \\ -3\frac{7}{8}\times\frac{2}{2} & = & -3\frac{14}{16} \\ \hline \end{array}$$

$$\begin{array}{r} 4\frac{19}{16} \\ -\ 3\frac{14}{16} \\ \hline 1\frac{5}{16} \end{array}$$

We borrow 1 from 5 to obtain $5\frac{3}{16} = 4 + 1\frac{3}{16} = 4 + \frac{19}{16}$

$\boxed{4-3=1} \rightarrow 1\frac{5}{16} \leftarrow \boxed{\frac{19}{16}-\frac{14}{16}=\frac{5}{16}}$

The plumber had to cut off $1\frac{5}{16}$ inches of pipe.

Practice Problem 6 Hillary and Sam purchased $6\frac{1}{4}$ gallons of paint to paint the inside of their house. They used $4\frac{2}{3}$ gallons of paint. How much paint was left over?

ALTERNATIVE METHOD: Add or Subtract Mixed Numbers as Improper Fractions Can mixed numbers be added and subtracted as improper fractions? Yes. Recall Example 5(a).

$$9\frac{1}{4} - 6\frac{5}{14} = 2\frac{25}{28}$$

If we write $9\frac{1}{4} - 6\frac{5}{14}$ using improper fractions, we have $\frac{37}{4} - \frac{89}{14}$. Now we build up each of these improper fractions so that they both have the LCD for their denominators.

$$\begin{array}{rcr} \frac{37}{4} & \boxed{\times \frac{7}{7}} & = \frac{259}{28} \\ -\frac{89}{14} & \boxed{\times \frac{2}{2}} & = -\frac{178}{28} \\ \hline & & \frac{81}{28} = 2\frac{25}{28} \end{array}$$

The same result is obtained as in Example 5(a). This method does not require borrowing. However, you do work with larger numbers. For more practice, see exercises 53–54.

3 Evaluating Fractional Expressions Using the Order of Operations

Recall that in Section 1.6 we discussed the order of operations when we were combining whole numbers. We will now encounter some similar problems involving fractions and mixed numbers. We will repeat here the four-step order of operations that we studied previously:

ORDER OF OPERATIONS

With grouping symbols:

Do first	**1.** Perform operations inside the parentheses.
↑	**2.** Simplify any expressions with exponents.
↓	**3.** Multiply or divide from left to right.
Do last	**4.** Add or subtract from left to right.

EXAMPLE 7 Evaluate. $\frac{3}{4} - \frac{2}{3} \times \frac{1}{8}$

Solution $\frac{3}{4} - \frac{2}{3} \times \frac{1}{8} = \frac{3}{4} - \frac{1}{12}$ First we must multiply $\frac{2}{3} \times \frac{1}{8}$.

$= \frac{9}{12} - \frac{1}{12}$ Now we subtract, but first we need to build $\frac{3}{4}$ to an equivalent fraction with a common denominator of 12.

$= \frac{8}{12}$ Now we can subtract $\frac{9}{12} - \frac{1}{12}$.

$= \frac{2}{3}$ Finally we reduce the fraction.

Practice Problem 7 Evaluate. $\frac{3}{5} - \frac{1}{15} \times \frac{10}{13}$

EXAMPLE 8 Evaluate. $\frac{2}{3} \times \frac{1}{4} + \frac{2}{5} \div \frac{14}{15}$

Solution

$$\frac{2}{3} \times \frac{1}{4} + \frac{2}{5} \div \frac{14}{15} = \frac{1}{6} + \frac{2}{5} \div \frac{14}{15}$$ First we multiply $\frac{2}{3} \times \frac{1}{4}$.

$$= \frac{1}{6} + \frac{2}{5} \times \frac{15}{14}$$ We express the division as a multiplication problem. We invert $\frac{14}{15}$ and multiply.

$$= \frac{1}{6} + \frac{3}{7}$$ Now we perform the multiplication.

$$= \frac{7}{42} + \frac{18}{42}$$ We obtain equivalent fractions with an LCD of 42.

$$= \frac{25}{42}$$ We add the two fractions.

***NOTE TO STUDENT:** Fully worked-out solutions to all of the Practice Problems can be found at the back of the text starting at page SP-1*

Practice Problem 8 Evaluate. $\frac{1}{7} \times \frac{5}{6} + \frac{5}{3} \div \frac{7}{6}$

Developing Your Study Skills

Problems with Accuracy

Strive for accuracy. Mistakes are often made because of human error rather than lack of understanding. Such mistakes are frustrating. A simple arithmetic or copying error can lead to an incorrect answer.

These five steps will help you cut down on errors.

1. Work carefully, and take your time. Do not rush through a problem just to get it done.
2. Concentrate on the problem. Sometimes problems become mechanical, and your mind begins to wander. You become careless and make a mistake.
3. Check your problem. Be sure that you copied it correctly from the book.
4. Check your computations from step to step. Check the solution to the problem. Does it work? Does it make sense?
5. Keep practicing new skills. Remember the old saying, "Practice makes perfect." An increase in practice results in an increase in accuracy. Many errors are due simply to lack of practice.

There is no magic formula for eliminating all errors, but these five steps will be a tremendous help in reducing them.

2.8 Exercises

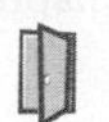
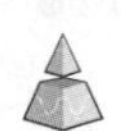

Student Solutions Manual | CD/ Video | PH Math Tutor Center | MathXL®Tutorials on CD | MathXL® | MyMathLab® | Interactmath.cor

Add or subtract. Express the answer as a mixed number. Simplify all answers.

1. $7\frac{1}{8} + 2\frac{5}{8}$ $9\frac{3}{4}$

2. $6\frac{3}{10} + 4\frac{1}{10}$ $10\frac{2}{5}$

3. $15\frac{3}{14} - 11\frac{1}{14}$ $4\frac{1}{7}$

4. $8\frac{3}{4} - 3\frac{1}{4}$ $5\frac{1}{2}$

5. $12\frac{1}{3} + 5\frac{1}{6}$ $17\frac{1}{2}$

6. $20\frac{1}{4} + 3\frac{1}{8}$ $23\frac{3}{8}$

7. $5\frac{4}{5} + 10\frac{3}{10}$ $16\frac{1}{10}$

8. $6\frac{3}{8} + 4\frac{1}{16}$ $10\frac{7}{16}$

9. $1 - \frac{3}{7}$ $\frac{4}{7}$

10. $1 - \frac{9}{11}$ $\frac{2}{11}$

11. $1\frac{5}{6} + \frac{7}{8}$ $2\frac{17}{24}$

12. $1\frac{2}{3} + \frac{13}{18}$ $2\frac{7}{18}$

13. $6\frac{1}{5} + 3\frac{1}{6}$ $9\frac{11}{30}$

14. $9\frac{5}{6} + 2\frac{2}{3}$ $12\frac{1}{2}$

15. $8\frac{1}{4} - 8\frac{4}{16}$ 0

16. $8\frac{11}{15} - 3\frac{3}{10}$ $5\frac{13}{30}$

17. $12\frac{1}{3} - 7\frac{2}{5}$ $4\frac{14}{15}$

18. $10\frac{10}{15} - 10\frac{2}{3}$ 0

19. $30 - 15\frac{3}{7}$ $14\frac{4}{7}$

20. $25 - 14\frac{2}{11}$ $10\frac{9}{11}$

21. $3 + 4\frac{2}{5}$ $7\frac{2}{5}$

22. $8 + 2\frac{3}{4}$ $10\frac{3}{4}$

23. $45 - 35\frac{2}{5}$ $9\frac{3}{5}$

24. $36 - 16\frac{5}{6}$ $19\frac{1}{6}$

Add or subtract. Express the answer as a mixed number. Simplify all answers.

25. $15\frac{4}{15} + 26\frac{8}{15}$ — $41\frac{4}{5}$

26. $22\frac{1}{8} + 14\frac{3}{8}$ — $36\frac{1}{2}$

27. $4\frac{1}{3} + 2\frac{1}{4}$ — $4\frac{4}{12} + 2\frac{3}{12} = 6\frac{7}{12}$

28. $6\frac{1}{8} + 7\frac{3}{4}$ — $6\frac{1}{8} + 7\frac{6}{8} = 13\frac{7}{8}$

29. $3\frac{3}{4} + 4\frac{5}{12}$ — $3\frac{9}{12} + 4\frac{5}{12} = 7\frac{14}{12} = 8\frac{1}{6}$

30. $11\frac{5}{8} + 13\frac{1}{2}$ — $11\frac{5}{8} + 13\frac{4}{8} = 24\frac{9}{8} = 25\frac{1}{8}$

31. $47\frac{3}{10} + 26\frac{5}{8}$ — $47\frac{12}{40} + 26\frac{25}{40} = 73\frac{37}{40}$

32. $34\frac{1}{20} + 45\frac{8}{15}$ — $34\frac{3}{60} + 45\frac{32}{60} = 79\frac{35}{60} = 79\frac{7}{12}$

33. $19\frac{5}{6} - 14\frac{1}{3}$ — $19\frac{5}{6} - 14\frac{2}{6} = 5\frac{3}{6} = 5\frac{1}{2}$

34. $22\frac{7}{9} - 16\frac{1}{4}$ — $22\frac{28}{36} - 16\frac{9}{36} = 6\frac{19}{36}$

35. $6\frac{1}{12} - 5\frac{10}{24}$ — $5\frac{26}{24} - 5\frac{10}{24} = \frac{16}{24} = \frac{2}{3}$

36. $4\frac{1}{12} - 3\frac{7}{18}$ — $3\frac{39}{36}$, $3\frac{14}{36}$, $\frac{25}{36}$

37. $\begin{array}{r} 12\dfrac{3}{20} \\ -\ 7\dfrac{7}{15} \\ \hline \end{array}$ $\begin{array}{r} 11\dfrac{69}{60} \\ -\ 7\dfrac{28}{60} \\ \hline 4\dfrac{41}{60} \end{array}$

38. $\begin{array}{r} 8\dfrac{5}{12} \\ -5\dfrac{9}{10} \\ \hline \end{array}$ $\begin{array}{r} 7\dfrac{85}{60} \\ -5\dfrac{54}{60} \\ \hline 2\dfrac{31}{60} \end{array}$

39. $\begin{array}{r} 12 \\ -\ 3\dfrac{7}{15} \\ \hline \end{array}$ $\begin{array}{r} 11\dfrac{15}{15} \\ -\ 3\dfrac{7}{15} \\ \hline 8\dfrac{8}{15} \end{array}$

40. $\begin{array}{r} 40 \\ -\ 6\dfrac{3}{7} \\ \hline \end{array}$ $\begin{array}{r} 39\dfrac{7}{7} \\ -\ 6\dfrac{3}{7} \\ \hline 33\dfrac{4}{7} \end{array}$

41. $\begin{array}{r} 120 \\ -\ 17\dfrac{3}{8} \\ \hline \end{array}$ $\begin{array}{r} 119\dfrac{8}{8} \\ -\ 17\dfrac{3}{8} \\ \hline 102\dfrac{5}{8} \end{array}$

42. $\begin{array}{r} 98 \\ -89\dfrac{15}{17} \\ \hline \end{array}$ $\begin{array}{r} 97\dfrac{17}{17} \\ -89\dfrac{15}{17} \\ \hline 8\dfrac{2}{17} \end{array}$

43. $\begin{array}{r} 3\dfrac{5}{8} \\ 2\dfrac{2}{3} \\ +7\dfrac{3}{4} \\ \hline \end{array}$ $\begin{array}{r} 3\dfrac{15}{24} \\ 2\dfrac{16}{24} \\ +7\dfrac{18}{24} \\ \hline 12\dfrac{49}{24} \end{array} = 14\dfrac{1}{24}$

44. $\begin{array}{r} 4\dfrac{2}{3} \\ 3\dfrac{4}{5} \\ +6\dfrac{3}{4} \\ \hline \end{array}$ $\begin{array}{r} 4\dfrac{40}{60} \\ 3\dfrac{48}{60} \\ +6\dfrac{45}{60} \\ \hline 13\dfrac{133}{60} \end{array} = 15\dfrac{13}{60}$

Applications

45. ***Mountain Biking*** Lee Hong rode his mountain bike through part of the Sangre de Cristo Mountains in New Mexico. On Wednesday he rode $20\frac{3}{4}$ miles. On Thursday he rode $22\frac{3}{8}$ miles. What was his total biking distance during those two days?
$43\frac{1}{8}$ miles

46. ***Hiking*** Ryan and Omar are planning an afternoon hike. Their map shows three loops measuring $2\frac{1}{8}$ miles, $1\frac{5}{6}$ miles, and $1\frac{2}{3}$ miles. If they hike all three loops, what will their total hiking distance be?
$5\frac{5}{8}$ miles

47. ***Food Purchase*** Renaldo bought $6\frac{3}{8}$ pounds of cheese for a party. The guests ate $4\frac{1}{3}$ pounds of the cheese. How much cheese did Renaldo have left over?
$2\frac{1}{24}$ pounds

48. ***Stock Market*** Shanna purchased stock in 1985 at \$$21\frac{3}{8}$ per share. When her son was ready for college, she sold the stock in 1999 at \$$93\frac{5}{8}$ per share. How much did she make per share for her son's tuition?
\$$72\frac{1}{4}$ per share

49. ***Basketball*** Nina and Julie are the two tallest basketball players on their high school team. Nina is $69\frac{3}{4}$ inches tall and Julie is $72\frac{1}{2}$ inches tall. How many inches taller is Julie than Nina?
$2\frac{3}{4}$ inches

50. ***Food Purchase*** Christos bought $1\frac{3}{4}$ pounds of French vanilla coffee beans. He also purchased $2\frac{5}{6}$ pounds of hazelnut coffee beans. How many total pounds of coffee beans did Christos buy?
$4\frac{7}{12}$ pounds

51. ***Food Purchase*** Lara needs 8 pounds of haddock for her dinner party. At the grocery store, haddock weighing $1\frac{3}{4}$ pounds and $2\frac{1}{6}$ pounds are placed on the scale.

(a) How many pounds of haddock are on the scale? $3\frac{11}{12}$ pounds

(b) How many more pounds of haddock does Lara need? $4\frac{1}{12}$ pounds

52. ***Medical Care*** A young man has been under a doctor's care to lose weight. His doctor wanted him to lose 46 pounds in the first three months. He lost $17\frac{5}{8}$ pounds the first month and $13\frac{1}{2}$ pounds the second month.

(a) How much did he lose during the first two months? $31\frac{1}{8}$ pounds

(b) How much would he need to lose in the third month to reach the goal? $14\frac{7}{8}$ pounds

To Think About

Use improper fractions and the Alternative Method as discussed in the text to perform each calculation.

53. $\frac{379}{8} + \frac{89}{5}$

$\frac{1895}{40} + \frac{712}{40} = \frac{2607}{40}$ or $65\frac{7}{40}$

54. $\frac{151}{6} - \frac{130}{7}$

$\frac{1057}{42} - \frac{780}{42} = \frac{277}{42}$ or $6\frac{25}{42}$

When adding or subtracting mixed numbers, it is wise to estimate your answer by rounding each mixed number to the nearest whole number.

55. Estimate your answer to $35\frac{1}{6} + 24\frac{5}{12}$ by rounding each mixed number to the nearest whole number. Then find the exact answer. How close was your estimate?

We estimate by adding 35 + 24 to obtain 59. The exact answer is $59\frac{7}{12}$. Our estimate is very close. We are off by only $\frac{7}{12}$.

56. Estimate your answer to $102\frac{5}{7} - 86\frac{2}{3}$ by rounding each mixed number to the nearest whole number. Then find the exact answer. How close was your estimate?

We estimate by subtracting 103 − 87 to obtain 16. The exact answer is $16\frac{1}{21}$. Our estimate is very close. We are off by only $\frac{1}{21}$.

Evaluate using the correct order of operation.

57. $\frac{6}{7} - \frac{4}{7} \times \frac{1}{3}$ $\frac{2}{3}$

58. $\frac{3}{5} - \frac{1}{3} \times \frac{6}{5}$ $\frac{1}{5}$

59. $\frac{1}{2} + \frac{3}{8} \div \frac{3}{4}$ 1

60. $\frac{3}{4} + \frac{1}{4} \div \frac{5}{3}$ $\frac{9}{10}$

61. $\frac{5}{7} \times \frac{7}{2} \div \frac{3}{2}$ $\frac{5}{3}$ or $1\frac{2}{3}$

62. $\frac{2}{7} \times \frac{3}{4} \div \frac{1}{2}$ $\frac{3}{7}$

63. $\frac{3}{5} \times \frac{1}{2} + \frac{1}{5} \div \frac{2}{3}$ $\frac{3}{5}$

64. $\frac{5}{6} \times \frac{1}{2} + \frac{2}{3} \div \frac{4}{3}$ $\frac{11}{12}$

65. $\left(\frac{3}{5} - \frac{3}{20}\right) \times \frac{4}{5}$

$\frac{9}{25}$

66. $\left(\frac{1}{3} + \frac{1}{6}\right) \times \frac{5}{11}$

$\frac{5}{22}$

67. $\left(\frac{4}{3}\right)^2 \div \frac{11}{9}$

$\frac{16}{11}$ or $1\frac{5}{11}$

68. $\left(\frac{5}{3}\right)^2 \div \frac{10}{7}$

$\frac{35}{18}$ or $1\frac{17}{18}$

69. $\frac{1}{4} \times \left(\frac{2}{3}\right)^2$ $\frac{1}{9}$

70. $\frac{5}{8} \times \left(\frac{2}{5}\right)^2$ $\frac{1}{10}$

71. $\frac{5}{6} \div \left(\frac{2}{3} + \frac{1}{6}\right)^2$

$\frac{6}{5}$ or $1\frac{1}{5}$

72. $\frac{4}{3} \div \left(\frac{3}{5} - \frac{3}{10}\right)^2$

$14\frac{22}{27}$ or $\frac{400}{27}$

Cumulative Review

Multiply.

73. 1200×400

480,000

74. 4050×2106

8,529,300

75. ***Home Decoration*** Suzanne is redecorating her family's house. She has a budget of \$6300 for painting, new flooring, and kitchen appliances. The painter charges her \$25/hour for 27 hours of work and \$520 for materials. The flooring contractor charges her \$30/hour for 8 hours of work and \$2972 for materials. How much will she have left over for appliances? \$1893

2.9 Solving Applied Problems Involving Fractions

Student Learning Objectives

After studying this section, you will be able to:

 **Solve real-life problems with fractions.**

 Solving Real-Life Problems with Fractions

All problem solving requires the same kind of thinking. In this section we will combine problem-solving skills with our new computational skills with fractions. Sometimes the difficulty is in figuring out what must be done. Sometimes it is in doing the computation. Remember that *estimating* is important in problem solving. We may use the following steps.

1. ***Understand the problem.***
   (a) Read the problem carefully.
   (b) Draw a picture if this helps you.
   (c) Fill in the Mathematics Blueprint.
2. ***Solve.***
   (a) Perform the calculations.
   (b) State the answer, including the unit of measure.
3. ***Check.***
   (a) Estimate the answer. Round fractions to the nearest whole number.
   (b) Compare the exact answer with the estimate to see if your answer is reasonable.

▲ 

EXAMPLE 1 In designing a modern offshore speedboat, the designing engineer has determined that one of the oak frames near the engine housing needs to be $26\frac{1}{8}$ inches long. At the end of the oak frame there will be $2\frac{5}{8}$ inches of insulation. Finally there will be a steel mounting that is $3\frac{3}{4}$ inches long. When all three items are assembled, how long will the oak frame and insulation and steel mounting extend?

Solution

1. ***Understand the problem.***
   We draw a picture to help us. See drawing on next page.
   Then we fill in the Mathematics Blueprint.

Mathematics Blueprint for Problem Solving

Gather the Facts	What Am I Asked to Do?	How Do I Proceed?	Key Points to Remember
Oak frame: $26\frac{1}{8}''$ Insulation: $2\frac{5}{8}''$ Steel mounting: $3\frac{3}{4}''$	Find the total length.	Add the lengths of the three items.	When adding mixed numbers, add the whole numbers first and then add the fractions.

2. *Solve and state the answer.*

Add the three amounts. $26\frac{1}{8} + 2\frac{5}{8} + 3\frac{3}{4}$

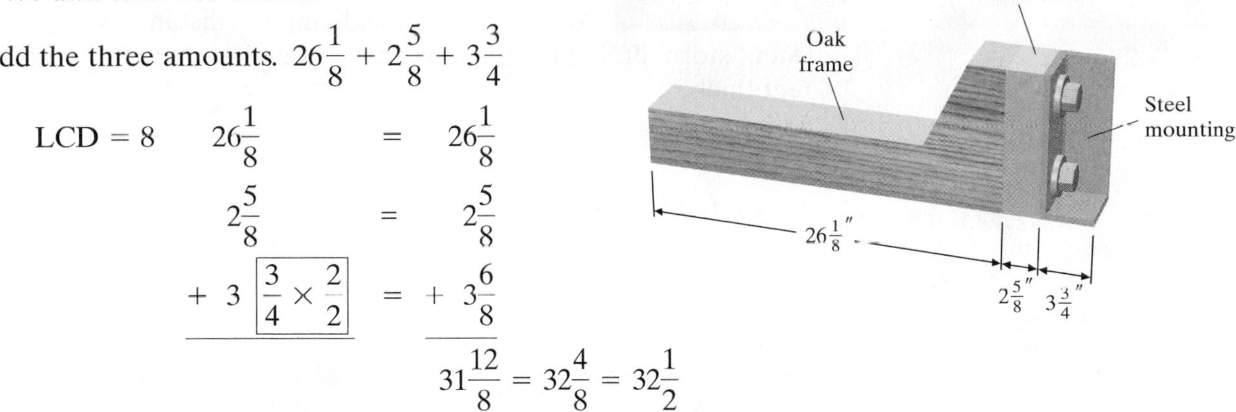

$$\begin{array}{lrcr} \text{LCD} = 8 & 26\frac{1}{8} & = & 26\frac{1}{8} \\ & 2\frac{5}{8} & = & 2\frac{5}{8} \\ & +\ 3\boxed{\frac{3}{4} \times \frac{2}{2}} & = & +\ 3\frac{6}{8} \\ \hline & & & 31\frac{12}{8} = 32\frac{4}{8} = 32\frac{1}{2} \end{array}$$

The entire assembly will be $32\frac{1}{2}$ inches.

3. *Check.* Estimate the sum by rounding each fraction to one nonzero digit.

$$\begin{array}{ll} \text{Thus} & 26\frac{1}{8} + 2\frac{5}{8} + 3\frac{3}{4} \\ \text{becomes} & 26 + 3 + 4 = 33 \end{array}$$

This is close to our answer, $32\frac{1}{2}$. Our answer seems reasonable.

One of the most important uses of estimation in mathematics is in the calculation of problems involving fractions. People find it easier to detect significant errors when working with whole numbers. However, the extra steps involved in the calculation with fractions and mixed numbers often distract our attention from an error that we should have detected.

Thus it is particularly critical to take the time to check your answer by estimating the results of the calculation with whole numbers. Be sure to ask yourself, is this answer reasonable? Does this answer seem realistic? Only by estimating our results with whole numbers will we be able to answer that question. It is this estimating skill that you will find more useful in your own life as a consumer and as a citizen.

Practice Problem 1 Nicole required the following amounts of gas for her farm tractor in the last three fill-ups: $18\frac{7}{10}$ gallons, $15\frac{2}{5}$ gallons, and $14\frac{1}{2}$ gallons. How many gallons did she need altogether?

NOTE TO STUDENT: Fully worked-out solutions to all of the Practice Problems can be found at the back of the text starting at page SP-1

The word *diameter* has two common meanings. First, it means a line segment that passes through the center of and intersects a circle twice. It has its endpoints on the circle. Second, it means the *length* of this segment.

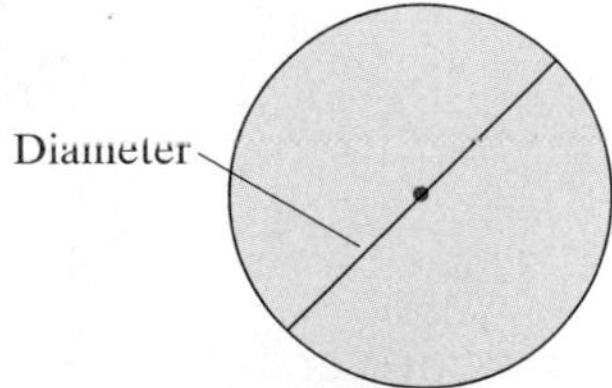

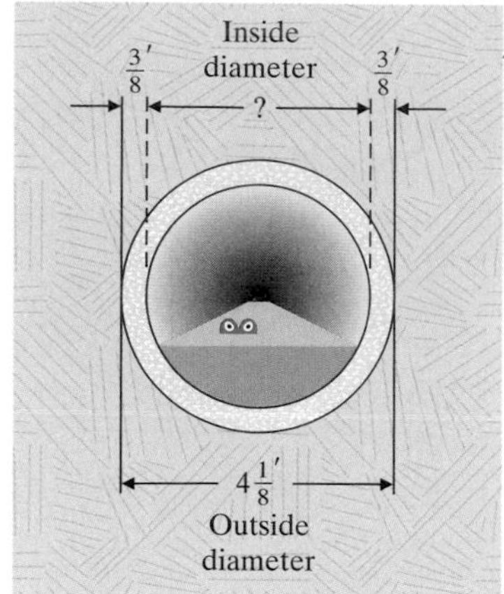

EXAMPLE 2 What is the inside diameter (distance across) of a cement storm drain pipe that has an outside diameter of $4\frac{1}{8}$ feet and is $\frac{3}{8}$ feet thick?

Solution

1. ***Understand the problem.*** Read the problem carefully. Draw a picture. The picture is in the margin on the left. Now fill in the Mathematics Blueprint.

Mathematics Blueprint for Problem Solving

Gather the Facts	What Am I Asked to Do?	How Do I Proceed?	Key Points to Remember
Outside diameter is $4\frac{1}{8}$ feet. Thickness is $\frac{3}{8}$ foot on both ends of the diameter.	Find the *inside* diameter of the pipe.	Add the two measures of thickness. Then subtract this total from the outside diameter.	Since the LCD = 8, all fractions must have this denominator.

2. ***Solve and state the answer.*** Add the two thickness measurements together.
Adding $\frac{3}{8} + \frac{3}{8} = \frac{6}{8}$ gives the total thickness of the pipe, $\frac{6}{8}$ foot. We will not reduce $\frac{6}{8}$ since the LCD is 8.
We subtract the total of the two thickness measurements from the outside diameter.

$$\begin{array}{rcr} 4\frac{1}{8} & = & 3\frac{9}{8} \\ -\ \frac{6}{8} & = & -\ \frac{6}{8} \\ \hline & & 3\frac{3}{8} \end{array}$$

We borrow 1 from 4 to get $3 + 1\frac{1}{8}$ or $3\frac{9}{8}$.

The inside diameter is $3\frac{3}{8}$ feet.

3. ***Check.*** We will work backward to check. We will use the exact values. If we have done our work correctly, $\frac{3}{8}$ foot + $3\frac{3}{8}$ feet + $\frac{3}{8}$ foot should add up to the outside diameter, $4\frac{1}{8}$ feet.

$$\frac{3}{8} + 3\frac{3}{8} + \frac{3}{8} \stackrel{?}{=} 4\frac{1}{8}$$

$$3\frac{9}{8} \stackrel{?}{=} 4\frac{1}{8}$$

$$4\frac{1}{8} = 4\frac{1}{8} \quad \checkmark$$

Our answer of $3\frac{3}{8}$ feet is correct.

▲ **Practice Problem 2** A poster is $12\frac{1}{4}$ inches long. We want a $1\frac{3}{8}$-inch border on the top and a 2-inch border on the bottom. What is the length of the inside portion of the poster?

***NOTE TO STUDENT:** Fully worked-out solutions to all of the Practice Problems can be found at the back of the text starting at page SP-1*

EXAMPLE 3 On Tuesday Michael earned $\$8\frac{1}{4}$ per hour working for eight hours. He also earned overtime pay, which is $1\frac{1}{2}$ times his regular rate of $\$8\frac{1}{4}$, for four hours on Tuesday. How much pay did he earn altogether on Tuesday?

Solution

1. ***Understand the problem.*** We draw a picture of the parts of Michael's pay on Tuesday.
   Michael's earnings on Tuesday are the sum of two parts:

   [Pay at regular pay rate] + [Pay at overtime pay rate] = [Total pay for the day]

   Now fill in the Mathematics Blueprint.

Mathematics Blueprint for Problem Solving

Gather the Facts	What Am I Asked to Do?	How Do I Proceed?	Key Points to Remember
He works eight hours at $\$8\frac{1}{4}$ per hour. He works four hours at the overtime rate, $1\frac{1}{2}$ times the regular rate.	Find his total pay for Tuesday.	Find out how much he is paid for regular time. Find out how much he is paid for overtime. Then add the two.	The overtime rate is $1\frac{1}{2}$ multiplied by the regular rate.

2. ***Solve and state the answer.*** Find his overtime pay rate.

$$1\frac{1}{2} \times 8\frac{1}{4} = \frac{3}{2} \times \frac{33}{4} = \frac{\$99}{8} \text{ per hour}$$

   We leave our answer as an improper fraction because we will need to multiply it by another fraction.
   How much was he paid for regular time? For overtime?

   For eight regular hours, he earned $8 \times 8\frac{1}{4} = \overset{2}{\cancel{8}} \times \frac{33}{\underset{1}{\cancel{4}}} = \$66$.

   For four overtime hours, he earned $\overset{1}{\cancel{4}} \times \frac{99}{\underset{2}{\cancel{8}}} = \frac{99}{2} = \$49\frac{1}{2}$.

   Now we add to find the total pay.

$$\begin{array}{rl} \$66 & \text{Pay at regular pay rate} \\ +\$49\frac{1}{2} & \text{Pay at overtime rate} \\ \hline \end{array}$$

   Michael earned $\$115\frac{1}{2}$ working on Tuesday. (This is the same as \$115.50, which we will use in Chapter 3.)

3. ***Check.*** We estimate his regular pay rate at \$8 per hour. We estimate his overtime pay rate at $1\frac{1}{2} \times 8 = \frac{3}{2} \times 8 = 12$ or \$12 per hour.

$$8 \text{ hours} \times \$8 \text{ per hour} = \$64 \text{ regular pay}$$
$$4 \text{ hours} \times \$12 \text{ per hour} = \$48 \text{ overtime pay}$$

Estimated sum. $\$64 + \$48 \approx \$60 + \$50 = \$110$
\$110 is close to our calculated value, $\$115\frac{1}{2}$, so our answer is reasonable. ✓

NOTE TO STUDENT: Fully worked-out solutions to all of the Practice Problems can be found at the back of the text starting at page SP-1

Practice Problem 3 A tent manufacturer uses $8\frac{1}{4}$ yards of waterproof duck cloth to make a regular tent. She uses $1\frac{1}{2}$ times that amount to make a large tent. How many yards of cloth will she need to make 6 regular tents and 16 large tents?

EXAMPLE 4 Alicia is buying some 8-foot boards for shelving. She wishes to make two bookcases, each with three shelves. Each shelf will be $3\frac{1}{4}$ feet long.

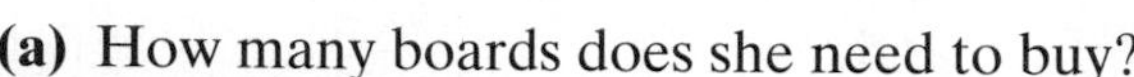

(a) How many boards does she need to buy?
(b) How many linear feet of shelving are actually needed to build the bookcases?
(c) How many linear feet of shelving will be left over?

Solution

1. ***Understand the problem.*** Draw a sketch of a bookcase. Each bookcase will have three shelves. Alicia is making two such bookcases. (Alicia's boards are for the shelves, not the sides.)
Now fill in the Mathematics Blueprint.

Mathematics Blueprint for Problem Solving

Gather the Facts	What Am I Asked to Do?	How Do I Proceed?	Key Points to Remember
She needs three shelves for each bookcase. Each shelf is $3\frac{1}{4}$ feet long. She will make two bookcases. Shelves are cut from 8-foot boards.	Find out how many boards to buy. Find out how many feet of board are needed for shelves and how many feet will be left over.	First find out how many $3\frac{1}{4}$-foot shelves she can get from one board. Then see how many boards she needs to make all six shelves.	Each time she cuts up an 8-foot board, she will get some shelves and some leftover wood.

2. ***Solve and state the answer.*** We want to know how many $3\frac{1}{4}$-foot boards are in an 8-foot board. By drawing a rough sketch, we would probably guess the answer is 2. To find exactly how many $3\frac{1}{4}$-foot long pieces are in 8 feet, we will use division.

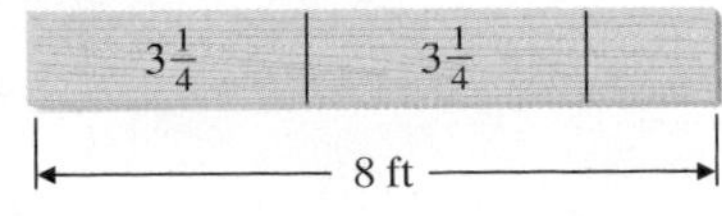

$$8 \div 3\frac{1}{4} = \frac{8}{1} \div \frac{13}{4} = \frac{8}{1} \times \frac{4}{13} = \frac{32}{13} = 2\frac{6}{13} \text{ boards}$$

She will get two shelves from each board, and some wood will be left over.

(a) How many boards does Alicia need to build two bookcases? For two bookcases, she needs six shelves. She will get two shelves out of each board. 6 ÷ 2 = 3. She will need three 8-foot boards.

(b) How many linear feet of shelving are actually needed to build the book cases?
She needs 6 shelves at $3\frac{1}{4}$ feet

$$6 \times 3\frac{1}{4} = \overset{3}{\cancel{6}} \times \frac{13}{\underset{2}{\cancel{4}}} = \frac{39}{2} = 19\frac{1}{2}$$

linear feet. A total of $19\frac{1}{2}$ linear feet of shelving is needed.

(c) How many linear feet of shelving will be left over?
Each time she uses one board she will have

$$8 - 3\frac{1}{4} - 3\frac{1}{4} = 8 - 6\frac{1}{2} = 1\frac{1}{2}$$

feet left over. Each of the three boards will have $1\frac{1}{2}$ feet left over.

$$3 \times 1\frac{1}{2} = 3 \times \frac{3}{2} = \frac{9}{2} = 4\frac{1}{2}$$

linear feet. A total of $4\frac{1}{2}$ linear feet of shelving will be left over.

3. ***Check.*** Work backward. See if you can check that with three 8-foot boards you

(a) can make the six shelves for the two bookcases.
(b) will use exactly $19\frac{1}{2}$ linear feet to make the shelves.
(c) will have exactly $4\frac{1}{2}$ linear feet left over.

The check is left to you.

Teaching Tip Many building and construction problems are similar to Example 4. After you have covered Example 4, ask the students if they can think of similar types of problems that they have encountered in their daily lives. Often you will get some interesting problems. If they have nothing to share, give them the following problem: "A man had three rooms with wall-to-wall carpeting. One room had $12\frac{1}{2}$ square yards, a second room had $13\frac{1}{4}$ square yards, and the third room had $15\frac{7}{8}$ square yards. The carpet-cleaning company charged $\$3\frac{1}{2}$ per square yard or any 3 rooms cleaned for \$159. The man could not determine which was a better buy. Can you?" The answer is that it is cheaper to get the $\$3\frac{1}{2}$-per-square-yard rate because then the cleaning job will only come to $\$145\frac{11}{16}$.

▲ **Practice Problem 4** Michael is purchasing 12-foot boards for shelving. He wishes to make two bookcases, each with four shelves. Each shelf will be $2\frac{3}{4}$ feet long.

(a) How many boards does he need to buy?
(b) How many linear feet of shelving are actually needed to build the bookcases?
(c) How many linear feet of shelving will be left over?

Another useful method for solving applied problems is called "Do a similar, simpler problem." When a problem seems difficult to understand because of the fractions, change the problem to an easier but similar problem.

Then decide how to solve the simpler problem and use the same steps to solve the original problem. For example:

How many gallons of water can a tank hold if its volume is $58\frac{2}{3}$ cubic feet? (1 cubic foot holds about $7\frac{1}{2}$ gallons.)

A similar, easier problem would be: "If 1 cubic foot holds about 8 gallons and a tank holds 60 cubic feet, how many gallons of water does the tank hold?"

The easier problem can be read more quickly and seems to make more sense. Probably we will see how to solve the easier problem right away: "I can find the number of gallons by multiplying 8×60." Therefore we can solve the first problem by multiplying $7\frac{1}{2} \times 58\frac{2}{3}$ to obtain the number of gallons of water. See the next example.

EXAMPLE 5 A fishing boat traveled $69\frac{3}{8}$ nautical miles in $3\frac{3}{4}$ hours. How many knots (nautical miles per hour) did the fishing boat average?

Solution

1. ***Understand the problem.*** Let us think of a simpler problem. If a boat traveled 70 nautical miles in 4 hours, how many knots did it average? We would divide distance by time.

$$70 \div 4 = \text{average speed}$$

Likewise in our original problem we need to divide distance by time.

$$69\frac{3}{8} \div 3\frac{3}{4} = \text{average speed}$$

Now fill in the Mathematics Blueprint.

Mathematics Blueprint for Problem Solving

Gather the Facts	What Am I Asked to Do?	How Do I Proceed?	Key Points to Remember
Distance is $69\frac{3}{8}$ nautical miles. Time is $3\frac{3}{4}$ hours.	Find the average speed of the boat.	Divide the distance in nautical miles by the time in hours.	You must change the mixed numbers to improper fractions before dividing.

2. ***Solve and state the answer.*** Divide distance by time to get speed in knots.

$$69\frac{3}{8} \div 3\frac{3}{4} = \frac{555}{8} \div \frac{15}{4} = \frac{\overset{37}{\cancel{555}}}{\underset{2}{\cancel{8}}} \cdot \frac{\overset{1}{\cancel{4}}}{\underset{1}{\cancel{15}}}$$

$$= \frac{37}{2} \cdot \frac{1}{1} = \frac{37}{2} = 18\frac{1}{2} \text{ knots}$$

The speed of the boat was $18\frac{1}{2}$ knots.

3. ***Check.***

We estimate $69\frac{3}{8} \div 3\frac{3}{4}$.

$$\text{Use } 70 \div 4 = 17\frac{1}{2} \text{ knots}$$

Our estimate is close to the calculated value.
Our answer is reasonable. ✓

Practice Problem 5 Alfonso traveled $199\frac{3}{4}$ miles in his car and used $8\frac{1}{2}$ gallons of gas. How many miles per gallon did he get?

NOTE TO STUDENT: Fully worked-out solutions to all of the Practice Problems can be found at the back of the text starting at page SP-1

Developing Your Study Skills

Why Study Mathematics?

Students often question the value of mathematics. They see little real use for it in their everyday lives. However, mathematics is often the key that opens the door to a better-paying job.

In our present-day technological world, many people use mathematics daily. Many vocational and professional areas—such as the fields of business, statistics, economics, psychology, finance, computer science, chemistry, physics, engineering, electronics, nuclear energy, banking, quality control, and teaching—require a certain level of expertise in mathematics. Those who want to work in these fields must be able to function at a given mathematical level. Those who cannot will not be able to enter these job areas.

So, whatever your field, be sure to realize the importance of mastering the basics of this course. It is very likely to help you advance to the career of your choice.

2.9 Exercises

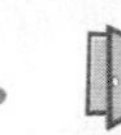

Student Solutions Manual | CD/ Video | PH Math Tutor Center | MathXL®Tutorials on CD | MathXL® | MyMathLab® | Interactmath.com

You may benefit from using the Mathematics Blueprint for Problem Solving when solving the following exercises.

Applications

▲ **1.** ***Geometry*** A triangle has three sides that measure $5\frac{1}{4}$ feet, $2\frac{5}{6}$ feet, and $8\frac{7}{12}$ feet. What is the perimeter (total distance around) of the triangle? $16\frac{2}{3}$ feet

2. ***Automobile Travel*** On Tuesday, Sally drove $10\frac{1}{2}$ miles while running errands. On Friday and Saturday, she had more errands to run and drove $6\frac{1}{3}$ miles and $12\frac{1}{4}$ miles, respectively. How many total miles did Sally drive this week while running errands? $29\frac{1}{12}$ miles

3. ***Storm Damage*** The coastal area of North Carolina experienced devastating erosion caused by fierce storms. The National Guard assisted by hauling sand to try to rebuild the dunes. During three days in September, the National Guard hauled $7\frac{5}{6}$ tons of sand on the first day, $8\frac{1}{8}$ tons on the second day, and $9\frac{1}{2}$ tons on the third. How many tons of sand were hauled over the three days? $25\frac{11}{24}$ tons

4. ***Plant Growth*** Lexi wonders if a tree in her front yard will grow to be 8 feet tall. When the tree was planted, it was $2\frac{1}{6}$ feet above the ground in height. It grew $1\frac{1}{4}$ feet the first year after being planted. How many more feet does it need to grow to reach a height of 8 feet? $4\frac{7}{12}$ feet

5. ***Carpentry*** A bolt extends through $\frac{3}{4}$ inch-thick plywood, two washers that are each $\frac{1}{16}$ inch thick, and a nut that is $\frac{3}{16}$ inch thick. The main body of the bolt must be $\frac{1}{2}$ inch longer than the sum of the thicknesses of plywood, washers, and nut. What is the minimum length of the bolt? $1\frac{9}{16}$ inches

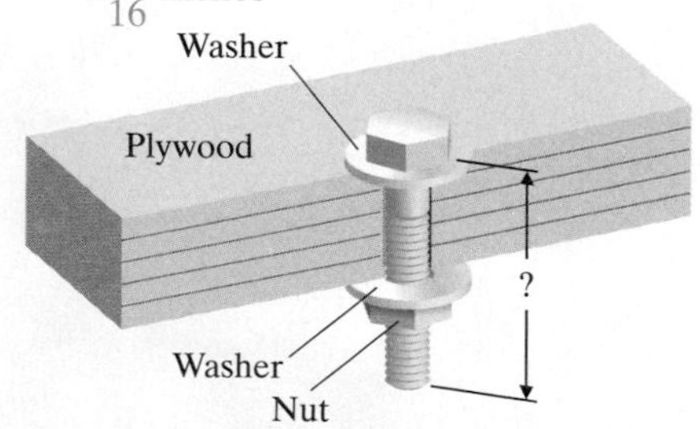

6. ***Carpentry*** A carpenter is using an 8-foot length of wood for a frame. The carpenter needs to cut a notch in the wood that is $4\frac{7}{8}$ feet from one end and $1\frac{2}{3}$ feet from the other end. How long does the notch need to be? $1\frac{11}{24}$ feet

7. ***Running a Marathon*** Hank is running the Boston Marathon, which is $26\frac{1}{5}$ miles long. At $6\frac{3}{4}$ miles from the start, he meets his wife, who is cheering him on. $9\frac{1}{2}$ miles further down the marathon course, he sees some friends from his running club volunteering at a water stop. Once he passes his friends, how many more miles does Hank have left to run? $9\frac{19}{20}$ miles

8. ***Carpentry*** Norman Olerud makes birdhouses as a hobby. He has a long piece of lumber that measures $14\frac{1}{4}$ feet. He needs to cut it into pieces that are $\frac{3}{4}$ foot long for the birdhouse floors. How many floors will he be able to cut from the long piece? 19 floors

9. ***Personal Finance*** Javier earned $\$10\frac{1}{2}$ per hour for 8 hours of work on Saturday. His manager asked his to stay for an additional 4 hours, for which he was paid $1\frac{1}{2}$ times the regular rate. How much did Javier earn on Saturday? $147

10. ***Food Purchase*** For a party of the British Literature Club using all "English foods," Nancy bought a $10\frac{2}{3}$-pound wheel of Stilton cheese, to go with the pears and the apples, at $\$8\frac{3}{4}$ per pound. How much did the wheel of Stilton cheese cost? $\$93\frac{1}{3}$

▲ **11.** ***Geometry*** How many gallons can a tank hold that has a volume of $36\frac{3}{4}$ cubic feet? (Assume that 1 cubic foot holds about $7\frac{1}{2}$ gallons.)
$275\frac{5}{8}$ gallons

▲ **12.** ***Geometry*** A tank can hold a volume of $7\frac{1}{4}$ cubic feet. If it is filled with water, how much does the water weigh? (Assume that 1 cubic foot of water weighs $62\frac{1}{2}$ pounds.)
$453\frac{1}{8}$ pounds

13. ***Titanic Disaster*** The night of the *Titanic* cruise ship disaster, the captain decided to run his ship at $22\frac{1}{2}$ knots (nautical miles per hour). The *Titanic* traveled at that speed for $4\frac{3}{4}$ hours before it met its tragic demise. How far did the *Titanic* travel at this excessive speed before the disaster? $106\frac{7}{8}$ nautical miles

14. ***Personal Finance*** William built a porch for his neighbor and got paid \$1200. He gave $\frac{1}{10}$ of this to his brother to pay back a debt. He used $\frac{1}{3}$ of it to pay bills and used $\frac{1}{6}$ to pay his helper. How much of the \$1200 did William have left?
\$480

15. ***Personal Finance*** Noriko earns \$660 per week. She has $\frac{1}{5}$ of her income withheld for federal taxes, $\frac{1}{15}$ of her income withheld for state taxes, and $\frac{1}{20}$ of her income withheld for medical coverage. How much per week is left for Noriko after these three deductions?
\$451 per week

16. ***Real Estate*** Brenda and Luis are saving for a down payment on a house. Their total take-home pay is \$870 per week. They have determined that $\frac{1}{5}$ of their weekly income is needed for car payments and insurance, $\frac{1}{10}$ is needed for groceries and monthly bills, and $\frac{1}{15}$ is needed for clothing and entertainment. How much per week is left to be saved for their down payment? \$551 per week

17. ***Making Clothing*** Caroline makes bandanas and sells them for $\$2\frac{1}{2}$. She has a long piece of fabric that measures $13\frac{1}{2}$ feet. Each bandana requires $\frac{2}{3}$ foot to make.

(a) How many bandanas can Caroline make from the large piece of fabric?
20 bandanas

(b) How much fabric is left over?
$\frac{1}{6}$ foot

(c) If Caroline sells all the bandanas, how much money will she make?
\$50

▲ **18.** ***Geometry*** Jay and Pam are building a dog run for their collie. The fenced-in area measures $38\frac{1}{3}$ feet by $52\frac{3}{4}$ feet.

(a) How many feet of coated fence wire will they need, if the wire is wrapped around the perimeter and $2\frac{1}{3}$ feet of wire is necessary to secure the end of the last wrap?
$184\frac{1}{2}$ feet

(b) How many feet of fence wire would they need if they made the fence exactly one-half as long and one-half as wide?
$93\frac{5}{12}$ feet

19. ***Food Purchase*** Cecilia bought a loaf of sourdough bread that was made by a local gourmet bakery. The label said that the bread, plus its fancy box, weighed $18\frac{1}{2}$ ounces in total. Of this, $1\frac{1}{4}$ ounces turned out to be the weight of the ribbon. The box weighed $3\frac{1}{8}$ ounces.

(a) How many ounces of bread did she actually buy?
$14\frac{1}{8}$ ounces of bread

(b) The box stated its net weight as $14\frac{3}{4}$ ounces. (This means that she should have found $14\frac{3}{4}$ ounces of gourmet sourdough bread in the box.) How much in error was this measurement?
$\frac{5}{8}$ of an ounce

20. ***Cooking*** Marnie has $12\frac{1}{2}$ cups of flour. She wants to make two pies, each requiring $1\frac{1}{4}$ cups of flour, and three cakes, each requiring $2\frac{1}{8}$ cups. How much flour will be left after Marnie makes the pies and cakes?
$3\frac{5}{8}$ cups

21. ***Coast Guard Boat Operation*** The largest Coast Guard boat stationed at San Diego can travel $160\frac{1}{8}$ nautical miles in $5\frac{1}{4}$ hours.
(a) At how many knots is the boat traveling?
$30\frac{1}{2}$ knots
(b) At this speed, how long would it take the Coast Guard boat to travel $213\frac{1}{2}$ nautical miles?
7 hours

22. ***Water Ski Boat*** Russ and Norma's Mariah water ski boat can travel $72\frac{7}{8}$ nautical miles in $2\frac{3}{4}$ hours.
(a) At how many knots is the boat traveling?
$26\frac{1}{2}$ knots
(b) At this speed, how long would it take their water ski boat to travel $92\frac{3}{4}$ nautical miles?
$3\frac{1}{2}$ hours

▲ 23. ***Farming*** A Kansas wheat farmer has a storage bin with a capacity of $6856\frac{1}{4}$ cubic feet.
(a) If a bushel of wheat is $1\frac{1}{4}$ cubic feet, how many bushels can the storage bin hold?
5485 bushels
(b) If a farmer wants to make a new storage bin $1\frac{3}{4}$ times larger, how many cubic feet will it hold?
$11{,}998\frac{7}{16}$ cubic feet
(c) How many bushels will the new bin hold?
$9598\frac{3}{4}$ bushels

▲ 24. ***Farming*** A wheat farmer from Texas has a storage bin with a capacity of $8693\frac{1}{3}$ cubic feet.
(a) If a bushel of wheat is $1\frac{1}{3}$ cubic feet, how many bushels can the storage bin hold?
6520 bushels
(b) If a farmer wants to make a new storage bin $1\frac{1}{3}$ times larger, how many cubic feet will it hold?
$11{,}951\frac{1}{9}$ cubic feet
(c) How many bushels will the new bin hold?
$8693\frac{1}{3}$ bushels

To Think About

Cooking *The Ortega family is having a birthday party for their 94-year-old grandmother. There are huge platters of cold cuts, salads, and cheese on the table. The table contains $9\frac{2}{3}$ lb of roast beef, $16\frac{1}{2}$ lb of honey-roasted ham, $4\frac{1}{4}$ lb of turkey, $12\frac{3}{4}$ lb of tuna salad, $10\frac{1}{3}$ lb of taco salad, $6\frac{8}{9}$ lb of corn salad, $10\frac{1}{8}$ lb of swiss cheese, $8\frac{1}{3}$ lb of cheddar cheese, and a wheel of brie cheese that weighs $6\frac{1}{5}$ lb.*

Grandpa Ortega remarked that this is exactly twice as much food in each category as they had at the party last year.

25. How many pounds of salad did they have at the party last year?
$14\frac{71}{72}$ lb

26. How many pounds of meat did they have at the party last year?
$15\frac{5}{24}$ lb

Cumulative Review

27. Add. $\begin{array}{r} 16{,}846 \\ 19{,}321 \\ +\ 8{,}078 \\ \hline 44{,}245 \end{array}$

28. Subtract. $\begin{array}{r} 209{,}364 \\ -\ 186{,}927 \\ \hline 22{,}437 \end{array}$

29. Multiply. $\begin{array}{r} 1683 \\ \times\ \ 27 \\ \hline 45{,}441 \end{array}$

30. Divide. $37\overline{)13{,}172}$ 356

Putting Your Skills to Work

Distances and Fractions in Colonial Times

In early colonial New England, fractions were used extensively in measuring local distances in rural areas. A man delivering supplies using a horse and carriage might very well need to calculate his total distances using fractions. The map pictured below measures the approximate distances between various villages in Essex County in Massachusetts. Use the map to answer the following questions.

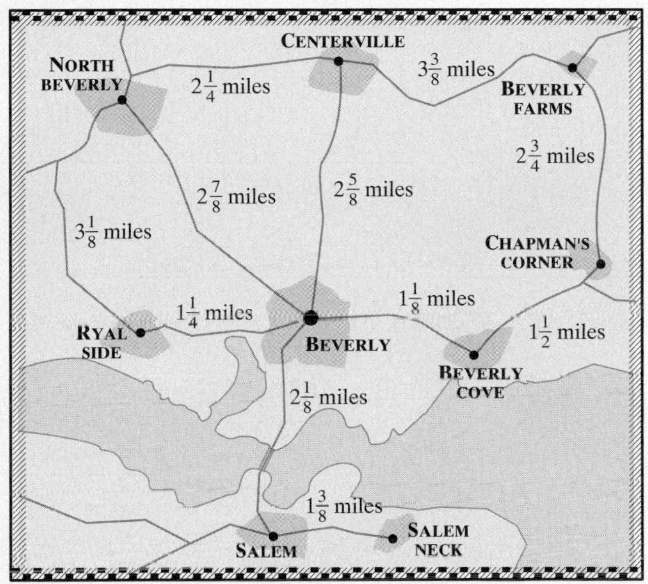

Problems for Individual Investigation

1. How many miles would it be for a horse and carriage to travel from Salem to Beverly and then on to Beverly Cove?

   $3\frac{1}{4}$ miles

2. How many miles would it be for a horse and carriage to travel from Centerville to North Beverly and then on to Ryal Side?

   $5\frac{3}{8}$ miles

Problems for Group Investigation and Cooperative Study

3. A delivery man leaves Beverly, delivers supplies in Centerville, delivers supplies in North Beverly, and then returns to Beverly by the shortest route. How many miles does he travel?

   $7\frac{3}{4}$ miles

4. Refer to problem 3 above. Suppose the man, after finishing in North Beverly, decides to come back to Beverly by first passing through Ryal Side. How many extra miles will he travel than if he came from North Beverly to Beverly by the shortest route?

   $1\frac{1}{2}$ miles

5. Mrs. Sarah Capen is traveling from Salem Neck to Salem and then on to Beverly. She said that her average speed with the horse and carriage is 8 miles per hour. How long will it take her to complete the trip?

   $\frac{7}{16}$ of an hour or $26\frac{1}{4}$ minutes

6. Mrs. Alice Conan has only 3 hours to travel from Ryal Side to North Beverly to pick up the mail, then to Centerville to pick up a package, and then to Beverly Farms to get her husband. Once they are in Beverly Farms, they will travel to Beverly by the shortest possible route. Can Mrs. Conan complete this entire trip in 3 hours if she has an average speed of 6 miles per hour with her horse and carriage? If so, how much extra time does she have left over?

   Yes, she will make it. She will have $38\frac{3}{4}$ minutes left over.

Chapter 2 Organizer

Topic	Procedure	Examples
Concept of a fractional part, p. 110.	The numerator is the number of parts selected. The denominator is the number of total parts.	What part of this sketch is shaded? $\frac{7}{10}$
Prime factorization, p. 118.	Prime factorization is the writing of a number as the product of prime numbers.	Write the prime factorization of 36. $36 = 4 \times 9$; $4 = 2 \times 2$, $9 = 3 \times 3$; $= 2 \times 2 \times 3 \times 3$
Reducing fractions, p. 120.	**1.** Factor numerator and denominator into prime factors. **2.** Divide out factors common to numerator and denominator.	Reduce. $\frac{54}{90}$ $\frac{54}{90} = \frac{\cancel{3} \times \cancel{3} \times 3 \times \cancel{2}}{\cancel{3} \times \cancel{3} \times \cancel{2} \times 5} = \frac{3}{5}$
Changing a mixed number to an improper fraction, p. 128.	**1.** Multiply whole number by denominator. **2.** Add product to numerator. **3.** Place sum over denominator.	Write as an improper fraction. $7\frac{3}{4} = \frac{4 \times 7 + 3}{4} = \frac{28 + 3}{4} = \frac{31}{4}$
Changing an improper fraction to a mixed number, p. 129.	**1.** Divide denominator into numerator. **2.** The quotient is the whole number. **3.** The fraction is the remainder over the divisor.	Change to a mixed number. $\frac{32}{5}$ $32 \div 5 = 6$ remainder 2 ($32 - 30 = 2$) $= 6\frac{2}{5}$
Multiplying fractions, p. 135.	**1.** Divide out common factors from the numerators and denominators whenever possible. **2.** Multiply numerators. **3.** Multiply denominators.	Multiply. $\frac{3}{7} \times \frac{5}{13} = \frac{15}{91}$ Multiply. $\frac{\overset{1}{\cancel{8}}}{\underset{1}{\cancel{8}}} \times \frac{\overset{2}{\cancel{16}}}{\underset{3}{\cancel{15}}} = \frac{2}{3}$
Multiplying mixed and/or whole numbers, p. 137.	**1.** Change any whole numbers to fractions with a denominator of 1. **2.** Change any mixed numbers to improper fractions. **3.** Use multiplication rule for fractions.	Multiply. $7 \times 3\frac{1}{4}$ $\frac{7}{1} \times \frac{13}{4} = \frac{91}{4} = 22\frac{3}{4}$
Dividing fractions, p. 142.	To divide two fractions, we invert the second fraction and multiply.	Divide. $\frac{3}{7} \div \frac{2}{9} = \frac{3}{7} \times \frac{9}{2} = \frac{27}{14}$ or $1\frac{13}{14}$
Dividing mixed numbers and/or whole numbers, p. 144.	**1.** Change any whole numbers to fractions with a denominator of 1. **2.** Change any mixed numbers to improper fractions. **3.** Use rule for division of fractions.	Divide. $8\frac{1}{3} \div 5\frac{5}{9} = \frac{25}{3} \div \frac{50}{9}$ $= \frac{\overset{1}{\cancel{25}}}{\underset{1}{\cancel{3}}} \times \frac{\overset{3}{\cancel{9}}}{\underset{2}{\cancel{50}}} = \frac{3}{2}$ or $1\frac{1}{2}$
Finding the least common denominator, p. 155.	**1.** Write each denominator as the product of prime factors. **2.** List all the prime factors that appear in both products. **3.** Form a product of those factors, using each factor the greatest number of times it appears in any denominator.	Find LCD of $\frac{1}{10}, \frac{3}{8}$, and $\frac{7}{25}$. $10 = 2 \times 5$ $8 = 2 \times 2 \times 2$ $25 = 5 \times 5$ LCD $= 2 \times 2 \times 2 \times 5 \times 5 = 200$

Topic	Procedure	Examples
Building fractions, p. 158.	**1.** Find how many times the original denominator can be divided into the new denominator. **2.** Multiply that value by numerator and denominator of original fraction.	Build $\frac{5}{7}$ to an equivalent fraction with a denominator of 42. First we find $7\overline{)42}$ = 6. Then we multiply the numerator and denominator by 6. $\frac{5}{7} \times \frac{6}{6} = \frac{30}{42}$
Adding or subtracting fractions with a common denominator, p. 164.	**1.** Add or subtract the numerators. **2.** Keep the common denominator.	Add. $\frac{3}{13} + \frac{5}{13} = \frac{8}{13}$ Subtract. $\frac{15}{17} - \frac{12}{17} = \frac{3}{17}$
Adding or subtracting fractions without a common denominator, p. 165.	**1.** Find the LCD of the fractions. **2.** Build up each fraction, if needed, to obtain the LCD in the denominator. **3.** Follow the steps for adding and subtracting fractions with the same denominator.	Add. $\frac{1}{4} + \frac{3}{7} + \frac{5}{8}$ LCD = 56 $\frac{1 \times 14}{4 \times 14} + \frac{3 \times 8}{7 \times 8} + \frac{5 \times 7}{8 \times 7}$ $= \frac{14}{56} + \frac{24}{56} + \frac{35}{56} = \frac{73}{56} = 1\frac{17}{56}$
Adding mixed numbers, p. 172.	**1.** Change fractional parts to equivalent fractions with LCD as a denominator, if needed. **2.** Add whole numbers and fractions separately. **3.** If improper fractions occur, change to mixed numbers and simplify.	Add. $6\frac{3}{4} + 2\frac{5}{8}$ $6\boxed{\frac{3}{4} \times \frac{2}{2}} = 6\frac{6}{8}$ $+2\frac{5}{8} = 2\frac{5}{8}$ $8\frac{11}{8} = 9\frac{3}{8}$
Subtracting mixed numbers, p. 173.	**1.** Change fractional parts to equivalent fractions with LCD as a denominator, if needed. **2.** If necessary, borrow from whole number to subtract fractions. **3.** Subtract whole numbers and fractions separately.	Subtract. $8\frac{1}{5} - 4\frac{2}{3}$ $8\boxed{\frac{1}{5} \times \frac{3}{3}} = 8\frac{3}{15} = 7\frac{18}{15}$ $-4\boxed{\frac{2}{3} \times \frac{5}{5}} = -4\frac{10}{15} = -4\frac{10}{15}$ $3\frac{8}{15}$
Order of Operations p. 175.	***Order of Operations*** With grouping symbols: Do first ↑ **1.** Perform operations inside the parentheses. **2.** Simplify any expressions with exponents. **3.** Multiply or divide from left to right. **4.** Add or subtract from left to right. ↓ Do last	$\frac{5}{6} \div \left(\frac{4}{5} - \frac{7}{15}\right)$ First combine numbers inside the parentheses. $\frac{5}{6} \div \left(\frac{12}{15} - \frac{7}{15}\right)$ Transform $\frac{4}{5}$ to equivalent fraction $\frac{12}{15}$. $\frac{5}{6} \div \frac{1}{3}$ Subtract the two fractions inside the parentheses. $\frac{5}{6} \times \frac{3}{1}$ Invert the second fraction and multiply. $\frac{5}{2}$ or $2\frac{1}{2}$ Simplify.

Procedure for Solving Applied Problems

Using the Mathematics Blueprint for Problem Solving, p. 180

In solving an applied problem with fractions, students may find it helpful to complete the following steps. You will not use the steps all of the time. Choose the steps that best fit the conditions of the problem.

1. ***Understand the problem.***

(a) Read the problem carefully.

(b) Draw a picture if this helps you to visualize the situation. Think about what facts you are given and what you are asked to find.

(c) It may help to write a similar, simpler problem to get started and to determine what operation to use.

(d) Use the Mathematics Blueprint for Problem Solving to organize your work. Follow these four parts.

1. Gather the facts (Write down specific values given in the problem.)
2. What am I asked to do? (Identify what you must obtain for an answer.)
3. How do I proceed? (What calculations need to be done.)
4. Key points to remember (Record any facts, warnings, formulas, or concepts you think will be important as you solve the problem.)

2. ***Solve and state the answer.***

(a) Perform the necessary calculations.

(b) State the answer, including the unit of measure.

3. ***Check.***

(a) Estimate the answer to the problem. Compare this estimate to the calculated value. Is your answer reasonable?

(b) Repeat your calculations.

(c) Work backward from your answer. Do you arrive at the original conditions of the problem?

EXAMPLE

A wire is $95\frac{1}{3}$ feet long. It is cut up into smaller, equal-sized pieces, each $4\frac{1}{3}$ feet long. How many pieces will there be?

1. ***Understand the problem.***

Draw a picture of the situation.

$95\frac{1}{3}$ ft wire $4\frac{1}{3}$ $4\frac{1}{3}$?

How will we find the number of pieces?

Now we will use a simpler problem to clarify the idea. A wire 100 feet long is cut up into smaller pieces each 4 feet long. How many pieces will there be? We readily see that we would divide 100 by 4. Thus in our original problem we should divide $95\frac{1}{3}$ feet by $4\frac{1}{3}$ feet. This will tell us the number of pieces. Now we fill in the Mathematics Blueprint.

Mathematics Blueprint for Problem Solving

Gather The Facts	What Am I Asked To Do?	How Do I Proceed?	Key Points To Remember
Wire is $95\frac{1}{3}$ feet. It is cut into equal pieces $4\frac{1}{3}$ feet long.	Determine how many pieces of wire there will be.	Divide $95\frac{1}{3}$ by $4\frac{1}{3}$.	Change mixed numbers to improper fractions before carrying out the division.

continues on next page

Procedure for Solving Applied Problems *(continued)*

2. ***Solve and state the answer.***

We need to divide $95\frac{1}{3} \div 4\frac{1}{3}$

$$\frac{286}{3} \div \frac{13}{3} = \frac{\overset{22}{\cancel{286}}}{\underset{1}{\cancel{3}}} \times \frac{\overset{1}{\cancel{3}}}{\underset{1}{\cancel{13}}} = \frac{22}{1} = 22$$

There will be 22 pieces of wire.

3. ***Check.***

Estimate. Rounded to the nearest ten, $95\frac{1}{3} \approx 100$.

Rounded to the nearest integer, $4\frac{1}{3} \approx 4$.

$$100 \div 4 = 25$$

This is close to our estimate. Our answer is reasonable. ✓

Chapter 2 Review Problems

Be sure to simplify all answers.

Section 2.1

Use a fraction to represent the shaded part of each object.

1. $\frac{3}{8}$

2. $\frac{5}{12}$

In exercises 3 and 4, draw a sketch to illustrate each fraction.

3. $\frac{4}{7}$ of an object Answers will vary.

4. $\frac{7}{9}$ of an object Answers will vary.

5. ***Quality Control*** An inspector looked at 31 parts and found 6 of them defective. What fractional part of these items was defective? $\frac{6}{31}$

6. ***Education*** The dean asked the 100 freshmen if they would be staying in the dorm over the holidays. A total of 87 said they would not. What fractional part of the freshmen said they would not? $\frac{87}{100}$

Section 2.2

Express each number as a product of prime factors.

7. 54 2×3^3

8. 120 $2^3 \times 3 \times 5$

9. 168 $2^3 \times 3 \times 7$

Determine which of the following numbers are prime. If a number is composite, express it as the product of prime factors.

10. 59 prime

11. 78 $2 \times 3 \times 13$

12. 167 prime

Reduce each fraction.

13. $\frac{12}{42}$ $\frac{2}{7}$

14. $\frac{13}{52}$ $\frac{1}{4}$

15. $\frac{21}{36}$ $\frac{7}{12}$

16. $\frac{26}{34}$ $\frac{13}{17}$

17. $\frac{168}{192}$ $\frac{7}{8}$

18. $\frac{51}{105}$ $\frac{17}{35}$

Section 2.3

Change each mixed number to an improper fraction.

19. $4\frac{3}{8}$ $\frac{35}{8}$

20. $15\frac{3}{4}$ $\frac{63}{4}$

21. $5\frac{2}{7}$ $\frac{37}{7}$

22. $6\frac{3}{5}$ $\frac{33}{5}$

Change each improper fraction to a mixed number.

23. $\frac{45}{8}$ $5\frac{5}{8}$

24. $\frac{100}{21}$ $4\frac{16}{21}$

25. $\frac{53}{7}$ $7\frac{4}{7}$

26. $\frac{74}{9}$ $8\frac{2}{9}$

27. Reduce and leave your answer as a mixed number.
$3\frac{15}{55}$ $3\frac{3}{11}$

28. Reduce and leave your answer as an improper fraction.
$\frac{234}{16}$ $\frac{117}{8}$

29. Change to a mixed number and then reduce. $\frac{132}{32}$ $4\frac{1}{8}$

Section 2.4

Multiply.

30. $\frac{4}{7} \times \frac{5}{11}$ $\frac{20}{77}$

31. $\frac{7}{9} \times \frac{21}{35}$ $\frac{7}{15}$

32. $12 \times \frac{3}{7} \times 0$ 0

33. $\frac{3}{5} \times \frac{2}{7} \times \frac{10}{27}$ $\frac{4}{63}$

34. $12 \times 8\frac{1}{5}$
$\frac{492}{5}$ or $98\frac{2}{5}$

35. $5\frac{1}{4} \times 4\frac{6}{7}$
$\frac{51}{2}$ or $25\frac{1}{2}$

36. $5\frac{1}{8} \times 3\frac{1}{5}$
$\frac{82}{5}$ or $16\frac{2}{5}$

37. $35 \times \frac{7}{10}$
$24\frac{1}{2}$

38. ***Stock Market*** In 1999, one share of stock cost $\$37\frac{5}{8}$. How much money did 18 shares cost?
$\$677\frac{1}{4}$

▲ **39.** ***Geometry*** Find the area of a rectangle that is $12\frac{1}{5}$ inches long and $3\frac{4}{7}$ inches wide.
$43\frac{4}{7}$ or $\frac{305}{7}$ square inches

Section 2.5

Divide, if possible.

40. $\frac{3}{7} \div \frac{2}{5}$
$\frac{15}{14}$ or $1\frac{1}{14}$

41. $\frac{3}{5} \div \frac{1}{10}$
6

42. $1200 \div \frac{5}{8}$
1920

43. $900 \div \frac{3}{5}$
1500

44. $5\frac{3}{4} \div 11\frac{1}{2}$ $\frac{1}{2}$

45. $\dfrac{20}{2\frac{1}{2}}$ 8

46. $0 \div 3\frac{7}{5}$ 0

47. $4\frac{2}{11} \div 3$ $\frac{46}{33}$ or $1\frac{13}{33}$

▲ **48.** ***Floor Carpeting*** Each roll of carpet covers $28\frac{1}{2}$ square yards. The community center has 342 square yards of flooring to carpet. How many rolls are needed?
12 rolls

49. There are 420 calories in $2\frac{1}{4}$ cans of grape soda. How many calories are in 1 can of soda?
$186\frac{2}{3}$ or $\frac{560}{3}$ calories

Section 2.6

Find the LCD for each pair of fractions.

50. $\frac{7}{14}$ and $\frac{3}{49}$ 98

51. $\frac{7}{40}$ and $\frac{11}{30}$ 120

52. $\frac{5}{18}, \frac{1}{6}, \frac{7}{45}$ 90

Build each fraction to an equivalent fraction with the specified denominator.

53. $\frac{3}{7} = \frac{?}{56}$ $\frac{24}{56}$

54. $\frac{11}{24} = \frac{?}{72}$ $\frac{33}{72}$

55. $\frac{9}{43} = \frac{?}{172}$ $\frac{36}{172}$

56. $\frac{17}{18} = \frac{?}{198}$ $\frac{187}{198}$

Section 2.7

Add or subtract.

57. $\frac{3}{7} - \frac{5}{14}$

$\frac{1}{14}$

58. $\frac{1}{2} + \frac{1}{3} + \frac{1}{4}$

$\frac{13}{12}$ or $1\frac{1}{12}$

59. $\frac{4}{7} + \frac{7}{9}$

$\frac{85}{63}$ or $1\frac{22}{63}$

60. $\frac{7}{8} - \frac{3}{5}$

$\frac{11}{40}$

61. $\frac{7}{30} + \frac{2}{21}$

$\frac{23}{70}$

62. $\frac{5}{18} + \frac{5}{12}$

$\frac{25}{36}$

63. $\frac{15}{16} - \frac{13}{24}$

$\frac{19}{48}$

64. $\frac{14}{15} - \frac{3}{25}$

$\frac{61}{75}$

Section 2.8

Evaluate using the correct order of operations.

65. $2 - \frac{3}{4}$

$\frac{5}{4}$ or $1\frac{1}{4}$

66. $6 - \frac{5}{9}$

$\frac{49}{9}$ or $5\frac{4}{9}$

67. $3 + 5\frac{2}{3}$

$8\frac{2}{3}$

68. $8 + 12\frac{5}{7}$

$20\frac{5}{7}$

69. $3\frac{3}{8} + 2\frac{3}{4}$

$\frac{49}{8}$ or $6\frac{1}{8}$

70. $5\frac{11}{16} - 2\frac{1}{5}$

$\frac{279}{80}$ or $3\frac{39}{80}$

71. $\frac{3}{5} \times \frac{1}{2} + \frac{2}{5} \div \frac{2}{3}$

$\frac{9}{10}$

72. $\left(\frac{3}{7} - \frac{1}{14}\right) \times \frac{8}{9}$

$\frac{20}{63}$

73. ***Jogging*** Bob jogged $1\frac{7}{8}$ miles on Monday, $2\frac{3}{4}$ miles on Tuesday, and $4\frac{1}{10}$ miles on Wednesday. How many miles did he jog on these three days?

$8\frac{29}{40}$ miles

74. ***Fuel Economy*** When it was new, Ginny Sue's car got $28\frac{1}{6}$ miles per gallon. It now gets $1\frac{5}{6}$ miles per gallon less. How far can she drive now if the car has $10\frac{3}{4}$ gallons in the tank?

$283\frac{1}{12}$ miles

Section 2.9

75. ***Cooking*** A recipe calls for $3\frac{1}{3}$ cups of sugar and $4\frac{1}{4}$ cups of flour. How much sugar and how much flour would be needed for $\frac{1}{2}$ of that recipe?

$1\frac{2}{3}$ cups sugar, $2\frac{1}{8}$ cups flour

76. ***Fuel Economy*** Rafael travels in a car that gets $24\frac{1}{4}$ miles per gallon. He has $8\frac{1}{2}$ gallons of gas in the gas tank. Approximately how far can he drive?

$206\frac{1}{8}$ miles

77. ***Construction*** How many lengths of pipe $3\frac{1}{5}$ inches long can be cut from a pipe 48 inches long?

15 lengths

78. ***Automobile Maintenance*** A car radiator holds $15\frac{3}{4}$ liters. If it contains $6\frac{1}{8}$ liters of antifreeze and the rest is water, how much is water?

$9\frac{5}{8}$ liters

79. ***Reading Speed*** Tim found that he can read 5 pages of his biology book in $32\frac{1}{2}$ minutes. He has three chapters to read over the weekend. The first is 12 pages, the second is 9 pages, and the third is 14 pages. How long will it take him?

$227\frac{1}{2}$ minutes or 3 hours and $47\frac{1}{2}$ minutes

80. ***Personal Finance*** Alicia earns $\$4\frac{1}{2}$ per hour for regular pay and $1\frac{1}{2}$ times that rate of pay for overtime. On Monday she worked eight hours at regular pay and three hours at overtime. How much did she earn on Monday?

$\$56\frac{1}{4}$

81. ***Stock Market*** Cheryl bought 80 shares of stock in 2001 at $\$12\frac{1}{4}$ a share. She sold all the shares in 2003 for \$18 each. How much did Cheryl make when she sold her shares?

\$460

82. ***Carpentry*** A 3-inch bolt passes through $1\frac{1}{2}$ inches of pine board, a $\frac{1}{16}$-inch washer, and a $\frac{1}{8}$-inch nut. How many inches does the bolt extend beyond the board, washer, and nut if the head of the bolt is $\frac{1}{4}$ inch long?

$1\frac{1}{16}$ inch

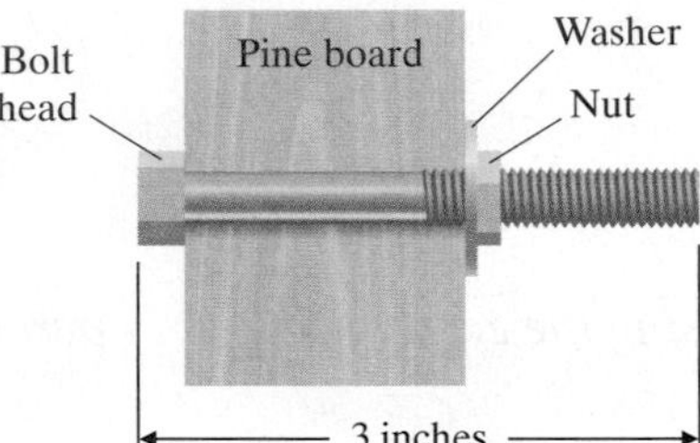

83. ***Budgeting*** Francine has a take-home pay of \$880 per month. She gives $\frac{1}{10}$ of it to her church, spends $\frac{1}{2}$ of it for rent and food, and spends $\frac{1}{8}$ of it on electricity, heat, and telephone. How many dollars per month does she have left for other things?

\$242

84. ***Cost of Auto Travel*** Manuel's new car used $18\frac{2}{5}$ gallons of gas on a 460-mile trip.

(a) How many miles can his car travel on 1 gallon of gas?

25 miles per gallon

(b) How much did his trip cost him in gasoline expense if the average cost of gasoline was $\$1\frac{1}{5}$ per gallon?

$\$22\frac{2}{25}$

Mixed Practice

Perform each calculation or each requested operation.

85. Reduce. $\frac{27}{63}$

$\frac{3}{7}$

86. $\frac{7}{15}+\frac{11}{25}$

$\frac{68}{75}$

87. $4\frac{1}{3}-2\frac{11}{12}$

$1\frac{5}{12}$

88. $\frac{36}{49}\times\frac{14}{33}$

$\frac{24}{77}$

89. $5\frac{2}{3}\div 1\frac{1}{6}$

$4\frac{6}{7}$ or $\frac{34}{7}$

90. $\left(\frac{4}{7}\right)^3$

$\frac{64}{343}$

91. $3\frac{7}{8}+2\frac{5}{16}$

$6\frac{3}{16}$

92. $5\frac{1}{2}\times 7\frac{5}{6}$

$43\frac{1}{12}$ or $\frac{517}{12}$

93. $120\div 3\frac{3}{5}$

$33\frac{1}{3}$ or $\frac{100}{3}$

How Am I Doing? Chapter 2 Test

Remember to use your Chapter Test Prep Video CD to see the worked-out solutions to the test problems you want to review.

Note to Instructor: The Chapter 2 Test file in the TestGen program provides algorithms specifically matched to these problems so you can easily replicate this test for additional practice or assessment purposes.

Solve.

1. Use a fraction to represent the shaded part of the object.

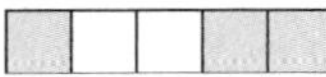

2. A basketball star shot at the hoop 388 times. The ball went in 311 times. Write a fraction that describes the part of the time that his shots went in.

Reduce each fraction.

3. $\frac{18}{42}$ **4.** $\frac{15}{70}$ **5.** $\frac{225}{50}$

6. Change to an improper fraction. $6\frac{4}{5}$

7. Change to a mixed number. $\frac{145}{14}$

Multiply.

8. $42 \times \frac{2}{7}$ **9.** $\frac{7}{9} \times \frac{2}{5}$ **10.** $2\frac{2}{3} \times 5\frac{1}{4}$

Divide.

11. $\frac{7}{8} \div \frac{5}{11}$ **12.** $\frac{12}{31} \div \frac{8}{13}$

13. $7\frac{1}{5} \div 1\frac{1}{25}$ **14.** $5\frac{1}{7} \div 3$

Find the least common denominator of each set of fractions.

15. $\frac{5}{12}$ and $\frac{7}{18}$ **16.** $\frac{3}{16}$ and $\frac{1}{24}$ **17.** $\frac{1}{4}, \frac{3}{8}, \frac{5}{6}$

18. Build the fraction to an equivalent fraction with the specified denominator. $\frac{5}{12} = \frac{?}{72}$

1. $\frac{3}{5}$

2. $\frac{311}{388}$

3. $\frac{3}{7}$

4. $\frac{3}{14}$

5. $\frac{9}{2}$

6. $\frac{34}{5}$

7. $10\frac{5}{14}$

8. 12

9. $\frac{14}{45}$

10. 14

11. $\frac{77}{40}$ or $1\frac{37}{40}$

12. $\frac{39}{62}$

13. $6\frac{12}{13}$ or $\frac{90}{13}$

14. $\frac{12}{7}$ or $1\frac{5}{7}$

15. 36

16. 48

17. 24

18. $\frac{30}{72}$

19. $\frac{13}{36}$

Evaluate using the correct order of operations.

19. $\frac{7}{9} - \frac{5}{12}$

20. $\frac{2}{15} + \frac{5}{12}$

21. $\frac{1}{4} + \frac{3}{7} + \frac{3}{14}$

22. $8\frac{3}{5} + 5\frac{4}{7}$

23. $18\frac{6}{7} - 13\frac{13}{14}$

24. $\frac{2}{9} \div \frac{8}{3} \times \frac{1}{4}$

25. $\left(\frac{1}{2} + \frac{1}{3}\right) \times \frac{7}{5}$

Answer each question.

▲ **26.** Erin needs to find the area of her kitchen so she knows how much tile to purchase. The room measures $16\frac{1}{2}$ feet by $9\frac{1}{3}$ feet. How many square feet is the kitchen?

27. A butcher has $18\frac{2}{3}$ pounds of steak that he wishes to place into packages that average $2\frac{1}{3}$ pounds each. How many packages can he make?

28. From central parking it is $\frac{9}{10}$ of a mile to the science building. Bob started at central parking and walked $\frac{1}{5}$ of a mile toward the science building. He stopped for coffee. When he finished, how much farther did he have to walk to reach the science building?

29. Robin jogged $4\frac{1}{8}$ miles on Monday, $3\frac{1}{6}$ miles on Tuesday, and $6\frac{3}{4}$ miles on Wednesday. How far did she jog on those three days?

30. Mr. and Mrs. Samuel visited Florida and purchased 120 oranges. They gave $\frac{1}{4}$ of them to relatives, ate $\frac{1}{12}$ of them in the hotel, and gave $\frac{1}{3}$ of them to friends. They shipped the rest home to Illinois.

(a) How many oranges did they ship?

(b) If it costs 24¢ for each orange to be shipped to Illinois, what was the total shipping bill?

31. A candle company purchased $48\frac{1}{8}$ pounds of wax to make specialty candles. It takes $\frac{5}{8}$ pound of wax to make one candle. The owners of the business plan to sell the candles for $12 each. The specialty wax cost them $2 per pound.

(a) How many candles can they make?

(b) How much does it cost to make one candle?

(c) How much profit will they make if they sell all of the candles?

20. $\frac{11}{20}$

21. $\frac{25}{28}$

22. $14\frac{6}{35}$

23. $4\frac{13}{14}$

24. $\frac{1}{48}$

25. $\frac{7}{6}$ or $1\frac{1}{6}$

26. 154 square feet

27. 8 packages

28. $\frac{7}{10}$ mile

29. $14\frac{1}{24}$ miles

30. (a) 40 oranges

(b) $\$9\frac{3}{5}$

31. (a) 77 candles

(b) $\$1\frac{1}{4}$

(c) $\$827\frac{3}{4}$

Cumulative Test for Chapters 1–2

One-half of this test is based on Chapter 1 material. The remainder is based on material covered in Chapter 2.

1. Write in words. 84,361,208

2. Add.
$$\begin{array}{r} 128 \\ 452 \\ 178 \\ 34 \\ +\ 77 \\ \hline \end{array}$$

3. Add.
$$\begin{array}{r} 156{,}200 \\ 364{,}700 \\ +198{,}320 \\ \hline \end{array}$$

4. Subtract.
$$\begin{array}{r} 5718 \\ -3643 \\ \hline \end{array}$$

5. Subtract.
$$\begin{array}{r} 1{,}000{,}361 \\ -\ \ 983{,}145 \\ \hline \end{array}$$

6. Multiply.
$$\begin{array}{r} 126 \\ \times\ 38 \\ \hline \end{array}$$

7. Multiply.
$$\begin{array}{r} 16{,}908 \\ \times\ \ 12 \\ \hline \end{array}$$

8. Divide. $7\overline{)32{,}606}$

9. Divide. $15\overline{)4631}$

10. Evaluate. 7^2

11. Round to the nearest thousand. 6,037,452

12. Perform the operations in their proper order. $6 \times 2^3 + 12 \div (4 + 2)$

13. Roone bought three shirts at \$26 and two pairs of pants at \$48. What was his total bill?

14. Leslie had a balance of \$64 in her checking account. She deposited \$1160. She made checks out for \$516, \$199, and \$203. What will be her new balance?

15. A supermarket survey found that of the 112 people that went grocery shopping on Friday morning, 83 were women. Write the fractions that describe the part of the shoppers that was women and the part that was men.

16. Reduce. $\dfrac{28}{52}$

17. Write as an improper fraction. $18\dfrac{3}{4}$

1. eighty-four million, three hundred sixty-one thousand, two hundred eight
2. 869
3. 719,220
4. 2075
5. 17,216
6. 4788
7. 202,896
8. 4658
9. 308 R11
10. 49
11. 6,037,000
12. 50
13. \$174
14. \$306
15. $\frac{83}{112}$ were women; $\frac{29}{112}$ were men
16. $\frac{7}{13}$
17. $\frac{75}{4}$

18. $14\frac{2}{7}$

19. $\frac{527}{48}$ or $10\frac{47}{48}$

20. $\frac{12}{35}$

21. 40

22. $\frac{61}{54}$ or $1\frac{7}{54}$

23. $\frac{71}{8}$ or $8\frac{7}{8}$

24. $\frac{113}{15}$ or $7\frac{8}{15}$

25. $\frac{5}{6}$

26. $2\frac{3}{4}$ pounds

27. $24\frac{3}{5}$ miles per gallon

28. $8\frac{1}{8}$ cups of sugar; $5\frac{5}{6}$ cups of flour

29. 60,000,000 miles

30. $160

18. Write as a mixed number. $\frac{100}{7}$

19. Multiply. $3\frac{7}{8} \times 2\frac{5}{6}$

20. Divide. $\frac{44}{49} \div 2\frac{13}{21}$

21. Find the least common denominator of $\frac{5}{8}$ and $\frac{7}{10}$.

Evaluate using the correct order of operations.

22. $\frac{7}{18} + \frac{20}{27}$

23. $2\frac{1}{8} + 6\frac{3}{4}$

24. $12\frac{1}{5} - 4\frac{2}{3}$

25. $\frac{1}{3} + \frac{5}{8} \div \frac{5}{4}$

Answer each question.

26. Marcos is on a special diet and exercise plan to lose weight. By the end of May, his goal is to have lost 15 pounds. In March he lost $5\frac{1}{2}$ pounds, and in April he lost $6\frac{3}{4}$ pounds. How many pounds must he lose in May to reach his goal?

27. Melinda traveled $221\frac{2}{5}$ miles on 9 gallons of gas. How many miles per gallon did her car get?

28. A recipe requires $3\frac{1}{4}$ cups of sugar and $2\frac{1}{3}$ cups of flour. Marcia wants to make $2\frac{1}{2}$ times that recipe. How many cups of each will she need?

29. A space probe travels at 28,356 miles per hour for 2142 hours. Estimate how many miles it travels.

30. To raise money for cancer research, the local YMCA sponsored a road race. Contributions totaled $960. One-sixth of this amount was used for refreshments and T-shirts for the participants. How much did the refreshments and T-shirts cost?

CHAPTER

3

Each year in our country about 72,900,000,000 gallons of gasoline are consumed by automobiles. How much of a savings would occur if half of the miles traveled were done in small, fuel-efficient automobiles? Can you use your mathematics knowledge to determine this saving? Turn to page 260 to find out.

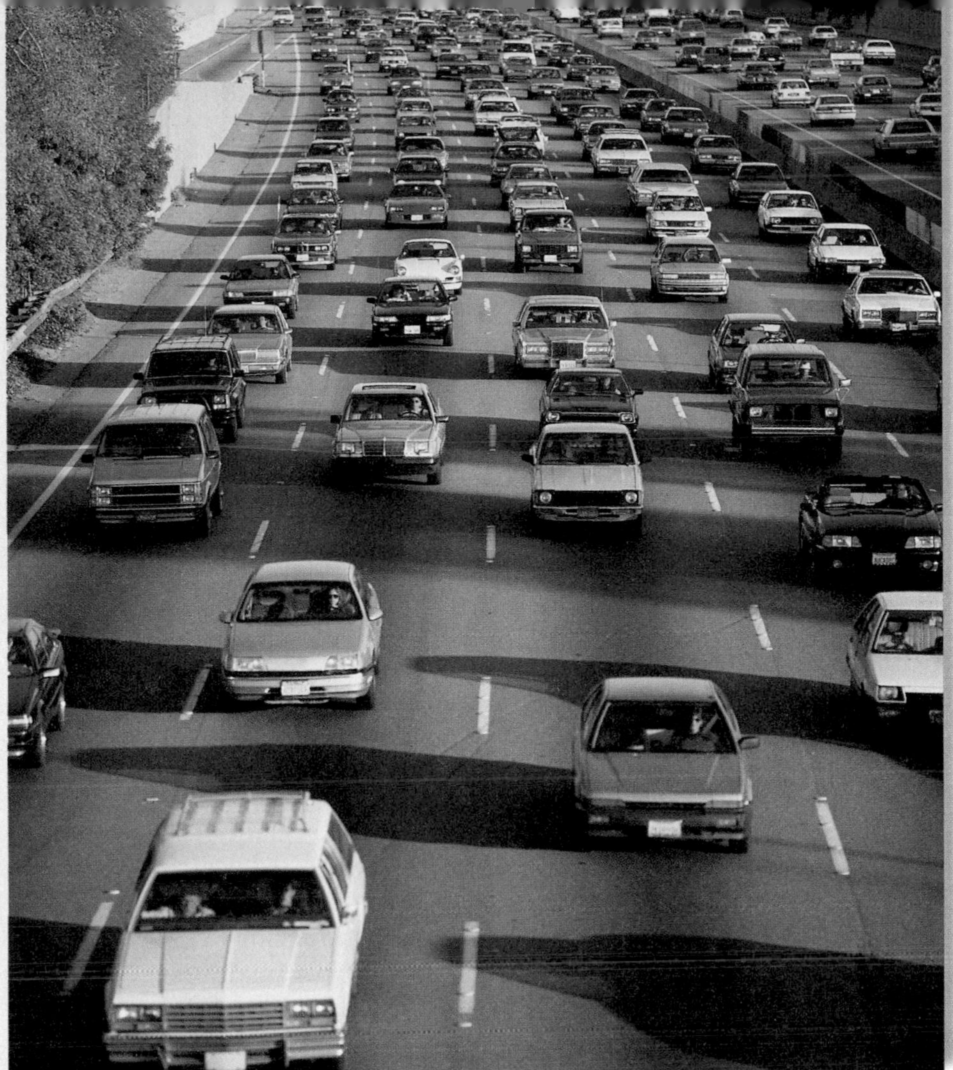

Decimals

3.1 Using Decimal Notation

Student Learning Objectives

After studying this section, you will be able to:

1. **Write a word name for a decimal fraction.**
2. **Change from fractional notation to decimal notation.**
3. **Change from decimal notation to fractional notation.**

1 Writing a Word Name for a Decimal Fraction

In Chapter 2 we discussed *fractions*—the set of numbers such as $\frac{1}{2}$, $\frac{2}{3}$, $\frac{1}{10}$, $\frac{6}{7}$, $\frac{18}{100}$, and so on. Now we will take a closer look at **decimal fractions**—that is, fractions with 10, 100, 1000, and so on, in the denominator, such as $\frac{1}{10}$, $\frac{18}{100}$, and $\frac{43}{1000}$.

Why, of all fractions, do we take special notice of these? Our hands have 10 digits. Our U.S. money system is based on the dollar, which has 100 equal parts, or cents. And the international system of measurement called the *metric system* is based on 10 and powers of 10.

As with other numbers, these decimal fractions can be written in different ways (forms). For example, the shaded part of the whole in the following drawing can be written:

in words (one-tenth)
in fractional form $\left(\frac{1}{10}\right)$
in decimal form (0.1)

All mean the same quantity, namely 1 out of 10 equal parts of the whole. We'll see that when we use decimal notation, computations can be easily done based on the old rules for whole numbers and a few new rules about where to place the decimal point. In a world where calculators and computers are commonplace, many of the fractions we encounter are decimal fractions. A decimal fraction is a fraction whose denominator is a power of 10.

$\frac{7}{10}$ is a decimal fraction. $\frac{89}{10^2} = \frac{89}{100}$ is a decimal fraction.

Decimal fractions can be written with numerals in two ways: fractional form or decimal form. Some decimal fractions are shown in decimal form below.

Fractional Form		Decimal Form
$\frac{3}{10}$	=	0.3
$\frac{59}{100}$	=	0.59
$\frac{171}{1000}$	=	0.171

Teaching Tip Students often wonder if they should write .45 rather than 0.45 because they think the 0 is unnecessary. Remind them that the extra zero in front of a decimal point reduces the chances for error. Also, many calculators display this extra 0.

The zero in front of the decimal point is not actually required. We place it there simply to make sure that we don't miss seeing the decimal point. A number written in decimal notation has three parts.

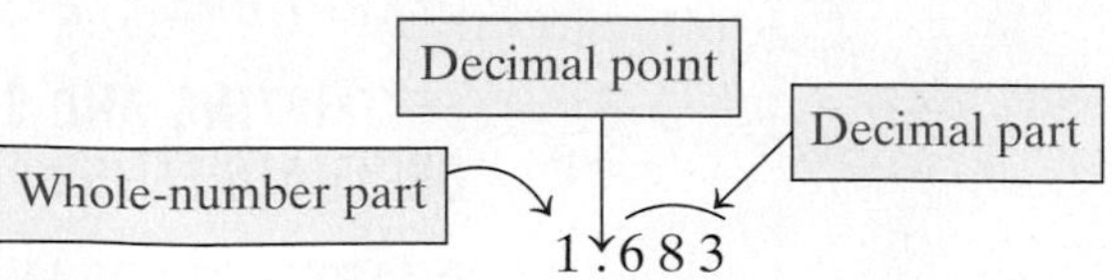

When a number is written in decimal form, the first digit to the right of the decimal point represents tenths, the next digit hundredths, the next digit thousandths, and so on. 0.9 means nine tenths and is equivalent to $\frac{9}{10}$. 0.51 means fifty-one hundredths and is equivalent to $\frac{51}{100}$.

Some decimals are larger than 1. For example, 1.683 means one and six hundred eighty-three thousandths. It is equivalent to $1\frac{683}{1000}$. Note that the word *and* is used to indicate the decimal point. A place-value chart is helpful.

Decimal Place Values

Hundreds	Tens	Ones	Decimal point	Tenths	Hundredths	Thousandths	Ten-thousandths
100	10	1	"and"	$\frac{1}{10}$	$\frac{1}{100}$	$\frac{1}{1,000}$	$\frac{1}{10,000}$
1	5	6	.	2	8	7	4

So, we can write 156.2874 in words as one hundred fifty-six and two thousand eight hundred seventy-four ten-thousandths. We say ten-thousandths because it is the name of the last decimal place on the right.

EXAMPLE 1 Write a word name for each decimal.

(a) 0.79 **(b)** 0.5308 **(c)** 1.6 **(d)** 23.765

Solution

(a) 0.79 = seventy-nine hundredths
(b) 0.5308 = five thousand three hundred eight ten-thousandths
(c) 1.6 = one and six tenths
(d) 23.765 = twenty-three and seven hundred sixty-five thousandths

Practice Problem 1 Write a word name for each decimal.

(a) 0.073 **(b)** 4.68 **(c)** 0.0017 **(d)** 561.78

***NOTE TO STUDENT:** Fully worked-out solutions to all of the Practice Problems can be found at the back of the text starting at page SP-1*

Sometimes, decimals are used where we would not expect them. For example, we commonly say that there are 365 days in a year, with 366 days in every fourth year (or leap year). However, this is not quite correct. In fact, from time to time further adjustments need to be made to the calendar to adjust for these inconsistencies. Astronomers know that a more accurate measure of a year is called a **tropical year** (measured from one equinox to the next). Rounded to the nearest hundred-thousandth, 1 tropical year = 365.24122 days. This is read "three hundred sixty-five and twenty-four thousand, one hundred twenty-two hundred-thousandths." This approximate value is a more accurate measurement of the amount of time it takes the earth to complete one orbit around the sun.

Note the relationship between fractions and their equivalent numbers' decimal forms.

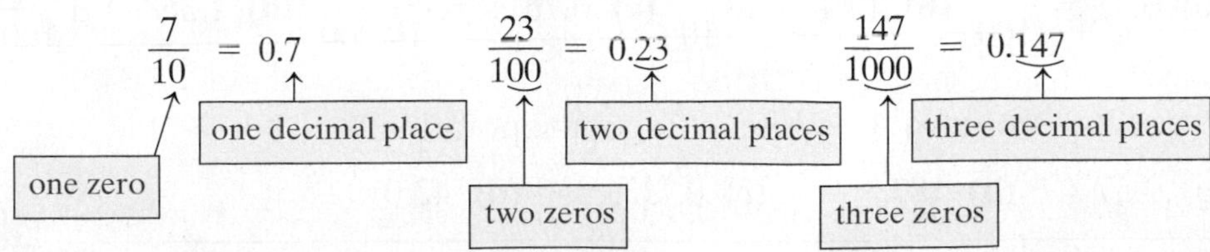

Teaching Tip Students often think that writing a word name for 4.36 is "Four point three six" instead of the correct answer, which is four and thirty-six hundredths. Explain to them the correct answer, and emphasize that if they gain the ability to write word names for decimals correctly, it will help them make out checks correctly.

Teaching Tip Additional coverage on balancing a checkbook can be found in the Consumer Finance appendix.

Decimal notation is commonly used with money. When writing a check, we often write the amount that is less than 1 dollar, such as 23¢, as $\frac{23}{100}$ dollar.

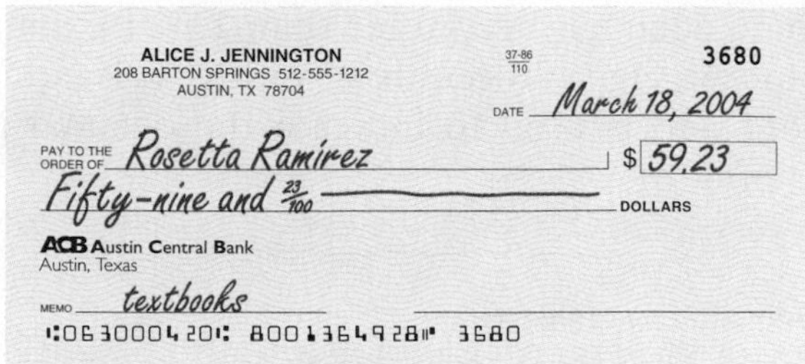

EXAMPLE 2 Write a word name for the amount on a check made out for $672.89.

Solution: six hundred seventy-two and $\frac{89}{100}$ dollars

Practice Problem 2 Write a word name for the amount of a check made out for $7863.04.

2 Changing from Fractional Notation to Decimal Notation

It is helpful to be able to write decimals in both decimal notation and fractional notation. First we illustrate changing a fraction with a denominator of 10, 100, or 1000 into decimal form.

EXAMPLE 3 Write as a decimal.

(a) $\frac{8}{10}$ **(b)** $\frac{74}{100}$ **(c)** $1\frac{3}{10}$ **(d)** $2\frac{56}{1000}$

Solution

(a) $\frac{8}{10} = 0.8$ **(b)** $\frac{74}{100} = 0.74$ **(c)** $1\frac{3}{10} = 1.3$ **(d)** $2\frac{56}{1000} = 2.056$

Note: In part (d), we need to add a zero before the digits 56. Since there are three zeros in the denominator, we need three decimal places in the decimal number.

Practice Problem 3 Write as a decimal.

(a) $\frac{9}{10}$ **(b)** $\frac{136}{1000}$ **(c)** $2\frac{56}{100}$ **(d)** $34\frac{86}{1000}$

NOTE TO STUDENT: Fully worked-out solutions to all of the Practice Problems can be found at the back of the text starting at page SP-1

3 Changing from Decimal Notation to Fractional Notation

EXAMPLE 4 Write in fractional notation.

(a) 0.51 **(b)** 18.1 **(c)** 0.7611 **(d)** 1.363

Solution

(a) $0.51 = \frac{51}{100}$ **(b)** $18.1 = 18\frac{1}{10}$ **(c)** $0.7611 = \frac{7611}{10{,}000}$ **(d)** $1.363 = 1\frac{363}{1000}$

Practice Problem 4 Write in fractional notation.

(a) 0.37 **(b)** 182.3 **(c)** 0.7131 **(d)** 42.019

When we convert from decimal form to fractional form, we reduce whenever possible.

EXAMPLE 5 Write in fractional notation. Reduce whenever possible.

(a) 2.6 **(b)** 0.38 **(c)** 0.525 **(d)** 361.007

Solution

(a) $2.6 = 2\frac{6}{10} = 2\frac{3}{5}$ **(b)** $0.38 = \frac{38}{100} = \frac{19}{50}$

(c) $0.525 = \frac{525}{1000} = \frac{105}{200} = \frac{21}{40}$

(d) $361.007 = 361\frac{7}{1000}$ (cannot be reduced)

Practice Problem 5 Write in fractional notation. Reduce whenever possible.

(a) 8.5 **(b)** 0.58 **(c)** 36.25 **(d)** 106.013

EXAMPLE 6 A chemist found that the concentration of lead in a water sample was 5 parts per million. What fraction would represent the concentration of lead?

Solution Five parts per million means 5 parts out of 1,000,000. As a fraction, this is $\frac{5}{1,000,000}$. We can reduce this by dividing numerator and denominator by 5. Thus

$$\frac{5}{1,000,000} = \frac{1}{200,000}.$$

The concentration of lead in the water sample is $\frac{1}{200,000}$.

Practice Problem 6 A chemist found that the concentration of PCBs in a water sample was 2 parts per billion. What fraction would represent the concentration of PCBs?

Teaching Tip You may want to show the class a few simple problems like 5 parts per thousand, or 15 parts per ten thousand, before doing the type of numbers contained in Example 6.

Developing Your Study Skills

Steps Toward Success In Mathematics

Mathematics is a building process, mastered one step at a time. The foundation of this process is formed by a few basic requirements. Those who are successful in mathematics realize the absolute necessity for building a study of mathematics on the firm foundation of these six minimum requirements.

1. Attend class every day.
2. Read the textbook.
3. Take notes in class.
4. Do assigned homework every day.
5. Get help immediately when needed.
6. Review regularly.

If you are in an online class or self-paced class, do some of your math assignment on five days during each week.

3.1 Exercises

Student Solutions Manual | CD/ Video | PH Math Tutor Center | MathXL®Tutorials on CD | MathXL® | MyMathLab® | Interactmath.com

Verbal and Writing Skills

1. Describe a decimal fraction and provide examples.
A decimal fraction is a fraction whose denominator is a power of 10. $\frac{23}{100}$ and $\frac{563}{1000}$ are decimal fractions

2. What word is used to describe the decimal point when writing the word name for a decimal that is greater than one?
The word that describes the decimal point is *and*.

3. What is the name of the last decimal place on the right for the decimal 132.45678?
hundred-thousandths

4. When writing \$82.75 on a check, we write 75¢ as $\frac{75}{100}$.

Write a word name for each decimal.

5. 0.57
fifty-seven hundredths

6. 0.78
seventy-eight hundredths

7. 3.8
three and eight tenths

8. 12.4
twelve and four tenths

9. 5.803
five and eight hundred three thousandths

10. 3.117
three and one hundred seventeen thousandths

11. 28.0037
twenty-eight and thirty-seven ten-thousandths

12. 54.0013
fifty-four and thirteen ten-thousandths

Write a word name as you would on a check.

13. \$124.20
one hundred twenty-four and $\frac{20}{100}$ dollars

14. \$510.31
five hundred ten and $\frac{31}{100}$ dollars

15. \$1236.08
one thousand, two hundred thirty-six and $\frac{8}{100}$ dollars

16. \$7652.02
seven thousand, six hundred fifty-two and $\frac{2}{100}$ dollars

17. \$12,015.45
twelve thousand fifteen and $\frac{45}{100}$ dollars

18. \$20,000.67
twenty thousand and $\frac{67}{100}$ dollars

Write in decimal notation.

19. seven tenths
0.7

20. six tenths
0.6

21. twelve hundredths
0.12

22. forty-five hundredths
0.45

23. four hundred eighty-one thousandths
0.481

24. twenty-two thousandths
0.022

25. two hundred eighty-six millionths
0.000286

26. one thousand three hundred eighteen millionths
0.001318

Write each fraction as a decimal.

27. $\frac{7}{10}$
0.7

28. $\frac{3}{10}$
0.3

29. $\frac{76}{100}$
0.76

30. $\frac{84}{100}$
0.84

31. $\frac{1}{100}$
0.01

32. $\frac{6}{100}$
0.06

33. $\frac{53}{1000}$
0.053

34. $\frac{328}{1000}$
0.328

35. $\frac{2403}{10,000}$
0.2403

36. $\frac{7794}{10,000}$
0.7794

37. $8\frac{3}{10}$
8.3

38. $3\frac{1}{10}$
3.1

39. $84\frac{13}{100}$
84.13

40. $52\frac{77}{100}$
52.77

41. $3\frac{529}{1000}$
3.529

42. $2\frac{23}{1000}$
2.023

43. $126\frac{571}{10,000}$
126.0571

44. $198\frac{333}{10,000}$
198.0333

Write in fractional notation. Reduce whenever possible.

45. 0.02
$\frac{2}{100} = \frac{1}{50}$

46. 0.05
$\frac{5}{100} = \frac{1}{20}$

47. 3.6
$3\frac{6}{10} = 3\frac{3}{5}$

48. 8.9
$8\frac{9}{10}$

49. 7.41
$7\frac{41}{100}$

50. 15.75
$15\frac{75}{100} = 15\frac{3}{4}$

51. 12.625
$12\frac{625}{1000} = 12\frac{5}{8}$

52. 29.875
$29\frac{875}{1000} = 29\frac{7}{8}$

53. 7.0615
$7\frac{615}{10,000} = 7\frac{123}{2000}$

54. 4.0016
$4\frac{16}{10,000} = 4\frac{1}{625}$

55. 8.0108
$8\frac{108}{10,000} = 8\frac{27}{2500}$

56. 7.0605
$7\frac{605}{10,000} = 7\frac{121}{2000}$

57. 235.1254
$235\frac{1254}{10,000} = 235\frac{627}{5000}$

58. 581.2406
$581\frac{2406}{10,000} = 581\frac{1203}{5000}$

59. 0.0187
$\frac{187}{10,000}$

60. 0.5405
$\frac{5405}{10,000} = \frac{1081}{2000}$

Applications

61. ***Cigarette Use*** The highest use of cigarettes in the United States takes place in Kentucky. In 2001 31,700 out of every 100,000 men age 18 or older who lived in Kentucky were smokers. That same year 30,100 out of every 100,000 women age 18 or older who lived in Kentucky were smokers. (a) What fractional part of the male population in Kentucky were smokers? (b) What fractional part of the female population in Kentucky were smokers? Be sure to express these fractions in reduced form. *(Source: U.S. Centers for Disease Control and Prevention)*
(a) $\frac{317}{1000}$ (b) $\frac{301}{1000}$

62. ***Cigarette Use*** The lowest use of cigarettes in the United States takes place in Utah. In 2001 14,600 out of every 100,000 men age 18 or older who lived in Utah were smokers. That same year 12,100 out of every 100,000 women age 18 or older who lived in Utah were smokers. (a) What fractional part of the male population in Utah were smokers? (b) What fractional part of the female population in Utah were smokers? Be sure to express these fractions in reduced form. *(Source: U.S. Centers for Disease Control and Prevention)*
(a) $\frac{73}{500}$ (b) $\frac{121}{1000}$

63. ***Bald Eagle Eggs*** American bald eagles have been fighting extinction due to environmental hazards such as DDT, PCBs, and dioxin. The problem is with the food chain. Fish or rodents consume contaminated food and/or water. Then the eagles ingest the poison, which in turn affects the durability of the eagles' eggs. It takes only 4 parts per million of certain chemicals to ruin an eagle egg; write this number as a fraction in lowest terms. (In 1994 the bald eagle was removed from the endangered species list.)
$\frac{4}{1,000,000} = \frac{1}{250,000}$

64. ***Turtle Eggs*** Every year turtles lay eggs on the islands of South Carolina. Unfortunately, due to illegal polluting, a lot of the eggs are contaminated. If the turtle eggs contain more than 2 parts per one hundred million of chemical pollutants, they will not hatch and the population will continue to head toward extinction. Write the preceding amount of chemical pollutants as a fraction in the lowest terms.
$\frac{2}{100,000,000} = \frac{1}{50,000,000}$

Cumulative Review

65. Add.
$$\begin{array}{r} 207 \\ 54 \\ 123 \\ 86 \\ +\ 55 \\ \hline 525 \end{array}$$

66. Subtract.
$$\begin{array}{r} 12,843 \\ -\ 11,905 \\ \hline 938 \end{array}$$

67. Round to the nearest *hundred*. 56,758
56,800

68. Round to the nearest *thousand*. 8,069,482
8,069,000

3.2 COMPARING, ORDERING, AND ROUNDING DECIMALS

Student Learning Objectives

After studying this section, you will be able to:

1. Compare decimals.
2. Place decimals in order from smallest to largest.
3. Round decimals to a specified decimal place.

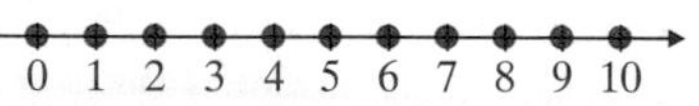

Teaching Tip This is a good time to stress the use of the number line. Remind the students that they can make any scale they want for a number line. It is just as easy to make a number line with units labeled 1.34, 1.35, 1.36, 1.37, 1.38, etc., as it is to make a number line with units labeled 1, 2, 3, 4, 5, 6, etc. This will help them whenever they compare decimals. The number line can be expanded to show decimal numbers between whole numbers.

1 Comparing Decimals

All of the numbers we have studied have a specific order. To illustrate this order, we can place the numbers on a **number line.** Look at the number line on the right. Each number has a specific place on it. The arrow points in the direction of increasing value. Thus, if one number is to the right of a second number, it is larger, or greater, than that number. Since 5 is to the right of 2 on the number line, we say that 5 is greater than 2. We write $5 > 2$.

Since 4 is to the left of 6 on the number line, we say that 4 is less than 6. We write $4 < 6$. The symbols ">" and "<" are called **inequality symbols.**

$a < b$ is read "a is less than b."

$a > b$ is read "a is greater than b."

We can assign exactly one point on the number line to each decimal number. When two decimal numbers are placed on a number line, the one farther to the right is the larger. Thus we can say that $3.4 > 2.7$ and $4.3 > 4.0$. We can also say that $0.5 < 1.0$ and $1.8 < 2.2$. Why?

0.5 1.8 2.2 2.7 3.4 4.3

0 1.0 2.0 3.0 4.0 5.0

To compare or order decimals, we compare each digit.

COMPARING TWO NUMBERS IN DECIMAL NOTATION

1. Start at the left and compare corresponding digits. If the digits are the same, move one place to the right.
2. When two digits are different, the larger number is the one with the larger digit.

EXAMPLE 1 Write an inequality statement with 0.167 and 0.166.

Solution The numbers in the tenths place are the same. They are both 1.

0.1 6 7 0.1 6 6

The numbers in the hundredths place are the same. They are both 6.

0.1 6 7 0.1 6 6

The numbers in the thousandths place differ.

0.1 6 7 0.1 6 6

Since $7 > 6$, we know that $0.167 > 0.166$.

Practice Problem 1 Write an inequality statement with 5.74 and 5.75.

NOTE TO STUDENT: Fully worked-out solutions to all of the Practice Problems can be found at the back of the text starting at page SP-1

Whenever necessary, extra zeros can be written to the right of the last digit—that is, to the right of the decimal point—without changing the value of the decimal. Thus

$$0.56 = 0.56000 \quad \text{and} \quad 0.7768 = 0.77680.$$

The zero to the left of the decimal point is optional. Thus 0.56 = .56. Both notations are used. You are encouraged to place a zero to the left of the decimal point so that you don't miss the decimal point when you work with decimals.

EXAMPLE 2 Fill in the blank with one of the symbols <, =, or >.

0.77 _____ 0.777

Solution We begin by adding a zero to the first decimal.

0.770 0.777

We see that the tenths and hundredths digits are equal. But the thousandths digits differ. Since $0 < 7$, we have $0.770 < 0.777$.

Practice Problem 2 Fill in the blank with one of the symbols <, =, or >.

0.894 _____ 0.89

Teaching Tip You may want to mention that in a mathematics class it does not make any difference whether you write 0.56 or 0.5600 since both numbers have exactly the same value. You might point out that in physics, chemistry, and engineering classes, however, the extra zeros to the right of the 56 are often used to indicate the level of accuracy of the measurement.

2 Placing Decimals in Order from Smallest to Largest

Which is the heaviest—a puppy that weighs 6.2 ounces, a puppy that weighs 6.28 ounces, or a puppy that weighs 6.028 ounces? Did you choose the puppy that weighs 6.28 ounces? You are correct.

You can place three or more decimals in order. If you are asked to order the decimals from smallest to largest, look for the smallest decimal and place it first.

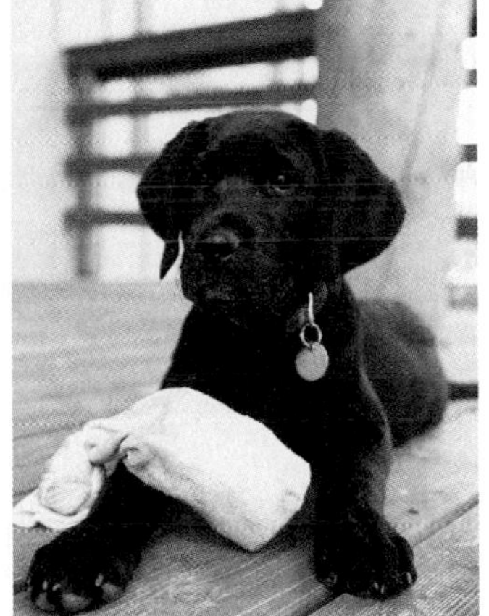

EXAMPLE 3 Place the following five decimal numbers in order from smallest to largest.

1.834, 1.83, 1.381, 1.38, 1.8

Solution First we add zeros to make the comparison easier.

1.834, 1.830, 1.381, 1.380, 1.800

Now we rearrange with smallest first.

1.380, 1.381, 1.800, 1.830, 1.834

Practice Problem 3 Place the following five decimal numbers in order from smallest to largest.

2.45, 2.543, 2.46, 2.54, 2.5

Teaching Tip You may want to ask the students if they are familiar with other rules for rounding off numbers. Particularly if you have students from other countries or cultures, you may see some unusual round-off rules. Remind them that this rule for rounding decimals is used uniformly in mathematics courses in the United States and that they will need to be able to use it.

3 Rounding Decimals to a Specified Decimal Place

Sometimes in calculations involving money, we see numbers like $386.432 and $29.5986. To make these useful, we usually round them to the nearest cent. $386.432 is rounded to $386.43. $29.5986 is rounded to $29.60. A general rule for rounding decimals follows.

ROUNDING DECIMALS

1. Find the decimal place (units, tenths, hundredths, and so on) for which rounding off is required.
2. If the first digit to the right of the given place value is less than 5, drop it and all digits to the right of it.
3. If the first digit to the right of the given place value is 5 or greater, increase the number in the given place value by one. Drop all digits to the right of this place.

EXAMPLE 4 Round 156.37 to the nearest tenth.

Solution

156.3 7

↑ We find the tenths place.

Note that 7, the next place to the right, is greater than 5. We round up to 156.4 and drop the digits to the right. The answer is 156.4.

Practice Problem 4 Round 723.88 to the nearest tenth.

NOTE TO STUDENT: Fully worked-out solutions to all of the Practice Problems can be found at the back of the text starting at page SP-1

EXAMPLE 5 Round to the nearest thousandth.

(a) 0.06358 **(b)** 128.37448

Solution

(a) 0.06 3 58

↑ We locate the thousandths place.

Note that the digit to the right of the thousandths place is 5. We round up to 0.064 and drop all the digits to the right.

(b) 128.37 4 48

↑ We locate the thousandths place.

Note that the digit to the right of the thousandths place is less than 5. We round to 128.374 and drop all the digits to the right.

Practice Problem 5 Round to the nearest thousandth.

(a) 12.92647 **(b)** 0.007892

Remember that rounding up to the next digit in a position may result in several digits being changed.

EXAMPLE 6 Round to the nearest hundredth. Fred and Linda used 203.9964 kilowatt hours of electricity in their house in May.

203.9 9 64

↑ We locate the hundredths place.

Solution Since the digit to the right of hundredths is greater than 5, we round up. This affects the next two positions. Do you see why? The result is 204.00 kilowatt hours. Notice that we have the two zeros to the right of the decimal place to show we have rounded to the nearest hundredth.

Practice Problem 6 Round to the nearest tenth. Last month the college gymnasium used 15,699.953 kilowatt hours of electricity.

Sometimes we round a decimal to the nearest whole number. For example, when writing figures on income tax forms, a taxpayer may round all figures to the nearest dollar.

EXAMPLE 7 To complete her income tax return, Marge needs to round these figures to the nearest whole dollar.

Medical bills $779.86	Taxes $563.49
Retirement contributions $674.38	Contributions to charity $534.77

Solution Round the amounts.

	Original Figure	***Rounded To Nearest Dollar***
Medical bills	779.86	780
Taxes	563.49	563
Retirement	674.38	674
Charity	534.77	535

Practice Problem 7 Round the following figures to the nearest whole dollar.

Medical bills $375.50	Taxes $981.39
Retirement contributions $980.49	Contributions to charity $817.65

CAUTION Why is it so important to consider only *one* digit to the right of the desired round-off position? What is wrong with rounding in steps? Suppose that Mark rounds 1.349 to the nearest tenth in steps. First he rounds 1.349 to 1.35 (nearest hundredth). Then he rounds 1.35 to 1.4 (nearest tenth). What is wrong with this reasoning?

1.349

1.30 1.31 1.32 1.33 1.34 1.35 1.36 1.37 1.38 1.39 1.40

To round 1.349 to the nearest tenth, we ask if 1.349 is closer to 1.3 or to 1.4. It is closer to 1.3. Mark got 1.4, so he is not correct. He "rounded in steps" by first moving to 1.35, thus increasing the error and moving in the wrong direction. To control rounding errors, we consider *only* the first digit to the right of the decimal place to which we are rounding.

3.2 Exercises

Student Solutions Manual

CD/Video

PH Math Tutor Center

MathXL®Tutorials on CD

MathXL®

MyMathLab®

Interactmath.com

Fill in the blank with one of the symbols $<$, $=$, or $>$.

1. 1.3 $\underline{>}$ 1.29
2. 2.6 $\underline{>}$ 2.58
3. 0.34 $\underline{=}$ 0.340
4. 72.54 $\underline{<}$ 72.56

5. 18.92 $\underline{<}$ 18.93
6. 0.460 $\underline{=}$ 0.46
7. 0.006 $\underline{>}$ 0.00063
8. 0.0037 $\underline{<}$ 0.036

9. 1.002 $\underline{<}$ 1.0021
10. 2.0056 $\underline{<}$ 2.006
11. 126.34 $\underline{>}$ 125.35
12. 406.78 $\underline{<}$ 407.75

13. 0.888 $\underline{<}$ 0.8888
14. 0.666 $\underline{<}$ 0.6666
15. 0.777 $\underline{>}$ 0.7077
16. 0.555 $\underline{>}$ 0.5505

17. $\frac{72}{1000}$ $\underline{=}$ 0.072
18. $\frac{54}{1000}$ $\underline{=}$ 0.054
19. $\frac{8}{10}$ $\underline{>}$ 0.08
20. $\frac{5}{100}$ $\underline{>}$ 0.005

Arrange each set of decimals from smallest to largest.

21. 12.6, 12.8, 12.65
12.6, 12.65, 12.8

22. 18.32, 18.038, 18.04
18.038, 18.04, 18.32

23. 0.0071, 0.05, 0.007
0.007, 0.0071, 0.05

24. 0.0025, 0.0052, 0.002
0.002, 0.0025, 0.0052

25. 5.2, 5.23, 5.3, 5.12
5.12, 5.2, 5.23, 5.3

26. 4.5, 4.67, 4.73, 4.6
4.5, 4.6, 4.67, 4.73

27. 26.034, 26.003, 26.04, 26.033
26.003, 26.033, 26.034, 26.04

28. 33.082, 33.02, 33.088, 33.079
33.02, 33.079, 33.082, 33.088

29. 18.006, 18.060, 18.066, 18.606, 18.065
18.006, 18.060, 18.065, 18.066, 18.606

30. 15.020, 15.002, 15.001, 15.018, 15.0019
15.001, 15.0019, 15.002, 15.018, 15.020

Round to the nearest tenth.

31. 5.67
5.7

32. 8.35
8.4

33. 28.98
29.0

34. 47.94
47.9

35. 578.064
578.1

36. 311.95
312.0

37. 2176.83
2176.8

38. 4082.74
4082.7

Round to the nearest hundredth.

39. 26.032
26.03

40. 47.071
47.07

41. 36.997
37.00

42. 24.999
25.00

43. 156.1749
156.17

44. 283.8441
283.84

45. 2786.706
2786.71

46. 4609.285
4609.29

Round to the nearest given place.

47. 1.06132 thousandths
1.061

48. 8.10263 thousandths
8.103

49. 0.05951 ten-thousandths
0.0595

50. 0.063148 ten-thousandths
0.0631

51. 5.00761238 hundred-thousandths
5.00761

52. 7.038972 hundred-thousandths
7.03897

53. 135.564 nearest whole number
136

54. 208.7302 nearest whole number
209

Round to the nearest dollar.

55. \$2536.85
\$2537

56. \$5319.62
\$5320

57. \$15,020.50
\$15,021

58. \$20,159.48
\$20,159

Round to the nearest cent.

59. \$56.9832
\$56.98

60. \$28.3951
\$28.40

61. \$5783.716
\$5783.72

62. \$3928.649
\$3928.65

Applications

63. ***Baseball*** During the 2003 baseball season, the winning percentages of the New York Yankees and the Toronto Blue Jays were 0.60869 and 0.52127, respectively. Round these values to the nearest thousandth.
0.609; 0.521

64. ***Employment Compensation*** During 2003 the employer costs in the United States for employee compensation per hour worked were \$29.4235 for union members. For nonunion members the figure was \$20.7896. Round these values to the nearest cent.
\$29.42; \$20.79

65. ***Astronomy*** The number of days in a year is 365.24122. Round this value to the nearest hundredth.
365.24

66. ***Mathematics History*** The numbers π and e are approximately equal to 3.14159 and 2.71828, respectively. We will be using π later in this textbook. You will encounter e in higher level mathematics courses. Round these values to the nearest hundredth.
3.14; 2.72

To Think About

67. Arrange in order from smallest to largest.

$$0.61, 0.062, \frac{6}{10}, 0.006, 0.0059,$$
$$\frac{6}{100}, 0.0601, 0.0519, 0.0612$$

$0.0059, 0.006, 0.0519, \frac{6}{100}, 0.0601, 0.0612, 0.062, \frac{6}{10}, 0.61$

68. Arrange in order from smallest to largest.

$$1.05, 1.512, \frac{15}{10}, 1.0513, 0.049,$$
$$\frac{151}{100}, 0.0515, 0.052, 1.051$$

$0.049, 0.0515, 0.052, 1.05, 1.051, 1.0513, \frac{15}{10}, \frac{151}{100}, 1.512$

69. A person wants to round 86.23498 to the nearest hundredth. He first rounds 86.23498 to 86.2350. He then rounds to 86.235. Finally, he rounds to 86.24. What is wrong with his reasoning?
You should consider only one digit to the right of the decimal place that you wish to round to. 86.23498 is closer to 86.23 than to 86.24.

70. ***Personal Finance*** Fred is checking the calculations on his monthly bank statement. An interest charge of \$16.3724 was rounded to \$16.38. An interest charge of \$43.7214 was rounded to \$43.73. What rule does the bank use for rounding off to the nearest cent?
The bank rounds up for any fractional part of a cent.

Cumulative Review

71. Add. $3\frac{1}{4} + 2\frac{1}{2} + 6\frac{3}{8}$ $12\frac{1}{8}$

72. Subtract. $27\frac{1}{5} - 16\frac{3}{4}$ $10\frac{9}{20}$

73. ***Car Travel*** Mary drove her Dodge Caravan on a trip. At the start of the trip, the odometer (which measures distance) read 46,381. At the end of the trip, it read 47,073. How many miles long was the trip?
692 miles

74. ***Boat Sales*** Don's New and Used Watercraft sold four boats one weekend for \$18,650, \$2490, \$835, and \$9845. Estimate the total amount of the sales.
\$31,800

3.3 Adding and Subtracting Decimals

Student Learning Objectives

After studying this section, you will be able to:

 **Add decimals.**

 **Subtract decimals.**

1 Adding Decimals

We often add decimals when we check the addition of our bill at a restaurant or at a store. We can relate addition of decimals to addition of fractions. For example,

$$\frac{3}{10} + \frac{6}{10} = \frac{9}{10} \qquad \text{and} \qquad 1\frac{1}{10} + 2\frac{8}{10} = 3\frac{9}{10}.$$

These same problems can be written more efficiently as decimals.

$$\begin{array}{r} 0.3 \\ +\ 0.6 \\ \hline 0.9 \end{array} \qquad \begin{array}{r} 1.1 \\ +\ 2.8 \\ \hline 3.9 \end{array}$$

The steps to follow when adding decimals are listed in the following box.

ADDING DECIMALS

1. Write the numbers to be added vertically and line up the decimal points. Extra zeros may be placed to the right of the decimal points if needed.
2. Add all the digits with the same place value, starting with the right column and moving to the left.
3. Place the decimal point of the sum in line with the decimal points of the numbers added.

Teaching Tip This is a good time to assign a classroom activity. Ask your class to add the following. 7.21 + 13.346 + 0.0008 + 9.6. The correct answer is 30.1568. Usually you will find that almost all of the students who made an error in adding these numbers did not insert the extra zeros to the right. Students will be quite convinced of the merit of this suggestion to add in the extra zeros to the right when they do Example 1(c) or any similar example.

EXAMPLE 1 Add.

(a) 2.8 + 5.6 + 3.2 **(b)** 158.26 + 200.07 + 315.98

(c) 5.3 + 26.182 + 0.0007 + 624

Solution

(a) $\begin{array}{r} \scriptstyle 1 \\ 2.8 \\ 5.6 \\ +\ 3.2 \\ \hline 11.6 \end{array}$ **(b)** $\begin{array}{r} \scriptstyle 11\ 2 \\ 158.26 \\ 200.07 \\ +\ 315.98 \\ \hline 674.31 \end{array}$ **(c)** $\begin{array}{r} 5.3000 \\ 26.1820 \\ 0.0007 \\ +\ 624.0000 \\ \hline 655.4827 \end{array}$

Extra zeros have been added to make the problem easier.
Note: The decimal point is understood to be to the right of the digit 4.

Practice Problem 1 Add.

(a) $\begin{array}{r} 9.8 \\ 3.6 \\ +\ 5.4 \\ \hline \end{array}$ **(b)** $\begin{array}{r} 300.72 \\ 163.75 \\ +\ 291.08 \\ \hline \end{array}$ **(c)** 8.9 + 37.056 + 0.0023 + 945

NOTE TO STUDENT: Fully worked-out solutions to all of the Practice Problems can be found at the back of the text starting at page SP-1

SIDELIGHT: Adding in Extra Zeros

When we add decimals like 3.1 + 2.16 + 4.007, we may write in zeros, as shown:

$$\begin{array}{r} 3.100 \\ 2.160 \\ +\ 4.007 \\ \hline 9.267 \end{array}$$

What are we really doing here? What is the advantage of adding these extra zeros?

"Decimals" means "decimal fractions." If we look at the number as fractions, we see that we are actually using the property of multiplying a fraction by 1 in order to obtain common denominators. Look at the problem this way:

$$\left.\begin{aligned} 3.1 &= 3\frac{1}{10} \\ 2.16 &= 2\frac{16}{100} \\ 4.007 &= 4\frac{7}{1000} \end{aligned}\right\}$$

The least common denominator is 1000. To obtain the common denominator for the first two fractions, we multiply.

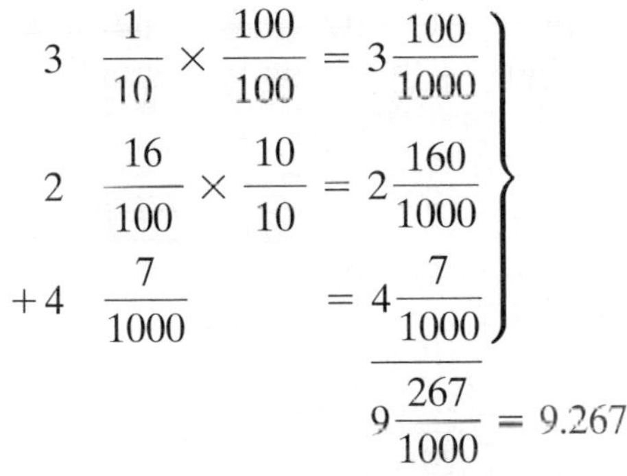

$$\left.\begin{aligned} 3\frac{1}{10} \times \frac{100}{100} &= 3\frac{100}{1000} \\ 2\frac{16}{100} \times \frac{10}{10} &= 2\frac{160}{1000} \\ +4\frac{7}{1000} \quad\quad &= 4\frac{7}{1000} \end{aligned}\right\}$$

$$9\frac{267}{1000} = 9.267$$

Once we obtain a common denominator, we can add the three fractions.

This is the answer we arrived at earlier using the decimal form for each number. Thus writing in zeros in a decimal fraction is really an easy way to transform fractions to equivalent fractions with a common denominator. Working with decimal fractions is easier than working with other fractions.

The final digit of most odometers measures tenths of a mile. The odometer reading shown in the odometer on the right is 38,516.2 miles.

Calculator

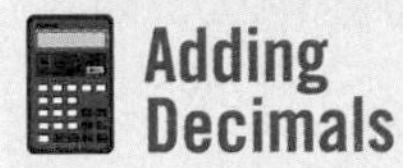

Adding Decimals

The calculator can be used to verify your work. You can use your calculator to add decimals. To find 23.08 + 8.53 + 9.31 enter:

23.08 [+] 8.53 [+]
9.31 [=]

Display:

[40.92]

EXAMPLE 2 Barbara checked her odometer before the summer began. It read 49,645.8 miles. She traveled 3852.6 miles that summer in her car. What was the odometer reading at the end of the summer?

Solution

$$\begin{array}{r} {}^{1\,1\quad 1} \\ 49{,}645.8 \\ +\ 3{,}852.6 \\ \hline 53{,}498.4 \end{array}$$

The odometer read 53,498.4 miles.

Practice Problem 2 A car odometer read 93,521.8 miles before a trip of 1634.8 miles. What was the final odometer reading?

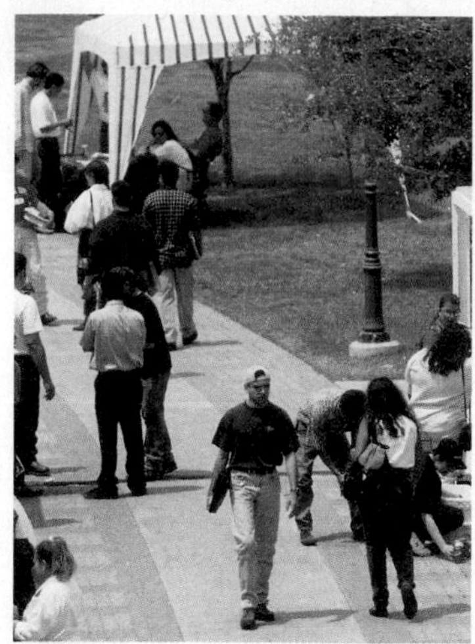

NOTE TO STUDENT: Fully worked-out solutions to all of the Practice Problems can be found at the back of the text starting at page SP-1

EXAMPLE 3 During his first semester at Tarrant County Community College, Kelvey deposited checks into his checking account in the amounts of $98.64, $157.32, $204.81, $36.07, and $229.89. What was the sum of his five checks?

Solution

$$\begin{array}{r} {\scriptstyle 2\,3\,2\;\,2} \\ \$\ \ 98.64 \\ 157.32 \\ 204.81 \\ 36.07 \\ +\ 229.89 \\ \hline \$726.73 \end{array}$$

Practice Problem 3 During the spring semester, Will deposited the following checks into his account: $80.95, $133.91, $256.47, $53.08, and $381.32. What was the sum of his five checks?

2 Subtracting Decimals

It is important to see the relationship between the decimal form of a mixed number and the fractional form of a mixed number. This relationship helps us understand why calculations with decimals are done the way they are. Recall that when we subtract mixed numbers with common denominators, sometimes we must borrow from the whole number.

$$\begin{array}{rcr} 5\frac{1}{10} & = & 4\frac{11}{10} \\ -\ 2\frac{7}{10} & = & -\ 2\frac{7}{10} \\ \hline & & 2\frac{4}{10} \end{array}$$

We could write the same problem in decimal form:

$$\begin{array}{r} {\scriptstyle 4\ 11} \\ \not{5}.\not{1} \\ -\ 2.7 \\ \hline 2.4 \end{array}$$

Subtraction of decimals is thus similar to subtraction of fractions (we get the same result), but it's usually easier to subtract with decimals than to subtract with fractions.

SUBTRACTING DECIMALS

1. Write the decimals to be subtracted vertically and line up the decimal points. Additional zeros may be placed to the right of the decimal point if not all numbers have the same number of decimal places.
2. Subtract all digits with the same place value, starting with the right column and moving to the left. Borrow when necessary.
3. Place the decimal point of the difference in line with the decimal point of the two numbers being subtracted.

EXAMPLE 4 Subtract.

(a) $\begin{array}{r} 84.8 \\ -\ 27.3 \\ \hline \end{array}$ **(b)** $\begin{array}{r} 1076.320 \\ -\ 983.518 \\ \hline \end{array}$

Solution

(a) $\begin{array}{r} 84.8 \\ -\ 27.3 \\ \hline 57.5 \end{array}$ (borrowing: 8 → 7, 4 → 14)

(b) $\begin{array}{r} 1076.320 \\ -\ 983.518 \\ \hline 92.802 \end{array}$ (borrowing: 1 0 → 10 → 9, 7 → 17, 6 → 5, 3 → 13, 2 → 1, 0 → 10)

Practice Problem 4 Subtract.

(a) $\begin{array}{r} 38.8 \\ -\ 26.9 \\ \hline \end{array}$ **(b)** $\begin{array}{r} 2034.908 \\ -\ 1986.325 \\ \hline \end{array}$

Teaching Tip Some students get very nervous about showing the borrowing steps as in Example 4. You will reduce their anxiety if you remind them that writing out the borrowing steps is always allowed and is very helpful for some students. They should never hesitate to add these steps if it will help them. If, however, they find that they can accurately subtract without writing the borrowing steps, then it is not necessary to write down the borrowing notation.

When the two numbers being subtracted do not have the same number of decimal places, write in zeros as needed.

Teaching Tip Students are usually surprised to realize how easy it is to get a common denominator when adding fractions or subtracting fractions that are written in decimal form. Remind them that the word *decimal* means *decimal fraction.* They usually remember that common denominators are needed for adding normal fractions but not decimal fractions. This concept is very interesting to students.

EXAMPLE 5 Subtract.

(a) 12 − 8.362 **(b)** 156.381 − 99.82

Solution

(a) $\begin{array}{r} 12.000 \\ -\ 8.362 \\ \hline 3.638 \end{array}$ (borrowing: 1 2 → 11, 0 → 10 → 9, 0 → 10 → 9, 0 → 10)

(b) $\begin{array}{r} 156.381 \\ -\ 99.820 \\ \hline 56.561 \end{array}$ (borrowing: 1 5 → 14 → 14, 6 → 5 → 15, 3 → 13)

Practice Problem 5 Subtract.

(a) 19 − 12.579 **(b)** 283.076 − 96.38

EXAMPLE 6 On Tuesday, Don Ling filled the gas tank in his car. The odometer read 56,098.5. He drove for four days. The next time he filled the tank, the odometer read 56,420.2. How many miles had he driven?

Solution

$\begin{array}{r} 56,420.2 \\ -56,098.5 \\ \hline 321.7 \end{array}$ (borrowing: 4 → 3, 2 → 1 → 11, 0 → 10 → 9, 2 → 12)

He had driven 321.7 miles.

Practice Problem 6 Abdul had his car oil changed when his car odometer read 82,370.9 miles. When he changed the oil again, the odometer read 87,160.1 miles. How many miles did he drive between oil changes?

EXAMPLE 7 Find the value of x if $x + 3.9 = 14.6$.

Solution Recall that the letter x is a variable. It represents a number that is added to 3.9 to obtain 14.6. We can find the number x if we calculate $14.6 - 3.9$.

$$\begin{array}{r} \overset{3}{\cancel{4}}\overset{16}{\cancel{6}} \\ 14.6 \\ -\ 3.9 \\ \hline 10.7 \end{array}$$

Thus $x = 10.7$.

Check. Is this true? If we replace x by 10.7, do we get a true statement?

$$x + 3.9 = 14.6$$
$$10.7 + 3.9 \stackrel{?}{=} 14.6$$
$$14.6 = 14.6 \checkmark$$

Practice Problem 7 Find the value of x if $x + 10.8 = 15.3$.

NOTE TO STUDENT:* *Fully worked-out solutions to all of the Practice Problems can be found at the back of the text starting at page SP-1

Developing Your Study Skills

Making a Friend in the Class

Attempt to make a friend in your class. You may find that you enjoy sitting together and drawing support and encouragement from each other. Exchange phone numbers so you can call each other whenever you get stuck in your study. Set up convenient times to study together on a regular basis, to do homework, and to review for exams.

You must not depend on a friend or fellow student to tutor you, do your work for you, or in any way be responsible for your learning. However, you will learn from each other as you seek to master the course. Studying with a friend and comparing notes, methods, and solutions can be very helpful. And it can make learning mathematics a lot more fun!

3.3 Exercises

Add.

1. 57.1 + 19.7
76.8

2. 78.3 + 29.4
107.7

3. 718.98 + 496.57
1215.55

4. 813.47 + 629.86
1443.33

5. 13.4 + 7.6 + 275.2
296.2

6. 176.5 + 8.4 + 22.5
207.4

7. 4.71 + 8.05
12.76

8. 9.284 + 5.77
15.054

9. 4.9637 + 28.12 + 3.645
36.7287

10. 7.0276 + 3.451 + 16.98
27.4586

11. 12. + 3.62 + 51.8
67.42

12. 13. + 4.52 + 63.7
81.22

13. 156.35 + 2.79 + 126.3 + 86
371.44

14. 172.49 + 3.52 + 138.6 + 77
391.61

15. 753.61 + 28.75 + 162.3 + 100.5 + 67
1112.16

16. 432.51 + 16.08 + 892.1 + 301.2 + 84
1725.89

Applications

In exercises 17 and 18, calculate the perimeter of each triangle.

17.

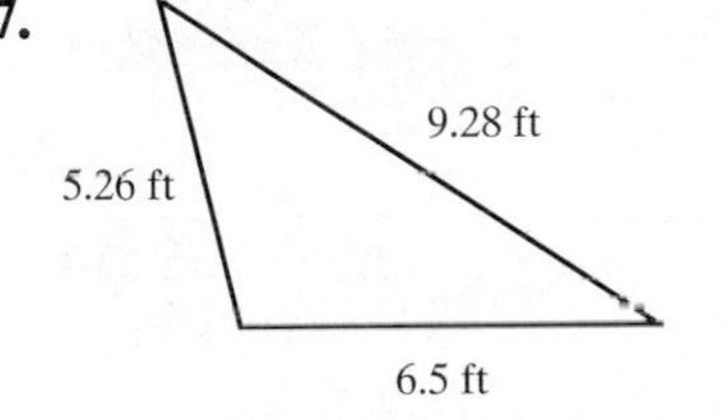

21.04 ft

18.

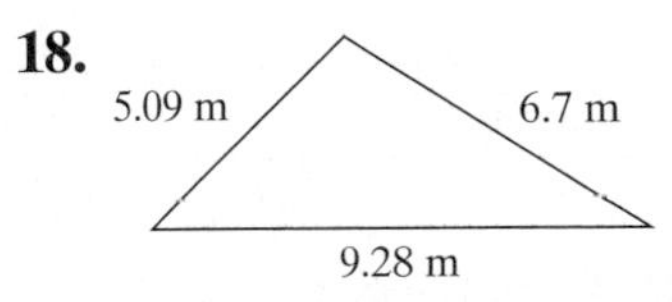

21.07 m

19. ***Weight Loss*** Lamar is losing weight by walking each evening after dinner. During the first week in February he lost 1.75 pounds. During the second, third, and fourth weeks, he lost 2.5 pounds, 1.55 pounds, and 2.8 pounds, respectively. How many total pounds did Lamar lose in February?
8.6 pounds

20. ***Health*** Teresa knows she needs to drink more water while at work. One day during her morning break she drank 7.15 ounces. At lunch she drank 12.45 ounces and throughout the afternoon she drank 10.75 ounces. How many total ounces of water did she drink?
30.35 ounces

21. ***Beach Vacation*** Mick and Keith have arrived in Miami and are going to the beach. They buy sunblock for $4.99, beverages for $12.50, sandwiches for $11.85, towels for $28.50, bottled water for $3.29, and two novels for $16.99. After they got what they needed, what was Mick and Keith's bill for their day at the beach?
$78.12

22. ***Consumer Mathematics*** Erin bought school supplies at the campus bookstore. She bought pens for $3.45, a calculator for $18.25, notebooks for $6.29 and pencils for $0.89. What was the total of Erin's bill?
$28.88

23. ***Truck Travel*** A truck odometer read 46,276.0 miles before a trip of 778.9 miles. What was the final odometer reading?
47,054.9

24. ***Car Travel*** Jane traveled 1723.1 miles. The car odometer at the beginning of the trip read 23,195.0 miles. What was the final odometer reading?
24,918.1

Personal Banking *In exercises 25 and 26, a portion of a bank checking account deposit slip is shown. Add the numbers to determine the total deposit. The line drawn between the dollars and the cents column serves as the decimal point.*

25.

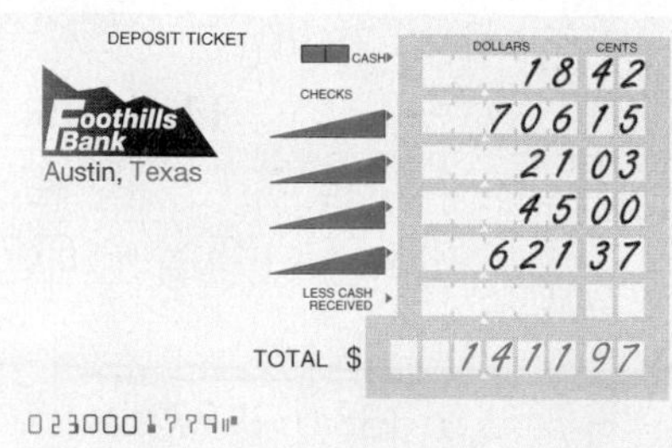

26.

Subtract.

27. 6.8 − 2.9
3.9

28. 3.6 − 2.8
0.8

29. 35.75 − 9.82
25.93

30. 84.33 − 8.09
76.24

31. 126 − 76.22
49.78

32. 209 − 81.54
127.46

33. 586.513 − 78.2
508.313

34. 243.967 − 84.2
159.767

35. 162.4 − 97.52
64.88

36. 119.6 − 23.91
95.69

37. 24.0079 − 19.3614
4.6465

38. 52.0708 − 41.9312
10.1396

39. 8 − 1.263
6.737

40. 12 − 7.981
4.019

41. 7362.14 − 6173.07
1189.07

42. 4986.71 − 3615.93
1370.78

43. 1.5 − 0.0365
1.4635

44. 2.8 − 0.07763
2.72237

Mixed Practice

Add or subtract.

45. 123.621 + 52.96
176.581

46. 241.983 + 75.48
317.463

47. 98.3 − 56.71
41.59

48. 79.2 − 45.93
33.27

49. 0.0763 + 2 + 3.16
5.2363

50. 0.0698 + 5 + 7.34
12.4098

51. 197.600 − 124.375
73.225

52. 382.700 − 291.927
90.773

Applications

53. ***BOA Constrictor*** The average length of an Emerald Tree boa constrictor is 1.8 meters. A native found one that measured 3.264 meters long. How much longer was this snake than the average Emerald Tree boa constrictor?
1.464 meters

54. ***Health*** At her 4-month checkup, baby Grace weighed 7.675 kilograms. When she was born, she weighed 3.7 kilograms. How much weight has Grace gained since she was born?
3.975 kilograms

55. ***Telescope*** A child's beginner telescope is priced at $79.49. The price of a certain professional telescope is $37,026.65. How much more does the professional telescope cost?
$36,947.16

56. ***Automobile Travel*** Tamika drove on a summer trip. When she began, the odometer read 26,052.3 miles. At the end of the trip, the odometer read 28,715.1 miles. How long was the trip?
2662.8 miles

57. ***Taxi Trip*** Malcolm took a taxi from John F. Kennedy Airport in New York to his hotel in the city. His fare was $47.70 and he tipped the driver $7.00. How much change did Malcolm get back if he gave the driver a $100 bill?
$45.30

58. ***Personal Banking*** Nathan took $150 out of the ATM. He bought shoes for $45.50, groceries for $38.72, and gas for $12.86. How much money does he have left?
$52.92

59. ***Electric Wire Construction*** An insulated wire measures 12.62 centimeters. The last 0.98 centimeter of the wire is exposed. How long is the part of the wire that is not exposed?

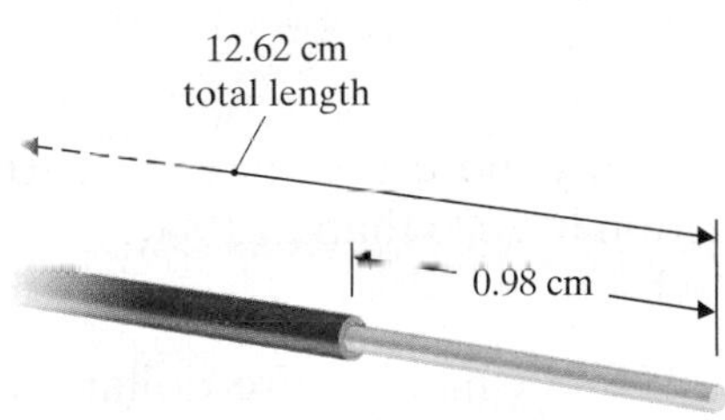

$$\begin{array}{r} 12.62 \\ -\ 0.98 \\ \hline 11.64 \end{array}$$ centimeters

60. ***Plumbing*** The outside radius of a pipe is 9.39 centimeters. The inside radius is 7.93 centimeters. What is the thickness of the pipe?

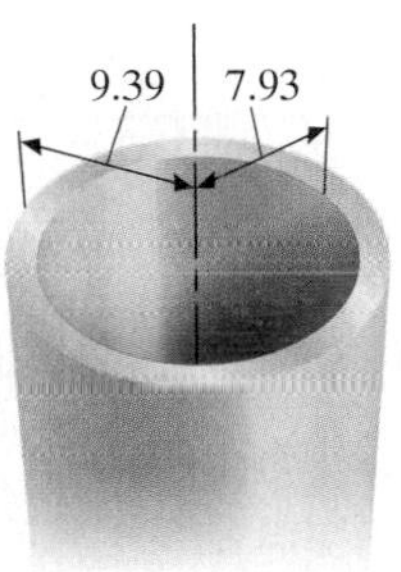

$$\begin{array}{r} 9.39 \\ -\ 7.93 \\ \hline 1.46 \end{array}$$ centimeters

61. ***Medical Research*** A cancer researcher is involved in an important experiment. She is trying to determine how much of an anticancer drug is necessary for a Stage I (nonhuman or animal) test. She pours 2.45 liters of the experimental anticancer formula in one container and 1.35 liters of a reactive liquid in another. She then pours the contents of one container into the other. If 0.85 liters is expected to evaporate during the process, how much liquid will be left?
2.95 liters

62. ***Rainforest*** Everyone is becoming aware of the rapid loss of the Earth's rainforests. Between 1981 and 2003, tropical South America lost a substantial amount of natural resources due to deforestation and development. In 1981, there were 797,100,000 hectares of rainforest. In 2003 there were 637,200,000 hectares. How much rainforest, in hectares, was destroyed? *(Source: United Nations Statistics Division)*
159,900,000 hectares

The federal water safety standard requires that drinking water contain no more than 0.015 milligram of lead per liter of water. (Source: Environmental Protection Agency)

63. ***Well Water Safety*** Carlos and Maria had the well that supplies their home analyzed for safety. A sample of well water contained 0.0089 milligram of lead per liter of water. What is the difference between their sample and the federal safety standard? Is it safe for them to drink the water?
0.0061 milligram; yes.

64. ***City Water Safety*** Fred and Donna use water provided by the city for the drinking water in their home. A sample of their tap water contained 0.023 milligram of lead per liter of water. What is the difference between their sample and the federal safety standard? Is it safe for them to drink the water?
0.008 milligram; no.

Income of Industries *The following table shows the income of the United States by industry. Use this table for exercises 65–68. Write each answer as a decimal and as a whole number. The table values are recorded in billions of dollars.*

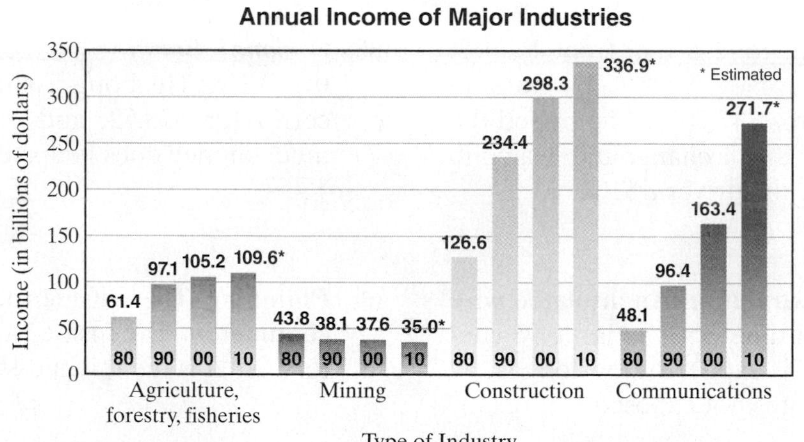

Source: Bureau of Labor Statistics

65. How many more dollars were earned in mining in 1980 than in 2000?
\$6.2 billion; \$6,200,000,000

66. How many more dollars were earned in construction in 2000 than in 1980?
\$171.7 billion; \$171,700,000,000

67. In 2010, how many more dollars will be earned in communications than in agriculture, forestry, and fisheries?
\$162.1 billion; \$162,100,000,000

68. In 1990, how many more dollars were earned in communication than in mining?
\$58.3 billion; \$58,300,000,000 dollars

To Think About

Mr. Jensen made up the following shopping list of items he needs and the cost of each item. Use the list to answer exercises 69 and 70.

can cranberry sauce $0.99
hot dog relish $0.79
ranch salad dressing $1.47
large can solid white tuna $2.29
can tomato soup $0.68
can sliced peaches $1.26
large jar tomato sauce $1.65
large box Cheerios $3.79
medium box Raisin Bran $2.63
medium jar peanut butter $2.19

69. ***Grocery Shopping*** Mr. Jensen goes to the store to buy the following items from his list: Raisin Bran, ranch salad dressing, sliced peaches, hot dog relish, and peanut butter. He has a ten-dollar bill. Round each item to the nearest ten cents and estimate the cost of buying these items by first rounding the cost of each item to the nearest ten cents. Does he have enough money to buy all of them? Find the exact cost of these items. How close was your estimate?
$8.40; yes, $8.34; very close: the estimate was off by 6¢

70. ***Grocery Shopping*** The next day the Jensens' daughter, Brenda, goes to the store to buy the following items from the list: Cheerios, tomato sauce, peanut butter, white tuna, tomato soup, and cranberry sauce. She has fifteen dollars. Estimate the cost of buying these items by first rounding the cost of each item to the nearest ten cents. Does she have enough money to buy all of them? Find the exact cost of these items. How close was your estimate?
$11.70; yes; $11.59; very close: the estimate was off by 11¢

Find the value of x.

71. $x + 7.1 = 15.5$
$x = 8.4$

72. $x + 4.8 = 23.1$
$x = 18.3$

73. $156.9 + x = 200.6$
$x = 43.7$

74. $210.3 + x = 301.2$
$x = 90.9$

75. $4.162 = x + 2.053$
$x = 2.109$

76. $7.076 = x + 5.602$
$x = 1.474$

Cumulative Review

Multiply.

77. 2536×8
20,288

78. 827×59
48,793

79. $\frac{22}{7} \times \frac{49}{50}$
$\frac{77}{25}$ or $3\frac{2}{25}$

80. $2\frac{1}{3} \times 3\frac{3}{4}$
$\frac{35}{4}$ or $8\frac{3}{4}$

3.4 Multiplying Decimals

Student Learning Objectives

After studying this section, you will be able to:

1. **Multiply a decimal by a decimal or a whole number.**
2. **Multiply a decimal by a power of 10.**

1 Multiplying a Decimal by a Decimal or a Whole Number

We learned previously that the product of two fractions is the product of the numerators over the product of the denominators. For example,

$$\frac{3}{10} \times \frac{7}{100} = \frac{21}{1000}$$

In decimal form this product would be written

$$\underbrace{0.3}_{\text{one decimal place}} \times \underbrace{0.07}_{\text{two decimal places}} = \underbrace{0.021}_{\text{three decimal places}}$$

MULTIPLICATION OF DECIMALS

1. Multiply the numbers just as you would multiply whole numbers.
2. Find the sum of the decimal places in the two factors.
3. Place the decimal point in the product so that the product has the same number of decimal places as the sum in step 2. You may need to write zeros to the left of the number found in step 1.

Now use these steps to do the preceding multiplication problem.

EXAMPLE 1 Multiply. 0.07 × 0.3

Solution

0.07	2 decimal places
× 0.3	1 decimal place
0.021	3 decimal places in product (2 + 1 = 3)

Practice Problem 1 Multiply 0.09 × 0.6

NOTE TO STUDENT: Fully worked-out solutions to all of the Practice Problems can be found at the back of the text starting at page SP-1

When performing the calculation, it is usually easier to place the factor with the smallest number of nonzero digits underneath the other factor.

EXAMPLE 2 Multiply.

(a) 0.38 × 0.26 **(b)** 12.64 × 0.572

Solution

(a)

0.38	2 decimal places
× 0.26	2 decimal places
228	
76	
0.0988	4 decimal places (2 + 2 = 4)

Note that we need to insert a zero before the 988.

(b)

12.64	2 decimal places
× 0.572	3 decimal places
2528	
8848	
6 320	
7.23008	5 decimal places (2 + 3 = 5)

Calculator

Multiplying Decimals

You can use your calculator to multiply a decimal by a decimal. To find 0.08 × 1.53 enter:

0.08 [×] 1.53 [=]

Display:

[0.1224]

Practice Problem 2 Multiply.

(a) 0.47×0.28 **(b)** 0.436×18.39

When multiplying decimal fractions by a whole number, you need to remember that a whole number has no decimal places.

EXAMPLE 3 Multiply. 5.261×45

Solution

5.261	3 decimal places
× 45	0 decimal places
26 305	
210 44	
236.745	3 decimal places (3 + 0 = 3)

Practice Problem 3 Multiply. 0.4264×38

▲ **EXAMPLE 4** Uncle Roger's rectangular front lawn measures 50.6 yards wide and 71.4 yards long. What is the area of the lawn in square yards?

Solution Since the lawn is rectangular, we will use the fact that to find the area of a rectangle we multiply the length by the width.

71.4	1 decimal place
× 50.6	1 decimal place
42 84	
3570 0	
3612.84	2 decimal places

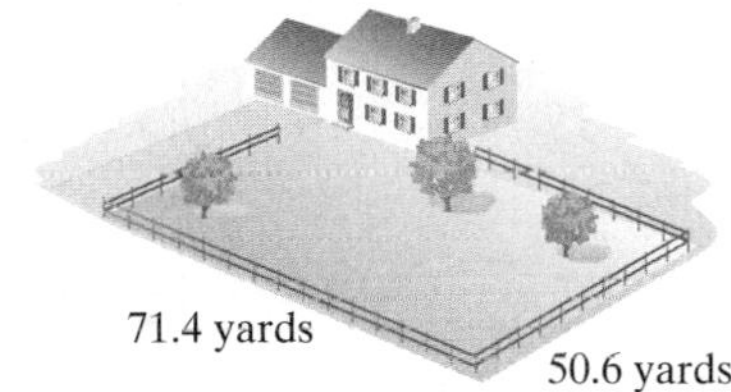

The area of the lawn is 3612.84 square yards.

▲ **Practice Problem 4** A rectangular computer chip measures 1.26 millimeters wide and 2.3 millimeters long. What is the area of the chip in square millimeters?

2 Multiplying a Decimal by a Power of 10

Observe the following pattern.

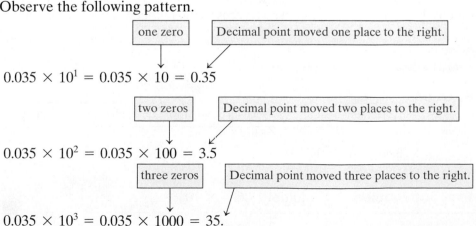

one zero — Decimal point moved one place to the right.

$0.035 \times 10^1 = 0.035 \times 10 = 0.35$

two zeros — Decimal point moved two places to the right.

$0.035 \times 10^2 = 0.035 \times 100 = 3.5$

three zeros — Decimal point moved three places to the right.

$0.035 \times 10^3 = 0.035 \times 1000 = 35.$

Teaching Tip Multiplication of a decimal by a power of 10 is a simple idea. However, it will be totally new to some students. Be sure to do an example such as this one: Multiply $123.8765 \times 100{,}000 = 12{,}387{,}650$ "in your head." Ask students what rule allowed them to do that problem. You will usually get several enthusiastic suggestions from the class. Students like the idea of this type of shortcut.

MULTIPLICATION OF A DECIMAL BY A POWER OF 10

To multiply a decimal by a power of 10, move the decimal point to the right the same number of places as the number of zeros in the power of 10.

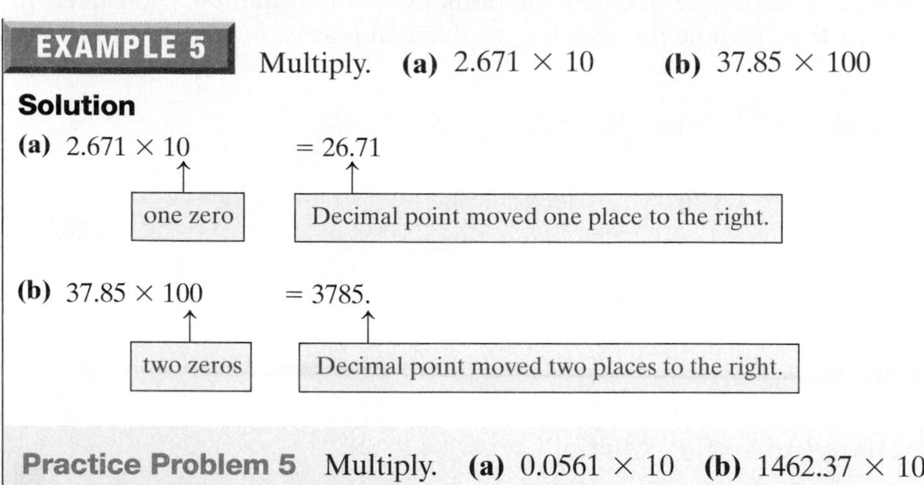

EXAMPLE 5 Multiply. **(a)** 2.671×10 **(b)** 37.85×100

Solution

(a) $2.671 \times 10 = 26.71$

one zero — Decimal point moved one place to the right.

(b) $37.85 \times 100 = 3785.$

two zeros — Decimal point moved two places to the right.

Practice Problem 5 Multiply. **(a)** 0.0561×10 **(b)** 1462.37×100

***NOTE TO STUDENT:* Fully worked-out solutions to all of the Practice Problems can be found at the back of the text starting at page SP-1**

Sometimes it is necessary to add extra zeros before placing the decimal point in the answer.

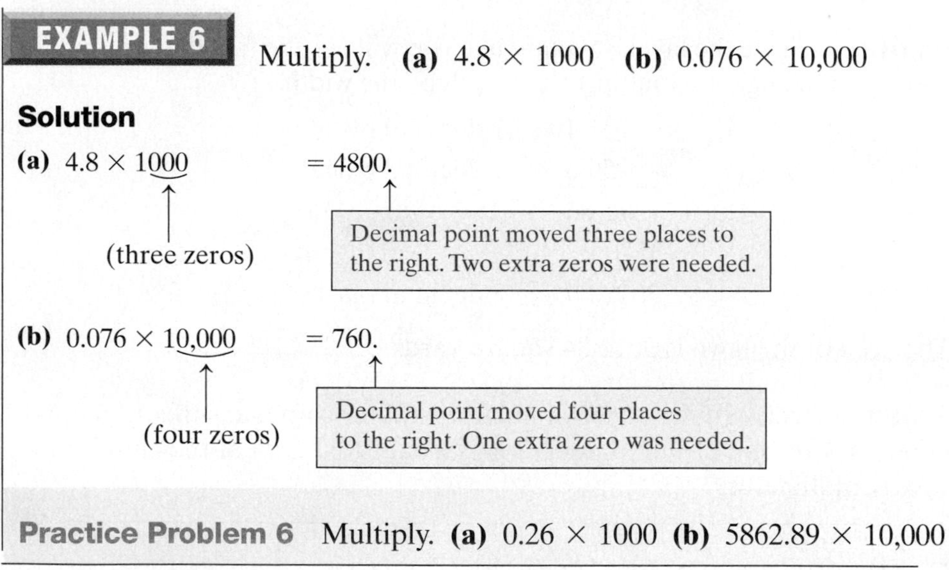

EXAMPLE 6 Multiply. **(a)** 4.8×1000 **(b)** $0.076 \times 10{,}000$

Solution

(a) $4.8 \times 1000 = 4800.$

(three zeros) — Decimal point moved three places to the right. Two extra zeros were needed.

(b) $0.076 \times 10{,}000 = 760.$

(four zeros) — Decimal point moved four places to the right. One extra zero was needed.

Practice Problem 6 Multiply. **(a)** 0.26×1000 **(b)** $5862.89 \times 10{,}000$

If the number that is a power of 10 is in exponent form, move the decimal point to the right the same number of places as the number that is the exponent.

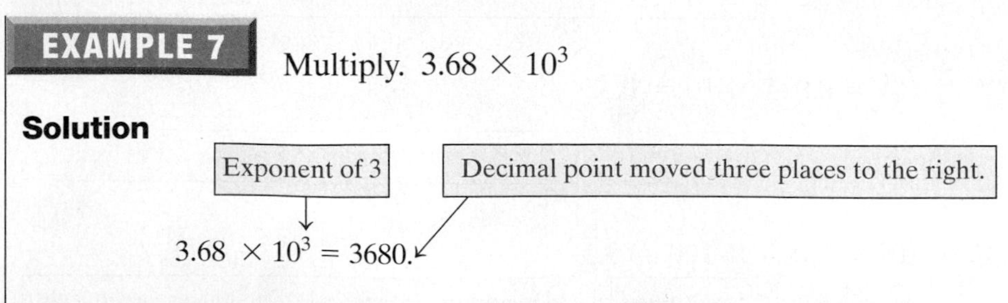

EXAMPLE 7 Multiply. 3.68×10^3

Solution

Exponent of 3 — Decimal point moved three places to the right.

$3.68 \times 10^3 = 3680.$

Practice Problem 7 Multiply. 7.684×10^4

SIDELIGHT: Moving the Decimal Point

Can you devise a quick rule to use when multiplying a decimal fraction by $\frac{1}{10}$, $\frac{1}{100}$, $\frac{1}{1000}$, and so on? How is it like the rules developed in this section? Consider a few examples

Original Problem	Change Fraction to Decimal	Decimal Multiplication	Observation
$86 \times \frac{1}{10}$	86×0.1	$\begin{array}{r} 86 \\ \times\ 0.1 \\ \hline 8.6 \end{array}$	Decimal point moved one place to the left.
$86 \times \frac{1}{100}$	86×0.01	$\begin{array}{r} 86 \\ \times\ 0.01 \\ \hline 0.86 \end{array}$	Decimal point moved two places to the left.
$86 \times \frac{1}{1000}$	86×0.001	$\begin{array}{r} 86 \\ \times\ 0.001 \\ \hline 0.086 \end{array}$	Decimal point moved three places to the left.

Teaching Tip You can expand on the discussion in Example 8 with the following example: "To convert from kilometers to centimeters, multiply by 100,000. How many centimeters are there in 3457.39 kilometers?" The answer is 345,739,000 centimeters. Ask the students to imagine how difficult the problem would have been had it been formulated in terms of changing miles to inches. This will help them to appreciate the advantages of the metric system.

Can you think of a way to describe a rule that you could use in solving this type of problem without going through all the foregoing steps?

You use multiplying by a power of 10 when you convert a larger unit of measure to a smaller unit of measure in the metric system.

EXAMPLE 8 Change 2.96 kilometers to meters.

Solution Since we are going from a larger unit of measure to a smaller one, we multiply. There are 1000 meters in 1 kilometer. Multiply 2.96 by 1000.

$$2.96 \times 1000 = 2960$$

2.96 kilometers is equal to 2960 meters.

1 kilometer

1000 meters

Practice Problem 8 Change 156.2 kilometers to meters.

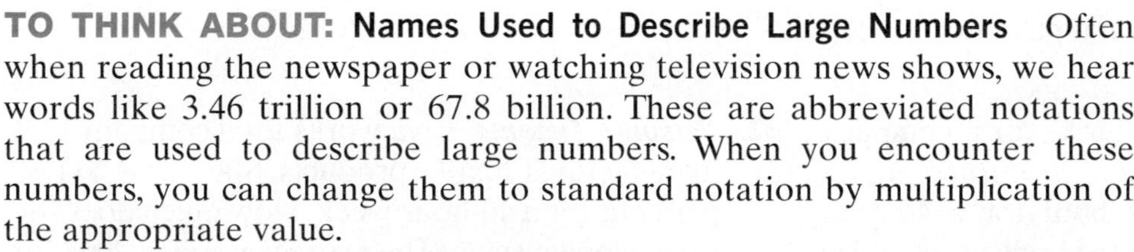

TO THINK ABOUT: Names Used to Describe Large Numbers Often when reading the newspaper or watching television news shows, we hear words like 3.46 trillion or 67.8 billion. These are abbreviated notations that are used to describe large numbers. When you encounter these numbers, you can change them to standard notation by multiplication of the appropriate value.

For example, if someone says that the population of China is 1.31 billion people, we can write 1.31 billion = 1.31 × 1 billion = 1.31 × 1,000,000,000 = 1,310,000,000. If someone says the population of Chicago is 2.92 million people, we can write

$$2.92 \text{ million} = 2.92 \times 1 \text{ million} = 2.92 \times 1{,}000{,}000 = 2{,}920{,}000.$$

3.4 Exercises

Student Solutions Manual | CD/Video | PH Math Tutor Center | MathXL®Tutorials on CD | MathXL® | MyMathLab® | Interactmath.com

Multiply.

Verbal and Writing Skills

1. Explain in your own words how to determine where to put the decimal point in the answer when you multiply 0.67 × 0.08.

Each factor has two decimal places. You add the number of decimal places to get four decimal places. You multiply 67 × 8 to obtain 536. Now you must place the decimal point four places to the left in your answer. The result is 0.0536.

2. Explain in your own words how to determine where to put the decimal point in the answer when you multiply 3.45 × 0.9.

The first factor has two decimal places. The second factor has one. You add the number of decimal places to get three decimal places. You multiply 345 × 9 to obtain 3105. Now you must place the decimal three places to the left in your answer. The result is 3.105.

3. Explain in your own words how to determine where to put the decimal point in the answer when you multiply 0.0078 × 100.

When you multiply a number by 100 you move the decimal point two places to the right. The answer is 0.78.

4. Explain in your own words how to determine where to put the decimal point in the answer when you multiply 5.0807 by 1000.

When you multiply a number by 1000 you move the decimal point three places to the right. The answer is 5080.7.

5. 0.6 × 0.2 — 0.12

6. 0.9 × 0.3 — 0.27

7. 0.12 × 0.5 — 0.06

8. 0.17 × 0.4 — 0.068

9. 0.0036 × 0.8 — 0.00288

10. 0.067 × 0.07 — 0.00469

11. 452 × 0.12 — 54.24

12. 316 × 0.24 — 75.84

13. 0.043 × 0.012 — 0.000516

14. 0.037 × 0.011 — 0.000407

15. 10.97 × 0.06 — 0.6582

16. 18.07 × 0.05 — 0.9035

17. 5167 × 0.19 — 981.73

18. 7986 × 0.32 — 2555.52

19. 2.163 × 0.008 — 0.017304

20. 1.892 × 0.007 — 0.013244

21. 0.7613 × 1009 — 768.1517

22. 0.6178 × 5004 — 3091.4712

23. 2350 × 3.6 — 8460

24. 3720 × 8.1 — 30,132

25. 4.57 × 11.8 — 53.926

26. 73.2 × 2.45 — 179.34

27. 0.001 × 6523.7 — 6.5237

28. 0.01 × 826.75 — 8.2675

Applications

29. ***Car Payments*** Kenny is making payments on his Ford Escort of $155.40 per month for the next 60 months. How much will he have spent in car payments after he sends in his final payment?

$9324

30. ***Food Purchase*** Each carton of ice cream contains 1.89 liters. Paul stocked his freezer with 25 cartons. How many total liters of ice cream did he buy?

47.25 liters

31. ***Personal Income*** Mei Lee works for a company that manufactures electric and electronic equipment and earns $14.70 per hour for a 40-hour week. How much does she earn in one week? (The average wage in 2002 for U.S. electric/electronic equipment manufacturers was $14.53 per hour for an average of 41.3 hours, for a total of $600.09 per week.) (*Source*: Bureau of Labor Statistics)

$588

32. ***Personal Income*** Elva works for a company that manufactures textile products. She earns $11.80 per hour for a 40-hour week. How much does she earn in one week? (The average wage in 2002 for U.S. textile mill industries was $11.35 per hour for an average of 40.6 hours, for a total of $460.81 per week. (*Source*: Bureau of Labor Statistics)

$472

▲ **33.** ***Geometry*** Ralph and Darlene are getting new carpet in their bedroom and need to find how many square feet they need to purchase. The dimensions of their rectangular bedroom are 15.5 feet and 19.2 feet. What is the area of the room in square feet?
297.6 square feet

▲ **34.** ***Geometry*** Sal is having his driveway paved by a company that charges by the square yard. Sal's driveway measures 8.6 yards by 17.5 yards. How many square yards is his driveway?
150.5 square yards

35. ***Student Loan*** Dwight is paying off a student loan at Westmont College with payments of $36.90 per month for the next 18 months. How much will he pay off during the next 18 months?
$664.20

36. ***Car Payments*** Marcia is making car payments to Highfield Center Chevrolet of $230.50 per month for 16 more months. How much will she pay for car payments in the next 16 months?
$3688.00

37. ***Fuel Efficiency*** Steve's car gets approximately 26.4 miles per gallon. His gas tank holds 19.5 gallons. Approximately how many miles can he travel on a full tank of gas?
514.8 miles

38. ***Fuel Efficiency*** Jim's 4×4 truck gets approximately 18.6 miles per gallon. His gas tank holds 19.5 gallons. Approximately how many miles can he travel on a full tank of gas? Compare this to your answer in exercise 37.
362.7 miles; Steve can travel 152.1 miles further than Jim on a tank of gas.

Multiply.

39. 2.86×10
28.6

40. 1.98×10
19.8

41. 52.0×100
5200

42. 83.0×100
8300

43. 22.615×1000
22,615

44. 34.105×1000
34,105

45. $5.60982 \times 10{,}000$
56,098.2

46. $1.27986 \times 10{,}000$
12,798.6

47. $17{,}561.44 \times 10^2$
1,756,144

48. 7163.241×10^2
716,324.1

49. 816.32×10^3
816,320

50. 763.49×10^4
7,634,900

Applications

51. ***Metric Conversion*** To convert from meters to centimeters, multiply by 100. How many centimeters are in 5.932 meters?
593.2 centimeters

52. ***Metric Conversion*** One meter is about 39.36 inches. About how many inches are in 100 meters?
3936 inches

53. ***Metric Conversion*** To convert from kilometers to meters, multiply by 1000. How many meters are in 2.98 kilometers?
2980 meters

54. ***Stock Market*** Jeremy bought 1000 shares of stock each worth $1.45. How much did Jeremy spend on the stock?
$1450

55. ***Personal Finance*** In June, Gabrielle receives $820.00 on her tax return. She decides to spend her money on family holiday gifts, six months early, so that she doesn't have to worry about it later. She spends $124.00 on a gift for her parents, $110.00 on a gift for her sister, $83.60 on a gift for her brother, $76.00 on a gift for her grandmother, and $44.60 for a gift for each of her four cousins. How much money does she have left over.
$248.00

56. ***Pet Cats*** Tomba is a beautiful orange tabby cat. When he was found by the side of the road, he was three weeks old and weighed 0.95 lb. At the age of three months, he weighed 2.85 lb. At the age of nine months, he weighed 6.30 lb; at one year, he weighed 11.7 lb. Today, Tomba the cat is $1\frac{1}{2}$ years old, and weighs 15.75 lb.

(a) How much weight did he gain?

(b) If the veterinarian wants him to lose 0.25 lb per week until he weighs 13.5 lb, how long will it take?

(a) 14.8 lb (b) nine weeks

▲ **57.** ***Geometry*** The college is purchasing new carpeting for the learning center. What is the price of a carpet that is 19.6 yards wide and 254.2 yards long if the cost is $12.50 per square yard?

$$\begin{array}{r} 254.2 \\ \times\ 19.6 \\ \hline 4982.32 \text{ square yards} \end{array} \qquad \begin{array}{r} 4982.32 \\ \times\ 12.5 \\ \hline \$62{,}279.00 \end{array}$$

58. ***Jewelry Store Operations*** A jewelry store purchased long lengths of gold chain, which will be cut and made into necklaces and bracelets. The store purchased 3220 grams of gold chain at $3.50 per gram.

(a) How much did the jewelry store spend?
$11,270

(b) If they sell a 28-gram gold necklace for $17.75 per gram, how much profit will they make on the necklace?
$399

To Think About

59. State in your own words a rule for mental multiplication by 0.1, 0.01, 0.001, 0.0001, and so on.
To multiply by numbers such as 0.1, 0.01, 0.001, and 0.0001, count the number of decimal places in this first number. Then, in the other number, move the decimal point to the left from its present position the same number of decimal places as was in the first number.

60. State in your own words a rule for mental multiplication by 0.2, 0.02, 0.002, 0.0002, and so on.
To multiply by numbers such as 0.2, 0.02, 0.002, and 0.0002, first double the second number. Then move the decimal point to the left using the rule stated in exercise 59.

Cumulative Review

Divide. Be sure to include any remainder as part of your answer.

61. $17\overline{)1462}$
86

62. $26\overline{)2418}$
93

63. $48\overline{)6099}$
127 R 3

64. $124\overline{)56{,}024}$
451 R 100

Hurricane Damage Relief *The relief costs for the five most devastating hurricanes to hit the United States are represented in the following bar graph. Use the bar graph to answer exercises 65–68.*

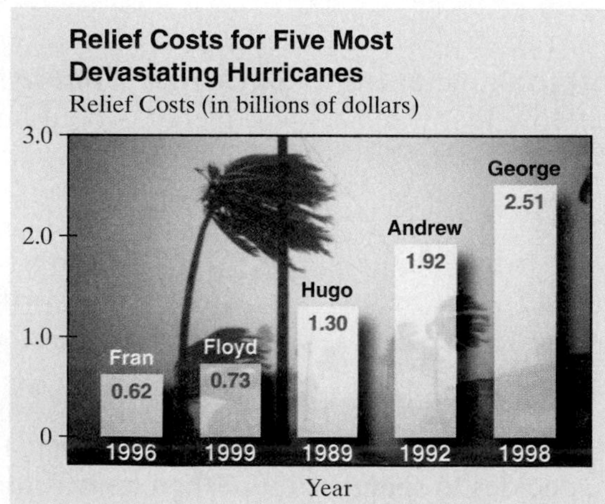

65. How much more were the relief costs for Hurricane Floyd than Hurricane Fran?
$0.11 billion or $110,000,000

66. How much more were the relief costs for Hurricane Hugo than Hurricane Floyd?
$0.57 billion or $570,000,000

67. FEMA has projected a contingency budget for 2005 that is $1.963 billion more than the relief budget for Hurricane George. What is that total amount of money?
$4.473 billion

68. FEMA has projected a contingency budget for 2006 that is $3.618 billion more than the relief budget for Hurricane Hugo. What is that total amount of money?
$4.918 billion

How Am I Doing? Sections 3.1–3.4

How are you doing with your homework assignments in Sections 3.1 to 3.4? Do you feel you have mastered the material so far? Do you understand the concepts you have covered? Before you go further in the textbook, take some time to do each of the following problems.

3.1

1. Write a word name for the decimal. 47.813

2. Express as a decimal. $\frac{567}{10{,}000}$

Write as a fraction or a mixed number. Reduce whenever possible.

3. 2.11 **4.** 0.525

3.2

5. Place the set of numbers in the proper order from smallest to largest. 1.6, 1.59, 1.61, 1.601

6. Round to the nearest tenth. 123.49268

7. Round to the nearest thousandth. 1.053458

8. Round to the nearest hundredth. 17.98523

3.3

Add.

9. $5.12 + 4.7 + 8.03 + 1.6$

10. $24.613 + 0.273 + 2.305$

Subtract.

11. $42.16 - 31.57$ **12.** $26 - 18.329$

3.4

Multiply.

13. 11.67×0.03 **14.** 4.7805×1000 **15.** 0.0003796×10^5

16. 0.768×0.085 **17.** 982×0.007 **18.** 0.00052×0.006

Now turn to page SA-7 for the answer to each of these problems. Each answer also includes a reference to the objective in which the problem is first taught. If you missed any of these problems, you should stop and review the Examples and Practice Problems in the referenced objective. A little review now will help you master the material in the upcoming sections of the text.

1. forty-seven and eight hundred thirteen thousandths
2. 0.0567
3. $2\frac{11}{100}$
4. $\frac{21}{40}$
5. 1.59, 1.6, 1.601, 1.61
6. 123.5
7. 1.053
8. 17.99
9. 19.45
10. 27.191
11. 10.59
12. 7.671
13. 0.3501
14. 4780.5
15. 37.96
16. 0.06528
17. 6.874
18. 0.00000312

3.5 Dividing Decimals

Student Learning Objectives

After studying this section, you will be able to:

 **Divide a decimal by a whole number.**

 **Divide a decimal by a decimal.**

Teaching Tip This is a good time to remind students that they will be very glad they learned these three words: *divisor, dividend, quotient.* They are referred to frequently in mathematics. You may want to put a diagram on the board as a reminder:

$$\begin{array}{r} \text{quotient} \\ \text{divisor}\overline{)\text{dividend}} \end{array}$$

1 Dividing a Decimal by a Whole Number

When you divide a decimal by a whole number, place the decimal point for the quotient directly above the decimal point in the dividend. Then divide as if the numbers were whole numbers.

To divide 26.8 by 4, we place the decimal point of our answer (the quotient) directly *above* the decimal point in the dividend.

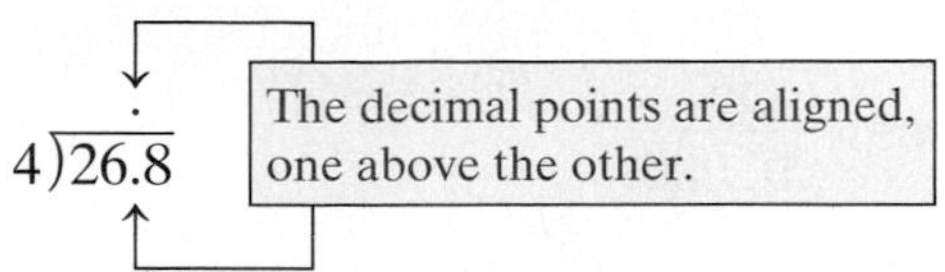

4)26.8
The decimal points are aligned, one above the other.

Then we divide as if we were dividing whole numbers.

$$\begin{array}{r} 6.7 \\ 4\overline{)26.8} \\ \underline{24} \\ 28 \\ \underline{28} \\ 0 \end{array}$$

The quotient is 6.7.

The quotient to a problem may have all digits to the right of the decimal point. In some cases you will have to put a zero in the quotient as a "place holder." Let's divide 0.268 by 4.

$$\begin{array}{r} 0.067 \\ 4\overline{)0.268} \\ \underline{24} \\ 28 \\ \underline{28} \\ 0 \end{array}$$

Note that we must have a zero after the decimal point in 0.067.

EXAMPLE 1 Divide.

(a) $9\overline{)0.3204}$ **(b)** $14\overline{)36.12}$

Solution

Note the zero *after* the decimal point.

(a)
$$\begin{array}{r} 0.0356 \\ 9\overline{)0.3204} \\ \underline{27} \\ 50 \\ \underline{45} \\ 54 \\ \underline{54} \\ 0 \end{array}$$

(b)
$$\begin{array}{r} 2.58 \\ 14\overline{)36.12} \\ \underline{28} \\ 81 \\ \underline{70} \\ 112 \\ \underline{112} \\ 0 \end{array}$$

Practice Problem 1 Divide.

(a) $7\overline{)1.806}$ **(b)** $16\overline{)0.0928}$

NOTE TO STUDENT: *Fully worked-out solutions to all of the Practice Problems can be found at the back of the text starting at page SP-1*

Some division problems do not yield a remainder of zero. In such cases, we may be asked to round off the answer to a specified place. To round off, we carry out the division until our answer contains a digit that is one place to the right of that to which we intend to round. Then we round our answer to the specified place. For example, to round to the nearest thousandth, we carry out the division to the ten-thousandths place. In some division problems, you will need to write in zeros at the end of the dividend so that this division can be carried out.

Teaching Tip It is helpful for most students to see a variety of division problems with decimals where the divisor is an integer and to master the ability to perform each division before going onward. Be sure to discuss carefully Examples 1 and 2 or similar problems before moving on to the case where the divisor is a decimal fraction.

EXAMPLE 2 Divide and round the quotient to the nearest thousandth.

$$12.67 \div 39$$

Solution We will carry out our division to the ten-thousandths place. Then we will round our answer to the nearest thousandth.

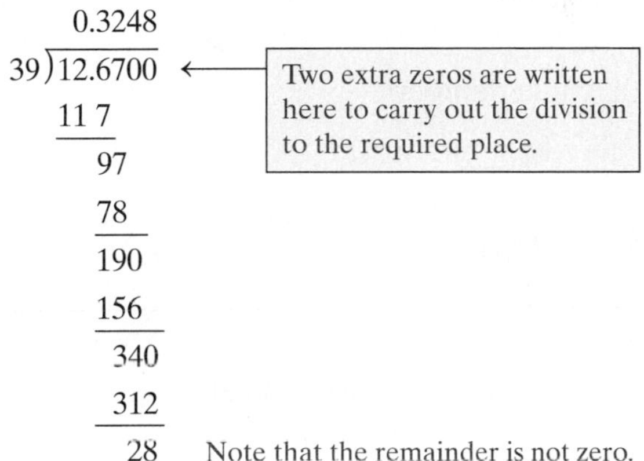

Now we round 0.3248 to 0.325. The answer is rounded to the nearest thousandth.

Practice Problem 2 Divide and round the quotient to the nearest hundredth. $23.82 \div 46$

EXAMPLE 3 Maria paid \$5.92 for 16 pounds of tomatoes. How much did she pay per pound?

Solution The cost of one pound of tomatoes equals the total cost, \$5.92, divided by 16 pounds. Thus we will divide.

```
     0.37     Maria paid
16)5.92      $0.37 per pound
   4 8       for the tomatoes.
   112
   112
     0
```

Practice Problem 3 Won Lin will pay off his auto loan for \$3538.75 over 19 months. If the monthly payments are equal, how much will he pay each month?

2 Dividing a Decimal by a Decimal

When the divisor is not a whole number, we can convert the division problem to an equivalent problem that has a whole number as a divisor. Think about the reasons why this procedure will work. We will ask you about it after you study Examples 4 and 5.

DIVIDING A DECIMAL BY A DECIMAL

1. Make the divisor a whole number by moving the decimal point to the right. Mark that position with a caret ($_\wedge$). Count the number of places the decimal point moved.
2. Move the decimal point in the dividend to the right the same number of places. Mark that position with a caret.
3. Place the decimal point of your answer directly above the caret marking the decimal point of the dividend.
4. Divide as with whole numbers.

EXAMPLE 4 **(a)** Divide. $0.08\overline{)1.632}$ **(b)** Divide. $1.352 \div 0.026$

Solution

(a) $0.08.\overline{)1.63.2}$ Move each decimal point two places to the right.

Place the decimal point of the answer directly above the caret.

$0.08_\wedge\overline{)1.63_\wedge 2}$ Mark the new position by a caret ($_\wedge$).

$$
\begin{array}{r}
20.4 \\
0.08_\wedge\overline{)1.63_\wedge 2} \\
\underline{16} \\
3\;2 \\
\underline{3\;2} \\
0
\end{array}
$$

The answer is 20.4.

Perform the division.

(b)
$$
\begin{array}{r}
52. \\
0.026_\wedge\overline{)1.352_\wedge} \\
\underline{1\;30} \\
52 \\
\underline{52} \\
0
\end{array}
$$

Move each decimal point three places to the right and mark the new position by a caret.

The answer is 52.

NOTE TO STUDENT: Fully worked-out solutions to all of the Practice Problems can be found at the back of the text starting at page SP-1

Practice Problem 4 Divide.

(a) $0.09\overline{)0.1008}$ **(b)** $1.702 \div 0.037$

TO THINK ABOUT: The Multiplicative Identity Why do we move the decimal point to the right in the divisor and the dividend? What rule allows us to do this? How do we know the answer will be valid? We are actually using the property that multiplication of a fraction by 1 leaves the fraction unchanged. This is called the *multiplication identity.* Let us examine Example 4(b) again. We will write 1.352 ÷ 0.026 as a fraction.

$$\frac{1.352}{0.026} \times 1$$ Multiplication of a fraction by 1 does not change the value of the fraction.

$$= \frac{1.352}{0.026} \times \frac{1000}{1000}$$ We know that $\frac{1000}{1000} = 1$.

$$= \frac{1352}{26}$$ Multiplication by 1000 can be done by moving the decimal point three places to the right.

$$= 52$$ Divide the whole numbers.

Thus in Example 4(b) when we moved the decimal point three places to the right in the divisor and the dividend, we were actually creating an equivalent fraction where the numerator and the denominator of the original fraction were multiplied by 1000.

Teaching Tip Sometimes students find that they need to work out some additional examples like Example 5 until they are sure of themselves. If you see a need for this, have them do 0.03154 ÷ 0.019 (the answer is 1.66) and 1.2999 ÷ 0.0007 (the answer is 1857).

EXAMPLE 5 Divide.

(a) $1.7\overline{)0.0323}$ **(b)** $0.0032\overline{)7.68}$

Solution

(a)
```
          0.019
1.7^)0.0^323
         17
         153
         153
           0
```
Move the decimal point in the divisor and dividend one place to the right and mark that position with a caret.

(b)
```
                2400.
0.0032^)7.6800^
          6 4
          1 28
          1 28
            000
```
Note that two extra zeros are needed in the dividend as we move the decimal point four places to the right.

Practice Problem 5 Divide.

(a) $1.8\overline{)0.0414}$ **(b)** $0.0036\overline{)8.316}$

EXAMPLE 6 **(a)** Find $2.9\overline{)431.2}$ rounded to the nearest tenth

(b) Find $2.17\overline{)0.08}$ rounded to the nearest thousandth.

Calculator

Dividing Decimals

You can use your calculator to divide a decimal by a decimal. To find $21.38\overline{)54.53}$ rounded to the nearest hundredth, enter:

54.53 [÷] 21.38 [=]

Display:

[2.5505145]

This is an approximation. Some calculators will round to eight digits. The answer rounded to the nearest hundredth is 2.55.

Teaching Tip You may want to ask the students if they are familiar with other rules for rounding off numbers. Particularly if you have students from other countries or cultures, you may see some unusual round-off rules. Remind them that this rule for rounding decimals is used uniformly in mathematics courses in the United States and that they will need to be able to use it.

NOTE TO STUDENT: Fully worked-out solutions to all of the Practice Problems can be found at the back of the text starting at page SP-1

Solution

(a)

```
          148.68
2.9∧)431.2∧00
     29
     141
     116
      25 2
      23 2
       2 0 0
       1 7 4
         2 60
         2 32
           28
```

Calculate to the hundredths place and round the answer to the nearest tenth.

The answer rounded to the nearest tenth is 148.7.

(b)

```
           0.0368
2.17∧)0.08∧0000
           6 51
           1 490
           1 302
            1880
            1736
             144
```

Calculate to the ten-thousandths place and then round the answer. Rounding 0.0368 to the nearest thousandth, we obtain 0.037.

Practice Problem 6 **(a)** Find $3.8\overline{)521.6}$ rounded to the nearest tenth. **(b)** Find $8.05\overline{)0.17}$ rounded to the nearest thousandth.

EXAMPLE 7 John drove his 1997 Cavalier 420.5 miles to Chicago. He used 14.5 gallons of gas on the trip. How many miles per gallon did his car get on the trip?

Solution To find miles per gallon we need to divide the number of miles, 420.5, by the number of gallons, 14.5.

```
          2 9.
14.5∧)420.5∧
      290
      1305
      1305
```

John's car achieved 29 miles per gallon on the trip to Chicago.

Practice Problem 7 Sarah rented a large truck to move to Boston. She drove 454.4 miles yesterday. She used 28.5 gallons of gas on the trip. How many miles per gallon did the rental truck get? Round to the nearest tenth.

EXAMPLE 8 Find the value of n if $0.8 \times n = 2.68$.

Solution Here 0.8 is multiplied by some number n to obtain 2.68. What is this number n? If we divide 2.68 by 0.8, we will find the value of n.

```
        3.35
0.8∧)2.6∧80
     2 4
       2 8
       2 4
         40
         40
```

Thus the value of n is 3.35.

Check. Is this true? Are we sure the value of $n = 3.35$?
We substitute the value of $n = 3.35$ into the equation to see if it makes the statement true.

$$0.8 \times n = 2.68$$
$$0.8 \times 3.35 \stackrel{?}{=} 2.68$$
$$2.68 = 2.68 \checkmark \quad \text{Yes, it is true.}$$

Practice Problem 8 Find the value of n if $0.12 \times n = 0.696$.

EXAMPLE 9 The level of sulfur dioxide emissions in the air has slowly been decreasing over the last 20 years, as can be seen in the accompanying bar graph. Find the average amount of sulfur dioxide emissions in the air over these five specific years.

Solution

First we take the sum of the five years.

$$\begin{array}{r} 9.37 \\ 9.30 \\ 8.68 \\ 7.37 \\ +\ 6.06 \\ \hline 40.78 \end{array}$$

Then we divide by five to obtain the average.

$$\begin{array}{r} 8.156 \\ 5\overline{)40.780} \\ \underline{40} \\ 7 \\ \underline{5} \\ 28 \\ \underline{25} \\ 30 \\ \underline{30} \end{array}$$

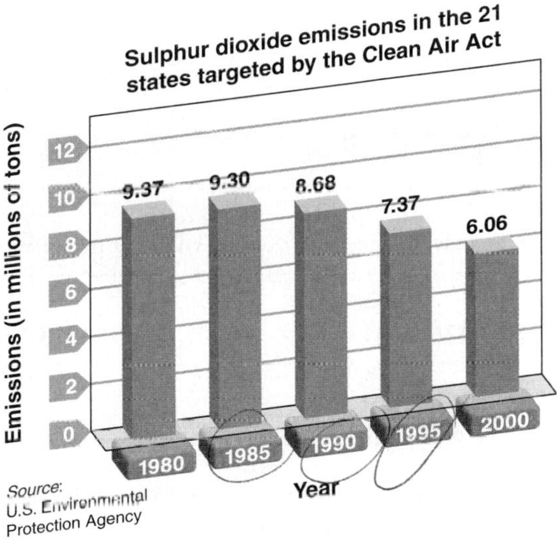

Thus the yearly average is 8.156 million tons of sulfur dioxide emissions in these 21 states.

Practice Problem 9 Use the accompanying bar graph to find the average level of sulfur dioxide for the three years: 1985, 1990, and 1995. By how much does the three-year average differ from the five-year average?

Developing Your Study Skills

Exam Time: How To Review

Reviewing adequately for an exam enables you to bring together the concepts you have learned over several sections. For your review, you will need to do the following:

1. Reread your textbook. Make a list of any terms, rules, or formulas you need to know for the exam. Be sure you understand them all.
2. Reread your notes. Go over returned homework and quizzes. Redo the problems you missed.
3. Practice some of each type of problem covered in the chapter(s) you are to be tested on. In fact, it is a good idea to construct a practice test of your own and then discuss it with a friend from class.
4. Use the end-of-chapter materials provided in your textbook. Read carefully through the Chapter Organizer. Do the Chapter Review Problems. Take the Chapter Test. When you are finished, check your answers. Redo any problems you missed.
5. Get help if any concepts give you difficulty.

3.5 Exercises

Student Solutions Manual | CD/Video | PH Math Tutor Center | MathXL®Tutorials on CD | MathXL® | MyMathLab® | Interactmath.com

Divide until there is a remainder of zero.

1. $6\overline{)12.6}$ 2.1 **2.** $8\overline{)17.28}$ 2.16 **3.** $4\overline{)71.32}$ 17.83 **4.** $6\overline{)83.16}$ 13.86

5. $7\overline{)73.64}$ 10.52 **6.** $8\overline{)168.48}$ 21.06 **7.** $0.6\overline{)81.9}$ 136.5 **8.** $0.5\overline{)32.15}$ 64.3

9. $0.362 \div 0.04$ 9.05 **10.** $0.7209 \div 0.09$ 8.01 **11.** $153.7 \div 2.9$ 53 **12.** $75.6 \div 3.6$ 21

13. $68.4 \div 3.8$ 18 **14.** $728 \div 5.6$ 130 **15.** $40.30 \div 0.31$ 130

Divide and round your answer to the nearest tenth.

16. $8\overline{)44}$ 5.5 **17.** $9\overline{)47.31}$ 5.3 **18.** $1.8\overline{)4.16}$ 2.3

19. $1.9\overline{)2.36}$ 1.2 **20.** $0.95\overline{)32.067}$ 33.8 **21.** $0.85\overline{)41.901}$ 49.3

Divide and round your answer to the nearest hundredth.

22. $4\overline{)263.82}$ 65.96 **23.** $5\overline{)471.03}$ 94.21 **24.** $1.7\overline{)20.8}$ 12.24

25. $1.8\overline{)24.41}$ 13.56 **26.** $29\overline{)4.073}$ 0.14 **27.** $36\overline{)6.125}$ 0.17

Divide and round your answer to the nearest thousandth.

28. $8\overline{)0.2019}$ 0.025 **29.** $7\overline{)0.5681}$ 0.081 **30.** $0.69\overline{)8.45}$ 12.246 **31.** $0.87\overline{)79.40}$ 91.264

Divide and round your answer to the nearest whole number.

32. $12\overline{)1396}$ 116 **33.** $19\overline{)2341}$ 123 **34.** $0.0024\overline{)0.2168}$ 90 **35.** $0.0046\overline{)0.981}$ 213

Applications

36. ***Travel in Mexico*** Rhett and Liza are traveling in Mexico, where distances on the highway are given in kilometers. There are approximately 1.6 kilometers in one mile. They see a sign that reads "Mexico City: 342 km." How many miles is it to Mexico City?
213.75 miles

37. ***Computer Payments*** The Miller family wants to use the latest technology to access the Internet from their home television system. The equipment needed to upgrade their existing equipment will cost $992.76. If the Millers make 12 equal monthly payments, how much will they pay per month?
$82.73

38. ***Lasagna Dinner*** Four students sit down to their weekly lasagna dinner. At one end of the table, there is a bottle containing 67.6 ounces of a popular soft drink. At the other end of the table is a bottle that contains 33.6 ounces of water.

(a) If the students share the soft drink and water equally, how many ounces of liquid will each student drink?

25.3 ounces

(b) At the last minute, another student is asked to join the group. How many ounces of liquid will each of the five students share?

20.24 ounces

39. ***Fuel Efficiency*** Wally owns a Plymouth Breeze that travels 360 miles on 13.2 gallons of gas. How many miles per gallon does it achieve? (Round your answer to the nearest tenth.)

approximately 27.3 miles per gallon

40. ***Costs of a Ski Trip*** The church youth group went on a ski trip. The ski resort charged the group $1200 for 32 lift tickets. How much was each ticket?

$37.50

41. ***Flower Sales*** Andrea makes Mother's Day bouquets each year for extra income. This year her goal is to make $300. If she sells each bouquet for $12.50, how many bouquets must she sell to reach her goal?

24 bouquets

42. ***Outdoor Deck Payments*** Demitri had a contractor build an outdoor deck for his back porch. He now has $1131.75 to pay off, and he agreed to pay $125.75 per month. How many more payments on the outdoor deck must he make?

9 payments

43. ***Wedding Reception Costs*** For their wedding reception, Sharon and Richard spent $1865.50 on food and drinks. If the caterer charged them $10.25 per person, how many guests did they have?

182 guests

44. ***Food Consumption***

(a) Using the chart below, find the average number of pounds of turkey consumed per person for the years 1980 and 1985.

8.65 lb

(b) What is the average increase in turkey consumption per person per year over this 25-year period?

0.292 lb

Year	U.S. Annual Per Capita Turkey Consumption (Boneless Weight)
1980	8.1 lb
1985	9.2 lb
1990	13.9 lb
1995	14.3 lb
2000	14.6 lb
2005*	15.4 lb

* estimated
Source: U.S. Department of Agriculture

45. ***Quality Inspection*** Yoshi is working as an inspector for a company that makes snowboards. A Mach 1 snowboard weighs 3.8 kilograms. How many of these snowboards are contained in a box in which the contents weigh 87.40 kilograms? If the box is labeled CONTENTS: 24 SNOWBOARDS, how great an error was made in packing the box?

23 snowboards; the error was in putting one less snowboard in the box than was required

Find the value of n.

46. $0.9 \times n = 0.3222$
0.358

47. $0.8 \times n = 5.768$
7.21

48. $1.7 \times n = 129.2$
76

49. $1.3 \times n = 1267.5$
975

50. $n \times 0.063 = 2.835$
45

51. $n \times 0.098 = 4.312$
44

To Think About

Multiply the numerator and denominator of each fraction by 10,000. Then divide the numerator by the denominator. Is the result the same if we divided the original numerator by the original denominator? Why?

yes; multiplying and dividing the numerator and denominator by 10,000 is the same as multiplying by $\frac{10{,}000}{10{,}000}$, which is 1

52. $\frac{3.8702}{0.0523}$

$\times \frac{10{,}000}{10{,}000} = \frac{38{,}702}{523} = 74$

53. $\frac{2.9356}{0.0716}$

$\times \frac{10{,}000}{10{,}000} = \frac{29{,}356}{716} = 41$

Cumulative Review

54. Add. $\frac{3}{8} + 2\frac{4}{5}$

$\frac{127}{40}$ or $3\frac{7}{40}$

55. Subtract. $2\frac{13}{16} - 1\frac{7}{8}$

$\frac{15}{16}$

56. Multiply. $3\frac{1}{2} \times 2\frac{1}{6}$

$\frac{91}{12}$ or $7\frac{7}{12}$

57. Divide. $4\frac{1}{3} \div 2\frac{3}{5}$

$\frac{5}{3}$ or $1\frac{2}{3}$

Radio Use in Europe *Use the chart at the right to answer the following questions.*

58. How many more radios per 1000 people are there in the United States than in the United Kingdom? 673

59. How many more radios per 1000 people are there in France than in Greece? 471

60. If you took an average of these five countries, how many radios are there for every 1000 people? Round to the nearest whole number if necessary. 1172

61. If you took an average of all the countries except the United States, how many radios are there for every 1000 people? Round to the nearest whole number if necessary. 936

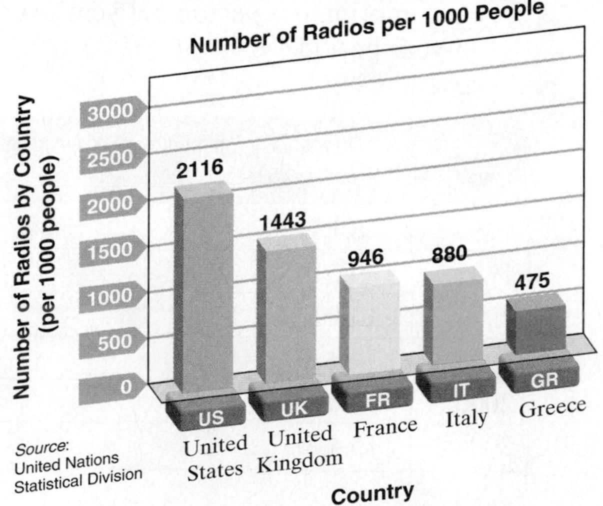

3.6 Converting Fractions to Decimals and the Order of Operations

1 Converting a Fraction to a Decimal

A number can be expressed in two equivalent forms: as a fraction or as a decimal.

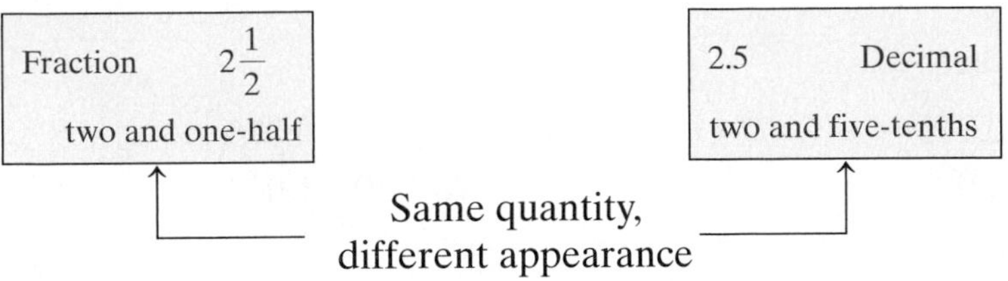

Every decimal in this chapter can be expressed as an equivalent fraction. For example,

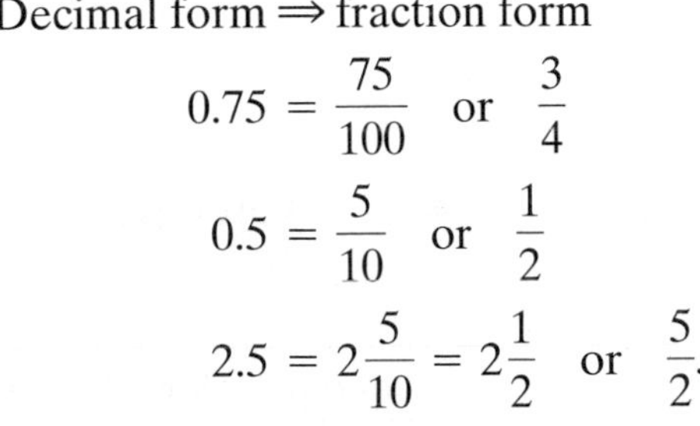

Decimal form ⇒ fraction form

$$0.75 = \frac{75}{100} \quad \text{or} \quad \frac{3}{4}$$

$$0.5 = \frac{5}{10} \quad \text{or} \quad \frac{1}{2}$$

$$2.5 = 2\frac{5}{10} = 2\frac{1}{2} \quad \text{or} \quad \frac{5}{2}.$$

And every fraction can be expressed as an equivalent decimal, as we will learn in this section. For example,

Fraction form ⇒ decimal form

$$\frac{1}{5} = 0.20 \quad \text{or} \quad 0.2$$

$$\frac{3}{8} = 0.375$$

$$\frac{5}{11} = 0.4545\ldots. \text{ (The "45" keeps repeating.)}$$

Some of these decimal equivalents are so common that people find it helpful to memorize them. You would be wise to memorize the following equivalents:

$$\frac{1}{2} = 0.5 \qquad \frac{1}{4} = 0.25 \qquad \frac{1}{5} = 0.2 \qquad \frac{1}{10} = 0.1.$$

We previously studied how to convert some fractions with a denominator of 10, 100, 1000, and so on to decimal form. For example, $\frac{3}{10} = 0.3$ and $\frac{7}{100} = 0.07$. We need to develop a procedure to write other fractions, such as $\frac{3}{8}$ and $\frac{5}{16}$, in decimal form.

> **CONVERTING A FRACTION TO AN EQUIVALENT DECIMAL**
>
> Divide the denominator into the numerator until
>
> **(a)** the remainder becomes zero, or
>
> **(b)** the remainder repeats itself, or
>
> **(c)** the desired number of decimal places is achieved.

Student Learning Objectives

After studying this section, you will be able to:

1. **Convert a fraction to a decimal.**
2. **Use the order of operations with decimals.**

Teaching Tip Sometimes a student will not see the difference in the three results discussed in Converting a Fraction to an Equivalent Decimal. You may want to give an example of each one right next to the rule.

(a) $\frac{5}{8} = 0.625$. The remainder becomes zero.

(b) $\frac{1}{3} = 0.333\ldots$. The remainder repeats itself.

(c) Rounded to the nearest thousandth, $\frac{13}{19} = 0.684$. The desired number of decimal places is achieved.

Teaching Tip Help students see the difference between the decimal 0.4787878... and the decimal 0.78787878... and the fact in general that a decimal can be a repeating decimal if all of its digits repeat like 1.23123123123... or if only some of its digits to the right of the decimal point repeat like 1.23232323.... Make sure the students put the bar over only the repeating digit or digits.

NOTE TO STUDENT: Fully worked-out solutions to all of the Practice Problems can be found at the back of the text starting at page SP-1

EXAMPLE 1 Write as an equivalent decimal.

(a) $\frac{3}{8}$ **(b)** $\frac{31}{40}$ of a second

Divide the denominator into the numerator until the remainder becomes zero.

Solution

(a)
$$\begin{array}{r} 0.375 \\ 8\overline{)3.000} \\ \underline{2\,4} \\ 60 \\ \underline{56} \\ 40 \\ \underline{40} \\ 0 \end{array}$$

(b)
$$\begin{array}{r} 0.775 \\ 40\overline{)31.000} \\ \underline{28\,0} \\ 3\,00 \\ \underline{2\,80} \\ 200 \\ \underline{200} \\ 0 \end{array}$$

Therefore, $\frac{3}{8} = 0.375$.

Therefore, $\frac{31}{40} = 0.775$ of a second.

Practice Problem 1 Write as an equivalent decimal.

(a) $\frac{5}{16}$ **(b)** $\frac{11}{80}$

Athletics' times in Olympic events, such as the 100-meter dash, are measured to the nearest hundredth of a second. Future Olympic athletics' times will be measured to the nearest thousandth of a second.

Decimals such as 0.375 and 0.775 are called **terminating decimals.** When converting $\frac{3}{8}$ to 0.375 or $\frac{31}{40}$ to 0.775, the division operation eventually yields a remainder of zero. Other fractions yield a repeating pattern. For example, $\frac{1}{3} = 0.3333\ldots$ and $\frac{2}{3} = 0.6666\ldots$ have a pattern of repeating digits. Decimals that have a digit or a group of digits that repeats are called **repeating decimals.** We often indicate the repeating pattern with a bar over the repeating group of digits:

$$0.3333\ldots = 0.\overline{3} \qquad 0.74\ 74\ 74\ \ldots = 0.\overline{74}$$
$$0.218\ 218\ 218\ \ldots = 0.\overline{218} \qquad 0.8942\ 8942\ \ldots = 0.\overline{8942}$$

If when converting fractions to decimal form the remainder repeats itself, we know that we have a repeating decimal.

EXAMPLE 2 Write as an equivalent decimal.

(a) $\frac{5}{11}$ **(b)** $\frac{13}{22}$ **(c)** $\frac{5}{37}$

Solution

(a) $11\overline{)5.0000}$ = 0.4545

```
      0.4545
 11 ) 5.0000
      4 4
        60  <-
        55     repeating
         50    remainders
         44
          60 <-
          55
           5
```

(b)

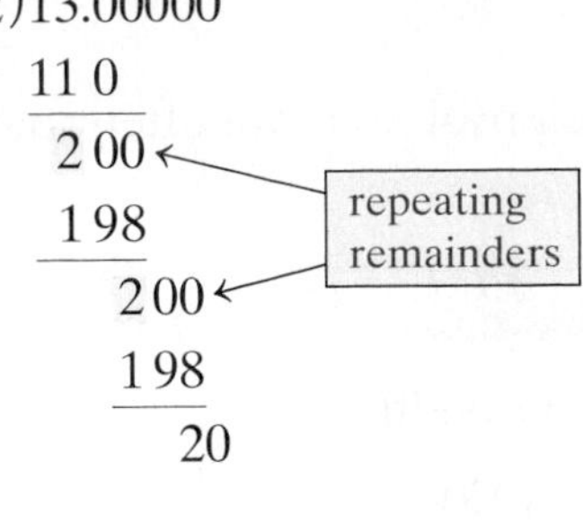

(c)

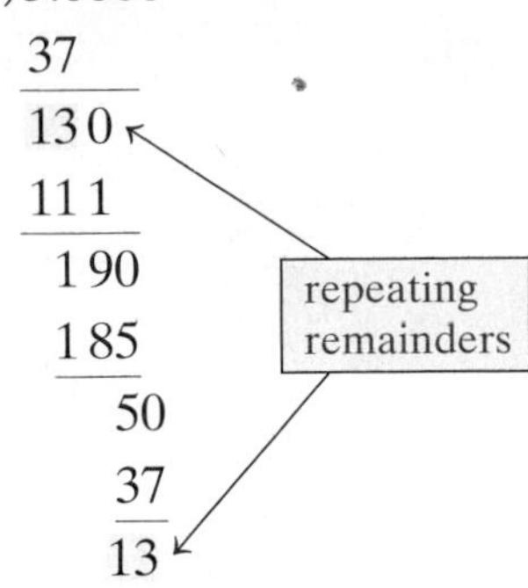

Thus $\frac{5}{11} = 0.4545\ldots = 0.\overline{45}$.

Thus $\frac{13}{22} = 0.5909090\ldots = 0.5\overline{90}$

Notice that the bar is over the digits 9 and 0 but *not* over the digit 5.

Thus $\frac{5}{37} = 0.135135\ldots = 0.\overline{135}$.

Practice Problem 2 Write as an equivalent decimal.

(a) $\frac{7}{11}$ **(b)** $\frac{8}{15}$ **(c)** $\frac{13}{44}$

EXAMPLE 3 Write as an equivalent decimal.

(a) $3\frac{7}{15}$ **(b)** $\frac{20}{11}$

Solution

(a) $3\frac{7}{15}$ means $3 + \frac{7}{15}$

```
      0.466
 15 ) 7.000
      60
      100
       90
      100
       90
       10
```

Thus $\frac{7}{15} = 0.4\overline{6}$ and $3\frac{7}{15} = 3.4\overline{6}$.

(b)

```
      1.818
 11 ) 20.000
      11
       90
       88
        20
        11
         90
         88
          2
```

Thus $\frac{20}{11} = 1.818181\ldots = 1.\overline{81}$.

Practice Problem 3 Write as an equivalent decimal.

(a) $2\frac{11}{18}$ **(b)** $\frac{28}{27}$

Calculator

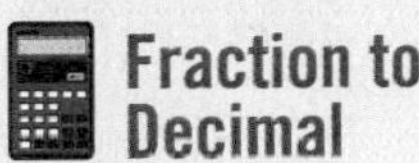

Fraction to Decimal

You can use a calculator to change $\frac{5}{8}$ to a decimal.

Enter:

5 [÷] 8 [=]

The display should read

[0.625]

Try the following.

(a) $\frac{17}{25}$ **(b)** $\frac{2}{9}$

(c) $\frac{13}{10}$ **(d)** $\frac{15}{19}$

Note: 0.78947368 is an approximation for $\frac{15}{19}$. Some calculators round to only eight places.

In some cases, the pattern of repeating is quite long. For example,

$$\frac{1}{7} = 0.142857142857\ldots = 0.\overline{142857}$$

Such problems are often rounded to a certain value.

EXAMPLE 4 Express $\frac{5}{7}$ as a decimal rounded to the nearest thousandth.

Solution

$$\begin{array}{r} 0.7142 \\ 7\overline{)5.0000} \\ \underline{4\,9} \\ 10 \\ \underline{7} \\ 30 \\ \underline{28} \\ 20 \\ \underline{14} \\ 6 \end{array}$$

Rounding to the nearest thousandth, we round 0.7142 to 0.714. (In repeating form, $\frac{5}{7} = 0.714285714285\ldots = 0.\overline{714285}$.)

Practice Problem 4 Express $\frac{19}{24}$ as a decimal rounded to the nearest thousandth.

***NOTE TO STUDENT:* Fully worked-out solutions to all of the Practice Problems can be found at the back of the text starting at page SP-1**

Recall that we studied placing two decimals in order in Section 3.2. If we are required to place a fraction and a decimal in order, it is usually easiest to change the fraction to decimal form and then compare the two decimals.

EXAMPLE 5 Fill in the blank with one of the symbols <, =, or >.

$$\frac{7}{16} \_\_\_ 0.43$$

Solution Now we divide to find the decimal equivalent of $\frac{7}{16}$.

$$\begin{array}{r} 0.4375 \\ 16\overline{)7.0000} \\ \underline{6\,4} \\ 60 \\ \underline{48} \\ 120 \\ \underline{112} \\ 80 \\ \underline{80} \\ 0 \end{array}$$

Now in the thousandths place 7 > 0, so we know

$$0.4375 > 0.4300.$$

Therefore, $\frac{7}{16} > 0.43$.

Practice Problem 5 Fill in the blank with one of the symbols <, =, or >.

$$\frac{5}{8} \_\_\_\_ 0.63$$

Using the Order of Operations with Decimals

The rules for order of operations that we discussed in Section 1.6 apply to operations with decimals.

ORDER OF OPERATIONS

Do first ↓
1. Perform operations inside parentheses.
2. Simplify any expressions with exponents.
3. Multiply or divide from left to right.
4. Add or subtract from left to right.

Do last

Sometimes exponents are used with decimals. In such cases, we merely evaluate using repeated multiplication.

$$(0.2)^2 = 0.2 \times 0.2 = 0.04$$
$$(0.2)^3 = 0.2 \times 0.2 \times 0.2 = 0.008$$
$$(0.2)^4 = 0.2 \times 0.2 \times 0.2 \times 0.2 = 0.0016$$

Teaching Tip If you sense that many of the class members are rusty on the rules for order of operations covered in Section 1.6, you may want to do a couple of simple examples with whole numbers first. You may want to try:

$$25 \times 2 - 15 + 3 \times 4$$

(The answer is 47.)

$$2^3 + 20 \div 10 \times 3 - 4.$$

(The answer is 10.)

EXAMPLE 6 Evaluate. $(0.3)^3 + 0.6 \times 0.2 + 0.013$

Solution First we need to evaluate $(0.3)^3 = 0.3 \times 0.3 \times 0.3 = 0.027$. Thus

$$(0.3)^3 + 0.6 \times 0.2 + 0.013$$
$$= 0.027 + 0.6 \times 0.2 + 0.013$$
$$= 0.027 + 0.12 + 0.013 \longleftarrow$$

When addends have a different number of decimal places, writing the problem in column form makes adding easier.

$$\begin{array}{r} 0.027 \\ 0.120 \\ +\ 0.013 \\ \hline 0.160 \end{array}$$

$$= 0.16$$

Practice Problem 6 Evaluate. $0.3 \times 0.5 + (0.4)^3 - 0.036$

In the next example all four steps of the rules for order of operations will be used.

EXAMPLE 7 Evaluate. $(8 - 0.12) \div 2^3 + 5.68 \times 0.1$

Solution

$(8 - 0.12) \div 2^3 + 5.68 \times 0.1$

$= 7.88 \div 2^3 + 5.68 \times 0.1$ First do subtraction inside the parentheses.

$= 7.88 \div 8 + 5.68 \times 0.1$ Simplify the expression with exponents.

$= 0.985 + 0.568$ From left to right do division and multiplication.

$= 1.553$ Add the final two numbers.

Practice Problem 7 Evaluate. $6.56 \div (2 - 0.36) + (8.5 - 8.3)^2$

***NOTE TO STUDENT:** Fully worked-out solutions to all of the Practice Problems can be found at the back of the text starting at page SP-1*

Developing Your Study Skills

Keep Trying

We live in a highly technical world, and you cannot afford to give up on the study of mathematics. Dropping mathematics may prevent you from entering certain career fields that you may find interesting. You may not have to take math courses as high-level as calculus, but such courses as intermediate algebra, finite math, college algebra, and trigonometry may be necessary. Learning mathematics can open new doors for you.

Learning mathematics is a process that takes time and effort. You will find that regular study and daily practice are necessary to strengthen your skills and to help you grow academically. This process will lead you toward success in mathematics. Then, as you become more successful, your confidence in your ability to do mathematics will grow.

3.6 EXERCISES

Student Solutions Manual | CD/Video | PH Math Tutor Center | MathXL®Tutorials on CD | MathXL® | MyMathLab® | Interactmath.com

Verbal and Writing Skills

1. 0.75 and $\frac{3}{4}$ are different ways to express the ___same quantity___.

2. To convert a fraction to an equivalent decimal, divide the ___denominator___ into the numerator.

3. Why is $0.\overline{8942}$ called a repeating decimal?
The digits 8942 repeat.

4. The order of operations for decimals is the same as the order of operations for whole numbers. Write the steps for the order of operations.
1. Perform operations inside parentheses.
2. Simplify any expressions with exponents.
3. Multiply or divide from left to right.
4. Add or subtract from left to right.

Write as an equivalent decimal. If a repeating decimal is obtained, use notation such as $0.\overline{7}$, $0.\overline{16}$, *or* $0.\overline{245}$.

5. $\frac{1}{4}$ 0.25

6. $\frac{3}{4}$ 0.75

7. $\frac{4}{5}$ 0.8

8. $\frac{2}{5}$ 0.4

9. $\frac{1}{8}$ 0.125

10. $\frac{3}{8}$ 0.375

11. $\frac{7}{20}$ 0.35

12. $\frac{3}{40}$ 0.075

13. $\frac{31}{50}$ 0.62

14. $\frac{23}{25}$ 0.92

15. $\frac{9}{4}$ 2.25

16. $\frac{14}{5}$ 2.8

17. $2\frac{1}{8}$ 2.125

18. $3\frac{13}{16}$ 3.8125

19. $1\frac{7}{16}$ 1.4375

20. $1\frac{1}{40}$ 1.025

21. $\frac{2}{3}$ $0.\overline{6}$

22. $\frac{5}{6}$ $0.8\overline{3}$

23. $\frac{5}{11}$ $0.\overline{45}$

24. $\frac{7}{11}$ $0.\overline{63}$

25. $3\frac{7}{12}$ $3.58\overline{3}$

26. $7\frac{1}{3}$ $7.\overline{3}$

27. $2\frac{5}{18}$ $2.2\overline{7}$

28. $6\frac{1}{6}$ $6.1\overline{6}$

Write as an equivalent decimal or a decimal approximation. Round your answer to the nearest thousandth if needed.

29. $\frac{4}{13}$ 0.308

30. $\frac{8}{17}$ 0.471

31. $\frac{19}{21}$ 0.905

32. $\frac{20}{21}$ 0.952

33. $\frac{7}{48}$ 0.146

34. $\frac{5}{48}$ 0.104

35. $\frac{35}{27}$ 1.296

36. $\frac{37}{23}$ 1.609

37. $\frac{21}{52}$
0.404

38. $\frac{1}{38}$
0.026

39. $\frac{17}{18}$
0.944

40. $\frac{5}{13}$
0.385

41. $\frac{22}{7}$
3.143

42. $\frac{17}{14}$
1.214

43. $3\frac{9}{19}$
3.474

44. $4\frac{11}{17}$
4.647

Fill in the blank with one of the symbols <, =, or >.

45. $\frac{7}{8}$ $\underline{<}$ 0.88

46. $\frac{3}{8}$ $\underline{<}$ 0.39

47. 0.573 $\underline{>}$ $\frac{9}{16}$

48. 0.689 $\underline{>}$ $\frac{11}{16}$

Applications

49. ***Fingernail Growth*** Your fingernails grow approximately $\frac{1}{8}$ inch each month. Write the thickness as a decimal.
0.125 inch

50. ***Hair Growth*** The hair on your head grows approximately $\frac{9}{20}$ inch a month. Write the thickness as a decimal.
0.45 inch

51. ***Mountain Climbing*** An ice climber had the local mountaineering shop install boot heaters to the back of his hiking boots. The installer drilled a hole $\frac{3}{8}$ inch in diameter. The hole for the boot heater should have been 0.5 inch in diameter. Is the hole too large or too small? By how much?
too small, 0.125 inch

52. ***Carpentry*** A master carpenter is re-creating a room for the set of a movie being filmed. He is using a burled maple veneer $\frac{7}{16}$ inch thick. The designer specified maple veneer 0.45 inch thick. Is the veneer he is using too thick or too thin? By how much?
too thin, 0.0125 inch

53. ***Safety Regulations*** Federal safety regulations specify that the slots between the bars on a baby's crib must not be more than $2\frac{3}{8}$ inches. One crib's slots measured 2.4 inches apart. Is this too wide? If so, by how much?
Yes; it is 0.025 inches too wide.

54. ***Manufacturing*** To manufacture a circuit board, Rick must program a computer to place a piece of thin plastic atop a circuit board. For the current to flow through the circuit, the top plastic piece must form a border of exactly $\frac{1}{16}$ inch with the circuit board. A few circuit boards were made with a border of 0.055 inch by accident. Is this border too small or too large? By how much?
too small; 0.0075 inch

Evaluate.

55. $2.4 + (0.5)^2 - 0.35$
$2.4 + 0.25 - 0.35 = 2.3$

56. $9.6 + 3.6 - (0.4)^2$
$9.6 + 3.6 - 0.16 = 13.04$

57. $2.3 \times 3.2 - 5 \times 0.8$
$7.36 - 4 = 3.36$

58. $9.6 \div 3 + 0.21 \times 6$
$3.2 + 1.26 = 4.46$

59. $12 \div 0.03 - 50 \times (0.5 + 1.5)^3$
$400 - 400 = 0$

60. $61.95 \div 1.05 - 2 \times (1.7 + 1.3)^3$
$59 - 54 = 5$

61. $(1.1)^3 + 2.6 \div 0.13 + 0.083$

$1.331 + 20 + 0.083 = 21.414$

62. $(1.1)^3 + 8.6 \div 2.15 - 0.086$

$1.331 + 4 - 0.086 = 5.245$

63. $(14.73 - 14.61)^2 \div (1.18 + 0.82)$

$0.0144 \div 2 = 0.0072$

64. $(32.16 - 32.02)^2 \div (2.24 + 1.76)$

$0.0196 \div 4 = 0.0049$

65. $(0.5)^3 + (3 - 2.6) \times 0.5$

0.325

66. $(0.6)^3 + (7 - 6.3) \times 0.07$

0.265

67. $(0.76 + 4.24) \div 0.25 + 8.6$

28.6

68. $(2.4)^2 + 3.6 \div (1.2 - 0.7)$

12.96

Evaluate.

69. $(1.6)^3 + (2.4)^2 + 18.666 \div 3.05 + 4.86$

$4.096 + 5.76 + 6.12 + 4.86 = 20.836$

70. $5.9 \times 3.6 \times 2.4 - 0.1 \times 0.2 \times 0.3 \times 0.4$

$50.976 - 0.0024 = 50.9736$

Write as a decimal. Round your answer to six decimal places.

71. $\dfrac{5236}{8921}$

0.586930

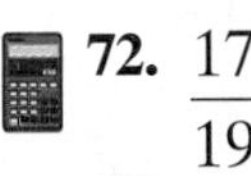
72. $\dfrac{17{,}359}{19{,}826}$

0.875567

To Think About

73. Subtract. $0.\overline{16} - 0.00\overline{16}$

(a) What do you obtain?

$$\begin{array}{r} 0.1616\overline{16} \\ -\ 0.0016\overline{16} \\ \hline 0.16 \end{array}$$

(b) Now subtract $0.\overline{16} - 0.01\overline{6}$. What do you obtain?

$$\begin{array}{r} 0.1616\overline{16} \\ -\ 0.0166\overline{66} \\ \hline 0.1449\overline{49} \end{array}$$

(c) What is different about these results?

(b) is a repeating and (a) is a nonrepeating decimal.

74. Subtract. $1.\overline{89} - 0.01\overline{89}$

(a) What do you obtain?

$$\begin{array}{r} 1.8989\overline{89} \\ -\ 0.0189\overline{89} \\ \hline 1.88 \end{array}$$

(b) Now subtract $1.\overline{89} - 0.18\overline{9}$. What do you obtain?

$$\begin{array}{r} 1.898989\overline{89} \\ -\ 0.189999\overline{99} \\ \hline 1.708989\overline{89} \end{array}$$

(c) What is different about these results?

(b) is a repeating and (a) is a nonrepeating decimal.

Cumulative Review

75. ***Boating Dock*** John and Nancy put in a new dock at the end of Tobey Lane. A pipe at the end of the dock supports the dock and is driven deep into the mud and sand at the bottom of Eel Pond. The pipe is 25 feet long. Half of the pipe is above the surface of the water at low tide. The pipe is driven $6\frac{3}{4}$ feet deep into the mud and sand. How deep is the water at the end of the dock at low tide?

$5\frac{3}{4}$ feet deep

76. ***Tidal Fluctuation*** Fisherman's Wharf in Digby, Nova Scotia has an average tidal range of $25\frac{4}{5}$ feet. These huge tidal ranges require considerable ingenuity in the design of docks and ramps for boats. If the water is $6\frac{1}{2}$ feet deep at low tide at the end of Fisherman's Wharf during an average low tide, how deep is the water at the same location during an average high tide. (*Source*: Nova Scotia Board of Tourism)

$32\frac{3}{10}$ feet

3.7 Estimating and Solving Applied Problems Involving Decimals

Student Learning Objectives

After studying this section, you will be able to:

1. **Estimate sums, differences, products, and quotients of decimals.**
2. **Solve applied problems using operations with decimals.**

1 Estimating Sums, Differences, Products, and Quotients of Decimals

When we encounter real-life applied problems, it is important to know if an answer is reasonable. A car may get 21.8 miles per gallon. However, a car will not get 218 miles per gallon. Neither will a car get 2.18 miles per gallon. To avoid making an error in solving applied problems, it is wise to make an estimate. The most useful time to make an estimate is at the end of solving the problem, in order to see if the answer is reasonable.

There are several different rules for estimating. Not all mathematicians agree what is the best method for estimating in each case. Most students find that a very quick and simple method to estimate is to round each number so that there is one nonzero digit. Then perform the calculation. We will use that approach in this section of the book. However, you should be aware that there are other valid approaches. Your instructor may wish you to use another method.

EXAMPLE 1 Estimate.

(a) $184{,}987.09 + 676{,}393.95$ **(b)** $0.00782 - 0.00358$

(c) 145.87×78.323 **(d)** $138.85 \div 5.887$

Solution In each case we will round to one nonzero digit to estimate.

(a) $184{,}987.09 + 676{,}393.95 \approx 200{,}000 + 700{,}000 = 900{,}000$

(b) $0.00782 - 0.00358 \approx 0.008 - 0.004 = 0.004$

(c) $145.87 \times 78.323 \approx$

$$\begin{array}{r} 100 \\ \times\ 80 \\ \hline 8000 \end{array}$$

Thus $145.87 \times 78.323 \approx 8000$

(d) $138.85 \div 5.887 \approx 6\overline{)100} = 16\frac{4}{6} \approx 17$

$$\begin{array}{r} 16 \\ 6\overline{)100} \\ \underline{6} \\ 40 \\ \underline{36} \\ 4 \end{array}$$

Thus $138.85 \div 5.887 \approx 17$

Here we round the answer to the nearest whole number.

Practice Problem 1 Round to one nonzero digit. Then estimate the result of the indicated calculation.

(a) $385.98 + 875.34$ **(b)** $0.0952 - 0.0579$

(c) 5876.34×0.087 **(d)** $46{,}873 \div 8.456$

NOTE TO STUDENT: Fully worked-out solutions to all of the Practice Problems can be found at the back of the text starting at page SP-1

Take a few minutes to review Example 1. Be sure you can perform these estimation steps. We will use this type of estimation to check our work in the applied problems in this section.

2 Solving Applied Problems Using Operations with Decimals

We use the basic plan of solving applied problems that we discussed in Section 1.8 and Section 2.9. Let us review how we analyze applied-problem situations.

1. *Understand the problem.*
2. *Solve and state the answer.*
3. *Check.*

In the United States for almost all jobs where you are paid an hourly wage, if you work more than 40 hours in one week, you should be paid overtime. The overtime rate is 1.5 times the normal hourly rate, for the extra hours worked in that week. The next problem deals with overtime wages.

EXAMPLE 2 A laborer is paid \$7.38 per hour for a 40-hour week and 1.5 times that wage for any hours worked beyond the standard 40. If he works 47 hours in a week, what will he earn?

Solution

1. ***Understand the problem.***

Mathematics Blueprint for Problem Solving

Gather the Facts	What Am I Asked to Do?	How Do I Proceed?	Key Points to Remember
He works 47 hours. He gets paid \$7.38 per hour for 40 hours. He gets paid $1.5 \times$ \$7.38 per hour for 7 hours.	Find the earnings of the laborer if he works 47 hours in one week.	Add the earnings of 40 hours at \$7.38 per hour to the earnings of 7 hours at overtime pay.	Multiply $1.5 \times$ \$7.38 to find the pay he earns for overtime.

2. ***Solve and state the answer.***

We want to compute his regular pay and his overtime pay and add the results.

Regular pay + Overtime pay = Total pay

Regular pay: Calculate his pay for 40 hours of work.

$$\begin{array}{r} 7.38 \\ \times \ \ 40 \\ \hline 295.20 \end{array}$$

He earns \$295.20 at \$7.38 per hour.

Overtime pay: Calculate his overtime pay rate. This is 7.38×1.5.

$$\begin{array}{r} 7.38 \\ \times\ 1.5 \\ \hline 3\ 690 \\ 7\ 38\ \\ \hline 11.070 \end{array}$$

He earns $11.07 per hour in overtime.

Calculate how much he earned doing 7 hours of overtime work.

$$\begin{array}{r} 11.07 \\ \times\quad 7 \\ \hline 77.49 \end{array}$$

For 7 hours overtime he earns $77.49.

Total pay: Add the two amounts.

$295.20	Regular 40-hour week earnings
+ 77.49	Overtime earnings
$372.69	Total earnings

The total earnings of the laborer for a 47-hour workweek will be $372.69.

3. *Check.*

Estimate his regular pay.

$$40 \times \$7 = \$280$$

Estimate his overtime rate of pay, and then his overtime pay.

$$1.5 \times \$7 = \$10.50$$
$$7 \times \$11 = \$77$$

Then add.

$$\begin{array}{r} \$280 \\ +\quad 77 \\ \hline \$357 \end{array}$$

Our estimate of $357 is close to our answer of $372.69. Our answer is reasonable. ✔

***NOTE TO STUDENT:** Fully worked-out solutions to all of the Practice Problems can be found at the back of the text starting at page SP-1*

Practice Problem 2 Melinda works for the phone company as a line repair technician. She earns $9.36 per hour. She worked 51 hours last week. If she gets time and a half for all hours worked above 40 hours per week, how much did she earn last week?

EXAMPLE 3 A chemist is testing 36.85 liters of cleaning fluid. She wishes to pour it into several smaller containers that each hold 0.67 liter of fluid. (a) How many containers will she need? (b) If each liter of this fluid costs $3.50, how much does the cleaning fluid in one container cost? (Round your answer to the nearest cent.)

Mathematics Blueprint for Problem Solving

Gather the Facts	What Am I Asked to Do?	How Do I Proceed?	Key Points to Remember
The total amount of cleaning fluid is 36.85 liters. Each small container holds 0.67 liter. Each liter of fluid costs \$3.50.	**(a)** Find out how many containers the chemist needs. **(b)** Find the cost of cleaning fluid in each small container.	**(a)** Divide the total, 36.85 liters, by the amount in each small container, 0.67 liter, to find the number of containers. **(b)** Multiply the cost of one liter, \$3.50, by the amount of liters in one container, 0.67.	If you are not clear as to what to do at any stage of the problem, then do a similar, simpler problem.

Solution **(a)** How many containers will the chemist need?

She has 36.85 liters of cleaning fluid and she wants to put it into several equal-sized containers each holding 0.67 liter. Suppose we are not sure what to do. Let's do a similar, simpler problem. If we had 40 liters of cleaning fluid and we wanted to put it into little containers each holding 2 liters, what would we do? Since the little containers would only hold 2 liters, we would need 20 containers. We know that $40 \div 2 = 20$. So we see that, in general, we divide the total number of liters by the amount in the small container. Thus $36.85 \div 0.67$ will give us the number of containers in this case.

$$\begin{array}{r} 55. \\ 0.67_{\wedge}\overline{)36.85_{\wedge}} \\ \underline{335} \\ 3\ 35 \\ \underline{3\ 35} \end{array}$$

The chemist will need 55 containers to hold this amount of cleaning fluid.

(b) How much does the cleaning fluid in each container cost? Each container will hold only 0.67 liter. If one liter costs \$3.50, then to find the cost of one container we multiply $0.67 \times \$3.50$.

$$\begin{array}{r} 3.50 \\ \underline{\times\ 0.67} \\ 2450 \\ \underline{2100} \\ 2.3450 \end{array}$$

We round our answer to the nearest cent. Thus each container would cost \$2.35.

Check.

(a) Is it really true that 55 containers each holding 0.67 liter will hold a total of 36.85 liters? To check we multiply.

$$\begin{array}{r} 55 \\ \times\ 0.67 \\ \hline 385 \\ 330 \\ \hline 36.85 \end{array} \checkmark$$

(b) One liter of cleaning fluid costs \$3.50. We would expect the cost of 0.67 liter to be less than \$3.50. \$2.35 is less than \$3.50. ✔

We use estimation to check more closely.

$$\begin{array}{rcr} \$3.50 & \longrightarrow & \$4.00 \\ \times\quad 0.67 & \longrightarrow & \times\quad 0.7 \\ \hline & & \$2.800 \end{array}$$

\$2.80 is fairly close to \$2.35. Our answer is reasonable. ✔

NOTE TO STUDENT: Fully worked-out solutions to all of the Practice Problems can be found at the back of the text starting at page SP-1

Practice Problem 3 A butcher divides 17.4 pounds of prime steak into small equal-sized packages. Each package contains 1.45 pounds of prime steak. **(a)** How many packages of steak will he have? **(b)** Prime steak sells for \$4.60 per pound. How much will each package of prime steak cost?

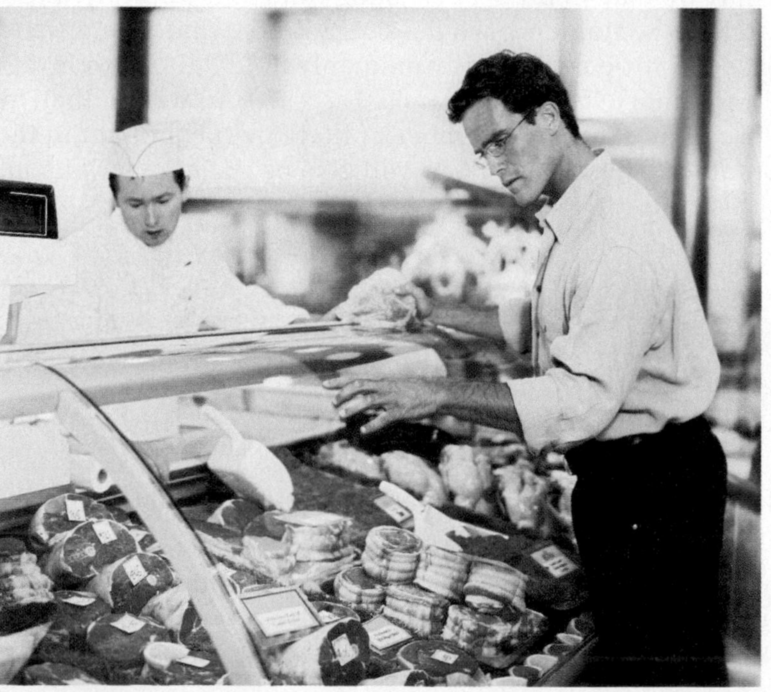

Developing Your Study Skills

Applications or Word Problems

Applications or word problems are the very life of mathematics! They are the reason for doing mathematics, because they teach you how to put into use the mathematical skills you have developed.

The key to success is practice. Make yourself do as many problems as you can. You may not be able to do them all correctly at first, but keep trying. If you cannot solve a problem, try another one. Ask for help from your teacher or the tutoring lab. Ask other classmates how they solved the problem. Soon you will see great progress in your own problem solving ability.

3.7 EXERCISES

Student Solutions Manual | CD/Video | PH Math Tutor Center | MathXL®Tutorials on CD | MathXL® | MyMathLab® | Interactmath.com

In exercises 1–10, first round each number to one nonzero digit. Then perform the calculation using the rounded numbers to obtain an estimate.

1. 238,598,980 + 487,903,870
700,000,000

2. 5,927,000 + 9,983,000
16,000,000

3. 56,789.345 − 33,875.125
30,000

4. 6949.45 − 1432.88
6000

5. 4832 × 0.532
2500

6. 68,976 × 0.875
63,000

7. 879.654 ÷ 56.82
15

8. 34.5684 ÷ 0.55
50

9. ***Canadian Tourists*** It is estimated that in 2003 a total of 15,627,350 tourists from Canada visited the United States. They spent a total of $3,922,464,850 in United States currency while visiting. Estimate the amount of money spent by each Canadian tourist. (*Source*: U.S. Dept of Commerce)
$200

10. ***Boat Sales*** Last year the sales of boats in Massachusetts totaled $865,987,273.45. If this represented a purchase of 55,872 boats, estimate the average price per boat.
$15,000

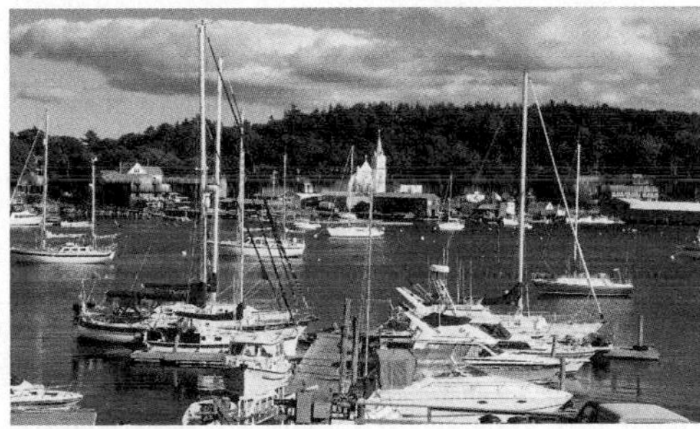

Applications

Estimate an answer to each of the following by rounding each number first, then perform the actual calculation.

11. ***Currency Conversion*** Carlos is taking a trip to Denmark. Before he leaves, he checks the newspaper and finds that every U.S. dollar is equal to 7.5 kroners (Danish currency). If Carlos takes $650 on his trip, how many kroners will he receive when he does the exchange?
4875 kroners

▲**12.** ***Geometry*** The dimensions of a football field, excluding the end zones, are 91.4 meters long and 48.75 meters wide. What is the area in square meters of the football field?
4455.75 square meters

▲**13.** ***Geometry*** Juan and Gloria are having their roof reshingled and need to determine its area in square feet. The dimensions of the roof are 48.3 feet by 56.9 feet. What is the area of the roof in square feet?
2748.27 square feet

14. ***Baby Formula*** A large can of infant formula contains 808 grams of powder. To prepare a bottle, 35.2 grams are needed. How many bottles can be prepared from the can? Round to the nearest whole number.
about 23 bottles

15. ***Cooking*** Hans is making gourmet chocolate in Switzerland. He has 11.52 liters of liquid white chocolate that will be poured into molds that hold 0.12 liter each. How many individual molds can Hans make with his 11.52 liters of liquid white chocolate?
96 molds

16. ***Food Purchase*** David bought MacIntosh apples and Anjou pears at the grocery store for a fruit salad. At the checkout counter, the apples weighed 2.7 pounds and the pears weighed 1.8 pounds. If the apples cost $1.29 per pound and the pears cost $1.49 per pound, how much did David spend on fruit? (Round your answer to the nearest cent.)
$6.17

17. ***Hawaii Rainfall*** One year in Mount Waialeale, Hawaii, considered the "rainiest place in the world," the yearly rainfall totaled 11.68 meters. The next year, the yearly rainfall on this mountain totaled 10.42 meters. The third year it was 12.67 meters. On average, how much rain fell on Mount Waialeale, Hawaii, per year?
11.59 meters

18. ***Auto Travel*** Emma and Jennie took a trip in their Ford Taurus from Saskatoon, Saskatchewan, to Calgary, Alberta, in Canada to check out the glacier lakes. When they left, their odometer read 54,089. When they returned home, the odometer read 55,401. They used 65.6 gallons of gas. How many miles per gallon did they get on the trip?
20 miles per gallon

19. ***Food Portions*** A jumbo bag of potato chips contains 18 ounces of chips. The recommended serving is 0.75 ounce. How many servings are in the jumbo bag?
24 servings

20. ***Telephone Costs*** Sylvia's telephone company offers a special rate of \$0.23 per minute on calls made to the Philippines during certain parts of the day. If Sylvia makes a 28.5-minute call to the Philippines at this special rate, how much will it cost?
\$6.56

21. ***Consumer Mathematics*** The local Police Athletic League raised enough money to renovate the local youth hall and turn it into a coffeehouse/activity center so that there is a safe place to hang out. The room that holds the Ping-Pong table needs 43.9 square yards of new carpeting. The entryway needs 11.3 square yards, and the stage/seating area needs 63.4 square yards. The carpeting will cost \$10.65 per square yard. What will be the total bill for carpeting these three areas of the coffeehouse?
\$1263.09

22. ***Painting Costs*** Kevin has a job as a house painter. One family needs its kitchen, family room, and hallway painted. The respective amounts needed are 2.7 gallons, 3.3 gallons, and 1.8 gallons. If paint costs \$7.40 per gallon, how much will Kevin need to spend on paint to do the job?
\$57.72

23. ***Overtime Pay*** Lucy earns \$8.50 per hour at the neighborhood café. She earns time and a half (1.5 times the hourly wage) for each hour she works on a holiday. Lucy worked eight hours each day for six days, then worked eight hours on New Year's Day. How much did she earn for that week?
\$510

24. ***Electrician's Pay*** An electrician is paid \$14.30 per hour for a 40-hour week. She is paid time and a half for overtime (1.5 times the hourly wage) for every hour more than 40 hours worked in the same week. If she works 48 hours in one week, what will she earn for that week?
\$743.60

25. ***Food Costs*** Belinda bought 1.5 pounds of fish for \$6.99 per pound, and 1.25 pounds of sliced turkey for \$4.59 per pound. How much total did she spend? Round to the nearest cent.
\$16.22

26. ***Consumer Mathematics*** At the beginning of each month, Perry withdraws \$80 for daily purchases. This month he spent \$22.75 on soda and snacks, \$3.25 on newspapers, and \$25.65 for bus fares. How much did Perry have left?
\$28.35

27. ***Car Payments*** Charlie borrowed \$11,500 to purchase a new car. His loan requires him to pay \$288.65 each month over the next 60 months (five years). How much will he pay over the five years? How much more will he pay back than the amount of the loan?
\$17,319; \$5819

28. ***House Payments*** Hector and Junita borrowed \$80,000 to buy their new home. They make monthly payments to the bank of \$450.25 to repay the loan. They will be making these monthly payments for the next 30 years. How much money will they pay to the bank in the next 30 years? How much more will they pay back than they borrowed?
\$162,090; \$82,090

29. ***Drinking Water Safety*** The EPA standard for safe drinking water is a maximum of 1.3 milligrams of copper per liter of water. A study was conducted on a sample of 7 liters of water drawn from Jeff Slater's house. The analysis revealed 8.06 milligrams of copper in the sample. Is the water safe or not? By how much?
yes; by 0.149 milligram per liter

30. ***Drinking Water Safety*** The EPA standard for safe drinking water is a maximum of 0.015 milligram of lead per liter of water. A study was conducted on 6 liters of water from West Towers Dormitory. The analysis revealed 0.0795 milligram of lead in the sample. Is the water safe or not? By how much?
yes; by 0.00175 milligram per liter

31. ***Jet Travel*** A jet fuel tank containing 17,316.8 gallons is being emptied at the rate of 126.4 gallons per minute. How many minutes will it take to empty the tank?
137 minutes

32. ***Monopoly Game*** In a New Jersey mall, the average price of a Parker Brothers Monopoly game is \$11.50. The Alfred Dunhill Company made a special commemorative set for \$25,000,000.00. Instead of plastic houses and hotels, you can buy and trade gold houses and silver hotels! How many regular Monopoly games could you purchase for the price of one special commemorative set?
2,173,913 games

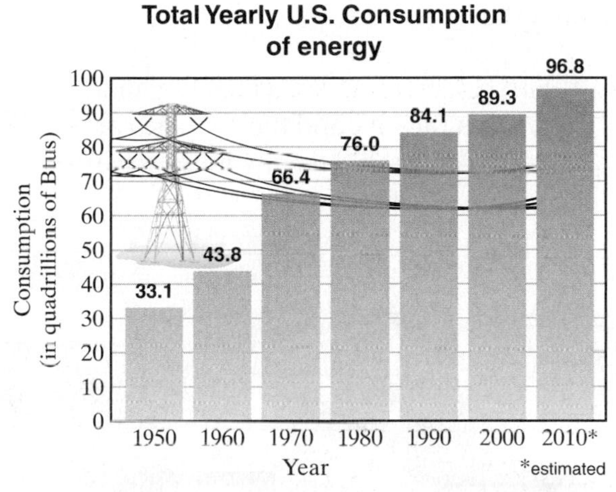

Source: U.S. Department of Energy

Energy Consumption *Use the preceding bar graph to answer exercises 33–36.*

33. How many more Btu were consumed in the United States during 2000 than in 1970?
22.9 quadrillion Btu

34. What was the greatest increase in consumption of energy in a 10-year period? When did it occur?
22.6 quadrillion Btu; from 1960 to 1970

35. What was the average consumption of energy per year in the United States for the years 1950, 1960, and 1970? Write your answer in quadrillion Btu and then write your answer in Btu. (Remember that a quadrillion is 1000 trillion)
approximately 47.8 quadrillion Btu;
47,800,000,000,000,000 Btu

36. What will be the average consumption of energy per year in the United States for the years 1990, 2000 and 2010? Write your answer in quadrillion Btu and then write your answer in Btu. (Remember that a quadrillion is 1000 trillion.)
approximately 90.1 quadrillion Btu;
90,100,000,000,000,000 Btu

Cumulative Review

Calculate.

37. $\frac{4}{7} + \frac{1}{2} \times \frac{2}{3}$
$\frac{19}{21}$

38. $\frac{3}{19} + \frac{5}{38} - \frac{2}{19}$
$\frac{7}{38}$

39. $\frac{7}{25} \times \frac{15}{42}$
$\frac{1}{10}$

40. $2\frac{2}{3} \div \frac{1}{3}$
8

Putting Your Skills to Work

The Mathematics of Fuel-Efficient Automobiles

According to the U.S. Federal Highway Administration, the automobiles driven in the United States currently consume 72,900,000,000 gallons of gasoline each year. The automobiles in the U.S. averaged 22.3 miles per gallon for fuel efficiency. (This is the figure for cars only—it does not include minivans, SUVs, and pickup trucks.) A lot of gasoline could be saved if some of these cars were converted to highly efficient hybrid automobiles. For example, the 2004 Toyota Prius has an EPA rating of 55 miles per gallon in combined highway/city driving. Use the above information in considering the following questions.

Problems for Individual Investigation and Analysis

1. If cars consume 72,900,000,000 gallons of gas per year in the U.S. and the average car gets 22.3 miles per gallon, how many miles are driven per year by cars in the U.S.?

1,625,670,000,000 miles

2. If all of these miles were driven by highly efficient hybrid cars that get 55 miles per gallon, how much gasoline would be used each year in the U.S. by these efficient cars? Round your answer to the nearest million gallons.

29,558,000,000 gallons

Problems for Group Analysis and Cooperative Investigation

3. It is probably not realistic to assume everyone would purchase a highly efficient hybrid automobile. However, it would be possible for half of the vehicles purchased in the U.S. to be this type of vehicle perhaps over the next twenty years. How much fuel would be used in a year by automobiles in the U.S. if half the mileage was driven by cars that obtained 22.3 miles per gallon and half the mileage was driven by cars that obtained 55 miles per gallon? Round your answer to the nearest million gallons.

51,229,000,000 gallons

4. If a person drives a car 20,000 miles each year and changes his car from a large SUV that gets 16 miles per gallon, to an efficient hydbrid that gets 55 miles per gallon, how much will that person save per year if gas is $1.95 per gallon? Round your answer to the nearest dollar.

$1728

Chapter 3 Organizer

<table>
<tr><th>Topic</th><th>Procedure</th><th>Examples</th></tr>
<tr><td>Word names for decimals, p. 205.</td><td>3 Hundreds
4 Tens
1 Ones
. Decimal point
6 Tenths
7 Hundredths
8 Thousandths
3 Ten-thousandths</td><td>The word name for 341.6783 is three hundred forty-one and six thousand seven hundred eighty-three ten-thousandths.</td></tr>
<tr><td>Writing a decimal as a fraction, p. 206.</td><td>1. Read the decimal in words.
2. Write it in fraction form.
3. Reduce if possible.</td><td>Write 0.36 as a fraction.
1. 0.36 is read "thirty-six hundredths."
2. Write the fractional form. $\frac{36}{100}$
3. Reduce. $\frac{36}{100} = \frac{9}{25}$</td></tr>
<tr><td>Determining which of two decimals is larger, p. 210.</td><td>1. Start at the left and compare corresponding digits. Write in extra zeros if needed.
2. When two digits are different, the larger number is the one with the larger digit.</td><td>Which is larger? 0.138 or 0.13
0.138 ? 0.130
$8 > 0$
So $0.138 > 0.130$.</td></tr>
<tr><td>Rounding decimals, p. 212.</td><td>1. Locate the place (units, tenths, hundredths, etc.) for which the round-off is required.
2. If the first digit to the right of the given place value is less than 5, drop it and all the digits to the right of it.
3. If the first digit to the right of the given place value is 5 or greater, increase the number in the given place value by one. Drop all digits to the right.</td><td>Round to the nearest hundredth. 0.8652
0.87
Round to the nearest thousandth. 0.21648
0.216</td></tr>
<tr><td>Adding and subtracting decimals, p. 216.</td><td>1. Write the numbers vertically and line up the decimal points. Extra zeros may be written to the right of the decimal points after the nonzero digits if needed.
2. Add or subtract all the digits with the same place value, starting with the right column, moving to the left. Use carrying or borrowing as needed.
3. Place the decimal point of the result in line with the decimal points of all the numbers added or subtracted.</td><td>Add. $36.3 + 8.007 + 5.26$
$\begin{array}{r} \overset{1}{3}6.300 \\ 8.007 \\ +\ 5.260 \\ \hline 49.567 \end{array}$
Subtract. $82.5 - 36.843$
$\begin{array}{r} \overset{7}{\not 8}\,\overset{11}{\not 2}.\overset{14}{\not 5}\,\overset{9}{\not 0}\,\overset{10}{\not 0} \\ -\ 36.843 \\ \hline 45.657 \end{array}$</td></tr>
<tr><td>Multiplying decimals, p. 226.</td><td>1. Multiply the numbers just as you would multiply whole numbers.
2. Find the sum of the decimal places in the two factors.
3. Place the decimal point in the product so that the product has the same number of decimal places as the sum in step 2. You may need to insert zeros to the left of the number found in step 1.</td><td>Multiply.
$\begin{array}{r} 0.2 \\ \times\ 0.6 \\ \hline 0.12 \end{array}$ $\begin{array}{r} 0.3174 \\ \times\ 0.8 \\ \hline 0.25392 \end{array}$
$\begin{array}{r} 0.0064 \\ \times\ 0.21 \\ \hline 64 \\ 128 \\ \hline 0.001344 \end{array}$ $\begin{array}{r} 1364 \\ \times\ 0.7 \\ \hline 954.8 \end{array}$</td></tr>
<tr><td>Multiplying a decimal by a power of 10, p. 227.</td><td>Move the decimal point to the right the same number of places as there are zeros in the power of 10. (Sometimes it is necessary to write extra zeros before placing the decimal point in the answer.)</td><td>Multiply.
$5.623 \times 10 = 56.23$
$0.0082 \times 1000 = 8.2$
$28.93 \times 10^2 = 2893$
$0.597 \times 10^4 = 5970$
$0.075 \times 10^6 = 75{,}000$</td></tr>
<tr><td>Dividing by a decimal, p. 236.</td><td>1. Make the divisor a whole number by moving the decimal point to the right. Mark that position with a caret (‸).
2. Move the decimal point in the dividend to the right the same number of places. Mark that position with a caret.
3. Place the decimal point of your answer directly above the caret in the dividend.
4. Divide as with whole numbers.</td><td>Divide.
(a) $0.06\overline{)0.162}$ (b) $0.003\overline{)85.8}$
(a) $\begin{array}{r} 2.7 \\ 0.06_\wedge\overline{)0.16_\wedge 2} \\ 12 \\ \hline 42 \\ 42 \\ \hline 0 \end{array}$
(b) $\begin{array}{r} 28\ 600. \\ 0.003_\wedge\overline{)85.800_\wedge} \\ 6 \\ \hline 25 \\ 24 \\ \hline 18 \\ 18 \\ \hline 0 \end{array}$</td></tr>
</table>

continues on next page

Topic	Procedure	Examples
Converting a fraction to a decimal, p. 243.	Divide the denominator into the numerator until **1.** the remainder is zero, or **2.** the decimal repeats itself, or **3.** the desired number of decimal places is achieved.	Find the decimal equivalent. **(a)** $\frac{13}{22}$ **(b)** $\frac{5}{7}$, rounded to the nearest ten-thousandth **(a)** $22\overline{)13.0000}$ = 0.5909 110 200 198 200 198 2 $\frac{13}{22} = 0.5\overline{90}$ or 0.5909090 . . . **(b)** $7\overline{)5.00000}$ = 0.71428 49 10 7 30 28 20 14 60 56 4 0.71428 rounded to the nearest ten thousandth is 0.7143.
Order of operations with decimal numbers, p. 247.	Same as order of operations of whole numbers. **1.** Perform operations inside parentheses. **2.** Simplify any expressions with exponents. **3.** Multiply or divide from left to right. **4.** Add or subtract from left to right.	Evaluate. $(0.4)^3 + 1.26 \div 0.12 - 0.12 \times (1.3 - 1.1)$ $= (0.4)^3 + 1.26 \div 0.12 - 0.12 \times 0.2$ $= 0.064 + 1.26 \div 0.12 - 0.12 \times 0.2$ $= 0.064 + 10.5 - 0.024$ $= 10.564 - 0.024$ $= 10.54$

Procedure for Solving Applied Problems

Using the Mathematics Blueprint for Problem Solving, p. 253

In solving a real-life problem with decimals, students may find it helpful to complete the following steps. You will not use all the steps all of the time. Choose the steps that best fit the conditions of the problem.

1. Understand the problem.

(a) Read the problem carefully.

(b) Draw a picture if it helps you visualize the situation. Think about what facts you are given and what you are asked to find.

(c) It may help to write a similar, simpler problem to get started and to determine what operation to use.

(d) Use the Mathematics Blueprint for Problem Solving to organize your work. Follow these four parts.

1. Gather the Facts (Write down specific values given in the problem.)
2. What Am I Asked to Do? (Identify what you must obtain for an answer.)
3. How Do I Proceed? (Determine what calculations need to be done.)
4. Key Points to Remember (Record any facts, warnings, formulas, or concepts you think will be important as you solve the problem.)

2. Solve and state the answer.

(a) Perform the necessary calculations.

(b) State the answer, including the unit of measure.

3. Check.

(a) Estimate the answer to the problem. Compare this estimate to the calculated value. Is your answer reasonable?

(b) Repeat your calculations.

(c) Work backward from your answer. Do you arrive at the original conditions of the problem?

EXAMPLE

▲Fred has a rectangular living room that is 3.5 yards wide and 6.8 yards long. He has a hallway that is 1.8 yards wide and 3.5 yards long. He wants to carpet each area using carpeting that costs $12.50 per square yard. What will the carpeting cost him?

Understand the problem.

It is helpful to draw a sketch.

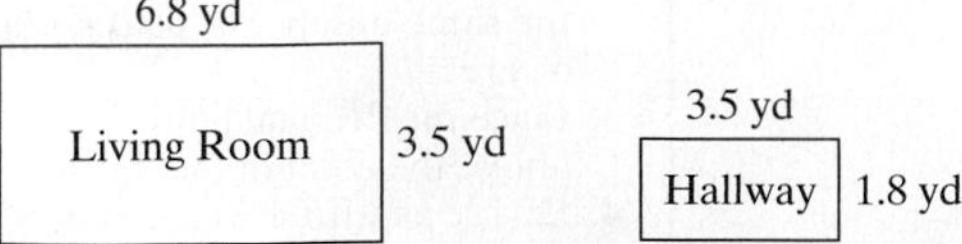

continues on next page

Procedure for Solving Applied Problems *(continued)*

Mathematics Blueprint for Problem Solving

Gather the Facts	What Am I Asked to Do?	How Do I Proceed?	Key Points to Remember
Living room: 6.8 yards by 3.5 yards Hallway: 3.5 yards by 1.8 yards Cost of carpet: \$12.50 per square yard	Find out what the carpeting will cost Fred.	Find the area of each room. Add the two areas. Multiply the total area by \$12.50.	Multiply the length by the width to get the area of the room. Remember, area is measured in square yards.

To find the area of each room, we multiply the dimensions for each room.

Living room $6.8 \times 3.5 = 23.80$ square yards

Hallway $3.5 \times 1.8 = 6.30$ square yards

Add the two areas.

$$\begin{array}{r} 23.80 \\ +\ 6.30 \\ \hline 30.10 \end{array}$$ square yards

Multiply the total area by the cost per square yard.

$30.1 \times 12.50 = \$376.25$

Estimate to check. You may be able to do some of this mentally.

$7 \times 4 = 28$ square yards $4 \times 2 = 8$ square yards

$$\begin{array}{r} 28 \\ +\ 8 \\ \hline 36 \end{array}$$ square yards

$36 \times 10 = \$360$ \$360 is close to \$376.25. ✓

Chapter 3 Review Problems

Section 3.1

Write a word name for each decimal.

1. 13.672
thirteen and six hundred seventy-two thousandths

2. 0.00084
eighty-four hundred-thousandths

Write as a decimal.

3. $\frac{7}{10}$
0.7

4. $\frac{81}{100}$
0.81

5. $1\frac{523}{1000}$
1.523

6. $\frac{79}{10{,}000}$
0.0079

Write as a fraction or a mixed number.

7. 0.17
$\frac{17}{100}$

8. 0.036
$\frac{9}{250}$

9. 34.24
$34\frac{6}{25}$

10. 1.00025
$1\frac{1}{4000}$

Section 3.2

Fill in the blank with <, =, or >.

11. $2\frac{9}{100}$ = 2.09

12. 0.716 > 0.706

13. $\frac{65}{100}$ < 0.655

14. 0.824 > 0.804

In exercises 15–18, arrange each set of decimal numbers from smallest to largest.

15. 0.981, 0.918, 0.98, 0.901
0.901, 0.918, 0.98, 0.981

16. 5.62, 5.2, 5.6, 5.26, 5.59
5.2, 5.26, 5.59, 5.6, 5.62

17. 0.754, 0.745, 0.7045, 0.704
0.704, 0.7045, 0.745, 0.754

18. 8.72, 8.2, 8.27, 8.702
8.2, 8.27, 8.702, 8.72

19. Round to the nearest tenth. 0.613
0.6

20. Round to the nearest hundredth. 19.2076
19.21

21. Round to the nearest ten-thousandth. 9.85215
9.8522

22. Round to the nearest dollar. $156.48
$156

Section 3.3

23. Add.

$$\begin{array}{r} 9.6 \\ 11.5 \\ 21.8 \\ +\ 34.7 \\ \hline 77.6 \end{array}$$

24. Add.

$$\begin{array}{r} 1.8 \\ 2.603 \\ 0.52 \\ +\ 1.716 \\ \hline 6.639 \end{array}$$

25. Subtract.

$$\begin{array}{r} 5.19 \\ -\ 1.296 \\ \hline 3.894 \end{array}$$

26. Subtract.

$$\begin{array}{r} 182.422 \\ -\ 68.55 \\ \hline 113.872 \end{array}$$

Section 3.4

In exercises 27–32, multiply.

27.

$$\begin{array}{r} 0.098 \\ \times\ 0.032 \\ \hline 0.003136 \end{array}$$

28.

$$\begin{array}{r} 126.83 \\ \times\ \quad 7 \\ \hline 887.81 \end{array}$$

29.

$$\begin{array}{r} 78 \\ \times\ 5.2 \\ \hline 405.6 \end{array}$$

30.

$$\begin{array}{r} 7053 \\ \times\ 0.34 \\ \hline 2398.02 \end{array}$$

31. 0.000613×10^3
0.613

32. 1.2354×10^5
123,540

33. ***Food Cost*** Roast beef was on sale for $3.49 per pound. How much would 2.5 pounds cost? Round to the nearest cent. $8.73

Section 3.5

In exercises 34–36, divide until there is a remainder of zero.

34. $0.07\overline{)0.0001806}$ 0.00258

35. $5.2\overline{)191.36}$ 36.8

36. $8\overline{)1863.2}$ 232.9

37. Divide and round your answer to the nearest tenth.

$1.3\overline{)746.75}$ 574.4

38. Divide and round your answer to the nearest thousandth.

$0.06\overline{)0.003539}$ 0.059

Section 3.6

Write as an equivalent decimal.

39. $\frac{5}{18}$
$0.2\overline{7}$

40. $\frac{7}{40}$
0.175

41. $1\frac{5}{6}$
$1.8\overline{3}$

42. $\frac{19}{16}$
1.1875

Write as a decimal rounded to the nearest thousandth.

43. $\frac{11}{14}$
0.786

44. $\frac{10}{29}$
0.345

45. $2\frac{5}{17}$
2.294

46. $3\frac{9}{23}$
3.391

Evaluate by doing the operations in proper order.

47. $2.3 \times 1.82 + 3 \times 5.12$
19.546

48. $0.03 + (1.2)^2 - 5.3 \times 0.06$
1.152

49. $(1.02)^3 + 5.76 \div 1.2 \times 0.05$
1.301208

50. $2.4 \div (2 - 1.6)^2 + 8.13$
23.13

Mixed Practice

Calculate.

51. $2398.26 - 1959.07$
439.19

52. $32.15 \times 0.02 \times 10^2$
64.3

53. $1.809 - 0.62 + 3.27$
4.459

54. $2.0792 \div 2.3$
0.904

55. $8 \div 0.4 + 0.1 \times (0.2)^2$
20.004

56. $(3.8 - 2.8)^3 \div (0.5 + 0.3)$
1.25

Applications

Section 3.7

Solve each problem.

57. ***Football Tickets*** At a large football stadium there are 2,600 people in line for tickets. In the first two minutes the computer is running slowly and tickets 228 people. Then the computer stops. For the next 2.5 minutes, the computer runs at medium speed and tickets 388 people per minute. For the next three minutes the computer runs at full speed and tickets 430 people per minute. Then the computer stops. How many people still have not received their tickets? 112 people

58. ***Fuel Efficiency*** Dan drove to the mountains. His odometer read 26,005.8 miles at the start, and 26,325.8 miles at the end of the trip. He used 12.9 gallons of gas on the trip. How many miles per gallon did his car get? (Round your answer to the nearest tenth.)
24.8 miles per gallon

59. ***Car Payments*** Robert is considering buying a car and making installment payments of $189.60 for 48 months. The cash price of the car is $6930.50. How much extra does he pay if he uses the installment plan instead of buying the car with one payment?
$2170.30

60. ***Comparing Job Salaries*** Mr. Zeno has a choice of working as an assistant manager at ABC Company at $315.00 per week or receiving an hourly salary of $8.26 per hour at the XYZ company. He learned from several previous assistant managers at both companies that they usually worked 38 hours per week. At which company will he probably earn more money? ABC Company

61. ***Drinking Water Safety*** The EPA standard for safe drinking water is a maximum of 0.002 milligram of mercury in one liter of water. The town wells at Winchester were tested. The test was done on 12 liters of water. The entire 12-liter sample contained 0.03 milligram of mercury. Is the water safe or not? By how much does it differ from the standard? no; by 0.0005 milligram per liter

62. ***Infant Head Size*** It is common for infants to have their heads measured during the first year of life. At two months, Will's head measured 40 centimeters. There are 2.54 centimeters in one inch. How many inches was this measurement? Round to the nearest hundredth.
15.75 inches

63. ***Geometry*** Dick Wright's new rectangular garden measures 18.3 feet by 9.6 feet. He needs to install wire fence on all four sides.

(a) How many feet of fence does he need? 55.8 feet

(b) The number of bags of wood chips Dick buys depends on the area of the garden. What is the area? 175.68 square feet

64. ***Geometry*** Bill Tupper's rectangular driveway needs to be resurfaced. It is 75.5 feet long and 18.5 feet wide. How large is the area of the driveway? 1396.75 square feet

65. ***Travel Distances*** The following strip map shows the distances in miles between several local towns in Pennsylvania. How much longer is the distance from Coudersport to Gaines, than the distance from Galeton to Wellsboro? 6.1 miles

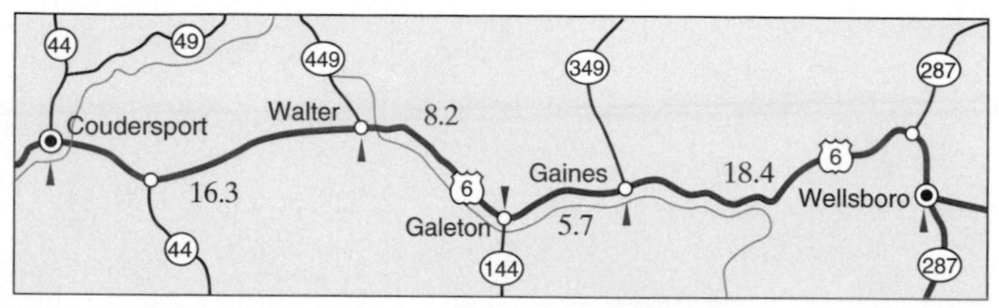

66. ***Geometry*** A farmer in Vermont has a field with an irregular shape. The distances are marked on the diagram. There is no fence but there is a path on the edge of the field. How long is the walking path around the field? 259.9 feet

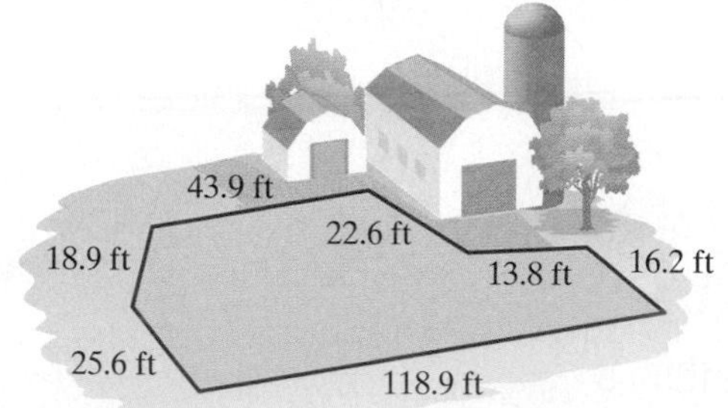

67. ***Car Payments*** Marcia and Greg purchased a new car. For the next five years they will be making monthly payments of $212.50. Their bank has offered to give them a loan at a smaller interest rate so that they would make monthly payments of only $199.50. The bank would charge them $285.00 to reissue their car loan. How much would it cost them to keep their original loan? How much would it cost them if they took the new loan from the bank? Should they make the change or keep the original loan? $12,750.00; $12,255.00; they should change to the new loan

Social Security Benefits *Use the following bar graph to answer exercises 68–73. Round all answers to the nearest cent.*

68. How much did the average monthly social security benefit increase from 1980 to 1990? $262

69. How much did the average monthly social security benefit increase from 1990 to 2000? $207

70. What was the average daily social security benefit in 1985? (Assume 30 days in a month.) $15.97

71. What was the average daily social security benefit in 1995? (Assume 30 days in a month.) $24.00

72. If the average daily social security benefit increases by the same amount from 2000 to 2015 as it did from 1985 to 2000, what will be the average daily social security benefit in 2015? $38.03

73. If the average daily social security benefit increases by the same amount from 2000 to 2010 as it did from 1990 to 2000, what will be the average daily social security benefit in 2010? $33.90

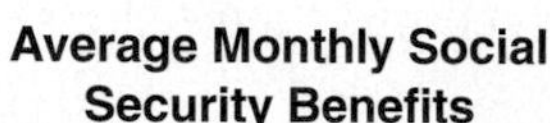

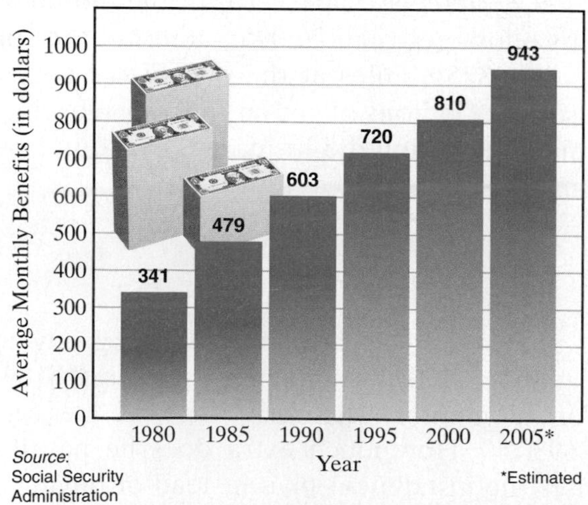

How Am I Doing? Chapter 3 Test

Remember to use your Chapter Test Prep Video CD to see the worked-out solutions to the test problems you want to review.

Note to Instructor: The Chapter 3 Test file in the TestGen program provides algorithms specifcally matched to these problems so you can easily replicate this test for additional practice or assessment purposes.

1. Write a word name for the decimal. 12.043

2. Write as a decimal. $\dfrac{3977}{10{,}000}$

In questions 3 and 4, write in fractional notation. Reduce whenever possible.

3. 7.15 **4.** 0.261

5. Arrange from smallest to largest. 2.19, 2.91, 2.9, 2.907

6. Round to the nearest hundredth. 78.6562

7. Round to the nearest ten-thousandth. 0.0341752

Add. **8.** $\begin{array}{r} 96.2 \\ 1.348 \\ +\ 2.15 \\ \hline \end{array}$ **9.** $17 + 2.1 + 16.8 + 0.04 + 1.59$

Subtract. **10.** $\begin{array}{r} 1.0075 \\ -\ 0.9096 \\ \hline \end{array}$ **11.** $72.3 - 1.145$

Multiply. **12.** $\begin{array}{r} 8.31 \\ \times\ 0.07 \\ \hline \end{array}$ **13.** 2.189×10^3

Divide. **14.** $0.08\overline{)0.01028}$ **15.** $0.69\overline{)32.43}$

Write as a decimal. **16.** $\dfrac{11}{9}$ **17.** $\dfrac{7}{8}$

In questions 18 and 19, perform the operations in the proper order.

18. $(0.3)^3 + 1.02 \div 0.5 - 0.58$

19. $19.36 \div (0.24 + 0.26) \times (0.4)^2$

20. Peter put 8.5 gallons of gas in his car. The price per gallon is $1.41. How much did Peter spend on gas? Round to the nearest cent.

21. Frank traveled from the city to the shore. His odometer read 42,620.5 miles at the start and 42,780.5 at the end of the trip. He used 8.5 gallons of gas. How many miles per gallon did his car achieve? Round to the nearest tenth.

22. The rainfall for March in Central City was 8.01 centimeters; for April, 5.03 centimeters; and for May, 8.53 centimeters. The normal rainfall for these three months is 25 centimeters. How much less rain fell during these three months than usual; that is, how does this year's figure compare with the figure for normal rainfall?

23. Wendy is earning $7.30 per hour in her new job as a teller trainee at the Springfield National Bank. She earns 1.5 times that amount for every hour over 40 hours she works in one week. She was asked to work 49 hours last week. How much did she earn last week?

1. twelve and forty-three thousandths

2. 0.3977

3. $7\frac{3}{20}$

4. $\frac{261}{1000}$

5. 2.19, 2.9, 2.907, 2.91

6. 78.66

7. 0.0342

8. 99.698

9. 37.53

10. 0.0979

11. 71.155

12. 0.5817

13. 2189

14. 0.1285

15. 47

16. $1.\overline{2}$

17. 0.875

18. 1.487

19. 6.1952

20. $11.99

21. 18.8 miles per gallon

22. 3.43 centimeters less

23. $390.55

Cumulative Test for Chapters 1–3

Approximately one-half of this test is based on Chapter 3 material. The remainder is based on material covered in Chapters 1 and 2.

1. Write in words. 38,056,954

2. Add.
$$\begin{array}{r} 156{,}028 \\ 301{,}579 \\ +\ 21{,}980 \\ \hline \end{array}$$

3. Subtract.
$$\begin{array}{r} 1{,}091{,}000 \\ -\ 1{,}036{,}520 \\ \hline \end{array}$$

4. Multiply.
$$\begin{array}{r} 589 \\ \times\ 67 \\ \hline \end{array}$$

5. Divide. $17\overline{)4386}$

6. Evaluate. $20 \div 4 + 2^5 - 7 \times 3$

7. Reduce. $\dfrac{18}{45}$

8. Add. $4\dfrac{1}{3} + 3\dfrac{1}{6}$

9. Subtract. $\dfrac{23}{35} - \dfrac{2}{5}$

10. Evaluate. $\dfrac{7}{10} \times \dfrac{5}{3} - \dfrac{5}{12} \times \dfrac{1}{2}$

11. Divide. $52 \div 3\dfrac{1}{4}$

12. Divide. $1\dfrac{3}{8} \div \dfrac{5}{12}$

13. Estimate. $58{,}216 \times 438{,}207$

14. Write as a decimal. $\dfrac{39}{1000}$

15. Arrange from smallest to largest. 2.1, 20.1, 2.01, 2.12, 2.11

16. Round to the nearest thousandth. 26.07984

17. Add.
$$\begin{array}{r} 1.9 \\ 2.36 \\ 15.2 \\ +\ 0.08 \\ \hline \end{array}$$

18. Subtract.
$$\begin{array}{r} 28.1 \\ -\ 14.982 \\ \hline \end{array}$$

19. Multiply. 56.8×0.02

20. Multiply. 0.1823×1000

21. Divide. $0.06\overline{)0.06348}$

22. Write as a decimal. $\dfrac{13}{16}$

23. Perform the operations in the correct order.

$1.44 \div 0.12 + (0.3)^3 + 1.57$

▲**24.** Dr. Bob Wells has a small square garden that measures 10.5 feet on each side.
(a) What is the area of this garden?
(b) What is the perimeter of this garden?

25. Sue's savings account balance is \$199.36. This month she earned interest of \$1.03. She deposited \$166.35 and \$93.50. She withdrew money three times, in the amounts of \$90.00, \$37.49, and \$137.18. What will her balance be at the start of next month?

26. Russ and Norma Camp borrowed some money from the bank to purchase a new car. They are paying off the car loan at the rate of \$320.50 per month. At the end of the loan period they will have paid \$19,230.00 to the bank. How many months will it take to pay off this car loan?

Answers

1. thirty-eight million, fifty-six thousand, nine hundred fifty-four
2. 479,587
3. 54,480
4. 39,463
5. 258
6. 16
7. $\dfrac{2}{5}$
8. $7\frac{1}{2}$
9. $\dfrac{9}{35}$
10. $\dfrac{23}{24}$
11. 16
12. $\dfrac{33}{10}$ or $3\dfrac{3}{10}$
13. 24,000,000,000
14. 0.039
15. 2.01, 2.1, 2.11, 2.12, 20.1
16. 26.080
17. 19.54
18. 13.118
19. 1.136
20. 182.3
21. 1.058
22. 0.8125
23. 13.597
24. (a) 110.25 square feet
 (b) 42 feet
25. \$195.57
26. 60 months

CHAPTER

4

We all know that too much fast food is not good for us. Many people eat fast food nearly every day, but they may be unaware of just how many calories they are consuming. Do you know many calories are in some common fast foods? How much exercise is necessary to burn off these calories? Try the Putting Your Skills to Work Problems on page 301 to find out.

Ratio and Proportion

4.1 Ratios and Rates

Student Learning Objectives

After studying this section, you will be able to:

1. **Use a ratio to compare two quantities with the same units.**
2. **Use a rate to compare two quantities with different units.**

1 Using a Ratio to Compare Two Quantities with the Same Units

Assume that you earn 13 dollars an hour and your friend earns 10 dollars per hour. The *ratio* 13 : 10 compares what you and your friend make. This ratio means that for every 13 dollars you earn, your friend earns 10. The *rate* you are paid is 13 dollars per hour, which compares 13 dollars to 1 hour. In this section we see how to use both ratios and rates to solve many everyday problems.

Suppose that we want to compare an object weighing 20 pounds to an object weighing 23 pounds. The ratio of their weights would be 20 to 23. We may also write this as $\frac{20}{23}$. A **ratio** is the comparison of two quantities that have the *same units*.

A commonly used video display for a computer has a horizontal dimension of 14 inches and a vertical dimension of 10 inches. The ratio of the horizontal dimension to the vertical dimension is 14 to 10. In reduced form we would write that as 7 to 5. We can express the ratio three ways.

We can write "the ratio of 7 to 5."
We can write 7 : 5 using a colon.
We can write $\frac{7}{5}$ using a fraction.

All three notations are valid ways to compare 7 to 5. Each is read as "7 to 5."

We always want to write a ratio in simplest form. A ratio is in **simplest form** when the two numbers do not have a common factor and both numbers are whole numbers.

EXAMPLE 1 Write in simplest form. Express your answer as a fraction.

(a) the ratio of 15 hours to 20 hours
(b) the ratio of 36 hours to 30 hours
(c) 125 : 150

Solution

(a) $\frac{15}{20} = \frac{3}{4}$ **(b)** $\frac{36}{30} = \frac{6}{5}$ **(c)** $\frac{125}{150} = \frac{5}{6}$

Notice that in each case the two numbers *do* have a common factor. When we form the fraction—that is, the ratio—we take the extra step of *reducing* the fraction. However, improper fractions *are not* changed to mixed numbers.

Teaching Tip You will need to remind students to reduce fractions to lowest terms when expressing ratios. They often forget this step or reduce only partially.

Practice Problem 1 Write in simplest form. Express your answer as a fraction.

(a) the ratio of 36 feet to 40 feet
(b) the ratio of 18 feet to 15 feet
(c) 220 : 270

***NOTE TO STUDENT:** Fully worked-out solutions to all of the Practice Problems can be found at the back of the text starting at page SP-1*

EXAMPLE 2 Martin earns \$350 weekly. However, he takes home only \$250 per week in his paycheck.

\$350.00 gross pay (what Martin earns)

45.00 withheld for federal tax
20.00 withheld for state tax
35.00 withheld for retirement
} (what is taken out of Martin's earnings)

\$250.00 take-home pay (what Martin has left)

(a) What is the ratio of the amount withheld for federal tax to gross pay?

(b) What is the ratio of the amount withheld for state tax to the amount withheld for federal tax?

Teaching Tip Some students will ask why the fractional form of the ratio is emphasized in this text. They may wonder why the colon (e.g., 4:9) ratio is not used more often. Explain that the fractional form is the most useful form to use when solving ratio or proportion problems.

Solution

(a) The ratio of the amount withheld for federal tax to gross pay is

$$\frac{45}{350} = \frac{9}{70}.$$

(b) The ratio of the amount withheld for state tax to the amount withheld for federal tax is

$$\frac{20}{45} = \frac{4}{9}.$$

Practice Problem 2 Recently President Burton conducted a survey of students at North Shore Community College who use the Internet. He wanted to determine how many of the students use the college Internet provider versus how many use AOL, MSN, or other commercial Internet providers. The results of his survey are shown in the circle graph.

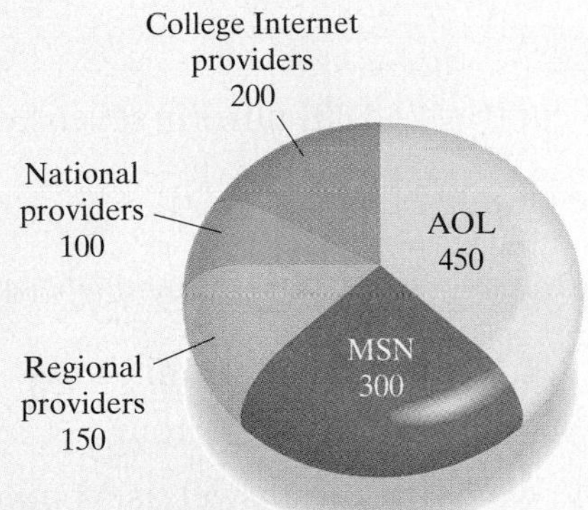

(a) Write the ratio of the number of students who use the college Internet provider to the number of students who use AOL.

(b) Write the ratio of the number of students who use MSN to the total number of students who use the Internet.

TO THINK ABOUT: Mach Numbers Perhaps you have heard statements like "a certain jet plane travels at Mach 2.2." What does that mean? A Mach number is a ratio that compares the velocity (speed) of an object to the velocity of sound. Sound travels at about 330 meters per second. The Mach number is written in decimal form.

What is the Mach number of a jet traveling at 690 meters per second?

$$\text{Mach number of jet} = \frac{690 \text{ meters per second}}{330 \text{ meters per second}} = \frac{69}{33} = \frac{23}{11}$$

Dividing this out, we obtain

$$2.09090909\ldots \text{ or } 2.\overline{09}$$

Rounded to the nearest tenth, the Mach number of the jet is 2.1.

Exercises 73 and 74 in Exercises 4.1 deal with Mach numbers.

 Using a Rate to Compare Two Quantities with Different Units

A **rate** is a comparison of two quantities with *different units*. Usually, to avoid misunderstanding, we express a rate as a reduced or simplified fraction with the units included.

EXAMPLE 3 Recently an automobile manufacturer spent $946,000 for a 48-second television commercial shown on a national network. What is the rate of dollars spent to seconds of commercial time?

Solution The rate is $\frac{946{,}000 \text{ dollars}}{48 \text{ seconds}} = \frac{59{,}125 \text{ dollars}}{3 \text{ seconds}}$.

Practice Problem 3 A farmer is charged a $44 storage fee for every 900 tons of grain he stores. What is the rate of the storage fee in dollars to tons of grain?

Often we want to know the rate for a single unit, which is the unit rate. A **unit rate** is a rate in which the denominator is the number 1. Often we need to divide the numerator by the denominator to obtain this value.

EXAMPLE 4 A car traveled 301 miles in seven hours. Find the unit rate.

Solution $\frac{301}{7}$ can be simplified. We find $301 \div 7 = 43$.

Thus

$$\frac{301 \text{ miles}}{7 \text{ hours}} = \frac{43 \text{ miles}}{1 \text{ hour}}$$

The denominator is 1. We write our answer as 43 miles/hour. The fraction line is read as the word *per*, so our answer here is read "43 miles per hour." *Per* means "for every," so a rate of 43 miles per hour means 43 miles traveled for every hour traveled.

Practice Problem 4 A car traveled 212 miles in four hours. Find the unit rate.

***NOTE TO STUDENT:** Fully worked-out solutions to all of the Practice Problems can be found at the back of the text starting at page SP-1*

EXAMPLE 5 A grocer purchased 200 pounds of apples for \$68. He sold the 200 pounds of apples for \$86. How much profit did he make per pound of apples?

Solution

$$\begin{array}{rl} \$86 & \text{selling price} \\ -\ 68 & \text{cost} \\ \hline \$18 & \text{profit} \end{array}$$

The rate that compares profit to pounds of apples sold is $\frac{18 \text{ dollars}}{200 \text{ pounds}}$. We will find $18 \div 200$.

$$\begin{array}{r} 0.09 \\ 200\overline{)18.00} \\ \underline{18\ 00} \\ 0 \end{array}$$

The unit rate of profit is \$0.09 per pound.

Practice Problem 5 A retailer purchased 120 nickel-cadmium batteries for flashlights for \$129.60. She sold them for \$170.40. What was her profit per battery?

EXAMPLE 6 Hamburger at a local butcher is packaged in large and extra-large packages. A large package costs \$7.86 for 6 pounds and an extra-large package is \$10.08 for 8 pounds.

(a) What is the unit rate in dollars per pound for each size package?

(b) How much per pound does a consumer save by buying the extra-large package?

Teaching Tip Ask students if they are familiar with unit pricing in a supermarket. Explain how unit pricing helps them decide which package is the best buy. (The largest size is not always the best buy.)

Solution

(a) $\frac{7.86 \text{ dollars}}{6 \text{ pounds}} = \$1.31/\text{pound}$ for the large package

$\frac{10.08 \text{ dollars}}{8 \text{ pounds}} = \$1.26/\text{pound}$ for the extra-large package

(b)
$$\begin{array}{r} \$1.31 \\ -\ 1.26 \\ \hline \$0.05 \end{array}$$
A person saves \$0.05/pound by buying the extra-large package.

Practice Problem 6 A 12-ounce package of Fred's favorite cereal costs \$2.04. A 20-ounce package of the same cereal costs \$2.80.

(a) What is the unit rate in cost per ounce of each size of cereal package?

(b) How much per ounce would Fred save by buying the larger size?

4.1 Exercises

Student Solutions Manual

CD/ Video

PH Math Tutor Center

MathXL®Tutorials on CD
MathXL®

MyMathLab®

Interactmath.com

Verbal and Writing Skills

1. A ____ratio____ is a comparison of two quantities that have the same units.

2. A rate is a comparison of two quantities that have ____different____ units.

3. The ratio 5 : 8 is read ____5 to 8____.

4. Marion compares the number of loaves of bread she bakes to the number of pounds of flour she needs to make the bread. Is this a ratio or a rate? Why?
a rate, it compares different units: loaves of bread to pounds of flour

Write in simplest form. Express your answer as a fraction.

5. 6:18
$\frac{1}{3}$

6. 8:20
$\frac{2}{5}$

7. 21:18
$\frac{7}{6}$

8. 50:35
$\frac{10}{7}$

9. 150:225
$\frac{2}{3}$

10. 360:480
$\frac{3}{4}$

11. 165 to 90
$\frac{11}{6}$

12. 135 to 120
$\frac{9}{8}$

13. 60 to 64
$\frac{15}{16}$

14. 33 to 57
$\frac{11}{19}$

15. 28 to 42
$\frac{2}{3}$

16. 21 to 98
$\frac{3}{14}$

17. 32 to 20
$\frac{8}{5}$

18. 90 to 54
$\frac{5}{3}$

19. 8 ounces to 12 ounces
$\frac{2}{3}$

20. 50 years to 85 years
$\frac{10}{17}$

21. 39 kilograms to 26 kilograms
$\frac{3}{2}$

22. 255 meters to 15 meters
$\frac{17}{1}$

23. \$82 to \$160
$\frac{41}{80}$

24. \$94 to \$140
$\frac{47}{70}$

25. 312 yards to 24 yards
$\frac{13}{1}$

26. 91 tons to 133 tons
$\frac{13}{19}$

27. $2\frac{1}{2}$ pounds to $4\frac{1}{4}$ pounds
$\frac{10}{17}$

28. $4\frac{1}{3}$ feet to $5\frac{2}{3}$ feet
$\frac{13}{17}$

Personal Finance

Use the following table to answer exercises 29–32.

ROBIN'S WEEKLY PAYCHECK

Total (Gross) Pay	Federal Withholding	State Withholding	Retirement	Insurance	Savings Contribution	Take-Home Pay
\$285	\$35	\$20	\$28	\$16	\$21	\$165

29. What is the ratio of take-home pay to total (gross) pay?
$\frac{165}{285} = \frac{11}{19}$

30. What is the ratio of retirement to insurance?
$\frac{28}{16} = \frac{7}{4}$

31. What is the ratio of federal withholding to take-home pay?

$\frac{35}{165} = \frac{7}{33}$

32. What is the ratio of retirement to total (gross) pay?

$\frac{28}{285}$

Useful Life of an Automobile *An automobile insurance company prepared the following analysis for its clients. Use this table for exercises 33–36.*

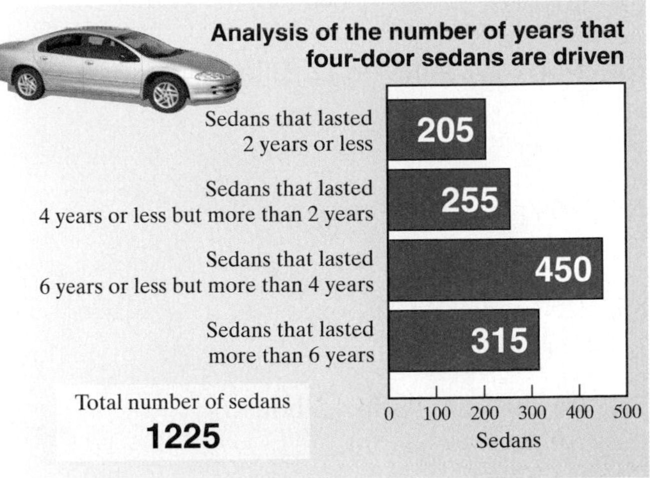

33. What is the ratio of sedans that lasted two years or less to the total number of sedans?

$\frac{205}{1225} = \frac{41}{245}$

34. What is the ratio of sedans that lasted more than six years to the total number of sedans?

$\frac{315}{1225} = \frac{9}{35}$

35. What is the ratio of the number of sedans that lasted six years or less but more than four years to the number of sedans that lasted two years or less?

$\frac{450}{205} = \frac{90}{41}$

36. What is the ratio of the number of sedans that lasted more than six years to the number of sedans that lasted four years or less but more than two years?

$\frac{315}{255} = \frac{21}{17}$

37. ***Basketball*** A basketball team scored a total of 704 points during one season. Of these, 44 points were scored by making one-point free throws. What is the ratio of free-throw points to total points?

$\frac{1}{16}$

38. ***Sales Tax*** When Michael bought his home theater sound system, he paid \$34 in tax. The total cost was \$714. What is the ratio of tax to total cost?

$\frac{1}{21}$

Write as a rate in simplest form.

39. \$40 for 16 magazines

$\frac{\$5}{2 \text{ magazines}}$

40. \$36 for 16 rolls of film

$\frac{\$9}{4 \text{ rolls of film}}$

41. \$170 for 12 bushes

$\frac{\$85}{6 \text{ bushes}}$

42. 98 pounds for 22 people

$\frac{49 \text{ pounds}}{11 \text{ people}}$

43. \$114 for 12 CDs

$\frac{\$19}{2 \text{ CDs}}$

44. \$150 for 12 house plants

$\frac{\$25}{2 \text{ plants}}$

45. 6150 revolutions for every 15 miles

$\frac{410 \text{ revolutions}}{1 \text{ mile}}$ or 410 rev/mile

46. 9540 revolutions for every 18 miles

$\frac{530 \text{ revolutions}}{1 \text{ mile}}$ or 530 rev/mile

47. $330,000 for 12 employees

$\frac{\$27,500}{1 \text{ employee}}$ or $27,500/employee

48. $156,000 for 24 people

$\frac{\$6500}{1 \text{ person}}$ or $6500/person

Write as a unit rate.

49. Earn $520 in 40 hours

$\frac{\$520}{40 \text{ hours}} = \$13/\text{hr}$

50. Earn $304 in 38 hours

$\frac{\$304}{38 \text{ hours}} = \$8/\text{hr}$

51. Travel 192 miles on 12 gallons of gas

$\frac{192 \text{ miles}}{12 \text{ gallons}} = 16 \text{ mi/gal}$

52. Travel 322 miles on 14 gallons of gas

$\frac{322 \text{ miles}}{14 \text{ gallons}} = 23 \text{ mi/gal}$

▲ **53.** 1800 people in 20 square miles

90 people/sq mi

▲ **54.** 6000 people in 15 square miles

400 people/sq mi

55. 2250 pencils in 18 boxes

125 pencils/box

56. 1122 books in 11 crates

102 books/crate

57. Travel 297 miles in 4.5 hours

66 mi/hour

58. Travel 374 miles in 5.5 hours

68 mi/hour

59. 475 patients for 25 doctors

19 patients/doctor

60. 375 trees planted on 15 acres

25 trees/acre

61. 60 eggs from 12 chickens

5 eggs/chickens

62. 78 children in 26 families

3 children/family

Applications

63. ***Stock Market*** $3870 was spent for 129 shares of Mattel stock. Find the cost per share.

$30/share

64. ***Stock Market*** $6150 was spent for 150 shares of Polaroid stock. Find the cost per share.

$41/share

65. ***Bookstore Profit*** A bookstore owner purchased 80 calendars for $760. He sold them for $1200. How much profit did he make per calendar?

$5.50 profit per calendar

66. ***Retail Profit*** A retailer purchased 24 compact disc players for $3380. She sold them for $4030. How much profit did she make per compact disc player? Round your answer to the nearest dollar.

$4030 - 3380 = \$650 \qquad \frac{\$650}{24 \text{ players}} = \$27/\text{player}$

67. ***Food Cost*** A 16-ounce box of dry pasta costs $1.28. A 24-ounce box of the same pasta costs $1.68.

(a) What is the cost per ounce of each box of pasta?

$0.08/oz small box; $0.07/oz large box

(b) How much does the educated consumer save by buying the larger box?

1¢ per ounce or $0.01 per ounce

(c) How much does the consumer save by buying 2 large boxes instead of 3 small boxes?

The consumer saves $0.48.

68. ***Food Cost*** A 16-ounce can of beef stew costs $2.88. A 26-ounce can of the same beef stew costs $4.16.

(a) What is the cost per ounce of each can of stew?

smaller can $\frac{\$2.88}{16 \text{ ounces}} = \$0.18/\text{oz}$ larger can $\frac{\$4.16}{26 \text{ ounces}} = \$0.16/\text{oz}$

(b) How much does the consumer save per ounce by buying the larger can?

2¢ per ounce or $0.02/oz

69. ***Moose Population Density*** Dr. Robert Tobey completed a count of two herds of moose in the central regions of Alaska. He recorded 3978 moose on the North Slope and 5520 moose on the South Slope. There are 306 acres on the North Slope and 460 acres on the South Slope.

(a) How many moose per acre were found on the North Slope?
13 moose

(b) How many moose per acre were found on the South Slope?
12 moose

(c) In which region are the moose more closely crowded together?
North Slope

70. ***Australia Population Density*** In Melbourne, Australia, 27,900 people live in the suburb of St. Kilda and 38,700 live in the suburb of Caulfield. The area of St. Kilda is 6500 acres. The area of Caulfield is 9200 acres. Round your answers to the nearest tenth.

(a) How many people per acre live in St. Kilda?
4.3 people

(b) How many people per acre live in Caulfield?
4.2 people

(c) Which suburb is more crowded?
St. Kilda

71. ***Stock Market***

(a) Mr. Jenson bought 525 shares of Zenith stock for $12,862.50. How much did he pay per share?
$24.50

(b) Mr. LeBlanc purchased 355 shares of Ford for $781. How much did he pay per share?
$2.20

(c) How much more per share did Mr. Jenson pay than Mr. LeBlanc?
$22.30

72. ***Baseball Statistics*** In 2002, Alex Rodriguez of the Texas Rangers hit 57 home runs with 624 "at-bats." In the same year, Jim Thome of the Cleveland Indians hit 52 home runs with 480 at-bats.

(a) What are the rates of at-bats per home run for Alex and for Jim? Round to the nearest tenth.
Alex Rodriguez, 10.9; Jim Thome, 9.2

(b) Which person hits homeruns more often?
Jim Thome

To Think About

For exercises 73 and 74, recall that the speed of sound is about 330 meters per second. (See the To Think About discussion on pages 271 and 272.) Round your answers to the nearest tenth.

73. ***Jet Speed*** A jet plane was originally designed to fly at 750 meters per second. It was modified to fly at 810 meters per second. By how much was its Mach number increased?
increased by Mach 0.2

74. ***Rocket Speed*** A rocket was first flown at 1960 meters per second. It proved unstable and unreliable at that speed. It is now flown at a maximum of 1920 meters per second. By how much was its Mach number decreased?
decreased by Mach 0.1

Cumulative Review

Calculate.

75. $2\frac{1}{4} + \frac{3}{8}$ $2\frac{5}{8}$

76. $\frac{5}{7} \div \frac{3}{21}$ 5

77. $\frac{3}{5} \times \frac{5}{8} - \frac{2}{3} \times \frac{1}{4}$ $\frac{5}{24}$

78. $3\frac{1}{16} - 2\frac{1}{24}$ $1\frac{1}{48}$

▲ **79.** ***Geometry*** A room 12 yards × 5.2 yards had a carpet installed. The bill was $764.40. What was the cost of the installed carpet per square yard?
$12.25/square yard

80. ***Electronic Game Store Profit*** An electronics superstore bought 1050 computer games for $23 each. How much did the store pay in all for these games? The store sold the games for $39 each. How much profit did the store make?
$24,150; $16,800

4.2 The Concept of Proportions

Student Learning Objectives

After studying this section, you will be able to:

1. **Write a proportion.**
2. **Determine whether a statement is a proportion.**

1 Writing a Proportion

A **proportion** states that two ratios or two rates are equal. For example, $\frac{5}{8} = \frac{15}{24}$ is a proportion and $\frac{7 \text{ feet}}{8 \text{ dollars}} = \frac{35 \text{ feet}}{40 \text{ dollars}}$ is also a proportion. A proportion can be read two ways. The proportion $\frac{5}{8} = \frac{15}{24}$ can be read "five eighths equals fifteen twenty-fourths," or it can be read "five *is to* eight *as* fifteen *is to* twenty-four."

EXAMPLE 1 Write the proportion 5 is to 7 as 15 is to 21.

Solution

$$\frac{5}{7} = \frac{15}{21}$$

Practice Problem 1 Write the proportion 6 is to 8 as 9 is to 12.

Teaching Tip Doing simple proportions like Example 2 will help students find the applications in Section 4.4 less difficult.

EXAMPLE 2 Write a proportion to express the following: If four rolls of wallpaper measure 300 feet, then eight rolls of wallpaper will measure 600 feet.

Solution When you write a proportion, order is important. Be sure that the similar units for the rates are in the same position in the fractions

$$\frac{4 \text{ rolls}}{300 \text{ feet}} = \frac{8 \text{ rolls}}{600 \text{ feet}}$$

NOTE TO STUDENT: Fully worked-out solutions to all of the Practice Problems can be found at the back of the text starting at page SP-1

Practice Problem 2 Write a proportion to express the following: If it takes two hours to drive 72 miles, then it will take three hours to drive 108 miles.

2 Determining Whether a Statement Is a Proportion

By definition, a proportion states that two ratios are equal. $\frac{2}{7} = \frac{4}{14}$ is a proportion because $\frac{2}{7}$ and $\frac{4}{14}$ are equivalent fractions. You might say that $\frac{2}{7} = \frac{4}{14}$ is a *true* statement. It is easy enough to see that $\frac{2}{7} = \frac{4}{14}$ is true. However, is $\frac{4}{14} = \frac{6}{21}$ true? Is $\frac{4}{14} = \frac{6}{21}$ a proportion? To determine whether a statement is a proportion, we use the equality test for fractions.

EQUALITY TEST FOR FRACTIONS

For any two fractions where $b \neq 0$ and $d \neq 0$,

$$\text{if and only if } \frac{a}{b} = \frac{c}{d}, \text{ then } a \times d = b \times c.$$

Thus, to see if $\frac{4}{14} = \frac{6}{21}$, we can multiply.

$$\frac{4}{14} \times \frac{6}{21} \qquad 14 \times 6 = 84 \qquad 4 \times 21 = 84$$

The cross products are equal.

$\frac{4}{14} = \frac{6}{21}$ is true. $\frac{4}{14} = \frac{6}{21}$ is a proportion.

This method is called finding **cross products.**

EXAMPLE 3 Determine which equations are proportions.

(a) $\frac{14}{18} \stackrel{?}{=} \frac{35}{45}$ **(b)** $\frac{16}{21} \stackrel{?}{=} \frac{174}{231}$

Solution

(a) $\frac{14}{18} \stackrel{?}{=} \frac{35}{45}$

$18 \times 35 = 630$

The cross products are equal.

$14 \times 45 = 630$

Thus $\frac{14}{18} = \frac{35}{45}$. This is a proportion.

(b) $\frac{16}{21} \stackrel{?}{=} \frac{174}{231}$

$21 \times 174 = 3654$

The cross products are not equal.

$16 \times 231 = 3696$

Thus $\frac{16}{21} \neq \frac{174}{231}$. This is not a proportion.

Practice Problem 3 Determine which of the following equations are proportions.

(a) $\frac{10}{18} \stackrel{?}{=} \frac{25}{45}$ **(b)** $\frac{42}{100} \stackrel{?}{=} \frac{22}{55}$

Teaching Tip In problems like this, remind students that they must complete the whole product on each side to verify that it is a proportion. However, to determine whether it is not a proportion, they can sometimes just check the final digits of the product. If these do not match, it's not a proportion. Show them this new example.

$$\frac{19}{23} = \frac{26}{29}$$

Explain that since the final digit of the cross products is not the same, the entire products are different and it is not a proportion. 551 is not equal to 598 simply because the final digit 1 is not equal to the final digit 8.

Proportions may involve fractions or decimals.

EXAMPLE 4 Determine which equations are proportions.

(a) $\frac{5.5}{7} \stackrel{?}{=} \frac{33}{42}$ **(b)** $\frac{5}{8\frac{3}{4}} \stackrel{?}{=} \frac{40}{72}$

Solution

(a) $\frac{5.5}{7} \stackrel{?}{=} \frac{33}{42}$

$7 \times 33 = 231$

The cross products are equal.

$5.5 \times 42 = 231$

Thus $\frac{5.5}{7} = \frac{33}{42}$. This is a proportion.

(b) $\dfrac{5}{8\frac{3}{4}} \stackrel{?}{=} \dfrac{40}{72}$

First we multiply $8\dfrac{3}{4} \times 40 = \dfrac{35}{\not{4}_1} \times \overset{10}{\not{40}} = 35 \times 10 = 350$

$8\dfrac{3}{4} \times 40 = 350$

$\dfrac{5}{8\frac{3}{4}} \times \dfrac{40}{72}$ The cross products are not equal. Thus $\dfrac{5}{8\frac{3}{4}} \neq \dfrac{40}{72}$. This is not a proportion.

$5 \times 72 = 360$

NOTE TO STUDENT: Fully worked-out solutions to all of the Practice Problems can be found at the back of the text starting at page SP-1

Practice Problem 4 Determine which equations are proportions.

(a) $\dfrac{2.4}{3} \stackrel{?}{=} \dfrac{12}{15}$ **(b)** $\dfrac{2\frac{1}{3}}{6} \stackrel{?}{=} \dfrac{14}{38}$

EXAMPLE 5 **(a)** Is the rate $\dfrac{\$86}{13 \text{ tons}}$ equal to the rate $\dfrac{\$79}{12 \text{ tons}}$?

(b) Is the rate $\dfrac{3 \text{ American dollars}}{2 \text{ British pounds}}$ equal to the rate $\dfrac{27 \text{ American dollars}}{18 \text{ British pounds}}$?

Solution

(a) We want to know whether $\dfrac{86}{13} = \dfrac{79}{12}$.

$13 \times 79 = 1027$

$\dfrac{86}{13} \times \dfrac{79}{12}$ The cross products are not equal. Thus the two rates are not equal. This is not a proportion.

$86 \times 12 = 1032$

(b) We want to know whether $\dfrac{3}{2} = \dfrac{27}{18}$.

$2 \times 27 = 54$

$\dfrac{3}{2} \times \dfrac{27}{18}$ The cross products are equal. Thus the two rates are equal. This is a proportion.

$3 \times 18 = 54$

Teaching Tip Some teachers like to use the statement "the proportion is false" for Example 5a, and the statement "the proportion is true" for Example 5b. That is an alternate way of describing the two possibilities.

Practice Problem 5

(a) Is the rate $\dfrac{1260 \text{ words}}{7 \text{ pages}}$ equal to the rate $\dfrac{3530 \text{ words}}{20 \text{ pages}}$?

(b) Is the rate $\dfrac{2 \text{ American dollars}}{11 \text{ French francs}}$ equal to the rate $\dfrac{16 \text{ American dollars}}{88 \text{ French francs}}$?

4.2 Exercises

Student Solutions Manual | CD/ Video | PH Math Tutor Center | MathXL®Tutorials on CD | MathXL® | MyMathLab® | Interactmath.com

Verbal and Writing Skills

1. A proportion states that two ratios or rates are equal.

2. Explain in your own words how we use the equality test for fractions to determine if a statement is a proportion. Give an example. Answers may vary. Check explanations and examples for accuracy.

Write a proportion.

3. 18 is to 9 as 2 is to 1.
$\frac{18}{9} = \frac{2}{1}$

4. 16 is to 12 as 4 is to 3.
$\frac{16}{12} = \frac{4}{3}$

5. 20 is to 36 as 5 is to 9.
$\frac{20}{36} = \frac{5}{9}$

6. 120 is to 15 as 160 is to 20
$\frac{120}{15} = \frac{160}{20}$

7. 220 is to 11 as 400 is to 20
$\frac{220}{11} = \frac{400}{20}$

8. $1\frac{1}{2}$ is to 6 as $7\frac{3}{4}$ is to 31.
$\frac{1\frac{1}{2}}{6} = \frac{7\frac{3}{4}}{31}$

9. $5\frac{1}{3}$ is to 16 as $7\frac{2}{3}$ is to 23.
$\frac{5\frac{1}{2}}{16} = \frac{7\frac{2}{3}}{23}$

10. 5.5 is to 10 as 11 is to 20.
$\frac{5.5}{10} = \frac{11}{20}$

11. 6.5 is to 14 as 13 is to 28.
$\frac{6.5}{14} = \frac{13}{28}$

Applications

Write a proportion.

12. ***Cooking*** When Jenny makes rice in her steamer, she mixes 2 cups of rice with 3 cups of water. To make 8 cups of rice, she needs 12 cups of water.
$\frac{2 \text{ cups rice}}{3 \text{ cups water}} = \frac{8 \text{ cups rice}}{12 \text{ cups water}}$

13. ***Cartography*** A cartographer (a person who makes maps) uses a scale of 3 inches to represent 40 miles. 27 inches would then represent 360 miles.
$\frac{3 \text{ inches}}{40 \text{ miles}} = \frac{27 \text{ inches}}{360 \text{ miles}}$

14. ***Reading Speed*** If Adam can read 18 pages of his biology book in two hours, he can read 45 pages in five hours.
$\frac{18 \text{ pages}}{2 \text{ hours}} = \frac{45 \text{ pages}}{5 \text{ hours}}$

15. ***Tips for Waitress*** If Sheila earned \$28 in tips for serving 6 tables, she should earn \$98 for serving 21 tables.
$\frac{\$28}{6 \text{ tables}} = \frac{\$98}{21 \text{ tables}}$

16. ***Food Cost*** If 20 pounds of pistachio nuts cost \$75, then 30 pounds will cost \$112.50.
$\frac{20 \text{ pounds}}{\$75} = \frac{30 \text{ pounds}}{\$112.50}$

17. ***Education*** If three credit hours at El Paso Community College costs \$525, then seven credit hours should cost \$1225.
$\frac{3 \text{ hours}}{\$525} = \frac{7 \text{ hours}}{\$1225}$

18. ***Education*** When Ridgewood Community College had 1200 students enrolled, 24 mathematics sections were offered. This year there are 1450 students enrolled, so 29 mathematics sections should be offered.
$\frac{1200 \text{ students}}{24 \text{ sections}} = \frac{1450 \text{ students}}{29 \text{ sections}}$

19. ***Teaching Ratio*** There are 3 teaching assistants for every 40 children in the elementary school. If we have 280 children, then we will have 21 teaching assistants.
$\frac{3 \text{ teaching assistants}}{40 \text{ children}} = \frac{21 \text{ teaching assistants}}{280 \text{ children}}$

▲ **20.** ***Lawn Care*** If 16 pounds of fertilizer cover 1520 square feet of lawn, then 19 pounds of fertilizer should cover 1805 square feet of lawn.

$$\frac{16 \text{ pounds}}{1520 \text{ square feet}} = \frac{19 \text{ pounds}}{1805 \text{ square feet}}$$

21. ***Restaurants*** When New City had 4800 people, it had three restaurants. Now New City has 11,200 people, so it should have seven restaurants. $\frac{4800 \text{ people}}{3 \text{ restaurants}} = \frac{11{,}200 \text{ people}}{7 \text{ restaurants}}$

Determine which equations are proportions.

22. $\frac{8}{6} \stackrel{?}{=} \frac{20}{15}$

$8 \times 15 \stackrel{?}{=} 6 \times 20$

$120 = 120$

It is a proportion.

23. $\frac{10}{25} \stackrel{?}{=} \frac{6}{15}$

$10 \times 15 \stackrel{?}{=} 25 \times 6$

$150 = 150$

It is a proportion.

24. $\frac{12}{7} \stackrel{?}{=} \frac{15}{9}$

$12 \times 9 \stackrel{?}{=} 7 \times 15$

$108 \neq 105$

It is not a proportion.

25. $\frac{11}{7} \stackrel{?}{=} \frac{20}{13}$

$11 \times 13 \stackrel{?}{=} 7 \times 20$

$143 \neq 140$

It is not a proportion.

26. $\frac{99}{100} \stackrel{?}{=} \frac{49}{50}$

$99 \times 50 = 100 \times 49$

$4950 \neq 4900$

It is not a proportion.

27. $\frac{17}{75} \stackrel{?}{=} \frac{22}{100}$

$17 \times 100 \stackrel{?}{=} 75 \times 22$

$1700 \neq 1650$

It is not a proportion.

28. $\frac{315}{2100} \stackrel{?}{=} \frac{15}{100}$

$315 \times 100 = 2100 \times 15$

$31{,}500 = 31{,}500$

It is a proportion.

29. $\frac{102}{120} \stackrel{?}{=} \frac{85}{100}$

$102 \times 100 \stackrel{?}{=} 120 \times 85$

$10{,}200 = 10{,}200$

It is a proportion.

30. $\frac{6}{14} \stackrel{?}{=} \frac{4.5}{10.5}$

$6 \times 10.5 \stackrel{?}{=} 14 \times 4.5$

$63 = 63$

It is a proportion.

31. $\frac{2.5}{4} \stackrel{?}{=} \frac{7.5}{12}$

$2.5 \times 12 \stackrel{?}{=} 4 \times 7.5$

$30 = 30$

It is a proportion.

32. $\frac{11}{12} \stackrel{?}{=} \frac{9.5}{10}$

$11 \times 10 \stackrel{?}{=} 12 \times 9.5$

$110 \neq 114$

It is not a proportion.

33. $\frac{3}{17} \stackrel{?}{=} \frac{4.5}{24.5}$

$3 \times 24.5 \stackrel{?}{=} 17 \times 4.5$

$73.5 \neq 76.5$

It is not a proportion.

34. $\frac{2}{4\frac{1}{3}} \stackrel{?}{=} \frac{6}{13}$

$2 \times 13 \stackrel{?}{=} 4\frac{1}{3} \times 6$

$26 = 26$

It is a proportion.

35. $\frac{2}{4\frac{3}{4}} \stackrel{?}{=} \frac{8}{19}$

$2 \times 19 \stackrel{?}{=} 4\frac{3}{4} \times 8$

$38 = 38$

It is a proportion.

36. $\frac{2\frac{1}{3}}{3} \stackrel{?}{=} \frac{7}{15}$

$2\frac{1}{3} \times 15 \stackrel{?}{=} 3 \times 7$

$35 \neq 21$

It is not a proportion.

37. $\frac{7\frac{1}{3}}{3} \stackrel{?}{=} \frac{23}{9}$

$7\frac{1}{3} \times 9 \stackrel{?}{=} 3 \times 23$

$66 \neq 69$

It is not a proportion.

38. $\frac{2.5}{\frac{1}{2}} \stackrel{?}{=} \frac{21}{5}$

$2.5 \times 5 \stackrel{?}{=} \frac{1}{2} \times 21$

$10.5 = 10.5$

It is a proportion.

39. $\frac{\frac{1}{4}}{2} \stackrel{?}{=} \frac{\frac{7}{20}}{2.8}$

$\frac{1}{4} \times 2.8 \stackrel{?}{=} 2 \times \frac{7}{20}$

$0.7 = 0.7$

It is a proportion.

40. $\frac{75 \text{ miles}}{5 \text{ hours}} \stackrel{?}{=} \frac{105 \text{ miles}}{7 \text{ hours}}$

$75 \times 7 \stackrel{?}{=} 5 \times 105$

$525 = 525$

It is a proportion.

41. $\frac{135 \text{ miles}}{3 \text{ hours}} \stackrel{?}{=} \frac{225 \text{ miles}}{5 \text{ hours}}$

$135 \times 5 \stackrel{?}{=} 3 \times 25$

$675 = 675$

It is a proportion.

42. $\frac{286 \text{ gallons}}{12 \text{ acres}} \stackrel{?}{=} \frac{429 \text{ gallons}}{18 \text{ acres}}$

$286 \times 18 \stackrel{?}{=} 12 \times 429$

$5148 = 5148$

It is a proportion.

43. $\frac{166 \text{ gallons}}{14 \text{ acres}} \stackrel{?}{=} \frac{249 \text{ gallons}}{21 \text{ acres}}$

$166 \times 21 \stackrel{?}{=} 14 \times 249$

$3486 = 3486$

It is a proportion.

44. $\frac{52 \text{ free throws}}{80 \text{ attempts}} \stackrel{?}{=} \frac{60 \text{ free throws}}{95 \text{ attempts}}$

$52 \times 95 \stackrel{?}{=} 80 \times 60$

$4940 \neq 4800$

It is not a proportion.

45. $\frac{21 \text{ homeruns}}{96 \text{ games}} \stackrel{?}{=} \frac{18 \text{ homeruns}}{81 \text{ games}}$

$21 \times 81 \stackrel{?}{=} 96 \times 18$

$1701 \neq 1728$

It is not a proportion.

46. ***Concert Audiences*** At the Michael W. Smith concert on Friday there were 9600 female fans and 8200 male fans. The concert on Saturday had 12,480 female fans and 10,660 male fans. Is the ratio of female fans to male fans the same for both nights of the concert? yes

47. ***Baseball Team Wins*** Since Harding High School opened, they have won 132 baseball games and have lost 22. Derry High School has won 160 games and lost 32 since it opened. Is the ratio of lost games to won games the same for both schools? no

48. ***Machine Operating Rate*** A machine folds 730 boxes in six hours. Another machine folds 1090 boxes in nine hours.

(a) Do they fold boxes at the same rate? no

(b) Which machine folds more boxes in 24 hours?

The machine that folds 730 boxes in six hours will fold more.

49. ***Speed of Vehicle*** A car traveled 550 miles in 15 hours. A bus traveled 230 miles in 6 hours.

(a) Did they travel at the same rate? no

(b) Which vehicle travels at a faster rate?

The bus is traveling at a faster rate.

▲ **50.** ***Television Screen Size*** A common size for a color television screen is 22 inches wide by 16 inches tall. Does a smaller color television screen that is 11 inches wide by 8.5 inches tall have the same ratio of width to length? no

▲ **51.** ***Driveway Size*** A common size for a driveway in suburban Wheaton, Illinois, is 75 feet long by 20 feet wide. Does a larger driveway that is 105 feet long by 28 feet wide have the same ratio of width to length? yes

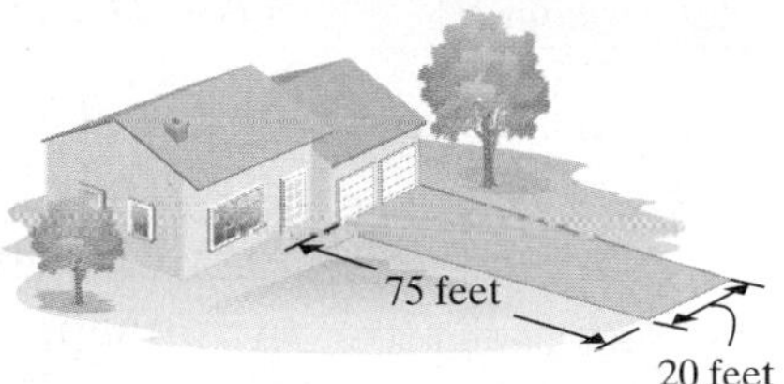

To Think About

52. Determine whether $\frac{63}{161} = \frac{171}{437}$

(a) by reducing each side to lowest terms.

$\frac{63}{161} = \frac{9}{23}$ $\frac{171}{437} = \frac{9}{23}$ yes

(b) by using the equality test for fractions. (This is the cross product method.)

$63 \times 437 \stackrel{?}{=} 161 \times 171$

$27{,}531 = 27{,}531$ yes

(c) Which method was faster? Why?

The equality test for fractions; for most students it is faster to multiply than to reduce fractions.

53. Determine whether $\frac{169}{221} = \frac{247}{323}$

(a) by reducing each side to lowest terms.

$\frac{169}{221} = \frac{13}{17}$ $\frac{247}{323} = \frac{13}{17}$ yes

(b) by using the equality test for fractions. (This is the cross product method.)

$169 \times 323 \stackrel{?}{=} 221 \times 247$

$54{,}587 = 54{,}587$ yes

(c) Which method was faster? Why?

The equality test for fractions; for most students it is faster to multiply than to reduce fractions.

Cumulative Review

Calculate.

54. $9.6 + 7.8 + 2.56 + 3.004 + 0.1765$ 23.1405

55. 5.92×3.04 17.9968

56. $29{,}366.215 - 28{,}963.807$

402.408

57. $7.03\overline{)181.374}$ 25.8

58. ***Walking Distance*** Susan has a goal of walking 20 miles this week. On Monday she walked $3\frac{1}{4}$ miles and on Tuesday she walked $4\frac{3}{8}$ miles. How many more miles does she need to walk to reach her goal? $12\frac{3}{8}$ miles

How Am I Doing? Sections 4.1–4.2

1. $\frac{13}{18}$

2. $\frac{1}{5}$

3. $\frac{9}{2}$

4. $\frac{11}{12}$

5. (a) $\frac{7}{24}$ (b) $\frac{11}{120}$

6. $\frac{3 \text{ flight attendants}}{100 \text{ passengers}}$

7. $\frac{31 \text{ gallons}}{42 \text{ square feet}}$

8. 30.5 miles per hour.

9. $29 per CD player

10. 160 cookies per pound of cookie dough

11. $\frac{13}{40} = \frac{39}{120}$

12. $\frac{116}{158} = \frac{29}{37}$

13. $\frac{33 \text{ nautical miles}}{2 \text{ hours}} = \frac{49.5 \text{ nautical miles}}{3 \text{ hours}}$

14. $\frac{3000 \text{ shoes}}{\$370} = \frac{7500 \text{ shoes}}{\$925}$

15. It is a proportion.

16. It is not a proportion.

17. It is a proportion.

18. It is not a proportion.

19. It is a proportion.

20. It is a proportion.

How are you doing with your homework assignments in Sections 4.1 to 4.2? Do you feel you have mastered the material so far? Do you understand the concepts you have covered? Before you go further in the textbook, take some time to do each of the following problems.

4.1

In questions 1–4, write each ratio in simplest form.

1. 13 to 18

2. 44 to 220

3. $72 to $16

4. 121 kilograms to 132 kilograms

5. Sam's take-home pay is $240 per week. $70 per week is withheld for federal taxes and $22 per week is withheld for state taxes.
   (a) Find the ratio of federal withholding to take-home pay.
   (b) Find the ratio of state withholding to take-home pay.

Write each rate in simplest form.

6. 9 flight attendants for 300 passengers

7. 620 gallons of water for each 840 square feet of lawn

Write as a unit rate. Round to the nearest tenth if necessary.

8. 122 miles are traveled in 4 hours. What is the rate in miles per hour?

9. 15 CD players are purchased for $435. What is the cost per CD player?

10. In a certain recipe, 2400 cookies are made with 15 pounds of cookie dough. How many cookies can be made with 1 pound of cookie dough?

4.2

Write a proportion.

11. 13 is to 40 as 39 is to 120

12. 116 is to 158 as 29 is to 37

13. If a speedboat can travel 33 nautical miles in 2 hours, then it can travel 49.5 nautical miles in 3 hours.

14. If the cost to manufacture 3000 athletic shoes is $370, then the cost to manufacture 7500 atheltic shoes is $925.

Determine whether each equation is a proportion.

15. $\frac{14}{31} = \frac{42}{93}$

16. $\frac{17}{33} = \frac{19}{45}$

17. $\frac{5.6}{3.2} = \frac{112}{64}$

18. $\frac{2\frac{1}{2}}{3\frac{1}{3}} = \frac{35}{46}$

19. The Pine Street Inn can produce 670 servings of a chicken dinner for homeless people at a cost of $1541. It will therefore cost $1886 to produce 820 servings of the same chicken dinner.

20. For every 30 flights that arrive at Logan Airport in December approximately 4 of them are more than 15 minutes late. Therefore, for every 3000 flights that arrive at Logan Airport in December approximately 400 of them will be more than 15 minutes late.

Now turn to page SA-9 for the answer to each of these problems. Each answer also includes a reference to the objective in which the problem is first taught. If you missed any of these problems, you should stop and review the Examples and Practice Problems in the referenced objective. A little review now will help you master the material in the upcoming sections of the text.

4.3 Solving Proportions

Student Learning Objectives

After studying this section, you will be able to:

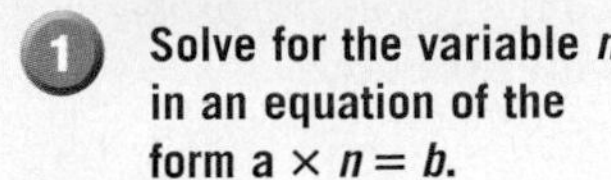
1. Solve for the variable n in an equation of the form $a \times n = b$.

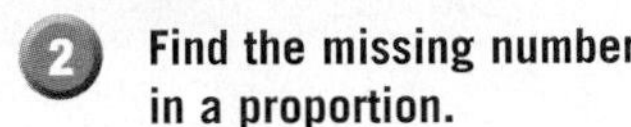
2. Find the missing number in a proportion.

Teaching Tip Some instructors prefer to introduce the notation $3n$ to indicate a product instead of $3 \times n$. Either is acceptable, but students who have never had algebra before may have trouble with the notation.

1 Solving for the Variable n in an Equation of the Form $a \times n = b$

Consider this expression: "3 times a number yields 15. What is the number?" We could write this as

$$3 \times \boxed{?} = 15$$

and guess that the number $\boxed{?} = 5$. There is a better way of solving this problem, a way that eliminates the guesswork. We will begin by using a **variable.** That is, we will use a letter to represent a number we do not yet know. We briefly used variables in Chapters 1–3. Now we use them more extensively.

Let the letter n represent the unknown number. We write

$$3 \times n = 15.$$

This is called an **equation.** An equation has an equal sign. This indicates that the values on each side of it are equivalent. We want to find the number n in this equation without guessing. We will not change the value of n in the equation if we divide both sides of the equation by 3. Thus if

$$3 \times n = 15,$$

we can say $$\frac{3 \times n}{3} = \frac{15}{3},$$

which is $$\frac{3}{3} \times n = 5$$

or $$1 \times n = 5.$$

Since $1 \times$ any number is the same number, we know that $n = 5$. Any equation of the form $a \times n = b$ can be solved in this way. We divide both sides of an equation of the form $a \times n = b$ by the number that is multiplied by n. (We do this because division is the inverse operation of multiplication. This method will not work for $3 + n = 15$, since here the 3 is added to n and not multiplied by n.)

EXAMPLE 1 Solve for n. **(a)** $16 \times n = 80$ **(b)** $24 \times n = 240$

Solution

(a) $16 \times n = 80$

$\frac{16 \times n}{16} = \frac{80}{16}$ Divide each side by 16.

$n = 5$ because $16 \div 16 = 1$ and $80 \div 16 = 5$.

(b) $24 \times n = 240$

$\frac{24 \times n}{24} = \frac{240}{24}$ Divide each side by 24.

$n = 10$ because $24 \div 24 = 1$ and $240 \div 24 = 10$.

Practice Problem 1 Solve for n.

(a) $5 \times n = 45$ **(b)** $7 \times n = 84$

NOTE TO STUDENT: Fully worked-out solutions to all of the Practice Problems can be found at the back of the text starting at page SP-1

The same procedure is followed if the variable n is on the right side of the equation.

Teaching Tip Remind students that $n = 6$ and $6 = n$ mean exactly the same thing. This is called the *symmetric property of equality.*

EXAMPLE 2 Solve for n. **(a)** $66 = 11 \times n$ **(b)** $143 = 13 \times n$

Solution

(a) $66 = 11 \times n$

$$\frac{66}{11} = \frac{11 \times n}{11} \quad \text{Divide each side by 11.}$$

$$6 = n$$

(b) $143 = 13 \times n$

$$\frac{143}{13} = \frac{13 \times n}{13} \quad \text{Divide each side by 13.}$$

$$11 = n$$

Practice Problem 2 Solve for n.

(a) $108 = 9 \times n$ **(b)** $210 = 14 \times n$

NOTE TO STUDENT: Fully worked-out solutions to all of the Practice Problems can be found at the back of the text starting at page SP-1

The numbers in the equations are not always whole numbers, and the answer to an equation is not always a whole number.

EXAMPLE 3 Solve for n. **(a)** $16 \times n = 56$ **(b)** $18.2 = 2.6 \times n$

Solution

(a) $16 \times n = 56$

$$\frac{16 \times n}{16} = \frac{56}{16} \quad \text{Divide each side by 16.}$$

$$n = 3.5$$

$$\begin{array}{r} 3.5 \\ 16\overline{)56.0} \\ \underline{48} \\ 8\,0 \\ \underline{8\,0} \\ 0 \end{array}$$

(b) $18.2 = 2.6 \times n$

$$\frac{18.2}{2.6} = \frac{2.6 \times n}{2.6} \quad \text{Divide each side by 2.6}$$

$$7 = n$$

$$\begin{array}{r} 7. \\ 2.6_{\wedge}\overline{)18.2_{\wedge}} \\ \underline{18\,2} \\ 0 \end{array}$$

Practice Problem 3 Solve for n.

(a) $15 \times n = 63$ **(b)** $39.2 = 5.6 \times n$

2 Finding the Missing Number in a Proportion

Sometimes one of the pieces of a proportion is unknown. We can use an equation such as $a \times n = b$ and solve for n to find the unknown quantity. Suppose we want to know the value of n in the proportion

$$\frac{5}{12} = \frac{n}{144}.$$

Since this is a proportion, we know that $5 \times 144 = 12 \times n$. Simplifying, we have

$$720 = 12 \times n.$$

Next we divide both sides by 12.

$$\frac{720}{12} = \frac{12 \times n}{12}$$

$$60 = n$$

We check to see if this is correct. Do we have a true proportion?

$$\frac{5}{12} \stackrel{?}{=} \frac{60}{144}$$

$$\frac{5}{12} \times \frac{60}{144} \qquad \begin{aligned} 12 \times 60 &= 720 \\ 5 \times 144 &= 720 \end{aligned}$$

The cross products are equal.

Thus $\frac{5}{12} = \frac{60}{144}$ is true. We have checked our answer.

TO SOLVE FOR A MISSING NUMBER IN A PROPORTION

1. Find the cross products.
2. Divide each side of the equation by the number multiplied by n.
3. Simplify the result.
4. Check your answer.

EXAMPLE 4 Find the value of n in $\frac{25}{4} = \frac{n}{12}$.

Solution

$$25 \times 12 = 4 \times n \quad \text{Find the cross products.}$$

$$300 = 4 \times n$$

$$\frac{300}{4} = \frac{4 \times n}{4} \quad \text{Divide each side by 4.}$$

$$75 = n$$

Check. Is this a proportion?

$$\frac{25}{4} \stackrel{?}{=} \frac{75}{12}$$

$$25 \times 12 \stackrel{?}{=} 4 \times 75$$

$$300 = 300 \quad \checkmark$$

It is a proportion. The answer $n = 75$ is correct.

Practice Problem 4 Find the value of n in $\frac{24}{n} = \frac{3}{7}$.

The answer to the next problem is not a whole number.

EXAMPLE 5 Find the value of n in $\frac{125}{2} = \frac{150}{n}$.

Solution

$125 \times n = 2 \times 150$ Find the cross products.

$125 \times n = 300$

$\frac{125 \times n}{125} = \frac{300}{125}$ Divide each side by 125.

$n = 2.4$

Check. $\frac{125}{2} \stackrel{?}{=} \frac{150}{2.4}$

$125 \times 2.4 \stackrel{?}{=} 2 \times 150$

$300 = 300$ ✓

NOTE TO STUDENT: Fully worked-out solutions to all of the Practice Problems can be found at the back of the text starting at page SP-1

Practice Problem 5 Find the value of n in $\frac{176}{4} = \frac{286}{n}$.

EXAMPLE 6 Find the value of n in

$$\frac{n}{20} = \frac{\frac{3}{4}}{5}.$$

Solution

$5 \times n = 20 \times \frac{3}{4}$ Find the cross products.

$5 \times n = 15$ Simplify.

$\frac{5 \times n}{5} = \frac{15}{5}$ Divide each side by 5.

$n = 3$

Check. Can you verify that this is a proportion?

Practice Problem 6 Find the value of n in $\frac{n}{30} = \frac{\frac{2}{3}}{4}$.

Teaching Tip Stress the importance of labeling the answers with the proper units in real-life problems. Students tend to avoid this, so remind them of the need to establish good habits now.

In real-life situations it is helpful to write the units of measure in the proportion. Remember, order is important. The same units should be in the same position in the fractions.

EXAMPLE 7 If 5 grams of a non-icing additive are placed in 8 liters of diesel fuel, how many grams n should be added to 12 liters of diesel fuel?

Solution We need to find the value of n in $\frac{n \text{ grams}}{12 \text{ liters}} = \frac{5 \text{ grams}}{8 \text{ liters}}$.

$$8 \times n = 12 \times 5$$
$$8 \times n = 60$$
$$\frac{8 \times n}{8} = \frac{60}{8}$$
$$n = 7.5$$

The answer is 7.5 grams. 7.5 grams of the additive should be added to 12 liters of the diesel fuel.

Check.

$$\frac{7.5 \text{ grams}}{12 \text{ liters}} \stackrel{?}{=} \frac{5 \text{ grams}}{8 \text{ liters}}$$

$$7.5 \times 8 \stackrel{?}{=} 12 \times 5$$

$$60 = 60 \quad \checkmark$$

Practice Problem 7 Find the value of n in $\dfrac{80 \text{ dollars}}{5 \text{ tons}} = \dfrac{n \text{ dollars}}{6 \text{ tons}}$.

Some answers will be exact values. In other cases we will obtain answers that are rounded to a certain decimal place. Recall that we sometimes use the $\approx$ symbol, which means "is approximately equal to."

EXAMPLE 8 Find the value of n in $\dfrac{141 \text{ miles}}{4.5 \text{ hours}} = \dfrac{67 \text{ miles}}{n \text{ hours}}$. Round to the nearest tenth.

Solution

$$141 \times n = 67 \times 4.5$$

$$141 \times n = 301.5$$

$$\frac{141 \times n}{141} = \frac{301.5}{141}$$

If we calculate to four decimal places, we have $n = 2.1382$. Rounding to the nearest tenth, $n \approx 2.1$.

The answer to the nearest tenth is $n = 2.1$. The check is up to you.

Practice Problem 8 Find the value of n in $\dfrac{264 \text{ meters}}{3.5 \text{ seconds}} = \dfrac{n \text{ meters}}{2 \text{ seconds}}$. Round to the nearest tenth.

TO THINK ABOUT: Proportions with Mixed Numbers or Fractions Suppose that the proportion contains many fractions or mixed numbers. Could you still follow all the steps? For example, find n when

$$\frac{n}{3\frac{1}{4}} = \frac{5\frac{1}{6}}{2\frac{1}{3}}$$

We have $2\frac{1}{3} \times n = 5\frac{1}{6} \times 3\frac{1}{4}$.

This can be written as

$$\frac{7}{3} \times n = \frac{31}{6} \times \frac{13}{4}$$

$$\boxed{\frac{7}{3} \times n = \frac{403}{24}} \qquad \text{equation (1)}$$

Now we divide each side of equation (1) by $\frac{7}{3}$. Why?

$$\frac{\frac{7}{3} \times n}{\frac{7}{3}} = \frac{\frac{403}{24}}{\frac{7}{3}}$$

Be careful here. The right-hand side means $\frac{403}{24} \div \frac{7}{3}$, which we evaluate by *inverting* the second fraction and multiplying.

$$\frac{403}{\cancelto{8}{24}} \times \frac{\cancelto{1}{3}}{7} = \frac{403}{56}$$

Thus $n = \frac{403}{56}$ or $7\frac{11}{56}$. Think about all the steps to solving this problem. Can you follow them? There is another way to do the problem. We could multiply each side of equation (1) by $\frac{3}{7}$.

$$\frac{7}{3} \times n = \frac{403}{24} \qquad \text{equation (1)}$$

$$\frac{3}{7} \times \frac{7}{3} \times n = \frac{3}{7} \times \frac{403}{24}$$

$$n = \frac{403}{56}$$

Why does this work? Now try exercises 56–59 in Exercises 4.3.

4.3 Exercises

Student Solutions Manual | CD/Video | PH Math Tutor Center | MathXL®Tutorials on CD | MathXL® | MyMathLab® | Interactmath.com

Verbal and Writing Skills

1. Suppose you have an equation of the form $a \times n = b$, where the letters a and b represent whole numbers and $a \neq 0$. Explain in your own words how you would solve the equation.

Divide each side of the equation by the number a. Calculate $\frac{b}{a}$. The value of n is $\frac{b}{a}$.

2. Suppose you have an equation of the form $\frac{n}{a} = \frac{b}{c}$, where a, b, and c represent whole numbers and $a, c \neq 0$. Explain in your own words how you would solve the equation.

Form the cross product $c \times n = a \times b$. Multiply $a \times b$. Then use the steps explained in exercise 1.

Solve for n.

3. $12 \times n = 132$ — $n = 11$

4. $14 \times n = 126$ — $n = 9$

5. $3 \times n = 16.8$ — $n = 5.6$

6. $2 \times n = 19.6$ — $n = 9.8$

7. $n \times 11.4 = 57$ — $n = 5$

8. $n \times 3.8 = 95$ — $n = 25$

9. $50.4 = 6.3 \times n$ — $n = 8$

10. $40.6 = 5.8 \times n$ — $n = 7$

11. $\frac{3}{4} \times n = 26$ (*Hint:* Divide each side by $\frac{3}{4}$.) — $n = 34\frac{2}{3}$

12. $\frac{5}{6} \times n = 32$ (*Hint:* Divide each side by $\frac{5}{6}$.) — $n = 38\frac{2}{5}$

Find the value of n. Check your answer.

13. $\frac{n}{20} = \frac{3}{4}$ — $n = 15$

14. $\frac{n}{28} = \frac{3}{7}$ — $n = 12$

15. $\frac{6}{n} = \frac{3}{8}$ — $n = 16$

16. $\frac{4}{n} = \frac{2}{7}$ — $14 = n$

17. $\frac{12}{40} = \frac{n}{25}$ — $7.5 = n$

18. $\frac{13}{30} = \frac{n}{15}$ — $6.5 = n$

19. $\frac{50}{100} = \frac{2.5}{n}$ — $n = 5$

20. $\frac{40}{160} = \frac{1.5}{n}$ — $n = 6$

21. $\frac{n}{6} = \frac{150}{12}$ — $n = 75$

22. $\frac{n}{22} = \frac{25}{11}$ — $n = 50$

23. $\frac{15}{4} = \frac{n}{6}$ — $22.5 = n$

24. $\frac{16}{10} = \frac{n}{9}$ — $14.4 = n$

25. $\frac{550}{n} = \frac{5}{3}$ — $n = 330$

26. $\frac{480}{n} = \frac{12}{11}$ — $n = 440$

Find the value of n. Round your answer to the nearest tenth when necessary.

27. $\frac{21}{n} = \frac{2}{3}$ — $n = 31.5$

28. $\frac{62}{n} = \frac{5}{4}$ — $n = 49.6$

29. $\frac{9}{26} = \frac{n}{52}$ — $n = 18$

30. $\frac{12}{8} = \frac{21}{n}$ — $n = 14$

31. $\frac{15}{12} = \frac{10}{n}$ — $n = 8$

32. $\frac{n}{18} = \frac{3.5}{1}$ — $n = 63$

33. $\frac{n}{36} = \frac{4.5}{1}$ — $n = 162$

34. $\frac{2.5}{n} = \frac{0.5}{10}$ — $n = 50$

35. $\frac{1.8}{n} = \frac{0.7}{12}$ — $n \approx 30.9$

36. $\frac{3}{4} = \frac{n}{3.8}$ — $n \approx 2.9$

37. $\frac{7}{8} = \frac{n}{4.2}$ — $n \approx 3.7$

38. $\frac{12.5}{16} = \frac{n}{12}$ — $n \approx 9.4$

39. $\frac{13.8}{15} = \frac{n}{6}$ — $n \approx 5.5$

40. $\frac{5}{n} = \frac{12\frac{1}{2}}{100}$ — $n = 40$

41. $\frac{3}{n} = \frac{6\frac{1}{4}}{100}$ — $n = 48$

Applications

Find the value of n. Round to the nearest hundredth when necessary.

42. $\frac{n \text{ grams}}{10 \text{ liters}} = \frac{7 \text{ grams}}{25 \text{ liters}}$ — $n = 2.8$

43. $\frac{n \text{ pounds}}{20 \text{ ounces}} = \frac{2 \text{ pounds}}{32 \text{ ounces}}$ — $n = 1.25$

44. $\frac{105 \text{ miles}}{2 \text{ hours}} = \frac{n \text{ miles}}{5 \text{ hours}}$ — $n = 262.5$

45. $\frac{128 \text{ miles}}{4 \text{ hours}} = \frac{80 \text{ miles}}{n \text{ hours}}$
$n = 2.5$

46. $\frac{50 \text{ gallons}}{12 \text{ acres}} = \frac{36 \text{ gallons}}{n \text{ acres}}$
$n = 8.64$

47. $\frac{32 \text{ meters}}{5 \text{ yards}} = \frac{24 \text{ meters}}{n \text{ yards}}$
$n = 3.75$

48. $\frac{10 \text{ miles}}{16.1 \text{ kilometers}} = \frac{n \text{ miles}}{7 \text{ kilometers}}$
$n \approx 4.35$

49. $\frac{3 \text{ inches}}{7.62 \text{ centimeters}} = \frac{n \text{ inches}}{10 \text{ centimeters}}$
$n \approx 3.94$

50. $\frac{12 \text{ quarters}}{3 \text{ dollars}} = \frac{87 \text{ quarters}}{n \text{ dollars}}$
$n = 21.75$

51. $\frac{35 \text{ dimes}}{3.5 \text{ dollars}} = \frac{n \text{ dimes}}{8 \text{ dollars}}$
$n = 80$

52. $\frac{2\frac{1}{2} \text{ acres}}{3 \text{ people}} = \frac{n \text{ acres}}{5 \text{ people}}$
$n = 4\frac{1}{6}$

53. $\frac{3\frac{1}{4} \text{ feet}}{8 \text{ pounds}} = \frac{n \text{ feet}}{12 \text{ pounds}}$
$n = 4\frac{7}{8}$

▲ **54.** ***Photography*** A photographic negative is 3.5 centimeters wide and 2.5 centimeters tall. If you want to make a color print that is 6 centimeters tall, how wide will the print be?
8.4 centimeters

▲ **55.** ***Photography*** A color photograph is 5 inches wide and 3 inches tall. If you want to make an enlargement of this photograph that is 6.6 inches tall, how wide will the enlargement be?
11 inches

To Think About

Study the "To Think About" example in the text. Then solve for n in exercises 56–59. Express n as a mixed number.

56. $\frac{n}{7\frac{1}{4}} = \frac{2\frac{1}{5}}{4\frac{1}{8}}$ $\quad n \times 4\frac{1}{8} = 15\frac{19}{20} \quad n = 3\frac{13}{15}$

57. $\frac{n}{2\frac{1}{3}} = \frac{4\frac{5}{6}}{3\frac{1}{9}}$ $\quad n \times 3\frac{1}{9} = 11\frac{5}{18} \quad n = 3\frac{5}{8}$

58. $\frac{9\frac{3}{4}}{n} = \frac{8\frac{1}{2}}{4\frac{1}{3}}$ $\quad n \times 8\frac{1}{2} = 42\frac{1}{4} \quad n = 4\frac{33}{34}$

59. $\frac{8\frac{1}{6}}{n} = \frac{5\frac{1}{2}}{7\frac{1}{3}}$ $\quad n \times 5\frac{1}{2} = 59\frac{8}{9} \quad n = 10\frac{8}{9}$

Cumulative Review

Evaluate by doing each operation in the proper order.

60. $4^3 + 20 \div 5 + 6 \times 3 - 5 \times 2$
$64 + 4 + 18 - 10 = 76$

61. $(3 + 1)^3 - 30 \div 6 - 144 \div 12$
47

62. Write a word name for the decimal 0.563.
five hundred sixty-three thousandths

63. Write thirty-four ten-thousandths in decimal notation. 0.0034

64. ***Profit on Cell Phones*** If a man purchases 156 cell phones for $32 each and sells half of them for $45 and half of them for $39, how much profit will he make? $1560

65. ***Soccer Team*** The North Bend Women's Soccer League has eight teams. If each team plays all the others twice, how many games will have been played? (Think carefully. This is a challenging question.) 56 games

4.4 Solving Applied Problems Involving Proportions

Student Learning Objectives

After studying this section, you will be able to:

 **Solve applied problems using proportions.**

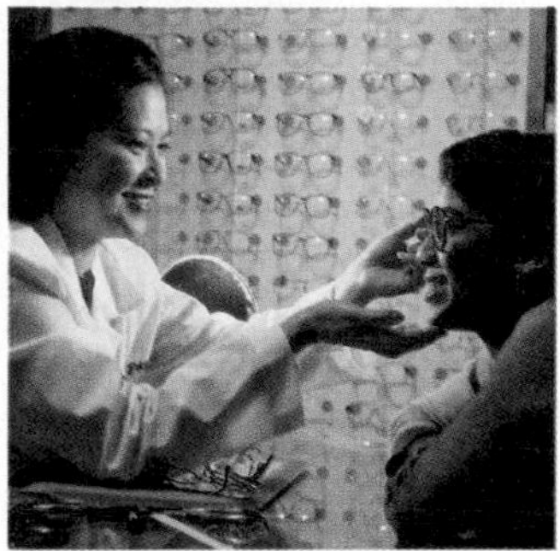

1 Solving Applied Problems Using Proportions

Let us examine a variety of applied problems that can be solved by proportions.

EXAMPLE 1 A company that makes eyeglasses conducted a recent survey using a quality control test. It was discovered that 37 pairs of eyeglasses in a sample of 120 pairs of eyeglasses were defective. If this rate remains the same each year, how many of the 36,000 pairs of eyeglasses made by this company each year are defective?

Solution

Mathematics Blueprint for Problem Solving

Gather the Facts	What Am I Asked to Do?	How Do I Proceed?	Key Points to Remember
Sample: 37 defective pairs in a total of 120 pairs 36,000 pairs were made by the company.	Find how many of the 36,000 pairs of eyeglasses are defective.	Set up a proportion comparing defective eyeglasses to total eyeglasses.	Make sure one fraction represents the sample and one fraction represents the total number of eyeglasses made by the company.

We will use the letter n to represent the number of defective eyeglasses in the total.

$$\underbrace{\frac{37 \text{ defective pairs}}{120 \text{ total pairs of eyeglasses}}}_{\text{We compare the sample}} \quad \text{to} \quad \underbrace{\frac{n \text{ defective pairs}}{36{,}000 \text{ total pairs of eyeglasses}}}_{\text{the total number}}$$

$$37 \times 36{,}000 = 120 \times n \quad \text{Form the cross products.}$$

$$1{,}332{,}000 = 120 \times n \quad \text{Simplify.}$$

$$\frac{1{,}332{,}000}{120} = \frac{120 \times n}{120} \quad \text{Divide each side by 120.}$$

$$11{,}100 = n$$

Thus, if the rate of defective eyeglasses holds steady, there are about 11,100 defective pairs of eyeglasses made by the company each year.

Practice Problem 1 Yesterday an automobile assembly line produced 243 engines, of which 27 were defective. If the same rate is true each day, how many of the 4131 engines produced this month are defective?

Teaching Tip This is a good time to stress the helpfulness of making an estimate in solving a proportion problem. Usually, a mistake will be obvious when compared with an estimate.

NOTE TO STUDENT: Fully worked-out solutions to all of the Practice Problems can be found at the back of the text starting at page SP-1

Looking back at Example 1, perhaps it occurred to you that the fractions in the proportion could be set up in an alternative way. **You can set up**

this problem in several different ways as long as the units are in correctly corresponding positions. It would be correct to set up the problem in the form

$$\frac{\text{defective pairs in sample}}{\text{total defective pairs}} = \frac{\text{total glasses in sample}}{\text{total glasses made by company}}$$

or

$$\frac{\text{total glasses in sample}}{\text{defective pairs in sample}} = \frac{\text{total glasses made by company}}{\text{total defective pairs}}.$$

But we **cannot** set up the problem this way.

$$\frac{\text{defective pairs in sample}}{\text{total glasses made by company}} = \frac{\text{total defective pairs}}{\text{total glasses in sample}}$$

This is *not* correct. Do you see why?

Teaching Tip Sometimes students ask why we do not find that Ted's car gets 35 miles per gallon first in Example 2. Respond by showing them how they would obtain the same answer. Point out that in real life, ratios may not reduce so nicely and that the general approach shown in Example 2 is more useful for solving everyday problems.

EXAMPLE 2 Ted's car can go 245 miles on 7 gallons of gas. Ted wants to take a trip of 455 miles. Approximately how many gallons of gas will this take?

Solution Let n = the unknown number of gallons.

$$\frac{245 \text{ miles}}{7 \text{ gallons}} = \frac{455 \text{ miles}}{n \text{ gallons}}$$

$$245 \times n = 7 \times 455 \quad \text{Form the cross products.}$$

$$245 \times n = 3185 \quad \text{Simplify.}$$

$$\frac{245 \times n}{245} = \frac{3185}{245} \quad \text{Divide both sides by 245.}$$

$$n = 13$$

Ted will need approximately 13 gallons of gas for the trip.

NOTE TO STUDENT: Fully worked-out solutions to all of the Practice Problems can be found at the back of the text starting at page SP-1

Practice Problem 2 Cindy's car travels 234 miles on 9 gallons of gas. How many gallons of gas will Cindy need to take a 312-mile trip?

EXAMPLE 3 In a certain gear, Alice's 18-speed bicycle has a gear ratio of three revolutions of the pedal for every two revolutions of the bicycle wheel. If her bicycle wheel is turning at 65 revolutions per minute, how many times must she pedal per minute?

Solution Let n = the number of revolutions of the pedal.

$$\frac{3 \text{ revolutions of the pedal}}{2 \text{ revolutions of the wheel}} = \frac{n \text{ revolutions of the pedal}}{65 \text{ revolutions of the wheel}}$$

$$3 \times 65 = 2 \times n \quad \text{Cross-multiply.}$$

$$195 = 2 \times n \quad \text{Simplify.}$$

$$\frac{195}{2} = \frac{2 \times n}{2} \quad \text{Divide both sides by 2.}$$

$$97.5 = n$$

Alice will pedal at the rate of 97.5 revolutions in 1 minute.

Practice Problem 3 Alicia must pedal at 80 revolutions per minute to ride her bicycle at 16 miles per hour. If she pedals at 90 revolutions per minute, how fast will she be riding?

EXAMPLE 4 Tim operates a bicycle rental center during the summer months on the island of Martha's Vineyard. He discovered that when the ferryboats brought 8500 passengers a day to the island, his center rented 340 bicycles a day. Next summer the ferryboats plan to bring 10,300 passengers a day to the island. How many bicycles a day should Tim plan to rent?

Teaching Tip Gear ratio problems often confuse students. Ask them why a bicycle might have a gear ratio of 3 to 2 as well as 4 to 5. Ask them which ratio would help them to ride at a higher rate of speed. Usually, this type of topic will generate a good class discussion.

Solution Two important cautions are necessary before we solve the proportion. We need to be sure that the bicycle rentals are directly related to the number of people on the ferryboat. (Presumably, people who fly to the island or who take small pleasure boats to the island also rent bicycles.)

Next we need to be sure that the people who represent the increase in passengers per day would be as likely to rent bicycles as the present number of passengers do. For example, if the new visitors to the island are all senior citizens, they are not as likely to rent bicycles as younger people. If we assume those two conditions are satisfied, then we can solve the problem as follows.

$$\frac{8500 \text{ passengers per day now}}{340 \text{ bike rentals per day now}} = \frac{10{,}300 \text{ passengers per day later}}{n \text{ bike rentals per day later}}$$

$$8500 \times n = 340 \times 10{,}300$$

$$8500 \times n = 3{,}502{,}000$$

$$\frac{8500 \times n}{8500} = \frac{3{,}502{,}000}{8500}$$

$$n = 412$$

If the two conditions are satisfied, we would predict 412 bicycle rentals.

Practice Problem 4 For every 4050 people who walk into Tom's Souvenir Shop, 729 make a purchase. Assuming the same conditions, if 5500 people walk into Tom's Souvenir Shop, how many people may be expected to make a purchase?

Wildlife Population Counting Biologists and others who observe or protect wildlife sometimes use the capture-mark-recapture method to determine how many animals are in a certain region. In this approach some animals are caught and tagged in a way that does not harm them. They are then released into the wild, where they mix with their kind.

It is assumed (usually correctly) that the tagged animals will mix throughout the entire population in that region, so that when they are recaptured in a future sample, the biologists can use them to make reasonable estimates about the total population. We will employ the capture-mark-recapture method in the next example.

EXAMPLE 5 A biologist catches 42 fish in a lake and tags them. She then quickly returns them to the lake. In a few days she catches a new sample of 50 fish. Of those 50 fish, 7 have her tag. Approximately how many fish are in the lake?

Solution

$$\frac{42 \text{ fish tagged in 1st sample}}{n \text{ fish in lake}} = \frac{7 \text{ fish tagged in 2nd sample}}{50 \text{ fish caught in 2nd sample}}$$

$$42 \times 50 = 7 \times n$$
$$2100 = 7 \times n$$
$$\frac{2100}{7} = \frac{7 \times n}{7}$$
$$300 = n$$

Assuming that no tagged fish died and that the tagged fish mixed throughout the population of fish in the lake, we estimate that there are 300 fish in the lake.

NOTE TO STUDENT: ***Fully worked-out solutions to all of the Practice Problems can be found at the back of the text starting at page SP-1***

Practice Problem 5 A park ranger in Alaska captures and tags 50 bears. He then releases them to range through the forest. Sometime later he captures 50 bears. Of the 50, 4 have tags from the previous capture. Estimate the number of bears in the forest.

Developing Your Study Skills

Getting Help

Getting the right kind of help at the right time can be a key ingredient in being successful in mathematics. When you have gone to class on a regular basis, taken careful notes, methodically read your textbook, and diligently done your homework—in other words, when you have made every effort possible to learn the mathematics—you may still find that you are having difficulty. If this is the case, then you need to seek help. Make an appointment with your instructor to find out what help is available to you. The instructor, tutoring services, a mathematics lab, videotapes, and computer software may be among the resources you can draw on.

Student Solutions Manual

CD/ Video

PH Math Tutor Center

MathXL®Tutorials on CD
MathXL®

MyMathLab®

Interactmath.com

Verbal and Writing Skills

1. Dan saw 12 people on the beach on Friday night. He counted 5 dogs on the beach at that time. On Saturday night he saw 60 people on the same beach. He is trying to estimate how many dogs might be on the beach Saturday night. He started by writing the equation

$$\frac{12 \text{ people}}{5 \text{ dogs}} = \qquad .$$

Explain how he should set up the rest of the proportion.

He should continue with people on the top of the fraction. That would be 60 people that he observed on Saturday night. He does not know the number of dogs, so this would be n. The proportion would be

$$\frac{12 \text{ people}}{5 \text{ dogs}} = \frac{60 \text{ people}}{n \text{ dogs}}.$$

2. Connie drove to the top of Mount Washington. As she drove up the roadway she observed 15 cars. On the same trip she observed 17 people walking. Later that afternoon she drove down the mountain. On that trip she observed 60 cars. She is trying to estimate how many people she might see walking. She started by writing the equation

$$\frac{15 \text{ cars}}{17 \text{ people}} = \qquad .$$

Explain how she should set up the rest of proportion.

She should continue with the number of cars observed on her trip down the mountain on the top of the fraction. That would be 60. She does not know the number of people, so this would be n. The proportion would be

$$\frac{15 \text{ cars}}{17 \text{ people}} = \frac{60 \text{ cars}}{n \text{ people}}.$$

Applications

3. ***Car Repairs*** An automobile dealership has found that for every 140 cars sold, 23 will be brought back to the dealer for major repairs. If the dealership sells 980 cars this year, approximately how many cars will be brought back for major repairs? 161 cars

4. ***Hotel Management*** The policy at the Colonnade Hotel is to have 19 desserts for every 16 people if a buffet is being served. If the Saturday buffet has 320 people, how many desserts must be available? 380 desserts

5. ***Consumer Product Use*** The directions on a bottle of bleach say to use $\frac{3}{4}$ cup bleach for every 1 gallon of soapy water. Ron needs a 4-gallon mixture to mop his floors. How many cups of bleach will he need?

3 cups

6. ***Food Preparation*** To make an 8-ounce serving of hot chocolate, $1\frac{1}{2}$ tablespoons of cocoa are needed. How much cocoa is needed to make 12 ounces of hot chocolate?

$2\frac{1}{4}$ tablespoons

7. ***Measurement*** There are approximately $1\frac{1}{2}$ kilometers in one mile. Approximately how many kilometers are in 5 miles?

$7\frac{1}{2}$ kilometers

8. ***Measurement*** There are approximately $2\frac{1}{2}$ centimeters in one inch. Approximately how many centimeters are in one foot?

30 centimeters

9. ***Exchange Rate*** When Douglas flew to Sweden, the exchange rate was 65 Swedish kronor for every 8 U.S. dollars. If Douglas brought 320 U.S. dollars for spending money, how many kronor did he receive?

2600 kronor

10. ***Exchange Rate*** One day in July 2003, one U.S. dollar was worth 0.622 British pounds. Hannah exchanged $240 U.S. dollars when she arrived in London. How many British pounds did she receive?

149.28 British pounds

Shadow Length *In exercises 11 and 12 two nearby objects cast shadows at the same time of day. The ratio of the height of one of the objects to the length of its shadow is equal to the ratio of the height of the other object to the length of its shadow.*

▲ **11.** A pro football offensive tackle who stands 6.5 feet tall casts a 5-foot shadow. At the same time, the football stadium, which he is standing next to, casts a shadow of 152 feet. How tall is the stadium? Round your answer to the nearest tenth. 197.6 feet

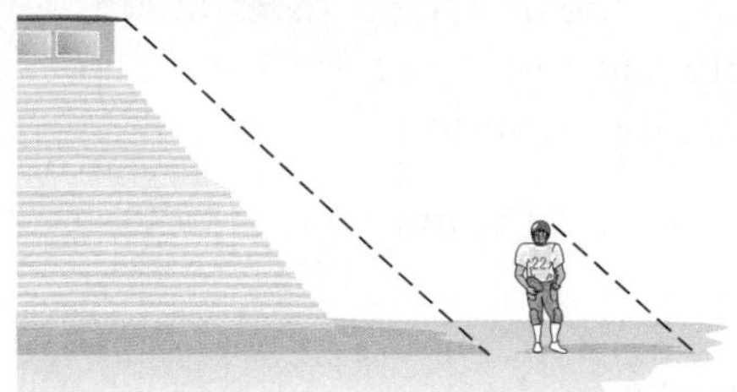

▲ **12.** In Copper Center, Alaska at 2:00 P.M., a boulder that is 7 feet high casts a shadow that is 11 feet long. At that same time, Melinda Tobey is standing by a tree that is on the bank of the river. She measured the tree and found it is exactly 22 feet tall. The tree has a shadow that crosses the entire width of the river. How wide is the river? Round your answer to the nearest tenth. 34.6 feet

13. ***Map Scale*** On a tour guide map of Madagascar, the scale states that 3 inches represent 125 miles. Two beaches are 5.2 inches apart on the map. What is the approximate distance in miles between the two beaches? Round your answer to the nearest mile.
217 miles

14. ***Map Scale*** On Dr. Jennings's map of Antarctica, the scale states that 4 inches represent 250 miles of actual distance. Two Antarctic mountains are 5.7 inches apart on the map. What is the approximate distance in miles between the two mountains? Round your answer to the nearest mile. 356 miles

15. ***Food Preparation*** In his lasagna recipe, Giovanni uses 2 cups of marinara sauce for every 7 people. How much sauce will he need to make lasagna for 62 people? Round to the nearest tenth. 17.7 cups

16. ***Food Preparation*** Jenny's blueberry pancake recipe uses 5 cups of blueberries for every 12 people. How many cups of blueberries does she need for the same recipe prepared for 28 people? Leave your answer as a mixed number.

$11\frac{2}{3}$ cups

17. ***Basketball*** During a basketball game against the Miami Heat, the Denver Nuggets made 17 out of 25 free throws attempted. If they attempt 150 free throws in the remaining games of the season, how many will they make if their success rate remains the same? 102 free throws

18. ***Baseball*** A baseball pitcher gave up 52 earned runs in 260 innings of pitching. At that rate, how many runs would he give up in a 9-inning game? (This decimal is called the pitcher's *earned run average.*)
1.8 runs

19. ***Fuel Efficiency*** In her Dodge Neon, Claire can drive 192 miles on 6 gallons of gas. During spring break, she plans to drive 600 miles. How many gallons of gas will she use? 18.75 gallons

20. ***Fuel Efficiency*** In her Nissan Pathfinder, Juanita can drive 75 miles on 5 gallons of gas. She drove 318 miles for a business trip. How many gallons of gas did she use? 21.2 gallons

21. ***Wildlife Population Counting*** An ornithologist is studying hawks in the Adirondack Mountains. She catches 24 hawks over a period of one month, tags them, and releases them back into the wild. The next month, she catches 20 hawks and finds that 12 are already tagged. Estimate the number of hawks in this part of the mountains. 40 hawks

22. ***Wildlife Population Counting*** In Kenya, a worker at the game preserve captures 26 giraffes, tags them, and then releases them back into the preserve. The next month, he captures 18 giraffes and finds that 6 of them have already been tagged. Estimate the number of giraffes on the preserve. 78 giraffes

23. ***Farming*** Bill and Shirley Grant are raising tomatoes to sell at the local co-op. The farm has a yield of 425 pounds of tomatoes for every 3 acres. The farm has 14 acres of good tomatoes. The crop this year should bring \$1.80 per pound for each pound of tomatoes. How much will the Grants get from the sale of the tomato crop?

\$3570

▲ **24.** ***Painting*** A paint manufacturer suggests 2 gallons of flat latex paint for every 750 square feet of wall. A painter is going to paint 7875 square feet of wall in a Dallas office building with paint that costs \$8.50 per gallon. How much will the painter spend for the paint?

\$178.50

25. ***Manufacturing Quality*** A company that manufactures computer chips expects 5 out of every 100 made to be defective. In a shipment of 5400 chips, how many are expected to be defective?

270 chips

26. ***Customer Satisfaction*** The editor of a small-town newspaper conducted a survey to find out how many customers are satisfied with their delivery service. Of the 100 people surveyed, 88 customers said they were satisfied. If 1700 people receive the newspaper, how many are satisfied?

1496 people

To Think About

Cooking *The following chart is used for several brands of instant mashed potatoes. Use this chart in answering exercises 27–30.*

TO MAKE	WATER	MARGARINE OR BUTTER	SALT (optional)	MILK	FLAKES
2 servings	2/3 cup	1 tablespoon	1/8 teaspoon	1/4 cup	2/3 cup
4 servings	1-1/3 cups	2 tablespoons	1/4 teaspoon	1/2 cup	1-1/3 cups
6 servings	2 cups	3 tablespoons	1/2 teaspoon	3/4 cup	2 cups
Entire box	5 cups	1/2 cup	1 teaspoon	2-1/2 cups	Entire box

27. How many cups of water and how many cups of milk are needed to make enough mashed potatoes for three people?

1 cup of water and $\frac{3}{8}$ cup of milk

28. How many cups of water and how many cups of milk are needed to make enough mashed potatoes for five people?

$1\frac{2}{3}$ cups of water and $\frac{5}{8}$ cup of milk

29. The instructions on the box say that in high altitude (above 5000 feet) the amount of water should be reduced by $\frac{1}{4}$. If you had to make enough mashed potatoes for eight people in a high-altitude city, how many cups of water and how many cups of milk would be needed?

2 cups of water and 1 cup of milk

30. The instructions on the box say that in high altitude (above 5000 feet) the amount of water should be reduced by $\frac{1}{4}$.

(a) If you had to make two boxes of mashed potatoes at a high altitude, how many cups of water and how many cups of milk would be used?

(b) How many servings would be obtained?

(a) $7\frac{1}{2}$ cups of water and 5 cups of milk (b) 30 servings

Baseball Salaries *During the 2002 playing season, Alex Rodriguez of the Texas Rangers hit 57 home runs and was paid an annual salary of $22,000,000. During the same season, Sammy Sosa of the Chicago Cubs hit 49 home runs and was paid an annual salary of $15,000,000. (Source: www.baseball-reference.com)*

31. Express the salary of each player as a unit rate in terms of dollars paid to home runs hit.
Alex Rodriguez, approximately $385,965 for each home run; Sammy Sosa, approximately $306,122 for each home run

32. Which player hit more home runs per dollar?
Sammy Sosa

Baseball Salaries *During the 2002 playing season, Randy Johnson of the Arizona Diamondbacks pitched in 35 games and was paid an annual salary of $13,350,000. During the same season, Mike Mussina of the New York Yankees pitched in 33 games and was paid $11,000,000. (Source: www.baseball-reference.com)*

33. Express the salary of each player as a unit rate in terms of dollars paid to number of games played.
Randy Johnson, approximately $381,429 per game; Mike Mussina, approximately $333,333 per game

34. Which player pitched more games per dollar?
Mike Mussina

Cumulative Review

35. Round to the nearest hundred. 56,179
56,200

36. Round to the nearest ten thousand. 196,379,910
196,380,000

37. Round to the nearest tenth. 56.148
56.1

38. Round to the nearest ten-thousandth. 2.74895
2.7490

▲ **39.** ***Sunglass Production*** An eyewear company makes very expensive carbon fiber sunglasses. The material is made into long sheets, and the basic eyeglass frame is punched out by a machine. It takes a section measuring $1\frac{3}{16}$ feet by $\frac{4}{5}$ foot to make one pair of glasses.

(a) How many square feet of this material are needed for one frame?
(a) $\frac{19}{20}$ of a square foot

(b) How many square feet of this material are needed for 1500 frames?
(b) 1425 square feet

Putting Your Skills to Work

The Mathematical Relationship Between Calories and Exercise

The following table gives the number of calories in a variety of fast foods. Under the table is a list of activities and the number of calories burned per hour during the activity. Use this information to answer the questions that follow.

Item	Calories
Dunkin Donuts—glazed donut	180
Dunkin Donuts—10-ounce coffee with cream	70
McDonald's—Big Mac	590
McDonalds's—medium fries	450
Wendy's—grilled chicken sandwich	300
Taco Bell—soft beef taco supreme	260
Pizza Hut—2 slices of pepperoni pan pizza	560
Starbucks—medium nonfat mocha	230

Source: www.dunkindonuts.com, www.mcdonalds.com, www.wendys.com, www.tacobell.com, www.pizzahut.com, www.starbucks.com

The following list gives the number of calories the average person* burns during 1 hour of the activity: jogging: 500; walking: 250; ballroom dancing: 200; cleaning house: 240; tae kwon do: 700; yard work: 350.

Problems for Individual Investigation

1. **(a)** Suppose you have a Dunkin Donuts glazed donut and 10-ounce coffee with cream every weekday morning. How many calories is this in one week? 1250 calories

   (b) How many hours of walking would it take to burn these calories? 5 hours

2. **(a)** If you ate a McDonald's Big Mac and medium fries every weekday for lunch, how many calories would you consume for lunch in one week? 5200 calories

   (b) How many hours of ballroom dancing would it take to burn these calories? 26 hours

*Number of calories per hour given is for a 155-lb person performing the activity at a moderate pace.

Problems for Group Investigation and Cooperative Study

In problems 3–6 assume 4 weeks in a month and 5 weekdays in one week.

3. Nora eats 2 slices of pepperoni pan pizza from Pizza Hut twice a week during her lunch break. Her friend, Gina, drinks a medium nonfat mocha from Starbucks each weekday morning before work. Who consumes more fast-food calories in one month? How many more calories does she consume? Gina; 120 calories

4. Barry walks for an hour 4 times a week. He also spends an hour doing yard work 3 times a week. His friend, Miguel, jogs twice a week for an hour and goes to tae kwon do class twice a week for an hour. Who burns more calories during these activities? How many more? Miguel; 350 calories more

5. Alison enjoys a soft beef taco supreme from Taco Bell every weekday for lunch but wants to make sure she burns the calories she consumes. Determine how many calories she consumes each month eating tacos. Choose 3 activities and give the number of hours she needs to spend on each to burn the calories.
   sample answer: 1 hour of ballroom dancing, 8 hours of walking, and 6 hours of jogging

6. **(a)** What is the ratio of the number of calories consumed in 1 hour of doing tae kwon do to the number of calories consumed in 1 hour of jogging? $\frac{7}{5}$

   (b) What is the ratio of the number of calories consumed in 1 hour of doing yard work to the number of calories consumed in walking? $\frac{7}{5}$

Chapter 4 Organizer

Topic	Procedure	Examples
Forming a ratio, p. 270.	A *ratio* is the comparison of two quantities that have the same units. A ratio is usually expressed as a fraction. The fractions should be in reduced form.	**1.** Find the ratio of 20 books to 35 books. $\frac{20}{35} = \frac{4}{7}$ **2.** Find the ratio in simplest form of 88 : 99. $\frac{88}{99} = \frac{8}{9}$ **3.** Bob earns \$250 each week, but \$15 is withheld for medical insurance. Find the ratio of medical insurance to total pay. $\frac{\$15}{\$250} = \frac{3}{50}$
Forming a rate, p. 272.	A *rate* is a comparison of two quantities that have different units. A rate is usually expressed as a fraction in reduced form.	A college has 2520 students with 154 faculty. What is the rate of students to faculty? $\frac{2520 \text{ students}}{154 \text{ faculty}} = \frac{180 \text{ students}}{11 \text{ faculty}}$
Forming a unit rate, p. 272.	A *unit rate* is a rate with a denominator of 1. Divide the denominator into the numerator to obtain the unit rate.	A car traveled 416 miles in 8 hours. Find the unit rate. $\frac{416 \text{ miles}}{8 \text{ hours}} = 52 \text{ miles/hour}$ Bob spread 50 pounds of fertilizer over 1870 square feet of land. Find the unit rate of square feet per pound. $\frac{1870 \text{ square feet}}{50 \text{ pounds}} = 37.4 \text{ square feet/pound}$
Writing proportions, p. 278.	A *proportion* is a statement that two rates or two ratios are equal. A proportion statement *a* is to *b* as *c* is to *d* can be written $\frac{a}{b} = \frac{c}{d}$.	Write a proportion for 17 is to 34 as 13 is to 26. $\frac{17}{34} = \frac{13}{26}$
Determining whether a relationship is a proportion, p. 278.	For any two fractions where $b \neq 0$ and $d \neq 0$, $\frac{a}{b} = \frac{c}{d}$ if and only if $a \times d = b \times c$. A proportion is a statement that two rates or two ratios are equal.	**1.** Is this a proportion? $\frac{7}{56} = \frac{3}{24}$ $7 \times 24 \stackrel{?}{=} 56 \times 3$ $168 = 168$ ✓ It is a proportion. **2.** Is this a proportion? $\frac{64 \text{ gallons}}{5 \text{ acres}} = \frac{89 \text{ gallons}}{7 \text{ acres}}$ $64 \times 7 \stackrel{?}{=} 5 \times 89$ $448 \neq 445$ It is not a proportion.
Solving a proportion, p. 285.	To solve a proportion where the value *n* is not known: **1.** Cross-multiply. **2.** Divide both sides of the equation by the number multiplied by *n*.	**1.** Solve for *n*. $\frac{17}{n} = \frac{51}{9}$ $17 \times 9 = 51 \times n$ Cross-multiply. $153 = 51 \times n$ Simplify. $\frac{153}{51} = \frac{51 \times n}{51}$ Divide by 51. $3 = n$

Topic	Procedure	Examples
Solving applied problems, p. 293.	**1.** Write a proportion with n representing the unknown value. **2.** Solve the proportion.	Bob purchased eight notebooks for \$19. How much would 14 notebooks cost? $\frac{8 \text{ notebooks}}{\$19} = \frac{14 \text{ notebooks}}{n}$ $8 \times n = 19 \times 14$ $8 \times n = 266$ $\frac{8 \times n}{8} = \frac{266}{8}$ $n = 33.25$ The 14 notebooks would cost \$33.25.

Chapter 4 Review Problems

Section 4.1

Write in simplest form. Express your answer as a fraction.

1. 88:40 $\frac{11}{5}$

2. 65:39 $\frac{5}{3}$

3. 28:35 $\frac{4}{5}$

4. 250:475 $\frac{10}{19}$

5. $2\frac{1}{3}$ to $4\frac{1}{4}$ $\frac{28}{51}$

6. 27 to 81 $\frac{1}{3}$

7. 280 to 651 $\frac{40}{93}$

8. 156 to 441 $\frac{52}{147}$

9. 26 tons to 65 tons $\frac{2}{5}$

Personal Finance *Bob earns \$480 per week and has \$60 withheld for federal taxes and \$45 withheld for state taxes.*

10. Write the ratio of federal taxes withheld to earned income.
$\frac{1}{8}$

11. Write the ratio of total withholdings to earned income.
$\frac{7}{32}$

Write as a rate in simplest form.

12. \$75 donated by every 6 people
$\frac{\$25}{2 \text{ people}}$

13. 44 revolutions every 121 minutes
$\frac{4 \text{ revolutions}}{11 \text{ minutes}}$

14. 188 vibrations every 16 seconds
$\frac{47 \text{ vibrations}}{4 \text{ seconds}}$

15. 20 cups of sugar for every 32 pies
$\frac{5 \text{ cups}}{8 \text{ pies}}$

In exercises 16–21, write as a unit rate. Round to the nearest tenth when necessary.

16. \$2125 was paid for 125 shares of stock. Find the cost per share.
\$17/share

17. ***Education*** \$1344 was paid for 12 credit-hours. Find the cost per credit-hour.
\$112/credit-hour

▲ **18.** \$742.50 was spent for 55 square yards of carpet. Find the cost per square yard.
\$13.50/square yard

19. ***DVD Cost*** Larry spent \$600 on 48 DVDs. Find the cost per DVD.
\$12.50/DVD

20. ***Food Costs*** A 4-ounce jar of instant coffee costs \$2.96. A 9-ounce jar of the same brand of instant coffee costs \$5.22.
(a) What is the cost per ounce of the 4-ounce jar? \$0.74
(b) What is the cost per ounce of the 9-ounce jar? \$0.58
(c) How much per ounce do you save by buying the larger jar? \$0.16

21. ***Food Costs*** A 12.5-ounce can of white tuna costs \$2.75. A 7.0-ounce can of the same brand of white tuna costs \$1.75.
(a) What is the cost per ounce of the large can? \$0.22
(b) What is the cost per ounce of the small can? \$0.25
(c) How much per ounce do you save by buying the larger can? \$0.03

Section 4.2

Write as a proportion.

22. 12 is to 48 as 7 is to 28
$\frac{12}{48} = \frac{7}{28}$

23. $1\frac{1}{2}$ is to 5 as 4 is to $13\frac{1}{3}$
$\frac{1\frac{1}{2}}{5} = \frac{4}{13\frac{1}{3}}$

24. 7.5 is to 45 as 22.5 is to 135
$\frac{7.5}{45} = \frac{22.5}{135}$

25. ***Bus Capacity*** If three buses can transport 138 passengers, then five buses can transport 230 passengers.
$\frac{3 \text{ buses}}{138 \text{ passengers}} = \frac{5 \text{ buses}}{230 \text{ passengers}}$

26. ***Cost of Products*** If 15 pounds cost \$4.50, then 27 pounds will cost \$8.10.
$\frac{15 \text{ pounds}}{\$4.50} = \frac{27 \text{ pounds}}{\$8.10}$

Determine whether each equation is a proportion.

27. $\frac{16}{48} = \frac{2}{12}$
It is not a proportion.

28. $\frac{20}{25} = \frac{8}{10}$
It is a proportion.

29. $\frac{24}{20} = \frac{18}{15}$
It is a proportion.

30. $\frac{84}{48} = \frac{14}{8}$
It is a proportion.

31. $\frac{37}{33} = \frac{22}{19}$
It is not a proportion.

32. $\frac{15}{18} = \frac{18}{22}$
It is not a proportion.

33. $\frac{84 \text{ miles}}{7 \text{ gallons}} = \frac{108 \text{ miles}}{9 \text{ gallons}}$
It is a proportion.

34. $\frac{156 \text{ revolutions}}{6 \text{ minutes}} = \frac{181 \text{ revolutions}}{7 \text{ minutes}}$
It is not a proportion.

Section 4.3

Solve for n.

35. $7 \times n = 161$
$n = 23$

36. $8 \times n = 42$
$n = 5\frac{1}{4}$ or 5.25

37. $442 = 20 \times n$
$n = 22\frac{1}{10}$ or 22.1

38. $663 = 39 \times n$
$n = 17$

Solve. Round to the nearest tenth when necessary.

39. $\frac{3}{11} = \frac{9}{n}$
$n = 33$

40. $\frac{2}{7} = \frac{12}{n}$
$n = 42$

✓ **41.** $\frac{n}{28} = \frac{6}{24}$
$n = 7$

42. $\frac{n}{32} = \frac{15}{20}$
$n = 24$

43. $\frac{2\frac{1}{4}}{9} = \frac{4\frac{3}{4}}{n}$
$n = 19$

44. $\frac{3\frac{1}{3}}{2\frac{2}{3}} = \frac{7}{n}$
$n = 5\frac{3}{5}$ or 5.6

45. $\frac{42}{50} = \frac{n}{6}$
$n \approx 5.0$

46. $\frac{38}{45} = \frac{n}{8}$
$n \approx 6.8$

47. $\frac{2.25}{9} = \frac{4.75}{n}$
$n = 19$

48. $\frac{3.5}{5} = \frac{10.5}{n}$
$n = 15$

49. $\frac{25}{7} = \frac{60}{n}$
$n = 16.8$

50. $\frac{60}{9} = \frac{31}{n}$
$n = 4.7$

51. $\frac{35 \text{ miles}}{28 \text{ gallons}} = \frac{15 \text{ miles}}{n \text{ gallons}}$
$n = 12$

52. $\frac{8 \text{ defective parts}}{100 \text{ perfect parts}} = \frac{44 \text{ defective parts}}{n \text{ perfect parts}}$
$n = 550$

Section 4.4

Solve using a proportion. Round your answer to the nearest hundredth when necessary.

53. ***Painting*** The school volunteers used 3 gallons of paint to paint two rooms. How many gallons would they need to paint 10 rooms of the same size?
15 gallons

54. ***Coffee Consumption*** Several recent surveys show that 49 out of every 100 adults in America drink coffee. If a computer company employs 3450 people, how many of those employees would you expect would drink coffee? Round to the nearest whole number.
1691 employees

55. ***Exchange Rate*** When Marguerite traveled as a child, the rate of French francs to American dollars was 24 francs to 5 dollars. How many francs did Marguerite receive for 420 dollars?
2016 francs

56. ***Exchange Rate*** When Melissa and Phil traveled to Switzerland, the rate of Swiss francs to U.S. dollars was 5 francs to 3.6 dollars. How many Swiss francs would they receive for 80 U.S. dollars?
111.11 Swiss francs

57. ***Map Scale*** Two cities located 225 miles apart appear 3 inches apart on a map. If two other cities appear 8 inches apart on the map, how many miles apart are the cities?
600 miles

58. ***Basketball*** In the first four basketball games of the season, Pedro made 14 free throws. There are 10 games left. How many free throws would you expect Pedro to make in these 10 games?
35 free throws

▲ 59. ***Shadow Length*** In the setting sun, a 6-foot man casts a shadow 16 feet long. At the same time a building casts a shadow of 320 feet. How tall is the building?
120 feet

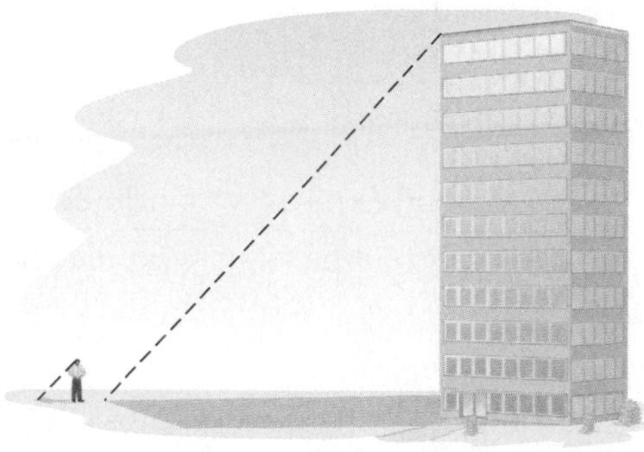

60. ***Gasoline Consumption*** During the first 680 miles of a trip, Johnny and Stephanie used 26 gallons of gas. They need to travel 200 more miles. Assume that the car will have the same rate of gas consumption.
(a) How many more gallons of gas will they need? 7.65 gallons
(b) If gas costs $1.85 per gallon, what will fuel cost them for the last 200 miles?
$14.15

▲ 61. ***Photography*** A film negative is 3.5 centimeters wide and 2.5 centimeters tall. If you want to make a color print that is 8 centimeters wide, how tall will the print be?
5.71 centimeters tall

62. The dosage of a certain medication is 3 grams for every 50 pounds of body weight. If a person weighs 125 pounds, how many grams of this medication should she take?
7.5 grams

63. Maria is making a brick walkway for the front of her new house. She needs 33 bricks to make a walkway section that is 2 feet long. How many bricks will she need to buy to make a walkway section that is 11 feet long?
technically 181.5 bricks, but in real life 182 bricks will be needed

64. Jeffrey found that he can read 20 pages on his business law textbook in 2.5 hours. He has set aside 6 hours this weekend to catch up on his reading. How many pages will he be able to read?
48 pages

▲ 65. When Carlos was painting his apartment he found that he used 3 gallons of paint to cover 500 square feet of wall space. He is planning to paint his sister's apartment. She said there are 1400 square feet of wall space that need to be painted. How many gallons of paint will Carlos need?
technically 8.4 gallons, but in real life 9 gallons of paint will be needed

66. From previous experience, the directors of a large running race know they need 2 liters of water for every 3 runners. This year 1250 runners will participate in the race. How many liters of water do they need?
Technically, 833.33 liters, but in real life, 834 liters will be needed.

▲ **67.** A scale model of a new church sanctuary has a length of 14 centimeters. When the church is built, the actual length will be 145 feet. In the scale model, the width measures 11 centimeters. What will be the actual width of the church sanctuary?
approximately 113.93 feet

68. Jeff checked the time on his new watch. In 40 days his watch gained 3 minutes. How much time will the watch gain in a year? (Assume it is not a leap year.)
approximately 27.38 minutes

69. Jean, the top soccer player for the Springfield Comets, scored a total of 68 goals during the season. During the season the team played 32 games, but Jean played in only 27 of them due to a leg injury. The league has been expanded and next season the team will play 34 games. If Jean scores goals at the same rate and is able to play in every game, how many goals might she be expected to score? Round your answer to the nearest whole number.
86 goals

70. Hank found out there were 345 calories in the 10-ounce chocolate milkshake that he purchased yesterday. Today he decided to order the 16-ounce milkshake. How many calories would you expect to be in the 16-ounce milkshake?
552 calories

71. A recent survey showed that 3 out of every 10 people in Massachusetts read the *Boston Globe*. In a Massachusetts town of 45,600 people, how many people would you expect read the *Boston Globe?*
13,680 people

72. Greg and Marcia are managing a boat for Eastern Whale Watching Tours this summer. For every 16 trips out to the ocean, the passengers spotted at least one whale during 13 trips. If Greg and Marcia send out 240 trips this month, how many trips will have the passengers spotting at least one whale?
195 trips

How Am I Doing? Chapter 4 Test

Note to Instructor: The Chapter 4 Test file in the TestGen program provides algorithms specifically matched to these problems so you can easily replicate this test for additional practice or assessment purposes.

1. $\frac{9}{26}$
2. $\frac{14}{37}$
3. $\frac{98 \text{ miles}}{3 \text{ gallons}}$
4. $\frac{140 \text{ square feet}}{3 \text{ pounds}}$
5. 3.8 tons/day
6. $8.28/hour
7. 245.45 feet/pole
8. $85.21/share
9. $\frac{17}{29} = \frac{51}{87}$
10. $\frac{2\frac{1}{2}}{10} = \frac{6}{24}$
11. $\frac{490 \text{ miles}}{21 \text{ gallons}} = \frac{280 \text{ miles}}{12 \text{ gallons}}$
12. $\frac{3 \text{ hours}}{180 \text{ miles}} = \frac{5 \text{ hours}}{300 \text{ miles}}$
13. It is not a proportion.
14. It is a proportion.
15. It is a proportion.
16. It is not a proportion.

Remember to use your Chapter Test Prep Video CD to see the worked-out solutions to the test problems you want to review.

Write as a ratio in simplest form.

1. 18:52

2. 70 to 185

Write as a rate in simplest form. Express your answer as a fraction.

3. 784 miles per 24 gallons

4. 2100 square feet per 45 pounds

Write as a unit rate. Round to the nearest hundredth when necessary.

5. 19 tons in five days

6. $57.96 for seven hours

7. 5400 feet per 22 telephone poles

8. $9373 for 110 shares of stock

Write as a proportion.

9. 17 is to 29 as 51 is to 87

10. $2\frac{1}{2}$ is to 10 as 6 is to 24

11. 490 miles is to 21 gallons as 280 miles is to 12 gallons

12. 3 hours is to 180 miles as 5 hours is to 300 miles

Determine whether each equation is a proportion.

13. $\frac{50}{24} = \frac{34}{16}$

14. $\frac{3\frac{1}{2}}{14} = \frac{5}{20}$

15. $\frac{32 \text{ smokers}}{46 \text{ nonsmokers}} = \frac{160 \text{ smokers}}{230 \text{ nonsmokers}}$

16. $\frac{\$0.74}{16 \text{ ounces}} = \frac{\$1.84}{40 \text{ ounces}}$

Solve. Round to the nearest tenth when necessary.

17. $\frac{n}{20} = \frac{4}{5}$

18. $\frac{8}{3} = \frac{60}{n}$

19. $\frac{2\frac{2}{3}}{8} = \frac{6\frac{1}{3}}{n}$

20. $\frac{4.2}{11} = \frac{n}{77}$

21. $\frac{45 \text{ women}}{15 \text{ men}} = \frac{n \text{ women}}{40 \text{ men}}$

22. $\frac{5 \text{ kg}}{11 \text{ pounds}} = \frac{32 \text{ kg}}{n \text{ pounds}}$

23. $\frac{n \text{ inches of snow}}{14 \text{ inches of rain}} = \frac{12 \text{ inches of snow}}{1.4 \text{ inches of rain}}$

24. $\frac{5 \text{ pounds of coffee}}{\$n} = \frac{1/2 \text{ pound of coffee}}{\$5.20}$

Solve using a proportion. Round your answer to the nearest hundredth when necessary.

25. Bob's recipe for pancakes calls for three eggs and will serve 11 people. If he wants to feed 22 people, how many eggs will he need?

26. A steel cable 42 feet long weighs 170 pounds. How much will 20 feet of this cable weigh?

27. If 9 inches on a map represents 57 miles, what distance does 3 inches represent?

28. John and Nancy found it would cost $240 per year to fertilize their front lawn of 4000 square feet. How much would it cost to fertilize 6000 square feet?

29. Tom Tobey knows that 1 mile is approximately 1.61 kilometers. While driving in Canada, a sign reads "Montreal 220 km." How many miles is Tom from Montreal? Round to the nearest tenth of a mile.

30. Stephen traveled 570 kilometers in 9 hours. At this rate, how far could he go in 11 hours?

31. During the first few games of the basketball season, Tyler shot 15 free throws and made 11 of them. If Tyler shoots free throws at the same rate as the first few games, he expects to shoot 120 more during the rest of the season. How many of these freer throws should he expect to make?

32. On the tri-city softball league, Lexi got seven hits in 34 times at bat last week. During the entire playing season, she was at bat 155 times. If she gets hits at the same rate all season as during last week's game, how many hits would she have for the entire season? Round to the nearest whole number.

17. $n = 16$

18. $n = 22.5$

19. $n = 19$

20. $n = 29.4$

21. $n = 120$

22. $n = 70.4$

23. $n = 120$

24. $n = 52$

25. 6 eggs

26. 80.95 pounds

27. 19 miles

28. $360

29. 136.7 miles

30. 696.67 kilometers

31. 88 free throws

32. 32 hits

Cumulative Test for Chapters 1–4

1.	twenty-six million, five hundred ninety-seven thousand, eighty-nine
2.	68
3.	$\frac{11}{32}$
4.	$\frac{27}{35}$
5.	65
6.	8.2584
7.	0.179586
8.	$\frac{1}{6}$
9.	16,145.5
10.	56.9
11.	362.278
12.	3.68
13.	0.15625
14.	It is a proportion.
15.	It is a proportion.
16.	$n = 3$
17.	$n = 2$
18.	$n = 64$
19.	$n = 9$
20.	$n \approx 3.4$
21.	$n = 21$
22.	8.33 inches
23.	\$396.67
24.	5 pounds
25.	214.5 gallons

Approximately one-half of this test is based on Chapter 4 material. The remainder is based on material covered in Chapters 1–3.

1. Write in words. 26,597,089

2. Divide. $23\overline{)1564}$

3. Combine. $\frac{1}{4} + \frac{1}{8} \times \frac{3}{4}$

4. Subtract. $2\frac{1}{5} - 1\frac{3}{7}$

5. Multiply. $20 \times 3\frac{1}{4}$

6. Subtract. $\begin{array}{r} 12.1 \\ -\ 3.8416 \\ \hline \end{array}$

7. Multiply. $\begin{array}{r} 0.8163 \\ \times\ \ 0.22 \\ \hline \end{array}$

8. Divide. $\frac{5}{12} \div \frac{15}{6}$

9. Multiply. 16.1455×10^3

10. Round to the nearest tenth. 56.8918

11. Add. $258.92 + 67.358$

12. Divide. $0.552 \div 0.15$

13. Change to a decimal. $\frac{5}{32}$

Determine whether each equation is a proportion.

14. $\frac{12}{17} = \frac{30}{42.5}$

15. $\frac{4\frac{1}{3}}{13} = \frac{2\frac{2}{3}}{8}$

Solve. Round to the nearest tenth when necessary.

16. $\frac{9}{2.1} = \frac{n}{0.7}$

17. $\frac{50}{20} = \frac{5}{n}$

18. $\frac{n}{56} = \frac{16}{14}$

19. $\frac{7}{n} = \frac{28}{36}$

20. $\frac{n}{11} = \frac{5}{16}$

21. $\frac{3\frac{1}{3}}{7} = \frac{10}{n}$

Solve. Round to the nearest hundredth when necessary.

22. Two cities that are located 300 miles apart appear 4 inches apart on a map. If two other cities are 625 miles apart, how far apart will they appear on the same map?

23. At her new waitressing job, Linnea earned \$34 in tips in 3 hours. Next week she is scheduled to work 35 hours. Assuming the same ratio, how much will she earn in tips next week? Round to the nearest cent.

24. Emily Robinson's lasagna recipe feeds 14 people and calls for 3.5 pounds of sausage. If she wants to feed 20 people, how much sausage does she need?

25. Loring Kerr in Nova Scotia produces his own maple syrup. He has found that 39 gallons of maple sap produce 2 gallons of maple syrup. How much sap is needed to produce 11 gallons of syrup?

CHAPTER

5

Compared to thirty years ago, today more high school students take a modern foreign language rather than an ancient language such as Latin. This is partially due to field trips to foreign countries and a greater emphasis on useful conversational skills in a foreign language. However, the increase is also due to a greater demand from business and health service sections of the United States that find a need for more employees who are bilingual. But is this demand being met in the local high schools? Can you use your mathematical skills to find an answer to this question? Turn to page 361 to find out.

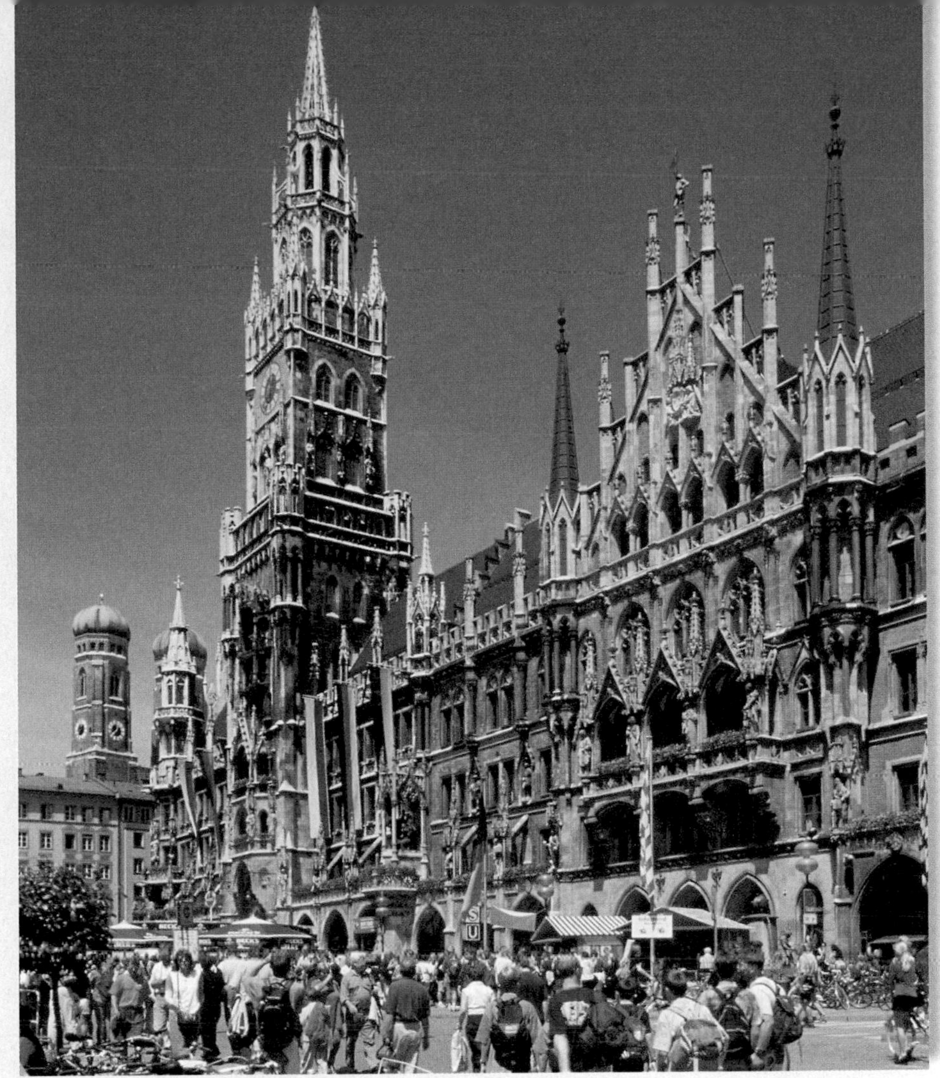

Percent

5.1 Understanding Percent

Student Learning Objectives

After studying this section, you will be able to:

1. **Write a fraction with a denominator of 100 as a percent.**
2. **Write a percent as a decimal.**
3. **Write a decimal as a percent.**

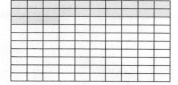

1 Writing a Fraction with a Denominator of 100 as a Percent

"My raise came through. I got a 6% increase!"

"The leading economic indicators show inflation rising at a rate of 1.3%."

"Mark McGwire and Babe Ruth each hit quite a few home runs. But I wonder who has the higher percentage of home runs per at-bat?"

We use percents often in our everyday lives. In business, in sports, in shopping, and in many areas of life, percentages play an important role. In this section we introduce the idea of percent, which means "*per centum*" or "per hundred." We then show how to use percentages.

In previous chapters, when we described parts of a whole, we used fractions or decimals. Using a percent is another way to describe a part of a whole. Percents can be described as ratios whose denominators are 100. The word **percent** means per 100. This sketch has 100 rectangles.

Of 100 rectangles, 23 are shaded. We can say that 23 percent of the whole is shaded. We use the symbol % for percent. It means "parts per 100." When we write 23 percent as 23%, we understand that it means 23 parts per one hundred, or, as a fraction, $\frac{23}{100}$.

EXAMPLE 1 Recently 100 college students were surveyed about their intentions for voting in the next presidential election. 39 students intended to vote for the Republican candidate, 28 students intended to vote for the Democratic candidate, and 22 students were undecided about which candidate to vote for. The remaining 11 students admitted that they were not planning to vote.

(a) What percent of the students intended to vote for the Democratic candidate?

(b) What percent of the students intended to vote for the Republican candidate?

(c) What percent of the students were undecided as to which candidate they would vote for?

(d) What percent of the students were not planning to vote?

Solution

(a) $\frac{28}{100} = 28\%$ **(b)** $\frac{39}{100} = 39\%$

(c) $\frac{22}{100} = 22\%$ **(d)** $\frac{11}{100} = 11\%$

Percent notation is often used in circle graphs or pie charts.

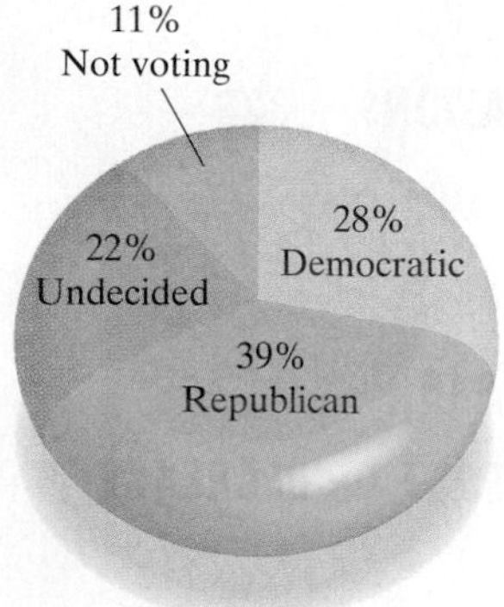

***NOTE TO STUDENT:** Fully worked-out solutions to all of the Practice Problems can be found at the back of the text starting at page SP-1*

Practice Problem 1 Write as a percent.

(a) 51 out of 100 students in the class were women.

(b) 68 out of 100 cars in the parking lot have front-wheel drive.

(c) 7 out of 100 students in the dorm quit smoking.

(d) 26 out of 100 students did not vote in class elections.

Some percents are larger than 100%. When you see expressions like 140% or 400%, you need to understand what they represent. Consider the following situations.

EXAMPLE 2

(a) Write $\frac{386}{100}$ as a percent.

(b) Twenty years ago, four car tires for a full-sized car cost \$100. Now the average price for four car tires for a full-sized car is \$270. Write the present cost as a percent of the cost 20 years ago.

Solution

(a) $\frac{386}{100} = 386\%$

(b) The ratio is $\frac{\$270 \text{ for four tires now}}{\$100 \text{ for four tires then}}$. $\quad \frac{270}{100} = 270\%$

The present cost of four car tires for a full-sized car is 270% of the cost 20 years ago.

Practice Problem 2

(a) Write $\frac{238}{100}$ as a percent.

(b) Last year 100 students tried out for varsity baseball. This year 121 students tried out. Write this year's number as a percent of last year's number.

Teaching Tip Students usually have trouble writing a fraction like $\frac{245}{100}$ as a percent. It is a good idea to put four simple examples on the board, ordered from smallest to largest, and show them in both fractional and decimal form. Try a display on the board like this:

FRACTIONAL FORM	PERCENT NOTATION FORM
$\frac{0.6}{100}$	0.6%
$\frac{7}{100}$	7%
$\frac{34}{100}$	34%
$\frac{245}{100}$	245%

Comparing one form to the other usually clarifies the idea for the student.

Some percents are smaller than 1%.

$\frac{0.7}{100}$ can be written as 0.7%. $\quad \frac{0.3}{100}$ can be written as 0.3%.

$\frac{0.04}{100}$ can be written as 0.04%.

EXAMPLE 3 Write as a percent.

(a) $\frac{0.9}{100}$ **(b)** $\frac{0.002}{100}$ **(c)** $\frac{0.07}{100}$

Solution

(a) $\frac{0.9}{100} = 0.9\%$ **(b)** $\frac{0.002}{100} = 0.002\%$ **(c)** $\frac{0.07}{100} = 0.07\%$

Practice Problem 3 Write as a percent.

(a) $\frac{0.5}{100}$ **(b)** $\frac{0.06}{100}$ **(c)** $\frac{0.003}{100}$

Remember: Whenever the denominator of a fraction is 100, the numerator is the percent.

2 Writing a Percent as a Decimal

Suppose we have a percent such as 59%. What would be the equivalent in decimal form? Using our definition of percent, 59% = $\frac{59}{100}$. This fraction could be written in decimal form as 0.59. In a similar way, we could write 21% as $\frac{21}{100}$ = 0.21. This pattern allows us to quickly change the form of a number from a percent to a fraction whose denominator is 100 to a decimal.

EXAMPLE 4 Write as a decimal.

(a) 38% **(b)** 6%

Solution

(a) $38\% = \frac{38}{100} = 0.38$ **(b)** $6\% = \frac{6}{100} = 0.06$

Practice Problem 4 Write as a decimal.

(a) 47% **(b)** 2%

NOTE TO STUDENT: Fully worked-out solutions to all of the Practice Problems can be found at the back of the text starting at page SP-1

The results of Example 4 suggest that **when you remove a percent sign (%) you are dividing by 100.** When we divide by 100 this moves the decimal point of a number two places to the left. Now that you understand this process we can abbreviate it with the following rule.

CHANGING A PERCENT TO A DECIMAL

1. Drop the % symbol.
2. Move the decimal point two places to the left.

EXAMPLE 5 Write as a decimal.

(a) 26.9% **(b)** 7.2% **(c)** 0.13% **(d)** 158%

Solution In each case, we drop the percent symbol and move the decimal point two places to the left.

(a) 26.9% = 0.269 = 0.269

(b) 7.2% = 0.072 = 0.072 Note that we need to add an extra zero to the left of the seven.

(c) 0.13% = 0.0013 = 0.0013 Here we added zeros to the left of the 1.

(d) 158% = 1.58 = 1.58

Practice Problem 5 Write as a decimal.

(a) 80.6% **(b)** 2.5% **(c)** 0.29% **(d)** 231%

3 Writing a Decimal as a Percent

In Example 4(a) we changed 38% to $\frac{38}{100}$ to 0.38. We can start with 0.38 and reverse the process. We obtain $0.38 = \frac{38}{100} = 38\%$. Study all the parts of Examples 4 and 5. You will see that the steps are reversible. Thus 0.38 = 38%, 0.06 = 6%, 0.70 = 70%, 0.269 = 26.9%, 0.072 = 7.2%, 0.0013 = 0.13%, and 1.58 = 158%.

In each part we are multiplying by 100. **To change a decimal number to a percent we are multiplying the number by 100.** In each part the decimal point is moved two places to the right. Then the percent symbol is written after the number.

Teaching Tip It's a good idea to stress the reversibility of the operation. Write a problem like this on the board and ask the class to complete it.

DECIMAL FORM	PERCENT FORM
0.58	—
—	62%
0.03	—
0.006	—
—	0.78%
—	321%

CHANGING A DECIMAL TO A PERCENT

1. Move the decimal point two places to the right.
2. Then write the % symbol at the end of the number.

EXAMPLE 6 Write as a percent.

(a) 0.47 **(b)** 0.08 **(c)** 6.31 **(d)** 0.055 **(e)** 0.001

Solution In each part we move the decimal point two places to the right and write the percent symbol at the end of the number.

(a) 0.47 = 47% **(b)** 0.08 = 8% **(c)** 6.31 = 631%
(d) 0.055 = 5.5% **(e)** 0.001 = 0.1%

Practice Problem 6 Write as a percent.

(a) 0.78 **(b)** 0.02 **(c)** 5.07 **(d)** 0.029 **(e)** 0.006

TO THINK ABOUT: The Meaning of Percent What is really happening when we change a decimal to a percent? Suppose that we wanted to change 0.59 to a percent.

$$0.59 = \frac{59}{100} \qquad \text{Definition of a decimal.}$$

$$= 59 \times \frac{1}{100} \qquad \text{Definition of multiplying fractions.}$$

$$= 59 \text{ percent} \qquad \text{Because "per 100" means percent.}$$

$$= 59\% \qquad \text{Writing the symbol for percent.}$$

Can you see why each step is valid? Since we know the reason behind each step, we know we can always move the decimal point two places to the right and write the percent symbol. See Exercises 5.1, exercises 81 and 82.

Calculator

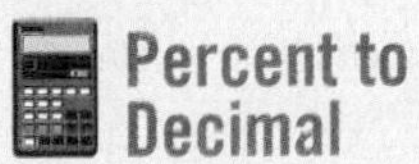

Percent to Decimal

You can use a calculator to change 52% to a decimal. Enter

52 [%]

The display should read

[0.52]

Try the following.

(a) 46% **(b)** 137%
(c) 9.3% **(d)** 6%

Note: The calculator divides by 100 when the percent key is pressed. If you do not have a [%] key then you can use the key-strokes [÷] [100] [=].

5.1 Exercises

Student Solutions Manual | CD/Video | PH Math Tutor Center | MathXL®Tutorials on CD | MathXL® | MyMathLab® | Interactmath.com

Verbal and Writing Skills

1. In this section we introduced percent, which means "per centum" or "per ___hundred___."

2. The number 1 written as a percent is ___100%___.

3. To change a percent to a decimal, move the decimal point ___two___ places to the ___left___. ___Drop___ the % symbol.

4. To change a decimal to a percent, move the decimal point ___two___ places to the ___right___. ___Write___ the % symbol at the end of the number.

Write as a percent.

5. $\frac{59}{100}$ 59% **6.** $\frac{67}{100}$ 67% **7.** $\frac{4}{100}$ 4% **8.** $\frac{7}{100}$ 7%

9. $\frac{80}{100}$ 80% **10.** $\frac{90}{100}$ 90% **11.** $\frac{245}{100}$ 245% **12.** $\frac{110}{100}$ 110%

13. $\frac{5.3}{100}$ 5.3% **14.** $\frac{8.3}{100}$ 8.3% **15.** $\frac{0.07}{100}$ 0.07% **16.** $\frac{0.019}{100}$ 0.019%

Applications

Write a percent to express each of the following.

17. 13 out of 100 loaves of bread had gone stale. 13%

18. 54 out of 100 dog owners have attended an obedience class. 54%

19. A survey showed that 8 out of 100 customers were dissatisfied. 8%

20. 2 out of 100 students majored in marine biology. 2%

Write as a decimal.

21. 51% 0.51 **22.** 42% 0.42 **23.** 7% 0.07 **24.** 6% 0.06 **25.** 20% 0.2 **26.** 40% 0.4

27. 43.6% 0.436 **28.** 81.5% 0.815 **29.** 0.03% 0.0003 **30.** 0.09% 0.0009 **31.** 0.72% 0.0072 **32.** 0.61% 0.0061

33. 1.25% 0.0125 **34.** 9.6% 0.096 **35.** 366% 3.66 **36.** 398% 3.98

Write as a percent.

37. 0.74 74% **38.** 0.66 66% **39.** 0.50 50% **40.** 0.40 40% **41.** 0.08 8% **42.** 0.03 3%

43. 0.563 56.3% **44.** 0.408 40.8% **45.** 0.002 0.2% **46.** 0.009 0.9% **47.** 0.0057 0.57% **48.** 0.0026 0.26%

49. 1.35 135% **50.** 1.86 186% **51.** 2.72 272% **52.** 3.04 304%

53. ***Income Taxes*** Robert Tansill paid $\frac{27}{100}$ of his income for federal income taxes. This means that 0.27 of his income was paid for federal taxes. Express this as a percent.
27%

54. ***Housing Costs*** Sally LeBlanc spends $\frac{37}{100}$ of her income for housing. This means that 0.37 of her income was spent for housing. Express this as a percent.
37%

55. ***Grade Distribution*** Prof Knutson gave $\frac{3}{10}$ of his students a grade of B for the semester. This means that 0.3 of his students will receive a B. Express this as a percent.
30%

56. ***Checking Account*** Anne Perkins puts $\frac{7}{10}$ of her income into her checking account each month. This means that 0.7 of her income goes into her checking account. Express this as a percent.
70%

Mixed Practice

Write as a percent.

57. 0.94 94%

58. 0.25 25%

59. 2.31 231%

60. 1.48 148%

61. $\frac{10}{100}$ 10%

62. $\frac{40}{100}$ 40%

63. 0.009 0.9%

64. 0.005 0.5%

Write as a decimal.

65. 62% 0.62

66. 49% 0.49

67. 138% 1.38

68. 210% 2.1

69. $\frac{0.3}{100}$ 0.003

70. $\frac{0.8}{100}$ 0.008

71. $\frac{80}{100}$ 0.8

72. $\frac{60}{100}$ 0.6

Applications

Write as a percent.

73. ***Presidential Election*** For every 100 people in Texas who voted in the 2000 presidential election, 59 voted for George W. Bush.
59%

74. ***Presidential Election*** For every 100 people in Tennessee who voted in the 2000 presidential election, 48 voted for Al Gore.
48%

The following are statements found in newspapers. In each case write the percent as a decimal.

75. ***House Value*** The value of the Sanchez's home increased by 115 percent during the last five years.
1.15

76. ***Charitable Donations*** According to the Gallup Poll, 1 percent of Americans receiving money from a federal tax cut plan to donate the money.
0.01

77. ***Home Value*** 0.6 percent of homes in the U.S. are valued at more than $1 million.
0.006

78. ***Vitamins*** Americans get 30 percent of their vitamin A from carrots.
0.3

79. ***Food Sales*** In 2002, Starbucks made 77 percent of its sales on beverages. Krispy Kreme made 90 percent of its sales from doughnuts.
0.77; 0.9

80. ***Crime Rates*** An FBI report showed a 0.2 percent decrease in the number of reported crimes in 2002. However, it showed a 3 percent increase in the number of robberies.
0.002; 0.03

To Think About

81. Suppose that we want to change 36% to 0.36 by moving the decimal point two places to the left and dropping the % symbol. Explain the steps to show what is really involved in changing 36% to 0.36. Why does the rule work?

36% = 36 percent = 36 "per one hundred" = $36 \times \frac{1}{100} = \frac{36}{100} = 0.36$. The rule is using the fact that 36% means 36 per one hundred.

82. Suppose that we want to change 10.65 to 1065%. Give a complete explanation of the steps.

$10.65 = 1065 \times \frac{1}{100}$ = 1065 "per one hundred" = 1065 percent = 1065%. We change 10.65 to $1065 \times \frac{1}{100}$ and use the idea that percent means "per one hundred."

Write the given value (a) as a decimal, (b) as a fraction with a denominator of 100, and (c) as a reduced fraction.

83. 1562%

(a) 15.62 (b) $\frac{1562}{100}$ (c) $\frac{781}{50}$

84. 3724%

(a) 37.24 (b) $\frac{3724}{100}$ (c) $\frac{931}{25}$

Cumulative Review

Write as a fraction in simplest form.

85. 0.56 $\frac{14}{25}$

86. 0.78 $\frac{39}{50}$

Write as a decimal.

87. $\frac{11}{16}$ 0.6875

88. $\frac{7}{8}$ 0.875

89. ***Ceramics Studio*** A very successful commercial ceramics studio makes beautiful vases for gift stores. In one corner of the warehouse, the storage area has 24 shelves. Three shelves have 246 vases each, seven shelves have 380 vases each, five shelves have 168 vases each, and nine shelves have 122 vases each. How many vases are there in this corner of the studio? 5336 vases

5.2 Changing Between Percents, Decimals, and Fractions

Student Learning Objectives

After studying this section, you will be able to:

1. Change a percent to a fraction.
2. Change a fraction to a percent.
3. Change a percent, a decimal, or a fraction to equivalent forms.

1 Changing a Percent to a Fraction

By using the definition of percent, we can write any percent as a fraction whose denominator is 100. Thus when we change a percent to a fraction, we remove the percent symbol and write the number over 100. To write a number over 100 means that we are dividing by 100. If possible, we then simplify the fraction.

EXAMPLE 1 Write as a fraction in simplest form.

(a) 37% **(b)** 75% **(c)** 2%

Solution

(a) $37\% = \frac{37}{100}$ **(b)** $75\% = \frac{75}{100} = \frac{3}{4}$ **(c)** $2\% = \frac{2}{100} = \frac{1}{50}$

Practice Problem 1 Write as a fraction in simplest form.

(a) 71% **(b)** 25% **(c)** 8%

NOTE TO STUDENT: Fully worked-out solutions to all of the Practice Problems can be found at the back of the text starting at page SP-1

In some cases, it may be helpful to write the percent as a decimal before you write it as a fraction in simplest form.

EXAMPLE 2 Write as a fraction in simplest form.

(a) 43.5% **(b)** 36.75%

Solution

(a) $43.5\% = 0.435$ Change the percent to a decimal.

$= \frac{435}{1000}$ Change the decimal to a fraction.

$= \frac{87}{200}$ Reduce the fraction.

(b) $36.75\% = 0.3675 = \frac{3675}{10{,}000} = \frac{147}{400}$

Teaching Tip Stress the fact that some percents have a simple fraction notation, while others have a rather large and unwieldy one.

Practice Problem 2 Write as a fraction in simplest form.

(a) 8.4% **(b)** 28.5%

√ If the percent is greater than 100%, the simplified fraction is usually changed to a mixed number.

Teaching Tip Some students will forget to reduce after they change the percent to a mixed fraction. Remind them that if they write $375\% = 3\frac{75}{100}$ the answer is not completely finished (and thus not correct) until it is reduced to $3\frac{3}{4}$.

EXAMPLE 3 Write as a mixed number. **(a)** 225% **(b)** 138%

Solution

(a) $225\% = 2.25 = 2\frac{25}{100} = 2\frac{1}{4}$ **(b)** $138\% = 1.38 = 1\frac{38}{100} = 1\frac{19}{50}$

Practice Problem 3 Write as a mixed number.

(a) 170% **(b)** 288%

Sometimes a percent is not a whole number, such as 9% or 10%. Instead, it contains a fraction, such as $9\frac{1}{12}\%$ or $9\frac{3}{8}\%$. Extra steps will be needed to write such a percent as a simplified fraction.

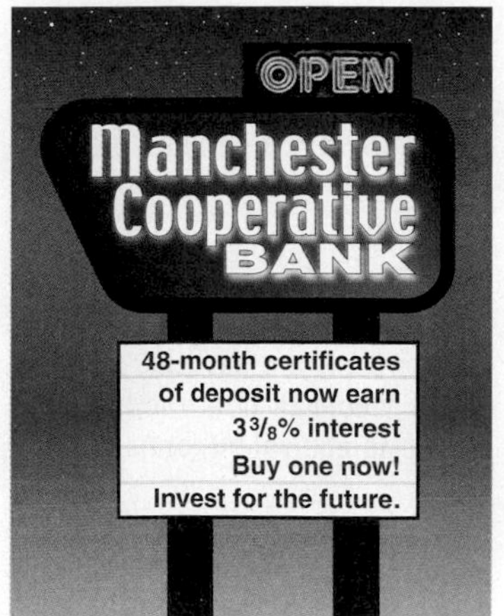

EXAMPLE 4 Convert $3\frac{3}{8}\%$ to a fraction in simplest form.

Solution

$$3\frac{3}{8}\% = \frac{3\frac{3}{8}}{100}$$ Change the percent to a fraction.

$$= 3\frac{3}{8} \div \frac{100}{1}$$ Write the division horizontally. $\frac{3\frac{3}{8}}{100}$ means $3\frac{3}{8}$ divided by 100.

$$= \frac{27}{8} \div \frac{100}{1}$$ Write $3\frac{3}{8}$ as an improper fraction.

$$= \frac{27}{8} \times \frac{1}{100}$$ Use the definition of division of fractions.

$$= \frac{27}{800}$$ Simplify.

***NOTE TO STUDENT:** Fully worked-out solutions to all of the Practice Problems can be found at the back of the text starting at page SP-1*

Practice Problem 4 Convert $7\frac{5}{8}\%$ to a fraction in simplest form.

EXAMPLE 5 In the fiscal 2002 budget of the United States, approximately $16\frac{7}{8}\%$ of the budget was designated for defense. (*Source:* U.S. Office of Management and Budget.) Write this percent as a fraction.

Solution

$$16\frac{7}{8}\% = \frac{16\frac{7}{8}}{100} = 16\frac{7}{8} \div 100 = \frac{135}{8} \times \frac{1}{100} = \frac{135}{800} = \frac{27}{160}$$

Thus we could say $\frac{27}{160}$ of the fiscal 2002 budget was designated for defense. That is, for every \$160 in the budget, \$27 was spent for defense.

Practice Problem 5 In the fiscal 2002 budget of the United States, approximately $22\frac{3}{8}\%$ was designated for social security. (*Source:* U.S. Office of Management and Budget.) Write this percent as a fraction.

Certain percents occur very often, especially in money matters. Here are some common equivalents that you may already know. If not, be sure to memorize them.

$$25\% = \frac{1}{4} \qquad 33\frac{1}{3}\% = \frac{1}{3} \qquad 10\% = \frac{1}{10}$$

$$50\% = \frac{1}{2} \qquad 66\frac{2}{3}\% = \frac{2}{3}$$

$$75\% = \frac{3}{4}$$

2 Changing a Fraction to a Percent

A convenient way to change a fraction to a percent is to write the fraction in decimal form first and then convert the decimal to a percent.

EXAMPLE 6 Write $\frac{3}{8}$ as a percent.

Solution We see that $\frac{3}{8} = 0.375$ by calculating $3 \div 8$.

$$\begin{array}{r} 0.375 \\ 8\overline{)3.000} \\ \underline{24} \\ 60 \\ \underline{56} \\ 40 \\ \underline{40} \\ 0 \end{array}$$

Thus $\frac{3}{8} = 0.375 = 37.5\%$.

Practice Problem 6 Write $\frac{5}{8}$ as a percent.

EXAMPLE 7 Write as a percent. **(a)** $\frac{7}{40}$ **(b)** $\frac{39}{50}$

Solution

(a) $\frac{7}{40} = 0.175 = 17.5\%$ **(b)** $\frac{39}{50} = 0.78 = 78\%$

Practice Problem 7 Write as a percent.

(a) $\frac{21}{25}$ **(b)** $\frac{7}{16}$

Changing some fractions to decimal form results in infinitely repeating decimals. In such cases, we usually round to the nearest hundredth of a percent.

Teaching Tip Remind students that if we want to write an exact value, we must say $\frac{2}{3} = 0.666666\ldots$ or use the bar notation over the 6. They must realize that $\frac{2}{3} \approx 0.67$ is a value rounded to the nearest hundredth. In contrast, $\frac{2}{3} \approx 0.66$ is a truncated value and is not an accurate approximation for our purposes.

EXAMPLE 8 Write as a percent. Round to the nearest hundredth of a percent.

(a) $\frac{1}{6}$ **(b)** $\frac{15}{33}$

Solution **(a)** We find that $\frac{1}{6} = 0.1666\ldots$ by calculating $1 \div 6$.

$$\begin{array}{r} 0.1666 \\ 6\overline{)1.0000} \\ \underline{6} \\ 40 \\ \underline{36} \\ 40 \\ \underline{36} \\ 40 \\ \underline{36} \\ 4 \end{array}$$

We will need a four place decimal so that we will obtain a percent to the nearest hundredth. If we round the decimal to the nearest ten-thousandth, we have $\frac{1}{6} \approx 0.1667$. If we change this to a percent, we have

$$\frac{1}{6} \approx 16.67\%.$$

This is correct to the nearest hundredth of a percent.

(b) By calculating $15 \div 33$, we see that $\frac{15}{33} = 0.45454545\ldots$. We will need a four-place decimal so that we will obtain a percent to the nearest hundredth. If we round to the nearest ten-thousandth, we have

$$\frac{15}{33} \approx 0.4545 = 45.45\%.$$

This rounded value is correct to the nearest hundredth of a percent.

Practice Problem 8 Write as a percent. Round to the nearest hundredth of a percent.

(a) $\frac{7}{9}$ **(b)** $\frac{19}{30}$

NOTE TO STUDENT: Fully worked-out solutions to all of the Practice Problems can be found at the back of the text starting at page SP-1

Recall that sometimes percents are written with fractions.

EXAMPLE 9 Express $\frac{11}{12}$ as a percent containing a fraction.

Solution We will stop the division after two steps and write the remainder in fraction form.

$$\begin{array}{r} 0.91 \\ 12\overline{)11.00} \\ \underline{10\,8} \\ 20 \\ \underline{12} \\ 8 \end{array}$$

This division tells us that we can write

$$\frac{11}{12} \text{ as } 0.91\frac{8}{12} \text{ or } 0.91\frac{2}{3}.$$

We now have a decimal with a fraction. When we express this decimal as a percent, we move the decimal point two places to the right. We do not write the decimal point in front of the fraction.

$$0.91\frac{2}{3} = 91\frac{2}{3}\%$$

Note that our answer in Example 9 is an *exact answer*. We have not rounded off or approximated in any way.

Practice Problem 9 Express $\frac{7}{12}$ as a percent containing a fraction.

3 Changing a Percent, a Decimal, or a Fraction to Equivalent Forms

We have seen so far that a fraction, a decimal, and a percent are three different forms (notations) for the same number. We can illustrate this in a chart.

EXAMPLE 10 Complete the following table of equivalent notations. Round decimals to the nearest ten-thousandth. Round percents to the nearest hundredth of a percent.

Fraction	Decimal	Percent
$\frac{11}{16}$		
	0.265	
		$17\frac{1}{5}\%$

Solution Begin with the first row. The number is written as a fraction. We will change the fraction to a decimal and then to a percent.

The fraction is changed to a decimal is changed to a percent.

$$\frac{11}{16} \longrightarrow 16\overline{)11.0000}\ \text{quotient } 0.6875 \longrightarrow 68.75\%$$

In the second row the number is written as a decimal. This can easily be written as a percent.

$$0.265 \longrightarrow 26.5\%$$

Now write 0.265 as a fraction and simplify.

$$0.265 \downarrow \frac{265}{1000} \longrightarrow \frac{53}{200}$$

In the third row the number is written as a percent. Proceed from right to left—that is, write the number as a decimal and then as a fraction.

$$\frac{17\frac{1}{5}}{100} \leftarrow 17\frac{1}{5}\%$$

$$\downarrow$$

$$\boxed{\frac{86}{5} \times \frac{1}{100}}$$

$$\downarrow$$

$$0.172 \longleftarrow \frac{86}{500} \quad \text{Divide.} \quad 500\overline{)86.000} = 0.172$$

and

$$\frac{43}{250} \longleftarrow \frac{86}{500}$$

Thus the completed table is as follows.

Fraction	Decimal	Percent
$\frac{11}{16}$	0.6875	68.75%
$\frac{53}{200}$	0.265	26.5%
$\frac{43}{250}$	0.172	$17\frac{1}{5}\%$

Practice Problem 10 Complete the following table of equivalent notations. Round decimals to the nearest ten-thousandth. Round percents to the nearest hundredth of a percent.

Fraction	Decimal	Percent
$\frac{23}{99}$	0.2323	23.23%
$\frac{129}{250}$	0.516	51.6%
$\frac{97}{250}$	0.388	$38\frac{4}{5}\%$

Calculator

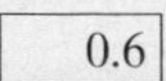

Fraction to Decimal

You can use a calculator to change $\frac{3}{5}$ to a decimal. Enter

3 [÷] 5 [=]

The display should read

[0.6]

Try the following.

(a) $\frac{17}{25}$ **(b)** $\frac{2}{9}$

(c) $\frac{13}{10}$ **(d)** $\frac{15}{19}$

Note: 0.78947368 is an approximation for $\frac{15}{19}$. Some calculators round to only eight places.

***NOTE TO STUDENT:** Fully worked-out solutions to all of the Practice Problems can be found at the back of the text starting at page SP-1*

ALTERNATIVE METHOD: Using Proportions to Convert from Fraction to Percent Another way to convert a fraction to a percent is to use a proportion. To change $\frac{7}{8}$ to a percent, write the proportion

$$\frac{7}{8} = \frac{n}{100}$$

$$7 \times 100 = 8 \times n \quad \text{Cross-multiply.}$$

$$700 = 8 \times n \quad \text{Simplify.}$$

$$\frac{700}{8} = \frac{8 \times n}{8} \quad \text{Divide each side by 8.}$$

$$87.5 = n \quad \text{Simplify.}$$

Thus $\frac{7}{8} = 87.5\%$. You will use this approach in Exercises 5.2, exercises 85 and 86.

5.2 EXERCISES

Student Solutions Manual | CD/Video | PH Math Tutor Center | MathXL®Tutorials on CD | MathXL® | MyMathLab® | Interactmath.com

Verbal and Writing Skills

1. Explain in your own words how to change a percent to a fraction.

Write the number in front of the percent symbol as the numerator of a fraction. Write the number 100 as the denominator of the fraction. Reduce the fraction if possible.

2. Explain in your own words how to change a fraction to a percent.

Change the fraction to a decimal by dividing the denominator into the numerator. Change the decimal to a percent by moving the decimal point two places to the right and add the % symbol.

Write as a fraction or as a mixed number.

3. 6% $\frac{3}{50}$

4. 8% $\frac{2}{25}$

5. 33% $\frac{33}{100}$

6. 47% $\frac{47}{100}$

7. 55% $\frac{11}{20}$

8. 35% $\frac{7}{20}$

9. 75% $\frac{3}{4}$

10. 25% $\frac{1}{4}$

11. 20% $\frac{1}{5}$

12. 40% $\frac{2}{5}$

13. 9.5% $\frac{19}{200}$

14. 6.5% $\frac{13}{200}$

15. 17.5% $\frac{7}{40}$

16. 77.5% $\frac{31}{40}$

17. 64.8% $\frac{81}{125}$

18. 12.2% $\frac{61}{500}$

19. 71.25% $\frac{57}{80}$

20. 38.75% $\frac{31}{80}$

21. 176% $\frac{176}{100} = 1\frac{19}{25}$

22. 228% $\frac{228}{100} = 2\frac{7}{25}$

23. 340% $\frac{340}{100} = 3\frac{2}{5}$

24. 420% $\frac{420}{100} = 4\frac{1}{5}$

25. 1200% $\frac{1200}{100} = 12$

26. 3600% $\frac{3600}{100} = 36$

27. $2\frac{1}{6}\%$ $\frac{\frac{13}{6}}{100} = \frac{13}{600}$

28. $7\frac{5}{6}\%$ $\frac{\frac{47}{6}}{100} = \frac{47}{600}$

29. $12\frac{1}{2}\%$ $\frac{\frac{25}{2}}{100} = \frac{1}{8}$

30. $37\frac{1}{2}\%$ $\frac{\frac{75}{2}}{100} = \frac{3}{8}$

31. $8\frac{4}{5}\%$ $\frac{\frac{44}{5}}{100} = \frac{11}{125}$

32. $9\frac{3}{5}\%$ $\frac{\frac{48}{5}}{100} = \frac{12}{125}$

Applications

33. ***Crime Rates*** Between 2001 and 2002, the change in the number of reported crimes in the Northeast region of the U.S. fell 3.3%. Write the percent as a fraction. (*Source*: Federal Bureau of Investigation)

$\frac{33}{1000}$

34. ***Crime Rates*** Between 2001 and 2002, the change in the number of reported crimes in the West region of the U.S. rose 2.9%. Write the percent as a fraction. (*Source*: Federal Bureau of Investigation)

$\frac{29}{1000}$

35. ***Federal Budget*** In the 2003 budget of the United States, approximately $2\frac{6}{11}\%$ was designated for veterans' benefits and Services. (*Source*: U.S. Office of Management and Budget.) Write this percent as a fraction.

$\frac{\frac{28}{11}}{100} = \frac{7}{275}$

36. ***Federal Budget*** In the 2003 budget of the United States, approximately $11\frac{1}{12}\%$ was designated for Medicare. (*Source:* U.S. Office of Management and Budget.) Write this percent as a fraction.

$\frac{\frac{133}{12}}{100} = \frac{133}{1200}$

Write as a percent. Round to the nearest hundredth of a percent when necessary.

37. $\frac{3}{4}$ 75% **38.** $\frac{1}{4}$ 25% **39.** $\frac{7}{10}$ 70% **40.** $\frac{9}{10}$ 90% **41.** $\frac{7}{20}$ 35% **42.** $\frac{11}{20}$ 55%

43. $\frac{7}{25}$ 28% **44.** $\frac{9}{25}$ 36% **45.** $\frac{11}{40}$ 27.5% **46.** $\frac{13}{40}$ 32.5% **47.** $\frac{18}{5}$ 360% **48.** $\frac{7}{4}$ 175%

49. $2\frac{1}{2}$ 250% **50.** $3\frac{3}{4}$ 375% **51.** $4\frac{1}{8}$ 412.5% **52.** $2\frac{5}{8}$ 262.5% **53.** $\frac{1}{3}$ 33.33% **54.** $\frac{2}{3}$ 66.67%

55. $\frac{3}{7}$ 42.86% **56.** $\frac{2}{7}$ 28.57% **57.** $\frac{17}{4}$ 425% **58.** $\frac{12}{5}$ 240% **59.** $\frac{26}{50}$ 52% **60.** $\frac{43}{50}$ 86%

Applications

Round to the nearest hundredth of a percent.

61. ***Human Brain*** The brain represents approximately $\frac{1}{40}$ of an average person's weight. Express this fraction as a percent. 2.5%

62. ***Monthly House Payments*** To calculate your maximum monthly house payment, a real estate agent multiplies your monthly income by $\frac{7}{25}$. Express this fraction as a percent. 28%

63. ***Mail Service*** The U.S. mail service estimates that people do not send as many personal letters through the mail as they did 20 years ago. In fact, in 2000 only $\frac{35}{3000}$ of all mail delivered consisted of personal letters. Express this fraction as a percent. 1.17%

64. ***Distribution of Labor Force*** People in age group 25–34 years made up $\frac{32}{145}$ of the civilian labor force in the year 2002 according to the U.S. Bureau of Labor Statistics. Express this fraction as a percent. 22.07%

Express as a percent containing a fraction. (See Example 9.)

65. $\frac{3}{8}$ $37\frac{1}{2}\%$ **66.** $\frac{5}{8}$ $62\frac{1}{2}\%$ **67.** $\frac{3}{40}$ $7\frac{1}{2}\%$ **68.** $\frac{11}{90}$ $12\frac{2}{9}\%$ **69.** $\frac{5}{6}$ $83\frac{1}{3}\%$

70. $\frac{1}{6}$ $16\frac{2}{3}\%$ **71.** $\frac{2}{9}$ $22\frac{2}{9}\%$ **72.** $\frac{8}{9}$ $88\frac{8}{9}\%$

Mixed Practice

In exercises 73–82, complete the table of equivalents. Round decimals to the nearest ten-thousandth. Round percents to the nearest hundredth of a percent.

	Fraction	Decimal	Percent
73.	$\frac{11}{12}$	0.9167	91.67%
75.	$\frac{14}{25}$	0.56	56%
77.	$\frac{1}{200}$	0.005	0.5%
79.	$\frac{5}{9}$	0.5556	55.56%
81.	$\frac{1}{32}$	0.0313	$3\frac{1}{8}\%$

	Fraction	Decimal	Percent
74.	$\frac{1}{12}$	0.0833	8.33%
76.	$\frac{17}{20}$	0.85	85%
78.	$\frac{17}{200}$	0.085	8.5%
80.	$\frac{7}{9}$	0.7778	77.78%
82.	$\frac{21}{800}$	0.0263	$2\frac{5}{8}\%$

83. Write $28\frac{15}{16}\%$ as a fraction.

$\frac{463}{16} \times \frac{1}{100} = \frac{463}{1600}$

84. Write $18\frac{7}{12}\%$ as a fraction.

$\frac{223}{12} \times \frac{1}{100} = \frac{223}{1200}$

Change each fraction to a percent by using a proportion.

85. $\frac{123}{800}$ $\frac{123}{800} = \frac{n}{100}$ $n = 15.375$ 15.375%

86. $\frac{417}{600}$ $\frac{417}{600} = \frac{n}{100}$ $n = 69.5$ 69.5%

To Think About

87. What can you say about a decimal number if it is equivalent to a percent that is less than 0.1%? How many zeros will it have to the right of the decimal point?

It will have at least two zeros to the right of the decimal point.

88. What can you say about a decimal number if it is equivalent to a percent that is less than 0.01%? How many zeros will it have to the right of the decimal point?

It will have at least three zeros to the right of the decimal point.

Cumulative Review

Solve for n.

89. $\frac{15}{n} = \frac{8}{3}$ $n = 5.625$

90. $\frac{32}{24} = \frac{n}{3}$ $n = 4$

91. ***Law Firm*** The law firm of Dewey, Cheatham & Howe was required to review 54 years of documents of one of its clients. The first file contained 10,041 documents. The second file contained 986 documents. The third file contained 4,283 documents. The last file contained 533,855 documents. How many total documents were there?

549,165 documents

▲ **92.** ***Restaurant Size*** A small neighborhood café has an area of 1800 square feet. A new bar and grill across the street is $2\frac{1}{2}$ times the size of the café. How many square feet is the bar and grill?

4500 square feet

5.3A Solving Percent Problems Using Equations

Student Learning Objectives

After studying this section, you will be able to:

1. Translate a percent problem into an equation.
2. Solve a percent problem by solving an equation.

1 Translating a Percent Problem into an Equation

In word problems like the ones in this section, we can translate from words to mathematical symbols and back again. After we have the mathematical symbols arranged in an *equation,* we solve the equation. When we find the values that make the equation true, we have also found the answer to our word problem.

To solve a percent problem, we express it as an equation with an unknown quantity. We use the letter n to represent the number we do not know. The following table is helpful when translating from a percent problem to an equation.

Word	Mathematical Symbol
of	Any multiplication symbol: $\times$ or ( ) or $\cdot$
is	$=$
what	Any letter; for example, n
find	$n =$

In Examples 1–5 we show how to translate words into an equation. Please do **not** solve the problem. Translate into an equation only.

EXAMPLE 1 Translate into an equation.

What is 5% of 19.00?

$$\downarrow \quad \downarrow\downarrow \quad \downarrow \quad \downarrow$$

Solution

$$n = 5\% \times 19.00$$

Practice Problem 1 Translate into an equation. What is 26% of 35?

NOTE TO STUDENT: Fully worked-out solutions to all of the Practice Problems can be found at the back of the text starting at page SP-1

Teaching Tip Practicing translation is an excellent way for students to develop confidence with percent problems. Encourage them not to skip the step of practicing translation.

EXAMPLE 2 Translate into an equation.

Find 0.6% of 400.

Solution Notice here that the words *what is* are missing. The word *find* is equivalent to *what is.*

Find 0.6% of 400.

$$\downarrow \quad \downarrow \quad \downarrow \quad \downarrow$$

$$n = 0.6\% \times 400$$

Practice Problem 2 Translate into an equation. Find 0.08% of 350.

The unknown quantity, n, does not always stand alone in an equation.

EXAMPLE 3 Translate into an equation.

(a) 35% of what is 60? **(b)** 7.2 is 120% of what?

Solution

(a) 35% of what is 60?

↓ ↓ ↓ ↓ ↓

$35\% \times n = 60$

(b) 7.2 is 120% of what?

↓ ↓ ↓ ↓ ↓

$7.2 = 120\% \times n$

Practice Problem 3 Translate into an equation.

(a) 58% of what is 400? **(b)** 9.1 is 135% of what?

EXAMPLE 4 Translate into an equation.

What percent of 50 is 10?

↓ ↓ ↓ ↓

$n \times 50 = 10$

Solution

We see here that the words *what percent* are represented by the letter *n*.

Practice Problem 4 Translate into an equation. What percent of 250 is 36?

EXAMPLE 5 Translate into an equation.

(a) 30 is what percent of 16? **(b)** What percent of 3000 is 2.6?

Solution

(a) 30 is what percent of 16?

↓ ↓ ↓ ↓ ↓

$30 = n \times 16$

(b) What percent of 3000 is 2.6?

↓ ↓ ↓ ↓ ↓

$n \times 3000 = 2.6$

Practice Problem 5 Translate into an equation.

(a) 50 is what percent of 20? **(b)** What percent of 2000 is 4.5?

2 Solving a Percent Problem by Solving an Equation

Teaching Tip Some students may insist that they do not need to use an equation to solve problems where the amount is not known. Remind them that it is good to do a few problems using an equation even though they do not need it, just so they will be better able to correctly identify which of the three types of percent problems they have encountered.

The percent problems we have translated are of three types. Consider the equation $60 = 20\% \times 300$. This problem has the form

$$\text{amount} = \text{percent} \times \text{base}.$$

Any one of these quantities—amount, percent, or base—may be unknown.

1. When *we do not know the amount,* we have an equation like
$$n = 20\% \times 300.$$
2. When *we do not know the base,* we have an equation like
$$60 = 20\% \times n.$$
3. When *we do not know the percent,* we have an equation like
$$60 = n \times 300.$$

We will study each type separately. It is not necessary to memorize the three types, but it is helpful to look carefully at the examples we give of each. In each example, do the computation in a way that is easiest for you. This may be using a pencil and paper, using a calculator, or, in some cases, doing the problem mentally.

Solving Percent Problems When the Amount Is Unknown In solving these equations we will need to change the percent number to decimal form.

EXAMPLE 6 What is 45% of 590?

Solution

$$n = 45\% \times 590 \quad \text{Translate into an equation.}$$
$$n = (0.45)(590) \quad \text{Change the percent to decimal form.}$$
$$n = 265.5 \quad \text{Multiply } 0.45 \times 590.$$

Practice Problem 6 What is 82% of 350?

EXAMPLE 7 Find 160% of 500.

Solution

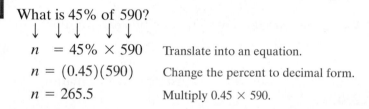

Find 160% of 500. When you translate, remember that the word *find* is equivalent to *what is.*

$$n = 160\% \times 500$$
$$n = (1.60)(500) \quad \text{Change the percent to decimal form.}$$
$$n = 800 \quad \text{Multiply 1.6 by 500.}$$

Practice Problem 7 Find 230% of 400.

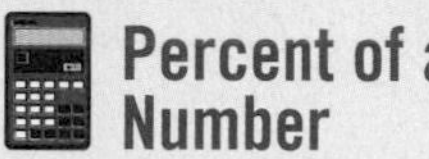

Calculator

Percent of a Number

You can use a calculator to find 12% of 48.
Enter

12 [%] [×] 48 [=]

The display should read

[5.76]

If your calculator does not have a percent key, use the keystrokes

0.12 [×] 48 [=]

What is 54% of 450?

NOTE TO STUDENT: Fully worked-out solutions to all of the Practice Problems can be found at the back of the text starting at page SP-1

EXAMPLE 8 When Rick bought a new Dodge Neon, he had to pay a sales tax of 5% on the cost of the car, which was \$12,000. What was the sales tax?

Solution This problem is asking

What is 5% of \$12,000?
↓ ↓ ↓ ↓ ↓

$$n = 5\% \times \$12{,}000$$

$$n = 0.05 \times 12{,}000$$

$$n = \$600$$

The sales tax was \$600.

Practice Problem 8 When Oprah bought an airplane ticket, she had to pay a tax of 8% on the cost of the ticket, which was \$350. What was the tax?

Solving Percent Problems When the Base Is Unknown If a number is multiplied by the letter n, this can be indicated by a multiplication sign, parentheses, a dot, or placing the number in front of the letter. Thus $3 \times n = 3(n) = 3 \cdot n = 3n$.

In this section we use equations like $3n = 9$ and $0.5n = 20$. To solve these equations we use the procedures developed in Chapter 4. We divide each side by the number multiplied by n.

In solving these equations we will need to change the percent number to decimal form.

EXAMPLE 9

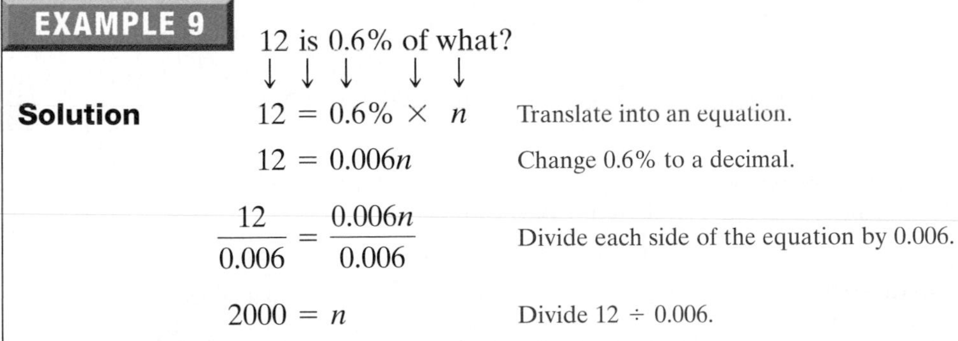

12 is 0.6% of what?
↓ ↓ ↓ ↓ ↓

Solution

$12 = 0.6\% \times n$ Translate into an equation.

$12 = 0.006n$ Change 0.6% to a decimal.

$\dfrac{12}{0.006} = \dfrac{0.006n}{0.006}$ Divide each side of the equation by 0.006.

$2000 = n$ Divide $12 \div 0.006$.

Practice Problem 9 32 is 0.4% of what?

Calculator

Finding the Percent

You can use a calculator to find a missing percent. What percent of 95 is 19?

1. Enter as a fraction. Enter the number after "is," and then the division key. Then enter the number after the word "of."

   19 [÷] 95

2. Change to a percent.

   19 [÷] 95 [×] 100 [=]

   The display should read

   [20]

   This means 20%.
   What percent of 625 is 250?

EXAMPLE 10 Dave and Elsie went out to dinner. They gave the waiter a tip that was 15% of the total bill. The tip the waiter received was \$6. What was the total bill (not including the tip)?

Solution This problem is asking

15% of what is \$6?

$$15\% \times n = 6$$

$$0.15n = 6$$

$$\frac{0.15n}{0.15} = \frac{6}{0.15} \qquad n = 40$$

The total bill for the meal (not including the tip) was \$40.

Practice Problem 10 The coach of the university baseball team said that 30% of the players on his team are left-handed. Six people on the team are left-handed. How many people are on the team?

Solving Percent Problems When the Percent Is Unknown In solving these problems, we notice that there is no % symbol in the problem. The percent is what we are trying to find. Therefore, our answer for this type of problem will always have a percent symbol.

EXAMPLE 11 What percent of 5000 is 3.8?

Solution

$n \times 5000 = 3.8$ Translate into an equation.

$5000n = 3.8$ Multiplication is commutative. $n \times 5000 = 5000 \times n$.

$\frac{5000n}{5000} = \frac{3.8}{5000}$ Divide each side by 5000.

$n = 0.00076$ Divide 3.8 by 5000.

$n = 0.076\%$ Express the decimal as a percent.

Practice Problem 11 What percent of 9000 is 4.5?

NOTE TO STUDENT: Fully worked-out solutions to all of the Practice Problems can be found at the back of the text starting at page SP-1

EXAMPLE 12 90 is what percent of 20?

Solution

$90 = n \times 20$ Translate into an equation.

$90 = 20n$ Multiplication is commutative. $n \times 20 = 20 \times n$.

$\frac{90}{20} = \frac{20n}{20}$ Divide each side by 20.

$4.5 = n$ Divide 90 by 20.

$450\% = n$ Express the decimal as a percent.

Practice Problem 12 198 is what percent of 33?

Teaching Tip In problems like Example 12, some students will want to reduce 90 divided by 20 before they divide, and others will divide directly. You might mention that either approach is OK as long as the student can do the problem fairly quickly.

EXAMPLE 13 In a recent basketball game for the New York Knicks, Patrick Ewing made 10 of his 24 shots. What percent of his shots did he make? (Round to the nearest tenth of a percent.)

Solution This is equivalent to

$$\underbrace{10}_{10}\ \underbrace{\text{is}}_{=}\ \underbrace{\text{what percent}}_{n}\ \underbrace{\text{of}}_{\times}\ \underbrace{24?}_{24}$$

$$10 = n \times 24$$

$$10 = 24n$$

$$\frac{10}{24} = \frac{24n}{24}$$

$$0.41666\ldots = n$$

To the nearest tenth of a percent we have

$$n = 41.7\%$$

Patrick Ewing made 41.7% of his shots in this game.

Practice Problem 13 In a basketball game for the Los Angeles Lakers, A.C. Green made 5 of his 16 shots. What percent of his shots did he make? (Round to the nearest tenth of a percent.)

5.3A Exercises

Student Solutions Manual

CD/ Video

PH Math Tutor Center

MathXL®Tutorials on CD
MathXL®

MyMathLab®

Interactmath.com

Verbal and Writing Skills

1. Give an example of a percent problem when we do not know the amount.
What is 20% of \$300?

2. Give an example of a percent problem when we do not know the base.
\$500 is 30% of what number?

3. Give an example of a percent problem when we do not know the percent.
20 baskets out of 25 shots is what percent?

4. When you encounter a problem like "What is 65% of \$600?" what type of percent problem is this? How would you solve such a problem?
This is a type called "a percent problem when we do not know the amount." We can translate this into an equation

$$n = 65\% \times 600$$
$$n = (0.65)(600)$$
$$n = 390$$

5. When you encounter a problem like "108 is 18% of what number?" what type of percent problem is this? How would you solve such a problem?
This is a type called "a percent problem when we do not know the base." We can translate this into an equation

$$108 = 18\% \times n$$
$$108 = 0.18n$$
$$\frac{108}{0.18} = \frac{0.18n}{0.18}$$
$$600 = n$$

6. When you encounter a problem like "What percent of 35 is 14?" what type of percent problem is this? How would you solve such a problem?
This is a type called "a percent problem when we do not know the percent." We can translate this into an equation

$$n \times 35 = 14$$
$$35n = 14$$
$$n = 0.4$$

The answer is 40%.

Translate into a mathematical equation in exercises 7–12. Use the letter n for the unknown quantity. Do **not** *solve, but rather just obtain the equation.*

7. What is 5% of 90?
$n = 5\% \times 90$

8. What is 9% of 65?
$n = 9\% \times 65$

9. 70% of what is 2?
$70\% \times n = 2$

10. 55% of what is 34?
$55\% \times n = 34$

11. 17 is what percent of 85?
$17 = n \times 85$

12. 24 is what percent of 144?
$24 = n \times 144$

Solve.

13. What is 20% of 140?
$n = 20\% \times 140 \quad n = 28$

14. What is 30% of 210?
$n = 30\% \times 210 \quad n = 63$

15. Find 40% of 140.
$n = 40\% \times 140 \quad n = 56$

16. Find 60% of 210.
$n = 60\% \times 210 \quad n = 126$

Applications

17. ***Sales Tax*** Rico bought a new television. The price before the 5% tax was added on was \$480. How much tax did Rico have to pay?
5% × \$480 = \$24

18. ***Check-Cashing Service*** The Center City check-cashing service charges a fee that is 7% of the amount of the check they cash. How much is the fee for cashing a check for \$240?
7% × \$240 = \$16.80

Solve.

19. 2% of what is 26?
$2\% \times n = 26 \quad n = 1300$

20. 3% of what is 18?
$3\% \times n = 18 \quad n = 600$

21. 52 is 4% of what?
$52 = 4\% \times n \quad n = 1300$

22. 36 is 6% of what?
$36 = 6\% \times n \quad n = 600$

Applications

23. ***Australia Tax*** In Australia, all general sales (except for food) have a hidden tax of 22% built into the final price. Walter is planning to purchase a camera while in Australia. He wants to know the before-tax price that the dealer is charging before he adds on the hidden tax of $33. Can you determine the amount of the before-tax price?
$22\% \times n = 33 \quad n = \$150$

24. ***Opinion Poll*** A newspaper states that 522 of its residents are in favor of building a new high school. This is 12% of the town's population. What is the population of the town?
$12\% \times n = 522$
$n = 4350$

Solve.

25. What percent of 200 is 168?
$n \times 200 = 168 \quad n = 84\%$

26. What percent of 300 is 135?
$n \times 300 = 135 \quad n = 45\%$

27. 56 is what percent of 200?
$56 = n \times 200 \quad 28\% = n$

28. 45 is what percent of 300?
$45 = n \times 300 \quad 15\% = n$

Applications

29. ***Basketball*** The total number of points scored in a basketball game was 120. The winning team scored 78 of those points. What percent of the points were scored by the winning team?
$78 = n \times 120 \quad n = 65\%$

30. ***Car Repairs*** Randy's bill for car repairs was $140. Of this amount, $28 was charged for labor and $112 was charged for parts. What percent of the bill was for labor?
$28 = n \times 140 \quad n = 20\%$

Mixed Practice

Solve.

31. 20% of 155 is what?
$20\% \times 155 = n \quad n = 31$

32. 60% of 215 is what?
$60\% \times 215 = n \quad n = 129$

33. 170% of what is 144.5?
$170\% \times n = 144.5 \quad n = 85$

34. 160% of what is 152?
$160\% \times n = 152 \quad n = 95$

35. 84 is what percent of 700?
$84 = n \times 700 \quad 12\% = n$

36. 72 is what percent of 900?
$72 = n \times 900 \quad 8\% = n$

37. Find 0.4% of 820.
$n = 0.4\% \times 820 \quad n = 3.28$

38. Find 0.3% of 540.
$n = 0.3\% \times 540 \quad n = 1.62$

39. What percent of 35 is 22.4?
$n \times 35 = 22.4 \quad n = 64\%$

40. What percent of 45 is 16.2?
$n \times 45 = 16.2 \quad n = 36\%$

41. 89 is 20% of what?
$89 = 20\% \times n \quad 445 = n$

42. 42 is 40% of what?
$42 = 40\% \times n \quad n = 105$

43. 8 is what percent of 1000?
$8 = n \times 1000 \quad 0.8\% = n$

44. 6 is what percent of 800?
$6 = n \times 800 \quad 0.75\% = n$

45. What is 16.5% of 240?
$n = 16.5\% \times 240 \quad n = 39.6$

46. What is 18.5% of 360?
$n = 18.5\% \times 360 \quad n = 66.6$

47. Scoring 44 problems out of 55 problems correctly on a test is what percent?
80%

48. Scoring 27 problems out of 45 problems correctly on a test is what percent?
60%

Applications

49. ***Movie Studio*** The special effects office at Magic Movie Studio fabricated 80 minimodels of office buildings for the disaster sequence of a new film. The director of photography discovered that on camera, only 68 of them really looked authentic. What percent of the models were acceptable? 85%

50. ***Equestrian Rider*** An Olympic equestrian rider practiced jumping over a water hazard. In 400 attempts, she and her horse touched the water 15 times. What percent of her jump attempts were not perfect? 3.75%

51. ***College Student Employment*** At Brookstone State College 44% of the freshman class is employed part-time. There are 1260 freshmen this year. How many of them are employed part-time? Round to the nearest whole number.
554 students

52. ***Student Health*** A recent study indicates that 15% of all middle school students do not eat a proper breakfast. If Pineridge Middle School has 420 students, how many do not eat a proper breakfast?
63 students

53. ***Swim Team*** The swim team at Stonybrook College has gone on to the state championships 24 times over the years. If that translates to 60% of the time in which the team has qualified for the finals, how many years has the swim team qualified for the finals? 40 years

54. ***Higher Education*** North Shore Community College found that 60% of its graduates go on for further education. Last year 570 of the graduates went on for further education. How many students graduated from the college last year? 950 students

55. Find 12% of 30% of $1600. $57.60

56. Find 90% of 15% of 2700. 364.5

To Think About

▲ **57.** ***Geometry*** A cruise ship has a rectangular swimming pool. The width of the pool is 30% of the length of the pool. The perimeter of the pool is 1040 feet. What is the width of the pool? What is the length of the pool?
width 120 feet; length 400 feet

▲ **58.** ***Geometry*** The back portion of the schoolyard is used as a playground for elementary children. The playground has a length that is 200% of the width. The area of the playground is 3200 square feet. What is the length and width of the playground?
width 40 feet; length 80 feet

Cumulative Review

Multiply or Divide.

59. $\begin{array}{r} 1.36 \\ \times\ 1.8 \\ \hline \end{array}$ 2.448

60. $\begin{array}{r} 5.06 \\ \times\ 0.82 \\ \hline \end{array}$ 4.1492

61. $0.06\overline{)170.04}$ 2834

62. $0.9\overline{)2.124}$ 2.36

5.3B Solving Percent Problems Using Proportions

Student Learning Objectives

After studying this section, you will be able to:

1. **Identify the parts of the percent proportion.**
2. **Use the percent proportion to solve percent problems.**

1 Identifying the Parts of the Percent Proportion

In Section 5.3A we showed you how to use an equation to solve a percent problem. Some students find it easier to use proportions to solve percent problems. We will show you how to use proportions in this section. The two methods work equally well. Using percent proportions allows you to see another of the many uses of the proportions that we studied in Chapter 4.

Suppose your math class of 25 students has 19 right-handed students and 6 left-handed students. You could say that $\frac{19}{25}$ of the class or 76% is right-handed. Consider the following relationship.

$$\frac{19}{25} = 76\%$$

This can be written as

$$\frac{19}{25} = \frac{76}{100}$$

Teaching Tip Some students dislike solving percent problems by the proportion method; other students find it very helpful.

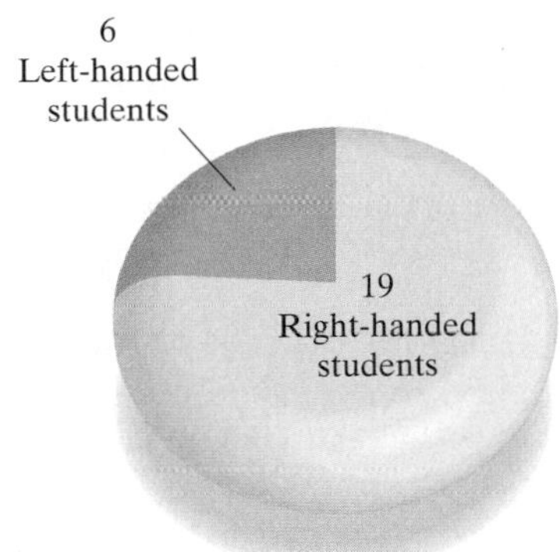

As a rule, we can write this relationship using the **percent proportion**

$$\frac{\text{amount}}{\text{base}} = \frac{\text{percent number}}{100}.$$

To use this equation effectively, we need to find the amount, base, and percent number in a word problem. The easiest of these three parts to find is the percent number. We use the letter p (a variable) to represent the **percent number.**

EXAMPLE 1 Identify the percent number p.

(a) Find 16% of 370 **(b)** 28% of what is 25?

(c) What percent of 18 is 4.5?

Solution

(a) Find 16% of 370.
The value of p is 16.

(b) 28% of what is 25?
The value of p is 28.

(c) $\underbrace{\text{What percent}}_{p}$ of 18 is 4.5?

We let p represent the unknown percent number.

Practice Problem 1 Identify the percent number p.

(a) Find 83% of 460. **(b)** 18% of what number is 90?

(c) What percent of 64 is 8?

***NOTE TO STUDENT:** Fully worked-out solutions to all of the Practice Problems can be found at the back of the text starting at page SP-1*

We use the letter b to represent the base number. The **base** is the entire quantity or the total involved. The number that is the base usually appears after the word *of.* The **amount,** which we represent by the letter a, is the part being compared to the whole.

EXAMPLE 2 Identify the base b and the amount a.

(a) 20% of 320 is 64. **(b)** 12 is 60% of what?

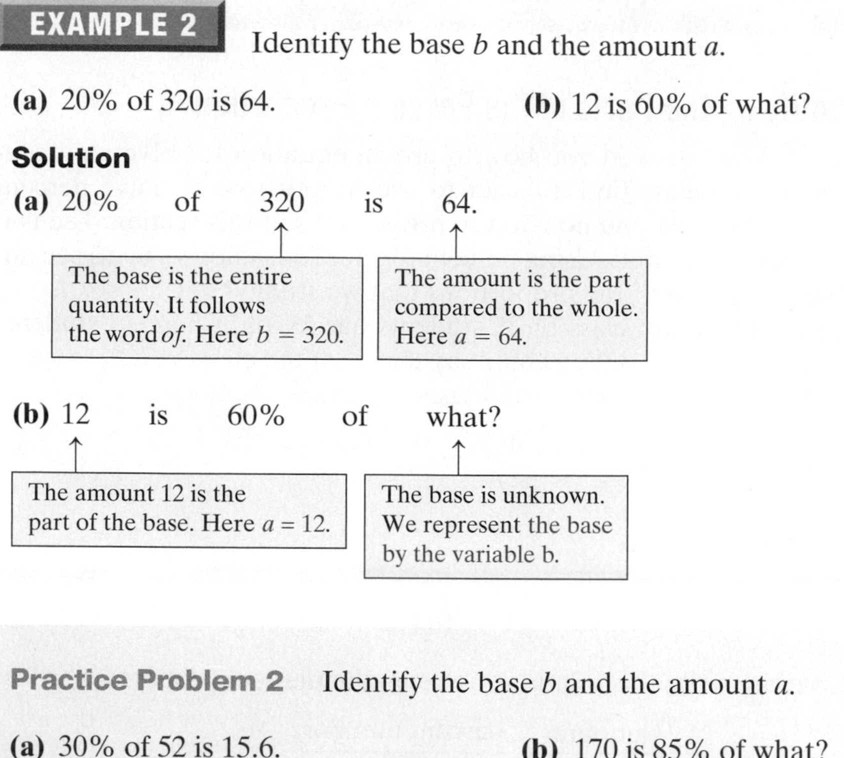

Solution
(a) 20% of 320 is 64.
The base is the entire quantity. It follows the word of. Here b = 320.
The amount is the part compared to the whole. Here a = 64.
(b) 12 is 60% of what?
The amount 12 is the part of the base. Here a = 12.
The base is unknown. We represent the base by the variable b.

NOTE TO STUDENT: Fully worked-out solutions to all of the Practice Problems can be found at the back of the text starting at page SP-1

Practice Problem 2 Identify the base b and the amount a.

(a) 30% of 52 is 15.6. **(b)** 170 is 85% of what?

When identifying p, b, and a in a problem, it is easiest to identify p and b first. The remaining quantity or variable is a.

Teaching Tip The key to getting students to use the proportion method successfully is seeing that they can correctly identify the three parts p, b, and a. Give this aspect special emphasis.

EXAMPLE 3 Find p, b, and a.

(a) What is 52% of 300? **(b)** What percent of 30 is 18?

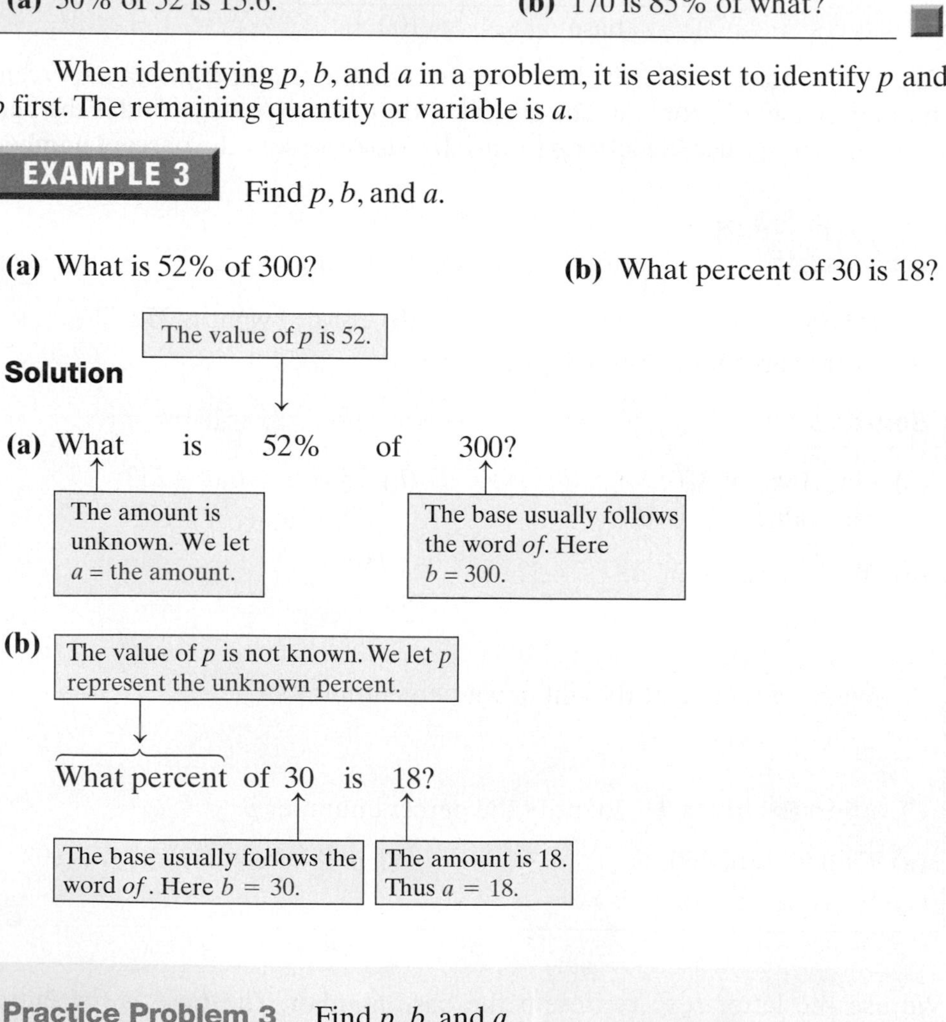

Practice Problem 3 Find p, b, and a.

(a) What is 18% of 240? **(b)** What percent of 64 is 4?

2 Using the Percent Proportion to Solve Percent Problems

When we solve the percent proportion, we will have enough information to state the numerical value for two of the three variables a, b, p in the equation

$$\frac{a}{b} = \frac{p}{100}.$$

We first identify those two values, and then substitute those values into the equation. Then we will use the skills that we acquired for solving proportions in Chapter 4 to find the value we do not know. Here and throughout the entire chapter we assume that $b \neq 0$.

When solving each problem it is a good idea to look at your answer and see if it is reasonable. Ask yourself, "Does my answer make sense?"

EXAMPLE 4 Find 260% of 40.

Solution The percent $p = 260$. The number that is the base usually appears after the word *of*. The base $b = 40$. The amount is unknown. We use the variable a. Thus

$$\frac{a}{b} = \frac{p}{100} \quad \text{becomes} \quad \frac{a}{40} = \frac{260}{100}.$$

If we reduce the fraction on the right-hand side, we have

$$\frac{a}{40} = \frac{13}{5}$$

$$5a = (40)(13) \quad \text{Cross multiply.}$$

$$5a = 520 \quad \text{Simplify.}$$

$$\frac{5a}{5} = \frac{520}{5} \quad \text{Divide each side of the equation by 5.}$$

$$a = 104$$

Thus 260% of 40 is 104.

Practice Problem 4 Find 340% of 70.

EXAMPLE 5 85% of what is 221?

Solution The percent $p = 85$. The base is unknown. We use the variable b. The amount a is 221. Thus

$$\frac{a}{b} = \frac{p}{100} \quad \text{becomes} \quad \frac{221}{b} = \frac{85}{100}.$$

If we reduce the fraction on the right-hand side, we have

$$\frac{221}{b} = \frac{17}{20}$$

$$(221)(20) = 17b \quad \text{Cross multiply.}$$

$$4420 = 17b \quad \text{Simplify.}$$

$$\frac{4420}{17} = \frac{17b}{17} \quad \text{Divide each side by 17.}$$

$$260 = b. \quad \text{Divide 4420 by 17.}$$

Thus 85% of 260 is 221.

***NOTE TO STUDENT:** Fully worked-out solutions to all of the Practice Problems can be found at the back of the text starting at page SP-1*

Practice Problem 5 68% of what is 476?

EXAMPLE 6 George and Barbara purchased some no-load mutual funds. The account manager charged a service fee of 0.2% of the value of the mutual funds. George and Barbara paid this fee, which amounted to \$53. When they got home they could not find the receipt that showed the exact value of the mutual funds that they purchased. Can you find the value of the mutual funds that they purchased?

Solution The basic situation here is that 0.2% of some number is \$53. This is equivalent to saying \$53 is 0.2% of what? If we want to answer the question "53 is 0.2% of what?", we need to identify a, b, and p.

The percent $p = 0.2$. The base is unknown. We use the variable b. The amount $a = 53$. Thus

$$\frac{a}{b} = \frac{p}{100} \quad \text{becomes} \quad \frac{53}{b} = \frac{0.2}{100}.$$

When we cross multiply, we obtain

$$(53)(100) = 0.2b$$

$$5300 = 0.2b$$

$$\frac{5300}{0.2} = \frac{0.2b}{0.2}$$

$$26{,}500 = b.$$

Thus \$53 is 0.2% of \$26,500. Therefore the value of the mutual funds was \$26,500.

Practice Problem 6 Everett Hatfield recently exchanged U.S. dollars to Canadian dollars for his company, Nova Scotia Central Trucking, Ltd. The bank charged a fee of 0.3% of the total U.S. dollars exchanged. The fee amounted to \$216 in U.S. money. How many U.S. dollars were exchanged?

EXAMPLE 7 What percent of 4000 is 160?

Solution The percent is unknown. We use the variable p. The base $b = 4000$. The amount $a = 160$. Thus

$$\frac{a}{b} = \frac{p}{100} \quad \text{becomes} \quad \frac{160}{4000} = \frac{p}{100}.$$

If we reduce the fraction on the left-hand side, we have

$$\frac{1}{25} = \frac{p}{100}$$

$$100 = 25p \quad \text{Cross multiply.}$$

$$\frac{100}{25} = \frac{25p}{25} \quad \text{Divide each side by 25.}$$

$$4 = p \quad \text{Divide 100 by 25.}$$

Thus 4% of 4000 is 160.

Practice Problem 7 What percent of 3500 is 105?

Developing Your Study Skills

Reading the Textbook

Homework time each day should begin with the careful reading of the section(s) assigned in your textbook. Much time and effort have gone into the selection of a particular text, and your instructor has chosen a book that will help you become successful in this mathematics class. Expensive textbooks can be a wise investment if you take advantage of them by reading them.

Reading a mathematics textbook is unlike reading many other types of books that you may use in your literature, history, psychology, or sociology courses. Mathematics texts are technical books that provide you with exercises to practice on. Reading a mathematics text requires slow and careful reading of each word, which takes time and effort.

Begin reading your textbook with a paper and pencil in hand. As you come across a new definition, or concept, underline it in the text and/or write it down in your notebook. Whenever you encounter an unfamiliar term, look it up and make a note of it. When you come to an example, work through it step by step. Be sure to read each word and to follow directions carefully.

Notice the helpful hints the author provides to guide you to correct solutions and prevent you from making errors. Take advantage of these pieces of expert advice.

Be sure that you understand what you are reading. Make a note of any of those things that you do not understand and ask your instructor about them. Do not hurry through the material. Learning mathematics takes time.

5.3B Exercises

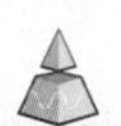

Student Solutions Manual | CD/Video | PH Math Tutor Center | MathXL®Tutorials on CD | MathXL® | MyMathLab® | Interactmath.com

Identify p, b, and a. Do not solve for the unknown.

	p	b	a
1. 75% of 660 is 495.	75	660	495
2. 65% of 820 is 532.	65	820	532
3. What is 22% of 60?	22	60	a
4. What is 35% of 95?	35	95	a
5. 49% of what is 2450?	49	b	2450
6. 38% of what is 2280?	38	b	2280
7. 30 is what percent of 50?	p	50	30
8. 50 is what percent of 250?	p	250	50

Solve using the percent proportion

$$\frac{a}{b} = \frac{p}{100}.$$

In exercises 9–14, the amount a is not known.

9. 40% of 70 is what?
$\frac{a}{70} = \frac{40}{100}$ $a = 28$

10. 80% of 90 is what?
$\frac{a}{90} = \frac{80}{100}$ $a = 72$

11. Find 280% of 70.
$\frac{a}{70} = \frac{280}{100}$ $a = 196$

12. Find 240% of 90.
$\frac{a}{90} = \frac{240}{100}$ $a = 216$

13. 0.7% of 8000 is what?
$\frac{a}{8000} = \frac{0.7}{100}$ $a = 56$

14. 0.8% of 9000 is what?
$\frac{a}{9000} = \frac{0.8}{100}$ $a = 72$

In exercises 15–20, the base b is not known.

15. 20 is 25% of what?
$\frac{20}{b} = \frac{25}{100}$ $b = 80$

16. 45 is 60% of what?
$\frac{45}{b} = \frac{60}{100}$ $b = 75$

17. 250% of what is 200?
$\frac{200}{b} = \frac{250}{100}$ $b = 80$

18. 120% of what is 90?
$\frac{90}{b} = \frac{120}{100}$ $b = 75$

19. 3000 is 0.5% of what?
$\frac{3000}{b} = \frac{0.5}{100}$ $b = 600{,}000$

20. 6000 is 0.4% of what?
$\frac{6000}{b} = \frac{0.4}{100}$ $b = 1{,}500{,}000$

In exercises 21–24, the percent p is not known.

21. 56 is what percent of 280?
$\frac{56}{280} = \frac{p}{100}$ $p = 20$

22. 70 is what percent of 1400?
$\frac{70}{1400} = \frac{p}{100}$ $p = 5$

23. What percent of 260 is 10.4?
$\frac{10.4}{260} = \frac{p}{100}$ $p = 4$

24. What percent of 350 is 10.5?
$\frac{10.5}{350} = \frac{p}{100}$ $p = 3$

Mixed Practice

25. 25% of 88 is what?

$\frac{a}{88} = \frac{25}{100}$ $a = 22$

26. 20% of 75 is what?

$\frac{a}{75} = \frac{20}{100}$ $a = 15$

27. 300% of what is 120?

$\frac{120}{b} = \frac{300}{100}$ $b = 40$

28. 200% of what is 120?

$\frac{120}{b} = \frac{200}{100}$ $b = 60$

29. 82 is what percent of 500?

$\frac{82}{500} = \frac{p}{100}$ $p = 16.4$ 16.4%

30. 75 is what percent of 600?

$\frac{75}{600} = \frac{p}{100}$ $p = 12.5$ 12.5%

31. Find 0.7% of 520.

$\frac{a}{520} = \frac{0.7}{100}$ $a = 3.64$

32. Find 0.4% of 650.

$\frac{a}{650} = \frac{0.4}{100}$ $a = 2.6$

33. What percent of 66 is 16.5?

$\frac{16.5}{66} = \frac{p}{100}$ $p = 25$ 25%

34. What percent of 49 is 34.3?

$\frac{34.3}{49} = \frac{p}{100}$ $p - 70$ 70%

35. 68 is 40% of what?

$\frac{68}{b} = \frac{40}{100}$ $b = 170$

36. 52 is 40% of what?

$\frac{40}{100} = \frac{52}{b}$ $b = 130$

Applications

When solving each applied problem, examine your answer and see if it is reasonable. Ask yourself, "Does my answer make sense?"

37. ***Pay Check Deposits*** Each time Nancy gets paid, 5% of her pay check is deposited into her savings account. Last week, $35 was deposited into her savings account. What was the amount of Nancy's pay check?

$700

38. ***Income Tax*** Last year Priscilla had 21% of her salary withheld for taxes. If the total amount withheld was $4200 for the year, what was her annual salary?

$20,000

39. ***Eating Out*** Ed and Suzie went out to eat at Pizzeria Uno. The dinner check was $26.00. They left a tip of $3.90. What percent of the check was the tip?

15%

40. ***Baseball*** During the baseball season, Damon was up to bat 60 times. Of these at-bats, 12 were home runs. What percent of Damon's at-bats resulted in a home run?

20%

41. ***Food Expiration Date*** The Super Shop and Save store had 120 gallons of milk placed on the shelf one night. During the next morning's inspection, the manager found that 15% of the milk had passed the expiration date. How many gallons of milk had passed the expiration date?

18 gallons

42. ***Police Arrests*** During June the Wenham police stopped 250 drivers for speed violations. It was found that 8% of the people who were stopped had outstanding warrants for their arrest. How many people had outstanding warrants for their arrest?

20 people

43. ***Car Purchase*** Victor purchased a used car for $9500. He made a down payment of 24% of the purchase price. How much was his down payment?

$2280

44. ***Education*** Trudy took a biology test with 40 problems. She got 8 of the problems wrong and 32 of the problems right. What percent of the test problems did she do incorrectly?

20%

To Think About

Texas Budget *Recently a estimate was made of the expenditures in Texas during 2003 for the following five categories: corrections, government and administration, interest and general debt, natural resources, and parks and recreation. The amount of money (in millions of dollars) expended in each of these five categories is given in the following pie chart.*

Round all answers to the nearest tenth.

Use the chart to answer questions 45–48.

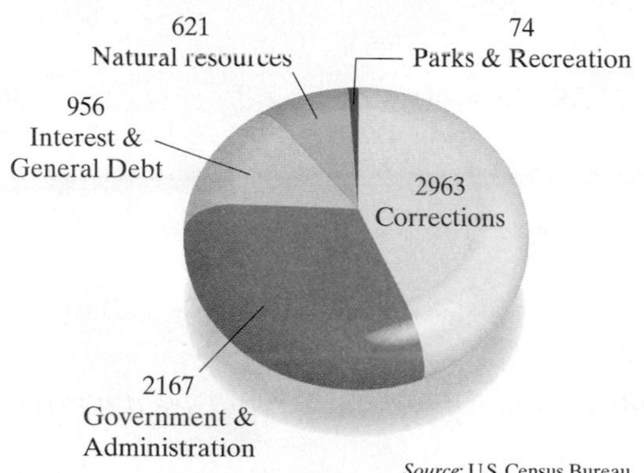

45. What percent of the total expenditures in these five categories was used for corrections?
43.7%

46. What percent of the total expenditures in these five categories was used for government and administration?
32%

47. Suppose that compared to 2003, expenditures in 2007 for corrections were 25% larger, expenditures for government and administration were 30% larger, and the other three categories remained at the same dollar figure. What percent of the total expenditures in 2007 in these five categories would be used for natural resources?
7.6%

48. Suppose that compared to 2003, expenditures in 2009 for corrections were 40% larger, expenditures for government and administration were 35% larger, expenditures for interest on general debt were 20% larger, and the other two categories remained at the same dollar figure. What percent of the total expenditures in 2009 would be used for interest on general debt?
12.9%

Cumulative Review

Simplify.

49. $\frac{4}{5} + \frac{8}{9}$ $1\frac{31}{45}$

50. $\frac{7}{13} - \frac{1}{2}$ $\frac{1}{26}$

51. $\left(2\frac{4}{5}\right)\left(1\frac{1}{2}\right)$ $4\frac{1}{5}$

52. $1\frac{2}{5} \div \frac{3}{4}$ $1\frac{13}{15}$

How Am I Doing? Sections 5.1–5.3

How are you doing with your homework assignments in Sections 5.1 to 5.3? Do you feel you have mastered the material so far? Do you understand the concepts you have covered? Before you go further in the textbook, take some time to do each of the following problems.

5.1

Write as a percent.

1. 0.17 **2.** 0.387 **3.** 1.34 **4.** 8.94

5. 0.006 **6.** 0.0004 **7.** $\frac{17}{100}$ **8.** $\frac{89}{100}$

9. $\frac{13.4}{100}$ **10.** $\frac{19.8}{100}$ **11.** $\frac{6\frac{1}{2}}{100}$ **12.** $\frac{1\frac{3}{8}}{100}$

5.2

Change to a percent. Round to the nearest hundredth of a percent when necessary.

13. $\frac{8}{10}$ **14.** $\frac{1}{40}$ **15.** $\frac{52}{20}$ **16.** $\frac{17}{16}$

17. $\frac{5}{7}$ **18.** $\frac{2}{7}$ **19.** $\frac{22}{23}$ **20.** $\frac{13}{19}$

21. $4\frac{2}{5}$ **22.** $2\frac{3}{4}$ **23.** $\frac{1}{300}$ **24.** $\frac{1}{400}$

Write as a fraction in simplified form.

25. 22% **26.** 53% **27.** 150% **28.** 160%

29. $6\frac{1}{3}\%$ **30.** $4\frac{2}{3}\%$ **31.** $51\frac{1}{4}\%$ **32.** $43\frac{3}{4}\%$

5.3

Solve. Round to the nearest hundredth when necessary.

33. Find 24% of 230. **34.** What is 78% of 62?

35. 68 is what percent of 72? **36.** What percent of 76 is 34?

37. 8% of what number is 240? **38.** 354 is 40% of what number?

Now turn to page SA-11 for the answer to each of these problems. Each answer also includes a reference to the objective in which the problem is first taught. If you missed any of these problems, you should stop and review the Examples and Practice Problems in the referenced objective. A little review now will help you master the material in the upcoming sections of the text.

1. 17%
2. 38.7%
3. 134%
4. 894%
5. 0.6%
6. 0.04%
7. 17%
8. 89%
9. 13.4%
10. 19.8%
11. $6\frac{1}{2}\%$
12. $1\frac{3}{8}\%$
13. 80%
14. 2.5%
15. 260%
16. 106.25%
17. 71.43%
18. 28.57%
19. 95.65%
20. 68.42%
21. 440%
22. 275%
23. 0.33%
24. 0.25%
25. $\frac{11}{50}$
26. $\frac{53}{100}$
27. $\frac{3}{2}$ or $1\frac{1}{2}$
28. $\frac{8}{5}$ or $1\frac{3}{5}$
29. $\frac{19}{300}$
30. $\frac{7}{150}$
31. $\frac{41}{80}$
32. $\frac{7}{16}$
33. 55.2
34. 48.36
35. 94.44%
36. 44.74%
37. 3000
38. 885

5.4 Solving Applied Percent Problems

Student Learning Objectives

After studying this section, you will be able to:

1. **Solve general applied percent problems.**
2. **Solve applied problems when percents are added.**
3. **Solve discount problems.**

1 Solving General Applied Percent Problems

In Sections 5.3A and 5.3B, we learned the three types of applied percent problems. Some problems ask you to find a percent of a number. Some problems give you an amount and a percent and ask you to find the base (or whole). Other problems give an amount and a base and ask you to find the percent. We will now see how the three types of percent problems occur in real life.

EXAMPLE 1 Of all the computers manufactured last month, an inspector found 18 that were defective. This is 2.5% of all the computers manufactured last month. How many computers were manufactured last month?

Solution

Method A Translate to an equation.

The problem is equivalent to: 2.5% of the number of computers is 18.
Let n = the number of computers.

$$\underbrace{2.5\%}_{2.5\%}\ \underbrace{\text{of}}_{\times}\ \underbrace{\text{the number of computers}}_{n}\ \underbrace{\text{is}}_{=}\ \underbrace{18}_{18}$$

$$0.025n = 18$$

$$\frac{0.025n}{0.025} = \frac{18}{0.025}$$

$$n = 720$$

720 computers were manufactured last month.

Method B Use the percent proportion $\frac{a}{b} = \frac{p}{100}$.

The percent $p = 2.5$. The base is unknown. We will use the variable b. The amount $a = 18$. Thus

$$\frac{a}{b} = \frac{p}{100} \quad \text{becomes} \quad \frac{18}{b} = \frac{2.5}{100}.$$

Using cross multiplication, we have

$$(18)(100) = 2.5b$$

$$1800 = 2.5b$$

$$\frac{1800}{2.5} = \frac{2.5b}{2.5}$$

$$720 = b.$$

720 computers were manufactured last month.

By either Method A or Method B, we obtain the same number of computers, 720.

Substitute 720 into the original problem to check.

2.5% of 720 computers are defective.

$$(0.025)(720) = 18 \checkmark$$

Practice Problem 1 4800 people, or 12% of all passengers holding tickets for American Airlines flights in one month, did not show up for their flights. How many people held tickets that month?

***NOTE TO STUDENT:** Fully worked-out solutions to all of the Practice Problems can be found at the back of the text starting at page SP-1*

EXAMPLE 2 How much sales tax will you pay on a color television priced at $299 if the sales tax is 5%?

Solution

Method A Translate to an equation.

$$\begin{array}{ccccc} \text{What} & \text{is} & 5\% & \text{of} & \$299? \\ \downarrow & \downarrow & \downarrow & \downarrow & \downarrow \\ n & = & 5\% & \times & 299 \end{array}$$

$$n = (0.05)(299)$$

$$n = 14.95 \qquad \text{The tax is \$14.95.}$$

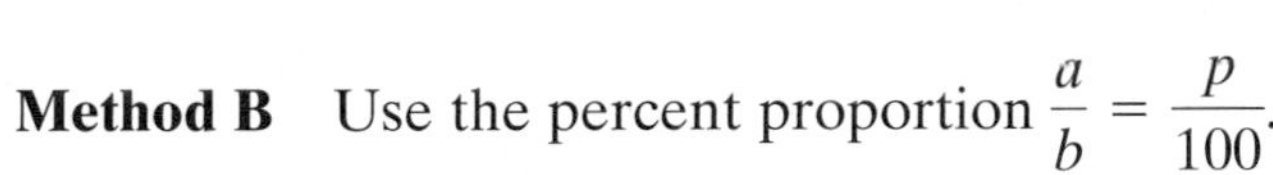

Method B Use the percent proportion $\frac{a}{b} = \frac{p}{100}$.

The percent $p = 5$. The base $b = 299$. The amount is unknown. We use the variable a. Thus

$$\frac{a}{b} = \frac{p}{100} \quad \text{becomes} \quad \frac{a}{299} = \frac{5}{100}.$$

If we reduce the fraction on the right-hand side, we have

$$\frac{a}{299} = \frac{1}{20}.$$

We then cross multiply to obtain

$$20a = 299$$

$$\frac{20a}{20} = \frac{299}{20}$$

$$a = 14.95 \qquad \text{The tax is \$14.95.}$$

Thus, by either method, the amount of the sales tax is $14.95.

Let's see if our answer is reasonable. Is 5% of $299 really $14.95? If we round $299 to one nonzero digit, we have $300. Thus we have 5% of 300 = 15. Since 15 is quite close to our value of $14.95, our answer seems reasonable.

Practice Problem 2 A salesperson rented a hotel room for $62.30 per night. The tax in her state is 8%. What tax does she pay for one night at the hotel? Round to the nearest cent.

EXAMPLE 3 A failing student attended class 39 times out of the 45 times the class met last semester. What percent of the classes did he attend? Round to the nearest tenth of a percent.

Solution

Method A Translate to an equation.

This problem is equivalent to:

$$\begin{array}{ccccc} 39 & \text{is} & \underbrace{\text{what percent}} & \text{of} & 45? \\ \downarrow & \downarrow & \downarrow & \downarrow & \downarrow \\ 39 & = & n & \times & 45 \end{array}$$

$$39 = 45n$$

$$\frac{39}{45} = \frac{45n}{45}$$

$$0.8666\ldots = n.$$

To the nearest tenth of a percent we have $n = 86.7\%$.

Method B Use the percent proportion $\frac{a}{b} = \frac{p}{100}$.

The percent is unknown. We use the variable p. The base b is 45. The amount a is 39. Thus

$$\frac{a}{b} = \frac{p}{100} \quad \text{becomes} \quad \frac{39}{45} = \frac{p}{100}.$$

When we cross multiply, we get

$$(39)(100) = 45p$$

$$3900 = 45p$$

$$\frac{3900}{45} = \frac{45p}{45}$$

$$86.666\ldots = p.$$

To the nearest tenth, the answer is 86.7%.

By using either method, we discover that the failing student attended approximately 86.7% of the classes.

Verify by estimating that the answer is reasonable.

NOTE TO STUDENT: Fully worked-out solutions to all of the Practice Problems can be found at the back of the text starting at page SP-1

Practice Problem 3 Of the 130 flights at Orange County Airport yesterday, only 105 of them were on time. What percent of the flights were on time? (Round to the nearest tenth of a percent.)

Now you have some experience solving the three types of percent problems in real-life applications. You can use either Method A or Method B to solve applied percent problems. In the following pages we will present more percent applications. We will not list all the steps of Method A or Method B. Most students will find after a careful study of Examples 1–3 that they do not need to write out all the steps of Method A or Method B when solving applied percent problems.

2 Solving Applied Problems when Percents Are Added

Percents can be added if the base (whole) is the same. For example, 50% of your salary added to 20% of your salary = 70% of your salary. 100% of your cost added to 15% of your cost = 115% of your cost. Problems like this are often called **markup problems.** If we add 15% of the cost of an item to the original cost, the markup is 15%. We will add percents in some applied situations.

The following example is interesting, but it is a little challenging. So please read it very carefully. A lot of students find it difficult at first.

EXAMPLE 4 Walter and Mary Ann are going out to a restaurant. They have a limit of \$63.25 to spend for the evening. They want to tip the waitress 15% of the cost of the meal. How much money can they afford to spend on the meal itself? (Assume there is no tax.)

Solution In some of the problems in this section, it may help you to use the Mathematics Blueprint. We will use it here for Example 4.

Mathematics Blueprint for Problem Solving

Gather the Facts	What Am I Asked to Do?	How Do I Proceed?	Key Points to Remember
They have a spending limit of \$63.25. They want to tip the waitress 15% of the cost of the meal.	Find the amount of money that the meal will cost.	Separate the \$63.25 into two parts: the cost of meal and the tip. Add these two parts to get \$63.25.	We are not taking 15% of \$63.25, but rather 15% of the cost of meal.

Let n = the cost of the meal. 15% of the cost = the amount of the tip. We want to add the percents of the meal.

$$\boxed{\text{Cost of meal } n} + \boxed{\text{tip of 15\% of the cost}} = \boxed{\$63.25}$$

$$100\% \text{ of } n + 15\% \text{ of } n = \$63.25$$

Note that 100% of n added to 15% of n is 115% of n.

$$115\% \text{ of } n = \$63.25$$

$$1.15 \times n = 63.25$$

$$\frac{1.15 \times n}{1.15} = \frac{63.25}{1.15} \quad \text{Divide both sides by 1.15.}$$

$$n = 55$$

They can spend up to \$55.00 on the meal itself.

Does this answer seem reasonable?

NOTE TO STUDENT: Fully worked-out solutions to all of the Practice Problems can be found at the back of the text starting at page SP-1

Practice Problem 4 Sue and Sam have \$46.00 to spend at a restaurant, including a 15% tip. How much can they spend on the meal itself? (Assume there is no tax.)

Solving Discount Problems

Frequently, we see signs urging us to buy during a sale when the list price is discounted by a certain percent.

The amount of a **discount** is the product of the discount rate and the list price.

$$\text{Discount} = \text{discount rate} \times \text{list price}$$

EXAMPLE 5 Jeff purchased a large screen color TV on sale at a 35% discount. The list price was \$430.00.

(a) What was the amount of the discount?

(b) How much did Jeff pay for the large screen color TV?

Solution

(a)
$$\begin{aligned} \text{Discount} &= \text{discount rate} \times \text{list price} \\ &= 35\% \times 430 \\ &= 0.35 \times 430 \\ &= 150.5 \end{aligned}$$

The discount was \$150.50.

(b) We subtract the discount from the list price to get the selling price.

\$430.00	list price
− \$150.50	discount
\$279.50	sellingprice

Jeff paid \$279.50 for the large screen color TV.

Practice Problem 5 Betty bought a car that lists for \$13,600 at a 7% discount.

(a) What was the discount? **(b)** What did she pay for the car?

5.4 Exercises

Student Solutions Manual | CD/Video | PH Math Tutor Center | MathXL®Tutorials on CD | MathXL® | MyMathLab® | Interactmath.com

Applications

Exercises 1–18 present the three types of percent problems. They are similar to Examples 1–3. Take the time to master exercises 1–18 before going on to the next ones. Round to the nearest hundredth when necessary.

1. ***Education*** No graphite was found in 4500 pencils shipped to Sureway School Supplies. This was 2.5% of the total number of pencils received by Sureway. How many pencils in total were in the order? 180,000 pencils

2. ***Track and Field*** A high-jumper on the track and field team hit the bar 58 times last week. This means that he did not succeed in 29% of his jump attempts. How many total attempts did he make last week? 200 attempts

3. ***Phone Bill*** Scott's phone bill averages $80.50 per month. This is 115% of his average monthly bill last year. What was his average monthly phone bill last year? $70

4. ***Salary Changes*** Renata now earns $9.50 per hour. This is 125% of what she earned last year. What did she earn per hour last year? $7.60 per hour

5. ***Airplane Flights*** In June 2003, ATA Airlines sold out 431 of its 6000 flights. What percent of the flights in June were sold out? 7.18%

6. ***Coffee Bar*** Every day this year, Sam ordered either cappuccino or espresso from the coffee bar downstairs. He had 85 espressos and 280 cappuccinos. What percent of the coffees were espressos? 23.29%

7. ***Sales Tax*** Elizabeth bought new towels and sheets for $65. How much tax did she pay if the sales tax is 6%? $3.90

8. ***Sales Tax*** Leon bought new clothes for his bank job. Before tax was added on, his total was $180. How much tax did he pay if the sales tax is 5%? $9

9. ***Mountain Bike*** Dean bought a new mountain bike. The sales tax in his state is 4%, and he paid $9.60 in tax. What was the price of the mountain bike before the tax? $240

10. ***Sales Tax*** Hiro bought some art work and paid $10.75 in tax. The sales tax in his state is 5%. What was the price of the art work? $215

11. ***Mortgage Payment*** Peter and Judy together earn $5060 per month. Their mortgage payment is $1265 per month. What percent of their household income goes toward paying the mortgage? 25%

12. ***Car Payment*** Jackie puts aside $65.50 per week for her monthly car payment. She earns $327.50 per week. What percent of her income is set aside for car payments. 20%

13. ***Charities*** The Children's Wish Charity raised 75% of its funds from sporting promotions. Last year the charity received $7,200,000 from its sporting promotions. What was the charity's total income last year? $9,600,000

14. ***Taxes*** Shannon paid $8400 in federal and state income taxes as a lab technician, which amounted to 28% of her annual income. What was her income last year? $30,000

15. ***Baseball*** 10,001 home runs have been hit in Boston's Fenway Park since it opened in 1912. Ted Williams of the Boston Red Sox hit 248 of these. What percent of the total home runs did Ted Williams hit? 2.48%

16. ***Baseball*** At one point during the 2003 baseball season, Nomar Garciaparra had been up to bat 480 times. Of these, 22 resulted in a home run. What percent of Garciaparra's at-bats resulted in a home run? 4.58%

17. ***Pediatrics*** In Flagstaff, Arizona, 0.9% of all babies walk before they reach the age of 11 months. If 24,000 babies were born in Flagstaff in the last 20 years, how many of them will have walked before they reached the age of 11 months? 216 babies

18. ***Basketball*** At a Boston Celtics basketball game scheduled for 8:30 P.M., 0.8% of the spectators were children under age 12. If 28,000 people showed up for the game, how many children under age 12 were in the stands? 224 children

Exercises 19–34 include percents that are added and discounts. Solve. Round to the nearest hundredth when necessary.

19. ***Sales Tax*** Henry has $800 total to spend on a new dining room table and chairs. If the sales tax is 5%, how much can he afford to spend on the table on chairs?
$761.90

20. ***Eating Out*** Belinda asked Martin out to dinner. She has $47.50 to spend. She wants to tip the waitress 15% of the cost of their meal. How much money can she afford to spend on the meal itself?
$41.30

21. ***Building Costs*** John and Chris Maney are building a new house. When finished, the house will cost $163,500. The price of the house is 9% higher than the price when the original plans were made. What was the price of the house when the original plans were made?
$150,000

22. ***SUV Purchase*** Dan and Connie Lacorazza purchased a new Honda Pilot. The purchase price was $25,440. The price was 6% higher than the price of a similar Honda Pilot three years ago. What was the price of the Honda Pilot three years ago?
$24,000

23. ***Manufacturing*** When a new computer case is made, there is some waste of the plastic material used for the front of the computer case. Approximately 3% of the plastic that is delivered is of poor quality and is thrown away. Furthermore, 8% of the plastic that is delivered is waste material that is thrown away as the front pieces are created by a giant stamping machine. If 20,000 pounds of plastic are delivered to make the front of computer cases each month, how many pounds are thrown away? 2200 pounds

24. ***Airport Operations*** In a recent survey of planes landing at Logan Airport in Boston, it was observed that 12% of the flights were delayed less than an hour and 7% of the flights were delayed an hour or more but less than two hours. If 9000 flights arrive at Logan Airport in a day, how many flights are delayed less than two hours?
1710 flights

25. ***Political Parties*** The Democratic National Committee has a budget of $33,000,000 to spend on the inauguration of the new president. 15% of the costs will be paid to personnel, 12% of the costs will go toward food, and 10% will go to decorations.

(a) How much money will go for personnel, food, and decorations?
$12,210,000 for personnel, food, and decorations

(b) How much will be left over to cover security, facility rental, and all other expenses?
$20,790,000 for security, facility rental, and all other expenses

26. ***Medical Research*** A major research facility has developed an experimental drug to treat Alzheimer's disease. Twenty percent of the research costs was paid to the staff. Sixteen percent of the research costs was paid to rent the building where the research was conducted. The rest of the money was used for research. The company has spent $6,000,000 in research on this new drug.

(a) How much was paid to cover the cost of staff and rental of the building? $2,160,000

(b) How much was left over for research?
$3,840,000

27. ***Clothes Purchase*** Melinda purchased a new blouse, jeans, and a sweater in Naperville, Illinois. All of the clothes were discounted 35%. Before the sale, the total purchase price would have been \$190 for these three items. How much did she pay for them with the discount?
\$123.50

28. ***Tire Purchase*** Juan went to purchase two new radial tires for his Honda Accord in Austin Texas. The set of two tires normally costs \$130. However, he bought them on sale at a discount of 30%. How much did he pay for the tires with the discount? \$91

29. ***Restaurant Discounts*** John and Stephanie went to purchase a discount coupon book for several local restaurants in their area of Philadelphia. The coupon book costs \$30 and is good for one year. It has coupons that are worth 15% off all meals served before 6 P.M. at 20 local restaurants. They usually spend \$400 a year dining at these restaurants. How many dollars would they save each year if they purchased this book? \$30

30. ***Shopping Club*** Anne wanted to join TJ's Discount Shopping Club in Tulsa, Oklahoma. The annual membership fee is \$45. With her membership card she will get a discount of 8% on any item in the store. She estimates that she would probably purchase around \$900 worth of store items each year. How much will she save over a period of one year if she purchases these items at the store but also pays the annual membership fee? \$27

31. ***Motorcycle Purchase*** Susie bought her first Harley-Davidson motorcycle in Manchester, New Hampshire. The list price was \$16,000, but the dealer gave her a discount of 8%.

(a) What was the discount?
\$1280

(b) How much did she pay for the motorcycle?
\$14,720

32. ***Sofa Purchase*** Jane bought a leather sofa that had been used as the floor model at LeHigh's Furniture in Jacksonville, Florida. The list price of the sofa was \$1200, but the dealer gave her a discount of 35%.

(a) What was the discount? \$420

(b) How much did she pay for the sofa? \$780

33. ***Antiques*** Old Tyme Antiques in Austin, Texas needs to clear space for a new shipment. They are offering 30% off all items in one area of the store. Angela collects antique clocks and finds a mantel clock priced at \$1110.

(a) What is the discount?
\$333

(b) How much will she pay for the mantel clock?
\$777

34. ***Swimming Pool*** The Harmelings are having a swimming pool and sauna installed in their backyard in Anaheim, California. They saw an advertisement that Swim & Spa was offering 15% off the total cost if you buy both items. When they went to the store, the sign read "10% off." Mrs. Harmeling and the store's owner agreed to average the discounts, and came up with a savings of 12.5%.

(a) If the total cost is \$13,000, how much do the Harmelings save with the 12.5% discount?
\$1625

(b) How much will they pay for the purchase?
\$11,375

Cumulative Review

35. Round to the nearest thousand. 1,698,481
1,698,000

36. Round to the nearest hundred. 2,452,399
2,452,400

37. Round to the nearest hundredth. 1.63474
1.63

38. Round to the nearest thousandth. 0.7995
0.800

39. Round to the nearest ten-thousandth. 0.055613
0.0556

40. Round to the nearest ten-thousandth. 0.079152
0.0792

5.5 Solving Commission, Percent of Increase or Decrease, and Interest Problems

Student Learning Objectives

After studying this section you will be able to:

1. **Solve commission problems.**
2. **Solve percent-of-increase or percent-of-decrease problems.**
3. **Solve simple interest problems.**

Teaching Tip Some students are familiar with commission sales and others are not. A few simple examples are always helpful. Ask your students if any of them have ever earned a commission on sales in one of their jobs.

NOTE TO STUDENT: Fully worked-out solutions to all of the Practice Problems can be found at the back of the text starting at page SP-1

1 Solving Commission Problems

If you work as a salesperson, your earnings may be in part or in total a certain percentage of the sales you make. The amount of money you get that is a percentage of the value of your sales is called your **commission.** It is calculated by multiplying the percentage (called the **commission rate**) by the value of the sales.

$$\text{Commission} = \text{commission rate} \times \text{value of sales}$$

EXAMPLE 1 A salesperson has a commission rate of 17%. She sells \$32,500 worth of goods in a department store in two months. What is her commission?

Solution

$$\begin{aligned} \text{Commission} &= \text{commission rate} \times \text{value of sales} \\ \text{Commission} &= 17\% \times \$32{,}500 \\ &= 0.17 \times 32{,}500 \\ &= 5525 \end{aligned}$$

Her commission is \$5525.00.

Does this answer seem reasonable? Check by estimating.

Practice Problem 1 A real estate salesperson earns a commission rate of 6% when he sells a \$156,000 home. What is his commission?

In some problems, the unknown quantity will be the commission rate or the value of sales. However, the same equation is used:

$$\text{Commission} = \text{commission rate} \times \text{value of sales}$$

2 Solving Percent-of-Increase or Percent-of-Decrease Problems

We sometimes need to find the percent by which a number increases or decreases. If a car costs \$7000 and the price decreases \$1750, we say that the percent of decrease is $\frac{1750}{7000} = 0.25 = 25\%$.

$$\textbf{Percent of decrease} = \frac{\text{amount of decrease}}{\text{original amount}}$$

Similarly, if a population of 12,000 people increases by 1920 people, we say that the percent of increase is $\frac{1920}{12{,}000} = 0.16 = 16\%$.

$$\textbf{Percent of increase} = \frac{\text{amount of increase}}{\text{original amount}}$$

Note that for these types of problems the base is always the *original amount.*

The most important thing to remember is that we must **first** find the amount of increase or decrease.

EXAMPLE 2 The population of Center City increased from 50,000 to 59,500. What was the percent of increase?

Solution For this problem as well as others in this section, you may find it helpful to use the Mathematics Blueprint.

Mathematics Blueprint for Problem Solving

Gather the Facts	What Am I Asked to Do?	How Do I Proceed?	Key Points to Remember
The population increased from 50,000 to 59,500.	We must find the percent of increase.	First subtract to find the amount of increase. Then divide the amount of increase by the original amount.	Always divide by the original amount.

Amount of increase

$$\begin{array}{r} 59{,}500 \\ -\ 50{,}000 \\ \hline 9{,}500 \end{array}$$

$$\text{Percent of increase} = \frac{\text{amount of increase}}{\text{original amount}} = \frac{9500}{50{,}000}$$
$$= 0.19 = 19\%$$

The percent of increase is 19%.

Practice Problem 2 A new car is sold for \$15,000. A year later its price had decreased to \$10,500. What is the percent of decrease?

Teaching Tip If students have trouble with percent-of-increase and percent-of-decrease problems, remind them that the denominator is always the original amount. Usually, students make mistakes in this type of problem because they have forgotten this fact.

Calculator

Interest

You can use a calculator to find simple interest. Find the interest on \$450 invested at 6.5% for 15 months. Notice the time is in months. Since the interest formula $I = P \times R \times T$, is in years, you need to change 15 months to years by dividing 15 by 12.

Enter

15 [÷] 12 [=]

Display

[1.25]

Leave this on the display and multiply as follows:

1.25 [×] 450 [×]

6.5 [%] [=]

The display should read

[36.5625]

which would round to \$36.56.

Try the following.

(a) \$9516 invested at 12% for 30 months

(b) \$593 borrowed at 8% for 5 months

3 Solving Simple Interest Problems

Interest is money paid for the use of money. If you deposit money in a bank, the bank uses that money and pays you interest. If you borrow money, you pay the bank interest for the use of that money. The **principal** is the amount deposited or borrowed. Interest is usually expressed as a percent rate of the principal. The **interest rate** is assumed to be per year, unless otherwise stated. The formula used in business to compute simple interest is

$$\text{Interest} = \text{principal} \times \text{rate} \times \text{time}$$
$$I = P \times R \times T$$

If the interest rate is *per year,* the time T *must* be in *years.*

EXAMPLE 3 Find the simple interest on a loan of \$7500 borrowed at 13% for one year.

Solution $I = P \times R \times T$

$P = \text{principal} = \$7500 \qquad R = \text{rate} = 13\% \qquad T = \text{time} = 1 \text{ year}$

$I = 7500 \times 13\% \times 1 = 7500 \times 0.13 = 975$

The interest is \$975.

***NOTE TO STUDENT:** Fully worked-out solutions to all of the Practice Problems can be found at the back of the text starting at page SP-1*

Practice Problem 3 Find the simple interest on a loan of \$5600 borrowed at 12% for one year.

Our formula is based on a yearly interest rate. Time periods of more than one year or a fractional part of a year are sometimes needed.

EXAMPLE 4 Find the simple interest on a loan of \$2500 that is borrowed at 9% for

(a) three years. **(b)** three months.

Solution

(a)

$$I = P \times R \times T$$

$$P = \$2500 \qquad R = 9\% \qquad T = 3 \text{ years}$$

$$I = 2500 \times 0.09 \times 3 = 225 \times 3 = 675$$

The interest for three years is \$675.

(b) Three months $= \frac{1}{4}$ year. The period must be in years to use the formula.

Since $T = \frac{1}{4}$ year, we have

$$I = 2500 \times 0.09 \times \frac{1}{4}$$

$$= 225 \times \frac{1}{4}$$

$$= \frac{225}{4} = 56.25$$

The interest for $\frac{1}{4}$ year is \$56.25.

Practice Problem 4 Find the simple interest on a loan of \$1800 that is borrowed at 11% for

(a) four years. **(b)** six months.

Many loans today are based on **compound interest.** This topic is covered in more advanced mathematics courses. The calculations for compound interest are tedious to do by hand. Usually people use a computer or a compound interest table to do compound interest problems.

5.5 Exercises

Student Solutions Manual | CD/ Video | PH Math Tutor Center | MathXL®Tutorials on CD | MathXL® | MyMathLab® | Interactmath.com

Applications

Exercises 1–18, are problems involving commissions, percent of increase or decrease, and simple interest.

1. ***Appliance Sales*** Walter works as an appliance salesman in a department store. Last month he sold \$170,000 worth of appliances. His commission rate is 2%. How much money did he earn in commission last month?
\$3400

2. ***Car Sales*** Susan works at the Acura dealership in Winchester. Last month she had car sales totaling \$230,000. Her commission rate is 3%. How much money did she earn in commission last month?
\$6900

3. ***Mobile Phone Sales*** Allison works in the local Verizon office selling mobile phones. She is paid \$300 per month plus 4% of her total sales in mobile phones. Last month she sold \$96,000 worth of mobile phones. What was her total income for the month?
\$4140

4. ***Stockbroker*** Matthew is a stockbroker. He is paid \$500 per month plus 0.5% of the total sales of stocks that he sells. Last month he sold \$340,000 worth of stock. What was his total income for the month?
\$2200

5. ***Airline Tickets*** Dawn is searching online for airline tickets. Two weeks ago the cost to fly from San Francisco to Minneapolis was \$275. The price now is \$330. What is the percent increase?
20%

6. ***Chandelier Purchase*** Wes is considering buying a chandelier for his dining room. The price tag reads \$500, but the store owner told Wes he would drop the price to \$450. What is the percent decrease?
10%

7. ***Monthly House Rental*** The rent for a three-bedroom home on Lake Sunapee in New Hampshire dropped from \$2000 to \$1200 per month. What percent decrease is this?
40%

8. ***Weight Loss*** Tay weighed 285 pounds two years ago. After careful supervision at a weight loss center, he reduced his weight to 171 pounds. What was the percent of decrease in his weight?
40%

9. ***CD Interest*** Phil placed \$2000 in a one-year CD at the bank. The bank is paying simple interest of 7% for one year on the CD. How much interest will Phil earn in one year?
\$140

10. ***Checking Account Interest*** Charlotte has a checking account that pays her simple interest of 1.2% on the average balance in her checking account. Last year her average balance was \$450. How much interest did she earn in her checking account?
\$5.40

11. ***Credit Card Expenses*** Melinda has a MasterCard account with Centerville Bank. She has to pay a monthly interest rate of 1.5% on the average daily balance of the amount she owes on her credit card. Last month her average daily balance was \$500. How much interest was she charged last month? (*Hint:* The formula $I = P \times R \times T$ can be used if the interest rate is *per month* and the time is in *months.*)
\$7.50

12. ***Student Loan*** Walter borrowed \$3000 for a student loan to finish college this year. Next year he will need to pay 7% simple interest on the amount he borrowed. How much interest will he need to pay next year?
\$210

13. ***House Construction Loan*** Abe had to borrow $12,000 for a house construction loan for three months. The interest rate was 16% per year. How much interest did he have to pay for borrowing the money for three months? $480

14. ***Small Business Loan*** Joan needed to borrow $8000 for two months to finance some renovations to her beauty salon. The interest rate she was charged was 18% per year. How much interest did she have to pay for borrowing the money for two months? $240

15. ***Life Insurance*** Robert sells life insurance for a major insurance company for a commission. Last year he sold $12,000,000 worth of insurance. He earned $72,000 in commissions. What was his commission rate? 0.6%

16. ***Medical Supplies*** Hillary sells medical supplies to doctors' offices for a major medical supply company. Last year she sold $9,000,000 worth of medical supplies. She works on a commission basis and last year she earned $63,000 in commissions. What was her commission rate? 0.7%

17. ***Furniture Sales*** Jennifer sells furniture for a major department store. Last year she was paid $48,000 for commissions. If her commission rate is 3%, what was the sales total of the furniture that she sold last year? $1,600,000

18. ***Auto Sales*** Michael sells used cars for Beltway Motors. Last year he was paid $42,000 in commissions. Beltway Motors pays the salespeople a commission rate of 6%. What was the sales total of the cars that Michael sold last year? $700,000

Mixed Applications

Exercises 19–36 are a variety of percent problems. They involve commissions, percent of increase or decreases, and simple interest. There are also some of each kind of percent problem encountered in the chapter. Unless otherwise directed, round to the nearest hundredth.

19. ***Entertainment Expenses*** Ted is trying to decrease his spending on entertainment. He earns $265 per week and is allowing himself to spend only 15% per week on movies, dining out, and so on. How much can Ted spend per week on entertainment? $39.75

20. ***Biology*** The maximum capacity of your lungs is 4.58 liters of air. In a typical breath, you breathe in 12% of the maximum capacity. How many liters of air do you breathe in a typical breath? approximately 0.55 liter

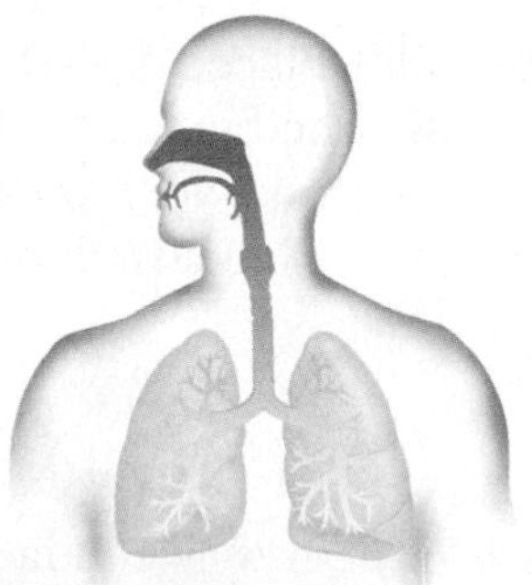

21. ***Girl Scout Cookies*** Of all the boxes of Girl Scout cookies sold, 25% are Thin Mints. If a Girl Scout troop sells 156 boxes of cookies, how many are Thin Mints?
39 boxes

22. ***Scotland*** 11% of the Scottish population have red hair. If the population of Scotland is 5,600,000, how many people have red hair?
616,000 people

23. ***Manufacturing Jobs*** The number of manufacturing jobs in the U.S. went from 17.32 million in July 2000 to 14.61 million in July 2003. What is the percent of decrease in the number of manufacturing jobs? (*Source:* U.S. Bureau of Labor Statistics) approximately 16%

24. ***Pediatrics*** Babies are born with 350 soft bones. When bone fusion is complete (at 20–25 years of age), the body has 206 bones. What is the percentage of decrease in the number of bones? approximately 41%

25. ***Sporting Goods*** A sporting goods store buys cross-training shoes for $40, and sells them for $72. What is the percentage of increase in the price of the shoes? 80%

26. ***Jewelry Costs*** A gift store buys earrings from an artist for $20 and sells them for $29. What is the percentage of increase in the price of the earrings? 45%

27. ***Savings Account*** Adam deposited $3700 in his savings account for one year. His savings account earns 2.3% interest annually. He did not add any more money within the year, and at the end of that time, he withdrew all funds.

(a) How much interest did he earn? $85.10

(b) How much money did he withdraw from the bank? $3785.10

28. ***Credit Card*** Nikki had $1258 outstanding on her MasterCard, which charges 2% monthly interest. At the end of this month Nikki paid off the loan.

(a) How much interest did Nikki pay for one month? $25.16

(b) How much did it cost to pay off the loan totally? $1283.16

29. ***Shopping Trip*** Bryce went shopping and bought a pair of sandals for $52, swimming trunks for $38, and sunglasses for $26. The tax in Bryce's city is 6%.

(a) What is the total sales tax? $6.96

(b) What is the total of the purchases? $122.96

30. ***Automobile Purchase*** Finn bought a used Honda Accord for $10,500. The tax in his state is 8%.

(a) What is the sales tax? $840

(b) What is the final price of the Honda Accord? $11,340

31. ***Property Taxes*** Smithville Kitchen Cabinetry Inc. is late in paying $9500 in property taxes to the city of Springfield. It will be assessed 14% interest for being late in property tax payment. Find the one total amount they need to pay off both the taxes and the interest charge. $10,830

32. ***Property Taxes*** Raymond and Elsie Ostram are late in paying $1600 in property taxes to the city of New Boston. They will be assessed 12% interest for being late in property tax payment. Find the one total amount they need to pay off both the taxes and the interest charge. $1792

33. ***Home Purchase*** Betty and Michael Bently purchased a new home for $349,000. They paid a down payment of 8% of the cost of the home. They took out a mortgage for the rest of the purchase price of the home.

(a) What was the amount of their down payment? $27,920

(b) What was the amount of their mortgage? $321,080

34. ***Home Purchase*** Marcia and Dan Perkins purchased a condominium for $188,000. They paid a down payment of 11% of the cost of the condominium. They took out a mortgage for the rest of the purchase price of the condominium.

(a) What was the amount of their down payment? $20,680

(b) What was the amount of their mortgage? $167,320

35. ***Interest Charges on a Mortgage*** Richard is making monthly mortgage payments of \$840 for his home mortgage. He noticed on his monthly statement that \$814 is used to pay off the interest charge. Only \$26 is used to pay off the principal. What percent of his monthly mortgage payment is used to pay off the interest charge? Round to the nearest tenth of a percent. 96.9%

36. ***Interest Charges on a Mortgage*** Alicia is making monthly mortgage payments of \$960 for her home mortgage. She noticed on her monthly statement that \$917 is used to pay off the interest charge. Only \$43 is used to pay off the principal. What percent of her monthly mortgage payment is used to pay off the interest charge? Round to the nearest tenth of a percent. 95.5%

Solve. Round to the nearest cent.

37. ***Sales Tax*** How much sales tax would you pay to purchase a new Honda Accord that costs \$18,456.82 if the sales tax rate is 4.6%? \$849.01

38. ***Living Room Set Purchase*** The Hartling family purchased a new living room set. The list price was \$1249.95. However, they got a discount of 29%. How much did they pay for the new living room set? \$887.46

Cumulative Review

Perform the following calculations using the correct order of operations.

39. $3(12 - 6) - 4(12 \div 3)$ 2

40. $7 + 4^3 \times 2 - 15$ 120

41. $\left(\frac{5}{2}\right)\left(\frac{1}{3}\right) - \left(\frac{2}{3} - \frac{1}{3}\right)^2$ $\frac{13}{18}$

42. $(6.8 - 6.6)^2 + 2(1.8)$ 3.64

Putting Your Skills to Work

Business and medical service areas in our country are finding a greater need for employees who are bilingual. The greatest demands are for Spanish and French speakers, but those who speak German, Italian, Japanese, and Russian are also in high demand. However, high schools are not keeping up with this increased demand. Although the number of students in high school who take a foreign language is increasing, the demand in society for bilingual people is increasing at a faster rate.

Study the chart below and then answer the questions that follow. Round answers to the nearest tenth when necessary.

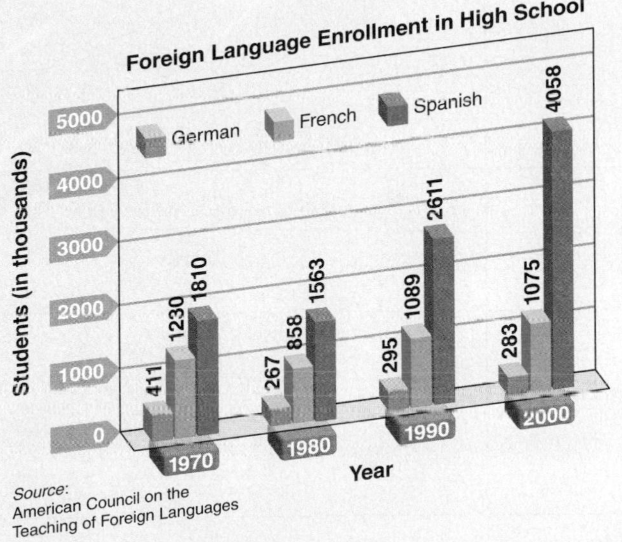

Problems for Individual Investigation

1. How many more students were learning Spanish in high school in 2000 than in 1970?
   2,248,000 students

2. How many more students were learning French in high school in 2000 than in 1980?
   217,000 students

Problems for Cooperative Study and Analysis

3. **(a)** What percent of foreign language students were studying French in 2000? 19.8%

   (b) How does that compare with the percent of foreign language students studying French in 1970?

   35.6% studied French in 1970. The percentage is less in 2000.

4. **(a)** What percent of foreign language students were studying German in 2000? 5.2%

   (b) How does that compare with the percent of foreign language students studying German in 1970?

   11.9% studied German in 1970. The percentage is less in 2000.

5. Recently the government estimated that 24% more students will need to study French in 2010 than in 2000 in order to meet projected employment needs. How many more students will need to take French in 2010 than took this language in 2000? 258,000 more students

6. Recently the government estimated that 6% more students will need to study Spanish in 2010 than in 2000 in order to meet projected employment needs. How many more students will need to take Spanish in 2010 than took this language in 2000?
   243,480 more students

Chapter 5 Organizer

Topic	Procedure	Examples
Converting a decimal to a percent, p. 315.	**1.** Move the decimal point two places to the right. **2.** Add the percent sign.	$0.19 = 19\%$ $0.516 = 51.6\%$ $0.04 = 4\%$ $1.53 = 153\%$ $0.006 = 0.6\%$
Converting a fraction with a denominator of 100 to a percent, p. 312.	**1.** Use the numerator only. **2.** Add the percent sign.	$\frac{29}{100} = 29\%$ $\quad \frac{5.6}{100} = 5.6\%$ $\frac{3}{100} = 3\%$ $\quad \frac{7\frac{1}{3}}{100} = 7\frac{1}{3}\%$ $\frac{231}{100} = 231\%$
Changing a fraction (whose denominator is not 100) to a percent, p. 321.	**1.** Divide the numerator by the denominator and obtain a decimal. **2.** Change the decimal to a percent.	$\frac{13}{50} = 0.26 = 26\%$ $\frac{1}{20} = 0.05 = 5\%$ $\frac{3}{800} = 0.00375 = 0.375\%$ $\frac{312}{200} = 1.56 = 156\%$
Changing a mixed number to a percent, p. 324.	**1.** Change the mixed number to a decimal. **2.** Transform the decimal to an equivalent percent.	$3\frac{1}{4} = 3.25 = 325\%$ $\quad 6\frac{4}{5} = 6.8 = 680\%$ $7\frac{3}{8} = 7.375 = 737.5\%$
Changing a percent to a decimal, p. 314.	**1.** Drop the percent sign. **2.** Move the decimal point two places to the left.	$49\% = 0.49$ $2\% = 0.02$ $0.5\% = 0.005$ $196\% = 1.96$ $1.36\% = 0.0136$
Changing a percent to a fraction, p. 319.	**1.** If the percent does not contain a decimal point, remove the % and write the number over a denominator of 100. Reduce the fraction if possible. **2.** If the percent contains a decimal point, change the percent to a decimal by removing the % and moving the decimal point two places to the left. Then write the decimal as a fraction and reduce if possible. **3.** If the percent contains a fraction, remove the % and write the number over a denominator of 100. If the numerator is a mixed number, change the numerator to an improper fraction. Next simplify by the "invert and multiply" rule. Then reduce the fraction if possible.	$25\% = \frac{25}{100} = \frac{1}{4}$ $38\% = \frac{38}{100} = \frac{19}{50}$ $130\% = \frac{130}{100} = \frac{13}{10}$ $5.8\% = 0.058$ $= \frac{58}{1000} = \frac{29}{500}$ $2.72\% = 0.0272$ $= \frac{272}{10{,}000} = \frac{17}{625}$ $7\frac{1}{8}\% = \frac{7\frac{1}{8}}{100}$ $= 7\frac{1}{8} \div \frac{100}{1}$ $= \frac{57}{8} \times \frac{1}{100} = \frac{57}{800}$

Topic	Procedure	Examples
Solving percent problems by translating to equations, p. 330.	**1.** Translate by replacing "of" with $\times$ "is" with $=$ "what" with n "find" with $n =$ **2.** Solve the resulting equation.	**(a)** What is 3% of 56? $n = 3\% \times 56$ $n = (0.03)(56)$ $n = 1.68$ **(b)** 16% of what is 208? $16\% \times n = 208$ $0.16n = 208$ $\frac{0.16n}{0.16} = \frac{208}{0.16}$ $n = 1300$ **(c)** What percent of 70 is 30? $n \times 70 = 30$ $70n = 30$ $\frac{70n}{70} = \frac{30}{70}$ $n = 0.428571\ldots$ $n \approx 42.86\%$.
Solving percent problems by using proportions, p. 339.	**1.** Identify the parts of the percent proportion. $a =$ the amount $b =$ the base (the whole; it usually appears after the word "of") $p =$ the percent number **2.** Write the percent proportion $\frac{a}{b} = \frac{p}{100}$ using the values obtained in step 1 and solve.	**(a)** What is 28% of 420? The percent $p = 28$. The base $b = 420$. The amount a is unknown. We use the variable a. $\frac{a}{b} = \frac{p}{100}$ becomes $\frac{a}{420} = \frac{28}{100}$ If we reduce the fraction on the right-hand side, we have $\frac{a}{420} = \frac{7}{25}$ $25a = (7)(420)$ $25a = 2940$ $\frac{25a}{25} = \frac{2940}{25}$ $a = 117.6$ Thus, 28% of 420 is 117.6. **(b)** 64% of what is 320? The percent $p = 64$. The base is unknown. We use the variable b. The amount $a = 320$. $\frac{a}{b} = \frac{p}{100}$ $\frac{320}{b} = \frac{64}{100}$ If we reduce the fraction on the right-hand side, we have $\frac{320}{b} = \frac{16}{25}$ $(320)(25) = 16b$ $8000 = 16b$ $\frac{8000}{16} = \frac{16b}{16}$ $500 = b$ Thus, 64% of 500 is 320.

(continues on next page)

Topic	Procedure	Examples
***(continued)* Solving percent problems by using proportions, p. 341.**	**1.** Identify the parts of the percent proportion. a = the amount b = the base (the whole; it usually appears after the word "of") p = the percent number **2.** Write the percent proportion $\frac{a}{b} = \frac{p}{100}$ using the values obtained in step 1 and solve.	**(c)** What percent of 140 is 105? The percent is unknown. The base b = 140, the amount a = 105. $\frac{a}{b} = \frac{p}{100}$ becomes $\frac{105}{140} = \frac{p}{100}$ If we reduce the fraction on the left side, we have $\frac{3}{4} = \frac{p}{100}$ $(100)(3) = 4p$ $300 = 4p$ $\frac{300}{4} = \frac{4p}{4}$ $75 = p$ Thus 105 is 75% of 140.
Solving discount problems, p. 350.	Discount = discount rate × list price	Carla purchased a color TV set that lists for \$350 at an 18% discount. **(a)** How much was the discount? **(b)** How much did she pay for the TV set? **(a)** Discount = (0.18)(350) = \$63 **(b)** 350 − 63 = 287 She paid \$287 for the color TV set.
Solving commission problems, p. 354.	Commission = commission rate × value of sales	A housewares salesperson gets a 16% commission on sales he makes. How much commission does he earn if he sells \$12,000 in housewares? Commission = (0.16)(12,000) = \$1920
Solving simple interest problems, p. 355.	Interest = principal × rate × time $I = P \times R \times T$ I = interest P = principal R = rate T = time	Hector borrowed \$3000 for 4 years at a simple interest rate of 12%. How much interest did he owe after 4 years? $I = P \times R \times T$ $I = (3000)(0.12)(4)$ $= (360)(4)$ $= 1440$ Hector owed \$1440 in interest.
Percent-of-increase or percent-of-decrease problems, p. 354.	Percent of increase or decrease = amount of increase or decrease ÷ base	A car that costs \$16,500 now cost only \$15,000 last year. What is the percent of increase? 16,500 − 15,000 = 1,500 increase $\frac{1500}{15,000} = 0.10$ Percent of increase = 10%

Chapter 5 Review Problems

Section 5.1

Write as a percent. Round to the nearest hundredth of a percent when necessary.

1. 0.62
62%

2. 0.43
43%

3. 0.372
37.2%

4. 0.529
52.9%

5. 1.05
105%

6. 2.1
210%

7. 2.52
252%

8. 4.37
437%

9. 1.036
103.6%

10. 1.052
105.2%

11. 0.006
0.6%

12. 0.002
0.2%

13. $\frac{12.5}{100}$
12.5%

14. $\frac{8.3}{100}$
8.3%

15. $\frac{4\frac{1}{12}}{100}$
$4\frac{1}{12}\%$

16. $\frac{3\frac{5}{12}}{100}$
$3\frac{5}{12}\%$

17. $\frac{317}{100}$
317%

18. $\frac{225}{100}$
225%

Section 5.2

Change to a percent. Round to the nearest hundredth of a percent when necessary.

19. $\frac{19}{25}$
76%

20. $\frac{13}{25}$
52%

21. $\frac{11}{20}$
55%

22. $\frac{9}{40}$
22.5%

23. $\frac{5}{11}$
45.45%

24. $\frac{4}{9}$
44.44%

25. $2\frac{1}{4}$
225%

26. $3\frac{3}{4}$
375%

27. $2\frac{7}{9}$
277.78%

28. $5\frac{5}{9}$
555.56%

29. $\frac{152}{80}$
190%

30. $\frac{200}{80}$
250%

31. $\frac{3}{800}$
0.38%

32. $\frac{5}{800}$
0.63%

Change to decimal form.

33. 32%
0.32

34. 68%
0.68

35. 82.7%
0.827

36. 59.6%
0.596

37. 236%
2.36

38. 177%
1.77

39. $32\frac{1}{8}\%$
0.32125

40. $26\frac{3}{8}\%$
0.26375

Change to fractional form.

41. 72%
$\frac{18}{25}$

42. 92%
$\frac{23}{25}$

43. 185%
$\frac{37}{20}$

44. 225%
$\frac{9}{4}$

45. 16.4%
$\frac{41}{250}$

46. 30.5%
$\frac{61}{200}$

47. $31\frac{1}{4}\%$
$\frac{5}{16}$

48. $43\frac{3}{4}\%$
$\frac{7}{16}$

49. 0.05%
$\frac{1}{2000}$

50. 0.06%
$\frac{3}{5000}$

Complete the following chart.

	Fraction	Decimal	Percent
51.	$\frac{3}{5}$	0.6	60%
52.	$\frac{7}{10}$	0.7	70%
53.	$\frac{3}{8}$	0.375	37.5%
54.	$\frac{9}{16}$	0.5625	56.25%
55.	$\frac{1}{125}$	0.008	0.8%
56.	$\frac{9}{20}$	0.45	45%

Section 5.3

Solve. Round to the nearest hundredth when necessary.

57. What is 20% of 85?
17

58. What is 25% of 92?
23

59. 18 is 20% of what number?
90

60. 70 is 40% of what number?
175

61. 50 is what percent of 130?
38.46%

62. 70 is what percent of 180?
38.89%

63. Find 162% of 60.
97.2

64. Find 124% of 80.
99.2

65. 92% of what number is 147.2?
160

66. 68% of what number is 95.2?
140

67. What percent of 70 is 14?
20%

68. What percent of 60 is 6?
10%

Sections 5.4 and 5.5

Solve. Round your answer to the nearest hundredth when necessary.

69. ***Education*** Professor Wonson found that 34% of his class is left-handed. He has 150 students in his class. How many are left-handed?
51 students

70. ***Truck Dealer*** A Vermont truck dealer found that 64% of all the trucks he sold had four-wheel drive. If he sold 150 trucks, how many had four-wheel drive? 96 trucks

71. ***Car Depreciation*** Today Yvonne's car has 61% of the value that it had two years ago. Today it is worth $6832. What was it worth two years ago?
$11,200

72. ***Administrative Expenses*** A charity organization spent 12% of its budget for administrative expenses. It spent $9624 on administrative expenses. What was the total budget?
$80,200

73. ***Rain in Seattle*** In Seattle it rained 20 days in February, 18 days in March, and 16 days in April. What percent of those three months did it rain? (Assume it was a leap year.)
60%

74. ***Job Applications*** Moorehouse Industries received 600 applications and hired 45 of the applicants. What percent of the applicants obtained a job?
7.5%

75. ***Appliance Purchase*** Nathan bought new appliances for $3670. The sales tax in his state is 5%. What did he pay in sales tax?
$183.50

76. ***Boat Purchase*** Chris and Annette bought a boat for $12,600. The sales tax is 6% in their state. What did they pay in sales tax?
$756

77. ***Budgets*** Joan and Michael budget 38% of their income for housing. They spend $684 per month for housing. What is their monthly income?
$1800

78. ***Real Estate Sales*** Beachfront property is very expensive. A real estate agent in South Carolina earned $26,000 in commissions. The property she sold was worth $650,000. What was her commission rate?
4%

79. ***Encyclopedia Sales*** Adam sold encyclopedias last summer to raise tuition money. He sold $83,500 worth of encyclopedias and was paid $5010 in commissions. What commission rate did he earn?
6%

80. ***Commission Sales*** Roberta earns a commission at the rate of 7.5%. Last month she sold $16,000 worth of goods. How much commission did she make last month?
$1200

81. ***Dining Room Set*** Irene purchased a dining room set at a 20% discount. The list price was $1595.

(a) What was the discount?
$319

(b) What did she pay for the set?
$1276

82. ***Laptop Computer*** A Dell laptop computer listed for $2125. This week, Lisa heard that the manufacturer is offering a rebate of 12%.

(a) What is the rebate?
$255

(b) How much will Lisa pay for the computer?
$1870

83. ***Construction Workers*** In 1998, the Big Dig project in Boston had 3243 construction workers. In 2000, there were 4618 construction workers. What was the percentage of increase in the number of workers?
42.40%

84. ***Cost of Gold*** In August 2003, one troy ounce of gold was worth $360.10. In August 2002, a troy ounce was worth $313.80. What was the percentage of increase in the value of one troy ounce of gold?
14.75%

85. ***Log Cabin*** Mark and Julie wanted to buy a prefabricated log cabin to put on their property in the Colorado Rockies. The price of the kit is listed at $24,000. At the after-holiday cabin sale, a discount of 14% was offered.

(a) What was the discount? $3360

(b) How much did they pay for the cabin? $20,640

86. ***Mutual Funds*** Sally invested $6000 in mutual funds earning 11% simple interest in one year. How much interest will she earn in

(a) six months? $330

(b) two years? $1320

87. ***College Loan*** Reed took out a college loan of $3000. He will be charged 8% simple interest on the loan.

(a) How much interest will be due on the loan in three months? $60

(b) How much interest will be due on the loan in three years? $720

How Am I Doing? Chapter 5 Test

Remember to use your Chapter Test Prep Video CD to see the worked-out solutions to the test problems you want to review.

Note to Instructor: The Chapter 5 Test file in the TestGen program provides algorithms specifically matched to these problems so you can easily replicate this test for additional practice or assessment purposes.

Write as a percent. Round to the nearest hundredth of a percent when necessary.

1. 0.57 **2.** 0.01

3. 0.008 **4.** 12.8

5. 3.56 **6.** $\frac{71}{100}$

7. $\frac{1.8}{100}$ **8.** $\frac{3\frac{1}{7}}{100}$

Change to a percent. Round to the nearest hundredth of a percent when necessary.

9. $\frac{19}{40}$ **10.** $\frac{27}{36}$

11. $\frac{225}{75}$ **12.** $1\frac{3}{4}$

Write as a percent.

13. 0.0825 **14.** 3.024

Write as a fraction in simplified form.

15. 152% **16.** $7\frac{3}{4}\%$

Solve. Round to the nearest hundredth if necessary.

17. What is 40% of 50?

18. 33.8 is 26% of what number?

19. What percent of 72 is 40?

20. Find 0.8% of 25,000.

1. 57%

2. 1%

3. 0.8%

4. 1280%

5. 356%

6. 71%

7. 1.8%

8. $3\frac{1}{7}\%$

9. 47.5%

10. 75%

11. 300%

12. 175%

13. 8.25%

14. 302.4%

15. $1\frac{13}{25}$

16. $\frac{31}{400}$

17. 20

18. 130

19. 55.56%

20. 200

21. 5000

22. 46%

23. 699.6

24. 20%

25. $6092

26. (a) $150.81

(b) $306.19

27. 89.29%

28. 23.24%

29. 12,000 registered voters

30. (a) $240

(b) $960

21. 16% of what number is 800?

22. 92 is what percent of 200?

23. 132% of 530 is what number?

24. What percent is 15 of 75?

Solve. Round to the nearest hundredth if necessary.

25. A real estate agent sells a house for $152,300. She gets a commission of 4% on the sale. What is her commission?

26. Julia and Charles bought a new dishwasher at a 33% discount. The list price was $457.

(a) What was the discount?
(b) How much did they pay for the dishwasher?

27. An inspector found that 75 out of 84 parts were not defective. What percent of the parts were not defective?

28. Last year Charlotte was the top player on the basketball team, scoring 185 points. This year, she scored 228 points. What is the percentage of increase in the number of points?

29. A total of 5160 people voted in the city election. This was 43% of the registered voters. How many registered voters are in the city?

30. Wanda borrowed $3000 at a simple interest rate of 16%.

(a) How much interest did she pay in six months?
(b) How much interest did she pay in two years?

Cumulative Test for Chapters 1–5

Approximately one-half of this test is based on Chapter 5 material. The remainder is based on material covered in Chapters 1–4.

Solve. Simplify your answer.

1. Add. $\begin{array}{r} 38 \\ 196 \\ +2007 \\ \hline \end{array}$

2. Subtract. $\begin{array}{r} 23{,}007 \\ -14{,}563 \\ \hline \end{array}$

3. Multiply. $\begin{array}{r} 126 \\ \times\ 42 \\ \hline \end{array}$

4. Divide. $36\overline{)3204}$

5. Add. $2\frac{1}{4} + 3\frac{1}{3}$

6. Subtract. $5\frac{2}{5} - 2\frac{7}{10}$

7. Multiply. $3\frac{1}{8} \times \frac{12}{5}$

8. Divide. $\frac{5}{12} \div 1\frac{3}{4}$

9. Round to the nearest thousandth. 77.1832

10. Add. $\begin{array}{r} 5.6 \\ 3.21 \\ 18.3 \\ +\ 7.008 \\ \hline \end{array}$

11. Multiply. $\begin{array}{r} 5.62 \\ \times\ 0.3 \\ \hline \end{array}$

12. Divide. $1.4\overline{)0.5152}$

▲ **13.** Write as a unit rate. 36 tiles in 9 square feet

14. Is this equation a proportion? $\frac{20}{25} = \frac{300}{375}$

15. Solve the proportion. $\frac{8}{2.5} = \frac{n}{7.5}$

16. A college has a ratio of 3 faculty members for every 19 students. The student body presently has 4263 students. How many faculty members are there? Round to the nearest whole number.

1. 2241

2. 8444

3. 5292

4. 89

5. $\frac{67}{12}$ or $5\frac{7}{12}$

6. $2\frac{7}{10}$

7. $\frac{15}{2}$ or $7\frac{1}{2}$

8. $\frac{5}{21}$

9. 77.183

10. 34.118

11. 1.686

12. 0.368

13. 4 tiles/square foot

14. yes

15. $n = 24$

16. 673 faculty members

In questions 17–30, round to the nearest hundredth when necessary.

Write as a percent.

17. 0.023

18. $\frac{46.8}{100}$

19. 1.98

20. $\frac{3}{80}$

In questions 21 and 22, write as a decimal.

21. 243%

22. $6\frac{3}{4}\%$

23. What percent of 214 is 38?

24. Find 1.7% of 6740.

25. 219 is 73% of what number?

26. 95% of 200 is what number?

27. While shopping for appliances, Cecilia sees a new washer and dryer for $680. A sign in the store reads "20% off." How much would she pay for the appliances?

28. A total of 896 freshmen were admitted to King Frederich College. Freshmen make up 28% of the student body. How big is the student body?

29. The air pollution level in Centerville is 8.86 parts per million. Ten years ago it was 7.96 parts per million. What is the percent of increase of the air pollution level?

30. Fred borrowed $1600 for two years. He was charged simple interest at a rate of 11%. How much interest did he pay?

17. 2.3%

18. 46.8%

19. 198%

20. 3.75%

21. 2.43

22. 0.0675

23. 17.76%

24. 114.58

25. 300

26. 190

27. $544

28. 3200 students

29. 11.31%

30. $352

CHAPTER

6

About fifty years ago a major change in the way land is measured occurred in Canada. All new deeds recording the amount of land area had to be changed from recording the area in acres to recording the area in hectares. Farmers, forestry personnel, and all land owners had to do some extensive calculations. How well do you think you would do if you had to change land records from acres to hectares? Turn to page 415 to find out.

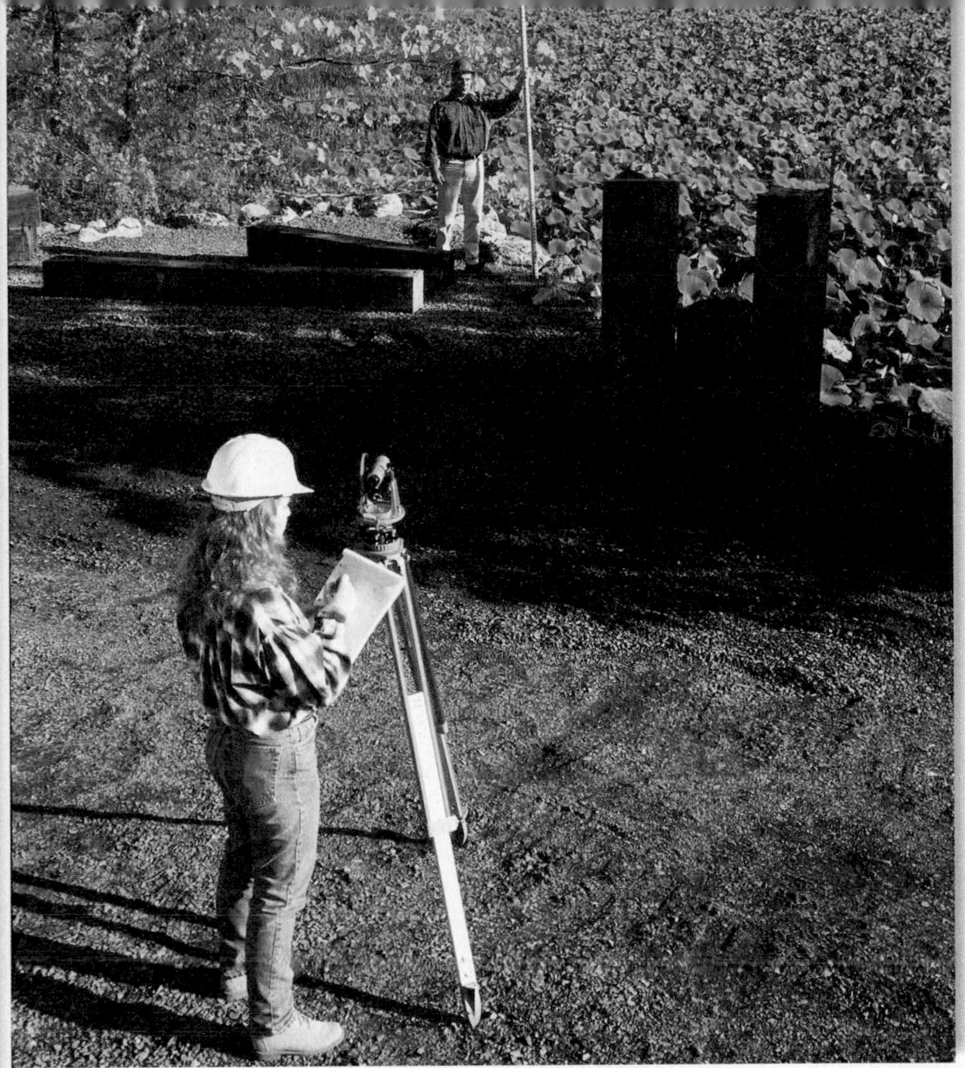

Measurement

6.1 AMERICAN UNITS

Student Learning Objectives

After studying this section, you will be able to:

1. **Identify the basic unit equivalencies in the American system.**
2. **Convert from one unit of measure to another.**

1 Identifying the Basic Unit Equivalencies in the American System

We often ask questions about measurements. How far is it to work? How much does this bottle hold? What is the weight of this box? How long will it be until exams? To answer these questions we need to agree on a unit of measure for each type of measurement.

At present there are two main systems of measurement, each with its own set of units: **the metric system** and the **American system.** Nearly all countries in the world use the metric system. In the United States, however, except in science, most measurements are made in American units.

The United States is using the metric system more and more frequently, however, and may eventually convert to metric units as the standard. But for now we need to be familiar with both systems. We cover American units in this section.

One of the most familiar measuring devices is a ruler that measures lengths as great as 1 foot. It is divided into 12 inches; that is,

$$1 \text{ foot} = 12 \text{ inches}.$$

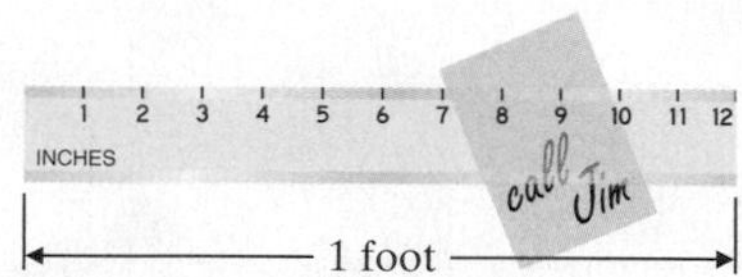

There are several other important relationships you need to know. Your instructor may require you to memorize the following facts.

Length
12 inches = 1 foot
36 inches = 1 yard
3 feet = 1 yard
5280 feet = 1 mile
1760 yards = 1 mile

Time
60 seconds = 1 minute
60 minutes = 1 hour
24 hours = 1 day
7 days = 1 week

Note that time is measured in the same units in both the metric and the American systems.

Weight
16 ounces = 1 pound
2000 pounds = 1 ton

Volume
8 fluid ounces = 1 cup
2 cups = 1 pint
2 pints = 1 quart
4 quarts = 1 gallon

We can choose to measure an object—say, a bridge—using a small unit (an inch), a larger unit (a foot), or a still larger unit (a mile). We may say that the bridge spans 7920 inches, 660 feet, or an eighth of a mile. Although we probably would not choose to express our measurement in inches because this is not a convenient measurement to work with for an object as long as a bridge, the bridge length is the same whatever unit of measurement we use.

Notice that the smaller the measuring unit, the larger the number of those units in the final measurement. The inch is the smallest unit in our example, and the inch measurement has the greatest number of units (7920). The mile is the largest unit, and it has the smallest number of units (an eighth equals 0.125). Whatever measuring system you use, and whatever you measure (length, volume, and so on), the smaller the unit of measurement you use, the greater the number of those units.

After studying the values in the length, time, weight, and volume tables, see if you can quickly do Example 1.

Teaching Tip An effective approach is to ask students to open their books and fill in the answers to Practice Problem 1. If they make several mistakes, or just cannot remember several facts, ask them to memorize the chart values showing the length, time, weight, and volume equivalencies. It wastes a lot of time looking these up when doing problems, so students should know them by heart.

EXAMPLE 1 Answer rapidly the following questions.

(a) How many inches in a foot?
(b) How many yards in a mile?
(c) How many seconds in a minute?
(d) How many hours in a day?
(e) How many pounds in a ton?
(f) How many cups in a pint?

Solution

(a) 12 **(b)** 1760 **(c)** 60 **(d)** 24 **(e)** 2000 **(f)** 2

Practice Problem 1 Answer rapidly the following questions.

(a) How many feet in a yard?
(b) How many feet in a mile?
(c) How many minutes in an hour?
(d) How many days in a week?
(e) How many ounces in a pound?
(f) How many pints in a quart?
(g) How many quarts in a gallon?

NOTE TO STUDENT: Fully worked-out solutions to all of the Practice Problems can be found at the back of the text starting at page SP-1

2 Converting from One Unit of Measure to Another

To convert or change one measurement to another, we simply multiply by 1 since multiplying by 1 does not change the value of a quantity. For example, to convert 180 inches to feet, we look for a name for 1 that has inches and feet.

$$1 = \frac{1 \text{ foot}}{12 \text{ inches}}$$

A ratio of measurements for which the measurement in the numerator is equivalent to the measurement in the denominator is called a **unit fraction.**

We now use the unit fraction $\frac{1 \text{ foot}}{12 \text{ inches}}$ to convert 180 inches to feet.

$$180 \cancel{\text{inches}} \times \frac{1 \text{ foot}}{12 \cancel{\text{inches}}} = \frac{180 \text{ feet}}{12} = 15 \text{ feet}$$

Notice that when we multiplied, the inches divided out. We are left with the unit feet.

What name for 1 (that is, unit fraction) should we choose if we want to change from feet to inches? Convert 4 feet to inches.

Teaching Tip The chief advantage of the "multiply by a fractional name for 1 and remove the common unit in the numerator of the first fraction and the denominator of the second fraction" method is that the students using it do not multiply by the wrong number. This method is very useful in science courses.

$$1 = \frac{12 \text{ inches}}{1 \text{ foot}}$$

$$4 \cancel{\text{feet}} \times \frac{12 \text{ inches}}{1 \cancel{\text{foot}}} = \frac{48 \text{ inches}}{1} = 48 \text{ inches}$$

When multiplying by a unit fraction, the unit we want to change to should be in the *numerator*. The unit we start with should be in the *denominator*. This unit will divide out.

EXAMPLE 2 Convert. 8800 yards to miles

Solution $8800 \text{ ~~yards~~} \times \dfrac{1 \text{ mile}}{1760 \text{ ~~yards~~}} = \dfrac{8800}{1760} \text{ miles} = 5 \text{ miles}$

NOTE TO STUDENT: Fully worked-out solutions to all of the Practice Problems can be found at the back of the text starting at page SP-1

Practice Problem 2 Convert. 15,840 feet to miles

Some conversions involve fractions or decimals. Whether you want to measure the area of a living room or the dimensions of a piece of property, it is helpful to be able to make conversions like these.

EXAMPLE 3 Convert.

(a) 26.48 miles to yards **(b)** $3\frac{2}{3}$ feet to yards

Solution

(a) $26.48 \text{ ~~miles~~} \times \dfrac{1760 \text{ yards}}{1 \text{ ~~mile~~}} = 46{,}604.8 \text{ yards}$

(b) $3\dfrac{2}{3} \text{ ~~feet~~} \times \dfrac{1 \text{ yard}}{3 \text{ ~~feet~~}} = \dfrac{11}{3} \times \dfrac{1}{3} \text{ yard} = \dfrac{11}{9} \text{ yards} = 1\dfrac{2}{9} \text{ yards}$

Practice Problem 3 Convert.

(a) 18.93 miles to feet **(b)** $16\frac{1}{2}$ inches to yards

EXAMPLE 4 Lynda's new car weighs 2.43 tons. How many pounds is that?

Solution $2.43 \text{ ~~tons~~} \times \dfrac{2000 \text{ pounds}}{1 \text{ ~~ton~~}} = 4860 \text{ pounds}$

Practice Problem 4 A package weighs 760.5 pounds. How many ounces does it weigh?

EXAMPLE 5 The chemistry lab has 34 quarts of weak hydrochloric acid. How many gallons of this acid are in the lab? (Express your answer as a decimal.)

Solution $34 \text{ ~~quarts~~} \times \dfrac{1 \text{ gallon}}{4 \text{ ~~quarts~~}} = \dfrac{34}{4} \text{ gallons} = 8.5 \text{ gallons}$

Practice Problem 5 19 pints of milk is the same as how many quarts? (Express your answer as a decimal.)

EXAMPLE 6 A window is 4 feet 5 inches wide. How many inches is that? 4 feet 5 inches means 4 feet and 5 inches. Change the 4 feet to inches and add the 5 inches.

Solution $4 \text{ \sout{feet}} \times \dfrac{12 \text{ inches}}{1 \text{ \sout{foot}}} = 48 \text{ inches}$

$48 \text{ inches} + 5 \text{ inches} = 53 \text{ inches}$ The window is 53 inches wide.

Practice Problem 6 A path through Dr. Sherf's property measures 26 yards 2 feet in length. How many feet long is the path?

EXAMPLE 7 The Charlotte all-night garage charges \$1.50 per hour for parking both day and night. A businessman left his car there for $2\frac{1}{4}$ days. How much was he charged?

Solution

1. ***Understand the problem.*** Here it might help to look at a simpler problem. If the businessman had left his car for two hours, we would multiply.

   The fraction bar means "per." $\rightarrow \dfrac{1.50 \text{ dollars}}{1 \text{ hour}} \times 2 \text{ hours} = 3.00 \text{ dollars or } \$3.00$

   Thus, if the businessman had left his car for two hours, he would have been charged \$3. We see that we need to multiply \$1.50 by the number of hours the car was in the garage to solve the problem.

   Since the original problem gave the time in days, not hours, we will need to change the days to hours.

2. ***Solve and state the answer.*** Now that we know that the way to solve the problem is to multiply by hours, we will begin. To make our calculations easier we will write $2\frac{1}{4}$ as 2.25. Change days to hours. Then multiply by \$1.50 per hour.

   $$2.25 \text{ \sout{days}} \times \frac{24 \text{ \sout{hours}}}{1 \text{ \sout{day}}} \times \frac{1.50 \text{ dollars}}{1 \text{ \sout{hour}}} = 81 \text{ dollars or } \$81$$

   The businessman was charged \$81.

3. ***Check.*** Is our answer in the desired unit? Yes. The answer is in dollars and we would expect it to be in dollars. ✓

   The check is up to you.

Practice Problem 7 A businesswoman parked her car at a garage for $1\frac{3}{4}$ days. The garage charges \$1.50 per hour. How much did she pay to park the car?

Teaching Tip It is important to carefully go over the type of problem that requires students to multiply by two unit fractions. You may also do an example of the type "The supermarket charges \$2.40 per pound for lean ground beef. If you bought 33 ounces, how much would you pay?" (The answer is \$4.95.)

ALTERNATIVE METHOD: Using Proportions How did people first come up with the idea of multiplying by a unit fraction? What mathematical principles are involved here? Actually, this is the same as solving a proportion. Consider Example 5, where we changed 34 quarts to 8.5 gallons by multiplying.

$$34 \text{ } \cancel{\text{quarts}} \times \frac{1 \text{ gallon}}{4 \text{ } \cancel{\text{quarts}}} = \frac{34}{4} \text{ gallons} = 8.5 \text{ gallons}$$

What we were actually doing is setting up the proportion:

1 gallon is to 4 quarts as *n* gallons is to 34 quarts.

$$\frac{1 \text{ gallon}}{4 \text{ quarts}} = \frac{n \text{ gallons}}{34 \text{ quarts}}$$

$$1 \text{ gallon} \times 34 \text{ quarts} = 4 \text{ quarts} \times n \text{ gallons} \qquad \text{Cross multiplying}$$

$$\frac{1 \text{ gallon} \times 34 \text{ } \cancel{\text{quarts}}}{4 \text{ } \cancel{\text{quarts}}} = \frac{\cancel{4} \text{ } \cancel{\text{quarts}} \times n \text{ gallons}}{\cancel{4} \text{ } \cancel{\text{quarts}}} \qquad \text{Dividing both sides of the equation by 4 quarts}$$

$$1 \text{ gallon} \times \frac{34}{4} = n \text{ gallons} \qquad \text{Simplifying}$$

$$8.5 \text{ gallons} = n \text{ gallons}$$

Thus the number of gallons is 8.5. Using proportions takes a little longer, so multiplying by a fractional name for 1 is the more popular method.

6.1 Exercises

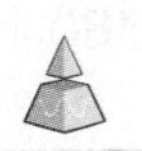

Student Solutions Manual | CD/Video | PH Math Tutor Center | MathXL®Tutorials on CD | MathXL® | MyMathLab® | Interactmath.com

Verbal and Writing Skills

1. Explain in your own words how you would use a unit fraction to change 23 miles to inches.

We know that each mile is 5280 feet. Each foot is 12 inches. So we know that one mile is $5280 \times 12 = 63{,}360$ inches. The unit fraction we want is $\frac{63{,}360 \text{ inches}}{1 \text{ mile}}$. So we multiply 23 miles $\times \frac{63{,}360 \text{ inches}}{1 \text{ mile}}$. The mile unit divides out. We obtain 1,457,280 inches. Thus 23 miles = 1,457,280 inches.

2. Explain in your own words how you would use a unit fraction to change 27 days to minutes.

We know that each day has 24 hours. Each hour has 60 minutes. Therefore we know that each day has $24 \times 60 = 1440$ minutes. The unit fraction we want is $\frac{1440 \text{ minutes}}{1 \text{ day}}$. So we multiply 27 days $\times \frac{1440 \text{ minutes}}{1 \text{ day}}$. The day unit divides out. We obtain 38,880 minutes. Thus 27 days = 38,880 minutes.

From memory, write the equivalent value

3. __1760__ yards = 1 mile

4. __5280__ feet = 1 mile

5. 1 ton = __2000__ pounds

6. 1 pound = __16__ ounces

7. __4__ quarts = 1 gallon

8. __2__ cups = 1 pint

9. 1 quart = __2__ pints

10. __60__ minutes = 1 hour

Convert. When necessary, express your answer as a decimal.

11. 21 feet = __7__ yards

12. 63 feet = __21__ yards

13. 108 inches = __9__ feet

14. 180 inches = __15__ feet

15. 12 feet = __144__ inches

16. 16 feet = __192__ inches

17. 10,560 feet = __2__ miles

18. 5 miles = __26,400__ feet

19. 7 miles = __12,320__ yards

20. 21,120 feet = __4__ miles

21. 7 gallons = __28__ quarts

22. 5 gallons = __20__ quarts

23. 48 quarts = __12__ gallons

24. 24 pints = __12__ quarts

25. 16 cups = __128__ fluid ounces

26. 40 fluid ounces = __5__ cups

27. $8\frac{1}{2}$ gallons = __68__ pints

28. $6\frac{1}{2}$ gallons = __52__ pints

29. 12 weeks = __84__ days

30. 7 weeks = __49__ days

31. 960 seconds = __16__ minutes

32. 1500 seconds = __25__ minutes

33. 8 ounces = __0.5__ pound

34. 12 ounces = __0.75__ pound

35. 12,500 pounds = __6.25__ tons

36. $4\frac{3}{4}$ tons = __9500__ pounds

37. 15 pints = __7.5__ cups

38. 23 pints = __11.5__ cups

39. 2.25 pounds = __36__ ounces

40. 4.25 pounds = __68__ ounces

41. 75 inches = __6.25__ feet

42. 87 inches = __7.25__ feet

Applications

43. ***Wheelchair Racer*** Candace Cable is a champion wheelchair racer. In Grandma's Marathon, she covered the 26.2-mile course in 1:46 (1 hour and 46 minutes). How many feet did she travel in the race?

$26.2 \text{ miles} \times \frac{5280 \text{ feet}}{1 \text{ mile}} = 138{,}336 \text{ feet}$

44. ***Marathon Record*** In April 2003, Paula Radcliffe of Great Britain set the women's world record for the marathon with a time of 2:15:25 (hours:minutes:seconds). How many seconds is that?

8125 seconds

45. ***Mountain Height*** Mount Kilimanjaro in Tanzania is 19,336 feet high. How many miles is that? Round to the nearest hundredth.

3.66 miles

46. ***Shot Put*** Curtis threw the shot put $39\frac{1}{2}$ feet at the high school track meet. How many inches is that?

474 inches

47. ***Food Purchase*** Judy is making a wild mushroom sauce for pasta tonight with a large group of friends. She bought 26 ounces of wild mushrooms at \$6.00 per pound. How much were the mushrooms?

$26 \cancel{\text{ounces}} \times \frac{1 \cancel{\text{pound}}}{16 \cancel{\text{ounces}}} \times \frac{\$6.00}{1 \cancel{\text{pound}}} = \$9.75$

48. ***Food Purchase*** Kurt is trying to eat a low-fat diet. He finds a store that sells 1% fat ground white-meat turkey breast at \$6.00 per pound. He buys one packet weighing 18 ounces and another weighing 22 ounces. How much does he pay?

$40 \cancel{\text{ounces}} \times \frac{1 \cancel{\text{pound}}}{16 \cancel{\text{ounces}}} \times \frac{\$6.00}{1 \cancel{\text{pound}}} = \$15.00$

▲ **49.** ***Geometry*** A window in Brianna and Nelson's house needs to be replaced. The rectangular window is 2 feet 3 inches wide and 3 feet 9 inches tall. Change each of the measurements to inches.

(a) Find the perimeter of the window in inches.

144 inches

(b) If the perimeter needs to be sealed with insulation that is \$0.85 per inch, how much will it cost to insulate the perimeter?

\$122.40

▲ **50.** ***Geometry*** The cellar in Jeff and Shelley's house needs to be sealed along the edge of the concrete floor. The rectangular floor measures 7 yards 2 feet wide and 12 yards 1 foot long. Change each of the measurements to feet.

(a) Find the perimeter of the cellar in feet.

120 feet

(b) If the perimeter (edge) of the cellar floor needs to be sealed with waterproof sealer that costs \$1.75 per foot, how much will it cost to seal the perimeter?

\$210

51. ***Heart Capacity*** Every day, your heart pumps 7200 quarts of blood through your body. How many cups is that?

28,800 cups

52. ***Peach Tree Growth*** A seedling peach tree grew for seven years until it produced its first fruit. How many hours was that if you assume that there are 365 days in a year? 61,320 hours

Estimating and Rounding

53. ***Boating Map*** A local boater's map shows a marker buoy 6 miles out in the ocean from the harbor at Woods Hole. Estimate the number of yards this distance is. (*Hint:* First round the number of yards in one mile to the nearest thousand yards. Then finish the calculation.)

≈12,000 yards

54. ***Plant Growth*** Dr. Russ Camp is examining a new experimental type of wheat in his laboratory at Gordon College. The plant is 618 days old. Estimate the age of the plant in months. (*Hint:* First round 618 to the nearest hundred. Then finish the calculation.)

≈20 months

55. ***Altitude of a Plane*** Greg Salzman is flying from Chicago. The pilot announced that the plane was flying at an altitude of 33,000 feet. The man seated next to Greg asked him to estimate how many miles high the plane was at that point. How should Greg answer the man? (*Hint:* First round the altitude to the nearest ten thousand feet. Then round the conversion equivalent for the number of feet in a mile to the nearest thousand feet. Then perform the calculation.) ≈6 miles

56. ***Plane Flight Time*** Melissa LaBelle is flying from Paris. The pilot announced that the plane had 3170 miles to go to complete the flight. Earlier he had said that the plane was flying at 640 miles per hour. Estimate the number of hours left in the flight. (*Hint:* First round the distance to the nearest thousand miles. Then round the speed to the nearest hundred. Then perform the calculation.) The pilot then announced that it would take 318 minutes to complete the flight. How close was our estimate?

≈5 hours. Our estimate was very close. It was only 18 minutes less than the Pilot's prediction.

To Think About

There are approximately 6080 feet in a nautical mile and 5280 feet in a land mile. On sea and in the air, distance and speed are often measured in nautical miles. The ratio of land miles to nautical miles is approximately 38 to 33.

57. ***Windjammer School*** A windjammer sailboat was used as a floating school. During one school year, the ship traveled 12,800 nautical miles. What would be the equivalent in land miles? Round to the nearest whole mile.

$12{,}800\ \cancel{\text{nautical miles}} \times \dfrac{38\text{ land miles}}{33\ \cancel{\text{nautical miles}}}$

$= 14{,}739\text{ land miles}$

58. ***Merchant Ship Travel*** A merchant ship took some tourists on board for extra income. The ship carried them the equivalent of 850 land miles. How many nautical miles did they travel? Round to the nearest whole mile.

$850\ \cancel{\text{land miles}} \times \dfrac{33\text{ nautical miles}}{38\ \cancel{\text{land miles}}} = 738\text{ nautical miles}$

Cumulative Review

59. ***House Payments*** Melinda and Robert are refinancing their house. They were making payments of \$560 per month for the next 20 years. They obtained a new mortgage for which their payments will be \$515 per month for the next 20 years. Starting with the date they began payments on the new mortgage, how much will they save with the new monthly payments during this entire period? \$10,800

60. ***Computer Disk Storage*** Michael has a computer disk that stores 650 MB of data. He stored three programs that each require 123 MB of storage space. He then stored two programs that each require 69 MB of storage space. What percent of his computer disk is still free for storage of future programs?

22%

61. ***Bicycle Trip*** Johnny and Phil took a bicycle trip covering 115 miles in five days last summer. This summer they hope to take a similar bicycle trip lasting seven days. How many miles are they likely to cover? 161 miles

62. ***Education*** The dean observed that 150 students enrolled in mathematics courses at the University of Hawaii during the last summer semester and that 12 of them dropped the course during the first two weeks. If this ratio has remained consistent over the past seven summers and if 1300 students enrolled in the summer mathematics courses during this time, how many of them dropped the course during the first two weeks of the semester? 104 students

6.2 Metric Measurements: Length

Student Learning Objectives

After studying this section, you will be able to:

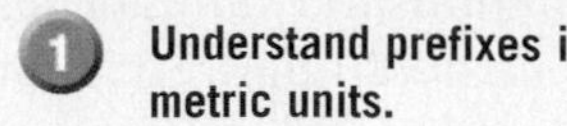
1. **Understand prefixes in metric units.**

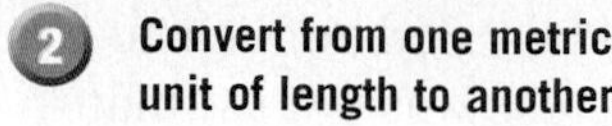
2. **Convert from one metric unit of length to another.**

1 Understanding Prefixes in Metric Units

The **metric system** of measurement is used in most industrialized nations of the world. As we move toward a global economy, it is important to be familiar with the metric system. The metric system is designed for ease in calculating and in converting from one unit to another.

In the metric system, the basic unit of length measurement is the **meter.** A meter is just slightly longer than a yard. To be more precise, the meter is approximately 39.37 inches long.

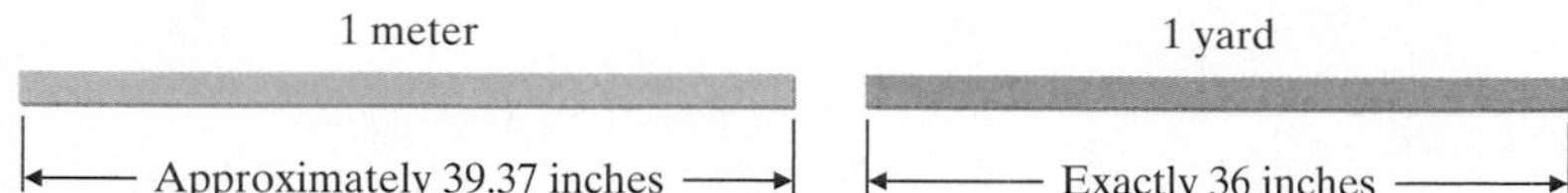

Units that are larger or smaller than the meter are based on the meter and powers of 10. For example, the unit *deka*meter is *ten* meters. The unit *deci*meter is *one-tenth* of a meter. The prefix *deka* means 10. What does the prefix *deci* mean? All the prefixes in the metric system are names for multiples of 10. A list of metric prefixes and their meanings follows.

Prefix	Meaning
kilo-	thousand
hecto-	hundred
deka-	ten
deci-	tenth
centi-	hundredth
milli-	thousandth

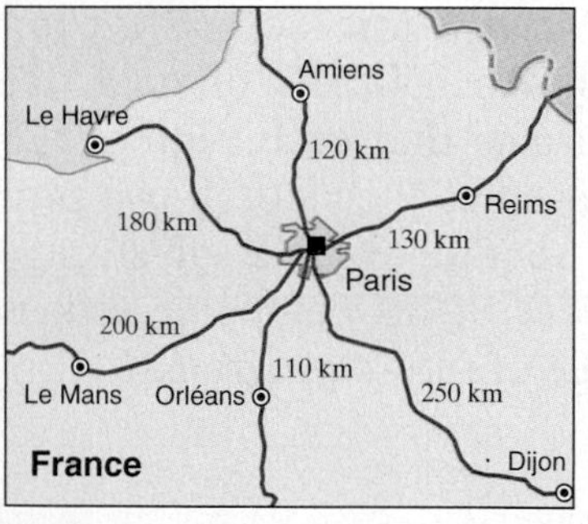

The most commonly used prefixes are *kilo-*, *centi-*, and *milli-*. *Kilo-* means thousand, so a *kilo*meter is a thousand meters. Similarly, *centi-* means one hundredth, so a *centi*meter is one hundredth of a meter. And *milli-* means one thousandth, so a *milli*meter is one thousandth of a meter.

The kilometer is used to measure distances much larger than the meter. How far did you travel in a car? The centimeter is used to measure shorter lengths. What are the dimensions of this textbook? The millimeter is used to measure very small lengths. What is the width of the lead in a pencil?

EXAMPLE 1 Write the prefixes that mean **(a)** thousand and **(b)** tenth.

Solution

(a) The prefix *kilo-* is used for thousand.
(b) The prefix *deci-* is used for tenth.

Practice Problem 1 Write the prefixes that mean **(a)** ten and **(b)** thousandth.

***NOTE TO STUDENT:** Fully worked-out solutions to all of the Practice Problems can be found at the back of the text starting at page SP-1*

2 Converting from One Metric Unit of Length to Another

How do we convert from one metric unit to another? For example, how do we change 5 kilometers into an equivalent number of meters?

Recall from Chapter 3 that when we multiply by 10 we move the decimal point one place to the right. When we divide by 10 we move the decimal point one place to the left. Let's see how we use this idea to change from one metric unit to another.

Changing from Larger Metric Units to Smaller Ones

When you change from one metric prefix to another by moving to the **right** on this prefix chart, move the decimal point to the **right** the same number of places.

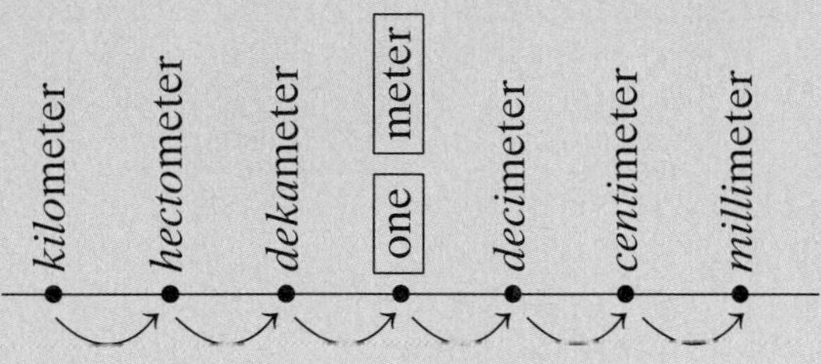

Thus 1 meter = 100 centimeters because we move two places to the right on the chart of prefixes and we also move the decimal point (1.00) two places to the right.

Now let us examine the four most commonly used metric measurements of length and their abbreviations.

COMMONLY USED METRIC LENGTHS

1 kilometer (km) = 1000 meters

1 meter (m) (the basic unit of length in the metric system)

1 centimeter (cm) = 0.01 meter

1 millimeter (mm) = 0.001 meter

Now let us see how we can change a measurement stated in a larger unit to an equivalent measurement stated in smaller units.

EXAMPLE 2

(a) Change 5 kilometers to meters. **(b)** Change 20 meters to centimeters.

Solution

(a) To go from *kilometer* to meter, we move three places to the right on the prefix chart. So we move the decimal point three places to the right.

5 kilometers = 5.000 meters (move three places) = 5000 meters

(b) To go from meter to *centimeter,* we move two places to the right on the prefix chart. Thus we move the decimal point two places to the right.

20 meters = 20.00 centimeters (move two places) = 2000 centimeters

Practice Problem 2

(a) Change 4 meters to centimeters.

(b) Change 30 centimeters to millimeters.

Changing from Smaller Metric Units to Larger Ones

When you change from one metric prefix to another by moving to the **left** on this prefix chart, move the decimal point to the **left** the same number of places.

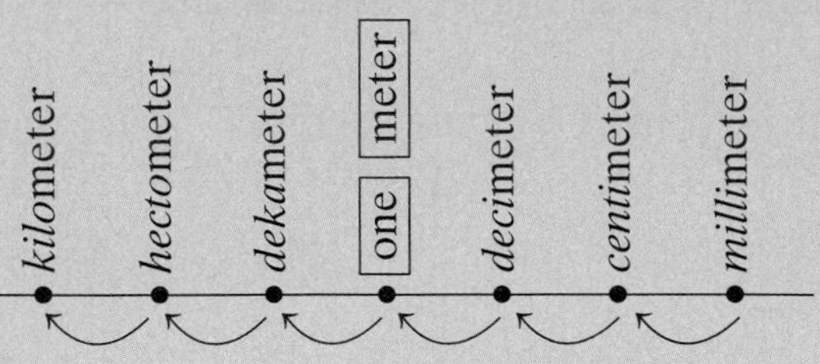

EXAMPLE 3

(a) Change 163 centimeters to meters.
(b) Change 56 millimeters to kilometers.

Solution

(a) To go from *centi*meter to meter, we move *two* places to the left on the prefix chart. Thus we move the decimal point two places to the left.

163 centimeters = 1.63. meters (move two places to the left)
= 1.63 meters

(b) To go from *milli*meter to *kilo*meter, we move six places to the left on the prefix chart. Thus we move the decimal point six places to the left.

56 millimeters = 0.000056. kilometer (move six places to the left)
= 0.000056 kilometer

NOTE TO STUDENT: Fully worked-out solutions to all of the Practice Problems can be found at the back of the text starting at page SP-1

Practice Problem 3

(a) Change 3 *milli*meters to meters.
(b) Change 47 *centi*meters to *kilo*meters.

Thinking Metric Long distances are customarily measured in kilometers. A kilometer is about 0.62 mile. It takes about 1.6 kilometers to make a mile. The following drawing shows the relationship between the kilometer and the mile.

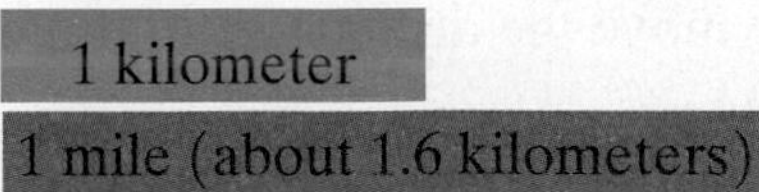

Many small distances are measured in centimeters. A centimeter is about 0.394 inch. It takes 2.54 centimeters to make an inch. You can get a good idea of their size by looking at a ruler marked in both inches and centimeters.

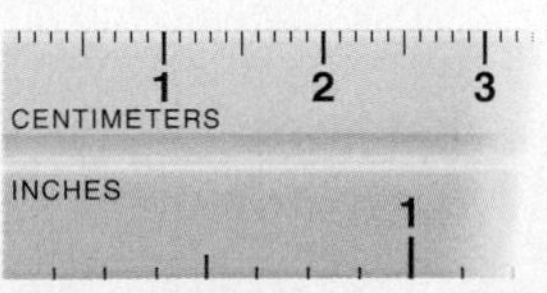

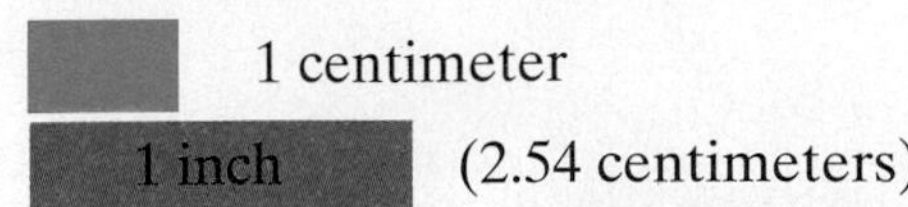

Try to visualize how many centimeters long or wide this book is. It is 21 centimeters wide and 27.5 centimeters long.

A millimeter is a very small unit of measurement, often used in manufacturing.

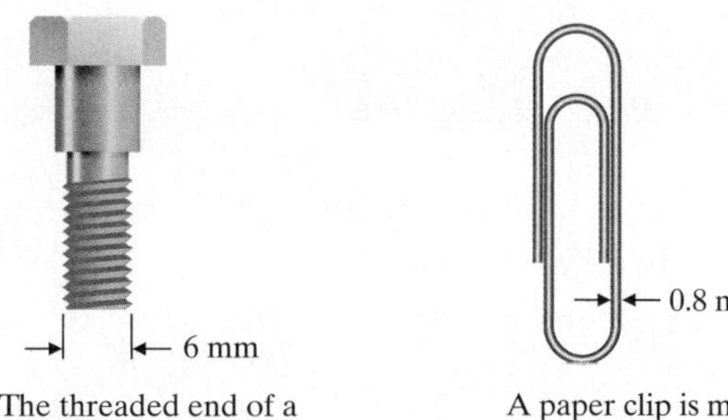

The threaded end of a bolt may be 6 mm wide.

A paper clip is made of wire 0.8 mm thick.

Now that you have an understanding of the size of these metric units, let's try to select the most convenient metric unit for measuring the length of an object.

EXAMPLE 4 Bob measured the width of a doorway in his house. He wrote down "73." What unit of measurement did he use?

(a) 73 kilometers **(b)** 73 meters **(c)** 73 centimeters

Solution The most reasonable choice is **(c)**, 73 centimeters. The other two units would be much too long. A meter is close to a yard, and a doorway would not be 73 yards wide! A kilometer is even larger than a meter.

Practice Problem 4 Joan measured the length of her car. She wrote down 3.8. Which unit of measurement did she use?

(a) 3.8 kilometers **(b)** 3.8 meters **(c)** 3.8 centimeters

Teaching Tip It is important to know the relative size of a meter, a centimeter, and a kilometer. Most students will benefit greatly from doing a few problems like Example 4.

The most frequent metric conversions in length are done between kilometers, meters, centimeters, and millimeters. Their abbreviations are km, m, cm, and mm, and you should be able to use them correctly.

EXAMPLE 5 Convert. **(a)** 982 cm to m **(b)** 5.2 m to mm

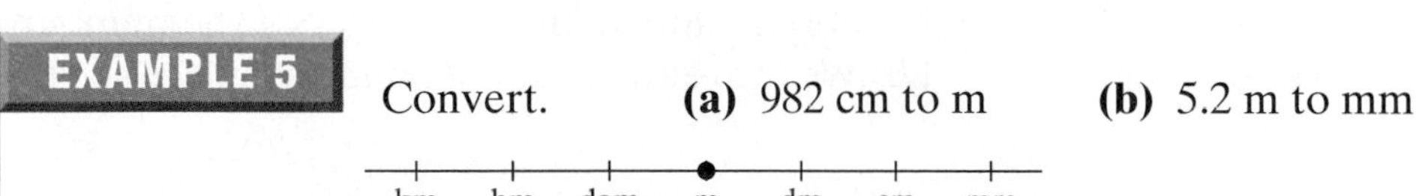

Solution In the first case, we move the decimal point to the left because we are going from a smaller unit to a larger unit.

(a) 982 cm = 9.82 m (two places to left)

= 9.82 m

In the last case, we need to move the decimal point to the right because we are going from a larger unit to a smaller unit.

(b) 5.2 m = 5200. mm (three places to right)

= 5200 mm

Practice Problem 5 Convert. **(a)** 375 cm to m **(b)** 46 m to mm

The other metric units of length are the hectometer, the dekameter, and the decimeter. These are not used very frequently, but it is good to understand how their lengths relate to the basic unit, the meter. A complete list of the metric lengths we have discussed appears in the following table.

Teaching Tip If students complain that they are getting all the meter prefixes mixed up, remind them that the kilometer, meter, and centimeter are the three most important and most frequently used metric measures. Units like decimeters and hectometers are rarely used and need not be memorized.

METRIC LENGTHS WITH ABBREVIATIONS

1 kilometer (km) = 1000 meters
1 hectometer (hm) = 100 meters
1 dekameter (dam) = 10 meters
1 meter (m)
1 decimeter (dm) = 0.1 meter
1 centimeter (cm) = 0.01 meter
1 millimeter (mm) = 0.001 meter

EXAMPLE 6 Convert.

(a) 426 decimeters to kilometers

(b) 9.47 hectometers to meters

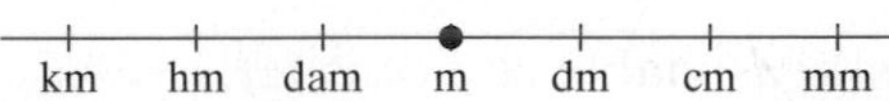

Solution

(a) We are converting from a smaller unit, dm, to a larger one, km. Therefore, there will be fewer kilometers than decimeters. (The number we get will be smaller than 426.) We move the decimal point four places to the left.

$$426 \text{ dm} = 0.0426. \text{ km} \quad \text{(four places to left)}$$
$$= 0.0426 \text{ km}$$

(b) We are converting from a larger unit, hm, to a smaller one, m. Therefore, there will be more meters than hectometers. (The number will be larger than 9.47.) We move the decimal point two places to the right.

$$9.47 \text{ hm} = 9.47. \text{ m} \quad \text{(two places to right)}$$
$$= 947 \text{ m}$$

NOTE TO STUDENT: Fully worked-out solutions to all of the Practice Problems can be found at the back of the text starting at page SP-1

Practice Problem 6 Convert.

(a) 389 millimeters to dekameters

(b) 0.48 hectometer to centimeters

When several metric measurements are to be added, we change them to a convenient common unit.

EXAMPLE 7 Add. 125 m + 1.8 km + 793 m

Solution First we change the kilometer measurement to a measurement in meters.

$$1.8 \text{ km} = 1800 \text{ m}$$

Then we add.

$$\begin{array}{r} 125 \text{ m} \\ 1800 \text{ m} \\ +\ 793 \text{ m} \\ \hline 2718 \text{ m} \end{array}$$

Practice Problem 7 Add. 782 cm + 2 m + 537 m

SIDELIGHT: Extremely Large Metric Distances

Is the biggest length in the metric system a kilometer? Is the smallest length a millimeter? No. The system extends to very large units and very small ones. Usually, only scientists use these units.

1 gigameter = 1,000,000,000 meters
1 megameter = 1,000,000 meters
1 kilometer = 1000 meters
1 meter
1 millimeter = 0.001 meter
1 micrometer = 0.000001 meter
1 nanometer = 0.000000001 meter

Teaching Tip The gigameter is useful in astronomy. Astronomers often refer to a basic measuring length of 1 astronomical unit (AU). The length of 1 astronomical unit is 150 gigameters.

For example, 26 megameters equals a length of 26,000,000 meters. A length of 31 micrometers equals a length of 0.000031 meter.

SIDELIGHT: Metric Measurements for Computers

A **byte** is the amount of computer memory needed to store one alphanumeric character. When referring to computers you may hear the following words: **kilobytes, megabytes,** and **gigabytes.** The following chart may help you.

Buy a new computer.
New 50 GB
Hard Drive
will store
all your data.

1 gigabyte (GB) = one billion bytes = 1,000,000,000 bytes
1 megabyte (MB) = one million bytes* = 1,000,000 bytes
1 kilobyte (KB) = one thousand bytes† = 1,000 bytes

TO THINK ABOUT: Converting Computer Measurements Before we move on to the exercises, see if you can use the large metric distance and use computer memory size to convert the following measurements.

1. 18 megameters = 18,000 kilometers
2. 26 millimeters = 26,000 micrometers
3. 17 nanometers = 0.000017 millimeter
4. 38 meters = 0.000038 megameter
5. 1.2 gigabytes = 1,200,000,000 bytes
6. 528 megabytes = 528,000,000 bytes
7. 78.9 kilobytes = 78,900 bytes
8. 24.9 gigabytes = 24,900,000,000 bytes

*Sometimes in computer science 1 megabyte is considered to be 1,048,576 bytes.
†Sometimes in computer science 1 kilobyte is considered to be 1024 bytes.

6.2 Exercises

Student Solutions Manual | CD/Video | PH Math Tutor Center | MathXL®Tutorials on CD | MathXL® | MyMathLab® | Interactmath.com

Verbal and Writing Skills

Write the prefix for each of the following.

1. Hundred hecto-

2. Hundredth centi-

3. Tenth deci-

4. Thousandth milli-

5. Thousand kilo-

6. Ten deka-

The following conversions involve metric units that are very commonly used. You should be able to perform each conversion without any notes and without consulting your text.

7. 46 centimeters = __460__ millimeters

8. 79 centimeters = __790__ millimeters

9. 2.61 kilometers = __2610__ meters

10. 8.3 kilometers = __8300__ meters

11. 1670 millimeters = __1.67__ meters

12. 2540 millimeters = __2.54__ meters

13. 7.32 centimeters = __0.0732__ meters

14. 9.14 centimeters = __0.0914__ meters

15. 2 kilometers = __200,000__ centimeters

16. 7 kilometers = __700,000__ centimeters

17. 78,000 millimeters = __0.078__ kilometer

18. 840 millimeters = __0.00084__ kilometer

Abbreviations are used in the following. Fill in the blanks with the correct values.

19. 35 mm = __3.5__ cm = __0.035__ m

20. 6300 mm = __630__ cm = __6.3__ m

21. 4.5 km = __4500__ m = __450,000__ cm

22. 6.8 km = __6800__ m = __680,000__ cm

Applications

23. ***Driving in Spain*** Amanda is driving in Spain and sees a road sign which gives the distance to the next city. Choose the most reasonable measurement.
(a) 24 m **(b)** 24 km **(c)** 24 cm (b)

24. ***College ID Card*** Wally measured the length of his college identification card. Choose the most reasonable measurement.
(a) 10 km **(b)** 10 m **(c)** 10 cm (c)

25. ***Compact Disc Case*** Eddie measured the width of a compact disc case. Choose the most appropriate measurement.
(a) 80.5 km **(b)** 80.5 m **(c)** 80.5 mm (c)

26. ***Botanical Gardens*** The Botanical Gardens are located in the center of the city. If it takes approximately 10 minutes to walk directly from the east entrance to the west entrance, which would be the most likely measurement of the width of the Gardens?
(a) 1.4 km **(b)** 1.4 m **(c)** 1.4 mm (a)

27. ***Book Cover*** Joan measured her math book so she could buy the right size book cover. Choose the most appropriate measurement for the length.
(a) 32 mm **(b)** 32 cm **(c)** 32 km (b)

28. ***Jewelry Box Construction*** Rich is making a jewelry box and must use very small screws. Which would be the most likely measurement of the screws?
(a) 8 cm **(b)** 8 m **(c)** 8 mm (c)

29. ***Street Length*** Brenda and Stanley live on Brookwood Street. It is a dead end street with about 12 houses on it. Which would be the most likely measurement of the length of the street?
(a) 0.5 km **(b)** 0.5 m **(c)** 0.5 cm (a)

30. ***Bookcase*** Gabe has a bookcase for all of his college textbooks. It is fairly full but one shelf has room for about 5 or 6 more textbooks. Which would be the most likely measurement of the available space on his bookcase?
(a) 32 km **(b)** 32 cm **(c)** 32 mm (b)

31. ***House Insulation*** Bob and Debbie found that one corner of their townhouse is chilly in the winter. Bob examined the wall and found one portion of the wall does not have insulation. He needed to install insulation between the wall board and the outer wall. What would be the most likely measurement of the thickness of the insulation?
(a) 11.9 mm **(b)** 11.9 cm **(c)** 11.9 m (b)

32. ***Baseball*** John and Stephanie got excellent tickets to see a Boston Red Sox game at Fenway Park. Which would be the most likely measurement of the distance from their seats to the pitcher's mound on the baseball field?
(a) 320 cm **(b)** 320 km **(c)** 320 m (c)

The following conversions involve metric units that are not used extensively. You should be able to perform each conversion, but it is not necessary to do it from memory.

33. 390 decimeters = __39__ meters

34. 270 decimeters = __27__ meters

35. 800 dekameters = __8000__ meters

36. 530 dekameters = __5300__ meters

37. 48.2 meters = __0.482__ hectometer

38. 435 hectometers = __43.5__ kilometers

Change to a convenient unit of measure and add.

39. 243 m + 2.7 km + 312 m
243 m + 2700 m + 312 m = 3255 m

40. 845 m + 5.79 km + 701 m
845 m + 5790 m + 701 m = 7336 m

41. 5.2 cm + 361 cm + 968 mm
5.2 cm + 361 cm + 96.8 cm = 463 cm

42. 4.8 cm + 607 cm + 834 mm
4.8 cm + 607 cm + 83.4 cm = 695.2 cm

43. 15 mm + 2 dm + 42 cm
1.5 cm + 20 cm + 42 cm = 63.5 cm

44. 8 dm + 21 mm + 38 cm
80 cm + 2.1 cm + 38 cm = 120.1 cm

45. ***Stereo Cabinet*** The outside casing of a stereo cabinet is built of plywood 0.95 centimeter thick attached to plastic 1.35 centimeters thick and a piece of mahogany veneer 2.464 millimeters thick. How thick is the stereo casing?
2.5464 cm or 25.464 mm

46. ***House Construction*** A plywood board is 2.2 centimeters thick. A layer of tar paper is 3.42 millimeters thick. A layer of false brick siding is 2.7 centimeters thick. A house wall consists of these three layers. How thick is the wall?
2.2 cm + 0.342 cm + 2.7 cm = 5.242 cm or 52.42 mm

Mixed Practice

47. 65 cm + 80 mm + 2.5 m
0.65 m + 0.08 m + 2.5 m = 3.23 m

48. 82 m + 471 cm + 0.32 km
82 m + 4.71 m + 320 m = 406.71 m

49. 46 m + 986 cm + 0.884 km
46 m + 9.86 m + 884 m = 939.86 m

50. 56.3 centimeters = ___0.563___ meter

51. 96.4 centimeters = ___0.964___ meter

Write true *or* false *for each statement.*

52. 0.001 kilometer = 1 meter true

53. 1 kilometer = 0.001 meter false

54. 1000 meters = 1 kilometer true

55. 10 millimeters = 1 centimeter true

56. An airport runway might be 2 kilometers long.
true

57. A man might be 2 meters tall.
true

58. A centimeter is larger than an inch. false

59. The glass in a drinking glass is usually about 2 millimeters thick. true

Applications

60. ***Trans-Siberian Train*** The world's longest train run is on the Trans-Siberian line in Russia, from Moscow to Nakhodka, a city on the Sea of Japan. The length of the run, which has 97 stops, measures 94,380,000 centimeters. The run takes 8 days, 4 hours, and 25 minutes to travel.
   (a) How many meters is the run?
   943,800 meters
   (b) How many kilometers is the run?
   943.8 kilometers

61. ***Peruvian Train*** The highest railroad line in the world is a track on the Morococha branch of the Peruvian State Railways at La Cima. The track is 4818 meters high.
   (a) How many centimeters high is the track?
   481,800 centimeters
   (b) How many kilometers high is the track?
   4.818 kilometers

62. ***Dam Construction*** The world's highest dam, the Rogun dam in Tajikistan, is 335 meters high.
   (a) How many kilometers high is the dam?
   0.335 kilometer
   (b) How many centimeters high is the dam?
   33,500 centimeters

63. ***Virus Size*** A typical virus of the human body measures just 0.000000254 centimeter in diameter. How many meters in diameter is a typical virus?
0.00000000254

To Think About

International Fishing *In countries outside the United States, very heavy items are measured in terms of* **metric tons.** *A metric ton is defined as 1000 kilograms (or 2205 pounds). Consider the following bar graph, which records the catch of fish (measured in millions of metric tons) by China for various years.*

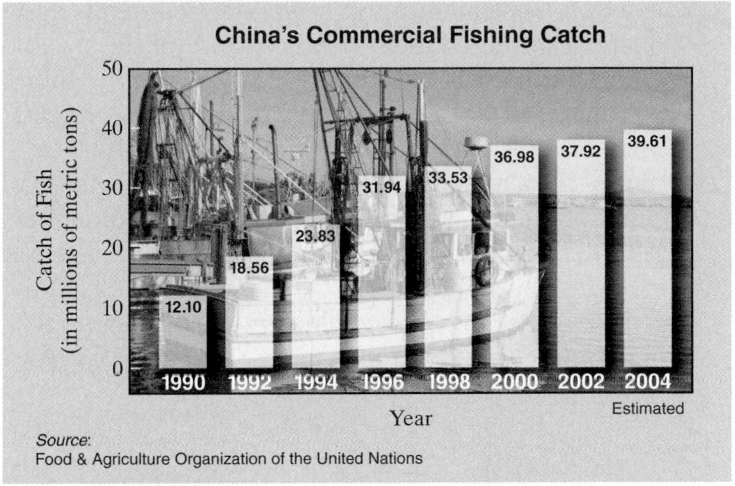

64. By how many metric tons did the commercial catch of fish by China increase from 1990 to 1994?
11,730,000 metric tons

65. By how many metric tons did the commercial catch of fish by China increase from 1996 to 2004?
7,670,000 metric tons

66. How many kilograms of fish were caught by China in 1998?
33,530,000,000 kilograms

67. How many kilograms of fish were caught by China in 1992?
18,560,000,000 kilograms

68. If the same *percent of increase* that occurred from 1994 to 2000 also occurs from 2000 to 2006, how many metric tons of fish will China catch in 2006?
approximately 57,390,000 metric tons

69. If the same *percent of increase* that occurred from 1990 to 2000 also occurs from 2000 to 2010, how many metric tons of fish will China catch in 2010?
approximately 113,020,000 metric tons

Cumulative Review

70. 57% of what number is 2850?
5000

71. Find 0.03% of 5900.
1.77

72. 13 is 25% of what number?
52

73. What is 75% of 20?
15

6.3 METRIC MEASUREMENTS: VOLUME AND WEIGHT

Student Learning Objectives

After studying this section, you will be able to:

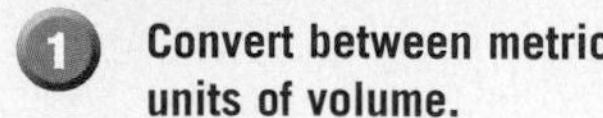

1 **Convert between metric units of volume.**

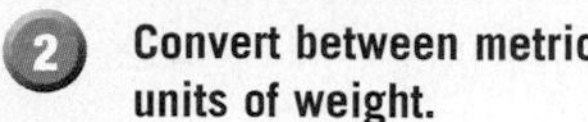

2 **Convert between metric units of weight.**

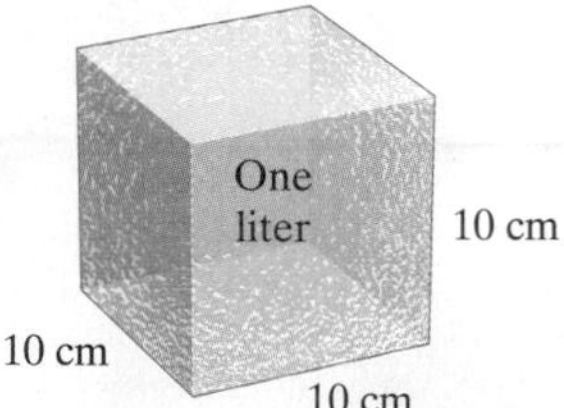

1 liter   2 liters

Teaching Tip It is helpful to emphasize the fact that a liter is just slightly larger than a quart. (1 liter is 1.057 quarts.)

1 Converting Between Metric Units of Volume

As products are distributed worldwide, more and more of them are being sold in metric units. Soft drinks come in 1-, 2-, or 3-liter bottles. Labels often contain these amounts in both American units and metric units. Get used to looking at these labels to gain a sense of the size of metric units of volume.

The basic metric unit for volume is the liter. A **liter** is defined as the volume of a box 10 cm × 10 cm × 10 cm, or 1000 cm^3. A cubic centimeter may be written as cc, so we sometimes see 1000 cc = 1 liter. A liter is slightly larger than a quart; 1 liter of liquid is 1.057 quarts of that liquid.

The most common metric units of volume are the milliliter, the liter, and the kiloliter. Often a capital letter L is used as an abbreviation for *liter*.

COMMON METRIC VOLUME MEASUREMENTS

1 kiloliter (kL) = 1000 liters
1 liter (L)
1 milliliter (mL) = 0.001 liter

We know that 1000 cc = 1 liter. Dividing each side of that equation by 1000, we get 1 cc = 1 mL.

Use this prefix chart as a guide when you change one metric prefix to another. Move the decimal point in the same direction and the same number of places.

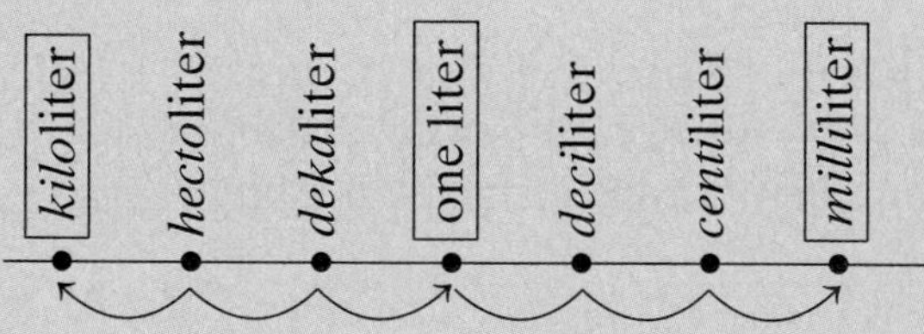

The prefixes for liter follow the pattern we have seen for meter. Note that *kilo-* is three places to the left of the liter and *milli-* is three places to the right. Because the kiloliter, the liter, and the milliliter are the most commonly used units of volume, we will focus exclusively on them in this text.

EXAMPLE 1 Convert.

(a) 3 L = _____ mL **(b)** 24 kL = _____ L **(c)** 0.084 L = _____ mL

Solution

(a) The prefix *milli-* is three places to the right. We move the decimal point three places to the right.

3 L = 3.000 = 3000 mL

(b) 24 kL = 24.000 = 24,000 L

(c) 0.084 L = 0.084 = 84 mL

Practice Problem 1 Convert.

(a) 5 L = _____ mL **(b)** 84 kL = _____ L **(c)** 0.732 L = _____ mL

NOTE TO STUDENT: Fully worked-out solutions to all of the Practice Problems can be found at the back of the text starting at page SP-1

EXAMPLE 2 Convert.

(a) 26.4 mL = ____ L **(b)** 5982 mL = ____ L **(c)** 6.7 L = ____ kL

Solution

(a) The unit L is three places to the left. We move the decimal three places to the left. 26.4 mL = 0.0264 L

(b) 5982 mL = 5.982 L

(c) 6.7 L = 0.0067 kL

Teaching Tip If your students are confused about the standard abbreviations of some of the units, call their attention to the chart of standard abbreviations at the beginning of Section 6.4. Remind them that it is a good idea to learn the standard abbreviations of these units as they go through the chapter.

Practice Problem 2 Convert.

(a) 15.8 mL = _____ L **(b)** 12,340 mL = _____ L

(c) 86.3 L = _____ kL

The cubic centimeter is often used in medicine. Recall 1 mL = 1 cm^3 or 1 cc.

EXAMPLE 3 Convert.

(a) 26 mL = _____ cm^3 **(b)** 0.82 L = _____ cm^3

Solution

(a) A milliliter and a cubic centimeter are equivalent. 26 mL = 26 cm^3

(b) We use the same rule to convert liters to cubic centimeters as we do to convert liters to milliliters. 0.82 L = 820 cm^3

Teaching Tip Students may wonder why there are two names for the same thing. Explain to them that the standard unit is the milliliter (mL) and that this is the one they are expected to learn. The cubic centimeter is used in nursing and a few other technical areas.

Practice Problem 3 Convert.

(a) 396 mL = _____ cm^3 **(b)** 0.096 L = _____ cm^3

2 Converting Between Metric Units of Weight

In a science class we make a distinction between weight and **mass.** Mass is the amount of material in an object. Weight is a measure of the pull of gravity on an object. The farther you are from the center of the earth, the less you weigh. If you were in an astronaut suit floating in outer space you would be weightless. The mass of your body, however, would not change. Throughout this book we will refer to *weight* only. The technical difference between weight and mass is not something we need to pay attention to in everyday life.

In the metric system the basic unit of weight is the gram. A **gram** is the weight of the water in a box that is 1 centimeter on each side. To get an idea of how small a gram is, we note that two small paper clips weigh about 1 gram. A gram is only about 0.035 ounce.

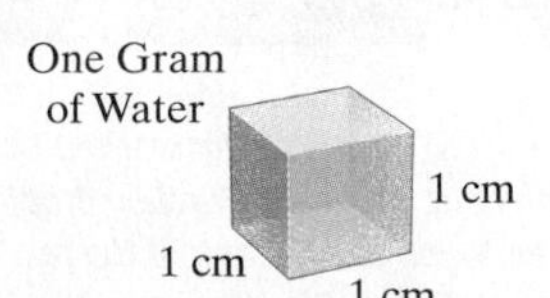

One kilogram is 1000 times larger than a gram. A kilogram weighs about 2.2 pounds. Some of the measures of weight in the metric system are shown in the following chart.

COMMON METRIC WEIGHT MEASUREMENTS

1 metric ton (t) = 1,000,000 grams
1 kilogram (kg) = 1000 grams
1 gram (g)
1 milligram (mg) = 0.001 gram

Use this prefix chart as a guide when you change one metric prefix to another. Move the decimal point in the same direction and the same number of places.

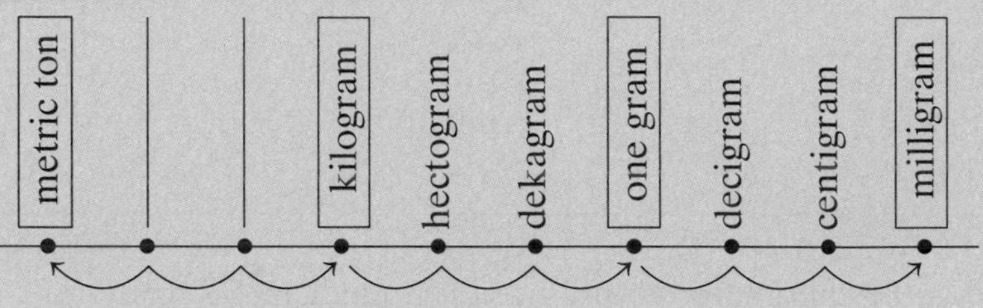

We will focus exclusively on the metric ton, kilogram, gram, and milligram. We convert weight measurements the same way we convert volume and length measurements. The metric ton is also called a **megagram.** *Mega* means "million" so a megagram is 1,000,000 grams.

EXAMPLE 4 Convert.

(a) 2 t = _____ kg **(b)** 0.42 kg = _____ g

Solution **(a)** 2 t = 2000 kg **(b)** 0.42 kg = 420 g

Practice Problem 4 Convert.

(a) 3.2 t = _____ kg **(b)** 7.08 kg = _____ g

A kilogram is slightly more than 2 pounds. A 3-week-old miniature pinscher weighs about 1 kilogram.

EXAMPLE 5 Convert.

(a) 283 kg = _____ t **(b)** 7.98 mg = _____ g

Solution **(a)** 283 kg = 0.283 t **(b)** 7.98 mg = 0.00798 g

Practice Problem 5 Convert.

(a) 59 kg = _____ t **(b)** 28.3 mg = _____ g

***NOTE TO STUDENT:** Fully worked-out solutions to all of the Practice Problems can be found at the back of the text starting at page SP-1*

EXAMPLE 6 If a chemical costs $0.03 per gram, what will it cost per kilogram?

Solution Since there are 1000 grams in a kilogram, a chemical that costs $0.03 per gram would cost 1000 times as much per kilogram.

$$1000 \times \$0.03 = \$30.00$$

The chemical would cost $30.00 per kilogram.

Practice Problem 6 If coffee costs $10.00 per kilogram, what will it cost per gram?

EXAMPLE 7 Select the most reasonable weight for Tammy's Toyota.

(a) 820 t **(b)** 820 g **(c)** 820 kg **(d)** 820 mg

Solution The most reasonable answer is **(c)** 820 kg. The other weight values are much too large or much too small. Since a kilogram is slightly more than 2 pounds, we see that this weight, 820 kg, most closely approximates the weight of a car.

Practice Problem 7 Select the most reasonable weight for Hank, starting linebacker for the college football team.

(a) 120 kg **(b)** 120 g **(c)** 120 mg

SIDELIGHT: Using Very Small Units

When dealing with very small particles or atomic elements, scientists sometimes use units smaller than a gram.

SMALL WEIGHT MEASUREMENTS

$$\begin{aligned} 1 \text{ milligram} &= 0.001 \text{ gram} \\ 1 \text{ microgram} &= 0.000001 \text{ gram} \\ 1 \text{ nanogram} &= 0.000000001 \text{ gram} \\ 1 \text{ picogram} &= 0.000000000001 \text{ gram} \end{aligned}$$

We could make the following conversions.

$$\begin{aligned} 2.6 \text{ picograms} &= 0.0026 \text{ nanogram} \\ 29.7 \text{ micrograms} &= 0.0297 \text{ milligram} \\ 58 \text{ nanograms} &= 58{,}000 \text{ picograms} \\ 58 \text{ nanograms} &= 0.058 \text{ microgram} \end{aligned}$$

See Exercises 6.3, exercises 63 and 64 for such conversion problems.

6.3 EXERCISES

Student Solutions Manual | CD/Video | PH Math Tutor Center | MathXL®Tutorials on CD | MathXL® | MyMathLab® | Interactmath.com

Verbal and Writing Skills

Write the metric unit that best represents each measurement.

1. one thousand liters
1 kL

2. one thousandth of a liter
1 mL

3. one thousandth of a gram
1 mg

4. one thousand grams
1 kg

5. one thousandth of a kilogram
1 g

6. one thousand milliliters
1 L

Perform each conversion.

7. 9 kL = 9000 L

8. 5 kL = 5000 L

9. 12 L = 12,000 mL

10. 25 L = 25,000 mL

11. 18.9 mL = 0.0189 L

12. 31.5 mL = 0.0315 L

13. 752 L = 0.752 kL

14. 493 L = 0.493 kL

15. 2.43 kL = 2,430,000 mL

16. 14.3 kL = 14,300,000 mL

17. 82 mL = 82 cm^3

18. 152 mL = 152 cm^3

19. 5261 mL = 0.005261 kL

20. 28,156 mL = 0.028156 kL

21. 74 L = 74,000 cm^3

22. 122 L = 122,000 cm^3

23. 216 g = 0.216 kg

24. 2940 g = 2.94 kg

25. 35 mg = 0.035 g

26. 13 mg = 0.013 g

27. 6328 mg = 6.328 g

28. 986 mg = 0.986 g

29. 2.92 kg = 2920 g

30. 14.6 kg = 14,600 g

31. 2.4 t = 2400 kg

32. 9500 kg = 9.5 t

Fill in the blanks with the correct values.

33. 7 mL = 0.007 L = 0.000007 kL

34. 18 mL = 0.018 L = 0.000018 kL

35. 128 cm^3 = 0.128 L = 0.000128 kL

36. 0.315 kL = 315 L = 315,000 cm^3

37. 0.033 kg = 33 g = 33,000 mg

38. 0.098 kg = 98 g = 98,000 mg

39. 2.58 metric tons = 2580 kg = 2,580,000 g

40. 7183 g = 7.183 kg = 0.007183 t

41. ***Jar Size*** Alice bought a jar of apple juice at the store. Choose the most reasonable measurement for its contents.
(a) 0.32 kL **(b)** 0.32 L **(c)** 0.32 mL (b)

42. ***Injection Dosage*** A nurse gave an injection of insulin to a diabetic patient. Choose the most reasonable measurement for the dose.
(a) 4 kL **(b)** 4 L **(c)** 4 mL (c)

43. ***Oil Tanker*** A ship owner purchased a new oil tanker. Choose the most reasonable measurement for its weight.
(a) 4000 t **(b)** 4000 kg **(c)** 4000 g (a)

44. ***College Textbooks*** Robert bought a new psychology textbook. Choose the most reasonable measurement for its weight.
(a) 0.49 t **(b)** 0.49 kg **(c)** 0.49 g (b)

Find a convenient unit of measure and add.

45. 83 L + 822 mL + 30.1 L
83 L + 0.822 L + 30.1 L = 113.922 L or 113,922 mL

46. 152 L + 473 mL + 77.3 L
152 L + 0.473 L + 77.3 L = 229.773 L or 229,773 mL

47. 20 g + 52 mg + 1.5 kg
20 g + 0.052 g + 1500 g = 1520.052 g or 1,520,052 mg

48. 2 kg + 42 mg + 120 g
2000 g + 0.042 g + 120 g = 2120.042 g or 2,120,042 mg

Mixed Practice

Write true *or* false *for each statement.*

49. 1 milliliter = 0.001 liter true

50. 1 kiloliter = 1000 liters true

51. Milk can be purchased in kiloliter jugs at the food store. false

52. Small amounts of medicine are often measured in liters. false

53. A reasonable weight for an adult man is 500,000 grams. false

54. A bottle of soda might contain 1000 mL.
true

55. A nickel weighs about 5 grams. true

56. A convenient size for a family purchase is 2 kilograms of ground beef. true

Applications

57. ***Food Costs*** Starbucks sells 453-gram bags of ground coffee for $8.99. Rhonda bought 3.624 kg. How much did she spend on coffee?
$71.92

58. ***Food Cost*** Randy's Premier Pizza needs to order tomato sauce from the distributor. The sauce comes in 4000-gram jars for $6.80. If Randy orders 56 kg, how much will he pay for the tomato sauce? $95.20

59. ***Biogenetic Research*** A very rare essence of an almost extinct flower found in the Amazon jungle of South America is extracted by a biogenetic company trying to copy and synthesize it. The company estimates that if the procedure is successful, the product will cost the company $850 per milliliter to produce. How much will it cost the company to produce 0.4 liter of the engineered essence? $340,000

60. ***Price of Gold*** In a recent year the price of gold was $9.40 per gram. At that price how much would a kilogram of gold cost?
$9400

61. ***Rocket Cargo Cost*** A government rocket is carrying cargo into space for a private company. The charge for the freight shipment is $22,450 per kilogram of the special cargo, and the cargo weighs 0.45 metric ton. How much will the private company have to pay?
$10,102,500

62. ***Coal Mining*** The individual coal-cutting record is 45.4 metric tons per person in one shift (six hours), by five Soviet miners. If the coal is sold for $2.00 per kilogram, what is the value of the coal?
$90,800

Convert.

63. 5632 picograms = 0.005632 micrograms

64. 0.076182 milligram = 76,182 nanograms

To Think About

Platinum Mining *The following chart shows the production in weight and value of platinum in the United States for various years. Use this chart to answer the following questions. Round to the nearest dollar when necessary.*

Production of Platinum Metal in the U.S.

Year	1990	1995	1997	1998	2000	2004
Production in kg	1810	1590	2610	3500	3860	4100*
Value in millions	$27	$21	$24	$33	$36	$38*

Source: U.S. Geological Survey Annual Reports

*estimated

65. How many more kilograms were produced in 2004 than in 1990? 2290 kg

66. How many fewer kilograms were produced in 1995 than in 1990? 220 kg

67. What was the value per kilogram of the platinum mined in 1997? What was the value per gram? approximately $9195.40; $9.20

68. What was the value per kilogram of the platinum mined in 2000? What was the value per gram? approximately $9326.42; $9.33

69. In what year was the value per kilogram of the platinum mined the lowest? 1997

70. In what year was the value per kilogram of the platinum mined the highest? 1990

Cumulative Review

71. 14 out of 70 is what percent? 20%

72. What is 23% of 250? 57.5

73. ***Home Theater Costs*** Marilyn bought a new home theater system for $4800. After a 10% discount was taken, a 5% sales tax was added. What did Marilyn pay for the theater system?
$4536

74. ***Commission Sales*** A salesperson earns a commission of 8%. She sold furniture worth $8960. How much commission did she earn?
$716.80

How Am I Doing? Sections 6.1–6.3

How are you doing with your homework assignments in Sections 6.1 to 6.3? Do you feel you have mastered the material so far? Do you understand the concepts you have covered? Before you go further in the textbook, take some time to do each of the following problems.

6.1 *Convert. When necessary, express your answer as a decimal rounded to the nearest hundredth.*

1. 52 in = ___ ft **2.** 24 qt = ___ gal **3.** 3 mi = ___ yd

4. 6400 lb = ___ tons **5.** 22 min = ___ sec **6.** 5 gal = ___ pt

7. Sam purchased some salmon steaks for an outdoor grill tonight. He picks up two salmon steaks and weighs them. They weigh 24 ounces. The purchase price on the display rack at the store as $5.50 per pound. How much will this purchase cost when Sam gets to the checkout counter?

6.2 *Perform each conversion.*

8. 6.75 km = ___ m **9.** 73.9 m = ___ cm **10.** 986 mm = ___ cm

11. 27 mm = ___ m **12.** 5296 mm = ___ cm **13.** 482 m = ___ km

Convert to meters and add.

14. 1.2 km + 192 m + 984 m **15.** 3862 cm + 9342 mm + 46.3 m

16. The wire that connects a motor to a control panel in Tokyo is 3.4 meters long. One portion of the wire measuring 78 centimeters has double insulation. Another portion of the wire measuring 128 centimeters has triple insulation. The remaining part of the wire has single insulation. How long is the portion of the wire that has single insulation?

6.3 *Perform each conversion.*

17. 5.66 L = ___ mL **18.** 7835 g = ___ kg **19.** 56.3 kg = ___ t

20. 4.8 kL = ___ L **21.** 568 mg = ___ g **22.** 8.9 L = ___ cm^3

23. A wholesaler sells peanut butter in canisters of 5000 g for $7.75. The Boston Rescue Mission needs to buy 75 kg to restock its pantry. How much will the mission spend on peanut butter if they buy from this wholesaler?

24. Old Italy Foods receives a special shipment of olive oil in 5000-gram cans. Old Italy Foods sells these cans for $32.50. If Ricardo's Restaurant needs 35 kg of olive oil and purchases it from Old Italy Foods, how much will it cost?

25. An antibacterial spray costs $36 per liter to produce. The Smiths purchased 200 milliliters of the spray. They know the owner of the company that produces the spray and he sold it to them at cost. How much did they pay for 200 milliliters?

26. Last year gold cost $11.20 per gram. Samuel purchased 0.5 kilogram of gold at that price? How much did he pay?

Now turn to page SA-12 for the answer to each of these problems. Each answer also includes a reference to the objective in which the problem is first taught. If you missed any of these problems, you should stop and review the Examples and Practice Problems in the referenced objective. A little review now will help you master the material in the upcoming sections of the test.

1. 4.33
2. 6
3. 5280
4. 3.2
5. 1320
6. 40
7. $8.25
8. 6750
9. 7390
10. 98.6
11. 0.027
12. 529.6
13. 0.482
14. 2376 m
15. 94.262 m
16. 1.34 meters or 134 centimeters
17. 5660
18. 7.835
19. 0.0563
20. 4800
21. 0.568
22. 8900
23. $116.25
24. $227.50
25. $7.20
26. $5600

6.4 Converting Units

Student Learning Objectives

After studying this section, you will be able to:

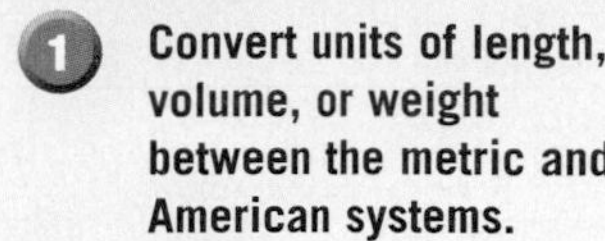

Convert units of length, volume, or weight between the metric and American systems.

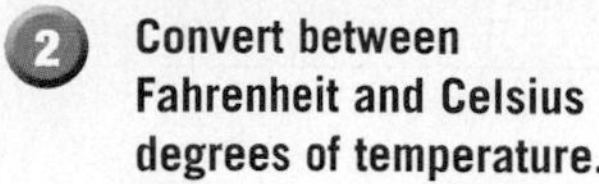

Convert between Fahrenheit and Celsius degrees of temperature.

Teaching Tip College math faculty do not agree on which, if any, of the topics of this section should be required in this course. In many schools the computer science, business and physical science departments assume that the math department is teaching this material. The majority of math faculty seem to feel it is better to introduce the topic but not require the memorization of all the equivalent measures.

1 Converting Units of Length, Volume, or Weight Between the Metric and American Systems

So far we've seen how to convert units when working *within* either the American or the metric system. Many people, however, work in *both* the metric and the American systems. If you study such fields as chemistry, electromechanical technology, business, X-ray technology, nursing, or computers, you will probably need to convert measurements between the two systems.

To convert between American units and metric units, it is helpful to have equivalent values. The most commonly used equivalents are listed in the following table. Most of these are approximate.

Equivalent Measures

	American to Metric	Metric to American
Units of length	1 mile ≈ 1.61 kilometers 1 yard ≈ 0.914 meter 1 foot ≈ 0.305 meter 1 inch = 2.54 centimeters*	1 kilometer ≈ 0.62 mile 1 meter ≈ 1.09 yards 1 meter ≈ 3.28 feet 1 centimeter ≈ 0.394 inch
Units of volume	1 gallon ≈ 3.79 liter 1 quart ≈ 0.946 liter	1 liter ≈ 0.264 gallon 1 liter ≈ 1.06 quarts
Units of weight	1 pound ≈ 0.454 kilogram 1 ounce ≈ 28.35 gram	1 kilogram ≈ 2.2 pounds 1 gram ≈ 0.0353 ounce

*exact value

Remember that to convert from one unit to another you multiply by a fraction that is equivalent to 1. Create a fraction from the equivalent measures table so that the unit in the denominator cancels the unit you are changing.

To change 5 miles to kilometers, we look in the table and find that 1 mile ≈ 1.61 kilometers. We will use the unit fraction

$$\frac{1.61 \text{ kilometers}}{1 \text{ mile}}$$

because we want to have miles in the denominator.

$$5 \cancel{\text{miles}} \times \frac{1.61 \text{ kilometers}}{1 \cancel{\text{mile}}} = 5 \times 1.61 \text{ kilometers} = 8.05 \text{ kilometers}$$

Thus 5 miles ≈ 8.05 kilometers.

Is this the only way to do the problem? No. To make the previous conversion, we also could have used the relationship that 1 kilometer ≈ 0.62 mile. Again, we want to have miles in the denominator, so we use

$$\frac{1 \text{ kilometer}}{0.62 \text{ mile}}$$

$$5 \cancel{\text{miles}} \times \frac{1 \text{ kilometers}}{0.62 \cancel{\text{mile}}} = \frac{5}{0.62} \approx 8.06 \text{ kilometers.}$$

Using this approach, we find that 5 miles ≈ 8.06 kilometers. This is not the same result we obtained before. The discrepancy between the results of the two conversions has occurred because the numbers given in the equivalent measures tables are approximations.

Often we have to make conversions in order to make comparisons. For example, is the 6 inches of attic insulation commonly used in the northeastern United States more or less than the 16 centimeters of attic insulation commonly used in Sweden? In order to find out we would have to convert 6 inches to centimeters. What unit fraction would we use? What result would we get?

EXAMPLE 1 Convert 3 feet to meters.

Solution

$$3\ \cancel{\text{feet}} \times \frac{0.305\text{ meter}}{1\ \cancel{\text{foot}}} = 0.915\text{ meter}$$

Practice Problem 1 Convert 7 feet to meters.

NOTE TO STUDENT: Fully worked-out solutions to all of the Practice Problems can be found at the back of the text starting at page SP-1

Unit abbreviations are quite common, so we will use them for the remainder of this section. We list them here for your reference.

American Measure (Alphabetical Order)	Standard Abbreviation
feet	ft
gallon	gal
inch	in.
mile	mi
ounce	oz
pound	lb
quart	qt
yard	yd

Metric Measure	Standard Abbreviation
centimeter	cm
gram	g
kilogram	kg
kilometer	km
liter	L
meter	m
millimeter	mm

EXAMPLE 2

(a) Convert 26 m to yd.
(b) Convert 1.9 km to mi.
(c) Convert 14 gal to L.
(d) Convert 2.5 L to qt.

Solution

(a) $26\ \cancel{\text{m}} \times \frac{1.09\text{ yd}}{1\ \cancel{\text{m}}} = 28.34\text{ yd}$

(b) $1.9\ \cancel{\text{km}} \times \frac{0.62\text{ mi}}{1\ \cancel{\text{km}}} = 1.178\text{ mi}$

(c) $14\ \cancel{\text{gal}} \times \frac{3.79\text{ L}}{1\ \cancel{\text{gal}}} = 53.06\text{ L}$

(d) $2.5\ \cancel{\text{L}} \times \frac{1.06\text{ qt}}{1\ \cancel{\text{L}}} = 2.65\text{ qt}$

Practice Problem 2

(a) Convert 17 m to yd.
(b) Convert 29.6 km to mi.
(c) Convert 26 gal to L.
(d) Convert 6.2 L to qt.

Some conversions require more than one step.

EXAMPLE 3 Convert 235 cm to ft. Round to the nearest hundredth of a foot.

Solution Our first fraction converts centimeters to inches. Our second fraction converts inches to feet.

$$235\ \cancel{\text{cm}} \times \frac{0.394\ \cancel{\text{in.}}}{1\ \cancel{\text{cm}}} \times \frac{1\text{ ft}}{12\ \cancel{\text{in.}}} = \frac{92.59}{12}\text{ ft}$$

$$\approx 7.72\text{ ft (rounded to the nearest hundredth)}$$

NOTE TO STUDENT: Fully worked-out solutions to all of the Practice Problems can be found at the back of the text starting at page SP-1

Practice Problem 3 Convert 180 cm to ft.

The same rules can be followed for a rate such as 50 miles per hour.

If the unit to be converted is in the numerator, be sure to express that unit in the denominator of the unit fraction. If the unit to be converted is in the denominator, be sure to express that unit in the numerator of the unit fraction.

Teaching Tip Remind students that changing rates from feet per second to miles per hour or miles per hour to kilometers per hour does not require a new skill. It is merely using the "multiply by a unit fraction" method that we have been using throughout the chapter.

EXAMPLE 4 Convert 100 km/hr to mi/hr.

Solution We need to multiply by a unit fraction. The fraction we multiply by must have kilometers in the denominator.

$$\frac{100\ \cancel{\text{km}}}{\text{hr}} \times \frac{0.62\text{ mi}}{1\ \cancel{\text{km}}} = 62\text{ mi/hr}$$

Thus 100 km/hr is approximately equal to 62 mi/hr.

Practice Problem 4 Convert 88 km/hr to mi/hr.

Sometimes we need more than one unit fraction to make the conversion of two rates. We will see how this is accomplished in Example 5.

EXAMPLE 5 A rocket carrying a communication satellite is launched from a rocket launch pad. It is traveling at 700 miles per hour. How many feet per second is the rocket traveling? Round to the nearest whole number.

Solution

$$\frac{700\ \cancel{\text{miles}}}{\cancel{\text{hr}}} \times \frac{5280\text{ ft}}{1\ \cancel{\text{mile}}} \times \frac{1\ \cancel{\text{hr}}}{60\ \cancel{\text{min}}} \times \frac{1\ \cancel{\text{min}}}{60\text{ sec}}$$

$$= \frac{700 \times 5280\text{ ft}}{60 \times 60\text{ sec}} = \frac{3{,}696{,}000\text{ ft}}{3600\text{ sec}} \approx 1027\text{ ft/sec}$$

The missile is traveling at 1027 feet per second.

Practice Problem 5 A Concorde jet in 2003 in a flight from Boston to Paris flew at 900 miles per hour. What was the speed of the Concorde jet in feet per second?

▲ SIDELIGHT: From Square Yards to Square Meters

Suppose we consider a rectangle that measures 2 yards wide by 4 yards long. The area would be 2 yards × 4 yards = 8 square yards. How could you change 8 square yards to square meters? Suppose that we look at 1 square yard. Each side is 1 yard long, which is equivalent to 0.9144 meter.

$$\text{Area} = 1 \text{ yard} \times 1 \text{ yard} \approx 0.9144 \text{ meter} \times 0.9144 \text{ meter}$$

$$\text{Area} = 1 \text{ square yard} \approx 0.8361 \text{ square meter}$$

Thus $1 \text{ yd}^2 \approx 0.8361 \text{ m}^2$. Therefore

$$8 \text{ yd}^2 \times \frac{0.8361 \text{ m}^2}{1 \text{ yd}^2} = 6.6888 \text{ m}^2.$$

8 square yards ≈ 6.6888 square meters.

Teaching Tip Students often have trouble with square units. They wonder why 1 square yard has 9 square feet instead of 3 square feet. Have them study the figure below to see the comparison. The problem is compounded with metric units. The key to understanding this topic is to always draw a sketch to illustrate the comparison. Encourage students to do this when they get to Exercises 6.4, problems 67 and 68. For classroom activity, ask students how many square centimeters are in one square meter. (Answer is 10.000 square centimeters.). Usually, every student who draws a quick sketch as he or she thinks, gets the right answer to this classroom problem.

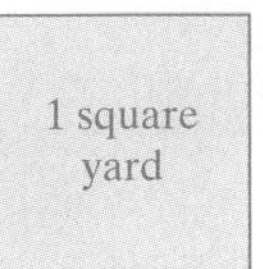

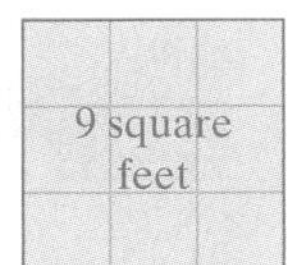

2 Converting Between Fahrenheit and Celsius Degrees of Temperature

In the metric system, temperature is measured on the **Celsius scale.** Water boils at 100° (100°C) and freezes at 0°(0°C) on the Celsius scale. In the **Fahrenheit system,** water boils at 212°(212°F) and freezes at 32°(32°F).

To convert Celsius to Fahrenheit, we can use the formula

$$F = 1.8 \times C + 32,$$

where C is the number of Celsius degrees and F is the number of Fahrenheit degrees.

EXAMPLE 6 When the temperature is 35°C, what is the Fahrenheit reading?

Solution

$$\begin{aligned} F &= 1.8 \times C + 32 \\ &= 1.8 \times 35 + 32 \\ &= 63 + 32 \\ &= 95 \end{aligned}$$

The temperature is 95°F.

Practice Problem 6 Convert 20°C to Fahrenheit temperature.

Calculator

Converting Temperatures

You can use your calculator to convert temperature readings between Fahrenheit and Celsius. To convert 30°C to Fahrenheit temperature, enter

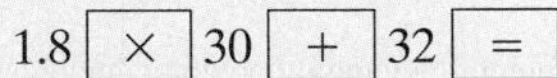

Display

86

The temperature is 86°F. To convert 82.4°F to Celsius temperature enter

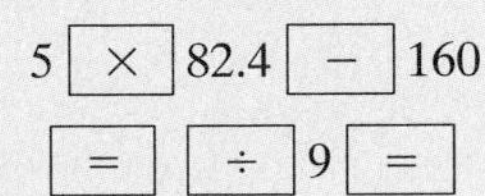

Display

28

The temperature is 28°C.

To convert Fahrenheit temperature to Celsius, we can use the formula

$$C = \frac{5 \times F - 160}{9},$$

where F is the number of Fahrenheit degrees and C is the number of Celsius degrees.

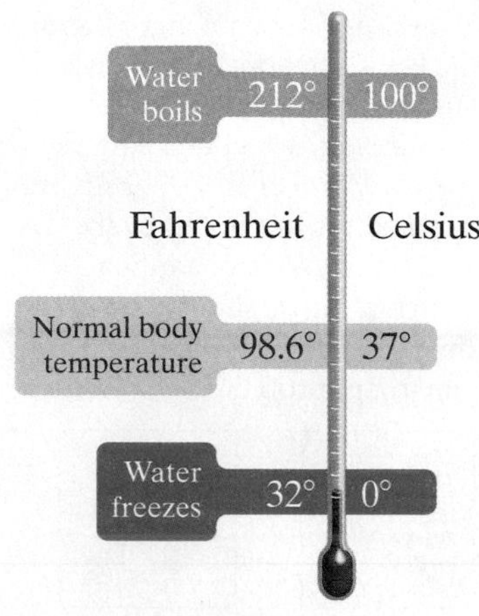

EXAMPLE 7 When the temperature is 50°F, what is the Celsius reading?

Solution

$$\begin{aligned} C &= \frac{5 \times F - 160}{9} \\ &= \frac{5 \times 50 - 160}{9} \\ &= \frac{250 - 160}{9} \\ &= \frac{90}{9} \\ &= 10 \end{aligned}$$

The temperature is 10°C.

***NOTE TO STUDENT:** Fully worked-out solutions to all of the Practice Problems can be found at the back of the text starting at page SP-1*

Practice Problem 7 When the temperature is 86°F, what is the Celsius reading?

6.4 EXERCISES

Student Solutions Manual | CD/ Video | PH Math Tutor Center | MathXL®Tutorials on CD | MathXL® | MyMathLab® | Interactmath.com

Verbal and Writing Skills

1. Which metric measure is approximately the same length as a yard? Which unit is larger?

The meter is approximately the same length as a yard. The meter is slightly longer.

2. Which metric measure of volume is approximately the same as quart? Which unit is larger?

The liter is approximately the same volume as a quart. The liter is slightly larger.

3. Which American measure is approximately twice the length of a centimeter?

The inch is approximately twice the length of a centimeter.

4. Which metric measure is approximately one-half of a pound?

The kilogram is approximately one-half of a pound?

Perform each conversion. Round to the nearest hundredth when necessary.

5. 7 ft to m

$7 \text{ ft} \times \frac{0.305 \text{ m}}{1 \text{ ft}} \approx 2.14 \text{ m}$

6. 11 ft to m

$11 \text{ ft} \times \frac{0.305 \text{ m}}{1 \text{ ft}} \approx 3.36 \text{ m}$

7. 9 in. to cm

$9 \text{ in.} \times \frac{2.54 \text{ cm}}{1 \text{ in.}} \approx 22.86 \text{ cm}$

8. 13 in. to cm

$13 \text{ in.} \times \frac{2.54 \text{ cm}}{1 \text{ in.}} \approx 33.02 \text{ cm}$

9. 14 m to yd

$14 \text{ m} \times \frac{1.09 \text{ yd}}{1 \text{ m}} \approx 15.26 \text{ yd}$

10. 200 m to yd

$200 \text{ m} \times \frac{1.09 \text{ yd}}{1 \text{ m}} \approx 218 \text{ yd}$

11. 30.8 yd to m

$30.8 \text{ yd} \times \frac{0.914 \text{ m}}{1 \text{ yd}} \approx 28.15 \text{ m}$

12. 42.5 yd to m

$42.5 \text{ yd} \times \frac{0.914 \text{ m}}{1 \text{ yd}} \approx 38.85 \text{ m}$

13. 82 mi to km

$82 \text{ mi} \times \frac{1.61 \text{ km}}{1 \text{ mi}} \approx 132.02 \text{ km}$

14. 68 mi to km

$68 \text{ mi} \times \frac{1.61 \text{ km}}{1 \text{ mi}} \approx 109.48 \text{ km}$

15. 16 m to yd

$16 \times \frac{1.09 \text{ yd}}{1 \text{ m}} \approx 17.44 \text{ yd}$

16. 12 m to yd

$12 \times \frac{1.09 \text{ yd}}{1 \text{ m}} \approx 13.08 \text{ yd}$

17. 17.5 cm to in.

$17.5 \text{ cm} \times \frac{0.394 \text{ in.}}{1 \text{ cm}} \approx 6.90 \text{ in.}$

18. 19.6 cm to in.

$19.6 \text{ cm} \times \frac{0.394 \text{ in.}}{1 \text{ cm}} \approx 7.72 \text{ in.}$

19. 200 m to ft

$200 \text{ m} \times \frac{3.28 \text{ ft}}{1 \text{ m}} \approx 656 \text{ ft}$

20. 400 m to ft

$400 \text{ m} \times \frac{3.28 \text{ ft}}{1 \text{ m}} \approx 1312 \text{ ft}$

21. 5 km to mi

$5 \text{ km} \times \frac{0.62 \text{ mi}}{1 \text{ km}} \approx 3.1 \text{ mi}$

22. 16 km to mi

$16 \text{ km} \times \frac{0.62 \text{ mi}}{1 \text{ km}} \approx 9.92 \text{ mi}$

23. 50 gal to L

$50 \text{ gal} \times \frac{3.79 \text{ L}}{1 \text{ gal}} \approx 189.5 \text{ L}$

24. 140 gal to L

$140 \text{ gal} \times \frac{3.79 \text{ L}}{1 \text{ gal}} \approx 530.6 \text{ L}$

25. 23 qt to L

$23 \text{ qt} \times \frac{0.946 \text{ L}}{1 \text{ qt}} \approx 21.76 \text{ L}$

26. 28 qt to L

$28 \text{ qt} \times \frac{0.946 \text{ L}}{1 \text{ qt}} \approx 26.49 \text{ L}$

27. 19 L to gal

$19 \text{ L} \times \frac{0.264 \text{ gal}}{1 \text{ L}} \approx 5.02 \text{ gal}$

28. 15 L to gal

$15 \text{ L} \times \frac{0.264 \text{ gal}}{1 \text{ L}} \approx 3.96 \text{ gal}$

29. 4.5 L to qt

$4.5 \text{ L} \times \frac{1.06 \text{ qt}}{1 \text{ L}} \approx 4.77 \text{ qt}$

30. 6.5 L to qt

$6.5 \text{ L} \times \frac{1.06 \text{ qt}}{1 \text{ L}} \approx 6.89 \text{ qt}$

31. 82 kg to lb

$82 \text{ kg} \times \frac{2.2 \text{ lb}}{1 \text{ kg}} \approx 180.4 \text{ lb}$

32. 45 kg to lb

$45 \text{ kg} \times \frac{2.2 \text{ lb}}{1 \text{ kg}} \approx 99 \text{ lb}$

33. 130 lb to kg

$130 \text{ lb} \times \frac{0.454 \text{ kg}}{1 \text{ lb}} \approx 59.02 \text{ kg}$

34. 155 lb to kg

$155 \text{ lb} \times \frac{0.454 \text{ kg}}{1 \text{ lb}} \approx 70.37 \text{ kg}$

35. 26 oz to g

≈ 737.1 g

36. 34 oz to g

≈ 963.9 g

37. 152 kg to lb

≈ 334.4 lb

38. 188 kg to lb

≈ 413.6 lb

Mixed Practice

Perform each conversion. Round to the nearest hundredth if necessary.

39. 126 g to oz

$126 \text{ g} \times \frac{0.0353 \text{ oz}}{1 \text{ g}} \approx 4.45 \text{ oz}$

40. 186 g to oz

$186 \text{ g} \times \frac{0.0353 \text{ oz}}{1 \text{ g}} \approx 6.57 \text{ oz}$

41. 1260 cm to ft

$1260 \text{ cm} \times \frac{0.394 \text{ in.}}{1 \text{ cm}} \times \frac{1 \text{ ft}}{12 \text{ in.}} \approx 41.37 \text{ ft}$

42. 85 cm to ft

$85 \text{ cm} \times \frac{0.394 \text{ in.}}{1 \text{ cm}} \times \frac{1 \text{ ft}}{12 \text{ in.}} \approx 2.791 \text{ ft}$

43. 55 km/hr to mi/hr

$\frac{55 \text{ km}}{\text{hr}} \times \frac{0.62 \text{ mi}}{1 \text{ km}} = 34.1 \text{ mi/hr}$

44. 120 km/hr to mi/hr

$\frac{120 \text{ km}}{\text{hr}} \times \frac{0.62 \text{ mi}}{1 \text{ km}} = 74.4 \text{ mi/hr}$

45. 400 ft/sec to mi/hr (Round to the nearest whole number.)

273 mi/hr

46. 300 ft/sec to mi/hr (Round to the nearest whole number.)

205 mi/hr

47. A wire that is 13 mm wide is how many inches wide?

$1.3 \text{ cm} \times \frac{0.394 \text{ in.}}{1 \text{ cm}} \approx 0.51 \text{ in.}$

48. A bolt that is 7 mm wide is how many inches wide?

$0.7 \text{ cm} \times \frac{0.394 \text{ in.}}{1 \text{ cm}} \approx 0.28 \text{ in.}$

49. 85°C to Fahrenheit

$1.8 \times 85 + 32 = 185°\text{F}$

50. 105°C to Fahrenheit

$1.8 \times 105 + 32 = 221°\text{F}$

51. 12°C to Fahrenheit

$1.8 \times 12 + 32 = 53.6°\text{F}$

52. 21°C to Fahrenheit

$1.8 \times 21 + 32 = 69.8°\text{F}$

53. 140°F to Celsius

$\frac{5 \times 140 - 160}{9} = 60°\text{C}$

54. 131°F to Celsius

$\frac{5 \times 131 - 160}{9} = 55°\text{C}$

55. 40°F to Celsius

$\frac{5 \times 40 - 160}{9} \approx 4.44°\text{C}$

56. 52°F to Celsius

$\frac{5 \times 52 - 160}{9} \approx 11.11°\text{C}$

Applications

Solve. Round to the nearest hundredth when necessary.

57. ***Speed Limit*** The speed limit on a New Zealand highway is 90 km/hr. Molly is driving at 65 mi/hr. Is Molly speeding?
yes

58. ***Spain Bike Trip*** John and Sandy Westphal traveled to Spain for a bike tour. The woman leading the tour told the participants they would be biking 75 km the first day, 83 km the second day, and 78 km the third day. How many miles did John and Sandy bike in those three days?
146.32 miles

59. ***Fuel Consumption*** Pierre had a Jeep imported into France. During a trip from Paris to Lyon, he used 38 liters of gas. The tank, which he filled before starting the trip, holds 15 gallons of gas. How many liters of gas were left in the tank when he arrived?
18.85 liters

60. ***Drinking Water*** It is recommended that each visitor to Death Valley, CA have 2 gallons of drinking water available. Rachel bought six 1-liter bottles of water. Does she have the recommended amount? Why?
no Six liters of water is only 1.584 gallons. This is less than 2 gallons.

61. ***Weight Records*** One of the heaviest males documented in medical records weighed 635 kg in 1978. What would have been his weight in pounds?
1397 pounds

62. ***Weight of a Child*** The average weight for a 7-year-old girl is 22.2 kilograms. What is the average weight in pounds?
48.84 lb

63. ***Australia Rock Climbing*** Tourists who visit Ayers Rock in central Australia in the summer begin climbing at four o'clock in the morning, when the temperature is 19° Celsius. They do this because after seven o'clock in the morning, the temperature can reach 45°C and can cause climbers to die of dehydration. What are equivalent Fahrenheit temperatures?
66.2°F at 4:00 A.M.; 113°F after 7:00 A.M.

64. ***Cooking*** A holiday turkey in Buenos Aires, Argentina, was roasted at 200° Celsius for 4 hours. What would have been the Fahrenheit roasting temperature in Joplin, Missouri?
392°F

65. ***Biology*** There are 96,550 km of blood vessels in the human body. How many miles of blood vessels is this?
59,861 miles

66. ***Earth Science*** The distance around the Earth at the equator (the circumference) is approximately 40,325 km. How many miles is this?
25,001.5 miles

Round to four decimal places.

▲ **67.** 28 square inches = ? square centimeters
$28 \times 2.54 \times 2.54 = 180.6448$ sq cm

▲ **68.** 36 square meters = ? square yards
$36 \times 1.09 \times 1.09 = 42.7716$ sq yd

To Think About

▲ **69.** ***Geometry*** Frederick Himlein is planning to carpet his rectangular living room, which measures 8 yards by 4 yards. He has found some carpet in New York City that he likes that costs $28 per square yard. While visiting family friends in Germany, his wife Gertrude found some carpet that costs $30 per square meter. The company has a New York City office and can sell it in America for the same price. Frederick says that the German carpet is much too expensive. How much would it cost to carpet the living room with the American carpet? How much would it cost to carpet the living room with the German carpet? How much difference in cost is there between these two choices? (Round to the nearest dollar.)
$896 for the American carpet; $802 for the German carpet; the German carpet is $94 cheaper

▲ **70.** ***Geometry*** Phillipe Bertoude is planning to carpet his rectangular family room, which measures 7 yards by 5 yards. His wife found some carpet in Boston that she likes, and it costs $24 per square yard. While Phillipe was visiting his father in Paris, he found some carpet that costs $26 per square meter. The company has a Boston office and can sell it in America for the same price. Phillipe told his wife that the carpet in Paris was a better buy. She does not agree. How much would it cost to carpet the family room with the French carpet? How much would it cost to carpet the family room with the American carpet? How much difference in cost is there between these two choices? (Round to the nearest dollar.)
$840 for the American carpet; $760 for the French carpet, the French carpet is $80 cheaper

Cumulative Review

Do the operations in the correct order.

71. $3^4 \times 2 - 5 + 12$ 169

72. $96 + 24 \div 4 \times 3$ 114

73. $\frac{1}{2} \cdot \frac{3}{4} - \frac{1}{5}\left(\frac{1}{2}\right)^2$ $\frac{13}{40}$

74. $\frac{5}{6} - \frac{1}{2}\left(\frac{5}{6} - \frac{1}{3}\right)$ $\frac{7}{12}$

6.5 Solving Applied Measurement Problems

 Solving Applied Problems Involving Metric and American Units

Once again, we will be using the Mathematics Blueprint for solving applied problems that we used previously.

Student Learning Objective

After studying this section, you will be able to:

1 Solve applied problems involving metric and American units.

EXAMPLE 1 A triangular support piece holds a solar panel. The sketch shows the dimensions of the triangle. Find the perimeter of this triangle. Express the answer in *feet*.

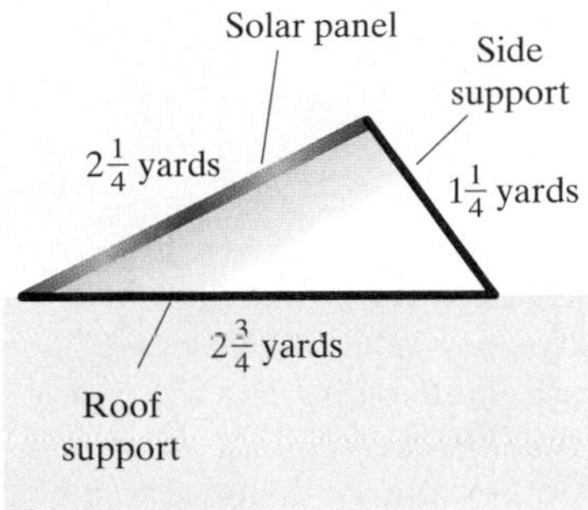

Solution

Mathematics Blueprint for Problem Solving

Gather the Facts	What Am I Asked to Do?	How Do I Proceed?	Key Points to Remember
The triangle has three sides: $2\frac{1}{4}$ yd, $1\frac{1}{4}$ yd, and $2\frac{3}{4}$ yd.	Find the perimeter.	Add the lengths of the three sides. Then change the answer from yards to feet.	Be sure to change the number of yards to an improper fraction before multiplying by 3 to obtain feet.

1. ***Understand the problem.*** The perimeter is the sum of the lengths of the sides. Remember to convert the yards to feet.
2. ***Solve and state the answer.***

Add the three sides.

$$\begin{array}{r} 2\frac{1}{4} \text{ yards} \\ 1\frac{1}{4} \text{ yards} \\ +\ 2\frac{3}{4} \text{ yards} \\ \hline 5\frac{5}{4} \text{ yards} \end{array} = 6\frac{1}{4} \text{ yards}$$

Convert $6\frac{1}{4}$ yards to feet using the fact that 1 yard = 3 feet. To make the calculation easier, we will change $6\frac{1}{4}$ to $\frac{25}{4}$.

$$6\frac{1}{4}\ \cancel{\text{yards}} \times \frac{3\text{ feet}}{1\ \cancel{\text{yard}}} = \frac{25}{4}\ \cancel{\text{yards}} \times \frac{3\text{ feet}}{1\ \cancel{\text{yard}}} = \frac{75}{4}\text{ feet} = 18\frac{3}{4}\text{ feet}$$

The perimeter of the triangle is $18\frac{3}{4}$ feet.

3. ***Check.*** We will check by estimating the answer.

$$2\frac{1}{4}\text{ yd} \approx 2\text{ yd} \qquad 1\frac{1}{4}\text{ yd} \approx 1\text{ yd} \qquad 2\frac{3}{4}\text{ yd} \approx 3\text{ yd}$$

Now we add the three sides, using our estimated values.

$$2 + 1 + 3 = 6\text{ yards} \qquad 6\text{ yards} = 18\text{ feet}$$

Our estimated answer, 18 feet, is close to our calculated answer, $18\frac{3}{4}$ feet. Thus our answer seems reasonable. ✓

Teaching Tip This is a good time to reinforce the concept of estimating your answer. Ask the students to estimate the answer for Practice Problem 1.

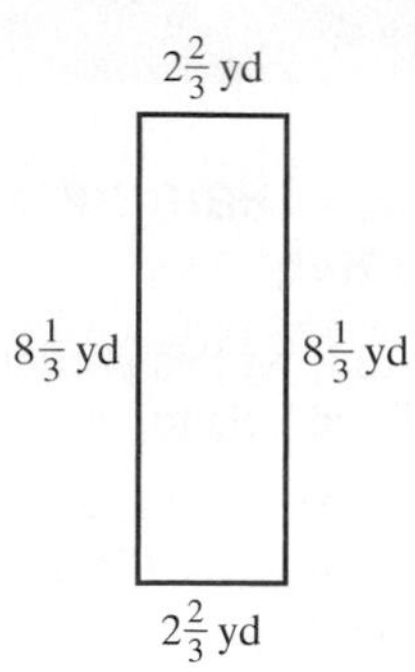

▲ **Practice Problem 1** Find the perimeter of the rectangle on the left. Express the answer in *feet*. ■

EXAMPLE 2 How many 210-*liter* gasoline barrels can be filled from a tank of 5.04 *kiloliters* of gasoline?

Solution

1. ***Understand the problem.***

Mathematics Blueprint for Problem Solving

Gather the Facts	What Am I Asked to Do?	How Do I Proceed?	Key Points to Remember
We have 5.04 kiloliters of gasoline. We are going to divide the gasoline into smaller barrels that hold 210 liters each.	Find out how many of these smaller 210-liter barrels can be filled.	We need to get all measurements in the same units. We choose to convert 5.04 kiloliters to liters. Then we divide that result by 210 to find out how many barrels can be filled.	To convert 5.04 kiloliters to liters, we move the decimal point three places to the right.

Teaching Tip If students have trouble visualizing Example 2, ask them to visualize how many times they can fill a 1-quart milk bottle from a 1-gallon milk jug. Draw a picture of each on the board. (Answer: 4 times.) Then ask them if they had 6 1-gallon jugs how much milk could be poured in 24 identical containers. (Answer: 1 quart of milk into each.)

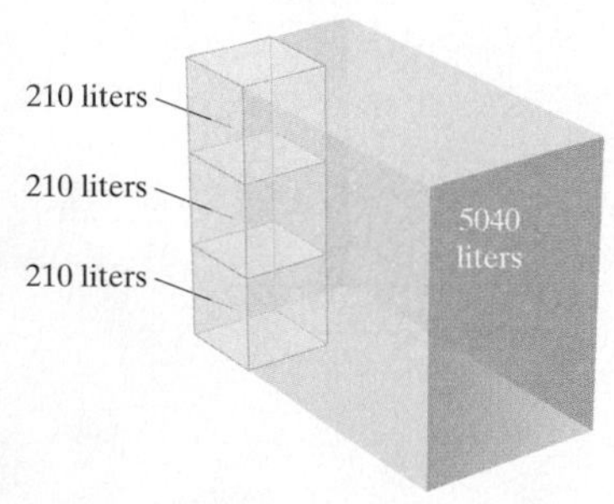

2. ***Solve and state the answer.*** First we convert 5.04 kiloliters to liters.

$$5.04 \text{ kiloliters} = 5040 \text{ liters}$$

Now we find out how many barrels can be filled. How many 210-liter barrels will 5040 liters fill? Visualize fitting 210-liter barrels into a big barrel that holds 5040 liters. (Use rectangular barrels.)

We need to divide

$$\frac{5040 \text{ liters}}{210 \text{ liters}} = 24.$$

Thus we can fill 24 of the 210-liter barrels.

3. ***Check.*** Estimate each value. First 5.04 kiloliters is approximately 5 kiloliters or 5000 liters. 210-liter barrels hold approximately 200 liters. How many times does 200 fit into 5000?

$$\frac{5000}{200} = 25$$

We estimate that 25 barrels can be filled. This is very close to our calculated value of 24 barrels. Thus our answer is reasonable. ✓

Practice Problem 2 A lab assistant must use 18.06 liters of solution to fill 42 jars. How many milliliters of the solution will go into each jar? ■

6.5 EXERCISES

Student Solutions Manual

CD/ Video

PH Math Tutor Center

MathXL®Tutorials on CD
MathXL®

MyMathLab®

Interactmath.com

Applications

Solve.

▲ **1.** Find the perimeter of the triangle. Express your answer in feet.

$14\frac{1}{3} + 8\frac{2}{3} + 13 = 36$ in. $= 3$ ft

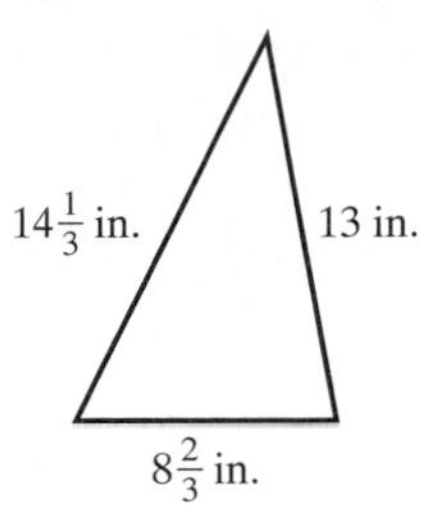

▲ **2.** Find the perimeter of the triangle. Express your answer in feet.

$27\frac{3}{4} + 22\frac{1}{4} + 10 = 60$ in. $= 5$ ft

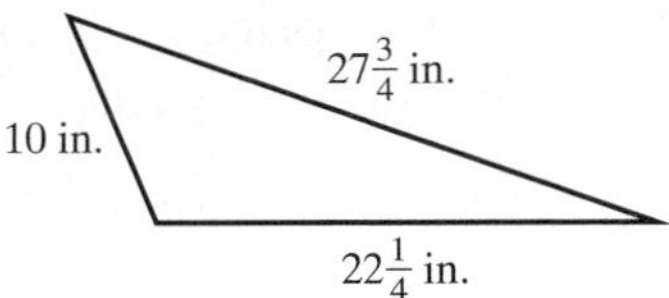

▲ **3.** ***Zoo Fencing*** The triangular giraffe enclosure at the zoo is being re-fenced. One side is 86 yards long and a second side is 77 yards long. If the zoo has 522 feet of fencing, how much will be left over for the third side? Express your answer in yards.

$\frac{522}{3} - 86 - 77 = 11$ yd

▲ **4.** ***Farm Fencing*** A farmer has 600 feet of fencing to fence in a triangular region. One side is to be 75 yards and a second 65 yards. How much fencing will the farmer have left for the third side? Express your answer in yards.

$75 + 65 = 140 \qquad 200 - 140 = 60$ yd

▲ **5.** ***Doorway Insulation*** A rectangular doorway measures 90 centimeters × 200 centimeters. Weatherstripping is applied on the top and the two sides. The weatherstripping costs $6.00 per meter. What did it cost to weatherstrip the door?

needed weatherstripping = 490 cm 490 × $0.06 = $29.40

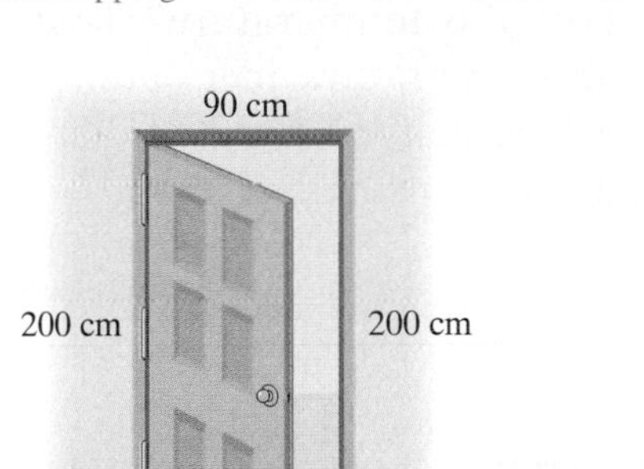
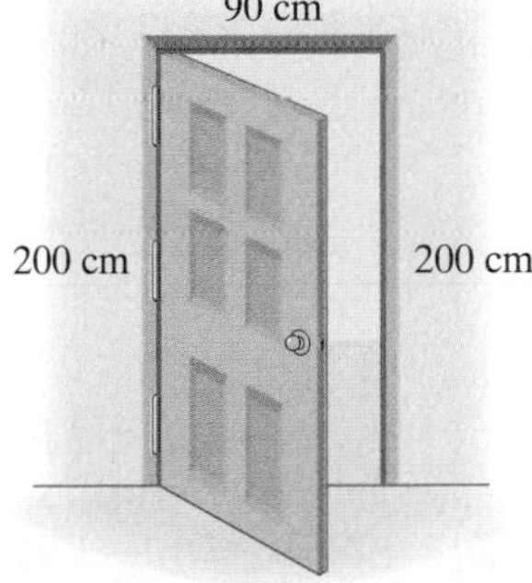

▲ **6.** ***Window Insulation*** A rectangular picture window measures 87 centimeters × 152 centimeters. Window insulation is applied along all four sides. The insulation costs $7.00 per meter. What will it cost to insulate the window?

Perimeter = 478 cm 478 × 0.07 = $33.46

	152 cm	
87 cm		87 cm
	152 cm	

7. ***Parking Space Dimensions*** A stretch of road 1.863 kilometers long has 230 parking spaces of equal length painted in white on the pavement. How many meters long is each parking space? 8.1 meters

8. ***Leather Horse Equipment*** A tack supply company, which makes saddles, fittings, bridles, and other equipment for horses and riders, has a length of braided leather 12.4 meters long. The leather must be cut into 4 equal pieces. How many centimeters long will each piece be? 310 centimeters per piece

9. ***Track and Field*** When Gary's father was in high school, he ran the 880 yd run. Gary is on the high school track team and is training for the 800 m race. Which distance is longer? By how much? 880 yd is 4.32 m longer than 800 m

10. ***Muscle Size*** The world record for the largest bicep is 30 in. How many centimeters in this? 76.2 cm

11. ***Gasoline Prices*** Don drove his car from Los Angeles, California to Tijuana, Mexico. The price of gasoline in California is \$1.65 per gallon. In Mexico, the price is \$0.46 per liter. Where is gasoline more expensive?

$\frac{\$0.46}{1 \text{ L}} \times \frac{3.79 \text{ L}}{1 \text{ gal}} \approx \$1.74/\text{gal}$ Gasoline is more expensive in Mexico.

12. ***Food Costs*** Kimberly lives in New Zealand, where bananas sell for \$0.80 per kilogram. While visiting her sister in Texas, she noticed that bananas were \$0.39 per pound. Where are bananas more expensive?

$\frac{\$0.39}{1 \text{ lb}} \times \frac{2.2 \text{ lb}}{1 \text{ kg}} \approx \$0.86/\text{kg}$ Bananas are more expensive in Texas.

13. ***Temperature*** While traveling in Scotland, Jenny noticed the high temperature one day was 25°C. What is the high temperature that day in Scotland in the Fahrenheit system? Jenny called her husband in Boston that day and found the high temperature in Boston that day was 86°F. How much hotter was Boston than Scotland that day? 77°F 9°F

14. ***Temperature*** The temperature in Dakar, Senegal, today is 33°C. The temperature on this date last year was 94°F. What is the difference in degrees Fahrenheit between the temperature in Dakar today and the temperature one year ago? The difference is 2.6°F.

15. ***Cooking*** A Swedish flight attendant is heating passenger meals for the new American airline he is working for. He is used to heating meals at 180°C. His co-worker tells him that the food must be heated at 350°F. What is the difference in temperature in degrees Fahrenheit between the two temperatures? Which temperature is hotter? The difference is 6°F. The temperature reading of 180°C is hotter.

16. ***Temperature*** The weather page in an American newspaper listed 117°F as the temperature one day in Kuwait City, Kuwait. This country uses the Celsius scale. How would a Kuwaiti report the temperature? That same day the high temperature in London was 28°C. How much cooler was London than Kuwait City on that day? 47.2°C 19.2°C

17. ***Travel in Mexico*** Sharon and James Hanson traveled from Arizona to Acapulco, Mexico. The last day of their trip, they traveled 520 miles and it took them eight hours. The maximum speed limit is 110 kilometers/hour.

(a) How many kilometers per hour did they average on the last day of the trip? about 105 kilometers per hour

(b) Did they break the speed limit? Probably not. We cannot be sure, since they could speed for a short time, but we have no evidence to indicate that they broke the speed limit.

18. ***Jet Travel*** A small corporate jet travels at 600 kilometers/hour for 1.5 hours. The pilot said the plane will be on time if it travels at 350 miles per hour.

(a) What was the jet's speed in miles per hour? 600 km/hr ≈ 372 mi/hr

(b) Will it arrive on time? yes

19. ***Concession Stand Sales*** The concession stand at the high school sells large 1-pint sodas. On an average night, two sodas are sold each minute. How many gallons of soda are sold per hour? 15 gallons

20. ***Leaky Faucet*** A leaky faucet drips 1 pint of water per hour. How many gallons of water is this per day? 3 gallons

21. ***Tax on Trucks*** Trucks in Sam's home state are taxed each year at \$0.03 per pound. Sam's empty truck weighs 1.8 tons. What is his annual tax? $1.8 \times 2000 \times 0.03 = \$108$

22. ***Banana Shipment*** 3.4 tons of bananas are offloaded from a ship that has just arrived from Costa Rica. The port taxes imported fruit at \$0.015 per pound. What is the tax on the entire shipment? \$102

23. ***Raisin Bran*** A box of raisin bran contains 14 ounces of cereal. Of this, 11 ounces are bran flakes and the rest is raisins. How many grams of raisins are in the box?

$14 - 11 = 3\text{ oz} \qquad 3\text{ oz} \times \dfrac{28.35\text{ g}}{1\text{ oz}} \approx 85.05\text{ g}$

24. ***Canned Fruit*** A can of peaches contains 16 ounces. Fred discovered that 5 ounces were syrup and the rest was fruit. How many grams of fruit were in the can?

$16 - 5 = 11\text{ oz} \qquad 11\text{ oz} \times \dfrac{28.35\text{ g}}{1\text{ oz}} \approx 311.85\text{ g}$

25. ***Tree Insecticide*** Carlos is mixing a fruit tree spray to retard the damage caused by beetles. He has 16 quarts of spray concentrate available. The old recipe he used in Mexico called for 11 liters of spray concentrate.

(a) How many extra quarts of spray concentrate does he have?

$11\text{ L} \times \dfrac{1.06\text{ qt}}{1\text{ L}} \approx 11.66\text{ qt} \quad 16 - 11.66 = 4.34\text{ qt}$

He has 4.34 quarts extra.

(b) If the spray costs \$2.89 per quart, how much will it cost him to prepare the old recipe? $11.66 \times 2.89 \approx \$33.70$

26. ***Motor Oil Consumption*** Maria bought 18 quarts of motor oil for \$1.39 per quart. Her uncle from Mexico is visiting this summer. He said he will use 12 liters of oil while driving his car this summer.

(a) How many extra quarts of oil did Maria buy?

$12\text{ L} \times \dfrac{1.06\text{ qt}}{1\text{ L}} \approx 12.72\text{ qt} \quad 18 - 12.72 = 5.28\text{ qt}$

She bought 5 qt extra.

(b) How much did this extra oil cost her?

$5 \times 1.39 = \$6.95$

27. ***Fuel Efficiency*** Jackson's small motorcycle gets 56 kilometers per liter. He drives 392 kilometers to Quebec City, Canada, and gas is \$0.78 per liter.

(a) How much does the gas used for the trip cost? \$5.46

(b) How many miles per gallon does Jackson's motorcycle get? about 132 miles per gallon

28. ***Fuel Efficiency*** Ryan's Volvo station wagon runs on diesel fuel. He gets 20 miles per gallon on the highway.

(a) If he drives 275 kilometers to Mexico City, and diesel fuel costs \$1.30 per gallon, how much does the fuel used for the trip cost him? Round to the nearest cent. \$11.08

(b) How many kilometers per liter does he get? approx. 8.50 kilometers per liter

29. ***Discharge Rate at a Dam*** The flow rate of a safety discharge pipe at a dam in Lowell, Massachusetts, is rated for a maximum of 240,000 gallons per hour. The inspector asked if this flow rate could have handled the floods of 1927. During the floods of 1927 the flow rate at the dam was measured as 440 pints per second. Could the safety discharge pipe safely handle a flow rate of 440 pints per second? Why? yes; 240,000 gal/hr is equivalent to $533\frac{1}{3}$ pt/sec

30. ***Water Reservoir*** The old main lines that run water from the Quabbin Reservoir to Boston have some leaks. It is estimated that the lines leak approximately 36,000 gallons per hour. A local newspaper said that the lines leak 80 pints per second. Did the newspaper have the correct information? Why? yes; 36,000 gal/hr is exactly equivalent to 80 pints per second

Cumulative Review

Solve for n.

31. $\frac{18}{27} = \frac{n}{15}$ $n = 10$

32. ***Geography of the Solar System*** The highest known mountain in the solar system is the Martian volcano Olympus Mons, which rises approximately three times higher than Mt. Everest. (Mt. Everest is 8846 meters high.)
   (a) How high is the Martian volcano in meters? 26,538 meters
   (b) How high is the Martian volcano in feet? (Round to the nearest foot.) 87,045 feet

33. ***Map Scale*** Dori is going to India. While reading his atlas he sees that Bombay is 6 inches from where he will begin his travel. The scale shows that 3 inches represents 7.75 miles on the ground. If he travels by train on a straight track, how far must the train go to take him to his destination? 15.5 miles

34. ***Scale Models*** Thompson has a scale model of a famous fishing schooner. Every 2 centimeters on the model represents an actual length of 7.5 yards. The model is 11 centimeters long. How long is the famous fishing schooner? 41.25 yards

Putting Your Skills to Work

Converting Acres to Hectares

When changing from the American System of Measurement to the Metric System of Measurement some special study is required with measurements of land area. In the American system land is usually measured in acres. An acre of land is defined as an area of 43,560 square feet. In the metric system land is usually measured in hectares. A hectare is defined as an area of 10,000 square meters. Sometimes people have difficulty visualizing how to change from one unit to the other. It may help to think of a hectare of land as approximately the amount of land in two football fields.

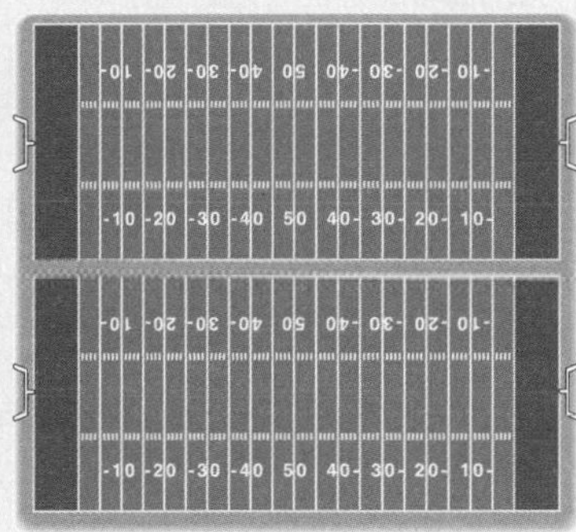

In recent years Canada has converted all units of measurement to the metric system. It is now necessary for all deeds for land owned in Canada to be recorded in hectares and not in acres. Possibly some time in the future a similar situation will occur in the United States.

Problems for Individual Investigation and Analysis

1. If 1 foot is approximately 0.305 meter, how many square meters are contained in 1 square foot? Approximately 0.093025 square meter in 1 square foot

2. If 1 acre = 43,560 square feet, how many square meters are in 1 acre? Approximately 4052.169 square meters in 1 acre

In problems 3–8 use the approximation that 1 acre is approximately 0.405 hectare.

3. A Nova Scotia farmer has 4 acres of land. The new deed for the land is measured in hectares. How many hectares does the farmer have? 1.62 hectares

4. The ramp for an exit from the Trans Canada Highway uses 3 hectares of land. How many acres of land are used? (Round to the nearest hundredth.) Approximately 7.41 acres of land

Problems for Group Investigation and Cooperative Learning

5. A rectangular field measures 340 feet by 590 feet. How many hectares of land are there in this field? (Round to nearest hundredth.) 1.87 hectares

6. A rectangular piece of land measures 200 yards by 340 yards. How many hectares of land are there in this field? (Round to nearest hundredth.) 5.69 hectares

7. Recently, one Prince Edward Island farmer told another that 2 hectares was the same as 5 acres. This is a good rough estimate, but it is not exact. If we wanted to be more accurate, what is the number of acres in 2 hectares? How great is the error in the estimation that 2 hectares is the same as 5 acres?
There are 4.94 acres in 2 hectares. The error is 0.06 acre.

8. Mr. and Mrs. Dubuzier purchased some land in Quebec. The deed was originally for 2.00 acres of land. The new deed (for the same amount of land) listed the area as 0.86 hectare of land. By how much is the deed in error?
It is 0.05 hectare in error. The new deed should actually be 0.81 hectare.

Chapter 6 Organizer

Topic	Procedure	Examples
Changing from one American unit to another, p. 375.	**1.** Find the equality statement that relates what you want to find and what you know. **2.** Form a unit fraction. The denominator will contain the units of the original measurement. **3.** Multiply by the unit fraction and simplify.	Convert 210 inches to feet. **1.** Use 12 inches = 1 foot. **2.** Unit fraction $= \dfrac{1 \text{ foot}}{12 \text{ inches}}$ **3.** $210 \cancel{\text{in.}} \times \dfrac{1 \text{ ft}}{12 \cancel{\text{in.}}} = \dfrac{210}{12} \text{ ft}$ $= 17.5 \text{ ft}$ Convert 86 yards to feet. **1.** Use 1 yard = 3 feet. **2.** Unit fraction $= \dfrac{3 \text{ feet}}{1 \text{ yard}}$ **3.** $86 \cancel{\text{yd}} \times \dfrac{3 \text{ ft}}{1 \cancel{\text{yd}}} = 86 \times 3 \text{ ft} = 258 \text{ ft}$
Changing from one metric unit to another, pp. 382, 392, 394.	When you change from one prefix to another by moving to the *left* in the prefix guide, move the decimal point to the *left* the same number of places. kilo- = 1000, hecto- = 100, deka- = 10, one unit = 1, deci- = 0.1, centi- = 0.01, milli- = 0.001 When you change from one prefix to another by moving to the *right* in the prefix guide, move the decimal point to the *right* the same number of places.	Change 7.2 meters to kilometers. **1.** Move three decimal places to the left. 0.007. 2 **2.** 7.2 m = 0.0072 km Change 196 centimeters to meters. **1.** Move two places to the left. 196. **2.** 196 cm = 1.96 m Change 17.3 liters to milliliters. **1.** Move three decimal places to the right. 17.300 **2.** 17.3 L = 17,300 mL
Changing from American units to metric units, p. 400.	**1.** From the list of approximate equivalent measures, pick an equality statement that begins with the unit in the original measurement. 1 mi ≈ 1.61 km 1 yd ≈ 0.914 m 1 ft ≈ 0.305 m 1 in. = 2.54 cm (exact) 1 gal ≈ 3.79 L 1 qt ≈ 0.946 L 1 lb ≈ 0.454 kg 1 oz ≈ 28.35 g **2.** Multiply by a unit fraction.	Convert 7 gallons to liters. **1.** 1 gal ≈ 3.79 L **2.** $7 \cancel{\text{gal}} \times \dfrac{3.79 \text{ L}}{1 \cancel{\text{gal}}} = 26.53 \text{ L}$ Convert 18 pounds to kilograms. **1.** 1 lb ≈ 0.454 kg **2.** $18 \cancel{\text{lb}} \times \dfrac{0.454 \text{ kg}}{1 \cancel{\text{lb}}} = 8.172 \text{ kg}$
Changing from metric units to American units, p. 400.	**1.** From the list of approximate equivalent measures, pick an equality statement that begins with the unit in the original measurement and ends with the unit you want. 1 km ≈ 0.62 mi 1 m ≈ 3.28 ft 1 m ≈ 1.09 yd 1 cm ≈ 0.394 in. 1 L ≈ 0.264 gal 1 L ≈ 1.06 qt 1 kg ≈ 2.2 lb 1 g ≈ 0.0353 oz **2.** Multiply by a unit fraction.	Convert 605 grams to ounces. **1.** 1 g ≈ 0.0353 oz **2.** $605 \cancel{\text{g}} \times \dfrac{0.0353 \text{ oz}}{1 \cancel{\text{g}}} = 21.3565 \text{ oz}$ Convert 80 km/hr to mi/hr. **1.** 1 km ≈ 0.62 mi **2.** $80 \dfrac{\cancel{\text{km}}}{\text{hr}} \times \dfrac{0.62 \text{ mi}}{1 \cancel{\text{km}}} = 49.6 \text{ mi/hr}$

Topic	Procedure	Examples
Changing from Celsius to Fahrenheit temperature, p. 403.	**1.** To convert Celsius to Fahrenheit, we use the formula $F = 1.8 \times C + 32.$ **2.** We replace C by the Celsius temperature. **3.** We calculate to find the Fahrenheit temperature.	Convert 65°C to Fahrenheit. **1.** $F = 1.8 \times C + 32$ **2.** $F = 1.8 \times 65 + 32$ **3.** $F = 117 + 32 = 149$ The temperature is 149°F.
Changing from Fahrenheit to Celsius temperature, p. 404.	**1.** To convert Fahrenheit to Celsius, we use the formula $C = \frac{5 \times F - 160}{9}.$ **2.** We replace F by the Fahrenheit temperature. **3.** We calculate to find the Celsius temperature.	Convert 50°F to Celsius. **1.** $C = \frac{5 \times F - 160}{9}$ **2.** $C = \frac{5 \times 50 - 160}{9}$ **3.** $C = \frac{250 - 160}{9} = \frac{90}{9} = 10$ The temperature is 10°C.

Chapter 6 Review Problems

Section 6.1

Convert. When necessary, express your answer as a decimal. Round to the nearest hundredth.

1. 33 ft = 11 yd

2. 27 ft = 9 yd

3. 5 mi = 8800 yd

4. 6 mi = 10,560 yd

5. 90 in. = 7.5 ft

6. 78 in. = 6.5 ft

7. 15,840 ft = 3 mi

8. 10,560 ft = 2 mi

9. 7 tons = 14,000 lb

10. 4 tons = 8000 lb

11. 8 oz = 0.5 lb

12. 12 oz = 0.75 lb

13. 15 gal = 60 qt

14. 21 gal = 84 qt

15. 31 pt = 15.5 qt

16. 27 pt = 13.5 qt

Section 6.2

Convert. Do not round.

17. 56 cm = 560 mm

18. 29 cm = 290 mm

19. 1763 mm = 176.3 cm

20. 2598 mm = 259.8 cm

21. 9.2 m = 920 cm

22. 7.4 m = 740 cm

23. 10,000 m = 10 km

24. 8200 m = 8.2 km

Change all units to meters and add.

25. 6.2 m + 121 cm + 0.52 m
7.93 m

26. 9.8 m + 673 cm + 0.48 m
17.01 m

27. 0.024 km + 1.8 m + 983 cm
35.63 m

28. 0.078 km + 5.5 m + 609 cm
89.59 m

Section 6.3

Convert. Do not round.

29. 17 kL = 17,000 L

30. 23 kL = 23,000 L

31. 196 kg = 196,000 g

32. 721 kg = 721,000 g

33. 778 mg = 0.778 g

34. 459 mg = 0.459 g

35. 3500 g = 3.5 kg

36. 12,750 g = 12.75 kg

37. 765 cm^3 = 765 mL

38. 423 cm^3 = 423 mL

39. 2.43 L = 2430 cm^3

40. 1.93 L = 1930 cm^3

Section 6.4

Perform each conversion. Round to the nearest hundredth.

41. 42 kg = 92.4 lb

42. 9 ft = 2.75 m

43. 45 mi = 72.45 km

44. 88 mi = 141.68 km

45. 14 cm = 5.52 in.

46. 18 cm = 7.09 in.

47. 20 lb = 9.08 kg

48. 30 lb = 13.62 kg

49. 12 yd = 10.97 m

50. 14 yd = 12.80 m

51. 80 km/hr = 49.6 mi/hr

52. 70 km/hr = 43.4 mi/hr

53. 12°C = 53.6° F

54. 32°C = 89.6° F

55. 221°F = 105° C

56. 185°F = 85° C

57. 32°F = 0° C

58. 212°F = 100° C

59. 13 L = 3.43 gallons

60. 27 quarts = 25.54 L

Section 6.5

Solve. Round to the nearest hundredth when necessary.

▲ **61.** ***Geometry*** Allison's living room measures 5 yd by 18 ft. What is the area of her living room on square feet? What is the area in square yards?
270 sq ft; 30 sq yd

▲ **62.** Find the perimeter of the triangle.
(a) Express your answer in feet. 17 ft
(b) Express your answer in inches. 204 in.

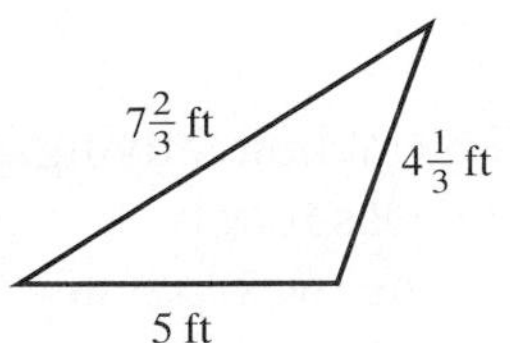

63. Find the perimeter of the rectangle.
(a) Express your answer in meters. 200 m
(b) Express your answer in kilometers 0.2 km

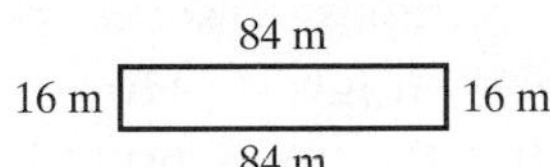

64. ***Food Cost*** The unit price on a box of cereal was \$0.14 per ounce. The net weight was 450 grams. How much did the cereal cost? \$2.22

65. ***Metric Conversion*** Lisa and Ray moved from Canada to the United States. For their new home, they need 15 meters of lumber to repair some stairs. They bought 45 feet at the lumberyard. Did they purchase enough lumber? How many feet extra or how short was this amount?
no; 4.2 feet short

66. ***Metric Conversion*** Lucia is from Mexico, where distances are measured in kilometers. While in San Diego, she rented a car and drove 70 mi/hr on the highway. In Mexico, she never drives faster than 100 km/hr. Was she driving faster than this in San Diego?
yes; she was driving 112.7 km/hr

67. ***Cooking*** A German cook wishes to bake a cake at 185° Celsius. The oven is set at 390° Fahrenheit. By how many degrees Fahrenheit is the oven temperature different from what is desired? Is the oven too hot or not hot enough?
25°F too hot

68. ***Flagpole*** A flagpole is 19 meters long. The bottom $\frac{1}{5}$ of it is coated with a special water seal before being placed in the ground. How many centimeters long is the portion that has the water seal?
380 centimeters

69. ***Track and Field*** A high school track member did a trial run of the 100 meter dash in 13.0 seconds. How many feet per second did he run?
approximately 25.2 ft/sec

70. ***Driving Speed*** Carlos was driving in Mexico City at 80 kilometers per hour. How many miles per hour was he driving?
49.6 miles per hour

71. ***Mountain Hike*** Marcia and Melissa went up the Mt. Washington hiking trail each carrying a backpack tent that weighed 2.2 kg, a canteen with water that weighed 1.4 kg, and some supplies that weighed 3.8 kg. They were told to carry less than 16 pounds while hiking. How close were they to the limit? Did they succeed in carrying less than the weight limit?
they are carrying 16.28 pounds; they are slightly over the weight limit

72. ***Gasoline Cost*** When buying gas in Canada, Greg Salzman was told by the attendant that in U.S. currency he was paying \$0.87 per liter. How much did the gas cost per gallon?
about \$3.30

73. ***Car Interior Measurements*** The Dodge Caravan that is made in Canada is built to accommodate a person who is 1.88 meters tall. A person who is taller than this will not be comfortable. Would a person who is 6 feet 2 inches tall be comfortable in this car?
yes

▲ **74.** ***Geometry*** The driveway of Sir Arthur Jensen in London is 4 meters wide and 12 meters long. How many square feet of sealer does he need to cover his driveway?
approximately 516.4 square feet

75. ***Food Cost*** While traveling through Berlin, Al Dundtreim purchased some powdered milk for \$1.23 per kilogram. How much would it have cost him to buy 4 pounds of powdered milk?
approximately \$2.23

How Am I Doing? Chapter 6 Test

Remember to use your Chapter Test Prep Video CD to see the worked-out solutions to the test problems you want to review.

Note to Instructor: The Chapter 6 Test file in the TestGen program provides algorithms specifically matched to these problems so you can easily replicate this test for additional practice or assessment purposes.

Convert. *Express your answer as a decimal rounded to the nearest hundredth when necessary.*

1. 1.6 tons = ____ lb

2. 19 ft = ____ in.

3. 21 gal = ____ qt

4. 36,960 ft = ____ mi

5. 1800 sec = ____ min

6. 3 cups = ____ qt

7. 8 oz = ____ lb

8. 5.5 yd = ____ ft

Perform each conversion. *Do not round.*

9. 9.2 km = ____ m

10. 9.88 cm = ____ m

11. 46 mm = ____ cm

12. 12.7 m = ____ cm

13. 0.936 cm = ____ mm

14. 46 L = ____ kL

15. 28.9 mg = ____ g

16. 983 g = ____ kg

17. 0.92 L = ____ mL

18. 9.42 g = ____ mg

Perform each conversion. *Round to the nearest hundredth when necessary.*

19. 42 mi = ____ km

20. 1.78 yd = ____ m

21. 9 cm = ____ in.

22. 30 km = ____ mi

23. 7.3 kg = ____ lb

24. 3 oz = ____ g

1. 3200

2. 228

3. 84

4. 7

5. 30

6. 0.75

7. 0.5

8. 16.5

9. 9200

10. 0.0988

11. 4.6

12. 1270

13. 9.36

14. 0.046

15. 0.0289

16. 0.983

17. 920

18. 9420

19. 67.62

20. 1.63

21. 3.55

22. 18.6

23. 16.06

24. 85.05

25. 56.85

26. 3.18

27. (a) 20 meters

(b) 21.8 yards

28. (a) 15°F

(b) yes

29. 82.5 gal/hr

30. (a) 300 km

(b) 14 miles

31. $5\frac{1}{4}$ lb

32. 104°F

Solve. Round to the nearest hundredth when necessary.

25. 15 gallons = ____ L

26. 3 L = ____ quarts

▲ **27.** A rectangular picture frame measures 3 m × 7 m.

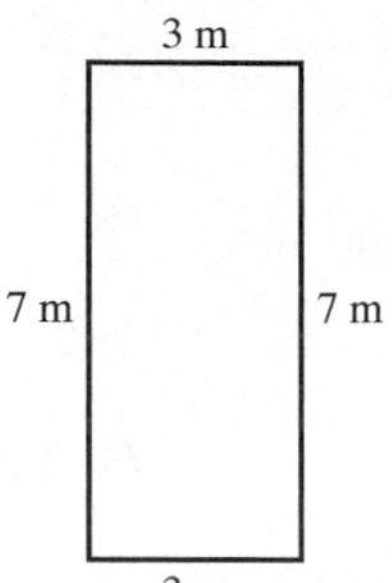

(a) What is the perimeter of the picture frame in meters?
(b) What is the perimeter of the picture frame in yards?

28. The temperature is 80°F today. Kristen's computer has a warning not to operate above 35°C.
(a) How many degrees Fahrenheit are there between the two temperatures?
(b) Can she use her computer today?

29. A pump is running at 5.5 quarts per minute. How many gallons per hour is this?

30. The speed limit on a Canadian road is 100 km/hr.
(a) How far can Samuel travel at this speed limit in three hours?
(b) If Samuel has to travel 200 miles, how much farther will he need to go after three hours of driving at 100 km/hr?

31. Rick bought 1 lb 6 oz of bananas, 2 lb 2 oz of grapes, and 1 lb 12 oz of plums. How many pounds of fruit did he buy?

32. The warmest day this year in Acapulco, Mexico, was 40°C. What was the Fahrenheit temperature?

Cumulative Test for Chapters 1–6

Approximately one-half of this test is based on Chapter 6 material. The remainder is based on material covered in Chapters 1–5.

Solve. Simplify your answer.

1. Subtract. $\begin{array}{r} 9824 \\ -\,3796 \\ \hline \end{array}$

2. Multiply. $\begin{array}{r} 608 \\ \times\,305 \\ \hline \end{array}$

3. Divide $32\overline{)8645}$

4. Add. $\frac{1}{7} + \frac{3}{14} + \frac{2}{21}$

5. Subtract. $3\frac{1}{8} - 1\frac{3}{4}$

6. Evaluate. $0.5 \times (9 - 3)^2 \div \frac{2}{3}$

7. Is this equation a proportion? $\frac{21}{35} = \frac{12}{20}$

8. Solve the proportion. $\frac{0.4}{n} = \frac{2}{30}$

9. A piece of wire 6.5 centimeters long weighs 68 grams. What will a 20-centimeter length of the same wire weigh? (Round to the nearest hundredth.)

10. What percent of 66 is 165?

11. Find 15% of 800.

12. 0.5% of what number is 100?

Convert. Express your answer as a decimal rounded to the nearest hundredth when necessary.

13. 38 qt = ____ gal

14. 2.5 tons = ____ lb

15. 7 pt = ____ qt

16. 25 feet = ____ in.

1. 6028

2. 185,440

3. 270 R 5

4. $\frac{19}{42}$

5. $1\frac{3}{8}$

6. 27

7. yes

8. $n = 6$

9. 209.23 grams

10. 250%

11. 120

12. 20,000

13. 9.5

14. 5000

15. 3.5

16. 300

Perform each conversion. Do not round.

17. 3.7 km = ____ m **18.** 62.8 g = ____ kg

Perform each conversion. Do not round answers to 19–21.

19. 0.79 L = ____ mL **20.** 5 cm = ____ m

21. 42 lb = ____ oz **22.** 50°F = ____ C

Perform each conversion. Round to the nearest hundredth when necessary.

23. 28 gal = ____ L **24.** 96 lb = ____ kg

25. 30 in. = ____ cm **26.** 9 mi = ____ km

▲ **27.** Find the perimeter in meters of this triangle.

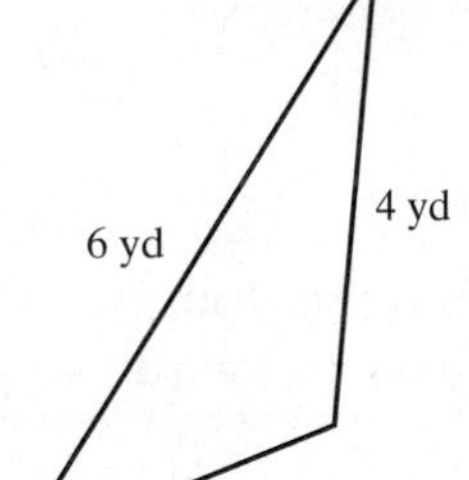

28. Change 15°C to Fahrenheit temperature. Now find the difference between 15°C and 15°F. Which figure represents the higher temperature?

29. Ricardo traveled on a Mexican highway at 100 km/hr for $1\frac{1}{2}$ hours. He needs to travel a total distance of 100 miles. How far does he still need to travel? (Express your answer in miles.)

30. Darla is making holiday placemats to sell at a craft fair. Each placemat requires 2.5 ft of cloth. The fabric store sells cloth by the yard. If Darla plans to make 80 placemats, how many yards of cloth should she buy?

17. 3700

18. 0.0628

19. 790

20. 0.05

21. 672

22. 10°

23. 106.12

24. 43.58

25. 76.2

26. 14.49

27. 11.88 meters

28. 59°F; the difference is 44°F; the 15°C temperature is higher.

29. 7 miles

30. Technically, she needs 66.6 yards, but in real life she should buy 67 yards

CHAPTER

7

Would you feel more crowded in Beijing than in New York City? Would you feel more crowded in Chicago than in Philadelphia? In mathematics, we measure population density to have an exact way to make these comparisons. Can you calculate them? Turn to page 509 to find out.

Geometry

7.1 Angles

Student Learning Objective

After studying this section, you will be able to:

 **Understand and use angles.**

Teaching Tip The word *angle* comes from the Latin word *angulus*, which means "corner." The idea that the number of degrees in a complete circle is 360 degrees is attributed by some scholars to ancient civilizations that believed that 60 was a "perfect number" and that 6 of these 60s made a "perfect circle."

 Understanding and Using Angles

Geometry is a branch of mathematics that deals with the properties of and relationships between figures in space. One of the simplest figures is a *line*. A **line** ↔ extends indefinitely, but a portion of a line, called a **line segment,** has a beginning and an end. A **ray** is a part of a line that has only one endpoint and goes on forever in one direction. An **angle** is made up of two rays that start at a common endpoint. The two line segments are called the **sides** of the angle. The point at which they meet is called the **vertex** of the angle.

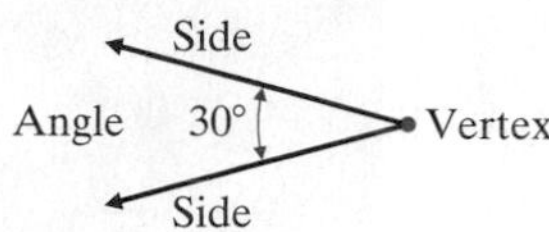

The "amount of opening" of an angle can be measured. Angles are commonly measured in **degrees.** In the preceding sketch the angle measures 30 degrees, or 30°. The symbol ° indicates degrees. If you fix one side of an angle and keep moving the other side, the angle measure will get larger and larger until eventually you have gone around in one complete revolution.

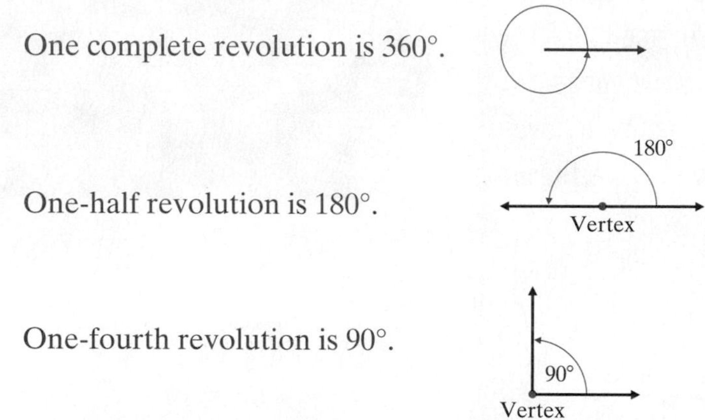

We call two lines **perpendicular** when they meet at an angle of 90°. A 90° angle is called a **right angle.** A 90° angle is often indicated by a small □ at the vertex. Thus when you see ∟ you know that the angle is 90° and also that the sides are perpendicular to each other. The following three angles are right angles.

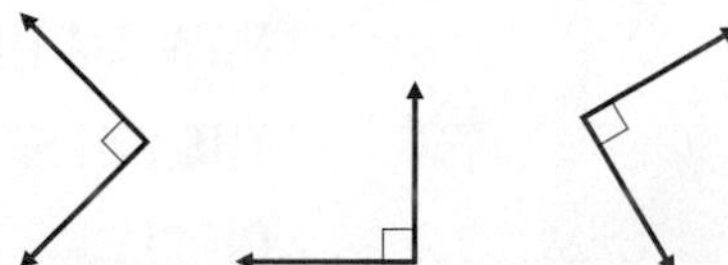

Often, to avoid confusion, angles are labeled with letters. Suppose we consider the angle with a vertex at point B. This angle can be called $\angle ABC$ or $\angle CBA$. Notice that when three letters are used, the middle letter is the vertex. This angle can also be called $\angle B$ or $\angle x$.

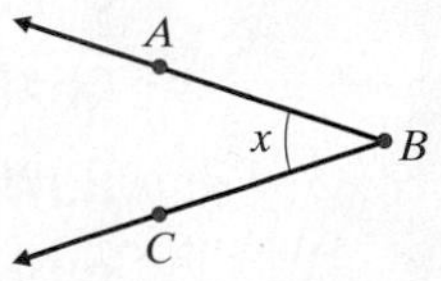

Now consider the following angles. We could label angle y as $\angle DEF$ or angle FED. However, we could not label it as $\angle E$ because this would be unclear. If we refer to $\angle E$, people would not know for sure whether we mean $\angle DEF$, $\angle FEG$, or $\angle DEG$. In such cases where there might be some confusion, the use of the three-letter label is always preferred.

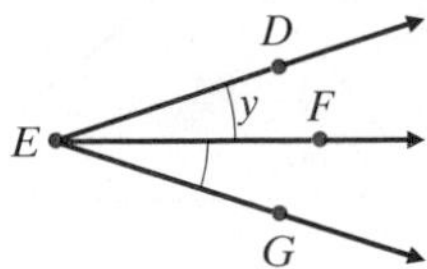

Certain types of angles are commonly encountered. It is important to learn their names. An angle that measures 180° is called a **straight angle.** Angle ABC in the following figure is a straight angle. As we mentioned previously, this is one-half of a revolution.

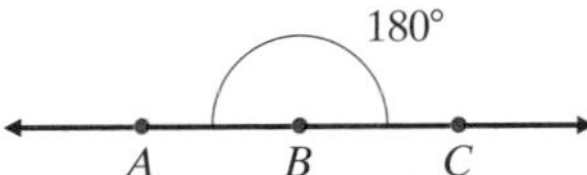

An angle whose measure is between but not including 0° and 90° is called an **acute angle.** $\angle DEF$ and $\angle GHJ$ are both acute angles.

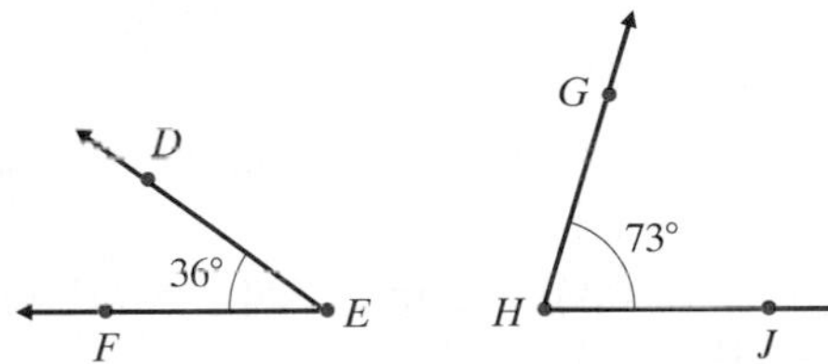

An angle whose measure is between but not including 90° and 180° is called an **obtuse angle.** $\angle ABC$ and $\angle JKL$ are both obtuse angles.

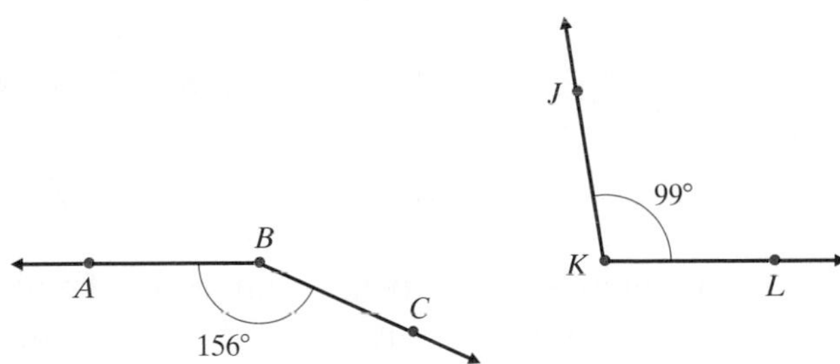

Surveyors make accurate measurements of angles so that *reliable* maps of land regions and buildings can be made.

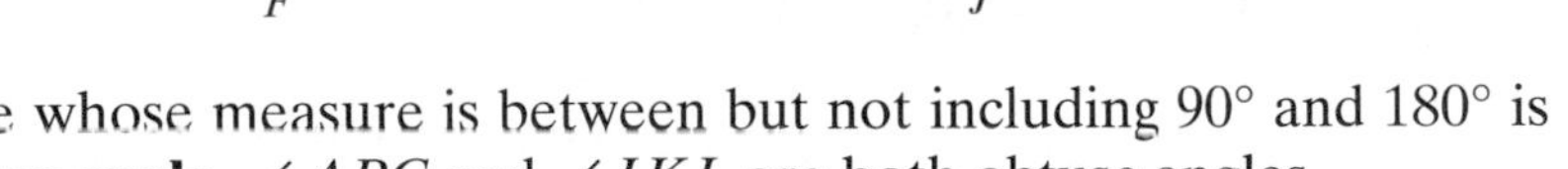

EXAMPLE 1 In the following sketch, determine which angles are acute, obtuse, right, or straight angles.

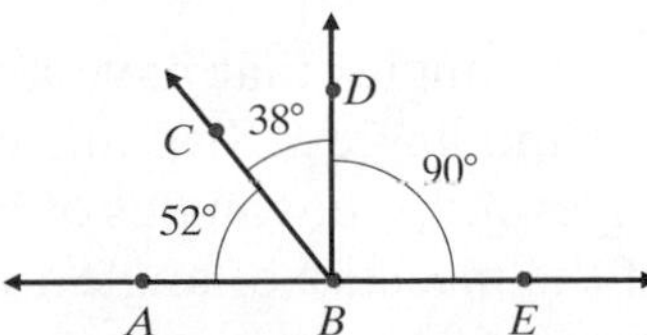

Solution $\angle ABC$ and $\angle CBD$ are acute angles, $\angle CBE$ is an obtuse angle, $\angle ABD$ and $\angle DBE$ are right angles, and $\angle ABE$ is a straight angle.

NOTE TO STUDENT: Fully worked-out solutions to all of the Practice Problems can be found at the back of the text starting at page SP-1

Practice Problem 1 In the following sketch, determine which angles are acute, obtuse, right, or straight angles.

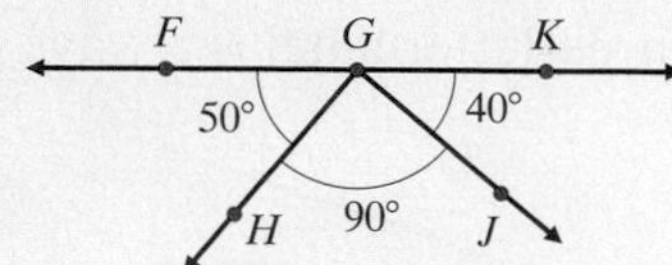

Two angles that have a sum of 90° are called **complementary angles.** We can therefore say that each angle is the **complement** of the other. Two angles that have a sum of 180° are called **supplementary angles.** In this case we say that each angle is the **supplement** of the other.

EXAMPLE 2 Angle A measures 39°.

(a) Find the complement of angle A.

(b) Find the supplement of angle A.

Solution

(a) Complementary angles have a sum of 90°. So the complement of angle A is $90° - 39° = 51°$.

(b) Supplementary angles have a sum of 180°. So the supplement of angle A is $180° - 39° = 141°$.

Practice Problem 2 Angle B measures 83°.

(a) Find the complement of angle B.

(b) Find the supplement of angle B.

Four angles are formed when two lines intersect. Think of how you have four angles if two straight streets intersect. The two angles that are opposite each other are called **vertical angles.** Vertical angles have the same measure. In the following sketch, angle x and angle z are vertical angles, and they have the same measure. Also, angle w and angle y are vertical angles, so they have the same measure.

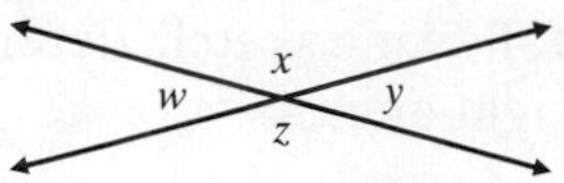

Now suppose we consider two angles that have a common side and a common vertex, such as angle w and angle x. Two angles that share a common side are called **adjacent** angles. Adjacent angles of intersecting lines are supplementary. If we know that the measure of angle x is 120°, then we also know that the measure of angle w is 60°.

EXAMPLE 3 In the following sketch, two lines intersect forming four angles. The measure of angle a is 55°. Find the measure of all the other angles.

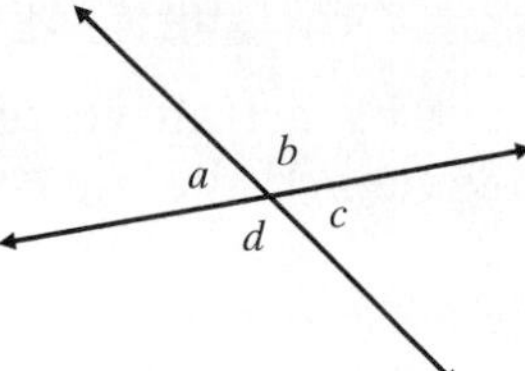

Solution Since $\angle a$ and $\angle c$ are vertical angles, we know that they have the same measure. Thus we know that $\angle c$ measures 55°.

Since $\angle a$ and $\angle b$ are adjacent angles of intersecting lines, we know that they are supplementary angles. Thus we know that $\angle b$ measures $180° - 55° = 125°$.

Finally, $\angle b$ and $\angle d$ are vertical angles, so we know that they have the same measure. Thus we know that $\angle d$ measures 125°.

Practice Problem 3 In the following sketch, two lines intersect forming four angles. The measure of angle y is 133°. Find the measure of all the other angles.

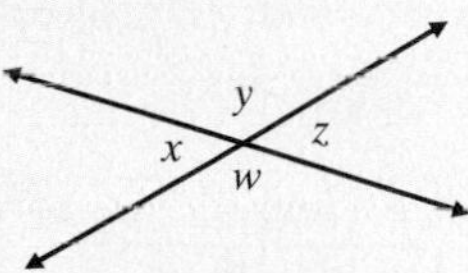

In mathematics there is a common notation for perpendicular lines. If line m is perpendicular to line n, we write $m \perp n$. **Parallel lines** never meet. If line p is parallel to line q, we write $p \parallel q$.

One more situation that is very important in geometry involves lines and angles. A line that intersects two or more lines at different points is called a **transversal.** In the following figure, line m is a transversal that intersects line n and line p. **Alternate interior angles** are two angles that are on opposite sides of the transversal and between the other two lines. In the figure, $\angle c$ and $\angle w$ are alternate interior angles.

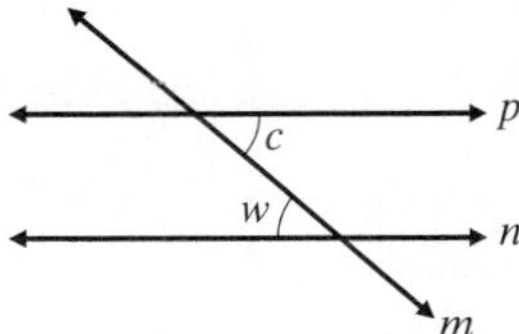

Corresponding angles are two angles that are on the same side of the transversal and are both above (or both below) the other two lines. In the following figure, angle a and angle b are corresponding angles.

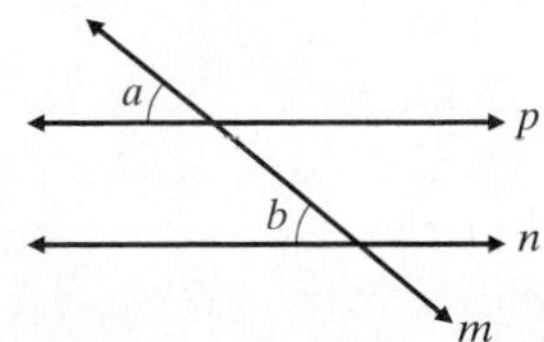

The most important case occurs when the two lines cut by the transversal are parallel. We will state this as follows:

PARALLEL LINES CUT BY A TRANSVERSAL

If two parallel lines are cut by a transversal, then the measures of **corresponding angles are equal** and the measures of **alternate interior angles** are equal.

EXAMPLE 4 In the following figure, $m \parallel n$ and the measure of $\angle a$ is 64°. Find the measures of $\angle b$, $\angle c$, $\angle d$, and $\angle e$.

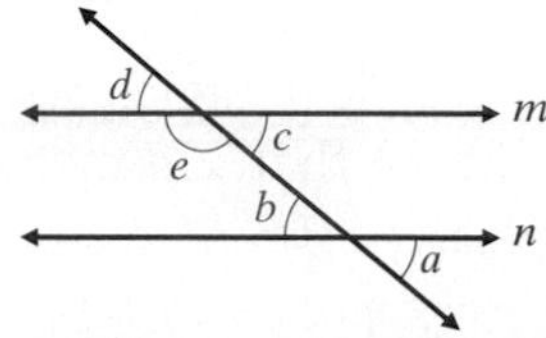

Solution

$\angle a = \angle b = 64°$. $\angle a$ and $\angle b$ are vertical angles.

$\angle b = \angle c = 64°$. $\angle b$ and $\angle c$ are alternate interior angles.

$\angle b = \angle d = 64°$. $\angle b$ and $\angle d$ are corresponding angles.

$\angle e = 180° - 64° = 116°$. $\angle e$ and $\angle d$ are adjacent angles of intersecting lines.

NOTE TO STUDENT: Fully worked-out solutions to all of the Practice Problems can be found at the back of the text starting at page SP-1

Practice Problem 4 In the following figure, $p \parallel q$ and the measure of $\angle x$ is 105°. Find the measures of $\angle w$, $\angle y$, $\angle z$, and $\angle v$.

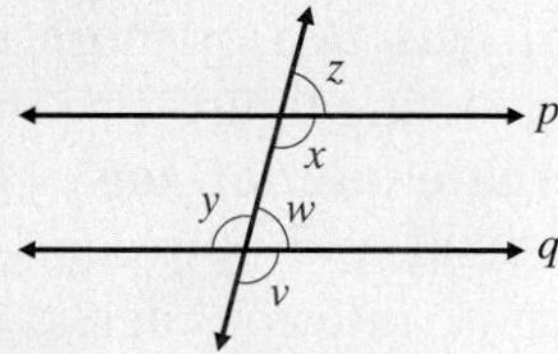

7.1 Exercises

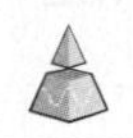

Student Solutions Manual | CD/Video | PH Math Tutor Center | MathXL®Tutorials on CD | MathXL® | MyMathLab® | Interactmath.com

Verbal and Writing Skills

In your own words, give a definition for each term.

1. acute angle
An acute angle is an angle whose measure is between 0° and 90°.

2. obtuse angle
An obtuse angle is an angle whose measure is between 90° and 180°.

3. complementary angles
Complementary angles are two angles whose measures have a sum of 90°.

4. supplementary angles
Supplementary angles are two angles whose measures have a sum of 180°.

5. vertical angles
When two lines intersect, the two angles that are opposite each other are called vertical angles.

6. adjacent angles
Two angles that are formed by intersecting lines and have a common side and a common vertex are called adjacent angles.

7. transversal
A transversal is a line that intersects two or more other lines at different points.

8. alternate interior angles
If a transversal intersects two lines, the two angles that are on opposite sides of the transversal and are between the other two lines are called alternate interior angles.

In exercises 9–14, two straight lines intersect at B, as shown in the following sketch.

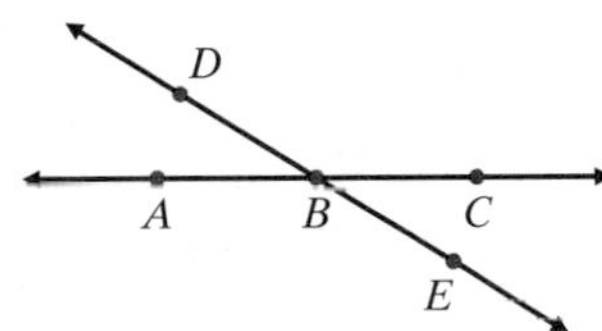

9. Name all the acute angles.
$\angle ABD$, $\angle CBE$

10. Name all the obtuse angles.
$\angle DBC$, $\angle ABE$

11. Name two pairs of angles that have the same measure.
$\angle ABD$ and $\angle CBE$; $\angle DBC$ and $\angle ABE$

12. Name two pairs of angles that are supplementary.
$\angle ABD$ and $\angle DBC$; $\angle ABE$ and $\angle CBE$

13. Name two pairs of angles that are complementary, if any exist.
There are no complementary angles.

14. Name two different straight angles.
$\angle ABC$ and $\angle DBE$

In exercises 15–22, find the measure of each angle, as shown in the following sketch. Assume that angle LOK is a right angle, and angle JOK is a straight angle.

15. $\angle LOJ$
90°

16. $\angle NOL$
65°

17. $\angle JON$
25°

18. $\angle NOM$
85°

19. $\angle JOM$
110°

20. $\angle MOK$
70°

21. $\angle NOK$
155°

22. $\angle KOJ$
180°

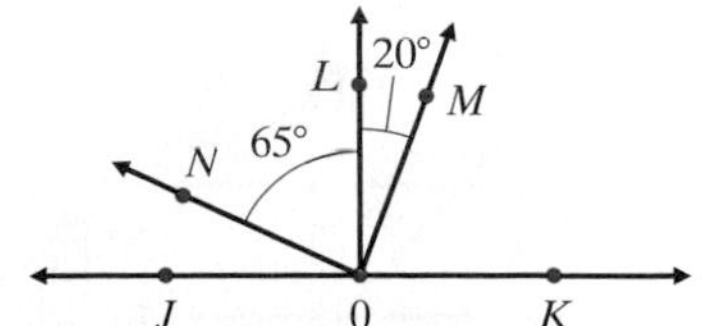

23. Find the complement of an angle that measures 31°. 59°

24. Find the complement of an angle that measures 86°. 4°

25. Find the supplement of an angle that measures 127°. 53°

26. Find the supplement of an angle that measures 8°. 172°

Find the measure of $\angle a$.

27. 34°

28. 16°

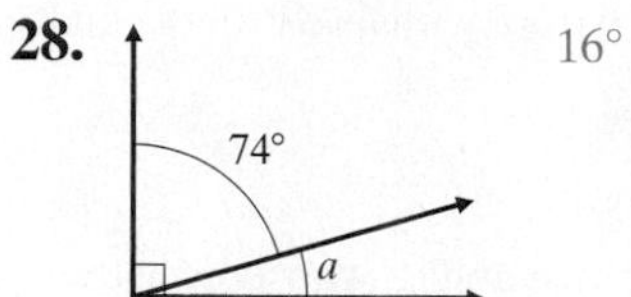

29. 35°

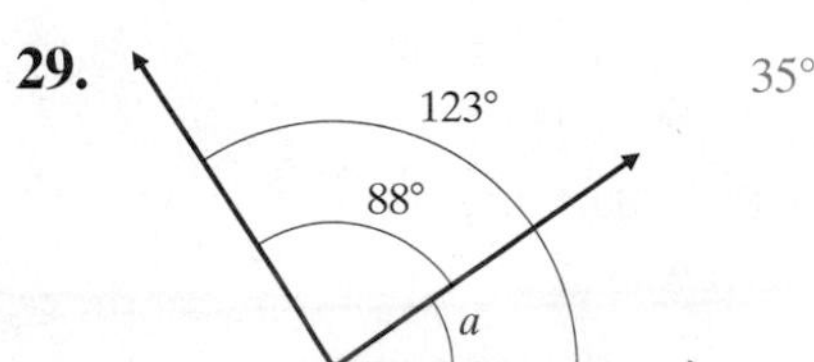

30. 44°

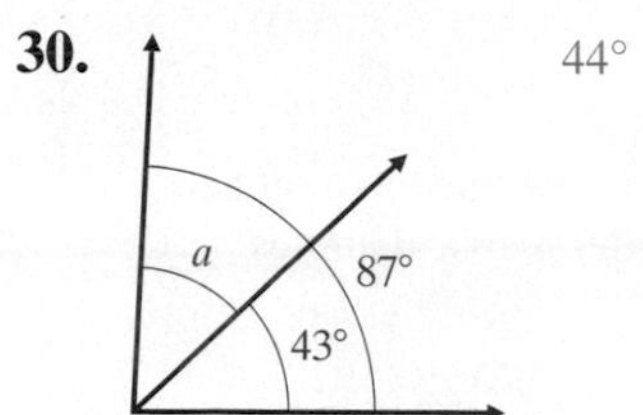

31. 25°

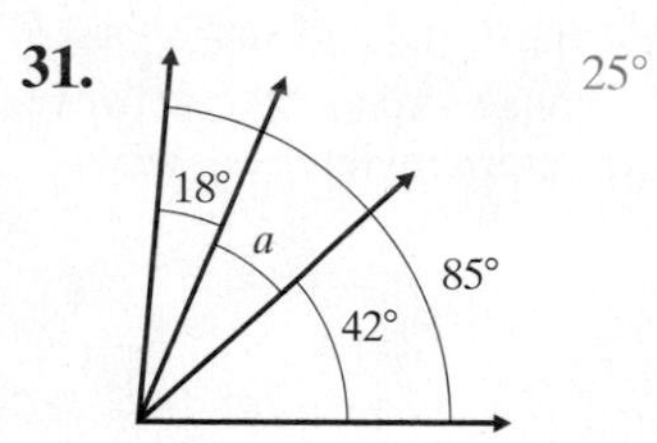

32. 45°

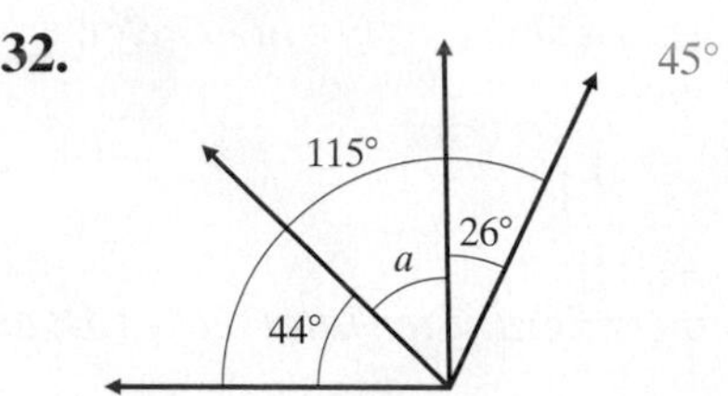

Find the measures of $\angle a$, $\angle b$, *and* $\angle c$.

33.

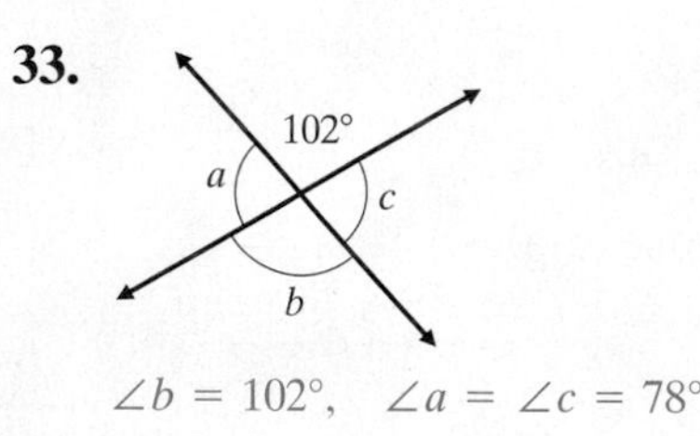

$\angle b = 102°$, $\angle a = \angle c = 78°$

34.

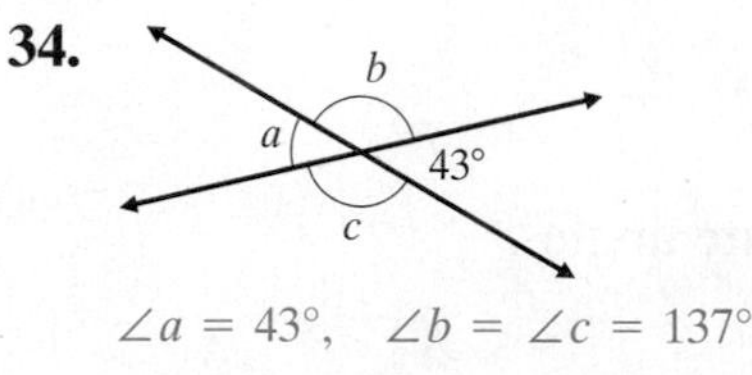

$\angle a = 43°$, $\angle b = \angle c = 137°$

35.

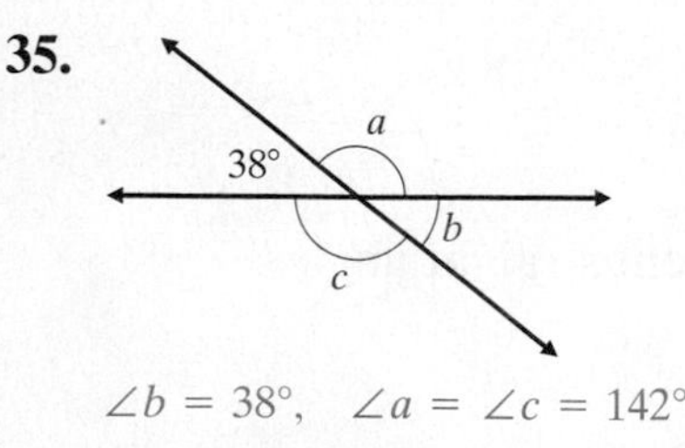

$\angle b = 38°$, $\angle a = \angle c = 142°$

36.

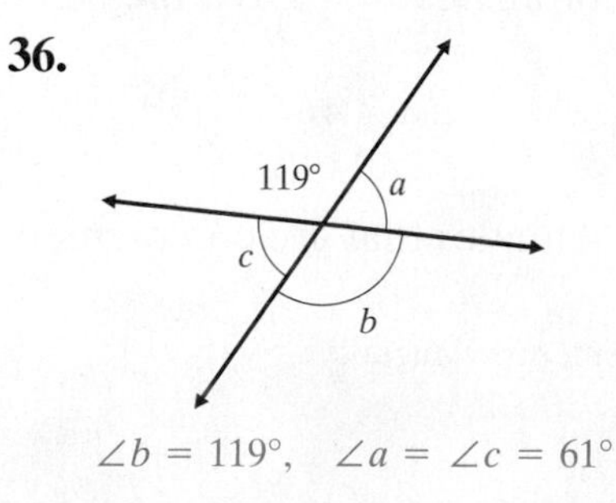

$\angle b = 119°$, $\angle a = \angle c = 61°$

Find the measures of $\angle a$, $\angle b$, *and* $\angle c$ *if we know that* $p \parallel q$.

37.

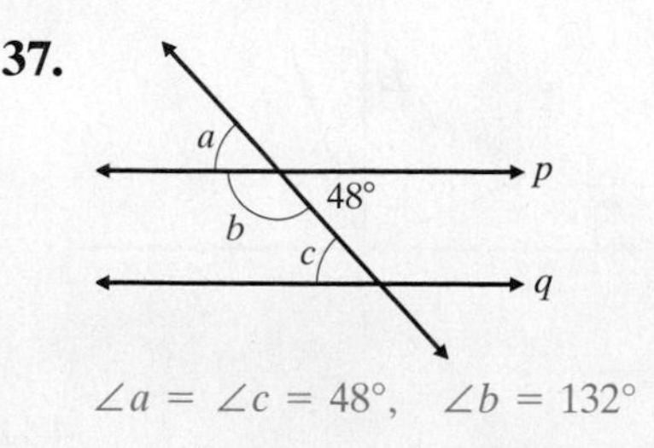

$\angle a = \angle c = 48°$, $\angle b = 132°$

38.

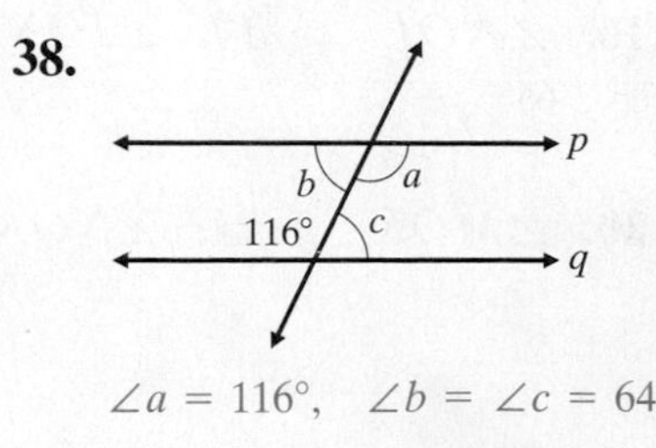

$\angle a = 116°$, $\angle b = \angle c = 64°$

Find the measures of $\angle a$, $\angle b$, $\angle c$, $\angle d$, $\angle e$, $\angle f$, and $\angle g$ if we know that $p \parallel q$.

39.

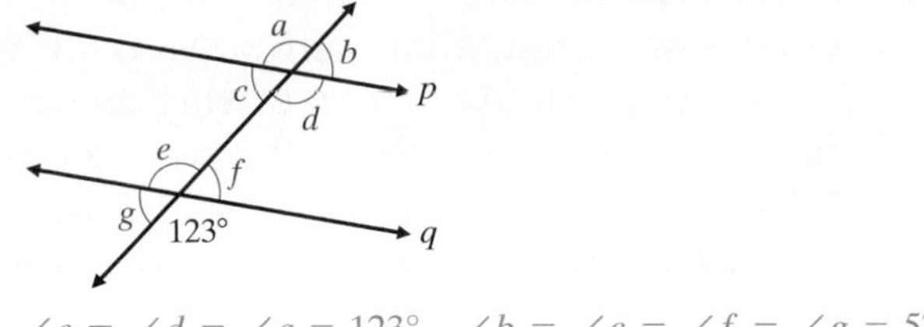

$\angle e = \angle d = \angle a = 123°$ $\angle b = \angle c = \angle f = \angle g = 57°$

40.

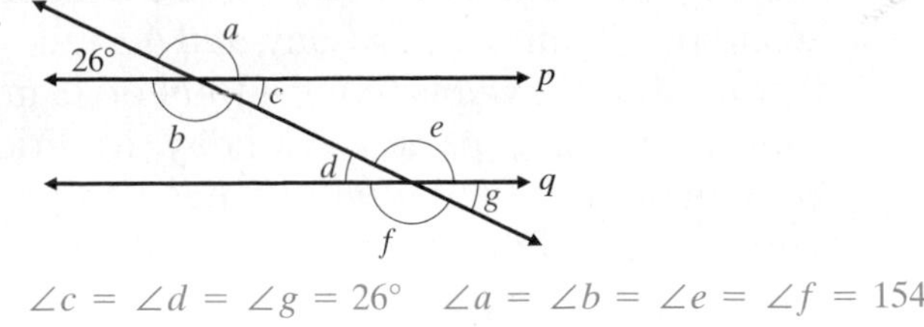

$\angle c = \angle d = \angle g = 26°$ $\angle a = \angle b = \angle e = \angle f = 154°$

Applications

41. ***Leaning Tower of Pisa*** The famous Leaning Tower of Pisa has an angle of inclination of 84°. Scientists are working now to move the tower slightly so that it does not lean so much. Find the angle x, at which the tower deviates from the normal upright position.

6°

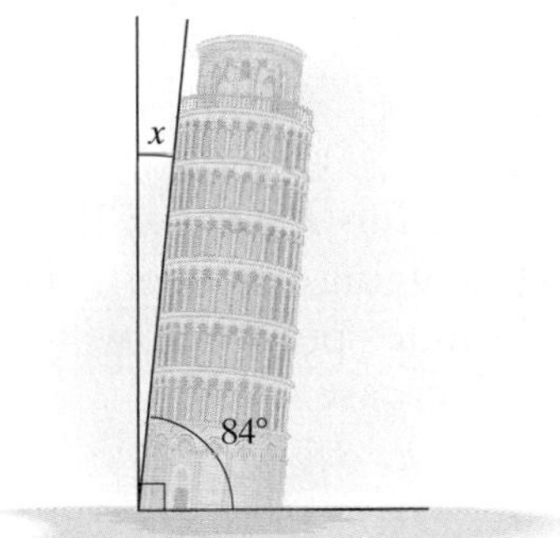

42. ***Pyramids of Monte Albán*** In Mexico the famous pyramids of Monte Albán are visited by thousands of tourists each month. The one most often climbed by tourists is steeper than the pyramids of Egypt and tourists find the climb very challenging. Find the angle x, which indicates the angle of inclination of the pyramid, based on the following sketch.

59°

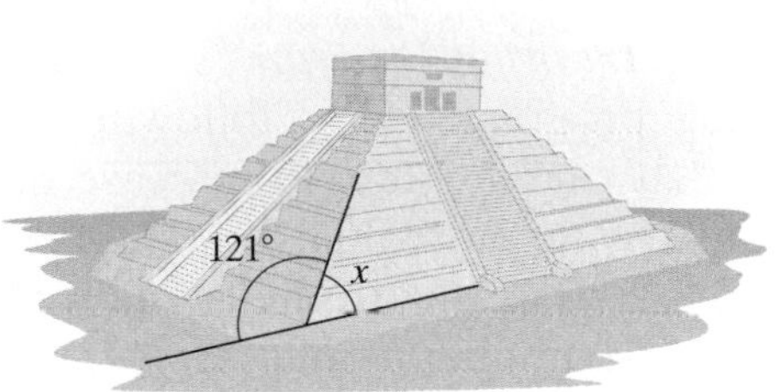

43. ***Course of Jet Plane*** A jetliner is flying 56° north of east when it leaves the airport at Dallas–Fort Worth. The control tower orders the plane to change course by turning to the right 7°. Describe the new course in terms of how many degrees north of east the plane is flying.

49° north of east

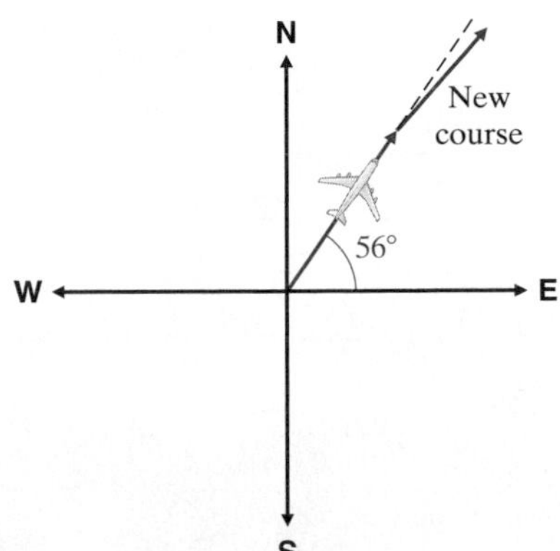

44. ***Course of Cruise Ship*** A cruise ship is leaving Bermuda and is heading on a course 72° north of west. The captain orders that the ship be turned to the left 9°. Describe the new course in terms of how many degrees north of west the ship is heading.

63° north of west

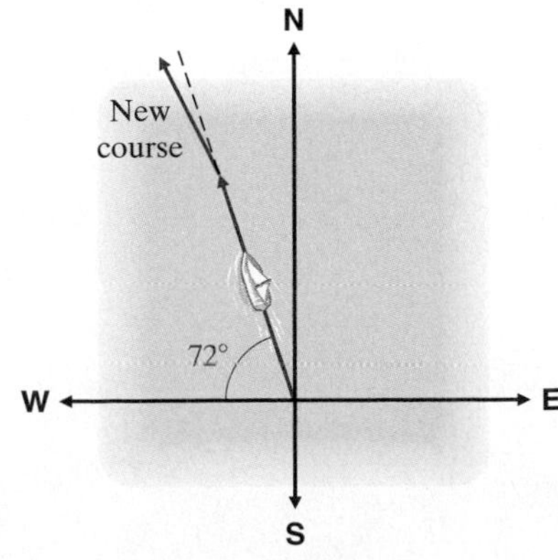

Cumulative Review

45. ***Jogging Workout*** Antonio jogged 3.2 miles on Monday, 5.8 miles on Tuesday, and 4.3 miles on Wednesday. He wants to jog a total of 23 miles over a five-day period, Monday to Friday. How many total miles will he need to jog on Thursday and Friday?
9.7 miles

46. ***Driving in Mexico*** While driving in Mexico, Greg saw a sign that said MEXICO CITY 34 KILOMETERS AHEAD. He is driving an American car with an odometer that reads in miles. How many miles farther does he need to drive? Round to the nearest tenth of a mile.
21.1 miles

47. ***Stock Market*** Dr. Finkelstein owns 1000 shares of IBM, 500 shares of America Online, 300 shares of General Electric, and 400 shares of Microsoft. As he was reviewing his stocks over the last year, he observed that IBM had increased in value \$4 per share, America Online had increased \$3 per share, and General Electric had increased \$1 per share. Unfortunately, Microsoft had lost value and was \$5 less per share. In total, how much had the value of Dr. Finkelstein's stock increased over the last year? \$3800

48. ***Boston Bridge*** In constructing a new suspension bridge over the Charles River in Boston, engineers have used 34 tons of cable on the first 800 feet of the bridge. How many additional tons of cable will be used to complete the remaining 1300 feet of the bridge? Round to the nearest tenth.
55.3 tons

49. ***Jogging Training Program*** Seth has started a jogging program. He is following a training plan that requires him to increase his weekly mileage by no more than 5%. This week he ran 24 miles. According to the training plan, what is the most number of miles he should run next week?
25.2 mi

50. ***Jogging Training Program*** If Seth increased his weekly jogging mileage from 24 miles per week to 27 miles per week, what would be the percent of increase?
12.5%

7.2 Rectangles and Squares

1 Finding the Perimeters of Rectangles and Squares

Geometry has a visual aspect that numbers and abstract ideas do not have. We can take pen in hand and draw a picture of a rectangle that represents a room with certain dimensions. We can easily visualize problems such as "What is the distance around the outside edges of the room (perimeter)?" or "How much carpeting will be needed for the room (area)?"

A rectangle is a four-sided figure like those shown here.

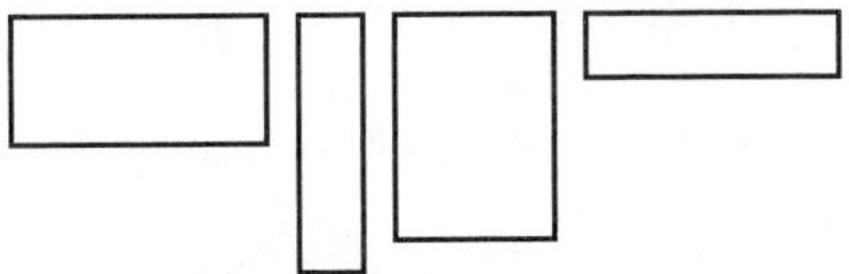

A rectangle has two interesting properties: (1) Any two adjoining sides are perpendicular and (2) the lengths of the opposite sides of a rectangle are equal. By "any two adjoining sides are perpendicular," we mean that any two sides that meet form an angle that measures 90°. We indicate the 90° angle with a small red box □ at each corner. When we say that "the lengths of the opposite sides of a rectangle are equal," we mean that the measure of one side is equal to the measure of the side opposite to it. Thus, we define a **rectangle** as a four-sided figure that has four right angles. If all four sides have the same length, then the rectangle is called a **square.**

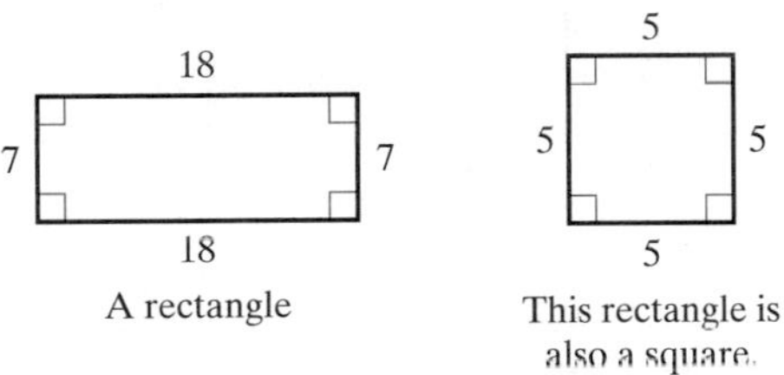

A rectangle

This rectangle is also a square.

A farmer owns some land in the Colorado mountains. It is in the shape of a rectangle. The **perimeter** of a rectangle is the sum of the lengths of all its sides. To find the perimeter of the rectangular field shown in the following figure, we add up the lengths of all the sides of the field.

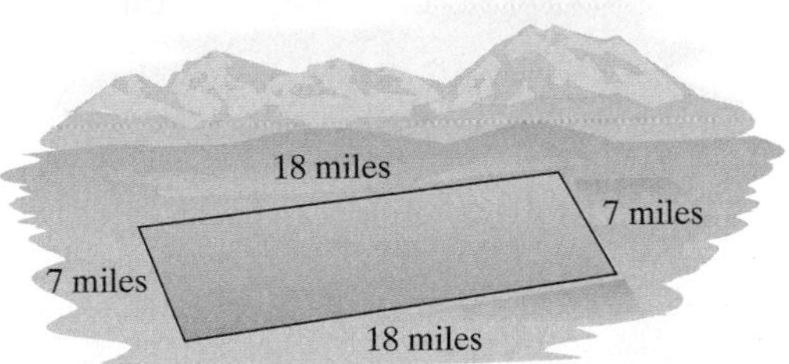

$$\begin{aligned}\text{Perimeter} &= 7 \text{ miles} + 18 \text{ miles} + 7 \text{ miles} + 18 \text{ miles}\\ &= 50 \text{ miles}\end{aligned}$$

Thus the perimeter of the field is 50 miles.

We could also use a formula to find the perimeter of a rectangle. In the formula we use letters to represent the measurements of the length and width of the rectangle. Let l represent the length, w represent the width, and P represent the perimeter. Note that the length is the longer side and the

Student Learning Objectives

After studying this section, you will be able to:

1. **Find the perimeters of rectangles and squares.**
2. **Find the perimeters of shapes made up of rectangles and squares.**
3. **Find the areas of rectangles and squares.**
4. **Find the areas of shapes made up of rectangles and squares.**

Teaching Tip The idea of perpendicular sides may be a little vague for students who have absolutely no background in geometry. A good class discussion question is, "Can you give me some examples from everyday life of perpendicular lines?" By discussing a couple of the class examples and sketching them on the blackboard, you can clear up any student confusion.

width is the shorter side. Since the perimeter is found by adding up the measurements all around the rectangle, we see that

$$\begin{aligned} P &= w + l + w + l \\ &= 2l + 2w. \end{aligned}$$

When we write $2l$ and $2w$, we mean 2 times l and 2 times w. We can use the formula to find the perimeter of the rectangle.

$$\begin{aligned} P &= 2l + 2w \\ &= (2)(18 \text{ mi}) + (2)(7 \text{ mi}) \\ &= 36 \text{ mi} + 14 \text{ mi} \\ &= 50 \text{ mi} \end{aligned}$$

Notice that we use parentheses ( ) here to indicate multiplication of 2×18 and 2×7.

Thus the perimeter can be found quickly by using the following formula.

The **perimeter (P) of a rectangle** is twice the length plus twice the width.

$$P = 2l + 2w$$

EXAMPLE 1 A helicopter has a 3-cm by 5.5-cm insulation pad near the control panel that is rectangular. Find the perimeter of the rectangle.

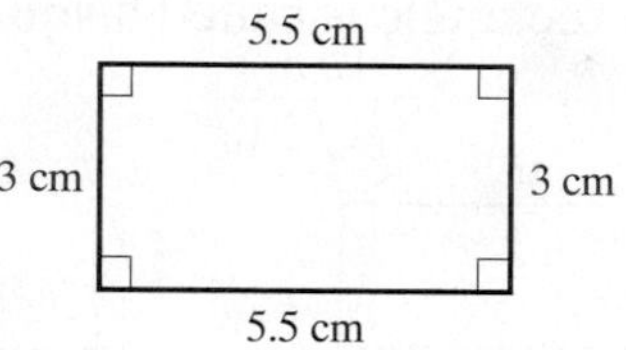

Solution

$$\text{Length} = l = 5.5 \text{ cm}$$
$$\text{Width} = w = 3 \text{ cm}$$

In the formula for the perimeter of a rectangle, we substitute 5.5 cm for l and 3 cm for w. Remember, $2l$ means 2 times l and $2w$ means 2 times w. Thus

$$\begin{aligned} P &= 2l + 2w \\ &= (2)(5.5 \text{ cm}) + (2)(3 \text{ cm}) \\ &= 11 \text{ cm} + 6 \text{ cm} = 17 \text{ cm}. \end{aligned}$$

***NOTE TO STUDENT:** Fully worked-out solutions to all of the Practice Problems can be found at the back of the text starting at page SP-1*

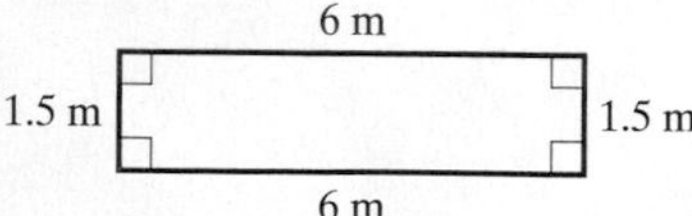

Practice Problem 1 Find the perimeter of the rectangle in the margin.

A square is a rectangle where all four sides have the same length. Since a rectangle is defined to have four right angles, all squares have four right angles. Some examples of squares are shown in the following figure.

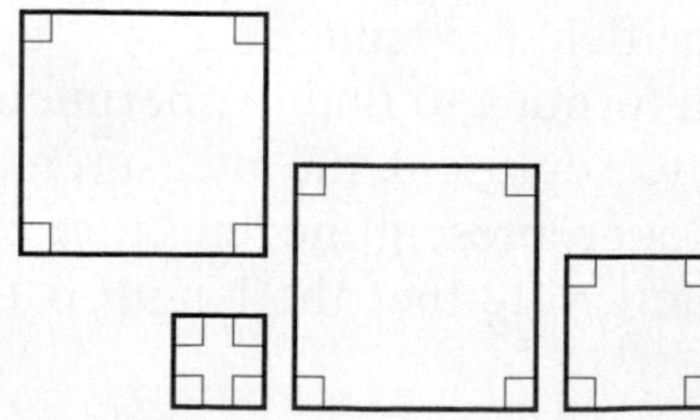

A square, then, is only a special type of rectangle. We can find the perimeter of a square just as we found the perimeter of a rectangle—by adding the measurements of all the sides of the square. Because the lengths of all sides are the same, the formula for the perimeter of a square is very simple. Let s represent the length of one side and P represent the perimeter. So, to find the perimeter, we multiply the length of a side by 4.

The **perimeter of a square** is four times the length of a side.

$$P = 4s$$

EXAMPLE 2 High Ridge Stables has a new sign at the highway entrance that is in the shape of a square, with each side measuring 8.6 yards. Find the perimeter of the sign.

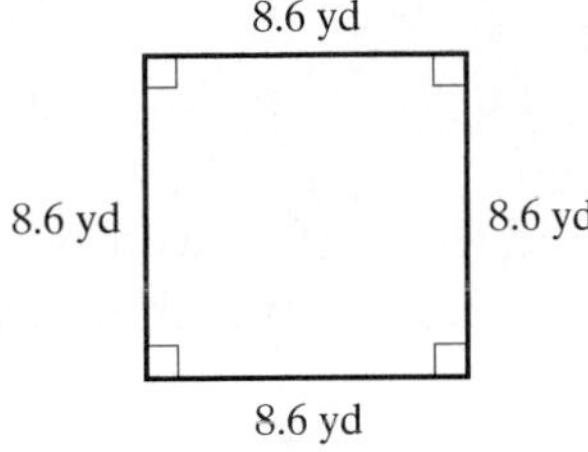

Solution

$$\text{Side} = s = 8.6 \text{ yd}$$
$$P = 4s$$
$$= (4)(8.6 \text{ yd})$$
$$= 34.4 \text{ yd}$$

Practice Problem 2 Find the perimeter of the square in the margin.

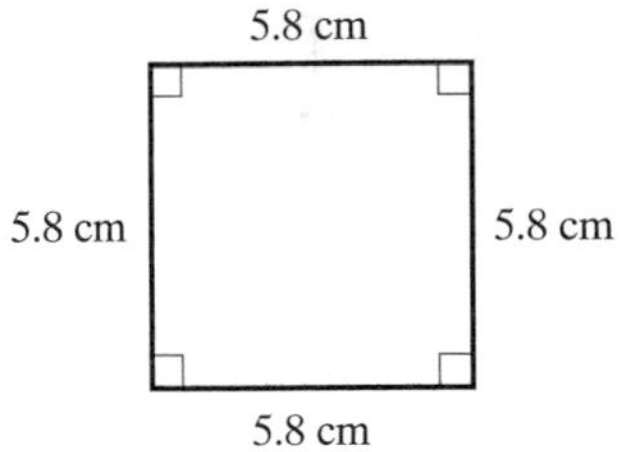

Since drawing the small red boxes sometimes makes drawings overly complicated, we will assume that all drawings in this chapter that appear to be rectangles and squares do in fact have four 90° angles.

2 Finding the Perimeters of Shapes Made Up of Rectangles and Squares

Some figures are a combination of rectangles and squares. To find the perimeter of the total figure, look only at the outside edges.

We can apply our knowledge to everyday problems. For example, by knowing how to find the perimeter of a rectangle, we can find out how many feet of picture framing a painting will need or how many feet of weather stripping will be needed to seal a doorway. Consider the following problem.

Teaching Tip Some students have significant difficulty visualizing an object that is made up of two or more geometric objects. Take the time to show how the object is "created" by putting two shapes together.

EXAMPLE 3 Find the cost of weather stripping needed to seal the edges of the hatch of a boat pictured at left. Weather stripping costs \$0.12 per foot.

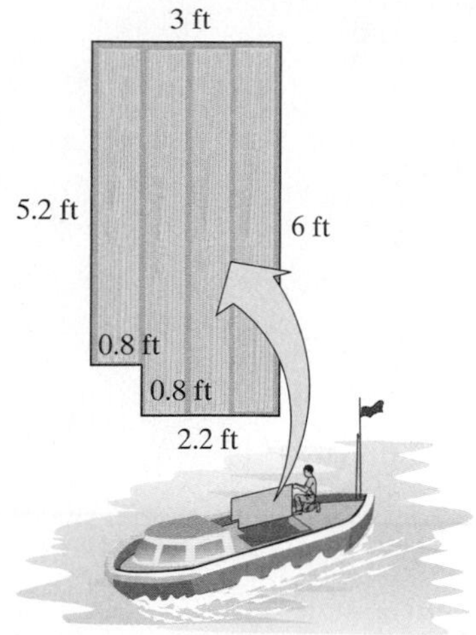

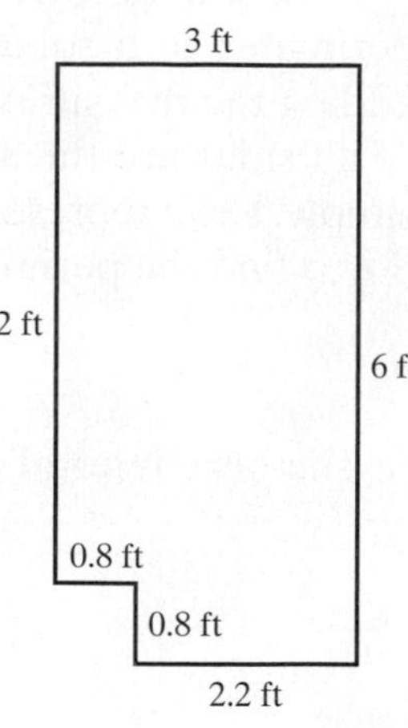

Solution First we need to find the perimeter of the hatch. The perimeter is the sum of all the edges.

$$\begin{array}{r} 3.0 \text{ ft} \\ 6.0 \text{ ft} \\ 2.2 \text{ ft} \\ 0.8 \text{ ft} \\ 0.8 \text{ ft} \\ +5.2 \text{ ft} \\ \hline 18.0 \text{ ft} \end{array}$$

The perimeter is 18 ft. Now we calculate the cost.

$$18.0 \cancel{\text{ft}} \times \frac{0.12 \text{ dollar}}{\cancel{\text{ft}}} = \$2.16 \text{ for weather stripping materials}$$

NOTE TO STUDENT: Fully worked-out solutions to all of the Practice Problems can be found at the back of the text starting at page SP-1

Practice Problem 3 Find the cost of weather stripping required to seal the edges of the hatch shown below. Weather seal stripping costs \$0.16 per foot.

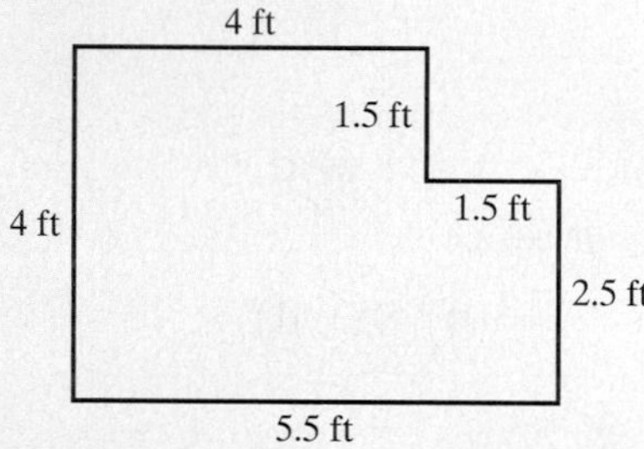

3 Finding the Areas of Rectangles and Squares

What do we mean by **area?** Area is the measure of the *surface inside* a geometric figure. For example, for a rectangular room, the area is the amount of floor in that room.

One *square meter* is the measure of a square that is 1 m long and 1 m wide.

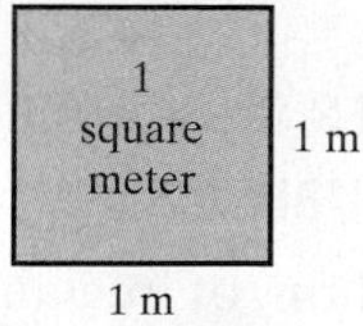

We can abbreviate *square meter* as m^2. In fact, all areas are measured in square meters, square feet, square inches, and so on (written as m^2, ft^2, $in.^2$, and so on).

We can calculate the area of a rectangular region if we know its length and its width. To find the area, *multiply* the length by the width.

> The **area (A) of a rectangle** is the length times the width.
>
> $$A = lw$$

EXAMPLE 4 Find the area of the rectangle shown below.

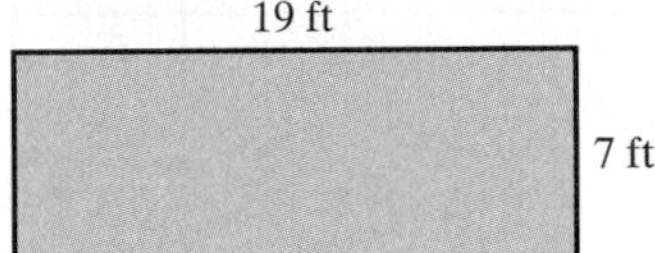

Solution Our answer must be in square feet because the measures of the length and width are in feet.

$$l = 19 \text{ ft} \qquad w = 7 \text{ ft}$$
$$A = (l)(w) = (19 \text{ ft})(7 \text{ ft}) = 133 \text{ ft}^2$$

The area is 133 square feet.

Practice Problem 4 Find the area of the rectangle shown below.

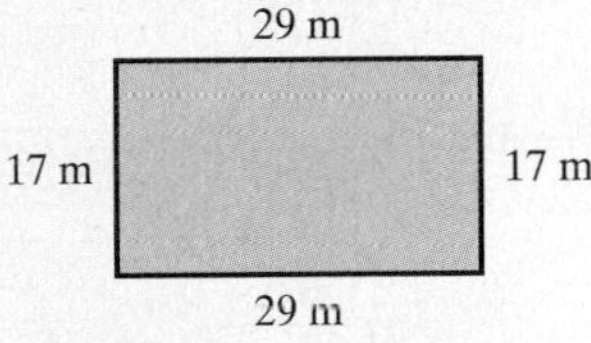

To find the area of a square, we multiply the length of one side by itself.

> The **area of a square** is the square of the length of one side.
>
> $$A = s^2$$

EXAMPLE 5 A square measures 9.6 in. on each side. Find its area.

Solution We know our answer will be measured in square inches. We will write this as in.^{2}.

$$\begin{aligned} A &= s^2 \\ &= (9.6)^2 \\ &= (9.6 \text{ in.})(9.6 \text{ in.}) \\ &= 92.16 \text{ in.}^2 \end{aligned}$$

Practice Problem 5 Find the area of a square computer chip that measures 11.8 mm on each side.

Teaching Tip You can raise the students' thinking level with the following question: "You have just seen how we multiplied 19 feet by 7 feet to obtain 133 square feet in Example 6. Suppose someone said to you, "Prove it." How could you convince someone that a rectangle that measures 19 feet by 7 feet really has an area of 133 square feet?" Student answers will vary. The most convincing proof to students is to take square tiles exactly 1 foot on each side and show that exactly 133 tiles fit inside the rectangle.

4 Finding the Areas of Shapes Made Up of Rectangles and Squares

EXAMPLE 6 Consider the shape shown below, which is made up of a rectangle and a square. Find the area of the shaded region.

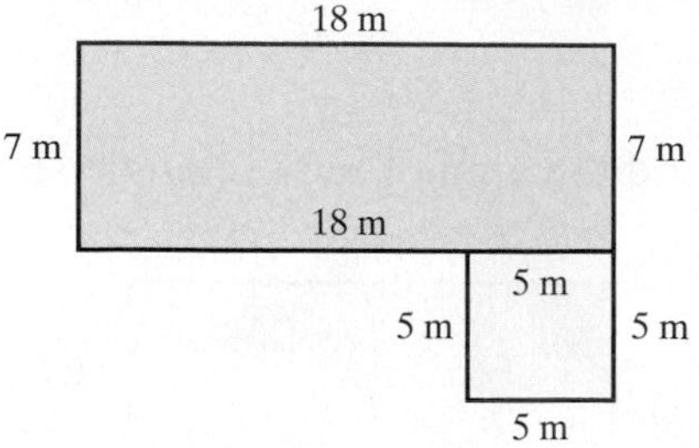

Solution The shaded region is made up of two separate regions. You can think of each separately, and calculate the area of each one. The total area is just the sum of the two separate areas.

Area of rectangle $= (7)(18) = 126 \text{ m}^2$ Area of square $= 5^2 = 25 \text{ m}^2$

The area of the rectangle	=	126 m^2
+ The area of the square	=	25 m^2
The total area is	=	151 m^2

NOTE TO STUDENT: Fully worked-out solutions to all of the Practice Problems can be found at the back of the text starting at page SP-1

Practice Problem 6 Find the area of the shaded region shown in figure below.

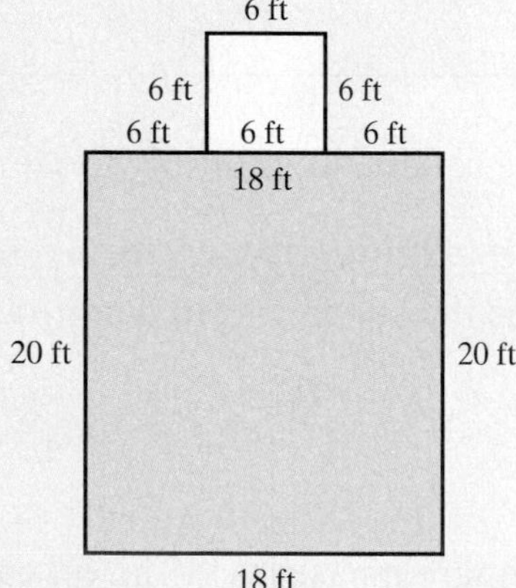

7.2 EXERCISES

Student Solutions Manual | CD/ Video | PH Math Tutor Center | MathXL®Tutorials on CD | MathXL® | MyMathLab® | Interactmath.com

Verbal and Writing Skills

1. A rectangle has two properties: (1) any two adjoining sides are ___perpendicular___ and (2) the lengths of opposite sides are ___equal___.

2. To find the perimeter of a figure, we ___add___ the lengths of all of the sides.

3. To find the area of a rectangle, we ___multiply___ the length by the width.

4. All area is measured in ___square___ units.

Find the perimeter of the rectangle or square.

5.

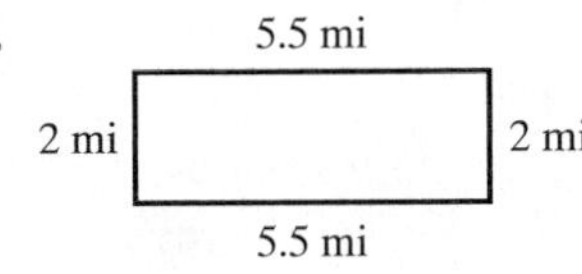

5.5 mi
2.0 mi
5.5 mi
2.0 mi
15 mi

6.

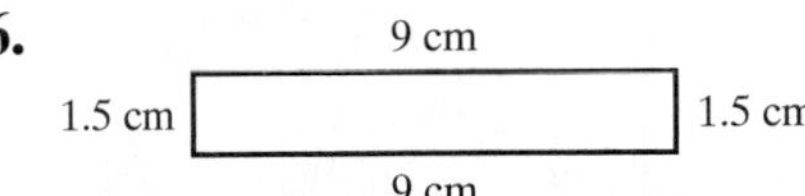

1.5 cm
9.0 cm
1.5 cm
9.0 cm
21 cm

7.

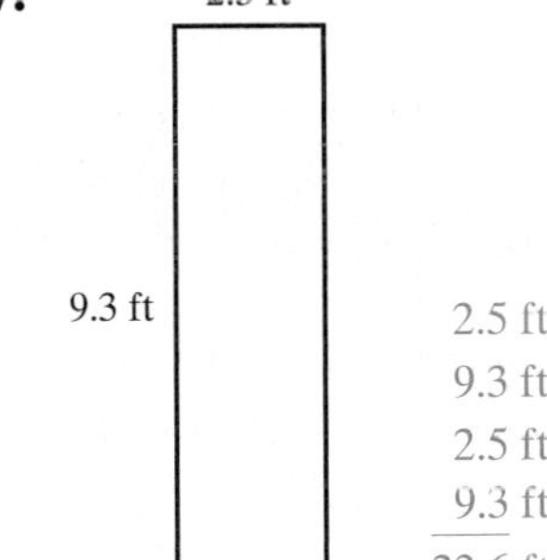

2.5 ft
9.3 ft
2.5 ft
9.3 ft
23.6 ft

8.

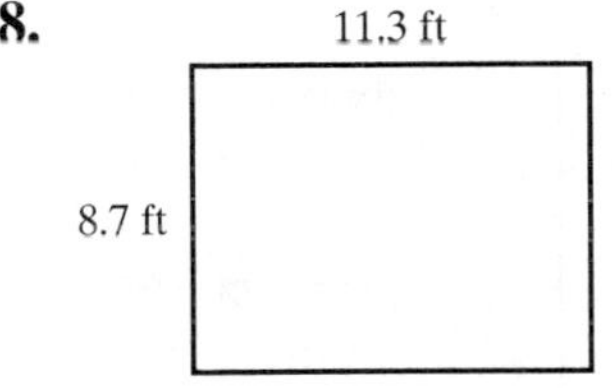

8.7 ft
11.3 ft
8.7 ft
11.3 ft
40 ft

9.

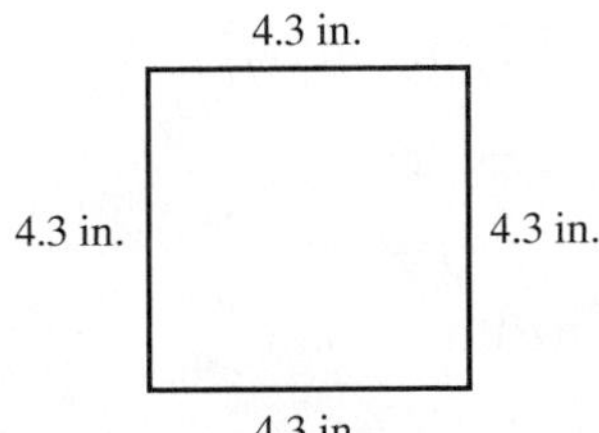

4×4.3 in. $= 17.2$ in.

10.

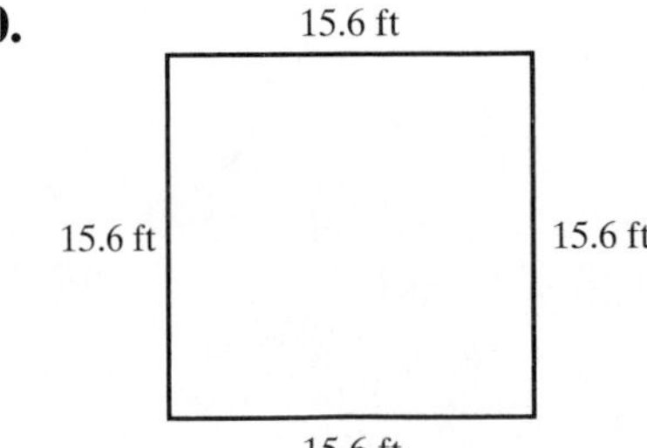

4×15.6 ft $= 62.4$ ft

11. Length $= 0.84$ mm, width $= 0.12$ mm

2×0.12 mm $+ 2 \times 0.84$ mm $= 1.92$ mm

12. Length $= 6.2$ in., width $= 1.5$ in.

2×1.5 in. $+ 2 \times 6.2$ in. $= 15.4$ in.

13. Length = width = 4.28 km
4 × 4.28 km = 17.12 km

14. Length = width = 9.63 cm
4 × 9.63 cm = 38.52 cm

15. Length = 3.2 ft, width = 48 in. (*Hint:* Make the units of length the same.)
2 × 3.2 ft + 2 × 4 ft = 14.4 ft
or 2 × 38.4 in. + 2 × 48 in. = 172.8 in.

16. Length = 8.5 ft, width = 30 in. (*Hint:* Make the units of length the same.)
2 × 8.5 ft + 2 × 2.5 ft = 22 ft
or 2 × 102 in. + 2 × 30 in. = 264 in.

Find the perimeter of the square. The length of the side is given.

17. 0.068 mm
4 × 0.068 mm = 0.272 mm

18. 0.097 mm
4 × 0.097 mm = 0.388 mm

19. 7.96 cm
4 × 7.96 cm = 31.84 cm

20. 6.32 cm
4 × 6.32 cm = 25.28 cm

Find the perimeter of each shape made up of rectangles and squares.

21.

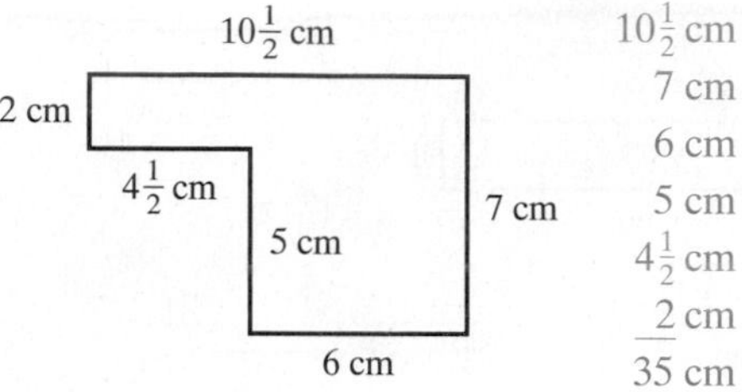

$10\frac{1}{2}$ cm
7 cm
6 cm
5 cm
$4\frac{1}{2}$ cm
2 cm
35 cm

22.

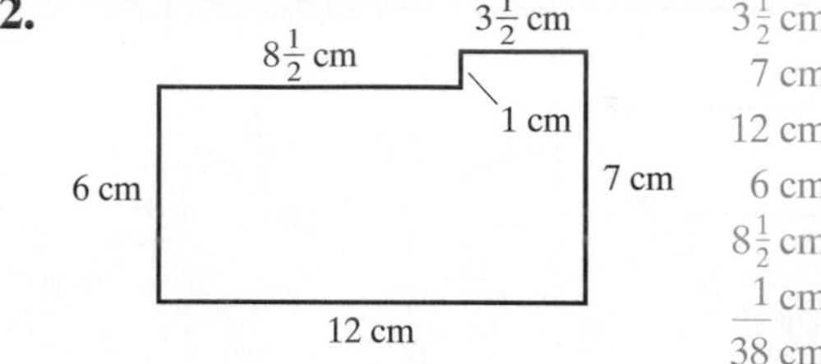

$3\frac{1}{2}$ cm
7 cm
12 cm
6 cm
$8\frac{1}{2}$ cm
1 cm
38 cm

23.

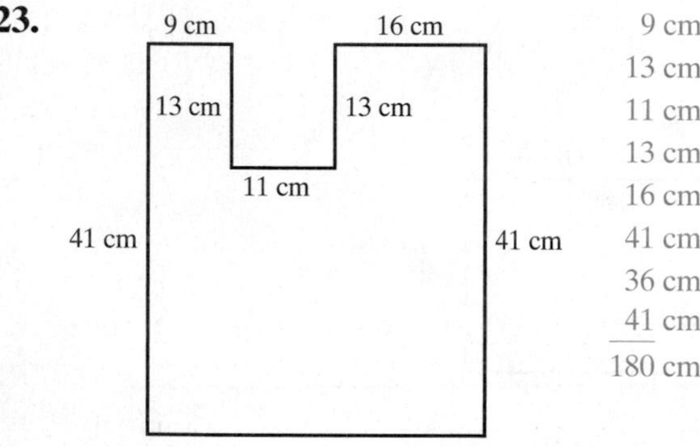

9 cm
13 cm
11 cm
13 cm
16 cm
41 cm
36 cm
41 cm
180 cm

24.

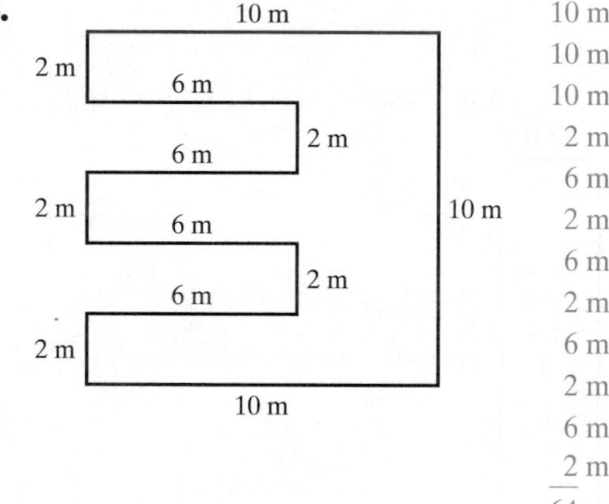

10 m
10 m
10 m
2 m
6 m
2 m
6 m
2 m
6 m
2 m
6 m
2 m
64 m

Find the area of the rectangle or square.

25. Length = width = 2.5 ft
$(2.5)^2 = 6.25\ \text{ft}^2$

26. Length = width = 5.1 m
$(5.1)^2 = 26.01\ \text{m}^2$

27. Length = 8 mi, width = 1.5 mi
8 mi × 1.5 mi = 12 mi^2

28. Length = 10.5 mi, width = 6 mi
10.5 mi × 6 mi = 63 mi^2

29. Length = 39 yd, width = 9 ft (*Hint:* Make the units of length the same.)
39 yd × 3 yd = 117 yd^2 or 117 ft × 9 ft = 1053 ft^2

30. Length = 57 yd, width = 15 ft (*Hint:* Make the units of length the same.)
57 yd × 5 yd = 285 yd^2 or 171 ft × 15 ft = 2565 ft^2

Mixed Practice

31.

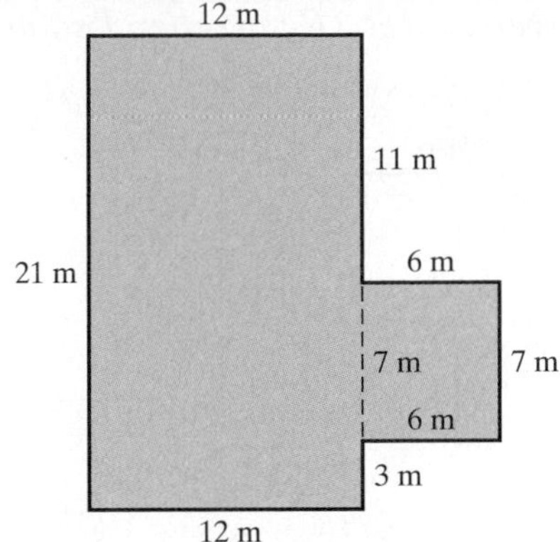

(a) Find the shaded area.
21 m × 12 m + 6 m × 7 m = 294 m^2

(b) Find the perimeter indicated by the black lines. Perimeter is 78 m

32.

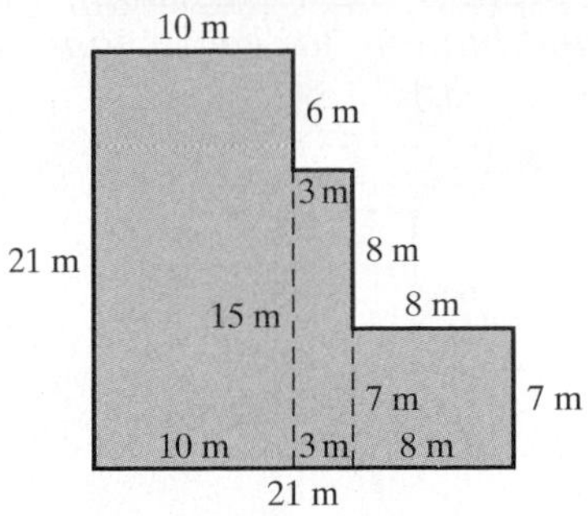

(a) Find the shaded area.
10 m × 6 m + 13 m × 8 m + 21 m × 7 m = 311 m^2

(b) Find the perimeter indicated by the black lines.
Perimeter is 84 m

Applications

Some of the following exercises will require that you find a perimeter. Others will require that you find an area. Read each problem carefully to determine which you are to find.

33. ***Indoor Fitness Area*** A hotel conference center is building an indoor fitness area measuring 220 ft × 50 ft. The flooring to cover the space is made of a special three-layered cushioned tile and costs $12.00 per square foot. How much will the new flooring cost?
$132,000

34. ***Soccer Team Warmup*** The high school soccer team is running the perimeter of their field to warm up for a game. The length of the field is 110 yards, and the width is 65 yards. If they run around the field 5 times, how many yards will they run? How many feet?
1750 yd; 5250 ft

35. ***Quilt at County Fair*** Arlene made a quilt that took first place at the county fair. The quilt measured 12.5 ft by 9.5 ft. She sewed a unique fringed border on each side of it. If the border material cost $1.35 per foot, how much did the border cost?
$59.40

36. ***Scuba Shop Sign*** Sammy's Scuba Shop is installing a new sign measuring 5.4 ft × 8.1 ft. The sign will be framed in purple neon light, which will cost $32.50 per foot. How much will it cost to frame the sign in purple neon light?
$877.50

37. A farmer has 16 feet of fencing. He constructs a rectangular garden whose sides are whole numbers. He uses all the fence to enclose the garden.

(a) How many possible shapes can the garden have?
1 × 7, 2 × 6, 3 × 5, 4 × 4 four possible shapes.

(b) What is the area of each possible garden?
7 sq ft, 12 sq ft, 15 sq ft, 16 sq ft

(c) Which shape has the largest area?
The square garden measuring 4 feet on a side

38. A farmer has 18 feet of fencing. She constructs a rectangular garden whose sides are whole numbers. She uses all the fence to enclose the garden.

(a) How many possible shapes can the garden have?
1 × 8, 2 × 7, 3 × 6, 4 × 5 there are four possible shapes

(b) What is the area of each possible garden?
8 sq ft, 14 sq ft, 18 sq ft, 20 sq ft

(c) Which shape has the largest area?
The rectangular garden measuring 4 feet by 5 feet

Installation of Carpeting *A family decides to have custom carpeting installed. It will cost \$14.50 per square yard. The binding, which runs along the outside edges of the carpet, will cost \$1.50 per yard. Find the cost of carpeting and binding for each room. Note that dimensions are given in feet. (Remember, 1 square yard equals 9 square feet.)*

39.

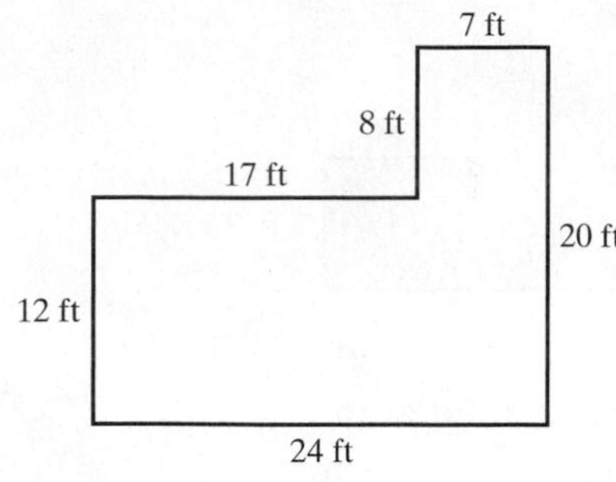

Area = 24 ft × 12 ft + 8 ft × 7 ft = 344 ft^2

Perimeter = 12 ft + 17 ft + 8 ft + 7 ft + 20 ft + 24 ft = 88 ft

$$\text{Cost} = 344\text{ ft}^2 \times \frac{\$14.50}{\text{yd}^2} \times \frac{1\text{ yd}^2}{9\text{ ft}^2} + 88\text{ ft} \times \frac{1\text{ yd}}{3\text{ ft}} \times \frac{1.50}{\text{yd}} = \$598.22$$

40.

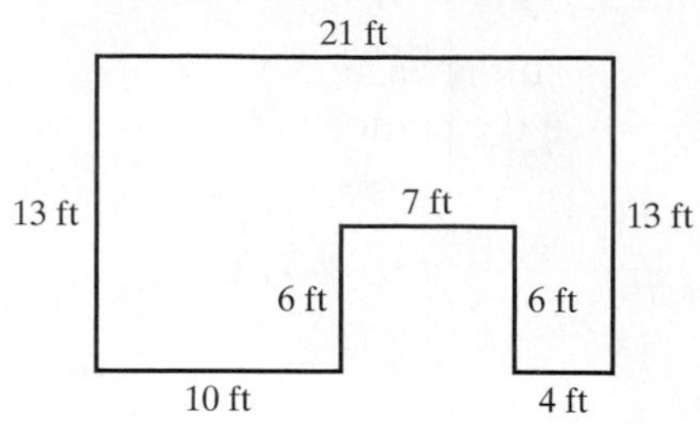

Area = 21 ft × 13 ft − 6 ft × 7 ft = 231 ft^2

Perimeter = 2 × 13 ft + 2 × 21 ft + 2 × 6 ft = 80 ft

$$\text{Cost} = 231\text{ ft}^2 \times \frac{1\text{ yd}^2}{9\text{ ft}^2} \times \frac{14.50}{\text{yd}^2} + 80\text{ ft} \times \frac{1\text{ yd}}{3\text{ ft}} \times \frac{1.50}{\text{yd}} = \$412.17$$

To Think About

41. ***Living Room Rug*** Phil and Melissa had a living room rug in their Nova Scotia home that was rectangular. It was 8 meters long and 3 meters shorter in width than length. It was badly damaged by the pets in the house. When Phil and Melissa replaced the rug, they could not find one the same size. They settled for a new rug that is 50 centimeters shorter in length and 40 centimeters shorter in width. How much smaller is the area of the new rug compared to the area of the old rug? Express your answer in square meters.
5.5 square meters less

42. ***Athletic Field*** The perimeter of a rectangular school athletic field in Mexico City is 600 meters. The length of the field is twice the width. The school principal would like to cut down some trees and increase the length of the field by 50 meters and the width of the field by 70 meters. How much larger will the area of the new field be compared to the area of the existing field?
22,500 square meters

Cumulative Review

43. Add. $156.8 + 27.2 + 39.3$
223.3

44. Subtract. $200.57 - 193.39$
7.18

45. Multiply. 1076×20.3
21,842.8

46. Divide. $12.3\overline{)19.384}$
approximately 1.5759

47. ***Linoleum in Cafe*** Cheryl owns a café and is having new linoleum put in the kitchen. The room measure $16\frac{1}{2}$ ft by $21\frac{1}{3}$ ft. What is the area of the room? If the linoleum costs \$0.75 per square foot, how much will it cost?
352 sq ft; \$264

7.3 Parallelograms, Trapezoids, and Rhombuses

Student Learning Objectives

After studying this section, you will be able to:

Find the perimeter and area of a parallelogram or a rhombus.

Find the perimeter and area of a trapezoid.

Finding the Perimeter and Area of a Parallelogram or a Rhombus

Parallelograms, rhombuses, and trapezoids are figures related to rectangles. Actually, they are in the same "family," the **quadrilaterals** (four-sided figures). For all these figures, the perimeter is the distance around the figure. But there is a different formula for finding the area of each.

A **parallelogram** is a four-sided figure in which both pairs of opposite sides are parallel. The opposite sides of a parallelogram are equal in length.

The following figures are parallelograms. Notice that the adjoining sides need not be perpendicular as in the rectangle.

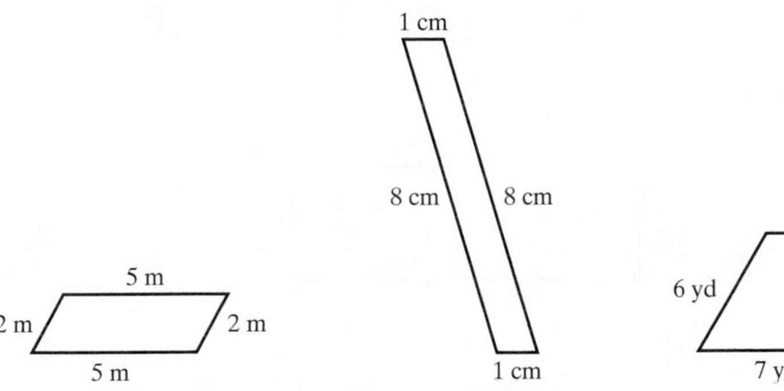

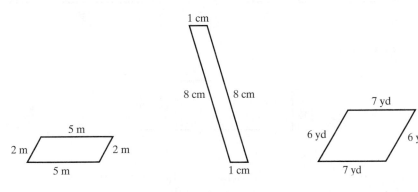

Teaching Tip Students must be clear on the meaning of *parallel lines*. Ask them to define what they mean by parallel lines. Usually, after two or three suggestions they will identify the key property that parallel lines are STRAIGHT LINES that are always the SAME DISTANCE apart.

The **perimeter** of a parallelogram is the distance around the parallelogram. It is found by adding the lengths of all the sides of the figure.

EXAMPLE 1 Find the perimeter.

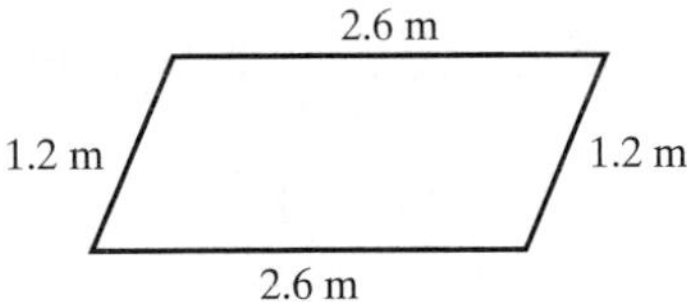

Solution $P = (2)(1.2 \text{ meters}) + (2)(2.6 \text{ meters})$
$= 2.4 \text{ meters} + 5.2 \text{ meters} = 7.6 \text{ meters}$

Practice Problem 1 Find the perimeter of the parallelogram in the margin.

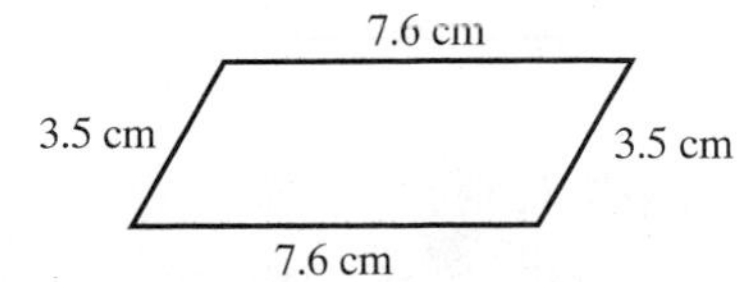

To find the **area** of a parallelogram, we multiply the base times the height. Any side of a parallelogram can be considered the **base.** The **height** is the shortest distance between the base and the side opposite the base. The height is a line segment that is perpendicular to the base. When we write the formula for area, we use the lengths of the base (b) and the height (h).

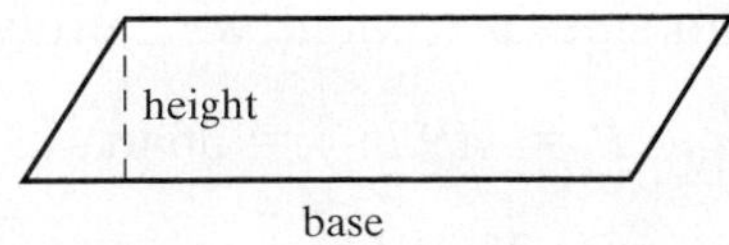

The **area of a parallelogram** is the base (b) times the height (h).

$$A = bh$$

Why is the area of a parallelogram equal to the base times the height? What reasoning leads us to that formula? Suppose that we cut off the triangular region on one side of the parallelogram and move it to the other side.

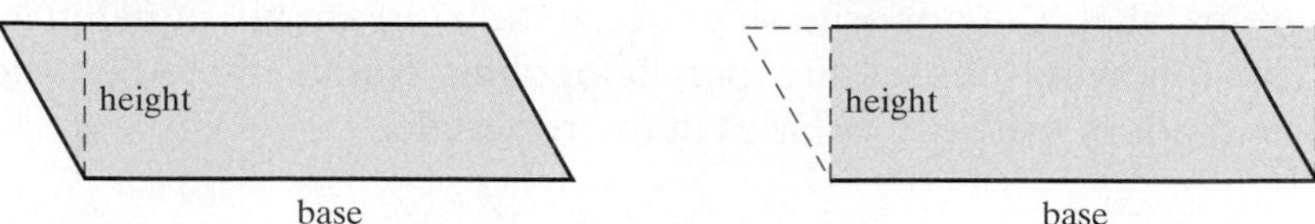

We now have a rectangle.

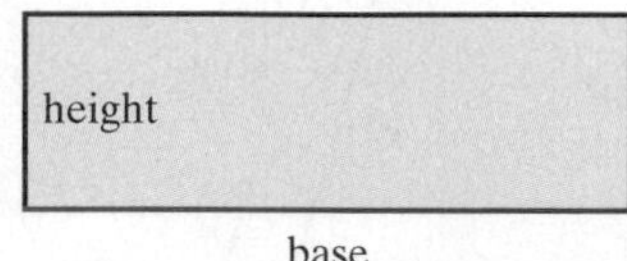

To find the area, we multiply the width by the length. In this case, $A = bh$. Thus finding the area of a parallelogram is like finding the area of a rectangle of length b and width h: $A = bh$.

EXAMPLE 2 Find the area of a parallelogram with base 7.5 m and height 3.2 m.

Solution

$$\begin{aligned} A &= bh \\ &= (7.5 \text{ m})(3.2 \text{ m}) \\ &= 24 \text{ m}^2 \end{aligned}$$

Practice Problem 2 Find the area of a parallelogram with base 10.3 km and height 1.5 km.

2 cm
2 cm Rhombus 2 cm
2 cm

A **rhombus** is a parallelogram with all four sides equal. The figure to the left, with each side of length 2 centimeters, is a rhombus. We will solve a problem involving a rhombus in Example 3.

EXAMPLE 3 A truck is manufactured with an iron brace welded to the truck frame. The brace is shaped like a rhombus. The brace has a base of 9 inches and a height of 5 inches. Find the perimeter and the area of this iron brace.

Solution Since all four sides are equal, we merely multiply

$$P = 4(9 \text{ in.}) = 36 \text{ in.}$$

The perimeter of this brace is 36 inches.

Since the rhombus is a special type of parallelogram, we can use the area formula for a parallelogram. In this case the base is 9 inches and the height is 5 inches.

$$A = bh = (9 \text{ in.})(5 \text{ in.}) = 45 \text{ in.}^2$$

Thus the area of the brace is 45 square inches.

Teaching Tip Be sure to mention *why* the formula for the area of a parallelogram is logical. This type of reasoning will benefit students throughout this chapter.

Practice Problem 3 An inlaid piece of cherry wood on the front of a hope chest is shaped like a rhombus. This piece has a base of 6 centimeters and a height of 4 centimeters. Find the perimeter and the area of this inlaid piece of cherry wood.

NOTE TO STUDENT: Fully worked-out solutions to all of the Practice Problems can be found at the back of the text starting at page SP-1

2 Finding the Perimeter and Area of a Trapezoid

A **trapezoid** is a four-sided figure with two parallel sides. The parallel sides are called **bases.** The lengths of the bases do not have to be equal. The adjoining sides do not have to be perpendicular.

Sometimes the trapezoid is sitting on a base. Then both bases are horizontal. But be careful. Sometimes the bases are vertical. You can recognize the bases because they are the two parallel sides. This becomes important when you use the formula for finding the area of a trapezoid.

Look at the following trapezoids. See if you can recognize the bases.

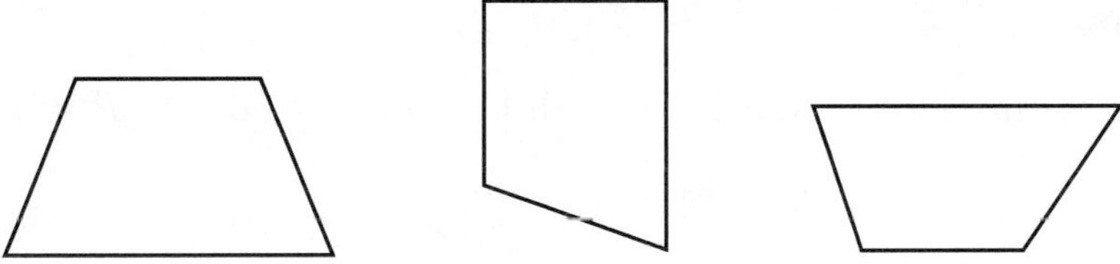

The perimeter of a trapezoid is the sum of the lengths of all of its sides.

EXAMPLE 4 Find the perimeter of the trapezoid on the right.

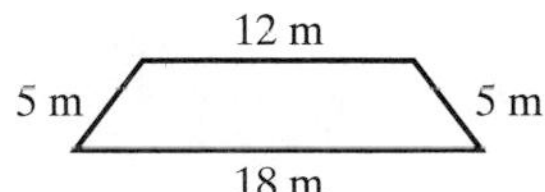

Solution

$$P = 18 \text{ m} + 5 \text{ m} + 12 \text{ m} + 5 \text{ m}$$
$$= 40 \text{ m}$$

Practice Problem 4 Find the perimeter of a trapezoid with sides of 7 yd, 15 yd, 21 yd, and 13 yd.

Remember, we often use parentheses as a way to group numbers together. The numbers inside parentheses should be combined first.

$(5)(7 + 2) = (5)(9)$ First we add numbers inside the parentheses.
$= 45$ Then we multiply.

The formula for the area of a trapezoid uses parentheses in this way.

The **height** of a trapezoid is the distance between the two parallel sides. The area of a trapezoid is one-half the height times the sum of the bases. (This means you add the bases *first.*)

Now this can be written $\frac{h}{2} \cdot (b + B)$ or $h\left(\frac{b + B}{2}\right)$ or $\frac{h(b + B)}{2}$.

Some students like to remember $h\left(\frac{b + B}{2}\right)$ because it is the height times the average of the bases.

The **area of a trapezoid** with a shorter base b, a longer base B, and height h is

$$A = \frac{h(b + B)}{2}.$$

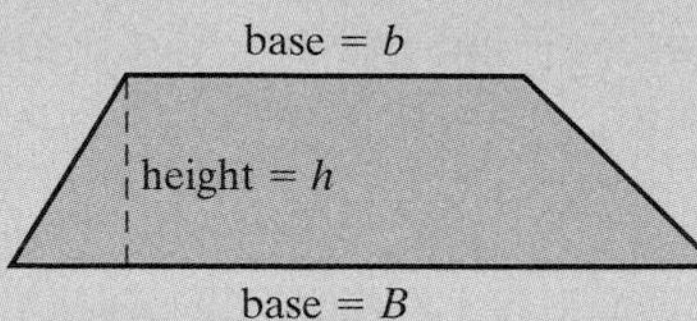

EXAMPLE 5 A roadside sign is in the shape of a trapezoid. It has a height of 30 ft, and the bases are 60 ft and 75 ft.

(a) What is the area of the sign?

(b) If 1 gallon of paint covers 200 ft^2, how many gallons of paint will be needed to paint the sign?

Solution

(a) We use the trapezoid formula with $h = 30$, $b = 60$, and $B = 75$.

$$A = \frac{h(b + B)}{2}$$

$$= \frac{(30 \text{ ft})(60 \text{ ft} + 75 \text{ ft})}{2}$$

$$= \frac{(30 \text{ ft})(135 \text{ ft})}{2} = \frac{4050}{2} \text{ ft}^2 = 2025 \text{ ft}^2$$

(b) Each gallon covers 200 ft^2, so we multiply the area by the fraction $\frac{1 \text{ gal}}{200 \text{ ft}^2}$. This fraction is equivalent to 1.

$$2025 \cancel{\text{ft}^2} \times \frac{1 \text{ gal}}{200 \cancel{\text{ft}^2}} = \frac{2025}{200} \text{ gal}$$

$$= 10.125 \text{ gal}$$

Thus 10.125 gallons of paint would be needed. In real life we would buy 11 gallons of paint.

Practice Problem 5 A corner parking lot is shaped like a trapezoid. The trapezoid has a height of 140 yd. The bases measure 180 yd and 130 yd.

(a) Find the area of the parking lot.

(b) If 1 gallon of sealant will cover 100 square yards of the parking lot, how many gallons are needed to cover the entire parking lot?

Some area problems involve two or more separate regions. Remember, areas can be added or subtracted.

EXAMPLE 6 Find the area of the following piece for inlaid woodwork made by a master carpenter. Since this shape is hard to cut, it is made of one trapezoid and one rectangle laid together.

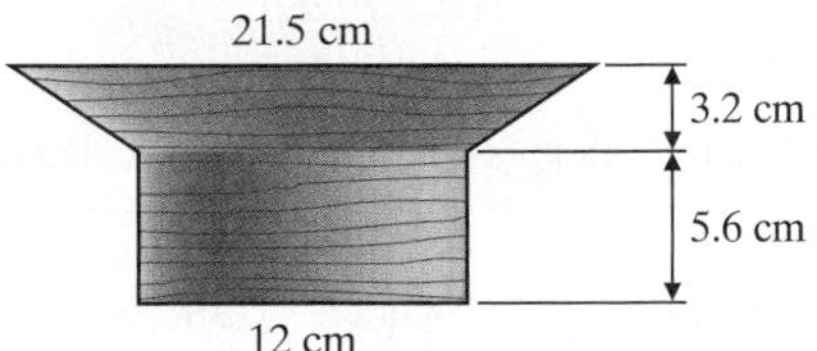

Solution We separate the area into two portions and find the area of each portion separately.

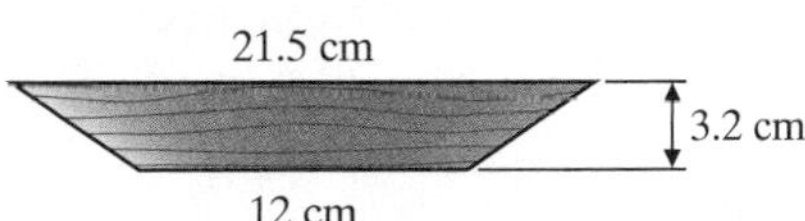

The area of the trapezoid is

$$\begin{aligned} A &= \frac{h(b + B)}{2} \\ &= \frac{(3.2 \text{ cm})(12 \text{ cm} + 21.5 \text{ cm})}{2} \\ &= \frac{(3.2 \text{ cm})(33.5 \text{ cm})}{2} \\ &= \frac{107.2}{2} \text{ cm}^2 \\ &= 53.6 \text{ cm}^2. \end{aligned}$$

The area of the rectangle is

$$\begin{aligned} A &= lw \\ &= (12 \text{ cm})(5.6 \text{ cm}) \\ &= 67.2 \text{ cm}^2. \end{aligned}$$

We now add each area.

$$\begin{array}{r} 67.2 \text{ cm}^2 \\ +53.6 \text{ cm}^2 \\ \hline 120.8 \text{ cm}^2 \end{array}$$

The total area of the piece for inlaid woodwork is 120.8 cm^2.

Practice Problem 6 Find the area of the piece for inlaid woodwork shown in the margin. The shape is made of one trapezoid and one rectangle.

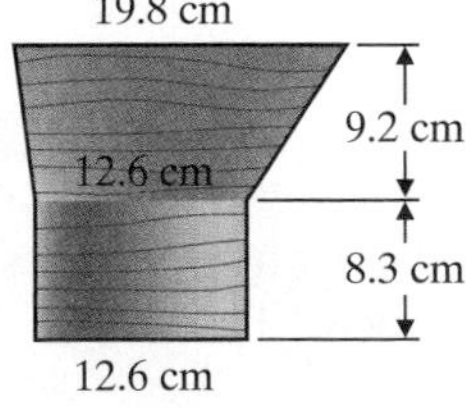

7.3 EXERCISES

Student Solutions Manual | CD/Video | PH Math Tutor Center | MathXL®Tutorials on CD | MathXL® | MyMathLab® | Interactmath.com

Verbal and Writing Skills

1. The perimeter of a parallelogram is found by ___adding___ the lengths of all the sides of the figure.

2. To find the area of a parallelogram, multiply the base times the ___height___.

3. The height of a parallelogram is a line segment that is ___perpendicular___ to the base.

4. The area of a trapezoid is one-half the height times the ___sum___ of the bases.

Find the perimeter of the parallelogram.

5. One side measures 2.8 m and a second side measures 17.3 m.
   $2 \times 17.3 \text{ m} + 2 \times 2.8 \text{ m} = 40.2 \text{ m}$

6. One side measures 4.6 m and a second side measures 20.5 m.
   $2 \times 4.6 \text{ m} + 2 \times 20.5 \text{ m} = 50.2 \text{ m}$

7. 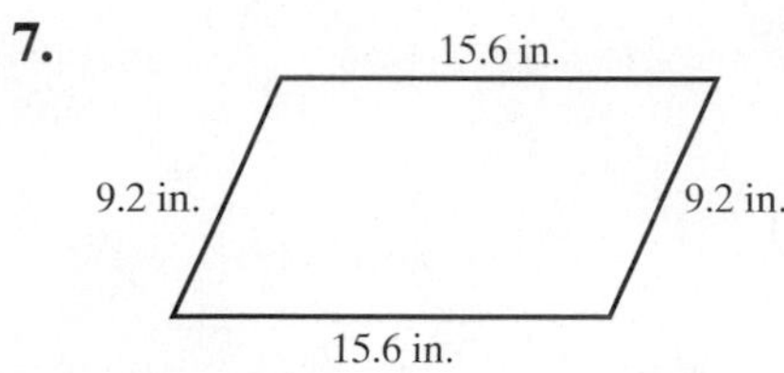

   $2 \times 9.2 \text{ in.} + 2 \times 15.6 \text{ in.} = 49.6 \text{ in.}$

8. 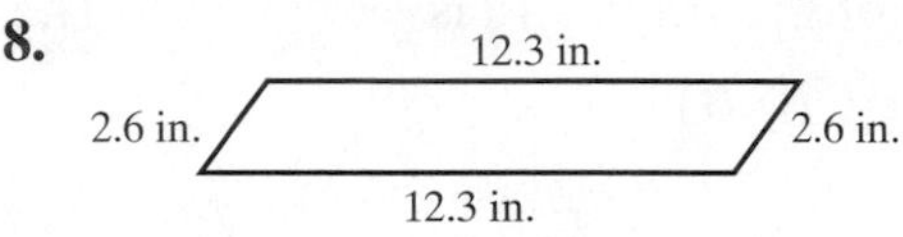

   $2 \times 2.6 \text{ in.} + 2 \times 12.3 \text{ in.} = 29.8 \text{ in.}$

Find the area of the parallelogram.

9. The base is 17.6 m and the height is 20.15 m.
   $17.6 \text{ m} \times 20.15 \text{ m} = 354.64 \text{ m}^2$

10. The base is 14.2 m and the height is 21.25 m.
    $14.2 \text{ m} \times 21.25 \text{ m} = 301.75 \text{ m}^2$

11. ***Music Theatre Seating*** The preferred seating area at the South Shore Music Theatre is in the shape of a parallelogram. Its base is 28 yd and its height is 21.5 yd. Find the area.
    602 yd^2

12. ***Courtyard*** A courtyard is shaped like a parallelogram. Its base is 126 yd and its height is 28 yd. Find its area.
    $126 \text{ yd} \times 28 \text{ yd} = 3528 \text{ yd}^2$

13. Find the perimeter and the area of a rhombus with height 6 meters and base 12 meters.
    $P = 48 \text{ m}$; $A = 72 \text{ m}^2$

14. Find the perimeter and the area of a rhombus with height 9 yards and base 14 yards.
    $P = 56 \text{ yd}$; $A = 126 \text{ yd}^2$

15. ***Kite*** Walter made his son Daniel a kite in the shape of a rhombus. The height of the kite is 1.5 feet. The length of the base of the kite is 2.4 feet. Find the perimeter and the area of the kite.
    $P = 9.6 \text{ ft}$; $A = 3.6 \text{ ft}^2$

16. ***State Park*** The lawn in front of Bradley Palmer State Park is constructed in the shape of a rhombus. The height of the lawn region is 17 feet. The length of the base is 25 feet. Find the perimeter and the area of this lawn.
    $P = 100 \text{ ft}$; $A = 425 \text{ ft}^2$

Find the perimeter of the trapezoid.

17.

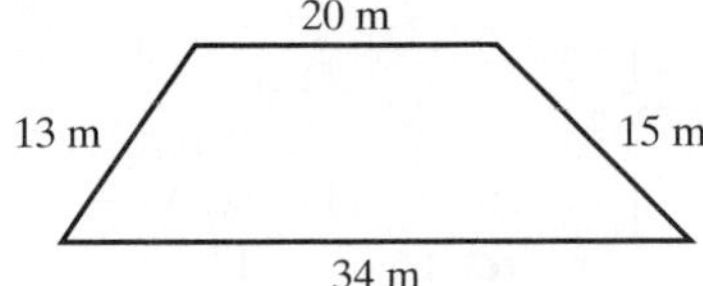

13 m + 20 m + 15 m + 34 m = 82 m

18.

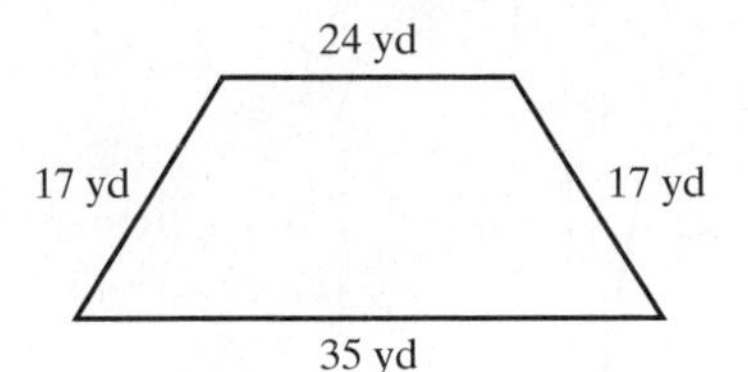

17 yd + 24 yd + 17 yd + 35 yd = 93 yd

19. The two bases are 55 ft and 135 ft. The other two sides are 80.5 ft and 75.5 ft.

55 ft + 135 ft + 80.5 ft + 75.5 ft = 346 ft

20. The two bases are 42 m and 53 m. The other two sides are 81.5 m and 94.5 m.

42 m + 53 m + 81.5 m + 94.5 m = 271 m

Find the area of the trapezoid.

21. The height is 12 yd and the bases are 9.6 yd and 10.2 yd.

$A = \frac{1}{2} \times 12 \text{ yd} \times (9.6 \text{ yd} + 10.2 \text{ yd}) = 118.8 \text{ yd}^2$

22. The height is 18 yd and the bases are 8.4 yd and 17.8 yd.

$A = \frac{1}{2} \times 18 \text{ yd} \times (8.4 \text{ yd} + 17.8 \text{ yd}) = 235.8 \text{ yd}^2$

23. ***Diving in Key West*** An underwater diving area for snorkelers and scuba divers in Key West, Florida, is designated by buoys and ropes, making the diving section into the shape of a trapezoid on the surface of the water. The trapezoid has a height of 265 meters. The bases are 300 meters and 280 meters. Find the area of the designated diving area.

$76{,}850 \text{ m}^2$

24. ***Provincial Park*** A provincial park in Canada is laid out in the shape of a trapezoid. The trapezoid has a height of 20 km. The bases are 24 km and 31 km. Find the area of the park.

$A = \frac{1}{2} \times 20 \text{ km} \times (24 \text{ km} + 31 \text{ km}) = 550 \text{ km}^2$

Mixed Practice

(a) Find the area of the entire shape made of trapezoids, parallelograms, squares, and rectangles.
(b) Name the object that is shaded in orange.
(c) Name the object that is shaded in yellow.

25.

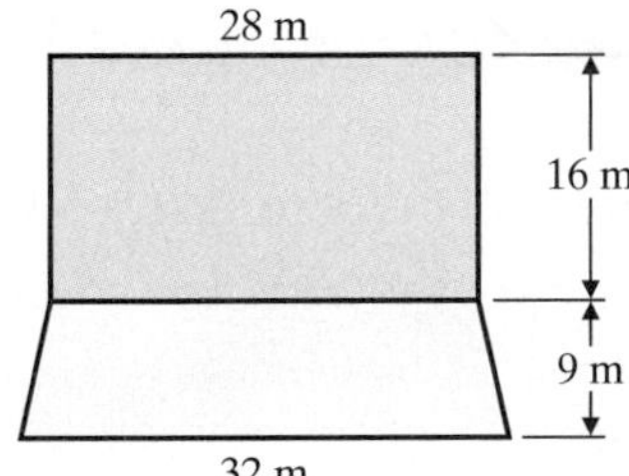

(a) $A = 28 \text{ m} \times 16 \text{ m} + \frac{1}{2} \times 9 \text{ m} \times (28 \text{ m} + 32 \text{ m})$

$= 718 \text{ m}^2$

(b) rectangle
(c) trapezoid

26.

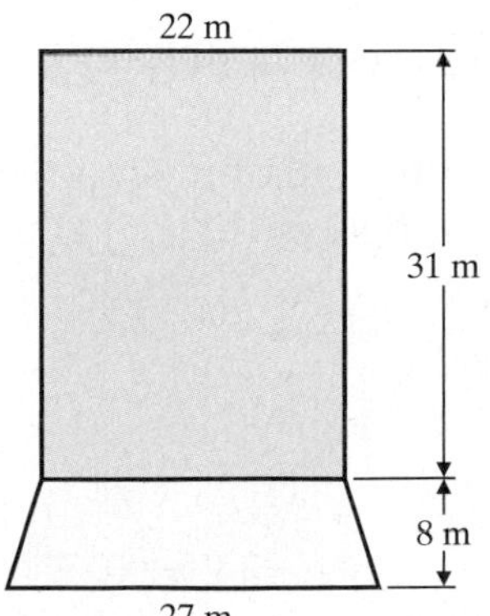

(a) $A = 22 \text{ m} \times 31 \text{ m} + \frac{1}{2} \times 8 \text{ m} \times (22 \text{ m} + 27 \text{ m})$

$= 878 \text{ m}^2$

(b) rectangle
(c) trapezoid

27.

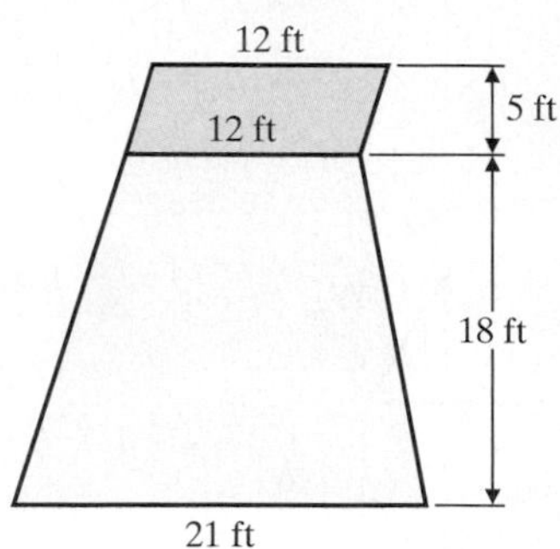

(b) parallelogram
(c) trapezoid

(a) $A = (12 \text{ ft} \times 5 \text{ ft}) + \frac{1}{2} \times 18 \text{ ft} \times (12 \text{ ft} + 21 \text{ ft})$

$= 357 \text{ ft}^2$

28.

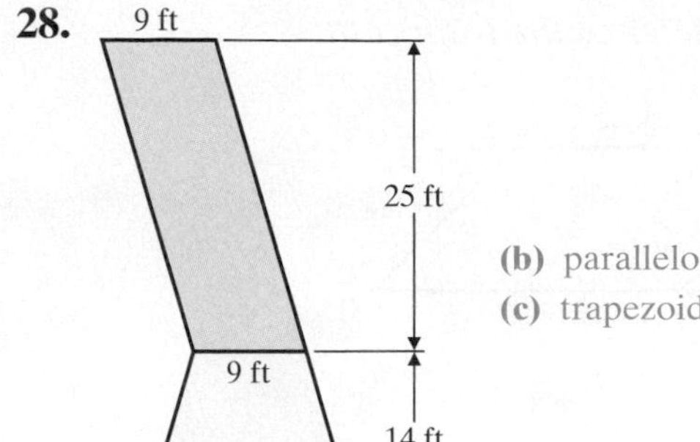

(b) parallelogram
(c) trapezoid

(a) $A = (9 \text{ ft} \times 25 \text{ ft}) + \frac{1}{2} \times 14 \text{ ft} \times (9 \text{ ft} + 17 \text{ ft})$

$= 407 \text{ ft}^2$

Applications

Carpeting in Conference Center *Each of the following shapes represents the lobby of a conference center. The lobby will be carpeted at a cost of \$22 per square yard. How much will the carpeting cost?*

29.

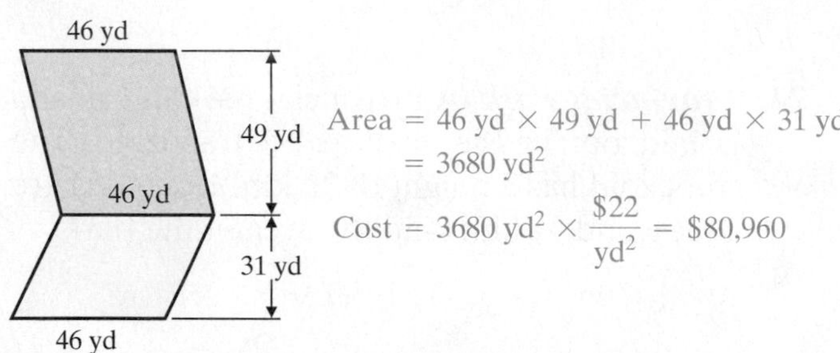

Area $= 46 \text{ yd} \times 49 \text{ yd} + 46 \text{ yd} \times 31 \text{ yd}$
$= 3680 \text{ yd}^2$

Cost $= 3680 \text{ yd}^2 \times \frac{\$22}{\text{yd}^2} = \$80{,}960$

30.

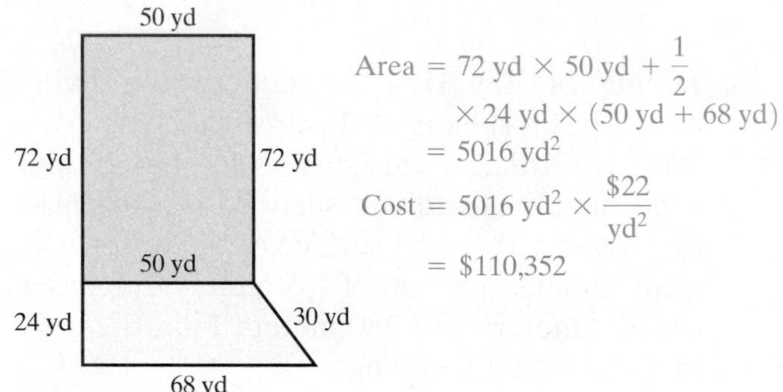

Area $= 72 \text{ yd} \times 50 \text{ yd} + \frac{1}{2} \times 24 \text{ yd} \times (50 \text{ yd} + 68 \text{ yd})$
$= 5016 \text{ yd}^2$

Cost $= 5016 \text{ yd}^2 \times \frac{\$22}{\text{yd}^2}$
$= \$110{,}352$

To Think About

31. See if you can find a formula that would give the area of a regular octagon. (A regular octagon is an eight-sided figure with all sides of equal length.) The dimensions of the rectangles and trapezoids are labeled on the sketch.

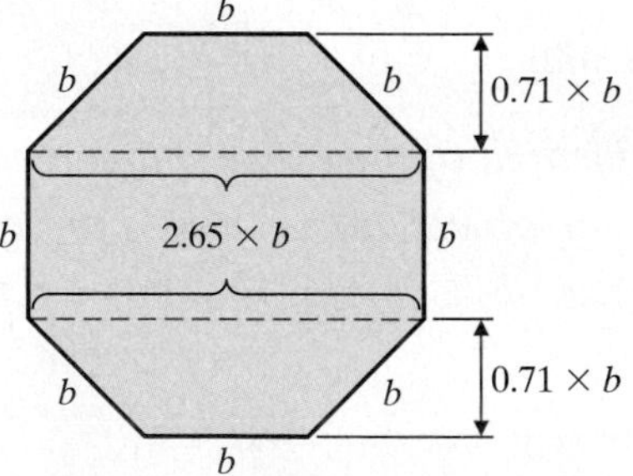

$2 \times \left[\frac{1}{2} \times 0.71 \times b \times (b + 2.65 \times b)\right] + 2.65 \times b \times b$

Area $= 5.2415 \times b^2$ sq units

32. See if you can find a formula for the area of a regular hexagon. Each side is b units long. (A regular hexagon is a six-sided figure with all sides of equal length.)

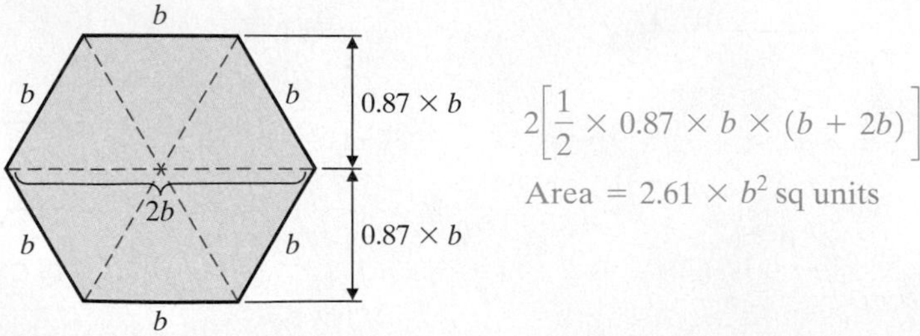

$2\left[\frac{1}{2} \times 0.87 \times b \times (b + 2b)\right]$

Area $= 2.61 \times b^2$ sq units

Cumulative Review *Complete each conversion.*

33. 10 yd = 30 ft **34.** 3 gal = 12 qt **35.** 18 m = 1800 cm **36.** 80 oz = 5 lb

 7.4 TRIANGLES

Student Learning Objectives

After studying this section, you will be able to:

 **Find the measures of angles in a triangle.**

 **Find the perimeter and the area of a triangle.**

1 Finding the Measures of Angles in a Triangle

A **triangle** is a three-sided figure with three angles. The prefix *tri*-means "three." Some triangles are shown.

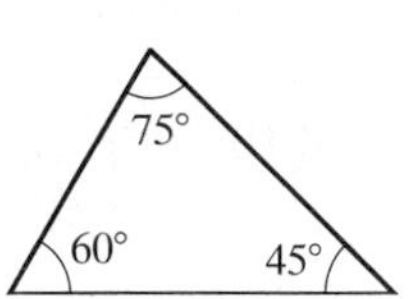

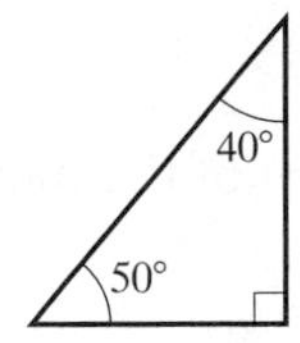

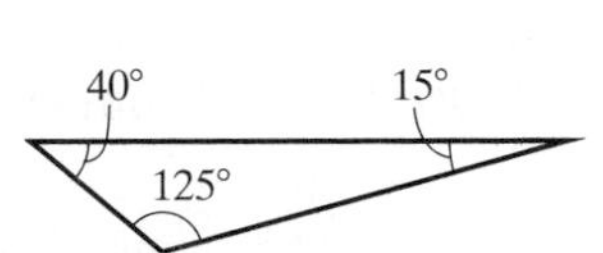

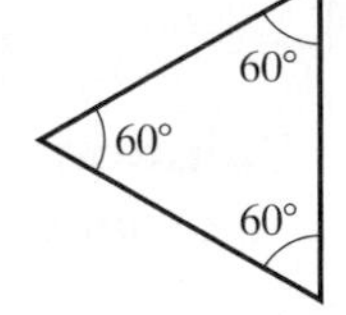

Although all triangles have three sides, not all triangles have the same shape. The shape of a triangle depends on the sizes of the angles and the lengths of the sides.

We will begin our study of triangles by looking at the angles. Although the sizes of the angles in triangles may be different, the sum of the angle measures of any triangle is always 180°.

> The sum of the measures of the angles in a triangle is 180°.

Why is this? Perhaps you are wondering why all the angles of a triangle have measures that add up to 180°.

Remember, a straight angle is 180°.

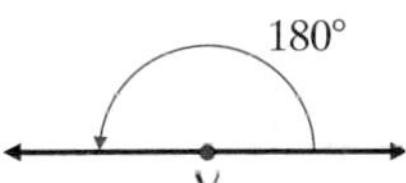

Suppose you take any triangle with $\angle A$, $\angle B$, and $\angle C$. Now cut off each side of the triangle.

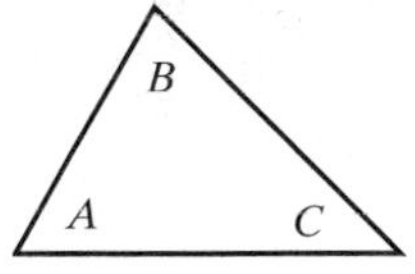

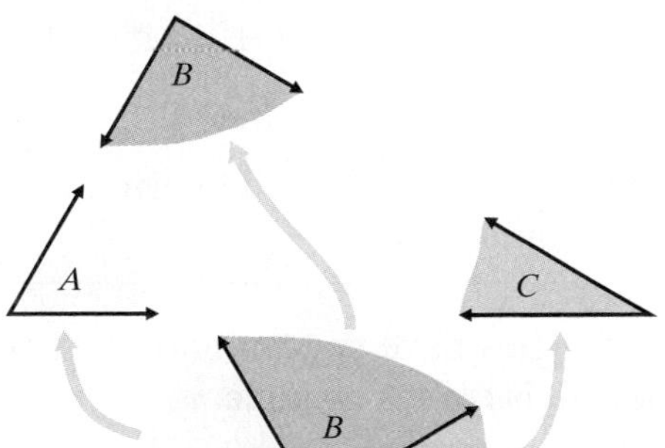

Now move the three angles so that they all have the same vertex and each pair of angles has an adjacent side. The three angles form a straight angle. Thus, the sum of the measures in a triangle is 180°.

We can use this fact to find the measure of an unknown angle in a triangle if we know the measures of the other two angles.

EXAMPLE 1 In the triangle below, angle A measures 35° and angle B measures 95°. Find the measure of angle C.

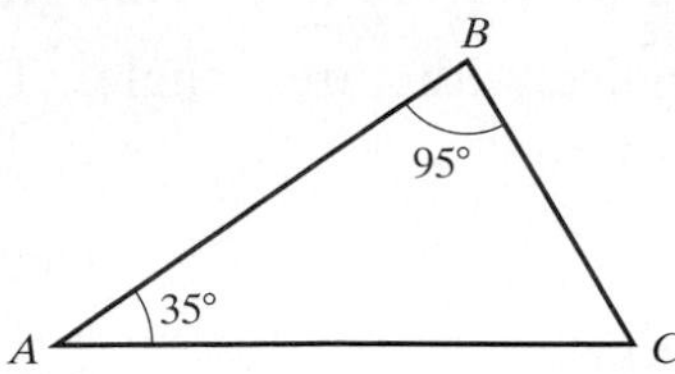

Solution We will use the fact that the sum of the measures of the angles of a triangle is 180°.

$$35 + 95 + x = 180$$
$$130 + x = 180$$

What number x when added to 130 equals 180? Since $130 + 50 = 180$, x must equal 50.

Angle C must equal 50°.

NOTE TO STUDENT: Fully worked-out solutions to all of the Practice Problems can be found at the back of the text starting at page SP-1

Practice Problem 1 In a triangle, angle B measures 125° and angle C measures 15°. What is the measure of angle A?

2 Finding the Perimeter and the Area of a Triangle

Recall that the perimeter of any figure is the sum of the lengths of its sides. Thus the perimeter of a triangle is the sum of the lengths of its three sides.

EXAMPLE 2 Find the perimeter of a triangular sail whose sides are 12 ft, 14 ft, and 17 ft.

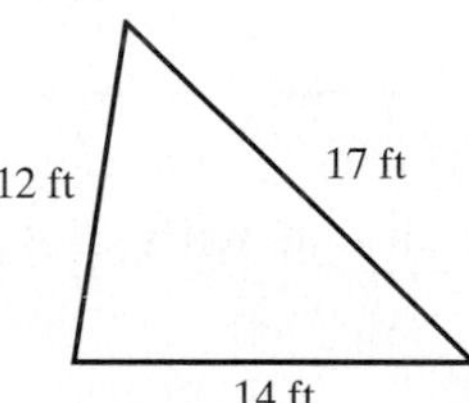

Solution $P = 12 \text{ ft} + 14 \text{ ft} + 17 \text{ ft} = 43 \text{ ft}$

Teaching Tip Emphasize to students that they must learn the terms *isosceles, equilateral,* and *right triangle.*

Practice Problem 2 Find the perimeter of a triangle whose sides are 10.5 m, 10.5 m, and 8.5 m.

Some triangles have special names. A triangle with at least two equal sides is called an **isosceles triangle.**

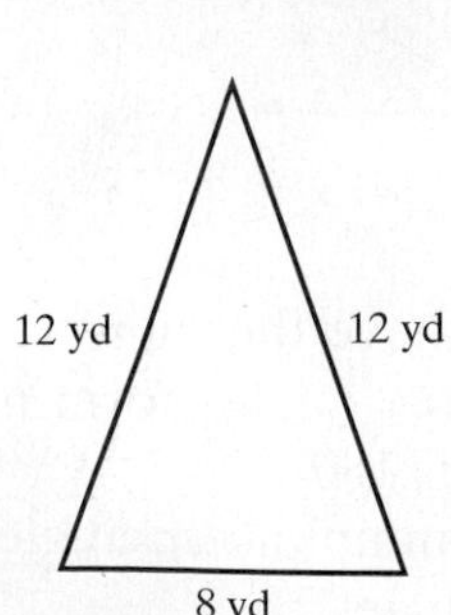

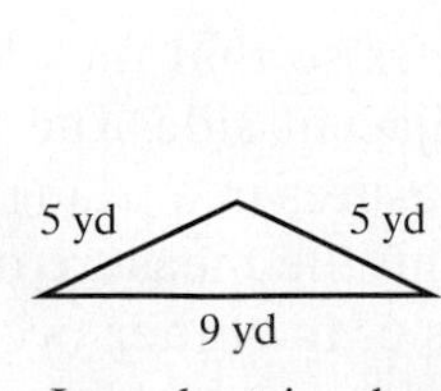

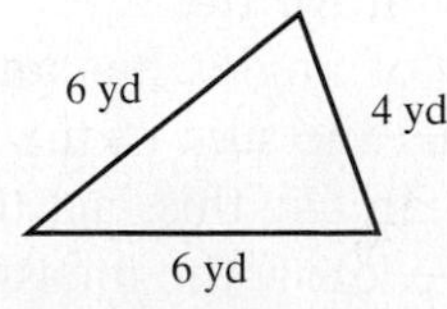

Isosceles triangles

A triangle with three equal sides is called an **equilateral triangle.** All angles in an equilateral triangle are exactly 60°.

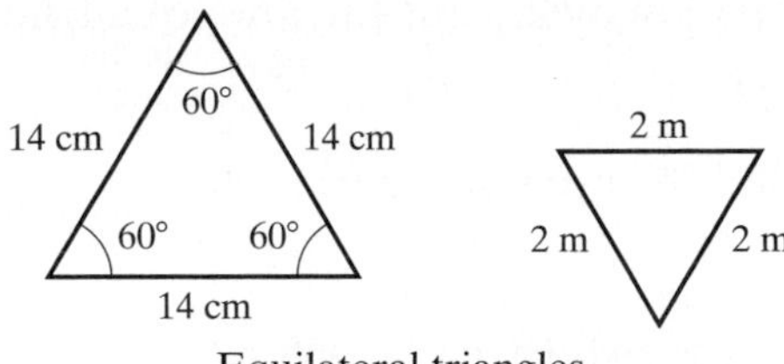

Equilateral triangles

A **scalene triangle** has no two sides of equal lengths and no two angles of equal measure.

A triangle with one 90° angle is called a **right triangle.**

The **height** of any triangle is the distance of a line drawn from a vertex perpendicular to the opposite side or an extension of the opposite side. The height may be one of the sides in a right triangle. The **base** of a triangle is perpendicular to the height.

Teaching Tip Students sometimes have trouble visualizing the height of a triangle if the height lies outside the triangle. Explain that the height lies outside the triangle only if one angle of the triangle is greater than 90°. Point out that height is always a perpendicular line and will be marked with a small square. The base will be the side the height is perpendicular to. If the height is outside the triangle, the base is only that part of the perpendicular that is the side of the triangle and does not include the extension.

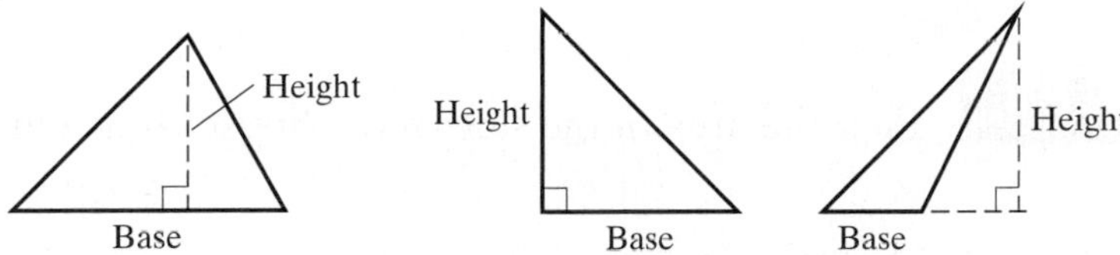

To find the area of a triangle, we need to be able to identify its height and base. The area of any triangle is half of the product of the base times the height of the triangle. The height is measured from the vertex above the base to that base.

The **area of a triangle** is the base times the height divided by 2.

$$A = \frac{bh}{2}$$

h = height

b = base

Where does the 2 come from in the formula $A = \frac{bh}{2}$? Why does this formula for the area of a triangle work? Suppose that we construct a triangle with base b and height h.

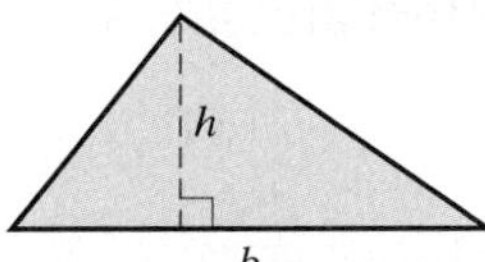

Now let us make an exact copy of the triangle and turn the copy around to the right exactly 180°. Carefully place the two triangles together. We now have a parallelogram of base b and height h. The area of a parallelogram is $A = bh$.

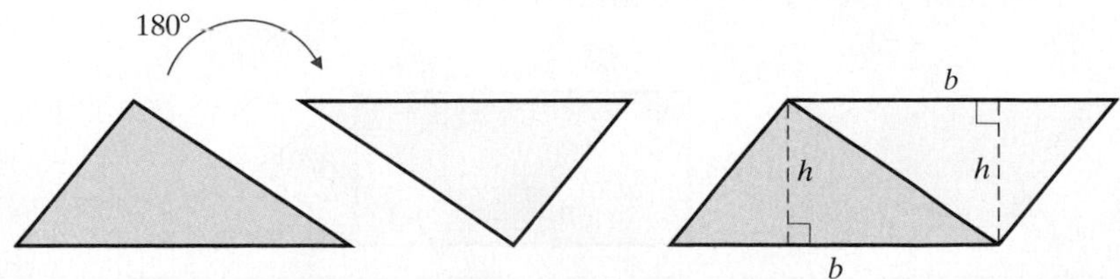

Because the parallelogram has area $A = bh$ and is made up of two triangles of identical shape and area, the area of one of the triangles is the area of the parallelogram divided by 2. Thus the area of a triangle is $A = \frac{bh}{2}$.

EXAMPLE 3 Find the area of the triangle.

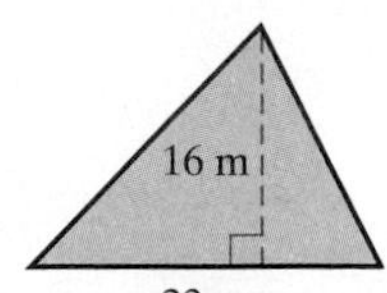

Solution

$$A = \frac{bh}{2} = \frac{(23\text{ m})(16\text{ m})}{2} = \frac{368\text{ m}^2}{2} = 184\text{ m}^2$$

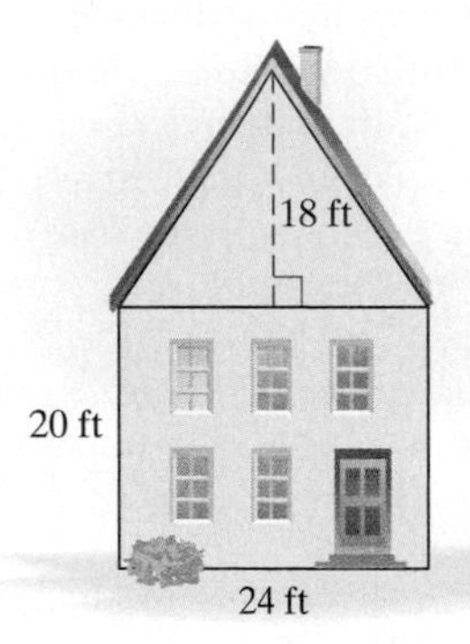

***NOTE TO STUDENT:** Fully worked-out solutions to all of the Practice Problems can be found at the back of the text starting at page SP-1*

Practice Problem 3 Find the area of the triangle in the margin.

In some geometric shapes, a triangle is combined with rectangles, squares, parallelograms, and trapezoids.

EXAMPLE 4 Find the area of the side of the house shown in the margin.

Solution Because the lengths of opposite sides of a rectangle are equal, the triangle has a base of 24 ft. Thus we can calculate its area.

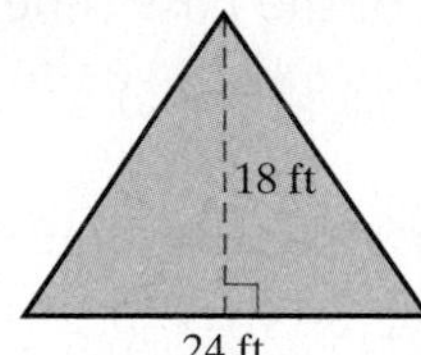

$$A = \frac{bh}{2} = \frac{(24\text{ ft})(18\text{ ft})}{2} = \frac{432\text{ ft}^2}{2} = 216\text{ ft}^2$$

The area of the rectangle is $A = lw = (24\text{ ft})(20\text{ ft}) = 480\text{ ft}^2$.

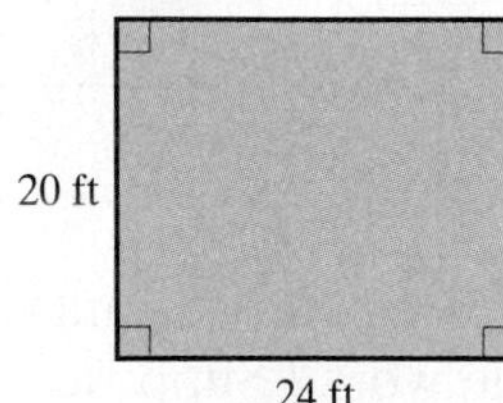

Now we find the sum of the two areas.

$$\begin{array}{r} 216\text{ ft}^2 \\ +480\text{ ft}^2 \\ \hline 696\text{ ft}^2 \end{array}$$

Thus the area of the side of the house is 696 square feet.

Practice Problem 4 Find the area of the figure.

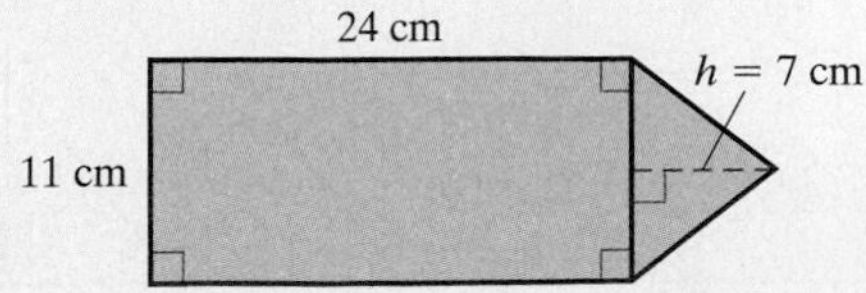

7.4 Exercises

Student Solutions Manual | CD/ Video | PH Math Tutor Center | MathXL®Tutorials on CD | MathXL® | MyMathLab® | Interactmath.com

Verbal and Writing Skills

1. A 90° angle is called a ___right___ angle.

2. The sum of the angle measures of a triangle is ___180°___.

3. Explain in your own words how you would find the measure of an unknown angle in a triangle if you knew the measures of the other two angles.
Add the measures of the two known angles and subtract that value from 180°.

4. If you were told that a triangle was an isosceles triangle, what could you conclude about the sides of that triangle?
You could conclude that two of the sides of the triangle are equal.

5. If you were told that a triangle was an equilateral triangle, what could you conclude about the sides of the triangle?
You could conclude that the lengths of all three sides of the triangle are equal.

6. How do you find the area of a triangle?
The area of a triangle is the base times the height divided by 2.

Write true or false for each statement.

7. Two lines that meet at a 90° angle are perpendicular.
true

8. A right triangle has two angles of 90°.
false

9. The sum of the angles of a triangle is 180°.
true

10. An isosceles triangle has two sides of equal length.
true

11. An equilateral triangle has one angle greater than 90°.
false

12. A scalene triangle has two angles of equal measures.
false

13. The measures of the angles of an equilateral triangle are all equal.
true

14. To find the perimeter of an equilateral triangle, you can multiply the length of one of the sides by 3.
true

Find the missing angle in the triangle.

15. Two angles are 36° and 74°.

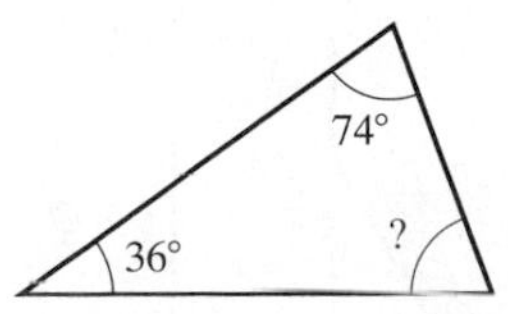

36° + 74° = 110° 180° − 110° = 70°

16. Two angles are 23° and 95°.

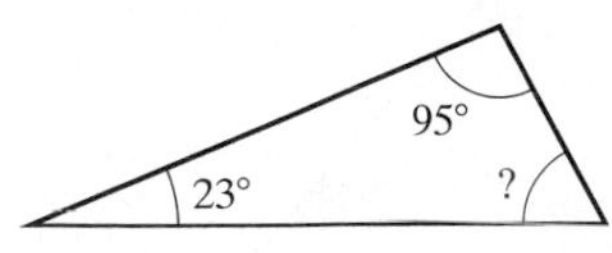

23° + 95° = 118° 180° − 118° = 62°

17. Two angles are 44.6° and 52.5°.
82.9°

18. Two angles are 136.2° and 17.5°.
26.3°

Find the perimeter of the triangle.

19. A scalene triangle whose sides are 18 m, 45 m, and 55 m $P = 18\text{ m} + 45\text{ m} + 55\text{ m} = 118\text{ m}$

20. A scalene triangle whose sides are 27 m, 44 m, and 23 m $P = 27\text{ m} + 44\text{ m} + 23\text{ m} = 94\text{ m}$

21. An isosceles triangle whose sides are 45.25 in., 35.75 in., and 35.75 in.
$P = 45.25\text{ in.} + 35.75\text{ in.} + 35.75\text{ in.} = 116.75\text{ in.}$

22. An isosceles triangle whose sides are 36.2 in., 47.65 in., and 47.65 in.
$P = 36.2\text{ in.} + 47.65\text{ in.} + 47.65\text{ in.} = 131.5\text{ in.}$

23. An equilateral triangle whose side measures $3\frac{1}{3}$ mi. $P = 3\left(3\frac{1}{3}\text{ mi}\right) = 10\text{ mi}$

24. An equilateral triangle whose side measures $4\frac{2}{3}$ mi. $P = 3\left(4\frac{2}{3}\right) = 14\text{ mi}$

Find the area of the triangle.

25.

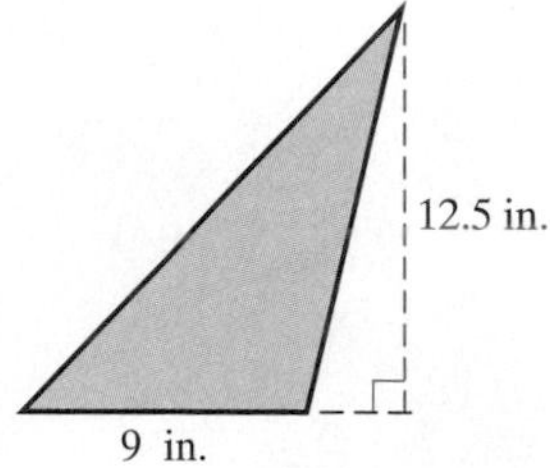

$A = \frac{1}{2} \times 9\text{ in.} \times 12.5\text{ in.} = 56.25\text{ in.}^2$

26.

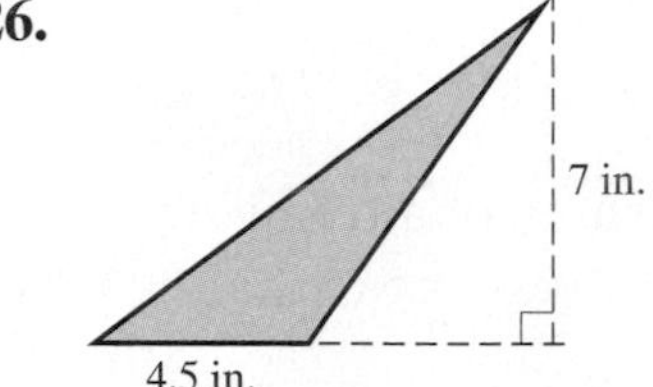

$A = \frac{1}{2} \times 4.5\text{ in.} \times 7\text{ in.} = 15.75\text{ in.}^2$

27. The base is 17.5 cm and the height is 9.5 cm.
$A = \frac{1}{2} \times 9.5\text{ cm} \times 17.5\text{ cm} = 83.125\text{ cm}^2$

28. The base is 3.6 cm and the height is 11.2 cm.
$A = \frac{1}{2} \times 3.6\text{ cm} \times 11.2\text{ cm} = 20.16\text{ cm}^2$

29. The base is $3\frac{1}{2}$ yd and the height is $4\frac{1}{3}$ yd.
$A = \frac{1}{2}\left(3\frac{1}{2}\text{ yd}\right)\left(4\frac{1}{3}\text{ yd}\right) = 7\frac{7}{12}\text{ yd}^2$

30. The base is $6\frac{1}{2}$ ft and the height is $5\frac{1}{3}$ ft.
$A = \frac{1}{2}\left(6\frac{1}{2}\right)\left(5\frac{1}{3}\right)$ $A = 17\frac{1}{3}\text{ ft}^2$

Mixed Practice

Find the area of the shaded region.

31.

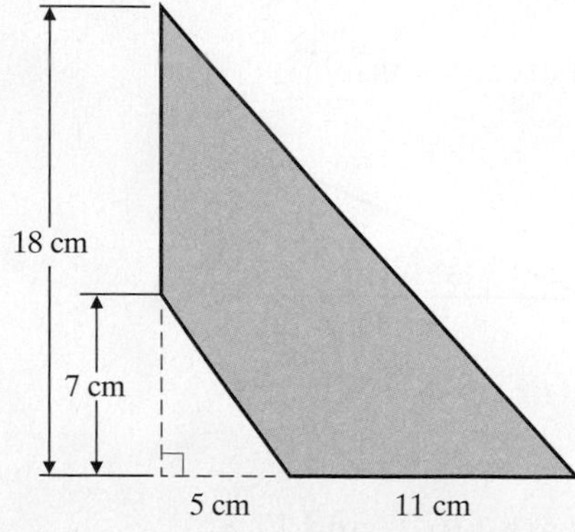

$A = \frac{1}{2} \times 16\text{ cm} \times 18\text{ cm} - \frac{1}{2} \times 15\text{ cm} \times 7\text{ cm} = 91.5\text{ cm}^2$

32.

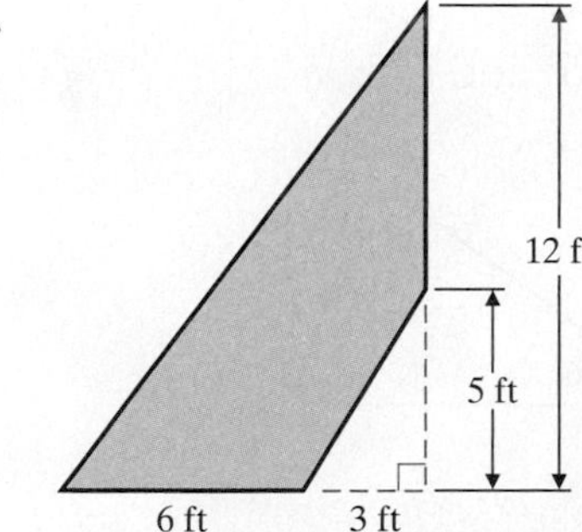

$A = \frac{1}{2} \times 9\text{ ft} \times 12\text{ ft} - \frac{1}{2} \times 3\text{ ft} \times 5\text{ ft} = 46.5\text{ ft}^2$

33.

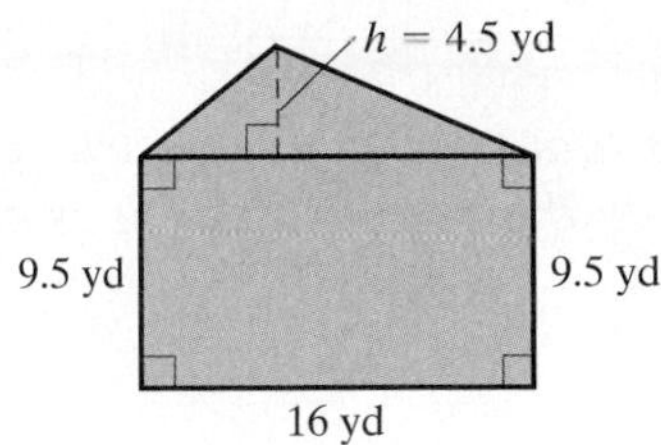

$A = 16 \text{ yd} \times 9.5 \text{ yd} + \frac{1}{2} \times 16 \text{ yd} \times 4.5 \text{ yd} = 188 \text{ yd}^2$

34.

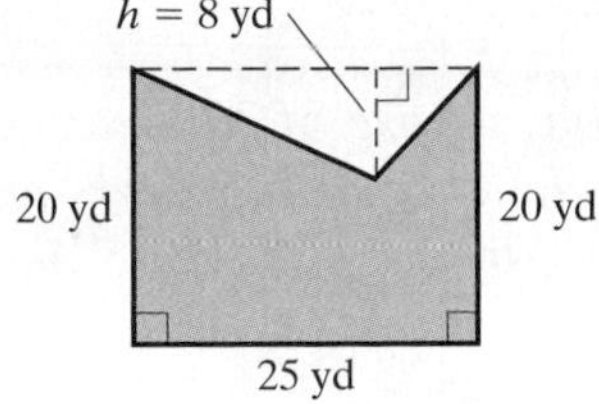

$A = 25 \text{ yd} \times 20 \text{ yd} - \frac{1}{2} \times 25 \text{ yd} \times 8 \text{ yd} = 400 \text{ yd}^2$

Applications

Area of Siding on a Building *Find the total area of all four vertical sides of the building.*

35.

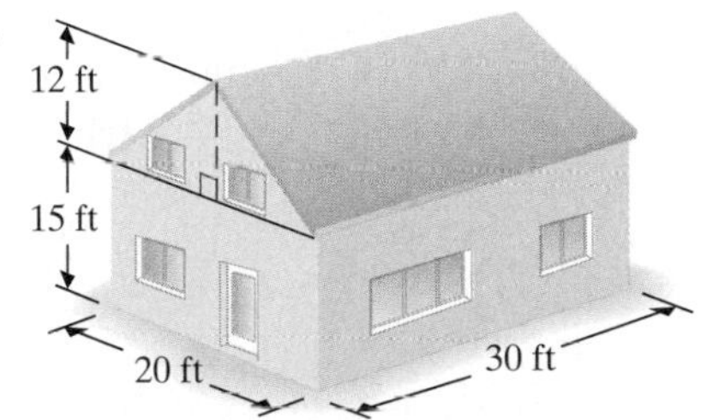

Area of each front and back

$= \frac{1}{2}(12 \text{ ft})(20 \text{ ft}) + (15 \text{ ft})(20 \text{ ft}) = 420 \text{ ft}^2.$

Area of each side $= (15 \text{ ft})(30 \text{ ft}) = 450 \text{ ft}^2.$

Total area $= 2(420 \text{ ft}^2) + 2(450 \text{ ft}^2) = 1740 \text{ ft}^2.$

36.

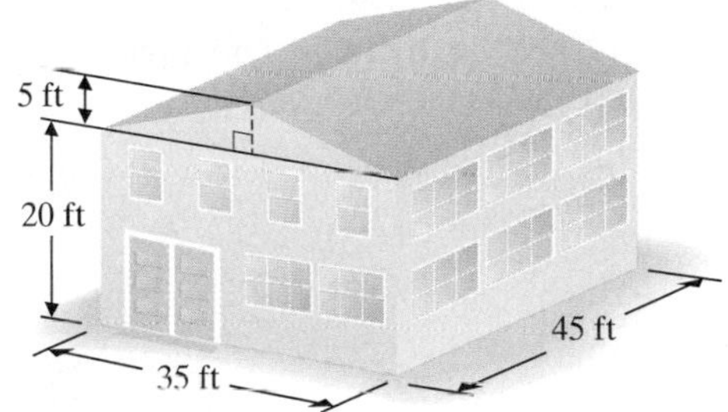

Area of each front and back

$= \frac{1}{2}(5 \text{ ft})(35 \text{ ft}) + (20 \text{ ft})(35 \text{ ft}) = 787.5 \text{ ft}^2.$

Area of each side $= (20 \text{ ft})(45 \text{ ft}) = 900 \text{ ft}^2.$

Total area $= 2(787.5 \text{ ft}^2) + 2(900 \text{ ft}^2) = 3375 \text{ ft}^2.$

Coating on Test Plane Wings *The top surface of the wings of a test plane must be coated with a special lacquer that costs \$90 per square meter. Find the cost to coat the shaded wing surface of the plane.*

37.

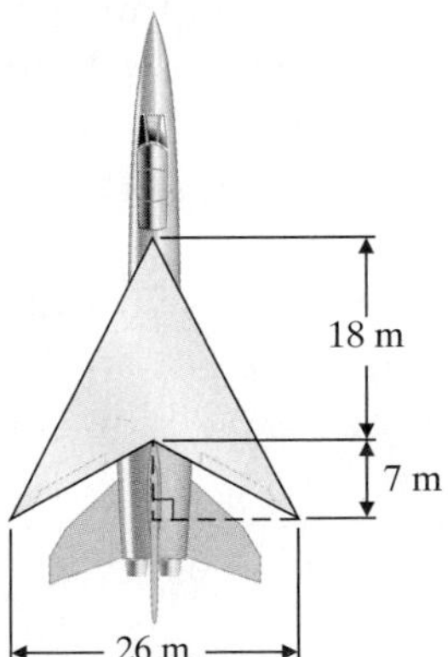

$A = 2 \times \frac{1}{2} \times 18 \text{ m} \times 13 \text{ m} = 234 \text{ m}^2$

$\text{Cost} = 234 \text{ m}^2 \times \frac{\$90}{\text{m}^2} = \$21{,}060$

38.

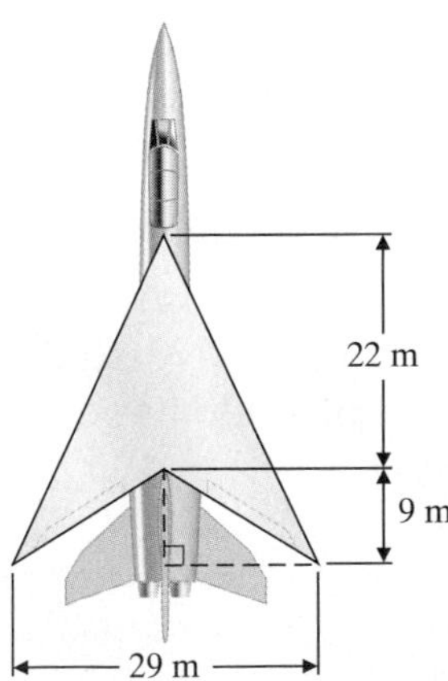

$A = 2 \times \frac{1}{2} \times 22 \text{ m} \times 14.5 \text{ m} = 319 \text{ m}^2$

$\text{Cost} = 319 \text{ m}^2 \times \frac{\$90}{\text{m}^2} = \$28{,}710$

To Think About

An equilateral triangle has a base of 20 meters and a height of h meters. Inside that triangle is constructed a second equilateral triangle of base 10 meters and a height of 0.5h meters. Inside the second triangle is constructed a third equilateral triangle of base 5 meters and a height of 0.25h meters.

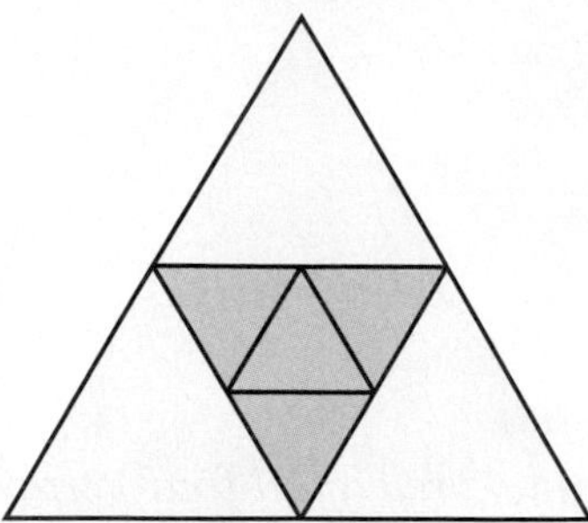

39. What percent of the area of the largest triangle is the area of the smallest triangle?
6.25%

40. What percent of the perimeter of the largest triangle is the perimeter of the smallest triangle?
25%

Cumulative Review

Solve for n. Round to the nearest hundredth.

41. $\frac{5}{n} = \frac{7.5}{18}$ $n = 12$

42. $\frac{n}{\frac{3}{4}} = \frac{7}{\frac{1}{8}}$ $n = 42$

43. ***Cruise Ship*** On the cruise ship *H.M.S. Salinora,* all restaurants on board must keep the ratio of waitstaff to patrons at 4 to 15. How many waitstaff should be serving if there are 2685 patrons dining at one time? 716

44. ***Transatlantic Plane Flight*** Recently, an airline found that after the transatlantic flight to Frankfurt, 68 people out of 300 passengers kept their in-flight magazines after being encouraged to take the magazines with them to read at their leisure. On a similar flight carrying 425 people, how many in-flight magazines would the airline expect to be taken? Round to the nearest whole number. 96 magazines

45. ***Education*** Dr. Hedman is teaching two sections of Biology 101. His evening section has 70 students and his day section has 135 students. On the first test, 14 students in his evening class got an A. The same proportion of students got an A in the day section. How many students in the day section earned an A? 27 students

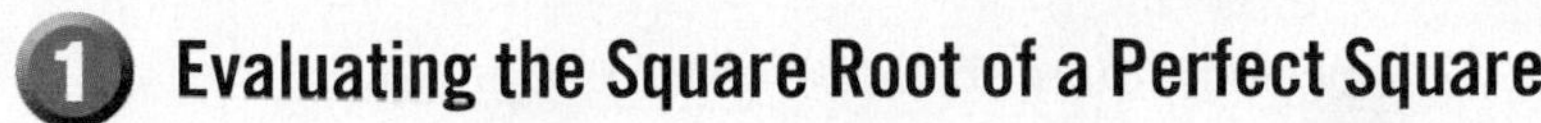

7.5 Square Roots

Student Learning Objectives

After studying this section, you will be able to:

1. Evaluate the square root of a perfect square.
2. Approximate the square root of a number that is not a perfect square.

1 Evaluating the Square Root of a Perfect Square

We know that by using the formula $A = s^2$ we can quickly find the area of a square with a side of 3 in. We simply square 3 in. That is, $A = (3 \text{ in.})(3 \text{ in.}) = 9 \text{ in.}^2$. Sometimes we want to ask another kind of question. If a square has an area of 64 in.^{2}, what is the length of its sides?

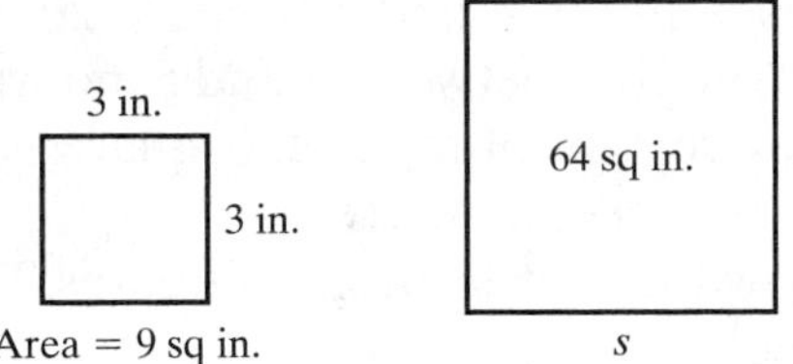

The answer is 8 in. Why? The skill we need to find a number when we are given the square of that number is called *finding the square root.* The square root of 64 is 8.

If a number is a product of two identical factors, then either factor is called a **square root.**

The square root of 64 is 8 because $(8)(8) = 64$.
The square root of 9 is 3 because $(3)(3) = 9$.

The symbol for finding the square root of a number is $\sqrt{\ }$. To write the square root of 64, we write $\sqrt{64} = 8$. Sometimes we speak of finding the square root of a number as *taking* the square root of the number, or we can say that we will *evaluate* the square root of the number. Thus to take the square root of 9, we write $\sqrt{9} = 3$; to evaluate the square root of 9, we write $\sqrt{9} = 3$.

EXAMPLE 1 Find. **(a)** $\sqrt{25}$ **(b)** $\sqrt{121}$

Solution

(a) $\sqrt{25} = 5$ because $(5)(5) = 25$.
(b) $\sqrt{121} = 11$ because $(11)(11) = 121$.

Practice Problem 1 Find. **(a)** $\sqrt{49}$ **(b)** $\sqrt{169}$

NOTE TO STUDENT: Fully worked-out solutions to all of the Practice Problems can be found at the back of the text starting at page SP-1

If square roots are added or subtracted, they must be evaluated *first,* then added or subtracted.

EXAMPLE 2 Find. $\sqrt{25} + \sqrt{36}$

Solution

$\sqrt{25} = 5$ because $(5)(5) = 25$.
$\sqrt{36} = 6$ because $(6)(6) = 36$.

Thus $\sqrt{25} + \sqrt{36} = 5 + 6 = 11$.

Practice Problem 2 Find. $\sqrt{49} - \sqrt{4}$

When a whole number is multiplied by itself, the number that is obtained is called a **perfect square.**

36 is a perfect square because $(6)(6) = 36$.
49 is a perfect square because $(7)(7) = 49$.

Teaching Tip Encourage students to fill in the chart of perfect squares and to memorize it. Remind them that if they can find those values quickly, they will find working with square roots much easier.

The numbers 20 or 48 are *not* perfect squares. There is no *whole number* that when squared—multiplied by itself—yields 20 or 48. Consider 20. $4^2 = 16$, which is less than 20. $5^2 = 25$, which is more than 20. We realize, then, that the square root of 20 is between 4 and 5 because 20 is between 16 and 25. Since there is no whole number between 4 and 5, no whole number squared equals 20. Since the square root of a perfect square is a whole number, we can say that 20 is *not* a perfect square.

It is helpful to know the first 15 perfect squares. Take a minute to complete the following table.

Number, n	1	2	3	4	5	6	7	8	9	10	11	12	13	14	15
Number Squared, n^2	1	4	9	16	25	36	49	64	81	100	121	144	169	196	225

EXAMPLE 3

(a) Is 81 a perfect square? **(b)** If so, find $\sqrt{81}$.

Solution

(a) Yes. 81 is a perfect square because $(9)(9) = 81$. **(b)** $\sqrt{81} = 9$

NOTE TO STUDENT: *Fully worked-out solutions to all of the Practice Problems can be found at the back of the text starting at page SP-1*

Practice Problem 3

(a) Is 144 a perfect square? **(b)** If so, find $\sqrt{144}$.

Approximating the Square Root of a Number That Is Not a Perfect Square

If a number is not a perfect square, we can only approximate its square root. This can be done by using a square root table such as the one that follows. Except for exact values such as $\sqrt{4} = 2.000$, all values are rounded to the nearest thousandth.

Number, n	Square Root of the Number, $\sqrt{n}$	Number, n	Square Root of the Number, $\sqrt{n}$
1	1.000	8	2.828
2	1.414	9	3.000
3	1.732	10	3.162
4	2.000	11	3.317
5	2.236	12	3.464
6	2.449	13	3.606
7	2.646	14	3.742

A square root table is located on page A-3. It gives you the square roots of whole numbers up to 200. Square roots can also be found with any calculator that has a square root key. Usually the key looks like this $\boxed{\sqrt{}}$ or this

$\boxed{\sqrt{x}}$. To find the square root of 8 on most calculators, enter the number 8 and press $\boxed{\sqrt{\ }}$ or $\boxed{\sqrt{x}}$. You will see displayed 2.8284271. On some calculators, you must enter the square root key first followed by the number. (Your calculator may display fewer or more digits.) Remember, no matter how many digits your calculator displays, when we find $\sqrt{8}$, we have only an **approximation.** It is not an exact answer. To emphasize this we use the $\approx$ notation to mean "is approximately equal to." Thus $\sqrt{8} \approx 2.828$.

EXAMPLE 4 Find approximate values using the square root table or a calculator. Round to the nearest thousandth.

(a) $\sqrt{2}$ **(b)** $\sqrt{12}$ **(c)** $\sqrt{7}$

Solution

(a) $\sqrt{2} \approx 1.414$ **(b)** $\sqrt{12} \approx 3.464$ **(c)** $\sqrt{7} \approx 2.646$

Practice Problem 4 Approximate to the nearest thousandth.

(a) $\sqrt{3}$ **(b)** $\sqrt{13}$ **(c)** $\sqrt{5}$

EXAMPLE 5 Approximate to the nearest thousandth of an inch the length of the side of a square that has an area of 6 in.^{2}.

Solution $\sqrt{6 \text{ in.}^2} \approx 2.449$ in.

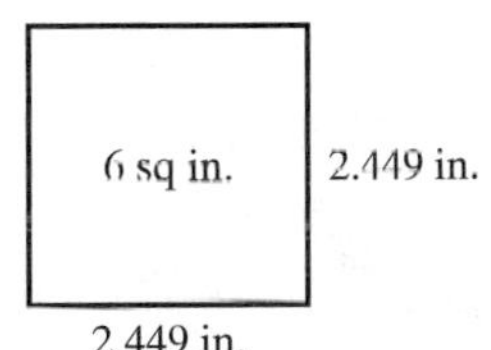

Thus, to the nearest thousandth of an inch, the side measures 2.449 in.

Practice Problem 5 Approximate to the nearest thousandth of a meter the length of the side of a square that has an area of 22 m^2.

Calculator

Square Roots

Locate the square root key $\boxed{\sqrt{\ }}$ on your calculator.

1. To find $\sqrt{289}$, enter

   289 $\boxed{\sqrt{\ }}$

   The display should read

   $\boxed{17}$

2. To find $\sqrt{194}$, enter

   194 $\boxed{\sqrt{\ }}$

   The display should read

   $\boxed{13.928388}$

   This is just an approximation of the actual square root. We will round the answer to the nearest thousandth.

   $\sqrt{194} \approx 13.928$

   Your calculator may require you to enter the square root key first and then the number.

Teaching Tip Since today many pocket calculators have a square root key, this might be a good time to encourage students who have a calculator to bring it to class, and to urge those who do not have a calculator to borrow one from a friend or relative for a few days. Square roots always seem less threatening to a student with a calculator.

7.5 EXERCISES

Student Solutions Manual | CD/ Video | PH Math Tutor Center | MathXL®Tutorials on CD | MathXL® | MyMathLab® | Interactmath.com

Verbal and Writing Skills

1. Why is $\sqrt{25} = 5$? $\sqrt{25} = 5$ because $(5)(5) = 25$.

2. $\sqrt{49}$ is read "the ___square root___ of 49".

3. 25 is a perfect square because its square root, 5, is a ___whole___ number.

4. Is 32 a perfect square? Why or why not?
32 is not a perfect square because no whole number when multiplied by itself equals 32.

5. How can you approximate the square root of a number that is not a perfect square?
To approximate the square root of a number that is not a perfect square, use the square root table or a calculator.

6. How would you find $\sqrt{0.04}$? $\sqrt{0.04} = 0.2$ since $(0.2)(0.2) = 0.04$

Find each square root. Do not use a calculator. Do not refer to a table of square roots.

7. $\sqrt{9}$ 3 **8.** $\sqrt{16}$ 4 **9.** $\sqrt{64}$ 8 **10.** $\sqrt{81}$ 9

13. $\sqrt{0}$ 0 **14.** $\sqrt{225}$ 15 **11.** $\sqrt{144}$ 12 **12.** $\sqrt{196}$ 14

15. $\sqrt{169}$ 13 **16.** $\sqrt{81}$ 9 **17.** $\sqrt{100}$ 10 **18.** $\sqrt{361}$ 19

In exercises 13–20, evaluate the square roots first, then add, subtract or multiply the results. Do not use a calculator or a square root table.

19. $\sqrt{49} + \sqrt{9}$
$7 + 3 = 10$

20. $\sqrt{25} + \sqrt{64}$
$5 + 8 = 13$

21. $\sqrt{100} + \sqrt{1}$
$10 + 1 = 11$

22. $\sqrt{0} + \sqrt{121}$
$0 + 11 = 11$

23. $\sqrt{225} - \sqrt{144}$
$15 - 12 = 3$

24. $\sqrt{169} - \sqrt{64}$
$13 - 8 = 5$

25. $\sqrt{169} - \sqrt{121} + \sqrt{36}$
$13 - 11 + 6 = 8$

26. $\sqrt{225} - \sqrt{100} + \sqrt{64}$
$15 - 10 + 8 = 13$

27. $\sqrt{4} \times \sqrt{121}$
$2 \times 11 = 22$

28. $\sqrt{225} \times \sqrt{9}$
$15 \times 3 = 45$

29. (a) Is 256 a perfect square? yes
(b) If so, find $\sqrt{256}$. 16

30. (a) Is 289 a perfect square? yes
(b) If so, find $\sqrt{289}$. 17

Use a table of square roots or a calculator with a square root key to approximate to the nearest thousandth.

31. $\sqrt{18}$
≈ 4.243

32. $\sqrt{24}$
≈ 4.899

33. $\sqrt{76}$
≈ 8.718

34. $\sqrt{82}$
9.055

35. $\sqrt{200}$
≈ 14.142

36. $\sqrt{194}$
≈ 13.928

Find the length of the side of the square. If the area is not a perfect square, approximate by using a square root table or a calculator with a square root key. Round to the nearest thousandth.

37. A square with area 34 m^2
$\sqrt{34 \text{ m}^2} \approx 5.831$ m

38. A square with an area of 54 m^2
$\sqrt{54 \text{ m}^2} \approx 7.348$ m

39. A square with area 136 m^2
$\sqrt{136 \text{ m}^2} \approx 11.662$ m

40. A square with area 180 m^2
$\sqrt{180 \text{ m}^2} \approx 13.416$ m

Mixed Practice

Evaluate the square roots first. Then combine the results. Use a calculator or square root table when needed. Round to the nearest thousandth.

41. $\sqrt{36} + \sqrt{20}$
10.472

42. $\sqrt{25} + \sqrt{30}$
10.477

43. $\sqrt{198} - \sqrt{49}$
7.071

44. $\sqrt{185} - \sqrt{64}$
5.601

Applications

Basketball Court *High school basketball is played on a standard rectangular court that measures 92 feet in length and 50 feet in width. Some middle schools have smaller basketball courts that measure 80 feet in length and 42 feet in width.*

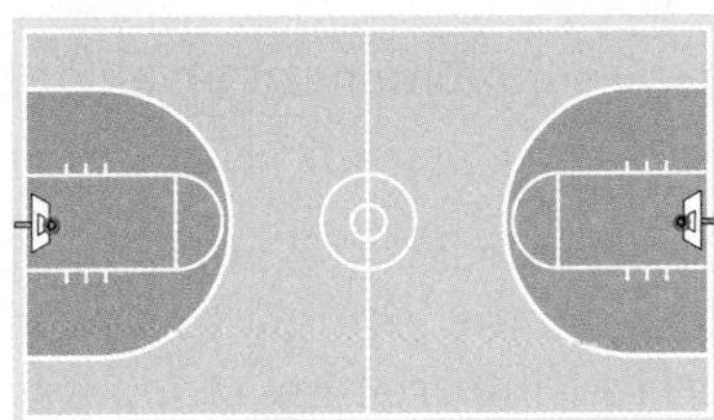

45. The diagonal of a standard high school basketball court measures $\sqrt{10{,}964}$ feet in length. Find the length of this diagonal to the nearest tenth of a foot. 104.7 ft

46. The diagonal of the smaller basketball court found in some middle schools measures $\sqrt{8164}$ feet in length. Find the length of this diagonal to the nearest tenth of a foot. 90.4 ft

47. ***Baseball*** The distance from second base to home plate on a professional baseball field is $\sqrt{16{,}200}$ ft. Find this distance to the nearest tenth of a foot. 127.3 ft

48. ***Room Measurements*** The diagonal of a rectangular room that measures 24 meters long and 12 meters wide is $\sqrt{720}$ meters long. Find the length of this diagonal to the nearest tenth of a meter. 26.8 m

To Think About

49. Find each square root.

(a) $\sqrt{4}$
2

(b) $\sqrt{0.04}$
0.2

(c) What pattern do you observe? Each answer is obtained from the previous answer by dividing by 10.

(d) Can you find $\sqrt{0.004}$ exactly? Why?
no; because 0.004 isn't a perfect square

50. Find each square root.

(a) $\sqrt{25}$
5

(b) $\sqrt{0.25}$
0.5

(c) What pattern do you observe?
same as 49(c)

(d) Can you find $\sqrt{0.025}$ exactly? Why?
no; because 0.025 isn't a perfect square

Using a calculator with a square root key, evaluate and round to the nearest thousandth.

51. $\sqrt{456} + \sqrt{322}$ 39.299

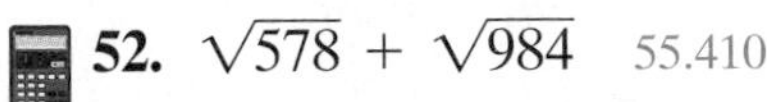

52. $\sqrt{578} + \sqrt{984}$ 55.410

Cumulative Review

53. ***Australia Zoo*** The Taronga Zoo in Sydney, Australia, has a viewing tank for its platypuses. The tank is 60 in. high and 80 in. wide. What is the area of the front of the rectangular tank?
4800 sq in.

54. ***Juggling and Running*** Ashrita Furman of Jamaica, New York, holds the world record in "joggling"—juggling and running at the same time—over 80.5 km. How many meters did he "joggle"? 80,500 meters

55. ***30 km Race*** North Shore Community College is hosting a 30-km running race. How many miles is the race? 18.6 mi

56. ***African Beetles*** Goliath beetles in equatorial Africa can weigh up to 98.9 grams. How many kilograms can they weigh? 0.0989 kilograms

How Am I Doing? Sections 7.1–7.5

1. 18°

2. 117°

3. $\angle b = 136°$
$\angle a = \angle c = 44°$

4. 18 m

5. 14 m

6. 23.04 sq cm

7. 2.43 sq cm

8. 25.6 yd

9. 79 ft

10. 351 sq in.

11. 171 sq in.

12. 97 sq m

How are you doing with your homework assignments in Sections 7.1 to 7.5? Do you feel you have mastered the material so far? Do you understand the concepts you have covered? Before you go further in the textbook, take some time to do each of the following problems.

7.1

1. Find the complement of an angle that is 72°.

2. Find the supplement of an angle that is 63°.

3. Find the measure of angle a, angle b, and angle c in the sketch to the right, which shows two intersecting straight lines.

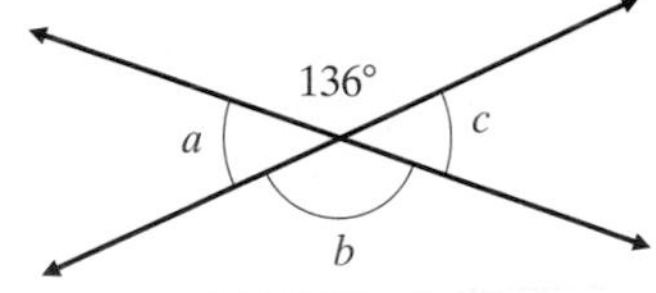

7.2

Find the perimeter of each rectangle or square.

4. Length = 6.5 m, width = 2.5 m

5. Length = width = 3.5 m

Find the area of each square or rectangle.

6. Length = width = 4.8 cm

7. Length = 2.7 cm, width = 0.9 cm

7.3

Find the perimeter.

8. A parallelogram with one side measuring 9.2 yd and another side measuring 3.6 yd.

9. A trapezoid with sides measuring 17 ft, 15 ft, $25\frac{1}{2}$ ft, and $21\frac{1}{2}$ ft.

Find the area.

10. A parallelogram with a base of 27 in. and a height of 13 in.

11. A trapezoid with a height of 9 in. and bases of 16 in. and 22 in.

12.

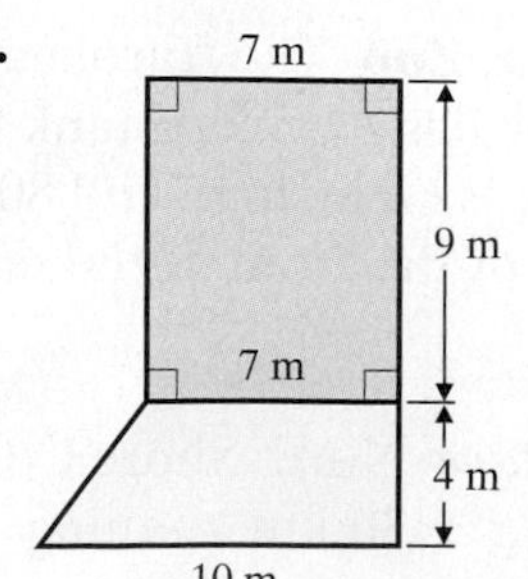

7.4

13. Find the third angle in the triangle if two angles are 39° and 118°.

13. 23°

14. Find the perimeter of the triangle whose sides measure $7\frac{1}{3}$ m, $4\frac{2}{3}$ m, and 3 m.

14. 15 m

15. Find the area of a triangle with a base of 16 m and a height of 9 m.

15. 72 sq m

Applications of 7.1 to 7.4

16. A college entrance has a sign shaped like this figure.

(a) How many square feet of paint are needed to cover the sign?

(b) How many feet of trim are needed to cover the edge (perimeter) of the sign?

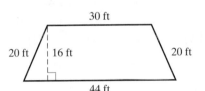

16. (a) 592 ft^2

(b) 114 ft

30 ft
20 ft
16 ft
20 ft
44 ft

7.5

Evaluate exactly.

17. $\sqrt{64}$

18. $\sqrt{4} + \sqrt{100}$

19. $\sqrt{169}$

20. $\sqrt{256}$

17. 8

18. 12

19. 13

20. 16

21. Approximate $\sqrt{46}$ using a square root table or a calculator with a square root key. Round to the nearest thousandth.

21. 6.782

Now turn to page SA-14 for the answer to each of these problems. Each answer also includes a reference to the objective in which the problem is first taught. If you missed any of these problems, you should stop and review the Examples and Practice Problems in the referenced objective. A little review now will help you master the material in the upcoming sections of the text.

7.6 The Pythagorean Theorem

Student Learning Objectives

After studying this section, you will be able to:

1. Find the hypotenuse of a right triangle given the length of each leg.
2. Find the length of a leg of a right triangle given the lengths of the hypotenuse and the other leg.
3. Solve applied problems using the Pythagorean Theorem.
4. Solve for the missing sides of special right triangles.

1 Finding the Hypotenuse of a Right Triangle Given the Length of Each Leg

The Pythagorean Theorem is a mathematical idea formulated long ago. It is as useful today as it was when it was discovered. The Pythagoreans lived in Italy about 2500 years ago. They studied various mathematical properties. They discovered that for any right triangle, the square of the hypotenuse equals the sum of the squares of the two legs of the triangle. This relationship is known as the **Pythagorean Theorem.** The side opposite the right angle is called the **hypotenuse;** the other two sides are called the legs of the right triangle.

The Pythagoreans discovered that this theorem could be used to find the length of the third side of any right triangle if the lengths of two of the sides was known. We still use this theorem today for the very same reason.

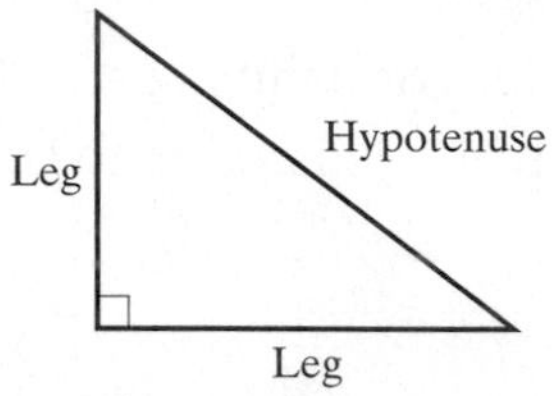

$$(\text{hypotenuse})^2 = (\text{leg})^2 + (\text{leg})^2$$

Note how the Pythagorean Theorem applies to the right triangle shown here.

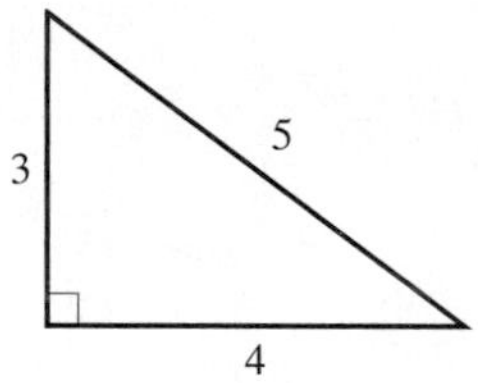

$$5^2 = 3^2 + 4^2$$
$$25 = 9 + 16$$
$$25 = 25 \checkmark$$

Teaching Tip Emphasize that it is important for students to know what side of a right triangle is the *hypotenuse*. Remind them that they will hear that word many times in mathematics and in other courses.

In a right triangle, the hypotenuse is the longest side. It is always opposite the largest angle, the right angle. The legs are the two shorter sides. When we know each leg of a right triangle, we use the following property.

$$\text{Hypotenuse} = \sqrt{(\text{leg})^2 + (\text{leg})^2}$$

EXAMPLE 1 Find the hypotenuse of a right triangle with legs of 5 in. and 12 in.

Solution

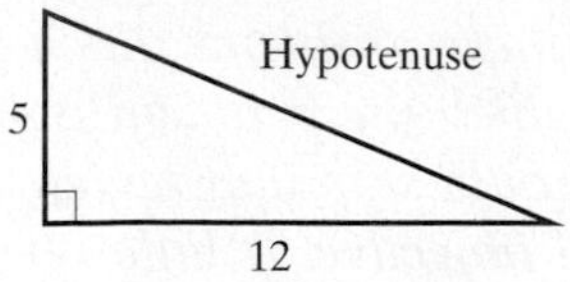

$$\text{Hypotenuse} = \sqrt{(5)^2 + (12)^2}$$

$= \sqrt{25 + 144}$ Square each value first.

$= \sqrt{169}$ Add together the two values.

$= 13$ in. Take the square root.

Teaching Tip This is a good time to remind students that they should keep reviewing the table of perfect squares they filled out in Section 7.5.

Practice Problem 1 Find the hypotenuse of a right triangle with legs of 8 m and 6 m.

NOTE TO STUDENT: Fully worked-out solutions to all of the Practice Problems can be found at the back of the text starting at page SP-1

Sometimes we cannot find the hypotenuse exactly. In those cases, we often approximate the square root by using a calculator or a square root table.

EXAMPLE 2 Find the hypotenuse of a right triangle with legs of 4 m and 5 m. Round to the nearest thousandth.

Solution

$$\text{Hypotenuse} = \sqrt{(4)^2 + (5)^2}$$
$$= \sqrt{16 + 25} \quad \text{Square each value first.}$$
$$= \sqrt{41}\text{ m} \quad \text{Add the two values together.}$$

Using the square root table or a calculator, we have the hypotenuse ≈ 6.403 m.

4
Hypotenuse
5

Practice Problem 2 Find to the nearest thousandth the hypotenuse of a right triangle with legs of 3 cm and 7 cm.

2 Finding the Length of a Leg of a Right Triangle Given the Lengths of the Hypotenuse and the Other Leg

When we know the hypotenuse and one leg of a right triangle, we find the length of the other leg by using the following property.

$$\text{Leg} = \sqrt{(\text{hypotenuse})^2 - (\text{leg})^2}$$

EXAMPLE 3 A right triangle has a hypotenuse of 15 cm and a leg of 12 cm. Find the length of the other leg.

Solution

$$\text{Leg} = \sqrt{(15)^2 - (12)^2}$$
$$= \sqrt{225 - 144} \quad \text{Square each value first.}$$
$$= \sqrt{81} \quad \text{Subtract.}$$
$$= 9\text{ cm} \quad \text{Find the square root.}$$

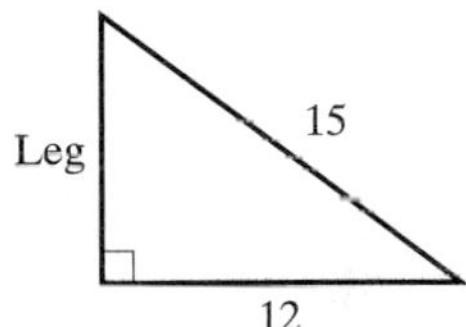

Practice Problem 3 A right triangle has a hypotenuse of 17 m and a leg of 15 m. Find the length of the other leg.

EXAMPLE 4 A sail for a sailboat is in the shape of a right triangle. The right triangle has a hypotenuse of 14 feet and a leg of 8 feet. Find the length of the other leg. Round to the nearest thousandth.

Solution

$$\text{Leg} = \sqrt{(14)^2 - (8)^2}$$
$$= \sqrt{196 - 64} \quad \text{Square each value first.}$$
$$= \sqrt{132}\text{ feet} \quad \text{Subtract the two numbers.}$$

Using a calculator or the square root table, we find that the leg ≈ 11.489 feet.

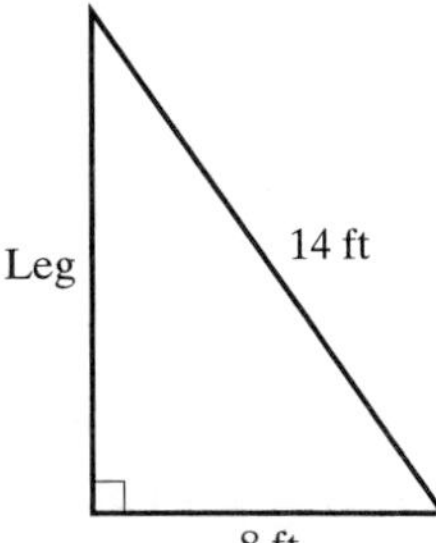

Practice Problem 4 A right triangle has a hypotenuse of 10 m and a leg of 5 m. Find the length of the other leg. Round to the nearest thousandth.

3 Solving Applied Problems Using the Pythagorean Theorem

Certain applied problems call for the use of the Pythagorean Theorem in the solution.

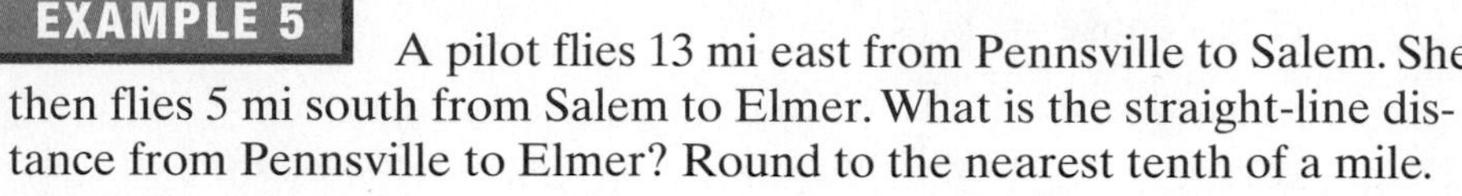

EXAMPLE 5 A pilot flies 13 mi east from Pennsville to Salem. She then flies 5 mi south from Salem to Elmer. What is the straight-line distance from Pennsville to Elmer? Round to the nearest tenth of a mile.

Solution

1. ***Understand the problem.*** It might help to draw a picture.

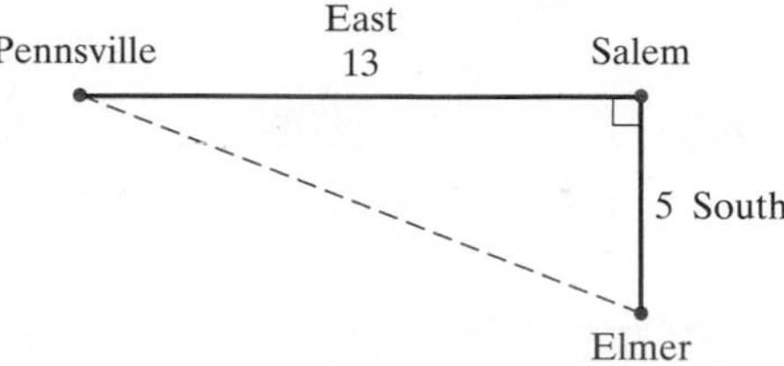

The distance from Pennsville to Elmer is the hypotenuse of the triangle.

2. ***Solve and state the answer.***

$$\text{Hypotenuse} = \sqrt{(\text{leg})^2 + (\text{leg})^2}$$
$$= \sqrt{(13)^2 + (5)^2}$$
$$= \sqrt{169 + 25}$$
$$= \sqrt{194}$$
$$\sqrt{194} \approx 13.928 \text{ mi}$$

Rounded to the nearest tenth, the distance is 13.9 mi.

3. ***Check.*** Work backward to check. Use the Pythagorean Theorem.

$13.9^2 \approx 13^2 + 5^2$ (We use $\approx$ because 13.9 is an approximate answer.)
$193.21 \approx 169 + 25$
$193.21 \approx 194$ ✓

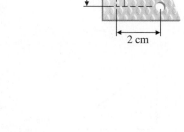

NOTE TO STUDENT: Fully worked-out solutions to all of the Practice Problems can be found at the back of the text starting at page SP-1

Practice Problem 5 Find the distance to the nearest thousandth between the centers of the holes in the triangular metal plate in the margin.

EXAMPLE 6 A 25-ft ladder is placed against a building at a point 22 ft from the ground. What is the distance of the base of the ladder from the building? Round to the nearest tenth.

Solution

1. ***Understand the problem.*** Draw a picture.

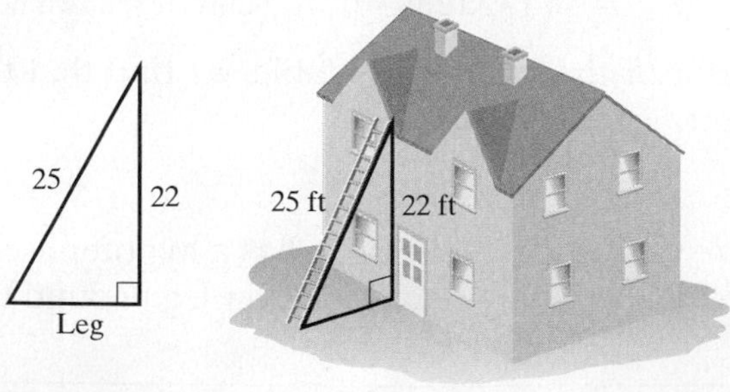

2. ***Solve and state the answer.***

$$\begin{aligned} \text{Leg} &= \sqrt{(\text{hypotenuse})^2 - (\text{leg})^2} \\ &= \sqrt{(25)^2 - (22)^2} \\ &= \sqrt{625 - 484} \\ &= \sqrt{141} \\ \sqrt{141} &\approx 11.874 \text{ ft} \end{aligned}$$

If we round to the nearest tenth, the ladder is 11.9 ft from the base of the house.

Practice Problem 6 A kite is out on 30 yd of string. The kite is directly above a rock. The rock is 27 yd from the boy flying the kite. How far above the rock is the kite? Round to the nearest tenth.

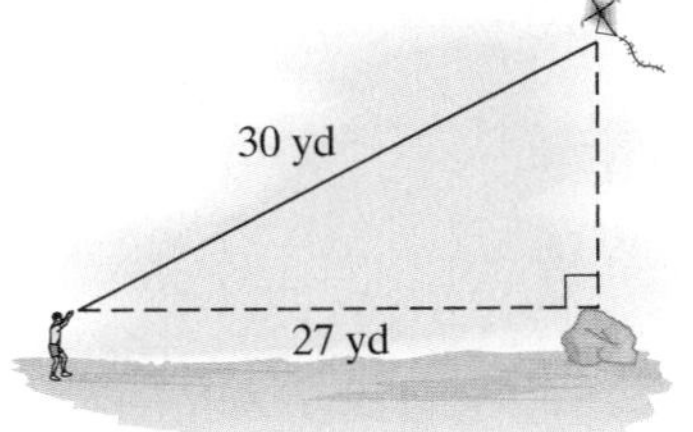

4 Solving for the Missing Sides of Special Right Triangles

If we use the Pythagorean Theorem and some other facts from geometry, we can find a relationship among the sides of two special right triangles. The first special right triangle is one that contains an angle that measures 30° and one that measures 60°. We call this the 30°–60°–90° right triangle.

Teaching Tip Instructors often assume that students have memorized the ratio of the sides of the 30–60–90 right triangle and the 45–45–90 right triangle. But many students have never heard these ratios before or may not remember them from high school. You may need to take extra time to explain this material to the class.

In a 30°–60°–90° triangle the length of the leg opposite the 30° angle is $\frac{1}{2}$ the length of the hypotenuse.

Notice that the hypotenuse of the first triangle is 10 m and the side opposite the 30° angle is exactly $\frac{1}{2}$ of that, or 5 m. The second triangle has a hypotenuse of 15 yd. The side opposite the 30° angle is exactly $\frac{1}{2}$ of that, or 7.5 yd.

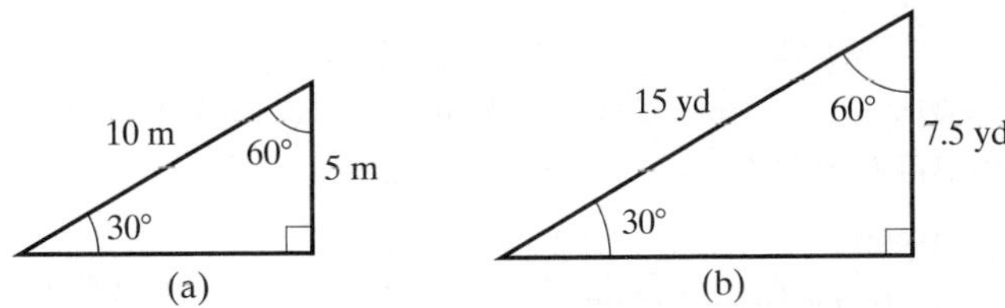

The second special right triangle is one that contains exactly two angles that each measure 45°. We call this the 45°–45°–90° right triangle.

In a 45°–45°–90° triangle the lengths of the sides opposite the 45° angles are equal. The length of the hypotenuse is equal to $\sqrt{2} \times$ the length of either leg.

We will use the decimal approximation $\sqrt{2} \approx 1.414$ with this property.

$$\begin{aligned} \text{Hypotenuse} &= \sqrt{2} \times 7 \\ &\approx 1.414 \times 7 \\ &\approx 9.898 \text{ cm} \end{aligned}$$

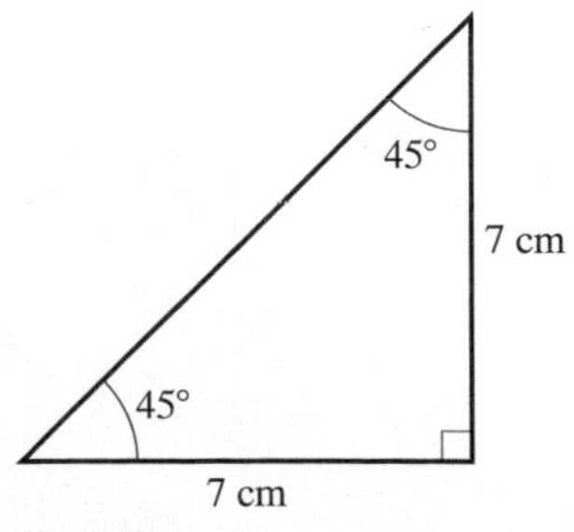

EXAMPLE 7 Find the requested sides of each special triangle. Round to the nearest tenth.

(a) Find the lengths of sides y and x. **(b)** Find the length of hypotenuse z.

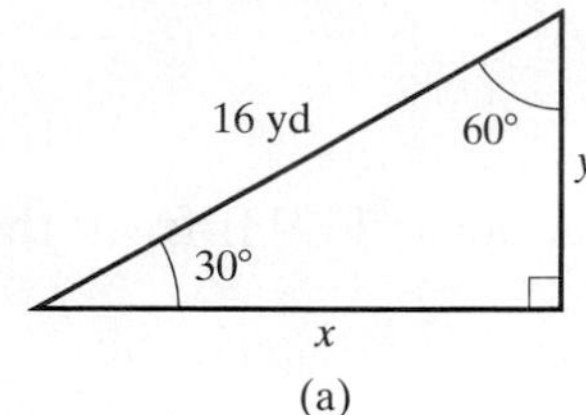

(a)

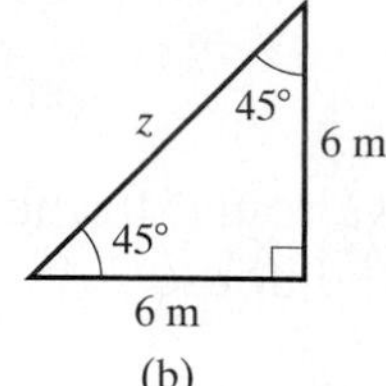

(b)

Solution

(a) In a 30°–60°–90° triangle the side opposite the 30° angle is $\frac{1}{2}$ of the hypotenuse.

$$\frac{1}{2} \times 16 = 8$$

Therefore, $y = 8$ yd.

When we know two sides of a right triangle, we find the third side using the Pythagorean Theorem.

$$\begin{aligned} \text{Leg} &= \sqrt{(\text{hypotenuse})^2 - (\text{leg})^2} \\ &= \sqrt{16^2 - 8^2} = \sqrt{256 - 64} \\ &= \sqrt{192} \approx 13.856 \text{ yd} \end{aligned}$$

Thus $x = 13.9$ yd rounded to the nearest tenth.

(b) In a 45°–45°–90° triangle we have the following.

$$\begin{aligned} \text{Hypotenuse} &= \sqrt{2} \times \text{leg} \\ &\approx 1.414(6) \\ &= 8.484 \text{ m} \end{aligned}$$

Rounded to the nearest tenth, the hypotenuse = 8.5 m.

***NOTE TO STUDENT:** Fully worked-out solutions to all of the Practice Problems can be found at the back of the text starting at page SP-1*

Practice Problem 7 Find the requested sides of each special triangle. Round to the nearest tenth.

(a) Find the lengths of sides y and x.

(b) Find the length of hypotenuse z.

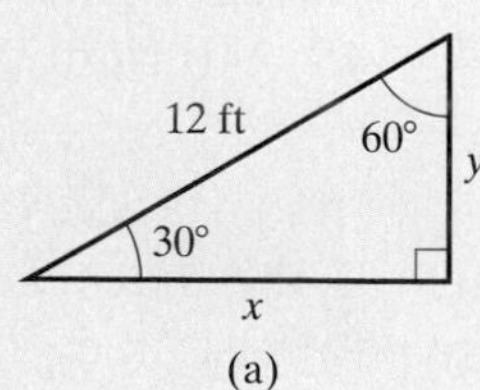

(a)

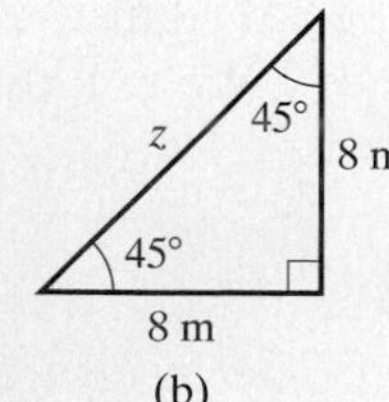

(b)

7.6 Exercises

Student Solutions Manual

CD/ Video

PH Math Tutor Center

MathXL®Tutorials on CD
MathXL®

MyMathLab®

Interactmath.com

Verbal and Writing Skills

1. Explain in your own words how to obtain the length of the hypotenuse of a right triangle if you know the length of each of the legs of the triangle.
Square the length of each leg and add those two results. Then take a square root of the remaining number.

2. Explain in your own words how to obtain the length of one leg of a right triangle if you know the length of the hypotenuse and the length of the other leg.
Square the length of the hypotenuse and square the length of the leg. Subtract the value of the leg squared from the value of the hypotenuse squared. Take the square root of this result.

Find the unknown side of the right triangle. Use a calculator or square root table when necessary and round to the nearest thousandth.

3.

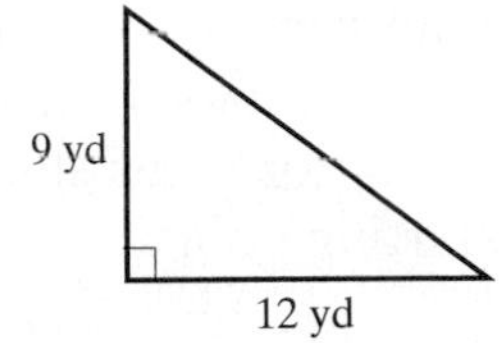

$h = \sqrt{9^2 + 12^2} = \sqrt{225} = 15$ yd

4.

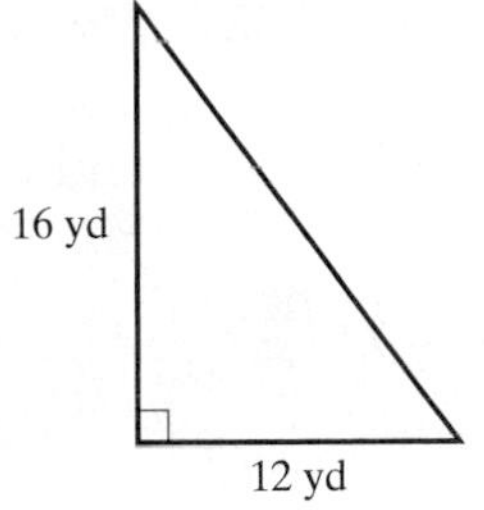

$h = \sqrt{16^2 + 12^2} = \sqrt{400} = 20$ yd

5.

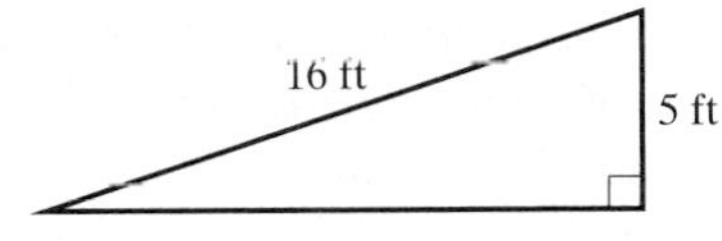

$l = \sqrt{16^2 - 5^2} \approx 15.199$ ft

6.

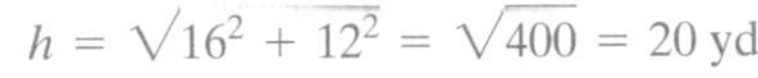

$l = \sqrt{21^2 - 7^2} \approx 19.799$ ft

Find the unknown side of the right triangle using the information given to the nearest thousandth.

7. leg = 11 m, leg = 3 m
$h = \sqrt{11^2 + 3^2} = \sqrt{130} \approx 11.402$ m

8. leg = 5 m, leg = 4 m
$h = \sqrt{5^2 + 4^2} = \sqrt{41} \approx 6.403$ m

9. leg = 10 m, leg = 10 m
$h = \sqrt{10^2 + 10^2} = \sqrt{200} \approx 14.142$ m

10. leg = 7 m, leg = 7 m
$h = \sqrt{7^2 + 7^2} = \sqrt{98} \approx 9.899$ m

11. hypotenuse = 10 ft, leg = 5 ft
$\text{leg} = \sqrt{10^2 - 5^2} = \sqrt{75} \approx 8.660$ ft

12. hypotenuse = 13 yd, leg = 11 yd
$\text{leg} = \sqrt{13^2 - 11^2} = \sqrt{48} \approx 6.928$ yd

13. hypotenuse = 14 yd, leg = 10 yd
$\text{leg} = \sqrt{14^2 - 10^2} = \sqrt{96} \approx 9.798$ yd

14. hypotenuse = 20 ft, leg = 14 ft
$\text{leg} = \sqrt{20^2 - 14^2} = \sqrt{204} \approx 14.283$ ft

15. leg = 12 m, leg = 9 m
$h = 15$ m

16. leg = 6 m, leg = 8 m
$h = 10$ m

17. hypotenuse = 16 ft, leg = 11 ft
$h \approx 11.619$ ft

18. hypotenuse = 18 ft, leg = 14 ft
$h \approx 11.314$ ft

Applications

Solve. Round to the nearest tenth.

19. ***Loading Ramp*** Find the length of this ramp to the back of a truck.

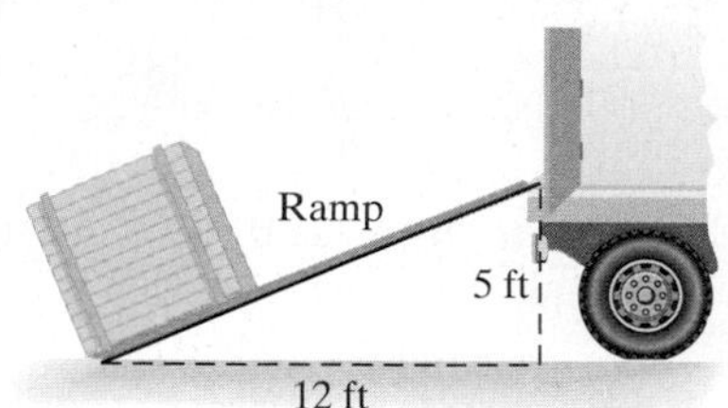

$h = \sqrt{12^2 + 5^2} = \sqrt{169} = 13$ ft

20. ***Telephone Pole*** Find the length of the guy wire supporting the telephone pole.

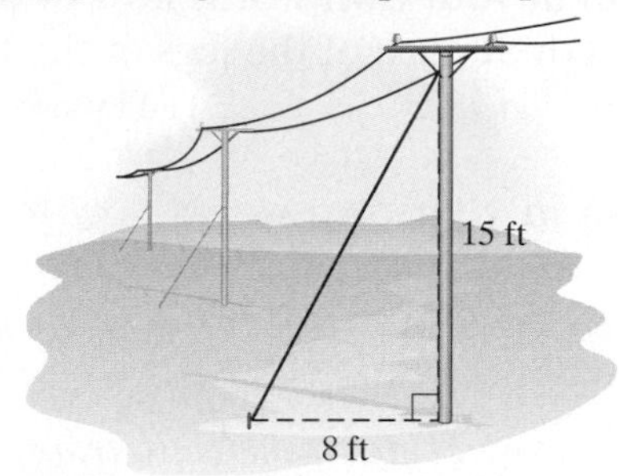

$h = \sqrt{15^2 + 8^2} = \sqrt{289} = 17$ ft

21. ***Construction of a Steel Plate*** A construction project requires a stainless steel plate with holes drilled as shown. Find the distance between the centers of the holes in this triangular plate.

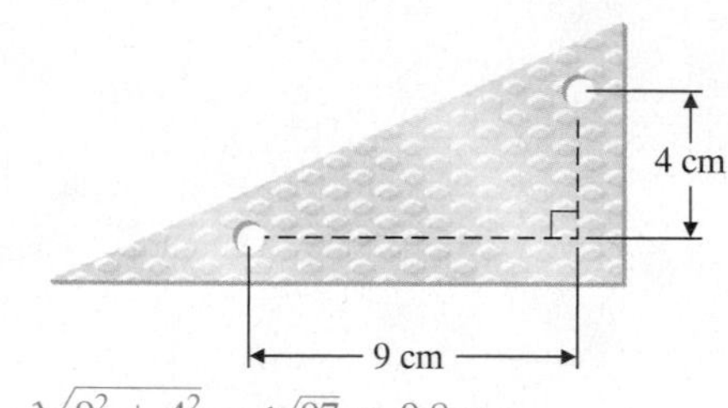

$h = \sqrt{9^2 + 4^2} = \sqrt{97} \approx 9.8$ cm

22. ***Running Out of Gas*** Juan runs out of gas in Los Lunas, New Mexico. He walks 4 mi west and then 3 mi south looking for a gas station. How far is he from his starting point?

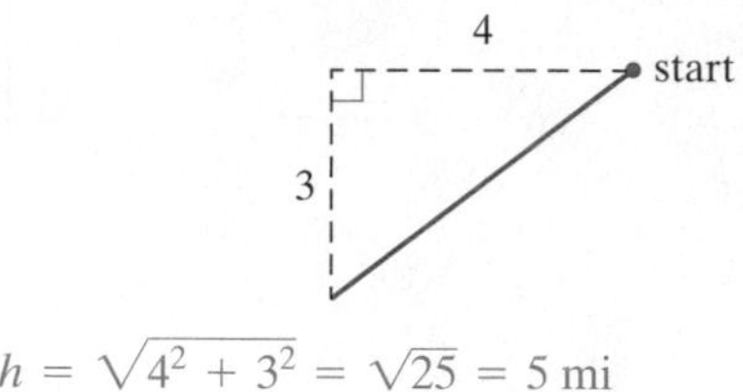

$h = \sqrt{4^2 + 3^2} = \sqrt{25} = 5$ mi

23. ***Kite Flying*** Barbara is flying her dragon kite on 32 yd of string. The kite is directly above the edge of a pond. The edge of the pond is 30 yd from where the kite is tied to the ground. How far is the kite above the pond?

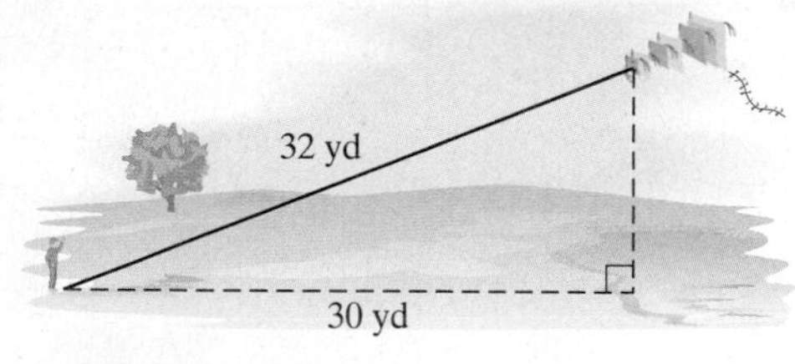

leg $= \sqrt{32^2 - 30^2} \approx 11.1$ yd

24. ***Ladder Distance*** A 20-ft ladder is placed against a college classroom building at a point 18 ft above the ground. What is the distance from the base of the ladder to the building?

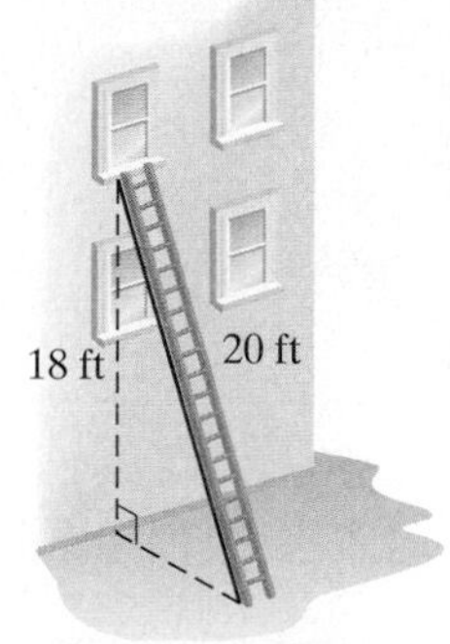

leg $= \sqrt{20^2 - 18^2} \approx 8.7$ ft

Using your knowledge of special right triangles, find the length of each leg. Round to the nearest tenth.

25.

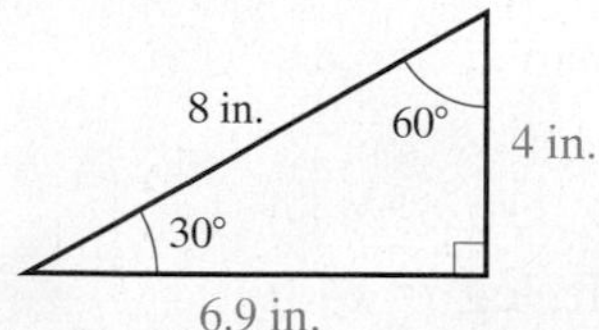

26.

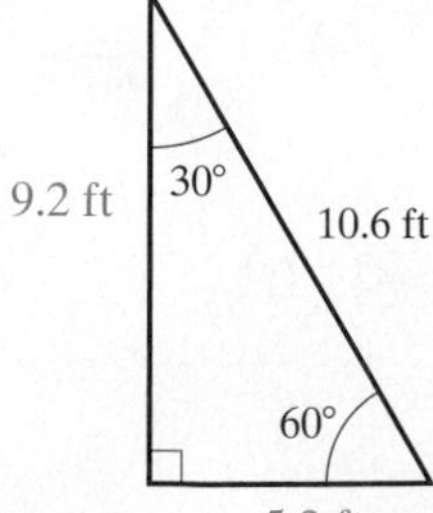

Using your knowledge of special right triangles, find the length of the hypotenuse. Round to the nearest tenth.

27.

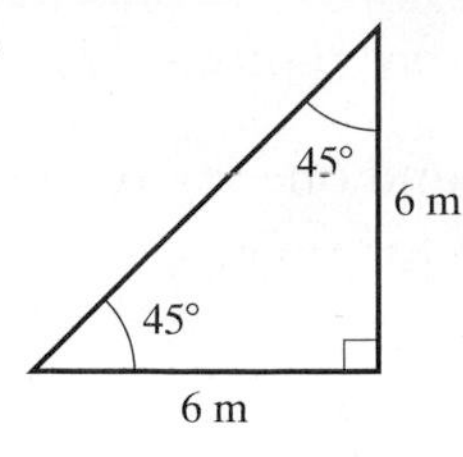

$6 \times \sqrt{2} \approx 8.5$ m

28.

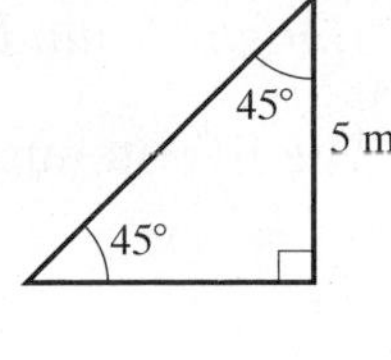

$5 \times \sqrt{2} \approx 7.1$ m

29.

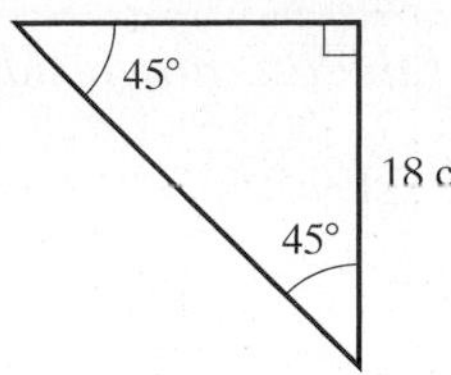

$18 \times \sqrt{2} \approx 25.5$ cm

30.

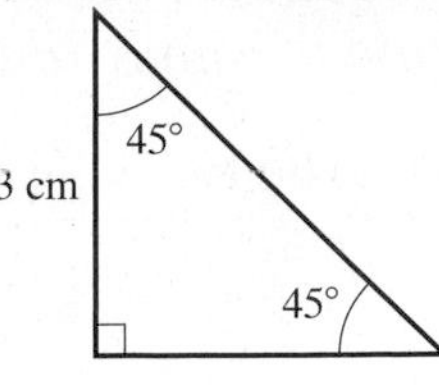

$3 \times \sqrt{2} \approx 4.2$ cm

To Think About

31. ***Flag Pole Construction*** A carpenter is going to use a wooden flagpole 10 in. in diameter, from which he will shape a rectangular base. The base will be 7 in. wide. The carpenter wishes to make the base as tall as possible, minimizing any waste. How tall will the rectangular base be? (Round to the nearest tenth.)

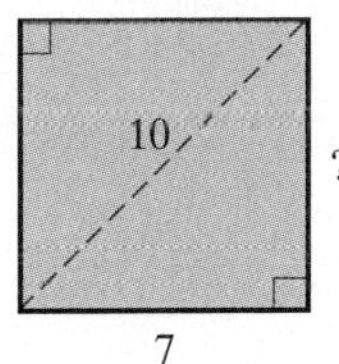

$\text{leg} = \sqrt{10^2 - 7^2} = \sqrt{51} \approx 7.1$ in.

32. ***Shortwave Antenna*** A 4-m shortwave antenna is placed on a garage roof that is 2 m above the lower part of the roof. The base of the garage roof is 16 m wide. How long is an antenna support from point A to point B?

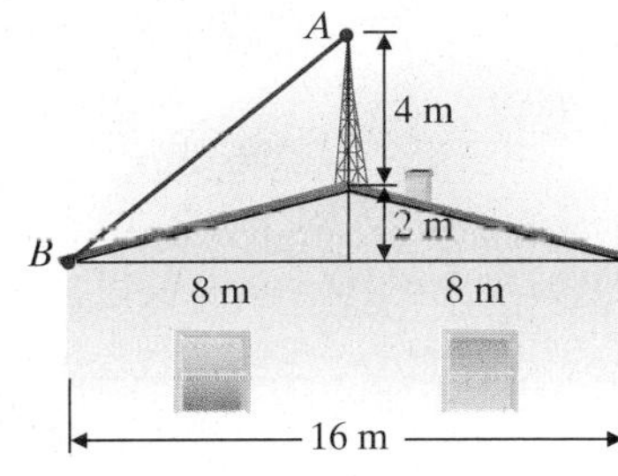

$h = \sqrt{8^2 + 6^2} = \sqrt{100} = 10$ m

33. ***Campus Walkway*** Natasha needs to walk from the campus library to her dormitory. The library is 0.4 mi directly east of the student common, and her dormitory is 0.25 mi directly south of the common. There is a straight walkway from her dormitory to the library. How long is this walkway? Round to the nearest hundredth.

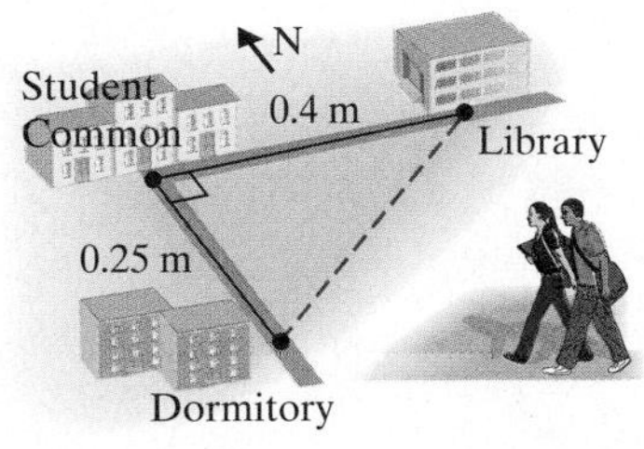

$h = \sqrt{(0.4)^2 + (0.25)^2} = \sqrt{0.2225} \approx 0.47$ mi

34. ***Picture Frame Construction*** Nancy makes pictures frames from strips of wood, plastic, and metal. She often measures the lengths of the frame's diagonals to be sure the corners are true right angles. If the dimensions of a frame are 7 in. by 9 in., how long would the diagonal measures be? (Round to the nearest hundredth.)

$h = \sqrt{7^2 + 9^2} = \sqrt{49 + 81} = \sqrt{130} \approx 11.40$ in.

Mixed Practice

Find the approximate value of the unknown side of the right triangle. Round to the nearest thousandth.

35. The two legs are 7 yd and 11 yd.
$h = \sqrt{7^2 + 11^2} = \sqrt{170} \approx 13.038$ yd

36. The hypotenuse is 45 yd and one leg is 43 yd.
$\text{leg} = \sqrt{45^2 - 43^2} = \sqrt{176} \approx 13.266$ yd

Cumulative Review

37. ***Land Area*** Find the area of a triangular piece of land with altitude 22 m and base 31 m.
341 m^2

38. ***Vegetable Garden*** Find the area of a backyard rectangular vegetable garden with length 14 m and width 12 m.
168 m^2

39. ***Lighthouse Window*** Find the area of a square window in a lighthouse that measures 21 inches on each side.
441 $in.^2$

40. ***Roof of a Building*** Find the area of the parallelogram-shaped roof of a building with a height of 48 yd and a base of 88 yd.
4224 yd^2

7.7 Circles

 Finding the Area and Circumference of a Circle

Every point on a circle is the same distance from the center of the circle, so the circle looks the same all around. In geometry we study the relationship between the parts of a circle and learn how to calculate the distance around a circle as well as the area of a circle.

A **circle** is a two-dimensional flat figure for which all points are at an equal distance from a given point. This given point is called the **center** of the circle.

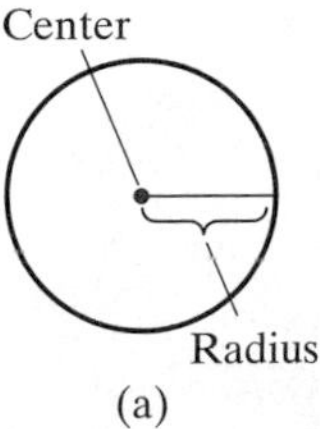

(a)

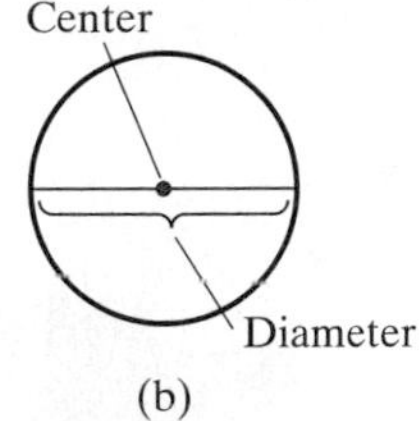

(b)

The **radius** is a line segment from the center to a point on the circle.

The **diameter** is a line segment across the circle that passes through the center with endpoints on the circle.

We often use the words **radius** and **diameter** to mean the length of those segments. Note that the plural of radius is radii. Clearly, then,

$$\text{diameter} = 2 \times \text{radius} \quad \text{or} \quad d = 2r.$$

We could also say that

$$\text{radius} = \text{diameter} \div 2 \quad \text{or} \quad r = \frac{d}{2}.$$

The distance around the circle is called the **circumference.**

There is a special number called **pi,** which we denote by the symbol π. π is the number we get when we divide the circumference of a circle by the diameter $\frac{C}{d} = \pi$. The value of π is approximately 3.14159265359. We can approximate π to any number of digits. For all work in this book we will use the following.

π is approximately 3.14, rounded to the nearest hundredth.

We find the **circumference** C of a circle by multiplying the length of the diameter d times π.

$$C = \pi d$$

Student Learning Objectives

After studying this section, you will be able to:

 **Find the area and circumference of a circle.**

 **Solve area problems containing circles and other geometric shapes.**

Teaching Tip You may want to point out that no one position on the rim of a circle is "more important" than another. Diplomats have been known to choose a circular table for their more difficult negotiations because every person's position at a circle has equal importance.

Circumference

EXAMPLE 1 Find the circumference of a quarter if we know the diameter is 2.4 cm. Use $\pi \approx 3.14$. Round to the nearest tenth.

Solution $C = \pi d = (3.14)(2.4 \text{ cm})$

$= 7.536 \text{ cm} \approx 7.5 \text{ cm}$ (rounded to the nearest tenth)

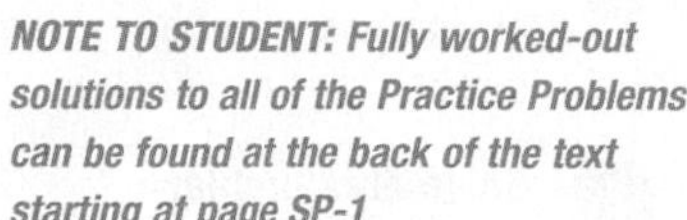

***NOTE TO STUDENT:** Fully worked-out solutions to all of the Practice Problems can be found at the back of the text starting at page SP-1*

Practice Problem 1 Find the circumference of a circle when the diameter is 9 m. Use $\pi \approx 3.14$. Round to the nearest tenth.

Teaching Tip Students sometimes wonder why we use 3.14 for pi throughout this book, when they may have used $\frac{22}{7}$ as an approximation for pi in high school. Explain that it is more convenient for mathematicians to use a decimal approximation since it can be easily adjusted to any desired level of accuracy by adding more decimal places.

An alternative formula is $C = 2\pi r$. Remember, $d = 2r$. We can use this formula to find the circumference if we are given the length of the radius.

When solving word problems involving circles, be careful. Ask yourself, "Is the radius given, or is the diameter given?" Then do the calculations accordingly.

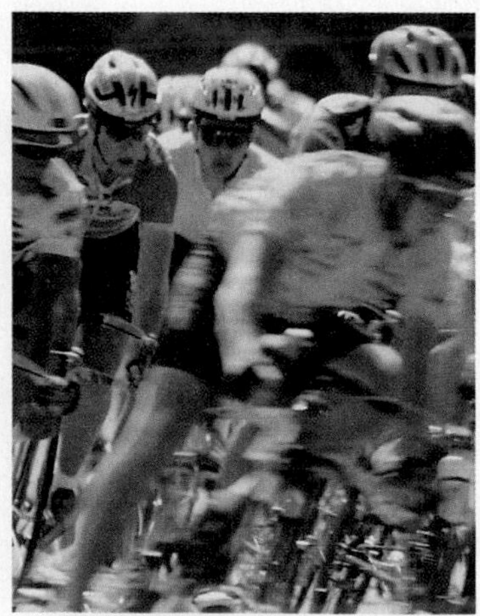

EXAMPLE 2 A bicycle tire has a diameter of 24 in. How many feet does the bicycle travel if the wheel makes one revolution?

Solution

1. ***Understand the problem.*** The distance the wheel travels when it makes 1 revolution is the circumference of the tire. Think of the tire unwinding.

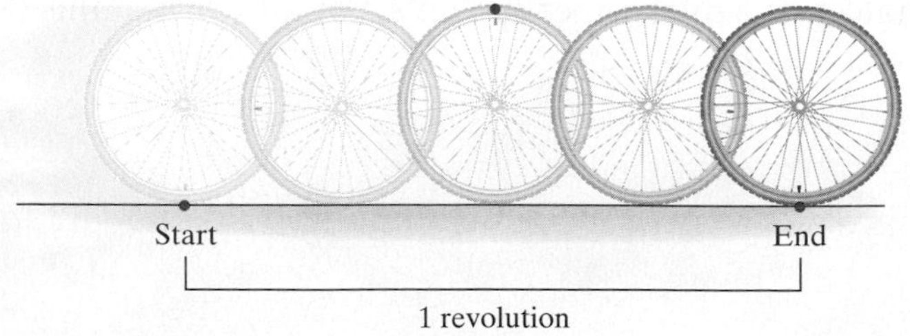

We are given the *diameter*. The diameter is given in *inches*. The answer should be in *feet*.

2. ***Solve and state the answer.*** Since we are given the diameter, we will use $C = \pi d$. We use 3.14 for π.

$$C = \pi d$$
$$= (3.14)(24 \text{ in.}) = 75.36 \text{ in.}$$

We will change 75.36 inches to feet.

$$75.36 \text{ in.} \times \frac{1 \text{ ft}}{12 \text{ in.}} = 6.28 \text{ ft}$$

When the wheel makes 1 revolution, the bicycle travels 6.28 ft.

3. *Check.* We estimate to check. Since $\pi \approx 3.14$, we will use 3 for π.

$$C \approx (3)(24 \text{ in.}) \times \frac{1 \text{ ft}}{12 \text{ in.}} \approx 6 \text{ ft} \checkmark$$

Practice Problem 2 A bicycle tire has a diameter of 30 in. How many feet does the bicycle travel if the wheel makes two revolutions?

The **area of a circle** is the product of π times the radius squared.

$$A = \pi r^2$$

EXAMPLE 3

(a) Estimate the area of a circle whose radius is 6 cm.

(b) Find the exact area of a circle whose radius is 6 cm. Use $\pi \approx 3.14$. Round to the nearest tenth.

Solution

(a) Since π is approximately equal to 3.14, we will use 3 for π to estimate the area.

$$\begin{aligned} A &= \pi r^2 \\ &\approx (3)(6 \text{ cm})^2 \\ &\approx (3)(6 \text{ cm})(6 \text{ cm}) \\ &\approx (3)(36 \text{ cm}^2) \\ &\approx 108 \text{ cm}^2 \end{aligned}$$

Thus our *estimated* area is 108 cm^2.

Teaching Tip When finding the area, some students try to multiply pi by the value of the radius before they have squared the radius. Remind them of the order of operations that we have used throughout the course. Numbers must be raised to a power first before an indicated multiplication is performed.

(b) Now let's compute a more exact area.

$$\begin{aligned} A &= \pi r^2 \\ &= (3.14)(6 \text{ cm})^2 \\ &= 3.14(6 \text{ cm})(6 \text{ cm}) \\ &= (3.14)(36 \text{ cm}^2) \end{aligned}$$

We *must* square the radius first before multiplying by 3.14.

$$\begin{aligned} &= 113.04 \text{ cm}^2 \\ &\approx 113.0 \text{ cm}^2 \text{ (rounded to the nearest tenth)} \end{aligned}$$

Our exact answer is close to the value 108 that we found in part (a).

Practice Problem 3 Find the area of a circle whose radius is 5 km. Use $\pi \approx 3.14$. Round to the nearest tenth. Estimate to check.

The formula for the area of a circle uses the length of the radius. If we are given the diameter, we can use the property that $r = \frac{d}{2}$.

EXAMPLE 4 Lexie Hatfield wants to buy a circular braided rug that is 8 ft in diameter. Find the cost of the rug at $35 a square yard.

Solution

1. ***Understand the problem.*** We are given the *diameter in feet*. We will need to find the radius. The cost of the rug is in *square yards*. We will need to change square feet to square yards.
2. ***Solve and state the answer.*** Find the radius.

$$\begin{aligned} r &= \frac{d}{2} \\ &= \frac{8 \text{ ft}}{2} \\ &= 4 \text{ ft} \end{aligned}$$

Use 3.14 for π.

$$\begin{aligned} A &= \pi r^2 \\ &= (3.14)(4 \text{ ft})^2 \\ &= (3.14)(4 \text{ ft})(4 \text{ ft}) \\ &= (3.14)(16 \text{ ft}^2) \\ &= 50.24 \text{ ft} \end{aligned}$$

Change square feet to square yards. Since 1 yd = 3 ft, $(1 \text{ yd})^2 = (3 \text{ ft})^2$. That is, $1 \text{ yd}^2 = 9 \text{ ft}^2$.

$$50.24 \cancel{\text{ft}^2} \times \frac{1 \text{ yd}^2}{9 \cancel{\text{ft}^2}} \approx 5.58 \text{ yd}^2$$

Find the cost.

$$\frac{\$35}{1 \cancel{\text{yd}^2}} \times 5.58 \cancel{\text{yd}^2} = \$195.30$$

3. ***Check.*** You may use a calculator to check.

Practice Problem 4 Dorrington Little wants to buy a circular pool cover that is 10 ft in diameter. Find the cost of the pool cover at $12 a square yard.

NOTE TO STUDENT: Fully worked-out solutions to all of the Practice Problems can be found at the back of the text starting at page SP-1

2 Solving Area Problems Containing Circles and Other Geometric Shapes

Several applied area problems have a circular region combined with another region.

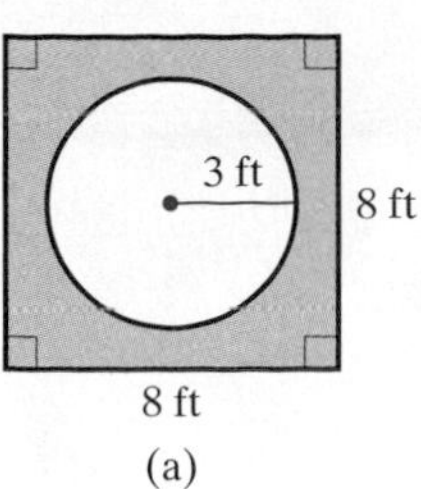

(a)

EXAMPLE 5 Find the area of the shaded region. Use $\pi \approx 3.14$. Round to the nearest tenth.

Solution We will subtract two areas to find the shaded region.

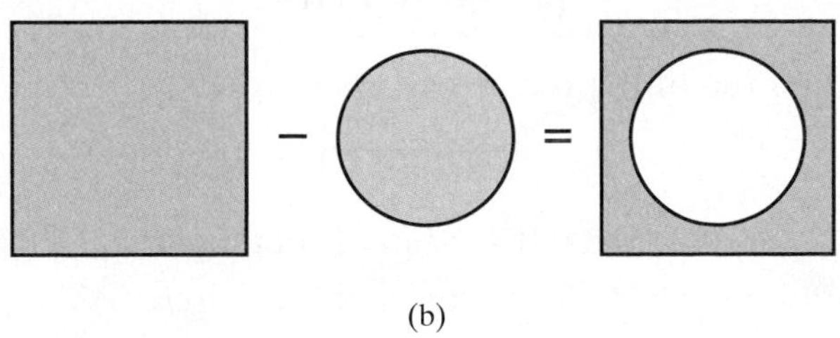

(b)

Area of the square − area of the circle = area of the shaded region

$$\begin{aligned} A &= s^2 \\ &= (8 \text{ ft})^2 \\ &= 64 \text{ ft}^2 \end{aligned} \qquad \begin{aligned} A &= \pi r^2 \\ &= (3.14)(3 \text{ ft})^2 \\ &= (3.14)(9 \text{ ft}^2) \\ &= 28.26 \text{ ft}^2 \end{aligned}$$

Area of the square		area of the circle		area of the shaded region
64 ft^2	−	28.26 ft^2	=	35.74 ft^2
				$\approx 35.7 \text{ ft}^2$
				(rounded to nearest tenth)

Practice Problem 5 Find the area of the shaded region in the margin. Use $\pi \approx 3.14$. Round to the nearest tenth.

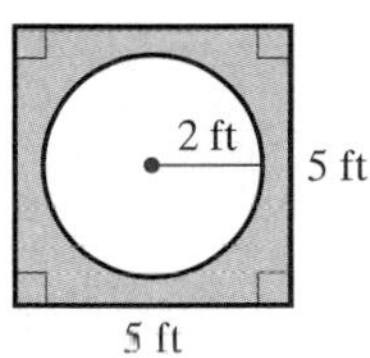

Many geometric shapes involve the semicircle. A **semicircle** is one-half of a circle. The area of a semicircle is therefore one-half of the area of a circle.

EXAMPLE 6 Find the area of the shaded region. Use $\pi \approx 3.14$. Round to the nearest tenth.

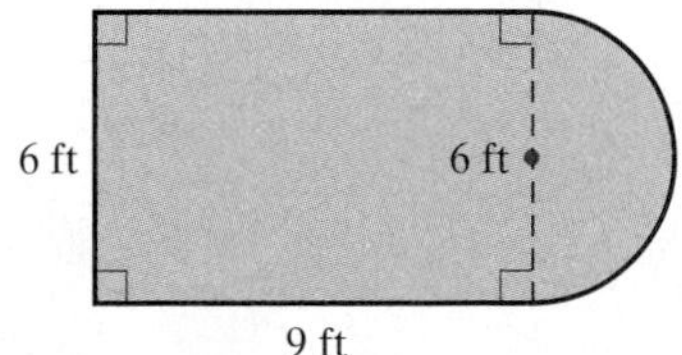

Teaching Tip Some students find these area problems with circles and rectangles very difficult. Emphasize the skill of looking at each part separately and then reasoning carefully to determine whether two areas should be added or subtracted.

Solution First we will find the area of the semicircle with diameter 6 ft.

$$r = \frac{d}{2} = \frac{6 \text{ ft}}{2} = 3 \text{ ft}$$

The radius is 3 ft. The area of a semicircle with radius 3 ft is

$$A_{\text{semicircle}} = \frac{\pi r^2}{2} = \frac{(3.14)(3\text{ ft})^2}{2} = \frac{(3.14)(9\text{ ft}^2)}{2}$$

$$= \frac{28.26\text{ ft}^2}{2} = 14.13\text{ ft}^2$$

Now we add the area of the rectangle.

$$A = lw = (9\text{ ft})(6\text{ ft}) = 54\text{ ft}^2$$

54.00 ft^2	area of rectangle
$+\ 14.13\text{ ft}^2$	area of semicircle
68.13 ft^2	total area

Rounded to the nearest tenth, area $= 68.1\text{ ft}^2$.

***NOTE TO STUDENT:** Fully worked-out solutions to all of the Practice Problems can be found at the back of the text starting at page SP-1*

Practice Problem 6 Find the area of the shaded region below. Use $\pi \approx 3.14$. Round to the nearest tenth.

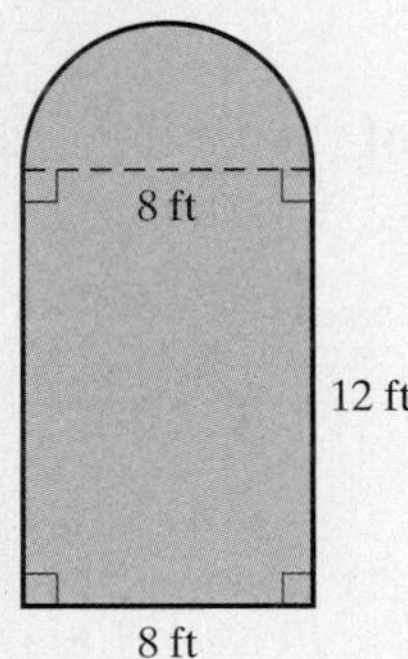

7.7 Exercises

Verbal and Writing Skills

1. The distance around a circle is called the circumference.

2. The radius is a line segment from the center to a point on the circle.

3. The diameter is two times the radius of the circle.

4. Explain in your own words how to *estimate* the area of a circle if you are given the diameter.
Divide the diameter by 2. Square the result, then multiply this by 3.

5. Explain in your own words how to find the circumference of a circle if you are given the radius.
You need to multiply the radius by 2, and then use the formula $C = \pi d$.

6. Explain in your own words how to find the area of a semicircle if you are given the radius.
You need to find the area of the circle with the given radius, and then divide by 2.

In all exercises, use $\pi \approx 3.14$. Round to the nearest hundredth.

Find the diameter of a circle if the radius has the value given.

7. $r = 29$ in.
2×29 in. $= 58$ in.

8. $r = 33$ in.
2×33 in. $= 66$ in.

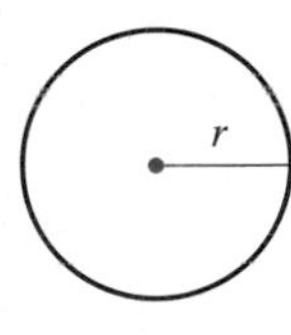

9. $r = 8\frac{1}{2}$ mm
$2 \times 8\frac{1}{2} = 17$ mm

10. $r = 10\frac{3}{4}$ ft
$2 \times 10\frac{3}{4} = 21\frac{1}{2}$ or 21.5 ft

Find the radius of a circle if the diameter has the value given.

11. $d = 45$ yd
45 yd $\div$ 2 = 22.5 yd

12. $d = 65$ yd
65 yd $\div$ 2 = 32.5 yd

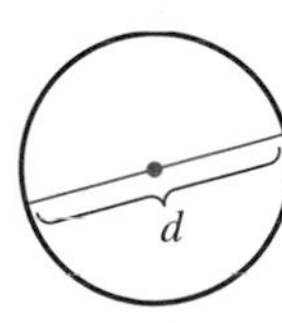

13. $d = 19.84$ cm
19.84 cm $\div$ 2 = 9.92 cm

14. $d = 27.06$ cm
27.06 cm $\div$ 2 = 13.53 cm

Find the circumference of the circle.

15. diameter = 32 cm
$C = 3.14 \times 32$ cm $= 100.48$ cm

16. diameter = 22 cm
$C = 3.14 \times 22$ cm $= 69.08$ cm

17. radius = 18.5 in.
$C = 2 \times 3.14 \times 18.5$ in. $= 116.18$ in.

18. radius = 24.5 in.
$C = 2 \times 3.14 \times 24.5$ in. $= 153.86$ in.

Travel Distance on a Bicycle *A bicycle wheel makes five revolutions. Determine how far the bicycle travels in feet.*

19. The diameter of the wheel is 32 in.
$C = (3.14)(32 \text{ in.}) = 100.48$ in.
$(100.48 \text{ in.})(5) \times \frac{1 \text{ ft}}{12 \text{ in.}} \approx 41.87$ ft

20. The diameter of the wheel is 24 in.
$C = (3.14)(24 \text{ in.}) = 75.36$ in.
$(75.36 \text{ in.})(5) \times \frac{1 \text{ ft}}{12 \text{ in.}} = 31.4$ ft

Find the area of each circle.

21. radius = 5 yd
$A = 3.14 \times (5 \text{ yd})^2 = 78.5 \text{ yd}^2$

22. radius = 7 yd
$A = 3.14 \times (7 \text{ yd})^2 = 153.86 \text{ yd}^2$

23. radius = 8.5 in.
$A = 3.14 \times (8.5 \text{ in.})^2 \approx 226.87 \text{ in.}^2$

24. radius = 12.5 in.
$A = 3.14 \times (12.5 \text{ in.})^2 \approx 490.63 \text{ in}$

25. diameter = 32 cm
$r = \frac{32}{2} = 16 \quad A = 3.14 \times (16 \text{ cm})^2 = 803.84 \text{ cm}^2$

26. diameter = 44 cm
$r = \frac{44}{2} = 22 \quad A = 3.14 \times (22 \text{ cm})^2 = 1519.76 \text{ cm}^2$

Water Sprinkler Distribution *A water sprinkler sends water out in a circular pattern. Determine how large an area is watered.*

27. The radius of watering is 12 ft.
$A = 3.14 \times (12 \text{ ft})^2 = 452.16 \text{ ft}^2$

28. The radius of watering is 8 ft.
$A = 3.14 \times (8 \text{ ft})^2 = 200.96 \text{ ft}^2$

Radio Signal Distribution *A radio station sends out radio waves in all directions from a tower at the center of the circle of broadcast range. Determine how large an area is reached.*

29. The diameter is 90 mi.
$r = \frac{90}{2} = 45 \quad A = 3.14 \times (45 \text{ mi})^2 = 6358.5 \text{ mi}^2$

30. The diameter is 120 mi.
$r = \frac{120}{2} = 60 \quad A = 3.14 \times (60 \text{ mi})^2 = 11{,}304 \text{ mi}^2$

Find the area of the shaded region.

31.

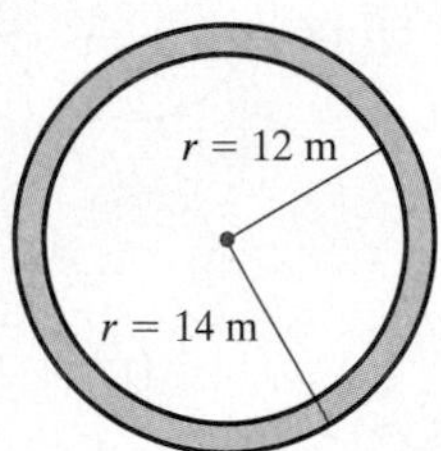

$A = 3.14 \times (14 \text{ m})^2 - 3.14 \times (12 \text{ m})^2$
$= 615.44 \text{ m} - 452.16 \text{ m} = 163.28 \text{ m}^2$

32.

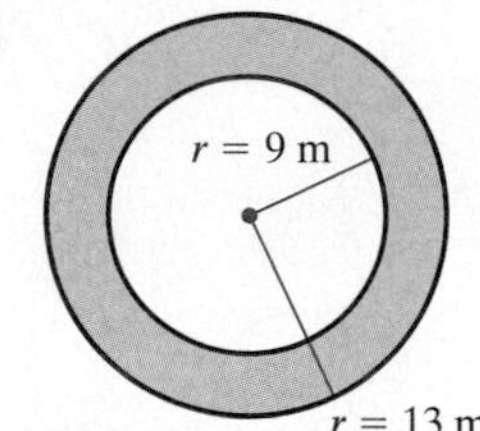

$A = 3.14 \times (13 \text{ m})^2 - 3.14 \times (9 \text{ m})^2$
$= 530.66 \text{ m} - 254.34 \text{ m} = 276.32 \text{ m}^2$

33.

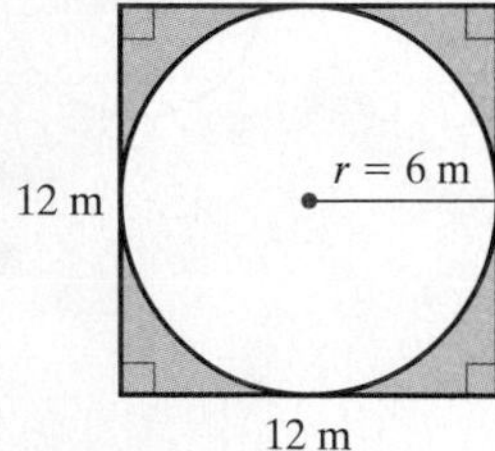

$A = 12 \text{ m} \times 12 \text{ m} - 3.14 \times (6 \text{ m})^2$
$= 144 \text{ m}^2 - 113.04 \text{ m}^2 = 30.96 \text{ m}^2$

34.

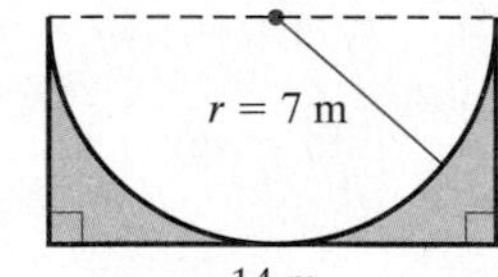

$A = 14 \text{ m} \times 7 \text{ m} - \frac{1}{2} \times 3.14 \times (7 \text{ m})^2$
$= 98 \text{ m}^2 - 76.93 \text{ m}^2 = 21.07 \text{ m}^2$

35.

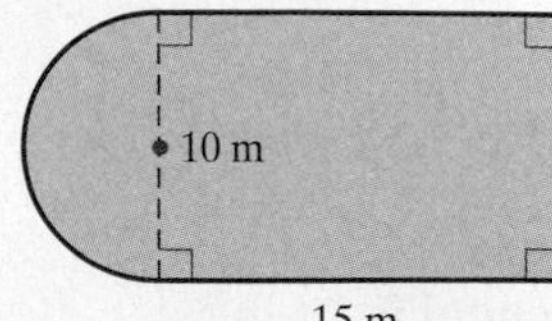

$r = \frac{10}{2} = 5 \quad A = \frac{1}{2} \times 3.14 \times (5 \text{ m})^2 + 15 \text{ m} \times 10 \text{ m}$
$= 39.25 \text{ m}^2 + 150 \text{ m}^2 = 189.25 \text{ m}^2$

36.

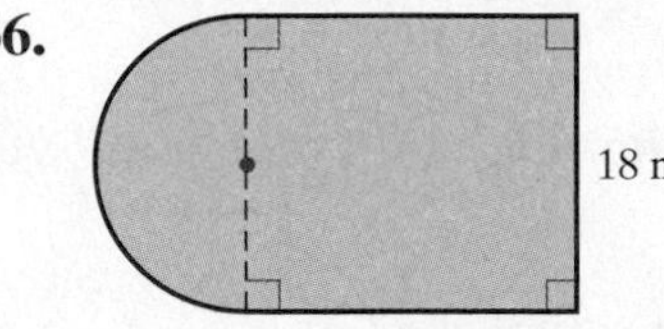

$r = \frac{18}{2} = 9 \quad A = \frac{1}{2} \times 3.14 \times (9 \text{ m})^2 + 20 \text{ m} \times 18 \text{ m}$
$= 127.17 \text{ m}^2 + 360 \text{ m}^2 = 487.17 \text{ m}^2$

Fertilizing a Playing Field *Find the cost of fertilizing a playing field at \$0.20 per square yard for the conditions stated.*

37. The rectangular part of the field is 120 yd long and the diameter of each semicircle is 40 yd.

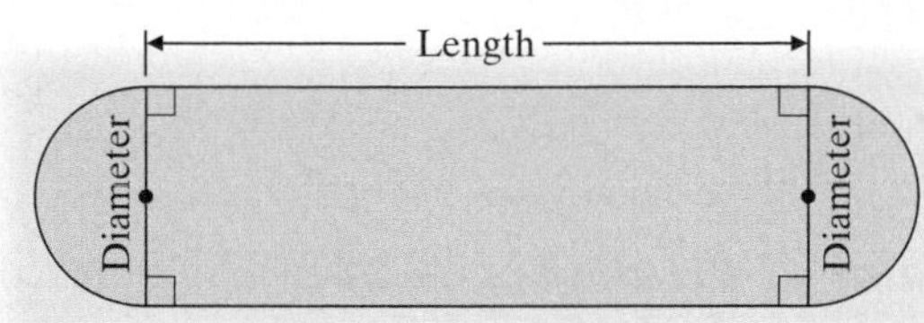

$r = \frac{40}{2}$ $A = 3.14 \times (20 \text{ yd})^2 + 120 \text{ yd} \times 40 \text{ yd}$

$= 1256 \text{ yd}^2 + 4800 \text{ yd}^2 = 6056 \text{ yd}^2$

$\text{Cost} = 6056 \text{ yd}^2 \times \frac{0.20}{\text{yd}^2} = \$1211.20$

38. The rectangular part of the field is 110 yd long and the diameter of each semicircle is 50 yd.

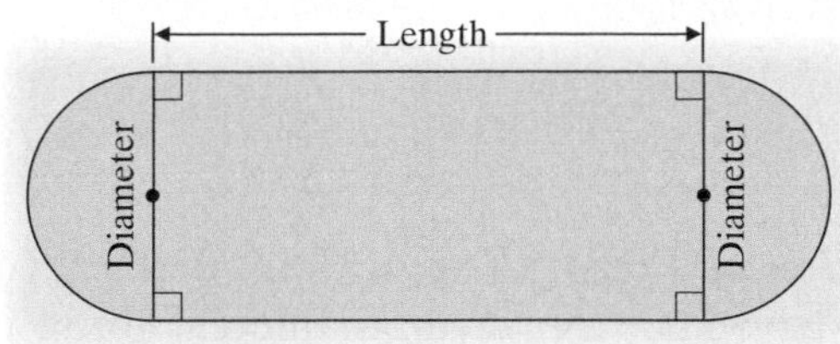

$r = \frac{50}{2}$ $A = 3.14 \times (25 \text{ yd})^2 + 110 \text{ yd} \times 50 \text{ yd}$

$= 7462.5 \text{ yd}^2$

$\text{Cost} = 7462.5 \text{ yd}^2 \times \frac{0.20}{\text{yd}} = \$1492.50$

Applications

Use $\pi \approx 3.14$ *in exercises 37–46. Round to the nearest hundredth.*

39. ***Manhole Cover*** A manhole cover has a diameter of 3 ft. What is the length of the brass grip-strip that encircles the cover, making it easier to manage?

9.42 ft

40. ***Ship Porthole*** A porthole window on a freighter ship has a diameter of 2 ft. What is the length of the insulating strip that encircles the window and keeps out wind and moisture?

6.28 ft

41. ***Truck Travel*** Jimmy's truck has tires with a radius of 30 inches. How many feet does his truck travel if the wheel makes nine revolutions?

141.3 feet

42. ***Car Travel*** Elena's car has tires with a radius of 14 in. How many feet does her car travel if the wheel makes 35 revolutions?

256.43 ft

43. ***Car Travel*** Lucy's car has tires with a radius of 16 in. How many complete revolutions do her wheels make in 1 mile? (*Hint:* First determine how many inches are in 1 mile.)

1 mile = 5280 ft × 12 in. = 63,360 in.

$C = 3.14 \times 32 \text{ in.} = 100.48 \text{ in.}$

$\frac{63{,}360 \text{ in.}}{100.48 \text{ in.}} = 630.57 \text{ revolutions}$

44. ***Car Travel*** Mickey's car has tires with a radius of 15 in. How many complete revolutions do his wheels make in 1 mile? (*Hint:* First determine how many inches are in 1 mile.)

1 mile = 5280 ft × 12 in. = 63,360 in.

$C = 3.14 \times 30 \text{ in.} = 94.2 \text{ in.}$

$\frac{63{,}360 \text{ in.}}{94.2 \text{ in.}} = 672.61 \text{ revolutions}$

45. ***Basketball Court*** In the center of a new basketball court, a circle with a diameter of 8 ft must be marked with tape before it is painted.

(a) How long will the tape be?

25.12 ft

(b) The circle is then painted. How large an area must be painted?

50.24 ft^2

46. ***Pizza Delivery*** Luca's Italian Pizza will make deliveries within a 2-mile radius of the restaurant.

(a) What is the area of the delivery area?

12.56 mi^2

(b) Giovanni drove once around the 2-mile radius of the restaurant. Since it is open country in West Texas, he was able to drive his Ford Explorer in this large circle on open land. How far did he drive on this circle?

12.56 mi

47. ***Mountain Radio Station Broadcast*** A radio station broadcasts from the top of a mountain. The signal can be heard 200 miles away in every direction. How many square miles are in the receiving range of the radio station?
125,600 mi^2

48. ***Distance of Sound Travel*** The sound of an explosion at a fireworks factory was heard 300 miles away in every direction. How many square miles are in the area where people heard the explosion?
282,600 mi^2

To Think About

49. ***Value of Pizza Slice*** A 15-in.-diameter pizza costs \$6.00. A 12-in.-diameter pizza costs \$4.00. The 12-in.-diameter pizza is cut into six slices. The 15-in.-diameter pizza is cut into eight slices.

(a) What is the cost per slice of the 15-in.-diameter pizza? How many square inches of pizza are in one slice?
\$0.75 per slice; ≈22.1 in.^{2}

(b) What is the cost per slice of the 12-in.-diameter pizza? How many square inches of pizza are in one slice?
≈\$0.67 per slice; 18.84 in.^{2}

(c) If you want more value for your money, which slice of pizza should you buy?
For 12-in. pizza, it is \$0.035 per in.^{2}; for 15-in. pizza, it is \$0.034 per in.^{2}; the 15-inch pizza is a better value

50. ***Value of Pizza Slice*** A 14-in.-diameter pizza costs \$5.50. It is cut into eight pieces. A 12.5 in. × 12.5 in. square pizza costs \$6.00. It is cut into nine pieces.

(a) What is the cost of one piece of the 14-in.-diameter pizza? How many square inches of pizza are in one piece?
≈\$0.69 a piece; ≈19.2 in.^{2}

(b) What is the cost of one piece of the 12.5 in. × 12.5 in. square pizza? How many square inches of pizza are in one piece?
≈\$0.67 per piece; ≈17.4 in.^{2}

(c) If you want more value for your money, which piece of pizza should you buy?
For 14-in.-diameter pizza, it is \$0.036 per in.^{2}; for square pizza it is \$0.038 per in.^{2}; the 14-inch round pizza is a better value

Cumulative Review

51. Find 16% of 87.
13.92

52. What is 0.5% of 60?
0.3

53. 12% of what number is 720?
6000

54. 19% of what number is 570?
3000

55. ***Grand Canyon Hike*** A total of 80 students started on the hike from the rim of the Grand Canyon to the river below. In the heat of the day, 40% of the students gave up and went back to the rim. How many students made it to the river?
48 students

56. ***Hotel Stay in Mexico*** Wally and Maureen Hersey stayed one night in a hotel room in Mexico. The room costs \$102 per day and they were charged a visitor tax of 7%. They also were charged \$14.50 for long-distance phone calls. What was the total amount due for the hotel stay?
\$123.64

7.8 Volume

Finding the Volume of a Rectangular Solid (Box)

How much grain can that shed hold? How much water is in the polluted lake? How much air is inside a basketball? These are questions of **volume.** In this section we compute the volume of several three-dimensional geometric figures: the rectangular solid (box), cylinder, sphere, cone, and pyramid.

We can start with a box 1 in. × 1 in. × 1 in.

This is a **cube** with a side of 1 inch.

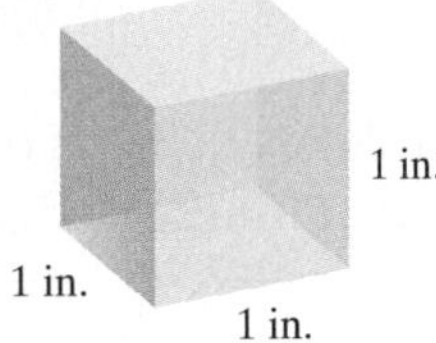

This box has a volume of 1 cubic inch (written 1 in.^3). We can use this as a **unit of volume.** Volume is measured in cubic units such as cubic meters (abbreviated m^3) or cubic feet (abbreviated ft^3). When we measure volume, we are measuring the space inside an object.

> The **volume of a rectangular solid** (box) is the length times the width times the height.
>
> $$V = lwh$$

Find the volume of a box of width 2 ft, length 3 ft, and height 4 ft.

Solution $V = lwh = (3\text{ ft})(2\text{ ft})(4\text{ ft}) = (6)(4)\text{ ft}^3 = 24\text{ ft}^3$

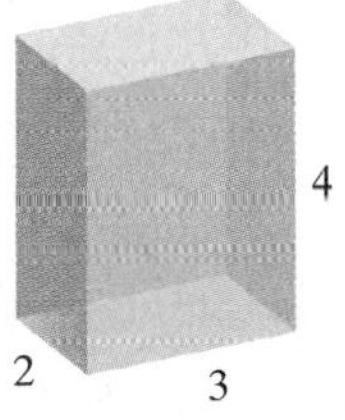

Practice Problem 1 Find the volume of a box of width 5 m, length 6 m, and height 2 m.

If all sides of the box are equal, then the formula is $V = s^3$ where s is the side of the cube.

2 Finding the Volume of a Cylinder

Cylinders are the shape we observe when we see a tin can or a tube.

> The **volume of a cylinder** is the area of its circular base, πr^2, times the height h.
>
> $$V = \pi r^2 h$$

We will continue to use 3.14 as an approximation for π, as we did in Section 7.7, in all volume problems requiring the use of π.

Student Learning Objectives

After studying this section, you will be able to:

1. **Find the volume of a rectangular solid (box).**
2. **Find the volume of a cylinder.**
3. **Find the volume of a sphere.**
4. **Find the volume of a cone.**
5. **Find the volume of a pyramid.**

Teaching Tip Depending on the ability level of your students and the goals of the course, you may find that you prefer to cover only some of the five types of volume problems in this section.

NOTE TO STUDENT: Fully worked-out solutions to all of the Practice Problems can be found at the back of the text starting at page SP-1

Teaching Tip The volume of a lot of unusually shaped solids can be found by using the general concept of "multiply the area of the base by the height." For example, draw a sketch of a solid whose base is a trapezoid or a parallelogram instead of a circle.

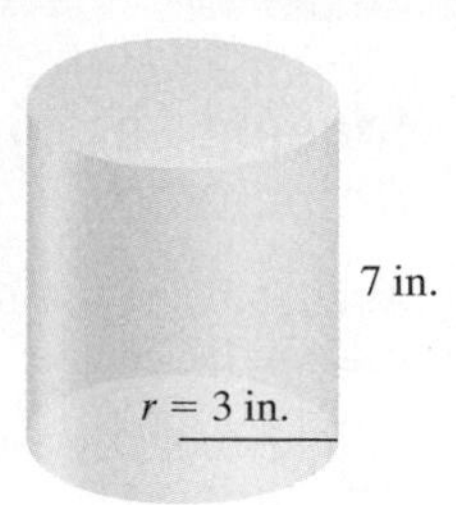

EXAMPLE 2 Find the volume of a cylinder of radius 3 in. and height 7 in. Round to the nearest tenth.

Solution $V = \pi r^2 h = (3.14)(3 \text{ in.})^2(7 \text{ in.}) = (3.14)(3 \text{ in.})(3 \text{ in.})(7 \text{ in.})$

Be sure to square the radius before doing any other multiplication.

$$V = (3.14)(9 \text{ in.}^2)(7 \text{ in.}) = (28.26 \text{ in.}^2)(7 \text{ in.})$$
$$= 197.82 \text{ in.}^3 \approx 197.8 \text{ in.}^3 \text{ rounded to the nearest tenth}$$

***NOTE TO STUDENT:** Fully worked-out solutions to all of the Practice Problems can be found at the back of the text starting at page SP-1*

Practice Problem 2 Find the volume of a cylinder of radius 2 in. and height 5 in. Round to the nearest tenth.

TO THINK ABOUT: Comparing Volume Formulas Take a minute to compare the formulas for the volumes of a rectangular solid and a cylinder. Do you see how they are similar? Consider the area of the base of each figure. In each case, what must you multiply the base area by to obtain the volume of the solid?

3 Finding the Volume of a Sphere

Have you ever considered how you would find the volume of the inside of a ball? How many cubic inches of air are inside a basketball? To answer these questions we need a volume formula for a *sphere*.

The **volume of a sphere** is 4 times π times the radius cubed divided by 3.

$$V = \frac{4\pi r^3}{3}$$

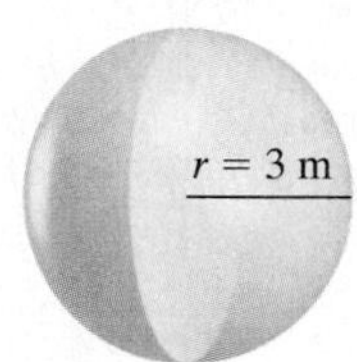

EXAMPLE 3 Find the volume of a sphere with radius 3 m. Round to the nearest tenth.

Solution $$V = \frac{4\pi r^3}{3} = \frac{(4)(3.14)(3 \text{ m})^3}{3}$$
$$= \frac{(4)(3.14)(3)(3)\cancel{(3)} \text{ m}^3}{\cancel{3}}$$
$$= (12.56)(9) \text{ m}^3 = 113.04 \text{ m}^3$$
$$\approx 113.0 \text{ m}^3 \text{ rounded to the nearest tenth}$$

Practice Problem 3 Find the volume of a sphere with radius 6 m. Round to the nearest tenth.

4 Finding the Volume of a Cone

We see the shape of a cone when we look at the sharpened end of a pencil or at an ice cream cone. To find the volume of a cone we use the following formula.

The **volume of a cone** is π times the radius of the base squared times the height divided by 3.

$$V = \frac{\pi r^2 h}{3}$$

EXAMPLE 4 Find the volume of a cone of radius 7 m and height 9 m. Round to the nearest tenth.

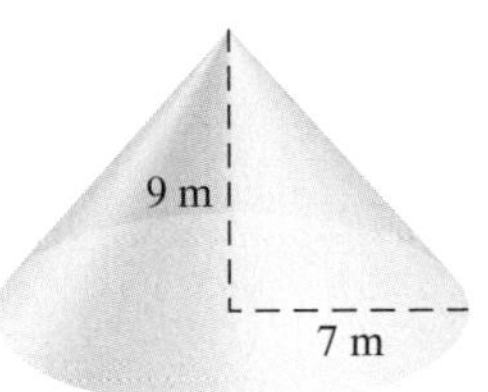

Solution

$$\begin{aligned} V &= \frac{\pi r^2 h}{3} \\ &= \frac{(3.14)(7\text{ m})^2(9\text{ m})}{3} \\ &= \frac{(3.14)(7\text{ m})(7\text{ m})(9\text{ m})}{3} \\ &= (3.14)(49)(3)\text{ m}^3 \\ &= (153.86)(3)\text{ m}^3 \\ &= 461.58\text{ m}^3 \\ &\approx 461.6\text{ m}^3 \text{ rounded to the nearest tenth} \end{aligned}$$

Practice Problem 4 Find the volume of a cone of radius 5 m and height 12 m. Round to the nearest tenth.

5 Finding the Volume of a Pyramid

You have seen pictures of the great pyramids of Egypt. These amazing stone structures are over 4000 years old.

The **volume of a pyramid** is obtained by multiplying the area B of the base of the pyramid by the height h and dividing by 3.

$$V = \frac{Bh}{3}$$

EXAMPLE 5 Find the volume of a pyramid with height 6 m, length of base 7 m, and width of base 5 m.

Solution The base is a rectangle.

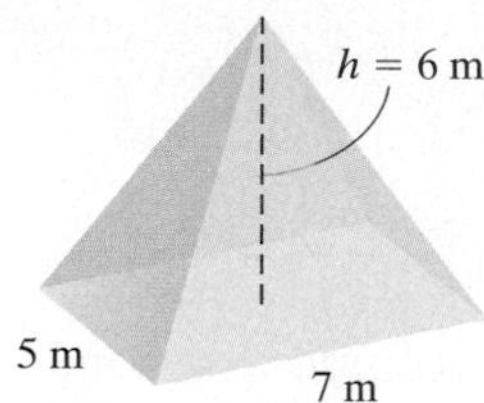

$$\text{Area of base} = (7 \text{ m})(5 \text{ m}) = 35 \text{ m}^2$$

Substituting the area of the base 35 m^2 and the height of 6 m, we have

$$V = \frac{Bh}{3} = \frac{(35 \text{ m}^2)(6 \text{ m})}{3}$$
$$= (35)(2) \text{ m}^3$$
$$= 70 \text{ m}^3$$

NOTE TO STUDENT: Fully worked-out solutions to all of the Practice Problems can be found at the back of the text starting at page SP-1

Practice Problem 5 Find the volume of a pyramid with the dimensions given.

(a) height 10 m, width 6 m, length 6 m

(b) height 15 m, width 7 m, length 8 m

7.8 Exercises

Student Solutions Manual

CD/ Video

PH Math Tutor Center

MathXL®Tutorials on CD
MathXL®

MyMathLab®
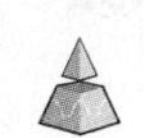
Interactmath.com

Verbal and Writing Skills

In this section, we have studied 6 volume formulas. They are $V = lwh$, $V = \pi r^2 h$, $V = \dfrac{4\pi r^3}{3}$, $V = s^3$, $V = \dfrac{\pi r^2 h}{3}$, *and* $V = \dfrac{Bh}{3}$.

For each of the following figures, (a) state the name of the figure and (b) the correct formula for its volume.

1. 

(a) sphere
(b) $V = \dfrac{4\pi r^3}{3}$

2. 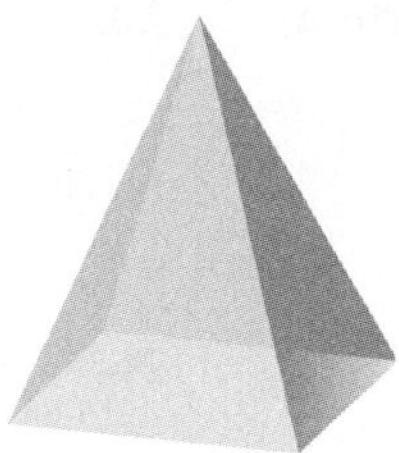

(a) pyramid
(b) $V = \dfrac{Bh}{3}$

3. 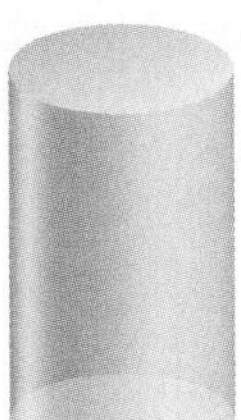

(a) cylinder
(b) $V = \pi r^2 h$

4. 

(a) box or rectangular solid
(b) $V = lwh$

5. 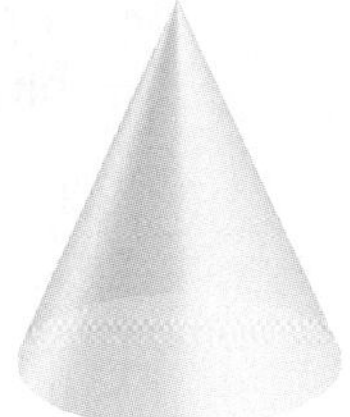

(a) cone
(b) $V = \dfrac{\pi r^2 h}{3}$

6. 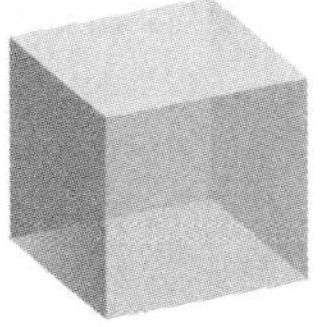

(a) cube
(b) $V = s^3$

Find the volume. Use $\pi \approx 3.14$. *Round to the nearest tenth when necessary.*

7. a rectangular solid with width = 12 mm, length = 30 mm, height = 1.5 mm
$V = 30 \text{ mm} \times 12 \text{ mm} \times 1.5 \text{ mm} = 540 \text{ mm}^3$

8. a rectangular solid with width = 14 mm, length = 20 mm, height = 2.5 mm
$V = 20 \text{ mm} \times 14 \text{ mm} \times 2.5 \text{ mm} = 700 \text{ mm}^3$

9. a cylinder with radius 3 m and height 8 m
$V = 3.14 \times (3 \text{ m})^2 \times 8 \text{ m} \approx 226.1 \text{ m}^3$

10. a cylinder with radius 2 m and height 7 m
$V = 3.14 \times (2 \text{ m})^2 \times 7 \text{ m} \approx 87.9 \text{ m}^3$

11. a cylinder with diameter 22 m and height 17 m
$V = 3.14 \times (11 \text{ m})^2 \times 17 \text{ m} \approx 6459.0 \text{ m}^3$

12. a cylinder with diameter 30 m and height 9 m
$V = 3.14 \times (15 \text{ m})^2 \times 9 \text{ m} = 6358.5 \text{ m}^3$

13. a sphere with radius 9 yd
$V = \dfrac{4 \times 3.14 \times (9 \text{ yd})^3}{3} \approx 3052.1 \text{ yd}^3$

14. a sphere with radius 12 yd
$V = \dfrac{4 \times 3.14 \times (12 \text{ yd})^3}{3} = 7234.6 \text{ yd}^3$

15. a pyramid with a base of 30 sq ft and a height of 70 feet.
700 ft^3

16. a pyramid with a base of 40 sq ft and a height of 90 feet.
1200 ft^3

17. a cube with side 0.6 cm
0.216 cm^3

18. a cube with side 0.8 cm
0.512 cm^3

19. a cone with a radius of 3 yd and a height of 7 yd
65.94 yd^3

20. a cone with a radius of 6 yd and a height of 4 yd
150.72 yd^3

Exercises 21 and 22 involve hemispheres. A hemisphere is exactly one half of a sphere.

21. Find the volume of a hemisphere with radius = 7 m.

$V = \frac{1}{2} \times \frac{4}{3} \times 3.14 \times (7\text{ m})^3 \approx 718.0\text{ m}^3$

22. Find the volume of a hemisphere with radius = 6 m.

$V = \frac{1}{2} \times \frac{4}{3} \times 3.14 \times (6\text{ m})^3 \approx 452.2\text{ m}^3$

Mixed Practice

Find the volume. Use $\pi \approx 3.14$. Round to the nearest tenth.

23. a cone with a height of 14 cm and a radius of 8 cm

$V = \frac{1}{3} \times 3.14 \times (8\text{ cm})^2 \times 14\text{ cm} = 937.8\text{ cm}^3$

24. a cone with a height of 12 cm and a radius of 9 cm

$V = \frac{1}{3} \times 3.14 \times (9\text{ cm})^2 \times 12\text{ cm} = 1017.4\text{ cm}^3$

25. a cone with a height of 10 ft and a radius of 5 ft

$V = \frac{1}{3} \times 3.14 \times (5\text{ ft})^2 \times 10\text{ ft} = 261.7\text{ ft}^3$

26. a cone with a height of 12 ft and a radius of 6 ft

$V = \frac{1}{3} \times 3.14 \times (6\text{ ft})^2 \times 12\text{ ft} = 452.2\text{ ft}^3$

27. a pyramid with a height of 10 m and a square base of 7 m on a side

$V = \frac{1}{3} \times 7\text{ m} \times 7\text{ m} \times 10\text{ m} = 163.3\text{ m}^3$

28. a pyramid with a height of 7 m and a square base of 3 m on a side

$V = \frac{1}{3} \times 3\text{ m} \times 3\text{ m} \times 7\text{ m} = 21\text{ m}^3$

29. a pyramid with a height of 10 m and a rectangular base measuring 8 m by 14 m

$V = \frac{1}{3} \times 14\text{ m} \times 8\text{ m} \times 10\text{ m} = 373.3\text{ m}^3$

30. a pyramid with a height of 5 m and a rectangular base measuring 6 m by 12 m

$V = \frac{1}{3} \times 12\text{ m} \times 6\text{ m} \times 5\text{ m} = 120\text{ m}^3$

Applications

Use $\pi \approx 3.14$ when necessary.

31. ***Sandbox*** The Essex Village neighborhood park has a great sandbox for young children that measures 20 yd × 18 yd. The parks committee will be adding 4 in. of new sand. How many cubic yards of sand will they need?

$4\text{ in.} = \frac{1}{9}\text{ yard} \quad V = 20\text{ yd} \times 18\text{ yd} \times \frac{1}{9}\text{ yd} = 40\text{ yd}^3$

32. ***Driveway Construction*** Mr. Bledsoe wants to put down a crushed-stone driveway to his summer camp. The driveway is 7 yd wide and 120 yd long. The crushed stone is to be 4 in. thick. How many cubic yards of stone will he need?

$4\text{ in.} = \frac{1}{9}\text{yd} \quad V = 120\text{ yd} \times 7\text{ yd} \times \frac{1}{9}\text{yd} \approx 93.3\text{ yd}^3$

Pipe Insulation *A collar of Styrofoam is made to insulate a pipe. Find the volume of the unshaded region (which represents the collar). The large radius R is to the outer rim. The small radius r is to the edge of the insulation.*

33. $r = 3$ in.
$R = 5$ in.
$h = 20$ in.

$V = 3.14 \times (5\text{ in.})^2 \times 20\text{ in.} - 3.14 \times (3\text{ in.})^2 \times 20\text{ in.} = 1004.8\text{ in.}^3$

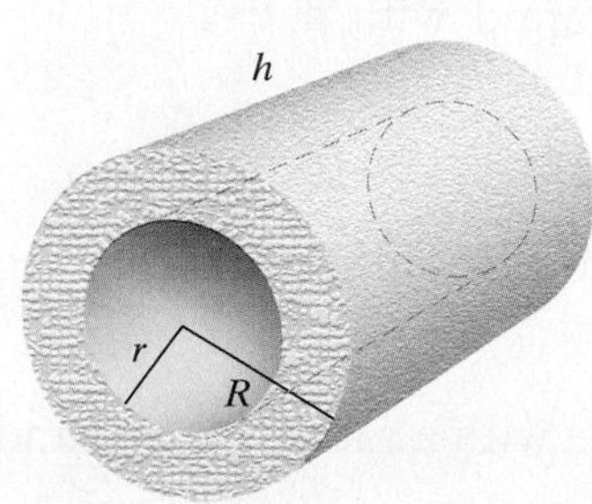

34. $r = 4$ in.
$R = 6$ in.
$h = 25$ in.

$V = 3.14 \times (6\text{ in.})^2 \times 25\text{ in.} - 3.14 \times (4\text{ in.})^2 \times 25\text{ in.} = 1570\text{ in.}^3$

35. ***Astronomy*** Jupiter has a radius of approximately 45,000 mi. Earth has a radius of approximately 3950 mi. Assuming both planets are spheres, what is the difference in volume between Earth and Jupiter?

$V_J \approx \frac{4}{3} \times 3.14 \times (45{,}000 \text{ mi})^3 = 381{,}510{,}000{,}000{,}000 \text{ mi}^3$

$V_E \approx \frac{4}{3} \times 3.14 \times (3950 \text{ mi})^3 = 258{,}023{,}743{,}333 \text{ mi}^3$

$381{,}510{,}000{,}000{,}000 \text{ mi}^3 - 258{,}023{,}743{,}333 \text{ mi}^3$

$= 381{,}251{,}976{,}256{,}667 \text{ mi}^3$

36. ***Tennis Ball*** A tennis ball has a diameter of 2.5 in. A baseball has a diameter of 2.9 in. What is the difference in volume between the baseball and the tennis ball? (Round your answer to the nearest tenth.)

$r_b = \frac{2.9 \text{ in.}}{2} \quad r_t = \frac{2.5 \text{ in.}}{2}$

$D = \frac{4}{3} \times 3.14 \times \left(\frac{2.9 \text{ in.}}{2}\right)^3 - \frac{4}{3} \times 3.14 \times \left(\frac{2.5 \text{ in.}}{2}\right)^3$

$\approx 12.8 \text{ in.}^3 - 8.2 \text{ in.}^3 = 4.6 \text{ in.}^3$

37. ***Shipping a Fragile Object*** A fragile glass box in the shape of a rectangular solid has width 6 in., length 18 in., and height 12 in. It is being shipped in a larger box of width 12 in., length 22 in., and height 16 in. All of the space between the glass box and shipping box will be packed with Styrofoam packing "peanuts." How many cubic inches of the shipping box will be Styrofoam peanuts?

$D = 12 \text{ in.} \times 22 \text{ in.} \times 16 \text{ in.} - 6 \text{ in.} \times 18 \text{ in.} \times 12 \text{ in.}$

$= 2928 \text{ in.}^3$

38. ***Storage Garage*** A storage garage rents space for people who need to store items on a monthly basis. The rooms measure 6 ft by 6 ft by 8 ft and rent for $45 per month. If you rented storage space for 1 month, how much would you pay per cubic foot? (Round to the nearest cent.)

$6 \text{ ft} \times 6 \text{ ft} \times 8 \text{ ft} = 288 \text{ ft}^3$

$\frac{\$45}{288 \text{ ft}^3} \approx \$0.16$ per cubic foot

39. ***Television Antenna Cone*** A special stainless steel cone sits on top of a cable television antenna. The cost of the stainless steel is $3.00 per cm³. The cone has a radius of 6 cm and a height of 10 cm. What is the cost of the stainless steel needed to make this *solid* steel cone?

$V = \frac{1}{3}(3.14)(6 \text{ cm})^2(10 \text{ cm}) = 376.8 \text{ cm}^3$

$376.8 \text{ cm}^2 \times \frac{\$3.00}{1 \text{ cm}^3} = \$1130.40$

40. ***Radar Nose Cone*** The nose cone of a passenger jet is used to receive and send radar. It is made of a special aluminum alloy that costs $4.00 per cm³. The cone has a radius of 5 cm and a height of 9 cm. What is the cost of the aluminum needed to make this *solid* nose cone?

$V = \frac{1}{3}(3.14)(5 \text{ cm})^2(9 \text{ cm}) = 235.5 \text{ cm}^3$

$235.5 \text{ cm}^3 \times \frac{\$4.00}{1 \text{ cm}^3} = \$942$

41. ***Root Beer Can*** The Old Smith Root Beer can was 13.5 cm high. The new can is 1.4 cm shorter. The new can has a diameter of 6.6 cm. What is the volume of the new can?

$V = 3.14 \times (3.3 \text{ cm})^2 \times 12.1 \text{ cm} \approx 413.8 \text{ cm}^3$

42. ***Hot Cocoa Mix*** Rhonda has 5 cylindrical-shaped tins of hot cocoa mix. Each tin is 6 in. high with a radius of 3 in. She wants to combine the contents of all 5 tins into one plastic container measuring 12 in. high with a radius of 5 in. Is there enough space for all 5 tins of cocoa mix?

yes

$V_t = 3.14 \times (3 \text{ in.})^2 \times 6 \text{ in.} = 169.56 \text{ in.}^3$

$5 \times 169.56 \text{ in.}^3 = 847.8 \text{ in.}^3$

$V_c = 3.14 \times (5 \text{ in.})^2 \times 12 \text{ in.} = 942 \text{ in.}^3$

43. ***Stone Pyramid*** Suppose that a stone pyramid has a rectangular base that measures 87 yd by 130 yd. Also suppose that the pyramid has a height of 70 yd. Find the volume.

$V = \frac{1}{3} \times 87 \text{ yd} \times 130 \text{ yd} \times 70 \text{ yd} = 263{,}900 \text{ yd}^3$

44. ***Stone Pyramid*** Suppose the pyramid in exercise 43 is made of solid stone. It is not hollow like the pyramids of Egypt. It is composed of layer after layer of cut stone. The stone weighs 422 lb per cubic yard. How many pounds does the pyramid weigh? How many tons does the pyramid weigh?

111,365,800 lb or 55,682.9 tons

To Think About

45. What is the formula for the volume of the given solid?

$V = \frac{1}{2}hbH$

46. You have seen that the volume of a rectangular solid or a cylinder is found by multiplying the area of the base of the solid by the height of the solid. Can you draw any other solids whose formulas might follow this same pattern?

Many possibilities. Two examples are

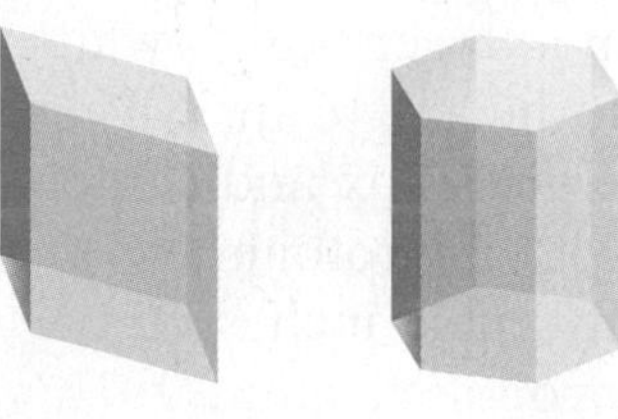

Cumulative Review

47. Add. $7\frac{1}{3} + 2\frac{1}{4}$

$9\frac{7}{12}$

48. Subtract. $9\frac{1}{8} - 2\frac{3}{4}$

$6\frac{3}{8}$

49. Multiply. $2\frac{1}{4} \times 3\frac{3}{4}$

$\frac{135}{16}$ or $8\frac{7}{16}$

50. Divide. $7\frac{1}{2} \div 4\frac{1}{5}$

$\frac{25}{14}$ or $1\frac{11}{14}$

51. Evaluate the following expression.

$$\left(\frac{5}{8} - \frac{1}{4}\right)^2 + \frac{7}{32}$$

$\frac{23}{64}$

52. Evaluate the following expression.

$$\left(6\frac{5}{6} + 2\frac{3}{4}\right) \times \frac{2}{3}$$

$\frac{115}{18}$ or $6\frac{7}{18}$

7.9 Similar Geometric Figures

Student Learning Objectives

After studying this section, you will be able to:

Find the corresponding parts of similar triangles.

Find the corresponding parts of similar geometric figures.

1 Finding the Corresponding Parts of Similar Triangles

In English, "similar" means that the things being compared are, in general, alike. But in mathematics, "similar" means that the things being compared are alike in a special way—they are *alike in shape,* even though they may be different in size. So photographs that are enlarged produce images *similar* to the original; a floor plan of a building is *similar* to the actual building; a model car is *similar* to the actual vehicle.

Two triangles with the same shape but not necessarily the same size are called **similar triangles.** Here are two pairs of similar triangles.

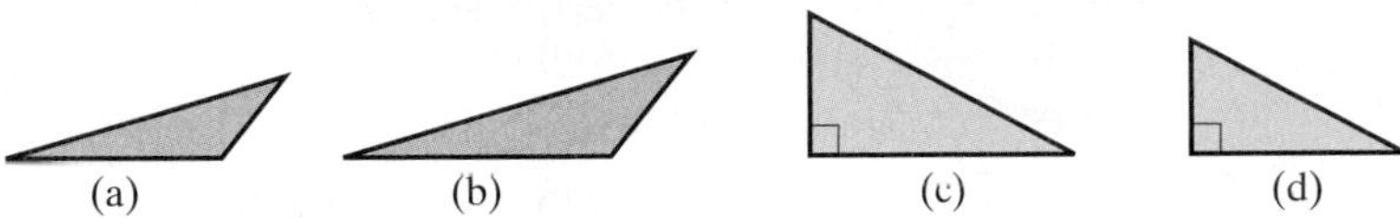

The two triangles on the right are similar. The smallest angle in the first triangle is angle A. The smallest angle in the second triangle is angle D. Both angles measure 36°. We say that angle A and angle D are **corresponding angles** in these similar triangles.

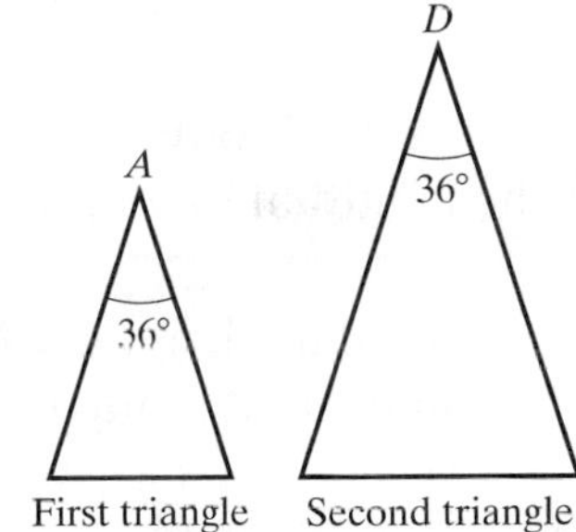

The **corresponding angles** of similar triangles are equal.

The following two triangles are similar. Notice the **corresponding sides.**

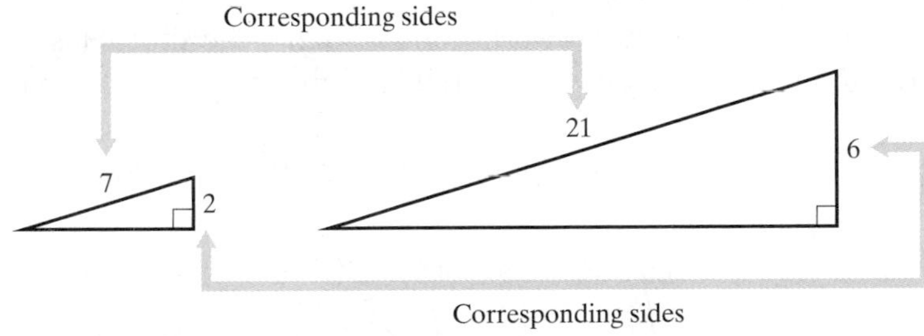

We see that the ratio of 7 to 21 is the same as the ratio of 2 to 6.

$$\frac{7}{21} = \frac{2}{6} \quad \text{is obviously true since} \quad \frac{1}{3} = \frac{1}{3}.$$

Teaching Tip Point out to students that in this section we will be solving a number of proportion problems. You may want to briefly review the steps for solving a proportion.

The corresponding sides of similar triangles have the same ratio.

We can use the fact that corresponding sides of similar triangles have the same ratio to find the lengths of the missing sides of triangles.

EXAMPLE 1 These two triangles are similar. Find the length of side n. Round to the nearest tenth.

Solution The ratio of 12 to 19 is the same as the ratio of 5 to n.

$$\frac{12}{19} = \frac{5}{n}$$

$$12n = (5)(19) \quad \text{Cross multiply.}$$

$$12n = 95 \quad \text{Simplify.}$$

$$\frac{12n}{12} = \frac{95}{12} \quad \text{Divide each side by 12.}$$

$$n = 7.91\overline{6}$$

$$\approx 7.9 \quad \text{Round to the nearest tenth.}$$

Side n is of length 7.9.

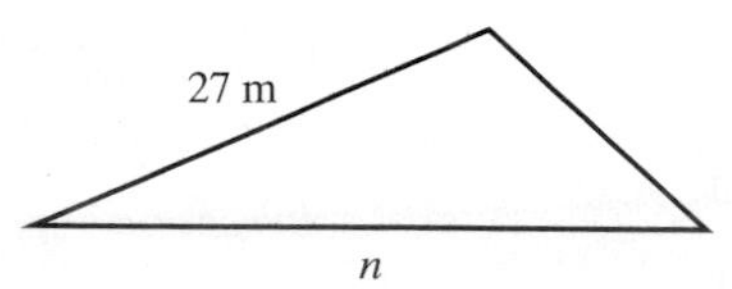

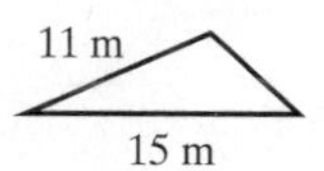

Practice Problem 1 The two triangles in the margin are similar. Find the length of side n. Round to the nearest tenth.

Similar triangles are not always oriented the same way. You may find it helpful to rotate one of the triangles so that the similarity is more apparent.

EXAMPLE 2 These two triangles are similar. Name the sides that correspond.

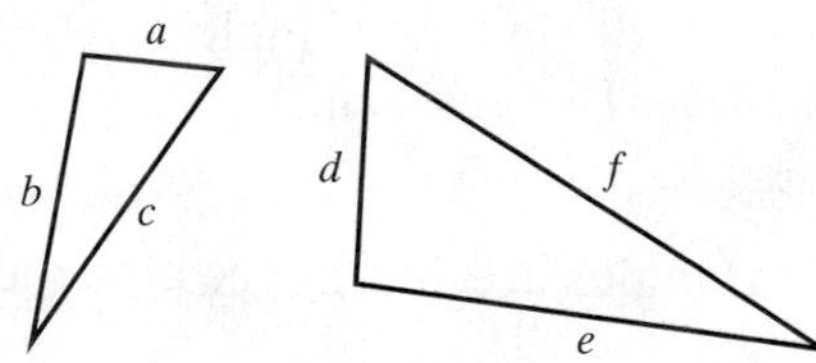

Solution First we turn the second triangle so that the shortest side is on the top, the intermediate side is to the left, and the longest side is on the right.

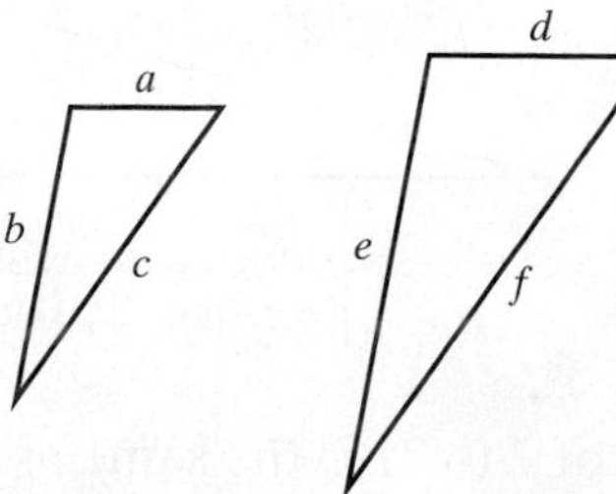

Now the shortest side of each triangle is on the top, the longest side of each triangle is on the right, and so on. We can see that

a corresponds to d.
b corresponds to e.
c corresponds to f.

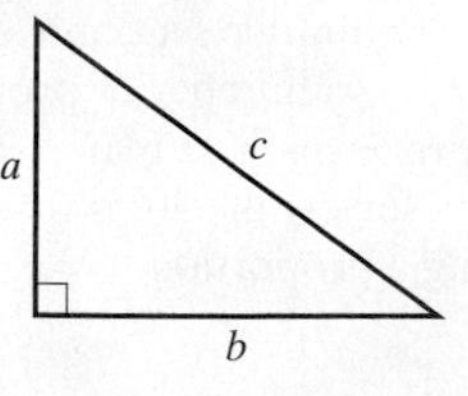

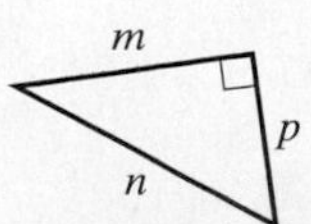

Practice Problem 2 The two triangles in the margin are similar. Name the sides that correspond.

The perimeters of similar triangles have the same ratio as the corresponding sides.

Similar triangles can be used to find distances or lengths that are difficult to measure.

EXAMPLE 3 A flagpole casts a shadow of 36 ft. At the same time, a nearby tree that is 3 ft tall has a shadow of 5 ft. How tall is the flagpole?

Solution

1. ***Understand the problem.*** The shadows cast by the sun shining on vertical objects at the same time of day form similar triangles. We draw a picture.

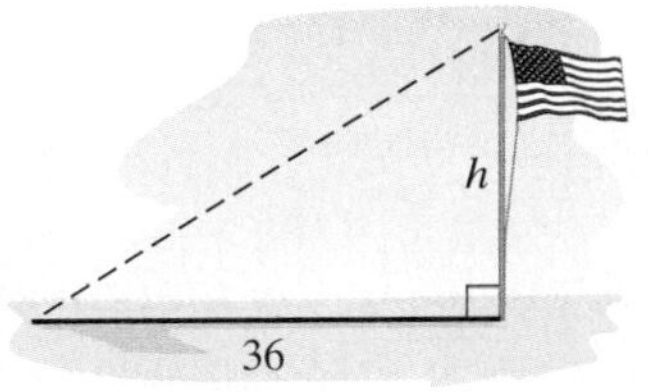

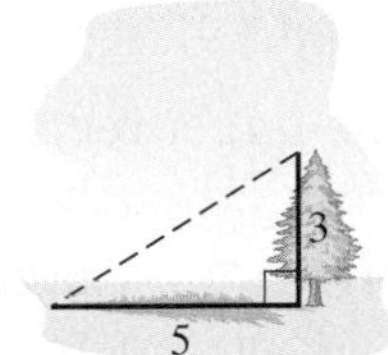

2. ***Solve and state the answer.*** Let h = the height of the flagpole. Thus we can say h is to 3 as 36 is to 5.

$$\frac{h}{3} = \frac{36}{5}$$
$$5h = (3)(36)$$
$$5h = 108$$
$$\frac{5h}{5} = \frac{108}{5}$$
$$h = 21.6$$

The flagpole is about 21.6 feet tall.

3. ***Check.*** The check is up to you.

Practice Problem 3 What is the height (h) of the side wall of the building in the margin if the two triangles are similar?

Teaching Tip Students often wonder where the idea of similar triangles came from. Tell them that the dimensions of the pyramids were determined by measuring and comparing similar triangles. Today models of proposed construction projects are made so that they are similar in proportion to the final objects.

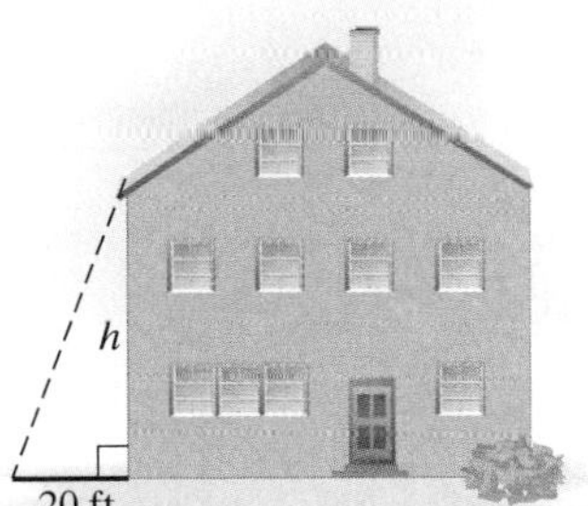

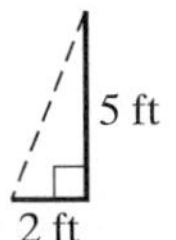

Finding the Corresponding Parts of Similar Geometric Figures

Geometric figures such as rectangles, trapezoids, and circles can be similar figures.

The corresponding sides of similar geometric figures have the same ratio.

EXAMPLE 4 The two rectangles shown here are similar because the corresponding sides of the two rectangles have the same ratio. Find the width of the larger rectangle.

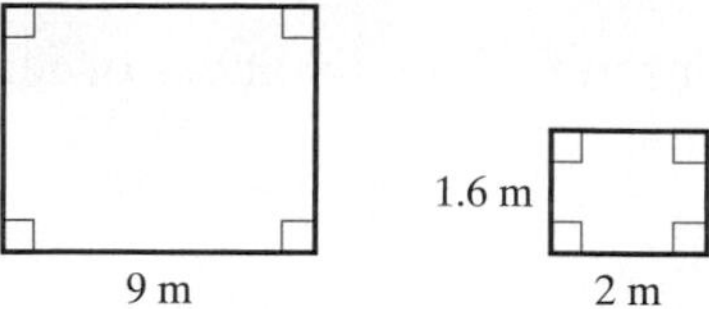

Solution Let w = the width of the larger rectangle.

$$\frac{w}{1.6} = \frac{9}{2}$$
$$2w = (1.6)(9)$$
$$2w = 14.4$$
$$\frac{2w}{2} = \frac{14.4}{2}$$
$$w = 7.2$$

The width of the larger rectangle is 7.2 meters.

Practice Problem 4 The two rectangles in the margin are similar. Find the width of the larger rectangle.

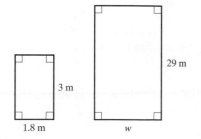

***NOTE TO STUDENT:** Fully worked-out solutions to all of the Practice Problems can be found at the back of the text starting at page SP-1*

The perimeters of similar figures—whatever the figures—have the same ratio as their corresponding sides. Circles are special cases. *All* circles are similar. The circumferences of two circles have the same ratio as their radii.

TO THINK ABOUT: Comparing Two Areas How would you find the relationship between the areas of two similar geometric figures? Consider the following two similar rectangles?

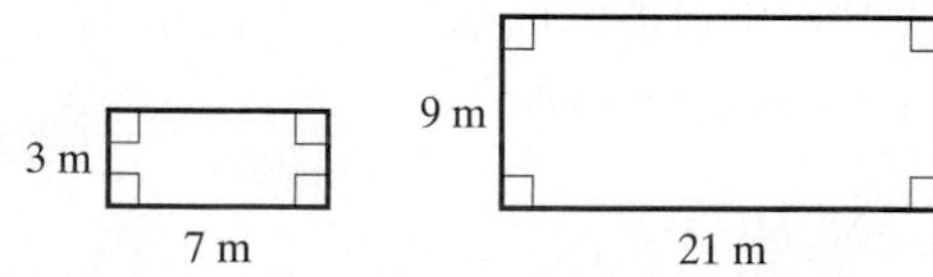

The area of the smaller rectangle is $(3\text{ m})(7\text{ m}) = 21\text{ m}^2$. The area of the larger rectangle is $(9\text{ m})(21\text{ m}) = 189\text{ m}^2$. How could you have predicted this result?

The ratio of small width to large width is $\frac{3}{9} = \frac{1}{3}$. The small rectangle has sides that are $\frac{1}{3}$ as large as the large rectangle. The ratio of the area of the small rectangle to the area of the large rectangle is $\frac{21}{189} = \frac{1}{9}$. Note that $\left(\frac{1}{3}\right)^2 = \frac{1}{9}$.

Thus we can develop the following principle: The areas of two similar figures are in the same ratio as the square of the ratio of two corresponding sides.

7.9 EXERCISES

Verbal and Writing Skills

1. Similar figures may be different in ___size___ but they are alike in ___shape___.

2. The corresponding sides of similar triangles have the same ___ratio___.

3. The perimeters of similar figures have the same ratio as their corresponding ___sides___.

4. You are given the lengths of the sides of a large triangle and the length of a corresponding side of a smaller, similar triangle. Explain in your own words how to find the perimeter of the smaller triangle.
Set up a proportion of the perimeter of the larger triangle to the perimeter of the smaller triangle equal to the side of the large triangle to the corresponding side of the small triangle. Use p for the perimeter of the smaller triangle. Substitute the numbers for the other quantities. Solve the proportion for p.

For each pair of similar triangles, find the missing side n. Round to the nearest tenth when necessary.

5.

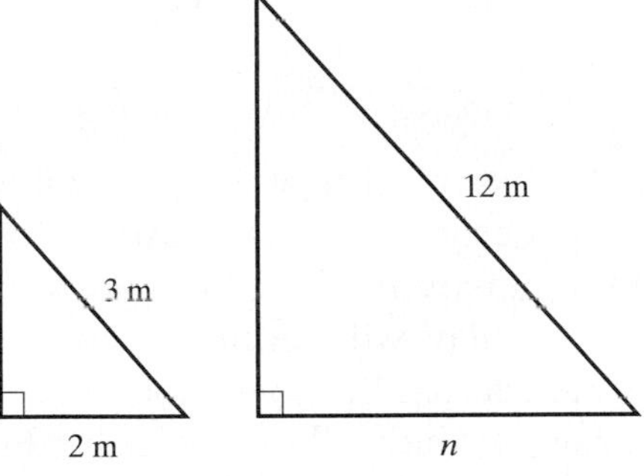

$\frac{n}{2} = \frac{12}{3}$ $n = 8$ m

6.

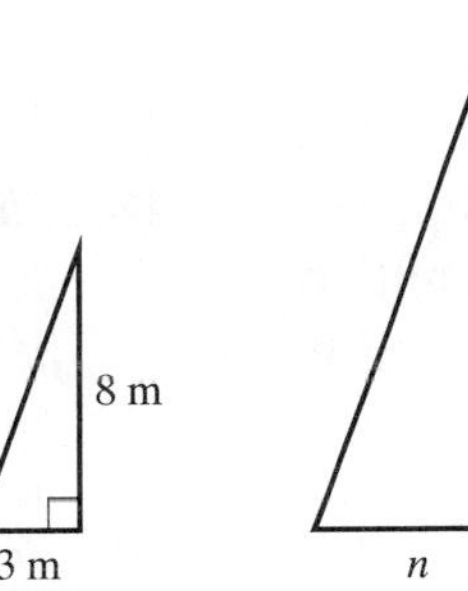

$\frac{n}{3} = \frac{24}{8}$ $n = 9$ m

7.

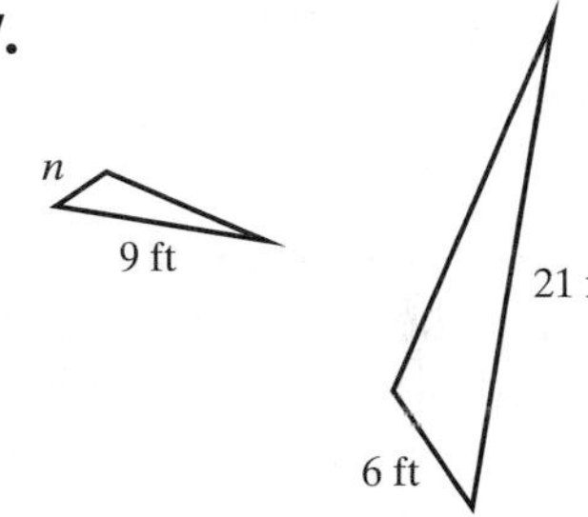

$\frac{n}{6} = \frac{9}{21}$ $n \approx 2.6$ ft

8.

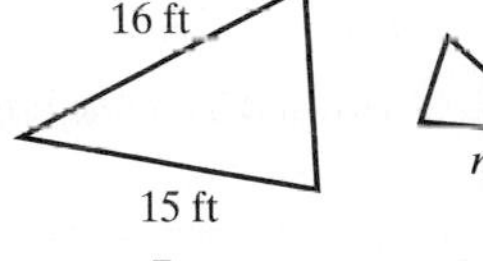

$\frac{n}{15} = \frac{7}{16}$ $n \approx 6.6$ ft

9.

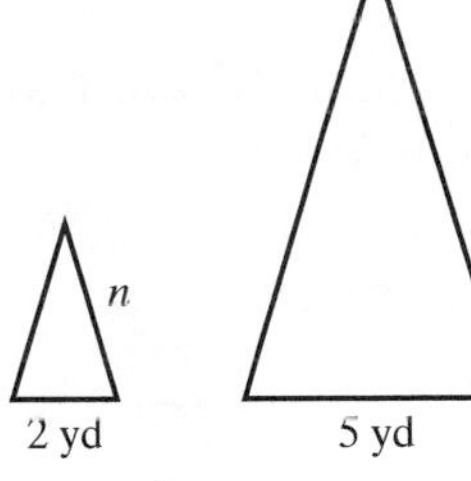

$\frac{n}{8.5} = \frac{2}{5}$ $n = 3.4$ yd

10.

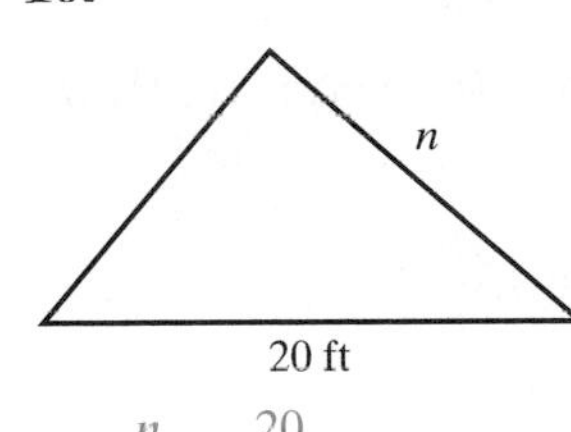

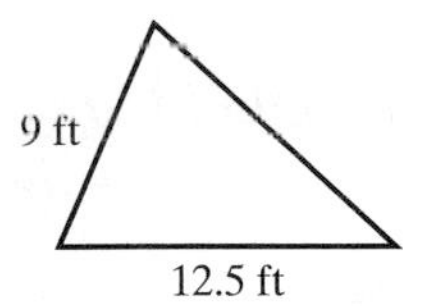

$\frac{n}{9} = \frac{20}{12.5}$ $n = 14.4$ ft

Each pair of triangles is similar. Determine the three pairs of corresponding sides in each case.

11.

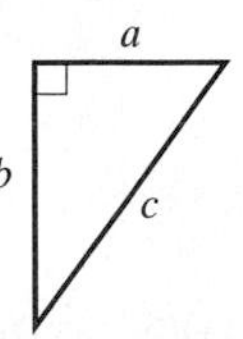

d e f

a corresponds to f
b corresponds to e
c corresponds to d

12.

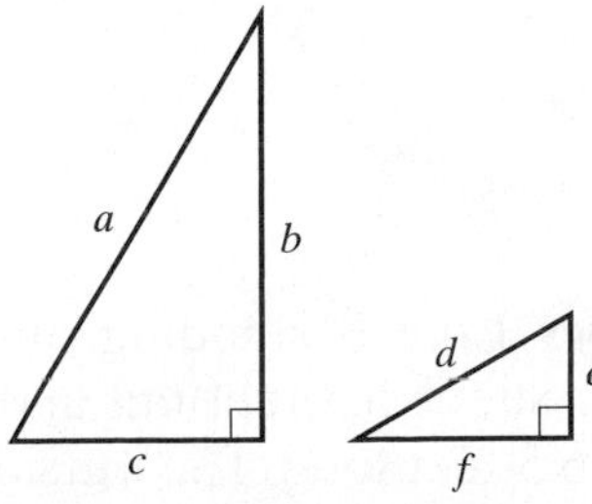

a corresponds to d
b corresponds to f
c corresponds to e

Applications

13. ***Sculptor*** A sculptor is designing her new triangular masterpiece. In her scale drawing, the shortest side of the triangular piece to be made measures 8 cm. The longest side of the drawing measures 25 cm. The longest side of the actual triangular piece to be sculpted must be 10.5 m long. How long will the shortest side of this piece be? Round to the nearest tenth.
3.4 m

14. ***Landscape Architect*** The zoo has hired a landscape architect to design the triangular lobby of the children's petting zoo. In his scale drawing, the longest side of the lobby is 9 cm. The shortest side of the lobby is 5 cm. The longest side of the actual lobby will be 30 m. How long will the shortest side of the actual lobby be? Round to the nearest tenth.
16.7 m

15. ***Photography*** Janice took a great photo of the entire family at this year's reunion. She brought it to a professional photography studio and asked that the 3-in.-by-5-in. photo be blown up to poster size, which is 3.5 feet tall. What is the smaller dimension (width) of the poster? 2.1 ft

16. ***Family Room Addition*** Rich and Anya are adding a new family room onto their home. On the blueprints, the room measures 2 in. by 3 in. The actual room is similar in shape with a length of 16 ft. What will the width of the family room be? Round to the nearest tenth.
10.7 ft

17. ***Blueprints of New House*** On the blueprints of their new home, Ben and Heather notice the bathtub measures 2 in. by 5 in. They know that the actual bathtub will be 5.5 ft long. How wide will the tub be?
2.2 ft

18. ***Theater Company Props*** A theater company's prop designer sends drawings of the props to the person in charge of construction. An upcoming play will require a large brick wall to stretch across the stage floor. In the designer's drawing, the wall measures $\frac{1}{4}$ ft by $\frac{3}{4}$ ft. If the length of the stage is 36 ft, how tall will the wall be? $\frac{n}{\frac{1}{4}} = \frac{36}{\frac{3}{4}}$ $\frac{3}{4}n = 9$ $n = 12$ ft

Length of Shadows *In exercises 19 and 20, a flagpole casts a shadow. At the same time, a nearby tree casts a shadow. Use the sketch to find the height n of each flagpole.*

19.

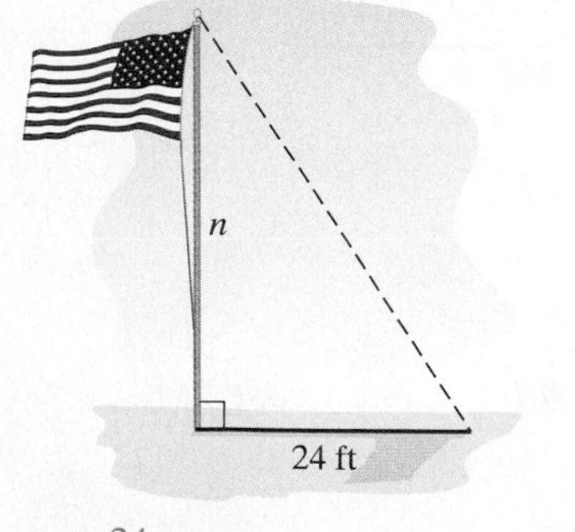

$\frac{n}{6} = \frac{24}{4}$ $n = 36$ ft

20.

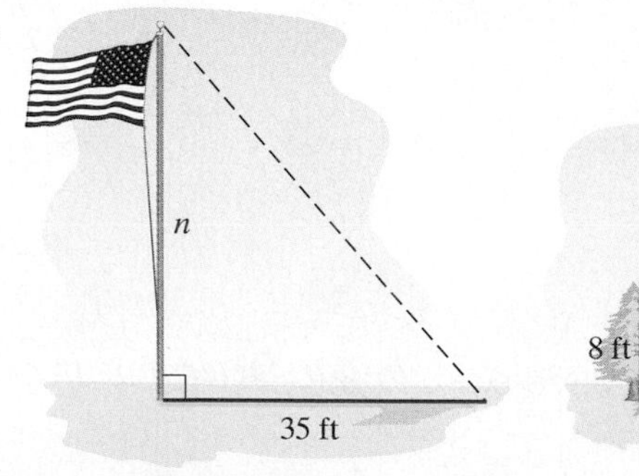

$\frac{n}{8} = \frac{35}{7}$ $n = 40$ ft

21. ***Length of Shadows*** Lola is standing outside the shopping mall. She is 5.5 feet tall and her shadow measures 6.5 feet long. The outside of the department store casts a shadow of 96 feet. How tall is the store? Round to the nearest foot. 81 ft

22. ***Length of Shadows*** Thomas is rock climbing in Utah. He is 6 feet tall and his shadow measures 8 feet long. The rock he wants to climb casts a shadow of 610 feet. How tall is the rock he is about to climb? 457.5 ft

Each pair of figures is similar. Find the missing side. Round to the nearest tenth when necessary.

23.

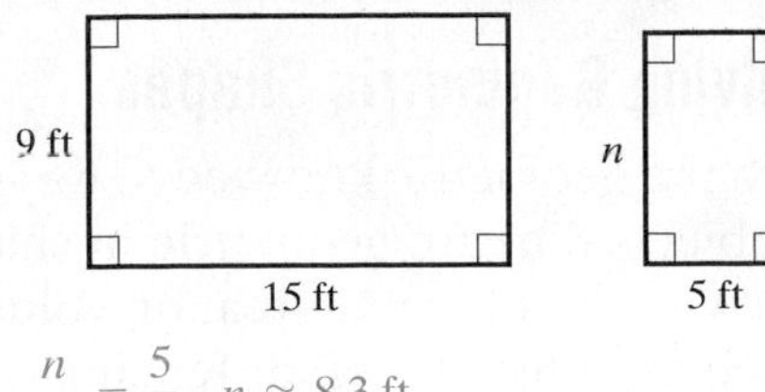

$\frac{n}{15} = \frac{5}{9}$ $n \approx 8.3$ ft

24.

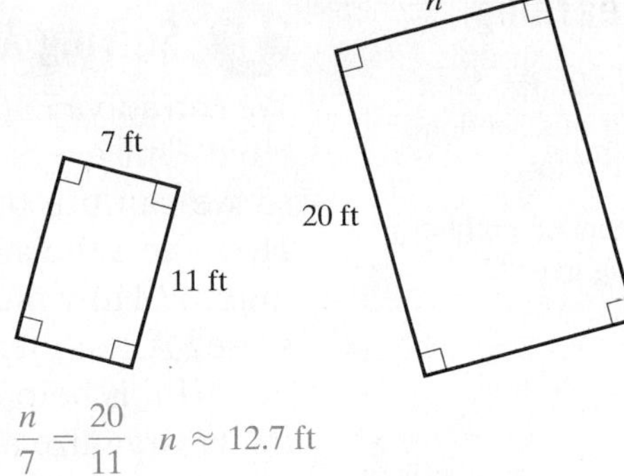

$\frac{n}{7} = \frac{20}{11}$ $n \approx 12.7$ ft

25.

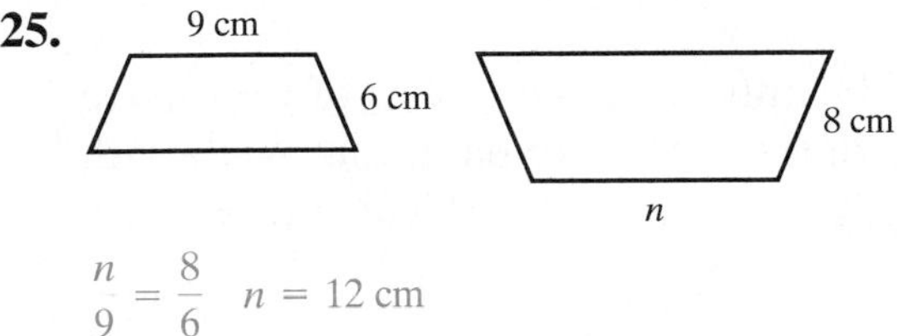

$\frac{n}{9} = \frac{8}{6}$ $n = 12$ cm

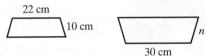

26. 22 cm, 10 cm

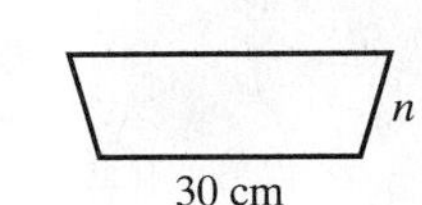

$\frac{n}{10} = \frac{30}{22}$ $n \approx 13.6$ cm

To Think About

Each pair of geometric figures is similar. Find the unknown area.

27.

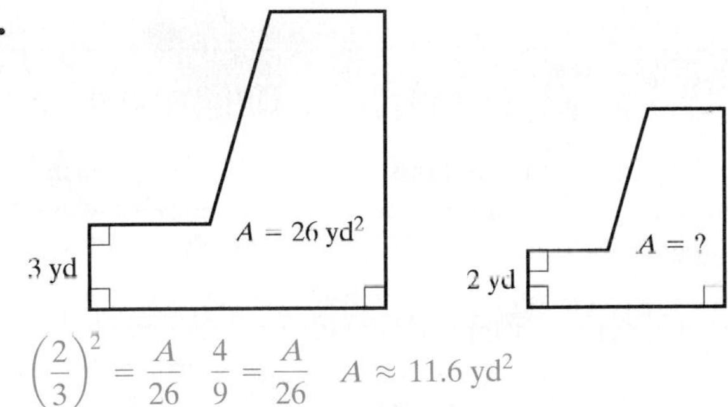

$\left(\frac{2}{3}\right)^2 = \frac{A}{26}$ $\frac{4}{9} = \frac{A}{26}$ $A \approx 11.6 \text{ yd}^2$

28.

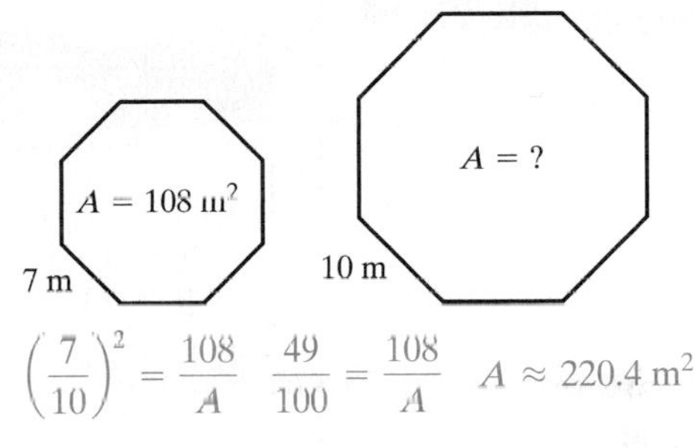

$\left(\frac{7}{10}\right)^2 = \frac{108}{A}$ $\frac{49}{100} = \frac{108}{A}$ $A \approx 220.4 \text{ m}^2$

Cumulative Review

Calculate. Use the correct order of operations.

29. $2 \times 3^2 + 4 - 2 \times 5$ 12

30. $100 \div (8 - 3)^2 \times 2^3$ 32

31. $(5)(9) - (21 + 3) \div 8$ 42

32. $108 \div 6 + 51 \div 3 + 3^3$ 62

7.10 Solving Applied Problems Involving Geometry

Student Learning Objective

After studying this section, you will be able to:

 **Solve applied problems involving geometric shapes.**

 Solving Applied Problems Involving Geometric Shapes

We can solve many real-life problems with the geometric knowledge we now have. Our everyday world is filled with objects that are geometric in shape, so we can use our knowledge of geometry to find length, area, or volume. How far is the automobile trip? How much framing, edging, or fencing is required? How much paint, siding, or roofing is required? How much can we store? All of these questions can be answered with geometry.

If it is helpful to you, use the Mathematics Blueprint for Problem Solving to organize a plan to solve these applied problems.

EXAMPLE 1 A professional painter can paint 90 ft^2 of wall space in 20 minutes. How long will it take the painter to paint four walls with the following dimensions: 14 ft × 8 ft, 12 ft × 8 ft, 10 ft × 7 ft, and 8 ft × 7 ft?

Solution

1. ***Understand the problem.***

Mathematics Blueprint for Problem Solving

Gather the Facts	What Am I Asked to Do?	How Do I Proceed?	Key Points to Remember
Painter paints four walls: 14 ft × 8 ft 12 ft × 8 ft 10 ft × 7 ft 8 ft × 7 ft Painter can paint 90 ft^2 in 20 minutes.	Find out how long it will take the painter to paint the four walls.	**(a)** Find the total area to be painted. **(b)** Then find how long it will take to paint the total area.	The area of each rectangular wall is length × width. To get the time, we set up a proportion.

2. ***Solve and state the answer.***

(a) Find the total area of the four walls.

Each wall is a rectangle. The first one is 8 ft wide and 14 ft long.

$$A = lw$$
$$= (14\text{ ft})(8\text{ ft}) = 112\text{ ft}^2$$

We find the areas of the other three walls.

$(12\text{ ft})(8\text{ ft}) = 96\text{ ft}^2 \quad (10\text{ ft})(7\text{ ft}) = 70\text{ ft}^2 \quad (8\text{ ft})(7\text{ ft}) = 56\text{ ft}^2$

The total area is obtained by adding.

$$\begin{array}{r} 112\text{ ft}^2 \\ 96\text{ ft}^2 \\ 70\text{ ft}^2 \\ +\ \ 56\text{ ft}^2 \\ \hline 334\text{ ft}^2 \end{array}$$

(b) Determine how long it will take to paint the four walls.
Now we set up a proportion. If 90 ft^2 can be done in 20 minutes, then 334 ft^2 can be done in t minutes.

$$\frac{90\text{ ft}^2}{20\text{ minutes}} = \frac{334\text{ ft}^2}{t\text{ minutes}}$$

$$\frac{90}{20} = \frac{334}{t}$$

$$90(t) = 334(20)$$

$$90t = 6680$$

$$\frac{90t}{90} = \frac{6680}{90}$$

$$t \approx 74 \qquad \text{We round our answer to the nearest minute.}$$

Thus the work can be done in approximately 74 minutes.

3. *Check.* Estimate the answer.

$$\begin{aligned} 14 \times 8 \approx 10 \times 8 &= 80\text{ ft}^2 \\ 12 \times 8 \approx 10 \times 8 &= 80\text{ ft}^2 \\ 10 \times 7 &= 70\text{ ft}^2 \\ 8 \times 7 &= 56\text{ ft}^2 \end{aligned}$$

If we estimate the sum of the number of square feet, we will have

$$80 + 80 + 70 + 56 = 286\text{ ft}^2.$$

Now since 60 minutes = 1 hour, we know that if you can paint 90 ft^2 in 20 minutes, you can paint 270 ft^2 in one hour. Our estimate of 286 ft^2 is slightly more than 270 ft^2, so we would expect that the answer would be slightly more than one hour.

Thus our calculated value of 74 minutes (1 hour 14 minutes) seems reasonable. ✓

Practice Problem 1 Mike rented an electric floor sander. It will sand 80 ft^2 of hardwood floor in 15 minutes. He needs to sand the floors in three rooms. The floor dimensions are 24 ft × 13 ft, 12 ft × 9 ft, and 16 ft × 3 ft. How long will it take him to sand the floors in all three rooms?

***NOTE TO STUDENT:** Fully worked-out solutions to all of the Practice Problems can be found at the back of the text starting at page SP-1*

EXAMPLE 2 Carlos and Rosetta want to put vinyl siding on the front of their home in West Chicago. The house dimensions are shown in the figure at right. The door dimensions are 6 ft × 3 ft. The windows measure 2 ft × 4 ft.

(a) Excluding windows and doors, how many square feet of siding will be needed?

(b) If the siding costs $2.25 per square foot, how much will it cost to side the front of the house?

Solution

1. ***Understand the problem.***

Mathematics Blueprint for Problem Solving

Gather the Facts	What Am I Asked to Do?	How Do I Proceed?	Key Points to Remember
House measures 19 ft × 25 ft. Windows measure 2 ft × 4 ft. Door measures 6 ft × 3 ft. Siding costs $2.25 per square foot.	Find the cost to put siding on the front of the house.	**(a)** Find the area of the entire front of the house by multiplying 19 ft by 25 ft. Find the area of one window by multiplying 2 ft by 4 ft. Find the area of the door by multiplying 6 ft by 3 ft. **(b)** Multiply shaded area by $2.25.	**(a)** To obtain the shaded area we must subtract the area of nine windows and one door from the area of the entire front. **(b)** We must multiply the resulting area by the cost of the siding per foot.

Teaching Tip After covering a couple of examples in Section 7.10, you may want to ask the students, "Can any of you think of some practical geometry problem like this that you or your friends have encountered in everyday life?" Often they will think of problems like the cost of paving a driveway, building a deck on the house, or finding the volume of the cylinders of a car to determine compression. Solving a real-life example from students' lives tends to impress on them that this section is really relevant.

2. ***Solve and state the answer.***

(a) Find the area of the front of the house (shaded area). We will find the area of the large rectangle representing the front of the house. Then we will subtract the area of the windows and the door.

$$\text{Area of each window} = (2\text{ ft})(4\text{ ft}) = 8\text{ ft}^2$$
$$\text{Area of 9 windows} = (9)(8\text{ ft})^2 = 72\text{ ft}^2$$
$$\text{Area of 1 door} = (6\text{ ft})(3\text{ ft}) = 18\text{ ft}^2$$
$$\text{Area of 9 windows} + 1\text{ door} = 90\text{ ft}^2$$
$$\text{Area of large rectangle} = (19\text{ ft})(25\text{ ft}) = 475\text{ ft}^2$$

Total area of front of house	475 ft^2
− Area of 9 windows and 1 door	− 90 ft^2
= Total area to be covered	385 ft^2

We see that 385 ft^2 of siding will be needed.

(b) Find the cost of the siding.

$$\text{Cost} = 385\ \text{ft}^2 \times \frac{\$2.25}{1\ \text{ft}^2} = \$866.25$$

The cost to put up siding on the front of the house will be \$866.25.

3. ***Check.*** We leave the check up to you.

Practice Problem 2 Below is a sketch of a roof of a commercial building.

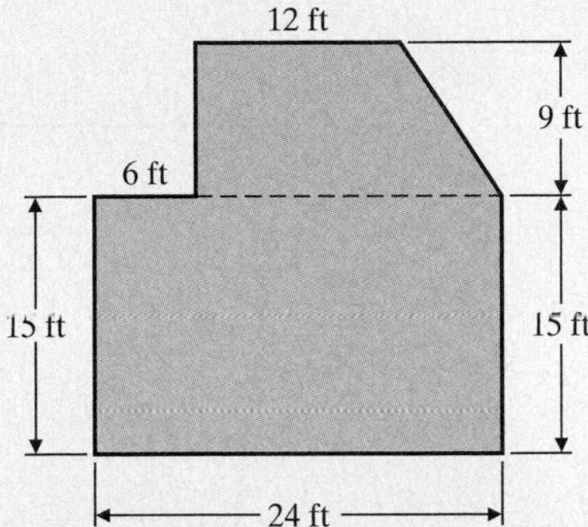

(a) What is the area of the roof?

(b) How much would it cost to install new roofing on the roof area shown if the roofing costs \$2.75 per square yard? (*Hint*: 9 square feet = 1 square yard.)

***NOTE TO STUDENT:** Fully worked-out solutions to all of the Practice Problems can be found at the back of the text starting at page SP-1*

7.10 Exercises

Student Solutions Manual

CD/ Video

PH Math Tutor Center

MathXL®Tutorials on CD

MathXL®

MyMathLab®

Interactmath.com

Applications

Round to the nearest tenth unless otherwise directed.

1. *Driving Distances* Monica drives to work each day from Bethel to Bridgeton. The following sketch shows the two possible routes.

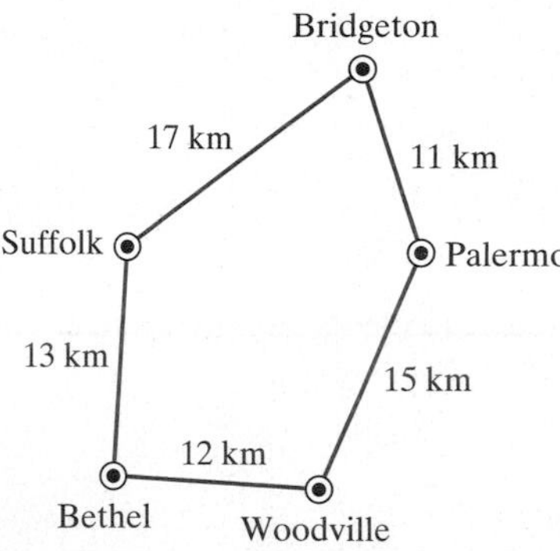

(a) How many kilometers is the trip if she drives through Suffolk? What is her average speed if this trip takes her 0.4 hour?

13 km + 17 km = 30 km $\quad \frac{30\text{ km}}{0.4\text{ hr}} = 75\text{ km/hr}$

(b) How many kilometers is the trip if she drives through Woodville and Palermo? What is her average speed if this trip takes her 0.5 hour?

15 km + 12 km + 11 km = 38 km $\quad \frac{38\text{ km}}{0.5\text{ hr}} = 76\text{ km/hr}$

(c) Over which route does she travel at a more rapid rate?

through Woodville and Palermo

2. *Driving Distances* Robert drives from work to either a convenience store and then home or to a supermarket and then home. The sketch shows the distances.

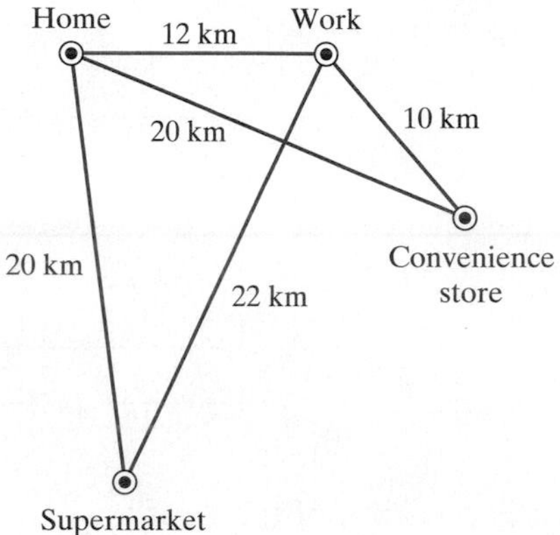

(a) How far does he travel if he goes from work to the supermarket and then home? How fast does he travel if the trip takes 0.6 hour?

22 km + 20 km = 42 km $\quad \frac{42\text{ km}}{0.6\text{ hr}} = 70\text{ km/hr}$

(b) How far does he travel if he goes from work to the convenience store and then home? How fast does he travel if the trip takes 0.5 hour?

10 km + 20 km = 30 km $\quad \frac{30\text{ km}}{0.5\text{ hr}} = 60\text{ km/hr}$

(c) Over which route does he travel at a more rapid rate?

via the supermarket

3. *Hanging Wallpaper* A professional wallpaper hanger can wallpaper 120 ft^2 in 15 minutes. She will be papering four walls in a house. They measure 7 ft × 10 ft, 7 ft × 14 ft, 6 ft × 10 ft, and 6 ft × 8 ft. How many minutes will it take her to paper all four walls?

$(7\text{ ft} \times 10\text{ ft}) + (7\text{ ft} \times 14\text{ ft}) + (6\text{ ft} \times 10\text{ ft}) + (6\text{ ft} \times 8\text{ ft}) = 276\text{ ft}^2$

$276\text{ ft}^2 \times \frac{15\text{ min}}{120\text{ ft}^2} = 34.5\text{ min}$

4. *Painting Apartment* Dave and Linda McCormick are painting the walls of their apartment in Seattle. When they worked together at Linda's mother's house, they were able to paint 80 ft^2 in 25 minutes. The living room they wish to paint has one wall that measures 16 feet by 7 feet, one wall that measures 14 feet by 7 feet, and two walls that measure 12 feet by 7 feet. How long will it take Dave and Linda together to paint the living room of their apartment? Round to the nearest minute.

118 minutes or 1 hour, 58 minutes

5. ***Painting Exterior of House*** The Crawfords need to paint the outside of their house. The front and back each measure 55 ft by 24 ft, and each side measures 32 ft by 24 ft. There are sixteen windows and two doors, measuring 4 ft by 2 ft and 7 ft by 3 ft, respectively. They need to calculate the total area to be painted so they know how much paint to purchase. What is the total area to be painted?
4006 ft^2

6. ***Carpeting a Conference Center*** A new conference center is ordering carpet for the main meeting room. The entire main meeting room measures 62 ft by 80 ft. The speaker's platform measures 15 ft by 20 ft and will not be carpeted. How many square feet of carpeting should be ordered?
4660 ft^2

7. ***Carpeting a Recreation Room*** The floor area of the recreation room at Yvonne's house is shown in the following drawing. How much will it cost to carpet the room if the carpet costs \$15 per square yard?

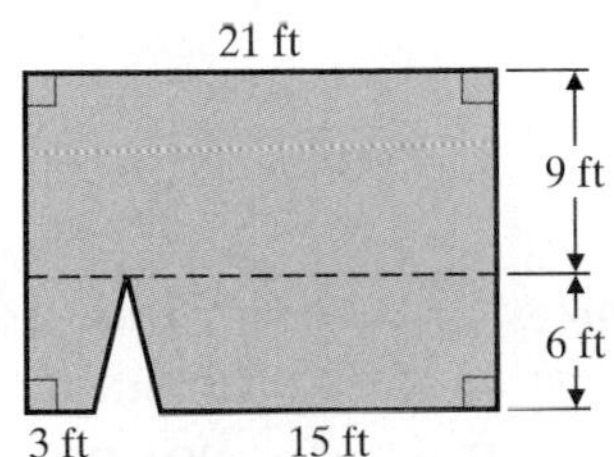

$A = 21 \text{ ft} \times 15 \text{ ft} - \frac{1}{2} \times 3 \text{ ft} \times 6 \text{ ft} = 306 \text{ ft}^2$

$\text{Cost} = 306 \text{ ft}^2 \times \frac{1 \text{ yd}^2}{9 \text{ ft}^2} \times \frac{\$15}{\text{yd}^2} = \$510$

8. ***Aluminum Siding on a Barn*** The side view of a barn is shown in the following diagram. The cost of aluminum siding is \$18 per square yard. How much will it cost to put siding on this side of the barn?

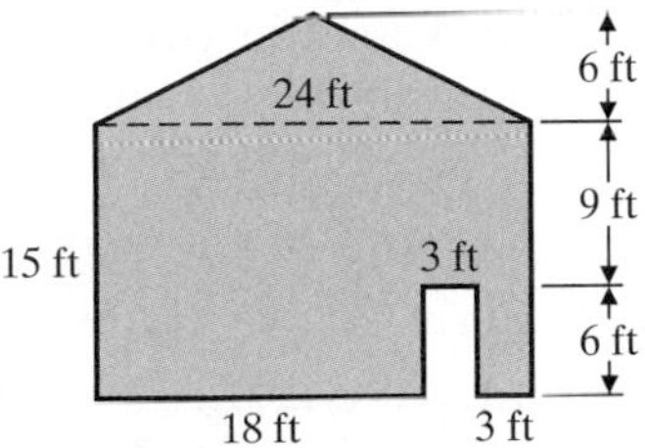

$A = \frac{1}{2} \times 24 \text{ ft} \times 6 \text{ ft} + 15 \text{ ft} \times 24 \text{ ft} - 3 \text{ ft} \times 6 \text{ ft} = 414 \text{ ft}^2$

$\text{Cost} = 414 \text{ ft}^2 \times \frac{1 \text{ yd}^2}{9 \text{ ft}^2} \times \frac{\$18}{\text{yd}^2} = \$828$

9. ***Gold Filling for a Tooth*** A dentist places a gold filling in the shape of a cylinder with a hemispherical top in a patient's tooth. The radius r of the filling is 1 mm. The height is 2 mm. Find the volume of the filling. If dental gold costs \$95 per cubic millimeter, how much did the gold cost for the filling?

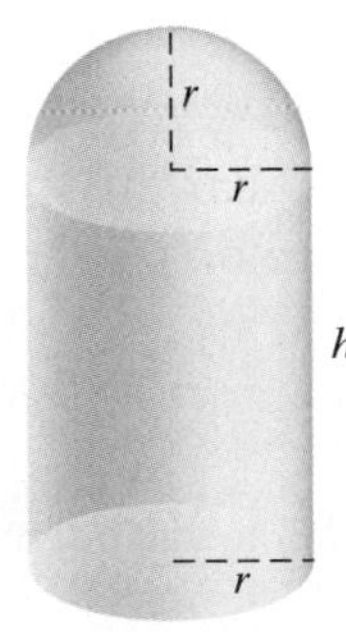

$V = \frac{1}{2} \times \frac{4}{3} \times 3.14 \times (1 \text{ mm})^3 + 3.14 \times 2 \text{ mm} \times (1 \text{ mm})^2 \quad V \approx 8.37 \text{ mm}^3$

$\text{Cost} = 8.37 \text{ mm}^3 \times \frac{\$95}{\text{mm}^3} = \$795.15$

10. ***City Sewer System*** Find the volume of a concrete connector for the city sewer system. A diagram of the connector is shown. It is shaped like a box with a hole of diameter 2 m. If it is formed using concrete that costs \$1.20 per cubic meter, how much will the necessary concrete cost?

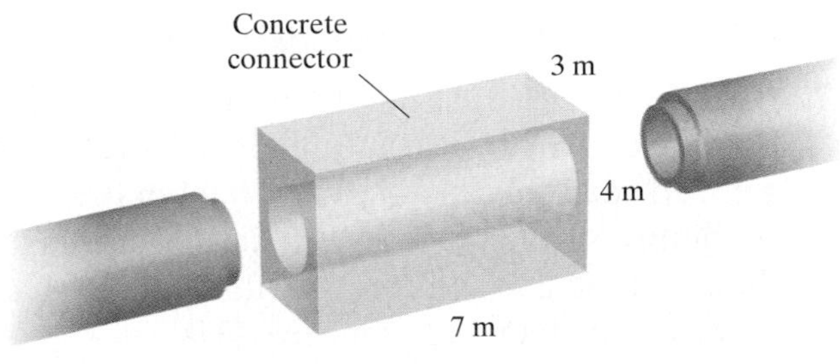

$r = \frac{2}{2} \text{m}$

$V = 7 \text{ m} \times 3 \text{ m} \times 4 \text{ m} - 3.14 \times (1 \text{ m})^2 \times 7 \text{ m} = 62.02 \text{ m}^3$

$\text{Cost} = 62.02 \text{ m}^3 \times \frac{\$1.20}{\text{m}^3} \approx \$74.42$

11. ***Satellite Orbit*** The Landstat satellite orbits the earth in an almost circular pattern. Assume that the radius of orbit (distance from center of Earth to the satellite) is 6500 km.

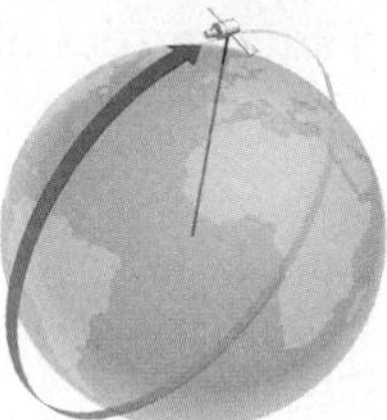

(a) How many kilometers long is one orbit of the satellite? (That is, what is the circumference of the orbit path?)
$C = 2 \times 3.14 \times 6500 \text{ km} = 40{,}820 \text{ km}$

(b) If the satellite goes through one orbit around the Earth in two hours, what is its speed in kilometers per hour?
$S = \dfrac{40{,}820 \text{ km}}{2 \text{ hr}}$ $S = 20{,}410 \text{ km/hr}$

12. ***City Park*** The North City Park is constructed in a shape that includes the region outside one-fourth of a circle. It is shaded in this sketch.

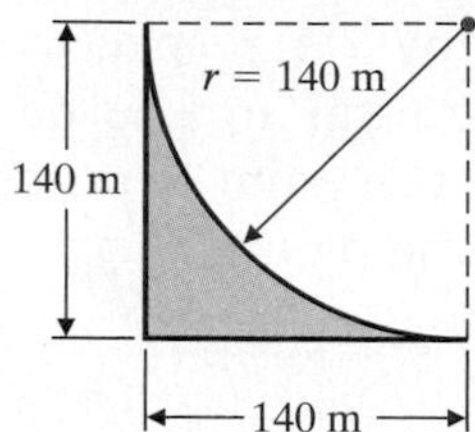

(a) What is the area of the park?
4214 m^2

(b) How much will it cost to resod the park with new grass at $4.00 per square meter?
$16,856

13. ***Valentine's Day Candy*** For Valentine's Day, a company makes decorative cylinder-shaped canisters and fills them with red hot candies. Each canister is 10 in. high and has a radius of 2 in. They want to make 400 canisters and need to determine how much candy to buy. What is the total number of cubic inches that will be filled with candy?
$V = 400 \times 3.14 \times (2)^2 \times 10 \approx 50{,}240 \text{ in.}^3$

14. ***Food Concessions at a State Fair*** Marcus is selling snow cones at the state fair. He has a cart with a refrigerated box measuring 2 ft by 3 ft by 2 ft. Each snow cone requires a spherical ball of shaved ice that measures 3 in. in diameter. If Marcus starts out with a full box of shaved ice, how many snow cones can be made?
1467 snow cones

15. ***Ranch Fencing*** Dr. Dobson's ranch in Colorado has a rectangular field that measures 456 feet by 625 feet. The fence surrounding the field needs to be replaced. If the fencing costs $4.57 per yard, what will it cost for materials to replace the fence? Round to the nearest cent.
$3293.45

16. ***Gasoline Storage Tank*** Boston Sunoco Dealerships owns a large gasoline storage tank in the shape of a cylinder in South Boston. The diameter of the tank is 122.9 feet. The height of the tank is 55.8 feet. How many gallons of gasoline will it hold if there are 7.5 gal in 1 ft^3? Round to the nearest tenth.
4,962,138.5 gallons

To Think About

17. ***Bicycle Race*** Jonathan Wells has entered a bike race in Stowe, Vermont. He is traveling at a speed such that his bicycle wheel makes 200 revolutions per minute. The bicycle tire has a diameter of 28 inches. How fast is Jonathan traveling in feet per minute? (Use $\pi = 3.14$ and round to the nearest hundredth.)
1465.33 feet per minute

18. ***Bicycle Race*** A car in front of Jonathan is traveling at 26 miles per hour. If Jonathan increases his speed by 20% from the speed he was traveling in exercise 17, will he be able to pass the car?
no; with this increase he will be traveling only 20 miles per hour

Cumulative Review

Divide.

19. $16\overline{)2048}$ 128

20. $42\overline{)12{,}936}$ 308

19. $1.3\overline{)0.325}$ 0.25

20. $0.52\overline{)2.5324}$ 4.87

Putting Your Skills to Work

The Mathematics of Population Density

Have you heard of Macau? This tiny region of China has a population of 460,000 and is only about one-tenth the size of Washington, D.C.! Just how crowded is this country and how does it compare to other countries or states of the United States? The topics of this chapter will help you answer the following questions.

The term *population density* refers to the number of people occupying an area in relation to the size of that area. A country's population density gives you an idea of how crowded it is. To find the population density, divide the population by the area.

Problems for Individual Investigation

1. Find the population density of Macau, whose population is 460,000 and area is 9.76 square miles. Round to the nearest whole number.
   47,131 people/sq mi

2. Find the population density of Libya, whose population is 5,368,600 and area is 676,367 square miles. Round to the nearest whole number.
   8 people/sq mi

3. The table below gives the population and area of 5 states. Make a guess which state has the highest and which has the lowest population density before doing any calculations.
   Answers will vary

4. Find the population density of each. Round to the nearest whole number. Were the guesses you made correct? Do any of these results surprise you?

Problems for Group Investigation and Cooperative Study

5. Recall that a square mile can be thought of as a square with each side measuring 1 mile. How many square yards are in one square mile? How many square feet are in one square mile?
   3,097,600 sq yd; 27,878,400 sq ft

6. Suppose the people of Macau were spread out evenly across the region. How many square yards would each person have? How many square feet?
   about 66 sq yd; about 592 sq ft

7. If the people of Rhode Island were spread out evenly across the state, how many square yards would each person have? How many square feet?
   about 3026 sq yd; about 27,234 sq ft

State	2002 Population	Area (sq mi)	Population Density
Alaska	643,786	571,951	1 person/sq mi
California	35,116,033	155,959	225 people/sq mi
Delaware	807,385	1954	413 people/sq mi
New Mexico	1,855,089	121,356	15 people/sq mi
Rhode Island	1,069,725	1045	1024 people/sq mi

Chapter 7 Organizer

Topic	Procedure	Examples
Perimeter of a rectangle, p. 436.	$P = 2l + 2w$	Find the perimeter of a rectangle with width = 3 m and length = 8 m. $P = (2)(8\text{ m}) + (2)(3\text{ m})$ $= 16\text{ m} + 6\text{ m} = 22\text{ m}$
Perimeter of a square, p. 437.	$P = 4s$	Find the perimeter of a square with side $s = 6$ m. $P = (4)(6\text{ m}) = 24\text{ m}$
Area of a rectangle, p. 439.	$A = lw$	Find the area of a rectangle with width = 2 m and length = 7 m. $A = (7\text{ m})(2\text{ m}) = 14\text{ m}^2$
Area of a square, p. 439.	$A = s^2$	Find the area of a square with a side of 4 m. $A = s^2 = (4\text{ m})^2 = 16\text{ m}^2$
Perimeter of parallelograms, trapezoids, and triangles, pp. 445, 447, 454.	Add up the lengths of all sides.	Find the perimeter of a triangle with sides of 3 m, 6 m, and 4 m. $3\text{ m} + 6\text{ m} + 4\text{ m} = 13\text{ m}$
Area of a parallelogram, p. 446.	$A = bh$ b = length of base h = height h b	Find the area of a parallelogram with a base of 12 m and a height of 7 m. $A = bh = (12\text{ m})(7\text{ m}) = 84\text{ m}^2$
Area of a trapezoid, p. 448.	$A = \frac{h(b + B)}{2}$ b = length of shorter base B = length of longer base h = height b h B	Find the area of a trapezoid whose height is 12 m and whose bases are 17 m and 25 m. $A = \frac{(12\text{ m})(17\text{ m} + 25\text{ m})}{2} = \frac{(12\text{ m})(42\text{ m})}{2}$ $= \frac{504\text{ m}^2}{2} = 252\text{ m}^2$
The sum of the measures of the three interior angles of a triangle is 180°, p. 453.	In a triangle, to find one missing angle if two are given: **1.** Add up the two known angles. **2.** Subtract the sum from 180°.	Find the missing angle if two known angles in a triangle are 60° and 70°. **1.** 60° + 70° = 130° **2.** 180° − 130° 50° The missing angle is 50°.
Area of a triangle, p. 455.	$A = \frac{bh}{2}$ b = base h = height h b	Find the area of the triangle whose base is 1.5 m and whose height is 3 m. $A = \frac{bh}{2} = \frac{(1.5\text{ m})(3\text{ m})}{2} = \frac{4.5\text{ m}^2}{2} = 2.25\text{ m}^2$
Evaluating square roots of numbers that are perfect squares, p. 461.	If a number is a product of two identical factors, then either factor is called a square root.	$\sqrt{0} = 0$ because $(0)(0) = 0$ $\sqrt{4} = 2$ because $(2)(2) = 4$ $\sqrt{100} = 10$ because $(10)(10) = 100$ $\sqrt{169} = 13$ because $(13)(13) = 169$

<table>
<tr><th>Topic</th><th>Procedure</th><th>Examples</th></tr>
<tr><td>Approximating the square root of a number that is not a perfect square, p. 462.</td><td>1. If a calculator with a square root key is available, enter the number and then press the $\boxed{\sqrt{x}}$ or $\boxed{\sqrt{\ }}$ key. The approximate value will be displayed.
2. If using a square root table, find the number n, then look for the square root of that number. The approximate value will be correct to the nearest thousandth.

Number, n | Square Root of That Number, $\sqrt{n}$
31 | 5.568
32 | 5.657
33 | 5.745
34 | 5.831</td><td>1. Find on a calculator.
(a) $\sqrt{13}$ (b) $\sqrt{182}$
Round to the nearest thousandth.
(a) 13 $\boxed{\sqrt{x}}$ 3.60555127 rounds to 3.606.
(b) 182 $\boxed{\sqrt{x}}$ 13.49073756 rounds to 13.491.
2. Find from a square root table.
(a) $\sqrt{31}$ (b) $\sqrt{33}$ (c) $\sqrt{34}$
To the nearest thousandth, the approximate values are as follows.
(a) $\sqrt{31} = 5.568$ (b) $\sqrt{33} = 5.745$
(c) $\sqrt{34} = 5.831$</td></tr>
<tr><td>Finding the hypotenuse of a right triangle when given the length of each leg, p. 468.</td><td>$\text{Hypotenuse} = \sqrt{(\text{leg})^2 + (\text{leg})^2}$
Leg
Leg
Hypotenuse</td><td>Find the hypotenuse of a triangle with legs of 9 m and 12 m.
$\text{hypotenuse} = \sqrt{(12)^2 + (9)^2}$
$= \sqrt{144 + 81} = \sqrt{225}$
$= 15$ m</td></tr>
<tr><td>Finding the leg of a right triangle when given the length of the other leg and the hypotenuse, p. 469.</td><td>$\text{Leg} = \sqrt{(\text{hypotenuse})^2 - (\text{leg})^2}$</td><td>Find the leg of a right triangle. The hypotenuse is 14 in. and the other leg is 12 in.
Round to nearest thousandth.
$\text{leg} = \sqrt{(14)^2 - (12)^2} = \sqrt{196 - 144} = \sqrt{52}$
Using a calculator or a square root table, the leg ≈ 7.211 in.</td></tr>
<tr><td>Solving applied problems involving the Pythagorean Theorem, p. 470.</td><td>1. Read the problem carefully.
2. Draw a sketch.
3. Label the two sides that are given.
4. If the hypotenuse is unknown, use
$\text{hypotenuse} = \sqrt{(\text{leg})^2 + (\text{leg})^2}$.
5. If one leg is unknown, use
$\text{leg} = \sqrt{(\text{hypotenuse})^2 - (\text{leg})^2}$.</td><td>A boat travels 5 mi south and then 3 mi east. How far is it from the starting point? Round to the nearest tenth.
$\text{hypotenuse} = \sqrt{(5)^2 + (3)^2}$
$= \sqrt{25 + 9} = \sqrt{34}$
Using a calculator or a square root table, the distance is approximately 5.8 mi.
5 miles south
3 miles east</td></tr>
<tr><td>The special 30°–60°–90° right triangle, p. 471.</td><td>The length of the leg opposite the 30° angle is $\frac{1}{2} \times$ the length of the hypotenuse.</td><td>Find y.
$y = \frac{1}{2}(26) = 13$ m
26 m
60°
y
30°</td></tr>
<tr><td>The special 45°–45°–90° right triangle, p. 471.</td><td>The sides opposite the 45° angles are equal. The hypotenuse is $\sqrt{2} \times$ the length of either leg.</td><td>Find z.
$z = \sqrt{2}(13) \approx (1.414)(13) = 18.382$ m
z
45°
13 m
45°
13 m</td></tr>
</table>

Topic	Procedure	Examples
Radius and diameter of a circle, p. 477.	r = radius d = diameter $r = \frac{d}{2}$ $d = 2r$	What is the radius of a circle with diameter 50 in. $r = \frac{50 \text{ in.}}{2} = 25 \text{ in.}$ What is the diameter of a circle with radius 16 in.? $d = (2)(16 \text{ in.}) = 32 \text{ in.}$
Pi, p. 477.	Pi is a decimal that goes on forever. It can be approximated by as many decimal places as needed. $\pi = \frac{\text{circumference of a circle}}{\text{diameter of same circle}}$	Use $\pi \approx 3.14$ for all calculations. Unless otherwise directed, round your final answer to the nearest tenth when any calculation involves π.
Circumference of a circle, p. 477.	$C = \pi d$	Find the circumference of a circle with diameter 12 ft. $C = \pi d = (3.14)(12 \text{ ft}) = 37.68$ ≈ 37.7 ft (rounded to the nearest tenth)
Area of a circle, p. 479.	$A = \pi r^2$ **1.** Square the radius first. **2.** Then multiply the result by 3.14.	Find the area of a circle with radius 7 ft. $A = \pi r^2 = (3.14)(7 \text{ ft})^2 = (3.14)(49 \text{ ft}^2)$ $= 153.86 \text{ ft}^2$ $\approx 153.9 \text{ ft}^2$ (rounded to the nearest tenth)
Volume of a rectangular solid (box), p. 487.	$V = lwh$	Find the volume of a box whose dimensions are 5 m by 8 m by 2 m. $V = (5 \text{ m})(8 \text{ m})(2 \text{ m}) = (40)(2) \text{ m}^3 = 80 \text{ m}^3$
Volume of a cylinder, p. 487.	r = radius h = height $V = \pi r^2 h$ **1.** Square the radius first. **2.** Then multiply the result by 3.14 and by the height. h r	Find the volume of a cylinder with radius 7 m and height 3 m. $V = \pi r^2 h = (3.14)(7 \text{ m})^2(3 \text{ m})$ $= (3.14)(49)(3) \text{ m}^3$ $= (153.86)(3) \text{ m}^3 = 461.58 \text{ m}^3$ $\approx 461.6 \text{ m}^3$ (rounded to the nearest tenth)
Volume of a sphere, p. 488.	$V = \frac{4\pi r^3}{3}$ r r = radius	Find the volume of a sphere of radius 3 m. $V = \frac{4\pi r^3}{3} = \frac{(4)(3.14)(3 \text{ m})^3}{3}$ $= \frac{(4)(3.14)(\overset{9}{\cancel{27}}) \text{ m}^3}{1}$ $= (12.56)(9) \text{ m}^3 = 113.04 \text{ m}^3$ $\approx 113.0 \text{ m}^3$ (rounded to the nearest tenth)
Volume of a cone, p. 489.	$V = \frac{\pi r^2 h}{3}$ r = radius h = height r	Find the volume of a cone of height 9 m and radius 7 m. $V = \frac{\pi r^2 h}{3} = \frac{(3.14)(7 \text{ m})^2(9 \text{ m})}{3}$ $= \frac{(3.14)(7^2)(\overset{3}{\cancel{9}}) \text{ m}^3}{\underset{1}{\cancel{3}}} = (3.14)(49)(3) \text{ m}^3$ $= (153.86)(3) \text{ m}^3 = 461.58 \text{ m}^3$ $\approx 461.6 \text{ m}^3$ (rounded to the nearest tenth)

Topic	Procedure	Examples
Volume of a pyramid, p. 489.	$V = \dfrac{Bh}{3}$ B = area of the base h = height **1.** Find the area of the base. **2.** Multiply this area by the height and divide the result by 3.	Find the volume of a pyramid whose height is 6 m and whose rectangular base is 10 m by 12 m. **1.** $B = (12\text{ m})(10\text{ m}) = 120\text{ m}^2$ **2.** $V = \dfrac{(120)(\overset{2}{\cancel{6}})\text{ m}^3}{\underset{1}{\cancel{3}}} = (120)(2)\text{ m}^3 = 240\text{ m}^3$
Similar figures, corresponding sides, p. 495.	The corresponding sides of similar figures have the same ratio.	Find n in the following similar figures. $\dfrac{n}{4} = \dfrac{9}{3}$ $3n = 36$ $n = 12\text{ m}$ n 9 m 4 m 3 m
Similar figures, corresponding perimeters, p. 498.	The perimeters of similar figures have the same ratio as the corresponding sides. For reasons of space, the procedure for the areas of similar figures is not given here but may be found in the text (see p. 500).	These two figures are similar. Find the perimeter of the larger figure. $\dfrac{6}{12} = \dfrac{29}{p}$ $6p = (12)(29)$ $6p = 348$ $\dfrac{6p}{6} = \dfrac{348}{6}$ $p = 58$ The perimeter of the larger figure is 58 m. 12 m 6 m 5 m 7 m 3 m 8 m

Chapter 7 Review Problems

Round to the nearest tenth when necessary. Use $\pi \approx 3.14$ in all calculations requiring the use of π.

Section 7.1

1. Find the complement of an angle of 76°. 14°

2. Find the supplement of an angle of 76°. 104°

3. Find the measures of $\angle a$, $\angle b$, and $\angle c$ in the following sketch.

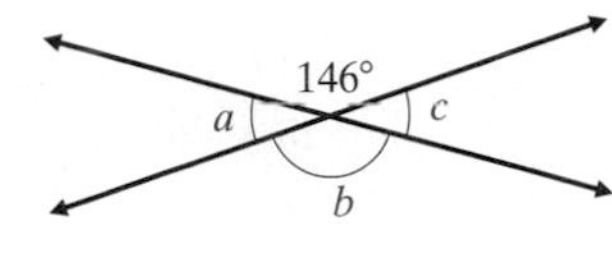

$\angle b = 146°$
$\angle a = \angle c = 34°$

4. Find $\angle s$, $\angle t$, $\angle u$, $\angle w$, $\angle x$, $\angle y$, and $\angle z$ in the following sketch if we know that line p is parallel to line q.

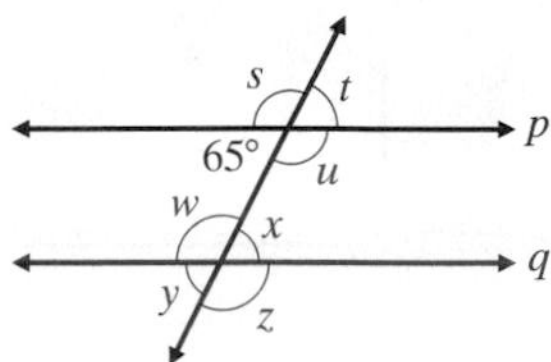

$\angle t = \angle x = \angle y = 65°$
$\angle s = \angle u = \angle w = \angle z = 115°$

Section 7.2

Find the perimeter of the square or rectangle.

5. length = 8.3 m, width = 1.6 m
$P = 19.8$ m

6. length = width = 2.4 yd
$P = 9.6$ yd

Find the area of the square or rectangle.

7. length = 5.9 cm, width = 2.8 cm
$A \approx 16.5 \text{ cm}^2$

8. length = width = 7.2 in.
$A \approx 51.8 \text{ in.}^2$

Find the perimeter of each object made up of rectangles and squares.

9. $P = 38$ ft

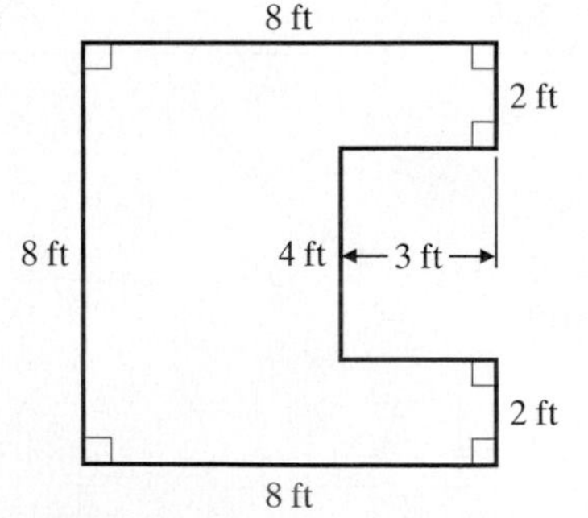

10. $P = 58$ ft

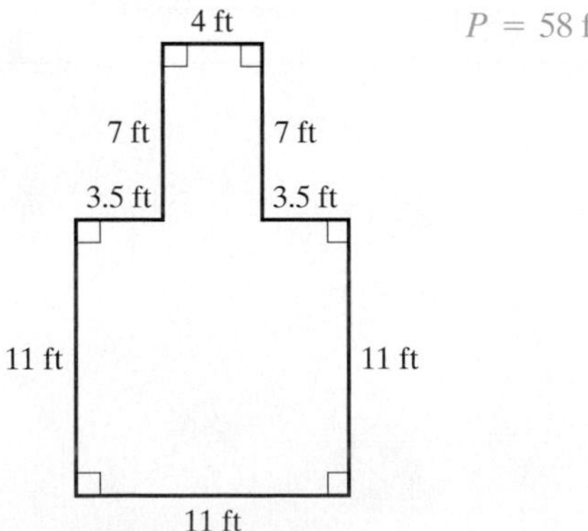

Find the area of each shaded region made up of rectangles and squares.

11.

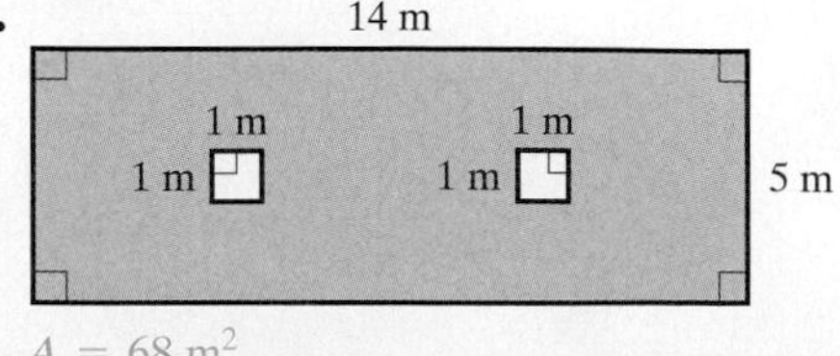

$A = 68 \text{ m}^2$

12.

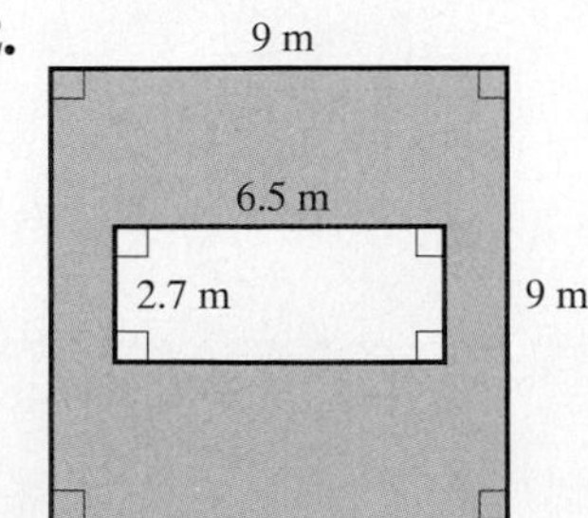

$A \approx 63.5 \text{ m}^2$

Section 7.3

Find the perimeter of the parallelogram or trapezoid.

13. Two sides of the parallelogram are 52 m and 20.6 m. $P = 145.2$ m

14. The sides of the trapezoid are 5 mi, 22 mi, 5 mi, and 30 mi. $P = 62$ mi

Find the area of the parallelogram or trapezoid.

15. The parallelogram has a base of 90 m and a height of 30 m. $A = 2700\text{ m}^2$

16. The trapezoid has a height of 36 yd and bases of 17 yd and 23 yd. $A = 720\text{ yd}^2$

Find the total area of each region made up of parallelograms, trapezoids, and rectangles

17.

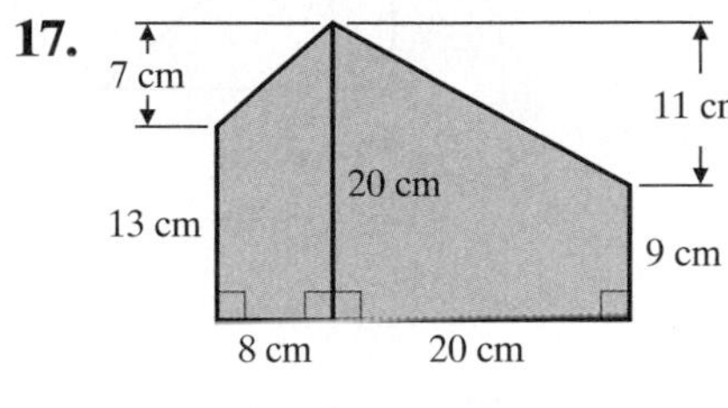

$A = 422\text{ cm}^2$

18.

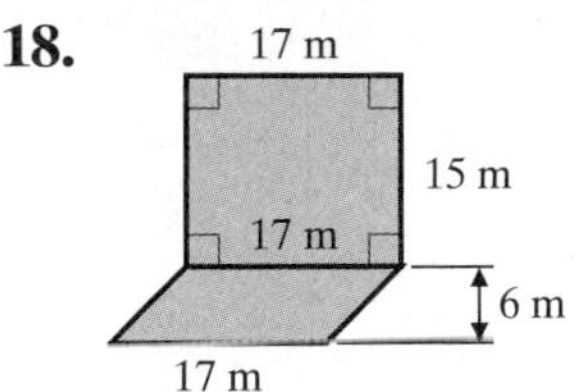

$A = 357\text{ m}^2$

Section 7.4

Find the perimeter of the triangle.

19. An isosceles triangle with one side 18 ft and the other two sides 21 ft. 60 ft

20. An equilateral triangle with each side 15.5 ft. 46.5 ft

Find the measure of the third angle in the triangle.

21. Two known angles are 15° and 12°. 153°

22. A right triangle with one angle of 35°. 55°

Find the area of the triangle.

23. base = 8.5 m, height = 12.3 m $A \approx 52.3\text{ m}^2$

24. base = 12.5 m, height = 9.5 m $A \approx 59.4\text{ m}^2$

Find the total area of each region made up of triangles and rectangles.

25.

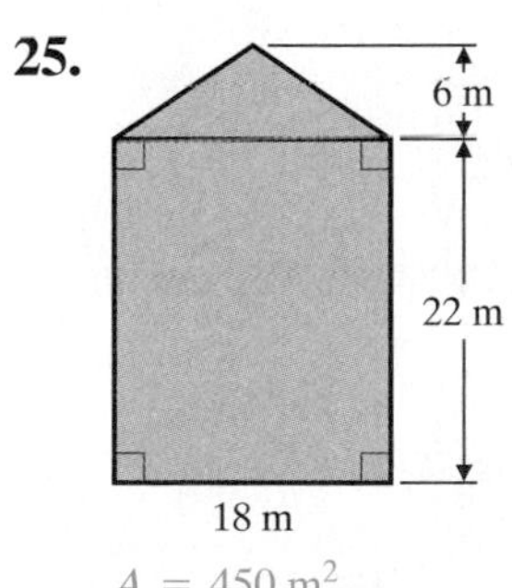

$A = 450\text{ m}^2$

26.

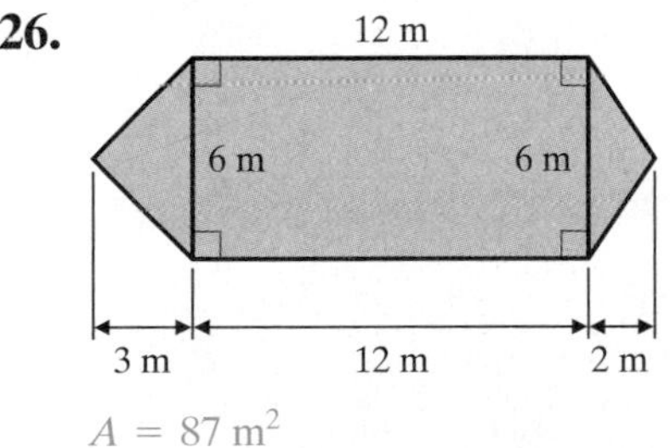

$A = 87\text{ m}^2$

Section 7.5

Evaluate exactly.

27. $\sqrt{81}$ 9

28. $\sqrt{64}$ 8

29. $\sqrt{121}$ 11

30. $\sqrt{169} - \sqrt{100}$ 3

31. $\sqrt{25} + \sqrt{9} + \sqrt{49}$ 15

Approximate using a square root table or a calculator with a square root key. Round to the nearest thousandth when necessary.

32. $\sqrt{35}$ ≈ 5.916 **33.** $\sqrt{48}$ ≈ 6.928 **34.** $\sqrt{165}$ ≈ 12.845 **35.** $\sqrt{180}$ ≈ 13.416

Section 7.6

Find the unknown side. If the answer cannot be obtained exactly, use a square root table or a calculator with a square root key. Round to the nearest hundredth when necessary.

36.

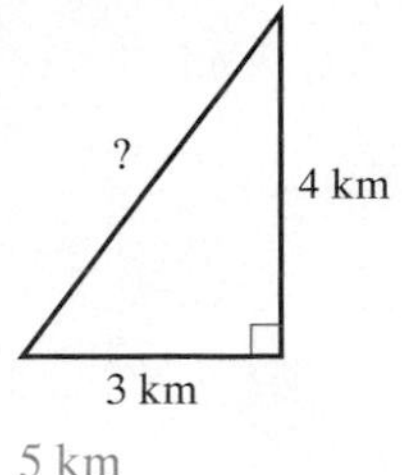

5 km

37.

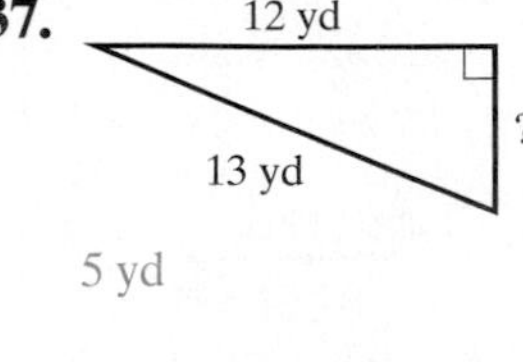

5 yd

38.

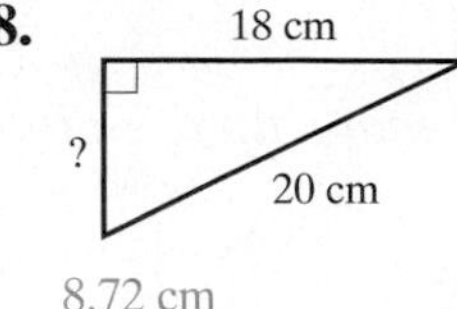

8.72 cm

39.

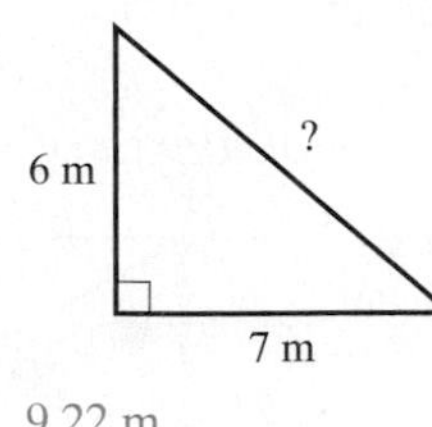

9.22 m

Round to the nearest tenth.

40. ***Construction of a Metal Plate*** Find the distance between the centers of the holes of a metal plate with the dimensions labeled in the following sketch. 6.4 cm

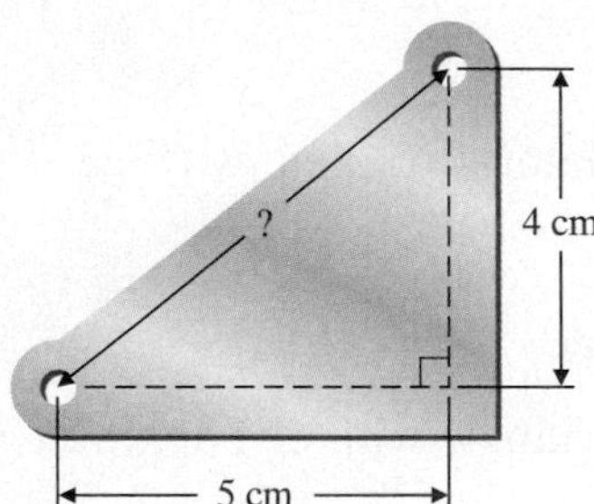

41. ***Building Ramp*** A building ramp has the following dimensions. Find the length of the ramp. 9.2 ft

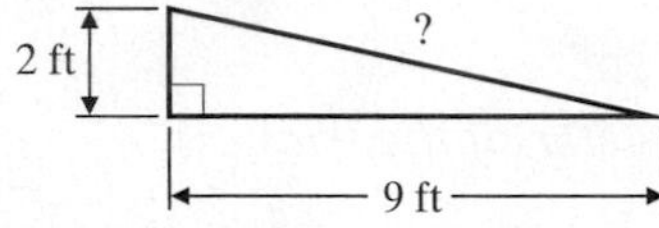

42. ***Shed Construction*** A shed is built with the following dimensions. Find the distance from the peak of the roof to the horizontal support brace. 6.3 ft

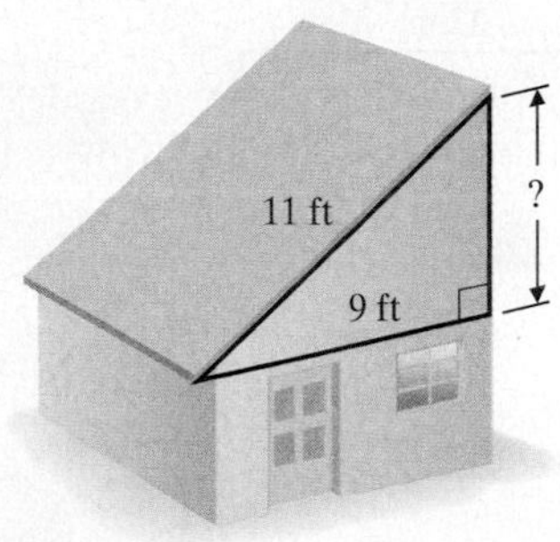

43. ***Replacing a Door*** Find the width of a replacement door if it is 6 ft tall and the diagonal measures 7 ft. 3.6 ft

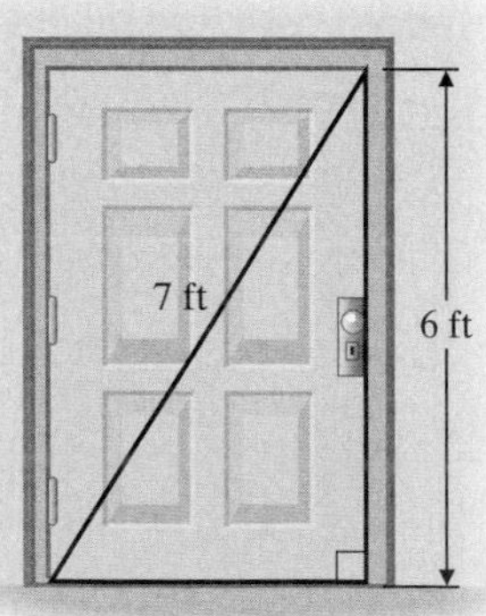

Section 7.7

44. What is the diameter of a circle whose radius is 53 cm? 106 cm

45. What is the radius of a circle whose diameter is 126 cm? 63 cm

46. Find the circumference of a circle with diameter 14 in. 44.0 in.

47. Find the circumference of a circle with radius 9 in. 56.5 in.

Find the area of each circle.

48. radius = 9 m $A \approx 254.3 \text{ m}^2$

49. diameter = 8.6 ft $A \approx 58.06 \text{ ft}^2$

Find the area of each shaded region made up of circles, semicircles rectangles, trapezoids, and parallelograms. Round your answer to the nearest tenth.

50.

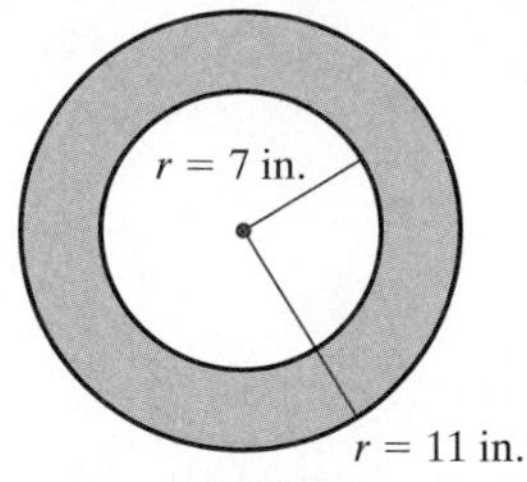

$A = 226.1 \text{ in.}^2$

51.

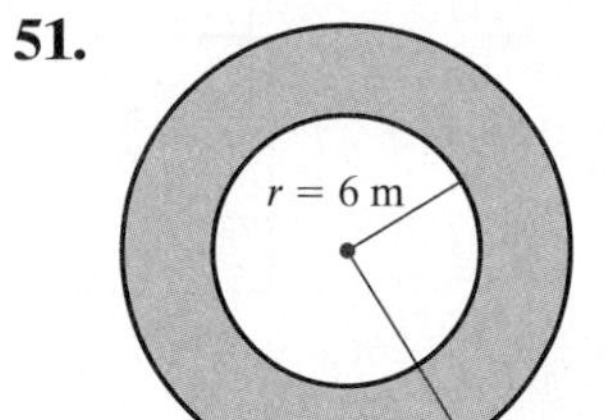

$A = 201.0 \text{ m}^2$

52.

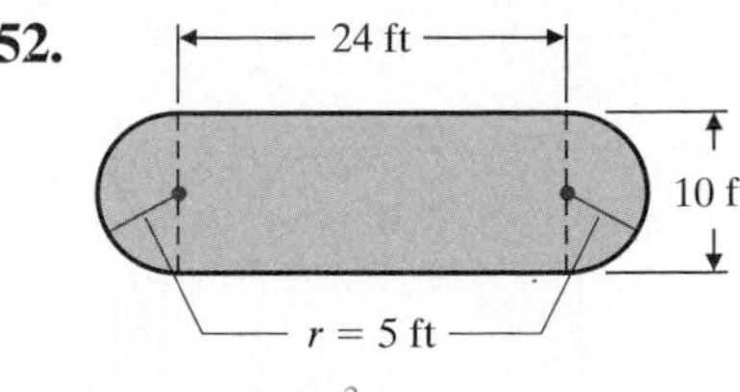

$A = 318.5 \text{ ft}^2$

53.

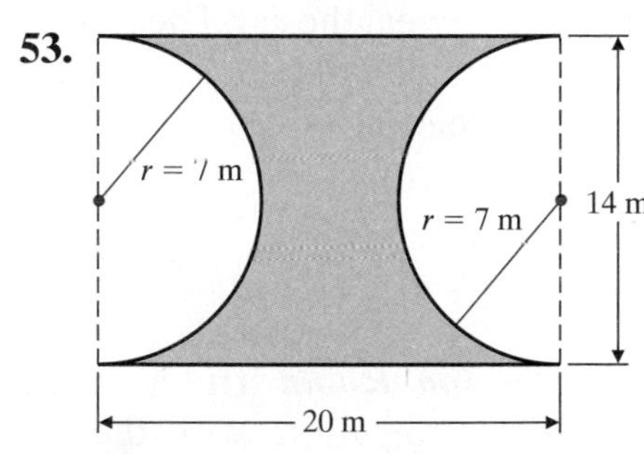

$A = 126.1 \text{ m}^2$

54.

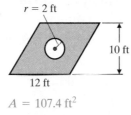

r = 2 ft

10 ft

12 ft

$A = 107.4 \text{ ft}^2$

55.

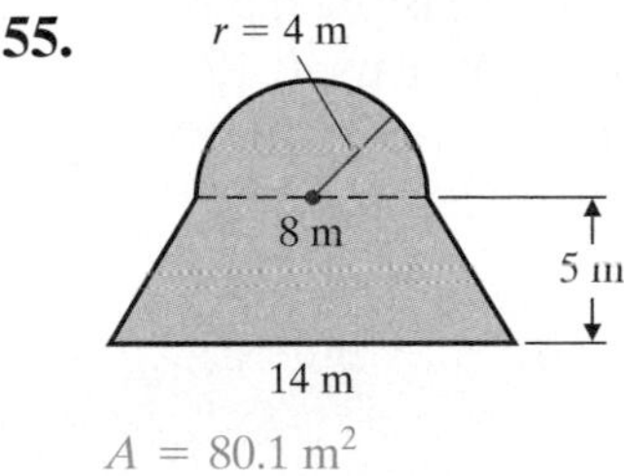

$A = 80.1 \text{ m}^2$

Section 7.8

In exercises 56–62, find the volume.

56. ***Moving Truck*** Find the volume of a moving truck's storage area measuring 12 ft by 6 ft by 6 ft. $V = 432 \text{ ft}^3$

57. ***Basketball*** Find the volume of a basketball with radius 4.75 inches. $V \approx 448.7 \text{ in.}^3$

58. ***Garbage Can*** Find the volume of a garbage can that is 3 ft high and has a radius of 1.5 ft. $V \approx 21.2 \text{ ft}^3$

59. ***Cup of Coffee*** Find the volume of a medium cup of coffee with a height of 5 in. and radius of 1.5 in. $V \approx 35.3 \text{ in.}^3$

60. Find the volume of a pyramid that is 15 m high and whose square base measures 7 m by 7 m. $V = 245 \text{ m}^3$

61. ***Sand Construction at the Beach*** Greg and Marcia took Lexi to play in the sand at the beach. Find the volume of a cone of sand Greg made which is 9 ft tall with a radius of 20 ft. $V = 3768 \text{ ft}^3$

62. ***Chemical Pollution*** A chemical has polluted a volume of ground in a cone shape. The depth of the cone is 30 yd. The radius of the cone is 17 yd. Find the volume of polluted ground. $V = 9074.6 \text{ yd}^3$

Section 7.9

Find n in each set of similar triangles.

63.

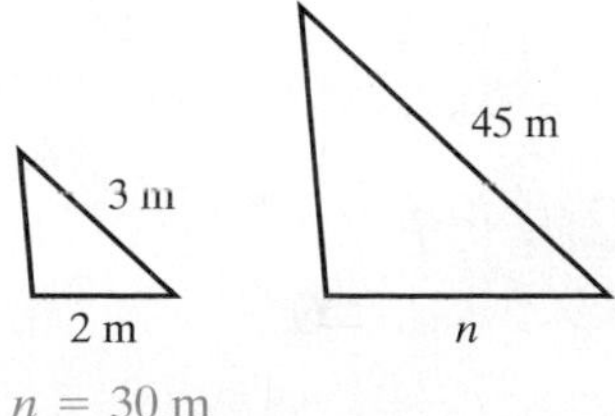

$n = 30$ m

64.

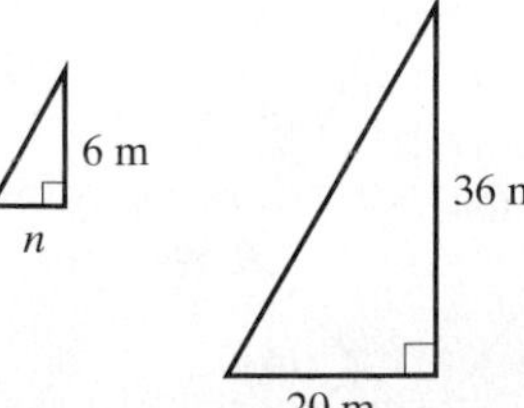

$n \approx 3.3$ m

Determine the perimeter of the unlabeled figure.

65.

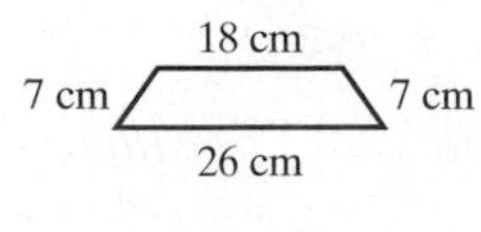

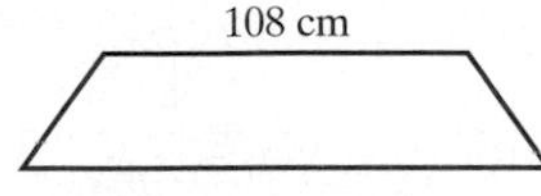

$P = 348$ cm

66.

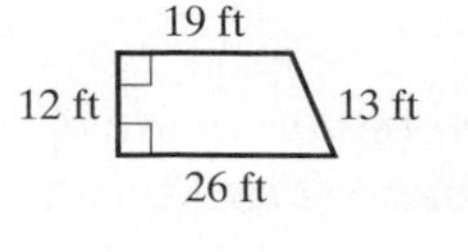

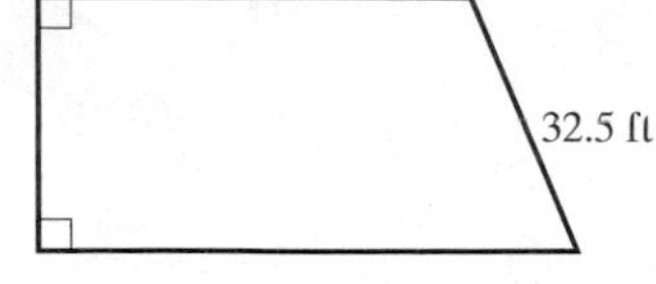

$P = 175$ ft

67. ***Banner Construction*** Anastasio is in charge of decorations for the "International Cars of the Future" show. He has designed a rectangular banner that will hang in front of the 1st prize-winning car, so that all he has to do is push a button and the banner will fly up into the ceiling space to reveal the car. The model of the banner used 12 square yards of fabric. The dimensions of the actual banner will be $3\frac{1}{2}$ times the length and the width of the model. How much fabric will the finished banner need? 147 square yards

Section 7.10

68. ***Chemistry Lab Tank*** A conical tank holds acid in a chemistry lab. The tank has a radius of 9 in. and a height of 24 in. How many cubic inches does the tank hold? The acid weighs 16 g per cubic inch. What is the weight of the acid if the tank is full?
$V \approx 2034.7$ in.^{3}; $W = 32{,}555.2$ g

69. ***Carpeting a Recreation Room*** The Wilsons are carpeting a recreation room with the dimensions shown. Carpeting costs \$8 per square yard. How much will the carpeting cost? \$736

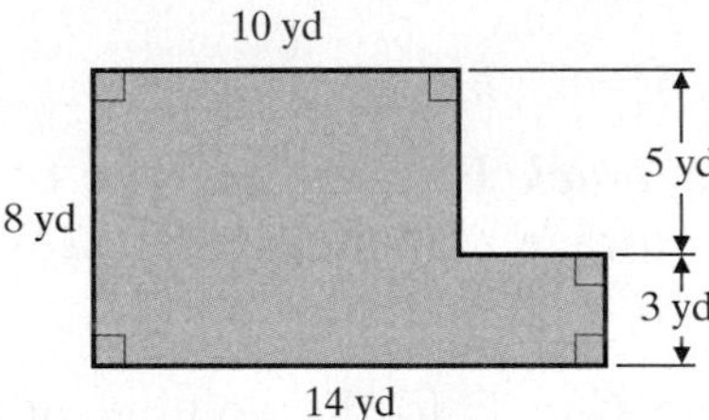

70. ***Driving Distances***

(a) In the following diagram, how many kilometers is it from Homeville to Seaview if you drive through Ipswich? How fast do you travel if it takes 0.5 hour to travel that way? 50 km; 100 km/hr

(b) How many kilometers is it from Homeville to Seaview if you drive through Acton and Westville? How fast do you travel if it takes 0.8 hour to travel that way? 56 km; 70 km/hr

(c) Over which route do you travel at a more rapid rate? through Ipswich

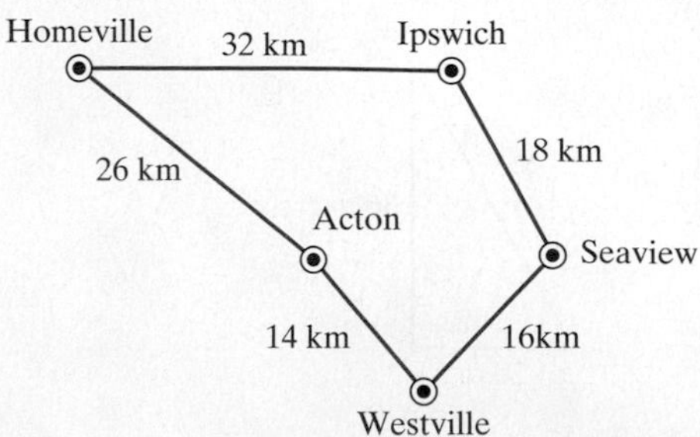

71. ***Silo Capacity*** A silo has a cylindrical shape with a hemispherical dome. It has dimensions as shown in the following figure.

(a) What is its volume in cubic feet?
$V \approx 21{,}873.2 \text{ ft}^3$

(b) If 1 cubic foot ≈ 0.8 bushel, how many bushels of grain will it hold?
$B \approx 17{,}498.6$ bushels

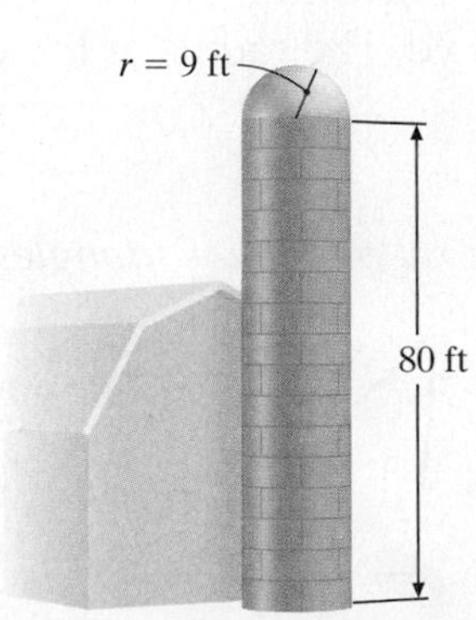

72. ***Farm Production*** During 2005, U.S. farms produced an estimated 2.757 billion bushels of soybeans. (*Source:* U.S. Department of Agriculture) Each bushel takes up 1.244 cubic feet of storage. How many cubic feet of storage was needed for the 2005 soybean crop?
3,429,708,000 cubic feet

73. ***Soybean Storage*** If all of the soybeans in exercise 72 were stored in a huge rectangular storage bin that is 10,000 feet wide and 20,000 feet long, how many feet high would the storage bin need to be? approximately 17.1 feet

74. ***Pet Aquarium*** The largest aquarium at PetsMart measures 2.25 ft by 4 ft by 2 ft. Water weighs about 62 pounds per cubic foot. How many pounds of water does the aquarium hold? There are about 8.6 pounds in one gallon. How many gallons of water does this aquarium hold? Round to the nearest whole gallon. 1116 lb; 130 gal

75. ***Pet Aquarium*** It is recommended that 1.5 inches of gravel be placed in the bottom of the aquarium in exercise 74. How many cubic inches of gravel are needed? Assume the base of the aquarium is 4 ft by 2 ft. 1728 in.^{3}

76. ***Pony Rides*** At the county fair a pony is tied to a 30-ft rope. The pony gives children rides by walking in a circle 5 times with the rope pulled taut. How many feet does the pony walk for each ride? 942 ft

77. ***Roller Rink*** Frank and Dolly manage a roller rink in Springfield. The floor of the rink consists of a large rectangle with a semicircle at each end. The rectangular part measures 16 yards by 20 yards. Each semicircle has a diameter of 16 yards. What is the area of the floor? approximately 521.0 square yards

78. ***Hardwood Flooring*** Frank and Dolly have decided to put down a new hardwood floor for the roller rink in exercise 77. The hardwood flooring will cost $50 per square yard to install (including both labor and materials). How much will it cost them to put down a new hardwood floor? $26,050

79. ***Flying a Kite*** A kite is flying exactly 30 feet above the edge of a pond. The person flying the kite is using exactly 33 feet of string. Assuming that the string is so tight that it forms a straight line, how far is the person standing from the edge of the pond? Round to the nearest tenth. 13.7 feet

80. ***Gas Tank Storage*** The Suburban Gas Company has a spherical gas tank. The diameter of the tank is 90 meters. Find the volume of the spherical tank.
381,510 cubic meters

81. ***Hot Water Tank*** Charlie and Ginny have a cylindrical hot water tank that is 5 feet high and has a diameter of 18 inches. How many cubic feet does the tank hold?
approximately 8.8 cubic feet

82. ***Hot Water Tank*** The tank in exercise 81 is filled with water. One cubic foot of water is about 7.5 gallons. How many gallons does the tank hold?
approximately 66 gallons

83. ***Lawn of a High School*** The front lawn at Central High School in Hanover is in the shape of a trapezoid. The bases of the trapezoid are 45 feet and 50 feet. The height of the trapezoid is 35 feet. What is the area of the front lawn? 1662.5 square feet

84. ***Fertilizer Cost*** The principal of the high school in exercise 83 has stated that the front lawn needs to be fertilized three times a year. The lawn company charges $0.50 per square foot to apply fertilizer. How much will it cost to have the front lawn fertilized three times a year? $2493.75

How Am I Doing? Chapter 7 Test

Note to Instructor: The Chapter 7 Test file in the TestGen program provides algorithms specifically matched to these problems so you can easily replicate this test for additional practice or assessment purposes.

1. $\angle b = 52°$, $\angle c = 128°$, $\angle e = 128°$

2. $P = 40$ yd

3. $P = 25.2$ ft

4. $P = 20$ m

5. $P = 80$ m

6. $P = 137$ m

7. $A = 180 \text{ yd}^2$

8. $A = 104.0 \text{ m}^2$

9. $A = 78 \text{ m}^2$

10. $A = 144 \text{ m}^2$

11. $A = 12 \text{ cm}^2$

12. 12

13. 13

14. 27°

15. 73°

16. 84°

Remember to use your Chapter Test Prep Video CD to see the worked-out solutions to the test problems you want to review.

1. In the following figure, lines m and n are parallel, and the measure of angle a is 52°. Find the measure of angle b, angle c, and angle e.

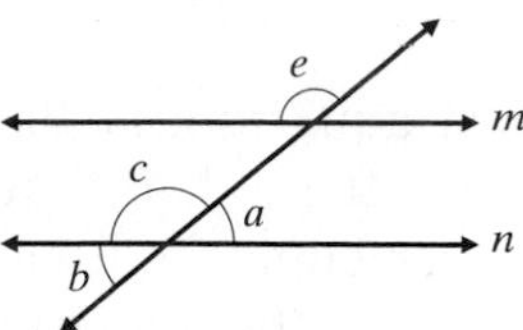

Find the perimeter.

2. a rectangle that measures 9 yd × 11 yd

3. a square with side 6.3 ft

4. a parallelogram with sides measuring 6.5 m and 3.5 m

5. a trapezoid with sides measuring 22 m, 13 m, 32 m, and 13 m

6. a triangle with sides measuring 58.6 m, 32.9 m, and 45.5 m

Find the area. Round to the nearest tenth.

7. a rectangle that measures 10 yd × 18 yd

8. a square 10.2 m on a side

9. a parallelogram with a height of 6 m and a base of 13 m

10. a trapezoid with a height of 9 m and bases of 7 m and 25 m

11. a triangle with a base of 4 cm and a height of 6 cm

Evaluate exactly.

12. $\sqrt{144}$ **13.** $\sqrt{169}$

14. Find the complement of an angle that measures 63°.

15. Find the supplement of an angle that measures 107°.

16. A triangle has an angle that measures 12.5° and another that measures 83.5°. What is the measure of the third angle?

Approximate using a square root table or a calculator with a square root key. Round to the nearest thousandth when necessary.

17. $\sqrt{54}$

18. $\sqrt{135}$

In exercises 19 and 20, find the unknown side. Use a calculator or a square root table to approximate square roots to the nearest thousandth.

19.

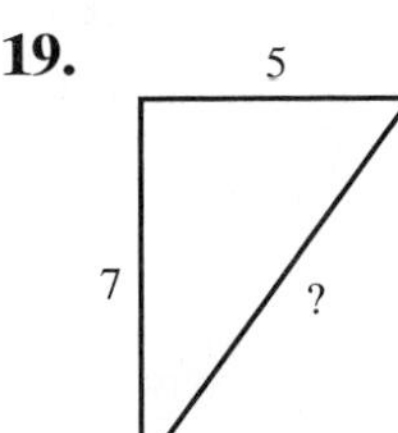

20.

26
24
?

In exercises 21–24, round to the nearest hundredth.

21. Find the distance between the centers of the holes drilled in a rectangular metal plate with the dimensions labeled in the following sketch.

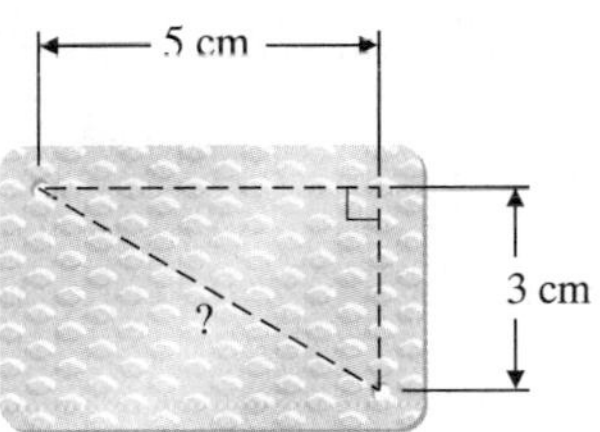

22. A 15-ft-tall ladder is placed so that it reaches 12 ft up on the wall of a house. How far is the base of the ladder from the wall of the house?

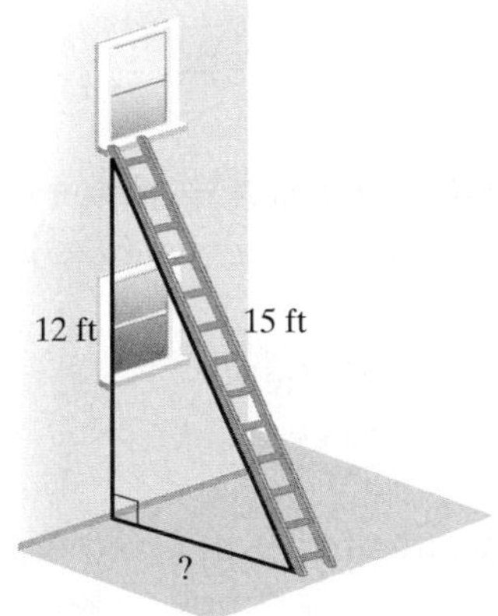

23. Find the circumference of a circle with diameter 18 ft.

24. Find the area of a circle with diameter 12 ft.

17. ≈ 7.348

18. ≈ 11.619

19. 8.602

20. 10

21. ≈ 5.83 cm

22. 9 ft

23. $C \approx 56.52$ ft

24. $A \approx 113.04\ \text{ft}^2$

25. $A \approx 107.4$ in.^{2}

26. $A \approx 144.3$ in.^{2}

27. $V = 700$ m^3

28. $V \approx 803.8$ m^3

29. $V \approx 113.0$ m^3

30. $V \approx 508.7$ ft^3

31. $V = 56$ m^3

32. $n = 46.8$ m

33. $n = 42$ ft

34. $A = 6456$ yd^2

35. \$2582.40

Find the shaded area of each region made up of circles, semicircles, rectangles, squares, trapezoids, and parallelograms. Round to the nearest tenth.

25.

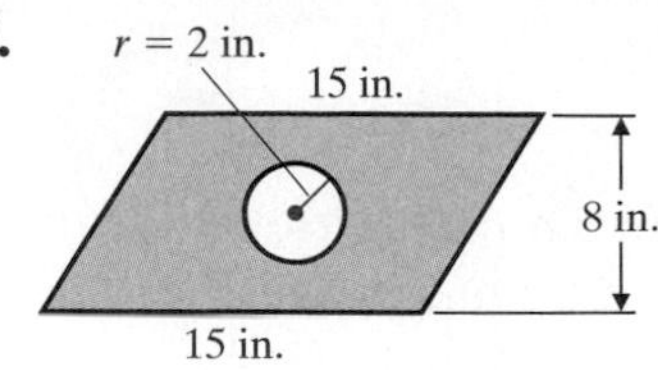

26.

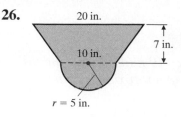

Find the volume. Round to the nearest tenth.

27. a rectangular box measuring 3.5 m by 20 m by 10 m

28. a cone with height 12 m and radius 8 m

29. a sphere of radius 3 m

30. a cylinder of height 2 ft and radius 9 ft

31. a pyramid of height 14 m and whose rectangular base measures 4 m by 3 m

Each pair of triangles is similar. Find the missing side n.

32.

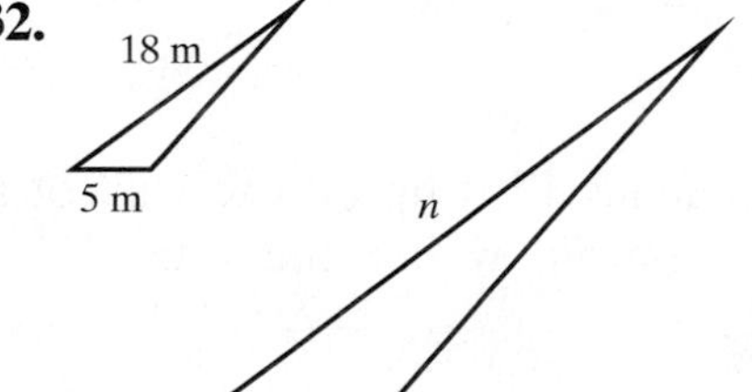

33.

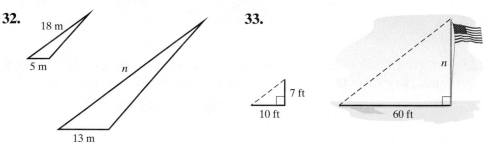

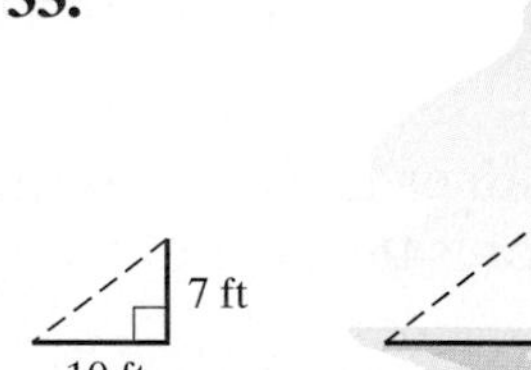

Solve. An athletic field has the dimensions shown in the figure below. Assume you are considering only the darker green shaded area.

34. What is the area of the athletic field?

35. How much will it cost to fertilize it at \$0.40 per square yard?

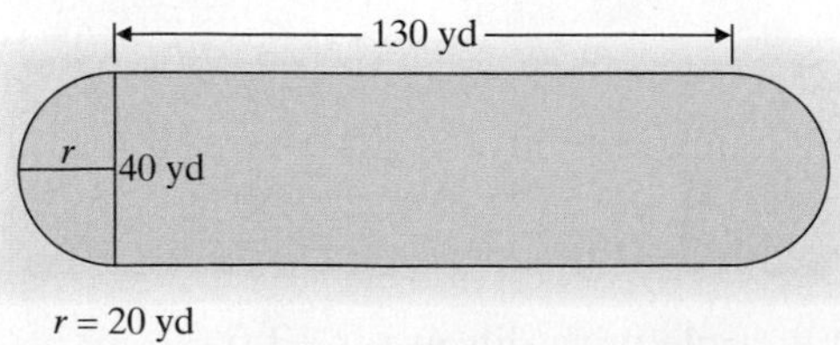

Cumulative Test for Chapters 1–7

Approximately one-half of this test is based on Chapter 7 material. The remainder is based on material covered in Chapters 1–6.

Solve. Simplify.

1. Add. $\begin{array}{r} 126{,}350 \\ 278{,}120 \\ +531{,}290 \\ \hline \end{array}$

2. Multiply. $\begin{array}{r} 163 \\ \times 205 \\ \hline \end{array}$

3. Subtract. $\dfrac{17}{18} - \dfrac{11}{12}$

4. Divide. $\dfrac{3}{7} \div 2\dfrac{1}{4}$

5. Round to the nearest hundredth. 56.1279

6. Multiply. $\begin{array}{r} 9.034 \\ \times \quad 0.8 \\ \hline \end{array}$

7. Multiply. 2.634×10^2

8. Divide. $0.021\overline{)1.743}$

9. Find n. $\dfrac{3}{n} = \dfrac{2}{18}$

10. There are seven teachers for every 100 students. If there are 56 teachers at the university, how many students would you expect?

11. Michael scored 18 baskets out of 24 shots on the court. What percent of his shots went into the basket?

12. 0.8% of what number is 16?

13. What is 15% of 120?

14. Convert 586 cm to m.

15. Convert 42 yd to in.

16. Ben traveled 88 km. How many miles did he travel? (1 kilometer $\approx$ 0.62 mile; 1 mile $\approx$ 1.61 kilometers.)

In questions 17–30, round to the nearest tenth when necessary. Use $\pi \approx 3.14$ in calculations involving π.

17. Find the perimeter of a rectangle of length 17 m and width 8 m

18. Find the perimeter of a trapezoid with sides of 86 cm, 13 cm, 96 cm, and 13 cm

19. Find the circumference of a circle with diameter 18 yd.

Find the area.

20. a triangle with base 1.2 cm and height 2.4 cm

21. a trapezoid with height 18 m and bases of 26 m and 34 m

22.

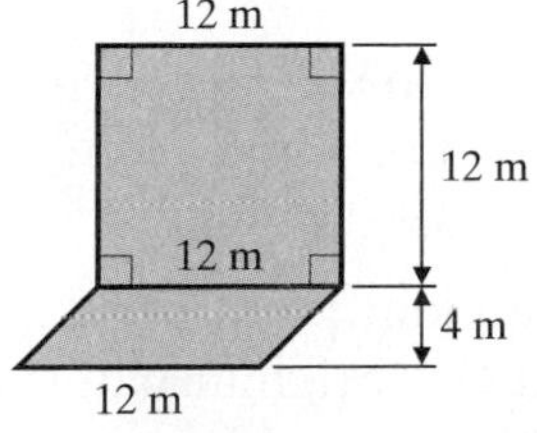

23.

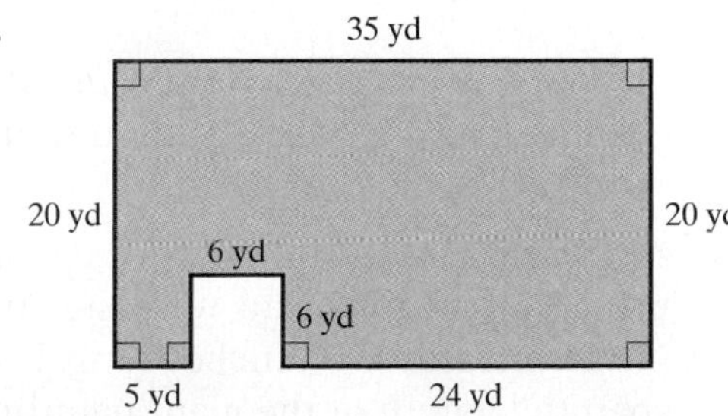

24. a circle with radius 4 m

1. 935,760
2. 33,415
3. $\frac{1}{36}$
4. $\frac{4}{21}$
5. 56.13
6. 7.2272
7. 263.4
8. 83
9. $n = 27$
10. 800 students
11. 75%
12. 2000
13. 18
14. 5.86 m
15. 1512 in.
16. 54.56 mi
17. $P = 50$ m
18. $P = 208$ cm
19. 56.5 yd
20. $A = 1.4 \text{ cm}^2$
21. $A = 540 \text{ m}^2$
22. $A = 192 \text{ m}^2$
23. $A = 664 \text{ yd}^2$
24. $A = 50.2 \text{ m}^2$

25. $V \approx 87.9 \text{ m}^3$

26. $V \approx 33.5 \text{ cm}^3$

27. $V = 3136 \text{ cm}^3$

28. $V = 2713.0 \text{ cm}^3$

29. $n = 38.6$ m

30. $n = 4.1$ ft

31. **(a)** 124 yd^2

(b) $992.00

32. 21

33. ≈ 7.550

34. 10.440 in.

35. 4.899 m

36. 13.9 mi

37. 32 paintbrushes

Find the volume.

25. a cylinder with height 7 m and radius 2 m

26. a sphere with diameter 4 cm

27. a pyramid with height 32 cm and a rectangular base 14 cm by 21 cm

28. A cone with height 18 m and radius 12 m

Find the value of n in each pair of similar figures. Round to the nearest tenth.

29. 30 m, n; 7 m, 9 m

30.

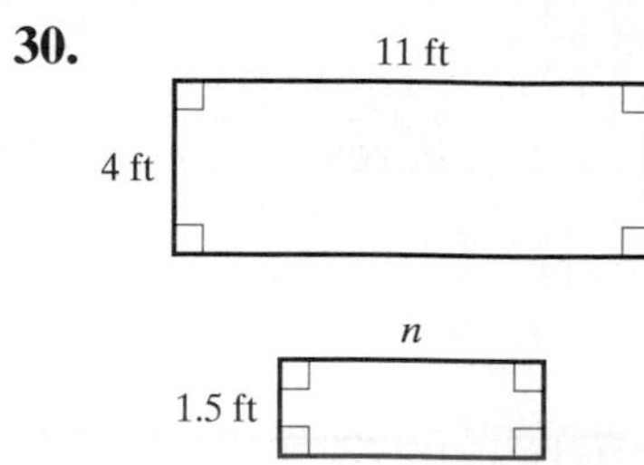

Solve.

31. Mary Ann and Wong Twan have a recreation room with the dimensions shown in the figure below. They wish to carpet it at a cost of $8.00 per square yard.

(a) How many square yards of carpet are needed? (Include the area of the rectangle, the triangle, and the square.)

(b) How much will it cost?

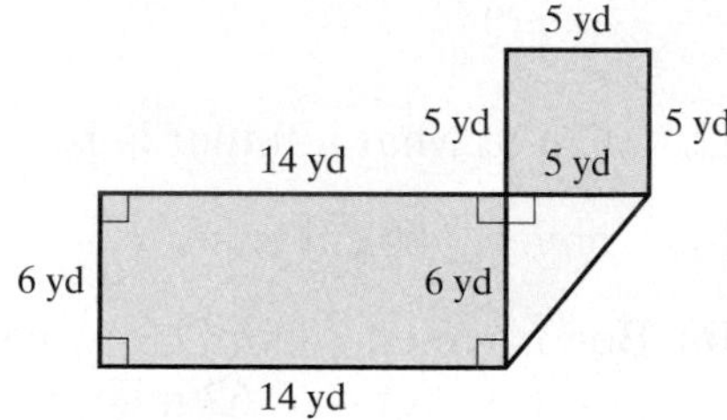

32. Evaluate exactly. $\sqrt{144} + \sqrt{81}$

33. Approximate to the nearest thousandth using a table or a calculator. $\sqrt{57}$

In questions 34 and 35, find the unknown side. Use a calculator or square root table if necessary. Round to the nearest thousandth.

34.

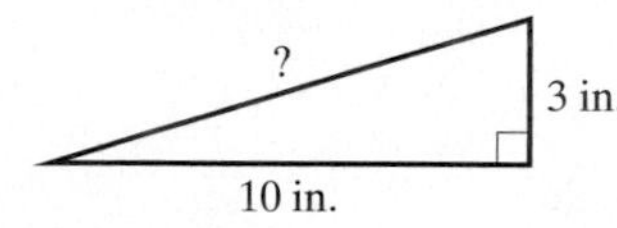

35.

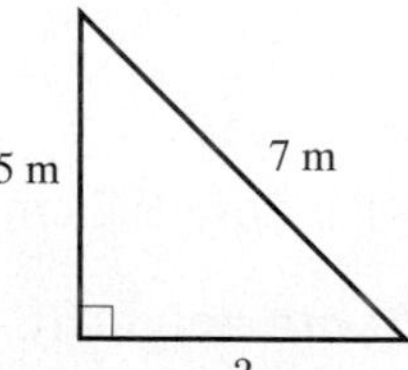

36. A boat travels 12 mi south and then 7 mi east. How far is it from its original starting point? Round to the nearest tenth of a mile.

37. Bo Sigvarson of Malmö, Sweden, constructed a giant paintbrush measuring 20 feet long and weighing 100 pounds. If the original paintbrush model measures $7\frac{1}{2}$ inches long, how many of them would you need to span the length of the giant paintbrush if you placed them end-to-end?

CHAPTER

8

While watching professional baseball on TV, you have probably heard the announcers talking about a player's batting average, fielding percentage, or slugging percentage. These are just a few of the statistics that suggest how valuable a player is. Do you know how these statistics are calculated? To find out, try the Putting Your Skills to Work questions on page 557.

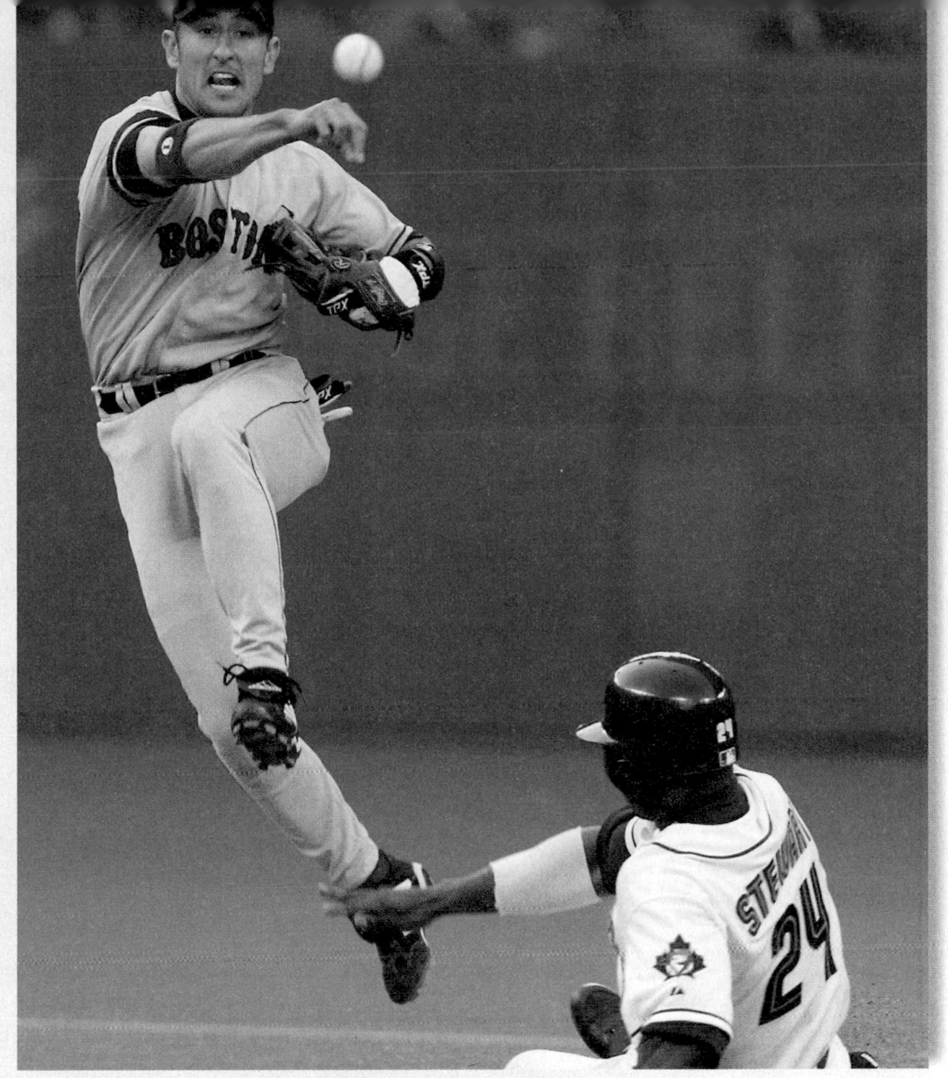

Statistics

8.1 Circle Graphs

Student Learning Objectives

After studying this section, you will be able to:

1. Read a circle graph with numerical values.
2. Read a circle graph with percentage values.

1 Reading a Circle Graph with Numerical Values

Statistics is that branch of mathematics that collects and studies data. Once the data are collected, they must be organized so that the information is easily readable. We use **graphs** to give a visual representation of the data that is easy to read. Graphs appeal to the eye. Their visual nature allows them to communicate information about the complicated relationships among statistical data. For this reason, newspapers often use graphs to help their readers quickly grasp information.

Circle graphs are especially helpful for showing the relationship of parts to a whole. The entire circle represents 100%; the pie-shaped pieces represent the subcategories. The following circle graph divides the 10,000 students at Westline College into five categories.

Distribution of Students at Westline College

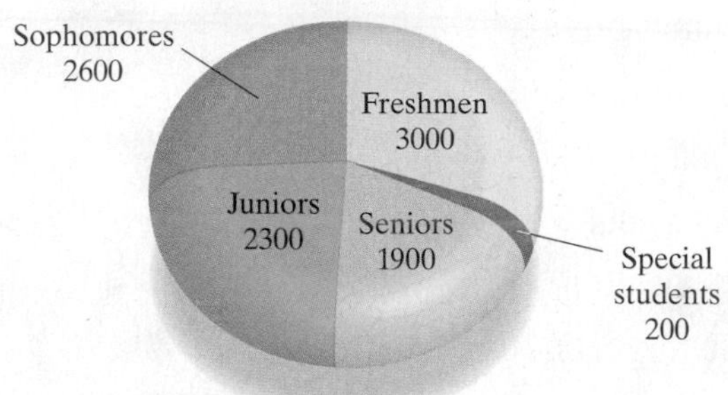

Teaching Tip With the advent of computer-generated graphics, students rarely have to construct circle graphs. Still, they often have to interpret them. This section of the text deals with interpretation only.

EXAMPLE 1 What is the largest category of students?

Solution The largest pie-shaped section of the circle is labeled "Freshmen." Thus the largest category is freshmen students.

Practice Problem 1 What is the smallest category of students?

EXAMPLE 2

(a) How many students are either sophomores or juniors?
(b) What percent of the students are sophomores or juniors?

Solution

(a) There are 2600 sophomores and 2300 juniors. If we add these two numbers, we have $2600 + 2300 = 4900$. Thus we see that there are 4900 students who are either sophomores or juniors.
(b) 4900 out of 10,000 are sophomores or juniors.

$$\frac{4900}{10{,}000} = 0.49 = 49\%$$

Practice Problem 2

(a) How many students are either freshmen or special students?
(b) What percent of the students are freshmen or special students?

***NOTE TO STUDENT:** Fully worked-out solutions to all of the Practice Problems can be found at the back of the text starting at page SP-1*

EXAMPLE 3 What is the ratio of freshmen to seniors?

Solution Number of freshmen ⟶ 3000
Number of seniors ⟶ 1900

Thus $\frac{3000}{1900} = \frac{30}{19}$.

The ratio of freshmen to seniors is $\frac{30}{19}$.

Practice Problem 3 What is the ratio of freshmen to sophomores?

Teaching Tip Emphasize that the order of the ratio is important. The ratio of freshmen to sophomores is not the same as the ratio of sophomores to freshmen. Remind students that the word following *to* is always the denominator of the ratio fraction.

EXAMPLE 4 What is the ratio of seniors to the total number of students?

Solution There are 1900 seniors. We find the total of all the students by adding the number of students in each section of the graph. There are 10,000 students. The ratio of seniors to the total number of students is

$$\frac{1900}{10{,}000} = \frac{19}{100}.$$

Practice Problem 4 What is the ratio of freshmen to the total number of students?

2 Reading a Circle Graph with Percentage Values

Together, the Great Lakes form the largest body of fresh water in the world. The total area of these five lakes is about 94,680 mi^2. The percentage of this total area taken up by each of the Great Lakes is shown in the circle graph below.

Teaching Tip Students sometimes get confused about when they can add percents. You can only add percents that are of the same value. Tell students that since these are all percents of the total area taken up by the five Great Lakes, it is all right to add them together.

Percentage of Area Occupied by Each of the Great Lakes

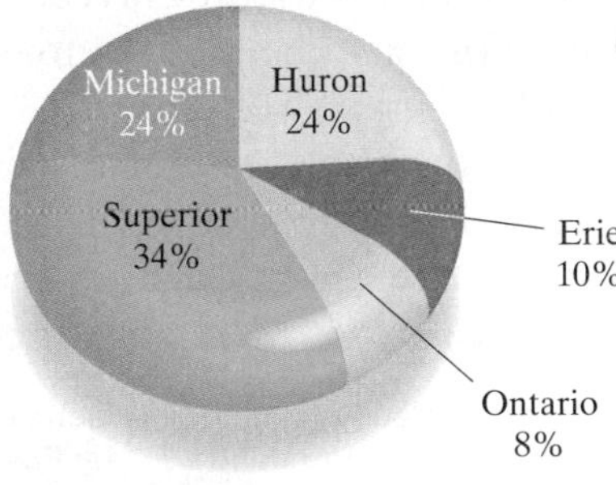

Source: U.S. Department of the Interior

EXAMPLE 5 Which of the Great Lakes occupies the largest area in square miles?

Solution The largest percent corresponds to the biggest area, which is occupied by Lake Superior. Lake Superior has the largest area in square miles.

Practice Problem 5 Which of the Great Lakes occupies the smallest area in square miles?

EXAMPLE 6 What percent of the total area is occupied either by Lake Erie or Lake Ontario?

Solution If we add 10% for Lake Erie and 8% for Lake Ontario, we get

$$10\% + 8\% = 18\%.$$

Thus 18% of the area is occupied by either Lake Erie or Lake Ontario.

NOTE TO STUDENT: Fully worked-out solutions to all of the Practice Problems can be found at the back of the text starting at page SP-1

Practice Problem 6 What percent of the total area is occupied by either Lake Superior or Lake Michigan?

EXAMPLE 7 How many of the total 94,680 mi^2 are occupied by Lake Michigan? Round to the nearest whole number.

Solution Remember that we multiply the percent times the base to obtain the amount. Here 24% of 94,680 is what is occupied by Lake Michigan.

$$(0.24)(94{,}680) = n$$
$$22{,}723.2 = n$$

Rounded to the nearest whole number, 22,723 mi^2 are occupied by Lake Michigan.

Practice Problem 7 How many of the total 94,680 mi^2 are occupied by Lake Superior? Round to the nearest whole number.

Sometimes a circle graph is used to investigate the distribution of one part of a larger group. For example, approximately 782,000 bachelor's degrees were awarded in the United States in 2003 to students majoring in the six most popular subject areas: business, social sciences, education, engineering, health sciences, and biological sciences. The circle graph shows how the degrees in these six subject areas were distributed.

Estimated Percentage of Bachelor's Degrees Earned in 2003 by Field

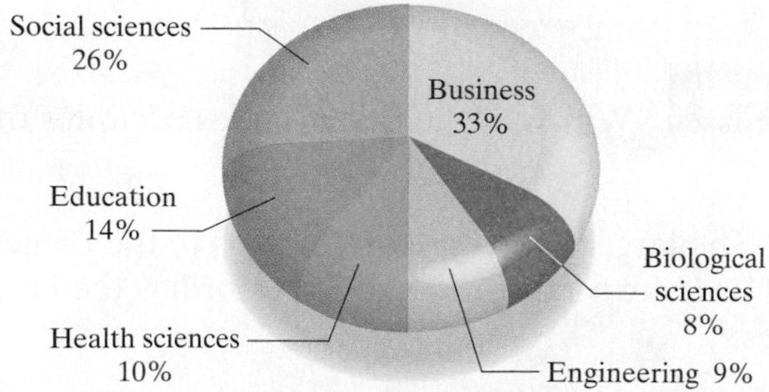

Source: U.S. National Center for Educational Statistics

EXAMPLE 8

(a) What percent of the bachelor's degrees represented in this circle graph are in the fields of health science or business?

(b) Of the approximately 782,000 degrees awarded in these six fields, how many were awarded in the field of engineering?

Solution

(a) We add 10% to 33% to obtain 43%. Thus 43% of the bachelor's degrees represented by this circle are in the fields of health science or business.

(b) We take 9% of the 782,000 people who obtained degrees in these six areas. Thus we have $(0.09)(782{,}000) = 70{,}380$. Approximately 70,380 degrees in engineering were awarded in 2003.

Practice Problem 8

(a) What percent of the bachelor's degrees represented in this circle graph are in the fields of social sciences or education? 26+14= 40

(b) How many bachelor's degrees in biological sciences were awarded in 2003?

8.1 EXERCISES

Student Solutions Manual | CD/Video | PH Math Tutor Center | MathXL®Tutorials on CD | MathXL® | MyMathLab® | Interactmath.com

Verbal and Writing Skills

Suppose that you must create a circle graph for 4000 students who attend Springfield Community College.

1. If 25% of the students live within five miles of the college, how would you determine how many students live within five miles of the college?
You multiply 25% × 4000, which is 0.25 × 4000 = 1000 students.

2. If 45% of the students live more than 8 miles from the college. How would you determine how many students live more than 8 miles from the college? You multiply 45% × 4000, which is 0.45 × 4000 = 1800 students

3. How would you construct a pie slice that describes the students who live within five miles of the college? You would divide the circle into quarters by drawing two perpendicular lines. Shade in one-quarter of the circle. Label this with the title "within five miles = 1000."

4. You plan to create a circle graph with three slices: one for those who live within five miles of the college, one for those who live within five and 8 miles of the college, and one for those who live more than 8 miles from the college. Explain how you would find how many students live more than 8 miles from the college. Subtract the total for answers 1 and 2 from the 4000.
4000 − 2800 = 1200. The answer would be 1200.

Applications

Monthly Budget *The following circle graph displays Bob and Linda McDonald's monthly $2700 family budget. Use the circle graph to answer exercises 5–14.*

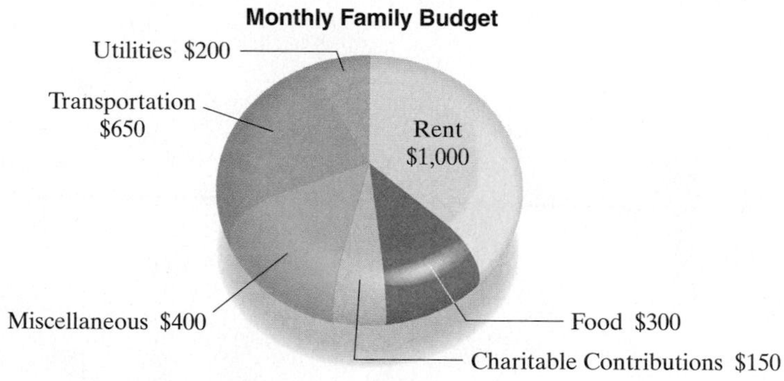

5. What category takes the largest amount of the budget? rent

6. Which two categories take the least amounts of the budget? charitable contributions and utilities

7. How much money is allotted each month for utilities? $200

8. How much money is allotted each month for transportation (this includes car payments, insurance, and gas)? $650

9. How much money in total is allotted each month for transportation and charitable contributions? $650 + $150 = $800

10. How much money is allotted for either food or rent? $300 + $1000 = $1300

11. What is the ratio of money spent for transportation to money spent on utilities?
$\frac{\$650}{\$200} = \frac{13}{4}$

12. What is the ratio of money spent on rent to money spent on miscellaneous items?
$\frac{\$1000}{\$400} = \frac{5}{2}$

13. What is the ratio of money spent on rent to the total amount of the monthly budget? $\frac{\$1000}{\$2700} = \frac{10}{27}$

14. What is the ratio of money spent on food to the total amount of the monthly budget? $\frac{\$300}{\$2700} = \frac{1}{9}$

Age Distribution *In 2002 there were approximately 288 million people in the United States. The following circle graph shows the age distribution of these people. Use the circle graph to answer exercises 15–24.*

Age of Americans in 2002

Older than 80 years 10 million

20 years or younger 81 million

80 years or younger but older than 60 37 million

60 years or younger but older than 40 78 million

40 years or younger but older than 20 82 million

Source: U.S. Census Bureau

15. What age group had the largest number of people? 40 years old or younger but older than 20

16. What age group had the second largest number of people? 20 years old or younger

17. How many people were 60 years old or younger but older than 40? 78 million

18. How many people were 80 years old or younger but older than 60? 37 million

19. How many people were older than 40 years in age? 125 million

20. How many people were older than 20 years in age? 207 million

21. What is the ratio of the number of people who were 40 years old or younger but older than 20 years old to the number of people who were 60 years old or younger but older than 40 years old? $\frac{41}{39}$

22. What is the ratio of the number of people who were 80 years old or younger but older than 60 years old to the number of people who were 60 years old or younger but older than 40 years old? $\frac{37}{78}$

23. What is the ratio of the number of people who are older than 40 years in age to the number of people who were 40 years old or younger? $\frac{125}{163}$

24. What is the ratio of the number of people who are older than 20 years in age to the number of people who were 20 years old or younger? $\frac{23}{9}$

Restaurant Preferences *In a survey, 1010 people were asked which aspect of dining out was most important to them. The results are shown in the circle graph below. Use the graph to answer exercises 25–30. Round all answers to nearest whole number.*

25. What percent of respondents feel either atmosphere or quick service is most important? 11% + 8% = 19%

26. What percent of respondents did not feel that great food was most important? 100% − 56% = 44%

27. Which two categories together make up approximately three-fourths of the circle graph? reasonable prices and great food

28. Of the 1010 respondents, how many responded that either quick service or reasonable prices was most important? 22% + 8% = 30%; 0.30 × 1010 = 303 people

What Consumers Want Most in a Restaurant

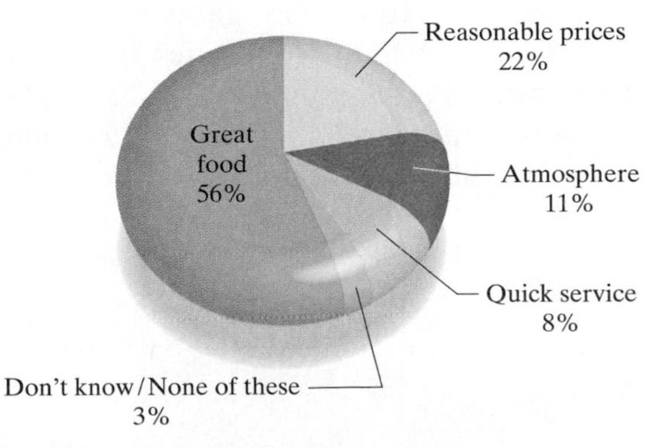

Source: National Restaurant Association

29. Of the 1010 people, how many more people felt that great food was more important than reasonable prices? 343 people

30. Of the 1010 people, how many more people felt that atmosphere was more important than quick service? 30 people

Religious Faith Distribution *Researchers estimate that the religious faith distribution of the 6,300,000,000 people in the world in 2003 was approximately that displayed in the following circle graph. Use the graph to answer exercises 31–38.*

Distribution of Religious Faith in the World Population of 2003

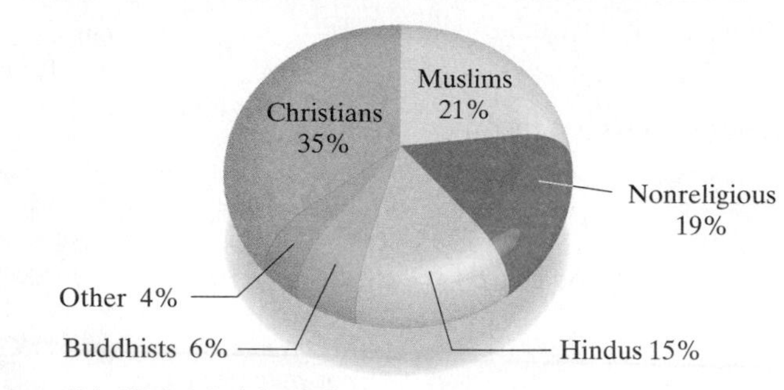

Source: United Nations Statistics Division

31. Approximately how many of the 6,300,000,000 people are Christians? 2,205,000,000

32. Approximately how many of the 6,300,000,000 people are Muslims? 1,323,000,000

33. What percent of the world's population is either Muslim or nonreligious? 40%

34. What percent of the world's population is either Hindu or Buddhist? 21%

35. What percent of the world's population is *not* Muslim? 79%

36. What percent of the world's population is *not* Christian? 65%

37. In 2003, it was estimated that there were 68,000,000 Anglicans in the world. What percent of the 6,300,000,000 people in the world were Anglican? Round to the nearest tenth of a percent. 1.1%

38. The circle graph contains a sector labeled Christians. What percent of the Christian sector represents Anglicans? Round to the nearest tenth of a percent. 3.1%

Cumulative Review

▲ **39.** ***Geometry*** Find the area of a right triangle with base 12 ft and height 20 ft. 120 ft^2

▲ **40.** ***Geometry*** Find the area of a parallelogram with base of 17 in. and height 12 in. 204 in.^2

▲ **41.** ***Paint Coverage*** How many gallons of paint will it take to cover the four sides of a barn with two sides that measure 7 yd by 12 yd and two sides that measure 7 yd by 20 yd? Assume that a gallon of paint covers 28 square yards. 16 gal

▲ **42.** ***Reflector Construction*** A circular reflector has a radius of 8 cm. How many grams of silver will it take to cover the reflector if each gram will cover 64 sq cm? Assume that the reflector is covered on one side only. (Use $\pi \approx 3.14$.) Round to the nearest whole number. about 3 g

8.2 Bar Graphs and Line Graphs

1 Reading and Interpreting a Bar Graph

Bar graphs are helpful for seeing changes over a period of time. Bar graphs or line graphs are especially helpful when the same type of data is repeatedly studied. The following bar graph shows the approximate population of California from 1940 to 2000.

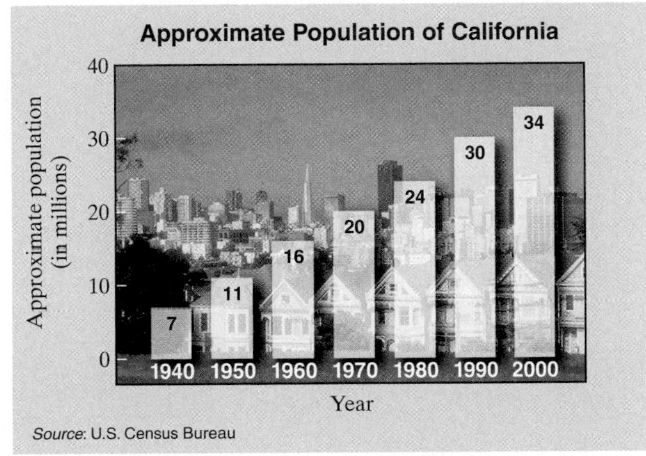

Student Learning Objectives

After studying this section, you will be able to:

1. **Read and interpret a bar graph.**
2. **Read and interpret a double-bar graph.**
3. **Read and interpret a line graph.**
4. **Read and interpret a comparison line graph.**

EXAMPLE 1 What was the approximate population of California in 2000?

Solution The bar for 2000 rises to 34. This represents 34 million; thus the approximate population was 34,000,000.

Practice Problem 1 What was the approximate population of California in 1980?

Teaching Tip Explain to students that the scale of these bar graphs is not always as simple as the one in this example. Some will have a scale of 12 million, 12.2 million, 12.4 million, etc. Students will encounter such examples as they read *USA Today*, *Time*, or *Newsweek*.

EXAMPLE 2 What was the increase in population from 1980 to 1990?

Solution The bar for 1980 rises to 24. Thus the approximate population is 24,000,000. The bar for 1990 rises to 30. Thus the approximate population is 30,000,000. To find the increase in population from 1980 to 1990, we subtract.

$$30{,}000{,}000 - 24{,}000{,}000 = 6{,}000{,}000$$

Practice Problem 2 What was the increase in population from 1940 to 1960?

NOTE TO STUDENT: Fully worked-out solutions to all of the Practice Problems can be found at the back of the text starting at page SP-1

2 Reading and Interpreting a Double-Bar Graph

Double-bar graphs are useful for making comparisons. For example, when a company is analyzing its sales, it may want to compare different years or different quarters. The following double-bar graph illustrates the sales of new cars at a Ford dealership for two different years, 2003 and 2004. The sales are recorded for each quarter of the year.

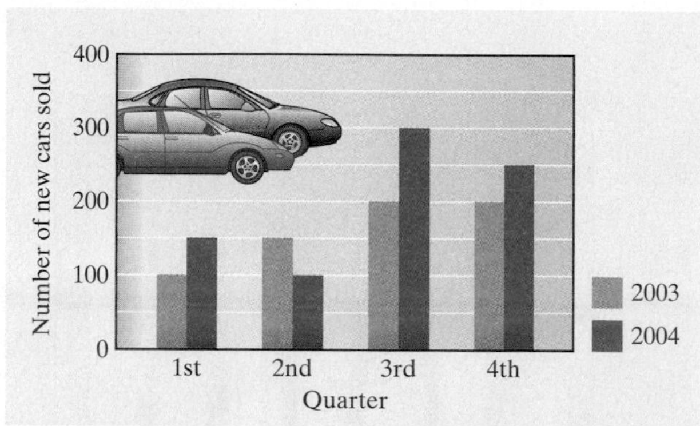

EXAMPLE 3 How many cars were sold in the second quarter of 2003?

Solution The bar rises to 150 for the second quarter of 2003. Therefore, 150 cars were sold.

***NOTE TO STUDENT:** Fully worked-out solutions to all of the Practice Problems can be found at the back of the text starting at page SP-1*

Practice Problem 3 How many cars were sold in the fourth quarter of 2004?

EXAMPLE 4 How many more cars were sold in the third quarter of 2004 than in the third quarter of 2003?

Solution From the double-bar graph, we see that 300 cars were sold in the third quarter of 2004 and that 200 cars were sold in the third quarter of 2003.

$$\begin{array}{r} 300 \\ -200 \\ \hline 100 \end{array}$$

Thus, 100 more cars were sold.

Practice Problem 4 How many fewer cars were sold in the second quarter of 2004 than in the second quarter of 2003?

3 Reading and Interpreting a Line Graph

A **line graph** is useful for showing trends over a period of time. In a line graph only a few points are actually plotted from measured values. The points are then connected by straight lines to show a trend. The intervening values between points may not lie exactly on the line. The following line graph shows the number of customers per month coming into a restaurant in a vacation community.

Teaching Tip Sometimes students try to interpret in between the points on a line graph. In the line graph for Example 5, they may say, "The number of customers was decreasing from the middle of March to the end of March." This, of course, is *not* valid. The graph is only valid for the one point plotted in March. Remind students that the purpose of the lines between the plotted points is to show the *direction of trends*, not to show intermediate values between the points.

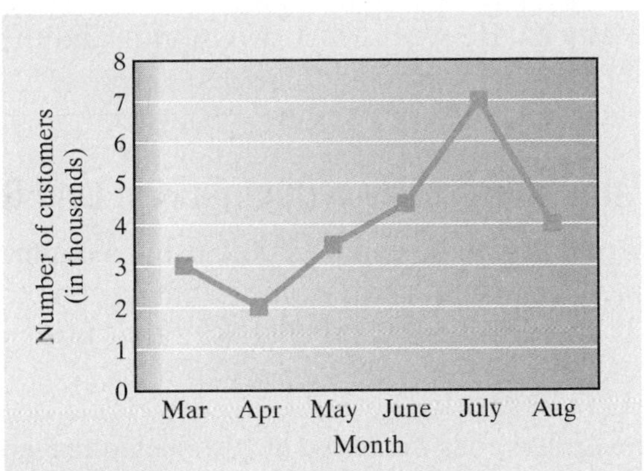

EXAMPLE 5 In which month did the smallest number of customers come into the restaurant?

Solution The lowest point on the graph occurs for the month of April. Thus the fewest number of customers came in April.

Practice Problem 5 In which month did the greatest number of customers come into the restaurant?

EXAMPLE 6

(a) Approximately how many customers came into the restaurant during the month of June?

(b) From May to June, did the number of customers increase or decrease?

Solution

(a) Notice that the dot is halfway between 4 and 5. This represents a value halfway between 4000 and 5000 customers. Thus we would estimate that 4500 customers came during the month of June.

(b) From May to June the line goes up, so the number of customers increased.

Practice Problem 6

(a) Approximately how many customers came into the restaurant during the month of May?

(b) From March to April, did the number of customers increase or decrease?

EXAMPLE 7 Between what two months was the increase in the number of customers the largest?

Solution The line from June to July goes upward at the steepest angle. This represents the largest increase. (You can check this by reading the numbers from the left axis.) Thus the greatest increase in attendance was between June and July.

Practice Problem 7 Between what two months did the biggest decrease occur?

NOTE TO STUDENT:* *Fully worked-out solutions to all of the Practice Problems can be found at the back of the text starting at page SP-1

4 Reading and Interpreting a Comparison Line Graph

Two or more sets of data can be compared by using a **comparison line graph.** A comparison line graph shows two or more line graphs together. A different style for each line distinguishes them. Note that using a blue line and a red line in the following graph makes it easy to read.

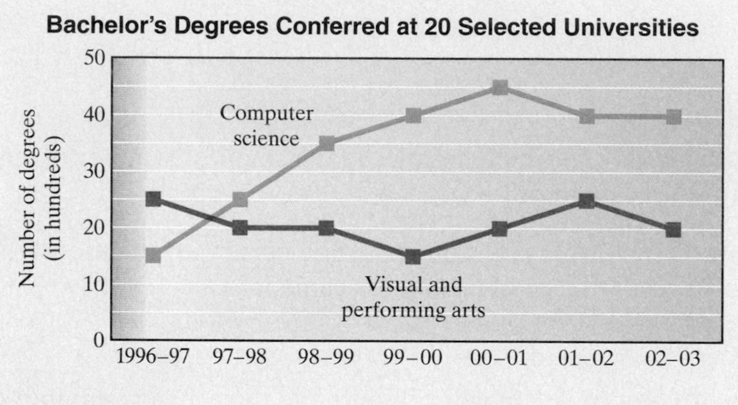

Teaching Tip When discussing the comparison line graph example, ask students if they remember seeing a similar graph in a newspaper or magazine in the last year. Students often have a wealth of interesting examples that they remember. Class discussion of this topic helps them to see how relevant this type of graph is to everyday life.

EXAMPLE 8 How many bachelor's degrees in computer science were awarded in the academic year 2002–2003?

Solution Because the dot corresponding to 2002–2003 is at 40 and the scale is in hundreds, we have $40 \times 100 = 4000$. Thus 4000 degrees were awarded in computer science in 2002–2003.

Practice Problem 8 How many bachelor's degrees in visual and performing arts were awarded in the academic year 2002–2003?

EXAMPLE 9 In what academic year were more degrees awarded in the visual and performing arts than in computer science?

Solution The only year when more bachelor's degrees were awarded in the visual and performing arts was the academic year 1996–1997.

Practice Problem 9 What was the first academic year in which more degrees were awarded in computer science than in the visual and performing arts?

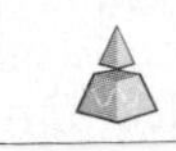

Student Solutions Manual | CD/Video | PH Math Tutor Center | MathXL®Tutorials on CD | MathXL® | MyMathLab® | Interactmath.com

Applications

Texas Population *The following bar graph shows the approximate population of Texas from 1940 to 2000. Use the graph to answer exercises 1–6.*

1. What was the approximate population in 2000?
21 million people

2. What was the approximate population in 1950?
8 million people

3. What was the approximate population in 1980?
14 million people

4. What was the approximate population in 1990?
17 million people

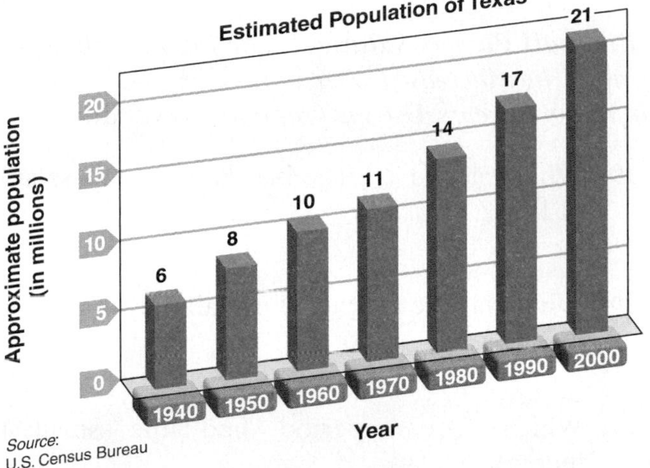

5. Between what years did the population of Texas increase by the smallest amount?
1960–1970

6. Between what years did the population of Texas increase by the largest amount?
1990–2000

Production of Coal and Oil *The following double-bar graph displays the production levels of oil and coal in the United States. Because oil is measured in barrels and coal is measured in tons, energy officials compare the production using British thermal units (Btu), which is a measure of how much heat is produced from the oil or coal. Use the following bar graph to answer exercises 7–18.*

7. How much coal was produced in 1990?
22 quadrillion Btu

8. How much oil was produced in 1995?
14 quadrillion Btu

9. What year had the highest production of oil?
1970

10. In what years was the production of oil and coal the same in terms of Btu?
in both 1980 and 1985

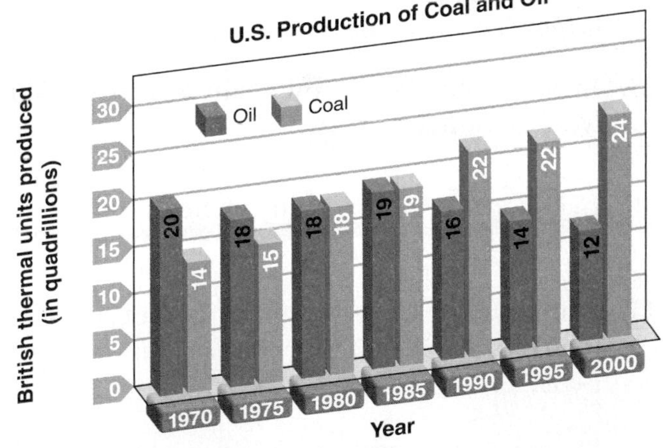

11. How much more oil was produced in 1975 than in 1995? 4 quadrillion Btu

12. How much more coal was produced in 2000 than in 1980? 6 quadrillion Btu

13. How much more coal was produced than oil in 1990? 6 quadrillion Btu

14. How much more oil was produced than coal in 1975? 3 quadrillion Btu

15. During what five-year period(s) did the biggest increase in coal production occur? from 1975 to 1980 and from 1985 to 1990

16. During what five year period did the biggest decrease in oil production occur? from 1985 to 1990

17. If the production of coal increases at the same rate from 2000 to 2020 as it did from 1980 to 2000, how much coal will be produced in 2020? 30 quadrillion Btu

18. If the production of oil decreases at the same rate from 2000 to 2015 as it did from 1985 to 2000, how much oil will be produced in 2015? 5 quadrillion Btu

Baseball Players Salaries *The following line graph shows how the average major league baseball player's salary has increased over a 10-year period.* Notice that we have only listed the average salary for odd-numbered years. Use the graph to answer exercises 19–24.*

19. What was the average baseball player's salary in 1993? $1,000,000

20. What was the average baseball player's salary in 2003? $2,400,000

21. Which 2-year period had the smallest increase? 1993 to 1995

22. Which 2-year period had the largest increase? 1999 to 2001

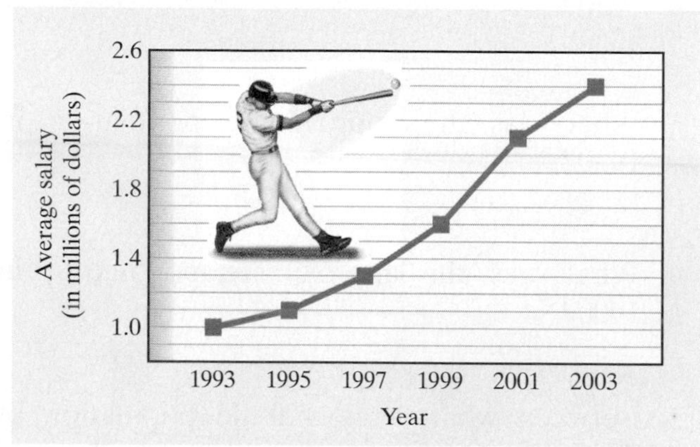

*Figures have been rounded.
(*Source:* Major League Baseball Players Association Web site at www.mlb.com)

23. Compare the average salary in 2003 to the average salary in 2001. By how much did the average salary increase from 2001 to 2003? $300,000

24. If the average salary continues to increase by the same amount as from 2001 to 2003, what will the average salary be in 2015? $4,200,000

Springfield Rainfall *The following comparison line graph indicates the rainfall for the last six months of two different years in Springfield. Use the graph to answer exercises 25–30.*

25. In September 2001, how many inches of rain were recorded? 2.5 in.

26. In October 2000 how many inches of rain were recorded? 2.5 in.

27. During what months was the rainfall of 2001 less than the rainfall of 2000? October, November, and December

28. During what months was the rainfall of 2001 greater than the rainfall of 2000? July, August, and September

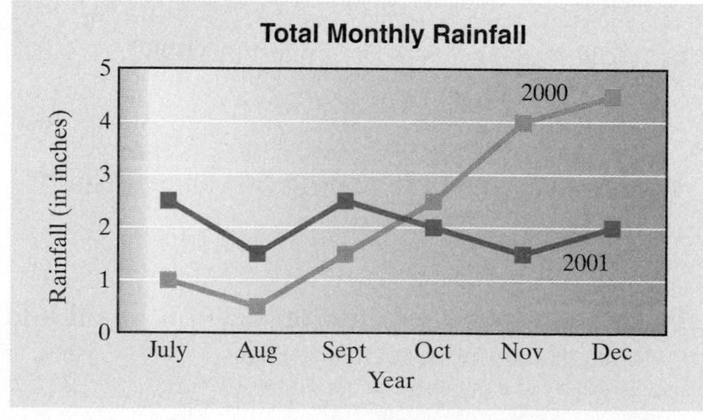

29. How many more inches of rain fell in November 2000 than in October 2000? 1.5 in.

30. How many more inches of rain fell in September 2000 than in August 2000? 1 in.

To Think About

Pizza Sales *The following table shows the number of pizzas sold at a pizza parlor near a college campus.*

Number of Pizzas Sold by Alfredo's Pizza Parlor	300	400	100	200	600
Month of the Year	Jan.	Feb.	Mar.	Apr.	May

31. Use the graph paper below to construct a line graph of the information in the table. Let the vertical scale (height) represent the number of pizzas sold. Let the horizontal scale (width) represent the month.

32. Is the biggest change on the graph an increase or a decrease in the number of pizzas sold per month? increase

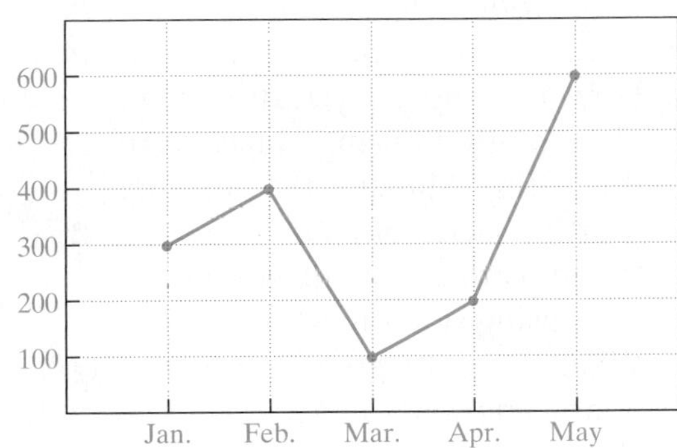

Cumulative Review

Do each calculation in the proper order.

33. $8 \times 0.5 + 2.5 - 5(0.5)$ 4

34. $(5 + 6)^2 - 18 \div 9 \times 3$ 115

35. $\frac{1}{5} + \left(\frac{1}{5} - \frac{1}{6}\right) \times \frac{2}{3}$ $\frac{2}{9}$

36. $150 \div 5 \div 6 \times 3 + 2 - 4$ 13

How Am I Doing? Sections 8.1–8.2

1. 14%

2. 2 people

3. 39%

4. 3,278,910 households

5. 66,671,170 households

6. 300 housing starts

7. 450 housing starts

8. during the second quarter of 2002

9. during the third quarter of 2003

10. 300

11. 50

12. Aug and Dec

13. Dec

14. Nov

15. (a) 20,000 sets

(b) 35,000 sets

How are you doing with your homework assignments in Sections 8.1 to 8.2? Do you feel you have mastered the material so far? Do you understand the concepts you have covered? Before you go further in the textbook, take some time to do each of the following problems.

8.1 *The following circle graph displays the sizes of U.S. households in 2002. There were a total of 109,297,000 households in 2002.*

Households by Size

2 people 35%
One person 26%
6 or more people 3%
5 people 6%
4 people 14%
3 people 16%

Source: www.census.gov

1. What percentage of households in 2002 had 4 people?

2. What household size comprises the largest percent of all households?

3. What percent of households had 3 or more people?

4. How many of the 109,297,000 households had 6 or more people?

5. How many households had 1 or 2 people?

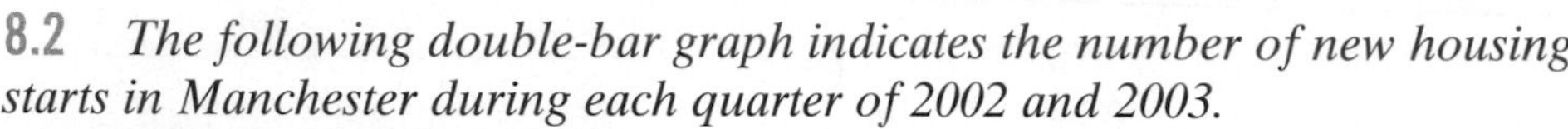

8.2 *The following double-bar graph indicates the number of new housing starts in Manchester during each quarter of 2002 and 2003.*

6. How many housing starts were in Manchester in the first quarter of 2003?

7. How many housing starts were there in Manchester in the fourth quarter of 2002?

8. When were the smallest number of housing starts in Manchester?

9. When were the greatest number of housing starts in Manchester?

10. How many more housing starts were there in the second quarter of 2003 than in the second quarter of 2002?

11. How many fewer housing starts were there in the fourth quarter of 2003 than in the fourth quarter of 2002?

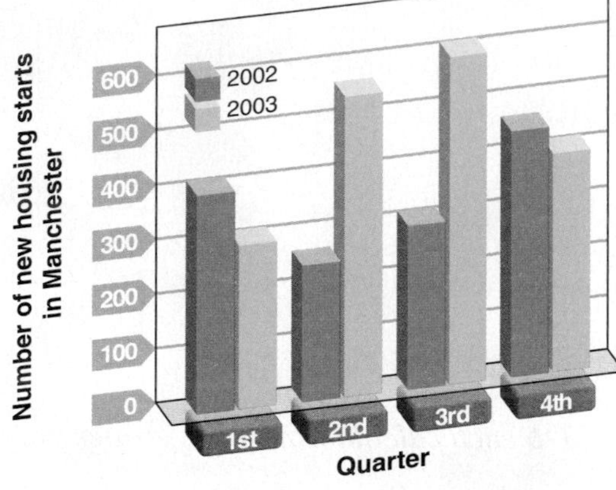

The line graph indicates sales and production of color television sets by a major manufacturer during the specified months.

12. During what months was the production of television sets the lowest?

13. During what month was the sales of television sets the highest?

14. What was the first month in which the production of television sets was lower than the sales of television sets?

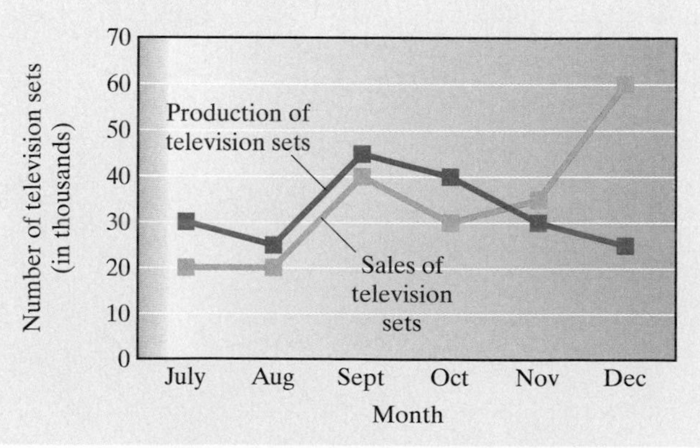

15. **(a)** How many television sets were sold in August? **(b)** November?

Now turn to page SA-15 for the answer to each of these problems. Each answer also includes a reference to the objective in which the problem is first taught. If you missed any of these problems, you should stop and review the Examples and Practice Problems in the referenced objective. A little review now will help you master the material in the upcoming sections of the text.

8.3 HISTOGRAMS

Student Learning Objectives

After studying this section, you will be able to:

1. Understand and interpret a histogram.
2. Construct a histogram from raw data.

1 Understanding and Interpreting a Histogram

In business or in higher education you are often asked to take data and organize them in some way. This section shows you the technique for making a *histogram*—a type of bar graph.

Suppose that a mathematics professor announced the results of a class test. The 40 students in the class scored between 50 and 99 on the test. The results are displayed on the following chart.

Scores on the Test	Class Frequency
50–59	4
60–69	6
70–79	16
80–89	8
90–99	6

The results in the table can be organized in a special type of bar graph known as a **histogram.** In a histogram the width of each bar is the same. The width represents the range of scores on the test. This is called a **class interval.** The height of each bar gives the class frequency of each class interval. The **class frequency** is the number of times a score occurs in a particular class interval. Be sure to notice that the bars touch each other. This is a main difference between the bar graph and the histogram. Use the histogram below to do Examples 1 and 2.

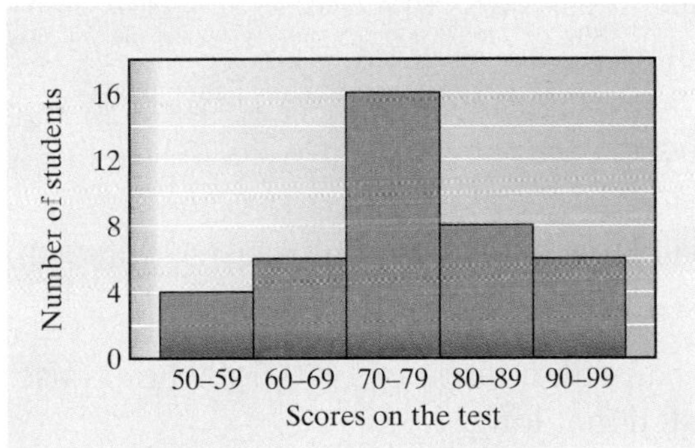

Teaching Tip Students will sometimes encounter bar graphs that have an abbreviated scale on the side. This indicates that the bar graph has been cut off. Remind them that the effect of the histogram may be exaggerated by a cut-off scale. This applies to all the types of graphs covered in Section 8.2 as well. If the left-hand scale does not start at zero at the bottom of the graph, the graph is said to have a cut-off scale.

EXAMPLE 1 How many students scored a B on the test if the professor considers a test score of 80–89 a B?

Solution Since the 80–89 bar rises to a height of 8, eight students scored a B on the test.

Practice Problem 1 How many students scored a D on the test if the professor considers a test score of 60–69 a D?

NOTE TO STUDENT: Fully worked-out solutions to all of the Practice Problems can be found at the back of the text starting at page SP-1

EXAMPLE 2 How many students scored less than 80 on the test?

Solution From the histogram, we see that there are three different bar heights to be included. Four tests were scored 50–59, six tests were scored 60–69, and 16 tests were scored 70–79. When we combine $4 + 6 + 16 = 26$, we can see that 26 students scored less than 80 on the test.

Practice Problem 2 How many students scored greater than 69 on the test?

NOTE TO STUDENT: Fully worked-out solutions to all of the Practice Problems can be found at the back of the text starting at page SP-1

The following histogram tells us about the length of life of 110 new light bulbs tested at a research center. The number of hours the bulbs lasted is indicated on the horizontal scale. The frequency of bulbs lasting that long is indicated on the vertical scale. Use this histogram to do Examples 3 and 4.

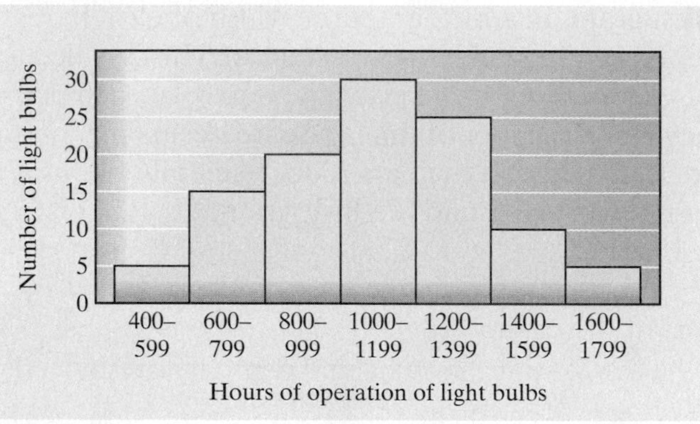

Teaching Tip When reviewing Examples 3 and 4 with the class, bring up the topic of quality control and the use of graphs to track quality maintenance and improvements in manufactured products.

EXAMPLE 3 How many light bulbs lasted between 1400 and 1599 hours?

Solution The bar with a range of 1400–1599 hours rises to 10. Thus 10 light bulbs lasted that long.

Practice Problem 3 How many light bulbs lasted between 800 and 999 hours?

EXAMPLE 4 How many light bulbs lasted less than 1000 hours?

Solution We see that there are three different bar heights to be included. Five bulbs lasted 400–599 hours, 15 bulbs lasted 600–799 hours, and 20 bulbs lasted 800–999 hours. We add $5 + 15 + 20 = 40$. Thus 40 light bulbs lasted less than 1000 hours.

Practice Problem 4 How many light bulbs lasted more than 1199 hours?

2 Constructing a Histogram from Raw Data

To construct a histogram, we start with *raw data,* data that have not yet been organized or interpreted. We perform the following steps.

1. Select data class intervals of equal width for the data.
2. Make a table with class intervals and a *tally* (count) of how many numbers occur in each interval. Add up the tally to find the class frequency for each class interval.
3. Draw the histogram.

First we will practice making the table. Later we will use the table to draw the histogram.

EXAMPLE 5 Each number in the following chart represents the number of kilowatt-hours of electricity used in a home during a one-month period. Create a set of class intervals for this data and then determine the frequency of each class interval.

770	520	850	900	1100
1200	1150	730	680	900
1160	590	670	1230	980

Solution

1. We select class intervals of equal width for the data. We choose intervals of 200. We might have chosen smaller or larger intervals, but we choose 200 because it gives us a convenient number of intervals to work with, as we will see.
2. We make a table. We write down the class intervals, then count (tally) how many numbers occur within each interval. Then we write the total. This is the class frequency.

Kilowatt-Hours Used (Class Interval)	Tally	Frequency
500–699	||||	4
700–899	|||	3
900–1099	|||	3
1100–1299	~~||||~~	5

Practice Problem 5 Each number in the following chart represents the weight in pounds of a new car.

2250	1760	2000	2100	1900
1640	1820	2300	2210	2390
2150	1930	2060	2350	1890

Teaching Tip You may find that students will ask for an additional example after you have reviewed the procedure in Examples 5 and 6. You may use the following: Thirty-one students in a math class recorded their height to the nearest inch. The following results were turned in: 55, 60, 66, 72, 79, 52, 75, 70, 60, 59, 54, 65, 71, 76, 67, 64, 56, 63, 68, 69, 60, 60, 65, 64, 63, 69, 68, 67, 71, 66, 70. These raw data can be arranged as follows:

Height Range in Inches	Frequency
48–53	1
54–59	4
60–65	10
66–71	12
72–77	3
78–83	1

Complete the following table to determine the frequency of each class interval for the preceding data.

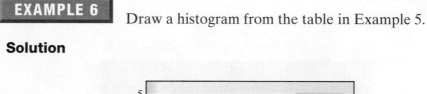

Weight in Pounds (Class Interval)	Tally	Frequency
1600–1799		
1800–1999		
2000–2199		
2200–2399		

***NOTE TO STUDENT:** Fully worked-out solutions to all of the Practice Problems can be found at the back of the text starting at page SP-1*

One of the purposes of a histogram is to give you a visual sense of how the data is distributed. For example, if you look at the raw data of Example 5 you may be left with the sense that the homes use very different amounts of electricity during a month. For most of us, looking at the raw data does not help us understand the situation. However, once we construct a histogram such as the one in Example 6, we are able to see patterns and trends of electricity use.

Draw a histogram from the table in Example 5.

Solution

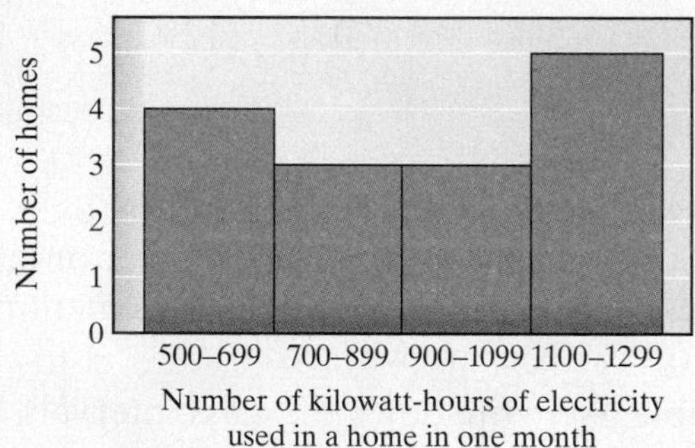

Practice Problem 6 Draw a histogram using the data from the table in Practice Problem 5.

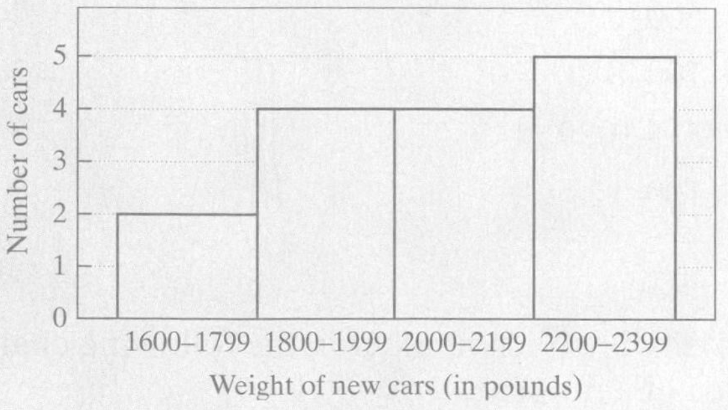

Note: Usually it is desirable for the class intervals to be of equal size. However, sometimes data is collected such that this is not the case. We will see this situation in Example 7. Here government data was collected with unequal class intervals.

EXAMPLE 7 Draw a histogram for the following table of data showing the number of people in the United States as of July 1, 2001, in each of five age categories.

Age Category	Number of People in the United States
19 or younger	80,701,000
20–34	59,288,000
35–54	84,207,000
55–64	25,308,000
65 or older	35,291,000

Source: U.S. Census Bureau

Solution

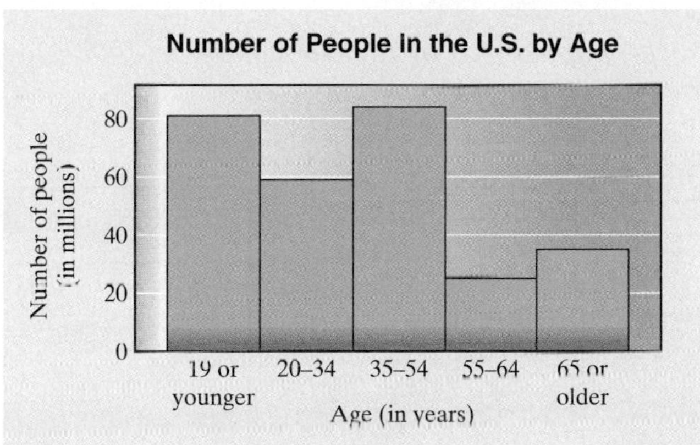

Practice Problem 7 Based on the preceding histogram, between what two age categories is there the greatest difference in population in the United States?

8.3 Exercises

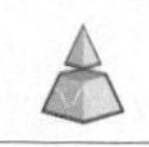

Student Solutions Manual | CD/Video | PH Math Tutor Center | MathXL®Tutorials on CD | MathXL® | MyMathLab® | Interactmath.com

Verbal and Writing Skills

1. Describe two main differences between the bar graph and the histogram.

The horizontal label for each item in a bar graph is usually a single number or a word title. For the histogram it is a class interval. The vertical bars have a space between them in the bar graph. For the histogram the vertical bars join each other.

2. Suppose you had test items 22, 24, 33, 44, 55, 66, 38, 48, and 60. If you were going to have a histogram with three class intervals, explain how you would pick the intervals.

You want the intervals to be of equal class width. The range from 22 to 66 is 44. Dividing 44 by 3, you get 15 to the nearest whole number. So you make each interval of width 15. Thus a good choice would be class widths of 33–37, 38–53, 54–69.

3. Explain in your own words what is meant by class frequency.

A class frequency is the number of times a score occurs in a particular class interval.

4. Jason made a histogram and it had the following class intervals: 300–400, 400–500, 500–600, 600–700. Explain why that is not a good choice of class intervals.

In the intervals that Jason made, the number 400 would be counted twice. Similarly, 500 and 600 would be counted twice. He should rather use 300–399, 400–499, 500–599, and 600–699.

Applications

City Population *The number of U.S. cities with populations of 75,000 or more in 2002 is depicted in the following histogram. Use the histogram to answer exercises 5–12.*

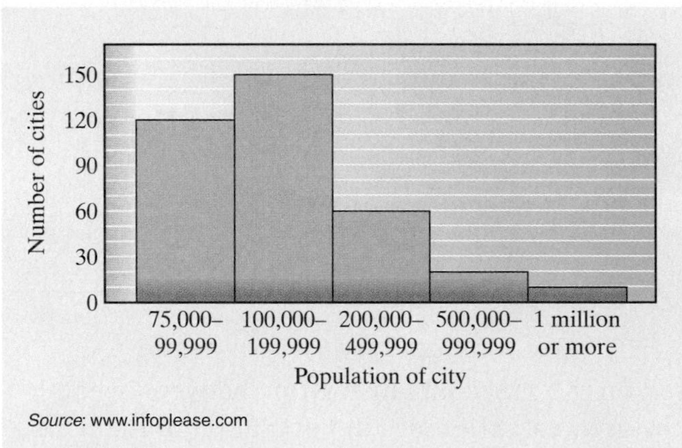

5. How many U.S. cities have a population of 100,000–199,999?

150 cities

6. How many U.S. cities have a population of 200,000–499,999?

60 cities

7. How many U.S. cities have a population of 1 million or more?

10 cities

8. How many U.S. cities have a population of 75,000–99,999?

120 cities

9. How many U.S. cities have a population of 500,000 or more?

$20 + 10 = 30$ cities

10. How many U.S. cities have a population of 75,000 or more?

$120 + 150 + 60 + 20 + 10 = 360$ cities

11. How many U.S. cities have between 75,000 and 199,999 people?

$120 + 150 = 270$ cities

12. How many U.S. cities have between 100,000 and 999,999 people?

$150 + 60 + 20 = 230$ cities

Book Sales *A large company comprising of three bookstores studied its yearly report to find out its customers' spending habits. The company sold a total of 70,000 books. The following histogram indicates the number of books sold in certain price ranges. Use the histogram to answer exercises 13–22.*

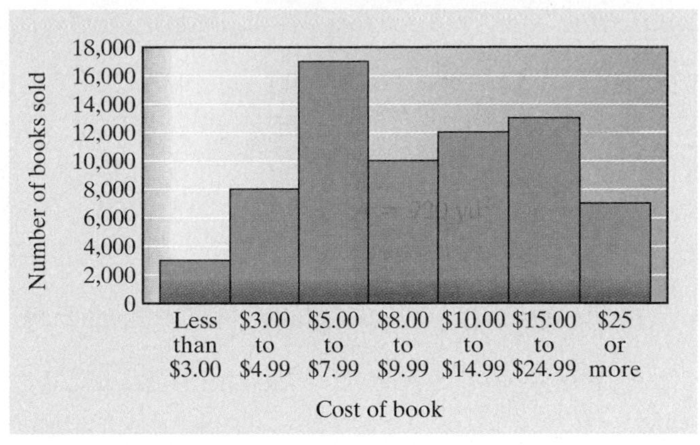

13. How many books priced at $3.00 to $4.99 were sold? 8000

14. How many books priced at $25.00 or more were sold? 7000

15. Which price category of books did the bookstore sell the most of? books costing $5.00–$7.99

16. What price category of books did the bookstore sell the least of? books costing less than $3.00

17. How many books priced at less than $8.00 were sold? 17,000 + 8000 + 3000 = 28,000 books

18. How many books priced at more than $9.99 were sold?
12,000 + 13,000 + 7000 = 32,000 books

19. How many books priced between $5.00 and $24.99 were sold?
17,000 + 10,000 + 12,000 + 13,000 = 52,000 books

20. How many books priced between $3.00 and $9.99 were sold?
8000 + 17,000 + 10,000 = 35,000 books

21. What percent of the 70,000 books sold were over $14.99?
$\frac{20{,}000}{70{,}000} = \frac{2}{7} \approx 28.6\%$

22. What percent of the 70,000 books sold were under $8.00? $\frac{28{,}000}{70{,}000} = \frac{2}{5} = 40\%$

Boston Temperature *The numbers in the following chart are the daily high temperatures in degrees Fahrenheit in Boston during February. In exercises 23–30, determine the frequencies of the class intervals for this data.*

23°	26°	30°	18°	42°	17°	19°
51°	42°	38°	36°	12°	18°	14°
20°	24°	26°	30°	18°	17°	16°
35°	38°	40°	33°	19°	22°	26°

	Temperature (Class Interval)	Tally	Frequency
23.	12°–16°	|||	3
24.	17°–21°	~~||||~~ |||	8
25.	22°–26°	~~||||~~ |	6
26.	27°–31°	||	2
27.	32°–36°	|||	3
28.	37°–41°	|||	3
29.	42°–46°	||	2
30.	47°–51°	|	1

31. Construct a histogram using the table prepared in exercises 19–26.

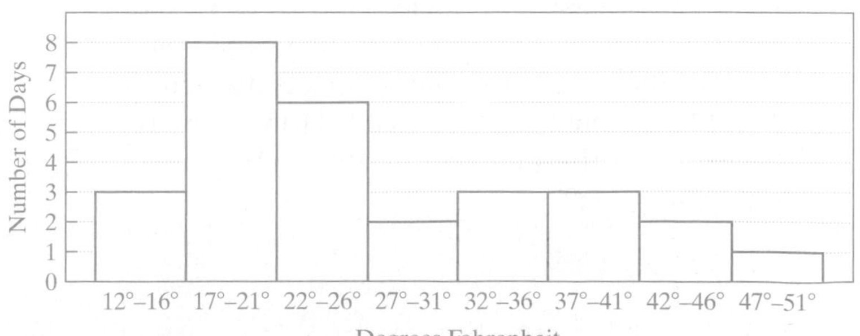

32. How many days in February was the temperature in Boston greater than 36°? 6 days

33. How many days in February was the temperature in Boston less than 27°? 17 days

Prescription Costs *Each number in the following chart is the cost of a prescription purchased by the Lin family this year. In exercises 34–41, determine the frequencies of the class intervals for this data.*

\$28.50	\$16.00	\$32.90	\$46.20	\$ 9.85
\$27.30	\$16.00	\$41.95	\$36.00	\$24.20
\$ 7.65	\$ 8.95	\$ 4.50	\$11.35	\$ 7.75
\$12.30	\$21.85	\$46.20	\$15.50	\$ 4.50

Purchase Price (Class Interval)	Tally	Frequency	Purchase Price (Class Interval)	Tally	Frequency
34. \$ 4.00–\$ 9.99	𝍸 I	6	35. \$10.00–\$15.99	III	3
36. \$16.00–\$21.99	III	3	37. \$22.00–\$27.99	II	2
38. \$28.00–\$33.99	II	2	39. \$34.00–\$39.99	I	1
40. \$40.00–\$45.99	I	1	41. \$46.00–\$51.99	II	2

42. Construct a histogram using the table constructed in exercises 34–41.

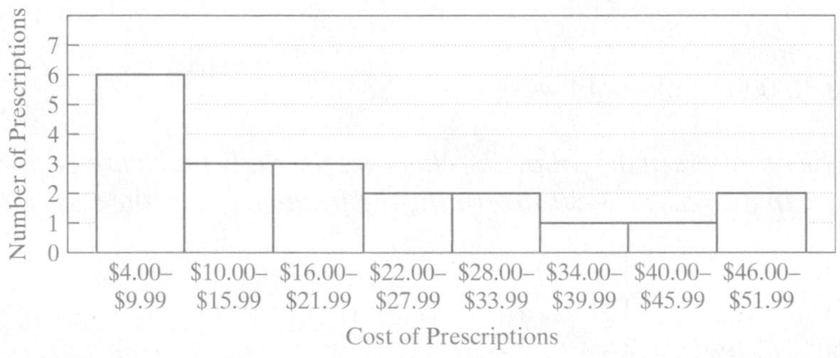

43. How many prescriptions cost less than \$22.00? 12 prescriptions

44. How many prescriptions cost more than \$33.99? 4 prescriptions

Cumulative Review

45. Solve for n. $\frac{126}{n} = \frac{36}{17}$ $n = 59.5$

46. Solve for n. $\frac{n}{18} = \frac{3.5}{9}$ $n = 7$

47. ***Volkswagen Sales*** At North Shore Volkswagen, 5 of the last 12 vehicles sold were Passat station wagons. By studying past years' sales reports, the sales manager expects to sell a total of 36 vehicles next month. How many of these should the manager expect to be Passat station wagons? 15

48. ***Snowfall on Mount Washington*** Tim and Judy Newitt worked as scientists on Mount Washington last year. They found that every 23 in. of snow corresponded to 2 in. of water. During the month of January they measured 150 in. of snow at the mountain weather observatory. How many inches of water does this correspond to? Round to the nearest tenth of an inch.
13.0 in.

8.4 Mean, Median, and Mode

1 Finding the Mean of a Set of Numbers

We often want to know the "middle value" of a group of numbers. In this section we learn that, in statistics, there is more than one way of describing this middle value: there is the *mean* of the group of numbers, there is the *median* of the group of numbers, and in most cases there is a *mode* of the group of numbers. In some situations it's more helpful to look at the mean; in others it's more helpful to look at the median, and in yet others the mode. We'll learn to tell which situations lend themselves to one or the other.

The **mean** of a set of values is the sum of the values divided by the number of values. The mean is often called the **average.**

The mean value is often rounded to a certain decimal-place accuracy.

EXAMPLE 1 Carl recorded the miles per gallon achieved by his car for the last two months. His results were as follows:

Week	1	2	3	4	5	6	7	8
Miles per Gallon	26	24	28	29	27	25	24	23

What is the mean miles-per-gallon figure for the last eight weeks? Round to the nearest whole number.

Solution

$$\begin{array}{l} \text{Sum of values} \longrightarrow \\ \text{Number of values} \longrightarrow \end{array} \quad \frac{26 + 24 + 28 + 29 + 27 + 25 + 24 + 23}{8}$$

$$= \frac{206}{8} = 25.75 \approx 26 \text{ rounded to the nearest whole number}$$

The mean miles-per-gallon rating is 26.

Practice Problem 1 Bob and Wally kept records of their phone bills for the last six months. Their bills were \$39.20, \$43.50, \$81.90, \$34.20, \$51.70, and \$48.10. Find the mean monthly bill. Round to the nearest cent.

2 Finding the Median of a Set of Numbers

If a set of numbers is arranged in order from smallest to largest, the **median** is that value that has the same number of values above it as below it.

If the numbers are not arranged in order, then the first step for finding the median is to put the numbers in order.

Student Learning Objectives

After studying this section, you will be able to:

1. **Find the mean of a set of numbers.**
2. **Find the median of a set of numbers.**
3. **Find the mode of a set of numbers.**

Teaching Tip In cases like Example 1, discuss with students what happens when we try to write answers like "The average is 23.45872 miles per gallon." Remind them that the accuracy of the answer is dependent on the accuracy of the data that are being averaged. Usually, we round our answer to the same number of significant digits that are in the items being averaged. In Example 2 we have two-digit accuracy in the miles-per-gallon ratings, so it is logical to round our answer to two significant digits.

***NOTE TO STUDENT:** Fully worked-out solutions to all of the Practice Problems can be found at the back of the text starting at page SP-1*

EXAMPLE 2 Find the median value of the following costs for a microwave oven: \$100, \$60, \$120, \$200, \$190, \$120, \$320, \$290, \$180.

Solution We must arrange the numbers in order from smallest to largest (or largest to smallest).

$$\underbrace{\$60, \$100, \$120, \$120}_{\text{four numbers}} \quad \underset{\text{middle number}}{\underset{\uparrow}{\$180}} \quad \underbrace{\$190, \$200, \$290, \$320}_{\text{four numbers}}$$

Thus \$180 is the median cost.

NOTE TO STUDENT: Fully worked-out solutions to all of the Practice Problems can be found at the back of the text starting at page SP-1

Practice Problem 2 Find the median value of the following weekly salaries: \$320, \$150, \$400, \$600, \$290, \$150, \$450.

If a list of numbers contains an even number of items, then of course there is no one middle number. In this situation we obtain the median by taking the average of the two middle numbers.

EXAMPLE 3 Find the median of the following numbers: 26, 31, 39, 33, 13, 16, 18, 38.

Solution First we place the numbers in order from smallest to largest.

$$\underbrace{13,\ 16,\ 18}_{\text{three numbers}} \quad \underset{\text{two middle numbers}}{\underset{\uparrow}{26, 31}} \quad \underbrace{33,\ 38,\ 39}_{\text{three numbers}}$$

The average (mean) of 26 and 31 is

$$\frac{26 + 31}{2} = \frac{57}{2} = 28.5.$$

Thus the median value is 28.5.

Teaching Tip After discussing the SIDELIGHT for a few minutes, ask students if they can think of some examples where the mean is used as an average for data but the figure is misleading. Usually, students offer several good suggestions. Among the best are the following:

(a) The mean salary for a family in poor and developing countries is misleadingly high because of a small class of well-paid government officials and leaders.

(b) A certain major university bragged that its "experienced" faculty had an average of 15 years of university teaching. However, the records showed that most of the faculty had taught only 1–3 years. In fact, about 35% of the faculty were near retirement and had taught (in most cases) over 35 years.

Practice Problem 3 Find the median value of the following numbers: 126, 105, 88, 100, 90, 118.

SIDELIGHT

When would someone want to use the mean, and when would someone want to use the median? Which is more helpful?

The mean, or average, is used more frequently. It is most helpful when the data are distributed fairly evenly, that is, when no one value is "much larger" or "much smaller" than the rest.

For example, suppose a company had employees with annual salaries of \$9000, \$11,000, \$14,000, \$15,000, \$17,000, and \$20,000. All the salaries fall within a fairly limited range. The mean salary

$$\frac{9000 + 11{,}000 + 14{,}000 + 15{,}000 + 17{,}000 + 20{,}000}{6} = \$14{,}333.33$$

gives us a reasonable idea of the typical salary.

However, suppose the company had six employees with salaries of \$9000, \$11,000, \$14,000, \$15,000, \$17,000, and \$90,000. Talking about the mean salary, which is $26,000, is deceptive. No one earns a salary very close to the mean salary. The typical worker in that company does not earn around $26,000. In this case, the median value is more appropriate. Here the median is $14,500. See exercises 31 and 32 in Exercises 8.4 for more on this.

3 Finding the Mode of a Set of Numbers

Another value that is sometimes used to describe a set of data is the mode. The **mode** of a set of data is the number or numbers that occur most often.

EXAMPLE 4 The following numbers are the weights of automobiles measured in pounds:

$$2345, 2567, 2785, 2967, 3105, 3105, 3245, 3546.$$

Find the mode of these weights.

Solution The value 3105 occurs twice, whereas each of the other values occurs just once. Thus the mode is 3105 pounds.

Practice Problem 4 The following numbers are the heights in inches of 10 male students in Basic Mathematics: 64, 66, 67, 69, 70, 71, 71, 73, 75, 76. Find the mode of these heights.

A set of numbers may have more than one mode.

EXAMPLE 5 The following numbers are finish times for 12 high school students who ran a distance of one mile. The finish times are measured in minutes.

$$290, 272, 268, 260, 290, 272, 330, 355, 368, 290, 370, 272$$

Find the mode of these finish times.

Solution First we need to arrange the numbers in order from smallest to largest and include all repeats.

$$260, 268, 272, 272, 272, 290, 290, 290, 330, 355, 368, 370$$

Now we can see that the value 272 occurs three times, as does the value 290. Thus the modes for these finish times are 272 minutes and 290 minutes.

Practice Problem 5 The following numbers are distances in miles that 16 students traveled to take classes at Massasoit Community College each day.

$$2, 5, 8, 3, 12, 15, 28, 8, 3, 14, 16, 31, 33, 27, 3, 28$$

Find the mode of these distances.

A set of numbers may have **no mode** at all. For example, the set of numbers 50, 60, 70, 80, 90 has no mode because each number occurs just once. The set of numbers 33, 33, 44, 44, 55, 55 has no mode because each number occurs twice. If all numbers occur the same number of times, there is no mode.

Student Solutions Manual

CD/ Video

PH Math Tutor Center

MathXL®Tutorials on CD
MathXL®

MyMathLab®

Interactmath.com

Verbal and Writing Skills

1. Explain the difference between a median and a mean.

The median of a set of numbers when they are arranged in order from smallest to largest is that value that has the same number of values above it as below it.

The mean of a set of values is the sum of the values divided by the number of values. The mean is most likely to be not typical of the value you would expect if there are many extremely low values or many extremely high values. The median is more likely to be typical of the value you would expect.

2. Explain why some sets of numbers have one mode, others two modes, and others no modes.

The mode of a set of data is the number or numbers that occur most often.

Some sets of data, like 5, 6, 6, 7, clearly have one mode. It is 6. Other sets of data, like 5, 5, 6, 7, 7, have two values that occur most often. There are two modes—5 and 7. Some sets of data, like 4, 5, 6, 7, 8, do not have any value that occurs more often than any other value. This set does not have a mode.

Applications

In exercises 3–12, find the mean. Round to the nearest tenth when necessary.

3. ***Coffeehouse Customers*** The numbers of customers who were served at Grinders Coffeehouse between 8:00 A.M. and 9:00 A.M. in the past seven days were as follows: 30, 29, 28, 35, 34, 37, 31. mean = 32

4. ***Pizza Delivery*** The numbers of pizzas delivered by Papa John's over the last 7 days were as follows: 28, 17, 18, 21, 24, 30, 30. mean = 24

5. ***Study Time*** Sam studied for the following number of hours during the past week: mean = 2.6 hours

S	M	T	W	Th	F	S
6	2	3	3.5	2.5	1	0

6. ***Temperature at Los Angeles*** The high temperature recorded in Los Angeles at the Airport in January 2004 is recorded as follows. Find the mean high temperature for these days. Mean = 61°F

Jan 1	Jan 2	Jan 3	Jan 4	Jan 5
63°F	55°F	59°F	61°F	67°F

Source: National Weather Service

7. ***Baseball*** The captain of the college baseball team achieved the following results:

	Game 1	Game 2	Game 3	Game 4	Game 5
Hits	0	2	3	2	2
Times at Bat	5	4	6	5	4

Find his batting average by dividing his total number of hits by the total times at bat.
$9 \div 24 = 0.375$

8. ***Bowling*** The captain of the college bowling team had the following results after practice:

	Practice 1	Practice 2	Practice 3	Practice 4
Score (Pins)	541	561	840	422
Number of Games	3	3	4	2

Find her bowling average by dividing the total number of pins scored by the total number of games. $2364 \div 12 = 197$ pins

9. ***Population of Guam*** The population on the island of Guam has increased significantly over the last 40 years. Find an approximate value for the mean population for this 40-year period from the following population chart: mean = 109,200

1960 Population	1970 Population	1980 Population	1990 Population	2000 Population
67,000	86,000	107,000	134,000	152,000

Source: U.S. Census Bureau

10. ***Population of Virgin Islands*** The population on the U.S. Virgin Islands went through a significant increase from 1970 to 1990 but since then has stayed relatively unchanged. A slight decrease is projected for 2010. Find the approximate value for the mean population from 1970 to 2010 by using all the values in the following population chart: mean = 96,400

1970 Population	1980 Population	1990 Population	2000 Population	2010* Population
63,000	98,000	104,000	109,000	108,000

Source: U.S. Census Bureau *estimated

11. ***Gas Used on a Trip*** Frank and Wally traveled to the West Coast during the summer. The number of miles they drove and the number of gallons of gas they used are recorded in the following chart.

	Day 1	Day 2	Day 3	Day 4
Miles Driven	276	350	391	336
Gallons of Gas	12	14	17	14

Find the average miles per gallon achieved by the car on the trip by dividing the total number of miles driven by the total number of gallons used. $1353 \div 57 = 23.7$ mi/gal

12. ***Gas Used on a Trip*** Cindy and Andrea traveled to Boston this fall. The number of miles they drove and the number of gallons of gas they used are recorded in the following chart.

	Day 1	Day 2	Day 3	Day 4
Miles Driven	260	375	408	416
Gallons of Gas	10	15	17	16

Find the average miles per gallon achieved by the car on the trip by dividing the total number of miles driven by the total number of gallons used. $1459 \div 58 = 25.2$ mi/gal

In exercises 13–24, find the median value.

13. 126, 232, 180, 195, 229 median = 195

14. 548, 554, 560, 539, 512 median = 548

15. 12.4, 11.6, 11.9, 12.1, 12.5, 11.9 median = 12

16. 8.2, 8.1, 7.8, 8.8, 7.9, 7.5 median = 8.0

17. ***Annual Salary*** The annual salaries of the employees of a local cable television office are \$17,000, \$11,600, \$23,500, \$15,700, \$26,700, and \$31,500. median = \$20,250

18. ***Family Income*** The annual incomes of six families are \$24,000, \$60,000, \$32,000, \$18,000, \$29,000, and \$35,000. median = \$30,500

19. ***Waiter Workload*** The number of tables Carl waited on at his job the past 9 days were 12, 10, 21, 25, 31, 18, 28. median = 21 tables

20. ***Walking Times*** The number of minutes Lucia spent walking the past 7 days were 28, 20, 45, 38, 50, 32, 40. median = 38 minutes

21. ***Phone Bills*** The phone bills for Dr. Price's cellular phone over the last seven months were as follows: \$109, \$207, \$420, \$218, \$97, \$330, and \$185. median = \$207

22. ***Compact Disc Prices*** The prices of the same compact disc sold at several different music stores or by mail order were as follows: \$15.99, \$11.99, \$5.99, \$12.99, \$14.99, \$9.99, \$13.99, \$7.99, and \$10.99. median = \$11.99

23. ***Grade Point Averages*** The grade point averages (GPA) for eight students were 1.8, 1.9, 3.1, 3.7, 2.0, 3.1, 2.0, and 2.4. median = 2.2

24. ***Food Purchases*** The numbers of pounds of smoked turkey breast purchased at a deli by the last eight customers were 1.2, 2.0, 1.7, 2.5, 2.4, 1.6, 1.5, and 2.3. median = 1.85 pounds

25. ***Number of Patents*** Every five years the U.S. Patent Office posts the approximate number of patents granted to industry during a specific year. This data is recorded in the following chart. Determine the median number of patents granted based on this data. median = 95,500

Year	1985	1990	1995	2000
Approximate Number of Patents Granted During That Year	72,000	90,000	101,000	157,00

26. ***Number of Accidents*** The National Safety Council each year posts the number of motor vehicle accidents that take place in the United States. This data is recorded in the following chart. Determine the median number of accidents that occur per year based on this data. median = 12,600,000

Year	1998	1999	2000	2001
Approximate Number of Motor Vehicle Accidents During That Year	12,700,000	11,400,000	13,400,000	12,500,000

Find the mean. Round to the nearest cent when necessary.

27. ***Business Owners Salary*** The salaries of eight small business owners in Big Rapids are \$30,000, \$74,500, \$47,890, \$89,000, \$57,645, \$78,090, \$110,370, and \$65,800.
mean = \$69,161.88

28. ***Price of Laptop Computer*** The prices of nine laptop computers with a Pentium IV chip are \$5679, \$6902, \$1530, \$2738, \$2999, \$4105, \$3655, \$5980, and \$4430. mean = \$4224.22

In exercises 29 and 30, find the median.

29. 2576, 8764, 3700, 5000, 7200, 4700, 9365, 1987
4850

30. 15.276, 21.375, 18.90, 29.2, 14.77, 19.02
18.96

31. ***Holiday Shopping*** It took Jenny 5 days to complete her holiday shopping. The amounts she spent during these five days were \$120.50, \$66.74, \$80.95, \$210.52, and \$45.00. Find the mean and the median.
mean = \$104.74, median = \$80.95

32. ***Property Taxes*** The amounts the Dayton family has paid in property taxes the past five years are \$1381, \$1405, \$1405, \$1520, \$1592. Find the mean and the median.
mean = \$1460.60, median = \$1405

Find the mode.

33. 60, 65, 68, 60, 72, 59, 80 The mode is 60.

34. 86, 84, 82, 87, 84, 88, 90 The mode is 84.

35. 121, 150, 116, 150, 121, 181, 117, 123
The two modes are 121 and 150.

36. 144, 143, 140, 141, 149, 144, 141, 150
The two modes are 141 and 144.

37. ***Bicycle Prices*** The last eight bicycles sold at the Skol Bike shop cost \$249, \$649, \$269, \$259, \$269, \$249, and \$269. The mode is \$269.

38. ***Color Television Prices*** The last seven color televisions sold at the local Circuit City cost \$315, \$430, \$315, \$330, \$430, \$315 and \$460.
The mode is \$315.

39. ***Internet Inquiries*** An Internet shopping site received the following numbers of inquiries over the last seven days: 869, 992, 482, 791, 399, 855, and 869. Find the mean, the median, and the mode.
mean = 751, median = 855, mode = 869

40. ***Commuter Passengers*** The numbers of passengers taking the Rockport to Boston train during the last seven days were 568, 388, 588, 688, 750, 900, and 388. Find the mean, the median, and the mode.
mean = 610, median = 588, mode = 388

Mixed Practice

41. ***Salary of Employees*** A local travel office has 10 employees. Their monthly salaries are $1500, $1700, $1650, $1300, $1440, $1580, $1820, $1380, $2900, and $6300.

(a) Find the mean. $2157

(b) Find the median. $1615

(c) Find the mode. There is no mode

(d) Which of these numbers best represents what the typical person earns? Why?
The median, because the mean is affected by the high amount $6300.

42. ***Track Running Times*** A college track star in California ran the 100-meter event in eight track meets. Her times were 11.7 seconds, 11.6 seconds, 12.0 seconds, 12.1 seconds, 11.9 seconds, 18 seconds, 11.5 seconds, and 12.4 seconds.

(a) Find the mean. 12.65 sec

(b) Find the median. 11.95 sec

(c) Find the mode. There is no mode

(d) Which of these numbers represents her typical running time? Why?
The median, because the long time of 18 seconds affects the mean.

43. ***Number of Phone Calls*** Sally made a record of the number of phone calls she received each night this last week.

Day of the Week	Sun	Mon	Tues	Wed	Thurs	Fri	Sat
Number of Phone Calls	23	3	2	3	7	10	11

(a) Find the mean. Round your answer to the nearest tenth. 8.4

(b) Find the median 7

(c) Find the mode 3

(d) Which of these three measures best represents the number of phone calls Sally receives on a typical night? Why?
The median is the most representative. On three nights she gets more calls than 7. On three nights she gets fewer calls than 7. On one night she got 7 calls. The mean is distorted a little because of the very large number of calls on Sunday night. The mode is artificially low because she gets so few calls on Monday and Wednesday and it just happened to be the same number, 3.

44. ***Number of Overnight Business Trips*** David has to travel as a representative for his computer software company. He made a record of the number of nights he had to spend away from home on business travel during the first seven months of the year.

Month of the Year	Jan	Feb	Mar	April	May	June	July
Number of Nights Spent Away from Home on Business	3	7	8	6	9	28	3

(a) Find the mean. Round to the nearest tenth. 9.1

(b) Find the median. 7

(c) Find the mode. 3

(d) Which of these three measures best represents the number of nights that David has to spend away from home on business? Why?
The median is the most representative. For three months he travels more than this. For three months he travels less than this. In one month he was away from home exactly 7 nights. The mean is distorted because of the very high value of 28 in June. The mode is artificially low because he is overnight for travel 3 nights in January and July. This is not representative of what usually happens.

To Think About

Grade Point Average *Finding the grade point average or GPA.*

At most universities and colleges, students are assigned grade point values for the grades they have earned. A student's grade point average is the average of all the grade point values for the credit hours taken. In most schools, the grade point values are $A = 4$, $B = 3$, $C = 2$, $D = 1$, and $F = 0$.

To find your GPA you need to multiply the number of credit hours of each course you have taken by the grade point value you received for the course. Add the results. Then divide by the total number of credit hours.

Suppose you took a six credit hour course and received a grade of B and a three credit hour course and receival a grade of A. Multiply $6(3) = 18$ as well as $3(4) = 12$. Add the results: $18 + 12 = 30$. Since you have a total of nine credit hours, you need to divide by 9. $30 \div 9 = 3.33333\ldots$. Rounded to the nearest tenth, your GPA is 3.3.

In exercises 45 and 46, find the GPA rounded to the nearest tenth.

45.

Number of Credit Hours	Grade
3	A
4	B
3	C
3	B

GPA = 3.0

46.

Number of Credit Hours	Grade
3	B
5	A
3	C
4	C

GPA = 2.9

Cumulative Review

Round to the nearest tenth. Use $\pi \approx 3.14$.

▲**47.** ***Geometry*** A triangular piece of insulation is located under the dash of a Ford Explorer. It has a base of 7 inches and a height of 5.5 inches. What is the area of this piece of insulation?
19.3 square inches

▲**48.** ***Geometry*** The Canaan Family Farm has two fields that are irrigated with a rotating sprinkler system. This system waters a circular area. The system is designed to deliver 2 gallons per hour for each square foot of the field. If each circular area has a radius of 40 feet, how many gallons per hour are needed to water these fields? 20,096 gallons per hour

▲**49.** ***Sign Costs*** Collette Camp has made a huge advertising sign in the shape of a rhombus. The sign has a base of 5 feet and a height of 4 feet. The sign is made out of aluminum that costs \$16.50 per square foot. How much did it cost for the aluminum used to make the sign?
\$330

▲**50.** ***Coffee Cost*** A medium coffee at Coffee Time costs \$1.98 and comes in a paper cup with a height of 7 in. and a radius of 1.5 in. What is the cost of the coffee per cubic inch? Round to the nearest cent.
about \$0.04 per cubic inch

Putting Your Skills to Work

The Mathematics of Baseball

Each year the Gold Glove Award is given to the top player in each of the eighteen baseball positions in both the American League (AL) and National League (NL). In the table below, 6 of the top shortstops for the 2003 baseball season are listed along with their number of at bats (AB), hits (H), total bases (TB), putouts (PO), assists (A), and errors (E) (for definitions of these terms, go to www.mlb.com).

Managers and coaches who take into account several factors decide the winners of the Gold Glove Awards. Three important statistics that are considered are listed below the table, as well as an example of how to calculate each. (Each statistic is rounded to the nearest thousandth.)

Player	League	AB	H	TB	PO	A	E	AVG	FPCT	SLG
Derek Jeter (New York Yankees)	AL	482	156	217	159	271	14	.324	.968	.450
Nomar Garciaparra (Boston Red Sox)	AL	658	198	345	216	456	20	.301	.971	.524
Alex Rodriguez (Texas Rangers)	AL	607	181	364	227	464	8	.298	.989	.600
Alex Gonzalez (Chicago Cubs)	NL	536	122	219	193	422	10	.228	.984	.409
Edgar Renteria (St. Louis Cardinals)	NL	587	194	282	191	439	16	.330	.975	.480
Orlando Cabrera (Montreal Expos)	NL	626	186	288	258	456	18	.297	.975	.460

Source: www.mlb.com

Batting Average (AVG): Divide a player's base hits by the number of at bats. Derek Jeter had 156 hits and 482 at bats. His batting average is $156/482 \approx .324$.

Fielding Percentage (FPCT): Divide the sum of putouts and assists, by the sum of putouts, assists, and errors. Nomar Garciaparra had 216 putouts, 456 assists, and 20 errors. His fielding average is $(216 + 456)/(216 + 456 + 20) \approx .971$.

Slugging Percentage (SLG): Divide the number of total bases by the number of at bats. Alex Gonzalez had 219 total bases and 536 at bats. His slugging percentage is $219/536 \approx .409$.

Problems for Individual Investigation

1. Calculate the batting average for each player. If the Gold Glove Award was based solely on batting average, who would win for the American League? Who would win for the National League?

Derek Jeter; Edgar Renteria

2. Calculate the slugging percentage for each player. If the Gold Glove Award was based solely on slugging percentage, who would win for the American League? Who would win for the National League?

Alex Rodriguez; Edgar Renteria

Problems for Group Investigation and Cooperative Study

3. Calculate the fielding percentage for each player. If the Gold Glove Award was based solely on fielding percentage, who would win for the American League? Who would win for the National League?

Alex Rodriguez; Alex Gonzalez

4. Based on your answers to questions 1–3, which shortstop in the American League would you award the Gold Glove Award to? Which shortstop in the National League would get the award? Explain why you chose these players.

Chapter 8 Organizer

Topic	Procedure	Examples
Circle graphs, p. 526.	The following circle graph describes the ages of the 200 men and women of the Glover City police force. 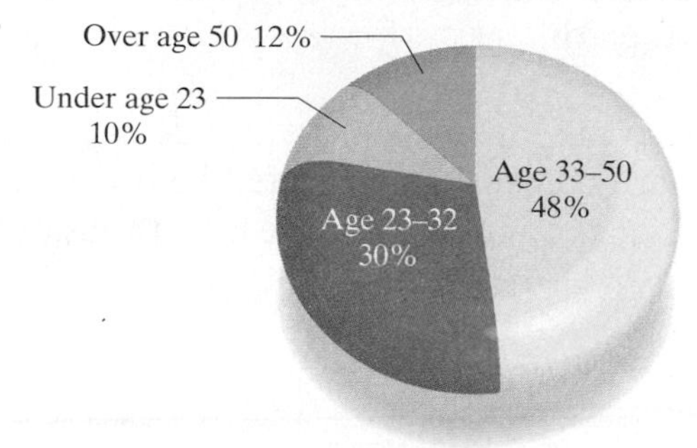	**1.** What percent of the police force is between 23 and 32 years old? 30% **2.** How many men and women in the police force are over 50 years old? 12% of 200 $= (0.12)(200) = 24$ people
Bar graphs and double-bar graphs, p. 533.	The following double-bar graph illustrates the sales of color television sets by a major store chain for 2003 and 2004 in three regions of the country. 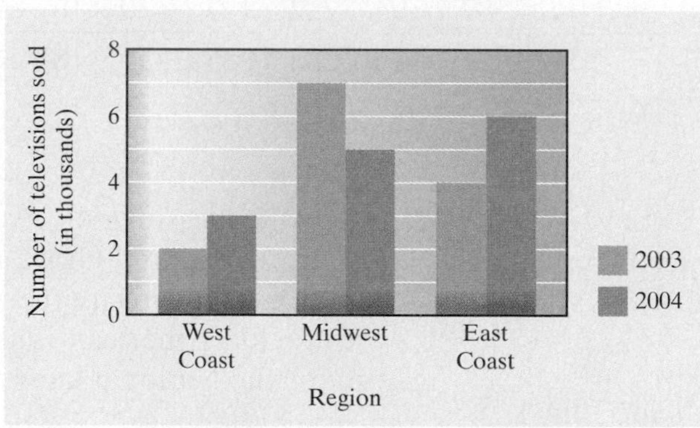	**1.** How many color television sets were sold by the chain on the East Coast in 2004? 6000 sets **2.** How many *more* color television sets were sold in 2004 than in 2003 on the West Coast? 3000 sets were sold in 2004; 2000 sets were sold in 2003. $\begin{array}{r} 3000 \\ -\ 2000 \\ \hline 1000 \end{array}$ sets more in 2004
Line graphs and comparison line graphs, pp. 535–536.	The following line graph indicates the number of visitors to Wetlands State Park during a four-month period in 2003 and 2004. 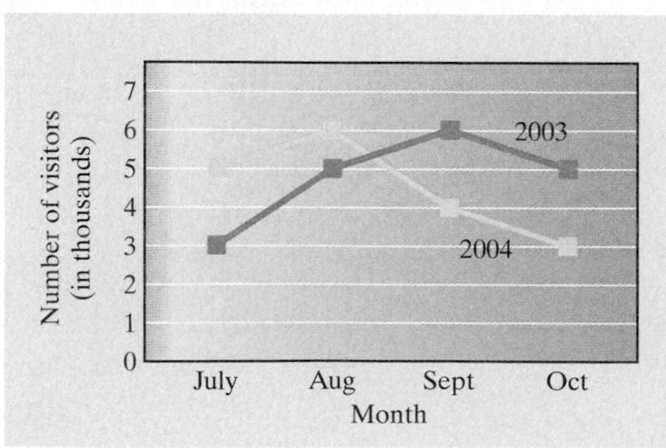	**1.** How many visitors came to the park in July 2003? 3000 visitors **2.** In what months were there more visitors in 2003 than in 2004? September and October **3.** The sharpest *decrease* in attendance took place between what two months? Between August 2004 and September 2004

Topic	Procedure	Examples
Histograms, p. 541.	The following histogram indicates the number of students in a math class who scored within each interval on a 15-point quiz. Number of students: 0, 4, 8, 12, 16, 20 Score: 0–3, 4–7, 8–11, 12–15	**1.** How many students had a score between 8 and 11? 20 students **2.** How many students had a score of less than 8? 12 + 6 = 18 students
Finding the mean, p. 549.	The *mean* of a set of values is the sum of the values divided by the number of values. The mean is often called the *average*.	**1.** Find the mean of 19, 13, 15, 25, and 18. $\frac{19 + 13 + 15 + 25 + 18}{5} = \frac{90}{5} = 18$ The mean is 18.
Finding the median, p. 549.	If a set of numbers is arranged in order from smallest to largest, the *median* is that value that has the same number of values above it as below it. How do we find the median? **1.** Arrange the numbers in order from smallest to largest. **2.** If there is an odd number of values, the middle value is the median. **3.** If there is an even number of values, the average of the two middle values is the median.	**1.** Find the median of 19, 29, 36, 15, and 20. First we arrange in order from smallest to largest: 15, 19, 20, 29, 36. 15, 19 (two numbers) 20 ↑ (middle number) 29, 36 (two numbers) The median is 20. **2.** Find the median of 67, 28, 92, 37, 81, and 75. First we arrange in order from smallest to largest: 28, 37, 67, 75, 81, 92. There is an even number of values 28, 37, 67, 75, 81, 92 ↑ two middle numbers $\frac{67 + 75}{2} = \frac{142}{2} = 71$ The median is 71.
Finding the mode, p. 551.	The *mode* of a set of values is the value that occurs most often. A set of values may have more than one mode or no mode.	**1.** Find the mode of 12, 15, 18, 26, 15, 9, 12, and 27. The modes are 12 and 15. **2.** Find the mode of 4, 8, 15, 21, and 23. There is no mode.

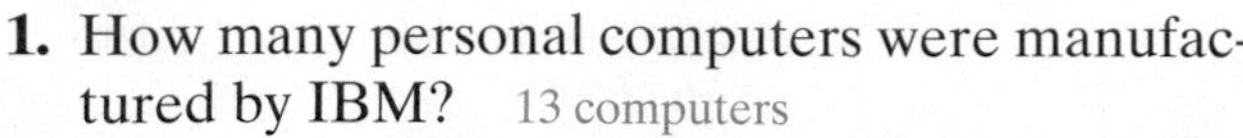

Chapter 8 Review Problems

Computer Manufacturers *A student found that there were a total of 140 personal computers owned by students in the dormitory. The following circle graph displays the distribution of manufacturers of these computers. Use the graph to answer exercises 1–8.*

Distribution of Computers by Manufacturer in a Dormitory

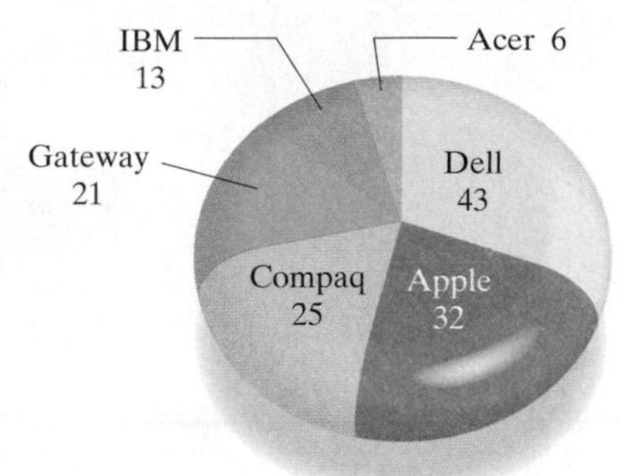

1. How many personal computers were manufactured by IBM? 13 computers

2. How many personal computers were manufactured by Apple? 32 computers

3. How many personal computers were manufactured by Dell or Compaq? 68 computers

4. How many personal computers were manufactured by Gateway or Acer? 27 computers

5. What is the ratio of the number of computers manufactured by IBM to the number of computers manufactured by Gateway? $\frac{13}{21}$

6. What is the ratio of the number of computers manufactured by Dell to the number of computers manufactured by Apple? $\frac{43}{32}$

7. What percent of the 140 computers are manufactured by Compaq? Round to the nearest tenth. approximately 17.9%

8. What percent of the computers are manufactured by Apple? Round to the nearest tenth. approximately 22.9%

College Majors *Bradford College offers majors in 6 areas: business, science, social science, language arts, education, and art. The distribution by category is displayed in the circle graph below. Use the graph to answer exercises 9–16.*

Majors at Bradford College

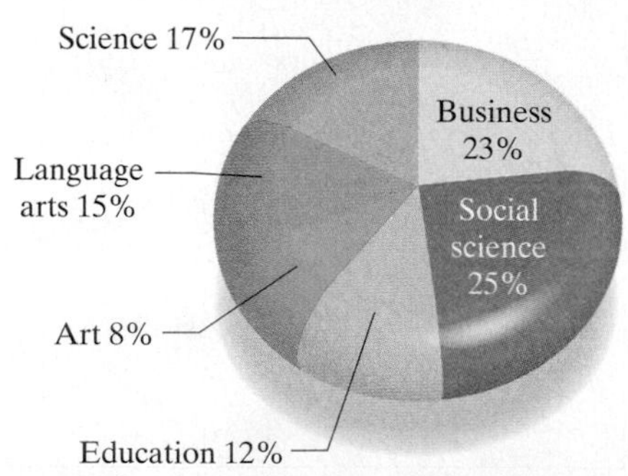

9. What percent of the students are majoring in either business or social science? 48%

10. What percent of the students are majoring in an area other than business? 77%

11. Which area has the least number of students? art

12. Which area has the second highest number of students? business

13. Which two areas together make up one-fifth of the graph? art and education

14. If Bradford College has 8000 students, how many of them are majoring in language arts? 1280 students

15. How many students are majoring in a science (either science or social science)? 3360 students

16. How many more students are majoring in business than education? 880 students

Milk Consumption of Children *The following double-bar graph illustrates the numbers of glasses of milk consumed each week by children in various age categories for the years 1990 and 2000. Use the graph to answer exercises 17–24.*

17. How many glasses of milk per week were consumed by children age 2–5 years in 1990? 36

18. How many glasses of milk per week were consumed by children age 6–9 years in 2000? 26

19. What age group saw the greatest decrease in milk consumption between 1990 and 2000? 10–13 years

20. What age group saw an increase in milk consumption between 1990 and 2000? 6–9 years

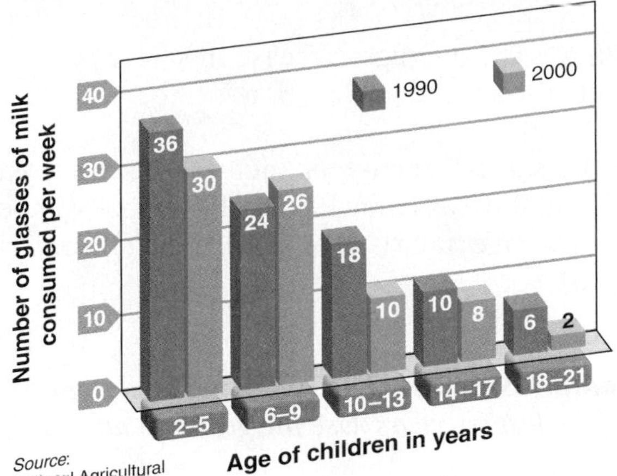

21. In 1990, how many fewer glasses of milk were consumed per week by children age 18–21 years than children age 10–13 years? 12 glasses

22. In 2000, how many more glasses of milk were consumed per week by children age 2–5 years than by children age 14–17 years? 22 glasses

23. What is the ratio of number of glasses of milk consumed each week by children age 10–13 years to that consumed by children age 18–21 years in 2000? $\frac{5}{1}$

24. What is the ratio of number of glasses of milk consumed each week by children age 2–5 years to that consumed by children aged 10–13 years in 1990? $\frac{2}{1}$

Education Salaries *The following double-bar graph shows the average salaries paid to public school classroom teachers and principals in the U.S. for selected years from 1980 to 2005. Use the bar graph to answer exercises 25–36.*

25. What was the average salary of a classroom teacher in 1990? \$31,000

26. What was the average salary of a principal in 2000? \$58,000

27. During what five-year period was there the greatest increase in salary for a principal? between 1985 and 1990

28. During what five-year period was there the greatest increase in salary for a classroom teacher? between 1980 and 1985

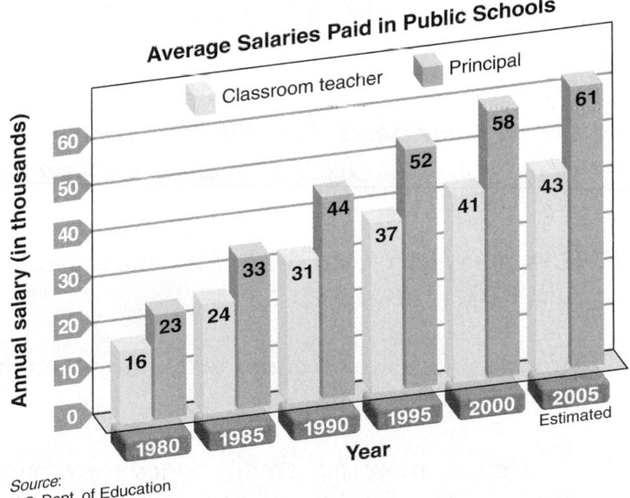

29. How much more did the average principal make per year than the average classroom teacher in 1985? \$9,000

30. How much more did the average principal make per year than the average classroom teacher in 1995? \$15,000

31. In what year was the difference between the average salary of a principal and the average salary of a classroom teacher become the greatest? 2000

32. In what year was the difference between the average salary of a principal and the average salary of a classroom teacher the smallest? 1980

33. What is the average increase in salary per five years for a principal? $7600

34. What is the average increase in salary per five years for a classroom teacher? $5400

35. If the same fifteen-year increase occurs from 1995 to 2010 as from 1980 to 1995 what will be the average salary of a classroom teacher in 2010? $58,000

36. If the same ten-year increase occurs from 2000 to 2010 as from 1990 to 2000, what will be the average salary of a principal in 2010? $72,000

Graduating Students *The following line graph shows the numbers of graduates of Williamston University during the last six years. Use the graph to answer exercises 37–44.*

37. How many Williamston University students graduated in 2000? 400 students

38. How many Williamston University students graduated in 1999? 500 students

39. How many Williamston University students graduated in 2001? 650 students

40. How many Williamston University students graduated in 1998? 450 students

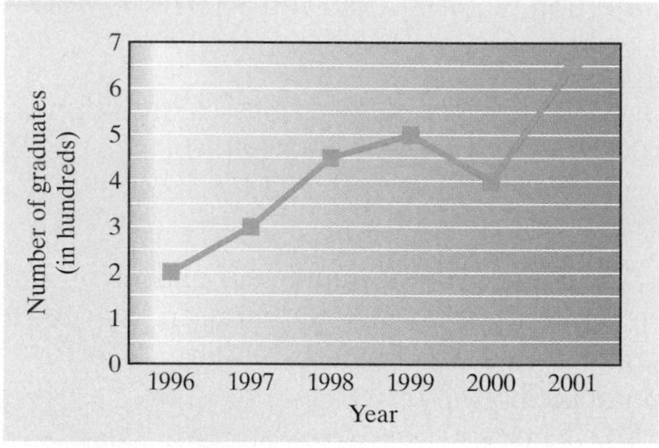

41. How many more Williamston University students graduated in 1997 than in 1996? 100 students

42. How many more Williamston University students graduated in 1999 than in 1998? 50 students

43. Between what two years did the number of graduates decline? 1999–2000

44. Between what two years did the number of graduates increase by the greatest amount? 2000–2001

Ice Cream Cone Sales *The following comparison line graph shows the number of ice cream cones purchased at the Junction Ice Cream Stand during a five-month period in 2002 and in 2003. Use this graph to answer exercises 45–52.*

45. How many ice cream cones were purchased in July 2003? 45,000 cones

46. How many ice cream cones were purchased in August 2002? 30,000 cones

47. How many more ice cream cones were purchased in May 2002 than in May 2003? 10,000 cones

48. How many more ice cream cones were purchased in August 2003 than in August 2002? 30,000 cones

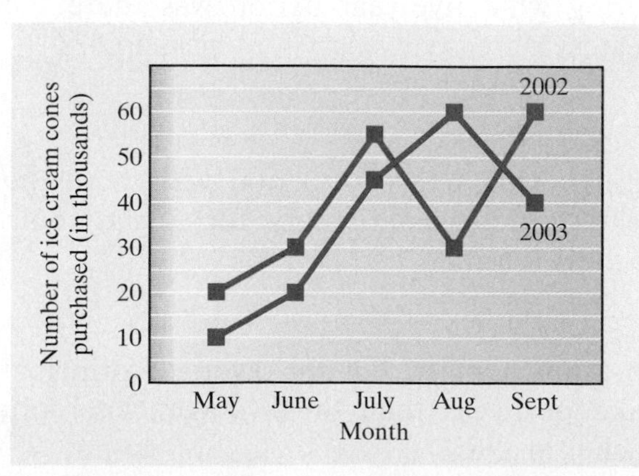

49. How many more ice cream cones were purchased in July 2002 than in June 2003?
35,000 cones

50. How many more ice cream cones were purchased in September 2002 than in August 2002? 30,000 cones

51. At the location of the ice cream stand, July 2002 was warm and sunny, whereas August 2002 was cold and rainy. Describe how the weather might have played a role in the trend shown on the graph from July to August 2002.
The sharp drop in the number of ice cream cones purchased from July 2002 to August 2002 is probably directly related to the weather. Since August was cold and rainy, significantly fewer people wanted ice cream during August.

52. At the location of the ice cream stand, June 2003 was cold and rainy, whereas July 2003 was warm and sunny. Describe how the weather might have played a role in the trend shown on the graph from June to July 2003.
The sharp increase in the number of ice cream cones purchased from June 2003 to July 2003 is probably directly related to the weather. Since June was cold and rainy and July was warm and sunny, significantly more people wanted ice cream during July.

Electronics Sales *The following comparison line graph shows the annual sales of cordless telephones and answering machines in the United States for selected years. Please use the graph to answer exercises 53–64.*

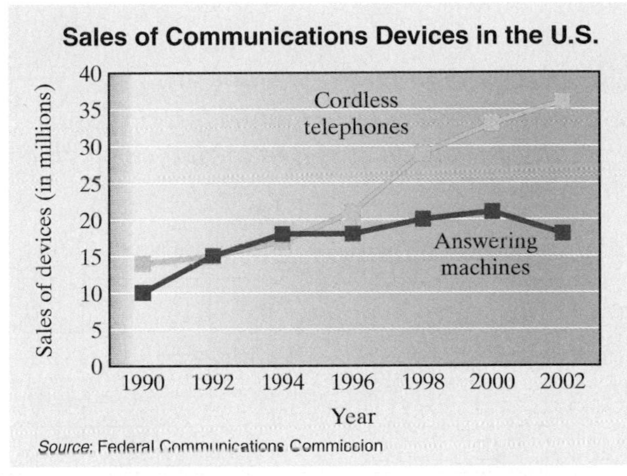

53. How many cordless telephones were sold in 1990? 14,000,000

54. How many answering machines were sold in 2002? 18,000,000

55. Between what two years was the increase in the sales of cordless telephones the greatest?
between 1996 and 1998

56. Between what two years was the increase in the sales of answering machines the greatest?
between 1990 and 1992

57. How many more answering machines than cordless telephones were sold in 1994?
1,000,000

58. How many more cordless telephones than answering machines were sold in 2000?
12,000,000

59. In what year were the sales of answering machines and cordless phones at the same level?
1992

60. In what year were the sales of answering machines and cordless phones the most significantly different? 2002

61. What was the average increase in sales per two-year period for cordless telephones?
≈3,666,667

62. What was the average increase in sales per two-year period for answering machines?
≈1,333,333

63. If the amount of increase from 1996 to 2000 continues in the four years from 2000 to 2004, what will be the sales of answering machines in 2004? 24,000,000

64. If the amount of increase from 1994 to 2000 continues in the next six years from 2000 to 2006, what will be the sales of cordless telephones in 2006? 49,000,000

Women's Shoe Sales *The following histogram shows the numbers of pairs of women's shoes sold at the grand opening of a new shoe store. Use the histogram to answer exercises 65–70.*

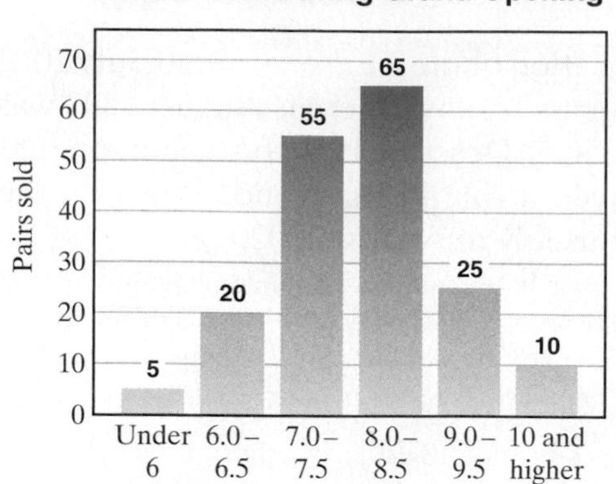

65. How many pairs sold were size 8–8.5? 65 pairs

66. How many pairs sold were size 10 or higher? 10 pairs

67. How many pairs sold were between size 7 and size 9.5? 145 pairs

68. Before the grand opening, the store had 200 pairs of women's shoes. What percent of these were sold during the grand opening? 90%

69. How many more pairs of size 8–8.5 were sold than size 6–6.5? 45 pairs

70. What is the ratio of pairs sold of size 9–9.5 to total pairs sold? 5 to 36 or $\frac{5}{36}$

Color Television Manufacturing *During the last 28 days, a major manufacturer produced 400 new color television sets each day. The manufacturer recorded the number of defective television sets produced each day. The results are shown in the following chart. In exercises 71–75, determine frequencies of the class intervals for this data.*

	Mon.	Tues.	Wed.	Thurs.	Fri.
Week 1	3	5	8	2	0
Week 2	13	6	3	4	1
Week 3	0	2	16	5	7
Week 4	12	10	17	5	4
Week 5	1	7	8	12	13
Week 6	14	0	3	closed	
	Mon.	Tues.	Wed.	Thurs.	Fri.

	Number of Defective Televisions Produced (Class Interval)	Tally	Frequency
71.	0–3	𝍸 𝍸	10
72.	4–7	𝍸 III	8
73.	8–11	III	3
74.	12–15	𝍸	5
75.	16–19	II	2

76. Construct a histogram using the table prepared in exercises 71–75.

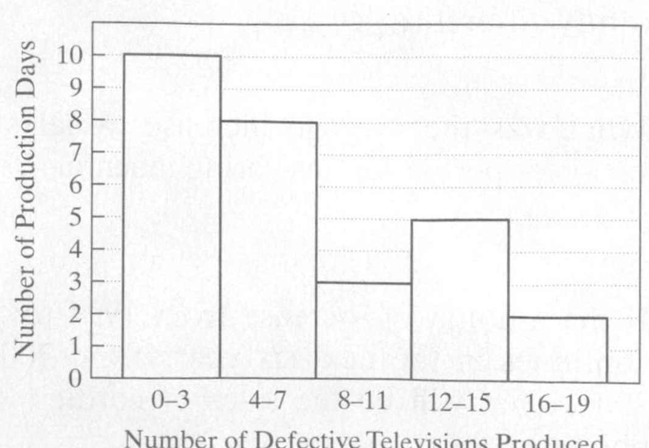

77. Based on the data of exercises 71–75, how often were between 0 and 7 defective television sets identified in the production? 18 times

Find the mean.

78. ***Temperatures in Los Angeles*** The maximum temperature readings in Los Angeles for the last seven days in July were: 86°, 83°, 88°, 95°, 97°, 100°, and 81°. 90°

79. ***Gas Heating Expenses*** The LeBlanc family's gas bills for January through June were \$145, \$162, \$95, \$67, \$43, and \$26. Round to the nearest cent. \$89.67

80. ***Visitors at Yellowstone National Park*** The approximate number of people who visited Yellowstone National Park from November of 2002 to March of 2003 was Nov 11,000; Dec 19,000; Jan 40,000, Feb 52,000, and March 23,000. Find the mean number of people who visited during a winter month. (*Source:* National Park Service) 29,000 people

81. ***Rental Car Employees*** The numbers of employees throughout the nation employed annually by Freedom Rent a Car for the last six years were 882, 913, 1017, 1592, 1778, and 1936. 1353 employees

Find the median.

82. ***Cost of Trucks*** The costs of eight trucks purchased by the highway department: \$28,500, \$29,300, \$21,690, \$35,000, \$37,000, \$43,600, \$45,300, \$38,600. \$36,000

83. ***Cost of Houses*** The costs of 10 houses recently purchased at Stillwater: \$98,000, \$150,000, \$120,000, \$139,000, \$170,000, \$156,000, \$135,000, \$144,000, \$154,000, \$126,000. \$141,500

In exercises 84 and 85, find the median and the mode.

84. ***San Diego Zoo*** The ages of the last 16 people who have passed through the entrance of the San Diego Zoo: 28, 30, 15, 54, 77, 79, 10, 8, 43, 38, 28, 31, 4, 7, 34, 35.

median = 30.5 years

mode = 28 years

85. ***Pizza Deliveries*** The daily numbers of deliveries made by the Northfield House of Pizza: 21, 16, 15, 3, 19, 24, 13, 18, 9, 31, 36, 25, 28, 14, 15, 26.

median = 18.5 deliveries

mode = 15 deliveries

86. ***Test Scores*** The scores on eight tests taken by Wong Yin in calculus last semester were 96, 98, 88, 100, 31, 89, 94, and 98. Which is a better measure of his usual score, the *mean* or the *median*? Why?

The median is better because the mean is skewed by the one low score, 31.

87. ***Sales of Cars*** The ten sales people at People's Dodge sold the following numbers of cars last month: 13, 16, 8, 4, 5, 19, 15, 18, 39, 12. Which is a better measure of the usual sales of these salespersons, the *mean* or the *median*? Why?

The median is better because the mean is skewed by the one high data item, 39.

88. Barbara made a record of the number of hours she uses the computer each day in her apartment.

Day of the Week	Sun	Mon	Tues	Wed	Thurs	Fri	Sat
Number of Hours Spent on the Computer	2	3	2	4	7	12	5

(a) Find the mean. Round your answer to the nearest tenth if necessary. 5

(b) Find the median. 4

(c) Find the mode. 2

(d) Which of these three measures best represents the number of hours she uses the computer on a typical day? Why?

The median is the most representative. On three days she uses the computer more than 4 hours and on three days she uses the computer less than 4 hours. One day she used it exactly 7 hours. The mean is distorted a little because of the very large number of hours on Friday. The mode is artificially low because she happened to use the computer only two hours on Sunday and Tuesday. All other days it was more than this.

How Am I Doing? Chapter 8 Test

Note to Instructor: The Chapter 8 Test file in the TestGen program provides algorithms specifically matched to these problems so you can easily replicate this test for additional practice or assessment purposes.

1. 37%

2. 21%

3. 12%

4. 90,000 automobiles

5. 81,000 automobiles

Remember to use your Chapter Test Prep Video CD to see the worked-out solutions to the test problems you want to review.

A state highway safety commission recently reported the results of inspecting 300,000 automobiles. The following circle graph depicts the percent of automobiles that passed and the percent that had one or more safety violations. Use this graph to answer questions 1–5.

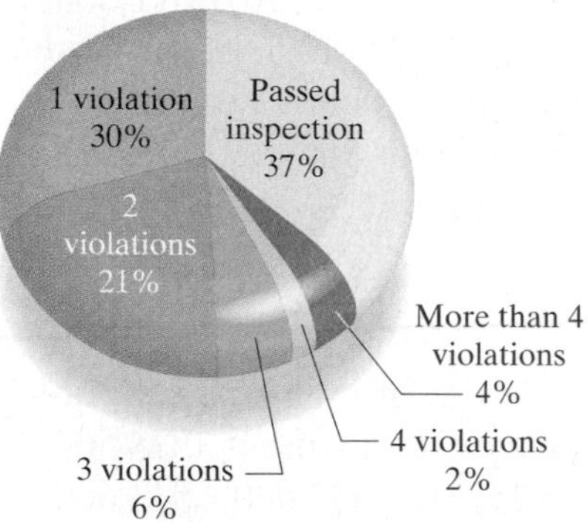

1. What percent of the automobiles passed inspection?

2. What percent of the automobiles had two safety violations?

3. What percent of the automobiles had more than two safety violations?

4. If 300,000 automobiles were inspected, how many of them had one safety violation?

5. If 300,000 automobiles were inspected, how many of them had either two violations or three violations?

The following double-bar graph shows how the average college tuition and fees costs per year have increased at public and private colleges since 1988. Use the graph to answer questions 6–11.*

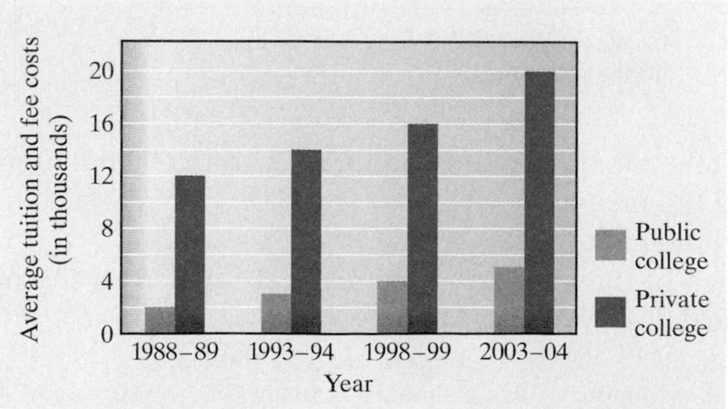

Source: The College Board
*Inflation has been taken into account. Dollar amounts are given in 2003 dollars.

6. What was the average cost per year of a private college in 1988–1989?

6. $12,000

7. What was the average cost per year of a public college in 1993–1994?

8. How much did the average cost per year of private colleges increase from 1988–1989 to 2003–2004?

7. $3,000

9. How much did the average cost per year of public colleges increase from 1993–1994 to 2003–2004?

8. $8,000

10. How much more did a year of private college cost than a year of public college in 1993–1994?

11. How much more did a year of private college cost than a year of public college in 2003–2004?

9. $2,000

A research study by 10 midwestern universities produced the following line graph. Use the graph to answer questions 12–16.

10. $11,000

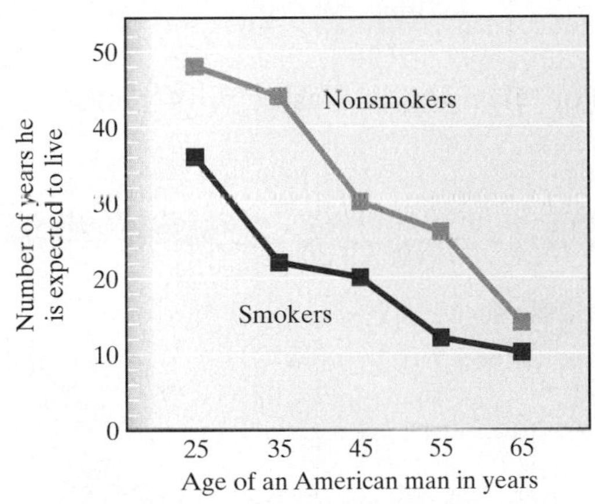

11. $15,000

12. 20 yr

12. Approximately how many more years is a 45-year-old American man expected to live if he smokes?

13. 26 yr

13. Approximately how many more years is a 55-year-old American man expected to live if he does not smoke?

14. 12 yr

14. According to this graph, approximately how much longer is a 25-year-old nonsmoker expected to live than a 25-year-old smoker?

15. According to this graph, at what age is the difference between the life expectancy of a smoker and a nonsmoker the greatest?

15. age 35

16. age 65

16. According to this graph, at what age is the difference between the life expectancy of a smoker and a nonsmoker the smallest?

The following histogram was prepared by a consumer research group. Use the histogram to answer questions 17–20.

17. 60,000 televisions

18. 25,000 televisions

19. 20,000 televisions

17. How many color television sets lasted 6–8 years?

20. 60,000 televisions

18. How many color television sets lasted 3–5 years?

19. How many color television sets lasted more than 11 years?

21. 14.5

20. How many color television sets lasted 9–14 years?

22. 14.5

A chemistry student had the following scores on ten quizzes in her chemistry class: 10, 16, 15, 12, 18, 17, 14, 10, 13, 20.

21. Find the mean quiz score.

23. 10

22. Find the median quiz score.

23. Find the mode quiz score.

24. Mean or Median

24. Which of the values mean, median, or mode best represents a typical quiz score?

Cumulative Test for Chapters 1–8

Approximately one-half of this test is based on Chapter 8 material. The remainder is based on material covered in Chapters 1–7.

1. Add. $1376 + 2804 + 9003 + 7642$

2. Multiply. 2008×37

3. Subtract. $7\frac{1}{5} - 3\frac{3}{8}$

4. Divide. $10\frac{3}{4} \div \frac{3}{8}$

5. Round to the nearest hundredth. 1796.4289

6. Subtract. $\begin{array}{r} 200.58 \\ -\ 127.93 \\ \hline \end{array}$

7. Divide. $52.0056 \div 0.72$

8. Find n. $\frac{7}{n} = \frac{35}{3}$

9. Of every 2030 cars manufactured, 3 have major engine defects. If the total number of these cars manufactured was 26,390, approximately how many had major engine defects?

10. What is 1.3% of 25?

11. 20% of what number is 12?

12. Convert 198 cm to m.

13. Convert 18 yd to ft.

▲ **14.** Find the area of a circle with radius of 3 in. Round to the nearest tenth. Use $\pi \approx 3.14$.

▲ **15.** Find the perimeter and area of a square with a side of 15 ft.

The following circle graph shows the ages of the 280 people who attended the last play at the Lexington Play House.

Attendees of Lexington Play House

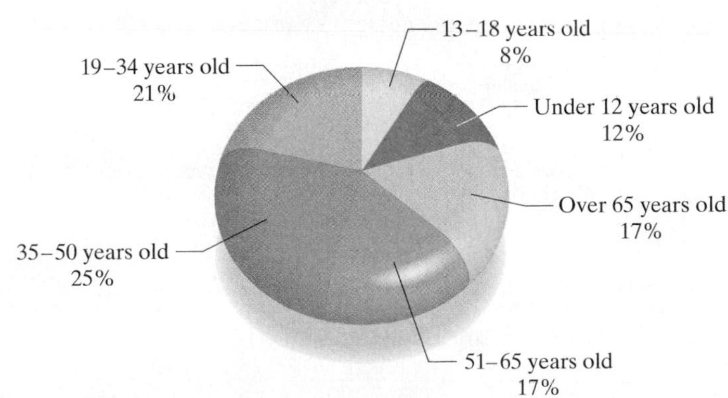

16. What percent were over 50 years old?

17. How many of the attendees were under 19 years old?

1. 20,825

2. 74,296

3. $\frac{153}{40}$ or $3\frac{33}{40}$

4. $\frac{86}{3}$ or $28\frac{2}{3}$

5. 1796.43

6. 72.65

7. 72.23

8. $n = 0.6$

9. 39 cars

10. 0.325

11. 60

12. 1.98 m

13. 54 ft

14. 28.3 in.^{2}

15. perimeter = 60 ft, area = 225 ft^2

16. 34%

17. 56 people

18. $3 million

19. $1 million

20. 16 in.

21. 1960 and 1970

22. 8 students

23. 16 students

24. $6.00

25. $4.95

26. $4.50

The following double-bar graph indicates the quarterly profits for Dedalon Corporation for 2002 and 2003.

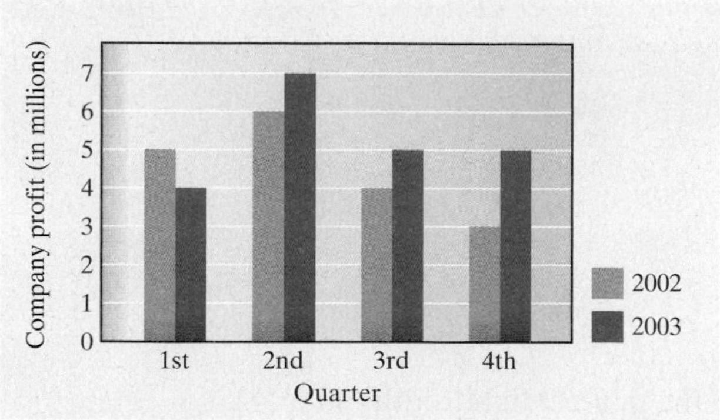

18. What was the quarterly profit for Dedalon Corporation in the fourth quarter of 2002?

19. How much greater was the profit of Dedalon Corporation in the second quarter of 2003 than in the second quarter of 2002?

The following comparison line graph depicts the annual rainfall in Dixville compared to the annual rainfall in Weston for five specific years.

20. How many inches of rain fell in Dixville in 1980?

21. In what years was the annual rainfall in Weston greater than the annual rainfall in Dixville?

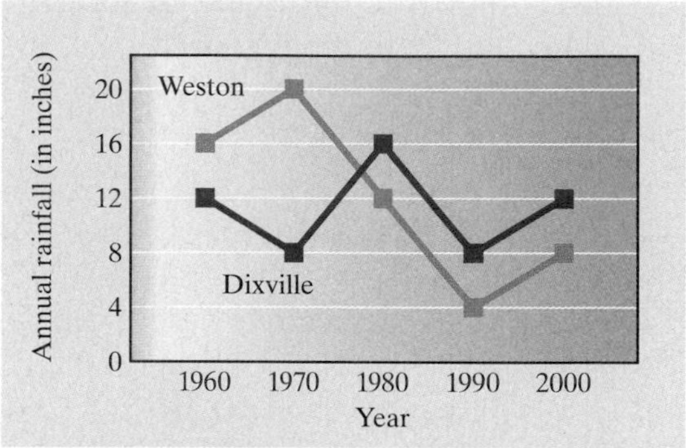

The following histogram depicts the number of students in the Basic Mathematics course who fall into various age groups.

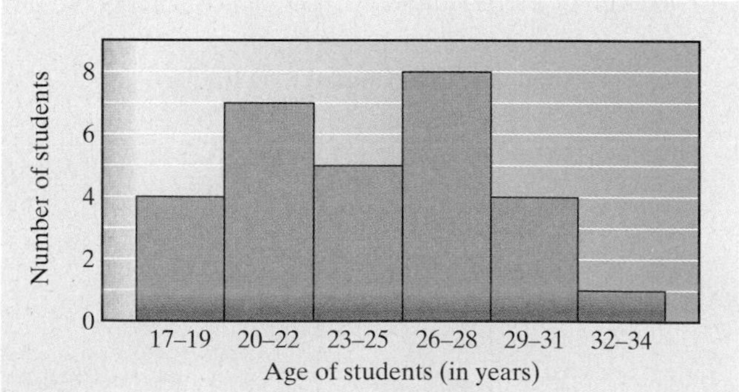

22. How many students are 26–28 years old?

23. How many students are under 26 years old?

The following are the hourly wages of eight employees of the Hamilton House of Pizza: $5.00, $4.50, $3.95, $4.90, $7.00, $12.15, $4.50, $6.00.

24. Find the mean hourly wage.

25. Find the median hourly wage.

26. Find the mode hourly wage.

CHAPTER

9

Each weekday, billions of dollars in stock-market transactions are made on Wall Street in New York City. This activity is the foundation of our national economy and greatly influences global markets as well. Do you know what kinds of transactions are made? To gain a better understanding of how the stock market operates, try the Putting Your Skills to Work problems on page 611.

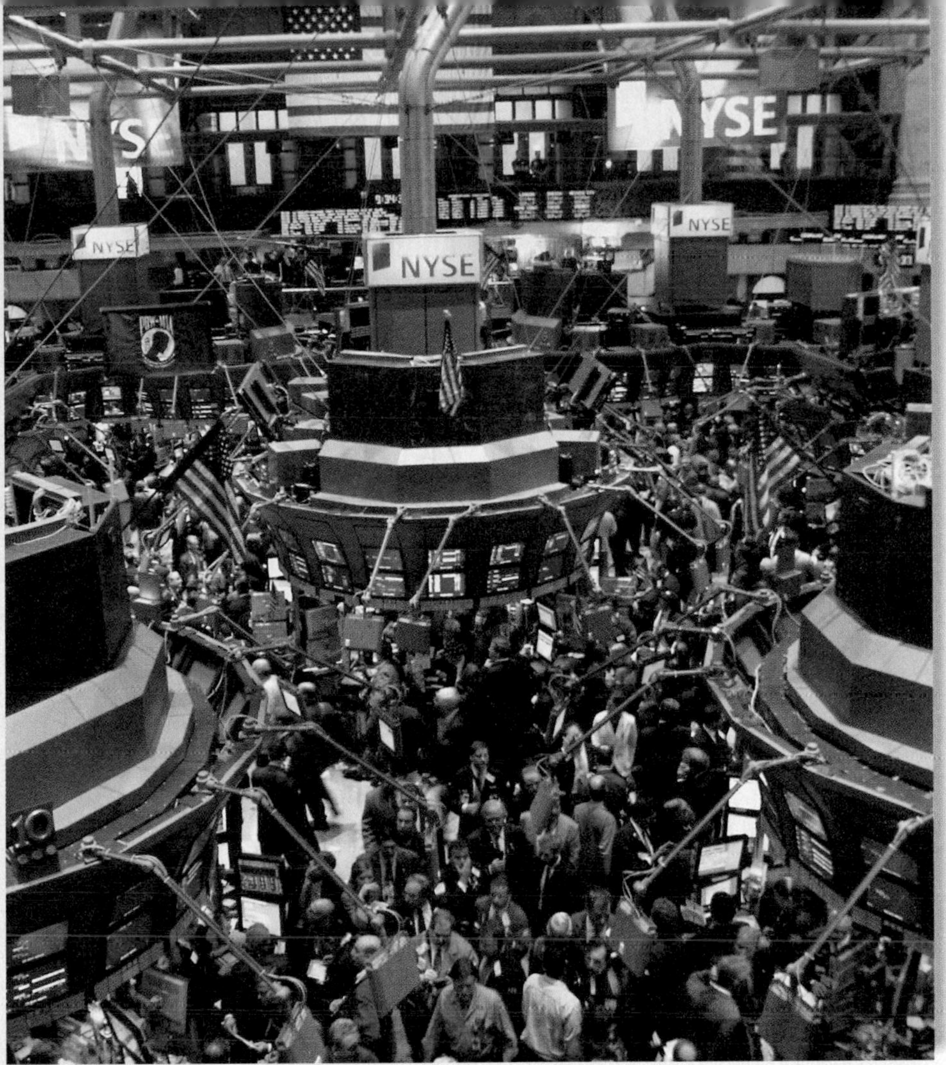

Signed Numbers

9.1 Adding Signed Numbers

Student Learning Objectives

After studying this section, you will be able to:

1. Add two signed numbers with the same sign.
2. Add two signed numbers with different signs.
3. Add three or more signed numbers.

1 Adding Two Signed Numbers with the Same Sign

In Chapters 1–8 we worked with whole numbers, fractions, and decimals. In this chapter we enlarge the set of numbers we work with to include numbers that are less than zero. Many real-life situations require using numbers that are less than zero. A debt that is owed, a financial loss, temperatures that fall below zero, and elevations that are below sea level can be expressed only in numbers that are less than zero, or negative numbers.

The following is a graph of the financial reports of four small airlines for the year. It shows **positive numbers**—those numbers that rise above zero—and **negative numbers**—those numbers that fall below zero. The positive numbers represent money gained. The negative numbers represent money lost.

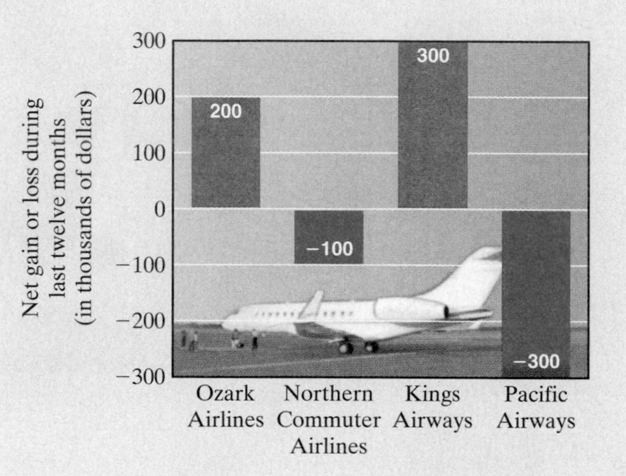

A value of $-100{,}000$ is shown for Northern Commuter Airlines. This means that Northern Commuter Airlines lost $100,000 during the year. A value of $-300{,}000$ is recorded for Pacific Airways. What does this mean?

Another way to picture positive and negative numbers is on a number line. A **number line** is a line on which each number is associated with a point. The numbers may be positive or negative, whole numbers, fractions, or decimals. Positive numbers are to the right of zero on the number line. Negative numbers are to the left of zero on the number line. Zero is neither positive nor negative.

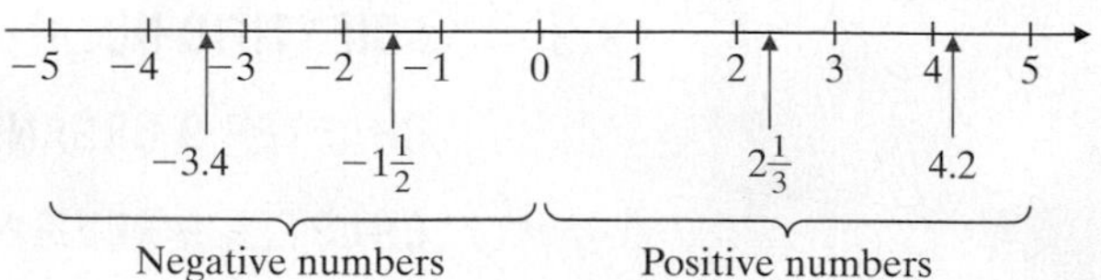

Positive numbers can be written with a plus sign—for example, $+2$—but this is not usually done. Positive 2 is usually written as 2. It is understood that the *sign* of the number is positive although it is not written. Negative numbers must always have the negative sign so that we know they are negative numbers. Negative 2 is written as -2. The *sign* of the number is negative. The set of positive numbers, negative numbers, and zero is called the set of **signed numbers.**

Order Signed numbers are named in **order** on the number line. Smaller numbers are to the left. Larger numbers are to the right. For any two numbers on the number line, the number on the left is less than the number on the right.

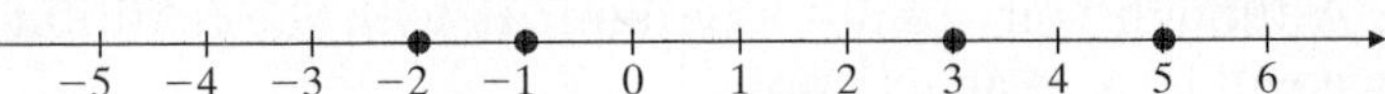

We use the symbol $<$ to mean "is less than." Thus the mathematical sentence $-2 < -1$ means "-2 is less than -1." We use the symbol $>$ to mean "is greater than." Thus the mathematical sentence $5 > 3$ means "5 is greater than 3."

EXAMPLE 1 In each case, replace the ? with $<$ or $>$.

(a) $-8 \ ? -4$ **(b)** $7 \ ? \ 1$ **(c)** $-2 \ ? \ 0$
(d) $-6 \ ? \ 3$ **(e)** $2 \ ? -5$

Solution

(a) Since -8 lies to the left of -4, we know that $-8 < -4$.
(b) Since 7 lies to the right of 1, we know that $7 > 1$.
(c) Since -2 lies to the left of 0, we know that $-2 < 0$.
(d) Since -6 lies to the left of 3, we know that $-6 < 3$.
(e) Since 2 lies to the right of -5, we know that $2 > -5$.

Practice Problem 1 In each case replace the ? with $<$ or $>$.

(a) $4 \ ? \ 2$ **(b)** $-3 \ ? -5$ **(c)** $0 \ ? -6$
(d) $-2 \ ? \ 1$ **(e)** $5 \ ? -7$

***NOTE TO STUDENT:** Fully worked-out solutions to all of the Practice Problems can be found at the back of the text starting at page SP-1*

Absolute Value Sometimes we are only interested in the distance a number is from zero. For example, the distance from 0 to $+3$ is 3. The distance from 0 to -3 is also 3. Notice that distance is always a positive number, regardless of which direction we travel on the number line. This distance is called the *absolute value.*

The **absolute value** of a number is the distance between that number and zero on the number line.

The symbol for absolute value is $| \ |$. When we write $|5|$, we are looking for the distance from 0 to 5 on the number line. Thus $|5| = 5$. This is read, "The absolute value of 5 is 5." $|-5|$ is the distance from 0 to -5 on the number line. Thus $|-5| = 5$. This is read, "The absolute value of -5 is 5."

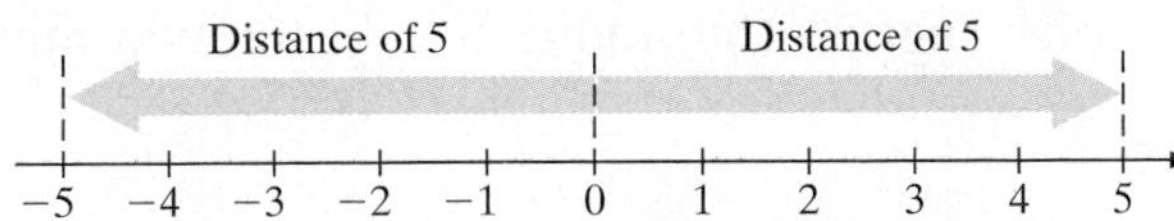

Other examples of absolute value are shown next.

$$|6| = 6 \qquad |-3| = 3$$

$$|7.2| = 7.2 \qquad \left|-\frac{1}{5}\right| = \frac{1}{5}$$

$$|0| = 0 \qquad |-26| = 26$$

When we find the absolute value of any nonzero number, we always get a positive value. We use the concept of absolute value to develop rules for adding signed numbers. We begin by looking at addition of numbers with the same sign. Although you are already familiar with the addition of positive numbers, we will look at an example.

Suppose that we earn \$52 one day and earn \$38 the next day. To learn what our two-day total is, we add the positive numbers. We earn

$$\$52 + \$38 = +\$90.$$

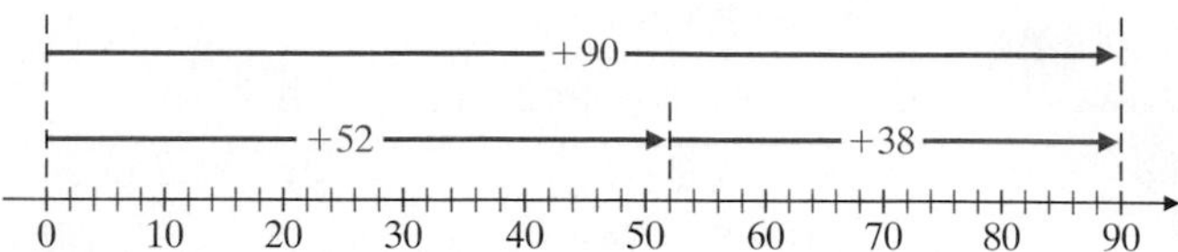

We can show this sum on a number line by drawing an arrow that starts at 0 and points 52 units to the right (because 52 is positive). At the end of this arrow we draw a second arrow that points 38 units to the right. Note that the second arrow ends at the sum, 90.

The money is coming in, and the plus sign records a gain. Notice that we added the numbers and that the sign of the sum is the same as the sign of the addends.

Now let's consider an example of addition of two negative numbers.

Suppose that we consider money spent as negative dollars. If we spend \$52 one day ($-\$52$) and we spend \$38 the next day ($-\$38$), we must add two negative numbers. What is our financial position?

$$-\$52 + (-\$38) = -\$90$$

We have spent \$90. The negative sign tells us the direction of the money: out! Notice that we added the numbers and that the sign of the sum is the same as the sign of the addends.

We can illustrate this sum on a number line by drawing a line that starts at 0 and points 52 units to the left (because -52 is negative). At the end of this arrow we add a second arrow that points 38 units to the left. This second arrow ends at the sum, -90.

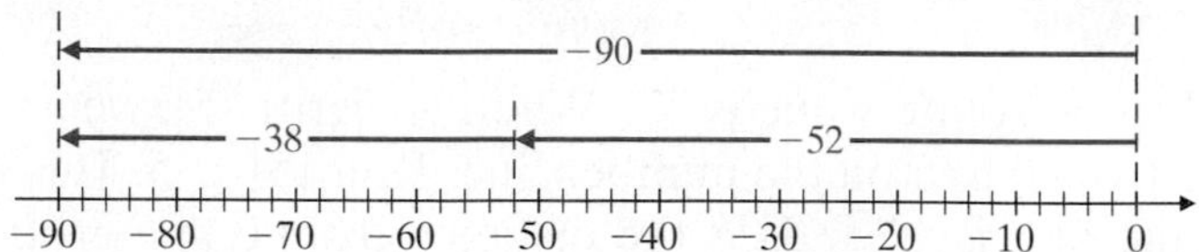

These examples suggest the addition rule for two numbers with the same sign.

ADDITION RULE FOR TWO NUMBERS WITH THE SAME SIGN

To add two numbers with the same sign:

1. Add the absolute value of the numbers.

2. Use the common sign in the answer.

EXAMPLE 2 Add. **(a)** $7 + 5$ **(b)** $-3.2 + (-5.6)$

Solution

(a) $\begin{array}{r} 7 \\ +\ 5 \\ \hline 12 \end{array}$ We add the absolute value of the numbers 7 and 5. The positive sign, although not written, is common to both numbers. The answer is a positive 12. (The + sign is not written.)

(b) $\begin{array}{r} -3.2 \\ +\ -5.6 \\ \hline -8.8 \end{array}$ We add the absolute value of the numbers 3.2 and 5.6.

We use a negative sign in our answer because we added two negative numbers.

Practice Problem 2 Add. **(a)** $9 + 14$ **(b)** $-4.5 + (-1.9)$

NOTE TO STUDENT: Fully worked-out solutions to all of the Practice Problems can be found at the back of the text starting at page SP-1

These rules can be applied to fractions as well.

EXAMPLE 3 Add. **(a)** $\frac{5}{18} + \frac{1}{3}$ **(b)** $-\frac{1}{7} + \left(-\frac{3}{5}\right)$

Solution

(a) The LCD = 18. The first fraction already has the LCD.

$$\begin{array}{rcr} \frac{5}{18} & = & \frac{5}{18} \\ + \frac{1}{3} \cdot \frac{6}{6} & = & +\frac{6}{18} \\ \hline & & \frac{11}{18} \end{array}$$

We add two positive numbers, so the answer is positive.

Teaching Tip Some students devote all their attention to finding the LCD and forget the sign of the fraction. Suggest that it is good practice to double-check the sign of any addition problem involving fractions.

(b) The LCD = 35.

$$\frac{1}{7} \cdot \frac{5}{5} = \frac{5}{35}$$

Because $\frac{1}{7} = \frac{5}{35}$ it follows that $-\frac{1}{7} = -\frac{5}{35}$.

$$\frac{3}{5} \cdot \frac{7}{7} = \frac{21}{35}$$

Because $\frac{3}{5} = \frac{21}{35}$ it follows that $-\frac{3}{5} = -\frac{21}{35}$. Thus

$$\begin{array}{r} -\frac{1}{7} \\ + \ -\frac{3}{5} \\ \hline \end{array} \quad \text{is equivalent to} \quad \begin{array}{r} -\frac{5}{35} \\ + \ -\frac{21}{35} \\ \hline -\frac{26}{35} \end{array}$$

We add two negative numbers, so the answer is negative.

It can also be written by adding $\left(\frac{-5}{35}\right) + \left(\frac{-21}{35}\right)$ to get $\frac{-26}{35}$. The negative sign can be placed in the numerator or in front of the fraction bar.

NOTE TO STUDENT: Fully worked-out solutions to all of the Practice Problems can be found at the back of the text starting at page SP-1

Practice Problem 3 Add. **(a)** $\frac{5}{12} + \frac{1}{4}$ **(b)** $-\frac{1}{6} + \left(-\frac{2}{7}\right)$

It is interesting to see how often negative numbers appear in statements of the Federal Budget. Observe the data in the following bar graph.

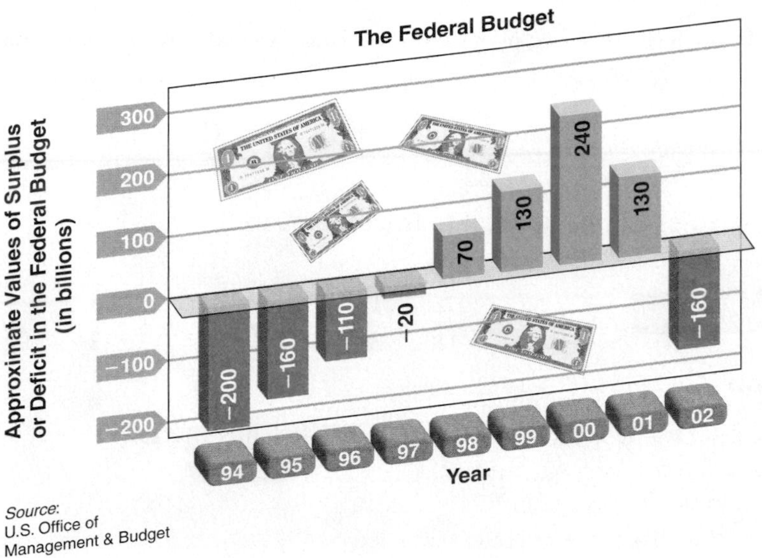

EXAMPLE 4 Find the total value of surplus or deficit for the two years 1995 and 1996.

Solution We add (−\$160 billion) + (−\$110 billion) to obtain −\$270 billion.

The total debt for these two years is \$270,000,000,000.

Practice Problem 4 Find the total value of surplus or deficit for the two years 1997 and 2002.

2 Adding Two Signed Numbers with Different Signs

Let's look at some real-life situations involving addition of signed numbers with different signs. Suppose that we earn \$52 one day and we spend \$38 the next day. If we combine (add) the two transactions, it would look like

$$\$52 + (-\$38) = +\$14.$$

On a number line we draw an arrow that starts at zero and points 52 units to the right. From the end of this arrow we draw an arrow that points 38 units to the left. (Remember, the arrow points to the left for a negative number.)

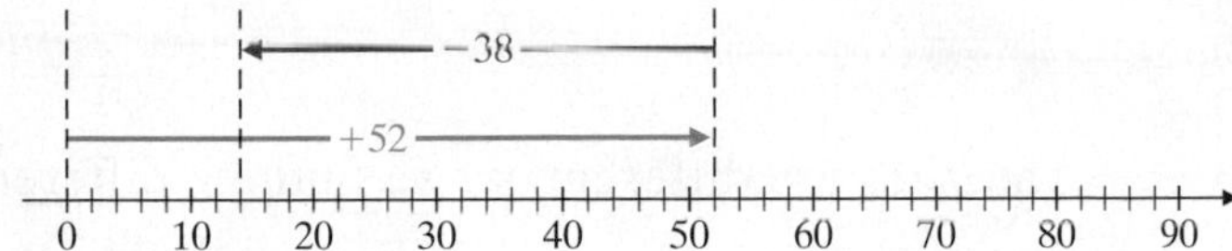

This is a situation with which we are familiar. What we actually do is subtract. That is, we take the difference between \$52 and \$38. Notice that the sign of the larger number is positive and that the sign of the answer is also positive.

Let's look at another situation. Suppose we earn \$38 and we spend \$52. The situation would look like this.

$$(-\$52) + \$38 = -\$14$$

On a number line we draw an arrow that starts at zero and points 52 units to the left. Again the arrow must point to the left to represent a negative number.

From the end of this arrow we draw an arrow that points 38 units to the right.

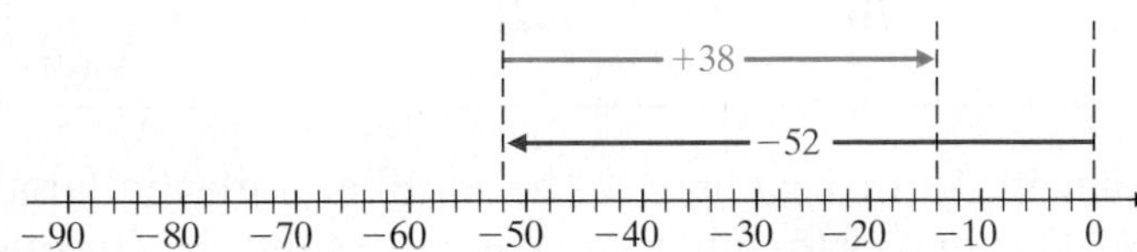

On our number line we end up at −14.

In our real-life situation we end up owing \$14, which is represented by a negative number. To find the sum, we actually find the difference between \$52 and \$38. Notice that if we do not account for sign, the larger number is 52. The sign of that number is negative and the sign of the answer is also negative. This suggests the addition rule for two numbers with different signs.

ADDITION RULE FOR TWO NUMBERS WITH DIFFERENT SIGNS

To add two numbers with different signs:

1. Subtract the absolute values of the numbers.
2. Use the sign of the number with the larger absolute value.

Teaching Tip After covering the addition of two numbers with the same sign and with different signs, ask students to do the following additional class exercises at their seats.

(a) $-12 + (-3) =$
(b) $45 + (-25) =$
(c) $36 + 10 =$
(d) $-18 + 24 =$
(e) $-30 + (-40) =$

After allowing enough time for the average student to do these, list the 5 correct answers (−15, 20, 46, 6, −70) and remind students of the rule involved in each case. This will help them to quickly think of which rule applies when they are doing several problems involving the addition of two signed numbers.

EXAMPLE 5 Add.

(a) $8 + (-10)$ **(b)** $-16.6 + 12.3$ **(c)** $\frac{3}{4} + \left(-\frac{2}{3}\right)$

Solution

(a)
$$\begin{array}{r} 8 \\ + -10 \\ \hline -2 \end{array}$$

-2 ← The signs are different, so we find the difference: $10 - 8 = 2$.

↑ The sign of the number with the larger absolute value is negative, so the answer is negative.

(b)

$$\begin{array}{r} -16.6 \\ +\ \ 12.3 \\ \hline -4.3 \end{array}$$

$\leftarrow$ The signs are different, so we find the difference.

The sign of the number with the larger absolute value is negative, so the answer is negative.

(c) $\dfrac{3}{4} + \left(-\dfrac{2}{3}\right) = \dfrac{9}{12} + \left(-\dfrac{8}{12}\right) = \dfrac{9 + (-8)}{12} = \dfrac{1}{12}$

The signs are different, so we find the difference. The sign of the number with the larger absolute value is positive, so the answer is positive.

***NOTE TO STUDENT:** Fully worked-out solutions to all of the Practice Problems can be found at the back of the text starting at page SP-1*

Practice Problem 5 Add.

(a) $7 + (-12)$ **(b)** $-20.8 + 15.2$ **(c)** $\dfrac{5}{6} + \left(-\dfrac{3}{4}\right)$

Notice that in **(b)** of Example 5, the number with the larger absolute value is on top. This makes the numbers easier to subtract. Because addition is commutative, we could have written **(a)** as

$$\begin{array}{r} -10 \\ +\ \ 8 \\ \hline \end{array}.$$

This makes the computation easier. If you are adding two numbers with different signs, place the number with the larger absolute value on top so that you can find the difference easily. As noted, the commutative property of addition holds for signed numbers.

COMMUTATIVE PROPERTY OF ADDITION

For any real numbers a and b,

$$a + b = b + a.$$

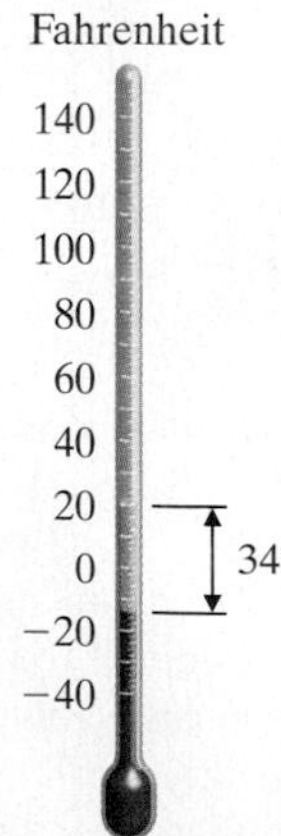

EXAMPLE 6 Last night the temperature dropped to −14°F. From that low, today the temperature rose 34°F. What was the highest temperature today?

Solution We want to add −14°F and 34°F. Because addition is commutative, it does not matter whether we add $-14 + 34$ or $34 + (-14)$.

$$\begin{array}{r} 34°\text{F} \\ +\ -14°\text{F} \\ \hline 20°\text{F} \end{array}$$

$\leftarrow$ The 34 is larger than 14. The difference between 34 and 14 is 20. The number with the larger absolute value is positive, so the answer is positive.

Practice Problem 6 Last night the temperature dropped to −19°F. From that low, today the temperature rose 28°F. What was the highest temperature today?

3 Adding Three or More Signed Numbers

We can add three or more numbers using these rules. Since addition is associative, we may group the numbers to be added in reverse order. That is, it does not matter which two numbers are added first.

EXAMPLE 7 Add. $24 + (-16) + (-10)$

Solution We can go from left to right and start with 24, or we can start with -16.

$$\begin{array}{lr} \textbf{Step 1} & 24 \\ & +\ -16 \\ \hline & 8 \\ \textbf{Step 2} & 8 \\ & +\ -10 \\ \hline & -2 \end{array} \qquad \text{or} \qquad \begin{array}{lr} \textbf{Step 1} & -16 \\ & +\ -10 \\ \hline & -26 \\ \textbf{Step 2} & -26 \\ & +\ 24 \\ \hline & -2 \end{array}$$

Practice Problem 7 Add. $36 + (-21) + (-18)$

ASSOCIATIVE PROPERTY OF ADDITION

For any three real numbers a, b, and c,

$$(a + b) + c = a + (b + c).$$

If there are many numbers to add, it may be easier to add the positive numbers and the negative numbers separately and then combine the results.

EXAMPLE 8 The results of a new company's operations over five months are listed in the following table. What is the company's overall profit or loss over the five-month period?

Net Operations
Profit/Loss Statement in Dollars

Month	Profit	Loss
January	30,000	
February		−50,000
March		−10,000
April	20,000	
May	15,000	

Solution First we will add separately the positive numbers and the negative numbers.

$$\begin{array}{r} 30{,}000 \\ 20{,}000 \\ +\,15{,}000 \\ \hline 65{,}000 \end{array} \qquad \begin{array}{r} -50{,}000 \\ +\ -10{,}000 \\ \hline -60{,}000 \end{array}$$

Teaching Tip Students are often interested to learn that negative numbers were used by the Chinese to some extent around 300 B.C. However, this idea did not spread rapidly to other parts of the world. It was not until the 1700s that most mathematics textbooks included operations with signed numbers. Some scientists feel that certain properties of electricity, magnetism, and astronomy would have been discovered earlier in the history of civilization if negative numbers had gained earlier acceptance.

Now we add the positive number 65,000 and the negative number −60,000.

$$\begin{array}{r} 65{,}000 \\ + \; -60{,}000 \\ \hline 5{,}000 \end{array}$$

The company had an overall profit of $5000 for the five-month period.

Practice Problem 8 The results of the next five months of operations for the same company are listed in the following table. What is the overall profit or loss over this five-month period?

Net Operations
Profit/Loss Statement in Dollars

Month	Profit	Loss
June		−20,000
July	30,000	
August	40,000	
September		−5,000
October		−35,000

NOTE TO STUDENT: *Fully worked-out solutions to all of the Practice Problems can be found at the back of the text starting at page SP-1*

9.1 EXERCISES

Student Solutions Manual | CD/Video | PH Math Tutor Center | MathXL®Tutorials on CD | MathXL® | MyMathLab® | Interactmath.com

Verbal and Writing Skills

1. Explain in your own words how to add two signed numbers if the signs are the same.
First find the absolute value of each number. Then add those two absolute values. Use the common sign in the answer.

2. Explain in your own words how to add two signed numbers if one number is positive and one number is negative.
First find the absolute value of each number. Then subtract those two absolute values. Use the sign of the number with the larger absolute value.

In each case, replace the ? with $<$ or $>$.

3. $-9 \ ? \ 2$ $<$ **4.** $-8 \ ? \ 6$ $<$ **5.** $-3 \ ? \ -5$ $>$ **6.** $-7 \ ? \ -14$ $>$

7. $5 \ ? \ -2$ $>$ **8.** $6 \ ? \ -3$ $>$ **9.** $-12 \ ? \ -10$ $<$ **10.** $-15 \ ? \ -13$ $<$

Simplify each absolute value expression.

11. $|7|$ 7 **12.** $|6|$ 6 **13.** $|-16|$ 16 **14.** $|-18|$ 18

Add each pair of signed numbers that have the same sign.

15. $-6 + (-11)$ -17 **16.** $-5 + (-13)$ -18 **17.** $-4.9 + (-2.1)$ -7 **18.** $-8.3 + (-3.7)$ -12

19. $8.9 + 7.6$ 16.5 **20.** $5.8 + 2.7$ 8.5 **21.** $\frac{1}{5} + \frac{2}{7}$ $\frac{17}{35}$ **22.** $\frac{2}{3} + \frac{1}{4}$ $\frac{11}{12}$

23. $-2\frac{1}{2} + \left(-\frac{1}{2}\right)$ -3 **24.** $-5\frac{1}{4} + \left(-\frac{3}{4}\right)$ -6

Add each pair of signed numbers that have different signs.

25. $14 + (-5)$ 9 **26.** $15 + (-6)$ 9 **27.** $-17 + 12$ -5 **28.** $-21 + 15$ -6

29. $-36 + 58$ 22 **30.** $-42 + 57$ 15 **31.** $-9.3 + 6.05$ -3.25 **32.** $-7.2 + 4.04$ -3.16

33. $\frac{1}{12} + \left(-\frac{3}{4}\right)$ $-\frac{2}{3}$ **34.** $\frac{1}{15} + \left(-\frac{3}{5}\right)$ $-\frac{8}{15}$

Mixed Practice

Add.

35. $\frac{7}{9} + \left(-\frac{2}{9}\right)$ $\frac{5}{9}$ **36.** $\frac{5}{12} + \left(-\frac{7}{12}\right)$ $-\frac{1}{6}$ **37.** $-18 + (-4)$ -22 **38.** $-34 + (-2)$ -36

39. $1.48 + (-2.2)$ -0.72 **40.** $3.72 + (-4.1)$ -0.38 **41.** $-125 + (-238)$ -363 **42.** $-514 + (-176)$ -690

43. $13 + (-9)$
4

44. $-18 + 7$
-11

45. $-2\frac{1}{4} + \left(-1\frac{5}{6}\right)$
$-4\frac{1}{12}$

46. $-3\frac{1}{2} + \left(-3\frac{5}{6}\right)$
$-7\frac{1}{3}$

47. $-7.56 + 13.8$
6.24

48. $-6.89 + 15.9$
9.01

49. $-5 + \left(-\frac{1}{2}\right)$
$-\frac{11}{2}$ or $-5\frac{1}{2}$

50. $\frac{5}{6} + (-3)$
$-\frac{13}{6}$ or $-2\frac{1}{6}$

Add.

51. $-20.5 + 18.1 + (-12.3)$ -14.7

52. $-9.8 + 14.7 + (-16.8)$ -11.9

53. $11 + (-9) + (-10) + 8$ 0

54. $(-13) + 8 + (-12) + 17$ 0

55. $-7 + 6 + (-2) + 5 + (-3) + (-5)$
-6

56. $-2 + 1 + (-12) + 7 + (-4) + (-1)$
-11

57. $\left(-\frac{1}{5}\right) + \left(-\frac{2}{3}\right) + \left(\frac{4}{25}\right)$ $-\frac{53}{75}$

58. $\left(-\frac{1}{7}\right) + \left(-\frac{5}{21}\right) + \left(\frac{3}{14}\right)$ $-\frac{1}{6}$

Applications

Profit and Loss Statements *Use signed numbers to represent the total profit or loss for a company after the following reports.*

59. A \$43,000 loss in February followed by a \$51,000 loss in March.
$-\$94,000$

60. A \$16,000 loss in May followed by a \$25,000 loss in June.
$-\$41,000$

61. A \$28,000 profit in July followed by a \$19,000 loss in August.
\$9000

62. An \$89,000 profit in April followed by a \$33,000 loss in May.
\$56,000

63. A \$35,000 loss in April, a \$17,000 profit in May, and a \$20,000 loss in June.
$-\$38,000$

64. A \$34,000 loss in October, a \$12,000 loss in November, and a \$15,500 profit in December.
$-\$30,500$

Solve.

65. ***Temperature Change*** Last night, the temperature was −1°F. In the morning, the temperature dropped 17°F. What was the new temperature?
−18°F

66. ***Temperature Change*** This morning the temperature was −12°F. This afternoon, the temperature rose 19°F. What was the new temperature? 7°F

67. ***Temperature Change*** This morning, the temperature was −5°F. This evening, the temperature rose 4°F. What was the new temperature?
−1°F

68. ***Temperature Change*** Last night the temperature was −7°F. The temperature dropped 15°F this afternoon. What was the new temperature?
−22°F

69. ***Stock Market*** The following list of signed numbers is the daily loss or gain of one share of Kraft Food, Inc. stock for the week of September 22–September 26, 2003: $-0.15, +0.20, +0.21, -0.24, -0.36$. What was the net loss or gain for the week?
-0.34

70. ***Stock Market*** The following list of signed numbers is the daily loss or gain of one share of Coca-Cola Enterprises stock for the week of September 22–September 26, 2003: $+0.16, +0.23, -0.18, +0.13, -0.43$. What was the net loss or gain for the week?
-0.09

71. ***Football*** In three plays of a football game, the quarterback threw passes that lost 8 yards, gained 13 yards, and lost 6 yards. What was the total gain or loss of the three plays?
loss of 1 yard (-1 yard)

72. ***Football*** In three plays of a football game, the quarterback threw passes that gained 20 yards, lost 13 yards, and gained 5 yards. What was the total gain or loss of the three plays?
gain of 12 yards ($+12$ yards)

To Think About

73. ***Credit Cards*** Jeffrey had a credit card balance of $-\$28$. He immediately made a payment of \$30. The next day the credit card company notified him that they had charged him \$15 because his account balance was negative. What was the credit card balance in Jeffrey's account after these actions took place?
$-\$13$

74. ***Credit Cards*** Susan had a credit card balance of $-\$39$. She immediately made a payment of \$100. Before the payment was credited, the credit card bank charged her \$0.79 in interest. What was the credit card balance in Susan's account after these actions took place?
\$60.21

75. ***Checking Accounts*** Bob examined his checking account register. He thought his balance was \$89.50. However, he forgot to subtract an ATM withdrawal of \$50.00. Since the ATM that he used was at a different bank, he was also charged \$2.50 for using the ATM. What was the actual balance in his checking account?
\$37.00

76. ***Checking Accounts*** Nancy examined her checking account register. She thought her balance was \$97.40. However, she forgot to subtract a check of \$95.00 that she had made out the previous week. She also forgot to subtract the monthly \$4.50 fee charged by her bank for having a checking account. What was the actual balance in her checking account?
$-\$2.10$

Cumulative Review

▲ **77.** ***Geometry*** Use $V = \dfrac{4\pi r^3}{3}$ to find the volume of a sphere of radius 6 feet. Use $\pi \approx 3.14$ and round to the nearest tenth.
904.3 ft^3

▲ **78.** ***Geometry*** Use $V = \dfrac{Bh}{3}$ to find the volume of a pyramid whose rectangular base measures 9 meters by 7 meters and whose height is 10 meters.
210 m^3

79. Find the median value. 62, 59, 60, 57, 60, 61
60

80. Find the mean. \$36, \$42, \$39, \$39, \$41, \$43
\$40

9.2 Subtracting Signed Numbers

Student Learning Objectives

After studying this section, you will be able to:

1. **Subtract one signed number from another.**
2. **Solve problems involving both addition and subtraction of signed numbers.**
3. **Solve simple applied problems that involve the subtraction of signed numbers.**

1 Subtracting One Signed Number from Another

We begin our discussion by defining the word **opposite.** The opposite of a positive number is a negative number with the same absolute value. For example, the opposite of 7 is -7.

The opposite of a negative number is a positive number with the same absolute value. For example, the opposite of -9 is 9. If a number is the opposite of another number, these two numbers are at an equal distance from zero on the number line.

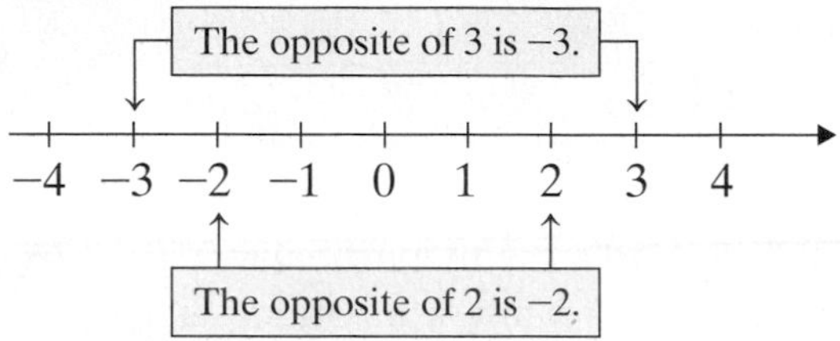

The opposite of 3 is −3.
−4 −3 −2 −1 0 1 2 3 4
The opposite of 2 is −2.

The sum of a number and its opposite is zero.

$$-3 + 3 = 0 \qquad 2 + (-2) = 0$$

We will use the concept of opposite to develop a way to subtract integers.

Let's think about how a checking account works. Suppose that you deposit \$25 and the bank adds a service charge of \$5 for a new checkbook. Your account looks like this:

$$\$25 + (-\$5) = \$20$$

Suppose instead that you deposit \$25 and the bank adds no charge. The next day, you write a check for \$5. The result of these two transactions is

$$\$25 - \$5 = \$20$$

Note that your account has the same amount of money (\$20) in both cases.

We see that adding a negative 5 to 25 is the same as subtracting a positive 5 from 25. That is, $25 + (-5) = 20$ and $25 - 5 = 20$.

Subtracting is equivalent to adding the opposite.

Subtracting	**Adding the Opposite**
$25 - 5 = 20$	$25 + (-5) = 20$
$19 - 6 = 13$	$19 + (-6) = 13$
$7 - 3 = 4$	$7 + (-3) = 4$
$15 - 5 = 10$	$15 + (-5) = 10$

We define a rule for the subtraction of signed numbers.

SUBTRACTION OF SIGNED NUMBERS

To subtract signed numbers, add the opposite of the second number to the first number.

Teaching Tip Some students will ask, "Why do we have to change subtraction to adding the opposite?" The most satisfying answer seems to be that if we get used to thinking of subtraction as adding the opposite, then we will be able to consider problems with indicated subtraction and addition such as $-5 - 3 - (-2)$ as addition problems like $-5 + (-3) + 2$. When we are able to do this, we can use the commutative property, which makes many problems easier to solve.

Thus, to do a subtraction problem, we first change it to an equivalent addition problem in which the first number does not change but the second number is replaced by its opposite. Then we follow the rules of *addition* for signed numbers.

EXAMPLE 1 Subtract. $-8 - (-2)$

Solution

$$-8 - (-2)$$

$$-8 + 2$$

Write the opposite of −2, which is 2.

Change subtraction to addition.

Now we use the rules of addition for two numbers with opposite signs.

$$-8 + 2 = -6$$

Practice Problem 1 Subtract. $-10 - (-5)$

NOTE TO STUDENT: Fully worked-out solutions to all of the Practice Problems can be found at the back of the text starting at page SP-1

EXAMPLE 2 Subtract. **(a)** $7 - 8$ **(b)** $-12 - 16$

Solution

(a)

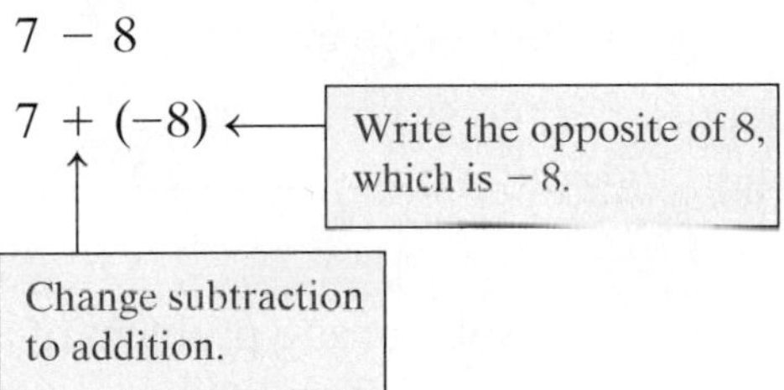

Now we use the rules of addition for two numbers with opposite signs.

$$7 + (-8) = -1$$

(b)

$$-12 - 16$$

$$-12 + (-16)$$

Write the opposite of 16, which is −16.

Change subtraction to addition.

Now we follow the rules of addition of two numbers with the same sign.

$$-12 + (-16) = -28$$

Practice Problem 2 Subtract. **(a)** $5 - 12$ **(b)** $-11 - 17$

Sometimes the numbers we subtract are fractions or decimals.

EXAMPLE 3 Subtract. **(a)** $5.6 - (-8.1)$ **(b)** $-\frac{6}{11} - \left(-\frac{1}{22}\right)$

Solution

(a) Change the subtraction to adding the opposite. Then add.

$$5.6 - (-8.1) = 5.6 + 8.1$$
$$= 13.7$$

(b) Change the subtraction to adding the opposite. Then add.

$$-\frac{6}{11} - \left(-\frac{1}{22}\right) = -\frac{6}{11} + \frac{1}{22}$$

$$= -\frac{6}{11} \cdot \frac{2}{2} + \frac{1}{22}$$ We see that the LCD = 22. We change $\frac{6}{11}$ to a fraction with a denominator of 22.

$$= -\frac{12}{22} + \frac{1}{22}$$ Add.

$$= -\frac{11}{22} \text{ or } -\frac{1}{2}$$

NOTE TO STUDENT: Fully worked-out solutions to all of the Practice Problems can be found at the back of the text starting at page SP-1

Practice Problem 3 Subtract. **(a)** $3.6 - (-9.5)$ **(b)** $-\frac{5}{8} - \left(-\frac{5}{24}\right)$

Remember that in performing subtraction of two signed numbers:

1. The first number does not change.
2. The subtraction sign is changed to addition.
3. We write the opposite of the second number.
4. We find the result of this addition problem.

Think of each subtraction problem as a problem of adding the opposite.

If you see $7 - 10$, think $7 + (-10)$.
If you see $-3 - 19$, think $-3 + (-19)$.
If you see $6 - (-3)$, think $6 + (+3)$.

EXAMPLE 4 Subtract.

(a) $6 - (+3)$ **(b)** $-\frac{1}{2} - \left(-\frac{1}{3}\right)$ **(c)** $2.7 - (-5.2)$

Solution

(a) $6 - (+3) = 6 + (-3) = 3$

(b) $-\frac{1}{2} - \left(-\frac{1}{3}\right) = -\frac{1}{2} + \frac{1}{3} = -\frac{3}{6} + \frac{2}{6} = -\frac{1}{6}$

(c) $2.7 - (-5.2) = 2.7 + 5.2 = 7.9$

Practice Problem 4 Subtract.

(a) $20 - (-5)$ **(b)** $-\frac{1}{5} - \left(-\frac{1}{2}\right)$ **(c)** $3.6 - (-5.5)$

Calculator

Negative Numbers

To enter a negative number on most scientific calculators, find the key marked [+/−]. To enter the number −3, press the key 3 and then the key [+/−]. The display should read

[−3]

To find $(-32) + (-46)$, enter

32 [+/−] [+] 46 [+/−] [=]

The display should read

[−78]

Try the following.

(a) $-756 + 184$
(b) $92 + (-51)$
(c) $-618 - (-824)$
(d) $-36 + (-10) - (-15)$

Note: The [+/−] key changes the sign of a number from + to − or − to +.

Solving Problems Involving Both Addition and Subtraction of Signed Numbers

EXAMPLE 5 Perform the following set of operations, working from left to right. $-8 - (-3) + (-5)$

Solution $-8 - (-3) + (-5) = -8 + 3 + (-5)$
$= -5 + (-5) = -10$

First we change subtracting a -3 to adding a 3.

Practice Problem 5 Perform the following set of operations.

$$-5 - (-9) + (-14)$$

Solving Simple Applied Problems That Involve the Subtraction of Signed Numbers

When we want to find the difference in altitude between two mountains, we subtract. We subtract the lower altitude from the higher altitude. Look at the illustration at the right. The difference in altitude between A and B is 3480 feet − 1260 feet = 2220 feet.

Land that is below sea level is considered to have a negative altitude. The Dead Sea is 1312 feet below sea level. Look at the following illustration. The difference in altitude between C and D is 2590 feet − (−1312 feet) = 3902 feet.

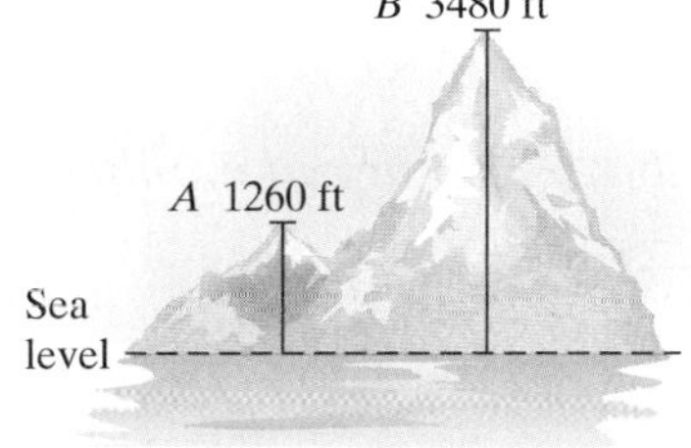

3480 ft − 1260 ft = 2220 ft

Teaching Tip Students find this example helpful. During the spring flooding season the Esconti River continued to rise for four days. The following data were recorded:

Day	Height of River (number of feet above normal)
Monday	−4
Tuesday	−2
Wednesday	+5
Thursday	+8

(a) What was the difference in height from Monday to Tuesday?

$-2 - (-4) = +2$ ft

(b) What was the difference in height from Tuesday to Wednesday?

$+5 - (-2) = +7$ ft

(c) What was the difference in height from Wednesday to Thursday?

$+8 - (+5) = +3$ ft

The three subtraction examples here seem to make sense to students because they know the result should be positive in each case, since the river is rising.

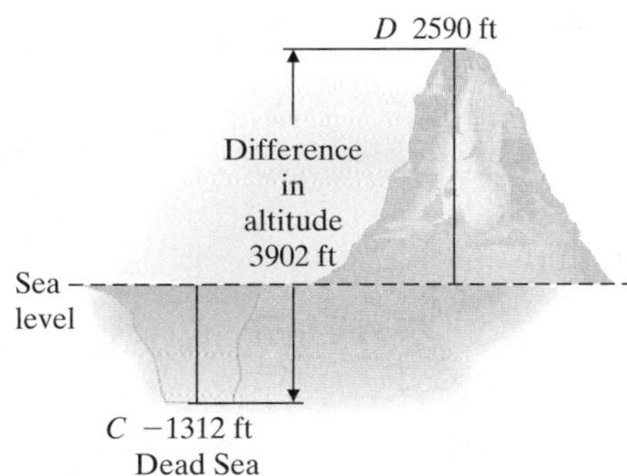

2590 ft − (−1312 ft) = 2590 ft + 1312 ft = 3902 ft

EXAMPLE 6 Find the difference in temperature between 38°F during the day in Anchorage, Alaska, and −26°F at night.

Solution We subtract the lower temperature from the higher temperature.

$$38 - (-26) = 38 + 26 = 64$$

The difference is 64°F.

Fahrenheit
140
120
100
80
60
40
20
0
−20
−40
64°

Practice Problem 6 Find the difference in temperature between 31°F during the day in Fairbanks, Alaska, and −37° at night.

9.2 Exercises

Student Solutions Manual | CD/Video | PH Math Tutor Center | MathXL®Tutorials on CD | MathXL® | MyMathLab® | Interactmath.com

Subtract the signed numbers by adding the opposite of the second number to the first number.

1. $-9 - (-3)$ -6 **2.** $-7 - (-5)$ -2 **3.** $-12 - (-7)$ -5 **4.** $-10 - (-2)$ -8 **5.** $3 - 9$ -6

6. $5 - 12$ -7 **7.** $-14 - 3$ -17 **8.** $-6 - 18$ -24 **9.** $-12 - (-10)$ -2 **10.** $-27 - (-12)$ -15

11. $46 - (-39)$ 85 **12.** $53 - (-28)$ 81 **13.** $12 - 30$ -18 **14.** $10 - 14$ -4 **15.** $-12 - (-15)$ 3

16. $-17 - (-30)$ 13 **17.** $150 - 210$ -60 **18.** $500 - 150$ 350 **19.** $300 - (-256)$ 556 **20.** $420 - (-300)$ 720

21. $-2.5 - 4.2$ -6.7 **22.** $-4.1 - 3.9$ -8 **23.** $4.2 - 10.7$ -6.5 **24.** $3.8 - 12.3$ -8.5

Mixed Practice

Subtract the signed numbers by adding the opposite of the second number to the first number.

25. $-10.9 - (-2.3)$ -8.6 **26.** $-6.8 - (-2.9)$ -3.9 **27.** $20.23 - (-12.71)$ 32.94 **28.** $13.92 - (-14.86)$ 28.78

29. $\frac{1}{4} - \left(-\frac{3}{4}\right)$ 1 **30.** $\frac{5}{7} - \left(-\frac{6}{7}\right)$ $\frac{11}{7}$ or $1\frac{4}{7}$ **31.** $-\frac{5}{6} - \frac{1}{3}$ $-\frac{7}{6}$ or $-1\frac{1}{6}$ **32.** $-\frac{2}{8} - \frac{1}{4}$ $-\frac{1}{2}$

33. $-7\frac{2}{5} - \left(-2\frac{1}{3}\right)$ $-5\frac{1}{15}$ **34.** $-8\frac{2}{3} - \left(-5\frac{1}{4}\right)$ $-3\frac{5}{12}$ **35.** $\frac{5}{9} - \frac{2}{7}$ $\frac{17}{63}$ **36.** $\frac{5}{11} - \frac{1}{5}$ $\frac{14}{55}$

Perform each set of operations, working from left to right.

37. $2 - (-8) + 5$ 15 **38.** $7 - (-3) + 9$ 19 **39.** $-5 - 6 - (-11)$ 0

40. $-3 - 12 - (-5)$ -10 **41.** $21 - (-15) - (-10)$ 46 **42.** $32 - (-12) - (-18)$ 62

43. $-16 - (-6) - 12$ -22 **44.** $-13 - (-4) - 15$ -24 **45.** $9 - 3 - 2 - 6$ -2

46. $12 - 5 - 4 - 8$ -5 **47.** $-2.4 - 7.1 + 1.3 - (-2.8)$ -5.4 **48.** $-4.3 - 6.5 + 2.1 - (-5.2)$ -3.5

Applications

Use your knowledge of signed numbers to answer exercises 49–52.

49. ***Altitude Change*** The highest point in California is Mt. Whitney at 14,494 feet. The lowest point is −282 feet in Death Valley. How far above Death Valley is Mt. Whitney?
14,776 ft

50. ***Altitude Change*** The highest point in Mexico is Volcan Pico de Orizaba at 18,689 feet. The lowest point is −33 feet at Laguna Salada. How far above Laguna Salada is Volcan Pico de Orizaba?
18,722 ft

51. ***Temperature Change*** Find the difference in temperature in Alta, Utah, between 23°F during the day and −19°F at night. 42°F

52. ***Temperature Change*** Find the difference in temperature in Fairbanks, Alaska, between 27°F during the day and −33°F at night. 60°F

53. ***Temperature Change*** In Thule, Greenland, yesterday, the temperature was −29°F. Today the temperature rose 16°F. What is the new temperature? −13°F

54. ***Height Change*** Find the difference in height between the top of a hill 642 feet high and a crack caused by an earthquake 57 feet below sea level. 699 ft

Profit and Loss Statements *A company's profit and loss statement in dollars for the last five months is shown in the following table.*

Month	Profit	Loss
January	18,700	
February		−34,700
March		−6,300
April	43,600	
May		−12,400

55. What is the change in the profit/loss status of the company from the first of January to the end of February? −\$16,000

56. What is the change in the profit/loss status of the company from the first of February to the end of March? −\$41,000

57. What is the change in the profit/loss status of the company from the first of March to the end of April? +\$37,300

58. What is the change in the profit/loss status of the company from the first of January to the end of May? +\$8900

59. In January 2000, the value of one share of a certain stock was $\$15\frac{1}{2}$. During the next three days, the value fell $\$1\frac{1}{2}$, rose $\$2\frac{3}{4}$, and fell $\$3\frac{1}{4}$. What was the value of one share at the end of the three days?
$\$13\frac{1}{2}$

60. In February 2000, the value of one share of a certain stock was \$28. During the next three days, the value rose $\$2\frac{1}{4}$, fell $\$5\frac{1}{2}$, and fell $\$1\frac{1}{2}$. What was the value of one share at the end of those three days?
$\$23\frac{1}{4}$

To Think About

61. Write a word problem about a bank and the calculation $50 - (-80) = 50 + 80 = 130$.
The bank finds that a customer has \$50 in his checking account. However, the bank must remove an erroneous debit of \$80 from the customer's account. When the bank makes the correction, what will the new balance be?

62. Write a word problem about a bank and the calculation $-100 - 50 = -100 + (-50) = -150$.
The bank finds that a customer has overdrawn his checking account by \$100. However, the bank must remove an erroneous credit of \$50 that should not have been deposited in the customer's account. When the bank makes the correction, what will the new balance be?

Cumulative Review

In exercises 63–64, perform the operations in the proper order.

63. $20 \times 2 \div 10 + 4 - 3$ 5

64. $2 + 3 \times (5 + 7) \div 9$ 6

▲ 65. ***Geometry*** A metalworker is making a copper sign. Find the area of the largest possible copper circle that can be made from a square piece of copper that measures 6 inches on each side. Use $\pi \approx 3.14$.
28.26 square inches

▲ 66. ***Geometry*** A triangular wooden sign is mounted in the laundry room of Dori and Elizabeth Little. The sign reads "Blessed are they who wash the clothes." The piece of wood is 7 inches high and has a base of 11 inches. Dori needs to replace the wood with a larger triangle with the same height but a new base of 14 inches. How much does the area of the piece of wood increase with this change?
10.5 square inches

9.3 Multiplying and Dividing Signed Numbers

Student Learning Objectives

After studying this section, you will be able to:

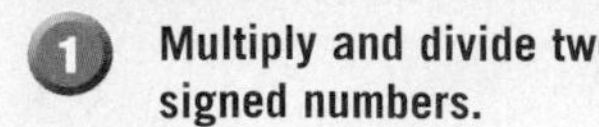

Multiply and divide two signed numbers.

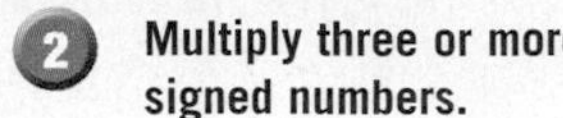

Multiply three or more signed numbers.

NOTE TO STUDENT: Fully worked-out solutions to all of the Practice Problems can be found at the back of the text starting at page SP-1

Teaching Tip Explain to students that multiplication of signed numbers and multiplication of other types of expressions in algebra are indicated by parentheses. They will see this notation in any math courses they take in the future.

Teaching Tip Students often find it helpful to consider a few examples from everyday life that illustrate multiplying a positive number by a negative number to obtain a negative number. Students seem to find the following illustrations helpful.

(a) Repeated plays where yardage is lost in a football game. Four plays losing 3 yards each is a loss of 12 yards. $4(-3) = -12$

(b) Repeated sales made by a company at a loss. Four sales made losing 9 dollars each is a loss of 36 dollars. $4(-9) = -36$

(c) Repeated dives by a submarine going farther and farther beneath the surface. The submarine captain gave four commands for the sub to dive 50 feet farther down. The total depth dropped was 200 feet. $4(-50) = -200$.

Repeatedly interjecting practical examples of different types tends to convince students that the sign rules are logical and not "artificial."

1 Multiplying and Dividing Two Signed Numbers

Recall the different ways we can indicate multiplication.

$$3 \times 5 \qquad 3 \cdot 5 \qquad (3)(5) \qquad 3(5)$$

It is common to use parentheses to mean multiplication.

EXAMPLE 1 Evaluate. **(a)** $(7)(8)$ **(b)** $3(12)$

Solution

(a) $(7)(8) = 56$ **(b)** $3(12) = 36$

Practice Problem 1 Evaluate. **(a)** $(6)(9)$ **(b)** $7(12)$

Now suppose we multiply a positive number times a negative number. What will happen? Let us look for a pattern.

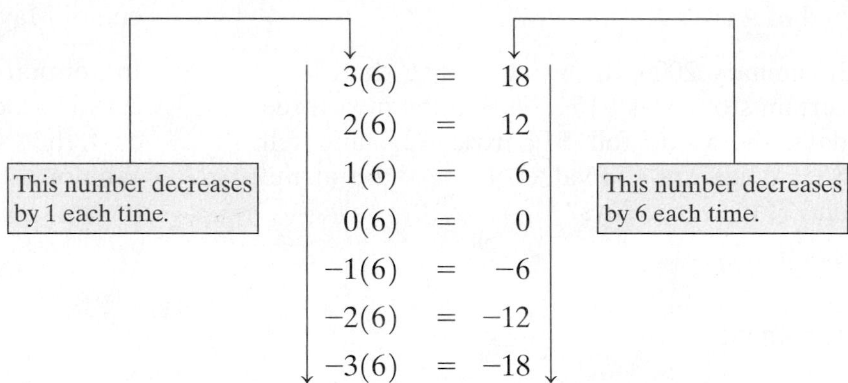

Our pattern suggests that when we multiply a positive number times a negative number, we get a negative number. Thus we will state the following rule.

MULTIPLICATION RULE FOR TWO NUMBERS WITH DIFFERENT SIGNS

To multiply two numbers with different signs, multiply the absolute values. The result is negative.

EXAMPLE 2 Multiply. **(a)** $2(-8)$ **(b)** $(-3)(25)$

Solution In each case, we are multiplying two signed numbers with different signs. We will always get a negative number for an answer.

(a) $2(-8) = -16$ **(b)** $(-3)(25) = -75$

Practice Problem 2 Multiply. **(a)** $(-8)(5)$ **(b)** $3(-60)$

A similar rule applies to division of two numbers when the signs are not the same.

DIVISION RULE FOR TWO NUMBERS WITH DIFFERENT SIGNS

To divide two numbers with different signs, divide the absolute values. The result is negative.

Teaching Tip An interesting example of division of two numbers that are different in sign is the following. Mount Washington recorded the following daily average Fahrenheit temperatures in January for a four-day period: −30°, −38°, −40°, and −32°. What was the average temperature over this four-day period? $-30 + (-38) + (-40) + (-32) = -140$. Now −140 divided by 4 yields −35. The average temperature was −35 degrees Fahrenheit.

EXAMPLE 3 Divide. **(a)** $-20 \div 5$ **(b)** $36 \div (-18)$

Solution

(a) $-20 \div 5 = -4$ **(b)** $36 \div (-18) = -2$

Practice Problem 3 Divide. **(a)** $-50 \div 25$ **(b)** $49 \div (-7)$

What happens if we multiply $(-2)(-6)$? What sign will we obtain? Let us once again look for a pattern.

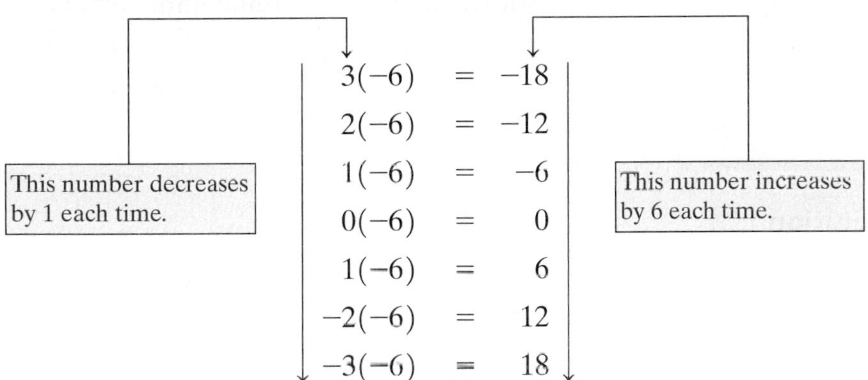

Our pattern suggests that when we multiply a negative number by a negative number, we get a positive number. A similar pattern occurs for division. Thus we are ready to state the following rule.

MULTIPLICATION AND DIVISION RULE FOR TWO NUMBERS WITH THE SAME SIGN

To multiply or divide two numbers with the same sign, multiply or divide the absolute values. The sign of the result is positive.

EXAMPLE 4 Multiply. **(a)** $-5(-6)$ **(b)** $\left(-\frac{1}{2}\right)\left(-\frac{3}{5}\right)$

Solution In each case, we are multiplying two numbers with the same sign. We will always obtain a positive number.

(a) $-5(-6) = 30$ **(b)** $\left(-\frac{1}{2}\right)\left(-\frac{3}{5}\right) = \frac{3}{10}$

Practice Problem 4 Multiply. **(a)** $-10(-6)$ **(b)** $\left(-\frac{1}{3}\right)\left(-\frac{2}{7}\right)$

Because division is related to multiplication, we find, just as in multiplication, that whenever we divide two numbers with the same sign, the result is a positive number.

EXAMPLE 5 Divide. **(a)** $(-50) \div (-2)$ **(b)** $(-9.9) \div (-3.0)$

Solution

(a) $(-50) \div (-2) = 25$ **(b)** $(-9.9) \div (-3.0) = 3.3$

NOTE TO STUDENT: Fully worked-out solutions to all of the Practice Problems can be found at the back of the text starting at page SP-1

Practice Problem 5 Divide.

(a) $-78 \div (-2)$ **(b)** $(-1.2) \div (-0.5)$

2 Multiplying Three or More Signed Numbers

When multiplying more than two numbers, multiply any two numbers first, then multiply the result by another number. Continue until each factor has been used.

Teaching Tip Remind students that multiplication is commutative. Thus to evaluate an expression like $12(-3)(5)\left(-\frac{1}{4}\right)$, they may prefer—although it is not necessary—to write the problem in the form $12\left(-\frac{1}{4}\right)(-3)(5)$ in order to facilitate cancellation.

EXAMPLE 6 Multiply. $5(-2)(-3)$

Solution

$$5(-2)(-3) = -10(-3) \quad \text{First multiply } 5(-2) = -10.$$
$$= 30 \quad \text{Then multiply } -10(-3) = 30.$$

Practice Problem 6 Multiply. $(-6)(3)(-4)$

EXAMPLE 7 Chemists have determined that a phosphate ion has an electrical charge of -3. If 10 phosphate ions are removed from a substance, what is the change in the charge of the remaining substance?

Solution Removing 10 ions from a substance can be represented by the number -10. We can use multiplication to determine the result of removing 10 ions with an electrical charge of -3.

$$(-3)(-10) = 30$$

Thus the change in charge would be $+30$.

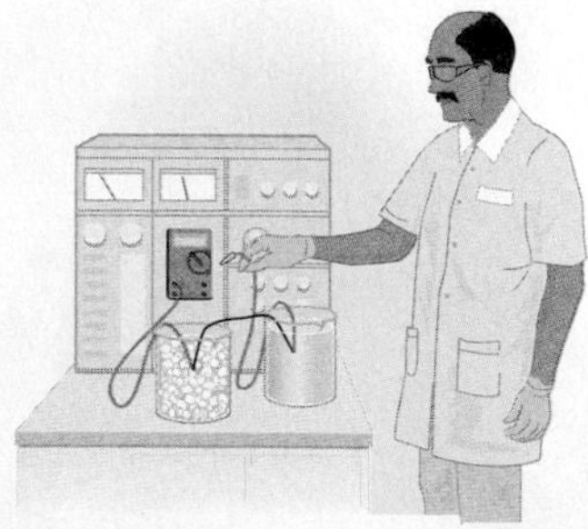

Practice Problem 7 An oxide ion has an electrical charge of -2. If 6 oxide ions are added to a substance, what is the change in the charge of the new substance?

EXAMPLE 8 Travis Tobey went outside his house to measure the temperature at 4:00 P.M. for seven days in October in Copper Center, Alaska. His temperature readings in degrees Fahrenheit were $-11°$, $-8°$, $-15°$, $-3°$, $5°$, $12°$, and $-1°$. Find the average temperature for this seven-day period.

Solution To find the average, we take the sum of the seven days of temperature readings and divide by seven.

$$\frac{-11 + (-8) + (-15) + (-3) + 5 + 12 + (-1)}{7} = \frac{-21}{7} = -3$$

The average temperature was $-3°$F.

Practice Problem 8 In March Tony Pitkin measured the morning temperature at his farm in North Dakota at 7:00 A.M. each day. For a six-day period the temperatures in degrees Fahrenheit were $17°$, $19°$, $2°$, $-4°$, $-3°$, and $-13°$. What was the average temperature at his farm over this six-day period?

9.3 Exercises

Student Solutions Manual | CD/ Video | PH Math Tutor Center | MathXL®Tutorials on CD | MathXL® | MyMathLab® | Interactmath.com

Verbal and Writing Skills

1. In your own words, state the rule for multiplying two signed numbers if the signs are the same.
To multiply two numbers with the same sign, multiply the absolute values. The sign of the result is positive.

2. In your own words, state the rule for multiplying two signed numbers if the signs are different.
To multiply two numbers with different signs, multiply the absolute values. The sign of the result is negative.

Multiply.

3. $(12)(3)$ 36

4. $(15)(5)$ 75

5. $(-20)(-3)$ 60

6. $(-30)(-6)$ 180

7. $(-20)(8)$ -160

8. $(-15)(12)$ -180

9. $(3)(-22)$ -66

10. $(4)(-34)$ -136

11. $(2.5)(-0.6)$ -1.5

12. $(8.5)(-0.3)$ -2.55

13. $(-12.5)(-2.25)$ 28.125

14. $(-7.35)(-10.5)$ 77.175

15. $\left(-\frac{2}{5}\right)\left(\frac{3}{7}\right)$ $-\frac{6}{35}$

16. $\left(-\frac{5}{12}\right)\left(-\frac{2}{3}\right)$ $\frac{5}{18}$

17. $\left(-\frac{4}{12}\right)\left(-\frac{3}{23}\right)$ $\frac{1}{23}$

18. $\left(-\frac{7}{9}\right)\left(\frac{3}{28}\right)$ $-\frac{1}{12}$

Divide.

19. $-64 \div 8$ -8

20. $-63 \div 7$ -9

21. $\frac{48}{-6}$ -8

22. $\frac{52}{-13}$ -4

23. $\frac{-120}{-20}$ 6

24. $\frac{-180}{-10}$ 18

25. $-25 \div (-5)$ 5

26. $-36 \div (-4)$ 9

27. $-\frac{4}{9} \div \left(-\frac{16}{27}\right)$ $\frac{3}{4}$

28. $-\frac{3}{20} \div \left(-\frac{6}{5}\right)$ $\frac{1}{8}$

29. $\dfrac{-\frac{4}{5}}{-\frac{7}{10}}$ $\frac{8}{7}$ or $1\frac{1}{7}$

30. $\dfrac{-\frac{26}{15}}{-\frac{13}{7}}$ $\frac{14}{15}$

31. $50.28 \div (-6)$ -8.38

32. $30.45 \div (-5)$ -6.09

33. $\frac{45.6}{-8}$ -5.7

34. $\frac{66.6}{-9}$ -7.4

35. $\frac{-21{,}000}{-700}$ 30

36. $\frac{-320{,}000}{-8000}$ 40

Mixed Practice

Multiply or divide.

37. $5(-9)$ -45

38. $6(-11)$ -66

39. $(-12)(-4)$ 48

40. $(-10)(-7)$ 70

41. $\frac{15}{-3}$ -5

42. $\frac{36}{-2}$ -18

43. $-30 \div (-3)$ 10

44. $-50 \div (-5)$ 10

45. $(-1.4)(2)$
-2.8

46. $(-1.6)(3)$
-4.8

47. $0.028 \div (-1.4)$
-0.02

48. $0.069 \div (-2.3)$
-0.03

49. $\left(-\frac{3}{5}\right)\left(-\frac{5}{7}\right)$
$\frac{3}{7}$

50. $\left(-\frac{4}{9}\right)\left(-\frac{9}{13}\right)$
$\frac{4}{13}$

51. $\left(-\frac{8}{5}\right) \div \left(-\frac{16}{15}\right)$
$\frac{3}{2}$

52. $\left(-\frac{9}{16}\right) \div \left(-\frac{3}{4}\right)$
$\frac{3}{4}$

Multiply.

53. $3(-6)(-4)$ 72

54. $3(-5)(-9)$ 135

55. $(-8)(4)(-6)$ 192

56. $(-7)(-3)(6)$ 126

57. $2(-8)(3)\left(-\frac{1}{3}\right)$
16

58. $7(-2)(-5)\left(\frac{1}{7}\right)$
10

59. $(-20)(6)(-30)(-5)$
$-18{,}000$

60. $(5)(-40)(-20)(-8)$
$-32{,}000$

61. $8(-3)(-5)(0)(-2)$ 0

62. $9(-6)(-4)(-3)(0)$ 0

Perform each set of operations. Simplify your answer. Work from left to right.

63. $\left(-\frac{2}{3}\right)\left(-\frac{3}{4}\right)\left(-\frac{5}{6}\right)$ $-\frac{5}{12}$

64. $\left(-\frac{2}{3}\right) \div \left(-\frac{2}{3}\right)\left(\frac{3}{5}\right)$ $\frac{3}{5}$

Applications

65. ***Stock Market*** Carla owns 90 shares of a stock whose value went down \$1.50 per share. She also owns 70 shares of a stock whose value went up \$0.80 per share. How much did Carla gain or lose?
She lost \$79.

66. ***Stock Market*** Vincent owns 50 shares of a stock whose value went down \$2.10 per share. He also owns 65 shares of a stock whose value went up \$3.20 per share. How much did Vincent gain or lose?
He gained \$103.

67. ***Temperature Records*** In Missoula, Montana, Bob recorded the following temperature readings in degrees Fahrenheit at 7 A.M. each morning for eight days in a row: $-12°, -14°, -3°, 5°, 8°, -1°, -10°$, and $-23°$. What was the average temperature?
$-6.25°$F

68. ***Temperature Records*** In Rhinelander, Wisconsin, Nancy recorded the following temperature readings in degrees Fahrenheit at 9 A.M. each morning for seven days in a row: $-8°, -5°, -18°, -22°, -6°, 3°$, and $7°$. What was the average temperature?
$-7°$F

69. ***Underwater Photography*** An underwater photographer made seven sets of underwater shots. He started at the surface, dropped down ten feet, and shot some pictures. Then he dropped ten more feet and shot more pictures. He continued this pattern until he had dropped seven times. How far beneath the surface was he at that point?
70 feet

70. ***Company Losses*** A new high-tech company reported losses of \$3 million for five quarters in a row. What was the amount of total losses after five quarters?
\$15 million

Electrical Charges *Ions are atoms or groups of atoms with positive or negative electrical charges. The charges of some ions are given in the following box.*

aluminum +3	chloride −1	magnesium +2
oxide −2	phosphate −3	silver +1

In exercises 71–76, find the total charge.

71. 17 magnesium ions 34

72. 8 phosphate ions −24

73. 4 silver ions and 2 chloride ions (*Hint:* multiply first, and then add.) 2

74. 3 oxide ions and 9 chloride ions (*Hint:* Multiply first, and then add.) −15

75. Eight chloride ions are removed from a substance. What is the change in the charge of the remaining substance? +8

76. Six oxide ions are removed from a substance. What is the change in the charge of the remaining substance? +12

Golf *A round of golf is nine holes. Pine Hills Golf Course is a par 3 course. This means that the expected number of strokes it takes to complete each hole is 3. The following table indicates points above or below par for each hole. For example, if someone shoots a 5 on one of the holes (that is, takes 5 strokes to complete the hole), this is called a double bogey and is 2 points over par.*

Birdie −1	Bogey +1
Eagle −2	Double Bogey +2

77. During a round of golf with his friend, Louis got two birdies, one eagle, and two double bogeys. He made par on the rest of the holes. How many points above or below par for the entire nine-hole course is this? (*Hint:* Multiply first, and then add.)
0; at par

78. Judy shot three birdies, two bogeys, and one double bogey. On the other three holes, she made par. How many points above or below par for the entire nine-hole course is this? (*Hint:* Multiply first, and then add.)
+1; one point above par

To Think About

79. Bob multiplied −12 times a mystery number that was not negative and he did not get a negative answer. What was the mystery number? 0

80. Fred divided a mystery number that was not negative by −3 and did not get a negative answer. What was the mystery number? 0

In exercises 81 and 82, the letters a, b, c, and d represent numbers that may be either positive or negative. The first three numbers are multiplied to obtain the fourth number. This can be stated by the equation $(a)(b)(c) = d$.

81. In the case where a, c, and d are all negative numbers, is the number b positive or negative?
b is negative

82. In the case where a is positive but b and d are negative, is the number c positive or negative?
c is positive

Cumulative Review

▲ **83.** ***Geometry*** Find the area of a parallelogram with height of 6 inches and base of 15 inches.
90 square inches

▲ **84.** ***Geometry*** Find the area of a trapezoid with height of 12 meters and bases of 18 meters and 26 meters.
264 square meters

How Am I Doing? Sections 9.1–9.3

How are you doing with your homework assignments in Sections 9.1 to 9.3? Do you feel you have mastered the material so far? Do you understand the concepts you have covered? Before you go further in the textbook, take some time to do each of the following problems.

9.1 *Add.*

1. $-7 + (-12)$ **2.** $-23 + 19$ **3.** $7.6 + (-3.1)$

4. $8 + (-5) + 6 + (-9)$ **5.** $\frac{5}{12} + \left(-\frac{3}{4}\right)$ **6.** $-\frac{5}{6} + \left(-\frac{1}{3}\right)$

7. $-2.8 + (-4.2)$ **8.** $-3.7 + 5.4$

9.2 *Subtract.*

9. $13 - 21$ **10.** $-26 - 15$ **11.** $\frac{5}{17} - \left(-\frac{9}{17}\right)$

12. $-19 - (-7)$ **13.** $-4.9 - (-6.3)$ **14.** $2.8 - 5.6$

15. $21 - (-21)$ **16.** $\frac{2}{3} - \left(-\frac{3}{5}\right)$

9.3 *Multiply or divide.*

17. $(-3)(-8)$ **18.** $-48 \div (-12)$ **19.** $-72 \div 9$

20. $(5)(-4)(2)(-1)\left(-\frac{1}{4}\right)$ **21.** $\frac{72}{-3}$ **22.** $\frac{-\frac{3}{4}}{-\frac{4}{5}}$

23. $(-8)(-2)(-4)$ **24.** $120 \div (-12)$

Mixed Practice

Perform the indicated operations. Simplify your answer.

25. $18 - (-6)$ **26.** $-7(-3)$

27. $28 \div (-4)$ **28.** $1.6 + (-1.8) + (-3.4)$

29. $2.9 - 3.5$ **30.** $-\frac{1}{3} + \left(-\frac{2}{5}\right)$

31. $\left(-\frac{5}{6}\right)\left(\frac{3}{7}\right)$ **32.** $\frac{3}{4} \div \left(-\frac{5}{6}\right)$

33. The temperature in Detroit in January was $-12°F$, $-8°F$, $3°F$, $7°F$, $-2°F$. What was the average temperature for those five days?

Now turn to page SA-18 for the answer to each of these problems. Each answer also includes a reference to the objective in which the problem is first taught. If you missed any of these problems, you should stop and review the Examples and Practice Problems in the referenced objective. A little review now will help you master the material in the upcoming sections of the text.

1. -19
2. -4
3. 4.5
4. 0
5. $-\frac{1}{3}$
6. $-\frac{7}{6}$ or $-1\frac{1}{6}$
7. -7
8. 1.7
9. -8
10. -41
11. $\frac{14}{17}$
12. -12
13. 1.4
14. -2.8
15. 42
16. $\frac{19}{15}$ or $1\frac{14}{15}$
17. 24
18. 4
19. -8
20. -10
21. -24
22. $\frac{15}{16}$
23. -64
24. -10
25. 24
26. 21
27. -7
28. -3.6
29. -0.6
30. $-\frac{11}{15}$
31. $-\frac{5}{14}$
32. $-\frac{9}{10}$
33. $-2.4°F$

9.4 ORDER OF OPERATIONS WITH SIGNED NUMBERS

Student Learning Objective

After studying this section, you will be able to:

Calculate with signed numbers using more than one operation.

1 Calculating with Signed Numbers Using More Than One Operation

The order of operations we discussed for whole numbers applies to signed numbers as well. The rules that we will use is this chapter are listed in the following box.

> **ORDER OF OPERATIONS FOR SIGNED NUMBERS**
>
> With grouping symbols:
>
> Do first **1.** Perform operations inside the parentheses.
>
> ↑ **2.** Simplify any expressions with exponents, and find any square roots.
>
> ↓ **3.** Multiply or divide from left to right.
>
> Do last **4.** Add or subtract from left to right.

EXAMPLE 1 Perform the indicated operations in the proper order.

$$-6 \div (-2)(5)$$

Solution Multiplication and division are of equal priority. So we work from left to right and divide first.

$$\underbrace{-6 \div (-2)}(5)$$
$$= \quad 3 \quad (5) = 15$$

NOTE TO STUDENT: Fully worked-out solutions to all of the Practice Problems can be found at the back of the text starting at page SP-1

Practice Problem 1 Perform the indicated operations in the proper order.

$$20 \div (-5)(-3)$$

Teaching Tip Remind students that keeping extra parentheses in the problem after the first step of operations, as in Example 2, will usually increase their accuracy as they do problems of this type.

Teaching Tip Encourage students to show the step of changing $3 - (-8)$ to $3 + 8$ when working out the problem.

EXAMPLE 2 Perform the indicated operations in the proper order.

(a) $7 + 6(-2)$ **(b)** $9 \div 3 - 16 \div (-2)$

Solution

(a) Multiplication and division must be done first. We begin with $6(-2)$.

$$7 + \underbrace{6(-2)}$$
$$= 7 + (-12) = -5$$

(b) There is no multiplication, but there is division, and we do that first.

$$\underbrace{9 \div 3} - \underbrace{16 \div (-2)}$$
$$= \quad 3 \quad - \quad (-8) \quad = 3 + 8 \quad \text{Transform subtraction to adding the opposite.}$$
$$= 11$$

Practice Problem 2 Perform the indicated operations in the proper order.

(a) $25 \div (-5) + 16 \div (-8)$ **(b)** $9 + 20 \div (-4)$

If a fraction has operations written in the numerator, in the denominator, or both, these operations must be done first. Then the fraction may be simplified or the division carried out.

EXAMPLE 3 Perform the indicated operations in the proper order.

$$\frac{7(-2) + 4}{8 \div (-2)(5)}$$

Solution

$$\frac{7(-2) + 4}{8 \div (-2)(5)} = \frac{-14 + 4}{(-4)(5)}$$ We perform the multiplication and division first in the numerator and the denominator, respectively.

$$= \frac{-10}{-20}$$ Simplify the fraction.

$$= \frac{1}{2}$$

Note that the answer is positive since a negative number divided by a negative number gives a positive result.

Practice Problem 3 Perform the indicated operations in the proper order.

$$\frac{9(-3) - 5}{2(-4) \div (-2)}$$

Some problems involve parentheses and exponents and will require additional steps.

EXAMPLE 4 Perform the indicated operations in the proper order.

$$4(6 - 9) + (-2)^3 + 3(-5)$$

Solution

$4(-3) + (-2)^3 + 3(-5)$ First we combine the numbers inside the parentheses.

$= 4(-3) + (-8) + 3(-5)$ Next we simplify the expression with exponents. We obtain $(-2)^3 = (-2)(-2)(-2) = -8$.

$= -12 + (-8) + (-15)$ Next we perform each multiplication.

$= -35$ Now we add three numbers.

Practice Problem 4 Perform the indicated operations in the proper order.

$$-2(-12 + 15) + (-3)^4 + 2(-6)$$

Be sure to use extra caution with problems involving fractions or decimals. It is easy to make an error in finding a common denominator or in placing a decimal point.

EXAMPLE 5 Perform the indicated operations in the proper order.

$$\left(\frac{1}{2}\right)^3 + 2\left(\frac{3}{4} - \frac{3}{8}\right) \div \left(-\frac{3}{5}\right)$$

Solution

$\left(\frac{1}{2}\right)^3 + 2\left(\frac{6}{8} - \frac{3}{8}\right) \div \left(-\frac{3}{5}\right)$ First find the LCD and write $\frac{3}{4} = \frac{6}{8}$.

$= \left(\frac{1}{2}\right)^3 + 2\left(\frac{3}{8}\right) \div \left(-\frac{3}{5}\right)$ Next we combine the two fractions inside the parentheses.

$= \frac{1}{8} + 2\left(\frac{3}{8}\right) \div \left(-\frac{3}{5}\right)$ Next we simplify $\left(\frac{1}{2}\right)^3 = \left(\frac{1}{2}\right)\left(\frac{1}{2}\right)\left(\frac{1}{2}\right) = \frac{1}{8}$.

$= \frac{1}{8} + \frac{3}{4} \div \left(-\frac{3}{5}\right)$ Next multiply: $\left(\frac{2}{1}\right)\left(\frac{3}{8}\right) = \frac{3}{4}$.

$= \frac{1}{8} + \frac{3}{4} \cdot \left(-\frac{5}{3}\right)$ To divide two fractions, invert and multiply the second fraction.

$= \frac{1}{8} + \left(-\frac{5}{4}\right)$ Multiply $\left(\frac{3}{4}\right)\left(-\frac{5}{3}\right) = -\frac{5}{4}$.

$= \frac{1}{8} + \left(-\frac{10}{8}\right)$ Change $-\frac{5}{4}$ to the equivalent $-\frac{10}{8}$.

$= -\frac{9}{8}$ Add the two fractions.

Teaching Tip Emphasize to students that problems like Example 5 can have the answer written three ways. The negative sign can be written before the fraction, before the 9, or before the 8. You may want to mention, however, that it is generally preferred not to have the negative sign in the denominator of the fraction.

NOTE TO STUDENT: Fully worked-out solutions to all of the Practice Problems can be found at the back of the text starting at page SP-1

Practice Problem 5 Perform the indicated operations in the proper order.

$$\left(\frac{1}{5}\right)^2 + 4\left(\frac{1}{5} - \frac{3}{10}\right) \div \frac{2}{3}$$

9.4 Exercises

Student Solutions Manual | CD/ Video | PH Math Tutor Center | MathXL®Tutorials on CD | MathXL® | MyMathLab® | Interactmath.com

Perform the indicated operations in the proper order.

1. $-8 \div (-4)(3)$ 6

2. $-20 \div (-5)(6)$ 24

3. $50 \div (-25)(4)$ -8

4. $60 \div (-20)(5)$ -15

5. $16 + 32 \div (-4)$ $16 + (-8) = 8$

6. $15 - (-18) \div 3$ $15 - (-6) = 21$

7. $24 \div (-3) + 16 \div (-4)$ $-8 + (-4) = -12$

8. $(-56) \div (-7) + 30 \div (-15)$ $8 + (-2) = 6$

9. $3(-4) + 5(-2) - (-3)$ $-12 + (-10) + 3 = -19$

10. $6(-3) + 8(-1) - (-2)$ $-18 + (-8) + 2 = -24$

11. $-54 \div 6(9)$ -81

12. $-72 \div 9(8)$ -64

13. $5 - 30 \div 3$ $5 - 10 = -5$

14. $8 - 70 \div 5$ $8 - 14 = -6$

15. $36 \div 12(-2)$ $3(-2) = -6$

16. $9(-6) + 6$ $-54 + 6 = -48$

17. $3(-4) + 6(-2) - 3$ $-12 + (-12) - 3 = -27$

18. $-6(7) + 8(-3) + 5$ $-42 + (-24) + 5 = -61$

19. $11(-6) - 3(12)$ $-66 - 36 = -102$

20. $10(-5) - 4(12)$ $-50 + (-48) = -98$

21. $16 - 4(8) + 18 \div (-9)$ $16 - 32 + (-2) = -18$

22. $20 - 3(-2) + (-20) \div (-5)$ $20 + 6 + 4 = 30$

In exercises 23–32, simplify the numerator and denominator first, using the proper order of operations. Then reduce the fraction if possible.

23. $\frac{8 + 6 - 12}{3 - 6 + 5}$ $\frac{2}{2} = 1$

24. $\frac{10 - 5 - 7}{8 + 3 - 9}$ $\frac{-2}{2} = -1$

25. $\frac{6(-2) + 4}{6 - 3 - 5}$ $\frac{-12 + 4}{-2} = 4$

26. $\frac{-8 + 2 - 6}{4(-6) - 12}$ $\frac{-12}{-24 - 12} = \frac{1}{3}$

27. $\frac{2(8) \div 4 - 5}{-35 \div (-7)}$ $\frac{4 - 5}{5} = -\frac{1}{5}$

28. $\frac{-25 \div 5 + (-3)(-4) - 7}{8 - (18 \div 3)}$ $\frac{0}{2} = 0$

29. $\frac{24 \div (-3) - (6 - 2)}{-5(4) + 8}$ $\frac{-8 - 4}{-20 + 8} = 1$

30. $\frac{6(-3) \div (-2) + 5}{32 \div (-8)}$ $\frac{14}{-4} = -\frac{7}{2}$

31. $\frac{12 \div 3 + (-2)(2)}{9 - 9 \div (-3)}$ $\frac{0}{9 + 3} = 0$

32. $\frac{8(-4) - 4}{(-42) \div (-6) + (12 - 10)}$ $\frac{-32 - 4}{7 + 2} = -4$

Perform the operations in the proper order.

33. $3(2 - 6) + 4^2$ 4

34. $-5(8 - 3) + 6^2$ 11

35. $12 \div (-6) + (7 - 2)^3$ 123

36. $-20 \div 2 + (6 - 3)^4$ 71

37. $\left(-1\frac{1}{2}\right)(-4) - 3\frac{1}{4} \div \frac{1}{4}$ -7

38. $(6)\left(-2\frac{1}{2}\right) + 2\frac{1}{3} \div \frac{1}{3}$ -8

39. $\left(\frac{3}{5} - \frac{2}{5}\right)^2 + \left(\frac{3}{2}\right)\left(-\frac{1}{5}\right)$ $-\frac{13}{50}$

40. $\left(\frac{5}{4} - \frac{3}{4}\right)^2 + \left(\frac{5}{8}\right)\left(-\frac{2}{3}\right)$ $-\frac{1}{6}$

41. $(1.2)^2 - 3.6(-1.5)$ 6.84

42. $(0.6)^2 - 5.2(-3.4)$ 18.04

Applications

Temperature Averages *The following chart gives the average monthly temperatures in Fahrenheit for two cities in Alaska.*

	Jan.	Feb.	Mar.	Apr.	May	June	July	Aug.	Sept.	Oct.	Nov.	Dec.
Barrow	−14	−20	−16	−2	19	33	39	38	31	14	−1	−13
Fairbanks	−13	−4	9	30	48	59	62	57	45	25	4	−10

Use the chart to solve. Round to the nearest tenth of a degree.

43. What is the average temperature in Barrow during December, January, and February?
$\frac{-13 + (-14) + (-20)}{3} \approx -15.7°\text{F}$

44. What is the average temperature in Fairbanks during December, January, and February?
$\frac{(-10) + (-13) + (-4)}{3} = -9°\text{F}$

45. What is the average temperature in Barrow from October through March?
$\frac{14 + (-1) + (-13) + (-14) + (-20) + (-16)}{6} \approx -8.3°\text{F}$

46. What is the average temperature in Fairbanks from October through March?
$\frac{25 + 4 + (-10) + (-13) + (-4) + 9}{6} \approx 1.8°\text{F}$

47. What is the average yearly temperature in Barrow?
$\frac{108}{12} = 9°\text{F}$

48. What is the average yearly temperature in Fairbanks?
$\frac{312}{12} = 26°\text{F}$

To Think About

49. In Hveragerdi, Iceland, yesterday the average temperature was −12°C. Today the average temperature rose 3°C. What is the new average temperature? If each of the next four days experience the same rise, what will the temperature be in four days?
−9°C; 3°C

50. In January 2005, Jeff Slater was at the research weather station on the top of Mount Washington. One day the low temperature was −15°F. For the next four days, the low temperature dropped 6 degrees each day. What was the low temperature at the end of those four days? −39°F

Cumulative Review

51. ***Metric Measure*** A telephone wire that is 3840 meters long is how long measured in kilometers? 3.84 kilometers

52. ***Metric Measure*** A container with 36.8 grams of protein contains how many milligrams of protein? 36,800 milligrams

▲ **53.** ***Geometry*** Find the area of a circle of radius 6 m. Use $\pi \approx 3.14$. 113.04 m^2

54. ***Junk Mail*** Janice receives a lot of mail-order catalogs and junk mail. In June she received 28 catalogs and 48 pieces of junk mail. In July she received exactly twice as many of each. In August she received exactly half of what she received in June. Over the three months, how many catalogs did she receive? In July and August, how many pieces of junk mail did she receive? 98 catalogs; 120 pieces of junk mail

9.5 Scientific Notation

Changing Numbers in Standard Notation to Scientific Notation

Scientists who frequently work with very large or very small measurements use a certain way to write numbers, called *scientific notation*. In our usual way of writing a number, which we call "standard notation" or "ordinary form," we would express the distance to the nearest star, Proxima Centauri, as 24,800,000,000,000 miles. In scientific notation, we more conveniently write this as

$$2.48 \times 10^{13} \text{ miles.}$$

For a very small number, like two millionths, the standard notation is

$$0.000002.$$

The same quantity in scientific notation is

$$2 \times 10^{-6}.$$

Notice that each number in scientific notation has two parts: (1) a number that is 1 or greater but less than 10, which is multiplied by (2) a power of 10. That power is either a whole number or the negative of a whole number. In this section we learn how to go back and forth between standard and scientific notation.

A positive number is in **scientific notation** if it is in the form $a \times 10^n$, where a is a number greater than (or equal to) 1 and less than 10, and n is an integer.

We begin our investigation of scientific notation by looking at large numbers. We recall from Section 1.6 that

$$10 = 10^1$$
$$100 = 10^2$$
$$1000 = 10^3.$$

The number of zeros tells us the number for the exponent. To write a number in scientific notation, we want the first number to be greater than or equal to 1 and less than 10, and the second number to be a power of 10.

Let's see how this can be done. Consider the following.

$$6700 = \underbrace{6.7}_{\text{greater than 1 and less than 10}} \times 1000 = 6.7 \times \underset{\uparrow \text{ a power of 10}}{10^3}$$

Let us look at two more cases.

$$530 = 5.3 \times 100 \qquad\qquad 156{,}000 = 1.56 \times 100{,}000$$
$$= \underbrace{5.3 \times 10^2} \qquad\qquad = \underbrace{1.56 \times 10^5}$$

These numbers are in scientific notation.

Now that we have seen some examples of numbers in scientific notation, let us think through the steps in the next example.

Student Learning Objectives

After studying this section, you will be able to:

1. **Change numbers in standard notation to scientific notation.**
2. **Change numbers in scientific notation to standard notation.**
3. **Add and subtract numbers in scientific notation.**

It is important to remember that all numbers *greater than* 10 always have a positive exponent when expressed in scientific notation.

Teaching Tip To demonstrate the usefulness of scientific notation, tell students that scientists and astronomers often refer to a measurement of one solar mass. This is 1.99×10^{30} kilograms. Ask students to try writing that number out without using scientific notation!

EXAMPLE 1 Write in scientific notation.

(a) 9826 **(b)** 163,457

Solution

(a)

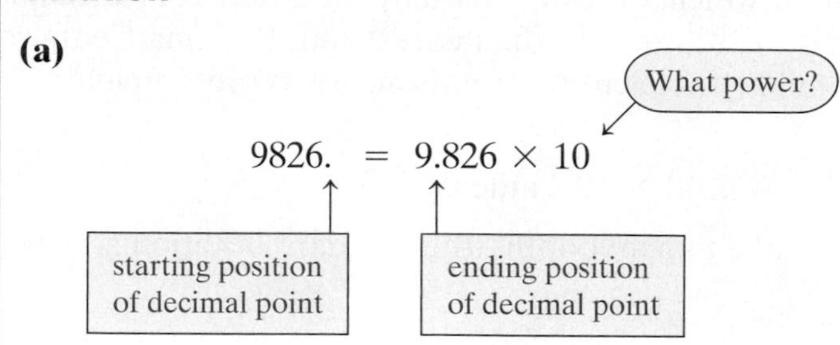

The decimal point moves 3 places to the left. We therefore use 3 for the power of 10.

$$9826 = 9.826 \times 10^3$$

(b)

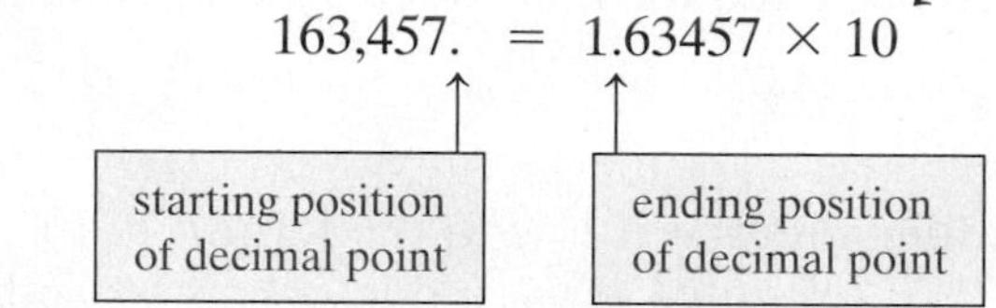

The decimal point moves 5 places to the left. We therefore use 5 for the power of 10.

$$163{,}457 = 1.63457 \times 10^5$$

Practice Problem 1 Write in scientific notation.

(a) 3729 **(b)** 506,936

Calculator

Standard to Scientific

Many calculators have a setting that will display numbers in scientific notation. Consult your manual on how to do this. To convert 154.32 into scientific notation, first change your setting to display in scientific notation. Often this is done by pressing

[SCI] or [2nd] [SCI].

Often SCI is then displayed on the calculator. Then enter:

154.32 [=]

Display:

[1.5432 02]

1.5432 02 means 1.5432×10^2.

Note that your calculator display may show the power of 10 in a different manner. Be sure to change your setting back to the regular display when you are done.

When changing to scientific notation, all zeros to the *right* of the final nonzero digit may be eliminated. This does not change the value of your answer.

Now we will look at numbers that are *less than* 1. In the introduction to this section we saw that 0.000002 can be written as 2×10^{-6}. How is it possible to have a negative exponent? Let's take another look at the powers of 10.

$$10^3 = 1000$$
$$10^2 = 100$$
$$10^1 = 10$$
$$10^0 = 1$$ Recall that any number to the zero power is 1.

Now, following this pattern, what would you expect the next number on the left of the equal sign to be? What would be the next number on the right of the equal sign? Each number on the right is one-tenth the number above it. We continue the pattern.

$$10^{-1} = 0.1$$
$$10^{-2} = 0.01$$
$$10^{-3} = 0.001$$

Thus you can see how it is possible to have negative exponents. We use negative exponents to write numbers that are less than 1 in scientific notation.

Let's look at 0.76. This number is less than 1. Recall that in our definition for scientific notation the first number must be greater than or equal to 1 and less than 10. To change 0.76 to such a number, we will have to move the decimal point.

$$0.76 = 7.6 \times 10^{-1}$$

The decimal point moves 1 place to the right. We therefore put a -1 for the power of 10. We use a negative exponent because the original number is less than 1. Let's look at two more cases.

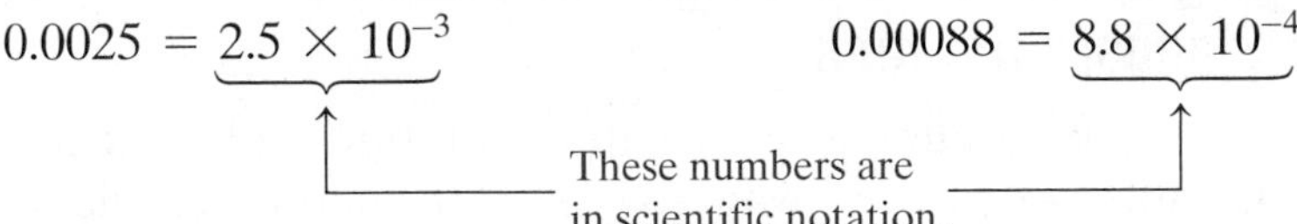

When we start with a positive number that is less than 1 and write it in scientific notation, we will get a result with 10 to a negative power. Think carefully through the steps of the following example.

EXAMPLE 2 Write in scientific notation.

(a) 0.036 **(b)** 0.72

Solution

(a) We change the given number to a number greater than or equal to 1 and less than 10. Thus we change 0.036 to 3.6.

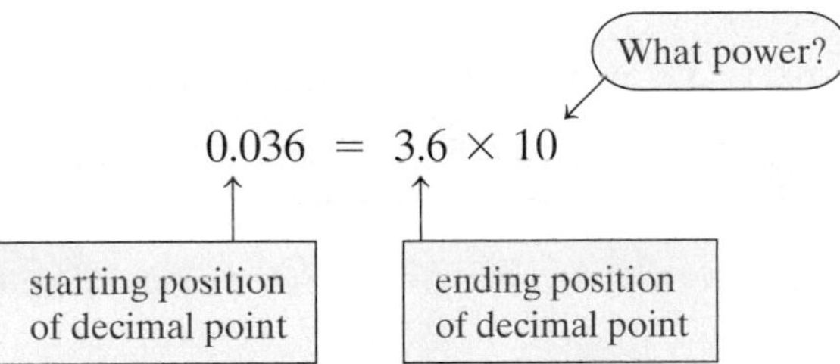

The decimal point moved 2 places to the right. Since the original number is less than 1, we use -2 for the power of 10.

$$0.036 = 3.6 \times 10^{-2}$$

(b)

What power?

$0.72 = 7.2 \times 10$

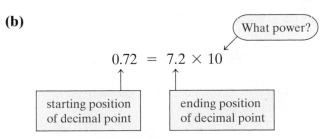

starting position of decimal point | ending position of decimal point

The decimal point moved 1 place to the right. Because the original number is less than 1, we use -1 for the power of 10.

$$0.72 = 7.2 \times 10^{-1}$$

Teaching Tip Whether changing from or to scientific notation, it is important to remember that all positive numbers less than 1 always have a negative exponent when expressed in scientific notation.

Practice Problem 2 Write in scientific notation.

(a) 0.076 **(b)** 0.982

***NOTE TO STUDENT:** Fully worked-out solutions to all of the Practice Problems can be found at the back of the text starting at page SP-1*

It is very helpful when writing numbers in scientific notation to remember these two concepts.

1. A number that is larger than 10 and written in scientific notation will always have a positive exponent as the power of 10.
2. A positive number that is smaller than 1 and written in scientific notation will always have a negative exponent as the power of 10.

2 Changing Numbers in Scientific Notation to Standard Notation

Often we are given a number in scientific notation and want to write it in standard notation—that is, we want to write it in what we consider "ordinary form." To do this, we reverse the process we've just used. If the number in scientific notation has a positive power of 10, we know the number is greater than 10. Therefore, we move the decimal point to the right to convert to standard notation.

EXAMPLE 3 Write in standard notation.

$$5.8671 \times 10^4$$

Solution

$5.8671 \times 10^4 = 58{,}671$ Move the decimal point four places to the right.

Practice Problem 3 Write in standard notation.

$$6.543 \times 10^3$$

NOTE TO STUDENT: Fully worked-out solutions to all of the Practice Problems can be found at the back of the text starting at page SP-1

In some cases, we will need to add zeros as we move the decimal point to the right.

EXAMPLE 4 Write in standard notation.

(a) 9.8×10^5 **(b)** 3×10^3

Solution

(a) $9.8 \times 10^5 = 980{,}000$ Move the decimal point five places to the right. Add four zeros.

(b) $3 \times 10^3 = 3000$ Move the decimal point three places to the right. Add three zeros.

Practice Problem 4 Write in standard notation.

(a) 4.3×10^5 **(b)** 6×10^4

If the number in scientific notation has a negative power of 10, we know the number is less than 1. Therefore, we move the decimal point to the left to convert to standard notation. In some cases, we need to add zeros as we move the decimal point to the left.

EXAMPLE 5 Write in standard notation.

(a) 2.48×10^{-3} **(b)** 1.2×10^{-4}

Solution

(a) $2.48 \times 10^{-3} = 0.00248$ Move the decimal point three places to the left. Add two zeros between the decimal point and 2.

(b) $1.2 \times 10^{-4} = 0.00012$ Move the decimal point four places to the left. Add three zeros between the decimal point and 1.

Practice Problem 5 Write in standard notation.

(a) 7.72×10^{-3} **(b)** 2.6×10^{-5}

3 Adding and Subtracting Numbers in Scientific Notation

Scientists calculate with numbers in scientific notation when they study galaxies (the huge) and when they study microbes (the tiny). To add or subtract numbers in scientific notation, the numbers must have the same power of 10.

Numbers in scientific notation may be added or subtracted if they have the same power of 10. We add or subtract the decimal part and leave the power of 10 unchanged.

EXAMPLE 6 Add. 5.89×10^{-20} meters + 3.04×10^{-20} meters

Solution

$$\begin{array}{r} 5.89 \times 10^{-20} \text{ meters} \\ +\ 3.04 \times 10^{-20} \text{ meters} \\ \hline 8.93 \times 10^{-20} \text{ meters} \end{array}$$

Teaching Tip This concept of adding numbers in scientific notation is very important. Students often miss the fact that each number must have the same power of 10 before the addition can be carried out.

Practice Problem 6 Add.

6.85×10^{22} kilograms + 2.09×10^{22} kilograms

EXAMPLE 7 Add. $7.2 \times 10^6 + 5.2 \times 10^5$

Solution Note that these two numbers have different powers of ten, so we cannot add them as written. We will change one number from scientific notation to another form.

Now $5.2 \times 10^5 = 520{,}000$ can be rewritten as 0.52×10^6. Now we can add.

$$\begin{array}{r} 7.2 \times 10^6 \\ +0.52 \times 10^6 \\ \hline 7.72 \times 10^6 \end{array}$$

Practice Problem 7 Subtract. $4.36 \times 10^5 - 3.1 \times 10^4$

9.5 Exercises

Student Solutions Manual | CD/Video | PH Math Tutor Center | MathXL®Tutorials on CD | MathXL® | MyMathLab® | Interactmath.com

Verbal and Writing Skills

1. Why does scientific notation use exponents with base 10?

Our number system is structured according to base 10. By making scientific notation also in base 10, the calculations are easier to perform.

2. Explain how to write 0.787 in scientific notation.

We need to move the decimal point one place to the right. This means we will need a negative exponent for 10. Since we moved the decimal point one place we will use 10^{-1}. Thus $0.787 = 7.87 \times 10^{-1}$.

3. What are the two parts of a number in scientific notation.

The first part is a number greater than or equal to 1 but smaller than 10. It has at least one nonzero digit. The second part is 10 raised to some integer power.

4. If $a \times 10^6$ is a number written in scientific notation, what are the possible values of a?

The value of a must be a number greater than or equal to 1 but smaller than 10.

Write in scientific notation.

5. 120
1.2×10^2

6. 340
3.4×10^2

7. 1900
1.9×10^3

8. 5200
5.2×10^3

9. 26,300
2.63×10^4

10. 78,100
7.81×10^4

11. 288,000
2.88×10^5

12. 238,000
2.38×10^5

13. 10,000
1×10^4

14. 100,000
1×10^5

15. 12,000,000
1.2×10^7

16. 28,000,000
2.8×10^7

17. 0.0931
9.31×10^{-2}

18. 0.0242
2.42×10^{-2}

19. 0.00279
2.79×10^{-3}

20. 0.00613
6.13×10^{-3}

21. 0.82
8.2×10^{-1}

22. 0.17
1.7×10^{-1}

23. 0.000016
1.6×10^{-5}

24. 0.000072
7.2×10^{-5}

25. 0.00000531
5.31×10^{-6}

26. 0.00000198
1.98×10^{-6}

27. 0.000008
8×10^{-6}

28. 0.00000007
7×10^{-8}

Write in standard notation.

29. 5.36×10^4
53,600

30. 2.19×10^4
21,900

31. 5.334×10^3
5334

32. 8.1215×10^4
81,215

33. 4.6×10^{12}
4,600,000,000,000

34. 3.8×10^{11}
380,000,000,000

35. 6.2×10^{-2}
0.062

36. 3.5×10^{-2}
0.035

37. 3.71×10^{-1}
0.371

38. 7.44×10^{-4}
0.000744

39. 9×10^{11}
900,000,000,000

40. 2×10^{12}
2,000,000,000,000

41. 3.862×10^{-8}
0.00000003862

42. 8.139×10^{-9}
0.000000008139

Mixed Practice

43. Write in scientific notation. 35,689
3.5689×10^4

44. Write in scientific notation. 76,371
7.6371×10^4

45. Write in standard notation. 5.2×10^{-2}
0.052

46. Write in standard notation. 7.9×10^{-2}
0.079

47. Write in scientific notation. 0.000398
3.98×10^{-4}

48. Write in scientific notation. 0.000976
9.76×10^{-4}

49. Write in standard notation. 1.88×10^6
1,880,000

50. Write in standard notation. 3.49×10^6
3,490,000

Applications

Write in scientific notation.

51. ***Light Speed*** In 1 year, light will travel 5,878,000,000,000 miles. 5.878×10^{12} miles

52. ***Forest*** The world's forests total 2,700,000,000 acres of wooded area. 2.7×10^{9} acres

▲ **53.** ***Red Blood Cells*** The average volume of a red blood cell is 0.000000000000092 liter.
9.2×10^{-14} liter

54. ***Duckweed*** The flowing duckweed of Australia has a small fruit that weighs 0.0000024 ounce.
2.4×10^{-6} ounce

Write in standard notation.

55. ***Locusts*** During the period August 15–25, 1875, a huge swarm of Rocky Mountain locusts invaded Nebraska. The swarm contained an estimated 1.25×10^{13} insects.
12,500,000,000,000 insects

56. ***National Debt*** On October 17, 2003, the United States' national debt was approximately 6.831×10^{12} dollars.
\$6,831,000,000,000

▲ **57.** ***Human Blood*** The diameter of a red corpuscle of human blood is about 7.5×10^{-5} centimeter.
0.000075 centimeter

▲ **58.** ***Mercury Drill*** Recently a tiny hole with an approximate diameter of 3.16×10^{-10} meter was produced by Doctors Heckl and Maddocks using a chemical method involving a mercury drill.
0.000000000316 meter

59. ***Volcano Eruption*** During the eruption of the Taupo volcano in New Zealand in 130 A.D., approximately 1.4×10^{10} tons of pumice were carried up in the air. 14,000,000,000 tons

60. ***Earth Size*** The number of grains of sand it would take to fill a sphere the size of the earth is 1×10^{23} 100,000,000,000,000,000,000,000

Add.

61. 3.38×10^{7} dollars + 5.63×10^{7} dollars
9.01×10^{7} dollars

62. 8.17×10^{9} atoms + 2.76×10^{9} atoms
1.093×10^{10} atoms

63. ***Planets*** The mass of the earth is 5.87×10^{21} tons, and the mass of Venus is 4.81×10^{21} tons. What is the total mass of the two planets? 1.068×10^{22} tons

64. ***Planets*** The masses of Mars, Pluto, and Mercury are 6.34×10^{20} tons, 1.76×10^{20} tons, and 3.21×10^{20} tons, respectively. What is the total mass of the three planets? 1.131×10^{21} tons

Subtract.

65. 4×10^8 feet $- 3.76 \times 10^7$ feet
$36.24 \times 10^7 = 3.624 \times 10^8$ feet

66. 9×10^{10} meters $- 1.26 \times 10^9$ meters
8.874×10^{10} meters

▲ **67.** ***Geography*** The two largest continents are Asia and Africa with areas of 1.76×10^7 square miles and 1.16×10^7 square miles, respectively. How much larger is Asia than Africa? 6.0×10^6 square miles

▲ **68.** ***Geography*** The total ocean area and land area of earth is approximately 1.39×10^8 square miles and 5.747×10^7 square miles, respectively. What is the total area of earth covered by land or ocean? 1.9647×10^8 square miles

To Think About

69. ***Planets*** Recently astronomers have discovered evidence of a planet named Epsilon Eridani b. This planet is in a different solar system and is 10.5 light-years from earth. A light-year is approximately 5.88 trillion miles. Find in scientific notation how many miles Epsilon Eridani b is from earth. 6.174×10^{13} miles

70. ***Planets*** Recently astronomers have discovered evidence of a planet named Cancri c. This planet is in a different solar system and is 41 light-years from earth. A light-year is approximately 5.88 trillion miles. Find in scientific notation how many miles Cancri c is from earth. 2.4108×10^{14} miles

Cumulative Review

Calculate.

71. 12.5×0.21 2.625

72. $0.53\overline{)0.13674}$ 0.258

▲ **73.** ***Radar Construction*** A radar mounting plate is constructed in the shape of a parallelogram. Two sides of the parallelogram are 9 feet and 13 feet. A rubber piece of safety bumper must be placed around all sides of the parallelogram. It costs \$4 per foot. How much will it cost to place the safety bumper around the plate? \$176

74. ***Mountain Climbing*** Michael is mountain climbing in the Rocky Mountains. He is 70 inches tall. At 2 P.M. his shadow measures 95 inches. He wants to climb to the top of a cliff that casts a shadow of 800 feet at 2 P.M. How tall is the cliff? Round to the nearest whole number. 589 feet

Putting Your Skills to Work

Signed Numbers on the Stock Market

When people buy shares of a company's stock, they become owners of a small piece of the company. The value of a share fluctuates throughout the day until the market closes. When the market closes at the end of each day, the value of one share at this time is called the closing price. Positive or negative news about a company's performance and economic indicators such as unemployment rates and inflation affect the value or price of the shares.

The chart below gives the closing price of one share of stock for four companies, on the first day of the month that the stock market was open, from June to November 2003. Use the information given to answer the questions below.

Company	6-2-03	7-1-03	8-1-03	9-2-03	10-1-03	11-3-03
Krispy Kreme	34.65	41.27	43.25	42.97	39.88	44.14
Six Flags, Inc.	7.94	6.77	4.65	5.42	5.45	6.10
Sears, Roebuck & Co.	31.98	33.70	40.62	45.70	45.76	53.25
Coca-Cola Enterprises	18.92	18.75	17.40	18.71	19.52	20.50

Problems for Individual Investigation

1. Using signed numbers, give the loss or gain of one share of Coca-Cola Enterprises stock each month from June 2, 2003 to November 3, 2003. −.017, −1.35, +1.31, +0.81, +0.98
2. Between June 2, 2003 and August 1, 2003, what was the change in value of one share of Sears, Roebuck & Co. stock? During the same time period, what was the change in value of one share of Coca-Cola Enterprises stock? +8.64; −1.52
3. Between August 1, 2003, and November 3, 2003, what was the change in value of one share of Six Flags, Inc. stock? If you bought 100 shares on August 1, and sold them on November 3, how money would you have made? +$1.45; $145

Problems for Group Investigation and Cooperative Study

4. Natalia had $1000 on July 1, 2003 to invest in Sears, Roebuck & Co. stock. How many shares was she able to buy? (Round to the nearest whole number.) She sold these shares on November 3, 2003. How much money did she make on this investment?
29 shares; $566.95
5. On July 1, 2003, David had $3000 to invest in the stock market. He spent $\frac{1}{3}$ of his money on Krispy Kreme stock, and the rest on Six Flags, Inc. stock. How many shares of each was he able to buy? (Round to the nearest whole number.) He sold all of these shares on September 2, 2003. Did he make or lose money? How much?
24 shares of Krispy Kreme; 295 shares of Six Flags, Inc.; he lost $357.45

Chapter 9 Organizer

Topic	Procedure	Examples
Absolute value, p. 573.	The absolute value of a number is the distance between the number and zero on the number line.	$\lvert -6\rvert = 6 \quad \lvert 3\rvert = 3 \quad \lvert 0\rvert = 0$
Adding signed numbers with the same sign, p. 574.	To add two numbers with the same sign: **1.** Add the absolute value of the numbers. **2.** Use the common sign in the answer.	$12 + 5 = 17$ $-6 + (-8) = -14$ $-5.2 + (-3.5) = -8.7$ $-\frac{1}{7} + \left(-\frac{3}{7}\right) = -\frac{4}{7}$
Adding signed numbers with different signs, p. 577.	To add two numbers with different signs: **1.** Subtract the absolute value of the numbers. **2.** Use the sign of the number with the larger absolute value.	$14 + (-8) = 6$ $-4 + 8 = 4$ $-3.2 + 7.1 = 3.9$ $\frac{5}{13} + \left(-\frac{8}{13}\right) = -\frac{3}{13}$
Subtracting signed numbers, p. 584.	To subtract signed numbers, add the opposite of the second number to the first number.	$-9 - (-3) = -9 + 3 = -6$ $5 - (-7) = 5 + 7 = 12$ $8 - 12 = 8 + (-12) = -4$ $-4 - 13 = -4 + (-13) = -17$ $-\frac{1}{12} - \left(-\frac{5}{12}\right) = -\frac{1}{12} + \frac{5}{12} = \frac{4}{12} = \frac{1}{3}$
Multiplying or dividing signed numbers with the same sign, p. 591.	To multiply or divide two numbers with the same sign, multiply or divide the absolute values. The sign of the result is positive.	$(0.5)(0.3) = 0.15$ $(-6)(-2) = 12$ $\frac{-20}{-2} = 10$ $\frac{-\frac{1}{3}}{-\frac{1}{7}} = \left(-\frac{1}{3}\right)\left(-\frac{7}{1}\right) = \frac{7}{3}$
Multiplying or dividing signed numbers with different signs, pp. 590–591.	To multiply or divide two numbers with different signs, multiply or divide the absolute values. The result is negative.	$7(-3) = -21$ $(-6)(4) = -24$ $(-36) \div (2) = -18$ $\frac{41.6}{-8} = -5.2$
Order of operations for signed numbers, p. 598.	With grouping symbols: Do first **1.** Perform operations inside the parentheses. ↑ **2.** Simplify any expressions with exponents, and find any square roots. ↓ **3.** Multiply or divide from left to right. Do last **4.** Add or subtract from left to right.	Perform the operations in the proper order. $-2(12 - 8) + (-3)^3 + 4(-6)$ $= -2(4) + (-3)^3 + 4(-6)$ $= -2(4) + (-27) + 4(-6)$ $= -8 + (-27) + (-24)$ $= -59$
Simplifying fractions with combined operations in numerator and denominator, p. 599.	**1.** Perform the operations in the numerator. **2.** Perform the operations in the denominator. **3.** Simplify the fraction.	Perform the operations in the proper order. $\frac{7(-4) - (-2)}{8 - (-5)}$ The numerator is $7(-4) - (-2) = -28 - (-2) = -26.$ The denominator is $8 - (-5) = 8 + 5 = 13.$ Thus the fraction becomes $-\frac{26}{13} = -2.$

Topic	Procedure	Examples
Writing a number in scientific notation, p. 603.	**1.** Move the decimal point to the position immediately to the right of the first nonzero digit. **2.** Count the number of decimal places you moved the decimal point. **3.** Multiply the result by a power of 10 equal to the number of places moved. If the number you started with is larger than 10, use a positive exponent. If the number you started with is less than 1, use a negative exponent.	Write in scientific notation. **(a)** 178 **(b)** 25,000,000 **(c)** 0.006 **(d)** 0.00001732 **(a)** $178 = 1.78 \times 10^2$ **(b)** $25{,}000{,}000 = 2.5 \times 10^7$ **(c)** $0.006 = 6 \times 10^{-3}$ **(d)** $0.00001732 = 1.732 \times 10^{-5}$
Changing from scientific notation to standard notation, p. 606.	**1.** If the exponent of 10 is *positive,* move the decimal point to the *right* as many places as the exponent shows. Insert extra zeros as necessary. **2.** If the exponent of 10 is *negative,* move the decimal point to the *left* as many places as the exponent shows. *Note:* Remember that numbers in scientific notation that have positive exponents are always greater than 10. Numbers in scientific notation that have negative exponents are always less than 1.	Write in standard notation. **(a)** 8×10^6 **(b)** 1.23×10^4 **(c)** 7×10^{-2} **(d)** 8.45×10^{-5} **(a)** $8 \times 10^6 = 8{,}000{,}000$ **(b)** $1.23 \times 10^4 = 12{,}300$ **(c)** $7 \times 10^{-2} = 0.07$ **(d)** $8.45 \times 10^{-5} = 0.0000845$

Chapter 9 Review Problems

Section 9.1

Add.

1. $-20 + 5$
-15

2. $-18 + 4$
-14

3. $-3.6 + (-5.2)$
-8.8

4. $10.4 + (-7.8)$
2.6

5. $-\frac{1}{5} + \left(-\frac{1}{3}\right)$
$-\frac{8}{15}$

6. $-\frac{2}{7} + \frac{5}{14}$
$\frac{1}{14}$

7. $20 + (-14)$
6

8. $-80 + 60$
-20

9. $(-82) + 50 + 35 + (-18)$
-15

10. $12 + (-7) + (-8) + 3$
0

Section 9.2

Subtract.

11. $25 - 36$
-11

12. $12 - 40$
-28

13. $12 - (-7)$
19

14. $14 - (-3)$
17

15. $-11.4 - 5.8$
-17.2

16. $-5.2 - 7.1$
-12.3

17. $-\frac{2}{5} - \left(-\frac{1}{3}\right)$
$-\frac{1}{15}$

18. $3\frac{1}{4} - \left(-5\frac{2}{3}\right)$
$\frac{107}{12}$ or $8\frac{11}{12}$

Perform the indicated operations from left to right.

19. $5 - (-2) - (-6)$
13

20. $-15 - (-3) + 9$
-3

21. $9 - 8 - 6 - 4$
-9

22. $-7 - 8 - (-3)$
-12

Section 9.3

Multiply or divide.

23. $\left(-\frac{2}{7}\right)\left(-\frac{1}{5}\right)$
$\frac{2}{35}$

24. $\left(-\frac{6}{15}\right)\left(\frac{5}{12}\right)$
$-\frac{1}{6}$

25. $(5.2)(-1.5)$
-7.8

26. $(-3.6)(-1.2)$
4.32

27. $-60 \div (-20)$
3

28. $-18 \div (-3)$
6

29. $\frac{-36}{4}$
-9

30. $\frac{-60}{12}$
-5

31. $\frac{-13.2}{-2.2}$
6

32. $\frac{48}{-3.2}$
-15

33. $\frac{-\frac{2}{5}}{\frac{4}{7}}$
$-\frac{7}{10}$

34. $\frac{-\frac{1}{3}}{-\frac{7}{9}}$
$\frac{3}{7}$

35. $3(-5)(-2)$
30

36. $(-2)(3)(-6)(-1)$
-36

Section 9.4

Perform the indicated operations in the proper order.

37. $8 - (-30) \div 6$
13

38. $26 + (-28) \div 4$
19

39. $2(-6) + 3(-4) - (-13)$
-11

40. $-49 \div (-7) + 3(-2)$
1

41. $36 \div (-12) + 50 \div (-25)$
-5

42. $21 - (-30) \div 15$
23

43. $50 \div 25(-4)$
-8

44. $-3.5 \div (-5) - 1.2$
-0.5

45. $2.5(-2) + 3.8$
-1.2

In exercises 46–49, simplify the numerator and denominator first, using the proper order of operations. Then reduce the fraction if possible.

46. $\frac{5 - 9 + 2}{3 - 5}$ 1

47. $\frac{4(-6) + 8 - 2}{15 - 7 + 2}$ $-\frac{9}{5}$ or $-1\frac{4}{5}$

48. $\frac{20 \div (-5) - (-6)}{(2)(-2)(-5)}$ $\frac{1}{10}$

49. $\frac{(22 - 4) \div (-2)}{-12 - 3(-5)}$ -3

Perform the operations in the proper order.

50. $-3 + 4(2 - 6)^2 \div (-2)$
-35

51. $-3(12 - 15)^2 + 2^5$
5

52. $-50 \div (-10) + (5 - 3)^4$
21

53. $\frac{2}{3} - \frac{2}{5} \div \left(\frac{1}{3}\right)\left(-\frac{3}{4}\right)$
$\frac{47}{30}$ or $1\frac{17}{30}$

54. $\left(\frac{2}{3}\right)^2 - \frac{3}{8}\left(\frac{8}{5}\right)$
$-\frac{7}{45}$

55. $(0.8)^2 - 3.2(-1.6)$
5.76

56. $1.4(4.7 - 4.9) - 12.8 \div (-0.2)$ 63.72

Section 9.5

Write in scientific notation.

57. 4160
4.16×10^3

58. 3,700,000
3.7×10^6

59. 200,000
2×10^5

60. 0.007
7.0×10^{-3}

61. 0.0000218
2.18×10^{-5}

62. 0.00000763
7.63×10^{-6}

Write in standard notation.

63. 1.89×10^4
18,900

64. 3.76×10^3
3760

65. 7.52×10^{-2}
0.0752

66. 6.61×10^{-3}
0.00661

67. 9×10^{-7}
0.0000009

68. 8×10^{-8}
0.00000008

69. 5.36×10^{-4}
0.000536

70. 1.98×10^{-5}
0.0000198

Add or subtract. Express your answer in scientific notation.

71. $5.26 \times 10^{11} + 3.18 \times 10^{11}$
8.44×10^{11}

72. $7.79 \times 10^{15} + 1.93 \times 10^{15}$
9.72×10^{15}

73. $3.42 \times 10^{14} - 1.98 \times 10^{14}$
1.44×10^{14}

74. $1.76 \times 10^{26} - 1.08 \times 10^{26}$
6.8×10^{25}

75. ***River*** There are 123,120,000,000,000 drops of water in a river 12 kilometers long, 270 meters wide, and 38 meters deep. Write this number in scientific notation. 1.2312×10^{14} drops

76. ***Earth Science*** The mass of the earth is approximately 5,983,000,000,000,000,000,000,000 kilograms. Since writing down all of these zeros is not fun, write this number in scientific notation. This can also be characterized as 5983 Yg (yottagrams). 5.983×10^{24} kilograms

77. ***Measurement*** How many feet are in 2,500,000,000 miles? Write this number in scientific notation. 1.32×10^{13} feet

78. ***Astronomy*** The distance from earth to the sun is approximately 93,000,000 miles. How many feet is that? Write your answer in scientific notation. 4.9104×10^{11} ft

79. ***Atomic Particles*** The mass of a proton is about 1.67 yg (yoctograms), and the mass of an electron is about 0.00091 yg. If *yocto-* means 0.000000000000000000000001, write the mass of a proton and that of an electron in grams in scientific notation.
1.67×10^{-24} grams; 9.1×10^{-28} grams

80. The prefix *femto-* means 10^{-15}. Write this number in standard form. 0.000000000000001

81. ***Moon*** The average distance to the moon is 384.4 Mm (megameters). If a megameter is equal to 10^6 meters, write out the number in meters. 384,400,000 meters

82. ***Football*** In three plays of a football game, the quarterback threw passes that lost 5 yards, gained 6 yards, and lost 7 yards. What was the total gain or loss of the three plays?
total loss 6 yards

83. ***Small Plane*** Fred is 6 feet tall and is standing at the lowest point of Death Valley, California, which is 282 feet below sea level. A small plane flies directly over Fred at an altitude of 2400 feet above sea level. Find the distance from the top of Fred's head to the plane. 2676 ft

84. ***Checking Account*** Max has overdrawn his checking account by \$18. His bank charged him \$20 for an overdraft fee. He quickly deposited \$40. What is his current balance? \$2

85. ***Temperature Statistics*** In Duluth, Minnesota, the high temperature in degrees Fahrenheit for five days during January was $-16°$, $-18°$, $-5°$, $3°$, and $-12°$. What was the average high temperature for these five days? $-9.6°$F

86. ***Golf*** Frank played golf with his friend Samuel. They played nine holes of golf on a special practice course. The expected number of strokes for each hole is 3. A birdie is 1 below par. An eagle is 2 below par. A bogie is 1 above par. A double bogey is 2 above par. Frank played on par for 1 hole and got two birdies, one eagle, four bogies, and one double bogey on the rest of the course. How many points above or below par was Frank on this nine-hole course? 2 points above par

How Am I Doing? Chapter 9 Test

Remember to use your Chapter Test Prep Video CD to see the worked-out solutions to the test problems you want to review.

Note to Instructor: The Chapter 9 Test file in the TestGen program provides algorithms specifically matched to these problems so you can easily replicate this test for additional practice or assessment purposes.

Add.

1. $-26 + 15$

2. $-31 + (-12)$

3. $12.8 + (-8.9)$

4. $-3 + (-6) + 7 + (-4)$

5. $-5\frac{3}{4} + 2\frac{1}{4}$

6. $-\frac{1}{4} + \left(-\frac{5}{8}\right)$

Subtract.

7. $-32 - 6$

8. $23 - 18$

9. $\frac{4}{5} - \left(-\frac{1}{3}\right)$

10. $-50 - (-7)$

11. $-2.5 - (-6.5)$

12. $-8.5 - 2.8$

13. $\frac{1}{12} - \left(-\frac{5}{6}\right)$

14. $-15 - (-15)$

Multiply or divide.

15. $(-20)(-6)$

16. $27 \div \left(-\frac{3}{4}\right)$

17. $-40 \div (-4)$

18. $(-9)(-1)(-2)(4)\left(\frac{1}{4}\right)$

19. $\frac{-39}{-13}$

20. $\dfrac{-\frac{3}{5}}{\frac{6}{7}}$

21. $(-12)(0.5)(-3)$

22. $96 \div (-3)$

1. -11

2. -43

3. 3.9

4. -6

5. $-\frac{7}{2}$ or $-3\frac{1}{2}$

6. $-\frac{7}{8}$

7. -38

8. 5

9. $\frac{17}{15}$ or $1\frac{2}{15}$

10. -43

11. 4

12. -11.3

13. $\frac{11}{12}$

14. 0

15. 120

16. -36

17. 10

18. -18

19. 3

20. $-\frac{7}{10}$

21. 18

22. -32

23. 17

24. 0.5

25. −8

26. −46

27. −7.2

28. −20

29. $-\frac{1}{7}$

30. $-\frac{1}{6}$

31. 8.054×10^4

32. 7×10^{-6}

33. 0.0000936

34. 72,000

35. −6.2°F

36. 2.72×10^{-4} meter

37. 186.9°F

Perform the indicated operations in the proper order.

23. $7 - 2(-5)$

24. $-2.5 - 1.2 \div (-0.4)$

25. $18 \div (-3) + 24 \div (-12)$

26. $-6(-3) - 4(3 - 7)^2$

27. $1.3 - 9.5 - (-2.5) + 3(-0.5)$

28. $-48 \div (-6) - 7(-2)^2$

29. $\dfrac{3 + 8 - 5}{(-4)(6) + (-6)(3)}$

30. $\dfrac{5 + 28 \div (-4)}{7 - (-5)}$

Write in scientific notation.

31. 80,540

32. 0.000007

Write in standard notation.

33. 9.36×10^{-5}

34. 7.2×10^4

Solve.

35. In Chicago the high temperatures in degrees Fahrenheit for five days during February were −14°, −8°, −5°, 7°, and −11°. What was the average high temperature for these five days?

▲**36.** A rectangular computer chip is 5.8×10^{-5} meter wide and 7.8×10^{-5} meter long. Find the perimeter of the chip and express your answer in scientific notation.

37. The lowest recorded temperature in Antarctica is −128.6°F in Vostock II on July 21, 1983. The highest recorded temperature in Antarctica is 58.3°F in Hope Bay on January 5, 1974. What is the difference between these temperatures?

Cumulative Test for Chapters 1–9

Approximately one-half of this test is based on Chapter 9 material. The remainder is based on material covered in Chapters 1–8.

Solve. Simplify your answers.

1. Subtract. $\begin{array}{r} 28{,}981 \\ -\ 16{,}598 \\ \hline \end{array}$

2. Divide. $36\overline{)4572}$

3. Add. $3\frac{1}{4} + 8\frac{2}{3}$

4. Multiply. $1\frac{5}{6} \times 2\frac{1}{2}$

5. Round to the nearest thousandth. 9.812456

6. Add. $5.82 + 38.964 + 0.571 + 9.305 + 8.8$

7. Multiply. 12.89×5.12

8. Find n. $\frac{n}{8} = \frac{56}{7}$

9. For every 156 parts manufactured, there are seven defects. If 2808 parts are manufactured, how many defects would you expect?

10. What is 0.8% of 38?

11. 10% of what number is 12?

12. Convert 94 kilometers to meters.

13. Convert 180 inches to yards.

▲ **14.** Find the perimeter of a rectangle that measures 12.5 ft $\times$ 10 ft.

▲ **15.** Find the area of a circle with radius 5 meters. Round to the nearest tenth. Use $\pi \approx 3.14$.

16. The following histogram depicts the ages of students at Wolfville College.

(a) How many students are between ages 23 and 25?

(b) How many students are older than 19 years?

(c) How many students are either less than 20 years or older than 25 years?

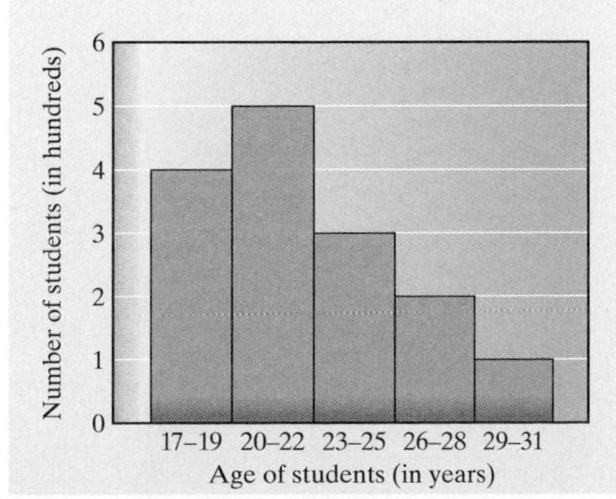

1. 12,383

2. 127

3. $\frac{143}{12}$ or $11\frac{11}{12}$

4. $\frac{55}{12}$ or $4\frac{7}{12}$

5. 9.812

6. 63.46

7. 65.9968

8. $n = 64$

9. 126 defects

10. 0.304

11. 120

12. 94,000 meters

13. 5 yards

14. 45 ft

15. 78.5 square meters

16. **(a)** 300 students

(b) 1100 students

(c) 700 students

17. 13

18. −4.7

19. $\frac{5}{12}$

20. −11

21. −5

22. −60

23. $\frac{4}{3}$ or $1\frac{1}{3}$

24. 18

25. 4

26. −2

27. $\frac{4}{15}$

28. 8.6972×10^4

29. 5.49×10^{-5}

30. 38,500,000

31. 0.00007

17. Evaluate. $\sqrt{36} + \sqrt{49}$

Add.

18. $-1.2 + (-3.5)$

19. $-\frac{1}{4} + \frac{2}{3}$

Subtract.

20. $7 - 18$

21. $-12 - (-7)$

Multiply or divide.

22. $(5)(-3)(-1)(-2)(2)$

23. $\dfrac{-\frac{4}{5}}{-\frac{21}{35}}$

Perform the indicated operations in the proper order.

24. $6 - 3(-4)$

25. $(-20) \div (-2) + (-6)$

26. $\dfrac{(-2)(-1) + (-4)(-3)}{1 + (-4)(2)}$

27. $\dfrac{(-11)(-8) \div 22}{1 - 7(-2)}$

Write in scientific notation.

28. 86,972

29. 0.0000549

Write in standard notation.

30. 3.85×10^7

31. 7×10^{-5}

CHAPTER

10

Traffic is a major problem in many parts of the United States. The wise use of traffic lights can do much to help the traffic flow. A linear equation can be used to determine the time that a light should show green for a particular number of vehicles. Can you use your mathematics to solve problems of this nature? Turn to page 666 to find out.

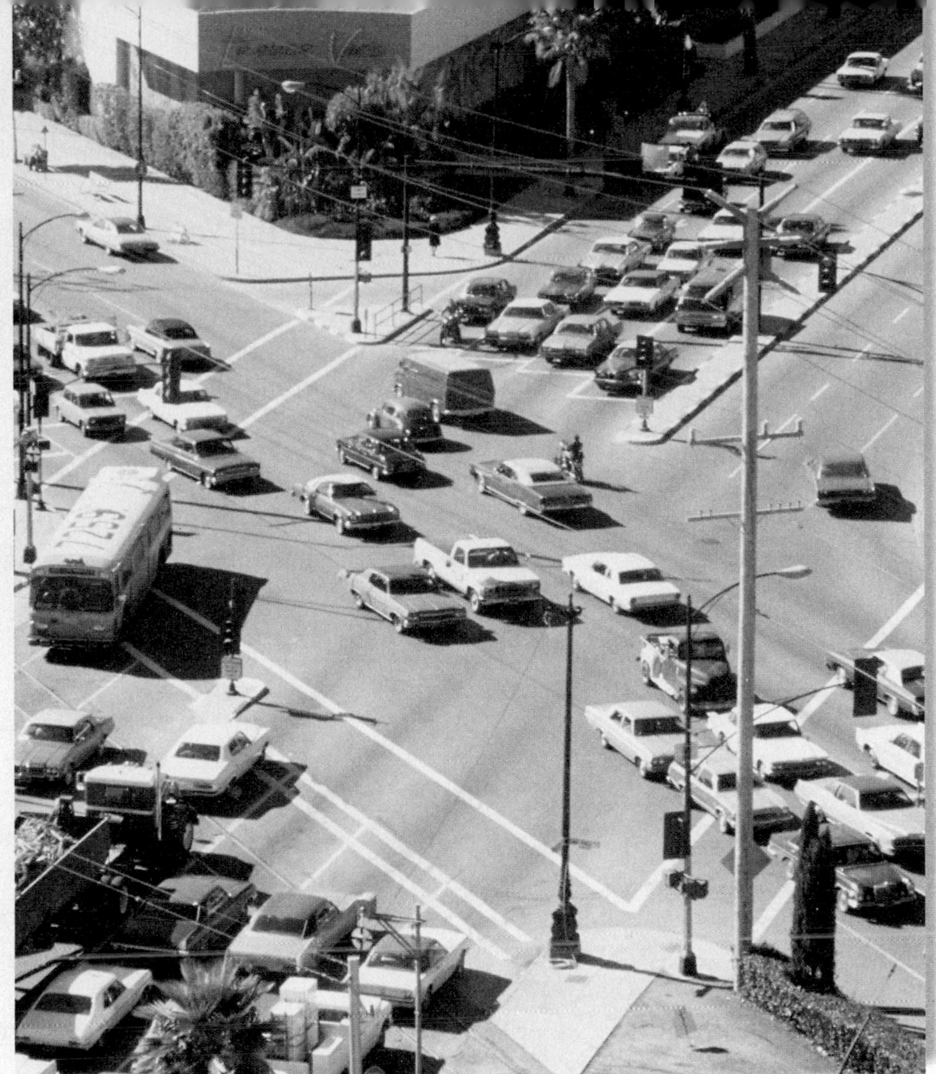

Introduction to Algebra

10.1 Variables and Like Terms

Student Learning Objectives

After studying this section, you will be able to:

1. Recognize the variable in an equation or a formula.
2. Combine like terms containing a variable.

1 Recognizing the Variable in an Equation or a Formula

In algebra we reason and solve problems by means of symbols. A **variable** is a symbol, usually a letter of the alphabet, that stands for a number. We can use the variable even though we may not know what number the variable stands for. We can find that number by following a logical order of steps. These are the rules of algebra.

We begin by taking a closer look at variables. In the formula for the area of a circle, the equation $A = \pi r^2$ contains two variables. r represents the value of the radius. A represents the value of the area. π is a known value. We often use the decimal approximation 3.14 for π.

EXAMPLE 1 Name the variables in each equation.

(a) $A = lw$ **(b)** $V = \dfrac{4\pi r^3}{3}$

Solution

(a) $A = lw$ The variables are A, l, and w.

(b) $V = \dfrac{4\pi r^3}{3}$ The variables are V and r.

NOTE TO STUDENT: Fully worked-out solutions to all of the Practice Problems can be found at the back of the text starting at page SP-1

Practice Problem 1 Name the variables.

(a) $A = \dfrac{bh}{2}$ **(b)** $V = lwh$

We have seen various ways to write multiplication. For example, three times n can be written as $3 \times n$. In algebra we usually do not write the multiplication symbol. We can simply write $3n$ to mean three times n. A number just to the left of a variable indicates that the number is multiplied by the variable. Thus $4ab$ means 4 times a times b.

A number just to the left of a set of parentheses also means multiplication. Thus $3(n + 8)$ means $3 \times (n + 8)$. So a product can be written with or without a multiplication sign.

Teaching Tip Example 2 is helpful because students may have seen formulas written both ways. It is important for them to know that the formula $A = b \times h$ is the same as the formula $A = bh$. You may want to give some additional examples, such as $F = 1.8 \times C + 32$, which is equivalent to $F = 1.8C + 32$, and $I = P \times R \times T$, which is equivalent to $I = PRT$.

EXAMPLE 2 Write the formula without a multiplication sign.

(a) $V = \dfrac{B \times h}{3}$ **(b)** $A = \dfrac{h \times (B + b)}{2}$

Solution

(a) $V = \dfrac{Bh}{3}$ **(b)** $A = \dfrac{h(B + b)}{2}$

Practice Problem 2 Write the formula without a multiplication sign.

(a) $p = 2 \times w + 2 \times l$ **(b)** $A = \pi \times r^2$

 Combining Like Terms Containing a Variable

Recall that when we work with measurements we combine like quantities. A carpenter, for example, might perform the following calculations.

$$20 \text{ m} - 3 \text{ m} = 17 \text{ m}$$
$$7 \text{ in.} + 9 \text{ in.} = 16 \text{ in.}$$

We cannot combine quantities that are not the same. We cannot add 5 yd + 7 gal. We cannot subtract 12 lb − 3 in.

Similarly, when using variables, we can add or subtract only when the same variable is used. For example, we can add $4a + 5a = 9a$, but we cannot add $4a + 5b$.

A **term** is a number, a variable, or a product of a number and one or more variables separated from other terms in an expression by a + sign or a − sign. In the expression $2x + 4y + (-1)$ there are three terms: $2x$, $4y$, and -1. **Like terms** have identical variables and identical exponents, so in the expression $3x + 4y + (-2x)$, the two terms $3x$ and $(-2x)$ are called *like terms*. The terms $2x^2y$ and $5xy$ are not like terms, since the exponents for the variable x are not the same. To combine like terms, you combine the numbers, called the **numerical coefficients,** that are directly in front of the terms by using the rules for adding signed numbers. Then you use this new number as the coefficient of the variable.

Teaching Tip Remind students that the word *term* is used extensively in mathematics, so it is best to make sure they grasp its definition now.

EXAMPLE 3 Combine like terms. $5x + 7x$

Solution We add $5 + 7 = 12$. Thus $5x + 7x = 12x$.

Practice Problem 3 Combine like terms. $9x + 2x$

When we combine signed numbers, we try to combine them mentally. Thus to combine $7 - 9$, we think $7 + (-9)$ and we write -2. In a similar way, to combine $7x - 9x$, we think $\boxed{7x + (-9x)}$ and we write $-2x$. Your instructor may ask you to write out this "think" step as part of your work. In the following example we show this extra step inside a $\boxed{}$ box. You should determine from your instructor whether he or she feels this step is necessary.

EXAMPLE 4 Combine like terms.

(a) $3x + 7x - 15x$ **(b)** $9x - 12x - 11x$

Solution

(a) $3x + 7x - 15x = \boxed{3x + 7x + (-15x)} = 10x + (-15x) = -5x$

(b) $9x - 12x - 11x = \boxed{9x + (-12x) + (-11x)} = -3x + (-11x)$
$= -14x$

Practice Problem 4 Combine like terms.

(a) $8x - 22x + 5x$ **(b)** $19x - 7x - 12x$

A variable without a numerical coefficient is understood to have a coefficient of 1.

$7x + y$ means $7x + 1y$. $5a - b$ means $5a - 1b$.

Teaching Tip Students often forget that a variable without a numerical coefficient is understood to have a coefficient of 1. In addition to Example 5, you may want to show them the following illustrations.

(a) $8w + w - 6w = 3w$

(b) $-13z - 18z - z = -32z$

***NOTE TO STUDENT:** Fully worked-out solutions to all of the Practice Problems can be found at the back of the text starting at page SP-1*

EXAMPLE 5 Combine like terms.

(a) $3x - 8x + x$ **(b)** $12x - x - 20.5x$

Solution

(a) $3x - 8x + x = 3x - 8x + 1x = -5x + 1x = -4x$

(b) $12x - x - 20.5x = 12x - 1x - 20.5x = 11.0x - 20.5x = -9.5x$

Practice Problem 5 Combine like terms.

(a) $9x - 12x + x$ **(b)** $5.6x - 8x - x$

Numbers cannot be combined with variable terms.

EXAMPLE 6 Combine like terms.

$$7.8 - 2.3x + 9.6x - 10.8$$

Solution In each case, we combine the numbers separately and the variable terms separately. It may help to use the commutative and associative properties first.

$$\begin{aligned} &7.8 - 2.3x + 9.6x - 10.8 \\ &= 7.8 - 10.8 - 2.3x + 9.6x \\ &= -3 + 7.3x \end{aligned}$$

Practice Problem 6 Combine like terms.

$$17.5 - 6.3x - 8.2x + 10.5$$

There may be more than one variable in a problem. Keep in mind, however, that only like terms may be combined.

EXAMPLE 7 Combine like terms.

(a) $5x + 2y + 8 - 6x + 3y - 4$ **(b)** $\frac{3}{4}x - 12 + \frac{1}{6}x + \frac{2}{3}$

Solution For convenience, we will rearrange the problem to place like terms next to each other. This is an optional step; you do not need to do this.

(a) $5x - 6x + 2y + 3y + 8 - 4 = -1x + 5y + 4 = -x + 5y + 4$

(b) $\frac{3}{4}x - 12 + \frac{1}{6}x + \frac{2}{3} = \frac{9}{12}x + \frac{2}{12}x - \frac{36}{3} + \frac{2}{3}$

$$= \frac{11}{12}x - \frac{34}{3}$$

The order of the terms in an answer is not important in this type of problem. The answer to Example 7(a) could have been $5y + 4 - x$ or $4 + 5y - x$. Often we give the answer with the letters in alphabetical order.

Practice Problem 7 Combine like terms.

(a) $2w + 3z - 12 - 5w - z - 16$ **(b)** $\frac{3}{5}x + 5 - \frac{7}{15}x - \frac{1}{3}$

10.1 Exercises

Student Solutions Manual | CD/Video | PH Math Tutor Center | MathXL®Tutorials on CD | MathXL® | MyMathLab® | Interactmath.com

Verbal and Writing Skills

1. In your own words, write a definition for the word *variable*.

A variable is a symbol, usually a letter of the alphabet, that stands for a number.

2. In your own words, define *like terms*.

Like terms are terms that have identical variables and identical exponents.

3. Why is it that you cannot combine like terms with a problem such as $3x^2y + 5xy^2$?

All the exponents for like terms must be the same. The exponent for x must be the same. The exponent for y must be the same. In this case x is raised to the second power in the first term but y is raised to the second power in the second term.

4. Why is it that you cannot combine like terms with a problem such as $7xy + 9x$?

All the variables must be the same. The first term has two variables x and y. The second term has only the variable x. In order to have like terms, the same exact variables must appear in each term.

Name the variables in each equation.

5. $G = 5xy$

G, x, y

6. $S = 3\pi r^3$

S, r

7. $p = \dfrac{4ab}{3}$

p, a, b

8. $p = \dfrac{7ab}{4}$

p, a, b

Write each equation without multiplication signs.

9. $r = 3 \times m + 5 \times n$

$r = 3m + 5n$

▲ **10.** $p = 2 \times w + 2 \times l$

$p = 2w + 2l$

11. $H = 2 \times a - 3 \times b$

$H = 2a - 3b$

12. $A = \dfrac{a \times b + a \times c}{3}$

$A = \dfrac{ab + ac}{3}$

Combine like terms.

13. $-16x + 26x$

$10x$

14. $-12x + 40x$

$28x$

15. $2x - 8x + 5x$

$-x$

16. $4x - 10x + 3x$

$-3x$

17. $-\dfrac{1}{2}x + \dfrac{3}{4}x + \dfrac{1}{12}x$

$\dfrac{1}{3}x$

18. $\dfrac{2}{5}x - \dfrac{2}{3}x + \dfrac{7}{15}x$

$\dfrac{1}{5}x$

19. $x + 3x + 8 - 7$

$4x + 1$

20. $5x - x + 9 + 16$

$4x + 25$

21. $1.3x + 10 - 2.4x - 3.6$ $\quad -1.1x + 6.4$

22. $3.8x + 2 - 1.9x - 3.5$ $\quad 1.9x - 1.5$

23. $16x + 9y - 11 + 21x$ $\quad 37x + 9y - 11$

24. $22x - 13y - 23 - 8x$ $\quad 14x - 13y - 23$

Mixed Practice

Combine like terms.

25. $-25 + \left(2\dfrac{1}{2}\right)x + 15 - \left(3\dfrac{1}{4}\right)x$

$-\dfrac{3}{4}x - 10$

26. $-38 - \left(3\dfrac{2}{3}\right)x + 46 + \left(2\dfrac{1}{3}\right)x$

$\left(-1\dfrac{1}{3}\right)x + 8$ or $-\dfrac{4}{3}x + 8$

27. $7a - c + 6b - 3c - 10a$
$-3a + 6b - 4c$

28. $10b + 3a - 8b - 5c + a$
$4a + 2b - 5c$

29. $\frac{1}{2}x + \frac{1}{7}y - \frac{3}{4}x + \frac{5}{21}y$ $\quad -\frac{1}{4}x + \frac{8}{21}y$

30. $\frac{1}{4}x + \frac{1}{3}y - \frac{7}{12}x - \frac{1}{2}y$ $\quad -\frac{1}{3}x - \frac{1}{6}y$

31. $7.3x + 1.7x + 4 - 6.4x - 5.6x - 10$
$-3x - 6$

32. $3.1x + 2.9x - 8 - 12.8x - 3.2x + 3$
$-10x - 5$

33. $-7.6n + 1.2 + 11.2m - 3.5n - 8.1m$
$-11.1n + 3.1m + 1.2$

34. $4.5n - 5.9m + 3.9 - 7.2n + 9m$
$-2.7n + 3.1m + 3.9$

To Think About

▲ **35.** **(a)** Find the perimeter of the triangle.
$12x + 1$
(b) If each side of the triangle is doubled, what is the new perimeter?
It is doubled to obtain $24x + 2$

$4x - 2$ $\quad 3x + 6$ $\quad 5x - 3$

▲ **36.** **(a)** Find the perimeter of this four-sided figure.
$6x + 16$
(b) If each side is doubled, what is the new perimeter?
It is doubled to obtain $12x + 32$

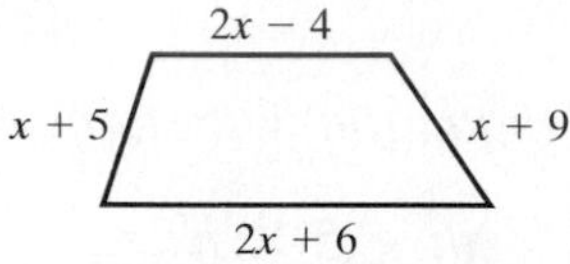

Cumulative Review

Solve for n.

37. $\frac{n}{6} = \frac{12}{15}$
$n = 4.8$

38. $\frac{n}{9} = \frac{36}{40}$
$n = 8.1$

39. $6n = 18$
$n = 3$

40. $5.5n = 46.75$
$n = 8.5$

41. $\frac{1}{4}n = 0.05$
$n = 0.2$

42. $\frac{1}{2}n = 6$
$n = 12$

43. ***Condominium Purchase*** John and Stephanie purchased a new condominium. They pay property taxes of \$2500 per year. In the town where they live property is taxed at the rate of \$1.25 for every \$100 in value. What is the value of the condominium that John and Stephanie purchased? \$200,000

44. ***Car Speed*** Carmela is driving her BMW at 55 miles per hour on route 1 in Maine. She enters a 45-miles-per-hour zone and brakes the car in such a way that the speed of the car slows by 2 miles per hour every three seconds. How long will it take until Carmela is driving the car at 45 miles per hour? 15 seconds

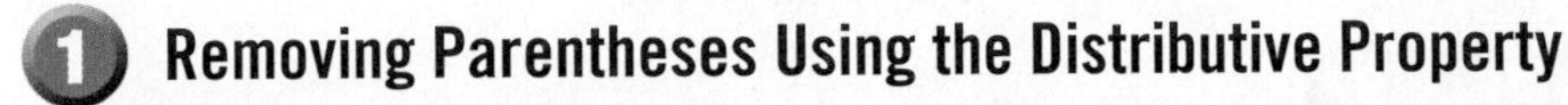

10.2 The Distributive Property

1 Removing Parentheses Using the Distributive Property

What do we mean by the word *property* in mathematics? A **property** is an essential characteristic. A property of addition is an essential characteristic of addition. In this section we learn about the distributive property and how to use this property to simplify expressions.

Sometimes we encounter expressions like $4(x + 3)$. We'd like to be able to simplify the expression so that no parentheses appear. Notice that this expression contains two operations, multiplication and addition. We will use the distributive property to distribute the 4 to both terms in the addition statement. That is,

$$4(x + 3) \quad \text{is equal to} \quad 4(x) + 4(3).$$

The numerical coefficient 4 can be "distributed over" the expression $x + 3$ by multiplying the 4 by each of the terms in the parentheses. The expression $4(x) + 4(3)$ can be simplified to $4x + 12$, which has no parentheses. Thus we can use the distributive property to remove the parentheses.

Using variables, we can write the distributive property two ways.

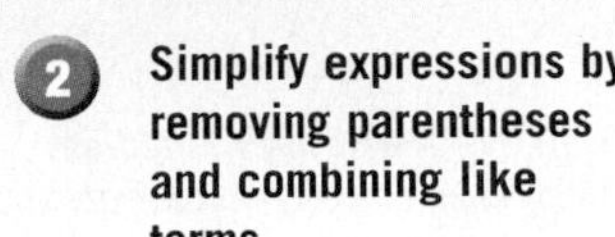

Student Learning Objectives

After studying this section, you will be able to:

1. **Remove parentheses using the distributive property.**
2. **Simplify expressions by removing parentheses and combining like terms.**

DISTRIBUTIVE PROPERTIES OF MULTIPLICATION OVER ADDITION

If a, b, and c are signed numbers, then

$$a(b + c) = ab + ac \quad \text{and} \quad (b + c)a = ba + ca.$$

A numerical example shows that the distributive property works.

$$(7)(4 + 6) = (7)(4) + (7)(6)$$
$$(7)(10) = 28 + 42$$
$$70 = 70$$

When the number is to the left of the parentheses, we use

$$a(b + c) = ab + ac.$$

There is also a distributive property of multiplication over subtraction: $a(b - c) = ab - ac$.

EXAMPLE 1 Simplify.

(a) $(8)(x + 5)$ **(b)** $(-3)(x + 3y)$ **(c)** $6(3a - 7b)$

Solution

(a) $(8)(x + 5) = 8x + (8)(5) = 8x + 40$

(b) $(-3)(x + 3y) = -3x + (-3)(3)(y) = -3x + (-9y) = -3x - 9y$

(c) $6(3a - 7b) = 6(3a) - (6)(7b) = 18a - 42b$

Practice Problem 1 Simplify.

(a) $(7)(x + 5)$ **(b)** $(-4)(x + 2y)$ **(c)** $5(6a - 2b)$

NOTE TO STUDENT: *Fully worked-out solutions to all of the Practice Problems can be found at the back of the text starting at page SP-1*

Sometimes the number is to the right of the parentheses, so we use

$$(b + c)a = ba + ca.$$

Teaching Tip You may want to warn students that although they need to be able to simplify both $5(x + 2y)$ and $(x + 2y)(5)$, it is the first type of problem that they will see most often.

EXAMPLE 2 Simplify. $(5x + y)(2)$

Solution

$$(5x + y)(2) = (5x)(2) + (y)(2) = 10x + 2y$$

Notice in Example 2 that we write our final answer with the numerical coefficient to the left of the variable. We would not leave $(y)(2)$ as an answer but would write $2y$.

NOTE TO STUDENT: Fully worked-out solutions to all of the Practice Problems can be found at the back of the text starting at page SP-1

Practice Problem 2 Simplify. $(x + 3y)(8)$

Sometimes the distributive property is used with three terms within the parentheses. When parentheses are used inside parentheses, the outside ( ) is often changed to a bracket [ ] notation.

Teaching Tip Warn students that it is very easy to make a sign error when doing problems of this type. After looking over Example 3 ask students to do the following exercises.

(a) $-9(-3 - 9x + 5y)$

(b) $12(2 - 7w + z)$

Ask them to check their work carefully. Answers are $27 + 81x - 45y$ and $24 - 84w + 12z$.

EXAMPLE 3 Simplify.

(a) $(-5)(x + 2y - 8)$ **(b)** $(1.5x + 3y + 7)(2)$

Solution

(a)
$$\begin{aligned}(-5)(x + 2y - 8) &= (-5)[x + 2y + (-8)] \\ &= -5(x) + (-5)(2y) + (-5)(-8) \\ &= -5x + (-10y) + 40 \\ &= -5x - 10y + 40\end{aligned}$$

(b)
$$\begin{aligned}(1.5x + 3y + 7)(2) &= (1.5x)(2) + (3y)(2) + (7)(2) \\ &= 3x + 6y + 14\end{aligned}$$

In this example every step is shown in detail. You may find that you do not need to write so many steps.

Practice Problem 3 Simplify.

(a) $(-5)(x + 4y + 5)$ **(b)** $(2.2x + 5.5y + 6)(3)$

The parentheses may contain four terms. The coefficients may be decimals or fractions.

EXAMPLE 4 Simplify. $\frac{2}{3}\left(x + \frac{1}{2}y - \frac{1}{4}z + \frac{1}{5}\right)$

Solution

$$\begin{aligned}&\frac{2}{3}\left(x + \frac{1}{2}y - \frac{1}{4}z + \frac{1}{5}\right) \\ &= \left(\frac{2}{3}\right)(x) + \left(\frac{2}{3}\right)\left(\frac{1}{2}y\right) + \left(\frac{2}{3}\right)\left(-\frac{1}{4}z\right) + \left(\frac{2}{3}\right)\left(\frac{1}{5}\right) \\ &= \frac{2}{3}x + \frac{1}{3}y - \frac{1}{6}z + \frac{2}{15}\end{aligned}$$

Practice Problem 4 Simplify. $\frac{3}{2}\left(\frac{1}{2}x - \frac{1}{3}y + 4z - \frac{1}{2}\right)$

2 Simplifying Expressions by Removing Parentheses and Combining Like Terms

After removing parentheses we may have a chance to combine like terms. The direction "simplify" means remove parentheses, combine like terms, and leave the answer in as simple and correct a form as possible.

EXAMPLE 5 Simplify. $2(x + 3y) + 3(4x + 2y)$

Solution

$2(x + 3y) + 3(4x + 2y) = 2x + 6y + 12x + 6y$ Use the distributive property.

$= 14x + 12y$ Combine like terms.

Practice Problem 5 Simplify. $3(2x + 4y) + 2(5x + y)$

Teaching Tip Caution students to be most careful of + and − signs in this type of problem. Ask them to simplify the following problems at their seats.

(a) Simplify.
$-3(2x - 3y) - 2(x + y)$

(b) Simplify.
$2(-5a + 2b) - 5(-3a - 3b)$

Students are often amazed at how easy it is to make a sign error and not obtain the correct answers of $-8x + 7y$ for problem a and $5a + 19b$ for problem b.

EXAMPLE 6 Simplify. $2(x - 3y) - 5(2x + 6)$

Solution

$2(x - 3y) - 5(2x + 6) = 2x - 6y - 10x - 30$ Use the distributive property.

$= -8x - 6y - 30$ Combine like terms.

Notice that in the final step of Example 6 only the x terms could be combined. There are no other like terms.

Practice Problem 6 Simplify. $-4(x - 5) + 3(-1 + 2x)$

10.2 Exercises

Student Solutions Manual

CD/ Video

PH Math Tutor Center

MathXL®Tutorials on CD

MathXL®

MyMathLab®

Interactmath.com

Verbal and Writing Skills

1. A ___variable___ is a symbol, usually a letter of the alphabet, that stands for a number.

2. What is the variable in the expression $5x + 9$? x

3. Identify the like terms in the expression $3x + 2y - 1 + x - 3y$. $3x$ and x; $2y$ and $-3y$

4. Explain the distributive property in your own words. Give an example.
The distributive property takes the multiplication of a number times a sum and distributes it to a multiplication by each value that was added. $3(x + 2y) = 3x + 3(2y) = 3x + 6y$

Simplify.

5. $9(3x - 2)$
$27x - 18$

6. $8(4x - 5)$
$32x - 40$

7. $(-2)(x + y)$
$-2x - 2y$

8. $(-5)(x + y)$
$-5x - 5y$

9. $-7(1.5x - 3y)$
$-10.5x + 21y$

10. $-4(3.5x - 6y)$
$-14x + 24y$

11. $(-3x + 7y)(-10)$
$30x - 70y$

12. $(-2x + 8y)(-12)$
$24x - 96y$

13. $(6a - 5b)(8)$
$48a - 40b$

14. $(7a - 11b)(6)$
$42a - 66b$

15. $(-5y - 6z)(-4)$
$20y + 24z$

16. $(-9y - 2z)(-5)$
$45y + 10z$

17. $4(p + 9q - 10)$
$4p + 36q - 40$

18. $6(2p - 7q + 11)$
$12p - 42q + 66$

19. $3\left(\frac{1}{5}x + \frac{2}{3}y - \frac{1}{4}\right)$
$\frac{3}{5}x + 2y - \frac{3}{4}$

20. $4\left(\frac{2}{3}x + \frac{1}{4}y - \frac{3}{8}\right)$
$\frac{8}{3}x + y - \frac{3}{2}$

21. $-15(-2a - 3.2b + 4.5)$
$30a + 48b - 67.5$

22. $-14(-5a + 1.4b - 2.5)$
$70a - 19.6b + 35$

23. $(8a + 12b - 9c - 5)(4)$
$32a + 48b - 36c - 20$

24. $(-7a + 11b - 10c - 9)(7)$
$-49a + 77b - 70c - 63$

25. $-2(1.3x - 8.5y - 5z + 12)$
$-2.6x + 17y + 10z - 24$

26. $-3(1.4x - 7.6y - 9z - 4)$
$-4.2x + 22.8y + 27z + 12$

27. $\frac{1}{2}\left(2x - 3y + 4z - \frac{1}{2}\right)$
$x - \frac{3}{2}y + 2z - \frac{1}{4}$

28. $\frac{1}{3}\left(-3x + \frac{1}{2}y + 2z - 3\right)$
$-x + \frac{1}{6}y + \frac{2}{3}z - 1$

29. $-\frac{1}{5}(20x - 30y + 5z)$
$-4x + 6y - z$

30. $-\frac{1}{4}(16x - 4y + 32z)$
$-4x + y - 8z$

Applications

▲ **31.** ***Geometry*** The perimeter of a rectangle is $p = 2(l + w)$. Write this formula without parentheses and without multiplication signs.
$p = 2l + 2w$

▲ **32.** ***Geometry*** The surface area of a rectangular solid is $S = 2(lw + lh + wh)$. Write this formula without parentheses and without multiplication signs. $S = 2lw + 2lh + 2wh$

▲ **33.** ***Geometry*** The area of a trapezoid is $A = \dfrac{h(B + b)}{2}$. Write this formula without parentheses and without multiplication signs.
$A = \dfrac{hB + hb}{2}$

▲ **34.** ***Geometry*** The surface area of a cylinder is $S = 2\pi r(h + r)$. Write this formula without parentheses and without multiplication signs.
$S = 2\pi rh + 2\pi r^2$

Simplify. Be sure to combine like terms.

35. $4(5x - 1) + 7(x - 5)$
$27x - 39$

36. $6(5x - 1) + 3(x - 4)$
$33x - 18$

37. $10(4a + 5b) - 8(6a + 2b)$
$-8a + 34b$

38. $12(3a + b) - 9(2a + 3b)$
$18a - 15b$

39. $1.5(x + 2.2y) + 3(2.2x + 1.6y)$
$8.1x + 8.1y$

40. $2.4(x + 3.5y) + 2(1.4x + 1.9y)$
$5.2x + 12.2y$

41. $2(3b + c - 2a) - 5(a - 2c + 5b)$
$-9a - 19b + 12c$

42. $3(-4a + c + 4b) - 4(2c + b - 6a)$
$12a + 8b - 5c$

To Think About

▲ **43.** Illustrate the distributive property by using the area of two rectangles.
$A = ab + ac$
$A = a(b + c)$

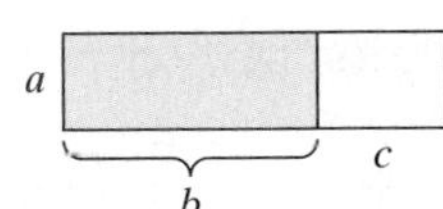

▲ **44.** Show that multiplication is distributive over subtraction by using the area of two rectangles.
$A = ab - ac$
$A = a(b - c)$

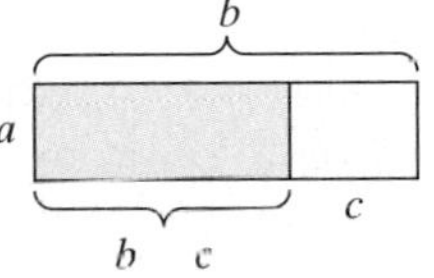

▲ **45.** Illustrate how the distributive property for three terms, $a(b + c + d) = ab + ac + ad$, could be used to discuss the area of three rectangles in the following figure.

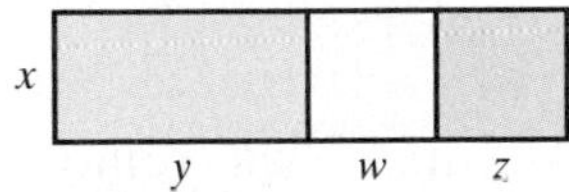

$A = xy + xw + xz$
$A = x(y + w + z)$

▲ **46.** Illustrate how the distributive property for three terms, $a(b + c + d) = ab + ac + ad$, could be used to discuss the area of three parallelograms in the following figure.

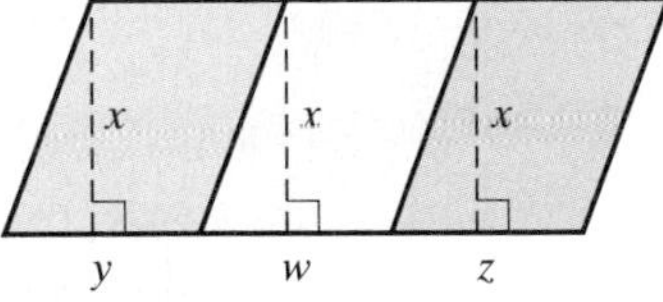

$A = xy + xw + xz$
$A = x(y + w + z)$

Cumulative Review

▲ **47.** ***Geometry*** Find the perimeter of a rectangular picture frame with a length of 8.5 inches and a width of 5 inches. 27 in.

▲ **48.** ***Geometry*** Find the perimeter of a triangle with sides measuring 6.4 feet, 5.2 feet, and 7.8 feet. 19.4 feet

10.3 Solving Equations Using the Addition Property

Student Learning Objective

After studying this section, you will be able to:

 **Solve equations using the addition property.**

1 Solving Equations Using the Addition Property

One of the most important skills in algebra is that of **solving an equation.** Starting from an equation with a variable whose value is unknown, we transform the equation into a simpler, equivalent equation by choosing a logical step. In the following sections we'll learn some of the logical steps for solving an equation successfully.

We begin with the concept of an equation. An **equation** is a mathematical statement that says that two expressions are equal. The statement $x + 5 = 13$ means that some number (x) plus 5 equals 13. Experience with the basic addition facts tells us that $x = 8$ since $8 + 5 = 13$. Therefore, 8 is the solution to the equation. The **solution** of an equation is that number which makes the equation true. What if we did not know that $8 + 5 = 13$? How could we find the value of x? Let's look at the first logical step in solving this equation.

The important thing to remember is that an equation is like a balance scale. Whatever you do to one side of the equation, you must do to the other side of the equation to maintain the balance. Our goal is to do this in such a way that we isolate the variable.

ADDITION PROPERTY OF EQUATIONS

You may add the same number to each side of an equation to obtain an equivalent equation.

Suppose we want to make the equation $x + 5 = 13$ into a simpler equation, such as $x =$ some number. What can we add to each side of the equation so that $x + 5$ becomes simply x? We can add the opposite of +5. Let's see what happens when we do this addition.

$$x + 5 = 13$$

$$x + 5 + (-5) = 13 + (-5)$$ Remember to add −5 to both sides of the equation.

$$x + 0 = 8$$

$$x = 8$$

We found the solution to the equation.

Notice that in the second step, the number 5 was removed from the left side of the equation. We know that we may add a number to both sides of the equation. But what number should we choose to add? We always add the opposite of the number we want to remove from one side of the equation.

EXAMPLE 1 Solve. $x - 9 = 3$

Solution We want to isolate the variable x.

$$x - 9 = 3$$ Think: "Add the opposite of −9 to both sides of the equation."

$$x - 9 + 9 = 3 + 9$$

$$x + 0 = 12$$

$$x = 12$$

To check the solution, we substitute 12 for x in the original equation.

$$x - 9 = 3$$
$$12 - 9 \stackrel{?}{=} 3$$
$$3 = 3 \quad \checkmark$$

It is always a good idea to check your solution, especially in more complicated equations.

Practice Problem 1 Solve. $x + 7 = -8$

NOTE TO STUDENT: Fully worked-out solutions to all of the Practice Problems can be found at the back of the text starting at page SP-1

When we solve an equation we want to isolate the variable. We do this by performing the same operation on both sides of the equation.

Equations can contain integers, decimals, or fractions.

EXAMPLE 2 Solve. Check your solution.

(a) $x + 1.5 = 4$ **(b)** $\frac{3}{8} = x - \frac{3}{4}$

Solution We use the addition property to solve these equations since each equation involves only addition or subtraction. In each case we want to isolate the variable.

(a) $x + 1.5 = 4$ Think: "Add the opposite of 1.5 to both sides of the equation."

$$x + 1.5 + (-1.5) = 4 + (-1.5)$$
$$x = 2.5$$

Check.

$x + 1.5 = 4$ Substitute 2.5 for x in the original equation.

$$2.5 + 1.5 \stackrel{?}{=} 4$$
$$4 = 4$$

(b) $\frac{3}{8} = x - \frac{3}{4}$ Note that here the variable is on the right-hand side.

$\frac{3}{8} + \frac{3}{4} = x - \frac{3}{4} + \frac{3}{4}$ Think: "Add the opposite of $-\frac{3}{4}$ to both sides of the equation."

$\frac{3}{8} + \frac{6}{8} = x + 0$ We need to change $\frac{3}{4}$ to $\frac{6}{8}$.

$$\frac{9}{8} \text{ or } 1\frac{1}{8} = x$$

Check.

$\frac{3}{8} = x - \frac{3}{4}$ Substitute $\frac{9}{8}$ for x in the original equation.

$$\frac{3}{8} \stackrel{?}{=} \frac{9}{8} - \frac{3}{4}$$
$$\frac{3}{8} \stackrel{?}{=} \frac{9}{8} - \frac{6}{8}$$
$$\frac{3}{8} = \frac{3}{8} \quad \checkmark$$

***NOTE TO STUDENT:** Fully worked-out solutions to all of the Practice Problems can be found at the back of the text starting at page SP-1*

Practice Problem 2 Solve. Check your solution.

(a) $y - 3.2 = 9$ **(b)** $\frac{2}{3} = x + \frac{1}{6}$

Sometimes you will need to use the addition property twice. If variables and numbers appear on both sides of the equation, you will want to get all of the variables on one side of the equation and all of the numbers on the other side of the equation.

EXAMPLE 3 Solve $2x + 7 = x + 9$. Check your solution.

Solution We want to remove the +7 on the left-hand side of the equation.

$$2x + 7 = x + 9$$

$$2x + 7 + (-7) = x + 9 + (-7)$$ Add −7 to both sides of the equation.

$$2x = x + 2$$ Now we need to remove the x on the right-hand side of the equation.

$$2x - x = x - x + 2$$ Add $-x$ to both sides of the equation.

$$x = 2$$

We would certainly want to check this solution. Solving the equation took several steps and we might have made a mistake along the way. To check, we substitute 2 for x in the original equation.

Check.

$$2x + 7 = x + 9$$
$$2(2) + 7 \stackrel{?}{=} 2 + 9$$
$$4 + 7 \stackrel{?}{=} 2 + 9$$
$$11 = 11 \checkmark$$

Practice Problem 3 Solve. $3x - 5 = 2x + 1$. Check your solution.

10.3 EXERCISES

Student Solutions Manual | CD/ Video | PH Math Tutor Center | MathXL®Tutorials on CD | MathXL® | MyMathLab® | Interactmath.com

Verbal and Writing Skills

1. An equation is a mathematical statement that says that two expressions are equal.

2. The solution of an equation is that number which makes the equation true.

3. To use the addition property, we add to both sides of the equation the opposite of the number we want to remove from one side of the equation.

4. To use the addition property to solve the equation $x - 8 = 9$, we add $+8$ to both sides of the equation.

Solve for the variable.

5. $y - 12 = 20$
   $y = 32$

6. $y - 14 = 27$
   $y = 41$

7. $x + 6 = 15$
   $x = 9$

8. $x + 8 = 12$
   $x = 4$

9. $x + 16 = -2$
   $x = -18$

10. $y + 12 = -8$
    $y = -20$

11. $14 + x = -11$
    $x = -25$

12. $10 + y = -9$
    $y = -19$

13. $-12 + x = 7$
    $x = 19$

14. $-20 + y = 10$
    $y = 30$

15. $5.2 = x - 4.6$
    $9.8 = x$

16. $3.1 = y - 5.4$
    $8.5 = y$

17. $x + 3.7 = -5$
    $x = -8.7$

18. $y + 6.2 = -3.1$
    $y = -9.3$

19. $x - 25.2 = -12$
    $x = 13.2$

20. $y - 29.8 = -15$
    $y = 14.8$

21. $\frac{4}{5} = x + \frac{2}{5}$
    $x = \frac{2}{5}$

22. $\frac{7}{10} = x + \frac{1}{10}$
    $\frac{3}{5} = x$

23. $x - \frac{3}{5} = \frac{2}{5}$
    $x = 1$

24. $y - \frac{2}{7} = \frac{6}{7}$
    $y = \frac{8}{7}$ or $1\frac{1}{7}$

25. $x + \frac{2}{3} = -\frac{5}{6}$
    $x = -\frac{3}{2}$ or $-1\frac{1}{2}$

26. $y + \frac{1}{2} = -\frac{3}{4}$
    $y = -\frac{5}{4}$ or $-1\frac{1}{4}$

27. $\frac{1}{5} + y = -\frac{2}{3}$
    $y = -\frac{13}{15}$

28. $\frac{5}{9} + y = -\frac{3}{18}$
    $y = -\frac{13}{18}$

Solve for the variable. You may need to use the addition property twice.

29. $3x - 5 = 2x + 9$
    $x = 14$

30. $5x + 1 = 4x - 3$
    $x = -4$

31. $5x + 12 = 4x - 1$
    $x = -13$

32. $8x - 3 = 7x - 12$
    $x = -9$

33. $7x - 9 = 6x - 7$
    $x = 2$

34. $4x - 8 = 3x + 12$
    $x = 20$

35. $18x + 28 = 17x + 19$
    $x = -9$

36. $15x + 41 = 14x + 27$
    $x = -14$

Mixed Practice

Solve for the variable.

37. $y - \frac{1}{2} = 6$ $y = \frac{13}{2}$ or $6\frac{1}{2}$

38. $x + 1.2 = -3.8$ $x = -5$

39. $5 = z + 13$ $-8 = z$

40. $-15 = -6 + x$ $-9 = x$

41. $-5.9 + y = -4.7$ $y = 1.2$

42. $z + \frac{2}{3} = \frac{7}{12}$ $z = -\frac{1}{12}$

43. $2x - 1 = x + 5$
$x = 6$

44. $3x - 8 = 2x - 15$
$x = -7$

45. $3.6x - 8 = 2.6x + 4$
$x = 12$

46. $5.4y + 3 = 4.4y - 1$
$y = -4$

47. $7x + 14 = 10x + 5$
$x = 3$

48. $9x + 5 = 3x - 13$
$x = -3$

To Think About

49. In the equation $x + 9 = 12$, we add -9 to both sides of the equation to solve. What would you do to solve the equation $3x = 12$? Why?
To solve the equation $3x = 12$, divide both sides of the equation by 3 so that x stands alone on one side of the equation.

50. In the equation $x - 7 = -12$, we add $+7$ to both sides of the equation to solve. What would you do to solve the equation $4x = -12$? Why?
To solve the equation $4x = -12$, divide both sides of the equation by 4 so that x stands alone on one side of the equation.

Cumulative Review

51. Combine like terms. $5x - y + 3 - 2x + 4y$
$3x + 3y + 3$

52. Simplify. $7(2x + 3y) - 3(5x - 1)$
$-x + 21y + 3$

53. ***U.S. Labor Force*** In 2001 there were 142 million people in the total labor force of the United States. Of this number, 6.7 million people were unemployed. (*Source:* Bureau of Labor Statistics) What percent of the total labor force was unemployed in 2001? Round to the nearest tenth of a percent. 4.7%

▲ 54. ***Geometry*** In the front of the Digitech Company building, there is a stone pyramid with a rectangular base. The base measures 3 feet wide and 4 feet long. The pyramid is 8 feet tall. What is the volume of this pyramid?
32 cubic feet

55. Find the mean. 18, 21, 24, 17, 21, 25 21

56. Find the median. 2.4, 2.2, 1.9, 1.8, 2.2, 2.7, 2.1
2.2

10.4 Solving Equations Using the Division or Multiplication Property

Student Learning Objectives

After studying this section, you will be able to:

1. Solve equations using the division property.
2. Solve equations using the multiplication property.

1 Solving Equations Using the Division Property

Recall that when we solve an equation using the addition property, we transform the equation to a simpler one where $x =$ some number. We use the same idea to solve the equation $3n = 75$. Think: "What can we do to the left side of the equation so that n stands alone?" If we divide the left side of the equation by 3, we will obtain $1 \cdot n = 25$. In this way n stands alone, since $1 \cdot n = n$. Remember, however, whatever you do to one side of the equation, you must do to the other side of the equation. Our goal once again is to isolate the variable.

$$3n = 75$$
$$\frac{3n}{3} = \frac{75}{3}$$
$$1 \cdot n = 25$$
$$n = 25$$

This is another important procedure used to solve equations.

DIVISION PROPERTY OF EQUATIONS

You may divide each side of an equation by the same nonzero number to obtain an equivalent equation.

Teaching Tip It has been a while since students have worked with proportion equations. You may want to give them a few exercises of this type to do at their seats. Solve for n for each of the following.

(a) $8n = 72$
(b) $5.5n = 55$
(c) $1.7n = 5.1$

The answers are $n = 9$, $n = 10$, and $n = 3$.

EXAMPLE 1 Solve for n. $6n = 72$

Solution

$6n = 72$	The variable n is multiplied by 6.
$\frac{6n}{6} = \frac{72}{6}$	Divide each side by 6.
$1 \cdot n = 12$	We have $72 \div 6 = 12$.
$n = 12$	Since $1 \cdot n = n$

Practice Problem 1 Solve for n. $8n = 104$

NOTE TO STUDENT: Fully worked-out solutions to all of the Practice Problems can be found at the back of the text starting at page SP-1

Sometimes the coefficient of the variable is a negative number. Therefore, in solving problems of this type, we need to divide each side of the equation by that negative number.

EXAMPLE 2 Solve for the variable. $-3n = 20$

Solution

$-3n = 20$	The coefficient of n is -3.
$\frac{-3n}{-3} = \frac{20}{-3}$	Divide each side of the equation by -3.
$n = -\frac{20}{3}$	Watch your signs!

Practice Problem 2 Solve for the variable. $-7n = 30$

Sometimes the coefficient of the variable is a decimal. In solving problems of this type, we need to divide each side of the equation by that decimal number.

EXAMPLE 3 Solve for the variable. $2.5y = 20$

Solution

$2.5y = 20$ The coefficient of y is 2.5.

$\dfrac{2.5y}{2.5} = \dfrac{20}{2.5}$ Divide each side of the equation by 2.5.

$$2.5_{\wedge}\overline{)20.0_{\wedge}} \quad 8$$

$y = 8$

To check, substitute 8 for y in the original equation.

$$2.5(8) \stackrel{?}{=} 20$$
$$20 = 20 \checkmark$$

It is always best to check the solution to equations involving decimals or fractions.

***NOTE TO STUDENT:** Fully worked-out solutions to all of the Practice Problems can be found at the back of the text starting at page SP-1*

Practice Problem 3 Solve for the variable. Check your solution.

$$3.2x = 16$$

2 Solving Equations Using the Multiplication Property

Sometimes the coefficient of the variable is a fraction, as in the equation $\frac{3}{4}x = 6$. Think: "What can we do to the left side of the equation so that x will stand alone?" Recall that when you multiply a fraction by its reciprocal, the product is 1. That is, $\frac{4}{3} \cdot \frac{3}{4} = 1$. We will use this idea to solve the equation. But remember, whatever you do to one side of the equation, you must do to the other side of the equation.

$$\frac{3}{4}x = 6$$

$$\frac{4}{3} \cdot \frac{3}{4}x = 6 \cdot \frac{4}{3}$$

$$1x = \frac{\overset{2}{\cancel{6}}}{1} \cdot \frac{4}{\cancel{3}}$$

$$x = 8$$

MULTIPLICATION PROPERTY OF EQUATIONS

You may multiply each side of an equation by the same nonzero number to obtain an equivalent equation.

EXAMPLE 4 Solve for the variable and check your solution.

(a) $\frac{5}{8}y = 1\frac{1}{4}$ **(b)** $1\frac{1}{2}z = 3$

Solution

(a)

$$\frac{5}{8}y = 1\frac{1}{4}$$

$$\frac{5}{8}y = \frac{5}{4}$$ Change the mixed number to a fraction. It will be easier to work with.

$$\frac{8}{5} \cdot \frac{5}{8}y = \frac{5}{4} \cdot \frac{8}{5}$$ Multiply both sides of the equation by $\frac{8}{5}$ because $\frac{8}{5} \cdot \frac{5}{8} = 1$.

$$1 \cdot y = 2$$

$$y = 2$$

Check.

$$\frac{5}{8}y = 1\frac{1}{4}$$

$$\frac{5}{8}(2) \stackrel{?}{=} 1\frac{1}{4}$$ Substitute 2 for y in the original equation.

$$\frac{10}{8} \stackrel{?}{=} 1\frac{1}{4}$$

$$1\frac{2}{8} \stackrel{?}{=} 1\frac{1}{4}$$

$$1\frac{1}{4} = 1\frac{1}{4} \checkmark$$

(b)

$$1\frac{1}{2}z = 3$$

$$\frac{3}{2}z = 3$$ Change the mixed number to a fraction.

$$\frac{2}{3} \cdot \frac{3}{2}z = 3 \cdot \frac{2}{3}$$ Multiply both sides of the equation by $\frac{2}{3}$. Why?

$$z = 2$$

It is always a good idea to check the solution to an equation involving fractions. We leave the check for this solution up to you.

Practice Problem 4 Solve for the variable and check your solution.

(a) $\frac{1}{6}y = 2\frac{2}{3}$ **(b)** $3\frac{1}{5}z = 4$

Hint: Remember to write all mixed numbers as improper fractions before solving linear equations. An alternate method that may also be used is to convert the mixed number to decimal form. Thus equations like $4\frac{1}{8}x = 12$ can first be written as $4.125x = 12$.

10.4 EXERCISES

Student Solutions Manual | CD/ Video | PH Math Tutor Center | MathXL®Tutorials on CD | MathXL® | MyMathLab® | Interactmath.com

Verbal and Writing Skills

1. How is an equation similar to a balance scale?
A sample answer is: To maintain the balance, whatever you do to one side of the scale, you need to do the exact same thing to the other side of the scale.

2. The division property states that we may divide each side of an equation by ___the same nonzero number___ to obtain an equivalent equation.

3. To change $\frac{3}{4}x = 5$ to a simpler equation, multiply both sides of the equation by ___$\frac{4}{3}$___.

4. Given the equation $1\frac{3}{5}y = 2$, we multiply both sides of the equation by $\frac{5}{8}$ to solve for y. Why?
The coefficient of the variable is $1\frac{3}{5} = \frac{8}{5}$. We multiply both sides of the equation by $\frac{5}{8}$ because $\frac{8}{5} \cdot \frac{5}{8} = 1$.

Solve for the variable.

5. $4x = 36$ $x = 9$

6. $8x = 56$ $x = 7$

7. $7y = -28$ $y = -4$

8. $5y = -45$ $y = -9$

9. $-9x = 16$
$x = -\frac{16}{9}$

10. $-7y = 22$
$y = -\frac{22}{7}$

11. $-12x = -144$
$x = 12$

12. $-11y = -121$
$y = 11$

13. $-64 = -4m$
$16 = m$

14. $-88 = -8m$
$11 = m$

15. $0.6x = 6$
$x = 10$

16. $0.5y = 50$
$y = 100$

17. $5.5z = 9.9$
$z = 1.8$

18. $3.5n = 7.7$
$n = 2.2$

19. $-0.5x = 6.75$
$x = -13.5$

20. $-0.4y = 6.88$
$y = -17.2$

Solve for the variable.

21. $\frac{5}{8}x = 5$ $x = 8$

22. $\frac{3}{4}x = 3$ $x = 4$

23. $\frac{2}{5}y = 4$ $y = 10$

24. $\frac{5}{6}y = 10$ $y = 12$

25. $\frac{3}{5}n = \frac{3}{4}$
$n = \frac{5}{4}$ or $1\frac{1}{4}$

26. $\frac{2}{3}z = \frac{1}{3}$
$z = \frac{1}{2}$

27. $\frac{3}{8}x = -\frac{3}{5}$
$x = -\frac{8}{5}$ or $-1\frac{3}{5}$

28. $\frac{4}{9}y = -\frac{4}{7}$
$y = -\frac{9}{7}$ or $-1\frac{2}{7}$

29. $\frac{1}{2}x = -2\frac{1}{4}$
$x = -\frac{9}{2}$ or $-4\frac{1}{2}$

30. $\frac{3}{4}y = -3\frac{3}{8}$
$y = -\frac{9}{2}$ or $-4\frac{1}{2}$

31. $\left(-2\frac{1}{2}\right)z = -20$
$z = 8$

32. $\left(-1\frac{3}{4}\right)z = -14$
$8 = z$

Mixed Practice

Solve for the variable.

33. $-40 = -5x$ $x = 8$

34. $-28 = -4x$ $x = 7$

35. $\frac{2}{3}x = -6$ $x = -9$

36. $\frac{4}{5}x = -8$ $x = -10$

37. $1.5x = 0.045$
$x = 0.03$

38. $1.6x = 0.064$
$x = 0.04$

39. $12 = -\frac{3}{5}x$
$x = -20$

40. $20 = -\frac{5}{6}x$
$x = -24$

To Think About

41. Mical solves an equation by multiplying both sides of the equation by 0.5 when he should have divided. He gets an answer of 4.5. What should the correct answer be?
18

42. Alexis solves an equation by multiplying both sides of the equation by $\frac{2}{3}$ when she should have multiplied both sides by the reciprocal of $\frac{2}{3}$. She gets an answer of 4. What should the correct answer be?
9

Cumulative Review

Combine like terms.

43. $6 - 3x + 5y + 7x - 12y$
$4x - 7y + 6$

44. Simplify. $-2(3a - 5b + c) + 5(-a + 2b - 5c)$
$-11a + 20b - 27c$

45. ***Cost to Produce Electricity*** It is estimated that the amount of electricity that could be produced for $26.10 in 1970 cost $140.50 to produce in 2000. (*Source*: U.S. Bureau of Labor Statistics) What is the percent of increase in the cost to produce electricity from 1970 to 2000? Round to the nearest tenth.
438.3%

46. ***World Oil Production*** It is estimated that in 2002, a total of 69.7 million barrels of oil were produced in the world. In that year the United Stated produced 5.7 million barrels of oil. (*Source*: U.S. Energy Administration) What percent of the world's oil production was produced in the United States in 2002? Round to the nearest tenth.
8.2%

47. ***Paper Product Consumption*** The local food warehouse club sells all kinds of paper products at wholesale prices. New England Camp Cherith needs a 30-day supply of paper plates, paper napkins, and paper towels for its campers. If the camp purchaser allows 5 plates, 7 napkins, and 4 paper towels per camper per day for 118 campers, and the same amount for each of its 32 staff members, how many of each paper product does the purchaser need to buy?
22,500 plates;
31,500 napkins;
18,000 towels

10.5 Solving Equations Using Two Properties

Student Learning Objectives

After studying this section, you will be able to:

1. **Use two properties to solve an equation.**
2. **Solve equations where the variable is on both sides of the equal sign.**
3. **Solve equations with parentheses.**

1 Using Two Properties to Solve an Equation

To solve an equation, we take logical steps to change the equation to a simpler equivalent equation. The simpler equivalent equation is x = some number. To do this, we use the addition property, the division property, or the multiplication property. In this section you will use more than one property to solve complex equations. Each time you use a property, you take a step toward solving the equation. At each step you try to isolate the variable. That is, you try to get the variable to stand alone.

EXAMPLE 1 Solve $3x + 18 = 27$. Check your solution.

Solution We want only x terms on the left and only numbers on the right. We begin by removing 18 from the left side of the equation.

$$3x + 18 + (-18) = 27 + (-18)$$ Add the opposite of 18 to both sides of the equation so that $3x$ stands alone.

$$3x = 9$$

$$\frac{3x}{3} = \frac{9}{3}$$ Divide both sides of the equation by 3 so that x stands alone.

$$x = 3$$ The solution to the equation is $x = 3$.

Check.

$$3(3) + 18 \stackrel{?}{=} 27$$ Substitute 3 for x in the original equation.

$$9 + 18 \stackrel{?}{=} 27$$

$$27 = 27 \ \checkmark$$

Practice Problem 1 Solve $5x + 13 = 33$. Check your solution.

Teaching Tip Stress that it is important to determine the correct number to add to each side of the equation. As an oral exercise after discussing Example 1 or a similar problem, ask students to decide, "What number should be added to each side of the equation in each of the following problems?"

(a) $-7x + 2 = 51$
(b) $-12x - 13 = 47$
(c) $20x + 3 = -103$

The answers are -2, $+13$, and -3.

NOTE TO STUDENT: Fully worked-out solutions to all of the Practice Problems can be found at the back of the text starting at page SP-1

Note that the variable is on the right-hand side in the next example.

EXAMPLE 2 Solve $-41 = 9x - 5$. Check your solution.

Solution We begin by removing -5 from the right-hand side of the equal sign.

$$-41 + 5 = 9x - 5 + 5$$ Add the opposite of -5 to both sides of the equation.

$$-36 = 9x$$

$$\frac{-36}{9} = \frac{9x}{9}$$ Divide both sides of the equation by 9.

$$-4 = x$$

The check is left up to you.

Practice Problem 2 Solve $-50 = 7x - 8$. Check your solution.

Solving Equations Where the Variable Is on Both Sides of the Equal Sign

Sometimes variables appear on both sides of the equation. When this occurs, we need to isolate the variable. This requires us to add a variable term to each side of the equation.

EXAMPLE 3 Solve. $8x = 5x - 21$

Solution We want to remove the $5x$ from the right-hand side of the equation so that all of the variables are on one side of the equation and all of the numbers are on the other side of the equation.

$$8x + (-5x) = 5x + (-5x) - 21$$ Add the opposite of $5x$ to both sides of the equation.

$$3x = -21$$

$$\frac{3x}{3} = \frac{-21}{3}$$ Divide both sides of the equation by 3.

$$x = -7$$

The check is left up to you.

Practice Problem 3 Solve. $4x = -8x + 42$

Suppose there is a variable term and a numerical term on each side of an equation. We want to collect all the variables on one side of the equation and all the numerical terms on the other side of the equation. To do this, we will have to use the addition property twice.

EXAMPLE 4 Solve. $2x + 9 = 5x - 3$

Solution We begin by collecting the numerical terms on the right-hand side of the equation.

$$2x + 9 + (-9) = 5x - 3 + (-9)$$ Add -9 to both sides of the equation.

$$2x = 5x - 12$$

Now we want to collect all the variable terms on the left-hand side of the equation.

$$2x + (-5x) = 5x + (-5x) - 12$$ Add $-5x$ to both sides of the equation.

$$-3x = -12$$

$$\frac{-3x}{-3} = \frac{-12}{-3}$$ Divide both sides of the equation by -3.

$$x = 4$$

Check:

$$2(4) + 9 \stackrel{?}{=} 5(4) - 3$$ Substitute 4 for x in the original equation.

$$8 + 9 \stackrel{?}{=} 20 - 3$$

$$17 = 17 \checkmark$$

Practice Problem 4 Solve. $4x - 7 = 9x + 13$

If there are like terms on one side of the equation, these should be combined first. Then proceed as before.

Teaching Tip Usually, about a quarter of the students will not take the time to collect like terms on one side of the equation before they add a value to each side of the equation. Emphasize that this step is very important because it makes working the problem easier and reduces the chance of making an error.

EXAMPLE 5 Solve for the variable. $-5 + 2y + 8 = 7y + 23$

Solution We begin by combining the like terms on the left-hand side of the equation.

$$-5 + 2y + 8 = 7y + 23$$

$$2y + 3 = 7y + 23 \quad \text{Combine } -5 + 8.$$

$$2y + (-7y) + 3 = 7y + (-7y) + 23 \quad \text{Remove the variables on the right-hand side of the equation.}$$

$$-5y + 3 = 23$$

$$-5y + 3 + (-3) = 23 + (-3) \quad \text{Remove the 3 on the left-hand side of the equation.}$$

$$-5y = 20$$

$$\frac{-5y}{-5} = \frac{20}{-5} \quad \text{Divide both sides of the equation by } -5.$$

$$y = -4$$

Check:

$$-5 + (2)(-4) + 8 \stackrel{?}{=} (7)(-4) + 23$$

$$-5 + (-8) + 8 \stackrel{?}{=} -28 + 23$$

$$-5 = -5 \quad \checkmark$$

***NOTE TO STUDENT:** Fully worked-out solutions to all of the Practice Problems can be found at the back of the text starting at page SP-1*

Practice Problem 5 Solve for the variable. $4x - 23 = 3x + 7 - 2x$

It is wise to check your answer when solving this type of linear equation. The chance of making a simple error with signs is quite high. Checking gives you a chance to detect this type of error.

Solving Equations with Parentheses

PROCEDURE TO SOLVE EQUATIONS

1. Remove any parentheses by using the distributive property.
2. Collect like terms on each side of the equation.
3. Add the appropriate value to both sides of the equation to get all numbers on one side.
4. Add the appropriate term to both sides of the equation to get all variable terms on the other side.
5. Divide both sides of the equation by the numerical coefficient of the variable term.
6. Check by substituting the solution back into the original equation.

You have probably noticed that steps 3 and 4 are interchangeable. You can do step 3 and then step 4, or you can do step 4 and then step 3.

If a problem contains one or more sets of parentheses, remove them using the distributive property. Then collect like terms on each side of the equation. Then solve.

EXAMPLE 6 Isolate the variable on the right-hand side. Then solve for x.

$$7x - 3(x - 4) = 9(x + 2)$$

Solution

$7x - 3x + 12 = 9x + 18$	Remove parentheses by using the distributive property.
$4x + 12 = 9x + 18$	Add like terms.
$4x + 12 + (-18) = 9x + 18 + (-18)$	Add -18 to each side.
$4x + (-6) = 9x$	Simplify.
$4x + (-4x) - 6 = 9x + (-4x)$	Add $-4x$ to each side. This isolates the variable on the right-hand side.
$-6 = 5x$	Simplify.
$\dfrac{-6}{5} = \dfrac{5x}{5}$	Divide each side of the equation by 5.
$-\dfrac{6}{5} = x$	We obtain a solution that is a fraction.

Practice Problem 6 Isolate the variable on the right-hand side. Then solve for x. $8(x - 3) + 5x = 15(x - 2)$

10.5 Exercises

Student Solutions Manual | CD/Video | PH Math Tutor Center | MathXL®Tutorials on CD | MathXL® | MyMathLab® | Interactmath.com

Verbal and Writing Skills

1. Explain how you would decide what to add to each side of the equation as you begin to solve $-5x - 6 = 29$ for x.

You want to obtain the x-term all by itself on one side of the equation. So you want to remove the -6 from the left-hand side of the equation. Therefore you would add the opposite of -6. This means you would add 6 to each side.

2. Explain how you would decide what to add to each side of the equation as you begin to solve $-11x = 4x - 45$ for x.

You want to obtain all the x-terms on the left side of the equation. So you want to remove the $4x$ from the right-hand side of the equation. This means you would add $-4x$ to each side.

Check to see whether the given answer is a solution to the equation.

3. Is 2 a solution to $3 - 4x = 5 - 3x$?

$3 - 4(2) \stackrel{?}{=} 5 - 3(2)$

$-5 \neq -1$

no

4. Is $x = 5$ a solution to $3x + 2 = -2x - 23$?

$3(5) + 2 \stackrel{?}{=} -2(5) - 23$

$17 \neq -33$

no

5. Is $x = \frac{1}{2}$ a solution to $8x - 2 = 10 - 16x$?

$8\left(\frac{1}{2}\right) - 2 \stackrel{?}{=} 10 - 16\left(\frac{1}{2}\right)$

$2 = 2$

yes

6. Is $x = \frac{1}{3}$ a solution to $12x - 7 = 3 - 18x$?

$12\left(\frac{1}{3}\right) - 7 \stackrel{?}{=} 3 - 18\left(\frac{1}{3}\right)$

$-3 = -3$

yes

Solve.

7. $15x - 10 = 35$ $x = 3$

8. $12x - 30 = 6$ $x = 3$

9. $6x - 9 = -12$ $x = -\frac{1}{2}$

10. $9x - 3 = -7$ $x = -\frac{4}{9}$

11. $-9x = 3x - 10$ $x = \frac{5}{6}$

12. $-3x = 7x + 14$ $x = -\frac{7}{5}$

13. $-12x - 3 = 21$ $x = -2$

14. $-15x - 2 = 13$ $x = -1$

15. $0.26 = 2x - 0.34$ $x = 0.3$

16. $0.78 = 3x - 0.12$ $x = 0.3$

17. $\frac{2}{3}x - 5 = 17$ $x = 33$

18. $\frac{3}{4}x + 2 = -10$ $x = -16$

19. $18 - 2x = 4x + 6$ $x = 2$

20. $3x + 4 = 7x - 12$ $x = 4$

21. $9 - 8x = 3 - 2x$ $x = 1$

22. $8 + x = 3x - 6$ $x = 7$

23. $2x - 7 = 3x + 9$ $x = -16$

24. $7 + 3x = 6x - 8$ $x = 5$

25. $1.2 + 0.3x = 0.6x - 2.1$ $x = 11$

26. $1.2 + 0.5y = -0.8 - 0.3y$ $y = -2.5$

27. $0.2x + 0.6 = -0.8 - 1.2x$ $x = -1$

28. $0.4x + 0.5 = -1.9 - 0.8x$ $x = -2$

29. $-10 + 6y + 2 = 3y - 26$ $y = -6$

30. $6 - 5x + 2 = 4x + 35$ $x = -3$

31. $12 + 4y - 7 = 6y - 9$ $y = 7$

32. $5 + 7x - 19 = 8x - 6$ $x = -8$

33. $-30 - 12y + 18 = -24y + 13 + 7y$ $y = 5$

34. $15 - 18y - 21 = 15y - 22 - 29y$ $y = 4$

35. $3(2x - 5) - 5x = 1$ $x = 16$

36. $4(2x - 1) - 7x = 9$ $x = 13$

37. $5(y - 2) = 2(2y + 3) - 16$
$y = 0$

38. $13 + 7(2y - 1) = 5(y + 6)$ $y = \frac{8}{3}$

39. $7x - 3(x - 6) = 2(x - 3) + 8$
$4x + 18 = 2x + 2$ $x = -8$

40. $5x - 6(x + 1) = 2x - 6$
$-x - 6 = 2x - 6$ $x = 0$

41. $5x + 9 = \frac{1}{3}(3x - 6)$ $x = -\frac{11}{4}$

42. $6x + 5 = \frac{1}{4}(8x - 4)$ $x = -\frac{3}{2}$

43. $-2x - 5(x + 1) = -3(2x + 5)$ $x = 10$

44. $-3x - 2(x + 1) = -4(x - 1)$ $x = -6$

To Think About

45. (a) Solve $7 + 3x = 6x - 8$ by collecting x-terms on the left. $-3x = -15$ $x = 5$
(b) Solve by collecting x-terms on the right.
$15 = 3x$ $5 = x$
(c) Which method is easier? Why?
For most students, collecting x-terms on the left is easier, since we usually write answers with x on the left. But either method is OK.

46. (a) Solve $8 + 5x = 2x - 6 + x$ by collecting x-terms on the left.
$8 + 5x = 3x - 6$ $2x = -14$ $x = -7$
(b) Solve by collecting x-terms on the right.
$14 = -2x$ $-7 = x$
(c) Which method is easier? Why?
For most students, collecting x-terms on the left is easier, since we usually write answers with x on the left. But either method is OK.

Cumulative Review

Use $\pi \approx 3.14$. Round to the nearest tenth.

47. ***Geometry*** A topographic globe has a radius of 46 centimeters. Find the volume of this sphere. 407,513.4 cubic centimeters

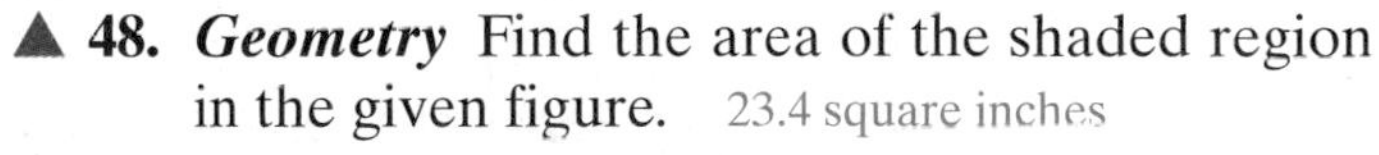

▲ **48.** ***Geometry*** Find the area of the shaded region in the given figure. 23.4 square inches

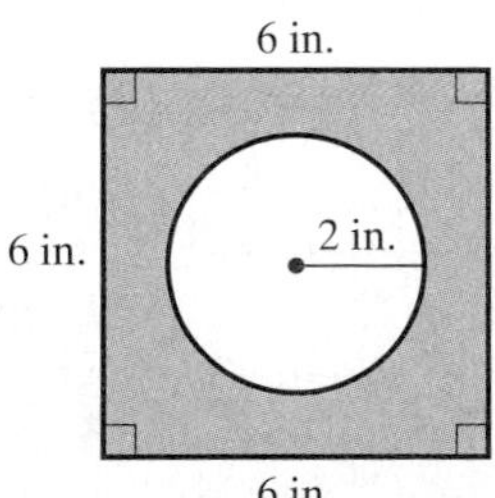

▲ **49.** ***Geometry*** The Pattersons are putting a circular swimming pool in their back yard. The pool will measure 12 feet in diameter. A square-shaped deck with each side measuring 18 feet will be built around the pool. What will the area of the deck be?
210.7 sq ft

▲ **50.** ***Slicing a Watermelon*** Challenge Problem: What is the greatest number of pieces that can result from slicing a spherical watermelon with five straight cuts? (*Hint:* Do not make all your cuts in the same way.)
22 pieces

How Am I Doing? Sections 10.1–10.5

1. $-17x$
2. y
3. $-3a + 2b$
4. $-12x - 4y + 10$
5. $16x - 2y - 6$
6. $4a - 12b + 3c$
7. $42x - 18y$
8. $-3a - 15b + 3$
9. $-3a - 6b + 12c + 10$
10. $x - 8y$
11. $-18x - 8y$
12. $-21x + 9y$
13. $x = 37$
14. $x = 3.5$
15. $x = \frac{7}{8}$
16. $x = 8$
17. $x = -8$
18. $x = 5$
19. $x = \frac{3}{2}$ or $1\frac{1}{2}$
20. $x = -12$
21. $x = 7$
22. $x = \frac{8}{5}$ or $1\frac{3}{5}$
23. $x = 3$
24. $x = \frac{9}{4}$ or $2\frac{1}{4}$
25. $x = 8$
26. $x = \frac{25}{6}$
27. $y = 7$
28. $x = 4$

How are you doing with your homework assignments in Sections 10.1 to 10.5? Do you feel you have mastered the material so far? Do you understand the concepts you have covered? Before you go further in the textbook, take some time to do each of the following problems.

10.1

Combine like terms.

1. $23x - 40x$

2. $-8y + 12y - 3y$

3. $6a - 5b - 9a + 7b$

4. $5x - y + 2 - 17x - 3y + 8$

5. $7x - 14 + 5y + 8 - 7y + 9x$

6. $4a - 7b + 3c - 5b$

10.2

Simplify.

7. $6(7x - 3y)$

8. $-3(a + 5b - 1)$

9. $-2(1.5a + 3b - 6c - 5)$

10. $5(2x - y) - 3(3x + y)$

11. $(9x + 4y)(-2)$

12. $(7x - 3y)(-3)$

10.3

Solve for the variable.

13. $5 + x = 42$

14. $x + 2.5 = 6$

15. $x - \frac{5}{8} = \frac{1}{4}$

16. $-12 = -20 + x$

10.4

Solve for the variable.

17. $7x = -56$

18. $5.4x = 27$

19. $\frac{3}{5}x = \frac{9}{10}$

20. $84 = -7x$

10.5

Solve for the variable.

21. $5x - 9 = 26$

22. $12 - 3x = 7x - 4$

23. $5(x - 1) = 7 - 3(x - 4)$

24. $3x + 7 = 5(5 - x)$

25. $5x - 18 = 2(x + 3)$

26. $8x - 5(x + 2) = -3(x - 5)$

27. $12 + 4y - 7 = 6y - 9$

28. $0.3x + 0.4 = 0.7x - 1.2$

Now turn to page SA-19 for the answer to each of these problems. Each answer also includes a reference to the objective in which the problem is first taught. If you missed any of these problems, you should stop and review the Examples and Practice Problems in the referenced objective. A little review now will help you master the material in the upcoming sections of the text.

10.6 Translating English to Algebra

Student Learning Objectives

After studying this section, you will be able to:

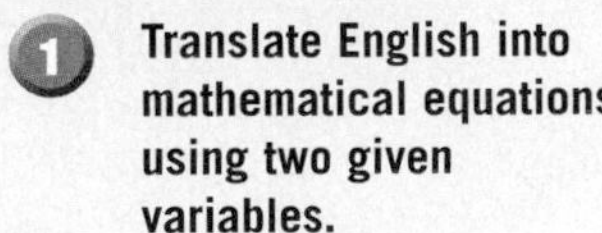

1. **Translate English into mathematical equations using two given variables.**
2. **Write algebraic expressions for several quantities using one given variable.**

1 Translating English into Mathematical Equations Using Two Given Variables

In the preceding section you learned how to solve an equation. We can use equations to solve applied problems, but before we can do that, we need to know how to write an equation that will represent the situation in a word problem.

In this section we practice *translating* to help you to write your own mathematical equations. That is, we translate English expressions into algebraic expressions. In the next section we'll apply this translation skill to a variety of word problems.

The following chart presents the mathematical symbols generally used in translating English phrases into equations.

The English Phrase:	Is Usually Represented by the Symbol:
greater than increased by more than added to sum of	$+$
less than decreased by smaller than fewer than shorter than difference of	$-$
multiplied by of product of times	$\times$
double	$2 \times$
triple	$3 \times$
divided by ratio of quotient of	$\div$
is was has costs equals represents amounts to	$=$

An English sentence describing the relationship between two or more quantities can often be translated into a short equation using variables. For example, if we say in English "Bob's salary is \$1000 greater than Fred's salary," we can express the mathematical relationship by the equation

$$b = 1000 + f$$

where b represents Bob's salary and f represents Fred's salary.

EXAMPLE 1 Translate the English sentence into an equation using variables. Use r to represent Roberto's weight and j to represent Juan's weight.

Roberto's weight is 42 pounds more than Juan's weight.

Solution Roberto's weight is 42 pounds more than Juan's weight.

Roberto's weight	is	42 pounds	more than	Juan's weight.
↓	↓	↓	↓	↓
r	$=$	42	$+$	j

The equation $r = 42 + j$ could also be written as

$$r = j + 42.$$

Both are correct translations because addition is commutative.

NOTE TO STUDENT: Fully worked-out solutions to all of the Practice Problems can be found at the back of the text starting at page SP-1

Practice Problem 1 Translate the English sentence into an equation using variables. Use t to represent Tom's height and a to represent Abdul's height.

Tom's height is 7 inches more than Abdul's height.

When translating the phrase "less than" or "fewer than," be sure that the number that appears before the phrase is the value that is subtracted. Words like "costs," "weighs," or "has the value of" are translated into an equal sign (=).

Teaching Tip After you have explained a few examples, have the class do the following exercises at their seats. In each case, let b = the number of articles Bob has and m = the number of articles Mike has.

(a) Bob has 12 more articles than Mike has.

(b) Mike has 7 fewer articles than Bob has.

(c) Mike has 5 more than double the number of articles Bob has.

(d) Bob has 15 fewer than triple the number of articles Mike has.

Answers:

(a) $b = 12 + m$;

(b) $m = b - 7$;

(c) $m = 5 + 2b$;

(d) $b = 3m - 15$

EXAMPLE 2 Translate the English sentence into an equation using variables. Use c to represent the cost of the chair in dollars and s to represent the cost of the sofa in dollars.

The chair costs $ 200 less than the sofa.

Solution The chair costs $ 200 *less than* the sofa.

$$c = s - 200$$

Note the order of the equation. We subtract 200 from s, so we have $s - 200$. It would be incorrect to write $200 - s$. Do you see why?

Practice Problem 2 Translate the English sentence into an equation using variables. Use n to represent the number of students in the noon class and m to represent the number of students in the morning class.

The noon class has 24 fewer students than the morning class.

In a similar way we can translate the phrases "more than" or "greater than," but since addition is commutative, we find it easier to write the mathematical symbols in the same order as the words in the English sentence.

EXAMPLE 3 Translate the English sentence into an equation using variables. Use f for the cost of a 14-foot truck and e for the cost of an 11-foot truck.

The daily cost of a 14-foot truck is 20 dollars more than the daily cost of an 11-foot truck.

Solution The 14-foot truck cost is 20 more than the 11-foot truck cost.

$$f = 20 + e$$

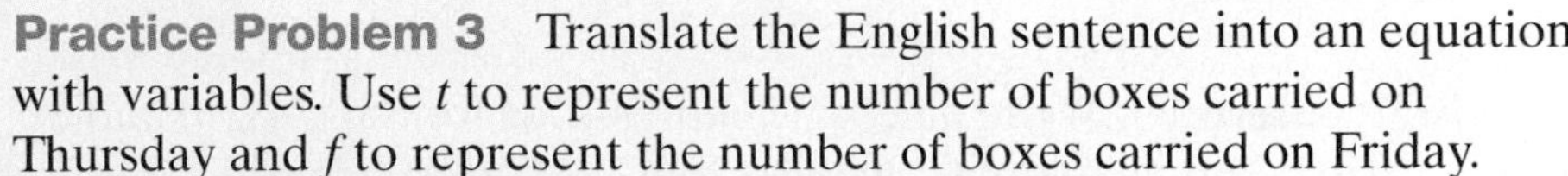

Practice Problem 3 Translate the English sentence into an equation with variables. Use t to represent the number of boxes carried on Thursday and f to represent the number of boxes carried on Friday.

On Thursday Adrianne carried five more boxes into the dorm than she did on Friday.

EXAMPLE 4 Translate the following English sentence into an equation using the variables indicated. *The length of the rectangle is 3 feet shorter than double the width.* Use l for the length and w for the width of the rectangle.

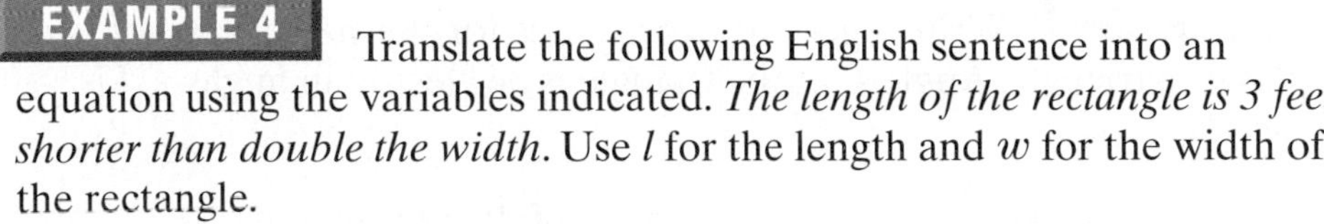

Solution The length of the rectangle is compared to the width. Therefore, we begin with the width. We have

$$w = \text{the width of the rectangle.}$$

Now the length of the rectangle is 3 feet shorter than double the width. Double the width is $2w$. If it is 3 feet shorter than the width, we will have to take away 3 from the $2w$. Therefore,

$$2w - 3 = \text{the length of the rectangle.}$$

So we have

$$l = 2w - 3.$$

Practice Problem 4 Translate the following English sentence into an equation using the variables indicated. The length of the rectangle is 7 feet longer than double the width. Use l for the length and w for the width of the rectangle.

Teaching Tip Present to the students this thought-stimulating problem. Let Bob's salary last year be x dollars. Describe in symbols:

(a) His salary was increased by \$400 and then he received a 10% raise.

(b) He received a 10% raise, and then he received a further increase of \$400.

(c) Which salary would be higher for Bob?

Answers:

(a) $1.10\,(x + 400)$;

(b) $1.10x + 400$;

(c) His salary is higher if he gets the \$400 raise first.

2 Writing Algebraic Expressions for Several Quantities Using One Given Variable

In each of the examples so far we have used two *different variables*. Now we'll learn how to write algebraic expressions for several quantities using the *same variable*. In the next section we'll use this skill to write and solve equations.

A mathematical expression that contains a variable is often called an **algebraic expression.**

EXAMPLE 5 Write algebraic expressions for Bob's salary and Fred's salary. Fred's salary is \$150 more than Bob's salary. Use the letter b.

Solution

Let b = Bob's salary.

Let $\underbrace{b + 150}_{\text{\$150 more than Bob's salary}}$ = Fred's salary.

Notice that Fred's salary is described in terms of Bob's salary. Thus it is logical to let Bob's salary be b and then to express Fred's salary as \$150 more than Bob's.

Practice Problem 5 Write algebraic expressions for Sally's trip and Melinda's trip. Melinda's trip is 380 miles longer than Sally's trip. Use the letter s.

NOTE TO STUDENT: Fully worked-out solutions to all of the Practice Problems can be found at the back of the text starting at page SP-1

▲ **EXAMPLE 6** Write algebraic expressions for the size of each of two angles of a triangle. Angle B of the triangle is 34° less than angle A. Use the letter A.

Solution

Let A = the number of degrees in angle A.

Let $\underbrace{A - 34}_{\text{34° less than angle } A.}$ = the number of degrees in angle B.

Practice Problem 6 Write algebraic expressions for the height of each of two buildings. Larson Center is 126 feet shorter than McCormick Hall. Use the letter m.

Teaching Tip After you have covered Examples 6 and 7, or similar class examples, have the students try the following exercises at their seats. In each case, express each quantity in terms of x.

(a) The Dodge truck weighs 300 pounds more than the Ford truck.

(b) The psychology class has 20 more than double the number of students in the music appreciation class.

(c) The first side of a triangle is 7 inches longer than the second. The third side is 2 inches shorter than double the second.

Answers:

(a) Dodge truck weight $= x + 300$, Ford truck weight $= x$;

(b) psychology class students $= 20 + 2x$, music appreciation class students $= x$;

(c) length of first side $= x + 7$, length of second side $= x$, length of third side $= 2x - 2$.

Often in algebra when we write expressions for one or two unknown quantities, we use the letter x.

▲ **EXAMPLE 7** Write algebraic expressions for the length of each of three sides of a triangle. The second side is 4 inches longer than the first. The third side is 7 inches shorter than triple the length of the first side. Use the letter x.

Solution Since the other two sides are described in terms of the first side, we start by writing an expression for the first side.

Let x = the length of the first side.

Let $x + 4$ = the length of the second side.

Let $3x - 7$ = the length of the third side.

▲ **Practice Problem 7** Write an algebraic expression for the length of each of three sides of a triangle. The second side is double the length of the first side. The third side is 6 inches longer than the first side. Use the letter x.

10.6 Exercises

Student Solutions Manual | CD/ Video | PH Math Tutor Center | MathXL®Tutorials on CD | MathXL® | MyMathLab® | Interactmath.com

Verbal and Writing Skills

Translate the English sentence into an equation using the variables indicated.

1. ***Weight Comparison*** Harry weighs 34 pounds more than Rita. Use h for Harry's weight and r for Rita's weight.
   $h = 34 + r$

2. ***Cereal Boxes*** The large cereal box contains 7 ounces more than the small box of cereal. Use l for the number of ounces in the large cereal box and s for the number of ounces in the small cereal box.
   $l = s + 7$

3. ***Jewelry*** The bracelet costs \$107 less than the necklace. Use b for the cost of the bracelet and n for the cost of the necklace.
   $b = n - 107$

4. ***Education*** There were 42 fewer students taking algebra in the fall semester than the spring semester. Use s to represent the number of students registered in spring and f to represent the number of students registered in fall.
   $f = s - 42$

5. ***Temperature Comparisons*** The temperature in New Delhi, India was 14°F more than the temperature in Athens, Greece. Use n for the temperature in New Delhi and a for the temperature in Athens.
   $n = a + 14$

6. ***Temperature Comparisons*** The temperature in Quebec City was 21°F less than the temperature in Rome. Use q for the temperature in Quebec City and r for the temperature in Rome.
   $q = r - 21$

▲ 7. ***Geometry*** The length of the rectangle is 7 meters longer than double the width. Use l for the length and w for the width of the rectangle.
   $l = 2w + 7$

▲ 8. ***Geometry*** The length of the rectangle is 8 meters shorter than double the width. Use l for the length and w for the width of the rectangle.
   $l = 2w - 8$

▲ 9. ***Geometry*** The length of the rectangle is 2 meters shorter than triple the width. Use l for the length and w for the width of the rectangle.
   $l = 3w - 2$

▲ 10. ***Geometry*** The length of the rectangle is 5 meters longer than triple the width. Use l for the length and w for the width of the rectangle.
   $l = 3w + 5$

11. ***Football*** During a college football game, Miami scored 10 points more than triple the number of points scored by Temple. Use m to represent the number of points Miami scored and t to represent the number of points Temple scored.
   $m = 3t + 10$

12. ***Football*** During a college football game, Texas scored 2 points more than double the number of points scored by Iowa State. Use t to represent the number of points Texas scored and I to represent the number of points Iowa State scored.
   $t = 2I + 2$

13. ***Television Viewing*** The combined number of hours Jenny and Sue watch television per week is 26 hours. Use j for the number of hours Jenny watches television and s for the number of hours Sue watches television.
   $j + s = 26$

14. ***Football Attendance*** The attendance at the Giants game was greater than the attendance at the Jets game. The difference in attendance between the two games was 7234 fans. Use g for the number of fans at the Giants game and j for the number of fans at the Jets game.
   $g - j = 7234$

15. ***Hourly Wage*** The product of your hourly wage and the amount of time worked is $500. Let h = the hourly wage and t = the number of hours worked.
$ht = 500$

16. ***Education*** The ratio of men to women at Central College is 5 to 3. Let m = the number of men and w = the number of women.
$\frac{m}{w} = \frac{5}{3}$

Write algebraic expressions for each quantity using the given variable.

17. ***Airfare*** The airfare from Chicago to San Diego was $135 more than the airfare from Chicago to Phoenix. Use the letter p.
p = cost of airfare to Phoenix;
$p + 135$ = cost of airfare to San Diego

18. ***Electronics Cost*** The cost of the television set was $212 more than the cost of the compact disc player. Use the letter p.
p = cost of disc player; $p + 212$ = cost of television

▲ 19. ***Geometry*** Angle A of the triangle is 46° less than angle B. Use the letter b.
b = number of degrees in angle B;
$b - 46$ = number of degrees in angle A

▲ 20. ***Geometry*** The top of the box is 38 centimeters shorter than the side of the box. Use the letter s.
s = length of the side of the box in centimeters;
$s - 38$ = length of the top of the box in centimeters

21. ***Mountain Height*** Mount Everest is 4430 meters taller than Mount Whitney. Use the letter w.
w = height of Mt. Whitney in meters;
$w + 4430$ = height of Mt. Everest in meters

22. ***Mountain Height*** Mount McKinley is 1802 meters taller than Mount Rainier. Use the letter r.
r = height of Mt. Rainier in meters;
$r + 1802$ = height of Mt. McKinley in meters

23. ***Reading Books*** During the summer, Nina read twice as many books as Aaron. Molly read five more books than Aaron. Use the letter a.
a = number of books Aaron read;
$2a$ = number of books Nina read;
$a + 5$ = number of books Molly read

24. ***Tip Salary*** Sam made $12 more in tips than Lisa one Friday night. Brenda made $6 less than Lisa. Use the letter l.
l = Lisa's tips;
$l + 12$ = Sam's tips;
$l - 6$ = Brenda's tips

▲ 25. ***Geometry*** The length of a box is 5 inches longer than its height. The width is triple the height. Use the letter h.
h = height;
$h + 5$ = length;
$3h$ = width

▲ 26. ***Geometry*** The height of a box is 7 inches longer than the width. The length is 1 inch shorter than double the width. Use the letter w.
w = width;
$w + 7$ = height;
$2w - 1$ = length

▲ 27. ***Triangles*** The second angle of a triangle is double the first. The third angle of the triangle is 14° smaller than the first. Use the letter x.
x = first angle;
$2x$ = second angle;
$x - 14$ = third angle

▲ 28. ***Triangles*** The second angle of a triangle is triple the first. The third angle of the triangle is 36° larger than the first. Use the letter x.
x = first angle;
$3x$ = second angle;
$x + 36$ = third angle

To Think About

29. ***Vehicle Speed*** A Caravan is traveling down the Massachusetts Turnpike at a speed s. A Lexus sedan is traveling 10 miles per hour faster and passes the Caravan. Both vehicles come to the top of a steep hill. As they go down the hill, both vehicles increase in speed by 8%. Write an expression for the speed of the Caravan going down the hill. Write an expression for the speed of the Lexus sedan going down the hill.

Speed in miles per hour of the Caravan $= 1.08s$;
Speed in miles per hour of the Lexus $= 1.08(s + 10)$

30. ***Planets*** Recently astronomers discovered evidence of two planets that are not part of our solar system. The first planet is 73 light-years from earth. It was given the designation HD 19994b. The second planet is directly behind the first one. It is 105 light-years from earth. The second planet has the designation HD 92788b. Suppose a space probe is sent out from earth in the direction of these two distant planets and it has traveled x light-years from earth. Write an expression for the distance from the probe to each planet.

distance from HD 19994b measured in light-years $= 73 - x$;
distance from HD 92788b measured in light-years $= 105 - x$

Cumulative Review

Perform the indicated operations in the proper order.

31. $-6 - (-7)(2)$

8

32. $5 - 5 + 8 - (-4) + 2 - 15$

-1

33. Solve for x. $-2(3x + 5) + 12 = 8$

$x = -1$

34. Solve for y. $3y - 4 - 5y = 10 - y$

$y = -14$

35. ***Basketball*** The following are recent statistics for some NBA players. Fill in the blanks to complete the season totals for three-point shots. (Round to three significant digits.)

Player	Team	3-Point Attempts	3-Point Shots Made	3-Point %
Dee Brown	Toronto	349	135	0.387
Tim Hardaway	Miami	311	112	0.360
Reggie Miller	Indiana	275	106	0.385

10.7 Solving Applied Problems

Student Learning Objectives

After studying this section, you will be able to:

1. Solve problems involving comparisons.
2. Solve problems involving geometric formulas.
3. Solve problems involving rates and percents.

 Solving Problems Involving Comparisons

To solve the following problem, we use the three steps for problem solving with which you are familiar, plus another step: *Write an equation.*

EXAMPLE 1 A 12-foot board is cut into two pieces. The longer piece is 3.5 feet longer than the shorter piece. What is the length of each piece?

Solution You may find it helpful to use the Mathematics Blueprint for Problem Solving to organize the data and make a plan for solving.

Mathematics Blueprint for Problem Solving

Gather the Facts	What Am I Asked to Do?	How Do I Proceed?	Key Points to Remember
The board is 12 feet long. It is cut into two pieces. One piece is 3.5 feet longer than the other.	Find the length of each piece.	Let x = length of shorter piece and use x to write an expression for the longer piece.	Make an equation by adding up the length of each piece to get 12 feet.

1. ***Understand the problem.***
   Draw a diagram.

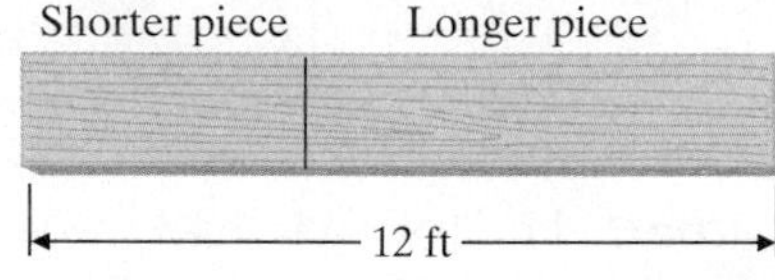

Since the longer piece is described in terms of the shorter piece, we let the variable represent the shorter piece. Let x = the length of the shorter piece. The longer piece is 3.5 feet longer than the shorter piece. Let $x + 3.5$ = the length of the longer piece. The sum of the two pieces is 12 feet. We write an equation.

2. ***Write an equation.***

$$x + (x + 3.5) = 12$$

3. ***Solve and state the answer.***

$$x + x + 3.5 = 12$$

$$2x + 3.5 = 12 \qquad \text{Collect like terms.}$$

$$2x + 3.5 + (-3.5) = 12 + (-3.5) \qquad \text{Add } -3.5 \text{ to each side.}$$

$$2x = 8.5 \qquad \text{Collect like terms.}$$

$$\frac{2x}{2} = \frac{8.5}{2} \qquad \text{Divide each side by 2.}$$

$$x = 4.25$$

The shorter piece is 4.25 feet long.

$$x + 3.5 = \text{the longer piece}$$

$$4.25 + 3.5 = 7.75$$

The longer piece is 7.75 feet long.

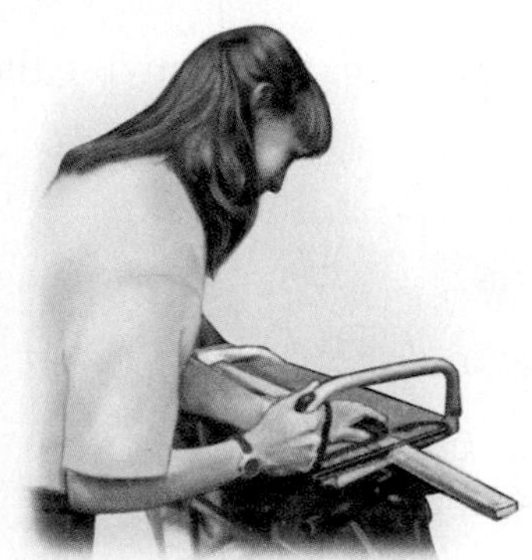

4. ***Check.*** We verify solutions to word problems by making sure that all the calculated values satisfy the original conditions. Do the two pieces add up to 12 feet?

$$4.25 + 7.75 \stackrel{?}{=} 12$$
$$12 = 12 \quad ✓$$

Is one piece 3.5 feet longer than the other?

$$7.75 \stackrel{?}{=} 3.5 + 4.25$$
$$7.75 = 7.75 \quad ✓$$

Practice Problem 1 An 18-foot board is cut into two pieces. The longer piece is 4.5 feet longer than the shorter piece. What is the length of each piece?

NOTE TO STUDENT: *Fully worked-out solutions to all of the Practice Problems can be found at the back of the text starting at page SP-1*

Sometimes three items are compared. Let a variable represent the quantity to which the other two quantities are compared. Then write an expression for the other two quantities.

EXAMPLE 2 Professor Jones is teaching 332 students in three sections of general psychology this semester. His noon class has 23 more students than his 8:00 A.M. class. His 2:00 P.M. class has 36 fewer students than his 8:00 A.M. class. How many students are in each class?

Solution

1. ***Understand the problem.*** Each class enrollment is described in terms of the enrollment in the 8:00 A.M. class.

   Let x = the number of students in the 8:00 A.M. class.

   The noon class has 23 more students than the 8:00 A.M. class.

   Let $x + 23$ = the number of students in the noon class.

   The 2:00 P.M. class has 36 fewer students than the 8:00 A.M. class.

   Let $x - 36$ = the number of students in the 2:00 P.M. class.

   The total enrollment for the three sections is 332.
   You can draw a diagram.

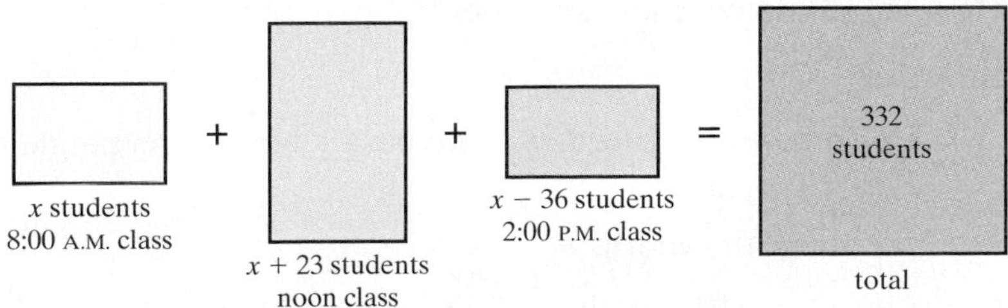

2. ***Write an equation.***

$$x + (x + 23) + (x - 36) = 332$$

Teaching Tip Have the students check the following problem answers. Central Freight dispatched 5 fewer trucks on Monday than on Tuesday. The company dispatched 9 more trucks on Wednesday than on Tuesday. The total number of trucks dispatched on the three days was 173 trucks. A student obtained the following answers: 53 trucks dispatched on Monday, 58 trucks on Tuesday, and 62 trucks on Wednesday. Was the student's answer correct? (Answer: No. The sum of the trucks does total 173. And the number of trucks on Monday is 5 fewer than on Tuesday. However, the number of trucks on Wednesday is not 9 more than the number on Tuesday.) Examples like this help students see the need for checking all parts of word problems to see if the answers they obtained are correct.

3. *Solve and state the answer.*

$$x + x + 23 + x - 36 = 332$$

$$3x - 13 = 332 \quad \text{Collect like terms.}$$

$$3x + (-13) + 13 = 332 + 13 \quad \text{Add 13 to each side.}$$

$$3x = 345 \quad \text{Simplify.}$$

$$\frac{3x}{3} = \frac{345}{3} \quad \text{Divide each side by 3.}$$

$$x = 115 \quad \text{8:00 A.M. class}$$

$$x + 23 = 115 + 23 = 138 \quad \text{noon class}$$

$$x - 36 = 115 - 36 = 79 \quad \text{2:00 P.M. class}$$

Thus there are 115 students in the 8:00 A.M. class, 138 students in the noon class, and 79 students in the 2:00 P.M. class.

4. *Check.* Do the numbers of students in the classes total 332?

$$115 + 138 + 79 \stackrel{?}{=} 332$$

$$332 = 332 \checkmark$$

Does the noon class have 23 more students than the 8:00 A.M. class?

$$138 \stackrel{?}{=} 23 + 115$$

$$138 = 138 \checkmark$$

Does the 2:00 P.M. class have 36 fewer students than the 8:00 A.M. class?

$$79 \stackrel{?}{=} 115 - 36$$

$$79 = 79 \checkmark$$

NOTE TO STUDENT: Fully worked-out solutions to all of the Practice Problems can be found at the back of the text starting at page SP-1

Practice Problem 2 The city airport had a total of 349 departures on Monday, Tuesday, and Wednesday. There were 29 more departures on Tuesday than on Monday. There were 16 fewer departures on Wednesday than on Monday. How many departures occurred on each day?

2 Solving Problems Involving Geometric Formulas

The following applied problems concern the geometric properties of two-dimensional figures. The problems involve perimeter or the measure of the angles in a triangle.

Recall that when we double something, we are multiplying by 2. That is, if something is x units, then double that value is $2x$. Triple that value is $3x$.

EXAMPLE 3 A farmer wishes to fence in a rectangular field with 804 feet of fence. The length is to be 3 feet longer than *double the width*. How long and how wide is the field?

Solution

1. *Understand the problem.* The perimeter of a rectangle is given by $P = 2w + 2l$.

Let w = the width.

The length is 3 feet longer than double the width.

$$\text{Length} = 3 + 2w$$

Thus $2w + 3$ = the length.

You may wish to draw a diagram and label the figures with the given facts.

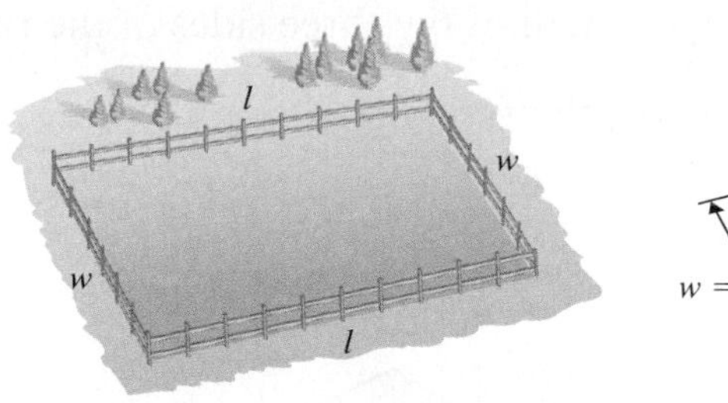

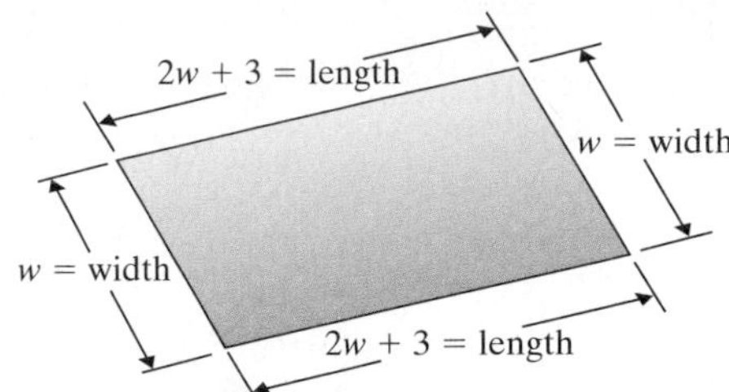

2. ***Write an equation.*** Substitute the given facts into the perimeter formula.

$$2w + 2l = P$$
$$2w + 2(2w + 3) = 804$$

3. ***Solve and state the answer.***

$2w + 2(2w + 3) = 804$	
$2w + 4w + 6 = 804$	Use the distributive property.
$6w + 6 = 804$	Collect like terms.
$6w + 6 + (-6) = 804 + (-6)$	Add −6 to each side.
$6w = 798$	Simplify.
$\frac{6w}{6} = \frac{798}{6}$	Divide each side by 6.
$w = 133$	

Teaching Tip Students sometimes forget that in perimeter problems there is an essential difference between triangle problems and rectangle problems. They often write $P = w + l$ for a rectangle instead of the correct formula $P = 2w + 2l$.

The width is 133 feet.
The length $= 2w + 3$. When $w = 133$, we have

$$(2)(133) + 3 = 266 + 3 = 269.$$

Thus the length is 269 feet.

4. ***Check.*** Is the length 3 feet longer than double the width?

$$269 \stackrel{?}{=} 3 + (2)(133)$$
$$269 \stackrel{?}{=} 3 + 266$$
$$269 = 269 \checkmark$$

Is the perimeter 804 feet?

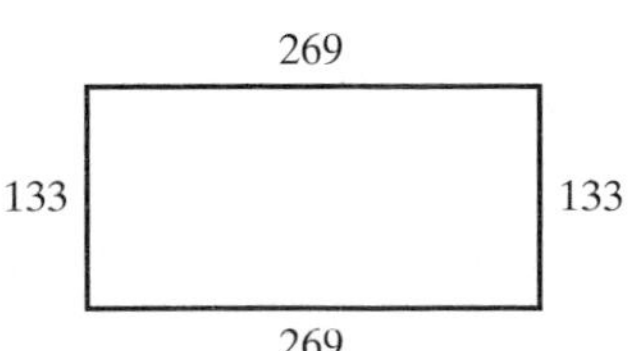

$$(2)(133) + (2)(269) \stackrel{?}{=} 804$$
$$266 + 538 \stackrel{?}{=} 804$$
$$804 = 804 \checkmark$$

▲ **Practice Problem 3** What are the length and width of a rectangular field that has a perimeter of 772 feet and a length that is 8 feet longer than double the width?

The perimeter of a triangular rug section is 21 feet. The second side is double the length of the first side. The third side is 3 feet longer than the first side. Find the length of the three sides of the rug.

Solution Let x = the length of the first side.

Let $2x$ = the length of the second side.

Let $x + 3$ = the length of the third side.

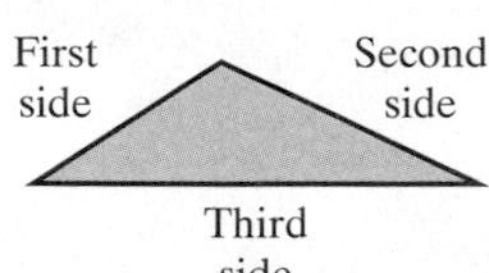

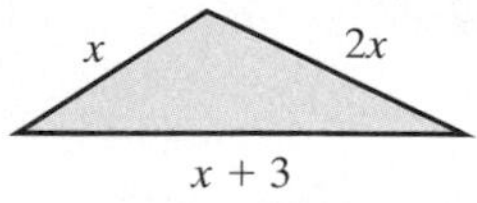

The distance around the three sides totals 21 feet. Thus

$$x + 2x + (x + 3) = 21 \quad \text{Use the perimeter formula.}$$
$$4x + 3 = 21 \quad \text{Collect like terms.}$$
$$4x + 3 + (-3) = 21 + (-3) \quad \text{Add } -3 \text{ to each side.}$$
$$4x = 18 \quad \text{Simplify.}$$
$$\frac{4x}{4} = \frac{18}{4} \quad \text{Divide each side by 4.}$$
$$x = 4.5$$

The first side is 4.5 feet long.

$$2x = (2)(4.5) = 9 \text{ feet}$$

The second side is 9 feet long.

$$x + 3 = 4.5 + 3 = 7.5 \text{ feet}$$

The third side is 7.5 feet long.

Check.

Do the three sides add up to a perimeter of 21 feet?

$$4.5 + 9 + 7.5 \stackrel{?}{=} 21$$
$$21 = 21 \checkmark$$

Is the second side double the length of the first side?

$$9 \stackrel{?}{=} (2)(4.5)$$
$$9 = 9 \checkmark$$

Is the third side 3 feet longer than the first side?

$$7.5 \stackrel{?}{=} 3 + 4.5$$
$$7.5 = 7.5 \checkmark$$

Practice Problem 4 The perimeter of a triangle is 36 meters. The second side is double the first side. The third side is 10 meters longer than the first side. Find the length of each side. Check your solutions.

***NOTE TO STUDENT:** Fully worked-out solutions to all of the Practice Problems can be found at the back of the text starting at page SP-1*

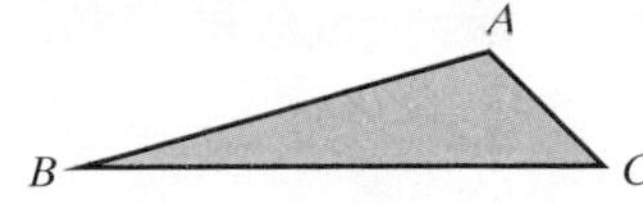

▲ **EXAMPLE 5** A triangle has three angles, A, B, and C. The measure of angle C is triple the measure of angle B. The measure of angle A is 105° larger than the measure of angle B. Find the measure of each angle. Check your answer.

Solution

Let x = the number of degrees in angle B.
Let $3x$ = the number of degrees in angle C.
Let $x + 105$ = the number of degrees in angle A.

The sum of the interior angles of a triangle is 180°. Thus we can write the following.

$$x + 3x + (x + 105) = 180$$
$$5x + 105 = 180$$
$$5x + 105 + (-105) = 180 + (-105)$$
$$5x = 75$$
$$\frac{5x}{5} = \frac{75}{5}$$
$$x = 15$$

Angle B measures 15°.

$$3x = (3)(15) = 45$$

Angle C measures 45°.

$$x + 105 = 15 + 105 = 120$$

Angle A measures 120°.

Check.
Do the angles total 180°?

$$15 + 45 + 120 \stackrel{?}{=} 180$$
$$180 = 180 \checkmark$$

Is angle C triple angle B?

$$45 \stackrel{?}{=} (3)(15)$$
$$45 = 45 \checkmark$$

Is angle A 105° larger than angle B?

$$120 \stackrel{?}{=} 105 + 15$$
$$120 = 120 \checkmark$$

Practice Problem 5 The measure of angle C of a triangle is triple the measure of angle A. The measure of angle B is 30° less than the measure of angle A. Find the measure of each angle.

3 Solving Problems Involving Rates and Percents

You can use equations to solve problems that involve rates and percents. Recall that the commission a salesperson earns is based on the total sales made. For example, a saleswoman earns $40 if she gets a 4% commission and she sells \$1000 worth of products. That is, 4% of \$1000 = \$40. Sometimes a salesperson earns a base salary. The commission will then be added to the base salary to determine the total salary. You can find the total salary if you know the amount of the sales. How would you find the amount of sales if the salary were known? We will use an equation.

Teaching Tip An additional example, similar to Example 6, that you may want to show your students is as follows. Roberta earned \$1960 last month. Her salary is based on a monthly salary of \$1000 plus 8% commission on her monthly sales. Find the amount of her monthly sales last month. (Answer: Her sales were \$12,000 for the month.)

EXAMPLE 6 This month's salary for an appliance saleswoman was \$3000. This includes her base monthly salary of \$1800 plus a 5% commission on total sales. Find the total sales for the month.

Solution

total salary of \$3000	=	base salary of \$1800	+	5% commission of total sales

Let s = the amount of total sales.
Then $0.05s$ = the amount of commission earned from the sales.

$$\begin{aligned} 3000 &= 1800 + 0.05s \\ 1200 &= 0.05s \\ \frac{1200}{0.05} &= \frac{0.05s}{0.05} \\ 24{,}000 &= s \end{aligned}$$

She sold \$24,000 worth of appliances.

Check.
Does 5% of \$24,000 added to \$1800 yield a salary of \$3000?

$$\begin{aligned} (0.05)(24{,}000) + 1800 &\stackrel{?}{=} 3000 \\ 1200 + 1800 &\stackrel{?}{=} 3000 \\ 3000 &= 3000 \quad \checkmark \end{aligned}$$

***NOTE TO STUDENT:** Fully worked-out solutions to all of the Practice Problems can be found at the back of the text starting at page SP-1*

Practice Problem 6 A salesperson at a boat dealership earns \$1000 a month plus a 3% commission on the total sales of the boats he sells. Last month he earned \$3250. What were the total sales of the boats he sold?

10.7 EXERCISES

Student Solutions Manual

CD/ Video

PH Math Tutor Center
MathXL®Tutorials on CD

MathXL®

MyMathLab®

Interactmath.com

Applications

Solve using an equation. Show what you let the variable equal.

1. ***Carpentry*** A 16-foot board is cut into two pieces. The longer piece is 5.5 feet longer than the shorter piece. What is the length of each piece?
x = length of shorter piece;
$x + 5.5$ = length of longer piece;
5.25 feet; 10.75 feet

2. ***Carpentry*** A 20-foot board is cut into two pieces. The longer piece is 4.5 feet longer than the shorter piece. What is the length of each piece?
x = length of shorter piece;
$x + 4.5$ = length of longer piece;
7.75 feet; 12.25 feet

3. ***Rugby*** During a rugby game, Japan scored 22 points less than France. A total of 80 points were scored. How many points did each team score?
x = number of points scored by France;
$x - 22$ = number of points scored by Japan;
France scored 51 points; Japan scored 29 points

4. ***Cross-Country*** In a cross-country race, St. Mark's scored 27 points less than Thayer. A total of 63 points were scored. How many points were scored by each team?
x = number of points scored by Thayer;
$x - 27$ = number of points scored by St. Mark's;
Thayer scored 45 points; St. Mark's scored 18 points

5. ***Car Wash*** The Business Club's thrice-yearly car wash serviced 398 cars this year. A total of 84 more cars participated in May than in November. A total of 43 fewer cars were washed in July than in November. How many cars were washed during each month?
x = the number of cars in November;
$x + 84$ = the number of cars in May;
$x - 43$ = the number of cars in July;
119 cars in November; 203 cars in May; 76 cars in July

6. ***Scrabble*** In the game Scrabble, wooden tiles with letters on them are placed on a board to spell words. There are three times as many A tiles as there are G tiles. The number of O tiles is two more than twice the number of G tiles. The total number of A, O, and G tiles is 20. How many tiles of each are there?
x = the number of G tiles;
$3x$ = the number of A tiles;
$2x + 2$ = the number of O tiles;
3 G tiles; 9 A tiles; 8 O tiles

Solve using an equation. Show what you let the variable equal. Check your answers.

7. ***Furniture*** A 12-foot solid cherry wood table-top is cut into two pieces to allow for an insert later on. Of the two original pieces, the shorter piece is 4.7 feet shorter than the longer piece. What is the length of each piece?
x = length of the shorter piece;
$x + 4.7$ = length of the longer piece;
The shorter piece is 3.65 feet long.
The longer piece is 8.35 feet long.

8. ***Painting Supplies*** An artist has created a huge painting 18 feet long. Her goal is to cut the canvas and have two pieces of the same painting. The longer piece of canvas will be 6.5 feet longer than the shorter piece. What will the length of each piece be?
x = length of the shorter piece;
$x + 6.5$ = length of the longer piece;
The shorter piece will be 5.75 feet long.
The longer piece will be 12.25 feet long.

▲ 9. ***Game Board*** The playing board of a new game has a perimeter of 76 inches. It was designed so that the length is 4 inches shorter than double the width. What are the dimensions of the playing board?
x = width of board;
$2x - 4$ = length of board;
width is 14 inches; length is 24 inches

▲ 10. ***Geometry*** The perimeter of a rectangle is 64 centimeters. The length is 4 centimeters less than triple the width. What are the dimensions of the rectangle?
x = width of rectangle;
$3x - 4$ = length of rectangle;
width is 9 centimeters; length is 23 centimeters

▲ **11.** ***Triangular Flag*** An unusual triangular wall flag at the United Nations has a perimeter of 199 millimeters. The second side is 20 millimeters longer than the first side. The third side is 4 millimeters shorter than the first side. Find the length of each side.

x = length of the first side;
$x + 20$ = length of the second side;
$x - 4$ = length of the third side;
61 millimeters; 81 millimeters; 57 millimeters

▲ **12.** ***Texas Oil Field*** There is a triangular piece of land adjoining an oil field in Texas, with a perimeter of 271 meters. The length of the second side is double the first side. The length of the third side is 15 meters longer than the first side. Find the length of each side.

x = length of the first side;
$2x$ = length of the second side;
$x + 15$ = length of the third side;
64 meters; 128 meters; 79 meters

▲ **13.** ***Puzzle*** A geometric puzzle has a triangular playing piece with a perimeter of 44 centimeters. The length of the second side is double the first side. The length of the third side is 12 centimeters longer than the first side. Find the length of each side.

x = length of the first side;
$2x$ = length of the second side;
$x + 12$ = length of the third side;
8 centimeters; 16 centimeters; 20 centimeters

▲ **14.** ***Triangular Pennant*** An unusual triangular pennant has a perimeter of 63 inches. The length of the first side is twice the length of the second side. The third side is 3 inches longer than twice the second side. Find the length of each side.

x = length of the second side;
$2x$ = length of the first side;
$2x + 3$ = length of the third side;
12 inches; 24 inches; 27 inches

▲ **15.** ***Geometry*** A triangle has three angles, A, B, and C. Angle B is triple angle A. Angle C is 40° larger than angle A. Find the measure of each angle.

x = number of degrees in angle A;
$3x$ = number of degrees in angle B;
$x + 40$ = number of degrees in angle C;
angle A measures 28°; angle B measures 84°; angle C measures 68°

▲ **16.** ***Geometry*** A triangle has three angles, A, B, and C. Angle C is double angle B. Angle A is 28° less than angle B. Find the measure of each angle.

$x - 28$ = number of degrees in angle A;
x = number of degrees in angle B;
$2x$ = number of degrees in angle C;
angle A measures 24°; angle B measures 52°; angle C measures 104°

17. ***Sales Commission*** A saleswoman at a car dealership earns $1200 per month plus a 5% commission on her total sales. If she earned $5000 last month, what was the amount of her sales?

x = total sales; $76,000

18. ***Sales Commission*** A salesman at a jewelry store earns $1000 per month plus a 6% commission on his total sales. If he earned $2200 last month, what was the amount of his sales?

x = total sales; $20,000

19. ***Real Estate*** A real estate agent charges $100 to place a rental listing plus 12% of the yearly rent. An apartment in Central City was rented by the agent for one year. She charged the landowner $820. How much did the apartment cost to rent for one year?

x = yearly rent; $6000

20. ***Real Estate*** A real estate agent charges $50 to place a rental listing plus 9% of the yearly rent. An apartment in the town of West Longmeadow was rented by the agent for one year. He charged the landowner $482. How much did the apartment cost to rent for one year?

x = yearly rent; $4800

21. ***Mural*** A community in South Florida has decided to paint a mural. Adults and children each have one section, but since there are more children interested in participating than adults, the children are awarded the larger piece of wall. If the wall is 32 feet long and the children's section is 6.2 feet longer than the adult section, what is the length of each section of wall?

x = length of the adult section;
$x + 6.2$ = length of the children section;
The adult section is 12.9 feet.
The children's section is 19.1 feet.

22. ***Children's Theater*** The new play at the children's theater attracted 321 people on opening night. There were 67 more children in attendance than adults. How many children attended? How many adults attended?

x = number of adults;
$x + 67$ = number of children;
127 adults and 194 children attended.

To Think About

23. ***Computer Software Costs*** The cost of three computer programs at a computer discount store is \$570.33. The cost of the second program is \$20 less than double the cost of the first program. The cost of the third program is \$17 more than triple the cost of the second program. How much does each program cost?

first program = \$70.37;
second program = \$120.74;
third program = \$379.22

24. ***Antique Car Club*** The cost of three vintage cars at an antique car club auction is \$45,000. The cost of the Model T is \$6000 less than double the cost of an Edsel. The cost of a GTO is \$3000 less than triple the cost of the Model T. How much does each vehicle cost?

The Edsel costs \$8000.
The Model T costs \$10,000.
The GTO costs \$27,000.

The following chart shows the percent of the gross domestic product of each country that is expended on healthcare.

Healthcare Expenditures as a Percent of GDP by Country

Country	1980	1990	1999	2000	2001
United States	8.7	11.9	13.0	13.1	13.9
Canada	7.1	9.0	9.1	9.2	9.7
Mexico	?	?	5.5	5.6	6.6
United Kingdom	5.6	6.0	7.2	7.3	7.6
Spain	5.4	6.7	?	?	7.5

Source: Organization for Economic Cooperation and Development, Paris, France. This means, for example, that the United States spent 8.7% of its gross domestic product on health care in 1980. By the year 2001 that figure had risen to 13.9% of the gross domestic product.

25. The value for Spain for 1999 and 2000 is not given. However, it is the same value for both years. The average value for Spain for these five listed years is 6.92. Find the value for Spain for 1999 and 2000. 7.5

26. The value for Mexico for 1980 and 1990 is not given. However, the value for 1990 is exactly 0.5 more than the value for 1980. The average value for Spain for these five listed years is 5.24. Find the value for Mexico for 1980 and 1990. 4.0 for 1980, 4.5 for 1990

Cumulative Review

27. What percent of 20 is 12? 60%

28. 38% of what number is 190? 500

29. Solve the proportion. $\frac{x}{12} = \frac{10}{15}$ $x = 8$

30. How many ounces are in 5 pounds? 80 oz

▲ **31.** ***Largest Pizza*** The largest pizza ever cooked was in Norwood, South Africa, and was 37.4 meters in diameter. Find the circumference of the pizza. Round to the nearest tenth. Use $\pi \approx 3.14$. 117.4 meters

Putting Your Skills to Work

Red Light, Green Light

According to the U.S. Census Bureau, almost 90% of Americans have access to a motor vehicle, and half of the nation's households have two or more vehicles. Is it any wonder then that traffic jams are becoming ever more common in our smaller cities and towns? One of the techniques that is being employed to assist in the flow of traffic is to modernize and synchronize traffic signals over heavily traveled routes.

Traffic lights that are lined within a traffic management network can be set for different scenarios adapted for different times of day and night and can be sequenced automatically through these scenarios by a central system. These scenarios can be programmed to favor home-to-office morning traffic and the office-to-home evening commute. Networks can be created for groups of traffic lights whose placement along a heavily traveled route requires specific synchronization to allow for peak traffic flow.

Bruce D. Greenshields of the Yale Bureau of Highway Traffic developed a formula. Greenshields' formula is commonly used to determine the amount of time that a traffic light at an intersection should remain green. The formula is as follows: $G = 2.1n + 3.7$, where G is the "green time" in seconds and n is the average number of vehicles that can travel in each lane per light cycle.

Problems for Individual Investigation and Study

Use Greenshield's formula for each of the following questions.

1. Determine how long a light should remain green to allow an average of 19 vehicles per lane for each cycle.
   $G = 2.1(19) + 3.7 = 43.6$ seconds

2. How many vehicles per lane, on average, would one expect to be able to traverse an intersection where the "green time" is 27 seconds?
   $27 = 2.1(n) + 3.7$
   $n = 11$ vehicles per lane average

3. Fire apparatus in many locations have strobe lights on top that cause the traffic lights in their path to turn green as they approach. If the traffic light is set to go green for 60 seconds when a strobe light is detected, how many vehicles per lane can pass through the green light in the opposite direction if it also gets the same green light?
   $60 = 2.1(n) + 3.7$
   an average of 26.8 or 27 cars per lane

Problems for Cooperative Group Activity and Investigation

Suppose that the normal light cycle for an inbound main road is 35 seconds. This means the light goes through a complete cycle of green, amber, and red in 35 seconds. During morning rush hour, the inbound light cycles are increased by 40%, so that there is a better flow of incoming traffic.

4. What is the new timing of the green cycle?
   $1.40(35) = 49$ seconds

5. How many vehicles per lane, on average, can travel past the traffic signal for each cycle?
   $49 = 2.1(n) + 3.7$
   $n = 21.57$ or 22 vehicles per cycle

6. During evening rush hour, the standard 35 second outgoing light cycles are increased by 55%, since it is important to clear vehicles from the city. What is the length of the green cycle for homebound traffic?
   $1.55(35) = 54.25$ seconds

7. How many vehicles per lane, on average, can travel past a traffic signal for each green cycle?
   $54.25 = 2.1(n) + 3.7$
   $n = 24$ vehicles per lane per cycle

Chapter 10 Organizer

Topic	Procedure	Examples
Combining like terms, p. 623.	If the terms are like terms, combine the numerical coefficients directly in front of the variables.	Combine like terms. **(a)** $7x - 8x + 2x = -1x + 2x = x$ **(b)** $3a - 2b - 6a - 5b = -3a - 7b$ **(c)** $a - 2b + 3 - 5a = -4a - 2b + 3$
The distributive properties, p. 627.	$a(b + c) = ab + ac$ and $(b + c)a = ba + ca$	$5(x - 4y) = 5x - 20y$ $3(a + 2b - 6) = 3a + 6b - 18$ $(-2x + y)(7) = -14x + 7y$
Problems involving parentheses and like terms, p. 629.	**1.** Remove the parentheses using the distributive property. **2.** Combine like terms.	Simplify. $2(4x - y) - 3(-2x + y) = 8x - 2y + 6x - 3y$ $= 14x - 5y$
Solving equations using the addition property, p. 632.	**1.** Add the appropriate value to both sides of the equation so that the variable is on one side and a number is on the other side of the equal sign. **2.** Check by substituting your answer back into the original equation.	Solve for x. $x - 2.5 = 7$ $x - 2.5 + 2.5 = 7 + 2.5$ $x + 0 = 9.5$ $x = 9.5$ Check. $x - 2.5 = 7$ $9.5 - 2.5 \stackrel{?}{=} 7$ $7 = 7$ ✓
Solving equations using the division property, p. 637.	**1.** Divide both sides of the equation by the numerical coefficient of the variable. **2.** Check by substituting your answer back into the original equation.	Solve for x. $-12x = 60$ $\frac{-12x}{-12} = \frac{60}{-12}$ $x = -5$ Check. $-12x = 60$ $(-12)(-5) \stackrel{?}{=} 60$ $60 = 60$ ✓
Solving equations using the multiplication property, p. 638.	**1.** Multiply both sides of the equation by the reciprocal of the numerical coefficient of the variable. **2.** Check by substituting your answer back into the original equation.	Solve for x. $\frac{3}{4}x = \frac{5}{8}$ $\frac{4}{3} \cdot \frac{3}{4}x = \frac{5}{8} \cdot \frac{4}{3}$ $x = \frac{5}{6}$ Check. $\left(\frac{3}{4}\right)\left(\frac{5}{6}\right) \stackrel{?}{=} \frac{5}{8}$ $\frac{5}{8} = \frac{5}{8}$ ✓
Solving equations using more than one step, p. 645.	**1.** Remove any parentheses by using the distributive property. **2.** Collect like terms on each side of the equation. **3.** Add the appropriate value to both sides of the equation to get all numbers on one side. **4.** Add the appropriate term to both sides of the equation to get all variable terms on the other side. **5.** Divide both sides of the equation by the numerical coefficient of the variable term. **6.** Check by substituting back into the original equation.	Solve for x. $5x - 2(6x - 1) = 3(1 + 2x) + 12$ $5x - 12x + 2 = 3 + 6x + 12$ $-7x + 2 = 15 + 6x$ $-7x + 2 + (-2) = 15 + (-2) + 6x$ $-7x = 13 + 6x$ $-7x + (-6x) = 13 + 6x + (-6x)$ $-13x = 13$ $x = -1$ Check. $5(-1) - 2[6(-1) - 1] \stackrel{?}{=} 3[1 + 2(-1)] + 12$ $-5 - 2[-6 - 1] \stackrel{?}{=} 3[1 - 2] + 12$ $-5 - 2[-7] \stackrel{?}{=} 3[-1] + 12$ $9 = 9$ ✓

Topic	Procedure	Examples
Translating an English sentence into an equation, p. 649.	When translating English into an equation: replace "greater than" by + and "less than" by −. See complete table on page 651.	Translate a comparison in English into an equation using two given variables. Use t to represent Thursday's temperature and w to represent Wednesday's temperature. The temperature Thursday was 12 degrees higher than the temperature on Wednesday. [Temperature on Thursday] [was] [12°] [higher than] [Temperature on Wednesday] ↓ ↓ ↓ ↓ ↓ t $=$ 12 $+$ w
Writing algebraic expressions for several quantities, p. 651.	**1.** Use a variable to describe the quantity that other quantities are described in terms of. **2.** Write an expression in terms of that variable for each of the other quantities.	▲ Write algebraic expressions for the size of each angle of a triangle. The second angle of a triangle is 7° less than the first angle. The third angle of a triangle is double the first angle. Use the letter x. Since two angles are described in terms of the first angle, we let the variable x represent that angle. Let $x =$ the number of degrees in the first angle. Let $x - 7 =$ the number of degrees in the second angle. Let $2x =$ the number of degrees in the third angle.
Solving applied problems using equations, p. 656.	**1.** *Understand the problem.* **(a)** Draw a sketch. **(b)** Choose a variable. **(c)** Represent other variables in terms of the first variable. **2.** *Write an equation.* **3.** *Solve the equation and state the answer.* **4.** *Check.*	▲ The perimeter of a field is 128 meters. The length of this rectangular field is 4 meters less than triple the width. Find the dimensions of the field. **1.** *Understand the problem.* Let $w =$ the width of the rectangle in meters. Let $3w - 4 =$ the length of the rectangle in meters. **2.** *Write an equation.* Perimeter $= 2(\text{width}) + 2(\text{length})$ $128 = 2(w) + 2(3w - 4)$ **3.** *Solve and state the answer.* $128 = 2w + 6w - 8$ $128 = 8w - 8$ $17 = w$ The width is 17 meters. $3w - 4 = 3(17) - 4 = 47$ The length is 47 meters. **4.** *Check.* Is the perimeter 128 meters? Does $17 + 47 + 17 + 47 = 128$? Yes. Is the length 4 less than triple the width? Is $47 = 3(17) - 4$? $47 = 47$ ✓

Chapter 10 Review Problems

Section 10.1

Combine like terms.

1. $-8a + 6 - 5a - 3$

$-13a + 3$

2. $\frac{3}{4}x + \frac{2}{3} + \frac{1}{8}x + \frac{1}{4}$

$\frac{7}{8}x + \frac{11}{12}$

3. $5x + 2y - 7x - 9y$

$-2x - 7y$

4. $3x - 7y + 8x + 2y$

$11x - 5y$

5. $5x - 9y - 12 - 6x - 3y + 18$

$-x - 12y + 6$

6. $7x - 2y - 20 - 5x - 8y + 13$

$2x - 10y - 7$

Section 10.2

Simplify.

7. $-3(5x + y)$

$-15x - 3y$

8. $-4(2x + 3y)$

$-8x - 12y$

9. $2(x - 3y + 4)$

$2x - 6y + 8$

10. $3(2x - 6y - 1)$

$6x - 18y - 3$

11. $-15\left(\frac{1}{3}a - \frac{2}{5}b - 2\right)$

$-5a + 6b + 30$

12. $-12\left(\frac{3}{4}a - \frac{1}{6}b - 1\right)$

$-9a + 2b + 12$

13. $5(1.2x + 3y - 5.5)$

$6x + 15y - 27.5$

14. $6(1.4x - 2y + 3.4)$

$8.4x - 12y + 20.4$

Simplify.

15. $2(x + 3y) - 4(x - 2y)$

$-2x + 14y$

16. $2(5x - y) - 3(x + 2y)$

$7x - 8y$

17. $-2(a + b) - 3(2a + 8)$

$-8a - 2b - 24$

18. $-4(a - 2b) + 3(5 - a)$

$-7a + 8b + 15$

Section 10.3

Solve for the variable.

19. $x - 3 = 9$

$x = 12$

20. $x + 8.3 = 20$

$x = 11.7$

21. $-8 = x - 12$

$x = 4$

22. $2.4 = x - 5$

$x = 7.4$

23. $3.1 + x = -9$

$x = -12.1$

24. $7 + x = 5.8$

$x = -1.2$

25. $x - \frac{3}{4} = 2$

$x = \frac{11}{4}$ or $2\frac{3}{4}$

26. $x + \frac{1}{2} = 3\frac{3}{4}$

$x = \frac{13}{4}$ or $3\frac{1}{4}$

27. $x + \frac{3}{8} = \frac{1}{2}$

$x = \frac{1}{8}$

28. $x - \frac{5}{6} = \frac{2}{3}$

$x = \frac{3}{2}$ or $1\frac{1}{2}$

29. $2x + 20 = 25 + x$

$x = 5$

30. $5x - 3 = 4x - 15$

$x = -12$

Section 10.4

Solve for the variable.

31. $8x = -20$

$x = -\frac{5}{2}$ or $-2\frac{1}{2}$

32. $-12y = 60$

$y = -5$

33. $1.5x = 9$

$x = 6$

34. $1.8y = 12.6$

$y = 7$

35. $-7.2x = 36$

$x = -5$

36. $6x = 1.5$

$x = 0.25$

37. $\frac{3}{4}x = 6$

$x = 8$

38. $\frac{2}{3}x = \frac{5}{9}$

$x = \frac{5}{6}$

Section 10.5

Solve for the variable.

39. $5x - 3 = 27$
$x = 6$

40. $8x - 5 = 19$
$x = 3$

41. $10 - x = -3x - 6$
$x = -8$

42. $7 - 2x = -4x - 11$
$x = -9$

43. $9x - 3x + 18 = 36$
$x = 3$

44. $4 + 3x - 8 = 12 + 5x + 4$
$x = -10$

45. $5(2x - 3) = -3 + 6x - 8x$
$x = 1$

46. $2(3x - 4) = 7 - 2x + 5x$
$x = 5$

47. $5 + 2y + 5(y - 3) = 6(y + 1)$
$y = 16$

48. $3 + 5(y + 4) = 4(y - 2) + 3$
$y = -28$

Section 10.6

Translate the English sentence into an equation using the variables indicated.

49. ***Vehicle Weight*** The weight of the truck is 3000 pounds more than the weight of the car. Use w for the weight of the truck and c for the weight of the car. $w = c + 3000$

50. ***Education*** The afternoon class had 18 fewer students than the morning class. Use a for the number of students in the afternoon class and m for the number of students in the morning class. $a = m - 18$

▲ **51.** ***Geometry*** The number of degrees in angle A is triple the number of degrees in angle B. Use A for the number of degrees in angle A and B for the number of degrees in angle B.
$A = 3B$

▲ **52.** ***Geometry*** The length of a rectangle is 3 inches shorter than double the width of the rectangle. Use w for the width of the rectangle in inches and l for the length of the rectangle in inches.
$l = 2w - 3$

Write an algebraic expression for each quantity using the given variable.

53. ***Salary Comparison*** Michael's salary is \$2050 more than Roberto's salary. Use the letter r.
r = Roberto's salary; $r + 2050$ = Michael's salary

▲ **54.** ***Geometry*** The length of the second side of a triangle is double the length of the first side of the triangle. Use the letter x.
x = length of first side; $2x$ = length of second side

55. ***Summer Employment*** During the summer, Carmen worked 12 more days than double the number of days Dennis worked. Use the letter d.
d = the number of days Dennis worked
$2d + 12$ = the number of days Carmen worked

56. ***Library*** The number of books in the new library is 450 more than double the number of books in the old library. Use the letter b.
b = number of books in old library;
$2b + 450$ = number of books in new library

Section 10.7

Solve using an equation. Show what you let the variable equal.

57. ***Plumbing*** A 60-ft length of pipe is divided into two pieces. One piece is 6.5 ft longer than the other. Find the length of each piece.
26.75 ft, 33.25 ft

58. ***Salary Comparison*** Two clerks work in a store. The new employee earns $28 less per week than a person hired six months ago. Together they earn $412 per week. What is the weekly salary of each person? $192, $220

59. ***Fast-Food Restaurant*** A local fast-food restaurant had twice as many customers in March as in February. It had 3000 more customers in April than in February. Over the three months 45,200 customers came to the restaurant. How many came each month?
Feb. = 10,550, Mar. = 21,100, Apr. = 13,550

60. ***Trip Distance*** Alfredo drove 856 mi during three days of travel. He drove 106 more miles on Friday than on Thursday. He drove 39 fewer miles on Saturday than on Thursday. How many miles did he drive each day?
Thurs. = 263 mi, Fri. = 369 mi, Sat. = 224 mi

▲ **61.** ***Geometry*** A rectangle has a perimeter of 72 in. The length is 3 in. less than double the width. Find the dimensions of the rectangle.
width = 13 in., length = 23 in.

▲ **62.** ***Geometry*** A rectangle has a perimeter of 180 m. The length is 2 m more than triple the width. Find the dimensions of the rectangle.
width = 22 m, length = 68 m

▲ **63.** ***Geometry*** A triangle has three angles, X, Y, and Z. Angle Y is double the measure of angle Z. Angle X is 12 degrees smaller than angle Z. Find the measure of each angle.
$X = 36°, Y = 96°, Z = 48°$

▲ **64.** ***Geometry*** A triangle has three angles labeled A, B, and C. Angle C is triple the measure of angle B. Angle A is 74 degrees larger than angle B. Find the measure of each angle.
$A = 95.2°$, $B = 21.2°$, $C = 63.6°$

▲ **65.** ***Football*** A regulation NFL football field is in the shape of a rectangle. The width of the field is 67 yards shorter than the length. The perimeter of the field is 346 yards. Find the width and length of the field. The width is 53 yards. The length is 120 yards.

▲ **66.** ***Basketball*** A regulation NBA basketball court is in the shape of a rectangle. The width of the court is 44 feet shorter than the length. The perimeter of the court is 288 feet. Find the width and length of the court.
The width is 50 feet.
The length is 94 feet.

67. ***Trip Distance*** Victor and Samuel drove 760 miles one way to visit their mother over Christmas vacation. It took two days for them to make the trip. On the first day they drove 88 more miles than they did on the second day. How many miles did they drive each day?
The first day they drove 424 miles. The second day they drove 336 miles.

68. ***Education*** During the second week of July, the North Lake Community College admissions office received 156 more applications than it did during the first week of July. During the third week of July, it received 142 fewer applications than it did during the first week of July. During these three weeks it received 800 applications. How many were received each week?
They received 262 the first week, 418 the second week, and 120 the third week.

69. ***Sales Commission*** Wayne receives a 4% commission on the used cars that he sells. Last week his total salary was $600. His base salary was $200. What was the cost of the cars he sold last month? $10,000

70. ***Sales Commission*** Megan receives an 8% commission on the furniture that she sells. Last month her total salary was $3050. Her base salary for the month was $1500. What was the cost of the furniture she sold last month?
$19,375

How Am I Doing? Chapter 10 Test

Note to Instructor: The Chapter 10 Test file in the TestGen program provides algorithms specifically matched to these problems so you can easily replicate this test for additional practice or assessment purposes.

1. $-6a$
2. $\frac{2}{15}x + \frac{9}{8}y$
3. $\frac{5}{8}a - \frac{2}{3}b$
4. $a - 8b$
5. $7x - 7z$
6. $-4x - 2y + 5$
7. $60x - 25y$
8. $2x - \frac{10}{3}y$
9. $-4.5a + 3b - 1.5c + 12$
10. $-11a + 14b$
11. $x = -8$
12. $x = -6.35$
13. $x = 15$
14. $x = -\frac{1}{2}$
15. $x = -5$
16. $x = -\frac{7}{10}$

Remember to use your Chapter Test Prep Video CD to see the worked-out solutions to the test problems you want to review.

Combine like terms.

1. $5a - 11a$

2. $\frac{1}{3}x + \frac{5}{8}y - \frac{1}{5}x + \frac{1}{2}y$

3. $\frac{1}{4}a - \frac{2}{3}b + \frac{3}{8}a$

4. $6a - 5b - 5a - 3b$

5. $7x - 8y + 2z - 9z + 8y$

6. $x + 5y - 6 - 5x - 7y + 11$

Simplify.

7. $5(12x - 5y)$

8. $4\left(\frac{1}{2}x - \frac{5}{6}y\right)$

9. $-1.5(3a - 2b + c - 8)$

10. $2(-3a + 2b) - 5(a - 2b)$

Solve for the variable.

11. $-5 - 3x = 19$

12. $x - 3.45 = -9.8$

13. $-5x + 9 = -4x - 6$

14. $8x - 2 - x = 3x - 9 - 10x$

15. $0.5x + 0.6 = 0.2x - 0.9$

16. $-\frac{5}{6}x = \frac{7}{12}$

Translate the English sentence into an equation using the variables indicated.

17. The second floor of Trah[illegible] Laboratory has 15 more classrooms than the first floor. Use s to represent the number of classrooms on the second floor and f to represent the number of classrooms on the first floor.

17. $s = f + 15$

18. The north field yields 15,000 fewer bushels of wheat than the south field. Use n to represent the number of bushels of wheat in the north field and s to represent the number of bushels of wheat in the south field.

18. $n = s - 15{,}000$

Write an algebraic expression for each quantity using the given variable.

▲ **19.** The first angle of a triangle is half the second angle. The third angle of the triangle is twice the second angle. Use the variable s.

19. $\frac{1}{2}s$ = measure of the first angle; s = measure of the second angle; $2s$ = measure of the third angle

▲ **20.** The length of a rectangle is 5 inches shorter than double the width. Use the letter w.

20. w = width; $2w - 5$ = length

Solve using an equation.

21. The number of acres of land in the old Smithfield farm is three times the number of acres of land in the Prentice farm. Together the two farms have 348 acres. How many acres of land are there on each farm?

21. Prentice farm = 87 acres; Smithfield farm = 261 acres

22. Sam earns \$1500 less per year than Marcia does. The combined income of the two people is \$46,500 per year. How much does each person earn?

22. Marcia = \$24,000; Sam = \$22,500

23. During the fall semester, 183 students registered for Introduction to Biology. The morning class has 24 fewer students than the afternoon class. The evening class has 12 more students than the afternoon class. How many students registered for each class?

23. 41 students in the morning class; 65 students in the afternoon class; 77 students in the evening class

▲ **24.** A rectangular field has a perimeter of 118 feet. The width is 8 feet longer than half the length. Find the dimensions of the rectangle.

24. width = 25 feet length = 34 feet

Cumulative Test for Chapters 1–10

Approximately one-half of this test is based on Chapter 10 material. The remainder is based on material covered in Chapters 1–9.

Solve. Simplify your answers.

1. Add. $456 + 89 + 123 + 79$

2. Multiply. $\begin{array}{r} 309 \\ \times\ 35 \\ \hline \end{array}$

3. Round to the nearest hundred. 45,678,934

4. Divide. $\dfrac{5}{12} \div \dfrac{1}{6}$

5. Multiply. $3\dfrac{1}{4} \times 2\dfrac{1}{2}$

6. Multiply. 9.3228×10^3

7. Subtract. $4182.7 - 3555.28$

8. Find n. $\dfrac{9}{n} = \dfrac{40.5}{72}$

9. What is 20% of 150?

10. 34% of what number is 1870?

11. Convert 345 millimeters to meters.

12. Convert 30 inches to feet.

▲**13.** Find the circumference of a circle with diameter 12 yards. Round to the nearest tenth. Use $\pi \approx 3.14$.

▲**14.** Find the area of a triangle that has a base of 13 meters and a height of 22 meters.

1. 747

2. 10,815

3. 45,678,900

4. $\frac{5}{2}$ or $2\frac{1}{2}$

5. $\frac{65}{8}$ or $8\frac{1}{8}$

6. 9322.8

7. 627.42

8. $n = 16$

9. 30

10. 5500

11. 0.345 meters

12. 2.5 feet

13. 37.7 yards

14. 143 square meters

Perform the indicated operations.

15. $4 - 8 + 12 - 32 - 7$

15. -31

16. $(6)(-3)(-4)(2)$

16. 144

Combine like terms.

17. $\frac{1}{2}a + \frac{1}{7}b + \frac{1}{4}a - \frac{3}{14}b$

17. $\frac{3}{4}a - \frac{1}{14}b$

18. $3x - 5y - 12 + x - 8y + 20$

18. $4x - 13y + 8$

Simplify.

19. $-8(2x - 3y + 5)$

19. $-16x + 24y - 40$

20. $2(3x - 4y) - 8(x + 2y)$

20. $-2x - 24y$

Solve for the variable.

21. $5x - 5 = 7x - 13$

21. $x = 4$

22. $7 - 9y - 12 = 3y + 5 - 8y$

22. $y = -\frac{5}{2}$ or $-2\frac{1}{2}$

23. $x - 2 + 5x + 3 = 183 - x$

23. $x = 26$

24. $9(2x + 8) = 20 - (x + 5)$

24. $x = -3$

Write an algebraic expression for each quantity, using the given variable.

25. The weight of the computer was 322 pounds more than the weight of the printer. Use the letter p.

25. p = weight of printer; $p + 322$ = weight of computer

26. The summer enrollment in algebra was 87 students less than the fall enrollment. Use the letter f.

26. f = enrollment during fall; $f - 87$ = enrollment during summer

Solve using an equation.

27. Barbara drove 1081 miles in three days. She drove 48 more miles on Friday than on Thursday. She drove 95 fewer miles on Saturday than on Thursday. How many miles did she drive each day?

27. Thursday = 376 miles; Friday = 424 miles; Saturday = 281 miles

▲ **28.** A rectangle has a perimeter of 98 feet. The length is 8 feet longer than double the width. Find each dimension.

28. width = $13\frac{2}{3}$ feet; length = $35\frac{1}{3}$ feet

Practice Final Examination

1. eighty-two thousand, three hundred sixty-seven
2. 30,333
3. 173
4. 34,103
5. 4212
6. 217,745
7. 158
8. 606
9. 116
10. 32 miles/gallon
11. $\frac{7}{15}$
12. $\frac{42}{11}$
13. $\frac{33}{20}$ or $1\frac{13}{20}$
14. $\frac{89}{15}$ or $5\frac{14}{15}$
15. $\frac{31}{14}$ or $2\frac{3}{14}$
16. 4
17. $\frac{14}{5}$ or $2\frac{4}{5}$
18. $\frac{22}{13}$ or $1\frac{9}{13}$
19. $6\frac{17}{20}$ miles
20. 5 packages

This examination is based on Chapters 1–10 of the book. There are 10 questions covering the content of each chapter.

Chapter 1

1. Write in words. 82,367

2. Add. $\begin{array}{r} 13{,}428 \\ +\ 16{,}905 \\ \hline \end{array}$

3. Add. $\begin{array}{r} 19 \\ 23 \\ 16 \\ 45 \\ +\ 70 \\ \hline \end{array}$

4. Subtract. $\begin{array}{r} 89{,}071 \\ -\ 54{,}968 \\ \hline \end{array}$

Multiply.

5. $\begin{array}{r} 78 \\ \times\ 54 \\ \hline \end{array}$

6. $\begin{array}{r} 2035 \\ \times\ 107 \\ \hline \end{array}$

In questions 7 and 8, divide. (Be sure to indicate the remainder if one exists.)

7. $7\overline{)1106}$

8. $26\overline{)15{,}756}$

9. Evaluate. Perform operations in the proper order.
$3^4 + 20 \div 4 \times 2 + 5^2$

10. Melinda traveled 512 miles in her car. The car used 16 gallons of gas on the entire trip. How many miles per gallon did the car achieve?

Chapter 2

11. Reduce the fraction. $\dfrac{14}{30}$

12. Change to an improper fraction. $3\dfrac{9}{11}$

13. Add. $\dfrac{1}{10} + \dfrac{3}{4} + \dfrac{4}{5}$

14. Add. $2\dfrac{1}{3} + 3\dfrac{3}{5}$

15. Subtract. $4\dfrac{5}{7} - 2\dfrac{1}{2}$

16. Multiply. $1\dfrac{1}{4} \times 3\dfrac{1}{5}$

17. Divide. $\dfrac{7}{9} \div \dfrac{5}{18}$

18. Divide. $\dfrac{5\frac{1}{2}}{3\frac{1}{4}}$

19. Lucinda jogged $1\frac{1}{2}$ miles on Monday, $3\frac{1}{4}$ miles on Tuesday, and $2\frac{1}{10}$ miles on Wednesday. How many miles in all did she jog over the three-day period?

20. A butcher has $11\frac{2}{3}$ pounds of steak. She wishes to place them in several equal-sized packages. Each package will hold $2\frac{1}{3}$ pounds of steak. How many packages can be made?

Chapter 3

21. Express as a decimal. $\frac{719}{1000}$

22. Write in reduced fractional notation. 0.86

23. Fill in the blank with <, =, or > 0.315 _____ 0.309

24. Round to the nearest hundredth. 506.3782

25. Add.

$$\begin{array}{r} 9.6 \\ 3.82 \\ 1.05 \\ +\ 7.3 \\ \hline \end{array}$$

26. Subtract.

$$\begin{array}{r} 3.61 \\ -\ 2.853 \\ \hline \end{array}$$

27. Multiply.

$$\begin{array}{r} 1.23 \\ \times\ 0.4 \\ \hline \end{array}$$

28. Divide. $0.24\overline{)0.8856}$

29. Write as a decimal. $\frac{13}{16}$

30. Evaluate by performing operations in proper order.
$0.7 + (0.2)^3 - 0.08(0.03)$

Chapter 4

31. Write a rate in simplest form to compare 7000 students to 215 faculty.

32. Is this a proportion? $\frac{12}{15} = \frac{17}{21}$

Solve the proportion. Round to the nearest tenth when necessary.

33. $\frac{5}{9} = \frac{n}{17}$

34. $\frac{3}{n} = \frac{7}{18}$

35. $\frac{n}{12} = \frac{5}{4}$

36. $\frac{n}{7} = \frac{36}{28}$

21. 0.719
22. $\frac{43}{50}$
23. >
24. 506.38
25. 21.77
26. 0.757
27. 0.492
28. 3.69
29. 0.8125
30. 0.7056
31. $\frac{1400 \text{ students}}{43 \text{ faculty}}$
32. no
33. $n \approx 9.4$
34. $n \approx 7.7$
35. $n = 15$
36. $n = 9$

37.	$3333.33
38.	9.75 inches
39.	$294.12
40.	1.6 pounds
41.	0.63%
42.	21.25%
43.	1.64
44.	17.3%
45.	302.4
46.	250
47.	4284
48.	$10,856
49.	4500 students
50.	34.3%
51.	4.25 gallons
52.	6500 pounds
53.	192 inches
54.	5600 meters

Solve using a proportion. Round to the nearest hundredth when necessary.

37. Bob earned $2000 for painting three houses. How much would he earn for painting five houses?

38. Two cities that are actually 200 miles apart appear 6 inches apart on the map. Two other cities are 325 miles apart. How far apart will they appear on the same map?

39. Roberta earned $68 last week on her part-time job. She had $5 withheld for federal income tax. Last year she earned $4000 on her part-time job. Assuming the same rate, how much was withheld for federal income tax last year?

40. Malaga's recipe feeds 18 people and calls for 1.2 pounds of butter. If she wants to feed 24 people, how many pounds of butter does she need?

Chapter 5

41. Write as a percent. 0.0063

42. Change $\frac{17}{80}$ to a percent.

Round to the nearest tenth when necessary.

43. Write as a decimal. 164%

44. What percent of 300 is 52?

45. Find 6.3% of 4800.

46. 145 is 58% of what number?

47. 126% of 3400 is what number?

48. Pauline bought a new car. She got an 8% discount. The car listed for $11,800. How much did she pay for the car?

49. A total of 1260 freshmen were admitted to Central College. This is 28% of the student body. How big is the student body?

50. There are 11.28 centimeters of water in the rain gauge this week. Last week the rain gauge held 8.40 centimeters of water. What is the percent of increase from last week to this week?

Chapter 6

Convert. Express your answers as a decimal rounded to the nearest hundredth when necessary.

51. 17 quarts = ______ gallons

52. 3.25 tons = ______ pounds

53. 16 feet = ______ inches

54. 5.6 kilometers = ______ meters

55. 69.8 grams = ______ kilogram

56. 2.48 milliliters = ______ liter

57. 12 miles = ______ kilometers

In questions 58 and 59, write in scientific notation.

58. 0.00063182

59. 126,400,000,000

60. Two metal sheets are 0.623 centimeter and 0.74 centimeter thick, respectively. An insulating foil is 0.0428 millimeter thick. When all three layers are placed tightly together, what is the total thickness? Express your answer in centimeters.

Chapter 7

Round to the nearest hundredth when necessary. Use $\pi \approx 3.14$ *when necessary.*

▲ 61. Find the perimeter of a rectangle that is 6 meters long and 1.2 meters wide.

▲ 62. Find the perimeter of a trapezoid with sides of 82 centimeters, 13 centimeters, 98 centimeters, and 13 centimeters.

▲ 63. Find the area of a triangle with base 6 feet and height 1.8 feet.

▲ 64. Find the area of a trapezoid with bases of 12 meters and 8 meters and a height of 7.5 meters.

▲ 65. Find the area of a circle with radius 6 meters.

▲ 66. Find the circumference of a circle with diameter 18 meters.

▲ 67. Find the volume of a cone with a radius of 4 centimeters and a height of 10 centimeters.

▲ 68. Find the volume of a rectangular pyramid with a base of 12 feet by 19 feet and a height of 2.7 feet.

▲ 69. Find the area of this object, consisting of a square and a triangle.

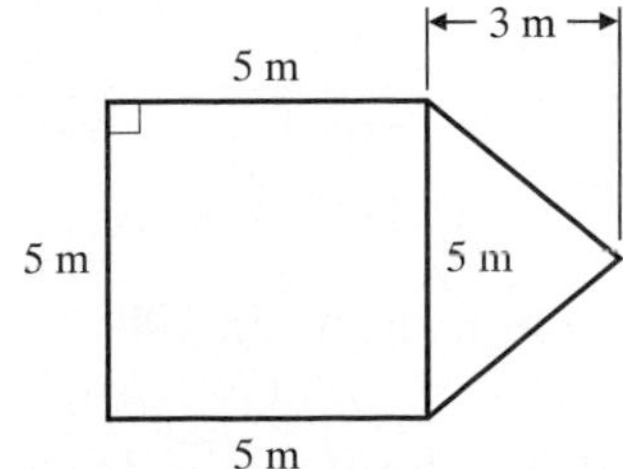

55. 0.0698 kilogram

56. 0.00248 liter

57. 19.32 kilometers

58. 6.3182×10^{-4}

59. 1.264×10^{11}

60. 1.36728 centimeters

61. 14.4 meters

62. 206 centimeters

63. 5.4 square feet

64. 75 square meters

65. 113.04 square meters

66. 56.52 meters

67. 167.47 cubic centimeters

68. 205.2 cubic feet

69. 32.5 square meters

70. $n = 32.5$

▲ **70.** In the following pair of similar triangles, find n.

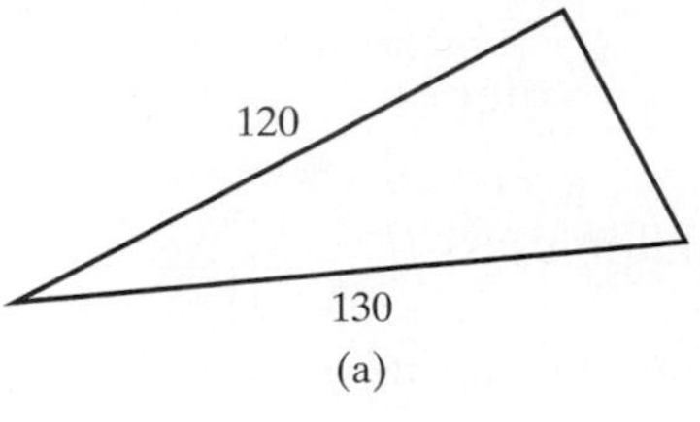

(a)

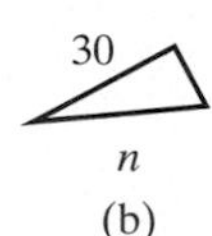

(b)

71. $8 million

Chapter 8

The following double-bar graph indicates the quarterly profits for Westar Corporation in 2003 and 2004.

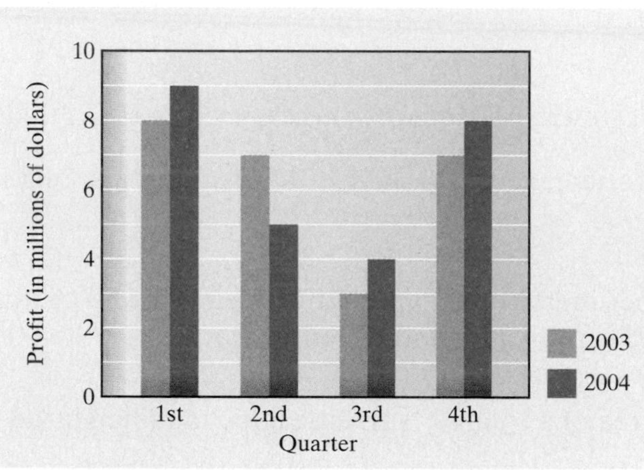

72. $1 million

71. What were the profits in the fourth quarter of 2004?

72. How much greater were the profits in the first quarter of 2004 than the profits in the first quarter of 2003?

The following line graph depicts the average annual temperature at West Valley for the years 1960, 1970, 1980, 1990, and 2000.

73. 50°F

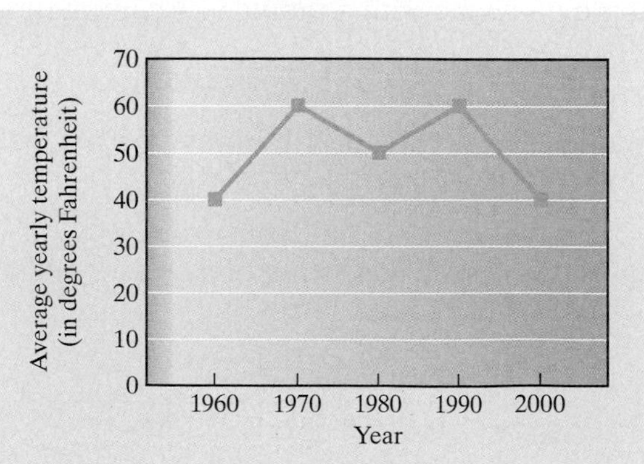

73. What was the average temperature in 1980?

74. from 1990 to 2000

74. In what 10-year period did the average temperature show the greatest decline?

The following histogram shows the number of students in each age category at Center City College.

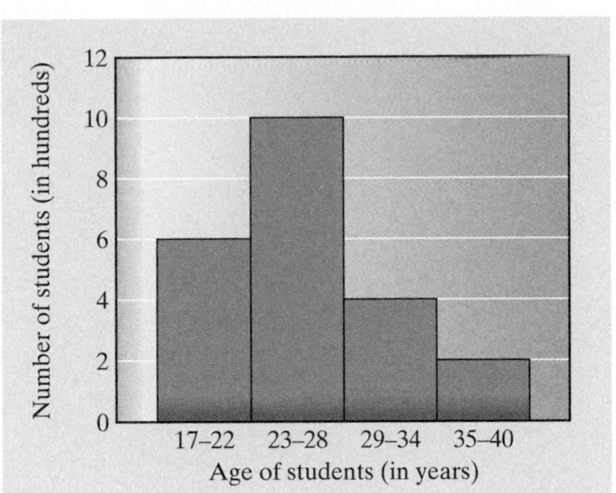

75. How many students are between 17 and 22 years old?

76. How many students are between 23 and 34 years old?

77. Find the *mean* and the *median* of the following. 8, 12, 16, 17, 20, 22. Round to the nearest hundredth.

78. Evaluate exactly. $\sqrt{49} + \sqrt{81}$

79. Approximate to the nearest thousandth using a calculator or the square root table. $\sqrt{123}$

▲ **80.** Find the unknown side of the right triangle.

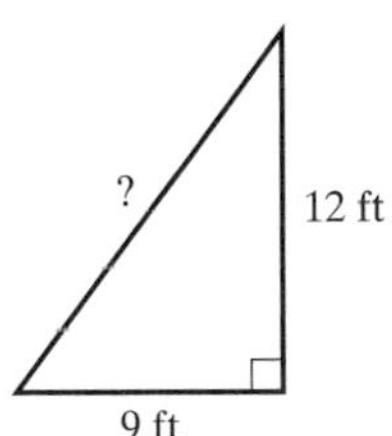

Chapter 9

Add.

81. $-8 + (-2) + (-3)$

82. $-\frac{1}{4} + \frac{3}{8}$

Subtract.

83. $9 - 12$

84. $-20 - (-3)$

85. Multiply. $(2)(-3)(4)(-1)$

86. Divide. $-\frac{2}{3} \div \frac{1}{4}$

75. 600 students

76. 1400 students

77. mean $\approx$ 15.83; median = 16.5

78. 16

79. 11.091

80. 15 feet

81. -13

82. $\frac{1}{8}$

83. -3

84. -17

85. 24

86. $-\frac{8}{3}$ or $-2\frac{2}{3}$

87. 4

88. 27

89. 8

90. 0.5 or $\frac{1}{2}$

91. $-3x - 7y$

92. $-7 - 4a - 17b$

93. $-2x + 6y + 10$

94. $-11x - 9y - 4$

95. $x = 2$

96. $x = -2$

97. $x = -\frac{1}{2}$ or $x = -0.5$

98. $x = -\frac{2}{5}$ or $x = -0.4$

99. 122 students are taking history; 110 students are taking math

100. The length is 37 meters. The width is 16 meters.

Perform the indicated operations in the proper order.

87. $(-16) \div (-2) + (-4)$

88. $12 - 3(-5)$

89. $7 - (-3) + 12 \div (-6)$

90. $\dfrac{(-3)(-1) + (-4)(2)}{(0)(6) + (-5)(2)}$

Chapter 10

Combine like terms.

91. $5x - 3y - 8x - 4y$

92. $5 + 2a - 8b - 12 - 6a - 9b$

Simplify.

93. $-2(x - 3y - 5)$

94. $-2(4x + 2) - 3(x + 3y)$

Solve for the variable.

95. $5 - 4x = -3$

96. $5 - 2(x - 3) = 15$

97. $7 - 2x = 10 + 4x$

98. $-3(x + 4) = 2(x - 5)$

Solve using an equation.

99. There are 12 more students taking history than math. There are twice as many students taking psychology as there are students taking math. There are 452 students taking these three subjects. How many are taking history? How many are taking math?

▲ **100.** A rectangle has a perimeter of 106 meters. The length is 5 meters longer than double the width. Find the length and width of the rectangle.

Basic College Mathematics Glossary

Absolute value of a number (9.1) The absolute value of a number is the distance between that number and zero on the number line. When we find the absolute value of a number, we use the $| \ |$ notation. To illustrate, $|-4| = 4$, $|6| = 6$, $|-20 - 3| = |-23| = 23$, $|0| = 0$.

Addends (1.2) When two or more numbers are added, the numbers being added are called addends. In the problem $3 + 4 = 7$, the numbers 3 and 4 are both addends.

Adjacent angles (7.1) Two angles that share a common side and a common vertex.

Algebraic expression (10.6) An algebraic expression consists of variables, numerals, and operation signs.

Altitude of a triangle (7.4) The height of a triangle.

Amount of a percent equation (5.3A) The product we obtain when we multiply a percent times a number. In the equation $75 = 50\% \times 150$, the amount is 75.

Angle (7.1) An **angle** is made up of two rays that start at a common endpoint.

Area (7.1) The measure of the surface inside a geometric figure. Area is measured in square units, such as square feet.

Associative property of addition (1.2) The property that tells us that when three numbers are added, it does not matter which two numbers are added first. An example of the associative property is $5 + (1 + 2) = (5 + 1) + 2$. Whether we add $1 + 2$ first and then add 5 to that, or add $5 + 1$ first and then add that result to 2, we will obtain the same result.

Associative property of multiplication (1.4) The property that tells us that when we multiply three numbers, it does not matter which two numbers we group together first to multiply; the result will be the same. An example of the associative property of multiplication is $2 \times (5 \times 3) = (2 \times 5) \times 3$.

Base (1.6) The number that is to be repeatedly multiplied in exponent form. When we write $16 = 2^4$, the number 2 is the base.

Base of a percent equation (5.3A) The quantity we take a percent of. In the equation $8 = 20\% \times 400$, the base is 400.

Billion (1.1) The number 1,000,000,000.

Borrowing (1.3) The renaming of a number in order to facilitate subtraction. When we subtract $42 - 28$, we rename 42 as 3 tens plus 12. This represents 3 tens and 12 ones. This renaming is called borrowing.

Box (7.8) A three-dimensional object whose every side is a rectangle. Another name for a box is a *rectangular solid.*

Building fraction property (2.6) For whole numbers a, b, and c, where neither b nor c equals zero,

$$\frac{a}{b} = \frac{a}{b} \times 1 = \frac{a}{b} \times \frac{c}{c} = \frac{a \times c}{b \times c}.$$

Building up a fraction (2.6) To make one fraction into an equivalent fraction by making the denominator and numerator larger numbers. For example, the fraction $\frac{3}{4}$ can be built up to the fraction $\frac{30}{40}$.

Caret (3.5) A symbol $\wedge$ used to indicate the new location of a decimal point when performing division of decimal fractions.

Celsius temperature (6.4) A temperature scale in which water boils at 100 degrees (100°C) and freezes at 0 degrees (0°C). To convert Celsius temperature to Fahrenheit, we use the helpful formula $F = 1.8 \times C + 32$.

Center of a circle (7.7) The point in the middle of a circle from which all points on the circle are an equal distance.

Centimeter (6.2) A unit of length commonly used in the metric system to measure small distances. 1 centimeter = 0.01 meter.

Circle (7.7) A two-dimensional figure for which all points are at an equal distance from a given point.

Circumference of a circle (7.7) The distance around the rim of a circle.

Commission (5.5) The amount of money a salesperson is paid that is a percentage of the value of the sales made by that salesperson. The commission is obtained by multiplying the commission rate times the value of the sales. If a salesman sells $120,000 of insurance and his commission rate is 0.5%, then his commission is $0.5\% \times 120{,}000 = \$600.00$

Common denominator (2.7) Two fractions have a common denominator if the same number appears in the denominator of each fraction. $\frac{3}{7}$ and $\frac{1}{7}$ have a common denominator of 7.

Commutative property of addition (1.2) The property that tells us that the order in which two numbers are added does not change the sum. An example of the commutative property of addition is $3 + 6 = 6 + 3$.

Commutative property of multiplication (1.4) The property that tells us that the order in which two numbers are multiplied does not change the value of the answer. An example of the commutative property of multiplication is $7 \times 3 = 3 \times 7$.

Composite number (2.2) A composite number is a whole number greater than 1 that can be divided by whole numbers other than itself. The number 6 is a composite number since it can be divided exactly by 2 and 3 (as well as by 1 and 6).

Cone (7.8) A three-dimensional object shaped like an ice-cream cone or the sharpened end of a pencil.

Cross-multiplying (4.3) If you have a proportion such as $\frac{n}{5} = \frac{12}{15}$, then to cross-multiply, you form products to obtain $n \times 15 = 5 \times 12$.

Cubic centimeter (6.3) A metric measurement of volume equal to 1 milliliter.

Cup (6.1) One of the smallest units of volume in the American system. 2 cups = 1 pint.

Cylinder (7.8) A three-dimensional object shaped like a tin can.

Debit (1.2) A debit in banking is the removing of money from an account. If you had a savings account and took $300 out of it on Wednesday, we would say that you had a debit of $300 from your account. Often a bank will add a service charge to your account and use the word *debit* to mean that it has removed money from your account to cover the charge.

Decimal fraction (3.1) A fraction whose denominator is a power of 10.

Decimal places (3.4) The number of digits to the right of the decimal point in a decimal fraction. The number 1.234 has three decimal places, while the number 0.129845 has six decimal places. A whole number such as 42 is considered to have zero decimal places.

Decimal point (3.1) The period that is used when writing a decimal fraction. In the number 5.346, the period between the 5 and the 3 is the decimal point. It separates the whole number from the fractional part that is less than 1.

Decimal system (1.1) Our number system is called the decimal system or base 10 system because the value of numbers written in our system is based on tens and ones.

Decimeter (6.2) A unit of length not commonly used in the metric system. 1 decimeter = 0.1 meter.

Degree (7.1) A unit used to measure an angle. A degree is $\frac{1}{360}$ of a complete revolution. An angle of 32 degrees is written as $32°$.

Dekameter (6.2) A unit of length not commonly used in the metric system. 1 dekameter = 10 meters.

Denominator (2.1) The number on the bottom of a fraction. In the fraction $\frac{2}{9}$ the denominator is 9.

Deposit (1.2) A deposit in banking is the placing of money in an account. If you had a checking account and on Tuesday you placed $124 into that account, we would say that you made a deposit of $124.

Diameter of a circle (7.7) A line segment across the circle that passes through the center of the circle. The diameter of a circle is equal to twice the radius of the circle.

Difference (1.3) The result of performing a subtraction. In the problem $9 - 2 = 7$ the number 7 is the difference.

Digits (1.1) The symbols 0, 1, 2, 3, 4, 5, 6, 7, 8, and 9 are called digits.

Discount (5.4) The amount of reduction in a price. The discount is a product of the discount rate times the list price. If the list price of a television is $430.00 and it has a discount rate of 35%, then the amount of discount is $35\% \times \$430.00 = \$150.50$. The price would be reduced by $150.50.

Distributive property of multiplication over addition (1.4) The property illustrated by $5 \times (4 + 3) = (5 \times 4) + (5 \times 3)$. In general, for any numbers a, b, and c, it is true that $a(b + c) = a \times b + a \times c$.

Dividend (1.5) The number that is being divided by another. In the problem $14 \div 7 = 2$, the number 14 is the dividend.

Divisor (1.5) The number that you divide into another number. In the problem $30 \div 5 = 6$, the number 5 is the divisor.

Earned run average (4.4) A ratio formed by finding the number of runs a pitcher would give up in a nine-inning game. If a pitcher has an earned run average of 2, it means that, on the average, he gives up two runs for every nine innings he pitches.

Equal fractions (2.2) Fractions that represent the same number. The fractions $\frac{3}{4}$ and $\frac{6}{8}$ are equal fractions.

Equality test of fractions (2.2) Two fractions $\frac{a}{b}$ and $\frac{c}{d}$ are equal if the product $a \times d = b \times c$. In this case, a, b, c, and d are whole numbers and b and $d \neq 0$.

Equations (10.3) Mathematical statements with variables that say that two expressions are equal, such as $x + 3 = -8$ and $2s + 5s = 34 - 4s$.

Equilateral triangle (7.4) A triangle with three equal sides.

Equivalent equations (10.3) Equations that have the same solution.

Equivalent fractions (2.2) Two fractions that are equal.

Expanded notation for a number (1.1) A number is written in expanded notation if it is written as a sum of hundreds, tens, ones, etc. The expanded notation for 763 is $700 + 60 + 3$.

Exponent (1.6) The number that indicates the number of times a factor occurs. When we write $8 = 2^3$, the number 3 is the exponent.

Factors (1.4) Each of the numbers that are multiplied. In the problem $8 \times 9 = 72$, the numbers 8 and 9 are factors.

Fahrenheit temperature (6.4) A temperature scale in which water boils at 212 degrees (212°F) and freezes at 32 degrees (32°F). To convert Fahrenheit temperature to Celsius, we use the formula $C = \dfrac{5 \times F - 160}{9}$.

Foot (6.1) American system unit of length. 3 feet = 1 yard. 12 inches = 1 foot.

Fundamental theorem of arithmetic (2.2) Every composite number has a unique product of prime numbers.

Gallon (6.1) A unit of volume in the American system. 4 quarts = 1 gallon.

Gigameter (6.2) A metric unit of length equal to 1,000,000,000 meters.

Gram (6.3) The basic unit of weight in the metric system. A gram is defined as the weight of the water in a box that is 1 centimeter on each side. 1 gram = 1000 milligrams. 1 gram = 0.001 kilogram.

Hectometer (6.2) A unit of length not commonly used in the metric system. 1 hectometer = 100 meters.

Height (7.3) The distance between two parallel sides in a four-sided figure such as a parallelogram or a trapezoid.

Height of a cone (7.8) The distance from the vertex of a cone to the base of the cone.

Height of a pyramid (7.8) The distance from the point on a pyramid to the base of the pyramid.

Height of a triangle (7.4) The distance of a line drawn from a vertex perpendicular to the other side, or an extension of the other side, of the triangle. This is sometimes called the *altitude of a triangle*.

Hexagon (7.3) A six-sided figure.

Hypotenuse (7.6) The side opposite the right angle in a right triangle. The hypotenuse is always the longest side of a right triangle.

Improper fraction (2.3) A fraction in which the numerator is greater than or equal to the denominator. The fractions $\frac{34}{29}$, $\frac{8}{7}$, and $\frac{6}{6}$ are all improper fractions.

Inch (6.1) The smallest unit of length in the American system. 12 inches = 1 foot.

Inequality symbol (3.2) The symbol that is used to indicate whether a number is greater than another number or less than another number. Since 5 is greater than 3, we would write this with a "greater than" symbol as follows: $5 > 3$. The statement "7 is less than 12" would be written as follows: $7 < 12$.

Interest (5.4) The money that is paid for the use of money. If you deposit money in a bank, the bank uses that money and pays you interest. If you borrow money, you pay the bank interest for the use of that money. Simple interest is determined by the formula $I = P \times R \times T$. Compound interest is usually determined by a table, a calculator, or a computer.

Invert a fraction (2.5) To invert a fraction is to interchange the numerator and the denominator. If we invert $\frac{5}{9}$, we obtain the fraction $\frac{9}{5}$. To invert a fraction is sometimes referred to as *to take the reciprocal of a fraction.*

Irreducible (2.2) A fraction that cannot be reduced (simplified) is called irreducible.

Isosceles triangle (7.4) A triangle with two sides equal.

Kilogram (6.3) The most commonly used metric unit of weight. 1 kilogram = 1000 grams.

Kiloliter (6.3) The metric unit of volume normally used to measure large volumes. 1 kiloliter = 1000 liters.

Kilometer (6.2) The unit of length commonly used in the metric system to measure large distances. 1 kilometer = 1000 meters.

Least common denominator (LCD) (2.6) The least common denominator (LCD) of two or more fractions is the smallest number that can be divided without remainder by each fraction's denominator. The LCD of $\frac{1}{3}$ and $\frac{1}{4}$ is 12. The LCD of $\frac{5}{6}$ and $\frac{4}{15}$ is 30.

Legs of a right triangle (7.6) The two shortest sides of a right triangle.

Length of a rectangle (7.2) Each of the longer sides of a rectangle.

Like terms (10.1) Like terms have identical variables with identical exponents. $-5x$ and $3x$ are like terms. $-7xyz$ and $-12xyz$ are like terms.

Line segment (7.3) A portion of a straight line that has a beginning and an end.

Liter (6.3) The standard metric measurement of volume. 1 liter = 1000 milliliters. 1 liter = 0.001 kiloliter.

Mean (8.4) The mean of a set of values is the sum of the values divided by the number of values. The mean of the numbers 10, 11, 14, 15 is 12.5. In everyday language, when people use the word *average,* they are usually referring to the mean.

Median (8.4) If a set of numbers is arranged in order from smallest to largest, the median is that value that has the same number of values above it as below it. The median of the numbers 3, 7, and 8 is 7. If the list contains an even number of items, we obtain the median by finding the mean of the two middle numbers. The median of the numbers 5, 6, 10, and 11 is 8.

Megameter (6.2) A metric unit of length equal to 1,000,000 meters.

Meter (6.2) The basic unit of length in the metric system. 1 meter = 1000 millimeters. 1 meter = 0.001 kilometer.

Metric ton (6.3) A metric unit of measurement for very heavy weights. 1 metric ton = 1,000,000 grams.

Microgram (6.3) A unit of weight equal to 0.000001 gram.

Micrometer (6.2) A metric unit of length equal to 0.000001 meter.

Mile (6.1) Largest unit of length in the American system. 5280 feet = 1 mile, 1760 yards = 1 mile.

Milligram (6.3) A metric unit of weight used for very, very small objects. 1 milligram = 0.001 gram.

Milliliter (6.3) The metric unit of volume normally used to measure small volumes. 1 milliliter = 0.001 liter.

Millimeter (6.2) A unit of length commonly used in the metric system to measure very small distances. 1 millimeter = 0.001 meter.

Million (1.1) The number 1,000,000.

Minuend (1.3) The number being subtracted from in a subtraction problem. In the problem $8 - 5 = 3$, the number 8 is the minuend.

Mixed number (2.3) A number created by the sum of a whole number greater than 1 and a proper fraction. The numbers $4\frac{5}{6}$ and $1\frac{1}{8}$ are both mixed numbers. Mixed numbers are sometimes referred to as *mixed fractions*.

Mode (8.4) The mode of a set of data is the number or numbers that occur most often.

Multiplicand (1.4) The first factor in a multiplication problem. In the problem $7 \times 2 = 14$, the number 7 is the multiplicand.

Multiplier (1.4) The second factor in a multiplication problem. In the problem $6 \times 3 = 18$, the number 3 is the multiplier.

Nanogram (6.3) A unit of weight equal to 0.000000001 gram.

Nanometer (6.2) A metric unit of length equal to 0.000000001 meter.

Negative numbers (9.1) All of the numbers to the left of zero on the number line. The numbers $-1.5, -16, -200.5, -4500$ are all negative numbers. All negative numbers are written with a negative sign in front of the digits.

Number line (1.7) A line on which numbers are placed in order from smallest to largest.

Numerator (2.1) The number on the top of a fraction. In the fraction $\frac{3}{7}$ the numerator is 3.

Numerical coefficients (10.1) The numbers in front of the variables in one or more terms. If we look at $-3xy + 12w$, we find that the numerical coefficient of the xy term is -3 while the numerical coefficient of the w term is 12.

Octagon (7.3) An eight-sided figure.

Odometer (1.8) A device on an automobile that displays how many miles the car has been driven since it was first put into operation.

Opposite of a signed number (9.2) The opposite of a signed number is a number that has the same absolute value. The opposite of -5 is 5. The opposite of 7 is -7.

Order of operations (1.6) An agreed-upon procedure to do a problem with several arithmetic operations in the proper order.

Ounce (6.1) Smallest unit of weight in the American system. 16 ounces = 1 pound.

Overtime (2.9) The pay earned by a person if he or she works more than a certain number of hours per week. In most jobs that pay by the hour, a person will earn $1\frac{1}{2}$ times as much per hour for every hour beyond 40 hours worked in one workweek. For example, Carlos earns $6.00 per hour for the first 40 hours in a week and overtime for each additional hour. He would earn $9.00 per hour for all hours he worked in that week beyond 40 hours.

Parallel lines (7.3) Two straight lines that are always the same distance apart.

Parallelogram (7.3) A four-sided figure with both pairs of opposite sides parallel.

Parentheses (1.4) One of several symbols used in mathematics to indicate multiplication. For example, (3)(5) means 3 multiplied by 5. Parentheses are also used as a grouping symbol.

Percent (5.1) The word *percent* means per one hundred. For example, 14 percent means $\frac{14}{100}$.

Percent of decrease (5.5) The percent that something decreases is determined by dividing the amount of decrease by the original amount. If a tape deck sold for $300 and its price was decreased by $60, the percent of decrease would be $\frac{60}{300} = 0.20 = 20\%$.

Percent of increase (5.5) The percent that something increases is determined by dividing the amount of increase by the original amount. If the population of a town was 5000 people and the population increased by 500 people, the percent of increase would be $\frac{500}{5000} = 0.10 = 10\%$.

Percent proportion (5.3B) The percent proportion is the equation $\frac{a}{b} = \frac{p}{100}$ where a is the amount, b is the base, and p is the percent number.

Percent symbol (5.1) A symbol that is used to indicate percent. To indicate 23 percent, we write 23%.

Perfect square (7.4) When a whole number is multiplied by itself, the number that is obtained is a perfect square. The numbers 1, 4, 9, 16, 25, 36, 49, 64, 81, and 100 are all perfect squares.

Perimeter (7.2) The distance around a figure.

Perpendicular lines (7.1) Lines that meet at an angle of 90 degrees.

Pi (7.7) Pi is an irrational number that we obtain if we divide the circumference of a circle by the diameter of a circle. It is represented by the symbol π. Accurate to eleven decimal places, the value of pi is given by 3.14159265359. For most work in this textbook, the value of 3.14 is used to approximate the value of pi.

Picogram (6.3) A unit of weight equal to 0.000000000001 gram.

Pint (6.1) Unit of volume in the American system. 2 pints = 1 quart.

Placeholder (1.1) The use of a digit to indicate a place. Zero is a placeholder in our number system. It holds a position and shows that there is no other digit in that place.

Place-value system (1.1) Our number system is called a place-value system because the placement of the digits tells the value of the number. If we use the digits 5 and 4 to write the number 54, the result is different than if we placed them in opposite order and wrote 45.

Positive numbers (9.1) All of the numbers to the right of zero on the number line. The numbers 5, 6.2, 124.186, 5000 are all positive numbers. A positive number such as +5 is usually written without the positive sign.

Pound (6.1) Basic unit of weight in the American system. 2000 pounds = 1 ton. 16 ounces = 1 pound.

Power of 10 (1.4) Whole numbers that begin with 1 and end in one or more zeros are called powers of 10. The numbers 10, 100, 1000, etc., are all powers of 10.

Prime factors (2.2) Factors that are prime numbers. If we write 15 as a product of prime factors, we have $15 = 5 \times 3$.

Prime number (2.2) A prime number is a whole number greater than 1 that can only be divided by 1 and itself. The first fifteen prime numbers are 2, 3, 5, 7, 11, 13, 17, 19, 23, 29, 31, 37, 41, 43, and 47. The list of prime numbers goes on forever.

Principal (5.4) The amount of money deposited or borrowed on which interest is computed. In the simple interest formula $I = P \times R \times T$, the P stands for the principal. (The other letters are I = interest, R = interest rate, and T = amount of time.)

Product (1.4) The answer in a multiplication problem. In the problem $3 \times 4 = 12$ the number 12 is the product.

Proper fraction (2.3) A fraction in which the numerator is less than the denominator. The fractions $\frac{3}{4}$ and $\frac{15}{16}$ are proper fractions.

Proportion (4.2) A statement that two ratios or two rates are equal. The statement $\frac{3}{4} = \frac{15}{20}$ is a proportion. The statement $\frac{5}{7} = \frac{7}{9}$ is false, and is therefore not a proportion.

Pyramid (7.8) A three-dimensional object made up of a geometric figure for a base and triangular sides that meet at a point. Some pyramids are shaped like the great pyramids of Egypt.

Pythagorean Theorem (7.6) A statement that for any right triangle the square of the hypotenuse equals the sum of the squares of the two legs of the triangle.

Quadrilateral (7.3) A four-sided geometric figure.

Quadrillion (1.1) The number 1,000,000,000,000,000.

Quart (6.1) Unit of volume in the American system. 4 quarts = 1 gallon.

Quotient (1.5) The answer after performing a division problem. In the problem $60 \div 6 = 10$ the number 10 is the quotient.

Radius of a circle (7.7) A line segment from the center of a circle to any point on the circle. The radius of a circle is equal to one-half the diameter of the circle.

Rate (4.1) A rate compares two quantities that have different units. Examples of rates are \$5.00 an hour and 13 pounds for every 2 inches. In fraction form, these two rates would be written as $\frac{\$5.00}{1\text{ hour}}$ and $\frac{13\text{ pounds}}{2\text{ inches}}$.

Ratio (4.1) A ratio is a comparison of two quantities that have the same units. To compare 2 to 3, we can express the ratio in three ways: the ratio of 2 to 3; 2 : 3; or the fraction $\frac{2}{3}$.

Ratio in simplest form (4.1) A ratio is in simplest form when the two numbers do not have a common factor.

Ray (7.1) A ray is a part of a line that has only one endpoint and goes on forever in one direction.

Rectangle (7.2) A four-sided figure that has four right angles.

Reduced fraction (2.2) A fraction for which the numerator and denominator have no common factor other than 1. The fraction $\frac{5}{7}$ is a reduced fraction. The fraction $\frac{15}{21}$ is not a reduced fraction because both numerator and denominator have a common factor of 3.

Regular hexagon (7.3) A six-sided figure with all sides equal.

Regular octagon (7.3) An eight-sided figure with all sides equal.

Remainder (1.5) When two numbers do not divide exactly, a part is left over. This part is called the remainder. For example, $13 \div 2 = 6$ with 1 left over; the 1 is the remainder.

Repeating decimals (3.6) Decimals that have a digit or a group of digits that repeat. The decimals 0.33333333333 . . . and 1.234234234234 . . . are repeating decimals. The pattern of repeating continues forever. Repeating decimals can be written in a form with a bar over the repeating digit(s). Thus the preceding decimals could be written as $0.\overline{3}$ and $1.\overline{234}$.

Right angle (7.1) and (7.4) An angle that measures 90 degrees.

Right triangle (7.4) A triangle with one 90-degree angle.

Rounding (1.7) The process of writing a number in an approximate form for convenience. The number 9756 rounded to the nearest hundred is 9800.

Sales tax (5.4) The amount of tax on a purchase. The sales tax for any item is a product of the sales tax rate times the purchase price. If an item is purchased for \$12.00 and the sales tax rate is 5%, the sales tax is $5\% \times 12.00 = \$0.60$.

Scientific notation (9.5) A positive number is written in scientific notation if it is in the form $a \times 10^n$ where a is a number greater than or equal to 1, but less than 10, and n is an integer. If we write 5678 in scientific notation, we have 5.678×10^3. If we write 0.00825 in scientific notation, we have 8.25×10^{-3}.

Semicircle (7.7) One-half of a circle. The semicircle usually includes the diameter of a circle connected to one-half the circumference of the circle.

Sides of an angle (7.1) The two line segments that meet to form an angle.

Signed numbers (9.1) All of the numbers on a number line. Numbers like -33, 2, 5, -4.2, 18.678, -8.432 are all signed numbers. A negative number always has a negative sign in front of the digits. A positive number such as $+3$ is usually written without the positive sign in front of it.

Similar triangles (7.9) Two triangles that have the same shape but are not necessarily the same size. The corresponding angles of similar triangles are equal. The corresponding sides of similar triangles have the same ratio.

Simple interest (5.4) The interest determined by the formula $I = P \times R \times T$ where I = the interest obtained, P = the principal or the amount borrowed or invested, R = the interest rate (usually on an annual basis), and T = the number of time periods (usually years).

Solution of an equation (10.3) A number is a solution of an equation if replacing the variable by the number makes the equation always true. The solution of $x - 5 = -20$ is the number -15.

Sphere (7.8) A three-dimensional object shaped like a perfectly round ball.

Square (7.2) A rectangle with all four sides equal.

Square root (7.5) The square root of a number is one of two identical factors of that number. The square root of 9 is 3. The square root of 121 is 11.

Square root sign (7.5) The symbol $\sqrt{\ }$. When we want to find the square root of 25, we write $\sqrt{25}$. The answer is 5.

Standard notation for a number (1.1) A number written in ordinary terms. For example, $70 + 2$ in standard notation is 72.

Subtrahend (1.3) The number being subtracted. In the problem $7 - 1 = 6$, the number 1 is the subtrahend.

Sum (1.2) The result of an addition of two or more numbers. In the problem $7 + 3 + 5 = 15$, the number 15 is the sum.

Term (10.1) A number, a variable, or a product of a number and one or more variables. $5x$, $2ab$, $-43cdef$ are three examples of terms, separated in an expression by a $+$ sign or a $-$ sign.

Terminating decimals (3.6) Every fraction can be written as a decimal. If the division process of dividing denominator into numerator ends with a remainder of zero, the decimal is a terminating decimal. Decimals such as 1.28, 0.007856, and 5.123 are terminating decimals.

Trapezoid (7.3) A four-sided figure with at least two parallel sides.

Triangle (7.4) A three-sided figure.

Trillion (1.1) The number 1,000,000,000,000.

Unit fraction (6.1) A fraction used to change one unit to another. For example, to change 180 inches to feet, we multiply by the unit fraction $\frac{1 \text{ foot}}{12 \text{ inches}}$. Thus we have

$$180 \text{ inches} \times \frac{1 \text{ foot}}{12 \text{ inches}} = 15 \text{ feet}.$$

Variable (10.1) A letter that is used to represent a number.

Vertex of a cone (7.8) The sharp point of a cone.

Vertex of an angle (7.1) The point at which two line segments meet to form an angle.

Volume (7.8) The measure of the space inside a three-dimensional object. Volume is measured in cubic units such as cubic feet.

Whole numbers (1.1) The whole numbers are the set of numbers 0, 1, 2, 3, 4, 5, 6, 7, 8, 9, 10, 11, 12, ... The set goes on forever. There is no largest whole number.

Width of a rectangle (7.2) Each of the shorter sides of a rectangle.

Word names for whole numbers (1.1) The notation for a number in which each digit is expressed by a word. To write 389 with a word name, we would write three hundred eighty-nine.

Zero (1.1) The smallest whole number. It is normally written 0.

Appendix A Consumer Finance Applications

A.1 Balancing a Checking Account

1 Calculating a Checkbook Balance

If you have a checking account, you should keep records of the checks written, ATM withdrawals, deposits, and other transactions on a check register. To find the amount of money in a checking account you subtract debits and add credits to the balance in the account. Debits are checks written, withdrawals made, or any other amount charged to a checking account. Credits include deposits made, as well as any other money credited to the account.

Student Learning Objectives

After studying this section, you will be able to:

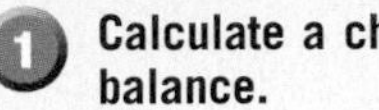

1. **Calculate a checkbook balance.**

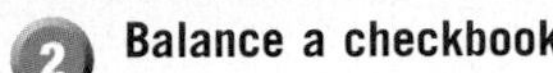

2. **Balance a checkbook.**

EXAMPLE 1 Jesse Holm had a balance of $1254.32 in his checking account before writing five checks and making a deposit. On 9/2 Jesse wrote check #243 to the Manor Apartments for $575, check #244 to the Electric Company for $23.41, and check #245 to the Gas Company for $15.67. Then on 9/3, he wrote check #246 to Jack's Market for $125.57, check #247 to Clothing Mart for $35.85, and made a $634.51 deposit. Record the checks and deposit in Jesse's check register and then find Jesse's ending balance.

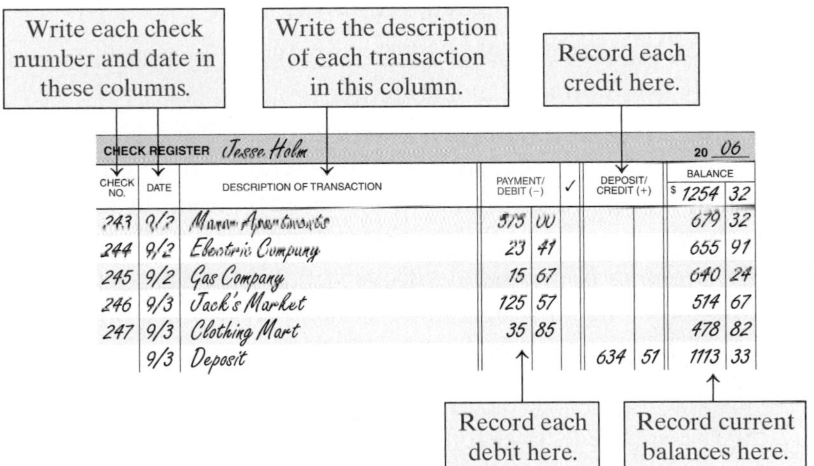

CHECK REGISTER Jesse Holm 20 06

CHECK NO.	DATE	DESCRIPTION OF TRANSACTION	PAYMENT/ DEBIT (−)	✓	DEPOSIT/ CREDIT (+)	BALANCE $ 1254 32
243	9/2	Manor Apartments	575 00			679 32
244	9/2	Electric Company	23 41			655 91
245	9/2	Gas Company	15 67			640 24
246	9/3	Jack's Market	125 57			514 67
247	9/3	Clothing Mart	35 85			478 82
	9/3	Deposit			634 51	1113 33

Solution To find the ending balance, we subtract each check written and add the deposit to the current balance. Then we record these amounts in the check register.

$$1254.32 - 575.00 = 679.32$$
$$679.32 - 23.41 = 655.91$$
$$655.91 - 15.67 = 640.24$$
$$640.24 - 125.57 = 514.67$$
$$514.67 - 35.85 = 478.82$$
$$478.82 + 634.51 = 1113.33$$

Jesse's balance is 1113.33.

Practice Problem 1 My Chung Nguyen had a balance of $1434.52 in her checking account before writing three checks and making a deposit. On 3/1 My Chung wrote check #144 to the Leland Mortgage Company for $908 and check #145 to the Phone Company for $33.21. Then on 3/2 she wrote check #146 to Sam's Food Market for $102.37 made a $524.41 deposit. Record the checks and deposit in My Chung's check register, and then find My Chung's ending balance.

NOTE TO STUDENT: Fully worked-out solutions to all of the Practice Problems can be found at the back of the text starting at page SP-1

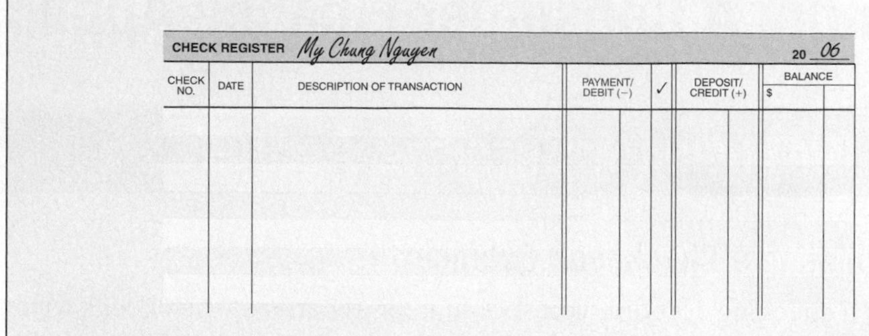

2 Balancing a Checkbook

The bank provides customers with bank statements each month. This statement lists the checks the bank paid, ATM withdrawals, deposits made, and all other debits and credits made to a checking account. It is very important to verify that these bank records match ours. We must make sure that we deducted all debits and added all credits in our check register. This is called **balancing a checkbook.** Balancing our checkbook allows us to make sure that the balance we think we have in our checkbook is correct. If our checkbook does not balance, we must look for any mistakes.

To balance a checkbook, proceed as follows.

1. First, *adjust the check register balance* so that it includes all credits and debits listed on the bank statement.
2. Then, *adjust the bank statement balance* so that it includes all credits and debits that may not have been received by the bank when the statement was printed. Checks that were written, but not received by the bank, are called **checks outstanding.**
3. Finally, *compare both balances* to verify that they are equal. If they are equal, the checking account balances. If they are not equal, we must find the error and make adjustments.

There are several ways to balance a checkbook. Most banks include a form you can fill out to assist you in this process.

EXAMPLE 2 Balance Jesse's checkbook using his check register and bank statement.

CHECK REGISTER *Jesse Holm* 20 06

CHECK NO.	DATE	DESCRIPTION OF TRANSACTION	PAYMENT/ DEBIT (−)	✓	DEPOSIT/ CREDIT (+)	BALANCE
						$ 1254 32
243	9/2	Manor Apartments	575 00	✓		679 32
244	9/2	Electric Company	23 41	✓		655 91
245	9/2	Gas Company	15 67	✓		640 24
246	9/3	Jack's Market	125 57	✓		514 67
247	9/3	Clothing Mart	35 85			478 82
	9/3	Deposit		✓	634 51	1113 33
248	9/12	College Bookstore	168 96	✓		944 37
	9/18	ATM	100 00	✓		844 37
249	9/25	Telephone Company	43 29	✓		801 08
250	9/30	Sports Emporium	40 00			761 08
	10/1	Deposit			530 90	1291 98

Bank Statement: JESSE HOLM 9/1/2006 to 9/30/2006

Beginning Balance $1254.32
Ending Balance $831.68

Checks cleared by the bank

#243	$575.00	#245	$15.67	#248	$168.96
#244	$23.41	#246*	$125.57	#249*	$43.29

* Indicates that the next check in the sequence is outstanding (hasn't cleared).

Deposits

9/3 $634.51

Other withdrawals

9/18 ATM $100.00 Service charge $5.25

Solution Follow steps 1–6 on the Checking Account Balance Form below.

CHECKING RECONCILEMENT This form is provided to assist you in balancing your checking account.

List checks outstanding* not charged to your checking account

CHECK NO.	AMOUNT	
247	35	85
250	40	00
TOTAL	75	85

* and ATM withdrawals

Period ending 9/30 ,2006

1. Check Register Balance		$ 1291.98
Subtract any charges listed on the bank statement which you have not previously deducted from your balance.	–	$ 5.25
Adjusted Check Register Balance		$ 1286.73
2. Enter the ending balance shown on the bank statement.		$ 831.68
3. Enter deposits made later than the ending date on the bank statement.	+	$ 530.90
	+	$
	+	$
TOTAL (Step 2 plus Step 3)		$ 1362.58
4. In your check register, **check off** all the checks paid. In the ares provided to the left, **list** numbers and amounts of all outstanding checks and ATM withdrawals.		
5. Subtract the total amount in Step 4.	–	$ 75.85
6. This adjusted bank balance should equal the adjusted Check Register Balance from Step 1.		$ 1286.73

The balances in steps **1** and **6** are equal, so Jesse's checkbook is balanced.

Practice Problem 2 Balance Anthony's checkbook using his check register, his bank statement, and the given Checking Account Balance Form.

***NOTE TO STUDENT:** Fully worked-out solutions to all of the Practice Problems can be found at the back of the text starting at page SP-1*

CHECK REGISTER *Anthony Maida* 20 06

CHECK NO.	DATE	DESCRIPTION OF TRANSACTION	PAYMENT/ DEBIT (–)		✓	DEPOSIT/ CREDIT (+)		BALANCE $ 1823	00
211	7/2	Apple Apartments	985	00				838	00
	7/9	ATM	101	50				736	50
212	7/10	Leland Groceries	98	87				637	63
213	7/21	The Gas Company	45	56				592	07
	7/21	Deposit				687	10	1279	17
214	7/28	Cellular for Less	59	98				1219	19
215	7/28	The Electric Company	89	75				1129	44
216	7/28	Sports World	129	99				999	45
217	7/28	Leland Groceries	205	99				793	46
	7/30	ATM	141	50				651	96
	8/1	Deposit				398	50	1050	46

Bank Statement: ANTHONY MAIDA 7/1/2006 to 8/1/2006

Beginning Balance $1823.00
Ending Balance $934.95

Checks cleared by the bank

#211	$985.00	#213	$45.56	#216	$129.99
#212	$98.87	#214*	$59.98		

* Indicates that the next check in the sequence is outstanding (hasn't cleared).

Deposits

7/21 $687.10

Other withdrawals

7/9 ATM	$101.50	Service charge	$3.50
7/30 ATM	$141.50	Check purchase	$9.25

CHECKING RECONCILEMENT This form is provided to assist you in balancing your checking account.

List checks outstanding* not charged to your checking account

CHECK NO.	AMOUNT	
TOTAL		

* and ATM withdrawals

Period ending ,20

1. Check Register Balance		$
Subtract any charges listed on the bank statement which you have not previously deducted from your balance.	–	$
Adjusted Check Register Balance		$
2. Enter the ending balance shown on the bank statement.		$
3. Enter deposits made later than the ending date on the bank statement.	+	$
	+	$
	+	$
TOTAL (Step 2 plus Step 3)		$
4. In your check register, **check off** all the checks paid. In the ares provided to the left, **list** numbers and amounts of all outstanding checks and ATM withdrawals.		
5. Subtract the total amount in Step 4.	–	$
6. This adjusted bank balance should equal the adjusted Check Register Balance from Step 1.		$

Note: To calculate the service charges, we add $3.50 + 9.25 = 12.75$. ■

A.1 Exercises

Student Solutions Manual

CD/ Video

PH Math Tutor Center

MathXL®Tutorials on CD

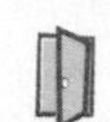
MathXL®

MyMathLab®

Interactmath.com

1. Shin Karasuda had a balance of $532 in his checking account before writing four checks and making a deposit. On 6/1 he wrote check #122 to the Mini Market for $124.95 and check #123 to Better Be Dry Cleaners for $41.50. Then on 6/9 he made a $384.10 deposit, wrote check #124 to Macy's Department Store for $72.98, and check #125 to Costco's for $121.55. Record the checks and deposit in Shin's check register, and then find the ending balance.

CHECK REGISTER *Shin Karasuda* 20 *06*

CHECK NO.	DATE	DESCRIPTION OF TRANSACTION	PAYMENT/ DEBIT (−)	✓	DEPOSIT/ CREDIT (+)	BALANCE
						$ 532 00
122	6/1	MiniMarket	124 95			407 05
123	6/1	BetterBe Dry Cleaners	41 50			365 55
	6/9	Deposit			384 10	749 65
124	6/9	Macy's Dept. Store	72 98			676 67
125	6/9	Costco's	121 55			555 12

2. Mary Beth O'Brian had a balance of $493 in her checking account before writing four checks and making a deposit. On 9/4 she wrote check #311 to the Ben's Garage for $213.45 and #312 to Food Mart for $132.50. Then on 9/5 she made a $387.50 check deposit, wrote check #313 to The Shoe Pavilion for $69.98, and check #314 to the Electric Company for $92.45. Record the checks and deposit in Mary Beth's check register, and then find the ending balance.

CHECK REGISTER *Mary Beth O'Brian* 20 *06*

CHECK NO.	DATE	DESCRIPTION OF TRANSACTION	PAYMENT/ DEBIT (−)	✓	DEPOSIT/ CREDIT (+)	BALANCE
						$ 493 00
311	9/4	Ben's Garage	213 45			279 55
312	9/4	Food Mart	132 50			147 05
	9/5	Deposit			387 50	534 55
313	9/5	Shoe Pavilion	69 98			464 57
314	9/5	Electric Company	92 45			372 12

3. The Harbor Beauty Salon had a balance of $2498.90 in its business checking account on 3/4. The manager made a deposit on the same day for $786 and wrote check #734 to the Beauty Supply Factory for $980. Then on 3/9 he wrote check #735 to the Water Department for $131.85 and check #736 to the Electric Company for $251.50. On 3/19 he made a $2614.10 deposit and wrote two payroll checks: #737 to Ranik Ghandi for $873 and #738 to Eduardo Gomez for $750. Record the checks and deposits in The Harbor Beauty Salon's check register, and then find the ending balance.

CHECK REGISTER *Harbor Beauty Salon* 20 *06*

CHECK NO.	DATE	DESCRIPTION OF TRANSACTION	PAYMENT/ DEBIT (−)	✓	DEPOSIT/ CREDIT (+)	BALANCE
						$ 2498 90
	3/4	Deposit			786 00	3284 90
734	3/4	Beauty Supply Factory	980 00			2304 90
735	3/9	Water Department	131 85			2173 05
736	3/9	Electric Company	251 50			1921 55
	3/19	Deposit			2614 10	4535 65
737	3/19	Ranik Ghandi	873 00			3662 65
738	3/19	Eduardo Gomez	750 00			2912 65

4. Joanna's Coffee Shop had a balance of $1108.50 in its business checking account on 1/7. The owner made a deposit on the same day for $963 and wrote check #527 to the Restaurant Supply Company for $492. Then on 1/11 she wrote check #528 to the Gas Company for $122.45 and check #529 to the Electric Company for $321.20. Then on 1/12 she made a $1518.20 deposit and wrote two payroll checks: #530 to Sara O'Conner for $579, and #531 to Niegan Raskin for $466. Record the checks and deposits in Joanna's Coffee Shop's check register, and then find the ending balance.

CHECK REGISTER *Joanna's Coffee Shop* 20 *06*

CHECK NO.	DATE	DESCRIPTION OF TRANSACTION	PAYMENT/ DEBIT (−)	✓	DEPOSIT/ CREDIT (+)	BALANCE
						$ 1108 50
	1/7	Deposit			963 00	2071 50
527	1/7	Restaurant Supply Company	492 00			1579 50
528	1/11	Gas Company	122 45			1457 05
529	1/11	Electric Company	321 20			1135 85
	1/12	Deposit			1518 20	2654 05
530	1/12	Sara O'Conner	579 00			2075 05
531	1/12	Niegan Raskin	466 00			1609 05

5. On 3/3 Justin Larkin had $321.94 in his checking account before he withdrew $101.50 at the ATM. On 3/7 he made a $601.90 deposit and wrote checks to pay the following bills: check #114 to the Third Street Apartments for $550, check #115 to the Cable Company for $59.50, check #116 to the Electric Company for $43.50, and check #117 to the Gas Company for $15.90. Does Justin have enough money left in his checking account to pay $99 for his car insurance?

CHECK REGISTER *Justin Larkin* 20 *06*

CHECK NO.	DATE	DESCRIPTION OF TRANSACTION	PAYMENT/DEBIT (−)	✓	DEPOSIT/CREDIT (+)	BALANCE $ 321.94
	3/3	ATM	101.50			220.44
	3/7	Deposit			601.90	822.34
114	3/7	Third Street Apartments	550.00			272.34
115	3/7	Cable Company	59.50			212.84
116	3/7	Electric Company	43.50			169.34
117	3/7	Gas Company	15.90			153.44

Yes, Justin can pay his car insurance.

6. On 5/2 Leon Jones had $423.54 in his checking account when he withdrew $51.50 at the ATM. On 5/9 he made a $601.80 deposit and wrote checks to pay the following bills: check #334 to Fasco Car Finance for $150.25, check #335 to A-1 Car Insurance for $89.20, check #336 to the Telephone Company for $33.40, and check #337 to the Apple Apartments for $615. Does Leon have enough money left in his checking account to pay $59 for his electric bill?

CHECK REGISTER *Leon Jones* 20 *06*

CHECK NO.	DATE	DESCRIPTION OF TRANSACTION	PAYMENT/DEBIT (−)	✓	DEPOSIT/CREDIT (+)	BALANCE $ 423.54
	5/2	ATM	51.50			372.04
	5/9	Deposit			601.80	973.84
334	5/9	Fasco Car Finance	150.25			823.59
335	5/9	A-1 Car Insurance	89.20			734.39
336	5/9	Telephone Company	33.40			700.99
337	5/9	Apple Apartments	615.00			85.99

Yes, Leon can pay his electric bill.

7. Balance the monthly statement for the Carson Maid Service on the Checking Account Balance Form shown below.

CHECK REGISTER *Carson Maid Service* 20 *06*

CHECK NO.	DATE	DESCRIPTION OF TRANSACTION	PAYMENT/DEBIT (−)	✓	DEPOSIT/CREDIT (+)	BALANCE $ 1721.50
102	4/3	A&R Cleaning Supplies	422.33			1299.17
103	4/9	Allison De Julio	510.50			788.67
104	4/9	Mai Vu	320.00			468.67
105	4/9	Jon Veldez	320.00			148.67
	4/11	Deposit			1890.00	2038.67
106	4/20	Mobil Gas Company	355.35			1683.32
107	4/25	Peterson Property Mangt. warehouse rent	525.00			1158.32
108	4/29	Jack's Garage	450.10			708.22
	5/2	Deposit			540.00	1248.22

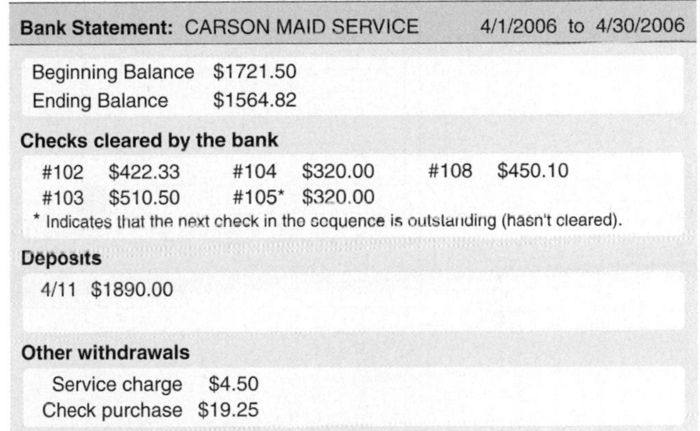

Bank Statement: CARSON MAID SERVICE 4/1/2006 to 4/30/2006

Beginning Balance $1721.50
Ending Balance $1564.82

Checks cleared by the bank

#102 $422.33 #104 $320.00 #108 $450.10
#103 $510.50 #105* $320.00

* Indicates that the next check in the sequence is outstanding (hasn't cleared).

Deposits

4/11 $1890.00

Other withdrawals

Service charge $4.50
Check purchase $19.25

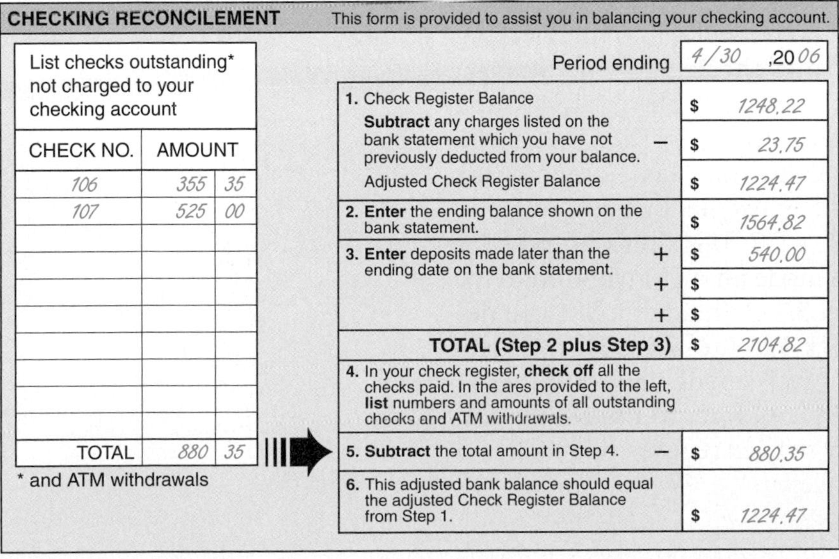

CHECKING RECONCILEMENT This form is provided to assist you in balancing your checking account.

List checks outstanding* not charged to your checking account

CHECK NO.	AMOUNT
106	355.35
107	525.00
TOTAL	880.35

* and ATM withdrawals

Period ending *4/30*, 20*06*

1. Check Register Balance		$ 1248.22
Subtract any charges listed on the bank statement which you have not previously deducted from your balance.	−	$ 23.75
Adjusted Check Register Balance		$ 1224.47
2. **Enter** the ending balance shown on the bank statement.		$ 1564.82
3. **Enter** deposits made later than the ending date on the bank statement.	+	$ 540.00
	+	$
	+	$
TOTAL (Step 2 plus Step 3)		$ 2104.82
4. In your check register, **check off** all the checks paid. In the ares provided to the left, **list** numbers and amounts of all outstanding checks and ATM withdrawals.		
5. **Subtract** the total amount in Step 4.	−	$ 880.35
6. This adjusted bank balance should equal the adjusted Check Register Balance from Step 1.		$ 1224.47

The account balances.

8. Balance the monthly statement for The Flower Shop on the Checking Account Balance Form shown below.

CHECK REGISTER *The Flower Shop* 20 *06*

CHECK NO.	DATE	DESCRIPTION OF TRANSACTION	PAYMENT/ DEBIT (−)	✓	DEPOSIT/ CREDIT (+)	BALANCE
						$ 3459 40
502	9/7	Whole Sale Flower Company	733 67			2725 73
503	9/7	Alexsandra Kruse	580 20			2145 53
504	9/7	Jose Sanchez	430 50			1715 03
505	9/7	Kamir Kosedag	601 90			1113 13
	9/11	Deposit			2654 00	3767 13
506	9/21	L&S Pottery	466 84			3300 29
507	9/24	Peterson Property Mangt. (warehouse rent)	985 00			2315 29
508	9/28	Barton Electric Company	525 60			1789 69
	10/2	Deposit			540 00	2329 69

Bank Statement: THE FLOWER SHOP 9/1/2005 to 9/30/2005

Beginning Balance $3459.40
Ending Balance $3220.28

Checks cleared by the bank

#502 $733.67 #504 $430.50 #508 $525.60
#503 $580.20 #505* $601.90

* Indicates that the next check in the sequence is outstanding (hasn't cleared).

Deposits

9/11 $2654.00

Other withdrawals

Service charge $4.75
Check purchase $16.50

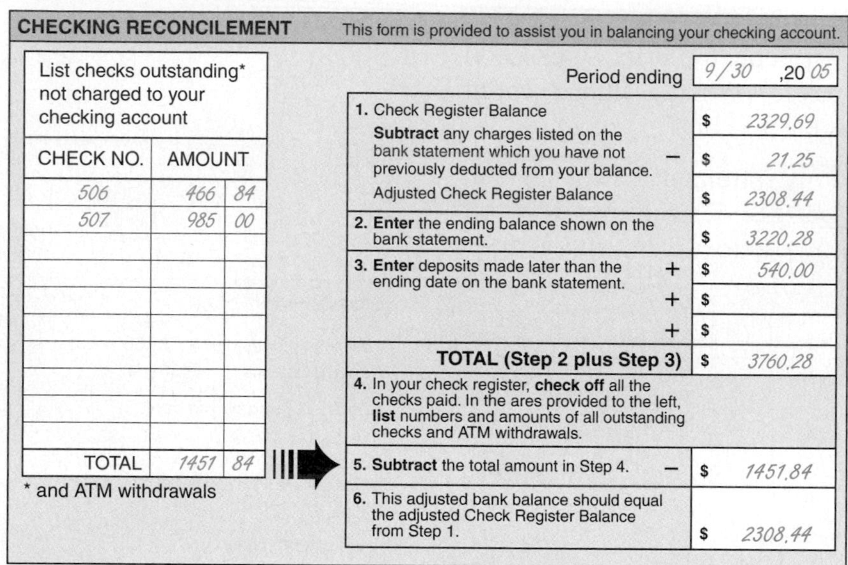

CHECKING RECONCILEMENT This form is provided to assist you in balancing your checking account.

List checks outstanding* not charged to your checking account

CHECK NO.	AMOUNT
506	466 84
507	985 00
TOTAL	1451 84

* and ATM withdrawals

Period ending 9/30 ,20 05

1. Check Register Balance — $ 2329.69
   Subtract any charges listed on the bank statement which you have not previously deducted from your balance. − $ 21.25
   Adjusted Check Register Balance — $ 2308.44
2. **Enter** the ending balance shown on the bank statement. — $ 3220.28
3. **Enter** deposits made later than the ending date on the bank statement. + $ 540.00
   + $
   + $
   TOTAL (Step 2 plus Step 3) — $ 3760.28
4. In your check register, **check off** all the checks paid. In the ares provided to the left, **list** numbers and amounts of all outstanding checks and ATM withdrawals.
5. **Subtract** the total amount in Step 4. − $ 1451.84
6. This adjusted bank balance should equal the adjusted Check Register Balance from Step 1. — $ 2308.44

The account balances.

9. On 2/1 Jeremy Sirk had a balance of $672.10 in his checking account. On the same day he deposited $735 of his paycheck into his checking account and wrote check #233 to Stanton Sporting Goods for $92.99, check #234 to the Garden Apartments for $680. Then on 2/20 he wrote check #235 to the Gas Company for $31.85, check #236 to the Cable Company for $51.50, check #237 to Ralph's Market for $173.98, and also made an ATM withdrawal for $101.50. Then on 3/1 he made an $814.10 deposit, wrote check #238 to State Farm Insurance for $98, and made another ATM withdrawal for $41.50. Record the checks and deposit in Jeremy's check register, and then find the ending balance.

CHECK REGISTER *Jeremy Sirk* 20 *06*

CHECK NO.	DATE	DESCRIPTION OF TRANSACTION	PAYMENT/ DEBIT (−)	✓	DEPOSIT/ CREDIT (+)	BALANCE
						$ 672 10
	2/1	Deposit			735 00	1407 10
233	2/1	Stanton Sporting Goods	92 99			1314 11
234	2/1	Garden Apartments	680 00			634 11
235	2/20	Gas Company	31 85			602 26
236	2/20	Cable Company	51 50			550 76
237	2/20	Ralph's Market	173 98			376 78
	2/20	ATM	101 50			275 28
	3/1	Deposit			814 10	1089 38
238	3/1	State Farm Insurance	98 00			991 38
	3/1	ATM	41 50			949 88

10. On 8/1 Shannon Mending had a balance of $525.90 in her checking account. Later that same day she deposited $588.23 into her checking account and wrote check #333 to Verizon Telephone Company for $33.20 and check #334 to Discount Car Insurance for $332.50. Then on 8/8 she wrote check #335 to the Walden Market for $21.35, made an ATM withdrawal for $81.50, and wrote check #336 to the Cable Company for $41.50. Then on 9/1 she made a $904.10 deposit, wrote check #337 to Next Day Dry Cleaners for $33.50, and check #338 to Marty's Dress Shop for $87.99. Record the checks and deposit in Shannon's check register, and then find the ending balance.

CHECK REGISTER *Shannon Mending* 20 *06*

CHECK NO.	DATE	DESCRIPTION OF TRANSACTION	PAYMENT/ DEBIT (−)	✓	DEPOSIT/ CREDIT (+)	BALANCE $ 525.90
	8/1	Deposit			588.23	1114.13
333	8/1	Verizon Telephone Company	33.20			1080.93
334	8/1	Discount Car Insurance	332.50			748.43
335	8/8	Walden Market	21.35			727.08
	8/8	ATM	81.50			645.58
336	8/8	Cable Company	41.50			604.08
	9/1	Deposit			904.10	1508.18
337	9/1	Next Day Dry Cleaners	33.50			1474.68
338	9/1	Marty's Dress Shop	87.99			1386.69

11. Refer to exercise #9 and use the bank statement and Checking Account Balance Form below to balance Jeremy's checking account.

Bank Statement: JEREMY SIRK 2/1/2005 to 2/28/2005

Beginning Balance $672.10
Ending Balance $293.88

Checks cleared by the bank

#233 $92.99 #236 $51.50
#234* $680.00 #237* $173.98

* Indicates that the next check in the sequence is outstanding (hasn't cleared).

Deposits

2/1 $735.00

Other withdrawals

2/20 ATM $101.50 Service charge $3.50
Check purchase $9.75

Jeremy's account balances.

CHECKING RECONCILEMENT This form is provided to assist you in balancing your checking account.

List checks outstanding* not charged to your checking account

CHECK NO.	AMOUNT
235	31.85
238	98.00
ATM	41.50
TOTAL	171.35

* and ATM withdrawals

Period ending 2/28, 20 05

1. Check Register Balance		$ 949.88
Subtract any charges listed on the bank statement which you have not previously deducted from your balance.	−	$ 13.25
Adjusted Check Register Balance		$ 936.63
2. Enter the ending balance shown on the bank statement.		$ 293.88
3. Enter deposits made later than the ending date on the bank statement.	+	$ 814.10
	+	$
	+	$
TOTAL (Step 2 plus Step 3)		$ 1107.98
4. In your check register, check off all the checks paid. In the ares provided to the left, list numbers and amounts of all outstanding checks and ATM withdrawals.		
5. Subtract the total amount in Step 4.	−	$ 171.35
6. This adjusted bank balance should equal the adjusted Check Register Balance from Step 1.		$ 936.63

12. Refer to exercise #10 and use the bank statement and Checking Account Balance Form below to balance Shannon's checking account.

Bank Statement: SHANNON MENDING 8/1/2005 to 8/31/2005

Beginning Balance $525.90
Ending Balance $923.83

Checks cleared by the bank

#333* $33.20 #336* $41.50
#335 $21.35

* Indicates that the next check in the sequence is outstanding (hasn't cleared).

Deposits

8/1 $588.23

Other withdrawals

8/8 ATM $81.50 Service charge $3.25
Check purchase $9.50

Shannon's account balances.

CHECKING RECONCILEMENT This form is provided to assist you in balancing your checking account.

List checks outstanding* not charged to your checking account

CHECK NO.	AMOUNT
334	332.50
337	33.50
338	87.99
TOTAL	453.99

* and ATM withdrawals

Period ending 8/31, 20 05

1. Check Register Balance		$ 1386.69
Subtract any charges listed on the bank statement which you have not previously deducted from your balance.	−	$ 12.75
Adjusted Check Register Balance		$ 1373.94
2. Enter the ending balance shown on the bank statement.		$ 923.83
3. Enter deposits made later than the ending date on the bank statement.	+	$ 904.10
	+	$
	+	$
TOTAL (Step 2 plus Step 3)		$ 1827.93
4. In your check register, check off all the checks paid. In the ares provided to the left, list numbers and amounts of all outstanding checks and ATM withdrawals.		
5. Subtract the total amount in Step 4.	−	$ 453.99
6. This adjusted bank balance should equal the adjusted Check Register Balance from Step 1.		$ 1373.94

A.2 Purchasing a Home

Student Learning Objectives

After studying this section you will be able to:

1. **Find the initial costs of buying a home.**
2. **Find the ongoing monthly costs of owning a home.**

Purchasing a home is a big investment and there are many questions we must consider. What will we pay for the *initial costs* to buy the home? Do we have enough money to cover all these costs? Can we afford the amount of the *house payment* plus all the *ongoing costs*? The interest rate on the loan is also an important factor, since it affects the amount of the house payment. Most home loans are either on a fixed-rate loan or a variable-rate loan. With a fixed-rate loan the payment stays the same for the entire life of the loan, whereas the payment can change on a variable-rate plan. We will discuss fixed-rate loans in this section, and define the terms used when buying or selling a home. Then we will see how we calculate the initial and ongoing cost of owning a home.

1 Finding the Initial Costs of Buying a Home

There are several initial costs to the buyer. By this we mean monies that must be paid before we purchase a home. They include a down payment, loan points, and escrow charges.

Down payment—This is the amount of money the buyer must put down on the home in order to purchase the home. The amount of the down payment is usually a percent of the purchase price. This percent can vary from 5% to 25% depending on the lender (bank or other loan companies) and the type of loan obtained.

down payment = percent × purchase price of home

Home loan (Mortgage)—The amount that is borrowed to purchase a home is called a home loan or mortgage.

home loan amount = purchase price − down payment

Teaching Tip Escrow charges include the cost for title insurance, loan processing, wire and money transfer fees, prepaid interest, taxes, and the first year's homeowner insurance. There are sometimes lawyer and appraisal fees, flood or PMI insurance (when the down payment is 5% or less), and several other fees that may be particular to the home purchase or state law. Most students are not aware of all these extra costs and what they mean. A brief discussion may help them understand why they need to think beyond the down payment when they purchase a home.

Loan fee (Points)—This is the amount that the buyer must pay the lender to process the loan. Lenders use the term **loan points** to describe the fee. The amount of points for a loan can vary. A loan point is the same as one percent. That is, *1 point is 1% of the home loan amount.* To find the loan fee we find the percent of the loan amount.

loan fee = percent × home loan amount

Escrow charges (Closing costs)—A closing agent, often from an escrow company, oversees the entire purchase process, and performs many other services necessary to complete the purchase. Escrow charges include all **closing costs** necessary for the purchase. Although escrow charges vary, we can estimate the total charges by calculating $1\frac{1}{2}\%$ of the purchase price.

escrow charges = $1\frac{1}{2}\%$ × purchase price of home

EXAMPLE 1 Victor purchased a home for \$160,000 and made a down payment of 15% of the purchase price. Find the amount of Victor's down payment and home loan.

Solution To find the down payment we calculate 15% of 160,000.

$$\text{down payment} = \text{percent} \times \text{purchase price of home}$$
$$= 0.15 \times 160{,}000$$
We must change 15% to a decimal: 15% = 0.15.
$$= 24{,}000$$

Victor made a \$24,000 down payment.

To find the amount of the home loan, we subtract the down payment from the purchase price.

$$\text{home loan amount} = \text{purchase price} - \text{down payment}$$
$$= 160{,}000 - 24{,}000 = 136{,}000$$

The home loan is for \$136,000.

Practice Problem 1 Alyssa purchased a home for \$110,000 and made a down payment of 10% of the purchase price. Find the amount of Alyssa's down payment and home loan.

***NOTE TO STUDENT:** Fully worked-out solutions to all of the Practice Problems can be found at the back of the text starting at page SP-1*

EXAMPLE 2 Victor paid a 2-point loan fee on his 136,000 home loan, and escrow charges on the 160,000 purchase price. Calculate the loan fee and escrow charges.

Solution

$$\text{loan fee} = \text{percent} \times \text{home loan amount}$$
$$= 2\% \times 136{,}000 = 0.02 \times 136{,}000$$
2 points = 2%; change 2% to a decimal: 0.02
$$= 2720$$

The loan fee is \$2720.

$$\text{escrow charges} = 1\tfrac{1}{2}\% \times \text{purchase price of home}$$
$$= 0.015 \times 160{,}000$$
Change $1\frac{1}{2}\%$ to a decimal: $1\frac{1}{2}\% = 1.5\% = 0.015$.
$$= 2400$$

The escrow charges are \$2400.

Caution: We must be sure to remember that the *loan fee* is based on the *loan amount* and the *escrow charges* are based on the *purchase price.*

Practice Problem 2 Alyssa paid a 3-point loan fee on her \$99,000 home loan, and escrow charges on the \$110,000 purchase price. Calculate the loan fee and escrow charges.

EXAMPLE 3 Refer to Examples 1 and 2 to complete the following. Find the total initial costs (down payment, loan fee, and escrow charges) Victor paid for the purchase of the $160,000 home purchase.

Solution We fill in the table with the correct amounts and then add to find the total initial costs.

Down payment	$24,000
Loan fee	$ 2720
Escrow charges	$ 2400
Total initial costs	$29,120

The total initial costs are $29,120.

NOTE TO STUDENT: Fully worked-out solutions to all of the Practice Problems can be found at the back of the text starting at page SP-1

Practice Problem 3 Refer to Practice Problems 1 and 2 to complete the following. Find the total initial costs (down payment, loan fee and escrow charges) Alyssa paid for the $110,000 home purchase.

Down payment	$
Loan fee	$
Escrow charges	$
Total initial costs	$

2 Finding the Ongoing Monthly Costs of Owning a Home

After considering the initial costs we must be sure we can afford the monthly ongoing costs. These expenses include the house payment (mortgage payment), property taxes, and homeowners insurance.

House payment—This is the amount we pay each month on our home loan. To estimate a house payment on a *fixed-rate* loan, we can multiply the amount of the home loan times a **payment factor.** To find the payment factor on the table below, locate the interest rate in the top row, and the number of years in the left column. The payment factor is where the interest rate column and the years row meet.

House payment = payment factor × home loan amount

Payment Factors						
	4%	5%	6%	7%	8%	9%
15 years	0.007397	0.007908	0.008439	0.008988	0.009557	0.010143
20 years	0.006059	0.006599	0.007164	0.007753	0.008364	0.008997
25 years	0.005278	0.005846	0.006443	0.007068	0.007718	0.008392
30 years	0.004774	0.005368	0.005996	0.006653	0.007338	0.008046

Property taxes and homeowner insurance—Homeowners must pay a tax on their property and obtain homeowners insurance. The amount of the tax and insurance varies depending on where your home is located. We can pay taxes and insurance in separate payments or include them as part of our house payment. Either way, we must determine if you can afford this extra cost. To determine the monthly cost, we divide the total yearly amount of each of these expenses by 12.

monthly tax payment = yearly property tax ÷ 12

monthly insurance payment = yearly premium ÷ 12

EXAMPLE 4 If Victor has a $136,000 home loan for 30 years at a fixed-rate of 7%, find his monthly house payment rounded to the nearest cent.

Solution We must find the payment factor on the table, and then multiply this amount times the amount of the home loan. To find the payment factor we locate the column labeled 7% and the row labeled 30 years, then we find where this column and row meet. The payment factor is 0.006653.

house payment = payment factor × home loan amount

= 0.006653 × 136,000

= 904.808 or $904.81 rounded to the nearest cent.

The monthly house payment is $904.81.

Practice Problem 4 If Alyssa has a $99,000 home loan for 25 years at a fixed-rate of 6%, find her monthly house payment rounded to the nearest cent.

EXAMPLE 5 The annual taxes and insurance for Victor's home are: taxes $2412; insurance $285. Find Victor's monthly tax and insurance payment.

Solution We divide each annual amount by 12 to get the monthly cost.

monthly tax payment = yearly property tax ÷ 12

= 2412 ÷ 12 = 201

Victor's monthly tax payment is $201.

monthly insurance payment = yearly premium ÷ 12

= 285 ÷ 12 = 23.75

Victor's monthly insurance payment is $23.75.

Practice Problem 5 The annual taxes and insurance for Alyssa's home are: taxes $1650; insurance $210. Find Alyssa's monthly tax and insurance payment.

EXAMPLE 6 Refer to Examples 4 and 5 to find Victor's total monthly payment for the mortgage, taxes, and insurance.

Solution We fill in the table with the correct amounts and then add to find the total monthly payment.

House payment	\$ 904.81
Monthly tax payment	\$ 201
Monthly insurance payment	\$ 23.75
Total monthly payment	\$1129.56

Victor's total monthly payment is \$1129.56.

NOTE TO STUDENT: Fully worked-out solutions to all of the Practice Problems can be found at the back of the text starting at page SP-1

Practice Problem 6 Refer to Practice Problems 4 and 5 to find Alyssa's total monthly payment for the mortgage, taxes and insurance.

House payment	\$
Monthly tax payment	\$
Monthly insurance payment	\$
Total monthly payment	\$

A.2 EXERCISES

Student Solutions Manual | CD/ Video | PH Math Tutor Center | MathXL®Tutorials on CD | MathXL® | MyMathLab® | Interactmath.com

Verbal and Writing Skills

Fill in the blank to complete each of the following formulas.

1. down payment = percent × purchase price of home

2. home loan amount = purchase price − down payment

3. loan fee = percent × home loan amount

4. escrow charges = $1\frac{1}{2}\%$ × purchase price of home

5. house payment = payment factor × home loan amount

6. monthly tax payment = yearly property tax ÷ 12

7. monthly insurance payment = yearly premium ÷ 12

8. Total initial buying costs = down payment + loan fee + escrow charges

Applications

Use the Payment Factor table on page A-10 as needed to do these exercises.

9. Sara May purchased a home for $250,000 and made a down payment of 20% of the purchase price. Find the amount of Sara May's down payment and home loan.
$50,000; $200,000

10. Jamel purchased a home for $200,000 and made a down payment of 15% of the purchase price. Find the amount of Jamel's down payment and home loan.
$30,000; $170,000

11. Stan paid a 3-point loan fee on his 144,000 home loan, and escrow charges on the 180,000 purchase price. Calculate the loan fee and escrow charges.
$4320; $2700

12. Kristina paid a 2-point loan fee on her $92,000 home loan, and escrow charges on the $115,000 purchase price. Calculate the loan fee and escrow charges.
$1840; $1725

13. If Lester has a $334,000 home loan for 30 years at a fixed-rate of 9%, find his monthly house payment rounded to the nearest cent.
$2687.36

14. If Amy has a $280,000 home loan for 25 years at a fixed-rate of 8%, find her monthly house payment rounded to the nearest cent.
$2161.04

15. The annual taxes and insurance for an office building are: taxes $6600; insurance $1920. Find monthly tax and insurance payments.
$550; $160

16. The annual taxes and insurance for a small warehouse are: taxes $4650; insurance $1410. Find monthly tax and insurance payments.
$387.50; $117.50

17. Find Mimi's initial buying costs on her home if her down payment is $25,000, loan fee is $2500 and escrow charges are $3750.
$31,250

18. Find Eric's initial buying costs on his vacation home if his down payment is $15,000, loan fee is $3000 and escrow charges are $2250.
$20,250

19. Find the total monthly payment for a home with the following monthly expenses: $1100 house payment, $250 property tax and $55 insurance payment.
$1405

20. Find the total monthly payment for the owner of a beach home with the following monthly expenses: $800 vacation home payment, $185 property tax and $45 insurance payment.
$1030

Mixed Practice

21. The Truman family purchased a mountain home for $320,000 with a 20% down payment and a loan with a 2-point fee. Find the total initial buying costs.

Down payment	$ 64,000
Loan fee	$ 5120
Escrow charges	$ 4800
Total initial costs	$ 73,920

22. An investor purchased a small office building for $550,000 with a 25% down payment and a loan with a 4-point fee. Find the total initial buying costs.

Down payment	$137,500
Loan fee	$ 16,500
Escrow charges	$ 8250
Total initial costs	$162,250

23. Jerry and Marian's monthly house payment is $985 and their *monthly* cost for taxes and insurance are: taxes $155; insurance $32. Find the total monthly payment for their home.
$1172

24. Ester and Jason's monthly house payment on their retirement home is $875 and their *monthly* cost for taxes and insurance are: taxes $95; insurance $29. Find the total monthly payment for their retirement home.
$999

25. The McMillan's have a mountain home with a view. Their home loan is $534,000 for a period of 25 years at a fixed-rate of 6%. Their *annual* taxes and insurance are: taxes $2100; insurance $1020. Find the total *monthly* payment. Round to the nearest cent.

House payment	$ 3440.56
Monthly tax payment	$ 175
Monthly insurance payment	$ 85
Total monthly payment	$ 3700.56

26. Sunhee Kim owns a beachfront home. Her home loan is $680,000 for a period of 30 years at a fixed-rate of 6%. Their *annual* taxes and insurance are: $3420; insurance $1590. Find the total *monthly* payment. Round to the nearest cent.

House payment	$ 4077.28
Monthly tax payment	$ 285
Monthly insurance payment	$ 132.50
Total monthly payment	$ 4494.78

27. The owner of Best Flavors Ice Creme owns the building where the store is located. The business loan on the building is $720,000 for a period of 20 years at a fixed-rate of 5%. The *annual* taxes and insurance are: taxes $4500; insurance $3600. Find the total *monthly* payment. Round to the nearest cent.

Building loan payment	$ 4751.28
Monthly tax payment	$ 375
Monthly insurance payment	$ 300
Total monthly payment	$ 5426.28

28. Huy Nguyen owns the building where his store, The Coffee House, is located. The business loan on the building is $690,000 for a period of 15 years at a fixed-rate of 4%. The *annual* taxes and insurance are: taxes $3420; insurance $1590. Find the total *monthly* payment. Round to the nearest cent.

Building loan payment	$ 5103.93
Monthly tax payment	$ 285
Monthly insurance payment	$ 132.50
Total monthly payment	$ 5521.43

A.3 Determining the Best Deal When Purchasing a Vehicle

When we buy a car there are several facts to consider in order to determine which car has the best price. The sale price offered by a car dealer or seller is only one factor we must consider—others include the interest rate on the loan and sale promotions such as: cash rebates, 0% interest, sales tax, and license fee. We must also consider the cost of options we choose such as extended warranties, sun roof, tinted glass, etc. In this section we will see how to calculate the total purchase price and determine the best deal when buying a vehicle.

Student Learning Objectives

After studying this section, you will be able to:

Find the true purchase price of a vehicle.

Find the total cost of a vehicle to determine the best deal.

1 Finding the True Purchase Price for a Vehicle

In most states there is a sales tax and a license or title fee that must be paid on all vehicles purchased. To find the true purchase price for a vehicle, we *add* these extra costs to the sale price. In addition, we must add to the sale price the cost of any extended warranties and extra options or accessories we buy.

$$\text{purchase price} = \text{sale price} + \text{sales tax} + \text{license fee} + \text{extended warranty (and other accessories)}$$

Sometimes lenders (banks and finance companies) require a **down payment.** The amount of the down payment is usually a percent of the purchase price. We subtract the down payment from the purchase price to find the amount we must finance.

$$\text{down payment} = \text{percent} \times \text{purchase price}$$

$$\text{amount financed} = \text{purchase price} - \text{down payment}$$

EXAMPLE 1 Daniel bought a truck that was on sale for \$28,999 in a city that has a 6% sales tax and a 2% license fee.

(a) Find the sales tax and license fee Daniel paid.

(b) Daniel also bought an extended warranty for \$1550. Find the purchase price of the truck.

Solution

(a) sales tax = 6% of sale price
= $0.06 \times 28{,}999$
sales tax = \$1739.94

license fee = 2% of sale price
= $0.02 \times 28{,}999$
license fee = \$579.98

(b) purchase price
= sales price + sales tax + license fee + extended warranty
= 28,999 + 1739.94 + 579.98 + 1550
purchase price = \$32,868.92

***NOTE TO STUDENT:** Fully worked-out solutions to all of the Practice Problems can be found at the back of the text starting at page SP-1*

Practice Problem 1 Huy Nguyen bought a van that was on sale for \$24,999 in a city that has a 7% sales tax and a 2% license fee.

(a) Find the sales tax and license fee Huy paid.

(b) Huy also bought an extended warranty for \$1275. Find the purchase price of the van.

EXAMPLE 2 The purchase price of a car Jerome plans to buy is \$19,999. In order to qualify for the loan on the car, Jerome must make a down payment of 20% of the purchase price.

(a) Find the down payment.

(b) Find the amount financed.

Solution

(a)
$$\begin{aligned} \text{down payment} &= \text{percent} \times \text{purchase price} \\ \text{down payment} &= 20\% \times 19{,}999 \\ &= 0.20 \times 19{,}999 \\ \text{down payment} &= \$3999.80 \end{aligned}$$

(b)
$$\begin{aligned} \text{amount financed} &= \text{purchase price} - \text{down payment} \\ &= 19{,}999 - 3999.80 \\ \text{amount financed} &= \$15{,}999.20 \end{aligned}$$

Practice Problem 2 The purchase price of a jeep Cheryl plans to buy is \$32,499. In order to qualify for the loan on the jeep, Cheryl must make a down payment of 15% of the purchase price. (a) Find the down payment. (b) Find the amount financed.

2 Finding the Total Cost of a Vehicle to Determine the Best Deal

When we borrow money to buy a car, we often pay interest on the loan and we must consider this extra cost when we calculate the **total cost** of the vehicle. If we know the amount of the car payment and the number of months it will take to pay off the loan, we can find the total cost of the car (the amount we borrowed plus interest) by multiplying the monthly payment amount times the number of months of the loan. Then we must add the down payment to that amount.

$$\text{Total Cost} = (\text{monthly payment} \times \text{number of months in loan}) + \text{down payment}$$

EXAMPLE 3 Marvin went to two dealerships to find the best deal on the truck he plans to purchase. From which dealership should Marvin buy the truck so that the *total cost* of the truck is the least expensive?

Dealership 1	Dealership 2
• Purchase price: \$39,999	• Purchase price: \$36,499
• Financing option: 0% financing with \$5000 down payment	• Financing option: 4% financing with no down payment
• Monthly payments: \$838.52 per month for 48 months	• Monthly payments: \$763.71 per month for 60 months

Solution First, we find the total cost of the truck at Dealership 1.

Total Cost = (monthly payment × number of months in loan) + down payment
= (838.52 × 48) + 5000 We multiply, then add.
= 45,248.96

The total cost of the truck at Dealership 1 is \$45,248.96.

Next, we find the total cost of the truck at Dealership 2.

Total cost = (monthly payment × number of months in loan) + down payment
= (763.71 × 60) + 0 There is no down payment.
= 45,822.60

The total cost of the truck at Dealership 2 is \$45,822.60.

We see that the best deal on the truck Marvin plans to buy is at Dealership 1.

Practice Problem 3 Phoebe went to two dealerships to find the best deal on the mini van she plans to purchase. From which dealership should Phoebe buy the mini van so that the *total cost* of the mini van is the least expensive?

Dealership 1	Dealership 2
• Purchase price: \$22,999	• Purchase price: \$24,299
• Financing option: 4% financing with no down payment	• Financing option: 0% financing with \$3000 down payment
• Monthly payments: \$453.21 per month for 60 months	• Monthly payments: \$479.17 per month for 48 months

Student Solutions Manual

CD/ Video

PH Math Tutor Center

MathXL®Tutorials on CD
MathXL®

MyMathLab®

Interactmath.com

1. A college student buys a car and pays a 6% sales tax on the $21,599 sale price. How much sales tax did the student pay?
$1259.94

2. A high school teacher buys a mini van and pays a 5% sales tax on the $26,800 sale price. How much sales tax did the teacher pay?
$1340

3. Francis is planning to buy a 4-door sedan that is on sale for $18,999. She must pay a 2% license fee. Find the license fee.
$379.98

4. Mai Vu saw an ad for a short bed truck that is on sale for $17,599. If she buys the truck she must pay a 2% license fee. Find the license fee.
$351.98

5. John must make a 10% down payment on the purchase price of the $42,450 sports car he is planning to buy. Find the down payment.
$4245

6. Kamir must make a 15% down payment on the purchase price of the $31,500 extended cab truck he is planning to buy. Find the down payment. $4725

7. Tabatha bought a mini van that was on sale for $24,899 in a city that has a 5% sales tax and a 2% license fee.

(a) Find the sales tax and license fee Tabatha paid. $1244.95; $497.98

(b) Tabatha also bought an extended warranty for $1100. Find the purchase price of the mini van. $27,741.93

8. Dante bought a truck that was on sale for $32,499 in a city that has a 7% sales tax and a 2% license fee.

(a) Find the sales tax and license fee Dante paid. $2274.93; $649.98

(b) Dante also bought an extended warranty for $1600. Find the purchase price of the truck. $37,023.91

9. Jeremiah bought a sports car that was on sale for $44,799 in a city that has a 7% sales tax and a 2% license fee.

(a) Find the sales tax and license fee Jeremiah paid. $3135.93; $895.98

(b) Jeremiah also bought an extended warranty for $2100. Find the purchase price of the sports car. $50,930.91

10. Dawn bought a 4-door sedan that was on sale for $31,899 in a city that has a 6% sales tax and a 2% license fee.

(a) Find the sales tax and license fee Dawn paid. $1913.94; $637.98

(b) Dawn also bought an extended warranty for $1300. Find the purchase price of the sedan. $35,750.92

11. The purchase price of an SUV that a soccer coach plans to buy is $49,999. In order to qualify for the loan on the SUV, the coach must make a down payment of 15% of the purchase price.

(a) Find the down payment. $7499.85

(b) Find the amount financed. $42,499.15

12. The purchase price of a flat bed truck a contractor plans to buy is $39,999. In order to qualify for the loan on the truck, the contractor must make a down payment of 10% of the purchase price.

(a) Find the down payment. $3999.90

(b) Find the amount financed. $35,999.10

13. Tammy went to two dealerships to find the best deal on the truck she plans to purchase. From which dealership should Tammy buy the truck so that the *total cost* of the truck is the least expensive?

Dealership 1	Dealership 2
• Purchase price: $35,999	• Purchase price: $32,499
• Financing option: 0% financing with $4000 down payment	• Financing option: 4% financing with $2000 down payment
• Monthly payments: $696.65 per month for 48 months	• Monthly payments: $603.58 per month for 60 months

Dealership 1

14. John went to two dealerships to find the best deal on a luxury SUV for his company to purchase. From which dealership should John buy the SUV so that the *total cost* of the SUV is the least expensive?

Dealership 1	Dealership 2
• Purchase price: $49,999	• Purchase price: $46,499
• Financing option: 0% financing with $5000 down payment	• Financing option: 4% financing with no down payment
• Monthly payments: $1010.39 per month for 48 months	• Monthly payments: $916.29 per month for 60 months

Dealership 1

To Think About

Natasha went to three car dealerships to check the prices of the same Ford 2-Door Coupe. The city where the dealerships are located has a sales tax of 5% and a license fee of 2%. All dealerships offer extended warranties that are 3 years/70,000 miles. Use the following information gathered by Natasha to answer Exercises 15 and 16.

Dealership 1—Ford Coupe	Dealership 2—Ford Coupe	Dealership 3—Ford Coupe
• $24,999 plus $2000 rebate	• $23,799; dealer pays sales tax	• $23,999
• extended warranty $1350	• extended warranty $1450	• Free extended warranty

15. **(a)** Which dealership offers the least expensive *purchase price*? State this amount. Dealership 3; $25,678.93

(b) Each dealership offers a *different interest rate* on a 60-month loan without a down payment, resulting in the following monthly payments:

Dealership 1	Dealership 2	Dealership 3
$480.65/month	$485.46/month	$496.44/month

From which dealership should Natasha buy the Ford Coupe so that the *total cost* of the car is the least expensive? State this amount. Dealership 1; $28,839

(c) Compare the results of parts (a) and (b). What conclusion can you make?
The least expensive purchase price does not guarantee the least expensive total cost. Many factors need to be considered to determine the best deal.

16. **(a)** Which dealership offers the most expensive *purchase price*? State this amount. Dealership 1; $26,098.93

(b) Each dealership offers a *different interest rate* on a 48-month loan without a down payment, resulting in the following monthly payments:

Dealership 1	Dealership 2	Dealership 3
$589.27/month	$594.07/month	$603.07/month

From which dealership should Natasha buy the Ford Coupe so that the *total cost* of the car is the most expensive? State this amount. Dealership 3; $28,947.36

(c) Compare the results of parts (a) and (b). What conclusion can you make?
The most expensive purchase price does not guarantee the most expensive total cost. Many factors need to be considered to determine the best deal.

Appendix B Tables

Table of Basic Addition Facts

+	0	1	2	3	4	5	6	7	8	9
0	0	1	2	3	4	5	6	7	8	9
1	1	2	3	4	5	6	7	8	9	10
2	2	3	4	5	6	7	8	9	10	11
3	3	4	5	6	7	8	9	10	11	12
4	4	5	6	7	8	9	10	11	12	13
5	5	6	7	8	9	10	11	12	13	14
6	6	7	8	9	10	11	12	13	14	15
7	7	8	9	10	11	12	13	14	15	16
8	8	9	10	11	12	13	14	15	16	17
9	9	10	11	12	13	14	15	16	17	18

Table of Basic Multiplication Facts

×	0	1	2	3	4	5	6	7	8	9	10	11	12
0	0	0	0	0	0	0	0	0	0	0	0	0	0
1	0	1	2	3	4	5	6	7	8	9	10	11	12
2	0	2	4	6	8	10	12	14	16	18	20	22	24
3	0	3	6	9	12	15	18	21	24	27	30	33	36
4	0	4	8	12	16	20	24	28	32	36	40	44	48
5	0	5	10	15	20	25	30	35	40	45	50	55	60
6	0	6	12	18	24	30	36	42	48	54	60	66	72
7	0	7	14	21	28	35	42	49	56	63	70	77	84
8	0	8	16	24	32	40	48	56	64	72	80	88	96
9	0	9	18	27	36	45	54	63	72	81	90	99	108
10	0	10	20	30	40	50	60	70	80	90	100	110	120
11	0	11	22	33	44	55	66	77	88	99	110	121	132
12	0	12	24	36	48	60	72	84	96	108	120	132	144

Table of Prime Factors

Number	Prime Factors	Number	Prime Factors	Number	Prime Factors	Number	Prime Factors
2	prime	**52**	$2^2 \times 13$	**102**	$2 \times 3 \times 17$	**152**	$2^3 \times 19$
3	prime	**53**	prime	**103**	prime	**153**	$3^2 \times 17$
4	2^2	**54**	2×3^3	**104**	$2^3 \times 13$	**154**	$2 \times 7 \times 11$
5	prime	**55**	5×11	**105**	$3 \times 5 \times 7$	**155**	5×31
6	2×3	**56**	$2^3 \times 7$	**106**	2×53	**156**	$2^2 \times 3 \times 13$
7	prime	**57**	3×19	**107**	prime	**157**	prime
8	2^3	**58**	2×29	**108**	$2^2 \times 3^3$	**158**	2×79
9	3^2	**59**	prime	**109**	prime	**159**	3×53
10	2×5	**60**	$2^2 \times 3 \times 5$	**110**	$2 \times 5 \times 11$	**160**	$2^5 \times 5$
11	prime	**61**	prime	**111**	3×37	**161**	7×23
12	$2^2 \times 3$	**62**	2×31	**112**	$2^4 \times 7$	**162**	2×3^4
13	prime	**63**	$3^2 \times 7$	**113**	prime	**163**	prime
14	2×7	**64**	2^6	**114**	$2 \times 3 \times 19$	**164**	$2^2 \times 41$
15	3×5	**65**	5×13	**115**	5×23	**165**	$3 \times 5 \times 11$
16	2^4	**66**	$2 \times 3 \times 11$	**116**	$2^2 \times 29$	**166**	2×83
17	prime	**67**	prime	**117**	$3^2 \times 13$	**167**	prime
18	2×3^2	**68**	$2^2 \times 17$	**118**	2×59	**168**	$2^3 \times 3 \times 7$
19	prime	**69**	3×23	**119**	7×17	**169**	13^2
20	$2^2 \times 5$	**70**	$2 \times 5 \times 7$	**120**	$2^3 \times 3 \times 5$	**170**	$2 \times 5 \times 17$
21	3×7	**71**	prime	**121**	11^2	**171**	$3^2 \times 19$
22	2×11	**72**	$2^3 \times 3^2$	**122**	2×61	**172**	$2^2 \times 43$
23	prime	**73**	prime	**123**	3×41	**173**	prime
24	$2^3 \times 3$	**74**	2×37	**124**	$2^2 \times 31$	**174**	$2 \times 3 \times 29$
25	5^2	**75**	3×5^2	**125**	5^3	**175**	$5^2 \times 7$
26	2×13	**76**	$2^2 \times 19$	**126**	$2 \times 3^2 \times 7$	**176**	$2^4 \times 11$
27	3^3	**77**	7×11	**127**	prime	**177**	3×59
28	$2^2 \times 7$	**78**	$2 \times 3 \times 13$	**128**	2^7	**178**	2×89
29	prime	**79**	prime	**129**	3×43	**179**	prime
30	$2 \times 3 \times 5$	**80**	$2^4 \times 5$	**130**	$2 \times 5 \times 13$	**180**	$2^2 \times 3^2 \times 5$
31	prime	**81**	3^4	**131**	prime	**181**	prime
32	2^5	**82**	2×41	**132**	$2^2 \times 3 \times 11$	**182**	$2 \times 7 \times 13$
33	3×11	**83**	prime	**133**	7×19	**183**	3×61
34	2×17	**84**	$2^2 \times 3 \times 7$	**134**	2×67	**184**	$2^3 \times 23$
35	5×7	**85**	5×17	**135**	$3^3 \times 5$	**185**	5×37
36	$2^2 \times 3^2$	**86**	2×43	**136**	$2^3 \times 17$	**186**	$2 \times 3 \times 31$
37	prime	**87**	3×29	**137**	prime	**187**	11×17
38	2×19	**88**	$2^3 \times 11$	**138**	$2 \times 3 \times 23$	**188**	$2^2 \times 47$
39	3×13	**89**	prime	**139**	prime	**189**	$3^3 \times 7$
40	$2^3 \times 5$	**90**	$2 \times 3^2 \times 5$	**140**	$2^2 \times 5 \times 7$	**190**	$2 \times 5 \times 19$
41	prime	**91**	7×13	**141**	3×47	**191**	prime
42	$2 \times 3 \times 7$	**92**	$2^2 \times 23$	**142**	2×71	**192**	$2^6 \times 3$
43	prime	**93**	3×31	**143**	11×13	**193**	prime
44	$2^2 \times 11$	**94**	2×47	**144**	$2^4 \times 3^2$	**194**	2×97
45	$3^2 \times 5$	**95**	5×19	**145**	5×29	**195**	$3 \times 5 \times 13$
46	2×23	**96**	$2^5 \times 3$	**146**	2×73	**196**	$2^2 \times 7^2$
47	prime	**97**	prime	**147**	3×7^2	**197**	prime
48	$2^4 \times 3$	**98**	2×7^2	**148**	$2^2 \times 37$	**198**	$2 \times 3^2 \times 11$
49	7^2	**99**	$3^2 \times 11$	**149**	prime	**199**	prime
50	2×5^2	**100**	$2^2 \times 5^2$	**150**	$2 \times 3 \times 5^2$	**200**	$2^3 \times 5^2$
51	3×17	**101**	prime	**151**	prime		

Table of Square Roots

Square Root Values Are Rounded to the Nearest Thousandth Unless the Answer Ends in .000

n	$\sqrt{n}$	n	$\sqrt{n}$	n	$\sqrt{n}$	n	$\sqrt{n}$	n	$\sqrt{n}$
1	1.000	**41**	6.403	**81**	9.000	**121**	11.000	**161**	12.689
2	1.414	**42**	6.481	**82**	9.055	**122**	11.045	**162**	12.728
3	1.732	**43**	6.557	**83**	9.110	**123**	11.091	**163**	12.767
4	2.000	**44**	6.633	**84**	9.165	**124**	11.136	**164**	12.806
5	2.236	**45**	6.708	**85**	9.220	**125**	11.180	**165**	12.845
6	2.449	**46**	6.782	**86**	9.274	**126**	11.225	**166**	12.884
7	2.646	**47**	6.856	**87**	9.327	**127**	11.269	**167**	12.923
8	2.828	**48**	6.928	**88**	9.381	**128**	11.314	**168**	12.961
9	3.000	**49**	7.000	**89**	9.434	**129**	11.358	**169**	13.000
10	3.162	**50**	7.071	**90**	9.487	**130**	11.402	**170**	13.038
11	3.317	**51**	7.141	**91**	9.539	**131**	11.446	**171**	13.077
12	3.464	**52**	7.211	**92**	9.592	**132**	11.489	**172**	13.115
13	3.606	**53**	7.280	**93**	9.644	**133**	11.533	**173**	13.153
14	3.742	**54**	7.348	**94**	9.695	**134**	11.576	**174**	13.191
15	3.873	**55**	7.416	**95**	9.747	**135**	11.619	**175**	13.229
16	4.000	**56**	7.483	**96**	9.798	**136**	11.662	**176**	13.266
17	4.123	**57**	7.550	**97**	9.849	**137**	11.705	**177**	13.304
18	4.243	**58**	7.616	**98**	9.899	**138**	11.747	**178**	13.342
19	4.359	**59**	7.681	**99**	9.950	**139**	11.790	**179**	13.379
20	4.472	**60**	7.746	**100**	10.000	**140**	11.832	**180**	13.416
21	4.583	**61**	7.810	**101**	10.050	**141**	11.874	**181**	13.454
22	4.690	**62**	7.874	**102**	10.100	**142**	11.916	**182**	13.491
23	4.796	**63**	7.937	**103**	10.149	**143**	11.958	**183**	13.528
24	4.899	**64**	8.000	**104**	10.198	**144**	12.000	**184**	13.565
25	5.000	**65**	8.062	**105**	10.247	**145**	12.042	**185**	13.601
26	5.099	**66**	8.124	**106**	10.296	**146**	12.083	**186**	13.638
27	5.196	**67**	8.185	**107**	10.344	**147**	12.124	**187**	13.675
28	5.292	**68**	8.246	**108**	10.392	**148**	12.166	**188**	13.711
29	5.385	**69**	8.307	**109**	10.440	**149**	12.207	**189**	13.748
30	5.477	**70**	8.367	**110**	10.488	**150**	12.247	**190**	13.784
31	5.568	**71**	8.426	**111**	10.536	**151**	12.288	**191**	13.820
32	5.657	**72**	8.485	**112**	10.583	**152**	12.329	**192**	13.856
33	5.745	**73**	8.544	**113**	10.630	**153**	12.369	**193**	13.892
34	5.831	**74**	8.602	**114**	10.677	**154**	12.410	**194**	13.928
35	5.916	**75**	8.660	**115**	10.724	**155**	12.450	**195**	13.964
36	6.000	**76**	8.718	**116**	10.770	**156**	12.490	**196**	14.000
37	6.083	**77**	8.775	**117**	10.817	**157**	12.530	**197**	14.036
38	6.164	**78**	8.832	**118**	10.863	**158**	12.570	**198**	14.071
39	6.245	**79**	8.888	**119**	10.909	**159**	12.610	**199**	14.107
40	6.325	**80**	8.944	**120**	10.954	**160**	12.649	**200**	14.142

Appendix C Scientific Calculators

This text *does not require* the use of a calculator. However, you may want to consider the purchase of an inexpensive scientific calculator. It is wise to ask your instructor for advice before you purchase any calculator for this course. It should be stressed that students are asked to avoid using a calculator for any of the exercises in which the calculations can be readily done by hand. The only problems in the text that really demand the use of a scientific calculator are marked with the symbol. Dependence on the use of the scientific calculator for regular exercises in the text will only hurt the student in the long run.

The Two Types of Logic Used in Scientific Calculators

Two major types of scientific calculators are popular today. The most common type employs a type of logic known as **algebraic** logic. The calculators manufactured by Casio, Sharp, and Texas Instruments as well as many other companies employ this type of logic. An example of calculation on such a calculator would be the following. To add 14 + 26 on an algebraic logic calculator, the sequence of buttons would be:

14 [+] 26 [=]

The second type of scientific calculator requires the entry of data in **Reverse Polish Notation (RPN).** Calculators manufactured by Hewlett-Packard and a few other specialized calculators use RPN. To add 14 + 26 on an RPN calculator, the sequence of buttons would be:

14 [enter] 26 [+]

Graphing scientific calculators such as the TI-82 and TI-83 have a large liquid display for viewing graphs. To perform the calculation on most graphing calculators, the sequence of buttons would be:

14 [+] 26 [enter]

Mathematicians and scientists do not agree on which type of scientific calculator is superior. However, the clear majority of college students own calculators that employ *algebraic* logic. Therefore this section of the text is explained with reference to the sequence of steps employed by an *algebraic* logic calculator. If you already own or intend to purchase a scientific calculator that uses RPN or a graphing calculator, you are encouraged to study the instruction booklet that comes with the calculator and practice the problems shown in the booklet. After this practice you will be able to solve the calculator problems discussed in this section.

Performing Simple Calculations

The following example will illustrate the use of the scientific calculator in doing basic arithmetic calculations.

NOTE TO STUDENT: Fully worked-out solutions to all of the Practice Problems can be found at the back of the text starting at page SP-1

EXAMPLE 1 Add. $156 + 298$

Solution We first enter the number 156, then press the [+] key, then enter the number 298, and finally press the [=] key.

156 [+] 298 [=] 454

Practice Problem 1 Add. $3792 + 5896$

EXAMPLE 2 Subtract. $1508 - 963$

Solution We first enter the number 1508, then press the [−] key, then enter the number 963, and finally press the [=] key.

1508 [−] 963 [=] 545

Practice Problem 2 Subtract. $7930 - 5096$

EXAMPLE 3 Multiply. 196×358

Solution 196 [×] 358 [=] 70168

Practice Problem 3 Multiply. 896×273

EXAMPLE 4 Divide. $2054 \div 13$

Solution 2054 [÷] 13 [=] 158

Practice Problem 4 Divide. $2352 \div 16$

Decimal Problems

Problems involving decimals can be readily done on the calculator. Entering numbers with a decimal point is done by pressing the decimal point key, the [·] key, at the appropriate time.

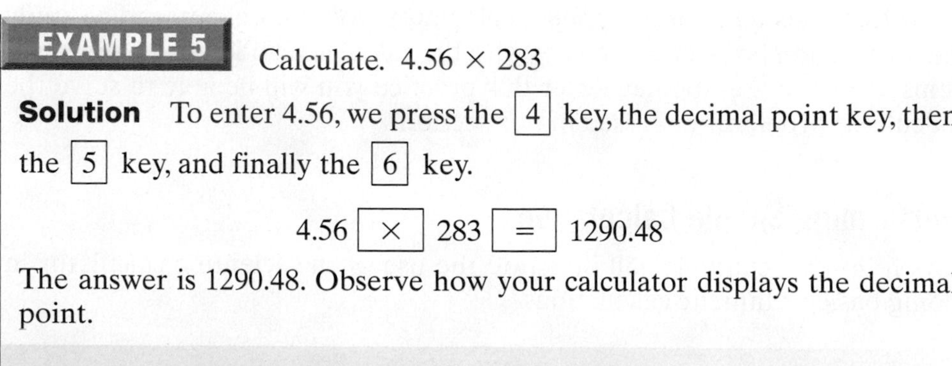

EXAMPLE 5 Calculate. 4.56×283

Solution To enter 4.56, we press the [4] key, the decimal point key, then the [5] key, and finally the [6] key.

4.56 [×] 283 [=] 1290.48

The answer is 1290.48. Observe how your calculator displays the decimal point.

Practice Problem 5 Calculate. 72.8×197

EXAMPLE 6 Add. 128.6 + 343.7 + 103.4 + 207.5

Solution 128.6 [+] 343.7 [+] 103.4 [+] 207.5 [=] 783.2

The answer is 783.2. Observe how your calculator displays the answer.

Practice Problem 6 Add. 52.98 + 31.74 + 40.37 + 99.82

Combined Operations

You must use extra caution concerning the order of mathematical operations when you are doing a problem on the calculator that involves two or more different operations.

Any scientific calculator with algebraic logic uses a priority system that has a clearly defined order of operations. It is the same order we use in performing arithmetic operations by hand. In either situation, calculations are performed in the following order:

1. First calculations within parentheses are completed.
2. Then numbers are raised to a power or a square root is calculated.
3. Then multiplication and division operations are performed from left to right.
4. Then addition and subtraction operations are performed from left to right.

This order is carefully followed on *scientific calculators* and *graphing calculators*. Small inexpensive calculators that do not have scientific functions often do not follow this order of operations.

The number of digits displayed in the answer varies from calculator to calculator. In the following examples, your calculator may display more or fewer digits than the answer we have listed.

EXAMPLE 7 Evaluate. $5.3 \times 1.62 + 1.78 \div 3.51$

Solution This problem requires that we multiply 5.3 by 1.62 and divide 1.78 by 3.51 first and then add the two results. If the numbers are entered directly into the calculator exactly as the problem is written, the calculator will perform the calculations in the correct order.

5.3 [×] 1.62 [+] 1.78 [÷] 3.51 [=] 9.09312251

Practice Problem 7 Evaluate. $0.0618 \times 19.22 - 59.38 \div 166.3$

The Use of Parentheses

In order to perform some calculations on the calculator the use of parentheses is helpful. These parentheses may or may not appear in the original problem.

EXAMPLE 8 Evaluate. $5 \times (2.123 + 5.786 - 12.063)$

Solution The problem requires that the numbers in the parentheses be combined first. By entering the parentheses on the calculator this will be accomplished.

5 [×] [(] 2.123 [+] 5.786 [−] 12.063 [)] [=] −20.77

Note: The result is a negative number.

***NOTE TO STUDENT:** Fully worked-out solutions to all of the Practice Problems can be found at the back of the text starting at page SP-1*

Practice Problem 8 Evaluate. $3.152 \times (0.1628 + 3.715 - 4.985)$

Negative Numbers

To enter a negative number, enter the number followed by the [+/−] button.

EXAMPLE 9 Evaluate. $(-8.634)(5.821) + (1.634)(-16.082)$

Solution The products will be evaluated first by the calculator. Therefore, parentheses are not needed as we enter the data.

8.634 [+/−] [×] 5.821 [+] 1.634 [×] 16.082 [+/−] [=] −76.536502

Note: The result is negative.

Practice Problem 9 Evaluate. $(0.5618)(-98.3) - (76.31)(-2.98)$

Scientific Notation

If you wish to enter a number in scientific notation, you should use the special scientific notation button. On most calculators it is denoted as [EXP] or [EE].

EXAMPLE 10 Multiply. $(9.32 \times 10^6)(3.52 \times 10^8)$

Solution 9.32 [EXP] 6 [×] 3.52 [EXP] 8 [=] 3.28064 15

This notation means the answer is 3.28064×10^{15}.

Practice Problem 10 Divide. $(3.76 \times 10^{15}) \div (7.76 \times 10^7)$

Raising a Number to a Power

All scientific calculators have a key for finding powers of numbers. It is usually labeled [y^x]. (On a few calculators the notation is [x^y] or sometimes [^].) To raise a number to a power on most scientific calculators, first you enter the base, then push the [y^x] key. Then you enter the exponent, then finally the [=] button.

Evaluate. $(2.16)^9$

Solution 2.16 $\boxed{y^x}$ 9 $\boxed{=}$ 1023.490369

Practice Problem 11 Evaluate. $(6.238)^6$

There is a special key to square a number. It is usually labeled $\boxed{x^2}$.

EXAMPLE 12 Evaluate. $(76.04)^2$

Solution 76.04 $\boxed{x^2}$ 5782.0816

Practice Problem 12 Evaluate. $(132.56)^2$

Finding Square Roots of Numbers

There is usually a key to approximate square roots on a scientific calculator labeled $\boxed{\sqrt{\ }}$. In this example we will need to use parentheses.

EXAMPLE 13 Evaluate. $\sqrt{5618 + 2734 + 3913}$

Solution $\boxed{(}$ 5618 $\boxed{+}$ 2734 $\boxed{+}$ 3913 $\boxed{)}$ $\boxed{\sqrt{\ }}$ 110.7474605

Practice Problem 13 Evaluate. $\sqrt{0.0782 - 0.0132 + 0.1364}$

On some calculators, you enter the square root key first and then enter the number. You will need to try this on your own calculator.

APPENDIX C EXERCISES

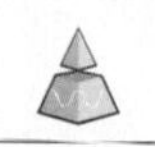

Student Solutions Manual | CD/Video | PH Math Tutor Center | MathXL®Tutorials on CD | MathXL® | MyMathLab® | Interactmath.com

Use your calculator to complete each of the following. Your answers may vary slightly because of the characteristics of individual calculators.

Complete the table.

To Do This Operation	Use These Keystrokes	Record Your Answer Here
1. 8963 + 2784	8963 [+] 2784 [=]	11,747
2. 15,308 − 7980	15308 [−] 7980 [=]	7328
3. 2631 × 134	2631 [×] 134 [=]	352,554
4. 70,221 ÷ 89	70221 [÷] 89 [=]	789
5. 5.325 − 4.031	5.325 [−] 4.031 [=]	1.294
6. 184.68 + 73.98	184.68 [+] 73.98 [=]	258.66
7. 2004.06 ÷ 7.89	2004.06 [÷] 7.89 [=]	254
8. 1.34 × 0.763	1.34 [×] 0.763 [=]	1.02242

Write down the answer and then show what problem you have solved.

9. 123.45 [+] 45.9876 [+] 8765.3 [=] 8934.7376
123.45 + 45.9876 + 8765.3

10. 0.0897 [×] 234.56 [×] 2.5428 [=] 53.50059337
0.0897 × 234.56 × 2.5428

11. 34 [÷] 8 [+] 12.56 [=] 16.81
$\frac{34}{8} + 12.56$

12. 458 [÷] 4 [−] 16.897 [=] 97.603
$\frac{458}{4} - 16.897$

Perform each calculation using your calculator.

13. 9.467 + 0.563 10.03

14. 0.347 + 23.457 23.804

15. 34.89 + 39.6 + 214.897 289.387

16. 12.567 + 48.31 + 189.38 250.257

17. 412,899 − 34,675 378,224

18. 87,456 − 2876 84,580

19. 3,567,089 − 2,876,805 690,284

20. 8,345,802 − 4,985,004 3,360,798

21. 234 × 4.567 1068.678

22. 1.9876 × 347 689.6972

23. 0.456 × 3.48 1.58688

24. 67,876 × 0.0946 6421.0696

25. 3458 ÷ 2.5 1383.2

26. 9764 ÷ 8 1220.5

27. 12.107524 ÷ 15.86 0.7634

28. 16.06513 ÷ 17.98 0.8935

Perform each calculation using your calculator.

29. $\begin{array}{r} 1.98 \\ 6.34 \\ +\ 7.71 \\ \hline \end{array}$ 16.03

30. $\begin{array}{r} 8.92 \\ 9.31 \\ +\ 7.79 \\ \hline \end{array}$ 26.02

31. $\begin{array}{r} \$103.91 \\ \$2653.82 \\ +\ \$9804.61 \\ \hline \end{array}$ \$12,562.34

32. $\begin{array}{r} \$3986.21 \\ \$4502.89 \\ +\ \$989.30 \\ \hline \end{array}$ \$9478.40

33. $\begin{array}{r} 368{,}781.5 \\ -\ 283{,}617.8 \\ \hline \end{array}$ 85,163.7

34. $\begin{array}{r} 571{,}809.6 \\ -\ 539{,}376.8 \\ \hline \end{array}$ 32,432.8

35. $\begin{array}{r} \$1{,}393{,}271.86 \\ -\ \$1{,}289{,}663.21 \\ \hline \end{array}$ \$103,608.65

36. $\begin{array}{r} \$8{,}571{,}300.76 \\ -\ \$4{,}098{,}789.39 \\ \hline \end{array}$ \$4,472,511.37

37. $\begin{array}{r} 345.34 \\ \times\ 45.7 \\ \hline \end{array}$ 15,782.038

38. $\begin{array}{r} 8954.34 \\ \times\ 425.4 \\ \hline \end{array}$ 3,809,176.236

39. $\begin{array}{r} 0.6314 \\ \times\ 3.96 \\ \hline \end{array}$ 2.500344

40. $\begin{array}{r} 0.0789 \\ \times\ 12.38 \\ \hline \end{array}$ 0.976782

41. $40.36\overline{)36202.92}$ 897

42. $52.98\overline{)172{,}608.84}$ 3258

43. $0.7613\overline{)17.12925}$ 22.5

44. $0.9854\overline{)3.59671}$ 3.65

Perform the following operations in the proper order using your calculator.

45. $4.567 + 87.89 - 2.45 \times 3.3$ 84.372

46. $4.891 + 234.5 - 0.98 \times 23.4$ 216.459

47. $7 \div 8 + 3.56$ 4.435

48. $9 \div 4.5 + 0.6754$ 2.6754

49. $(9.34)(0.345) + 98.345$ 101.5673

50. $(0.628)(398) + 34.4581$ 284.4021

51. $\dfrac{(95.34)(0.9874)}{381.36}$ 0.24685

52. $\dfrac{(0.8759)(45.87)}{183.48}$ 0.218975

53. $2.56 + 8.98 \times 3.14$ 30.7572

54. $1.62 + 3.81 - 5.23 \times 6.18$ −26.8914

55. $(-4.23)(1.863) - 5.998$ −13.87849

56. $12.34 - (26.314)(-1.856)$ 61.178784

57. $5.62(5 \times 3.16 - 18.12)$ −13.0384

58. $9.356(4.8 - 7.2 - 15.94)$ −171.58904

59. $(3.42 \times 10^8)(0.97 \times 10^{10})$ 3.3174×10^{18}

60. $(6.27 \times 10^{20})(1.35 \times 10^3)$ 8.4645×10^{23}

61. $\dfrac{(2.16 \times 10^3)(1.37 \times 10^{14})}{6.39 \times 10^5}$ $4.630985915 \times 10^{11}$

62. $\dfrac{(3.84 \times 10^{12})(1.62 \times 10^5)}{7.78 \times 10^8}$ 7.9958869×10^8

63. $\dfrac{2.3 + 5.8 - 2.6 - 3.9}{5.3 - 8.2}$ -0.5517241379

64. $\dfrac{(2.6)(-3.2) + (5.8)(-0.9)}{2.614 + 5.832}$ -1.60312574

65. $\sqrt{253.12}$ 15.90974544

66. $\sqrt{0.0713}$ 0.2670205985

67. $\sqrt{5.6213 - 3.7214}$ 1.378368601

68. $\sqrt{3417.2 - 2216.3}$ 34.6540041

69. $(1.78)^3 + 6.342$ 11.981752

70. $(2.26)^8 - 3.1413$ 677.4204134

71. $\sqrt{(6.13)^2 + (5.28)^2}$ 8.090444982

72. $\sqrt{(0.3614)^2 + (0.9217)^2}$ 0.9900206311

73. $\sqrt{56 + 83} - \sqrt{12}$ 8.325724507

74. $\sqrt{98 + 33} - \sqrt{17}$ 7.322417517

Find an approximate value. Round to five decimal places.

75. $\dfrac{7}{18} + \dfrac{9}{13}$ 1.08120

76. $\dfrac{5}{22} + \dfrac{1}{31}$ 0.25953

77. $\dfrac{7}{8} + \dfrac{3}{11}$ 1.14773

78. $\dfrac{9}{14} + \dfrac{5}{19}$ 0.90602

Solutions to Practice Problems

Chapter 1

1.1 Practice Problems

1. **(a)** 3182 = 3000 + 100 + 80 + 2
   (b) 520,890 = 500,000 + 20,000 + 800 + 90
   (c) 709,680,059 = 700,000,000 + 9,000,000 + 600,000 + 80,000 + 50 + 9
2. **(a)** 492 **(b)** 80,427
3. **(a)** 7 **(b)** 9 **(c)** 4,000
   (d) 900,000 for the first 9; 9 for the last 9
4. two hundred sixty-seven million, three hundred fifty-eight thousand, nine hundred eighty-one
5. **(a)** two thousand, seven hundred thirty-six
   (b) nine hundred eighty thousand, three hundred six
   (c) twelve million, twenty-one
6. The world population on January 1, 2004 was six billion, three hundred ninety-three million, six hundred forty-six thousand, five hundred twenty-five.
7. **(a)** 803 **(b)** 30,229
8. **(a)** 13,000 **(b)** 88,000 **(c)** 10,000

1.2 Practice Problems

1. **(a)** $\begin{array}{r} 7 \\ +5 \\ \hline 12 \end{array}$ **(b)** $\begin{array}{r} 9 \\ +4 \\ \hline 13 \end{array}$ **(c)** $\begin{array}{r} 3 \\ +0 \\ \hline 3 \end{array}$

2. $\begin{array}{r} 7 \\ 6 \\ 5 \\ 8 \\ +2 \\ \hline 28 \end{array}$ (7; 6 + 5 = 11; 8 + 2 = 10)

3. $\begin{array}{r} 1 \\ 7 \\ 2 \\ 9 \\ +3 \\ \hline 22 \end{array}$ (1 + 9 = 10; 7 + 3 = 10)

4. $\begin{array}{r} 8246 \\ +1702 \\ \hline 9948 \end{array}$

5. $\begin{array}{r} {}^{1} \\ 56 \\ +36 \\ \hline 92 \end{array}$

6. $\begin{array}{r} {}^{21} \\ 789 \\ 63 \\ +297 \\ \hline 1149 \end{array}$

7. **(a)** $\begin{array}{r} {}^{111} \\ 127 \\ 9876 \\ +\ 342 \\ \hline 10{,}345 \end{array}$ **(b)** Check by adding in opposite order. $\begin{array}{r} {}^{111} \\ 342 \\ 9876 \\ +\ 127 \\ \hline 10{,}345 \end{array}$ same

8. $\begin{array}{r} {}^{1\,2\,1\,2} \\ 18{,}316 \\ 24{,}789 \\ +22{,}965 \\ \hline 66{,}070 \end{array}$ total women

9. $\begin{array}{r} 1000 \\ 2000 \\ 1000 \\ +2000 \\ \hline 6000 \end{array}$ ft

1.3 Practice Problems

1. **(a)** $\begin{array}{r} 9 \\ -6 \\ \hline 3 \end{array}$ **(b)** $\begin{array}{r} 12 \\ -5 \\ \hline 7 \end{array}$ **(c)** $\begin{array}{r} 17 \\ -8 \\ \hline 9 \end{array}$ **(d)** $\begin{array}{r} 14 \\ -0 \\ \hline 14 \end{array}$ **(e)** $\begin{array}{r} 18 \\ -9 \\ \hline 9 \end{array}$

2. $\begin{array}{r} 7695 \\ -3481 \\ \hline 4214 \end{array}$

3. $\begin{array}{r} {}^{2\ 14} \\ \not{3}\ \not{4} \\ -1\ 6 \\ \hline 1\ 8 \end{array}$

4. $\begin{array}{r} {}^{8\ 13} \\ 6\ \not{9}\ \not{3} \\ -4\ 2\ 6 \\ \hline 2\ 6\ 7 \end{array}$

5. $\begin{array}{r} {}^{16} \\ {}^{8\ 9\ \not{6}\ 10} \\ \not{9}\ \not{0}\ \not{7}\ \not{0} \\ -5\ 8\ 8\ 6 \\ \hline 3\ 1\ 8\ 4 \end{array}$

6. **(a)** $\begin{array}{r} 8964 \\ -\ 985 \\ \hline 7979 \end{array}$ **(b)** $\begin{array}{r} 50{,}000 \\ -32{,}508 \\ \hline 17{,}492 \end{array}$

7. **(a)** Subtraction $\begin{array}{r} 9763 \\ -5732 \\ \hline 4031 \end{array}$ Checking by addition $\begin{array}{r} 5732 \\ +4031 \\ \hline 9763 \end{array}$ IT CHECKS

8. **(a)** $\begin{array}{r} 284{,}000 \\ -\ 96{,}327 \\ \hline 187{,}673 \end{array}$ Checking by addition $\begin{array}{r} 96{,}327 \\ +187{,}673 \\ \hline 284{,}000 \end{array}$ IT CHECKS

   (b) $\begin{array}{r} 8{,}526{,}024 \\ -6{,}397{,}518 \\ \hline 2{,}128{,}506 \end{array}$ Checking by addition $\begin{array}{r} 6{,}397{,}518 \\ +2{,}128{,}506 \\ \hline 8{,}526{,}024 \end{array}$ IT CHECKS

9. **(a)** $x = 5$ **(b)** $x = 12$
10. **(a)** $\begin{array}{r} 23{,}667{,}947 \\ -14{,}227{,}799 \\ \hline 9{,}440{,}148 \end{array}$ **(b)** $\begin{array}{r} 11{,}198{,}655 \\ -\ 9{,}579{,}677 \\ \hline 1{,}618{,}978 \end{array}$
11. **(a)** From the bar graph:

2003 sales	114
2002 sales	− 78
Sales increase	36

    (b) From the bar graph:

Springfield	91
Riverside	− 78
	13 more homes

    (c)

2003 sales	271	2004 sales	284
2002 sales	− 240	2003 sales	− 271
	31		13

    Therefore, the greatest increase in sales occurred from 2002 to 2003.

1.4 Practice Problems

1. (a) 64 **(b)** 42 **(c)** 40 **(d)** 63 **(e)** 81

2.
$$\begin{array}{r} 3021 \\ \times\ \ 3 \\ \hline 9063 \end{array}$$

3.
$$\begin{array}{r} \scriptstyle 2 \\ 43 \\ \times\ 8 \\ \hline 344 \end{array}$$

4.
$$\begin{array}{r} \scriptstyle 5\,6 \\ 579 \\ \times\ \ 7 \\ \hline 4053 \end{array}$$

5. (a) $1267 \times 10 = 12{,}670$
(b) $1267 \times 1000 = 1{,}267{,}000$
(c) $1267 \times 10{,}000 = 12{,}670{,}000$
(d) $1267 \times 1{,}000{,}000 = 1{,}267{,}000{,}000$

6. (a) $9 \times 6 \times 10{,}000 = 54 \times 10{,}000 = 540{,}000$
(b) $15 \times 4 \times 100 = 60 \times 100 = 6{,}000$
(c) $270 \times 800 = 27 \times 8 \times 10 \times 100 = 216 \times 1000 = 216{,}000$

7.
$$\begin{array}{r} 323 \\ \times\ 32 \\ \hline 646 \\ 969\ \ \\ \hline 10{,}336 \end{array}$$

8.
$$\begin{array}{r} 385 \\ \times\ 69 \\ \hline 3465 \\ 2310\ \ \\ \hline 26{,}565 \end{array}$$

9.
$$\begin{array}{r} 34 \\ \times 20 \\ \hline 680 \end{array}$$

10.
$$\begin{array}{r} 130 \\ \times\ 50 \\ \hline 6500 \end{array}$$

11.
$$\begin{array}{r} 923 \\ \times 675 \\ \hline 4615 \\ 6461\ \ \\ 5538\ \ \ \ \\ \hline 623{,}025 \end{array}$$

12. $25 \times 4 \times 17 = (25 \times 4) \times 17 = 100 \times 17 = 1700$

13.
$$\begin{aligned} 8 \times 4 \times 3 \times 25 &= 8 \times 3 \times (4 \times 25) \\ &= 24 \times 100 \\ &= 2400 \end{aligned}$$

14.
$$\begin{array}{r} 17348 \\ \times\ \ 378 \\ \hline 138784 \\ 121436\ \ \\ 52044\ \ \ \ \\ \hline 6{,}557{,}544 \end{array}$$

The total sales of cars is $6,557,544.

15. Area = 5 yards × 7 yards = 35 square yards.

1.5 Practice Problems

1. (a) $4\overline{)36}$ quotient 9 **(b)** $5\overline{)25}$ quotient 5 **(c)** $9\overline{)72}$ quotient 8 **(d)** $6\overline{)30}$ quotient 5

2. (a) $\frac{7}{1} = 7$ **(b)** $\frac{9}{9} = 1$ **(c)** $\frac{0}{5} = 0$

(d) $\frac{12}{0}$ cannot be done **(e)** $\frac{7}{0}$ cannot be done

3. $6\overline{)45}$ = 7 R3
$$\begin{array}{r} 45 \\ 42 \\ \hline 3 \end{array}$$
Check
$$\begin{array}{r} 6 \\ \times 7 \\ \hline 42 \\ +\ 3 \\ \hline 45 \end{array}$$

4. $6\overline{)129}$ = 21 R 3
$$\begin{array}{r} 129 \\ 12\ \\ \hline 9 \\ 6 \\ \hline 3 \end{array}$$
Check
$$\begin{array}{r} 21 \\ \times\ 6 \\ \hline 126 \\ +\ \ \ 3 \text{ Remainder} \\ \hline 129 \end{array}$$

5. $8\overline{)4237}$ = 529 R 5
$$\begin{array}{r} 4237 \\ 40\ \ \ \\ \hline 23\ \ \\ 16\ \ \\ \hline 77\ \\ 72\ \\ \hline 5 \end{array}$$

6. $32\overline{)229}$ = 7 R 5
$$\begin{array}{r} 229 \\ 224 \\ \hline 5 \end{array}$$

7. $33\overline{)42183}$ = 1278 R 9
$$\begin{array}{r} 42183 \\ 33\ \ \ \ \\ \hline 91\ \ \ \\ 66\ \ \ \\ \hline 258\ \ \\ 231\ \ \\ \hline 273\ \\ 264\ \\ \hline 9 \end{array}$$

8. $128\overline{)3227}$ = 25 R 27
$$\begin{array}{r} 3227 \\ 256\ \\ \hline 667 \\ 640 \\ \hline 27 \end{array}$$

9. $7\overline{)117{,}964}$ = 16,852 *Check*
$$\begin{array}{r} 16{,}852 \\ \times\ \ \ \ \ \ 7 \\ \hline 117{,}964 \end{array}$$

The cost of one car is $16,852.

10. $14\overline{)5138}$ = 367 The average speed was 367 mph.

1.6 Practice Problems

1. (a) $12 \times 12 \times 12 \times 12 = 12^4$. 12 is the base, 4 is the exponent.
(b) $2 \times 2 \times 2 \times 2 \times 2 \times 2 = 2^6$. 2 is the base, 6 is the exponent.

2. (a) $12^2 = 12 \times 12 = 144$
(b) $6^3 = 6 \times 6 \times 6 = 216$
(c) $2^6 = 2 \times 2 \times 2 \times 2 \times 2 \times 2 = 64$
(d) $1^{10} = 1 \times 1 \times 1 \times 1 \times 1 \times 1 \times 1 \times 1 \times 1 \times 1 = 1$

3. (a) $(7)(7)(7) + (8)(8) = 343 + 64 = 407$
(b) $(9)(9) + 1 = 81 + 1 = 82$
(c) $5^4 + 5 = (5)(5)(5)(5) + 5 = 625 + 5 = 630$

4.
$$\begin{aligned} 7 + 4^3 \times 3 &= 7 + 64 \times 3 && \text{Exponents} \\ &= 7 + 192 && \text{Multiply} \\ &= 199 && \text{Add} \end{aligned}$$

5. $37 - 20 \div 5 + 2 - 3 \times 4$
$= 37 - 4 + 2 - 3 \times 4$ Divide
$= 37 - 4 + 2 - 12$ Multiply
$= 33 + 2 - 12$ Subtract
$= 35 - 12$ Add
$= 23$ Subtract

6. $4^3 - 2 + 3^2$
$= 4 \times 4 \times 4 - 2 + 3 \times 3$ Evaluate exponents.
$= 64 - 2 + 9$ $4^3 = 64$ and $3^2 = 9$.
$= 62 + 9$ Subtract
$= 71$ Add

7. $(17 + 7) \div 6 \times 2 + 7 \times 3 - 4$
$= 24 \div 6 \times 2 + 7 \times 3 - 4$ Combine inside parentheses
$= 4 \times 2 + 7 \times 3 - 4$ Divide
$= 8 + 7 \times 3 - 4$ Multiply
$= 8 + 21 - 4$ Multiply
$= 29 - 4$ Add
$= 25$ Subtract

8. $5^2 - 6 \div 2 + 3^4 + 7 \times (12 - 10)$
$= 5^2 - 6 \div 2 + 3^4 + 7 \times 2$ Combine inside parentheses
$= 25 - 6 \div 2 + 81 + 7 \times 2$ Exponents
$= 25 - 3 + 81 + 7 \times 2$ Divide
$= 25 - 3 + 81 + 14$ Multiply
$= 22 + 81 + 14$ Subtract
$= 117$ Add

1.7 Practice Problems

1. 65,528 Locate the thousands round-off place.
65,528 The first digit to the right is 5 or more. We will increase the thousands digit by 1.
66,000 All digits to the right of thousands are replaced by zero.

2. 172,963 = 170,000 to the nearest ten thousand.

3. **(a)** 53,282 = 53,280 to the nearest ten. The digit to the right of tens was less than 5.
(b) 164,485 = 164,000 to the nearest thousand. The digit to the right of thousands was less than 5.
(c) 1,365,273 = 1,400,000 to the nearest hundred thousand. The digit to the right of hundred thousands was greater than 5.

4. **(a)** 935,682 = 936,000. The digit to the right of thousands was greater than 5.
(b) 935,682 = 900,000. The digit to the right of hundred thousands was less than 5.
(c) 935,682 = 1,000,000. The digit to the right of millions was greater than 5.

5. 9,460,000,000,000,000 = 9,500,000,000,000,000 meters to the nearest hundred trillion.

6.

Actual Sum	Estimated Sum	
3456	3000	
9876	10000	
5421	5000	
+ 1278	+ 1000	
20,031	19,000	Close to the actual sum

7.

\$697	\$700
35	40
+ 19	+ 20
	\$760

We estimate that the total cost is \$760. (The exact answer is \$751, so we can see that our answer is quite close.)

8. 10,000 + 10,000 + 20,000 + 60,000 = 100,000
This is significantly different from 81,358, so we would suspect that an error has been made. In fact, Ming did make an error. The exact sum is actually 101,358!

9. 30,000,000 − 20,000,000 = 10,000,000
We estimate that 10,000,000 more people lived in California than in Florida.

10. $9000 \times 7000 = 63{,}000{,}000$
We estimate the product to be 63,000,000.

11. $40\overline{)80{,}000}$ = 2,000 Our estimate is 2,000.

12. $60\overline{)2{,}000{,}000}$ = 33,333 R 20 Our estimate is \$33,333 for one truck.

1.8 Practice Problems

Practice Problem 1

1. Understand the problem.

Mathematics Blueprint for Problem Solving

Gather the Facts	What Am I Asked to Do?	How Do I Proceed?	Key Points to Remember
The deductions are \$135, \$28, \$13, and \$34.	Find out the total amount of deductions.	I must add the four deductions to obtain the total.	Watch out! Gross pay of \$1352 is not needed to solve the problem.

2. Solve and state the answer:
$135 + 28 + 13 + 34 = 210$
The total amount taken out of Diane's paycheck is \$210.

3. Check. Estimate the answer to see if it is reasonable.

Practice Problem 2

1. Understand the problem.

Mathematics Blueprint for Problem Solving			
Gather the Facts	**What Am I Asked to Do?**	**How Do I Proceed?**	**Key Points to Remember**
Gore had 50,999,897 votes. Bush had 50,456,002 votes.	Find out by how many votes Gore beat Bush.	I must subtract the amounts.	Gore is a Democrat and Bush is a Republican.

2. Solve and state the answer:

$$\begin{array}{r} 50{,}999{,}897 \\ -\ \underline{50{,}456{,}002} \\ 543{,}895 \end{array}$$

Gore beat Bush by 543,895 votes.

3. ***Check.*** Estimate the answer to see if it is reasonable.

4. Bush had more electoral college votes.

Practice Problem 3

1. Understand the problem.

Mathematics Blueprint for Problem Solving			
Gather the Facts	**What Am I Asked to Do?**	**How Do I Proceed?**	**Key Points to Remember**
1 gallon is 1,024 fluid drams.	Find out how many fluid drams in 9 gallons.	I need to multiply 1024 by 9.	I must use fluid drams as my measure in my answer.

2. Solve and state the answer:

$$\frac{1024 \text{ fluid drams}}{1 \cancel{\text{gal}}} \times 9 \cancel{\text{gal}} = 9{,}216 \text{ fluid drams.}$$

There would be 9,216 fluid drams in 9 gallons.

3. ***Check.*** Estimate the answer to see if it is reasonable.

Practice Problem 4

1. Understand the problem.

Mathematics Blueprint for Problem Solving			
Gather the Facts	**What Am I Asked to Do?**	**How Do I Proceed?**	**Key Points to Remember**
Donna bought 45 shares of stock. She paid $1620 for them.	Find out the cost per share of stock.	I need to divide $1620 by 45.	Use dollars as the unit in the answer.

2. Solve and state the answer:

$$\begin{array}{r} 36 \\ 45\overline{)1620} \\ \underline{135} \\ 270 \\ \underline{270} \\ 0 \end{array}$$

Donna paid $36 per share for the stock.

3. ***Check.*** Estimate the answer to see if it is reasonable.

Practice Problem 5

1. Understand the problem.

Mathematics Blueprint for Problem Solving			
Gather the Facts	**What Am I Asked to Do?**	**How Do I Proceed?**	**Key Points to Remember**
50 tables at $200 each. 180 chairs at $40 each. 6 carts at $65 each.	Find the total cost.	**1.** Multiply the number of purchases by each price. **2.** Add the total costs of each of the items.	There are three types of items involved in the total purchase.

2. Solve and state the answer:

50 tables at $200 each	$= 50 \times 200 =$	$10,000	cost of tables
180 chairs at $40 each	$= 180 \times 40 =$	7,200	cost of chairs
6 carts at $65 each	$= 6 \times 65 =$	390	cost of carts
		$17,590	total cost

The total purchase was $17,590.

3. ***Check.*** Estimate the answer to see if it is reasonable.

Practice Problem 6

1. Understand the problem.

Mathematics Blueprint for Problem Solving			
Gather the Facts	**What Am I Asked to Do?**	**How Do I Proceed?**	**Key Points to Remember**
Old balance: $498 New deposits: $607 $163 Interest: $36 Withdrawls: $ 19 $158 $582 $ 74	Find her new balance after the transactions.	**(a)** Add the new deposits and interest to the old balance. **(b)** Add the withdrawals. **(c)** Subtract the results from steps (a) and (b).	Deposits and interest are added and withdrawals are subtracted from savings accounts.

2. Solve and state the answer:

$$\textbf{(a)}\ \begin{array}{r} \$498 \\ 607 \\ 163 \\ +\ \ 36 \\ \hline \$1304 \end{array} \qquad \textbf{(b)}\ \begin{array}{r} \$\ 19 \\ 158 \\ 582 \\ +\ 74 \\ \hline \$833 \end{array} \qquad \textbf{(c)}\ \begin{array}{r} \$1304 \\ +\ \ 833 \\ \hline \$471 \end{array}$$

Her balance this month is $471.

3. ***Check.*** Estimate the answer to see if it is reasonable.

Practice Problem 7

1. Understand the problem.

Mathematics Blueprint for Problem Solving			
Gather the Facts	**What Am I Asked to Do?**	**How Do I Proceed?**	**Key Points to Remember**
Odometer reading at end of trip: 51,118 miles Odometer reading at start of trip: 50,698 miles Used on trip: 12 gallons of gas	Find the number of miles per gallon that the car obtained on the trip.	**(a)** Subtract the two odometer readings. **(b)** Divide that number by 12.	The gas tank was full at the beginning of the trip. 12 gallons fill the tank at the end of the trip.

2. Solve and state the answer:

$$\begin{array}{rl} 51{,}118 & \text{odometer at end of trip} \\ -\,50{,}698 & \text{odometer at start of trip} \\ \hline 420 & \text{miles traveled on trip} \end{array} \qquad \frac{420 \text{ miles}}{12 \text{ gallons of gas used}}$$

$$= 12\overline{)420} \text{ quotient } 35; \quad \begin{array}{r} 36 \\ \hline 60 \\ 60 \end{array}$$

35 miles per gallon on the trip

3. ***Check.*** Estimate the answer to see if it is reasonable.

Chapter 2

2.1 Practice Problems

1. (a) Four parts of twelve are shaded. The fraction is $\frac{4}{12}$.

(b) Three parts out of six are shaded. The fraction is $\frac{3}{6}$.

(c) Two parts of three are shaded. The fraction is $\frac{2}{3}$.

2. (a) $\frac{4}{5}$ of the object is shaded.

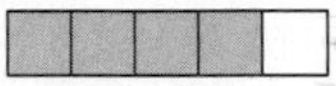

(b) $\frac{3}{7}$ of the group is shaded.

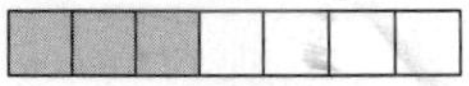

3. (a) $\frac{9}{17}$ represents 9 players out of 17.

(b) The total class is $382 + 351 = 733$.

The fractional part that is men is $\frac{382}{733}$.

(c) $\frac{7}{8}$ of a yard of material.

4. Total number of defective items $1 + 2 = 3$. Total number of items $7 + 9 = 16$. A fraction that represents the portion of the items that were defective is $\frac{3}{16}$.

2.2 Practice Problems

1. (a) $18 = 9 \times 2$
$= 3 \times 3 \times 2$
$= 3^2 \times 2$

(b) $72 = 9 \times 8$
$= 3 \times 3 \times 2 \times 2 \times 2$
$= 3^2 \times 2^3$

(c) $400 = 10 \times 40$
$= 5 \times 2 \times 5 \times 8$
$= 5 \times 2 \times 5 \times 2 \times 2 \times 2$
$= 5^2 \times 2^4$

2. (a) $\frac{30 \div 6}{42 \div 6} = \frac{5}{7}$

(b) $\frac{60 \div 12}{132 \div 12} = \frac{5}{11}$

3. (a) $\frac{120}{135} = \frac{2 \times 2 \times 2 \times \cancel{3} \times \cancel{5}}{3 \times 3 \times \cancel{3} \times \cancel{5}} = \frac{8}{9}$

(b) $\frac{715}{880} = \frac{\cancel{5} \times \cancel{11} \times 13}{\cancel{5} \times \cancel{11} \times 2 \times 2 \times 2 \times 2} = \frac{13}{16}$

4. (a) $\frac{84}{108} \stackrel{?}{=} \frac{7}{9}$
$84 \times 9 \stackrel{?}{=} 108 \times 7$
$756 = 756$ Yes

(b) $\frac{3}{7} \stackrel{?}{=} \frac{79}{182}$
$3 \times 182 \stackrel{?}{=} 7 \times 79$
$546 \neq 553$ No

2.3 Practice Problems

1. (a) $4\frac{3}{7} = \frac{7 \times 4 + 3}{7} = \frac{28 + 3}{7} = \frac{31}{7}$

(b) $6\frac{2}{3} = \frac{3 \times 6 + 2}{3} = \frac{18 + 2}{3} = \frac{20}{3}$

(c) $19\frac{4}{7} = \frac{7 \times 19 + 4}{7} = \frac{133 + 4}{7} = \frac{137}{7}$

2. (a) $4\overline{)17}$ quotient 4; $\begin{array}{r} 16 \\ \hline 1 \end{array}$ so $\frac{17}{4} = 4\frac{1}{4}$

(b) $5\overline{)36}$ quotient 7; $\begin{array}{r} 35 \\ \hline 1 \end{array}$ so $\frac{36}{5} = 7\frac{1}{5}$

(c) $27\overline{)116}$ quotient 4; $\begin{array}{r} 108 \\ \hline 8 \end{array}$ so $\frac{116}{27} = 4\frac{8}{27}$

(d) $13\overline{)91}$ quotient 7; $\begin{array}{r} 91 \\ \hline 0 \end{array}$ so $\frac{91}{13} = 7$

3. $\frac{51}{15} = \frac{3 \times 17}{3 \times 5} = \frac{\overset{1}{\cancel{3}} \times 17}{\underset{1}{\cancel{3}} \times 5} = \frac{17}{5}$

4. $\frac{16}{80} = \frac{1}{5}$ so therefore we find that

$3\frac{16}{80} = 3\frac{1}{5}$.

5. $\frac{1001}{572} = 1\frac{429}{572}$

Now the fraction $\frac{429}{572} = \frac{3 \times \overset{1}{\cancel{11}} \times \overset{1}{\cancel{13}}}{2 \times 2 \times \underset{1}{\cancel{11}} \times \underset{1}{\cancel{13}}} = \frac{3}{4}$.

Thus $\frac{1001}{572} = 1\frac{429}{572} = 1\frac{3}{4}$.

2.4 Practice Problems

1. (a) $\frac{6}{7} \times \frac{3}{13} = \frac{18}{91}$ **(b)** $\frac{1}{5} \times \frac{11}{12} = \frac{11}{60}$

2. $\frac{\overset{5}{\cancel{55}}}{\underset{9}{\cancel{72}}} \times \frac{\overset{2}{\cancel{16}}}{\underset{3}{\cancel{33}}} = \frac{5}{9} \times \frac{2}{3} = \frac{10}{27}$

3. (a) $7 \times \frac{5}{13} = \frac{7}{1} \times \frac{5}{13} = \frac{35}{13}$ or $2\frac{9}{13}$

(b) $\frac{13}{16} \times 8 = \frac{13}{\underset{2}{\cancel{16}}} \times \frac{\overset{1}{\cancel{8}}}{1} = \frac{13}{2}$ or $6\frac{1}{2}$

4. $\frac{3}{\underset{1}{\cancel{8}}} \times \overset{12{,}300}{\cancel{98{,}400}} = \frac{3}{1} \times 12{,}300 = 36{,}900$

There are 36,900 ft^2 in the wetland area.

5. (a) $2\frac{1}{6} \times \frac{4}{7} = \frac{13}{\underset{3}{\cancel{6}}} \times \frac{\overset{2}{\cancel{4}}}{7} = \frac{26}{21}$ or $1\frac{5}{21}$

(b) $10\frac{2}{3} \times 13\frac{1}{2} = \frac{\overset{16}{\cancel{32}}}{\underset{1}{\cancel{3}}} \times \frac{\overset{9}{\cancel{27}}}{\underset{1}{\cancel{2}}} = \frac{144}{1} = 144$

(c) $\frac{3}{5} \times 1\frac{1}{3} \times \frac{5}{8} = \frac{\overset{1}{\cancel{3}}}{\underset{1}{\cancel{5}}} \times \frac{\overset{1}{\cancel{4}}}{\underset{1}{\cancel{3}}} \times \frac{\overset{1}{\cancel{5}}}{\underset{2}{\cancel{8}}} = \frac{1}{2}$

(d) $3\frac{1}{5} \times 2\frac{1}{2} = \frac{\overset{8}{\cancel{16}}}{\underset{1}{\cancel{5}}} \times \frac{\overset{1}{\cancel{5}}}{\underset{1}{\cancel{2}}} = \frac{8}{1} = 8$

6. Area $= lw = 1\frac{1}{5} \times 4\frac{5}{6} = \frac{\overset{1}{\cancel{6}}}{5} \times \frac{29}{\underset{1}{\cancel{6}}} = \frac{29}{5} = 5\frac{4}{5}\ \text{m}^2$

7. Since $8 \cdot 10 = 80$ and $9 \cdot 9 = 81$,

we know that $\frac{8}{9} \cdot \frac{10}{9} = \frac{80}{81}$.

Therefore $x = \frac{10}{9}$.

2.5 Practice Problems

1. (a) $\frac{7}{13} \div \frac{3}{4} = \frac{7}{13} \times \frac{4}{3} = \frac{28}{39}$

(b) $\frac{16}{35} \div \frac{24}{25} = \frac{\overset{2}{\cancel{16}}}{\underset{7}{\cancel{35}}} \times \frac{\overset{5}{\cancel{25}}}{\underset{3}{\cancel{24}}} = \frac{10}{21}$

2. (a) $\frac{3}{17} \div \frac{6}{1} = \frac{\overset{1}{\cancel{3}}}{17} \times \frac{1}{\underset{2}{\cancel{6}}} = \frac{1}{34}$

(b) $\frac{14}{1} \div \frac{7}{15} = \frac{\overset{2}{\cancel{14}}}{1} \times \frac{15}{\underset{1}{\cancel{7}}} = 30$

3 (a) $1 \div \frac{11}{13} = \frac{1}{1} \times \frac{13}{11} = \frac{13}{11}$ or $1\frac{2}{11}$

(b) $\frac{14}{17} \div 1 = \frac{14}{17} \times \frac{1}{1} = \frac{14}{17}$

(c) $\frac{3}{11} \div 0$ Division by zero cannot be done. This problem cannot be done.

(d) $0 \div \frac{9}{16} = \frac{0}{1} \times \frac{16}{9} = \frac{0}{9} = 0$

4. (a) $1\frac{1}{5} \div \frac{7}{10} = \frac{6}{\underset{1}{\cancel{5}}} \times \frac{\overset{2}{\cancel{10}}}{7} = \frac{12}{7}$ or $1\frac{5}{7}$

(b) $2\frac{1}{4} \div 1\frac{7}{8} = \frac{9}{4} \div \frac{15}{8} = \frac{\overset{3}{\cancel{9}}}{\underset{1}{\cancel{4}}} \times \frac{\overset{2}{\cancel{8}}}{\underset{5}{\cancel{15}}} = \frac{6}{5}$ or $1\frac{1}{5}$

5. (a) $\frac{5\frac{2}{3}}{7} = 5\frac{2}{3} \div 7 = \frac{17}{3} \times \frac{1}{7} = \frac{17}{21}$

(b) $\frac{1\frac{2}{5}}{2\frac{1}{3}} = 1\frac{2}{5} \div 2\frac{1}{3} = \frac{7}{5} \div \frac{7}{3} = \frac{\overset{1}{\cancel{7}}}{5} \times \frac{3}{\underset{1}{\cancel{7}}} = \frac{3}{5}$

6. $x \div \frac{3}{2} = \frac{22}{36}$

$x \cdot \frac{2}{3} = \frac{22}{36}$

$\frac{11}{12} \cdot \frac{2}{3} = \frac{22}{36}$ Thus $x = \frac{11}{12}$.

7. $19\frac{1}{4} \div 14 = \frac{\overset{11}{\cancel{77}}}{4} \times \frac{1}{\underset{2}{\cancel{14}}} = \frac{11}{8}$ or $1\frac{3}{8}$

Each piece will be $1\frac{3}{8}$ feet long.

2.6 Practice Problems

1. The multiples of 14 are 14, 28, 42, 56, 70, 84, ...
The multiples of 21 are 21, 42, 63, 84, 105, 126, ...
42 is the least common multiple of 14 and 21.

2. The multiples of 10 are 10, 20, 30, 40 ...
The multiples of 15 are 15, 30, 45 ...
30 is the least common multiple of 10 and 15.

3. 54 is a multiple of 6. We know that $6 \times 9 = 54$.
The least common multiple of 6 and 54 is 54.

4. (a) The LCD of $\frac{3}{4}$ and $\frac{11}{12}$ is 12.
12 can be divided by 4 and 12.

(b) The LCD of $\frac{1}{7}$ and $\frac{8}{35}$ is 35.
35 can be divided by 7 and 35.

5. The LCD of $\frac{3}{7}$ and $\frac{5}{6}$ is 42.
42 can be divided by 7 and 6.

6. (a) $14 = 2 \times 7$
$10 = 2 \times 5$
The LCD $= 2 \times 5 \times 7 = 70$.

(b) $15 = 3 \times 5$
$50 = 2 \times 5 \times 5$
The LCD $= 2 \times 3 \times 5 \times 5 = 150$.

(c) $16 = 2 \times 2 \times 2 \times 2$
$12 = 2 \times 2 \times 3$
The LCD $= 2 \times 2 \times 2 \times 2 \times 3 = 48$.

7. $49 = 7 \times 7$
$21 = 7 \times 3$
$7 = 7 \times 1$
LCD $= 7 \times 7 \times 3 = 147$.

8. (a) $\frac{3}{5} \times \frac{8}{8} = \frac{24}{40}$

(b) $\frac{7}{11} \times \frac{4}{4} = \frac{28}{44}$

(c) $\frac{2}{7} = \frac{2}{7} \times \frac{4}{4} = \frac{8}{28}$

$\frac{3}{4} = \frac{3}{4} \times \frac{7}{7} = \frac{21}{28}$

9. (a) The LCD is 60 because it can be divided by 20 and 15.

(b) $\frac{3}{20} \times \frac{3}{3} = \frac{9}{60}$ $\frac{11}{15} \times \frac{4}{4} = \frac{44}{60}$

10. (a) LCD of 64 and 80 is 320.

(b) $\frac{5}{64} = \frac{25}{320}$ $\frac{3}{80} = \frac{12}{320}$

2.7 Practice Problems

1. $\frac{3}{17} + \frac{12}{17} = \frac{15}{17}$

2. (a) $\frac{1}{12} + \frac{5}{12} = \frac{6}{12} = \frac{1}{2}$

(b) $\frac{13}{15} + \frac{7}{15} = \frac{20}{15} = \frac{4}{3}$ or $1\frac{1}{3}$

3. (a) $\frac{5}{19} - \frac{2}{19} = \frac{3}{19}$ **(b)** $\frac{21}{25} - \frac{6}{25} = \frac{15}{25} = \frac{3}{5}$

4. $\frac{2}{15} = \frac{2}{15}$

$\frac{2}{5} \times \frac{3}{3} = \frac{6}{15}$

$\frac{8}{15}$

5. LCD $= 48$ $\frac{5}{12} \times \frac{4}{4} = \frac{20}{48}$ $\frac{5}{16} \times \frac{3}{3} = \frac{15}{48}$

$\frac{5}{12} + \frac{5}{16} = \frac{20}{48} + \frac{15}{48} = \frac{35}{48}$

6. LCD $= 48$

$\frac{3}{16} \times \frac{3}{3} = \frac{9}{48}$

$\frac{1}{8} \times \frac{6}{6} = \frac{6}{48}$

$\frac{1}{12} \times \frac{4}{4} = \frac{4}{48}$

$\frac{19}{48}$

7. LCD = 96 $\quad \frac{9}{48} \times \frac{2}{2} = \frac{18}{96}$

$\frac{5}{32} \times \frac{3}{3} = \frac{15}{96} \quad \frac{9}{48} - \frac{5}{32} = \frac{18}{96} - \frac{15}{96} = \frac{3}{96} = \frac{1}{32}$

8. $\frac{9}{10} \times \frac{2}{2} = \frac{18}{20}$

$-\frac{1}{4} \times \frac{5}{5} = -\frac{5}{20}$

$\frac{13}{20}$

There is $\frac{13}{20}$ gallon left.

9. The LCD of $\frac{3}{10}$ and $\frac{23}{25}$ is 50.

$\frac{3}{10} \times \frac{5}{5} = \frac{15}{50} \quad$ Now rewriting: $\quad x + \frac{15}{50} = \frac{46}{50}$

$\frac{23}{25} \times \frac{2}{2} = \frac{46}{50} \quad \frac{31}{50} + \frac{15}{50} = \frac{46}{50}$

So, $x = \frac{31}{50}$

10. $\frac{15}{16} + \frac{3}{40}$

$\frac{15}{16} \times \frac{40}{40} = \frac{600}{640} \quad \frac{3}{40} \times \frac{16}{16} = \frac{48}{640}$

Thus $\frac{15}{16} + \frac{3}{40} = \frac{600}{640} + \frac{48}{640} = \frac{648}{640} = \frac{81}{80}$ or $1\frac{1}{80}$

2.8 Practice Problems

1. $5\frac{1}{12}$

$+9\frac{5}{12}$

$14\frac{6}{12} = 14\frac{1}{2}$

2. $6\frac{1}{4} = 6\frac{5}{20}$

$+2\frac{2}{5} = +2\frac{8}{20}$

$8\frac{13}{20}$

3. LCD = 12 $\quad 7\boxed{\frac{1}{4} \times \frac{3}{3}} = 7\frac{3}{12}$

$+3\boxed{\frac{5}{6} \times \frac{2}{2}} \quad +3\frac{10}{12}$

$10\frac{13}{12} = 10 + 1\frac{1}{12} = 11\frac{1}{12}$

4. LCD = 12 $\quad 12\frac{5}{6} = 12\frac{10}{12}$

$-7\frac{5}{12} = -7\frac{5}{12}$

$5\frac{5}{12}$

5. (a) LCD = 24 $\quad 9\boxed{\frac{1}{8} \times \frac{3}{3}} = 9\frac{3}{24} = 8\frac{27}{24}$

$-3\boxed{\frac{2}{3} \times \frac{8}{8}} = -3\frac{16}{24} = -3\frac{16}{24}$

$5\frac{11}{24}$

Borrow 1 from 9:

$9\frac{3}{24} = 8 + 1 + \frac{3}{24} = 8\frac{27}{24}$

(b) $18 = 17\frac{18}{18}$

$-6\frac{7}{18} = -6\frac{7}{18}$

$11\frac{11}{18}$

6. $6\frac{1}{4} = 6\frac{3}{12} = 5\frac{15}{12}$

$-4\frac{2}{3} = -4\frac{8}{12} = -4\frac{8}{12}$

$1\frac{7}{12}$

They had $1\frac{7}{12}$ gallons left over.

7. $\frac{3}{5} - \frac{1}{15} \times \frac{10}{13}$

$= \frac{3}{5} - \frac{2}{39} \quad$ LCD $= 5 \cdot 39 = 195$

$= \frac{117}{195} - \frac{10}{195}$

$= \frac{107}{195}$

8. $\frac{1}{7} \times \frac{5}{6} + \frac{5}{3} \div \frac{7}{6} = \frac{1}{7} \times \frac{5}{6} + \frac{5}{3} \times \frac{6}{7}$

$= \frac{5}{42} + \frac{10}{7} \quad$ LCD = 42

$= \frac{5}{42} + \frac{60}{42} = \frac{65}{42}$ or $1\frac{23}{42}$

2.9 Practice Problems

Practice Problem 1

1. Understand the problem.

Mathematics Blueprint for Problem Solving

Gather the Facts	What Am I Asked to Do?	How Do I Proceed?	Key Points to Remember
Gas amounts: $18\frac{7}{10}$ gal, $15\frac{2}{5}$ gal, $14\frac{1}{2}$ gal	Find out how many gallons of gas she bought altogether.	Add the three amounts.	When adding mixed numbers, the LCD is needed for the fractions.

2. Solve and state the answer:

LCD = 10

$$18\frac{7}{10} = 18\frac{7}{10}$$
$$15\frac{2}{5} = 15\frac{4}{10}$$
$$14\frac{1}{2} = 14\frac{5}{10}$$
$$47\frac{16}{10} = 48\frac{6}{10}$$
$$= 48\frac{3}{5}\text{ gallons}$$

Total is $48\frac{3}{5}$ gallons.

3. ***Check.*** Estimate the answer to see if it is reasonable.

Practice Problem 2

1. Understand the problem.

Mathematics Blueprint for Problem Solving			
Gather the Facts	**What Am I Asked to Do?**	**How Do I Proceed?**	**Key Points to Remember**
Poster: $12\frac{1}{4}$ in. Top border: $1\frac{3}{8}$ in. Bottom border: 2 in.	Find the length of the inside portion of the poster.	**(a)** Add the two border lengths. **(b)** Subtract this total from the poster length.	When adding mixed numbers, the LCD is needed for the fractions.

2. Solve and state the answer:

(a) $1\frac{3}{8} + 2 = 3\frac{3}{8}$

(b) $12\frac{1}{4} = 12\frac{2}{8} = 11\frac{10}{8}$

$-\ 3\frac{3}{8} = -\ 3\frac{3}{8} = -\ 3\frac{3}{8}$

$8\frac{7}{8}$

The inside portion is $8\frac{7}{8}$ inches.

3. ***Check.*** Estimate the answer to see if it is reasonable.

Practice Problem 3

1. Understand the problem.

Mathematics Blueprint for Problem Solving			
Gather the Facts	**What Am I Asked to Do?**	**How Do I Proceed?**	**Key Points to Remember**
Regular tent uses $8\frac{1}{4}$ yards. Large tent uses $1\frac{1}{2}$ times the regular. She makes 6 regular and 16 large tents.	Find out how many yards of cloth will be needed to make the tents.	Find the amount used for regular tents, and the amount used for large tents. Then add the two.	Large tents use $1\frac{1}{2}$ times the regular amount.

2. Solve and state the answer:

We multiply $6 \times 8\frac{1}{4}$ for regular tents and $16 \times 1\frac{1}{2} \times 8\frac{1}{4}$ for large tents. Then add total yardage.

$$6 \times 8\frac{1}{4}\text{(regular tents)} = \overset{3}{\cancel{6}} \times \frac{33}{\underset{2}{\cancel{4}}} = \frac{99}{2} = 49\frac{1}{2} \text{ yards.}$$

(Large tents)

$$16 \times 1\frac{1}{2} \times 8\frac{1}{4} = \overset{\overset{2}{\cancel{8}}}{\cancel{16}} \times \frac{3}{\underset{1}{\cancel{2}}} \times \frac{33}{\underset{1}{\cancel{4}}} = \frac{198}{1} = 198 \text{ yards.}$$

Total yards for all tents $198 + 49\frac{1}{2} = 247\frac{1}{2}$ yards.

3. ***Check.*** Estimate the answer to see if it is reasonable.

Practice Problem 4

1. Understand the problem.

Mathematics Blueprint for Problem Solving			
Gather the Facts	**What Am I Asked to Do?**	**How Do I Proceed?**	**Key Points to Remember**
He purchases 12-foot boards. Each shelf is $2\frac{3}{4}$ ft. He needs four shelves for each bookcase and he is making two bookcases.	**(a)** Find out how many boards he needs to buy. **(b)** Find out how many feet of shelving are actually needed. **(c)** Find out how many feet will be left over.	Find out how many $2\frac{3}{4}$-ft. shelves he can get from one board. Then see how many boards he needs to make all eight shelves.	There will be three answers to this problem. Don't forget to calculate the leftover wood.

2. Solve and state answer:

We want to know how many $2\frac{3}{4}$-ft. shelves are in a 12-ft. board.

$$12 \div 2\frac{3}{4} = \frac{12}{1} \div \frac{11}{4} = \frac{12}{1} \times \frac{4}{11} = \frac{48}{11} = 4\frac{4}{11} \text{ shelves}$$

He will get 4 shelves from each board with some left over.

(a) For two bookcases, he needs eight shelves. He gets four shelves out of each board. $8 \div 4 = 2$. He will need two 12-ft. boards.

(b) He needs 8 shelves at $2\frac{3}{4}$ feet.

$$8 \times 2\frac{3}{4} = 8 \times \frac{11}{4} = 22 \text{ feet.}$$

He actually needs 22 feet of shelving.

(c)
  24 feet of shelving bought
− 22 feet of shelving used
  2 feet of shelving left over.

3. ***Check.*** Estimate the answer to see if it is reasonable.

Practice Problem 5

1. Understand the problem.

Mathematics Blueprint for Problem Solving			
Gather the Facts	**What Am I Asked to Do?**	**How Do I Proceed?**	**Key Points to Remember**
Distance is $199\frac{3}{4}$ miles. He uses $8\frac{1}{2}$ gallons of gas	Find out how many miles per gallon he gets.	Divide the distance by the number of gallons.	Change mixed numbers to improper fractions before dividing.

2. Solve and state the answer:

$$199\frac{3}{4} \div 8\frac{1}{2} = \frac{799}{4} \div \frac{17}{2}$$
$$= \frac{\overset{47}{\cancel{799}}}{\underset{2}{\cancel{4}}} \times \frac{\overset{1}{\cancel{2}}}{\underset{1}{\cancel{17}}}$$
$$= \frac{47}{2} = 23\frac{1}{2}$$

He gets $23\frac{1}{2}$ miles per gallon.

3. *Check.* Estimate the answer to see if it is reasonable.

Chapter 3

3.1 Practice Problems

1. (a) 0.073 seventy-three thousandths
(b) 4.68 four and sixty-eight hundredths
(c) 0.0017 seventeen ten-thousandths
(d) 561.78 five hundred sixty-one and seventy-eight hundredths

2. seven thousand eight hundred sixty-three and $\frac{4}{100}$ dollars

3. (a) $\frac{9}{10} = 0.9$ **(b)** $\frac{136}{1000} = 0.136$ **(c)** $2\frac{56}{100} = 2.56$
(d) $34\frac{86}{1000} = 34.086$

4. (a) $0.37 = \frac{37}{100}$ **(b)** $182.3 = 182\frac{3}{10}$
(c) $0.7131 = \frac{7131}{10{,}000}$ **(d)** $42.019 = 42\frac{19}{1000}$

5. (a) $8.5 = 8\frac{5}{10} = 8\frac{1}{2}$ **(b)** $0.58 = \frac{58}{100} = \frac{29}{50}$
(c) $36.25 = 36\frac{25}{100} = 36\frac{1}{4}$ **(d)** $106.013 = 106\frac{13}{1000}$

6. $\frac{2}{1{,}000{,}000{,}000} = \frac{1}{500{,}000{,}000}$

The concentration of PCBs is $\frac{1}{500{,}000{,}000}$.

3.2 Practice Problems

1. Since 4 < 5,

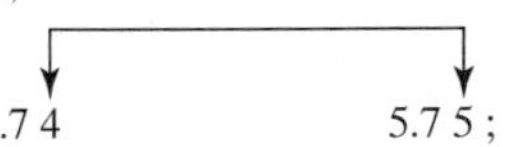

5.7 4 5.7 5; therefore 5.74 < 5.75.

2. 0.894 > 0.890, so 0.894 > 0.89

3. 2.45, 2.543, 2.46, 2.54, 2.5

It is helpful to add extra zeros and to place the decimals that begin with 2.4 in a group and the decimals that begin with 2.5 in the other.

2.450, 2.460, 2.543, 2.540, 2.500

In order, we have from smallest to largest

2.450, 2.460, 2.500, 2.540, 2.543.

It is OK to leave the extra terminal zeros in our answer.

4. 723.88 — Since the digit to right of tenths is greater than 5, we round up.
723.9

5. (a) 12.92 6 47 — Since the digit to right of thousandths is less than 5, we drop the digits 4 and 7.
12.926

(b) 0.00 7 892 — Since the digit to right of thousandths is greater than 5, we round up.
0.008

6. 15,699.953 — Since the digit to right of tenths is five, we round up.
15,700.0

7.

		Rounded to Nearest Dollar
Medical bills	375.50	376
Taxes	981.39	981
Retirement	980.49	980
Charity	817.65	818

3.3 Practice Problems

1. (a)
$$\begin{array}{r} \overset{1}{9}.8 \\ 3.6 \\ +\ 5.4 \\ \hline 18.8 \end{array}$$
(b)
$$\begin{array}{r} \overset{1}{3}\overset{1}{0}\overset{1}{0}.72 \\ 163.75 \\ +\ 291.08 \\ \hline 755.55 \end{array}$$
(c)
$$\begin{array}{r} \overset{2}{}8.9000 \\ 37.0560 \\ 0.0023 \\ +\ 945.0000 \\ \hline 990.9583 \end{array}$$

2.
$$\begin{array}{r} 9\overset{1}{3},5\overset{1}{2}1.8 \\ +\ 1{,}634.8 \text{ miles} \\ \hline 95{,}156.6 \text{ miles} \end{array}$$

3.
$$\begin{array}{r} \$\ 80.95 \\ 133.91 \\ 256.47 \\ 53.08 \\ +\ 381.32 \\ \hline \$905.73 \end{array}$$

4. (a)
$$\begin{array}{r} 3\overset{7}{8}.\overset{18}{8} \\ -\ 26.9 \\ \hline 11.9 \end{array}$$
(b)
$$\begin{array}{r} 2034.908 \\ -\ 1986.325 \\ \hline 48.583 \end{array}$$

5. (a)
$$\begin{array}{r} 19.000 \\ -\ 12.579 \\ \hline 6.421 \end{array}$$
(b)
$$\begin{array}{r} 283.076 \\ -\ 96.380 \\ \hline 186.696 \end{array}$$

6.
$$\begin{array}{r} 87{,}160.1 \\ -\ 82{,}370.9 \\ \hline 4{,}789.2 \text{ miles} \end{array}$$

7.
$$\begin{array}{r} 15.3 \\ -\ 10.8 \\ \hline 4.5 \end{array}$$
x is 4.5.

3.4 Practice Problems

1.
$$\begin{array}{rl} 0.09 & \text{2 decimal places} \\ \times\ 0.6 & +\ \text{1 decimal place} \\ \hline 0.054 & \text{3 decimal places in product} \end{array}$$

2. (a)
$$\begin{array}{rl} 0.47 & \text{2 decimal places} \\ \times\ 0.28 & +\ \text{2 decimal places} \\ \hline 376 & \\ 94 & \\ \hline 0.1316 & \text{4 decimal places in product} \end{array}$$

(b)
$$\begin{array}{rl} 0.436 & \text{3 decimal places} \\ \times\ 18.39 & +\ \text{2 decimal places} \\ \hline 8.01804 & \text{5 decimal places in product} \end{array}$$

3.
$$\begin{array}{rl} 0.4264 & \text{4 decimal places} \\ \times\ 38 & +\ \text{0 decimal place} \\ \hline 16.2032 & \text{4 decimal places in product} \end{array}$$

4. (a) Area = length × width
$$\begin{array}{r} 1.26 \\ \times\ 2.3 \\ \hline 2.898 \text{ square millimeters} \end{array}$$

5. (a) $0.0561 \times 10 = 0.561$ Decimal point moved one place to the right.
(b) $1462.37 \times 100 = 146{,}237$ Decimal point moved two places to the right.

6. (a) $0.26 \times \underline{1000} = 260$ Decimal point moved three places to the right. One extra zero needed.

(b) $5862.89 \times \underline{10{,}000} = 58{,}628{,}900$ Decimal point moved four places to the right. Two extra zeros needed.

7. (a) $7.684 \times 10^4 = 76{,}840$ Decimal point moved four places to the right. One extra zero needed.

8. $156.2 \times 1000 = 156{,}200$ meters

3.5 Practice Problems

1. (a)

0.258
7)1.806
14
40
35
56
56
0

(b)

0.0058
16)0.0928
80
128
128
0

2. 0.517 = 0.52 to the nearest hundredth

46)23.820
230
82
46
360
322
38

3. $186.25 each month

19)$3538.75
19
163
152
118
114
4 7
3 8
95
95
0

4. (a)

1.12
0.09∧)0.10∧08
9
10
9
18
18
0

(b)

46.
0.037∧)1.702∧
1.48
222
222
0

5. (a)

0.023
1.8∧)0.0∧414
36
54
54
0

(b)

2310.
0.0036∧)8.3160∧
72
111
108
36
36
0

6. (a)

137.26
3.8∧)521.6∧00
38
141
114
27 6
26 6
1 00
76
24 0
22 8
1 2

The answer rounded to the nearest tenth is 137.3.

(b)

0.0211
8.05∧)0.17∧0000
16 10
900
805
950
805
145

The answer rounded to the nearest thousandth is 0.021.

7.

15.9
28.5∧)454.4∧0
285
169 4
142 5
269 0
256 5
12 5

The truck got approximately 16 miles per gallon.

8.

5.8
0.12∧)0.69∧6
60
9 6
9 6
0

n is 5.8.

9. Find the sum of levels for the years 1985, 1990, and 1995.

9.30
8.68
+ 7.37
25.35

Then divide by three to obtain the average.

8.45
3)25.35
24
13
12
15
15
0

The three-year average is 8.45 million tons. The five-year average was found to be 8.156 in Example 9. Find the difference between the averages.

8.450
− 8.156
0.294

The three-year average differs from the five-year average by 0.294 million tons.

3.6 Practice Problems

1. (a)

$$\begin{array}{r} 0.3125 \\ 16\overline{)5.0000} \\ \underline{48} \\ 20 \\ \underline{16} \\ 40 \\ \underline{32} \\ 80 \\ \underline{80} \\ 0 \end{array}$$

(b)

$$\begin{array}{r} 0.1375 \\ 80\overline{)11.0000} \\ \underline{80} \\ 300 \\ \underline{240} \\ 600 \\ \underline{560} \\ 400 \\ \underline{400} \\ 0 \end{array}$$

2. (a) $0.6363 = 0.\overline{63}$

$$\begin{array}{r} 0.6363 \\ 11\overline{)7.0000} \\ \underline{66} \\ 40 \\ \underline{33} \\ 70 \\ \underline{66} \\ 40 \\ \underline{33} \\ 7 \end{array}$$

(b) $0.533 = 0.5\overline{3}$

$$\begin{array}{r} 0.533 \\ 15\overline{)8.000} \\ \underline{75} \\ 50 \\ \underline{45} \\ 50 \\ \underline{45} \\ 5 \end{array}$$

(c) $0.295454 = 0.29\overline{54}$

$$\begin{array}{r} 0.295454 \\ 44\overline{)13.000000} \\ \underline{88} \\ 420 \\ \underline{396} \\ 240 \\ \underline{220} \\ 200 \\ \underline{176} \\ 240 \\ \underline{220} \\ 200 \\ \underline{176} \\ 24 \end{array}$$

3. (a)

$$2\frac{11}{18} = 2 + \frac{11}{18}$$
$$= 2 + 0.6\overline{1}$$
$$= 2.6\overline{1}$$

$0.611 = 0.6\overline{1}$

$$\begin{array}{r} 0.611 \\ 18\overline{)11.000} \\ \underline{108} \\ 20 \\ \underline{18} \\ 20 \\ \underline{18} \\ 2 \end{array}$$

(b) $1.03703 = 1.\overline{037}$

$$\begin{array}{r} 1.03703 \\ 27\overline{)28.00000} \\ \underline{27} \\ 100 \\ \underline{81} \\ 190 \\ \underline{189} \\ 100 \\ \underline{81} \\ 19 \end{array}$$

4. $0.7916 = 0.792$ rounded to the nearest thousandth

$$\begin{array}{r} 0.7916 \\ 24\overline{)19.0000} \\ \underline{168} \\ 220 \\ \underline{216} \\ 40 \\ \underline{24} \\ 160 \\ \underline{144} \\ 16 \end{array}$$

5. Divide to find the decimal equivalent of $\frac{5}{8}$.

$$\begin{array}{r} 0.625 \\ 8\overline{)5.000} \\ \underline{48} \\ 20 \\ \underline{16} \\ 40 \\ \underline{40} \\ 0 \end{array}$$

In the hundredths place $2 < 3$, so we know

$$0.6\underline{2}5 < 0.6\underline{3}0.$$

Therefore, $\frac{5}{8} < 0.63$.

6.
$$0.3 \times 0.5 + (0.4)^3 - 0.036 = 0.3 \times 0.5 + 0.064 - 0.036$$
$$= 0.15 + 0.064 - 0.036$$
$$= 0.214 - 0.036$$
$$= 0.178$$

7. $6.56 \div (2 - 0.36) + (8.5 - 8.3)^2$

$= 6.56 \div (1.64) + (0.2)^2$	Parentheses
$= 6.56 \div 1.64 + 0.04$	Exponents
$= 4 + 0.04$	Divide
$= 4.04$	Add

3.7 Practice Problems

Practice Problem 1

(a) $385.98 + 875.34 \approx 400 + 900 = 1300$

(b) $0.0952 - 0.0579 \approx 0.09 - 0.05 = 0.04$

(c) $5876.34 \times 0.087 \approx$

$$\begin{array}{r} 6000 \\ \underline{\times\ 0.09} \\ 540.00 \end{array}$$

(d) $46{,}873 \div 8.456 \approx$

$$\begin{array}{r} 6250 \\ 8\overline{)50{,}000} \\ \underline{48} \\ 20 \\ \underline{16} \\ 40 \\ \underline{40} \\ 0 \end{array}$$

Practice Problem 2

1. Understand the problem.

Mathematics Blueprint for Problem Solving			
Gather the Facts	**What Am I Asked to Do?**	**How Do I Proceed?**	**Key Points to Remember**
She worked 51 hours. She gets paid \$9.36 per hour for 40 hours. She gets paid time-and-a-half for 11 hours.	Find the amount Melinda earned working 51 hours last week.	Add the earnings of 40 hours at \$9.36 per hour to the earnings of 11 hours at overtime pay.	Overtime pay is time-and-a-half, which is $1.5 \times \$9.36$.

2. Solve and state the answer:

(a) Calculate regular earnings for 40 hours.

$$\begin{array}{r} \$9.36 \\ \times \quad 40 \\ \hline \$374.40 \end{array}$$

(b) Calculate overtime pay rate.

$$\begin{array}{r} \$9.36 \\ \times \quad 1.5 \\ \hline \$14.04 \end{array}$$

(c) Calculate overtime earnings for 11 hours.

$$\begin{array}{r} \$14.04 \\ \times \quad 11 \\ \hline \$154.44 \end{array}$$

(d) Add the two amounts.

$$\begin{array}{rl} \$374.40 & \text{Regular earnings} \\ -\ 154.44 & \text{Overtime earnings} \\ \hline \$528.84 & \text{Total earnings} \end{array}$$

Melinda earned \$528.84 last week.

3. ***Check.*** $40 \times \$9 = \$360$ $10 \times \$14 = \$140$

$$\begin{array}{r} \$360 \\ +\ 140 \\ \hline \$500 \end{array}$$

The answer is reasonable.

Practice Problem 3

1. Understand the problem.

Mathematics Blueprint for Problem Solving			
Gather the Facts	**What Am I Asked to Do?**	**How Do I Proceed?**	**Key Points to Remember**
The total amount of steak is 17.4 pounds. Each package contains 1.45 pounds. Each package costs \$4.60 per pound.	**(a)** Find out how many packages of steak the butcher will have. **(b)** Find the cost of each package.	**(a)** Divide the total, 17.4, by the amount in each package, 1.45, to find the number of packages. **(b)** Multiply the cost of one pound, \$4.60, by the amount in one package, 1.45.	There will be two answers to this problem.

2. Solve and state the answer.

(a)

$$\begin{array}{r} 12. \\ 1.45_\wedge\overline{)17.40_\wedge} \\ 14\ 5 \\ \hline 290 \\ 290 \\ \hline 0 \end{array}$$

The butcher will have 12 packages of steak.

(b)

$$\begin{array}{r} \$4.60 \\ \times\ 1.45 \\ \hline \$6.67 \end{array}$$

Each package will cost \$6.67.

3. ***Check.***

(a)

$$\begin{array}{r} 11. \\ 1.5_\wedge\overline{)17.0_\wedge} \\ 15 \\ \hline 2\ 0 \end{array}$$

(b) $\$5 \times 1 = \$5$

The answers are reasonable.

Chapter 4

4.1 Practice Problems

1. **(a)** $\frac{36}{40} = \frac{9}{10}$ **(b)** $\frac{18}{15} = \frac{6}{5}$ **(c)** $\frac{220}{270} = \frac{22}{27}$

2. **(a)** $\frac{200}{450} = \frac{4}{9}$ **(b)** $\frac{300}{1200} = \frac{1}{4}$

3. $\frac{44 \text{ dollars}}{900 \text{ tons}} = \frac{11 \text{ dollars}}{225 \text{ tons}}$

4. $\frac{212 \text{ miles}}{4 \text{ hours}} = \frac{53 \text{ miles}}{1 \text{ hour}}$ 53 miles/hour

5.

selling price	\$170.40
− purchase price	− 129.60
= profit	\$ 40.80

She made a profit of \$40.80 on 120 batteries.

$$\begin{array}{r} .34 \\ 120\overline{)40.80} \\ \underline{360} \\ 480 \\ \underline{480} \\ 0 \end{array}$$

Her profit was \$0.34 per battery.

6. **(a)** $\frac{\$2.04}{12 \text{ ounces}} = \$0.17/\text{ounce}$ $\frac{\$2.80}{20 \text{ ounces}} = \$0.14/\text{ounce}$

(b) Fred saves \$0.03 per ounce by buying the larger size.

4.2 Practice Problems

1. 6 is to 8 as 9 is to 12.

$\frac{6}{8} = \frac{9}{12}$

2. $\frac{2 \text{ hours}}{72 \text{ miles}} = \frac{3 \text{ hours}}{108 \text{ miles}}$

3. **(a)** $\frac{10}{18} \stackrel{?}{=} \frac{25}{45}$

$18 \times 25 = 450$ ↑

$\frac{10}{18} \times \frac{25}{45}$ The cross products are equal.

↓ $10 \times 45 = 450$

Thus $\frac{10}{18} = \frac{25}{45}$. This is a proportion.

(b) $\frac{42}{100} \stackrel{?}{=} \frac{22}{55}$

$100 \times 22 = 2200$ ↑

$\frac{42}{100} \times \frac{22}{55}$ The cross products are not equal.

↓ $42 \times 55 = 2310$

Thus $\frac{42}{100} \neq \frac{22}{55}$. This is a not proportion.

4. **(a)** $\frac{2.4}{3} \stackrel{?}{=} \frac{12}{15}$

$3 \times 12 = 36$ ↑

$\frac{2.4}{3} \times \frac{12}{15}$ The cross products are equal.

↓ $2.4 \times 15 = 36$

Thus $\frac{2.4}{3} = \frac{12}{15}$. This is a proportion.

(b) $\frac{2\frac{1}{3}}{6} \stackrel{?}{=} \frac{14}{38}$

$2\frac{1}{3} \times 38 = \frac{7}{3} \times \frac{38}{1} = \frac{266}{3} = 88\frac{2}{3}$

$\frac{2\frac{1}{3}}{6} \times \frac{14}{38}$

$6 \times 14 = 84$

The cross products are not equal.

$2\frac{1}{3} \times 38 = 88\frac{2}{3}$

Thus $\frac{2\frac{1}{3}}{6} \neq \frac{14}{38}$. This is not a proportion.

5. **(a)** $\frac{1260}{7} \stackrel{?}{=} \frac{3530}{20}$

$7 \times 3530 = 24{,}710$ ↑

$\frac{1260}{7} \times \frac{3530}{20}$ The cross products are not equal.

↓ $1260 \times 20 = 25{,}200$

The rates are not equal.

(b) $\frac{2}{11} \stackrel{?}{=} \frac{16}{88}$

$11 \times 16 = 176$ ↑

$\frac{2}{11} \times \frac{16}{88}$ The cross products are equal.

↓ $2 \times 88 = 176$

The rates are equal.

4.3 Practice Problems

1. **(a)** $5 \times n = 45$

$\frac{5 \times n}{5} = \frac{45}{5}$

$n = 9$

(b) $7 \times n = 84$

$\frac{7 \times n}{7} = \frac{84}{7}$

$n = 12$

2. **(a)** $108 = 9 \times n$

$\frac{108}{9} = \frac{9 \times n}{9}$

$12 = n$

(b) $210 = 14 \times n$

$\frac{210}{14} = \frac{14 \times n}{14}$

$15 = n$

3. **(a)** $15 \times n = 63$

$\frac{15 \times n}{15} = \frac{63}{15}$

$n = 4.2$

$$\begin{array}{r} 4.2 \\ 15\overline{)63.0} \\ \underline{60} \\ 30 \\ \underline{30} \\ 0 \end{array}$$

(b) $39.2 = 5.6 \times n$

$\frac{39.2}{5.6} = \frac{5.6 \times n}{5.6}$

$7 = n$

$$\begin{array}{r} 7. \\ 5.6_\wedge\overline{)39.2_\wedge} \\ \underline{39\ 2} \\ 0 \end{array}$$

4. $\frac{24}{n} = \frac{3}{7}$

$24 \times 7 = n \times 3$

$168 = n \times 3$

$\frac{168}{3} = \frac{n \times 3}{3}$

$56 = n$

5. $\dfrac{176}{4} = \dfrac{286}{n}$

$176 \times n = 286 \times 4$
$176 \times n = 1144$
$\dfrac{176 \times n}{176} = \dfrac{1144}{176}$
$n = 6.5$

$$\begin{array}{r} 6.5 \\ 176\overline{)1144.0} \\ \underline{1056} \\ 88\,0 \\ \underline{88\,0} \\ 0 \end{array}$$

6. $\dfrac{n}{30} = \dfrac{\frac{2}{3}}{4}$

$30 \times \dfrac{2}{3} = n \times 4$
$20 = n \times 4$
$\dfrac{20}{4} = \dfrac{n \times 4}{4}$
$5 = n$

7. $80 \times 6 = 5 \times n$
$480 = 5 \times n$
$\dfrac{480}{5} = \dfrac{5 \times n}{5}$
$96 = n$

The answer is 96 dollars.

8. $264 \times 2 = 3.5 \times n$
$528 = 3.5 \times n$
$\dfrac{528}{3.5} = \dfrac{3.5 \times n}{3.5}$
$150.9 \approx n$

The answer is 150.9 meters.

4.4 Practice Problems

1. $\dfrac{27 \text{ defective engines}}{243 \text{ engines produced}} = \dfrac{n \text{ defective engines}}{4131 \text{ engines produced}}$

$27 \times 4131 = 243 \times n$
$111{,}537 = 243 \times n$
$\dfrac{111{,}537}{243} = \dfrac{243 \times n}{243}$
$459 = n$

Thus we estimate that 459 engines are defective.

2. $\dfrac{9 \text{ gallons of gas}}{234 \text{ miles traveled}} = \dfrac{n \text{ gallons of gas}}{312 \text{ miles traveled}}$

$9 \times 312 = 234 \times n$
$2808 = 234 \times n$
$\dfrac{2808}{234} = \dfrac{234 \times n}{234}$
$12 = n$

She will need 12 gallons of gas.

3. $\dfrac{80 \text{ revolutions per minute}}{16 \text{ miles per hour}} = \dfrac{90 \text{ revolutions per minute}}{n \text{ miles per hour}}$

$80 \times n = 16 \times 90$
$80 \times n = 1440$
$\dfrac{80 \times n}{80} = \dfrac{1440}{80}$
$n = 18$

Alicia will be riding 18 miles per hour.

4. $\dfrac{4050 \text{ walk-in}}{729 \text{ purchase}} = \dfrac{5500 \text{ walk-in}}{n \text{ purchase}}$

$4050 \times n = 729 \times 5500$
$4050 \times n = 4{,}009{,}500$
$\dfrac{4050 \times n}{4050} = \dfrac{4{,}009{,}500}{4050}$
$n = 990$

Tom will expect 990 people to make a purchase in his store.

5. $\dfrac{50 \text{ bears tagged in 1st sample}}{n \text{ bears in forest}} = \dfrac{4 \text{ bears tagged in 2nd sample}}{50 \text{ bears caught in 2nd sample}}$

$50 \times 50 = n \times 4$
$2500 = n \times 4$
$\dfrac{2500}{4} = \dfrac{n \times 4}{4}$
$625 = n$

We estimate that there are 625 bears in the forest.

Chapter 5

5.1 Practice Problems

1. (a) $\dfrac{51}{100} = 51\%$ (b) $\dfrac{68}{100} = 68\%$
(c) $\dfrac{7}{100} = 7\%$ (d) $\dfrac{26}{100} = 26\%$

2. (a) $\dfrac{238}{100} = 238\%$ (b) $\dfrac{121}{100} = 121\%$

3. (a) $\dfrac{0.5}{100} = 0.5\%$ (b) $\dfrac{0.06}{100} = 0.06\%$
(c) $\dfrac{0.003}{100} = 0.003\%$

4. (a) $47\% = 0.47$ (b) $2\% = 0.02$

5. (a) $80.6\% = 0.806$ (b) $2.5\% = 0.025$
(c) $0.29\% = 0.0029$ (d) $231\% = 2.31$

6. (a) $0.78 = 78\%$ (b) $0.02 = 2\%$
(c) $5.07 = 507\%$ (d) $0.029 = 2.9\%$
(e) $0.006 = 0.6\%$

5.2 Practice Problems

1. (a) $71\% = \dfrac{71}{100}$ (b) $25\% = \dfrac{25}{100} = \dfrac{1}{4}$
(c) $8\% = \dfrac{8}{100} = \dfrac{2}{25}$

2. (a) $8.4\% = 0.084 = \dfrac{84}{1000} = \dfrac{21}{250}$
(b) $28.5\% = 0.285 = \dfrac{285}{1000} = \dfrac{57}{200}$

3. (a) $170\% = 1.70 = 1\dfrac{7}{10}$ (b) $288\% = 2.88 = 2\dfrac{88}{100} = 2\dfrac{22}{25}$

4. $7\dfrac{5}{8}\% = 7\dfrac{5}{8} \div 100$
$= \dfrac{61}{8} \times \dfrac{1}{100}$
$= \dfrac{61}{800}$

5. $22\dfrac{3}{8}\% = 22\dfrac{3}{8} \div 100$
$= 22\dfrac{3}{8} \times \dfrac{1}{100}$
$= \dfrac{179}{8} \times \dfrac{1}{100}$
$= \dfrac{179}{800}$

6. $\dfrac{5}{8}$ $\quad \begin{array}{r} 0.625 \\ 8\overline{)5.000} \end{array}$ $\quad 62.5\%$

7. (a) $\dfrac{21}{25} = 0.84 = 84\%$ (b) $\dfrac{7}{16} = 0.4375 = 43.75\%$

8. (a) $\dfrac{7}{9} = 0.7777\overline{7} \approx 0.7778 = 77.78\%$
(b) $\dfrac{19}{30} = 0.6333\overline{3} \approx 0.6333 = 63.33\%$

9. $\frac{7}{12}$ If we divide

$$\begin{array}{r} 0.58 \\ 12\overline{)7.00} \\ \underline{60} \\ 100 \\ \underline{96} \\ 4 \end{array}$$

Thus $\frac{7}{12} = 58\frac{1}{3}\%$.

10.

Fraction	Decimal	Percent
$\frac{23}{99}$	0.2323	23.23%
$\frac{129}{250}$	0.516	51.6%
$\frac{97}{250}$	0.388	$38\frac{4}{5}\%$

5.3A Practice Problems

1. What is 26% of 35?
$n = 26\% \times 35$
2. Find 0.08% of 350.
$n = 0.08\% \times 350$
3. **(a)** $58\% \times n = 400$ **(b)** $9.1 = 135\% \times n$
4. What percent of 250 is 36?

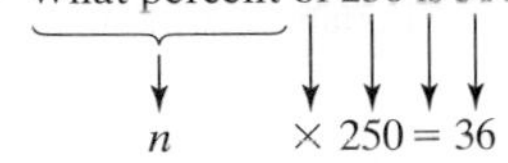

$n \times 250 = 36$

5. **(a)** $50 = n \times 20$ **(b)** $n \times 2000 = 4.5$
6. What is 82% of 350?
$n = 82\% \times 350$
$n = 0.82(350)$
$n = 287$
7. **(a)** $n = 230\% \times 400$
$n = (2.30)(400)$
$n = 920$
8. The problem asks: What is 8% of \$350?
$n = 8\% \times 350$
$n = \$28$ tax
9. $32 = 0.4\% \times n$
$32 = 0.004n$
$\frac{32}{0.004} = \frac{0.004n}{0.004}$
$8000 = n$
10. The problem asks: 30% of what is 6?
$30\% \times n = 6$
$0.30n = 6$
$\frac{0.30n}{0.30} = \frac{6}{0.30}$
$n = 20$
11. What percent of 9000 is 4.5?

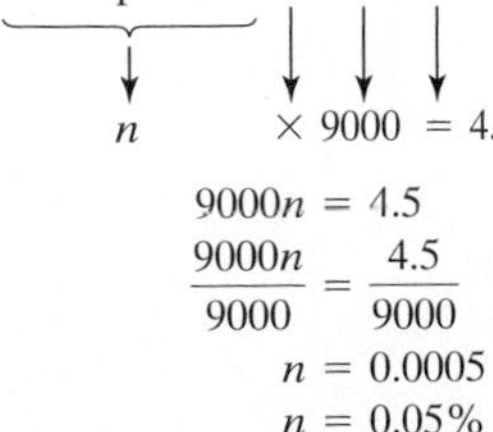

$n \times 9000 = 4.5$

$9000n = 4.5$
$\frac{9000n}{9000} = \frac{4.5}{9000}$
$n = 0.0005$
$n = 0.05\%$

12. $198 = n \times 33$
$\frac{198}{33} = \frac{n \times 33}{33}$
$6 = n$
Now express n as a percent: 600%
13. The problem asks: 5 is what percent of 16?
$5 = n \times 16$
$\frac{5}{16} = \frac{n \times 16}{16}$
$0.3125 = n$
Now express n as a percent rounded to tenths: 31.3%

5.3B Practice Problems

1. **(a)** Find 83% of 460.
$p = 83$
(b) 18% of what number is 90?
$p = 18$
(c) What percent of 64 is 8?
The percent is unknown. Use the variable p.
2. **(a)** $b = 52, a = 15.6$ **(b)** Base $= b, a = 170$
3. **(a)** What is 18% of 240?
Percent $p = 18$
Base $b = 240$
Amount is unknown; use the variable a.
(b) What percent of 64 is 4?
Percent is unknown; use the variable p.
Base $b = 64$
Amount $a = 4$
4. Percent $p = 340$
Base $b = 70$
Variable a
$\frac{a}{70} = \frac{340}{100}$
$100a = (70)(340)$
$a = 238$
Thus 340% of 70 is 238.
5. 68% of what is 476?
Percent $p = 68$
Base is unknown; use base $= b$.
Amount $a = 476$
$\frac{a}{b} = \frac{p}{100}$ becomes $\frac{476}{b} = \frac{68}{100}$
If we reduce the right-hand fraction, we have
$\frac{476}{b} = \frac{17}{25}$
$(476)(25) = 17b$ Using cross multiplication.
$11{,}900 = 17b$ Simplify.
$\frac{11{,}900}{17} = \frac{17b}{17}$ Divide each side by 17.
$700 = b$ Result of $11{,}900 \div 17$.
Thus 68% of 700 is 476.
6. Percent $p = 0.3$
Variable b
Amount $a = 216$
$\frac{216}{b} = \frac{0.3}{100}$
$0.3b = (216)(100)$
$b = 72{,}000$
Thus 216 is 0.3% of 72,000.
Therefore \$72,000 was exchanged.
7. Variable p
Base $b = 3500$
Amount $a = 105$
$\frac{105}{3500} = \frac{p}{100}$
$\frac{3}{100} = \frac{p}{100}$
$100p = (3)(100)$
$p = 3$
Thus 3% of 3500 is 105.

5.4 Practice Problems

1. Let n = number of people with reserved airline tickets.
   12% of $n = 4800$
   $0.12 \times n = 4800$
   $\frac{0.12 \times n}{0.12} = \frac{4800}{0.12}$
   $n = 40{,}000$
   40,000 people held airline tickets that month.
2. The problem asks: What is 8% of $62.30?
   $n = 0.08 \times 62.30$
   $n = 4.984$
   The tax is $4.98.
3. The problem asks: 105 is what percent of 130?
   $105 = n \times 130$
   $\frac{105}{130} = n$
   $0.8077 \approx n$
   Now express n as a percent rounded to tenths: 80.8%.
   Thus 80.8% of the flights were on time.
4. 100% Cost of meal + tip of 15% = $46.00
   Let n = Cost of meal
   100% of n + 15 of %n = $46.00
   115% of $n = 46.00$
   $1.15 \times n = 46.00$
   $n = 40.00$
   They can spend $40.00 on the meal itself.
5. **(a)** 7% of $13,600 is the discount.
   $0.07 \times 13{,}600$ = the discount
   $952 is the discount.
   (b) $13,600 list price
   − 952 discount
   $12,648 Amount Betty paid for the car.

5.5 Practice Problems

1. Commission = commission rate × value of sales
   Commission = 6% × $156,000
   = 0.06 × 156,000
   = 9360
   His commission is $9360.
2. 15,000
   −10,500
   4500 the amount of decrease
   Percent of decrease $= \frac{\text{amount of decrease}}{\text{original amount}} = \frac{\$4500}{\$15{,}000}$
   = 0.30 = 30%
   The percent of decrease is 30%.
3. $I = P \times R \times T$
   $P = \$5600 \quad R = 12\% \quad T = 1$ year
   $I = 5600 \times 12\% \times 1$
   $= 5600 \times 0.12$
   $= 672$
   The interest is $672.
4. **(a)** $I = P \times R \times T$
   $= 1800 \times 0.11 \times 4$
   $= \$792$
   (b) $I = P \times R \times T$
   $= 1800 \times 0.11 \times \frac{1}{2}$
   $= \$99$

Chapter 6

6.1 Practice Problems

1. **(a)** 3 **(b)** 5280 **(c)** 60 **(d)** 7 **(e)** 16
   (f) 2 **(g)** 4
2. $15{,}840 \text{ ft} \times \frac{1 \text{ mile}}{5280 \text{ ft}} = \frac{15{,}840}{5280}$ miles = 3 miles
3. **(a)** $18.93 \text{ miles} \times \frac{5280 \text{ feet}}{1 \text{ mile}} = 99{,}950.4$ feet
   (b) $16\frac{1}{2} \text{ inches} \times \frac{1 \text{ yard}}{36 \text{ inches}} = 16\frac{1}{2} \times \frac{1}{36}$ yard
   $= \frac{\overset{11}{33}}{2} \times \frac{1}{\underset{12}{36}}$ yard $= \frac{11}{24}$ yard
4. $760.5 \text{ lb} \times \frac{16 \text{ oz}}{1 \text{ lb}} = 760.5 \times 16 \text{ oz} = 12{,}168$ oz
5. $19 \text{ pints} \times \frac{1 \text{ quart}}{2 \text{ pints}} = \frac{19}{2}$ quarts = 9.5 quarts
6. **Step 1:** $26 \text{ yards} \times \frac{3 \text{ feet}}{1 \text{ yard}} = 26 \times 3$ feet = 78 feet
   Step 2: 78 feet + 2 feet = 80 feet
   The path is 80 feet long.
7. **Step 1:** $1\frac{3}{4} \text{ days} \times \frac{24 \text{ hours}}{1 \text{ day}} = \frac{7}{4} \times \frac{24}{1}$ hours = 42 hr parked
   Step 2: $42 \text{ hours} \times \frac{1.50 \text{ dollars}}{1 \text{ hour}} = 63.00$ dollars
   She paid $63.00.

6.2 Practice Problems

1. **(a)** deka- **(b)** milli-
2. **(a)** 4 meters = 4.00‸ centimeters = 400 cm
   (b) 30 centimeters = 30.0‸ millimeters = 300 mm
3. **(a)** 3 mm = ‸003. m = 0.003 m
   (b) 47 cm = ‸00047. km = 0.00047 km
4. The car length would logically be choice (b) 3.8 meters. (A meter is close to a yard and 3.8 yards seems reasonable)
5. **(a)** 3.75 m **(b)** 46,000 mm
6. **(a)** 389 mm = 0.0389 dam (four places to left)
   (b) 0.48 hm = 4800 cm (four places to right)
7. 782 cm = 7.82 m
   2 m = 2.00 m
   537 m = 537.00 m
   546.82 m

6.3 Practice Problems

1. **(a)** 5 L = 5000 mL **(b)** 84 kL = 84,000 L
   (c) 0.732 L = 732 mL
2. **(a)** 15.8 mL = 0.0158 L **(b)** 12,340 mL = 12.34 L
   (c) 86.3 L = 0.0863 kL
3. **(a)** 396 mL = 396 cm^3
   (because 1 milliliter = 1 cubic centimeter)
   (b) 0.096 L = 96 cm^3
4. **(a)** 3.2 t = 3200 kg **(b)** 7.08 kg = 7080 g
5. **(a)** 59 kg = 0.059 t **(b)** 28.3 mg = 0.0283 g
6. A gram is $\frac{1}{1000}$ of a kilogram. If the coffee costs $10.00 per kilogram, then 1 gram would be $\frac{1}{1000}$ of $10.
   $\frac{1}{1000} \times \$10 = \frac{\$10.00}{1000} = \$0.01$
   The coffee costs 1¢ per gram.
7. **(a)** 120 kg (A kg is slightly more than 2 lb.)

6.4 Practice Problems

1. **(a)** $7 \text{ ft} \times \frac{0.305 \text{ m}}{1 \text{ ft}} = 2.135$ m
2. **(a)** $17 \text{ m} \times \frac{1.09 \text{ yd}}{1 \text{ m}} = 18.53$ yard
   (b) $29.6 \text{ km} \times \frac{0.62 \text{ mi}}{1 \text{ km}} = 18.352$ mi
   (c) $26 \text{ gal} \times \frac{3.79 \text{ L}}{1 \text{ gal}} = 98.54$ L
   (d) $6.2 \text{ L} \times \frac{1.06 \text{ qt}}{1 \text{ L}} = 6.572$ qt

3. $180 \text{ cm} \times \frac{0.394 \text{ in.}}{1 \text{ cm}} \times \frac{1 \text{ ft}}{12 \text{ in.}} = 5.91 \text{ ft}$

4. $\frac{88 \text{ km}}{\text{hr}} \times \frac{0.62 \text{ mi}}{1 \text{ km}} = 54.56 \text{ mi/hr}$

5. $\frac{900 \text{ miles}}{\text{hr}} \times \frac{5280 \text{ ft}}{1 \text{ mile}} \times \frac{1 \text{ hr}}{60 \text{ min}} \times \frac{1 \text{ min}}{60 \text{ sec}}$

$= \frac{900 \times 5280 \text{ ft}}{60 \times 60 \text{ sec}} = \frac{4{,}752{,}000 \text{ ft}}{3600 \text{ sec}}$

$= 1320 \text{ ft/sec}$

The jet is traveling at 1320 feet per second.

6. $F = 1.8 \times C + 32$
$= 1.8 \times 20 + 32$
$= 36 + 32$
$= 68$

The temperature is 68°F.

7. $C = \frac{5 \times F - 160}{9}$
$= \frac{5 \times 86 - 160}{9}$
$= \frac{430 - 160}{9}$
$= \frac{270}{9}$
$= 30$

The temperature is 30°C.

Practice Problem 2

1. Understand the problem.

Mathematics Blueprint for Problem Solving

Gather the Facts	What Am I Asked to Do?	How Do I Proceed?	Key Points to Remember
He must use 18.06 liters of solution. He has 42 jars to fill.	Find out how many milliliters of solution will go into each jar.	We need to convert 18.06 liters to milliliters, and then divide that result by 42.	To convert 18.06 liters to milliliters, we move the decimal point three places to the right.

2. Solve and state the answer:

(a) 18.06 L = 18,060 mL

(b) $\frac{18{,}060 \text{ mL}}{42 \text{ jars}} = 430 \text{ mL/jar}$

Thus, 430 mL of a solution will go into each jar.

3. ***Check.*** 18.06 L is approximately 18 L, or 18,000 mL.

$\frac{18{,}000}{40} = 450$ and is reasonable.

6.5 Practice Problems

Practice Problem 1

Step 1:
$2\frac{2}{3}$ yd
$8\frac{1}{3}$ yd
$2\frac{2}{3}$ yd
$+\ 8\frac{1}{3}$ yd
22 yd

Step 2: $22 \text{ yd} \times \frac{3 \text{ ft}}{1 \text{ yd}} = 66 \text{ ft}$

The perimeter is 66 ft.

Chapter 7

7.1 Practice Problems

1. $\angle FGH$ and $\angle KGJ$ are acute angles, $\angle HGK$ and $\angle FGJ$ are obtuse angles, $\angle HGJ$ is a right angle, and $\angle FGK$ is a straight angle.
2. The complement of angle B is $90° - 83° = 7°$.
The supplement of angle B is $180° - 83° = 97°$.
3. $\angle y$ and $\angle w$ are vertical angles and so have the same measure. Thus $\angle w = 133°$. $\angle y$ and $\angle z$ are adjacent angles, so we know they are supplementary. Thus $\angle z$ measures $180° - 133° = 47°$. Finally, $\angle x$ and $\angle z$ are vertical angles, so we know they have the same measure. Thus $\angle x$ measures 47°.
4. $\angle z = 180° - 105° = 75°$ ($\angle x$ and $\angle z$ are adjacent angles).
$\angle x = \angle y = 105°$ ($\angle x$ and $\angle y$ are alternate interior angles).
$\angle v = \angle x = 105°$ ($\angle v$ and $\angle x$ are corresponding angles).
$\angle w = 180° - 105° = 75°$ ($\angle w$ and $\angle v$ are adjacent angles).

7.2 Practice Problems

1. $P = 2l + 2w$
$= 2(6) + 2(1.5)$
$= 12 + 3 = 15 \text{ m}$
2. $P = 4S$
$= 4(5.8) = 23.2 \text{ cm}$
3. $P = 4 + 4 + 5.5 + 2.5 + 1.5 + 1.5 = 19 \text{ ft}$
$\text{Cost} = 19 \text{ ft} \times \frac{0.16 \text{ dollar}}{1 \text{ ft}} = \$3.04$
4. $A = lw = (29)(17) = 493 \text{ m}^2$
5. $A = (s)^2$
$= (11.8)^2$
$= 139.24 \text{ mm}^2$
6. Area of rectangle $= (18)(20) = 360 \text{ ft}^2$
Area of square $= 6^2 = 36 \text{ ft}^2$
Total area $= 396 \text{ ft}^2$

7.3 Practice Problems

1. $P = (2)(7.6) + (2)(3.5)$
$= 15.2 + 7.0 = 22.2$ cm

2. $A = bh$
$= (10.3)(1.5)$
$= 15.45 \text{ km}^2$

3. $P = 4(6 \text{ cm}) = 24$ cm
$A = bh$
$= (4 \text{ cm})(6 \text{ cm}) = 24 \text{ cm}^2$

4. $P = 7 + 15 + 21 + 13 = 56$ yd

5. (a) $A = \dfrac{h(b + B)}{2} = \dfrac{140(180 + 130)}{2} = 21{,}700 \text{ yd}^2$
(b) $21{,}700 \text{ yd}^2 \times \dfrac{1 \text{ gallon}}{100 \text{ yd}^2} = 217$ gallons
Thus 217 gallons of sealer are needed.

6. The area of the trapezoid is
$A = \dfrac{(9.2)(12.6 + 19.8)}{2}$
$= \dfrac{(9.2)(32.4)}{2} = \dfrac{298.08}{2}$
$= 149.04 \text{ cm}^2$.
The area of the rectangle is
$A = (8.3)(12.6) = 104.58 \text{ cm}^2$
Total area $= 253.62 \text{ cm}^2$

7.4 Practice Problems

1. The sum of the angles in a triangle is 180°. The two given angles total $125° + 15° = 140°$. Thus $180° - 140° = 40°$. Angle A must be 40°.

2. $P = 10.5 + 10.5 + 8.5 = 29.5$ m

3. $A = \dfrac{bh}{2} = \dfrac{(38)(13)}{2} = \dfrac{494}{2} = 247 \text{ m}^2$

4. Area of rectangle $= (11)(24) = 264 \text{ cm}^2$
Area of triangle $= \dfrac{(11)(7)}{2} = \dfrac{77}{2} = 38.5 \text{ cm}^2$
Total area $= 302.5 \text{ cm}^2$

7.5 Practice Problems

1. (a) $\sqrt{49} = 7$ because $(7)(7) = 49$.
(b) $\sqrt{169} = 13$ because $(13)(13) = 169$.

2. $\sqrt{49} = 7$ because $(7)(7) = 49$.
$\sqrt{4} = 2$ because $(2)(2) = 4$.
Thus $\sqrt{49} - \sqrt{4} = 7 - 2 = 5$.

3. (a) Yes. 144 is a perfect square because $(12)(12) = 144$.
(b) $\sqrt{144} = 12$

4. (a) $\sqrt{3} \approx 1.732$ (b) $\sqrt{13} \approx 3.606$ (c) $\sqrt{5} \approx 2.236$

5. $\sqrt{22 \text{ m}^2} \approx 4.690$ m
Thus, to the nearest thousandth of an inch, the side measures 4.690 m.

7.6 Practice Problems

1. Hypotenuse $= \sqrt{(8)^2 + (6)^2}$
$= \sqrt{64 + 36}$ Square each value first.
$= \sqrt{100}$ Add together the two values.
$= 10$ m Take the square root.

2. Hypotenuse $= \sqrt{(3)^2 + (7)^2}$
$= \sqrt{9 + 49}$ Square each value first.
$= \sqrt{58}$ cm Add the two values together
Using the square root table or a calculator, we have the hypotenuse ≈ 7.616 m.

3. Leg $= \sqrt{(17)^2 - (15)^2}$
$= \sqrt{289 - 225}$ Square each value first.
$= \sqrt{64}$ Subtract.
$= 8$ m Find the square root.

4. Leg $= \sqrt{(10)^2 - (5)^2}$
$= \sqrt{100 - 25}$ Square each value first.
$= \sqrt{75}$ m Subtract the two numbers.
Using a calculator or the square root table, we find that the leg ≈ 8.660 m.

5. 1. Understand the problem.
We are given a picture.
The distance between the centers of the holes is the hypotenuse of the triangle.
2. Solve and state the answer.
Hypotenuse $= \sqrt{(\text{leg})^2 + (\text{leg})^2}$
$= \sqrt{(2)^2 + (5)^2}$
$= \sqrt{4 + 25}$
$= \sqrt{29}$
$\sqrt{29} \approx 5.385$ cm
Rounded to the nearest thousandth, the distance is 5.385 cm.
3. *Check.*
Work backward to check. Use the Pythagorean Theorem.
$5.385^2 \approx 2^2 + 5^2$? (We use $\approx$ because 5.385 is an approximate answer.)
$28.998 \approx 4 + 25$
$28.998 \approx 29$ ✓

6. 1. Understand the problem.
We are given a picture.
2. Solve and state the answer.
Leg $= \sqrt{(\text{hypotenuse})^2 - (\text{leg})^2}$
$= \sqrt{(30)^2 - (27)^2}$
$= \sqrt{900 - 729}$
$= \sqrt{171}$
$\sqrt{171} \approx 13.1$ yd
If we round to the nearest tenth, the kite is 13.1 yd above the rock.

7. (a) In a 30°–60°–90° triangle the side opposite the 30° angle is $\frac{1}{2}$ of the hypotenuse.
$$\frac{1}{2} \times 12 = 6$$
Therefore, $y = 6$ ft.
When we know two sides of a right triangle, we find the third side using the Pythagorean Theorem.
Leg $= \sqrt{(\text{hypotenuse})^2 - (\text{leg})^2}$
$= \sqrt{(12)^2 - (6)^2} = \sqrt{144 - 36}$
$= \sqrt{108} \approx 10.4$ ft
$x = 10.4$ ft rounded to the nearest tenth.
(b) In a 45°–45°–90° triangle we have the following:
Hypotenuse $= \sqrt{2} \times \text{leg}$
$\approx 1.414(8)$
$= 11.312$ m
Rounded to the nearest tenth, the hypotenuse $= 11.3$ m.

7.7 Practice Problems

1. $C = \pi d$
$= (3.14)(9)$
$= 28.26$
≈ 28.3 m to nearest tenth

2. (a) $C = \pi d$
$= (3.14)(30)$
$= 94.2$ in.
(b) Change 94.2 in. to ft.
$94.2 \text{ in.} \times \dfrac{1 \text{ ft}}{12 \text{ in.}} = 7.85$ ft
(c) When the wheel makes 2 revolutions, the bicycle travels $7.85 \times 2 = 15.7$ ft.

3. $A = \pi r^2$
$= (3.14)(5^2)$
$= (3.14)(25)$
$= 78.5 \text{ km}^2$

4. $r = \frac{d}{2} = \frac{10}{2} = 5 \text{ ft}$

$A = \pi r^2$
$= (3.14)(5)^2$
$= 78.5 \text{ ft}^2$

Change 78.5 ft² to yd²

$78.5 \text{ ft}^2 \times \frac{1 \text{ yd}^2}{9 \text{ ft}^2} \approx 8.7222 \text{ yd}^2$

Find the cost: $\frac{\$12}{1 \text{ ft}^2} \times 8.7222 \text{ yd}^2 = \$104.67.$

The cost of the pool cover is $104.67.

5. Area of square − area of circle = shaded area

s^2	−	πr^2	= shaded area	
5^2	−	$(3.14)(2^2)$		
25	−	$(3.14)(4)$		
25.00	−	12.56	≈	12.4 ft² (rounded to nearest tenth)

6. $r = \frac{d}{2} = \frac{8}{2} = 4 \text{ ft}$

$A \text{ semicircle} = \frac{\pi r^2}{2}$
$= \frac{(3.14)(4)^2}{2}$
$= 25.12 \text{ ft}^2$

$A \text{ rectangle} = lw = (12)(8) = 96 \text{ ft}^2.$

$$\begin{array}{r} 25.12 \text{ ft}^2 \\ +\ 96.00 \text{ ft}^2 \\ \hline 121.12 \text{ ft}^2 \end{array}$$

The total area is approximately 121.1 ft².

7.8 Practice Problems

1. $V = lwh$
$= (6)(5)(2)$
$= (30)(2)$
$= 60 \text{ m}^3$

2. $V = \pi r^2 h$
$= (3.14)(2)^2(5)$
$= (3.14)(4)(5)$
$= 62.8 \text{ in.}^3$

3. $V = \frac{4\pi r^3}{3} = \frac{(4)(3.14)(6^3)}{3} = \frac{(4)(3.14)(6)(6)\overset{2}{\cancel{(6)}}}{\underset{1}{\cancel{3}}}$

$= (12.56)(36)(2) = 904.32$
$\approx 904.3 \text{ m}^3$ rounded to nearest tenth

4. $V = \frac{\pi r^2 h}{3}$
$= \frac{(3.14)(5)^2(12)}{3}$
$= 314 \text{ m}^3$

5. (a) $V = \frac{Bh}{3}$
$= \frac{(6)(6)(10)}{3} = \frac{(36)(10)}{3}$
$= \frac{\overset{12}{\cancel{(36)}}(10)}{\underset{1}{\cancel{3}}} = 120 \text{ m}^3$

(b) $V = \frac{Bh}{3}$
$= \frac{(7)(8)(15)}{3} = \frac{(7)(8)\overset{5}{\cancel{(15)}}}{\underset{1}{\cancel{3}}}$
$= (56)(5) = 280 \text{ m}^3$

7.9 Practice Problems

1. $\frac{11}{27} = \frac{15}{n}$
$11n = (27)(15)$
$11n = 405$
$\frac{11n}{11} = \frac{405}{11}$
$n = 36.\overline{81}$
$= 36.8$ meters measured to the nearest tenth

2. a corresponds to p, b corresponds to m, c corresponds to n

3. $\frac{h}{5} = \frac{20}{2}$
$2h = 100$
$h = 50 \text{ ft}$

4. $\frac{3}{29} = \frac{1.8}{w}$
$3w = (1.8)(29)$
$3w = 52.2$
$\frac{3w}{3} = \frac{52.2}{3}$
$w = 17.4$ The width is 17.4 meters.

7.10 Practice Problems

Practice Problem 1

1. Understand the problem.

Mathematics Blueprint for Problem Solving			
Gather the Facts	**What Am I Asked to Do?**	**How Do I Proceed?**	**Key Points to Remember**
Mike needs to sand three rooms: 24 ft × 13 ft 12 ft × 9 ft 16 ft × 3 ft He can sand 80 ft² in 15 min.	Find out how long it will take him to sand all three rooms.	**(a)** Find the total area to be sanded. **(b)** Then find out how long it will take him to sand the total area.	Area = length × width To get the total time, multiply the total area by the fraction: $\frac{15 \text{ min.}}{80 \text{ ft}^2}$

2. Solve and state the answer:
$24 \times 13 = 312 \text{ ft}^2$ room 1
$12 \times 9 = 108 \text{ ft}^2$ room 2
$16 \times 3 = 48 \text{ ft}^3$ room 3
Total area $= 468 \text{ ft}^2$

$468 \text{ ft}^2 \times \dfrac{15 \text{ min}}{80 \text{ ft}^2} = 87.75$ min. It will take Mike 87.75 min. to sand the rooms.

3. ***Check.*** Estimate the answer to see if it is reasonable.

Practice Problem 2

1. Understand the problem.

Mathematics Blueprint for Problem Solving			
Gather the Facts	**What Am I Asked to Do?**	**How Do I Proceed?**	**Key Points to Remember**
The trapezoid has a height of 9 ft. The bases are 18 ft and 12 ft. The rectangular portion measures 24 ft × 15 ft. Roofing costs $2.75 per square yard.	**(a)** Find the area of the roof. **(b)** Find the cost to install new roofing.	**(a)** Find the area of the entire roof. Change square feet to square yards. **(b)** Multiply by $2.75.	9 square feet = 1 square yard

2. Solve and state the answer:

(a) Area of trapezoid $= \dfrac{1}{2}h(b + B)$

$= \dfrac{1}{2}(9)(12 + 18)$

$= 135 \text{ ft}^2$

Area of rectangle $= lw$

$= (15)(24)$

$= 360 \text{ ft}^2$

Total Area $= 135 \text{ ft}^2 + 360 \text{ ft}^2 = 495 \text{ ft}^2$
Change square feet to square yards.

$495 \text{ ft}^2 \times \dfrac{1 \text{ yd}^2}{9 \text{ ft}^2} = 55 \text{ yd}^2$

The area of the roof is 55 yd^2.

(b) Cost $= 55 \text{ yd}^2 \times \dfrac{\$2.75}{1 \text{ yd}^2} = \$151.25$

The cost to install new roofing would be $151.25.

3. ***Check.*** Estimate the answer to see if they seem reasonable.

Chapter 8

8.1 Practice Problems

1. The smallest category of students is special students.

2. (a) 3000 freshmen + 200 special students = 3200
There are 3200 students who are either freshmen or special students.

(b) 3200 out of 10,000 are freshmen or special students.
$\dfrac{3200}{10{,}000} = 0.32 = 32\%$

3. There are 3000 freshmen but only 2600 sophomores. The ratio of freshmen to sophomores is $\dfrac{3000}{2600} = \dfrac{15}{13}$.

4. The ratio of freshmen to the total number of students is $\dfrac{3000}{10{,}000} = \dfrac{3}{10}$.

5. Lake Ontario occupies the smallest area with 8%.

6. The percent of the total area occupied by either Lake Superior or Lake Michigan is: 34% + 24% = 58%.

7. Lake Superior has 34% of the area. 34% of $94{,}680 \text{ mi}^2$ is $(10.34)(94{,}680) = 32{,}191 \text{ mi}^2$.

8. (a) 26% + 14% = 40%
(b) 8% of 782,000 = (0.08)(782,000) = 62,560

8.2 Practice Problems

1. The bar rises to 24. The approximate population was 24,000,000.

2. 16,000,000 − 7,000,000 = 9,000,000. The population increased by 9,000,000.

3. The bar rises to 250. The number of new cars sold in the fourth quarter of 2004 was 250.

4. 150 − 100 = 50. Thus, 50 fewer cars were sold.

5. The greatest number of customers came in July, since the highest point of the graph occurs in July.

6. (a) 3500 **(b)** The number decreased.

7. The sharpest line of decrease is from July to August. The line slopes downward most steeply between those two months. Thus the biggest decrease in customers is between July and August.

8. Because the dot corresponding to 2002–2003 is at 20 and the scale is in hundreds, we have 20 × 100 = 2000. Thus, 2000 degrees in visual and performing arts were awarded.

9. The computer science line goes above the visual and performing arts line first in 1997–1998. Thus the first academic year with more degrees in computer science was 1997–1998.

8.3 Practice Problems

1. The 60–69 bar rises to a height of 6. Thus six tests would have a D grade.

2. From the histogram, 16 tests were 70–79, 8 tests were 80–89, and 6 tests were 90–99. When we combine 16 + 8 + 6 = 30, we can see that 30 students scored greater than 69 on the test.

3. The 800–999 bar rises to a height of 20. Thus 20 light bulbs lasted between 800 and 999 hours.

4. From the histogram, 25 bulbs lasted 1200–1399 hours, 10 lasted 1400–1599 hours, and 5 lasted 1600–1799 hours. When we combine 25 + 10 + 5 = 40, we can see that 40 light bulbs lasted more than 1199 hours.

5. Weight in Pounds

(Class Interval)	Tally	Frequency
1600–1799	||	2
1800–1999	||||	4
2000–2199	||||	4
2200–2399	~~||||~~	5

6. 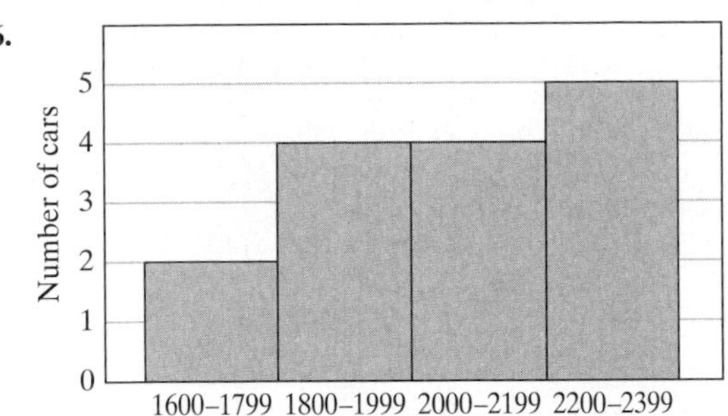

7. The greatest difference occurs between the 55–64 age category and the 35–54 category.

8.4 Practice Problems

1. $\dfrac{\$39.20 + \$43.50 + \$81.90 + \$34.20 + \$51.70 + \$48.10}{6} = \$49.77$

The mean monthly phone bill is \$49.77.

2. \$150, \$150, \$290 (three numbers)   \$320 (middle number)   \$400, \$450, \$600 (three numbers)

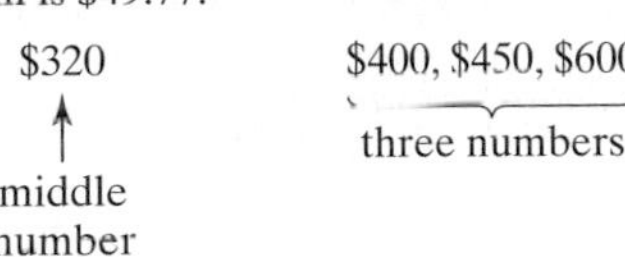

Thus, \$320 is the median salary.

3. 88, 90 (two numbers)   100, 105 (two middle numbers)   118, 126 (three numbers)

$\dfrac{100 + 105}{2} = \dfrac{205}{2} = 102.5$

The median is 102.5.

4. The value 71 occurs twice. The mode is 71 inches.
5. The value 3 occurs three times. The mode is 3.

Chapter 9

9.1 Practice Problems

1. (a) 4 lies to the right of 2, so $4 > 2$.
   (b) -3 lies to the right of -5, so $-3 > -5$.
   (c) 0 lies to the right of -6, so $0 > -6$.
   (d) -2 lies to the left of 1, so $-2 < 1$.
   (e) 5 lies to the right of -7, so $5 > -7$.

2. (a) $\begin{array}{r} 9 \\ +\ 14 \\ \hline 23 \end{array}$ (b) $\begin{array}{r} -\ 4.5 \\ +-\ 1.9 \\ \hline -\ 6.4 \end{array}$

3. (a) $\begin{array}{rcl} \dfrac{5}{12} & = & \dfrac{5}{12} \\ +\dfrac{1}{4}\times\dfrac{3}{3} & = & +\dfrac{3}{12} \\ \hline & & \dfrac{8}{12} = \dfrac{2}{3} \end{array}$

(b) The LCD = 42.

$\dfrac{1}{6}\times\dfrac{7}{7} = \dfrac{7}{42}$

Because $\dfrac{1}{6} = \dfrac{7}{42}$ it follows that $-\dfrac{1}{6} = -\dfrac{7}{42}$.

$\dfrac{2}{7}\times\dfrac{6}{6} = \dfrac{12}{42}$

Because $\dfrac{2}{7} = \dfrac{12}{42}$ it follows that $-\dfrac{2}{7} = -\dfrac{12}{42}$.

Thus

$\begin{array}{r} -\dfrac{1}{6} \\ +-\dfrac{2}{7} \\ \hline \end{array}$ is equivalent to $\begin{array}{r} -\dfrac{7}{42} \\ +-\dfrac{12}{42} \\ \hline -\dfrac{19}{42} \end{array}$

4. Add ($-\$20$ million) + ($-\$160$ million) to obtain $-\$180$ million. The total debt for these two years is \$180,000,000.

5. (a) $\begin{array}{r} 7 \\ +-12 \\ \hline -5 \end{array}$ (b) $\begin{array}{r} -20.8 \\ +\ \ 15.2 \\ \hline -\ 5.6 \end{array}$

   (c) $\dfrac{5}{6} + \left(-\dfrac{3}{4}\right) = \dfrac{10}{12} + \left(-\dfrac{9}{12}\right) = \dfrac{10 + (-9)}{12} = \dfrac{1}{12}$

6. $\begin{array}{r} 28°F \\ +-19°F \\ \hline 9°F \end{array}$

7. $\begin{array}{r} 36 \\ +-21 \\ \hline 15 \end{array}$ Then we add $\begin{array}{r} 15 \\ +-18 \\ \hline -3 \end{array}$

8. $\begin{array}{r} \$30,000 \\ +\ \$40,000 \\ \hline \$70,000 \end{array}$ $\begin{array}{r} \$20,000 \\ -\ \$5,000 \\ \hline +-\ \$35,000 \\ \hline -\ \$60,000 \end{array}$ $\begin{array}{r} \$70,000 \\ +-\ \$60,000 \\ \hline \$10,000 \end{array}$

The company had an overall profit of \$10,000 in the five-month period.

9.2 Practice Problems

1. $-10 - (-5) = -10 + 5 = -5$
2. (a) $5 - 12 = 5 + (-12) = -7$
   (b) $-11 - 17 = -11 + (-17) = -28$
3. (a) $3.6 - (-9.5) = 3.6 + 9.5 = 13.1$
   (b) $-\dfrac{5}{8} - \left(-\dfrac{5}{24}\right) = -\dfrac{5}{8} + \dfrac{5}{24}$
   $= -\dfrac{5}{8}\times\dfrac{3}{3} + \dfrac{5}{24}$
   $= -\dfrac{15}{24} + \dfrac{5}{24}$
   $= -\dfrac{10}{24}$ or $-\dfrac{5}{12}$
4. (a) $20 - (-5) = 20 + 5 = 25$
   (b) $-\dfrac{1}{5} - \left(-\dfrac{1}{2}\right) = -\dfrac{1}{5} + \dfrac{1}{2} = -\dfrac{2}{10} + \dfrac{5}{10} = \dfrac{3}{10}$
   (c) $3.6 - (-5.5) = 3.6 + 5.5 = 9.1$
5. $-5 - (-9) + (-14) = -5 + 9 + (-14) = 4 + (-14) = -10$
6. $31 - (-37) = 31 + 37 = 68$ The difference is 68°F.

9.3 Practice Problems

1. (a) $(6)(9) = 54$ (b) $(7)(12) = 84$
2. (a) $(-8)(5) = -40$ (b) $3(-60) = -180$

3. **(a)** $-50 \div 25 = -2$ **(b)** $49 \div (-7) = -7$

4. **(a)** $-10(-6) = 60$ **(b)** $\left(-\frac{1}{3}\right)\left(-\frac{2}{7}\right) = \frac{2}{21}$

5. **(a)** $-78 \div (-2) = 39$ **(b)** $(-1.2) \div (-0.5) = 2.4$

6. $(-6)(3)(-4) = (-18)(-4) = 72$

7. $(-2)(6) = -12$

Thus the change in charge would be -12.

8. $\dfrac{17 + 19 + 2 + (-4) + (-3) + (-13)}{6} = \dfrac{18}{6} = 3$

The average temperature was 3°.

9.4 Practice Problems

1. $20 \div (-5)(-3)$
$= (-4)\ (-3)$
$= 12$

2. **(a)** $25 \div (-5) + 16 \div (-8)$
$= (-5) + (-2)$
$= -7$

(b) $9 + 20 \div (-4)$
$= 9 + (-5)$
$= 4$

3. $\dfrac{9(-3) - 5}{2(-4) \div (-2)} = \dfrac{-27 - 5}{-8 \div (-2)} = \dfrac{-32}{+4} = -8$

4. $-2(-12 + 15) + (-3)^4 + 2(-6)$
$= -2(3) + (-3)^4 + 2(-6)$
$= -2(3) + 81 + 2(-6)$
$= -6 + 81 + (-12)$
$= 75 + (-12)$
$= 63$

5. $\left(\frac{1}{5}\right)^2 + 4\left(\frac{1}{5} - \frac{3}{10}\right) \div \frac{2}{3}$
$= \left(\frac{1}{5}\right)^2 + 4\left(\frac{2}{10} - \frac{3}{10}\right) \div \frac{2}{3}$
$= \left(\frac{1}{5}\right)^2 + 4\left(-\frac{1}{10}\right) \div \frac{2}{3}$
$= \frac{1}{25} + 4\left(-\frac{1}{10}\right) \div \frac{2}{3}$
$= \frac{1}{25} + \left(-\frac{2}{5}\right) \div \frac{2}{3}$
$= \frac{1}{25} + \left(-\frac{3}{5}\right)$
$= \frac{1}{25} + \left(-\frac{15}{25}\right)$
$= -\frac{14}{25}$

9.5 Practice Problems

1. **(a)** 3.729×10^3 **(b)** 5.06936×10^5

2. **(a)** 7.6×10^{-2} **(b)** 9.82×10^{-1}

3. $6.543 \times 10^3 = 6543$

4. **(a)** 430,000 **(b)** 60,000

5. **(a)** $7.72 \times 10^{-3} = 0.00772$ **(b)** $2.6 \times 10^{-5} = 0.000026$

6. 6.85×10^{22} kilograms
$+2.09 \times 10^{22}$ kilograms
8.94×10^{22} kilograms

7. $3.1 \times 10^4 - 31{,}000$

But $31{,}000 = 0.31 \times 10^5$

4.36×10^5
-0.31×10^5
4.05×10^5

Chapter 10

10.1 Practice Problems

1. **(a)** The variables are A, b and h.
(b) The variables are V, l, w, and h.

2. **(a)** $p = 2w + 2l$ **(b)** $A = \pi r^2$

3. We add $9 + 2 = 11$, therefore $9x + 2x = 11x$.

4. **(a)** $8x - 22x + 5x = 8x + (-22x) + 5x = (-14x) + 5x = -9x$
(b) $19x - 7x - 12x = 19x + (-7x) + (-12x)$
$= 12x + (-12x) = 0$

5. **(a)** $9x - 12x + x = 9x + (-12x) + 1x = -3x + 1x = -2x$
(b) $5.6x - 8x - x = 5.6x - 8x - 1x = -2.4x - 1x = -3.4x$

6. $17.5 - 6.3x - 8.2x + 10.5 =$
$17.5 + 10.5 + (-6.3x) + (-8.2x) = 28 - 14.5x$

7. **(a)** $2w + 3z - 12 - 5w - z - 16 =$
$2w + (-5w) + 3z + (-1z) + (-12) + (-16)$
$= -3w + 2z - 28$
(b) $\frac{3}{5}x + 5 - \frac{7}{15}x - \frac{1}{3} = \frac{9}{15}x - \frac{7}{15}x + \frac{15}{3} - \frac{1}{3} = \frac{2}{15}x + \frac{14}{3}$

10.2 Practice Problems

1. **(a)** $(7)(x + 5) = 7(x) + 7(5) = 7x + 35$
(b) $(-4)(x + 2y) = -4(x) + (-4)(2y) = -4x - 8y$
(c) $5(6a - 2b) = 5(6a) - 5(2b) = 30a - 10b$

2. $(x + 3y)(8) = (x)(8) + (3y)(8) = 8x + 24y$

3. **(a)** $(-5)(x + 4y + 5) = (-5)(1x) + (-5)(4y) + (-5)(5)$
$= -5x - 20y - 25$
(b) $(3)(2.2x + 5.5y + 6) = (3)(2.2x) + (3)(5.5y) + (3)(6)$
$= 6.6x + 16.5y + 18$

4. $\left(\frac{3}{2}\right)\left(\frac{1}{2}x - \frac{1}{3}y + 4z - \frac{1}{2}\right)$
$= \left(\frac{3}{2}\right)\left(\frac{1}{2}\right)(x) + \left(\frac{3}{2}\right)\left(-\frac{1}{3}\right)(y) + \left(\frac{3}{2}\right)(4)(z) + \left(\frac{3}{2}\right)\left(-\frac{1}{2}\right)$
$= \frac{3}{4}x - \frac{1}{2}y + 6z - \frac{3}{4}$

5. $3(2x + 4y) + 2(5x + y) = 6x + 12y + 10x + 2y = 16x + 14y$

6. $-4(x - 5) + 3(-1 + 2x)$
$= -4x + 20 - 3 + 6x$
$= 2x + 17$

10.3 Practice Problems

1.
$x + 7 = -8$
$x + 7 + (-7) = -8 + (-7)$
$x + 0 = -15$
$x = -15$

2. **(a)**
$y - 3.2 = 9$
$y - 3.2 + 3.2 = 9.0 + 3.2$
$y = 12.2$

Check.
$y - 3.2 = 9$
$(12.2) - 3.2 \stackrel{?}{=} 9$
$9 = 9$ ✓

(b)
$\frac{2}{3} = x + \frac{1}{6}$
$\frac{4}{6} + \left(-\frac{1}{6}\right) = x + \frac{1}{6} + \left(-\frac{1}{6}\right)$
$\frac{3}{6} = x$
$\frac{1}{2} = x$

Check.
$\frac{2}{3} = x + \frac{1}{6}$
$\frac{2}{3} \stackrel{?}{=} \left(\frac{1}{2}\right) + \frac{1}{6}$
$\frac{2}{3} \stackrel{?}{=} \frac{3}{6} + \frac{1}{6}$
$\frac{2}{3} = \frac{4}{6}$ ✓

3.
$3x - 5 = 2x + 1$
$3x - 5 + 5 = 2x + 1 + 5$
$3x = 2x + 6$
$3x + (-2x) = 2x + (-2x) + 6$
$x = 6$

Check.
$3x - 5 = 2x + 1$
$3(6) - 5 \stackrel{?}{=} 2(6) + 1$
$18 - 5 \stackrel{?}{=} 12 + 1$
$13 = 13$ ✓

10.4 Practice Problems

1. $8n = 104$

$\frac{8n}{8} = \frac{104}{8}$

$n = 13$

2. $-7n = 30$

$\frac{-7n}{-7} = \frac{30}{-7}$

$n = -\frac{30}{7}$

3. $3.2x = 16$

$\frac{3.2x}{3.2} = \frac{16}{3.2}$

$x = 5$

Check.

$3.2(5) \stackrel{?}{=} 16$

$16 = 16$ ✓

4. $\frac{1}{6}y = 2\frac{2}{3}$

$\left(\frac{1}{6}\right)y = \frac{8}{3}$

$\left(\frac{6}{1}\right)\left(\frac{1}{6}\right)y = \left(\frac{8}{3}\right)\left(\frac{6}{1}\right)$

$y = 16$

Check.

$\frac{1}{6}y = 2\frac{2}{3}$

$\left(\frac{1}{6}\right)(16) \stackrel{?}{=} 2\frac{2}{3}$

$\frac{16}{6} \stackrel{?}{=} \frac{8}{3}$

$\frac{8}{3} = \frac{8}{3}$ ✓

(b) $\left(3\frac{1}{5}\right)z = 4$

$\frac{16}{5}z = 4$

$\left(\frac{5}{16}\right)\left(\frac{16}{5}\right)z = (4)\left(\frac{5}{16}\right)$

$z = \frac{5}{4}$

Check.

$3\frac{1}{5}z = 4$

$\left(3\frac{1}{5}\right)\left(\frac{5}{4}\right) \stackrel{?}{=} 4$

$\left(\frac{16}{5}\right)\left(\frac{5}{4}\right) \stackrel{?}{=} 4$

$4 = 4$ ✓

10.5 Practice Problems

1. $5x + 13 = 33$

$5x + 13 + (-13) = 33 + (-13)$

$5x = 20$

$\frac{5x}{5} = \frac{20}{5}$

$x = 4$

Check.

$5(4) + 13 \stackrel{?}{=} 33$

$20 + 13 \stackrel{?}{=} 33$

$33 = 33$ ✓

2. $-50 = 7x - 8$

$-50 + 8 = 7x - 8 + 8$

$-42 = 7x$

$\frac{-42}{7} = \frac{7x}{7}$

$-6 = x$

Check.

$-50 \stackrel{?}{=} 7(-6) - 8$

$-50 \stackrel{?}{=} -42 - 8$

$-50 = -50$ ✓

3. $4x = -8x + 42$

$4x + 8x = -8x + 8x + 42$

$12x = 42$

$\frac{12x}{12} = \frac{42}{12}$

$x = \frac{7}{2}$

4. $4x - 7 = 9x + 13$

$4x - 7 + 7 = 9x + 13 + 7$

$4x = 9x + 20$

$4x + (-9x) = 9x + (-9x) + 20$

$-5x = 20$

$\frac{-5x}{-5} = \frac{20}{-5}$

$x = -4$

5. $4x - 23 = 3x + 7 - 2x$

$4x - 23 = x + 7$

$4x + (-1x) - 23 = (-1x) + x + 7$

$3x - 23 = 7$

$3x - 23 + 23 = 7 + 23$

$3x = 30$

$\frac{3x}{3} = \frac{30}{3}$

$x = 10$

6. $8(x - 3) + 5x = 15(x - 2)$

$8x - 24 + 5x = 15x - 30$

$13x - 24 = 15x - 30$

$13x - 24 + (30) = 15x - 30 + (30)$

$13x + 6 = 15x$

$13x + (-13x) + 6 = 15x + (-13x)$

$6 = 2x$

$\frac{6}{2} = \frac{2x}{2}$

$3 = x$

10.6 Practice Problems

1. Tom's height is 7 inches more than Abdul's height

Tom's height	is	7 inches	more than	Abdul's height
↓	↓	↓	↓	↓
t	=	7	+	a

2. The noon class has 24 fewer students than the morning class

The noon class	has	24 fewer students than		the morning class
↓	↓			
n	=	m	−	24

3. On Thursday carried 5 more than Friday. $t = 5 + f$

4. Let w = width of the rectangle. $l = 2w + 7$ is the length of the rectangle.

5. Let s = length in miles of Sally's trip

$s + 380$ = length in miles of Melinda's trip

380 miles longer then Sally's trip.

6. Let m = height in feet of McCormick Hall

$m - 126$ = height in feet of Larson Center

126 feet shorter than McCormick Hall.

7. Let x = length in inches of the first side of the triangle

$2x$ = length in inches of the second side of the triangle

$x + 6$ = length in inches of the third side of the triangle

10.7 Practice Problems

1. Let x = length in feet of shorter piece of board

$x + 4.5$ = length in feet of longer piece of board

$x + x + 4.5 = 18$

$2x + 4.5 = 18$

$2x + 4.5 + (-4.5) = 18 + (-4.5)$

$2x = 13.5$

$\frac{2x}{2} = \frac{13.5}{2}$

$x = 6.75$

The shorter piece is 6.75 feet long.

$x + 4.5 = 6.75 + 4.5 = 11.25$

The longer piece is 11.25 feet long.

Check.

$6.75 + 11.25 \stackrel{?}{=} 18$

$18 = 18$ ✓

2. Let

$$\begin{aligned} x &= \text{the number of departures on Monday} \\ x + 29 &= \text{the number of departures on Tuesday} \\ x - 16 &= \text{the number of departures on Wednesday} \\ x + x + 29 + x - 16 &= 349 \\ 3x + 13 &= 349 \\ 3x + 13 + (-13) &= 349 + (-13) \\ 3x &= 336 \\ \frac{3x}{3} &= \frac{336}{3} \\ x &= 112 \\ x + 29 &= 112 + 29 = 141 \\ x - 16 &= 112 - 16 = 96 \end{aligned}$$

There were 112 departures on Monday, 141 departures on Tuesday, and 96 departures on Wednesday.

Check.

$112 + 141 + 96 \stackrel{?}{=} 349$

$349 = 349$ ✓

$141 \stackrel{?}{=} 112 + 29$

$141 = 141$ ✓

$96 \stackrel{?}{=} 112 - 16$

$96 = 96$ ✓

3. Let

$$\begin{aligned} w &= \text{the width of the field measured in feet} \\ 2w + 8 &= \text{the length of the field measured in feet} \\ 2(\text{width}) + 2(\text{length}) &= \text{perimeter} \\ 2(w) + 2(2w + 8) &= 772 \\ 2w + 4w + 16 &= 772 \\ 6w + 16 &= 772 \\ 6w + 16 + (-16) &= 772 + (-16) \\ 6w &= 756 \\ \frac{6w}{6} &= \frac{756}{6} \\ w &= 126 \end{aligned}$$

The width of the field is 126 feet.

$2w + 8 = 2(126) + 8 = 252 + 8 = 260$

The length of the field is 260 feet.

Check.

$260 \stackrel{?}{=} 2(126) + 8$

$260 \stackrel{?}{=} 252 + 8$

$260 = 260$ ✓

$2(126) + 2(260) \stackrel{?}{=} 772$

$252 + 520 \stackrel{?}{=} 772$

$772 = 772$ ✓

4. Let

$$\begin{aligned} x &= \text{the length in meters of the first side of the triangle} \\ 2x &= \text{the length in meters of the second side of the triangle} \\ x + 10 &= \text{the length in meters of the third side of the triangle} \\ x + 2x + x + 10 &= 36 \\ 4x + 10 &= 36 \\ 4x + 10 + (-10) &= 36 + (-10) \\ 4x &= 26 \\ \frac{4x}{4} &= \frac{26}{4} \\ x &= 6.5 \end{aligned}$$

The first side of the triangle is 6.5 meters long.

$2x = 2(6.5) = 13$

The second side of the triangle is 13 meters long.

$x + 10 = 6.5 + 10 = 16.5$

The third side of the triangle is 16.5 meters long.

Check.

$6.5 + 13 + 16.5 \stackrel{?}{=} 36$

$36 = 36$ ✓

$13 \stackrel{?}{=} 2(6.5)$

$13 = 13$ ✓

$16.5 \stackrel{?}{=} 10 + 6.5$

$16.5 = 16.5$

5. Let

$$\begin{aligned} x &= \text{the measure of angle } A \text{ in degrees} \\ 3x &= \text{the measure of angle } C \text{ in degrees} \\ x - 30 &= \text{the measure of angle } B \text{ in degrees} \\ x + 3x + x - 30 &= 180 \\ 5x - 30 &= 180 \\ 5x - 30 + 30 &= 180 + 30 \\ 5x &= 210 \\ \frac{5x}{5} &= \frac{210}{5} \\ x &= 42 \end{aligned}$$

Angle A measures 42 degrees.

$3x = 3(42) = 126$

Angle C measures 126 degrees.

$x - 30 = 42 - 30 = 12$

Angle B measures 12 degrees.

Check.

$42 + 126 + 12 \stackrel{?}{=} 180$

$180 = 180$ ✓

$126 \stackrel{?}{=} 3(42)$

$126 = 126$ ✓

$12 \stackrel{?}{=} 42 - 30$

$12 = 12$ ✓

6. Let

$$\begin{aligned} s &= \text{the total amount of sales of the boats in dollars} \\ 0.03s &= \text{the amount of the commission earned on sales of } s \text{ dollars} \\ 3250 &= 1000 + 0.03s \\ 2250 &= 0.03s \\ \frac{2250}{0.03} &= \frac{0.03s}{0.03} \\ 75{,}000 &= s \end{aligned}$$

Therefore he sold \$75,000 worth of boats for the month.

Check.

$(0.03)(75{,}000) + 1000 \stackrel{?}{=} 3250$

$2250 + 1000 \stackrel{?}{=} 3250$

$3250 = 3250$ ✓

Appendix A.1 Balancing a Checking Account

Practice Problems

1.

CHECK REGISTER *My Chung Nguyen* 20 *06*

CHECK NO.	DATE	DESCRIPTION OF TRANSACTION	PAYMENT/ DEBIT (−)	✓	DEPOSIT/ CREDIT (+)	BALANCE
						\$ 1434 52
144	3/1	Leland Mortgage Company	908 00			526 52
145	3/1	Phone Company	33 21			493 31
146	3/2	Sam's Food Market	102 37			390 94
	3/2	Deposit			524 41	915 35

To find the ending balance we subtract each check written and add the deposit to the current balance. Then we record these amounts in the check register.

$$\begin{array}{r} 1434.52 \\ -\ 908.00 \\ \hline 526.52 \end{array} \quad \begin{array}{r} 526.52 \\ -\ 33.21 \\ \hline 493.31 \end{array} \quad \begin{array}{r} 493.31 \\ -\ 102.37 \\ \hline 390.94 \end{array} \quad \begin{array}{r} 390.94 \\ +\ 524.41 \\ \hline 915.35 \end{array}$$

My Chung's balance is \$915.35

2.

CHECKING RECONCILEMENT This form is provided to assist you in balancing your checking account.

List checks outstanding* not charged to your checking account

CHECK NO.	AMOUNT	
215	89	75
217	205	99
TOTAL	295	74

* and ATM withdrawals

Period ending 8/1 ,2006

1. Check Register Balance		\$ 1050.46
Subtract any charges listed on the bank statement which you have not previously deducted from your balance.	−	\$ 12.75
Adjusted Check Register Balance		\$ 1037.71
2. **Enter** the ending balance shown on the bank statement.		\$ 934.95
3. **Enter** deposits made later than the ending date on the bank statement.	+	\$ 398.50
	+	\$
	+	\$
TOTAL (Step 2 plus Step 3)		\$ 1333.45
4. In your check register, **check off** all the checks paid. In the ares provided to the left, **list** numbers and amounts of all outstanding checks and ATM withdrawals.		
5. **Subtract** the total amount in Step 4.	−	\$ 295.74
6. This adjusted bank balance should equal the adjusted Check Register Balance from Step 1.		\$ 1037.71

The balances in steps **1** and **6** are equal, so Anthony's checkbook is balanced.

Appendix A.2 Purchasing a Home

Practice Problems

1. To find the down payment we calculate 10% of 110,000.

down payment = percent × purchase price of home
= 0.10 × 110,000 We must change 10% to a decimal: 10% = 0.10.
= 11,000

Alyssa made a \$11,000 down payment.

To find the amount of the home loan, we subtract the down payment from the purchase price.

home loan amount = purchase price − down payment
= 110,000 11,000 = 99,000

The home loan is for \$99,000.

2. loan fee = percent × home loan amount
= 3% × 99,000 = 0.03 × 99,000
= 2970

{ 3 points = 3%; Change 3% to the decimal 0.03. }

The loan fee is \$2970.

escrow charges = $1\frac{1}{2}$% × purchase price of home
= 0.015 × 110,000 Change $1\frac{1}{2}$% to a decimal: $1\frac{1}{2}$% = 1.5% = 0.015.
= 1650

The escrow charges are \$1650.

3. We fill in the table with the correct amounts and then add to find the total initial costs.

Down payment	\$11,000
Loan Fee	\$2970
Escrow charges	\$1650
Total Initial Costs	\$15,620

The total initial costs are \$15,620.

4. We must find the payment factor on the table, and then multiply this amount times the amount of the home loan. To find the payment factor we locate the column labeled 6% and the row labeled 25 years, then we find where this column and row meet. The payment factor is 0.006443.

house payment = payment factor × home loan amount
= 0.006443 × 99,000
= 637.857 or \$637.86 rounded to the nearest cent.

The monthly house payment is \$637.86.

5. We divide each annual amount by 12 to get the monthly cost.

monthly tax payment = yearly property tax ÷ 12
= 1650 ÷ 12 = 137.5

Alyssa's monthly tax payment is \$137.50.

monthly insurance payment = yearly premium ÷ 12
= 210 ÷ 12 = 17.5

Alyssa's monthly insurance payment is \$17.50.

6. We fill in the table with the correct amounts and then add to find the total monthly payment.

House payment	\$637.86
Monthly tax payment	\$137.50
Monthly insurance payment	\$17.50
Total monthly payment	\$792.86

Alyssa's total monthly payment is \$792.86.

Appendix A.3 Purchasing a Vehicle

Practice Problems

1. (a) sales tax = 7% of sale price
= 0.07 × 24,999
sales tax = \$1749.93
license fee = 2% of sale price
= 0.02 × 24,999
license fee = \$499.98

(b) purchase price = sales price + sales tax + license fee + extended warranty
= 24,999 + 1749.93 + 499.98 + 1275
purchase price = \$28,523.91

2. (a) down payment = percent × purchase price.
down payment = 15% × 32,499
= 0.15 × 32,499
down payment = \$4874.85

(b) amount financed = purchase price − down payment
= 32,499 − 4874.85
amount financed = \$27,624.15

3. First, we find the total cost of the mini van at Dealer 1.

Total Cost = (monthly payment × number of months in loan) + down payment
= (453.21 × 60) + 0 There is no down payment.
= 27,192.60

The total cost of the mini van at Dealership 1 is \$27,192.60.

Next, we find the cost of the mini van at Dealer 2.

Total Cost = (monthly payment × number of months in loan) + down payment
= (479.17 × 48) + 3000 We multiply, then add.
= 26,000.16

The total cost of the mini van at Dealership 2 is \$26,000.16.

We see that the best deal on the mini van Phoebe plans to buy is at Dealership 2.

Appendix C Practice Problems

1. 3792 [+] 5896 [=] 9688
2. 7930 [−] 5096 [=] 2834
3. 896 [×] 273 [=] 244608
4. 2352 [÷] 16 [=] 147
5. 72.8 [×] 197 [=] 14341.6
6. 52.98 [+] 31.74 [+] 40.37 [+] 99.82 [=] 224.91
7. 0.0618 [×] 19.22 [−] 59.38 [÷] 166.3 [=] 0.830730455
8. 3.152 [×] [(] 0.1628 [+] 3.715 [−] 4.985 [)] [=] −3.4898944
9. 0.5618 [×] 98.3 [+/−] [−] 76.31 [×] 2.98 [+/−] [=] 172.17886
10. 3.76 [EXP] 15 [÷] 7.76 [EXP] 7 [=] 48453608.25
11. 6.238 [y^x] 6 [=] 58921.28674
12. 132.56 [x^2] 17572.1536
13. [(] 0.0782 [−] 0.0132 [+] 0.1364 [)] [$\sqrt{\ }$] 0.448776113

Answers to Selected Exercises

Chapter 1

1.1 Exercises

1. 6000 + 700 + 30 + 1 **3.** 100,000 + 8000 + 200 + 70 + 6 **5.** 20,000,000 + 3,000,000 + 700,000 + 60,000 + 1000 + 300 + 40 + 5 **7.** 100,000,000 + 3,000,000 + 200,000 + 60,000 + 700 + 60 + 8 **9.** 671 **11.** 9863 **13.** 40,885 **15.** 706,200 **17.** (a) 7 (b) 50,000 **19.** (a) 2 (b) 200,000 **21.** fifty-three **23.** nine thousand, three hundred four **25.** thirty-six thousand, one hundred eighteen **27.** one hundred five thousand, two hundred sixty-one **29.** fourteen million, two hundred three thousand, three hundred twenty-six **31.** four billion, three hundred two million, one hundred fifty-six thousand, two hundred **33.** 1561 **35.** 27,382 **37.** 100,079,826 **39.** one thousand, nine hundred sixty-five **41.** 9 million or 9,000,000 **43.** 33 million or 33,000,000 **45.** 930,000 **47.** 52,566,000 **49.** (a) 5 (b) 2 **51.** (a) 8 (b) 9 **53.** 613,001,033,208,003 **55.** three quintillion, six hundred eighty-two quadrillion, nine hundred sixty-eight trillion, nine billion, nine hundred thirty-one million, nine hundred sixty thousand, seven hundred forty-seven **57.** You would obtain 2 E 20. This is 200,000,000,000,000,000,000 in standard form.

1.2 Exercises

1. Answers may vary. A sample is: You can change the order of the addends without changing the sum.

3.

+	3	5	4	8	0	6	7	2	9	1
2	5	7	6	10	2	8	9	4	11	3
7	10	12	11	15	7	13	14	9	16	8
5	8	10	9	13	5	11	12	7	14	6
3	6	8	7	11	3	9	10	5	12	4
0	3	5	4	8	0	6	7	2	9	1
4	7	9	8	12	4	10	11	6	13	5
1	4	6	5	9	1	7	8	3	10	2
8	11	13	12	16	8	14	15	10	17	9
6	9	11	10	14	6	12	13	8	15	7
9	12	14	13	17	9	15	16	11	18	10

5. 23 **7.** 26 **9.** 57 **11.** 99 **13.** 4579 **15.** 9994 **17.** 13,861 **19.** 117,240 **21.** 121 **23.** 1143 **25.** 10,130 **27.** 11,579,426 **29.** 1,135,280,240 **31.** 2,303,820 **33.** 283 **35.** 20,909 **37.** \$707 **39.** \$20,625 **41.** 468 feet **43.** 121,100,000 square miles **45.** 4,095,622 **47.** (a) 1134 students (b) 1392 students **49.** 202 miles **51.** 434 feet **53.** (a) \$9553 (b) \$7319 (c) \$13,047 **55.** 1161 **57.** Answers may vary. A sample is: You could not group the addends in groups that sum to 10s to make column addition easier. **58.** seventy-six million, two hundred eight thousand, nine hundred forty-one. **59.** one hundred twenty-one million, three hundred seventy-four **60.** 8,724,396 **61.** 9,051,719 **62.** 28,387,018

1.3 Exercises

1. In subtraction the minuend minus the subtrahead equals the difference. To check the problem we add the subtrahead and the difference to see if we get the minuend. If we do, the answer is correct. **3.** We know that 1683 + 1592 = 32?5. Therefore if we add 8 tens and 9 tens we get 17 tens, which is 1 hundred and 7 tens. Thus the ?, should be replaced by 7. **5.** 5 **7.** 6 **9.** 16 **11.** 9 **13.** 7 **15.** 6 **17.** 3 **19.** 9

21. 21 $\begin{array}{r} 26 \\ +\ 21 \\ \hline 47 \end{array}$ **23.** 12 $\begin{array}{r} 73 \\ +\ 12 \\ \hline 85 \end{array}$ **25.** 124 $\begin{array}{r} 65 \\ +\ 124 \\ \hline 189 \end{array}$ **27.** 321 $\begin{array}{r} 548 \\ +\ 321 \\ \hline 869 \end{array}$ **29.** 2321 $\begin{array}{r} 572 \\ +\ 2321 \\ \hline 2893 \end{array}$ **31.** 143,235 $\begin{array}{r} 12{,}600 \\ +\ 143{,}235 \\ \hline 155{,}835 \end{array}$

33. 553,101 $\begin{array}{r} 433{,}201 \\ +\ 553{,}101 \\ \hline 986{,}302 \end{array}$ **35.** $\begin{array}{r} 19 \\ +\ 110 \\ \hline 129 \end{array}$ Correct **37.** $\begin{array}{r} 3215 \\ +\ 5781 \\ \hline 8996 \end{array}$ Incorrect Correct answer: 5381 **39.** $\begin{array}{r} 5020 \\ +\ 1020 \\ \hline 6040 \end{array}$ Incorrect Correct answer: 1010 **41.** $\begin{array}{r} 33{,}846 \\ +\ 13{,}023 \\ \hline 46{,}869 \end{array}$ Incorrect Correct answer: 14,023 **43.** 46 **45.** 92

47. 384 **49.** 718 **51.** 10,630 **53.** 34,092 **55.** 7447 **57.** 908,930 **59.** $x = 5$ **61.** $x = 29$ **63.** $x = 27$ **65.** 182 votes **67.** 6,225,255 **69.** \$762 **71.** 1,416,920 people **73.** 2,004,796 people **75.** 320,317 people **77.** 93,154 people **79.** 29 homes **81.** 80 homes **83.** between 2002 and 2003 **85.** Willow Creek and Irving **87.** It is true if a and b represent the same number, for example, if $a = 10$ and $b = 10$. **89.** \$550 **91.** 8,466,084 **92.** two hundred ninety-six thousand, three hundred eight **93.** 218 **94.** 1,174,750

1.4 Exercises

1. (a) You can change the order of the factors without changing the product. (b) You can group the factors in any way without changing the product.

3.

x	6	2	3	8	0	5	7	9	12	4
5	30	10	15	40	0	25	35	45	60	20
7	42	14	21	56	0	35	49	63	84	28
1	6	2	3	8	0	5	7	9	12	4
0	0	0	0	0	0	0	0	0	0	0
6	36	12	18	48	0	30	42	54	72	24
2	12	4	6	16	0	10	14	18	24	8
3	18	6	9	24	0	15	21	27	36	12
8	48	16	24	64	0	40	56	72	96	32
4	24	8	12	32	0	20	28	36	48	16
9	54	18	27	72	0	45	63	81	108	36

5. 96 **7.** 70 **9.** 522 **11.** 693 **13.** 3432 **15.** 18,306 **17.** 36,609 **19.** 31,308 **21.** 100,208 **23.** 942,808 **25.** 1560 **27.** 2,715,800 **29.** 482,000 **31.** 372,560,000 **33.** 8460 **35.** 63,600 **37.** 56,000,000 **39.** 6168 **41.** 7884 **43.** 5696 **45.** 15,175 **47.** 44,153 **49.** 69,312 **51.** 148,567 **53.** 823,823 **55.** 1,240,064 **57.** 89,496 **59.** 217,980 **61.** 1,435,083 **63.** 720,000 **65.** 10,000 **67.** 90,600 **69.** 70 **71.** 308 **73.** 18,432 **75.** 1600 **77.** $x = 0$ **79.** 760 square feet **81.** 195 square feet **83.** \$1200 **85.** \$3192 **87.** 612 miles **89.** \$6790 **91.** \$7,372,300,000 **93.** 198 **95.** 62 **97.** $x = 8$ **99.** $x = 9$ **101.** No, it would not always be true. In our number system 62 = 60 + 2. But in roman numerals IV ≠ I + V. The digit system in roman numerals involves subtraction. Thus (XII) × (IV) ≠ (XII × I) + (XII × V). **103.** 6756 **104.** 1249 **105.** \$139 **106.** \$59 **107.** 1829 people **108.** \$136,387,280,400

1.5 Exercises

1. (a) When you divide a nonzero number by itself, the result is one. (b) When you divide a number by 1, the result is that number. (c) When you divide zero by a nonzero number, the result is zero. (d) You cannot divide a number by zero. Division by zero is undefined. **3.** 7 **5.** 3 **7.** 5 **9.** 4 **11.** 3 **13.** 7 **15.** 9 **17.** 9 **19.** 7 **21.** 4 **23.** 0 **25.** undefined **27.** 0 **29.** 1 **31.** 4 R 5 **33.** 9 R 4 **35.** 25 R 3 **37.** 21 R 7 **39.** 32 **41.** 37 **43.** 322 R 1 **45.** 127 R 1 **47.** 753 **49.** 1122 R 1 **51.** 2056 R 2 **53.** 2562 R 3 **55.** 30 R 5 **57.** 5 R 7 **59.** 7 **61.** 418 R 8 **63.** 48 R 12 **65.** 327 **67.** 210 R 8 **69.** 14 R 2 **71.** 4 R 4 **73.** 16 **75.** 37 **77.** 61,693 runs per day **79.** 154 guests **81.** \$21,053 **83.** \$12,140 **85.** 165 sandwiches **87.** (a) 41,808 km (b) 8192 km **89.** a and b must represent the same number. For example, if $a = 12$, then $b = 12$. **91.** 5400 **92.** 1,038,490 **93.** 406,195 **94.** 66,844

How Am I Doing? Sections 1.1–1.5

1. seventy-eight million, three hundred ten thousand, four hundred thirty-six. (obj. 1.1.3) **2.** 30,000 + 8000 + 200 + 40 + 7 (obj. 1.1.1) **3.** 5,064,122 (obj. 1.1.2) **4.** 2,747,000 (obj. 1.1.4) **5.** 2,583,000 (obj. 1.1.4) **6.** 244 (obj. 1.2.4) **7.** 50,570 (obj. 1.2.4) **8.** 14,729 (obj. 1.2.4) **9.** 3750 (obj. 1.3.3) **10.** 76,311 (obj. 1.3.3) **11.** 1,981,652 (obj. 1.3.3) **12.** 108 (obj. 1.4.1) **13.** 1,280,000 (obj. 1.4.4) **14.** 18,606 (obj. 1.4.2) **15.** 6734 (obj. 1.4.4) **16.** 331,420 (obj. 1.4.4) **17.** 10,605 (obj. 1.5.2) **18.** 7376 R1 (obj. 1.5.2) **19.** 26 (obj. 1.5.3) **20.** 139 (obj. 1.5.3)

1.6 Exercises

1. 5^3 means 5 × 5 × 5. $5^3 = 125$. **3.** base

5. To ensure consistency we
1. perform operations inside parentheses
2. simplify any expressions with exponents
3. multiply or divide from left to right
4. add or subtract from left to right

7. 6^4 **9.** 5^6 **11.** 8^4 **13.** 9^1 **15.** 16 **17.** 64 **19.** 36 **21.** 10,000 **23.** 1 **25.** 64 **27.** 243 **29.** 225 **31.** 343 **33.** 256 **35.** 1 **37.** 625 **39.** 1,000,000 **41.** 169 **43.** 9 **45.** 2401 **47.** 17 **49.** 225 **51.** 520 **53.** $56 - 4 = 52$ **55.** $27 - 5 = 22$ **57.** $48 \div 8 + 4 = 6 + 4 = 10$ **59.** $3 \times 100 - 50 = 300 - 50 = 250$ **61.** $100 + 3 \times 5 = 100 + 15 = 115$ **63.** $20 \div 20 = 1$ **65.** $950 \div 5 = 190$ **67.** $60 - 17 = 43$ **69.** $9 + 16 \div 4 = 9 + 4 = 13$ **71.** $42 - 4 \div 4 = 42 - 1 = 41$ **73.** $100 - 9 \times 4 = 100 - 36 = 64$ **75.** $25 + 4 + 27 = 56$ **77.** $8 \times 3 \times 1 \div 2 = 24 \div 2 = 12$ **79.** $144 - 0 = 144$ **81.** $16 \times 6 \div 3 = 96 \div 3 = 32$ **83.** $16 - 3 \times 4 + 1 = 16 - 12 + 1 = 5$ **85.** $3 + 9 \times 6 + 4 = 3 + 54 + 4 = 61$ **87.** $32 \div 2 \times 16 = 16 \times 16 = 256$ **89.** $9 \times 6 \div 9 + 4 \times 3 = 6 + 12 = 18$ **91.** $36 + 81 = 117$ **93.** $1200 - 8(3) \div 6 = 1200 - 4 = 1196$ **95.** $50 + 20 - 9 = 61$ **97.** $250 \div 25 - 9 = 10 - 9 = 1$ **99.** $6 + (6)^2 - 20 = 6 + 36 - 20 = 22$ **101.** 86,164 seconds **103.** (a) 3 (b) 2,000,000 **104.** 200,765,909 **105.** two hundred sixty-one million, seven hundred sixty-three thousand, two **106.** 1460 feet of fencing will be needed. 120,000 square feet of grass must be planted

1.7 Exercises

1. Locate the rounding place. If the digit to the right of the rounding place is 5 or greater than 5, round up. If the digit to the right of the rounding place is less than 5, round down. **3.** 80 **5.** 70 **7.** 170 **9.** 7440 **11.** 1670 **13.** 200 **15.** 2800 **17.** 7700 **19.** 8000 **21.** 2000 **23.** 28,000 **25.** 800,000 **27.** 15,000,000 stars **29.** (a) 163,000 (b) 163,300 **31.** (a) 3,700,000 square miles; 9,600,000 square kilometers (b) 3,710,000 square miles; 9,600,000 square kilometers

33. $\begin{array}{r} 600 \\ 300 \\ +\ 100 \\ \hline 1000 \end{array}$ **35.** $\begin{array}{r} 40 \\ 70 \\ 100 \\ +\ 20 \\ \hline 230 \end{array}$ **37.** $\begin{array}{r} 200{,}000 \\ 50{,}000 \\ +\ 9{,}000 \\ \hline 259{,}000 \end{array}$ **39.** $\begin{array}{r} 600{,}000 \\ -\ 100{,}000 \\ \hline 500{,}000 \end{array}$ **41.** $\begin{array}{r} 800{,}000 \\ 80{,}000 \\ \hline 720{,}000 \end{array}$ **43.** $\begin{array}{r} 30{,}000{,}000 \\ -\ 20{,}000{,}000 \\ \hline 10{,}000{,}000 \end{array}$ **45.** $\begin{array}{r} 60 \\ \times\ 50 \\ \hline 3000 \end{array}$ **47.** $\begin{array}{r} 1000 \\ \times\ 8 \\ \hline 8000 \end{array}$

49. $\begin{array}{r} 600{,}000 \\ \times\ 300 \\ \hline 180{,}000{,}000 \end{array}$ **51.** $\begin{array}{r} 200 \\ 30\overline{)6000} \end{array}$ **53.** $\begin{array}{r} 5{,}000 \\ 40\overline{)200{,}000} \end{array}$ **55.** $\begin{array}{r} 5{,}000 \\ 800\overline{)4{,}000{,}000} \end{array}$ **57.** Incorrect $\begin{array}{r} 400 \\ 500 \\ 900 \\ +\ 200 \\ \hline 2000 \end{array}$ **59.** Incorrect $\begin{array}{r} 100{,}000 \\ 50{,}000 \\ +\ 40{,}000 \\ \hline 190{,}000 \end{array}$

61. Correct $\begin{array}{r} 300{,}000 \\ -\ 90{,}000 \\ \hline 210{,}000 \end{array}$ **63.** Incorrect $\begin{array}{r} 80{,}000{,}000 \\ -\ 50{,}000{,}000 \\ \hline 30{,}000{,}000 \end{array}$ **65.** Incorrect $\begin{array}{r} 200 \\ \times\ 20 \\ \hline 4000 \end{array}$ **67.** Correct $\begin{array}{r} 6000 \\ \times\ 70 \\ \hline 420{,}000 \end{array}$ **69.** $\begin{array}{r} 2000 \\ 40\overline{)80{,}000} \end{array}$ Correct **71.** $\begin{array}{r} 500 \\ 400\overline{)200{,}000} \end{array}$ Correct

73. 2,400 square feet **75.** 5,300,000 **77.** 30,000 pizzas **79.** 200,000,000 passengers **81.** 590,000 − 270,000 = 320,000 square miles **83.** (a) 400,000 hours (b) 20,000 days **85.** 83 **86.** 27 **87.** 28 **88.** 66 **89.** 367,763 **90.** 87

1.8 Exercises

1. $8800 **3.** 1560 bagels **5.** 6¢ per ounce **7.** $64 **9.** 5 hours; 300 minutes **11.** $20,382 **13.** 25,231; 466 **15.** 800,000 people **17.** $192 **19.** $1482 **21.** $16,405 **23.** 25 miles per gallon **25.** There are 54 oak trees, 108 maple trees, and 756 pine trees. In total there are 936 trees. **27.** 118 **29.** 217 **31.** $14,734,000,000 **33.** $69,603,000,000 **35.** 343 **36.** 21 **37.** 4788 **38.** 258 **39.** 802 **40.** 23,285 **41.** 526,196,000 **42.** 3,400,603,025

Putting Your Skills to Work

1. 74 **2.** 1821 **3.** 330,000 **4.** 350,000 **5.** 989 **6.** 624

Chapter 1 Review Problems

1. three hundred seventy-six **2.** fifteen thousand, eight hundred two **3.** One hundred nine thousand, two hundred seventy-six **4.** four hundred twenty-three million, five hundred seventy-six thousand, fifty-five **5.** 4000 + 300 + 60 + 4 **6.** 20,000, + 7000 + 900 + 80 + 6 **7.** 40,000,000 + 2,000,000 + 100,000 + 60,000 + 6000 + 30 + 7 **8.** 1,000,000 + 300,000 + 5000 + 100 + 20 + 8 **9.** 924 **10.** 5302 **11.** 1,328,828 **12.** 45,092,651 **13.** 115 **14.** 300 **15.** 690 **16.** 150 **17.** 400 **18.** 1598 **19.** 1007 **20.** 125,423 **21.** 14,703 **22.** 10,582 **23.** 17 **24.** 6 **25.** 27 **26.** 171 **27.** 159 **28.** 3167 **29.** 63,146 **30.** 3026 **31.** 6,236,011 **32.** 5,332,991 **33.** 144 **34.** 0 **35.** 240 **36.** 300 **37.** 62,100 **38.** 84,312,000 **39.** 83,200,000 **40.** 536,000,000 **41.** 1856 **42.** 1752 **43.** 4050 **44.** 13,680 **45.** 25,524 **46.** 24,096 **47.** 87,822 **48.** 268,513 **49.** 543,510 **50.** 255,068 **51.** 150,000 **52.** 240,000 **53.** 7,200,000 **54.** 7,500,000 **55.** 2,000,000,000 **56.** 12,000,000,000 **57.** 2 **58.** 5 **59.** 0 **60.** 12 **61.** 7 **62.** 0 **63.** 7 **64.** 7 **65.** undefined **66.** 4 **67.** 7 **68.** 6 **69.** 125 **70.** 125 **71.** 258 **72.** 309 **73.** 25,874 **74.** 3064 **75.** 36,958 **76.** 36,921 **77.** 15,986 R 2 **78.** 35,783 R 4 **79.** 7 R 21 **80.** 4 R 37 **81.** 31 R 15 **82.** 14 R 11 **83.** 38 R 30 **84.** 60 R 22 **85.** 95 **86.** 258 **87.** 54 **88.** 19 **89.** 13^2 **90.** 21^3 **91.** 8^5 **92.** 10^6 **93.** 64 **94.** 81 **95.** 128 **96.** 125 **97.** 49 **98.** 81 **99.** 216 **100.** 64 **101.** 8 **102.** 11 **103.** 22 **104.** 66 **105.** 107 **106.** 76 **107.** 17 **108.** 26 **109.** 86 **110.** 1280 **111.** 5900 **112.** 15,310 **113.** 42,640 **114.** 12,000 **115.** 23,000 **116.** 676,000 **117.** 202,000 **118.** 4,600,000 **119.** 10,000,000 **120.** $\begin{array}{r} 600 \\ 600 \\ 900 \\ +\ 900 \\ \hline 3000 \end{array}$ **121.** $\begin{array}{r} 30{,}000 \\ 7000 \\ +\ 70000 \\ \hline 107{,}000 \end{array}$ **122.** $\begin{array}{r} 4{,}000{,}000 \\ -\ 3{,}000{,}000 \\ \hline 1{,}000{,}000 \end{array}$ **123.** $\begin{array}{r} 30{,}000 \\ -\ 20{,}000 \\ \hline 10{,}000 \end{array}$ **124.** $\begin{array}{r} 1000 \\ \times\ 6000 \\ \hline 6{,}000{,}000 \end{array}$ **125.** $\begin{array}{r} 3{,}000{,}000 \\ \times\ 900 \\ \hline 2{,}700{,}000{,}000 \end{array}$

126. $\begin{array}{r} 4000 \\ 20\overline{)80{,}000} \end{array}$ **127.** $\begin{array}{r} 20{,}000 \\ 400\overline{)8{,}000{,}000} \end{array}$ **128.** 216 cans **129.** 175 words **130.** 7020 people **131.** $59,470 **132.** 10,301 feet **133.** $7028 **134.** $1356 **135.** $74 **136.** $278 **137.** 25 miles per gallon **138.** $5041 **139.** $2031 **140.** 40,500,000 tons **141.** 21,400,000 tons from 1990 to 1995 **142.** 93,400,000 tons **143.** 2284 **144.** 7867 **145.** 11,088 **146.** 129 **147.** 29 **148.** $747 **149.** (a) 330 square feet (b) 74 feet

How Am I Doing? Chapter 1 Test

1. forty-four million, seven thousand, six hundred thirty-five (obj. 1.1.3) **2.** 20,000 + 6000 + 800 + 50 + 9 (obj. 1.1.1) **3.** 3,581,076 (obj. 1.1.2) **4.** 831 (obj. 1.2.4) **5.** 1491 (obj. 1.2.4) **6.** 318,977 (obj. 1.2.4) **7.** 8067 (obj. 1.3.3) **8.** 172,858 (obj. 1.3.3) **9.** 5,225,768 (obj. 1.3.3) **10.** 378 (obj. 1.4.1) **11.** 4320 (obj. 1.4.4) **12.** 192,992 (obj. 1.4.4) **13.** 129,437 (obj. 1.4.2) **14.** 3014 R 1 (obj. 1.5.2) **15.** 2358 (obj. 1.5.2) **16.** 352 (obj. 1.5.3) **17.** 14^3 (obj. 1.6.1) **18.** 64 (obj. 1.6.1) **19.** 23 (obj. 1.6.2) **20.** 50 (obj. 1.6.2) **21.** 79 (obj. 1.6.2) **22.** 94,800 (obj. 1.7.1) **23.** 6,460,000 (obj. 1.7.1) **24.** 5,300,000 (obj. 1.7.1) **25.** 150,000,000,000 (obj. 1.7.2) **26.** 16,000 (obj. 1.7.2) **27.** $2148 (obj. 1.8.1) **28.** 467 feet (obj. 1.8.1) **29.** $127 (obj. 1.8.2) **30.** $292 (obj. 1.8.2) **31.** 748,000 square feet (obj. 1.8.2) **32.** 46 feet (obj. 1.8.2)

Chapter 2

2.1 Exercises

1. fraction **3.** denominator **5.** N: 3; D: 5 **7.** N: 7, D: 8 **9.** N: 1; D: 17 **11.** $\frac{1}{3}$ **13.** $\frac{7}{9}$ **15.** $\frac{3}{4}$ **17.** $\frac{3}{7}$ **19.** $\frac{2}{5}$ **21.** $\frac{7}{10}$ **23.** $\frac{5}{8}$ **25.** $\frac{4}{7}$ **27.** $\frac{7}{8}$ **29.** $\frac{3}{5}$ **31.** 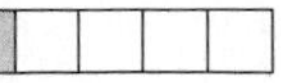 **33.** 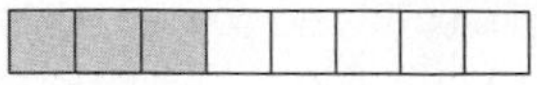 **35.** 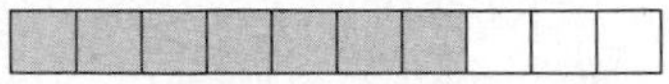 **37.** $\frac{31}{95}$ **39.** $\frac{209}{750}$ **41.** $\frac{89}{211}$ **43.** $\frac{7}{29}$ **45.** $\frac{9}{14}$ **47.** (a) $\frac{90}{195}$ (b) $\frac{22}{195}$ **49.** The amount of money each of six business owners gets if the business has a profit of \$0. **51.** 241 **52.** 13,216 **53.** 146,188 **54.** 1258 R 4 **55.** 181

2.2 Exercises

1. 11, 19, 41, 5 **3.** composite number **5.** $56 = 2 \times 2 \times 2 \times 7$ **7.** 3×5 **9.** 5×7 **11.** 7^2 **13.** 2^6 **15.** 5×11 **17.** $3^2 \times 7$ **19.** 3×5^2 **21.** 2×3^3 **23.** $2^3 \times 3 \times 5$ **25.** $2^3 \times 23$ **27.** prime **29.** 3×19 **31.** prime **33.** 2×31 **35.** prime **37.** prime **39.** 11×11 **41.** 3×43 **43.** $\frac{18 \div 9}{27 \div 9} = \frac{2}{3}$ **45.** $\frac{32 \div 16}{48 \div 16} = \frac{2}{3}$ **47.** $\frac{30 \div 6}{48 \div 6} = \frac{5}{8}$ **49.** $\frac{210 \div 10}{310 \div 10} = \frac{21}{31}$ **51.** $\frac{3 \times 1}{3 \times 5} = \frac{1}{5}$ **53.** $\frac{2 \times 3 \times 11}{2 \times 2 \times 2 \times 11} = \frac{3}{4}$ **55.** $\frac{2 \times 3 \times 5}{3 \times 3 \times 5} = \frac{2}{3}$ **57.** $\frac{3 \times 3 \times 3}{3 \times 3 \times 5} = \frac{3}{5}$ **59.** $\frac{3 \times 11}{3 \times 12} = \frac{11}{12}$ **61.** $\frac{9 \times 7}{9 \times 12} = \frac{7}{12}$ **63.** $\frac{11 \times 8}{11 \times 11} = \frac{8}{11}$ **65.** $\frac{3 \times 50}{4 \times 50} = \frac{3}{4}$ **67.** $\frac{11 \times 20}{13 \times 20} = \frac{11}{13}$ **69.** $3 \times 33 \stackrel{?}{=} 11 \times 9$; $99 = 99$; yes **71.** no **73.** no **75.** yes **77.** yes **79.** $\frac{3}{4}$ **81.** $\frac{16}{19}$ **83.** $\frac{2}{7}$ **85.** $\frac{17}{45}$ **87.** $\frac{8}{45}$ **89.** 164,050 **90.** 1296 **91.** 960,000 **92.** \$1,194,700,000

2.3 Exercises

1. (a) Multiply the whole number by the denominator of the fraction. (b) Add the numerator of the fraction to the product formed in step (a). (c) Write the sum found in step (b) over the denominator of the fraction. **3.** $\frac{14}{3}$ **5.** $\frac{17}{7}$ **7.** $\frac{83}{9}$ **9.** $\frac{32}{3}$ **11.** $\frac{65}{3}$ **13.** $\frac{55}{6}$ **15.** $\frac{121}{6}$ **17.** $\frac{131}{12}$ **19.** $\frac{79}{10}$ **21.** $\frac{201}{25}$ **23.** $\frac{65}{12}$ **25.** $\frac{494}{3}$ **27.** $\frac{131}{15}$ **29.** $\frac{138}{25}$ **31.** $1\frac{1}{3}$ **33.** $2\frac{3}{4}$ **35.** $2\frac{1}{2}$ **37.** $3\frac{3}{8}$ **39.** 5 **41.** $9\frac{5}{9}$ **43.** $23\frac{1}{3}$ **45.** $3\frac{3}{16}$ **47.** $9\frac{1}{3}$ **49.** $17\frac{1}{2}$ **51.** 13 **53.** 14 **55.** 6 **57.** $36\frac{7}{11}$ **59.** $2\frac{3}{4}$ **61.** $4\frac{1}{6}$ **63.** $15\frac{1}{4}$ **65.** 4 **67.** $\frac{12}{5}$ **69.** $\frac{13}{2}$ **71.** $2\frac{88}{126} = 2\frac{44}{63}$ **73.** $2\frac{20}{280} = 2\frac{1}{14}$ **75.** $1\frac{212}{296} = 1\frac{53}{74}$ **77.** $\frac{1082}{3}$ yards **79.** $50\frac{1}{3}$ acres **81.** $141\frac{3}{8}$ pounds **83.** No, 101 is prime and is not a factor of 5687. **85.** 260,247 **86.** 16,000,000,000 **87.** 300 **88.** 37 full cartons are needed. There are 5 books in the carton that is not full.

2.4 Exercises

1. $\frac{21}{55}$ **3.** $\frac{15}{52}$ **5.** 1 **7.** $\frac{35}{54}$ **9.** $\frac{5}{12}$ **11.** $\frac{21}{8}$ or $2\frac{5}{8}$ **13.** $\frac{24}{7}$ or $3\frac{3}{7}$ **15.** $\frac{10}{3}$ or $3\frac{1}{3}$ **17.** $\frac{1}{6}$ **19.** 3 **21.** $\frac{22}{9}$ or $2\frac{4}{9}$ **23.** 31 **25.** 0 **27.** $3\frac{7}{8}$ **29.** $\frac{55}{12}$ or $4\frac{7}{12}$ **31.** $\frac{69}{50}$ or $1\frac{19}{50}$ **33.** $\frac{154}{3}$ or $51\frac{1}{3}$ **35.** $\frac{8}{5}$ or $1\frac{3}{5}$ **37.** $\frac{7}{9}$ **39.** $15\frac{1}{6}$ or $\frac{91}{6}$ **41.** $x = \frac{9}{5}$ **43.** $x = \frac{8}{9}$ **45.** $37\frac{11}{12}$ square miles **47.** 1560 miles **49.** 1629 grams **51.** 22 students **53.** 377 companies **55.** $1\frac{8}{9}$ miles **57.** (a) $\frac{1}{3}$ of the garden (b) 40 square feet **59.** The step of dividing the numerator and denominator by the same number allows us to work with smaller numbers when we do the multiplication. Also, this allows us to avoid the step of having to simplify the fraction in the final answer. **61.** 529 cars **62.** 368 calls **63.** 5040 miles **64.** 173,040 gallons

2.5 Exercises

1. Think of a simple problem like $3 \div \frac{1}{2}$. One way to think of it is how many $\frac{1}{2}$'s can be placed in 3? For example, how many $\frac{1}{2}$-pound rocks could be put in a bag that holds 3 pounds of rocks? The answer is 6. If we inverted the first fraction by mistake, we would have $\frac{1}{3} \times \frac{1}{2} = \frac{1}{6}$. We know this is wrong since there are obviously several $\frac{1}{2}$-pound rocks in a bag that holds 3 pounds of rocks. The answer $\frac{1}{6}$ would make no sense. **3.** $\frac{21}{16}$ or $1\frac{5}{16}$ **5.** $\frac{9}{2}$ or $4\frac{1}{2}$ **7.** $\frac{3}{2}$ or $1\frac{1}{2}$ **9.** $\frac{4}{3}$ or $1\frac{1}{3}$ **11.** 1 **13.** $\frac{9}{49}$ **15.** $\frac{4}{5}$ **17.** $\frac{3}{44}$ **19.** $\frac{27}{7}$ or $3\frac{6}{7}$ **21.** 0 **23.** undefined **25.** 10 **27.** $\frac{7}{32}$ **29.** $\frac{3}{4}$ **31.** $\frac{13}{9}$ or $1\frac{4}{9}$ **33.** 2 **35.** 5000 **37.** $\frac{1}{250}$ **39.** $\frac{7}{40}$ **41.** $\frac{32}{3}$ or $10\frac{2}{3}$ **43.** $\frac{7}{18}$ **45.** 2 **47.** 4 **49.** $\frac{15}{7}$ or $2\frac{1}{7}$ **51.** 3 **53.** $\frac{63}{32}$ or $1\frac{31}{32}$ **55.** $\frac{5}{12}$ **57.** $\frac{30}{19}$ or $1\frac{11}{19}$ **59.** 0 **61.** $\frac{7}{44}$ **63.** $\frac{42}{5}$ or $8\frac{2}{5}$ **65.** $x = \frac{7}{5}$ **67.** $x = \frac{4}{7}$ **69.** $2\frac{1}{4}$ gallons **71.** $37\frac{1}{2}$ miles per hour **73.** 58 students **75.** 100 large styrofoam cups **77.** It took six drill attempts.

79. We estimate by dividing 15 ÷ 5, which is 3. The exact value is $2\frac{26}{31}$, which is very close. Our answer is off by only $\frac{5}{31}$.

81. thirty-nine million, five hundred seventy-six thousand, three hundred four **82.** 500,000 + 9000 + 200 + 70 **83.** 1099

84. 87,595,631

How Am I Doing? Sections 2.1–2.5

1. $\frac{3}{8}$ (obj. 2.1.1) **2.** $\frac{8}{69}$ (obj. 2.1.3) **3.** $\frac{5}{124}$ (obj. 2.1.3) **4.** $\frac{1}{6}$ (obj. 2.2.2) **5.** $\frac{1}{3}$ (obj. 2.2.2) **6.** $\frac{1}{7}$ (obj. 2.2.2) **7.** $\frac{7}{8}$ (obj. 2.2.2) **8.** $\frac{4}{11}$ (obj. 2.2.2) **9.** $\frac{11}{3}$ (obj. 2.3.1) **10.** $\frac{55}{9}$ (obj. 2.3.1) **11.** $24\frac{1}{4}$ (obj. 2.3.2) **12.** $5\frac{4}{5}$ (obj. 2.3.2) **13.** $2\frac{2}{17}$ (obj. 2.3.2) **14.** $\frac{5}{44}$ (obj. 2.4.1) **15.** $\frac{2}{3}$ (obj. 2.4.1) **16.** $\frac{407}{6}$ or $67\frac{5}{6}$ (obj. 2.4.3) **17.** 1 (obj. 2.5.1) **18.** $\frac{1}{2}$ (obj. 2.5.1) **19.** $\frac{69}{13}$ or $5\frac{4}{13}$ (obj. 2.5.3) **20.** $\frac{14}{3}$ or $4\frac{2}{3}$ (obj. 2.5.2)

Test on Sections 2.1–2.5

1. $\frac{23}{32}$ **2.** $\frac{85}{113}$ **3.** $\frac{1}{2}$ **4.** $\frac{7}{15}$ **5.** $\frac{4}{11}$ **6.** $\frac{25}{31}$ **7.** $\frac{3}{4}$ **8.** $\frac{7}{3}$ or $2\frac{1}{3}$ **9.** $\frac{43}{12}$ **10.** $\frac{33}{8}$ **11.** $6\frac{3}{7}$ **12.** $8\frac{1}{4}$ **13.** $\frac{21}{88}$ **14.** $\frac{9}{7}$ or $1\frac{2}{7}$ **15.** 15 **16.** $16\frac{1}{2}$ or $\frac{33}{2}$ **17.** $\frac{161}{12}$ or $13\frac{5}{12}$ **18.** $\frac{100}{21}$ or $4\frac{16}{21}$ **19.** $\frac{16}{21}$ **20.** $5\frac{1}{3}$ or $\frac{16}{3}$ **21.** 7 **22.** $\frac{56}{5}$ or $11\frac{1}{5}$ **23.** $\frac{63}{8}$ or $7\frac{7}{8}$ **24.** 14 **25.** $\frac{8}{3}$ or $2\frac{2}{3}$ **26.** $\frac{23}{8}$ or $2\frac{7}{8}$ **27.** $\frac{13}{16}$ **28.** $\frac{1}{14}$ **29.** $\frac{9}{32}$ **30.** $\frac{13}{15}$ **31.** $45\frac{15}{16}$ square feet **32.** 4 cups **33.** $46\frac{7}{8}$ miles **34.** 16 full packages; $\frac{3}{8}$ lb. left over **35.** 51 computers **36.** 8000 homes **37.** 16 hours **38.** 6 tents; 7 yards left over **39.** 41 days

2.6 Exercises

1. 24 **3.** 100 **5.** 60 **7.** 36 **9.** 147 **11.** 10 **13.** 28 **15.** 35 **17.** 18 **19.** 60 **21.** 32 **23.** 90 **25.** 60 **27.** 105 **29.** 90 **31.** 6 **33.** 60 **35.** 132 **37.** 84 **39.** 120 **41.** 3 **43.** 35 **45.** 20 **47.** 14 **49.** 96 **51.** 63 **53.** $\frac{21}{36}$ and $\frac{20}{36}$ **55.** $\frac{25}{80}$ and $\frac{68}{80}$ **57.** $\frac{144}{160}$ and $\frac{152}{160}$ **59.** LCD = 35; $\frac{14}{35}$ and $\frac{9}{35}$ **61.** LCD = 24; $\frac{5}{24}$ and $\frac{9}{24}$ **63.** LCD = 30; $\frac{21}{30}$ and $\frac{25}{30}$ **65.** LCD = 60; $\frac{16}{60}$ and $\frac{25}{60}$ **67.** LCD = 36; $\frac{10}{36}, \frac{11}{36}, \frac{21}{36}$ **69.** LCD = 56; $\frac{3}{56}, \frac{49}{56}, \frac{40}{56}$ **71.** LCD = 63; $\frac{5}{63}, \frac{12}{63}, \frac{56}{63}$ **73.** (a) LCD = 16 (b) $\frac{3}{16}, \frac{12}{16}, \frac{6}{16}$ **75.** 178 R 3 **76.** 31,636 **77.** 25 **78.** between 1980 and 1985 **79.** between 1985 and 1990 **80.** more than \$7469 a year **81.** more than \$7903 a year **82.** \$24,135 **83.** \$21,847

2.7 Exercises

1. $\frac{7}{9}$ **3.** $1\frac{2}{9}$ **5.** $\frac{1}{12}$ **7.** $\frac{17}{44}$ **9.** $\frac{5}{6}$ **11.** $\frac{9}{20}$ **13.** $\frac{7}{8}$ **15.** $\frac{23}{20}$ or $1\frac{3}{20}$ **17.** $\frac{37}{100}$ **19.** $\frac{26}{175}$ **21.** $\frac{31}{24}$ or $1\frac{7}{24}$ **23.** $\frac{27}{40}$ **25.** $\frac{1}{4}$ **27.** $\frac{8}{35}$ **29.** $\frac{5}{12}$ **31.** $\frac{11}{60}$ **33.** $\frac{1}{4}$ **35.** $\frac{2}{3}$ **37.** $\frac{1}{8}$ **39.** 0 **41.** $\frac{5}{12}$ **43.** 1 **45.** $\frac{11}{30}$ **47.** $1\frac{7}{15}$ **49.** $\frac{3}{14}$ **51.** $\frac{5}{33}$ **53.** $\frac{8}{15}$ **55.** $1\frac{5}{12}$ cups **57.** $1\frac{1}{2}$ pounds **59.** $\frac{19}{60}$ of the book report **61.** 16 chocolates **63.** $\frac{4}{21}$ of the membership **65.** $\frac{3}{17}$ **66.** $\frac{3}{23}$ **67.** $8\frac{3}{14}$ **68.** $\frac{101}{7}$ **69.** $26\frac{1}{8}$ **70.** $2\frac{8}{9}$

2.8 Exercises

1. $9\frac{3}{4}$ **3.** $4\frac{1}{7}$ **5.** $17\frac{1}{2}$ **7.** $16\frac{1}{10}$ **9.** $\frac{4}{7}$ **11.** $2\frac{17}{24}$ **13.** $9\frac{11}{20}$ **15.** 0 **17.** $4\frac{4}{15}$ **19.** $14\frac{4}{7}$ **21.** $7\frac{2}{5}$ **23.** $9\frac{3}{5}$ **25.** $41\frac{4}{5}$ **27.** $6\frac{7}{12}$ **29.** $8\frac{1}{6}$ **31.** $73\frac{37}{40}$ **33.** $5\frac{1}{2}$ **35.** $\frac{2}{3}$ **37.** $4\frac{41}{60}$ **39.** $8\frac{8}{15}$ **41.** $102\frac{5}{8}$ **43.** $14\frac{1}{24}$ **45.** $43\frac{1}{8}$ miles **47.** $2\frac{1}{24}$ pounds **49.** $2\frac{3}{4}$ inches **51.** (a) $3\frac{11}{12}$ pounds (b) $4\frac{1}{12}$ pounds **53.** $\frac{2607}{40}$ or $65\frac{7}{40}$

55. We estimate by adding 35 + 24 to obtain 59. The exact answer is $59\frac{7}{12}$. Our estimate is very close. We are off by only $\frac{7}{12}$. **57.** $\frac{2}{3}$

59. 1 **61.** $\frac{5}{3}$ or $1\frac{2}{3}$ **63.** $\frac{3}{5}$ **65.** $\frac{9}{25}$ **67.** $\frac{16}{11}$ or $1\frac{5}{11}$ **69.** $\frac{1}{9}$ **71.** $\frac{6}{5}$ or $1\frac{1}{5}$ **73.** 480,000 **74.** 8,529,300 **75.** \$1893

2.9 Exercises

1. $16\frac{2}{3}$ feet **3.** $25\frac{11}{24}$ tons **5.** $1\frac{9}{16}$ inches **7.** $9\frac{19}{20}$ miles **9.** \$147 **11.** $275\frac{5}{8}$ gallons **13.** $106\frac{7}{8}$ nautical miles

15. \$451 per week **17.** (a) 20 bandanas (b) $\frac{1}{6}$ ft (c) \$50 **19.** (a) $14\frac{1}{8}$ ounces of bread (b) $\frac{5}{8}$ ounce

21. (a) $30\frac{1}{2}$ knots (b) 7 hours **23.** (a) 5485 bushels (b) $11{,}998\frac{7}{16}$ cubic feet (c) $9598\frac{3}{4}$ bushels **25.** $14\frac{71}{72}$ lbs **27.** 44,245 **28.** 22,437 **29.** 45,441 **30.** 356

Putting Your Skills to Work

1. $3\frac{1}{4}$ miles **2.** $5\frac{3}{8}$ miles **3.** $7\frac{3}{4}$ miles **4.** $1\frac{1}{2}$ miles **5.** $\frac{7}{16}$ of an hour or $26\frac{1}{4}$ minutes

6. Yes, she will make it. She will have $38\frac{3}{4}$ minutes left over.

Chapter 2 Review Problems

1. $\frac{3}{8}$ **2.** $\frac{5}{12}$ **3.** answers will vary **4.** answers will vary **5.** $\frac{6}{31}$ **6.** $\frac{87}{100}$ **7.** 2×3^3 **8.** $2^3 \times 3 \times 5$ **9.** $2^3 \times 3 \times 7$ **10.** prime **11.** $2 \times 3 \times 13$ **12.** prime **13.** $\frac{2}{7}$ **14.** $\frac{1}{4}$ **15.** $\frac{7}{12}$ **16.** $\frac{13}{17}$ **17.** $\frac{7}{8}$ **18.** $\frac{17}{35}$ **19.** $\frac{35}{8}$ **20.** $\frac{63}{4}$ **21.** $\frac{37}{7}$ **22.** $\frac{33}{5}$ **23.** $5\frac{5}{8}$ **24.** $4\frac{16}{21}$ **25.** $7\frac{4}{7}$ **26.** $8\frac{2}{9}$ **27.** $3\frac{3}{11}$ **28.** $\frac{117}{8}$ **29.** $4\frac{1}{8}$ **30.** $\frac{20}{77}$ **31.** $\frac{7}{15}$ **32.** 0 **33.** $\frac{4}{63}$ **34.** $\frac{492}{5}$ or $98\frac{2}{5}$ **35.** $\frac{51}{2}$ or $25\frac{1}{2}$ **36.** $\frac{82}{5}$ or $16\frac{2}{5}$ **37.** $\frac{49}{2}$ or $24\frac{1}{2}$ **38.** $\$677\frac{1}{4}$ **39.** $\frac{305}{7}$ or $43\frac{4}{7}$ square inches **40.** $\frac{15}{14}$ or $1\frac{1}{14}$ **41.** 6 **42.** 1920 **43.** 1500 **44.** $\frac{1}{2}$ **45.** 8 **46.** 0 **47.** $\frac{46}{33}$ or $1\frac{13}{33}$ **48.** 12 rolls **49.** $\frac{560}{3}$ or $186\frac{2}{3}$ calories **50.** 98 **51.** 120 **52.** 90 **53.** $\frac{24}{56}$ **54.** $\frac{33}{72}$ **55.** $\frac{36}{172}$ **56.** $\frac{187}{198}$ **57.** $\frac{1}{14}$ **58.** $\frac{13}{12}$ or $1\frac{1}{12}$ **59.** $\frac{85}{63}$ or $1\frac{22}{63}$ **60.** $\frac{11}{40}$ **61.** $\frac{23}{70}$ **62.** $\frac{25}{36}$ **63.** $\frac{19}{48}$ **64.** $\frac{61}{75}$ **65.** $\frac{5}{4}$ or $1\frac{1}{4}$ **66.** $\frac{49}{9}$ or $5\frac{4}{9}$ **67.** $8\frac{2}{3}$ **68.** $20\frac{5}{7}$ **69.** $\frac{49}{8}$ or $6\frac{1}{8}$ **70.** $\frac{279}{80}$ or $3\frac{39}{80}$ **71.** $\frac{9}{10}$ **72.** $\frac{20}{63}$ **73.** $8\frac{29}{40}$ miles **74.** $283\frac{1}{12}$ miles **75.** $1\frac{2}{3}$ cups sugar; $2\frac{1}{8}$ cups flour **76.** $206\frac{1}{8}$ miles **77.** 15 lengths **78.** $9\frac{5}{8}$ liters **79.** $227\frac{1}{2}$ minutes or 3 hours and $47\frac{1}{2}$ min. **80.** $\$56\frac{1}{4}$ **81.** \$460 **82.** $1\frac{1}{16}$ inch **83.** \$242 **84.** (a) 25 miles per gallon (b) $\$22\frac{2}{25}$ **85.** $\frac{3}{7}$ **86.** $\frac{68}{75}$ **87.** $1\frac{5}{12}$ **88.** $\frac{24}{77}$ **89.** $\frac{34}{7}$ or $4\frac{6}{7}$ **90.** $\frac{64}{243}$ **91.** $6\frac{3}{16}$ **92.** $\frac{517}{12}$ or $43\frac{1}{12}$ **93.** $\frac{100}{3}$ or $33\frac{1}{3}$

How Am I Doing? Chapter 2 Test

1. $\frac{3}{5}$ (obj. 2.1.1) **2.** $\frac{311}{388}$ (obj. 2.1.3) **3.** $\frac{3}{7}$ (obj. 2.2.2) **4.** $\frac{3}{14}$ (obj. 2.2.2) **5.** $\frac{9}{2}$ (obj. 2.2.2) **6.** $\frac{34}{5}$ (obj. 2.3.1) **7.** $10\frac{5}{14}$ (obj. 2.3.2) **8.** 12 (obj. 2.4.2) **9.** $\frac{14}{45}$ (obj. 2.4.1) **10.** 14 (obj. 2.4.3) **11.** $\frac{77}{40}$ or $1\frac{37}{40}$ (obj. 2.5.1) **12.** $\frac{39}{62}$ (obj. 2.5.1) **13.** $\frac{90}{13}$ or $6\frac{12}{13}$ (obj. 2.5.3) **14.** $\frac{12}{7}$ or $1\frac{5}{7}$ (obj. 2.5.3) **15.** 36 (obj. 2.6.2) **16.** 48 (obj. 2.6.2) **17.** 24 (obj. 2.6.2) **18.** $\frac{30}{72}$ (obj. 2.6.3) **19.** $\frac{13}{36}$ (obj. 2.7.2) **20.** $\frac{11}{20}$ (obj. 2.7.2) **21.** $\frac{25}{28}$ (obj. 2.7.2) **22.** $14\frac{6}{35}$ (obj. 2.8.1) **23.** $4\frac{13}{14}$ (obj. 2.8.2) **24.** $\frac{1}{48}$ (obj. 2.8.3) **25.** $\frac{7}{6}$ or $1\frac{1}{6}$ (obj. 2.8.3) **26.** 154 square feet (obj. 2.9.1) **27.** 8 packages (obj. 2.9.1) **28.** $\frac{7}{10}$ mile (obj. 2.9.1) **29.** $14\frac{1}{24}$ miles (obj. 2.9.1) **30.** (obj. 2.9.1) (a) 40 oranges (b) $\$9\frac{3}{5}$ **31.** (obj. 2.9.1) (a) 77 candles (b) $\$1\frac{1}{4}$ (c) $\$827\frac{3}{4}$

Cumulative Test for Chapters 1–2

1. eighty-four million, three hundred sixty-one thousand, two hundred eight **2.** 869 **3.** 719,220 **4.** 2075 **5.** 17,216 **6.** 4788 **7.** 202,896 **8.** 4658 **9.** 308 R 11 **10.** 49 **11.** 6,037,000 **12.** 50 **13.** \$174 **14.** \$306 **15.** $\frac{83}{112}$ were women; $\frac{29}{112}$ were men **16.** $\frac{7}{13}$ **17.** $\frac{75}{4}$ **18.** $14\frac{2}{7}$ **19.** $\frac{527}{48}$ or $10\frac{47}{48}$ **20.** $\frac{12}{35}$ **21.** 40 **22.** $\frac{61}{54}$ or $1\frac{7}{54}$ **23.** $\frac{71}{8}$ or $8\frac{7}{8}$ **24.** $\frac{113}{15}$ or $7\frac{8}{15}$ **25.** $\frac{5}{6}$ **26.** $2\frac{3}{4}$ pounds **27.** $24\frac{3}{5}$ miles per gallon **28.** $8\frac{1}{8}$ cups of sugar; $5\frac{5}{6}$ cups of flour **29.** 60,000,000 miles **30.** \$160

Chapter 3

3.1 Exercises

1. A decimal fraction is a fraction whose denominator is a power of 10. $\frac{23}{100}$ and $\frac{563}{1000}$ are decimal fractions. **3.** hundred-thousandths **5.** fifty-seven hundredths **7.** three and eight tenths **9.** five and eight hundred three thousandths **11.** twenty-eight and thirty-seven ten-thousandths **13.** one hundred twenty-four and $\frac{20}{100}$ dollars **15.** one thousand, two hundred thirty-six and $\frac{8}{100}$ dollars **17.** twelve thousand fifteen and $\frac{45}{100}$ dollars **19.** 0.7 **21.** 0.12 **23.** 0.481 **25.** 0.000286 **27.** 0.7 **29.** 0.76 **31.** 0.01

33. 0.053 **35.** 0.2403 **37.** 8.3 **39.** 84.13 **41.** 3,529 **43.** 126.0571 **45.** $\frac{1}{50}$ **47.** $3\frac{3}{5}$ **49.** $7\frac{41}{100}$ **51.** $12\frac{5}{8}$ **53.** $7\frac{123}{2000}$ **55.** $8\frac{27}{2500}$ **57.** $235\frac{627}{5000}$ **59.** $\frac{187}{10,000}$ **61.** (a) $\frac{317}{1000}$ (b) $\frac{301}{1000}$ **63.** $\frac{1}{250,000}$ **65.** 525 **66.** 938 **67.** 56,800 **68.** 8,069,000

3.2 Exercises

1. > **3.** = **5.** < **7.** > **9.** < **11.** > **13.** < **15.** > **17.** = **19.** > **21.** 12.6, 12.65, 12.8 **23.** 0.007, 0.0071, 0.05 **25.** 5.12, 5.2, 5.23, 5.3 **27.** 26.003, 26.033, 26.034, 26.04 **29.** 18.006, 18.060, 18.065, 18.066, 18.606 **31.** 5.7 **33.** 29.0 **35.** 578.1 **37.** 2176.8 **39.** 26.03 **41.** 37.00 **43.** 156.17 **45.** 2786.71 **47.** 1.061 **49.** 0.0595 **51.** 5.00761 **53.** 136 **55.** $2537 **57.** $15,021 **59.** $56.98 **61.** $5783.72 **63.** 0.609; 0.521 **65.** 365.24 **67.** 0.0059, 0.006, 0.0519, $\frac{6}{100}$, 0.0601, 0.0612, 0.062, $\frac{6}{10}$, 0.61 **69.** You should consider only one digit to the right of the decimal place that you wish to round to. 86.23498 is closer to 86.23 than to 86.24. **71.** $12\frac{1}{8}$ **72.** $10\frac{9}{20}$ **73.** 692 miles **74.** $31,800

3.3 Exercises

1. 76.8 **3.** 1215.55 **5.** 296.2 **7.** 12.76 **9.** 36.7287 **11.** 67.42 **13.** 371.44 **15.** 1112.16 **17.** 21.04 ft. **19.** 8.6 pounds **21.** $78.12 **23.** 47,054.9 **25.** $1411.97 **27.** 3.9 **29.** 25.93 **31.** 49.78 **33.** 508.313 **35.** 64.88 **37.** 4.6465 **39.** 6.737 **41.** 1189.07 **43.** 1.4635 **45.** 176.581 **47.** 41.59 **49.** 5.2363 **51.** 73.225 **53.** 1.464 meters **55.** $36,947.16 **57.** $45.30 **59.** 11.64 centimeters **61.** 2.95 liters **63.** 0.0061 milligram; yes **65.** $6.2 billion; $6,200,000,000 **67.** $162.1 billion; $162,100,000,000 **69.** $8.40; yes; $8.34; very close the estimate was off by 6¢. **71.** $x = 8.4$ **73.** $x = 43.7$ **75.** $x = 2.109$ **77.** 20,288 **78.** 48,793 **79.** $\frac{77}{25}$ or $3\frac{2}{25}$ **80.** $\frac{35}{4}$ or $8\frac{3}{4}$

3.4 Exercises

1. Each factor has two decimal places you add the number of decimal places to get 4 decimal places. Multiply 67×8 to get 536. Place the decimal point 4 places to the left to obtain the result, 0.0536 **3.** When you multiply a number by 100, move the decimal point two places to the right. The answer is 0.78. **5.** 0.12 **7.** 0.06 **9.** 0.00288 **11.** 54.24 **13.** 0.000516 **15.** 0.6582 **17.** 981.73 **19.** 0.017304 **21.** 768.1517 **23.** 8460 **25.** 53.926 **27.** 6.5237 **29.** $9324 **31.** $588 **33.** 297.6 square feet **35.** $664.20 **37.** 514.8 miles **39.** 28.6 **41.** 5200 **43.** 22,615 **45.** 56,098.2 **47.** 1,756,144 **49.** 816.320 **51.** 593.2 centimeters **53.** 2980 meters **55.** $248.00 **57.** $62,279.00 **59.** To multiply by numbers such as 0.1, 0.01, 0.001, and 0.0001, count the number of decimal places in this first number. Then, in the other number, move the decimal point to the left from its present position the same number of decimal places as was in the first number. **61.** 86 **62.** 93 **63.** 127 R 3 **64.** 451 R 100 **65.** $0.11 billion or $110,000,000 **66.** $0.57 billion or $570,000,000 **67.** $4.473 billion **68.** $4.918 billion

How Am I Doing? Sections 3.1–3.4

1. forty-seven and eight hundred thirteen thousandths (obj. 3.1.1) **2.** 0.0567 (obj. 3.1.2) **3.** $2\frac{11}{100}$ (obj. 3.1.3) **4.** $\frac{21}{40}$ (obj. 3.1.3) **5.** 1.59, 1.6, 1.601, 1.61 (obj. 3.2.2) **6.** 123.5 (obj. 3.2.3) **7.** 1.053 (obj. 3.2.3) **8.** 17.99 (obj. 3.2.3) **9.** 19.45 (obj. 3.3.1) **10.** 27.191 (obj. 3.3.1) **11.** 10.59 (obj. 3.3.2) **12.** 7.671 (obj. 3.3.2) **13.** 0.3501 (obj. 3.4.1) **14.** 4780.5 (obj. 3.4.2) **15.** 37.96 (obj. 3.4.2) **16.** 0.06528 (obj. 3.4.1) **17.** 6.874 (obj. 3.4.1) **18.** 0.00000312 (obj. 3.4.1)

3.5 Exercises

1. 2.1 **3.** 17.83 **5.** 10.52 **7.** 136.5 **9.** 9.05 **11.** 53 **13.** 18 **15.** 130 **17.** 5.3 **19.** 1.2 **21.** 49.3 **23.** 94.21 **25.** 13.56 **27.** 0.17 **29.** 0.081 **31.** 91.264 **33.** 123 **35.** 213 **37.** $82.73 **39.** approximately 27.3 miles per gallon **41.** 24 bouquets **43.** 182 guests **45.** 23 snowboards. The error was in putting one less snowboard in the box than was required. **47.** 7.21 **49.** 975 **51.** 44 **53.** 41 **54.** $\frac{127}{40}$ or $3\frac{7}{40}$ **55.** $\frac{15}{16}$ **56.** $\frac{91}{12}$ or $7\frac{7}{12}$ **57.** $\frac{5}{3}$ or $1\frac{2}{3}$ **58.** 673 **59.** 471 **60.** 1172 **61.** 936

3.6 Exercises

1. same quantity **3.** The digits 8942 repeat. **5.** 0.25 **7.** 0.8 **9.** 0.125 **11.** 0.35 **13.** 0.62 **15.** 2.25 **17.** 2.125 **19.** 1.4375 **21.** $0.\overline{6}$ **23.** $0.\overline{45}$ **25.** $3.8\overline{3}$ **27.** $2.2\overline{7}$ **29.** 0.308 **31.** 0.905 **33.** 0.146 **35.** 1.296 **37.** 0.404 **39.** 0.944 **41.** 3.143 **43.** 3.474 **45.** < **47.** > **49.** 0.125 inch **51.** too small 0.125 inch **53.** yes; it is 0.025 inches too wide. **55.** 2.3 **57.** 3.36 **59.** 0 **61.** 21.414 **63.** 0.0072 **65.** 0.325 **67.** 28.6 **69.** 20.836 **71.** 0.586930 **73.** (a) 0.16 (b) $0.1449\overline{49}$ (c) b is a repeating decimal and a is a nonrepeating decimal **75.** $5\frac{3}{4}$ feet deep **76.** $32\frac{3}{10}$ feet

3.7 Exercises

1. 700,000,000 **3.** 30,000 **5.** 2500 **7.** 15 **9.** $200 **11.** 4875 kroners **13.** 2748.27 square feet **15.** 96 molds **17.** 11.59 meters **19.** 24 servings **21.** $1263.09 **23.** $510 **25.** $16.22 **27.** $17,319; $5819 **29.** yes; by 0.149 milligram per liter **31.** 137 minutes **33.** 22.9 quadrillion Btu **35.** approximately 47.8 quadrillion Btu; 47,800,000,000,000,000 Btu **37.** $\frac{19}{21}$ **38.** $\frac{7}{38}$ **39.** $\frac{1}{10}$ **40.** 8

Putting Your Skills to Work

1. 1,625,670,000,000 miles **2.** 29,558,000,000 gallons **3.** 51,229,000,000 gallons **4.** $1728 would be saved each year.

Chapter 3 Review Problems

1. thirteen and six hundred seventy-two thousandths **2.** eighty-four hundred-thousandths **3.** 0.7 **4.** 0.81 **5.** 1.523 **6.** 0.0079 **7.** $\frac{17}{100}$ **8.** $\frac{9}{250}$ **9.** $34\frac{6}{25}$ **10.** $1\frac{1}{4000}$ **11.** = **12.** > **13.** < **14.** > **15.** 0.901, 0.918, 0.98, 0.981 **16.** 5.2, 5.26, 5.59, 5.6, 5.62 **17.** 0.704, 0.7045, 0.745, 0.754 **18.** 8.2, 8.27, 8.702, 8.72 **19.** 0.6 **20.** 19.21 **21.** 9.8522 **22.** $156 **23.** 77.6 **24.** 6.639 **25.** 3.894 **26.** 113.872 **27.** 0.003136 **28.** 887.81 **29.** 405.6 **30.** 2398.02 **31.** 0.613 **32.** 123.540 **33.** $8.73 **34.** 0.00258 **35.** 36.8 **36.** 232.9 **37.** 574.4 **38.** 0.059 **39.** $0.2\overline{7}$ **40.** 0.175 **41.** $1.8\overline{3}$ **42.** 1.1875 **43.** 0.786 **44.** 0.345 **45.** 2.294 **46.** 3.391 **47.** 19.546 **48.** 1.152 **49.** 1.301208 **50.** 23.13 **51.** 439.19 **52.** 64.3 **53.** 4.459 **54.** 0.904 **55.** 20.004 **56.** 1.25 **57.** 112 people **58.** 24.8 miles per gallon **59.** $2170.30 **60.** ABC company **61.** no; by 0.0005 milligram per liter **62.** 15.75 inches **63.** (a) 55.8 feet (b) 175.68 square feet **64.** 1396.75 square feet **65.** 6.1 miles **66.** 259.9 feet **67.** $12,750.00; $12,255.00 they should change to the new loan **68.** $262 **69.** $207 **70.** $15.97 **71.** $24.00 **72.** $38.03 **73.** $33.90

How Am I Doing? Chapter 3 Test

1. twelve and forty-three thousandths (obj. 3.1.1) **2.** 0.3977 (obj. 3.1.2) **3.** $7\frac{3}{20}$ (obj. 3.1.3) **4.** $\frac{261}{1000}$ (obj. 3.1.3) **5.** 2.19, 2.9, 2.907, 2.91 (obj. 3.2.2) **6.** 78.66 (obj. 3.2.3) **7.** 0.0342 (obj. 3.2.3) **8.** 99.698 (obj. 3.3.1) **9.** 37.53 (obj. 3.3.1) **10.** 0.0979 (obj. 3.3.2) **11.** 71.155 (obj. 3.3.2) **12.** 0.5817 (obj. 3.4.1) **13.** 2189 (obj. 3.4.2) **14.** 0.1285 (obj. 3.5.2) **15.** 47 (obj. 3.5.2) **16.** $1.\overline{2}$ (obj. 3.6.1) **17.** 0.875 (obj. 3.6.1) **18.** 1.487 (obj. 3.6.2) **19.** 6.1952 (obj. 3.6.2) **20.** $11.99 (obj. 3.7.2) **21.** 18.8 miles per gallon (obj. 3.7.2) **22.** 3.43 centimeters (obj. 3.7.2) **23.** $390.55 (obj. 3.7.2)

Cumulative Test for Chapters 1–3

1. thirty-eight million, fifty-six thousand, nine hundred fifty-four **2.** 479,587 **3.** 54,480 **4.** 39,463 **5.** 258 **6.** 16 **7.** $\frac{2}{5}$ **8.** $7\frac{1}{2}$ **9.** $\frac{9}{35}$ **10.** $\frac{23}{24}$ **11.** 16 **12.** $\frac{33}{10}$ or $3\frac{3}{10}$ **13.** 24,000,000,000 **14.** 0.039 **15.** 2.01, 2.1, 2.11, 2.12, 20.1 **16.** 26.080 **17.** 19.54 **18.** 13.118 **19.** 1.136 **20.** 182.3 **21.** 1.058 **22.** 0.8125 **23.** 13.597 **24.** (a) 110.25 square feet (b) 42 feet **25.** $195.57 **26.** 60 months

Chapter 4

4.1 Exercises

1. ratio **3.** 5 to 8 **5.** $\frac{1}{3}$ **7.** $\frac{7}{6}$ **9.** $\frac{2}{3}$ **11.** $\frac{11}{6}$ **13.** $\frac{15}{16}$ **15.** $\frac{2}{3}$ **17.** $\frac{8}{5}$ **19.** $\frac{2}{3}$ **21.** $\frac{3}{2}$ **23.** $\frac{41}{80}$ **25.** $\frac{13}{1}$ **27.** $\frac{10}{17}$ **29.** $\frac{165}{285} = \frac{11}{19}$ **31.** $\frac{35}{165} = \frac{7}{33}$ **33.** $\frac{205}{1225} = \frac{41}{245}$ **35.** $\frac{450}{205} = \frac{90}{41}$ **37.** $\frac{1}{16}$ **39.** $\frac{\$5}{\text{2 magazines}}$ **41.** $\frac{\$85}{\text{6 bushes}}$ **43.** $\frac{\$38}{\text{4 CD's}}$ **45.** $\frac{\text{410 revolutions}}{\text{1 mile}}$ or 410 revolutions/mile **47.** $\frac{\$27{,}500}{\text{1 employee}}$ or $27,500/employee **49.** $13/hour **51.** 16 miles/gallon **53.** 90 people/sq mi **55.** 125 pencils/box **57.** 66 mi/hr **59.** 19 patients/doctor **61.** 5 eggs/chicken **63.** $30/share **65.** $5.50 profit per calendar **67.** (a) $0.08/oz small box $0.07/oz large box (b) 1¢ per ounce (c) The consumer saves $0.48. **69.** (a) 13 moose (b) 12 moose (c) North slope **71.** (a) $24.50 (b) $2.20 (c) $22.30 **73.** increased by Mach 0.3 **75.** $2\frac{5}{8}$ **76.** 5 **77.** $\frac{5}{24}$ **78.** $1\frac{1}{48}$ **79.** $12.25/sq. yard **80.** $24,150; $16,800

4.2 Exercises

1. equal **3.** $\frac{18}{9} = \frac{2}{1}$ **5.** $\frac{20}{36} = \frac{5}{9}$ **7.** $\frac{220}{11} = \frac{400}{20}$ **9.** $\frac{5\frac{1}{3}}{16} = \frac{7\frac{2}{3}}{23}$ **11.** $\frac{6.5}{14} = \frac{13}{28}$ **13.** $\frac{\text{3 inches}}{\text{40 miles}} = \frac{\text{27 inches}}{\text{360 miles}}$ **15.** $\frac{\$28}{\text{6 tables}} = \frac{\$98}{\text{21 tables}}$ **17.** $\frac{\text{3 hours}}{\$525} = \frac{\text{7 hours}}{\$1225}$ **19.** $\frac{\text{3 teaching assistants}}{\text{40 children}} = \frac{\text{21 teaching assistants}}{\text{280 children}}$ **21.** $\frac{\text{4800 people}}{\text{3 restaurants}} = \frac{\text{11,200 people}}{\text{7 restaurants}}$ **23.** It is a proportion. **25.** It is not a proportion. **27.** It is not a proportion. **29.** It is a proportion. **31.** It is a proportion. **33.** It is not a proportion. **35.** It is a proportion. **37.** It is not a proportion. **39.** It is a proportion. **41.** It is a proportion. **43.** It is a proportion. **45.** It is not a proportion. **47.** no **49.** (a) no (b) The bus is traveling at a faster rate. **51.** yes **53.** (a) yes (b) yes (c) the equality test for fractions **54.** 23.1405 **55.** 17.9968 **56.** 402.408 **57.** 25.8 **58.** $12\frac{3}{8}$ miles

How Am I Doing? Sections 4.1–4.2

1. $\frac{13}{18}$ (obj. 4.1.1) **2.** $\frac{1}{5}$ (obj. 4.1.1) **3.** $\frac{9}{2}$ (obj. 4.1.1) **4.** $\frac{11}{12}$ (obj. 4.1.1) **5.** (a) $\frac{7}{24}$ (b) $\frac{11}{120}$ (obj. 4.1.2)
6. $\frac{\text{3 flight attendants}}{\text{100 passengers}}$ (obj. 4.1.2) **7.** $\frac{\text{31 gallons}}{\text{42 square feet}}$ (obj. 4.1.2) **8.** 30.5 miles per hour (obj. 4.1.2) **9.** $29 per CD player (obj. 4.1.2)
10. 160 cookies per pound of cookie dough (obj. 4.1.2) **11.** $\frac{13}{40} = \frac{39}{120}$ (obj. 4.2.1) **12.** $\frac{116}{158} = \frac{29}{37}$ (obj. 4.2.1)
13. $\frac{\text{33 nautical miles}}{\text{2 hours}} = \frac{\text{49.5 nautical miles}}{\text{3 hours}}$ (obj. 4.2.1) **14.** $\frac{\text{3000 shoes}}{\$370} = \frac{\text{7500 shoes}}{\$925}$ (obj. 4.2.1) **15.** It is a proportion. (obj. 4.2.2)
16. It is not a proportion. (obj. 4.2.2) **17.** It is a proportion. (obj. 4.2.2) **18.** It is not a proportion. (obj. 4.2.2)
19. It is a proportion. (obj. 4.2.2) **20.** It is a proportion. (obj. 4.2.2)

4.3 Exercises

1. Divide each side of the equation by the number a. Calculate $\frac{b}{a}$. The value of n is $\frac{b}{a}$. **3.** $n = 11$ **5.** $n = 5.6$ **7.** $n = 5$ **9.** $n = 8$
11. $n = 34\frac{2}{3}$ **13.** $n = 15$ **15.** $n = 16$ **17.** $n = 7.5$ **19.** $n = 5$ **21.** $n = 75$ **23.** $n = 22.5$ **25.** $n = 330$ **27.** $n = 31.5$
29. $n = 18$ **31.** $n = 8$ **33.** $n = 162$ **35.** $n \approx 30.9$ **37.** $n \approx 3.7$ **39.** $n \approx 5.5$ **41.** $n = 48$ **43.** $n = 1.25$ **45.** $n = 2.5$
47. $n = 3.75$ **49.** $n \approx 3.94$ **51.** $n = 80$ **53.** $n = 4\frac{7}{8}$ **55.** 11 inches **57.** $n = 3\frac{5}{8}$ **59.** $n = 10\frac{8}{9}$ **60.** 76 **61.** 47
62. five hundred sixty-three thousandths **63.** 0.0034 **64.** $1560 **65.** 56 games

4.4 Exercises

1. He should continue with people on the top of the fraction. That would be 60 people he observed on Saturday night. He does not know the number of dogs, so this would be n. The proportion would be:

$$\frac{\text{12 people}}{\text{5 dogs}} = \frac{\text{60 people}}{n \text{ dogs}}.$$

3. 161 cars **5.** 3 cups **7.** $7\frac{1}{2}$ kilometers **9.** 2600 kronor **11.** 197.6 feet **13.** 217 miles **15.** 17.7 cups
17. 102 free throws **19.** 18.75 gallons **21.** 40 hawks **23.** $3570 **25.** 270 chips **27.** 1 cup of water and $\frac{3}{8}$ cup of milk
29. 2 cups of water and 1 cup of milk **31.** Alex Rodriguez, approximately $385,965 for each home run. Sammy Sosa, approximately $306,122 for each home run. **33.** Randy Johnson, approximately $381,429 per game. Mike Mussina, approximately $333,333 per game. **35.** 56,200
36. 196,380,000 **37.** 56.1 **38.** 2.7490 **39.** (a) $\frac{19}{20}$ of a square foot (b) 1425 square feet

Putting Your Skills to Work

1. (a) 1250 calories (b) 5 hours **2.** (a) 5200 calories (b) 26 hours **3.** Gina; 120 calories **4.** Miguel; 350 calories more
5. Sample answer; 1 hour of ballroom dancing, 8 hours of walking, and 6 hours of jogging **6.** (a) $\frac{7}{5}$ (b) $\frac{7}{5}$

Chapter 4 Review Problems

1. $\frac{11}{5}$ **2.** $\frac{5}{3}$ **3.** $\frac{4}{5}$ **4.** $\frac{10}{19}$ **5.** $\frac{28}{51}$ **6.** $\frac{1}{3}$ **7.** $\frac{40}{93}$ **8.** $\frac{52}{147}$ **9.** $\frac{2}{5}$ **10.** $\frac{1}{8}$ **11.** $\frac{7}{32}$ **12.** $\frac{\$25}{\text{2 people}}$ **13.** $\frac{\text{4 revolutions}}{\text{11 minutes}}$
14. $\frac{\text{47 vibrations}}{\text{4 seconds}}$ **15.** $\frac{\text{5 cups}}{\text{8 pies}}$ **16.** $17/share **17.** $112/credit-hour **18.** $13.50/square yard **19.** $12.50/DVD
20. (a) $0.74 (b) $0.58 (c) $0.16 **21.** (a) $0.22 (b) $0.25 (c) $0.03 **22.** $\frac{12}{48} = \frac{7}{28}$ **23.** $\frac{1\frac{1}{2}}{5} = \frac{4}{13\frac{1}{3}}$ **24.** $\frac{7.5}{45} = \frac{22.5}{135}$
25. $\frac{\text{3 buses}}{\text{138 passengers}} = \frac{\text{5 buses}}{\text{230 passengers}}$ **26.** $\frac{\text{15 pounds}}{\$4.50} = \frac{\text{27 pounds}}{\$8.10}$ **27.** It is not a proportion. **28.** It is a proportion.
29. It is a proportion. **30.** It is a proportion. **31.** It is not a proportion. **32.** It is not a proportion. **33.** It is a proportion.
34. It is not a proportion. **35.** $n = 23$ **36.** $n = 5\frac{1}{4}$ or 5.25 **37.** $n = 22.1$ or $22\frac{1}{10}$ **38.** $n = 17$ **39.** $n = 33$ **40.** $n = 42$
41. $n = 7$ **42.** $n = 24$ **43.** $n = 19$ **44.** $n = 5\frac{3}{5}$ or 5.6 **45.** $n \approx 5.0$ **46.** $n \approx 6.8$ **47.** $n = 19$ **48.** $n = 15$
49. $n = 16.8$ **50.** $n = 4.7$ **51.** $n = 12$ **52.** $n = 550$ **53.** 15 gallons **54.** 1691 employees **55.** 2016 francs

56. 111.11 swiss francs **57.** 600 miles **58.** 35 free throws **59.** 120 feet **60.** (a) 7.65 gallons (b) \$14.15 **61.** 5.71 centimeters tall **62.** 7.5 grams **63.** 182 bricks **64.** 48 pages **65.** 9 gallons **66.** 834 liters **67.** approximately 113.93 feet **68.** approximately 27.38 minutes **69.** 86 goals **70.** 552 calories **71.** 13,680 people **72.** 195 trips

How Am I Doing? Chapter 4 Test

1. $\frac{9}{26}$ (obj. 4.1.1) **2.** $\frac{14}{37}$ (obj. 4.1.1) **3.** $\frac{98 \text{ miles}}{3 \text{ gallons}}$ (obj. 4.1.2) **4.** $\frac{140 \text{ square feet}}{3 \text{ pounds}}$ (obj. 4.1.2) **5.** 3.8 tons/day (obj. 4.1.2) **6.** \$8.28/hour (obj. 4.1.2) **7.** 245.45 feet/pole (obj. 4.1.2) **8.** \$85.21/share (obj. 4.1.2) **9.** $\frac{17}{29} = \frac{51}{87}$ (obj. 4.2.1) **10.** $\frac{2\frac{1}{2}}{10} = \frac{6}{24}$ (obj.4.2.1) **11.** $\frac{490 \text{ miles}}{21 \text{ gallons}} = \frac{280 \text{ miles}}{12 \text{ gallons}}$ (obj. 4.2.1) **12.** $\frac{3 \text{ hours}}{180 \text{ miles}} = \frac{5 \text{ hours}}{300 \text{ miles}}$ (obj. 4.2.1) **13.** It is not a proportion. (obj. 4.2.2) **14.** It is a proportion. (obj. 4.2.2) **15.** It is a proportion. (obj. 4.2.2) **16.** It is not a proportion. (obj. 4.2.2) **17.** $n = 16$ (obj. 4.3.2) **18.** $n = 22.5$ (obj. 4.3.2) **19.** $n = 19$ (obj. 4.3.2) **20.** $n = 29.4$ (obj. 4.3.2) **21.** $n = 120$ (obj. 4.3.2) **22.** $n = 70.4$ (obj. 4.3.2) **23.** $n = 120$ (obj. 4.3.2) **24.** $n = 52$ (obj. 4.3.2) **25.** 6 eggs (obj. 4.4.1) **26.** 80.95 pounds (obj. 4.4.1) **27.** 19 miles (obj. 4.4.1) **28.** \$360 (obj. 4.4.1) **29.** 136.7 miles (obj. 4.4.1) **30.** 696.67 kilometers (obj. 4.4.1) **31.** 88 free throws (obj. 4.4.1) **32.** 32 hits (obj. 4.4.1)

Cumulative Test for Chapters 1–4

1. twenty-six million, five hundred ninety-seven thousand, eighty-nine **2.** 68 **3.** $\frac{11}{32}$ **4.** $\frac{27}{35}$ **5.** 65 **6.** 8.2584 **7.** 0.179856 **8.** $\frac{1}{6}$ **9.** 16,145.5 **10.** 56.9 **11.** 326.278 **12.** 3.68 **13.** 0.15625 **14.** It is a proportion **15.** It is a proportion **16.** $n = 3$ **17.** $n = 2$ **18.** $n = 64$ **19.** $n = 9$ **20.** $n \approx 3.4$ **21.** $n = 21$ **22.** 8.33 inches **23.** \$396.67 **24.** 5 pounds **25.** 214.5 gallons

Chapter 5

5.1 Exercises

1. hundred **3.** two; left; Drop **5.** 59% **7.** 4% **9.** 80% **11.** 245% **13.** 5.3% **15.** 0.07% **17.** 13% **19.** 8% **21.** 0.51 **23.** 0.07 **25.** 0.2 **27.** 0.436 **29.** 0.0003 **31.** 0.0072 **33.** 0.0125 **35.** 3.66 **37.** 74% **39.** 50% **41.** 8% **43.** 56.3% **45.** 0.2% **47.** 0.57% **49.** 135% **51.** 272% **53.** 27% **55.** 30% **57.** 94% **59.** 231% **61.** 10% **63.** 0.9% **65.** 0.62 **67.** 1.38 **69.** 0.003 **71.** 0.8 **73.** 59% **75.** 1.15 **77.** 0.006 **79.** 0.77; 0.9 **81.** 36% = 36 percent = 36 "per one hundred" $= 36 \times \frac{1}{100} = \frac{36}{100} = 0.36$. The rule is using the fact that 36% means 36 per one hundred. **83.** (a) 15.62 (b) $\frac{1562}{100}$ (c) $\frac{781}{50}$ **85.** $\frac{14}{25}$ **86.** $\frac{39}{50}$ **87.** 0.6875 **88.** 0.875 **89.** 5336 vases

5.2 Exercises

1. Write the number in front of the percent symbol as the numerator of a fraction. Write the number 100 as the denominator of the fraction. Reduce the fraction if possible. **3.** $\frac{3}{50}$ **5.** $\frac{33}{100}$ **7.** $\frac{11}{20}$ **9.** $\frac{3}{4}$ **11.** $\frac{1}{5}$ **13.** $\frac{19}{200}$ **15.** $\frac{7}{40}$ **17.** $\frac{81}{125}$ **19.** $\frac{57}{80}$ **21.** $1\frac{19}{25}$ **23.** $3\frac{2}{5}$ **25.** 12 **27.** $\frac{13}{600}$ **29.** $\frac{1}{8}$ **31.** $\frac{11}{125}$ **33.** $\frac{33}{1000}$ **35.** $\frac{7}{275}$ **37.** 75% **39.** 70% **41.** 35% **43.** 28% **45.** 27.5% **47.** 360% **49.** 250% **51.** 412.5% **53.** 33.33% **55.** 42.86% **57.** 425% **59.** 52% **61.** 2.5% **63.** 1.17% **65.** $37\frac{1}{2}\%$ **67.** $7\frac{1}{2}\%$ **69.** $83\frac{1}{3}\%$ **71.** $22\frac{2}{9}\%$

	Fraction	Decimal	Percent
73.	$\frac{11}{12}$	0.9167	91.67%
75.	$\frac{14}{25}$	0.56	56%
77.	$\frac{1}{200}$	0.005	0.5%
79.	$\frac{5}{9}$	0.5556	55.56%
81.	$\frac{1}{32}$	0.0313	$3\frac{1}{8}\%$

83. $\frac{463}{1600}$ **85.** 15.375% **87.** It will have at least two zeros to the right of the decimal point. **89.** $n = 5.625$ **90.** $n = 4$ **91.** 549,165 documents **92.** 4500 square feet

5.3A Exercises

1. What is 20% of \$300? **3.** 20 baskets out of 25 shots is what percent?
5. This is "a percent problem when we do not know the base."
Translated into an equation: $108 = 18\% \times n$

$$108 = 0.18n$$
$$\frac{108}{0.18} = \frac{0.18n}{0.18}$$
$$600 = n.$$

7. $n = 5\% \times 90$ **9.** $70\% \times n = 2$ **11.** $17 = n \times 85$ **13.** 28 **15.** 56 **17.** \$24 **19.** 1300 **21.** 1300 **23.** \$150 **25.** 84% **27.** 28% **29.** 65% **31.** 31 **33.** 85 **35.** 12% **37.** 3.28 **39.** 64% **41.** 445 **43.** 0.8% **45.** 39.6 **47.** 80% **49.** 85% **51.** 554 students **53.** 40 years **55.** \$57.60 **57.** width = 120 feet, length = 400 feet **59.** 2.448 **60.** 4.1492 **61.** 2834 **62.** 2.36

5.3B Exercises

	p	b	a
1.	75	660	495
3.	22	60	a
5.	49	b	2450
7.	p	50	30

9. 28 **11.** 196 **13.** 56 **15.** 80 **17.** 80 **19.** 600,000 **21.** 20 **23.** 4 **25.** 22 **27.** 40 **29.** 16.4% **31.** 3.64 **33.** 25% **35.** 170 **37.** \$700 **39.** 15% **41.** 18 gallons **43.** \$2280 **45.** 43.7% **47.** 7.6% **49.** $1\frac{31}{45}$ **50.** $\frac{1}{26}$ **51.** $4\frac{1}{5}$ **52.** $1\frac{13}{15}$

How Am I Doing? Sections 5.1–5.3

1. 17% (obj. 5.1.3) **2.** 38.7% (obj. 5.1.3) **3.** 134% (obj. 5.1.3) **4.** 894% (obj. 5.1.3) **5.** 0.6% (obj. 5.1.3) **6.** 0.04% (obj. 5.1.3) **7.** 17% (obj. 5.1.1) **8.** 89% (obj. 5.1.1) **9.** 13.4% (obj. 5.1.1) **10.** 19.8% (obj. 5.1.1) **11.** $6\frac{1}{2}\%$ (obj. 5.1.1) **12.** $1\frac{3}{8}\%$ (obj. 5.1.1) **13.** 80% (obj. 5.2.2) **14.** 2.5% (obj. 5.2.2) **15.** 260% (obj. 5.2.2) **16.** 106.25% (obj. 5.2.2) **17.** 71.43% (obj. 5.2.2) **18.** 28.57% (obj. 5.2.2) **19.** 95.65% (obj. 5.2.2) **20.** 68.42% (obj. 5.2.2) **21.** 440% (obj. 5.2.2) **22.** 275% (obj. 5.2.2) **23.** 0.33% (obj. 5.2.2) **24.** 0.25% (obj. 5.2.2) **25.** $\frac{11}{50}$ (obj. 5.2.1) **26.** $\frac{53}{100}$ (obj. 5.2.1) **27.** $\frac{3}{2}$ or $1\frac{1}{2}$ (obj. 5.2.1) **28.** $\frac{8}{5}$ or $1\frac{3}{5}$ (obj. 5.2.1) **29.** $\frac{19}{300}$ (obj. 5.2.1) **30.** $\frac{7}{150}$ (obj. 5.2.1) **31.** $\frac{41}{80}$ (obj. 5.2.1) **32.** $\frac{7}{16}$ (obj. 5.2.1) **33.** 55.2 (obj. 5.3.2) **34.** 48.36 (obj. 5.3.2) **35.** 94.44% (obj. 5.3.2) **36.** 44.74% (obj. 5.3.2) **37.** 3000 (obj. 5.3.2) **38.** 885 (obj. 5.3.2)

5.4 Exercises

1. 180,000 pencils **3.** \$70 **5.** 7.18% **7.** \$3.90 **9.** \$240 **11.** 25% **13.** \$9,600,000 **15.** 2.48% **17.** 216 babies **19.** \$761.90 **21.** \$150,000 **23.** 2200 pounds **25.** \$12,210,000 for personnel, food and decorations; \$20,790,000 for security, facility rental, and all other expenses **27.** \$123.50 **29.** \$30 **31.** (a) \$1280 (b) \$14,720 **33.** (a) \$333 (b) \$777 **35.** 1,698,000 **36.** 2,452,400 **37.** 1.63 **38.** 0.800 **39.** 0.0556 **40.** 0.0792

5.5 Exercises

1. \$3400 **3.** \$4140 **5.** 20% **7.** 40% **9.** \$140 **11.** \$7.50 **13.** \$480 **15.** 0.6% **17.** \$1,600,000 **19.** \$39.75 **21.** 39 boxes **23.** approximately 16% **25.** 80% **27.** (a) \$85.10 (b) \$3785.10 **29.** (a) \$6.96 (b) \$122.96 **31.** \$10,830 **33.** (a) \$27,920 (b) \$321,080 **35.** 96.9% **37.** \$849.01 **39.** 2 **40.** 120 **41.** $\frac{13}{18}$ **42.** 3.64

Putting Your Skills to Work

1. 2,248,000 students **2.** 217,000 students **3.** (a) 19.8% (b) 35.6% studied French in 1970. The percentage is less in 2000. **4.** (a) 5.2% (b) 11.9% studied German in 1970. The percentage is less in 2000. **5.** 258,000 more students **6.** 243,480 more students

Chapter 5 Review Problems

1. 62% **2.** 43% **3.** 37.2% **4.** 52.9% **5.** 105% **6.** 210% **7.** 252% **8.** 437% **9.** 103.6% **10.** 105.2% **11.** 0.6% **12.** 0.2% **13.** 12.5% **14.** 8.3% **15.** $4\frac{1}{12}\%$ **16.** $3\frac{5}{12}\%$ **17.** 317% **18.** 225% **19.** 76% **20.** 52% **21.** 55% **22.** 22.5% **23.** 45.45% **24.** 44.44% **25.** 225% **26.** 375% **27.** 277.78% **28.** 555.56% **29.** 190% **30.** 250% **31.** 0.38% **32.** 0.63% **33.** 0.32 **34.** 0.68 **35.** 0.827 **36.** 0.596 **37.** 2.36 **38.** 1.77 **39.** 0.32125 **40.** 0.26375 **41.** $\frac{18}{25}$ **42.** $\frac{23}{25}$ **43.** $\frac{37}{20}$ **44.** $\frac{9}{4}$ **45.** $\frac{41}{250}$ **46.** $\frac{61}{200}$ **47.** $\frac{5}{16}$ **48.** $\frac{7}{16}$ **49.** $\frac{1}{2000}$ **50.** $\frac{3}{5000}$

	Fraction	Decimal	Percent
51.		0.6	60%
52.		0.7	70%
53.	$\frac{3}{8}$	0.375	
54.	$\frac{9}{16}$	0.5625	
55.	$\frac{1}{125}$		0.8%
56.	$\frac{9}{20}$		45%

57. 17 **58.** 23 **59.** 90 **60.** 175 **61.** 38.46% **62.** 38.89% **63.** 97.2 **64.** 99.2 **65.** 160 **66.** 140 **67.** 20% **68.** 10% **69.** 51 students **70.** 96 trucks **71.** $11,200 **72.** $80,200 **73.** 60% **74.** 7.5% **75.** $183.50 **76.** $756 **77.** $1800 **78.** 4% **79.** 6% **80.** $1200 **81.** (a) $319 (b) $1276 **82.** (a) $255 (b) $1870 **83.** 42.40% **84.** 14.75% **85.** (a) $3360 (b) $20,640 **86.** (a) $330 (b) $1320 **87.** (a) $60 (b) $720

How Am I Doing? Chapter 5 Test

1. 57% (obj. 5.1.3) **2.** 1% (obj. 5.1.3) **3.** 0.8% (obj. 5.1.3) **4.** 1280% (obj. 5.1.3) **5.** 356% (obj. 5.1.3) **6.** 71% (obj. 5.1.1) **7.** 1.8% (obj. 5.1.1) **8.** $3\frac{1}{7}$% (obj. 5.1.1) **9.** 47.5% (obj. 5.2.2) **10.** 75% (obj. 5.2.2) **11.** 300% (obj. 5.2.2) **12.** 175% (obj. 5.2.2) **13.** 8.25% (obj. 5.2.3) **14.** 302.4% (obj. 5.2.3) **15.** $1\frac{13}{25}$ (obj. 5.2.3) **16.** $\frac{31}{400}$ (obj. 5.2.3) **17.** 20 (obj. 5.3.2) **18.** 130 (obj. 5.3.2) **19.** 55.56% (obj. 5.3.2) **20.** 200 (obj. 5.3.2) **21.** 5000 (obj. 5.3.2) **22.** 46% (obj. 5.3.2) **23.** 699.6 (obj. 5.3.2) **24.** 20% (obj. 5.3.2) **25.** $6092 (obj. 5.5.1) **26.** (a) $150.81 (obj. 5.4.3) (b) $306.19 **27.** 89.29% (obj. 5.4.1) **28.** 23.24% (obj. 5.4.1) **29.** 12,000 registered voters (obj. 5.4.1) **30.** (a) $240 (b) $960 (obj. 5.5.3)

Cumulative Test for Chapters 1–5

1. 2241 **2.** 8444 **3.** 5292 **4.** 89 **5.** $\frac{67}{12}$ or $5\frac{7}{12}$ **6.** $2\frac{7}{10}$ **7.** $\frac{15}{2}$ or $7\frac{1}{2}$ **8.** $\frac{5}{21}$ **9.** 77.183 **10.** 34.118 **11.** 1.686 **12.** 0.368 **13.** 4 tiles/square foot **14.** yes **15.** $n = 24$ **16.** 673 faculty members **17.** 2.3% **18.** 46.8% **19.** 198% **20.** 3.75% **21.** 2.43 **22.** 0.0675 **23.** 17.76% **24.** 114.58 **25.** 300 **26.** 190 **27.** $544 **28.** 3200 students **29.** 11.31% **30.** $352

Chapter 6

6.1 Exercises

1. We know that each mile is 5280 feet. Each foot is 12 inches. So we know that one mile is $5280 \times 12 = 63{,}360$ inches. The unit fraction we want is $\frac{63{,}360 \text{ inches}}{1 \text{ mile}}$. So we multiply 23 miles $\times \frac{63{,}360 \text{ inches}}{1 \text{ mile}}$. The mile unit divides out. We obtain 1,457,280 inches. Thus 23 miles = 1,457,280 inches. **3.** 1760 **5.** 2000 **7.** 4 **9.** 2 **11.** 7 **13.** 9 **15.** 144 **17.** 2 **19.** 12,320 **21.** 28 **23.** 12 **25.** 128 **27.** 68 **29.** 84 **31.** 16 **33.** 0.5 **35.** 6.25 **37.** 7.5 **39.** 36 **41.** 6.25 **43.** 138,336 feet **45.** 3.66 miles **47.** $9.75 **49.** (a) 144 inches (b) $122.40 **51.** 28,800 cups **53.** ≈ 12,000 yards **55.** ≈ 6 miles **57.** 14,739 land miles **59.** $10,800 **60.** 22% **61.** 161 miles **62.** 104 students

6.2 Exercises

1. hecto- **3.** deci- **5.** kilo- **7.** 460 **9.** 2610 meters **11.** 1.67 **13.** 0.0732 **15.** 200,000 **17.** 0.078 **19.** 3.5; 0.035 **21.** 4500; 450,000 **23.** b **25.** c **27.** b **29.** a **31.** b **33.** 39 **35.** 8000 **37.** 0.482 **39.** 3255 m **41.** 463 cm **43.** 63.5 cm **45.** 2.5464 cm or 25.464 mm **47.** 3.23 m **49.** 939.86 m **51.** 0.964 **53.** false **55.** true **57.** true **59.** true **61.** (a) 481,800 cm (b) 4,818 km **63.** 0.00000000254 **65.** 7,670,000 metric tons **67.** 18,560,000,000 kilograms **69.** approximately 113,020,000 metric tons **70.** 5000 **71.** 1.77 **72.** 52 **73.** 15

6.3 Exercises

1. 1 kL **3.** 1 mg **5.** 1 g **7.** 9000 **9.** 12,000 **11.** 0.0189 **13.** 0.752 **15.** 2,430,000 **17.** 82 **19.** 0.005261 **21.** 74,000 **23.** 0.216 **25.** 0.035 **27.** 6.328 **29.** 2920 **31.** 2400 **33.** 0.007; 0.000007 **35.** 0.128; 0.000128 **37.** 33; 33,000 **39.** 2580; 2,580,000 **41.** (b) **43.** (a) **45.** 83 L + 0.822 L + 30.1 L = 113.922 L or 113,922 mL **47.** 20 g + 0.052 g + 1500 g = 1520.052 g or 1,520,052 mg **49.** true **51.** false **53.** false **55.** true **57.** $71.92 **59.** $340,000 **61.** $10,102,500 **63.** 0.005632 **65.** 2290 kg **67.** approximately $9195.40; $9.20 **69.** 1997 **71.** 20% **72.** 57.5 **73.** $4536 **74.** $716.80

How Am I Doing? Sections 6.1–6.3

1. 4.33 ft (obj. 6.1.2) **2.** 6 (obj. 6.1.2) **3.** 5280 (obj. 6.1.2) **4.** 3.2 (obj. 6.1.2) **5.** 1320 (obj. 6.1.2) **6.** 40 (obj. 6.1.2) **7.** $8.25 (obj. 6.1.2) **8.** 6750 (obj. 6.2.2) **9.** 7390 (obj. 6.2.2) **10.** 98.6 (obj. 6.2.2) **11.** 0.027 (obj. 6.2.2) **12.** 529.6 (obj. 6.2.2)

13. 0.482 (obj. 6.2.2) **14.** 2376 m (obj. 6.2.2) **15.** 94.262 m (obj. 6.2.2) **16.** 1.34 meters or 134 centimeters (obj. 6.2.2) **17.** 5660 (obj. 6.3.1) **18.** 7.835 (obj. 6.3.2) **19.** 0.0563 (obj. 6.3.2) **20.** 4800 (obj. 6.3.1) **21.** 0.568 (obj. 6.3.2) **22.** 8900 (obj. 6.3.1) **23.** $116.25 (obj. 6.3.2) **24.** $227.50 (obj. 6.3.2) **25.** $7.20 (obj. 6.3.1) **26.** $5600 (obj. 6.3.2)

6.4 Exercises

1. The meter is approximately the same length as a yard. The meter is slightly longer. **3.** The inch is approximately twice the length of a centimeter. **5.** 2.14 m **7.** 22.86 cm **9.** 15.26 yd **11.** 28.15 m **13.** 132.02 km **15.** 17.44 yd **17.** 6.90 in **19.** 656 ft **21.** 3.1 mi **23.** 189.5 L **25.** 21.76 L **27.** 5.02 gal **29.** 4.77 qt **31.** 180.4 lb **33.** 59.02 kg **35.** 737.1 g **37.** 334.4 lb **39.** 4.45 oz **41.** 41.37 ft **43.** 34.1 mi/hr **45.** 273 mi/hr **47.** 0.51 in. **49.** 185°F **51.** 53.6°F **53.** 60°C **55.** 4.44°C **57.** yes **59.** 18.85 liters **61.** 1397 lb **63.** 66.2°F at 4 A.M. 113°F after 7 A.M. **65.** 59,861 miles **67.** 180.6448 sq cm **69.** $896 for the American carpet $802 for the German carpet The German carpet is cheaper. **71.** 169 **72.** 114 **73.** $\frac{13}{40}$ **74.** $\frac{7}{12}$

6.5 Exercises

1. 3 ft **3.** 11 yd **5.** $29.40 **7.** 8.1 meters **9.** 880 yd is 4.32 m longer than 880 m. **11.** $1.74/gal gasoline is more expensive in Mexico. **13.** 77°F; 9°F **15.** The difference is 6°F. The temperature reading of 180°C is hotter. **17.** (a) about 105 kilometers per hour (b) Probably not. We cannot be sure, but we have no evidence to indicate that they broke the speed limit. **19.** 15 gallons **21.** $108 **23.** 85.05 g **25.** (a) 4.34 quarts extra (b) $33.70 **27.** (a) $5.46 (b) about 132 miles per gallon **29.** yes; 240,000 gal/hr is equivalent to $533\frac{1}{3}$ pt/sec **31.** $n = 10$ **32.** (a) 26,538 meters (b) 87,045 feet **33.** 15.5 miles **34.** 41.25 yards

Putting Your Skills to Work

1. approximately 0.093025 square meter in 1 square foot **2.** approximately 4052.169 square meters in 1 acre **3.** 1.62 hectares **4.** approximately 7.41 acres of land **5.** 1.87 hectares **6.** 5.69 hectares **7.** There are 4.94 acres in 2 hectares. The error is 0.06 acre. **8.** It is 0.05 hectare in error. The new deed should actually be 0.81 hectare.

Chapter 6 Review Problems

1. 11 **2.** 9 **3.** 8800 **4.** 10,560 **5.** 7.5 **6.** 6.5 **7.** 3 **8.** 2 **9.** 14,000 **10.** 8000 **11.** 0.5 **12.** 0.75 **13.** 60 **14.** 84 **15.** 15.5 **16.** 13.5 **17.** 560 **18.** 290 **19.** 176.3 **20.** 259.8 **21.** 920 **22.** 740 **23.** 10 **24.** 8.2 **25.** 7.93 m **26.** 17.01 m **27.** 35.63 m **28.** 89.59 m **29.** 17,000 **30.** 23,000 **31.** 196,000 **32.** 721,000 **33.** 0.778 **34.** 0.459 **35.** 3.5 **36.** 12.75 **37.** 765 **38.** 423 **39.** 2430 **40.** 1930 **41.** 92.4 **42.** 2.75 **43.** 72.45 **44.** 141.68 **45.** 5.52 **46.** 7.09 **47.** 9.08 **48.** 13.62 **49.** 10.97 **50.** 12.80 **51.** 49.6 **52.** 43.4 **53.** 53.6° **54.** 89.6° **55.** 105° **56.** 85° **57.** 0° **58.** 100° **59.** 3.43 **60.** 25.54 **61.** 270 sq ft; 30 sq yd **62.** (a) 17 ft (b) 204 in **63.** (a) 200 m (b) 0.2 km **64.** $2.22 **65.** no; 4.2 feet short **66.** yes; 112.7 km/hr **67.** 25°F. Too hot **68.** 380 cm **69.** approximately 25.2 ft/sec **70.** 49.6 mph **71.** They are carrying 16.28 pounds. They are slightly over the weight limit. **72.** about $3.30 **73.** yes **74.** approximately 516.4 square feet **75.** approximately $2.23

How Am I Doing? Chapter 6 Test

1. 3200 (obj. 6.1.2) **2.** 228 (obj. 6.1.2) **3.** 84 (obj. 6.1.2) **4.** 7 (obj. 6.1.2) **5.** 30 (obj. 6.1.2) **6.** 0.75 (obj. 6.1.2) **7.** 0.5 (obj. 6.1.2) **8.** 16.5 (obj. 6.1.2) **9.** 9200 (obj. 6.2.2) **10.** 0.0988 (obj. 6.2.2) **11.** 4.6 (obj. 6.2.2) **12.** 1270 (obj. 6.2.2) **13.** 9.36 (obj. 6.2.2) **14.** 0.046 (obj. 6.3.1) **15.** 0.0289 (obj. 6.3.2) **16.** 0.983 (obj. 6.3.2) **17.** 920 (obj. 6.3.1) **18.** 9420 (obj. 6.3.2) **19.** 67.62 (obj. 6.4.1) **20.** 1.63 (obj. 6.4.1) **21.** 3.55 (obj. 6.4.1) **22.** 18.6 (obj. 6.4.1) **23.** 16.06 (obj. 6.4.1) **24.** 85.05 (obj. 6.4.1) **25.** 56.85 (obj. 6.4.1) **26.** 3.18 (obj. 6.4.1) **27.** (a) 20 meters (obj. 6.5.1) (b) 21.8 yards **28.** (a) 15°F (obj. 6.5.1) (b) yes (obj. 6.4.2) **29.** 82.5 gal/hr (obj. 6.5.1) **30.** (a) 300 km (obj. 6.5.1) (b) 14 miles **31.** $5\frac{1}{4}$ lb (obj. 6.5.1) **32.** 104°F (obj. 6.4.2) (obj. 6.5.1)

Cumulative Test for Chapters 1–6

1. 6028 **2.** 185,440 **3.** 270 R5 **4.** $\frac{19}{42}$ **5.** $1\frac{3}{8}$ **6.** 27 **7.** yes **8.** $n = 6$ **9.** 209.23 grams **10.** 250% **11.** 120 **12.** 20,000 **13.** 9.5 **14.** 5000 **15.** 3.5 **16.** 300 **17.** 3700 **18.** 0.0628 **19.** 790 **20.** 0.05 **21.** 672 **22.** 10° **23.** 106.12 **24.** 43.58 **25.** 76.2 **26.** 14.49 **27.** 11.88 meters **28.** 59°F; the difference is 44°F; the 15°C temperature is higher. **29.** 7 miles **30.** Technically, she needs 66.6 yards, but in real life, she should buy 67 yards.

Chapter 7

7.1 Exercises

1. An acute angle is an angle whose measure is between 0° and 90°. **3.** Complementary angles are two angles that have a sum of 90°. **5.** When two lines intersect, the two angles that are opposite each other are called vertical angles. **7.** A transversal is a line that intersects two or more other lines at different points. **9.** $\angle ABD$, $\angle CBE$ **11.** $\angle ABD$ and $\angle CBE$; $\angle DBC$ and $\angle ABE$ **13.** There are no complementary angles. **15.** 90° **17.** 25° **19.** 110° **21.** 155° **23.** 59° **25.** 53° **27.** 34° **29.** 35° **31.** 25° **33.** $\angle b = 102°$; $\angle c = \angle a = 78°$ **35.** $\angle b = 38°$; $\angle a = \angle c = 142°$ **37.** $\angle a = \angle c = 48°$; $\angle b = 132°$ **39.** $\angle e = \angle d = \angle a = 123°$; $\angle b = \angle c = \angle f = \angle g = 57°$ **41.** 6° **43.** 49° north of east **45.** 9.7 miles **46.** 21.1 miles **47.** $3800 **48.** 55.3 tons **49.** 25.2 mi **50.** 12.5%

7.2 Exercises

1. perpendicular; equal **3.** multiply **5.** 15 mi **7.** 23.6 ft **9.** 17.2 in. **11.** 1.92 mm **13.** 17.12 km **15.** 14.4 ft or 172.8 in. **17.** 0.272 mm **19.** 31.84 cm **21.** 35 cm **23.** 180 cm **25.** 6.25 ft^2 **27.** 12 mi^2 **29.** 117 yd^2 or 1053 ft^2 **31.** (a) 294 m^2 (b) 78 m **33.** \$132,000 **35.** \$59.40 **37.** (a) $1 \times 7,\ 2 \times 6,\ 3 \times 5,\ 4 \times 4$ (b) 7 sq ft, 12 sq ft, 15 sq ft, 16 sq ft (c) Square garden measuring 4 ft on a side. **39.** \$598.22 **41.** 5.5 m^2 less **43.** 223.3 **44.** 7.18 **45.** 21,842.8 **46.** approximately 1.5759 **47.** 352 sq ft; \$264

7.3 Exercises

1. adding **3.** perpendicular **5.** 40.2 m **7.** 49.6 in. **9.** 354.64 m^2 **11.** 602 square yards **13.** $P = 48$ m; $A = 72\ m^2$ **15.** $P = 9.6$ ft; $A = 3.6\ ft^2$ **17.** 82 m **19.** 55 ft + 135 ft + 80.5 ft + 75.5 ft = 346 ft **21.** 118.8 yd^2 **23.** 76,850 square meters **25.** (a) 718 m^2 (b) rectangle (c) trapezoid **27.** (a) 357 ft^2 (b) parallelogram (c) trapezoid **29.** \$80,960 **31.** $5.2415 \times b^2$ sq units **33.** 30 **34.** 12 **35.** 1800 **36.** 5

7.4 Exercises

1. right **3.** Add the measures of the two known angles and subtract that value from 180°. **5.** You could conclude that the lengths of all three sides of the triangle are equal. **7.** true **9.** true **11.** false **13.** true **15.** 70° **17.** 82.9° **19.** 118 m **21.** 116.75 in. **23.** 10 mi **25.** 56.25 $in.^2$ **27.** 83.125 cm^2 **29.** $7\frac{7}{12}$ yd^2 **31.** 91.5 ft^2 **33.** 188 yd^2 **35.** 3375 ft^2 **37.** \$21,060 **39.** 6.25% **41.** $n = 12$ **42.** $n = 42$ **43.** 716 **44.** 96 magazines **45.** 27 students

7.5 Exercises

1. $\sqrt{25} = 5$ because $(5)(5) = 25$ **3.** whole **5.** Use the square root table or a calculator. **7.** 3 **9.** 8 **11.** 12 **13.** 0 **15.** 13 **17.** 10 **19.** 10 **21.** 11 **23.** 3 **25.** 8 **27.** 22 **29.** (a) yes (b) 16 **31.** $\approx$4.243 **33.** $\approx$8.718 **35.** $\approx$14.142 **37.** $\approx$5.831 m **39.** $\approx$11.662 m **41.** 10.472 **43.** 7.071 **45.** 104.7 ft **47.** 127.3 ft **49.** (a) 2 (b) 0.2 (c) Each answer is obtained from the previous answer by dividing by 10. (d) no; because 0.004 isn't a perfect square. **51.** 39.299 **53.** 4800 sq in **54.** 80,500 meters **55.** 18.6 mi **56.** 0.0989 kilograms

How Am I Doing? Sections 7.1–7.5

1. 18° (obj. 7.1.1) **2.** 117° (obj. 7.1.1) **3.** $\angle b = 136°$; $\angle a = \angle c = 44°$ (obj.7.1.1) **4.** 18 m (obj.7.2.1) **5.** 14 m (obj.7.2.1) **6.** 23.04 sq cm (obj. 7.2.3) **7.** 2.43 sq cm (obj. 7.2.3) **8.** 25.6 yd (obj. 7.3.1) **9.** 79 ft (obj. 7.3.2) **10.** 351 sq in. (obj. 7.3.1) **11.** 171 sq in. (obj. 7.3.2) **12.** 97 sq m (obj. 7.3.2) **13.** 23° (obj.7.4.1) **14.** 15 m (obj. 7.4.2) **15.** 72 sq m (obj. 7.4.2) **16.** (a) 592 ft^2 (b) 114 ft (obj. 7.4.2) **17.** 8 (obj. 7.5.1) **18.** 12 (obj. 7.5.1) **19.** 13 (obj. 7.5.1) **20.** 16 (obj. 7.5.1) **21.** 6.782 (obj. 7.5.2)

7.6 Exercises

1. Square the length of each leg and add those two results. Then take the square root of the remaining number. **3.** 15 yd **5.** 15.199 ft **7.** $\approx$11.402 m **9.** $\approx$14.142 m **11.** $\approx$8.660 ft **13.** $\approx$9.798 yd **15.** 15 m **17.** $\approx$ 11.619 ft **19.** 13 ft **21.** 9.8 cm **23.** $\approx$11.1 yd **25.** 6.9 in.; 4 in. **27.** 8.5 m **29.** 25.5 cm **31.** $\approx$7.1 in. **33.** $\approx$0.47 mi **35.** $\approx$13.038 yd **37.** 341 m^2 **38.** 168 m^2 **39.** 441 $in.^2$ **40.** 4224 yd^2

7.7 Exercises

1. circumference **3.** radius **5.** Multiply the radius by 2 and then use $c = \pi d$. **7.** 58 in. **9.** 17 mm **11.** 22.5 yd **13.** 9.92 cm **15.** 100.48 cm **17.** 116.18 in. **19.** 41.87 ft **21.** 78.5 yd^2 **23.** $\approx$226.87 $in.^2$ **25.** 803.84 cm^2 **27.** 452.16 ft^2 **29.** 6358.5 mi^2 **31.** 163.28 m^2 **33.** 30.96 m^2 **35.** 189.25 m^2 **37.** \$1211.20 **39.** 9.42 ft **41.** 141.3 feet **43.** 630.57 revolutions **45.** (a) 25.12 ft (b) 50.24 ft^2 **47.** 125,600 square miles **49.** (a) $\approx$22.1 $in.^2$ (b) 18.84 $in.^2$ (c) For 12 in.; \$0.035 per $in.^2$; for 15 in.; \$0.034 per $in.^2$ **51.** 13.92 **52.** 0.3 **53.** 6000 **54.** 3000 **55.** 48 students **56.** \$123.64

7.8 Exercises

1. (a) sphere (b) $V = \dfrac{4\pi r^3}{3}$ **3.** (a) cylinder (b) $V = \pi r^2 h$ **5.** (a) cone (b) $V = \dfrac{\pi r^2 h}{3}$ **7.** 540 mm^3 **9.** 226.1 m^3 **11.** 6459.0 m^3 **13.** 3052.1 yd^3 **15.** 700 ft^3 **17.** 0.216 cm^3 **19.** 65.94 yd^3 **21.** 718.0 m^3 **23.** 937.8 cm^3 **25.** 261.7 ft^3 **27.** 163.3 m^3 **29.** 373.3 m^3 **31.** 40 yd^3 **33.** 1004.8 $in.^3$ **35.** 381,251,976,256.667 mi^3 **37.** 2928 $in.^3$ **39.** \$1130.40 **41.** 413.8 cm^3 **43.** 263,900 yd^3 **45.** $V = \frac{1}{2}hbH$ **47.** $9\frac{7}{12}$ **48.** $6\frac{3}{8}$ **49.** $\frac{135}{16}$ or $8\frac{7}{16}$ **50.** $\frac{25}{14}$ or $1\frac{11}{14}$ **51.** $\frac{23}{64}$ **52.** $6\frac{7}{18}$ or $\frac{115}{18}$

7.9 Exercises

1. size; shape **3.** sides **5.** 8 m **7.** 2.6 ft **9.** 3.4 yd **11.** *a* corresponds to *f*, *b* corresponds to *e*, *c* corresponds to *d* **13.** 3.4 m **15.** 2.1 ft **17.** 2.2 ft **19.** 36 ft **21.** 81 ft **23.** 8.3 ft **25.** 12 cm **27.** 11.6 yd^2 **29.** 12 **30.** 32 **31.** 42 **32.** 62

7.10 Exercises

1. (a) 75 km/hr (b) 76 km/hr (c) through Woodville and Palemo **3.** 34.5 min **5.** 4006 ft^2 **7.** \$510 **9.** \$795.15 **11.** (a) 40.820 km (b) 20.410 km/hr **13.** $\approx$50.240 $in.^3$ **15.** \$3293.45 **17.** 1465.33 feet/min **19.** 128 **20.** 308 **21.** 0.25 **22.** 4.87

Putting Your Skills to Work

1. 47,131 people/sq mi **2.** 8 people/sq mi **3.** answers will vary **4.** 1 person/sq mi—Alaska; 225 people/sq mi—California; 413 people/sq mi—Delaware; 15 people/sq mi—New Mexico; 1024 people/sq mi—Rhode Island **5.** 3,097,600 sq yd; 27,878,400 sq ft **6.** about 66 sq yd; about 592 sq ft **7.** about 3026 sq yd; about 27,234 sq ft

Chapter 7 Review Problems

1. 14° **2.** 104° **3.** $\angle b = 146°$, $\angle a = \angle c = 34°$ **4.** $\angle t = \angle x = \angle y = 65°$, $\angle s = \angle u = \angle w = \angle z = 115°$ **5.** 19.8 m **6.** 9.6 yd **7.** 16.5 cm^2 **8.** 51.8 $in.^2$ **9.** 38 ft **10.** 58 ft **11.** 68 m^2 **12.** 63.5 m^2 **13.** 145.2 m **14.** 62 mi **15.** 2700 m^2 **16.** 720 yd^2 **17.** 422 cm^2 **18.** 357 m^2 **19.** 60 ft **20.** 46.5 ft **21.** 153° **22.** 55° **23.** 52.3 m^2 **24.** 59.4 m^2 **25.** 450 m^2 **26.** 87 m^2 **27.** 9 **28.** 8 **29.** 11 **30.** 3 **31.** 15 **32.** 5.916 **33.** 6.928 **34.** 12.845 **35.** 13.416 **36.** 5 km **37.** 5 yd **38.** 8.72 cm **39.** 9.22 m **40.** 6.4 cm **41.** 9.2 ft **42.** 6.3 ft **43.** 3.6 ft **44.** 106 cm **45.** 63 cm **46.** 44.0 in. **47.** 56.5 in. **48.** 254.3 m^2 **49.** 58.06 ft^2 **50.** 226.1 $in.^2$ **51.** 201.0 m^2 **52.** 318.5 ft^2 **53.** 126.1 m^2 **54.** 107.4 ft^2 **55.** 80.1 m^2 **56.** 432 ft^3 **57.** 448.7 $in.^3$ **58.** 21.2 ft^3 **59.** 35.3 $in.^3$ **60.** 245 m^3 **61.** 3768 ft^3 **62.** 9074.6 yd^3 **63.** 30 m **64.** 3.3 m **65.** 348 cm **66.** 175 ft **67.** 147 sq yd **68.** 32,555.2 g **69.** \$736 **70.** (a) 50 km; 100 km/hr (b) 56 km; 70 km/hr (c) through Ipswich **71.** (a) $\approx$21,873.2 ft^3 (b) $\approx$17,498.6 bushels **72.** 3,429,708,000 ft^3 **73.** 17.1 ft **74.** 116 lb; 130 gal **75.** 1728 $in.^3$ **76.** 942 ft **77.** $\approx$521.0 yd^2 **78.** \$26,050 **79.** 13.7 ft **80.** 381,510 m^3 **81.** $\approx$8.8 ft^3 **82.** $\approx$66 gallons **83.** 1662.5 ft^2 **84.** \$2493.75

How Am I Doing? Chapter 7 Test

1. $\angle b = 52°$; $\angle c = 128°$; $\angle e = 128°$; (obj. 7.1.1) **2.** 40 yd (obj. 7.2.1) **3.** 25.2 ft (obj. 7.2.1) **4.** 20 m (obj. 7.3.1) **5.** 80 m (obj. 7.3.2) **6.** 137 m (obj. 7.4.2) **7.** 180 yd^2 (obj. 7.2.1) **8.** 104.0 m^2 (obj. 7.2.1) **9.** 78 m^2 (obj. 7.3.1) **10.** 144 m^2 (obj. 7.3.2) **11.** 12 cm^2 (obj. 7.4.2) **12.** 12 (obj. 7.5.1) **13.** 13 (obj. 7.5.1) **14.** 27° (obj. 7.1.1) **15.** 73° (obj. 7.1.1) **16.** 84° (obj. 7.4.1) **17.** $\approx$7.348 (obj. 7.5.2) **18.** $\approx$11.619 (obj. 7.5.2) **19.** 8.602 (obj. 7.6.2) **20.** 10 (obj. 7.6.2) **21.** 5.83 cm (obj. 7.6.3) **22.** 9 ft (obj. 7.6.3) **23.** 56.52 ft (obj. 7.7.1) **24.** 113.04 ft^2 (obj. 7.7.1) **25.** 107.4 $in.^2$ (obj. 7.7.2) **26.** 144.3 $in.^2$ (obj. 7.7.2) **27.** 700 m^3 (obj. 7.8.1) **28.** 803.8 m^3 (obj. 7.8.4) **29.** 113.0 m^3 (obj. 7.8.3) **30.** 508.7 ft^3 (obj. 7.8.2) **31.** 56 m^3 (obj. 7.8.5) **32.** 46.8 m (obj. 7.9.1) **33.** 42 ft (obj. 7.9.1) **34.** 6456 yd^2 (obj. 7.10.1) **35.** \$2582.40 (obj. 7.10.1)

Cumulative Test for Chapters 1–7

1. 935,760 **2.** 33,415 **3.** $\frac{1}{36}$ **4.** $\frac{4}{21}$ **5.** 56.13 **6.** 7.2272 **7.** 263.4 **8.** 83 **9.** 27 **10.** 800 students **11.** 75% **12.** 2000 **13.** 18 **14.** 5.86 m **15.** 1512 in. **16.** 54.56 mi **17.** 50 m **18.** 208 cm **19.** 56.5 yd **20.** 1.4 cm^2 **21.** 540 m^2 **22.** 192 m^2 **23.** 664 yd^2 **24.** 50.2 m^2 **25.** 87.9 m^3 **26.** 33.5 cm^3 **27.** 3136 cm^3 **28.** 2713.0 cm^3 **29.** 39.6 m **30.** 4.1 ft. **31.** (a) 124 yd^2 (b) \$992.00 **32.** 21 **33.** $\approx$7.550 **34.** 10.440 in **35.** 4.899 m **36.** 13.9 mi **37.** 32 paint brushes

Chapter 8

8.1 Exercises

1. Multiply 25% × 4000, which is 0.25 × 4000 = 1000 students **3.** You would divide the circle into quarters by drawing two perpendicular lines. Shade in one quarter of the circle. Label this with the title "within five miles 1000." **5.** rent **7.** \$200 **9.** \$800 **11.** $\frac{13}{4}$ **13.** $\frac{10}{27}$ **15.** 40 years or younger but older than 20 **17.** 78 million **19.** 125 million **21.** $\frac{41}{39}$ **23.** $\frac{125}{163}$ **25.** 19% **27.** reasonable prices and great food **29.** 343 people **31.** 2,205,000,000 **33.** 40% **35.** 79% **37.** 1.1% **39.** 120 ft^2 **40.** 204 $in.^2$ **41.** 16 gal **42.** about 3 g

8.2 Exercises

1. 21 million people **3.** 14 million people **5.** 1960–1970 **7.** 22 quadrillion Btu **9.** 1970 **11.** 4 quadrillion Btu **13.** 6 quadrillion Btu **15.** from 1975 to 1980 and from 1985 to 1990 **17.** 30 quadrillion Btu **19.** \$1,000,000 **21.** 1993 to 1995 **23.** \$300,000 **25.** 2.5 in. **27.** October, November, and December **29.** 1.5 in.

31.

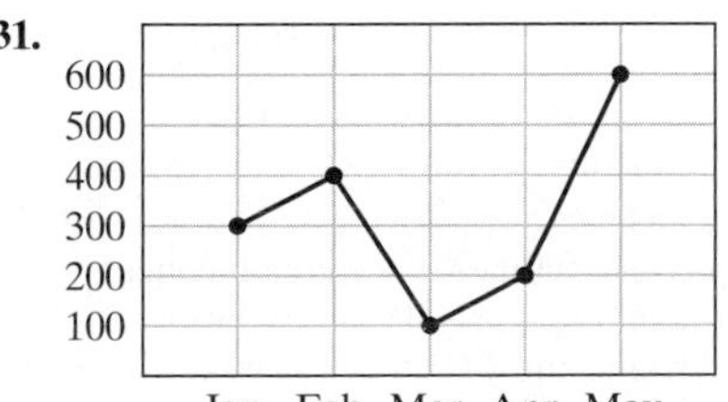

33. 4 **34.** 115 **35.** $\frac{2}{9}$ **36.** 13

How Am I Doing? Sections 8.1–8.2

1. 14% (obj. 8.1.2) **2.** 2 people (obj. 8.1.2) **3.** 39% (obj. 8.1.2) **4.** 3,278,910 households (obj. 8.1.2) **5.** 66,671,170 households (obj. 8.1.2) **6.** 300 housing starts (obj. 8.2.2) **7.** 450 housing starts (obj. 8.2.2) **8.** during the second quarter of 2002 (obj. 8.2.2) **9.** during the third quarter of 2003 (obj. 8.2.2) **10.** 300 (obj. 8.2.2) **11.** 50 (obj. 8.2.2) **12.** Aug. and Dec. (obj. 8.2.4) **13.** Dec. (obj. 8.2.4) **14.** Nov. (obj. 8.2.4) **15.** (a) 20,000 sets (obj. 8.2.4) (b) 35,000 sets (obj. 8.2.4)

8.3 Exercises

1. The horizontal label for each item in a bar graph is usually a single number or a word title. For the histogram it is a class interval. The vertical bars have a space between them in the bar graph. For the histogram, the vertical bars join each other. **3.** A class frequency is the number of times a score occurs in a particular class interval. **5.** 150 cities **7.** 10 cities **9.** 30 cities **11.** 270 cities **13.** 8000 **15.** books costing \$5.00–\$7.99 **17.** 28,000 books **19.** 52,000 books **21.** 28.6%

	Tally	Frequency
23.	|||	3
25.	~~||||~~ |	6
27.	|||	3
29.	||	2

31.

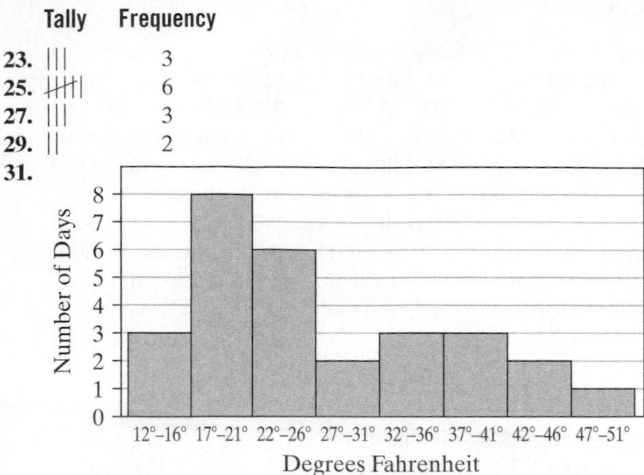

33. 17 days

	Tally	Frequency
35.	|||	3
37.	||	2
39.	|	1
41.	||	2

43. 12 prescriptions **45.** $n = 59.5$ **46.** $n = 7$ **47.** 15 **48.** 13.0 in.

8.4 Exercises

1. The median of a set of numbers when they are arranged in order from smallest to largest is that value that has the same number of values above it as below it. The mean of a set of values is the sum of the values divided by the number of values. The mean is most likely to be not typical of the values you would expect if there are many extremely low values or many extremely high values. The median is more likely to be typical of the value you would expect. **3.** 32 **5.** 2.6 hours **7.** 0.375 **9.** 109,200 **11.** 23.7 mi/gal **13.** 195 **15.** 12 **17.** \$20,250 **19.** 21 **21.** \$207 **23.** 2.2 **25.** 95,500 **27.** \$69,161.88 **29.** 4850 **31.** mean = \$104.74; median = \$80.95 **33.** 60 **35.** 121 and 150 **37.** \$269 **39.** mean = 751, median = 855, mode = 869 **41.** (a) \$2157 (b) \$1615 (c) There is no mode (d) The median because the mean is affected by the high amount, \$6300. **43.** (a) 8.4 (b) 7 (c) 3 (d) The median. On three nights she gets more calls than 7. On 3 nights she gets fewer calls than 7. On one night she got 7 calls. The mean is distorted a little because of the very large number of calls on Sunday night. The mode is artificially low because she gets so few calls on Monday and Wednesday and it just happened to be the same number, 3. **45.** GPA = 3.0 **47.** 19.3 square inches **48.** 20,096 gallons per hour **49.** \$330 **50.** about \$0.04 per cubic inch

Putting Your Skills to Work

1. Derek Jeter, Edgar Renteria **2.** Alex Rodriguez, Edgar Renteria **3.** Alex Rodriguez, Alex Gonzalez

Chapter 8 Review Problems

1. 13 computers **2.** 32 computers **3.** 68 computers **4.** 27 computers **5.** $\frac{13}{21}$ **6.** $\frac{43}{32}$ **7.** ≈17.9% **8.** ≈22.9% **9.** 48% **10.** 77% **11.** art **12.** business **13.** art and education **14.** 1280 students **15.** 3360 students **16.** 880 students **17.** 36 **18.** 26 **19.** 10–13 years **20.** 6–9 years **21.** 12 glasses **22.** 22 glasses **23.** $\frac{5}{1}$ **24.** $\frac{2}{1}$ **25.** \$31,000 **26.** \$58,000 **27.** between 1985 and 1990 **28.** between 1980 and 1985 **29.** \$9000 **30.** \$15,000 **31.** 2000 **32.** 1980 **33.** \$7600 **34.** \$5,400 **35.** \$58,000 **36.** \$72,000 **37.** 400 students **38.** 500 students **39.** 650 students **40.** 450 students **41.** 100 students **42.** 50 students **43.** 1999–2000 **44.** 2000–2001 **45.** 45,000 cones **46.** 30,000 cones **47.** 10,000 cones **48.** 30,000 cones **49.** 35,000 cones **50.** 30,000 cones **51.** The sharp drop in the number of ice cream cones purchased from July 2002 to August 2002 is probably directly related to the weather. Since August was cold and rainy significantly fewer people wanted ice cream during August. **52.** The sharp increase in the number of ice cream cones purchased from June 2003 to July 2003 is probably directly related to the weather. Since June was cold and rainy and July was warm and sunny, significantly more people wanted ice cream during July. **53.** 14,000,000 **54.** 18,000,000 **55.** between 1996 and 1998 **56.** between 1990 and 1992 **57.** 1,000,000 **58.** 12,000,000 **59.** 1992 **60.** 2002 **61.** ≈3,666,667 **62.** ≈1,333,333 **63.** 24,000,000 **64.** 49,000,000 **65.** 65 pairs **66.** 10 pairs **67.** 145 pairs **68.** 90% **69.** 45 pairs **70.** 5 to 36; or $\frac{5}{36}$

	Number of Defective Televisions (Class Intervals)	Tally	Frequency
71.	0–3	~~||||~~ ~~||||~~	10
72.	4–7	~~||||~~ |||	8
73.	8–11	|||	3
74.	12–15	~~||||~~	5
75.	16–19	||	2

76.

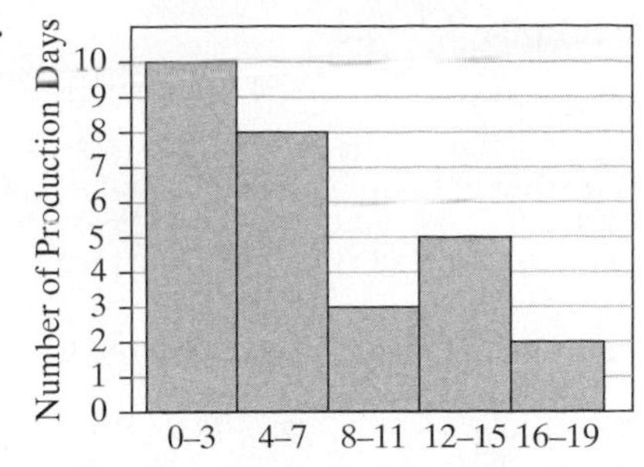

77. 18 times **78.** 90° **79.** \$89.67 **80.** 29,000 people **81.** 1353 employees **82.** \$36,000 **83.** \$141,500 **84.** median = 30.5 years; mode = 28 years **85.** median = 18.5 deliveries, mode = 15 deliveries **86.** The median is better because the mean is skewed by the one low score, 31. **87.** The median is better because the mean is skewed by the one high data item, 39. **88.** (a) 5 (b) 4 (c) 2 (d) The mean is the most representative. On three days she uses the computer more than 4 hours and on three days she uses the computer less than 4 hours. One day she used it exactly 7 hours. The mean is distorted a little because of the very large number of hours on Friday. The mode is artificially low because she happened to use the computer only two hours on Sunday and Tuesday. All other days it was more than this.

How Am I Doing? Chapter 8 Test

1. 37% (obj. 8.1.2) **2.** 21% (obj. 8.1.2) **3.** 12% (obj. 8.1.2) **4.** 90,000 automobiles (obj. 8.1.2) **5.** 81,000 automobiles (obj. 8.1.2) **6.** \$12,000 (obj. 8.2.2) **7.** \$3000 (obj. 8.2.2) **8.** \$8000 (obj. 8.2.2) **9.** \$2000 (obj. 8.2.2) **10.** \$11,000 (obj. 8.2.2) **11.** \$15,000 (obj. 8.2.2) **12.** 20 yr (obj. 8.2.4) **13.** 26 yr (obj. 8.2.4) **14.** 12 yr (obj. 8.2.4) **15.** age 35 (obj. 8.2.4) **16.** age 65 (obj. 8.2.4) **17.** 60,000 televisions (obj. 8.3.1) **18.** 25,000 televisions (obj. 8.3.1) **19.** 20,000 televisions (obj. 8.3.1) **20.** 60,000 televisions (obj. 8.3.1) **21.** 14.5 (obj. 8.4.1) **22.** 14.5 (obj. 8.4.2) **23.** 10 (obj. 8.4.3) **24.** mean or median (obj. 8.4.3)

Cumulative Test for Chapters 1–8

1. 20,825 **2.** 74,296 **3.** $\frac{153}{40}$ or $3\frac{33}{40}$ **4.** $\frac{86}{3}$ or $28\frac{2}{3}$ **5.** 1796.43 **6.** 72.65 **7.** 72.23 **8.** $n = 0.6$ **9.** 39 cars **10.** 0.325 **11.** 60 **12.** 1.98 m **13.** 54 ft **14.** 28.3 in.^{2} **15.** perimeter = 60 ft; area = 225 ft^2 **16.** 34% **17.** 56 people **18.** \$3 million **19.** \$1 million **20.** 16 in. **21.** 1960 and 1970 **22.** 8 students **23.** 16 students **24.** \$6.00 **25.** \$4.95 **26.** \$4.50

Chapter 9

9.1 Exercises

1. First, find the absolute value of each number. Then add those two absolute values. Use the common sign in the answer. **3.** < **5.** > **7.** > **9.** < **11.** 7 **13.** 16 **15.** −17 **17.** −7 **19.** 16.5 **21.** $\frac{17}{35}$ **23.** −3 **25.** 9 **27.** −5 **29.** 22 **31.** −3.25 **33.** $-\frac{2}{3}$ **35.** $\frac{5}{9}$ **37.** −22 **39.** −0.72 **41.** −363 **43.** 4 **45.** $-4\frac{1}{12}$ **47.** 6.24 **49.** $-\frac{11}{2}$ or $-5\frac{1}{2}$ **51.** −14.7 **53.** 0 **55.** −6 **57.** $-\frac{53}{75}$ **59.** −\$94,000 **61.** \$9000 **63.** −\$38,000 **65.** −18°F **67.** −1°F **69.** −0.34 **71.** −1 yard **73.** −\$13 **75.** \$37.00 **77.** 904.3 ft^3 **78.** 210 m^3 **79.** 60 **80.** \$40

9.2 Exercises

1. −6 **3.** −5 **5.** −6 **7.** −17 **9.** −2 **11.** 85 **13.** −18 **15.** 3 **17.** −60 **19.** 556 **21.** −6.7 **23.** −6.5 **25.** −8.6 **27.** 32.94 **29.** 1 **31.** $-\frac{7}{6}$ or $-1\frac{1}{6}$ **33.** $-\frac{76}{15}$ or $-5\frac{1}{15}$ **35.** $\frac{17}{63}$ **37.** 15 **39.** 0 **41.** 46 **43.** −22 **45.** −2 **47.** −5.4 **49.** 14,776 ft **51.** 42°F **53.** −13°F **55.** −\$16,000 **57.** +\$37,300 **59.** $\$13\frac{1}{2}$ **61.** The bank finds that a customer has \$50 in his checking account. However, the bank must remove an erroneous debit of \$80 from the customer's account. When the bank makes the correction, what will the new balance be? **63.** 5 **64.** 6 **65.** 28.26 square inches **66.** 10.5 square inches

9.3 Exercises

1. To multiply two numbers with the same sign, multiply the absolute values. The sign of the result is positive. **3.** 36 **5.** 60 **7.** −160 **9.** −66 **11.** −1.5 **13.** 28.125 **15.** $-\frac{6}{35}$ **17.** $\frac{1}{23}$ **19.** −8 **21.** −8 **23.** 6 **25.** 5 **27.** $\frac{3}{4}$ **29.** $\frac{8}{7}$ or $1\frac{1}{7}$ **31.** −8.38 **33.** −5.7 **35.** 30 **37.** −45 **39.** 48 **41.** −5 **43.** 10 **45.** −2.8 **47.** −0.02 **49.** $\frac{3}{7}$ **51.** $\frac{3}{2}$ **53.** 72 **55.** 192 **57.** 16 **59.** −18,000 **61.** 0 **63.** $-\frac{5}{12}$ **65.** She lost \$79. **67.** −6.25°F **69.** 70 feet **71.** 34 **73.** 2 **75.** +8 **77.** 0; at par **79.** 0 **81.** b is negative **83.** 90 square inches **84.** 264 square meters

How Am I Doing? Sections 9.1–9.3

1. −19 (obj. 9.1.1) **2.** −4 (obj. 9.1.2) **3.** 4.5 (obj. 9.1.2) **4.** 0 (obj. 9.1.3) **5.** $-\frac{1}{3}$ (obj. 9.1.2) **6.** $-\frac{7}{6}$ or $-1\frac{1}{6}$ (obj. 9.1.1)

7. −7 (obj. 9.1.1) **8.** 1.7 (obj. 9.1.2) **9.** −8 (obj. 9.2.1) **10.** −41 (obj. 9.2.1) **11.** $\frac{14}{17}$ (obj. 9.2.1) **12.** −12 (obj. 9.2.1)

13. 1.4 (obj. 9.2.1) **14.** −2.8 (obj. 9.2.1) **15.** 42 (obj. 9.2.1) **16.** $\frac{19}{15}$ or $1\frac{14}{15}$ (obj. 9.2.1) **17.** 24 (obj. 9.3.1) **18.** 4 (obj. 9.3.1)

19. −8 (obj. 9.3.1) **20.** −10 (obj. 9.3.2) **21.** −24 (obj. 9.3.1) **22.** $\frac{15}{16}$ (obj. 9.3.1) **23.** −64 (obj. 9.3.2) **24.** −10 (obj. 9.3.1)

25. 24 (obj. 9.2.1) **26.** 21 (obj. 9.3.1) **27.** −7 (obj. 9.3.1) **28.** −3.6 (obj. 9.1.3) **29.** −0.6 (obj. 9.2.1) **30.** $-\frac{11}{15}$ (obj. 9.1.1)
31. $-\frac{5}{14}$ (obj. 9.3.1) **32.** $-\frac{9}{10}$ (obj. 9.3.1) **33.** −2.4°F (obj. 9.3.2)

9.4 Exercises

1. 6 **3.** −8 **5.** 8 **7.** −12 **9.** −19 **11.** −81 **13.** −5 **15.** −6 **17.** −27 **19.** −102 **21.** −18 **23.** 1 **25.** 4
27. $-\frac{1}{5}$ **29.** 1 **31.** 0 **33.** 4 **35.** 123 **37.** −7 **39.** $-\frac{13}{50}$ **41.** 6.84 **43.** −15.7°F **45.** −8.3°F **47.** 9°F
49. −9°C; 3°C **51.** 3.84 kilometers **52.** 36,800 milligrams **53.** 113.04 m^2 **54.** 98 catalogs; 120 pieces of junk mail

9.5 Exercises

1. Our number system is structured according to base 10. By making scientific notation also in base 10, the calculations are easier to perform.
3. The first part is a number greater than or equal to 1 but smaller than 10. It has at least one nonzero digit. The second part is 10 raised to some integer power. **5.** 1.2×10^2 **7.** 1.9×10^3 **9.** 2.63×10^4 **11.** 2.88×10^5 **13.** 1×10^4 **15.** 1.2×10^7 **17.** 9.31×10^{-2}
19. 2.79×10^{-3} **21.** 8.2×10^{-1} **23.** 1.6×10^{-5} **25.** 5.31×10^{-6} **27.** 8×10^{-6} **29.** 53,600 **31.** 5334
33. 4,600,000,000,000 **35.** 0.062 **37.** 0.371 **39.** 900,000,000,000 **41.** 0.00000003862 **43.** 3.5689×10^4 **45.** 0.052
47. 3.98×10^{-4} **49.** 1,880,000 **51.** 5.878×10^{12} miles **53.** 9.2×10^{-14} **55.** 12,500,000,000,000 insects **57.** 0.000075 centimeter
59. 14,000,000,000 tons **61.** 9.01×10^7 dollars **63.** 1.068×10^{22} tons **65.** 3.624×10^8 feet **67.** 6.0×10^6 square miles
69. 6.174×10^{13} miles **71.** 2.625 **72.** 0.258 **73.** $176 **74.** 589 feet

Putting Your Skills to Work

1. −0.017, −1.35, +1.31, +0.81, +0.98 **2.** +8.64; −1.52 **3.** +$1.45; $145 **4.** 29 shares; $566.95
5. 24 shares of Krispy Kreme; 295 shares of Six Flags, Inc.; he lost $357.45.

Chapter 9 Review Problems

1. −15 **2.** −14 **3.** −8.8 **4.** 2.6 **5.** $-\frac{8}{15}$ **6.** $\frac{1}{14}$ **7.** 6 **8.** −20 **9.** −15 **10.** 0 **11.** −11 **12.** −28 **13.** 19

14. 17 **15.** −17.2 **16.** −12.3 **17.** $-\frac{1}{15}$ **18.** $\frac{107}{12}$ or $8\frac{11}{12}$ **19.** 13 **20.** −3 **21.** −9 **22.** −12 **23.** $\frac{2}{35}$ **24.** $-\frac{1}{6}$

25. −7.8 **26.** 4.32 **27.** 3 **28.** 6 **29.** −9 **30.** −5 **31.** 6 **32.** −15 **33.** $-\frac{7}{10}$ **34.** $\frac{3}{7}$ **35.** 30 **36.** −36 **37.** 13

38. 19 **39.** −11 **40.** 1 **41.** −5 **42.** 23 **43.** −8 **44.** −0.5 **45.** −1.2 **46.** 1 **47.** $-\frac{9}{5}$ or $-1\frac{4}{5}$ **48.** $\frac{1}{10}$ **49.** −3

50. −35 **51.** 5 **52.** 21 **53.** $\frac{47}{30}$ or $1\frac{17}{30}$ **54.** $-\frac{7}{45}$ **55.** 5.76 **56.** 63.72 **57.** 4.16×10^3 **58.** 3.7×10^6 **59.** 2×10^5

60. 7×10^{-3} **61.** 2.18×10^{-5} **62.** 7.63×10^{-6} **63.** 18,900 **64.** 3760 **65.** 0.0752 **66.** 0.00661 **67.** 0.0000009
68. 0.00000008 **69.** 0.000536 **70.** 0.0000198 **71.** 8.44×10^{11} **72.** 9.72×10^{15} **73.** 1.44×10^{14} **74.** 6.8×10^{25}
75. 1.2312×10^{14} drops **76.** 5.983×10^{24} kilograms **77.** 1.32×10^{13} feet **78.** 4.9104×10^{11} ft
79. 1.67×10^{-24} grams; 9.1×10^{-28} grams **80.** 0.000000000000001 **81.** 384,400,000 meters **82.** total loss 6 yards **83.** 2676 ft
84. $2 **85.** −9.6°F **86.** 2 points above par

How Am I Doing? Chapter 9 Test

1. −11 (obj. 9.1.2) **2.** −43 (obj. 9.1.1) **3.** 3.9 (obj. 9.1.2) **4.** −6 (obj. 9.1.3) **5.** $-3\frac{1}{2}$ (obj. 9.1.2) **6.** $-\frac{7}{8}$ (obj. 9.1.1)

7. −38 (obj. 9.2.1) **8.** 5 (obj. 9.2.1) **9.** $\frac{17}{15}$ or $1\frac{2}{15}$ (obj. 9.2.1) **10.** −43 (obj. 9.2.1) **11.** 4 (obj. 9.2.1) **12.** −11.3 (obj. 9.2.1)

13. $\frac{11}{12}$ (obj. 9.2.1) **14.** 0 (obj. 9.2.1) **15.** 120 (obj. 9.3.1) **16.** −36 (obj. 9.3.1) **17.** 10 (obj. 9.3.1) **18.** −18 (obj. 9.3.2)

19. 3 (obj. 9.3.1) **20.** $-\frac{7}{10}$ (obj. 9.3.1) **21.** 18 (obj. 9.3.2) **22.** −32 (obj. 9.3.1) **23.** 17 (obj. 9.4.1) **24.** 0.5 (obj. 9.4.1)

25. −8 (obj. 9.4.1) **26.** −46 (obj. 9.4.1) **27.** −7.2 (obj. 9.4.1) **28.** −20 (obj. 9.4.1) **29.** $-\frac{1}{7}$ (obj. 9.4.1) **30.** $-\frac{1}{6}$ (obj. 9.4.1)
31. 8.054×10^4 (obj. 9.5.1) **32.** 7×10^{-6} (obj. 9.5.1) **33.** 0.0000936(obj. 9.5.2) **34.** 72,000 (obj. 9.5.2) **35.** −6.2°F (obj. 9.2.3)
36. 2.72×10^{-4} meters (obj. 9.5.3) **37.** 186.9°F (obj. 9.2.2)

Cumulative Test for Chapters 1–9

1. 12,383 **2.** 127 **3.** $\frac{143}{12}$ or $11\frac{11}{12}$ **4.** $\frac{55}{12}$ or $4\frac{7}{12}$ **5.** 9.812 **6.** 63.46 **7.** 65.9968 **8.** $n = 64$ **9.** 126 defects **10.** 0.304 **11.** 120 **12.** 94,000 meters **13.** 5 yards **14.** 45 ft **15.** 78.5 square meters **16.** (a) 300 students (b) 1100 students (c) 700 students **17.** 13 **18.** -4.7 **19.** $\frac{5}{12}$ **20.** -11 **21.** -5 **22.** -60 **23.** $\frac{4}{3}$ or $1\frac{1}{3}$ **24.** 18 **25.** 4 **26.** -2 **27.** $\frac{4}{15}$ **28.** 8.6972×10^4 **29.** 5.49×10^{-5} **30.** 38,500,000 **31.** 0.00007

Chapter 10

10.1 Exercises

1. A variable is a symbol, usually a letter of the alphabet, that stands for a number.
3. All the exponents for like terms must be the same. The exponent for x must be the same. The exponent for y must be the same. In this case, x is raised to the second power in the first term but y is raised to the second power in the second term. **5.** G, x, y **7.** p, a, b **9.** $r = 3m + 5n$ **11.** $H = 2a - 3b$ **13.** $10x$ **15.** $-x$ **17.** $\frac{1}{3}x$ **19.** $4x + 1$ **21.** $-1.1x + 6.4$ **23.** $37x + 9y - 11$ **25.** $-\frac{3}{4}x - 10$ **27.** $-3a + 6b - 4c$ **29.** $-\frac{1}{4}x + \frac{8}{21}y$ **31.** $-3x - 6$ **33.** $-11.1n + 3.1m + 1.2$ **35.** (a) $12x + 1$ (b) It is doubled to obtain $24x + 2$. **37.** $n = 4.8$ **38.** $n = 8.1$ **39.** $n = 3$ **40.** $n = 8.5$ **41.** $n = 0.2$ **42.** $n = 12$ **43.** \$200,000 **44.** 15 seconds

10.2 Exercises

1. variable **3.** $3x$ and x; $2y$ and $-3y$ **5.** $27x - 18$ **7.** $-2x - 2y$ **9.** $-10.5x + 21y$ **11.** $30x - 70y$ **13.** $48a - 40b$ **15.** $20y + 24z$ **17.** $4p + 36q - 40$ **19.** $\frac{3}{5}x + 2y - \frac{3}{4}$ **21.** $30a + 48b - 67.5$ **23.** $32a + 48b - 36c - 20$ **25.** $-2.6x + 17y + 10z - 24$ **27.** $x - \frac{3}{2}y + 2z - \frac{1}{4}$ **29.** $-4x + 6y - z$ **31.** $p = 2l + 2w$ **33.** $A = \frac{hB + hb}{2}$ **35.** $27x - 39$ **37.** $-8a + 34b$ **39.** $8.1x + 8.1y$ **41.** $-9a - 19b + 12c$ **43.** $A = a(b + c) = ab + ac$ **45.** $A = x(y + w + z) = xy + xw + xz$ **47.** 27 in. **48.** 19.4 feet

10.3 Exercises

1. equation **3.** opposite **5.** $y = 32$ **7.** $x = 9$ **9.** $x = -18$ **11.** $x = -25$ **13.** $x = 19$ **15.** $9.8 = x$ **17.** $x = -8.7$ **19.** $x = 13.2$ **21.** $x = \frac{2}{5}$ **23.** $x = 1$ **25.** $x = -\frac{3}{2}$ or $-1\frac{1}{2}$ **27.** $y = -\frac{13}{15}$ **29.** $x = 14$ **31.** $x = -13$ **33.** $x = 2$ **35.** $x = -9$ **37.** $y = \frac{13}{2}$ or $6\frac{1}{2}$ **39.** $-8 = z$ **41.** $y = 1.2$ **43.** $x = 6$ **45.** $x = 12$ **47.** $x = 3$
49. To solve the equation $3x = 12$, divide both sides of the equation by 3 so that x stands alone on one side of the equation.
51. $3x + 3y + 3$ **52.** $-x + 21y + 3$ **53.** 4.7% **54.** 32 cubic feet **55.** 21 **56.** 2.2

10.4 Exercises

1. A sample answer is: To maintain the balance, whatever you do to one side of the scale, you need to do the exact same thing to the other side of the scale.
3. $\frac{4}{3}$ **5.** $x = 9$ **7.** $y = -4$ **9.** $x = -\frac{16}{9}$ **11.** $x = 12$ **13.** $16 = m$ **15.** $x = 10$ **17.** $z = 1.8$ **19.** $x = -13.5$ **21.** $x = 8$ **23.** $y = 10$ **25.** $n = \frac{5}{4}$ or $1\frac{1}{4}$ **27.** $x = -\frac{8}{5}$ or $-1\frac{3}{5}$ **29.** $x = -\frac{9}{2}$ or $-4\frac{1}{2}$ **31.** $z = 8$ **33.** $x = 8$ **35.** $x = -9$ **37.** $x = 0.03$ **39.** $x = -20$ **41.** 18 **43.** $4x - 7y + 6$ **44.** $-11a + 20b - 27c$ **45.** 438.3% **46.** 8.2%
47. 22,500 plates; 31,500 napkins; 18,000 towels

10.5 Exercises

1. You want to obtain the x-term all by itself on one side of the equation. So you want to remove the -6 from the left side of the equation. Therefore you would add the opposite of -6. This means you would add 6 to each side. **3.** no **5.** yes **7.** $x = 3$ **9.** $x = -\frac{1}{2}$ **11.** $x = \frac{5}{6}$ **13.** $x = -2$ **15.** $x = 0.3$ **17.** $x = 33$ **19.** $x = 2$ **21.** $x = 1$ **23.** $x = -16$ **25.** $x = 11$ **27.** $x = -1$ **29.** $y = -6$ **31.** $y = 7$ **33.** $y = 5$ **35.** $x = 16$ **37.** $y = 0$ **39.** $x = -8$ **41.** $x = -\frac{11}{4}$ **43.** $x = 10$
45. (a) $x = 5$ (b) $5 = x$ (c) For most students, collecting x-terms on the left is easier, since we usually write answers with x on the left. But either method is OK. **47.** 407,513.4 cubic centimeters **48.** 23.4 square inches **49.** 210.96 sq ft **50.** 22 pieces

How Am I Doing? Sections 10.1–10.5

1. $-17x$ (obj. 10.1.2) **2.** y (obj. 10.1.2) **3.** $-3a + 2b$ (obj. 10.1.2) **4.** $-12x - 4y + 10$ (obj. 10.1.2) **5.** $16x - 2y - 6$ (obj. 10.1.2) **6.** $4a - 12b + 3c$ (obj. 10.1.2) **7.** $42x - 18y$ (obj. 10.2.1) **8.** $-3a - 15b + 3$ (obj. 10.2.1) **9.** $-3a - 6b + 12c + 10$ (obj. 10.2.1) **10.** $x - 8y$ (obj. 10.2.2) **11.** $-18x - 12y$ (obj. 10.5.1) **12.** $-21x + 9y$ (obj. 10.2.1) **13.** $x = 37$ (obj. 10.5.2)

14. $x = 3.5$ (obj. 10.3.1) **15.** $x = \frac{7}{8}$ (obj. 10.3.1) **16.** $x = 8$ (obj. 10.3.1) **17.** $x = -8$ (obj. 10.4.1) **18.** $x = 5$ (obj. 10.4.1)

19. $x = \frac{3}{2}$ or $1\frac{1}{2}$ (obj. 10.4.2) **20.** $x = -12$ (obj. 10.4.1) **21.** $x = 7$ (obj. 10.5.1) **22.** $x = \frac{8}{5}$ or $1\frac{3}{5}$ (obj. 10.5.2)

23. $x = 3$ (obj. 10.5.3) **24.** $x = \frac{9}{4}$ or $2\frac{1}{4}$ (obj. 10.5.3) **25.** $x = 8$ (obj. 10.5.3) **26.** $x = \frac{25}{6}$ (obj. 10.5.3) **27.** $y = 7$ (obj. 10.5.2)

28. $x = 4$ (obj. 10.5.2)

10.6 Exercises

1. $h = 34 + r$ **3.** $b = n - 107$ **5.** $n = a + 14$ **7.** $l = 2w + 7$ **9.** $l = 3w - 2$ **11.** $m = 3t + 10$ **13.** $j + s = 26$ **15.** $ht = 500$ **17.** p = cost of airfare to Phoenix; $p + 135$ = cost of airfare to San Diego **19.** b = number of degrees in angle B; $b - 46$ = number of degrees in angle A **21.** w = height of Mt Whitney; $w + 4430$ = height of Mt. Everest **23.** a = number of books Aaron read; $2a$ = number of books Nina read; $a + 5$ = number of books Molly read **25.** h = height; $h + 5$ = length; $3h$ = width **27.** x = 1st angle; $2x$ = 2nd angle; $x - 14$ = 3rd angle **29.** Caravan's speed = $1.08s$; Lexus' speed = $1.08(s + 10)$ **31.** 8 **32.** -1 **33.** $x = -1$ **34.** $y = -14$ **35.** From top to bottom in the table: 0.387,311,106

10.7 Exercises

1. x = length of shorter piece; $x + 5.5$ = length of longer piece; 5.25 feet; 10.75 feet **3.** x = number of points scored by France $x - 22$ = number of points scored by Japan; France scored 51 points, Japan scored 29 points **5.** x = the number of cars in November; $x + 84$ = the number of cars in May; $x - 43$ = the number of cars in July; 119 cars in November; 203 cars in May; 76 cars in July **7.** x = length of shorter piece; $x + 4.7$ = length of the longer piece; the shorter piece is 3.65 feet long; the longer piece is 8.35 feet long **9.** x = width; $2x - 4$ = length; width is 14 inches; length is 24 inches **11.** x = length of the first side; $x + 20$ = length of the second side; $x - 4$ = length of the third side; 61 millimeters; 81 millimeters; 57 millimeters **13.** x = length of the first side; $2x$ = length of the second side; $x + 12$ = length of the third side; 8 centimeters; 16 centimeters; 20 centimeters **15.** x = number of degrees in angle A; $3x$ = number of degrees in angle B; $x + 40$ = number of degrees in angle C; angle A measures 28°; angle B measures 84°; angle C measures 68° **17.** x = total sales; $76,000 **19.** x = yearly rent; $6000 **21.** x = length of the adult section; $x + 6.2$ = length of the children section; the adult section is 12.9 feet; the children section is 19.1 feet **23.** first program = $70.37; second program = $120.74; third program = $379.22 **25.** 7.5 **27.** 60% **28.** 500 **29.** $x = 8$ **30.** 80 oz **31.** 117.4 meters

Putting Your Skills to Work

1. $G = 2.1(19) + 3.7 = 43.6$ seconds **2.** $n = 11$ vehicles per lane average **3.** 26.8 or 27 cars per lane average **4.** 49 seconds **5.** $n = 21.57$ or 22 vehicles per cycle **6.** 54.25 seconds **7.** $n = 24$ vehicles per lane per cycle

Chapter 10 Review Problems

1. $-13a + 3$ **2.** $\frac{7}{8}x + \frac{11}{12}$ **3.** $-2x - 7y$ **4.** $11x - 5y$ **5.** $-x - 12y + 6$ **6.** $2x - 10y - 7$ **7.** $-15x - 3y$

8. $-8x - 12y$ **9.** $2x - 6y + 8$ **10.** $6x - 18y - 3$ **11.** $-5a + 6b + 30$ **12.** $-9a + 2b + 12$ **13.** $6x + 15y - 27.5$ **14.** $8.4x - 12y + 20.4$ **15.** $-2x + 14y$ **16.** $7x - 8y$ **17.** $-8a - 2b - 24$ **18.** $-7a + 8b + 15$ **19.** $x = 12$ **20.** $x = 11.7$

21. $x = 4$ **22.** $x = 7.4$ **23.** $x = -12.1$ **24.** $x = -1.2$ **25.** $x = \frac{11}{4}$ or $2\frac{3}{4}$ **26.** $x = \frac{13}{4}$ or $3\frac{1}{4}$ **27.** $x = \frac{1}{8}$ **28.** $x = \frac{3}{2}$ or $1\frac{1}{2}$

29. $x = 5$ **30.** $x = -12$ **31.** $x = -\frac{5}{2}$ or $-2\frac{1}{2}$ **32.** $y = -5$ **33.** $x = 6$ **34.** $y = 7$ **35.** $x = -5$ **36.** $x = 0.25$

37. $x = 8$ **38.** $x = \frac{5}{6}$ **39.** $x = 6$ **40.** $x = 3$ **41.** $x = -8$ **42.** $x = -9$ **43.** $x = 3$ **44.** $x = -10$ **45.** $x = 1$

46. $x = 5$ **47.** $y = 16$ **48.** $y = -28$ **49.** $w = c + 3000$ **50.** $a = m - 18$ **51.** $A = 3B$ **52.** $l = 2w - 3$ **53.** r = Roberto's salary; $r + 2050$ = Michael's salary **54.** x = length of first side; $2x$ = length second side **55.** d = the number of days Dennis worked; $2d + 12$ = the number of days Carmen worked **56.** b = number of books in old library; $2b + 450$ = number of books in new library **57.** x = length of the shorter piece; $x + 6.5$ = length of the longer piece; 26.75 feet; 33.25 feet **58.** x = the old employee's salary; $x - 28$ = the new employee's salary; $192; $220 **59.** x = number of customers in February; $2x$ = number of customers in March; $x + 3000$ = number of customers in April; Feb = 10,550; Mar. = 21,100; Apr. = 13,550 **60.** x = number of miles on Thursday; $x + 106$ = number of miles on Friday; $x - 39$ = number of miles on Saturday; Thurs. = 263 miles; Fri. = 369 miles; Sat. = 224 miles **61.** x = width; $2x - 3$ = length; width = 13 inches; length = 23 inches **62.** x = width; $3x + 2$ = length; width = 22 meters; length = 68 meters **63.** $X = 36°$; $Y = 96°$; $Z = 48°$ **64.** x = number of degrees in angle B; $x + 74$ = number of degrees in angle A; $3x$ = number of degrees in angle C; angle A measures 95.2°; angle B measures 21.2°; angle C measures 63.6° **65.** width = 53 yards; length = 120 yards **66.** x = length; $x - 44$ = width; the width is 50 feet. The length is 94 feet. **67.** $x + 88$ = number of miles on the first day; x = number of miles on the second day; the first day they drove 424 miles. The second day they drove 336 miles. **68.** x = number of applications in the first week; $x + 156$ = number of applications in the second week; $x - 142$ = number of applications in the third week; they received 262 the first week, 418 the second week, and 120 the third week. **69.** $10,000 **70.** x = total sales; $19,375

How Am I Doing? Chapter 10 Test

1. $-6a$ (obj. 10.1.2) **2.** $\frac{2}{15}x + \frac{9}{8}y$ (obj. 10.1.2) **3.** $\frac{5}{8}a - \frac{2}{3}b$ (obj. 10.1.2) **4.** $a - 8b$ (obj. 10.1.2) **5.** $7x - 7z$ (obj. 10.1.2)

6. $-4x - 2y + 5$ (obj. 10.1.2) **7.** $60x - 25y$ (obj. 10.2.1) **8.** $2x - \frac{10}{3}y$ (obj. 10.2.1) **9.** $-4.5a + 3b - 1.5c + 12$ (obj. 10.2.1)

10. $-11a + 14b$ (obj. 10.2.2) **11.** $x = -8$ (obj. 10.5.1) **12.** $x = -6.35$ (obj. 10.3.1) **13.** $x = 15$ (obj. 10.5.1)

14. $x = -\frac{1}{2}$ (obj. 10.5.1) **15.** $x = -5$ (obj. 10.5.1) **16.** $x = -\frac{7}{10}$ (obj. 10.5.1) **17.** $s = f + 15$ (obj. 10.6.1)
18. $n = s - 15{,}000$ (obj. 10.6.1) **19.** $\frac{1}{2}s$ = measure of the first angle; s = measure of the second angle; $2s$ = measure of the third angle (obj. 10.6.2) **20.** w = width; $2w - 5$ = length (obj. 10.6.2)
21. Prentice farm = 87 acres; Smithfield farm = 261 acres (obj. 10.7.1) **22.** Marcia = \$24,000; Sam = \$22,500 (obj. 10.7.3)
23. 41 students in the morning class, 65 students in the afternoon class, 77 students in the evening class (obj. 10.7.1)
24. width = 25 feet; length = 34 feet (obj. 10.7.2)

Cumulative Test for Chapters 1–10

1. 747 **2.** 10,815 **3.** 45,678,900 **4.** $\frac{5}{2}$ or $2\frac{1}{2}$ **5.** $\frac{65}{8}$ or $8\frac{1}{8}$ **6.** 9322.8 **7.** 627.42 **8.** $n = 16$ **9.** 30 **10.** 5500
11. 0.345 meters **12.** 2.5 feet **13.** 37.7 yard **14.** 143 square meters **15.** −31 **16.** 144 **17.** $\frac{3}{4}a - \frac{1}{14}b$ **18.** $4x - 13y + 8$
19. $-16x + 24y - 40$ **20.** $-2x - 24y$ **21.** $x = 4$ **22.** $y = -\frac{5}{2}$ or $-2\frac{1}{2}$ **23.** $x = 26$ **24.** $x = -3$
25. p = weight of printer; $p + 322$ = weight of computer **26.** f = enrollment during fall; $f - 87$ = enrollment during summer
27. Thursday = 376 miles; Friday = 424 miles; Saturday = 281 miles **28.** width = $13\frac{2}{3}$ feet; length = $35\frac{1}{3}$ feet

Practice Final Examination

1. eighty-two thousand, three hundred sixty-seven **2.** 30,333 **3.** 173 **4.** 34,103 **5.** 4212 **6.** 217,745 **7.** 158 **8.** 606
9. 116 **10.** 32 miles/gallon **11.** $\frac{7}{15}$ **12.** $\frac{42}{11}$ **13.** $\frac{33}{20}$ or $1\frac{13}{20}$ **14.** $\frac{89}{15}$ or $5\frac{14}{15}$ **15.** $\frac{31}{14}$ or $2\frac{3}{14}$ **16.** 4 **17.** $\frac{14}{5}$ or $2\frac{4}{5}$
18. $\frac{22}{13}$ or $1\frac{9}{13}$ **19.** $6\frac{17}{20}$ miles **20.** 5 packages **21.** 0.719 **22.** $\frac{43}{50}$ **23.** > **24.** 506.38 **25.** 21.77 **26.** 0.757
27. 0.492 **28.** 3.69 **29.** 0.8125 **30.** 0.7056 **31.** $\frac{1400 \text{ students}}{43 \text{ faculty}}$ **32.** no **33.** $n \approx 9.4$ **34.** $n \approx 7.7$ **35.** $n = 15$
36. $n = 9$ **37.** \$3333.33 **38.** 9.75 inches **39.** \$294.12 **40.** 1.6 pounds **41.** 0.63% **42.** 21.25% **43.** 1.64 **44.** 17.3%
45. 302.4 **46.** 250 **47.** 4284 **48.** \$10,856 **49.** 4500 students **50.** 34.3% **51.** 4.25 gallons **52.** 6500 pounds
53. 192 inches **54.** 5600 meters **55.** 0.0698 kilograms **56.** 0.00248 liter **57.** 19.32 kilometers **58.** 6.3182×10^{-4}
59. 1.264×10^{11} **60.** 1.36728 centimeters **61.** 14.4 meters **62.** 206 centimeters **63.** 5.4 square feet **64.** 75 square meters
65. 113.04 square meters **66.** 56.52 meters **67.** 167.47 cubic centimeters **68.** 205.2 cubic feet **69.** 32.5 square meters
70. $n = 32.5$ **71.** \$8 million **72.** \$1 million dollars **73.** 50°F **74.** from 1990 to 2000 **75.** 600 students **76.** 1400 students
77. mean ≈ 15.83; median = 16.5 **78.** 16 **79.** 11.091 **80.** 15 feet **81.** −13 **82.** $\frac{1}{8}$ **83.** −3 **84.** −17 **85.** 24
86. $-\frac{8}{3}$ or $-2\frac{2}{3}$ **87.** 4 **88.** 27 **89.** 8 **90.** $\frac{1}{2}$ or 0.5 **91.** $-3x - 7y$ **92.** $-7 - 4a - 17b$ **93.** $-2x + 6y + 10$
94. $-11x - 9y - 4$ **95.** $x = 2$ **96.** $x = 2$ **97.** $x = -\frac{1}{2}$ or −0.5 **98.** $x = -\frac{2}{5}$ or −0.4
99. 122 students are taking history; 110 students are taking math **100.** The length is 37 meters. The width is 16 meters.

Appendix A.1 Balancing a Checking Account

Exercises

1. \$555.12 **3.** \$2912.65 **5.** \$153.44; Yes, Justin can pay his car insurance. **7.** The account balances. **9.** \$949.88
11. Jeremy's account balances.

Exercises Appendix A.2 Purchasing a Home

Verbal and Writing

1. percent × purchase price of home **3.** percent × home loan amount **5.** payment factor × home loan amount
7. yearly premium ÷ 12

Applications

9. \$50,000; \$200,000 **11.** \$4320; \$2700 **13.** \$2687.36 **15.** \$550; \$160 **17.** \$31,250 **19.** \$1405 **21.** \$73,920 **23.** \$1172
25. \$3700.56 **27.** \$5426.28

Exercises Appendix A.3 Purchasing a Vehicle

A.3 Exercises

1. \$1259.94 **3.** \$379.98 **5.** \$4245 **7.** (a) \$1244.95; \$497.98 (b) \$27,741.93 **9.** (a) \$3135.93; \$895.98 (b) \$50,930.91
11. (a) \$7499.85 (b) \$42,499.15 **13.** Dealership 1 **15.** (a) Dealership 3; \$25,678.93; (b) Dealership 1; \$28,839 (c) The most expensive purchase price does not guarantee the most expensive total cost. Many factors need to be considered to determine the best deal.

Appendix C Scientific Calculators

Exercises

1. 11,747 **3.** 352,554 **5.** 1.294 **7.** 254 **9.** 8934.7376; 123.45 + 45.9876 + 8765.3 **11.** 16.81; $\frac{34}{8} + 12.56$ **13.** 10.03 **15.** 289.387 **17.** 378,224 **19.** 690,284 **21.** 1068.678 **23.** 1.58688 **25.** 1383.2 **27.** 0.7634 **29.** 16.03 **31.** $12,562.34 **33.** 85,163.7 **35.** $103,608.65 **37.** 15,782.038 **39.** 2.500344 **41.** 897 **43.** 22.5 **45.** 84.372 **47.** 4.435 **49.** 101.5673 **51.** 0.24685 **53.** 30.7572 **55.** −13.87849 **57.** −13.0384 **59.** 3.3174×10^{18} **61.** $4.630985915 \times 10^{11}$ **63.** −0.5517241379 **65.** 15.90974544 **67.** 1.378368601 **69.** 11.981752 **71.** 8.090444982 **73.** 8.325724507 **75.** 1.08120 **77.** 1.14773

Applications Index

G

H

I

M

Subject Index

H

I

K

L

M

N

O

P

T

U

V

W

Y

Z

Photo Credits

TABLE OF CONTENTS **p. vii (a)** Photo courtesy of Kate Scannell/World Vision **p. vii (b)** R.P. Kingston/Index Stock Imagery, Inc. **p. viii (a)** EyeWire Collection/Getty Images/Photodisc **p. viii (b)** Kim Sayer © Dorling Kindersley **p. ix (a)** Hans Peter Merton/Robert Harding World Imagery **p. ix (b)** Blair Seitz/Photo Researchers, Inc. **p. ix (c)** Liu Liqun/Corbis Bettmann **p. x (a)** AP Wide World Photos **p. x (b)** Lynsey Addario/Corbis/Bettmann **p. xi** Charles E. Rotkin/Corbis Bettmann

CHAPTER 1 **CO** Photo courtesy of Kate Scannell/World Vision USA, Inc. **p. 29** Bob Winsett/Corbis Bettmann **p. 56** Jose L. Pelaez/Corbis/Stock Market **p. 64** Jeff Greenberg/PhotoEdit **p. 70** Stocktrek/Corbis/Stock Market **p. 81** Jean Miele/Corbis/Stock Market **p. 86** Rob & Sas/Corbis/Stock Market **p. 95** Ariel Skelley/Corbis/Stock Market

CHAPTER 2 **CO** A.R.P. Kingston, Index Stock Imagery, Inc. **p. 114** Rob Lewine/Corbis/Stock Market **p. 135** John Paul Endress/Corbis/Stock Market **p. 137** Bob Daemmrich/Stock Boston **p. 166** Stefan Lawrence/Image State/International Stock Photography Ltd. **p. 184** Index Stock Imagery, Inc. **p. 186** Bill Stanton/International Stock Photography Ltd.

CHAPTER 3 **CO** EyeWire Collection/Getty Images Inc.-Photodisc **p. 211** Dale C. Spartas/Corbis/Bettmann **p. 218** Daemmrich/Stock Boston **p. 253** Richard Bickel/Corbis/Bettmann **p. 254** A. Ramey/Woodfin Camp & Associates **p. 256** Stockbyte **p. 257** Catherine Ursillo/Photo Researchers, Inc. **p. 258** David R. Frazier Photolibrary/Photo Researchers, Inc. **p. 260** Andy Sacks/Getty Images Inc.-Stone Allstock

CHAPTER 4 **CO** Kim Sayer © Dorling Kindersley Media Library **p. 280** Lester Lefkowitz/Corbis/Stock Market **p. 293** Charles Gupton/Stock Boston **p. 294** PeterSaloutos/Corbis/Bettmann **p. 296** Greig Cranna/Stock Boston **p. 301** Michael Newman/PhotoEdit

CHAPTER 5 **CO** Hans Peter Merton/Robert Harding World Imagery **p. 313** Robert Rathe/Stock Boston **p. 320** Hillary Wilkes/Image State/International Stock Photography Ltd. **p. 333** Mike Mazzaschi/Stock Boston **p. 340** SuperStock, Inc. **p. 347** Michael Newman/PhotoEdit

CHAPTER 6 **CO** Blair Seitz/Photo Researchers, Inc. **p. 376** William Taufic/Corbis/Stock Market **p. 394** Gerard Lacz/Peter Arnold, Inc. **p. 398** Frank Grant/International Stock Photography Ltd. **p. 400** Tom McHugh/Photo Researchers, Inc.

CHAPTER 7 **CO** Liu Linqun/Corbis/Bettmann **p. 427** Marc Romanelli/Getty Images Inc./Image Bank **p. 436** Clint Clemens/International Stock Photography, Inc. **p. 449** Rosenthal/SuperStock, Inc. **p. 470** Tom McCarthy/PhotoEdit **p. 478** (top) Tom McCarthy/PhotoEdit (side) SuperStock, Inc. **p. 489** Dick Durrance/Woodfin Camp & Associates **p. 502** John Elk III/Stock Boston **p. 504** David Young-Wolff/PhotoEdit

CHAPTER 8 **CO** AP Wide World Photos **p. 527** Daemmrich/Stock Boston **p. 551** Ed Simpson/Getty Images Inc./Stone Allstock

CHAPTER 9 **CO** Lynsey Addario/Corbis Bettmann **p. 580** Gary Landsman/Corbis/Stock Market **p. 593** Michael Giannechini/Photo Researchers, Inc. **p. 611** David Karp/Bloomberg News/Landov LLC

CHAPTER 10 **CO** Charles E. Rotkin/Corbis Bettmann **p. 662** Jeffry W. Myers/Stock Boston